Principles of Cell and Molecular Biology

Second Edition

Principles of Cell and Molecular Biology

Lewis J. Kleinsmith *The University of Michigan*

Valerie M. Kish *University of Richmond*

HarperCollinsCollegePublishers

18314260

Acquisitions Editor: Susan McLaughlin
Developmental Editor: Karen Trost
Project Editor: Ellen MacElree
Design Manager: Lucy Krikorian
Text Designer: Robin Hoffmann
Cover Designer: John Callahan
Cover Photograph: Reproduced from *The Journal of Cell Biology*, Volume
    105, October 1987, pp. 1613–1622 by copyright permission of The Rocke-
    feller University Press. Courtesy of D. Lansing Taylor.
Art Studio: J. B. Woolsey
Photo Researcher: Joann DeSimone
Electronic Production Manager: Su Levine
Desktop Administrator: Laura Leever
Manufacturing Manager: Willie Lane
Electronic Page Makeup: Interactive Composition Corporation
Printer and Binder: R. R. Donnelley & Sons Company
Cover Printer: The Lehigh Press, Inc.

Principles of Cell and Molecular Biology, Second Edition

Library of Congress Cataloging-in-Publication Data

Kleinsmith, Lewis J.
    [Principles of cell biology]
    Principles of cell and molecular biology / Lewis J. Kleinsmith.
    Valerie M. Kish.—2nd ed.
        p.   cm.
    First ed. published under the title: Principles of cell biology.
    Includes bibliographical references and index.
    ISBN 0-06-500404-3
    1. Cytology.   2. Molecular bilogy   I. Kish, Valerie M.
II. Title.
QH581.2.K54   1995
574.87—dc20
                                                    94-22311
                                                        CIP

95 96 97 9 8 7 6 5 4 3 2

Brief

Contents

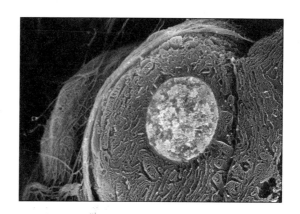

Detailed Contents

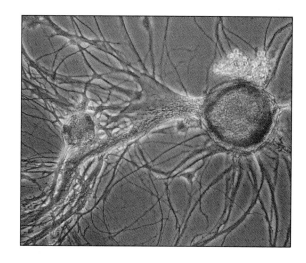

Chapter 9
Chloroplasts and the Capturing of Energy Derived from Sunlight   354

Part 3

Formation of Specialized Cells 671

Chapter 15
Gametes, Fertilization, and Early Development 673

Preface

The past two decades have witnessed an explosive growth in our understanding of the properties of living cells. As a consequence, the scientific literature is now growing so quickly that it is almost a full-time job to keep abreast of the major developments relating to cell organization and behavior. This enormous profusion of information presented us with a great challenge when we confronted the task of writing an introductory textbook that was both modest in length and readily comprehensible to students encountering the field of cell and molecular biology for the first time. In addressing this challenge, we have tried to create a textbook with a scope defined by three central goals. First, the primary goal of this book is to introduce students to the *fundamental principles* that guide cellular organization and function. Second, we think it is important for students to understand some of the crucial *scientific evidence* that has led to the formulation of these central concepts. And finally, we have sought to accomplish the two preceding goals in a book of *manageable length* that can be *easily read and understood* by a beginning cell biology student in the time allotted for a typical course. To accomplish the third objective, we obviously needed to be selective both in the types of examples chosen to illustrate key concepts and in the quantity of scientific evidence included. By exercising this selectivity, we have attempted to create a readable story that is accessible and interesting to introductory students rather than a compendium of the most recent, detailed facts about cell biology.

CHANGES IN THE SECOND EDITION

The second edition of our textbook bears the slightly altered title, *Principles of Cell and Molecular Biology*. This change from the first edition, entitled *Principles of Cell Biology*, was introduced to acknowledge the rapid strides that are being made in unraveling the molecular mechanisms that underlie cell functions. The book still focuses on the organization and functional activities of cells and their organelles; the new title simply reflects our growing understanding of the molecular components that are involved. In addition to its new title, the second edition includes the following changes:

• The scientific content of every chapter has been updated to reflect the numerous experimental advances in cell and molecular biology that have occurred since the first edition was written.

• Every chapter has been thoroughly rewritten to improve the readability of the text for the beginning cell biology student.

• The topic of muscle contraction has been integrated into the chapter on the Cytoskeleton and Cell Motility, and the topic of hormone signaling has been integrated into the chapter on the Cell Surface and Cellular Communication. As a result, we have been able to reduce the total number of chapters in the book from twenty to eighteen.

• An outline of concept headings and corresponding page references is provided at the beginning of each chapter to provide students with an introductory overview and quick reference.

• Every chapter is now subdivided into a series of small conceptual sections, each introduced by a Sentence Heading that summarizes the concept to be described.

• The illustration program has been completely revised. All of the line drawings are either new or have been extensively altered, and a large number of the micrographs are also new.

• Full color has been introduced in all line drawings to illustrate concepts more clearly.

• The narrative summary at the end of each chapter has been replaced by a bulleted Summary of Principal Points that is designed to help students identify the major issues covered in the chapter.

• The list of suggested readings at the end of each chapter has been thoroughly updated to reflect the most recent advances and now includes references through 1994.

• A Glossary containing a brief definition of every boldfaced term has been added at the end of the book. Each definition is followed by a specific page reference directing the student to the place in the book where the term is first discussed.

ORGANIZATION AND COVERAGE

Principles of Cell and Molecular Biology is divided into three major sections devoted to the discussion of background concepts, cell organization, and specialized cell types, respectively. This arrangement is designed to make the book adaptable to both single-semester and full-year courses, as well as to courses that differ in terms of the amount of biology background required of students. From comments made by reviewers of the first edition, we are aware that great variability exists in the sequence in which individual topics are covered in cell biology courses taught at different institutions. Because the sequence of chapters in this book represents only one of several possible ways of presenting the subject matter, we have tried to organize the chapters into clusters that can be read in several different sequences, using frequent page cross-references to provide appropriate background information if students read the chapters in a different order.

Part 1 (Chapters 1 through 4) is intended to provide students with a common set of background concepts that are essential to the understanding of the material discussed later in the book. Students with strong backgrounds in introductory biology may be able to simply use this material as a quick review. Chapter 1 provides an overview of the structural organization of prokaryotic and eukaryotic cells, including an introduction to the major types of organelles and the functions they carry out. The chapter then summarizes the types of molecules found within cells and concludes with a brief introduction to viruses. The evolutionary origin of cells, which was covered in Chapter 1 in the first edition, has been moved to Chapter 14 so that the topic can be discussed in greater depth.

Chapter 2 focuses on two underlying phenomena that are essential to the activities of all living cells: energy transformations and the use of enzymes to catalyze chemical reactions. Chapter 3 introduces the roles played by DNA and RNA in the flow of genetic information, and Chapter 4 summarizes some of the more important experimental techniques employed in the study of cells. Chapter 4 is intended primarily as a reference and background tool that students can consult as needed when various techniques are employed in experiments described later in the book.

After becoming acquainted with the background concepts introduced in the first four chapters, the student is ready to explore the core material covered in Part 2 (Chapters 5 through 14). This central section of the book deals with the general topic of how cells utilize a variety of organelles and specialized molecular components to carry out basic cell functions. We have chosen to begin this part of the book with a discussion of membrane architecture and function because virtually every important cellular activity involves events that occur within, on, attached to, or through membranes. Chapter 5 provides an introduction to the topics of membrane structure and membrane transport, followed by a detailed description of the plasma membrane and the cell surface in Chapter 6. Because of their close association with the cell surface, the topics of cell-cell signaling and the extracellular matrix are also included in Chapter 6. In Chapter 7 we cover the endoplasmic reticulum, Golgi complex, and lysosomes, describing the roles played by these membrane systems in cell secretion, endocytosis, and membrane biogenesis.

The next two chapters concentrate on cellular energetics. In Chapter 8 we describe how cells utilize glycolysis and respiration to transform energy derived from food into useful chemical forms. Throughout this discussion, heavy emphasis is placed on the role played by mitochondria and the chemiosmotic coupling mechanism. Chapter 9 then builds upon these concepts by showing how chloroplasts and related photosynthetic membranes utilize chemiosmotic coupling to capture energy derived from sunlight. The discussions of mitochondria and chloroplasts in Chapters 8 and 9 provide students with numerous opportunities to explore the relationship between organelle architecture and function.

The next three chapters are devoted to the transcription, translation, and transmission of genetic information. Chapter 10 builds on the background concepts provided in Chapter 3 by discussing how the transcription of genetic information is achieved and regulated and relates this information to the morphology of the nucleus. Chapter 11 discusses the biogenesis and structure of ribosomes and describes their role in the translation of genetic information. Chapter 12 concludes this sequence by describing prokaryotic and eukaryotic cell cycles, and the way in which genetic information is transmitted to newly forming cells during cell division.

In Chapter 13 we describe the cytoskeleton and cell motility, beginning with a description of muscle contraction and then discussing the general roles played by actin filaments, microtubules, and intermediate filaments. Chapter 14 discusses the evolutionary origin of cells and their major organelles, with heavy emphasis on our current understanding of the genetic systems of mitochondria and chloroplasts.

Because Chapters 1 through 14 focus mainly on principles of cell organization that are broadly applicable to "typical" cells, it is easy to lose sight of the fact that many cells found in multicellular organisms are programmed to perform a limited array of highly specialized functions. This differentiation is reflected in the unique shapes, sizes, subcellular organization, and molecular composition of the cells involved. In Part 3 we describe several examples of specialized cell types to show how a cell's basic architectural plan can be modified for the purpose of carrying out specialized functions. In Chapter 15 we introduce this section of the book by discussing the formation of gametes, their role in fertilization, and the events during early development that control cell differentiation. Chapter 16 concentrates on the properties of lymphocytes and their role in immune responses, and Chapter 17 describes neurons

and the phenomenon of synaptic signaling. Finally, we conclude in Chapter 18 with a description of cancer cells, whose abnormalities are beginning to provide important insights into the mechanism of growth control in normal cells.

PEDAGOGICAL FEATURES

The primary objective of this book is to make the fundamental concepts of cell biology and their experimental foundations readily accessible to students. To assist students in this endeavor, we have incorporated a variety of pedagogical tools:

• In a feature that is new to this edition, each chapter opens with an outline of concept headings that provides an introductory overview of the concepts to be discussed. Page references indicate where each concept is covered within the chapter.

• To assist students in identifying key concepts, we have utilized **boldface** type to highlight the most important terms in each chapter. All boldface terms are defined in the Glossary located at the back of the book.

• *Italics* are employed throughout the text to highlight newly introduced terms that are less important than boldfaced terms and thus do not appear in the Glossary. Occasionally, italics are also employed to highlight important phrases or sentences.

• The illustrations accompanying the text are of several different types. Some are composites of data taken from the original scientific literature, while others are interpretative diagrams designed to illustrate particular concepts. We have also included a large number of micrographs to round out discussions of morphology and to aid in tying together biochemical data with structural observations. As an added feature, the magnification of each micrograph has been transposed to a bar with the length noted above it, thereby allowing size comparisons to be made more easily.

• Each chapter is concluded by a bulleted Summary of Principal Points that can be used by the student to review the major concepts after the chapter has been read.

• Because cell biology is a rapidly advancing field whose ideas are continually being reevaluated in the light of current scientific evidence, the original scientific literature is the place to search for the most up-to-date information on any given subject. Therefore, at the end of each chapter we have provided a suggested reading list of books and articles that can be used by students as a bridge to the literature. Although space constraints have prevented us from including many of the classical papers used in preparing the text, we have incorporated enough recent articles to make it relatively easy for students to trace back through the literature to the earlier important experiments.

• A Glossary of brief definitions, another new feature, is included at the end of the book as a quick reference tool. The Glossary contains every boldfaced term from the text and includes specific page references that direct the student to the location in the book where the term is first discussed.

• Because a properly designed index is an extremely powerful tool, we have tried to compile an index that is pertinent, detailed without being cumbersome, and easy to use. As an added feature, pages on which key terms are defined appear in boldface type in the index.

SUPPLEMENTARY MATERIALS

A study guide/solutions manual and an instructor's manual, both written by Christine Case, are available with the text. Also available are full color transparencies.

ACKNOWLEDGMENTS

We are pleased to thank those colleagues who graciously consented to contribute micrographs to this endeavor, as well as the authors and publishers who have kindly granted permission to reproduce copyrighted material. Additionally, we are indebted to the students and reviewers whose comments have resulted in a much clearer, more accurate, and concisely written text. In particular, we wish to thank the following reviewers: Michael Adams, Eastern Connecticut State University; Howard Arnott, University of Texas—Arlington; L. Rao Ayyagari, Lindenwood College (MO); Robert Bast, Cleveland State University; Gerald Bergtrom, University of Wisconsin, Milwaukee; Foster Billheimer, California University of Pennsylvania; David Borst, Illinois State University; G. Benjamin Bouck, The University of Illinois, Chicago; Lawrence Bradford, Benedictine College (KS); Jim Bradley, Auburn University; Doug Carmichael, Pittsburg State University (KS); Keith Carson, Old Dominion University; Christine Case, Skyline College; Chen-Ho Chen, South Dakota State University; Catherine Chia, University of Nebraska—Lincoln; Randy Cohen, California State University, Northridge; Alan Comer, The University of North Carolina at Asheville; David Gale Davis, University of Alabama; Douglas Dennis, James Madison University; John Denny, University of Texas—San Antonio; Don Downer, Mississippi State University; Ernest DuBrul, University of Toledo; Dave Estervig, University of Wisconsin—Eau Claire; Richard Falk, University of California—Davis; John Fessler, University of California, Los Angeles; James Forbes, Hampton University (VA); John Freeman, The University of South Alabama; Allan Harrelson, University of Missouri; Marcia Harrison, Marshall University (WV); Thomas Herbert, University of Miami; Martinez Hewlett, University of Arizona; Peter Heywood, Brown University; M. E. S. Hudspeth, Northern Illinois University; Janice Knepper, Villanova University; Robert Koch, California State University—Fullerton; A. Krishna Kumaran, Marquette University; Cran Lucas, Louisiana State University at Shreveport; Thomas MacRae, Dalhousie University; Brian Masters, Towson State University; Hugh Miller, East Tennessee State University; Rathin Mitra, West Liberty State College; Robert Morris, Widener University; Donald Mykles, Colorado State University; James Nestler, Walla Walla College; Robert Paoletti, King's College (PA); Howard Petty, Wayne State University (MI); Robert Riley, Frostburg State (MD); Thomas Roberts, Florida State University; Edmund Samuel, Southern Connecticut State University; John Smarrelli, Loyola University of Chicago; Douglas Smith, Clarion University of Pennsylvania; Robert Smith, Skyline College; Richard Storey, Colorado College; Russell Stullken, Augusta College; Bruce Telzer, Pomona College; John Tyson, Virginia Polytechnic Institute and State University; David Vickers, University of Central Florida; and Douglas Walton, College of St. Scholastica.

We are also grateful for the guidance and technical expertise provided by individuals at HarperCollins who helped make this book a reality. We thank Glyn Davies, editor-in-chief, who provided unflagging support through the entire process. Susan McLaughlin, our acquisitions editor, consistently offered enthusiastic support and provided us with crucial guidance and information as we moved through the development and production processes. We are especially indebted to our developmental editor, Karen Trost, who played a major role in helping us revise the first edition. Her careful and thoughtful reading of the manuscript provided important insights that significantly enhanced the clarity of the text and art program. We also thank Ellen MacElree, our project editor, who successfully kept track of the manuscript and large illustration program during the production process. Finally, we thank John D. Woolsey and the artists at J/B Woolsey Associates, the studio responsible for the new look of our art program.

Finally, we would like to acknowledge the ongoing support of our families, whose patience and encouragement during these long days and nights have been paramount in making this book a reality.

LEWIS J. KLEINSMITH

VALERIE M. KISH

Part 1

Background Concepts

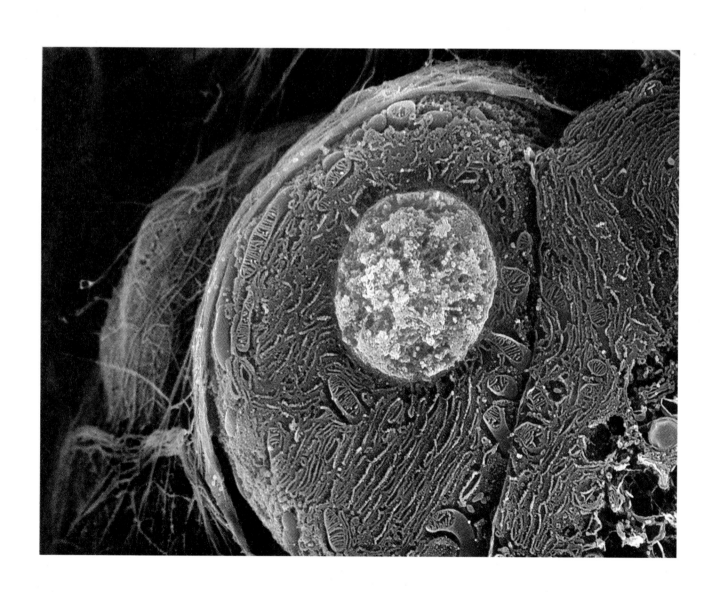

C h a p t e r   1

Prologue: Cells and Their Molecules

The cell is the fundamental unit of life, the underlying building block from which all organisms are constructed. The properties of cells, which are the smallest units exhibiting the characteristics of life, define both the potential capabilities and the inherent limitations of all living organisms. In the past several decades a wide variety of powerful new experimental approaches have been utilized to investigate the intricacies of cell organization and function. As a result, a revolution has occurred in our understanding of how cells are constructed and how they carry out the activities required for maintaining life. These experimental studies and the information obtained from them comprise the subject matter of this book.

To set the stage for the journey that is to follow, we will begin with a brief overview of the properties of cells and the molecules from which they are constructed. Students who have recently taken a course in introductory biology can use this background material as a brief review.

AN OVERVIEW OF CELL ORGANIZATION

The Discovery of Cells Led to the Formulation of the Cell Theory

Most cells are invisible to the naked eye, so scientists did not know of their existence prior to the invention of the microscope. When the ability to cast and grind magnifying lenses was first perfected in the early seventeenth century, it triggered a scientific and intellectual revolution. Pointed at the sky, such lenses opened the universe to our vision; pointed at small objects, these lenses opened the microscopic world to our sight. Robert Hooke, who served as Curator of Instruments for the Royal Society of London, was one of the first scientists to build his own microscope. In 1665 Hooke used one of his microscopes to examine thin slices of cork cut with a razor blade, leading him to discover that cork is constructed from tiny "little boxes," which he named *cells*.

What Hooke really had seen was the cell walls of dead plant material, with little evidence of internal cell structure (Figure 1-1). One of the limitations inherent in Hooke's work was that his microscope could only magnify objects about 30 times, making it difficult to learn much about the internal organization of cells. This obstacle was soon overcome, however, by Antonie van Leeuwenhoek, a Dutch storekeeper who produced hand-polished lenses that could magnify objects almost 300-fold. Van Leeuwenhoek's superior microscopes allowed him to discover blood cells, sperm cells, and the one-celled organisms present in pond water.

During the 150 years that followed these pioneering observations, many other cell types were discovered using light microscopy, but the biological significance

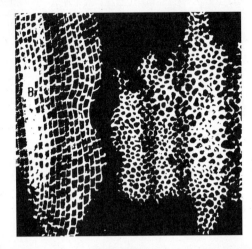

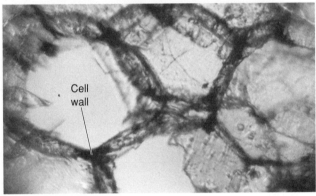

Figure 1-1 *First Views of Cell Structure*   (Top) *The structure of cork as drawn by Robert Hooke in 1665.* (Bottom) *An original slice of elder pith cut by van Leeuwenhoek in 1674, viewed through a surviving Leeuwenhoek microscope. The cell walls are clearly visible, but no internal cell structure can be seen. Bottom photograph courtesy of B. J. Ford.*

of cells remained unclear. Then in 1839 the German biologists Mathias Schleiden and Theodor Schwann integrated the growing body of information on the universal occurrence of cells into one of the first great unifying theories of biology, the **cell theory.** This theory had two important facets. First, it stated that in spite of the enormous diversity of living organisms on earth, all organisms are composed of cells. And second, it proposed that all living cells are structurally similar to one another. In 1855 the German physiologist Rudolf Virchow added a third principle to the cell theory when he concluded that all cells arise from the division of preexisting cells (an idea summarized in Virchow's famous Latin phrase, *omnis cellula e cellula,* which means "all cells from cells").

Considering that these three generalizations were based on relatively crude microscopic observations and virtually no biochemical data, it is remarkable how accurate they have turned out to be. In the century since the cell theory was first proposed, our tools for examining both the architectural organization and chemical activities of cells have been revolutionized, and have led to an understanding of cell structure and function whose extraordinary detail would have astounded Schleiden and Schwann. Although this detailed information has revealed that some differences do exist among cells, it has also been verified that cells are remarkably similar to one another both in the functions they carry out and in the kinds of cell structures they utilize to perform these functions.

Cells Perform Four Basic Functions

Although the cell theory was originally based on the observation that different kinds of cells resemble one another when viewed microscopically, the functional similarities between cells have turned out to be even more pronounced than their structural similarities. If one examines the properties of many different cell types, it soon becomes apparent that cells share the following functional characteristics:

1. Cells maintain a selective barrier, called the *plasma membrane,* which separates the inside of the cell from the external environment. By regulating the passage of materials into and out of cells, the plasma membrane ensures that optimum conditions for living processes prevail within the cell interior. Membrane barriers are also employed to subdivide the cell into multiple compartments specialized for different activities.

2. Cells utilize genetic information to guide the synthesis of most of the cell's components. This genetic information is stored in molecules of DNA and is duplicated prior to cell division so that each newly formed cell inherits a complete set of genetic instructions.

3. Cells contain catalysts called *enzymes,* which speed up chemical reactions involved in the synthesis and breakdown of organic molecules. The sum of all these reactions is referred to as *metabolism.* Metabolism converts foodstuffs into molecules that are needed by the cell, breaks down molecules that are no longer required, and traps energy in useful chemical forms which in turn provide the power for energy-requiring activities.

4. Cells almost always exhibit some type of motility. Mechanisms for moving components from one location to another within the cell are virtually universal. In many cell types, such mechanisms also permit movement of the cell as a whole.

The preceding functions are carried out with the aid of specialized subcellular structures known as **organelles.** In the following sections we will introduce the cell's main organelles and briefly state the functions

in which they are involved. Each organelle will then be described in depth later in the book.

Membranes and Walls Create Barriers and Compartments

When cells derived from different organisms are examined under the microscope, variations in subcellular structure are often observed. These differences are most apparent when the architecture of "typical" animal, plant, and bacterial cells is compared (Figures 1-2, 1-3, and 1-4). But in spite of this variation, cells always share certain structural features in common. Prominent among these shared features is the **plasma membrane,** a structure measuring 7–8 nanometers (nm) in thickness that constitutes the outer membrane boundary of all living cells. The plasma membrane regulates the flow of materials into and out of cells, creating an intracellular ("within the cell") milieu that differs significantly from the environment that typically surrounds a cell. Several different mechanisms are employed by the plasma membrane for regulating the flow of materials. First, membranes are selectively permeable; that is, some molecules can diffuse through the plasma membrane more readily than others. In addition, since membranes can actively transport specific molecules against a concentration gradient, the plasma membrane can establish either a lower or higher concentration of particular substances inside the cell than out. Finally, material can be moved into and out of cells by membrane vesicles that pinch off from the plasma membrane and enter the cell and by intracellular membrane vesicles that fuse with the plasma membrane and expel their contents to the cell exterior.

Besides its role in regulating the flow of materials into and out of the cell, the plasma membrane also plays an important role in cell-to-cell communication, transmitting signals from the cell exterior to the cell interior. In plant and bacterial cells, a rigid **cell wall** is usually found directly outside the plasma membrane (see Figures 1-3 and 1-4). The rigidity and strength of the cell wall allows it to mold the shape of the cell it encloses and to protect the cell from adverse environmental conditions. Cell walls are not essential features of cell structure, however, for they can be removed experimentally without destroying the enclosed cells. Such cells without walls, termed **protoplasts,** are more fragile and susceptible to breakage than are walled cells.

In addition to the presence of the plasma membrane at the outer cell surface, membranes are also employed to partition the cell interior into multiple compartments. This type of compartmentalization is almost exclusive to animal and plant cells, although internal membranes are occasionally observed in bacteria as well. In the cells of animals and plants, an elaborate system of interconnected membrane channels and vesicles known as the **endoplasmic reticulum (ER)** plays an important role in transporting newly synthesized proteins

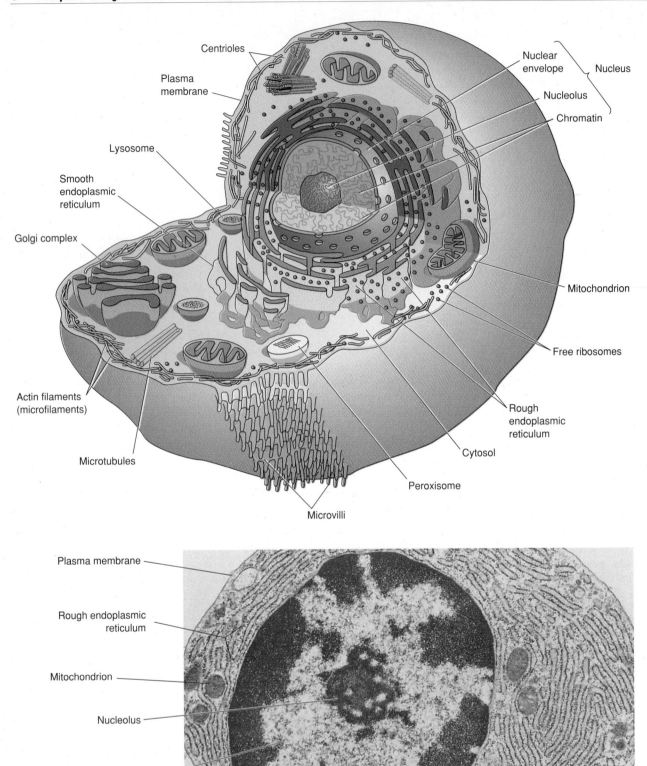

Figure 1-2 *Structure of a Typical Animal Cell*   (Top) *Three-dimensional model showing the distribution of major organelles in a "typical" animal cell. Animal cells of differing types vary in the degree to which the various organelles are present.* (Bottom) *Thin-section electron micrograph of a plasma cell. Micrograph courtesy of D. Fawcett.*

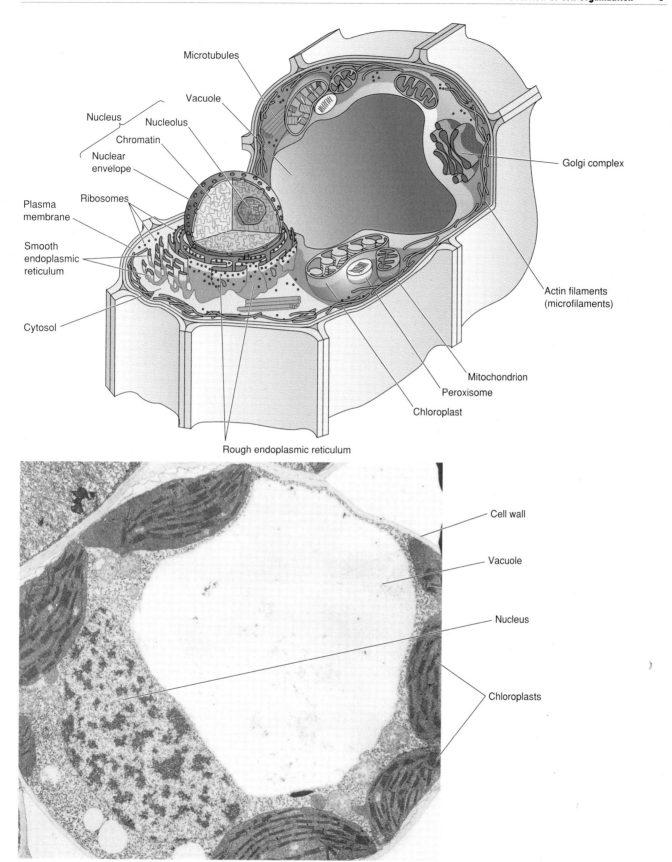

Figure 1-3 *Structure of a Typical Plant Cell* (Top) *Three-dimensional model showing the distribution of major organelles in a "typical" plant cell. Plant cells of differing types vary in the degree to which the various organelles are present.* (Bottom) *Thin-section electron micrograph of a leaf cell.*

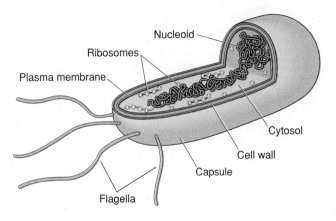

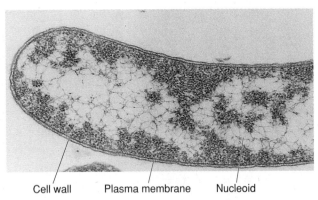

Figure 1-4 *Structure of a Typical Bacterial Cell* (Top) *Three-dimensional model showing the distribution of organelles in a typical bacterium.* (Bottom) *Thin-section electron micrograph of a Gram-negative bacterium. Micrograph courtesy of G. J. Brewer.*

to various destinations within the cell. Closely associated with the endoplasmic reticulum is the **Golgi complex,** a stack of membranes involved in processing newly synthesized proteins and packaging them into membrane vesicles for storage or secretion.

Several other kinds of membrane-bound organelles also occur in animals and plants. **Lysosomes** are small membrane-enclosed structures that serve a digestive function, breaking down foreign materials and intracellular constituents that are no longer needed by the cell. **Peroxisomes** carry out certain kinds of oxidation reactions, and **vacuoles,** which are especially prominent in plant cells, are large vesicles that function as storage compartments and in maintaining water balance. In addition to the preceding membrane-enclosed organelles, we will see shortly that membranes surround the genetic material and energy-transforming organelles of both plant and animal cells.

The Nucleus and Ribosomes Function in the Flow of Genetic Information

In spite of the basic similarities that have been observed in the cells of all organisms, some prominent structural

differences do exist. One of the more conspicuous differences involves the way in which the genetic material is organized; this differing organization has led to the division of the cellular world into two major cell types: prokaryotic cells and eukaryotic cells. **Eukaryotic** cells are defined as cells in which the bulk of the genetic information is contained in a **nucleus** surrounded by a double-membrane envelope, while the genetic material of **prokaryotic** cells is not membrane-enclosed. Prokaryotic cells include all forms of bacteria, while eukaryotic cells encompass other single-celled organisms (e.g., algae, fungi, and protozoa) as well as the cells of multicellular plants and animals.

The nucleus of eukaryotic cells is largely occupied by a mass of intertwined **chromatin** fibers, which contain the DNA molecules that store most of the cell's genetic information. A small portion of the nuclear DNA is localized in a small spherical structure known as the **nucleolus,** which contains DNA information involved in the formation of ribosomes. A fluidlike material called the **nucleoplasm** fills the spaces around the chromatin fibers and nucleoli. During cell division the chromatin fibers and nucleoli condense into compact structures known as **chromosomes.** The outer boundary of the nucleus is formed by two concentric membranes that together form the **nuclear envelope.** Occasional connections are observed between the outer membrane of the nuclear envelope and the endoplasmic reticulum. The most distinctive structural feature of the nuclear envelope is the presence of numerous *nuclear pores* that interrupt the continuity of the membranes. Nuclear pores help to regulate the flow of materials between the nucleus and the **cytoplasm** (the region outside the nuclear envelope), and serve as attachment points for chromatin fibers.

The organization of the genetic material is much simpler in prokaryotic cells than in eukaryotic cells. In prokaryotes the genetic material is packaged into a single DNA-containing chromosome that is folded into a compact structure which occupies a region of the cell known as the **nucleoid.** Although the nucleoid is visibly distinct from the rest of the cell, it is not enclosed by a membrane, nor are any nucleoli present.

In both prokaryotes and eukaryotes, the genetic information encoded in the cell's DNA guides the synthesis of specific protein molecules. This process of protein synthesis occurs on small cytoplasmic granules known as **ribosomes.** Prokaryotic ribosomes are slightly smaller than eukaryotic cytoplasmic ribosomes, but the two are basically similar in structure and function. Eukaryotic ribosomes are found free in the cytoplasm and attached to membranes of the endoplasmic reticulum; in addition, small numbers of ribosomes resembling those of prokaryotes occur in mitochondria and chloroplasts.

Mitochondria and Chloroplasts Play Key Roles in Metabolism

Metabolic reactions occur virtually everywhere in the cell, but those that are most central to the flow of energy are localized predominantly in the cytoplasm. Reactions involved in the initial breakdown of energy-rich nutrients occur in the **cytosol,** which is the fluidlike portion of the cytoplasm that surrounds the cytoplasmic organelles. Once this initial process is complete, a second set of reactions occurs in which most of the energy contained in the nutrients is released and used to drive the formation of the energy-rich molecule ATP. In eukaryotes, these latter reactions occur in specialized cytoplasmic organelles called **mitochondria** (see Figures 1-2 and 1-3). Found only in eukaryotic cells, mitochondria are as large as typical bacteria and are enclosed by two membranes. The inner of the two membranes is folded into a series of *cristae* that project into the internal cavity, or *matrix,* of the mitochondrion. In addition to being the site of many chemical reactions, the matrix also contains DNA and ribosomes, suggesting that mitochondria exhibit a certain degree of genetic autonomy. In prokaryotic cells, which do not have mitochondria, the reactions that typically take place in mitochondria occur instead in close association with the plasma membrane.

In addition to obtaining energy from the breakdown of energy-rich nutrients, some organisms are capable of trapping energy from sunlight and converting it into chemical energy. These *photosynthetic* organisms include both unicellular and multicellular plants (eukaryotes), as well as photosynthetic bacteria (prokaryotes). In photosynthetic eukaryotes, photosynthesis takes place in specialized organelles called **chloroplasts** (see Figure 1-3). Although chloroplasts are usually larger than mitochondria, the structure of the two organelles exhibits some fundamental similarities. Like mitochondria, chloroplasts are enclosed by two membranes surrounding an internal compartment, in this case designated the *stroma.* Within the stroma are the *thylakoid membranes,* which contain chlorophyll and other light-absorbing pigments. Like the matrix space of mitochondria, the stroma of chloroplasts contains both DNA and ribosomes, indicating that chloroplasts too exhibit a degree of genetic autonomy.

Cytoskeletal Filaments Function in Cell Motility

As noted previously, motility of some type is a characteristic of virtually all cells. Eukaryotic cells have developed an elaborate network of cytoplasmic filaments whose purpose is to move components within the cell as well as the cell as a whole. This network of filaments, called the **cytoskeleton,** consists of three distinct components:

1. The largest filaments, known as **microtubules,** are relatively rigid hollow structures measuring 25 nm in diameter; they contain the protein *tubulin* and occur widely in eukaryotic cells. Microtubules are used in the construction of the mitotic spindle that moves the chromosomes during cell division, and also form the core of motile appendages known as **cilia** and **flagella.**

2. Smaller filaments known as **actin filaments** (or *microfilaments*) also generate movements within the cytoplasm of eukaryotic cells. Actin filaments measure about 6 nm in diameter and are constructed from *actin,* a protein that occurs in the contractile fibers of muscle cells as well. Actin filaments generate bulk movements of the cytoplasm known as cytoplasmic streaming, as well as a crawling type of locomotion of the cell as a whole. Actin filaments also have been implicated in movements of the plasma membrane, especially those occurring when the cytoplasm is divided during cell division.

3. The third type of filament found in the cytoskeleton is called the **intermediate filament** because its diameter is in between that of actin filaments and microtubules. The protein composition of intermediate filaments varies, depending on the cell type. The role of intermediate filaments in motility, if any, is yet to be established, but they are known to play an important structural support and anchoring role in the cytoskeleton.

Microtubules, actin filaments, and intermediate filaments are not present in prokaryotic cells. Instead, the movement of prokaryotic cells is made possible by flagella that differ from eukaryotic flagella in both structure and chemical makeup. Prokaryotic flagella are thinner in diameter and constructed from a single structural protein, while eukaryotic flagella are larger structures containing nine pairs of microtubules surrounding two inner microtubules (the so-called "9 + 2" arrangement).

Prokaryotic and Eukaryotic Cells Differ in Their Structural Complexity

Although many of the functions carried out by prokaryotic and eukaryotic cells are quite similar, we have now seen that several structural features allow these two broad classes of cells to be differentiated from each other. The defining difference is the presence of a nuclear envelope surrounding the genetic material of eukaryotes but not prokaryotes. This difference is not the only one, however (Table 1-1). Eukaryotic cells are typically larger and more structurally complex than prokaryotic cells. Among the organelles contributing to this complexity are the aforementioned endoplasmic reticulum, Golgi complex, lysosomes, peroxisomes, mitochondria, and chloroplasts. Other features that distinguish eukaryotes are the presence of nucleoli, microtubules, actin filaments, intermediate filaments, and the 9 + 2 arrangement of their cilia and flagella.

Table 1-1   Comparison of Prokaryotic and Eukaryotic Cells

Feature	Prokaryotes	Eukaryotes	
		Animal	Plant
Plasma membrane	Yes	Yes	Yes
Cell wall	Yes	No	Yes
Ribosomes	Yes	Yes	Yes
Endoplasmic reticulum	No	Yes	Yes
Golgi complex	No	Yes	Yes
Lysosomes	No	Yes	Yes
Peroxisomes	No	Yes	Yes
Nuclear envelope	No	Yes	Yes
Nucleolus	No	Yes	Yes
Mitochondria	No	Yes	Yes
Chloroplasts	No	No	Yes
9 + 2 cilia/flagella	No	Yes	Yes
Microtubules	No	Yes	Yes
Actin filaments	No	Yes	Yes
Intermediate filaments	No	Yes	Yes

Each of the organelles that has been introduced in this chapter will be described in detail later in the book. At that time, comparisons will be made to show how prokaryotic and eukaryotic cells accomplish the same basic functions despite their differences in structure.

Cells Are Limited as to How Small or Large They Can Be

One of the most striking features of a cell is its ability to create an extraordinarily complex organizational structure in the confines of a very tiny space. Typical prokaryotic cells usually range in size from 1 to 10 μm in diameter, whereas average eukaryotic cells have diameters of 3–30 μm (Figure 1-5). Several factors limit the maximum size a cell can attain. A major constraint involves the quantitative relationship between cell volume and cell surface area. Although cells are not always spherical, a rough approximation of this quantitative relationship can be obtained using the formulas for calculating the surface area and volume of a sphere (Figure 1-6). Such calculations show that as a spherical cell increases in size, its volume increases much more rapidly than does its surface area. Thus, as cells get larger, a point is eventually reached where the surface area of the plasma membrane is not adequate for exchange of nutrients and wastes between the cell interior and the surrounding environment. Some cells combat this problem by increasing the surface area of the plasma membrane. For example, in cells lining the intestine, which are responsible for absorbing nutrients from the gastrointestinal tract, a series of fingerlike projections of the plasma membrane called **microvilli** have evolved (see Figure 13-37).

Another limitation on cell size is imposed by the need to maintain adequate concentrations of critical intracellular substances. Many important molecules are present in a relatively small number of copies per cell. As the size of the cell increases, these molecules face the problem of becoming diluted in the increasing volume of cytoplasm. Since the volume of the average eukaryotic cell is about a thousandfold greater than that of the typical prokaryotic cell, this problem has been especially pronounced for eukaryotic cells. This may explain why eukaryotic cells have evolved intracellular membrane systems that divide the cytoplasm into smaller compartments specialized for particular metabolic functions. By creating organelle compartments with smaller volumes and localizing particular kinds of molecules in each compartment, eukaryotic cells have been able to minimize the difficulties associated with increasing cell size.

A related problem faced by large cells is the need for enough genetic material to guide all the events taking place in the larger cell volume. For this reason the development of exceptionally large cells is sometimes associated with an increase in the amount of DNA per chromosome, the number of chromosome copies per cell, or the number of nuclei per cell. Such modifications allow certain eukaryotic cells, such as the amoeba with its multiple nuclei, to attain diameters of several millimeters or more, which is ten times greater than that of typical eukaryotic cells. Much larger sizes are reached in a few unusual cases, such as the egg cell of a hen or ostrich, but these situations are atypical because these eggs consist largely of stored food. The metabolically active portion of the egg, destined to become the future embryo, is no more than a tiny speck on the surface of the yolk.

Limitations also exist on how small a cell can be. Based on our current understanding of how essential functions are carried out by living cells, it can be estimated that a minimum of about 500–1000 different enzymes and other proteins are required in order for a cell to live and reproduce independently. This is about the amount of protein found in *mycoplasmas,* which are the smallest cells known. These primitive prokaryotic cells measure only 0.2–0.3 μm in diameter, yet are capable of a completely independent existence. Although it is conceivable that smaller cells will yet be discovered, it is clear that the lower limits of cell size are being approached in the mycoplasmas.

MOLECULAR COMPOSITION OF CELLS

The properties of living cells ultimately reflect the characteristics of the molecules from which they are constructed. Hence, before proceeding to a detailed discussion of how cells function, we need to briefly de-

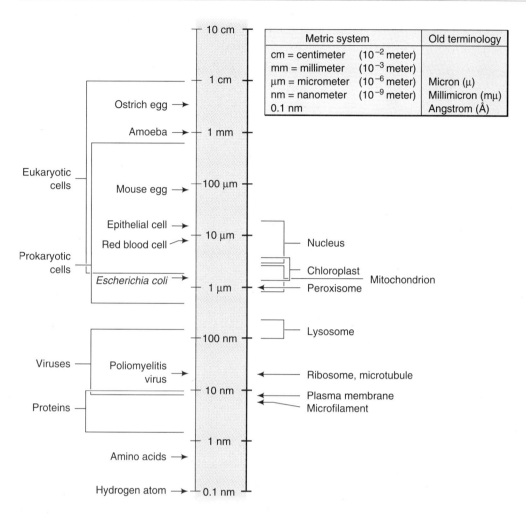

Figure 1-5 *Dimensions of Typical Cells, Organelles, Viruses, and Molecules* *Although typical prokaryotic cells range between 1 and 10 μm in diameter, a recently discovered prokaryote (the bacterium* E. fishelsoni) *is much larger, measuring up to 80 μm by 600 μm. For convenience in reading the original research literature, some older measurement units such as the micron and angstrom are included in this figure along with the modern metric terminology.*

scribe the kinds of molecules found in cells. Of the 92 natural elements that are potentially available for use in biological molecules, a relatively small number are of prime importance. *Carbon, hydrogen, oxygen, nitrogen, phosphorus,* and *sulfur* are crucial elements because they are employed in constructing the large *macromolecules* that comprise most of the cell's dry mass. In addition, about two dozen other elements are used by cells in smaller quantities for a variety of specialized purposes (Table 1-2).

Water Is the Most Abundant Substance in Cells

Of all the molecules found in cells, water is the most abundant, typically accounting for 70 to 80 percent of a cell's total mass (Figure 1-7). Water is the medium in which all cellular activities take place, and so it is a crucial factor in shaping intracellular events. Because water is so familiar to us in everyday life, we tend not to ap-

preciate how exotic a substance it is. Most other substances the size of the water molecule are gases at ambient temperature, but water is a liquid. Why does water exhibit this unusual behavior, and what does it tell us about the nature of the water molecule?

The unusual properties displayed by water arise from the fact that the electrons of the water molecule are not equally shared between the oxygen and hydrogen atoms, but are instead more closely associated with oxygen. Water is therefore a **polar** molecule—that is, a molecule in which one region displays a partial negative charge and another region displays a partial positive charge. This inherent polarity allows the hydrogen atom of one water molecule to bind weakly to the oxygen atom of another water molecule, forming a **hydrogen bond.** Each molecule of water can form as many as four hydrogen bonds with its neighbors, creating a loosely associated three-dimensional network of water molecules (Figure 1-8).

Spherical cells

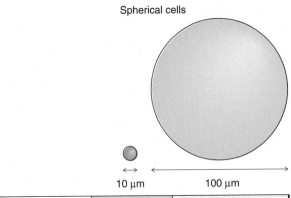

10 μm   100 μm

Radius	5 μm	50 μm
Surface area ($4\pi r^2$)	314 μm²	31,400 μm²
Volume ($\frac{4}{3}\pi r^3$)	524 μm³	524,000 μm³
Surface area / Volume	0.6	0.06

Figure 1-6 *Relationship between Cell Volume and Cell Surface Area* *This example involving two hypothetical spherical cells shows what happens when the diameter of a cell is increased tenfold. The volume of the cell increases proportionally to the cube of its radius, while the increase in surface area is proportional only to the square of the radius. Therefore the ratio of surface area to cell volume decreases as cells become larger.*

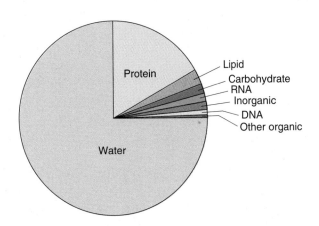

Figure 1-7 *Molecular Composition of a Typical Cell Expressed in Terms of Relative Mass* *This type of diagram emphasizes the high water content of a typical cell. Next to water, protein is by far the most abundant molecule in cells.*

Hydrogen bonding between neighboring water molecules gives water a unique set of properties for a molecule of its size. Among these properties, several are of special importance for living organisms: (1) Hydrogen bonding between water molecules endows water with a *high heat capacity,* where heat capacity refers to the amount of heat energy that must be absorbed to raise the temperature of water by a given amount. When water is heated, a significant fraction of the absorbed energy is employed to break the hydro-

Table 1-2   Chemical Elements Used by Living Organisms

Constituents of Macromolecules

Carbon (C) Hydrogen (H) Nitrogen (N) Oxygen (O) Phosphorus (P) Sulfur (S)	All are employed in constructing macromolecules; H and O are present in water molecules; C and H are present in all organic molecules

Other Major Elements

Calcium (Ca) Chlorine (Cl) Magnesium (Mg) Potassium (K) Sodium (Na)	Ions required for various enzyme reactions; Na⁺ is main extracellular cation; K⁺ is main cellular cation; Cl⁻ is main cellular and extracellular anion; Ca²⁺ is main component of bone

Essential Trace Elements

Cobalt (Co) Copper (Cu) Iron (Fe) Manganese (Mn) Zinc (Zn)	Metal ions required in trace amounts for the activity of various enzymes

Trace Elements Used by Some Organisms

Aluminum (Al) Arsenic (As) Boron (B) Bromine (Br) Chromium (Cr) Fluorine (F) Gallium (Ga) Iodine (I) Molybdenum (Mo) Selenium (Se) Silicon (Si) Tin (Sn) Vanadium (V)	Required by certain organisms in trace amounts for various purposes

gen bonds between neighboring water molecules and therefore does not contribute to elevating the water temperature. Hence it takes a relatively large input of heat to change the temperature of water. This phenomenon helps to stabilize cells against temperature fluctuations induced by transient changes in environmental temperature. (2) Hydrogen bonding between water molecules makes water more cohesive as a liquid, which gives water a *high surface tension.* This surface tension is responsible for the capillary action that pulls water upward through water-filled channels in woody plants. (3) Hydrogen bonding between water molecules is also responsible for the *high boiling point* of water relative to other substances of comparable molecular weight. If it were not for this property, water would boil away at the temperatures normally maintained in living cells.

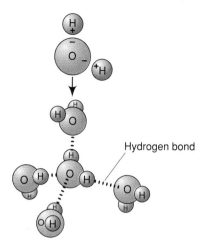

Figure 1-8 *Hydrogen Bonding in Water* (Top) *The electrons of the two covalent bonds in water are closer to the oxygen atom than to the hydrogens, imparting a partial negative charge to the oxygen and a partial positive charge to the hydrogens.* (Bottom) *This polarity promotes the formation of hydrogen bonds between adjacent water molecules in which the hydrogen atom of one water molecule interacts weakly with the oxygen atom of an adjacent water molecule.*

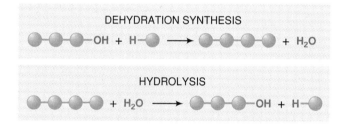

Figure 1-9 *Role of Water as a Product and Reactant in the Formation and Breakdown of Macromolecules* *The colored circles represent chemical building blocks used in the construction of macromolecules.* (Top) *Macromolecules are built up by dehydration reactions in which water is a product.* (Bottom) *Macromolecules are broken down by hydrolysis reactions in which water is a reactant.*

Water Acts as Both Chemical Reactant and Solvent

Several of life's most crucial processes depend upon the direct participation of water molecules. A prime example is the involvement of water in the formation and breakdown of the macromolecules that make up the bulk of a cell's structural and functional machinery. These macromolecules are synthesized by **dehydration** reactions, in which water is a product, and are broken down by **hydrolysis** reactions, in which water is a reactant (Figure 1-9).

In addition to its role as a reactant, water functions as a solvent for the thousands of molecules and ions that are required for cells to be able to carry out their essential activities. The polarity of the water molecule

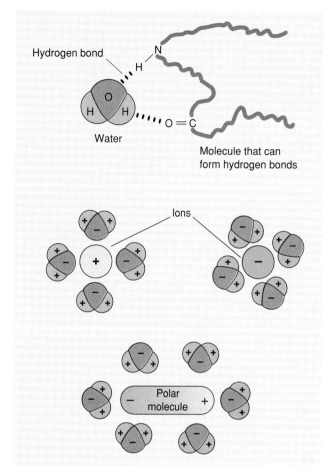

Figure 1-10 *Role of Water as a Solvent* (Top) *The ability of the water molecule to form hydrogen bonds helps it to solubilize other substances to which it can bond.* (Middle) *The polarity of the water molecule allows its positively and negatively charged regions to interact with ions.* (Bottom) *The polarity of the water molecule also facilitates its interaction with polar molecules.*

makes it an excellent solvent for ions and polar organic molecules, and its capacity to form hydrogen bonds makes water an excellent solvent for substances containing chemical groups that can form hydrogen bonds with the water molecule (Figure 1-10). This ability to solubilize inorganic ions as well as a broad spectrum of organic molecules makes water unique as a solvent.

It is often assumed that the water inside cells behaves like a normal liquid, but some evidence suggests that this view may be too simplistic. In one set of experiments designed to address this issue, Samuel Horowitz and Philip Paine introduced a small amount of a water-containing gel into the cytoplasm of amphibian egg cells. Radioactive sucrose molecules were then injected into the cells. After a brief waiting period, the cells were frozen and various regions of the cytoplasm and nucleus were removed by microdissection. Radioactivity measurements revealed that the concentrations of radioactive sucrose in the nucleus and in the water-containing gel were similar to what would be expected if sucrose had been dissolved in pure water. However, the concentration of radioactive sucrose in the cytoplasm was roughly 30 percent lower, suggesting that the solvent properties of cytoplasmic water differ from those of normal water or the water contained in the nucleus.

Measurements on the behavior of water inside other kinds of cells and organelles have tended to confirm the conclusion that cytoplasmic water often differs from normal water in its ability to dissolve solutes. Although the reasons for this unusual behavior are not thoroughly understood, the interaction of ions and water with the surfaces of large cellular macromolecules is believed to contribute to localized changes in the structure and behavior of water. Thus water inside cells may exhibit different properties depending on whether it is bound to macromolecular surfaces, in the close vicinity of macromolecular surfaces, or further away from such surfaces. The existence of these differing states of water may exert a significant influence on reactions taking place in different regions of the cell.

Most Biological Molecules Are Constructed from Sugars, Fatty Acids, Nucleotides, and Amino Acids

Aside from water, the molecules from which cells are constructed all share one feature in common: the presence of a carbon skeleton. Up until the early nineteenth century it was thought that these carbon-containing or **organic molecules** were too complex to be understood and that living systems perform some type of transcendental chemistry beyond the reach of laboratory experimentation. This mistaken impression was finally corrected in 1828, when the German chemist Friedrich Wöhler achieved the first synthesis of an organic compound (urea) outside a living organism.

In the years since this pioneering discovery, thousands of organic compounds have been synthesized in the laboratory, and the study of such compounds has greatly enhanced our understanding of how cells function. Contrary to the expectations of the early nineteenth-century scientists, the chemistry of living organisms has not turned out to be staggeringly complex. In fact, in spite of the enormous complexity of living organisms, most of the molecules that make up living cells have been found to be constructed from only four different kinds of chemical building blocks: **sugars, fatty acids, nucleotides,** and **amino acids.** These four building blocks are utilized for the same basic purposes by all cells, prokaryotic as well as eukaryotic.

One of the primary uses of these substances is in the construction of larger molecules. The complexity of the living cell requires that many of its constituent molecules be of immense size; therefore cells manufacture large molecules called **macromolecules** whose molecular weights typically fall in the range of several thousand to a million daltons or more. These macromolecules are all **polymers;** that is, they are constructed by linking together smaller chemical units called **monomers.** In the following sections, we will see how cells create macromolecules of enormous complexity and diversity using a relatively small number of different monomers.

Sugars Are the Building Blocks of Polysaccharides

Sugars belong to a class of organic molecules called **carbohydrates,** whose members are characterized by a 1:2:1 ratio of carbon, hydrogen, and oxygen. Simple sugars or **monosaccharides** are the smallest carbohydrates. Monosaccharides may contain as few as three carbons (*trioses*), but five- and six-carbon sugars (*pentoses* and *hexoses*) are more common in living organisms. The most widely occurring six-carbon sugar in nature is **glucose,** a key source of energy and a building block for larger carbohydrates. The glucose molecule occurs in the form of a ring as well as a linear chain, a pattern that is common for most sugars containing five or more carbons. The ring form of glucose occurs in two configurations, called α-glucose and β-glucose, which differ in the orientation of the hydroxyl group attached to the first carbon (Figure 1-11). In aqueous solution the linear and two ring forms of glucose are all in equilibrium with one another, but the ring forms predominate.

In addition to glucose, several other hexoses play important roles in the construction of biological macromolecules (Figure 1-12). Although these sugars all contain six carbon atoms, the positions of the hydroxyl (−OH) or aldehyde (HC=O) groups are different. Five-carbon sugars are also important building blocks of macromolecules; as we will see later in the chapter, the pentoses *ribose* and *deoxyribose* are utilized in the construction of nucleic acids, the polymers in which genetic information is stored.

When monosaccharides are joined together to form long chains of sugar molecules, the resulting products are called **polysaccharides.** The most commonly encountered polysaccharides in living organisms are cellulose, starch, and glycogen (Figure 1-13). During the synthesis of these polysaccharides, a molecule of water is released as each monosaccharide is added to the growing sugar chain. **Starch** and **glycogen** are formed in plants and animals, respectively, by the linking together of α-glucose molecules. The bond joining adjacent sugars is an *α(1 → 4) glycosidic linkage* between the number 1 carbon atom of one α-glucose molecule and the number 4 carbon of the adjoining glucose. Branching of the chain is generated by occasional *α(1 → 6) glycosidic linkages* between a number 1 carbon atom and a number 6 carbon atom. The main struc-

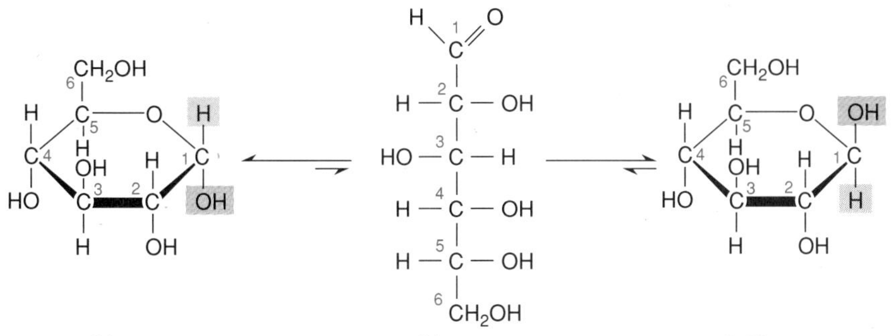

Figure 1-11 *The Structure of Glucose* *The chain and ring forms of glucose are in equilibrium with each other, but the ring forms predominate. The numbers used in identifying the six different carbon atoms are in color. Note that α-glucose differs from β-glucose in the orientation of the hydroxyl group attached to the first carbon atom.*

tural difference between starch and glycogen is that glycogen tends to be more extensively branched. Both of these polysaccharides function in the storage of glucose, releasing it for use by the organism when conditions demand. Glycogen occurs mainly in the liver and muscle tissue of animals, while starch is found in the chloroplasts and starch-storing organelles *(amyloplasts)* of plants.

In contrast to starch and glycogen, the polysaccharide **cellulose** is constructed from β-glucose rather than α-glucose and so is held together by *β(1 → 4) glycosidic linkages.* This difference in the orientation of a single hydroxyl group in glucose generates a polysaccharide whose properties are markedly different from those of starch and glycogen. Cellulose is a major constituent of the cell wall of plants, where its great mechanical strength contributes to many of the physical properties we associate with wood. Although cellulose, like starch, is constructed entirely from glucose units, humans cannot employ cellulose as a food source because we lack the enzyme required to break the β(1 → 4) bond. Since humans do produce an enzyme that breaks the α(1 → 4) bond, we can obtain nutrition from vegetables such as corn and potatoes, which contain large amounts of starch, but not from plant products like wood, paper, or grass, which consist mainly of cellulose. Cows and other grazing animals that are able to use grass as a food source can do so because their gastrointestinal tract contains bacteria that digest the cellulose for them, releasing free glucose.

While starch, glycogen, and cellulose are each constructed from a single type of sugar molecule, other polysaccharides are composed of more than one kind of sugar. Amino- and acetyl-substituted sugar derivatives are commonly used in such polysaccharides, as are sugars containing sulfate or phosphate. The most prominent polysaccharides in this category are the *glycosaminoglycans,* which are formed by alternating two different sugar derivatives. Such "mixed" polysaccharides serve a variety of important functions, especially as constituents of extracellular materials (page 224).

Fatty Acids Are Employed in the Construction of Many Types of Lipids

We have just seen that polysaccharides are macromolecules constructed by linking sugar molecules together to form long polymers. In contrast to this chemically precise definition of a polysaccharide, lipids cannot be defined in terms of the chemical building blocks used in their construction, nor are they macromolecules created by the sequential addition of building blocks to form polymers. Instead, **lipids** are defined as those biological molecules that are readily soluble in nonpolar organic solvents but only slightly soluble in water. Such water-insoluble substances are said to be **hydrophobic** ("water-fearing"). Because they are defined on the basis of solubility characteristics rather than chemical structure, it is not surprising to find that lipids are a rather heterogeneous group of molecules. When lipids are extracted from cells by exposing them to an organic solvent such as acetone, chloroform, or ether, the extract is found to contain at least six major groups of substances: *fatty acids, triacylglycerols, phospholipids, glycolipids, steroids,* and *terpenes* (Figure 1-14).

1. Fatty acids. The simplest group of lipids are the fatty acids, which serve as an energy source for cells and also function as building blocks for more complex lipids. Fatty acids are long carbon chains containing an acidic carboxyl ($-COO^-$) group at one end. Most biologically important fatty acids have an even number of carbon atoms, ranging up to a maximum of around twenty-four. **Unsaturated** fatty acids contain double bonds, whereas **saturated** fatty acids do not.

2. Triacylglycerols (also called **triglycerides** or **fats**). The triacylglycerols are a group of related lipids

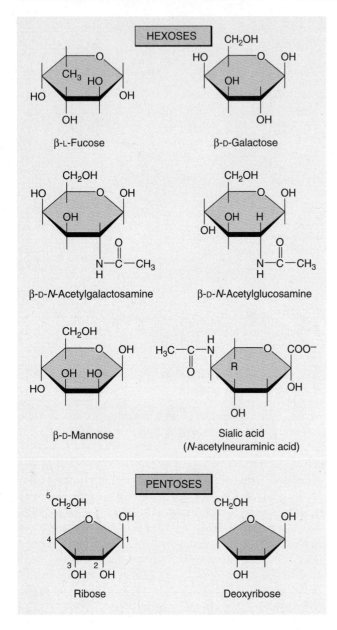

Figure 1-12 *Some Biologically Important Hexoses and Pentoses* *In drawing the structures of these monosaccharides, many of the hydrogen atoms have been left out for clarity.*

that are constructed by joining various kinds of fatty acids to *glycerol,* which is a reduced derivative of a three-carbon sugar. During the formation of triacylglycerols, a molecule of water is produced for each of the three fatty acids joined to glycerol. The fatty acids attached to the three carbons of glycerol may be of the same or different lengths, and may be saturated or unsaturated. Triacylglycerols function predominantly as a means of storing fatty acids, which represent an extremely concentrated energy source. In animals, triacylglycerols are stored in fat cells as a reserve for future energy needs. In such cells, fat droplets containing triacylglycerols occupy almost the entire volume of the cytoplasm.

3. Phospholipids. In contrast to the extremely low water solubility of triacylglycerols, phospholipids contain water-soluble phosphate groups incorporated into the lipid backbone. This produces an **amphipathic** molecule—that is, a molecule with both **hydrophilic** (water-soluble) and **hydrophobic** (water-insoluble) regions. Phospholipids are constructed by linking a phosphate group either to *glycerol* or to a long-chain alcohol derivative called *sphingosine.* Fatty acids are also involved in the construction of both types of phospholipid. Since phospholipids are major contributors to the structure of cellular membranes, they will be described in more detail in the chapter devoted to membranes (Chapter 5).

4. Glycolipids. Glycolipids are derivatives of sphingosine or glycerol that contain a carbohydrate group instead of a phosphate group. Carbohydrate groups, like phosphate groups, are water soluble. Therefore the presence of the carbohydrate group in glycolipids gives them an amphipathic nature. Glycolipids are important constituents of certain types of membranes, especially those found in plant cells and cells of the nervous system.

5. Steroids. Steroids are derivatives of a four-membered ring compound called *phenanthrene* (see Figure 1-14), which makes them structurally distinct from the other lipids discussed thus far. Although some steroids may be linked to fatty acids, this is not typical. Steroids such as cortisone, testosterone, estrogen, and progesterone function as hormones. Other important steroids include vitamin D, which regulates the absorption of calcium from the intestine, and cholesterol, a constituent of animal cell membranes and the main precursor for the synthesis of other steroids.

6. Terpenes. Lipids constructed from the five-carbon compound *isoprene* are called terpenes. Isoprene and its derivatives are joined together in various combinations to produce substances such as vitamin A, coenzyme Q, and carotenoids. Despite their underlying similarities in chemical structure, such molecules differ significantly in the functions they carry out.

As is evident from the foregoing discussion, lipids exhibit a variety of chemical structures, each suited for a particular purpose. Fatty acids are a commonly employed building block in the formation of many, but not all, of these lipids. The principal feature unifying this diverse group of compounds is not their chemical structures but their hydrophobic nature, a characteristic that imposes a certain set of physical properties. Probably the most important property in terms of the origin and evolution of cells is that hydrophobic molecules are ideally suited for creating barriers between aqueous environments. Hence lipids have been involved in the

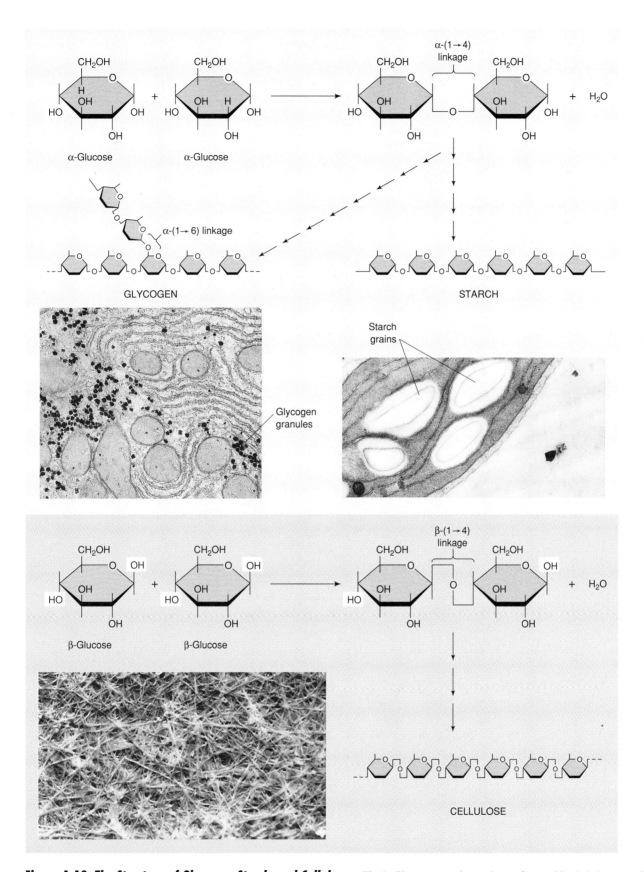

Figure 1-13 *The Structure of Glycogen, Starch, and Cellulose*   (Top) *Glycogen and starch are formed by joining together α-glucose molecules. The micrographs show glycogen granules in a salamander liver, and starch grains in a soybean leaf chloroplast. (Bottom) Cellulose is created by joining together β-glucose molecules. The micrograph shows cellulose fibrils in the wall of a higher plant. The metabolic pathways used by cells to synthesize these three polysaccharides require an input of energy and therefore contain more steps than are included in this diagram. Micrographs courtesy of D. W. Fawcett and Photo Researchers Inc. (top left), B. J. Ford (top right), and K. Mühlenthaler (bottom).*

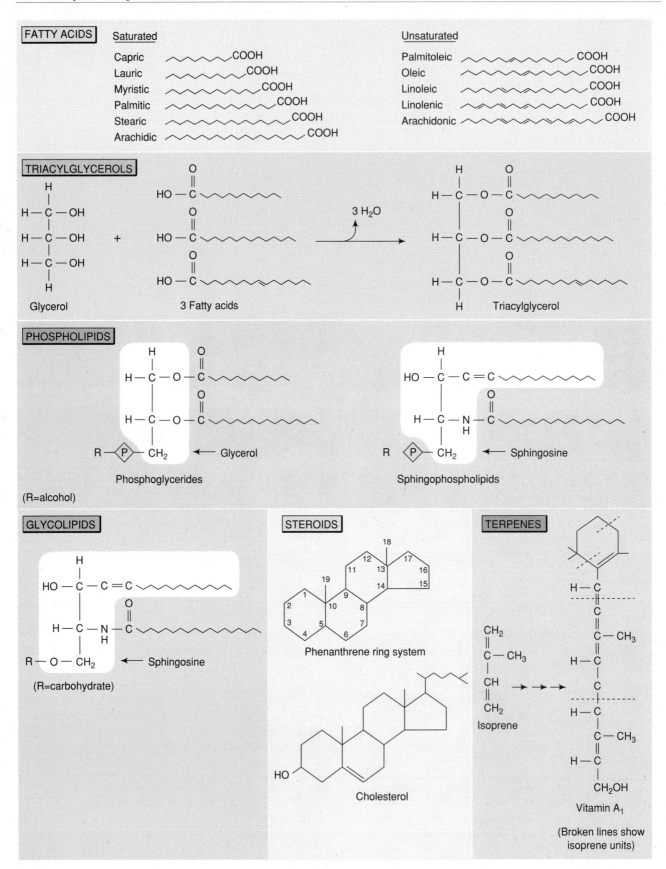

Figure 1-14 *The Six Major Classes of Lipids   In each category, one or more representative examples is illustrated. The zigzag lines represent linear chains of carbon atoms. The reaction by which triacylglycerols are synthesized is included to show how fatty acids are used as building blocks to create more complex lipids. The metabolic pathway used by cells to synthesize triacylglycerols requires an input of energy and therefore contains more steps than are included in this diagram.*

creation of the membrane systems required for the evolution of cells as we know them.

Nucleotides Are the Building Blocks of Nucleic Acids

Nucleic acids serve as vehicles for the storage and transmission of genetic information. This critically important information is encoded within the linear sequence of **nucleotides,** which are the monomer building blocks that are joined together to form a nucleic acid molecule. Each nucleotide is composed of three components: a *five-carbon sugar,* a *nitrogenous base,* and a *phosphate group.*

1. Five-carbon sugars (pentoses). Two different five-carbon sugars are employed in the construction of nucleic acids, **deoxyribose** and **ribose.** The only difference between these two sugars is the absence of oxygen on the 2'-carbon atom of the deoxyribose molecule (see Figure 1-12). Cells contain two major classes of nucleic acids referred to as **deoxyribonucleic acid (DNA)** and **ribonucleic acid (RNA)** that differ in the sugar they contain. DNA contains deoxyribose, whereas RNA utilizes ribose instead.

2. Nitrogenous bases. The nitrogenous bases employed in the construction of nucleic acids are members of two classes of nitrogen-containing ring compounds known as **purines** and **pyrimidines** (Figure 1-15). **Adenine** and **guanine** are derivatives of the two-membered purine ring, whereas **cytosine, thymine,** and **uracil** are derived from the single-ringed pyrimidine. The capital letters **A, G, C, T,** and **U** are commonly used to refer to adenine, guanine, cytosine, thymine, and uracil, respectively. DNA and RNA each utilize four of the five bases. A, G, and C are present in both DNA and RNA, but the fourth base is T in DNA and U in RNA. Chemically modified forms of these bases, as well as other more unusual bases, also occur occasionally in DNA and RNA. The nitrogenous bases are the key components of nucleic acid structure because the sequence of these bases is employed to encode genetic information.

3. Phosphate group. A phosphate group is the third component of a nucleotide. It is linked to the five-carbon sugar and plays a key role in joining nucleotides together to form nucleic acids. The phosphate group is a strong acid, which explains why DNA and RNA are referred to as nucleic "acids."

The way in which a five-carbon sugar, a nitrogenous base, and a phosphate group are linked together to form a nucleotide is illustrated at the top of Figure 1-16. To generate DNA or RNA, nucleotides are joined together by **phosphodiester bonds** between the phosphate group on the 5' carbon of one sugar and the

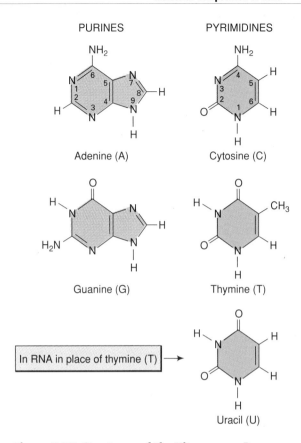

PURINES   PYRIMIDINES

Adenine (A)   Cytosine (C)

Guanine (G)   Thymine (T)

In RNA in place of thymine (T) →

Uracil (U)

Figure 1-15 *Structures of the Nitrogenous Bases Employed in the Construction of Nucleic Acids* *The bases A, C, G, and T are utilized in the construction of DNA, whereas A, C, G, and U are used for RNA.*

hydroxyl group on the 3' carbon of an adjacent sugar (see Figure 1-16, *bottom*). The *polynucleotide* chains formed by this process have an inherent directionality; one end of the chain is called the *5' end* because its 5' carbon contains a free phosphate group; similarly, the other end of the chain is called the *3' end* because its 3' carbon contains a free hydroxyl group. RNA chains are typically hundreds or thousands of bases long, and DNA chains can even be millions of bases long. The final three-dimensional structure of DNA and RNA molecules involves interactions between nitrogenous bases both within and between polynucleotide chains. Because such interactions are intimately associated with the process of information flow, we will delay further consideration of this topic until Chapter 3, where the subject of genetic information is introduced.

Amino Acids Are the Building Blocks of Proteins

Proteins are the most abundant and functionally diverse macromolecules in the cell, accounting for two-thirds or more of the total dry weight of most cells and performing a variety of essential functions (Table 1-3). Each type of protein is created by joining together a unique combination of monomer building blocks referred to as

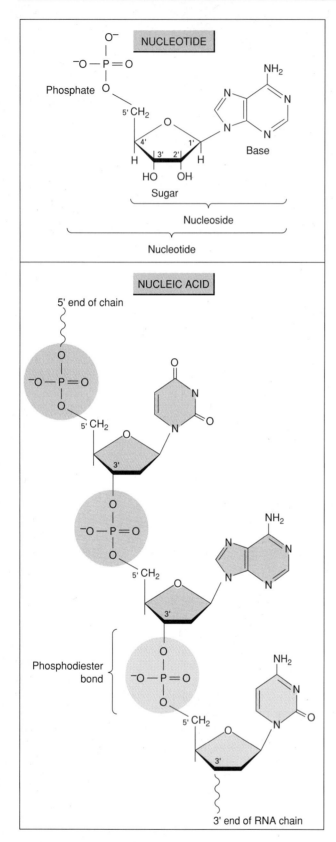

Figure 1-16 *The Structure of Nucleotides and Nucleic Acids* (Top) *Structure of a typical nucleotide, adenosine monophosphate.* (Bottom) *In nucleic acids, phosphodiester bonds join adjacent nucleotides together. Only a small segment of a nucleic acid (in this case RNA) is illustrated.*

Table 1-3   Some Functions of Protein Molecules

Function	Examples
Structural materials	Collagen, keratin
Hormones	Insulin, growth hormone
Motility	Actin, myosin
Solute transport	Na$^+$-K$^+$ pump
Nutrient storage	Casein, ferritin
Signal transduction	Acetylcholine receptor
Gene regulation	*Lac* repressor
Osmotic regulation	Serum albumin
Electron transfer	Cytochromes
Immune defense	Antibodies
Toxins	Diphtheria and cholera toxins
Enzymes (catalysis)	Oxidoreductases
	Transferases
	Hydrolases Defined in
	Lyases Chapter 2
	Isomerases
	Ligases

amino acids. The general structure of an amino acid is illustrated in Figure 1-17, which shows that each amino acid contains a centrally located carbon atom called the *α-carbon* atom to which four groups are attached: a basic amino group (—NH$_3^+$), an acidic carboxyl group (—COO$^-$), a hydrogen atom, and a group of varying chemical structure called a *side chain* (designated by the letter R). The four groups attached to the α-carbon can be arranged in two different mirror-image spatial orientations called the L and D configurations. Only L-amino acids are employed in the synthesis of proteins, but D-amino acids are utilized by cells for a few special purposes, such as construction of the bacterial cell wall (page 249).

Twenty different amino acids are employed in the construction of protein molecules. The differences between these amino acids are localized in their side

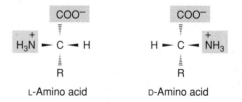

Figure 1-17 *The Structure of an Amino Acid* *The wedge-shaped bonds project above the plane of the page, and the dashed bonds project behind the page. Only L-amino acids are employed in the construction of proteins. Under the pH conditions that prevail within cells, the amino group binds a proton and is positively charged, whereas the carboxyl group loses a proton and is negatively charged. The structures of the 20 different side chains designated by the letter R are provided in Figure 1-18.*

chains, whose differing properties allow amino acids to be divided into four general categories: (1) *nonpolar* (hydrophobic) amino acids; (2) *polar* (hydrophilic) amino acids containing hydroxyl, sulfhydryl, or amide groups; (3) *acidic* amino acids; and (4) *basic* amino acids (Figure 1-18). Amino acids can be linked together by a reaction in which the carboxyl group of one amino acid is joined to the amino group of another amino acid, splitting out a molecule of water and forming a **peptide bond.** This type of reaction is illustrated in Figure 1-19, which shows two amino acids joining together to form a *dipeptide.* As more amino acids are added to form longer chains, the products are called **polypeptides.** Most polypeptides retain a free amino group at one end (called the amino terminus or **N-terminus**) and a free carboxyl group at the other end (called the carboxyl terminus or **C-terminus**). **Protein** molecules consist of one or more polypeptide chains, each containing from a few dozen to hundreds or even thousands of amino acids. As we will see in Chapter 11, the process by which amino acids are joined together by cells to form polypeptide chains is considerably more complex than the simple reaction illustrated in Figure 1-19.

An enormous number of different proteins can be made using the 20 amino acids found in cells. To illustrate, let us consider how many different polypeptide chains 200 amino acids long could theoretically be constructed. Since 20 possible amino acids can be located at each of the 200 positions in the polypeptide chain, the total number of different polypeptides is 20 multiplied by itself 200 times, or 20^{200}. This number is truly gigantic compared to the roughly 10^4 different kinds of proteins that occur in a typical eukaryotic cell. Thus the potential diversity inherent in protein structure has barely been tapped by living organisms.

The enormous structural diversity that is possible among protein molecules helps to explain why proteins can carry out so many different functions. The structural complexity of protein molecules also makes it difficult to describe the structures of individual proteins and compare them with one another. To simplify the task, discussions of protein structure are usually subdivided into four levels of organization known as *primary, secondary, tertiary,* and *quaternary* structure.

Each Protein Has a Unique Amino Acid Sequence Known as Its Primary Structure

Primary structure refers to the linear sequence of amino acids in a protein molecule. The first protein to have its complete amino acid sequence determined was insulin, a hormone that contains 51 amino acids (Figure 1-20). In 1956 Frederick Sanger reported a series of pioneering experiments in which he determined the sequence of the insulin molecule by cleaving it into smaller fragments and analyzing the amino acid arrange-

ment within the individual fragments. Sanger's techniques paved the way for the sequencing of hundreds of other proteins and ultimately led to the design of machines that can determine amino acid sequences automatically. A more recent approach for determining protein sequences has emerged from the discovery that nucleotide base sequences in DNA code for the amino acid sequences of protein molecules. It is now often easier to determine a DNA base sequence than to purify and analyze the amino acid sequence of a protein molecule. Once a DNA base sequence has been determined, the amino acid sequence encoded by that DNA segment can be inferred.

What have we learned from examining the amino acid sequences of protein molecules? First, it has been discovered that each type of protein is characterized by its own unique sequence of amino acids. The proteins insulin and hemoglobin, for example, are very different molecules because their amino acid sequences are markedly different. However, smaller differences in amino acid sequence can be tolerated without altering a protein's function. The insulin molecules of cows, humans, sheep, dogs, and rats differ in amino acid sequence in several locations, yet these proteins all perform the same function and are all properly referred to as insulin molecules (see Figure 1-20). Although the sequence of the human protein cytochrome *c* differs in almost 50 percent of its amino acids from the cytochrome *c* of fungi, these two cytochromes are similar enough to carry out the same crucial function in cellular energy metabolism.

The preceding observations seem to imply that wide variations in amino acid sequence can be tolerated without adverse effects on protein function, but this is not always the case. It depends on where the substituted amino acids are located. For example, in individuals suffering from sickle-cell anemia, one of the polypeptide chains in the hemoglobin molecule differs from normal hemoglobin by only one amino acid out of 146, but this single alteration has dire effects on the properties of hemoglobin.

Although a protein's primary structure determines the type of function to be carried out, it should not be assumed that a linear chain of amino acids is all that is required to create a functional protein molecule. Before a protein can carry out its proper functions, the amino acid chain must become folded and organized into its proper three-dimensional shape. This process involves several distinct levels of organization known as secondary, tertiary, and quaternary structure.

The Secondary Structure of Protein Molecules Involves α Helices and β Sheets

During the early 1950s, theoretical considerations and X-ray diffraction studies led Linus Pauling to conclude

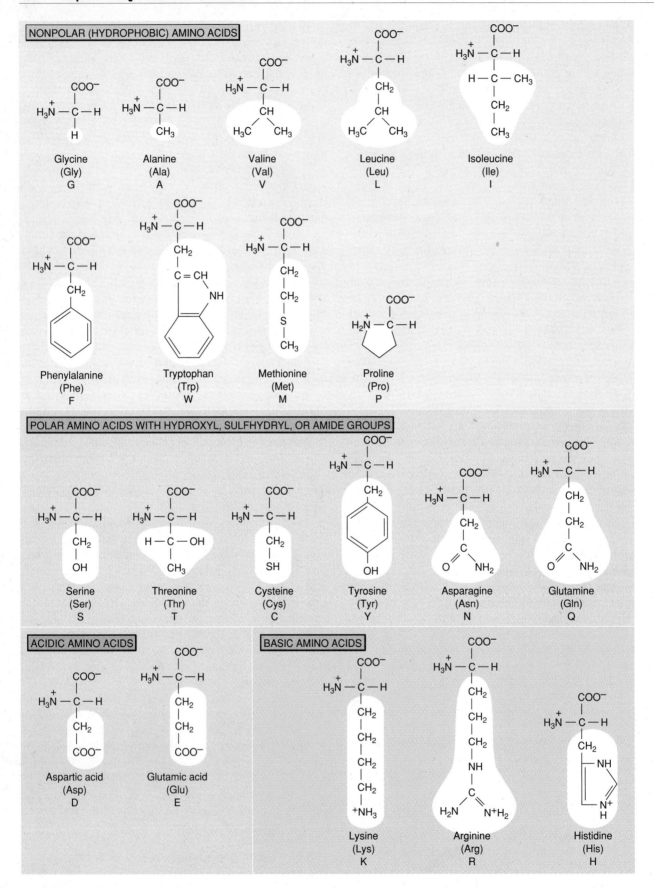

Figure 1-18 *Structures of the Twenty Amino Acids Employed in the Construction of Proteins* *The three-letter and single-letter abbreviations beneath each name are commonly employed when listing amino acid sequences. The amino acids are grouped into four categories based on whether their side chains are nonpolar, polar, acidic, or basic.*

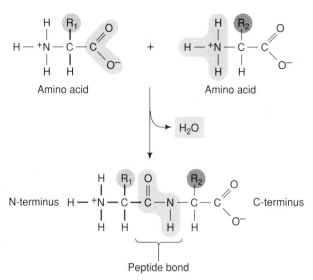

Figure 1-19 *The Dehydration Reaction for Creating a Peptide Bond between Two Amino Acids   Although the reaction for creating a peptide bond is illustrated here as a single step, cells utilize a complex, multistep pathway that provides both the energy needed to drive peptide-bond formation and the information that specifies the sequence of amino acids.*

that polypeptide chains are folded into two types of **secondary structure,** both of which result from hydrogen bonding between the —C=O group of one peptide bond and the —NH group of another. In one type of secondary structure, called an **alpha helix (α helix),** these hydrogen bonds cause a region of the polypeptide chain to become coiled into a spiral (Figure 1-21, *left*). An α helix is built from a continuous stretch of the polypeptide chain in which the —C=O group of each amino acid is hydrogen bonded to the —NH group of the fourth amino acid along the helix.

In the other type of secondary structure, called a

beta sheet (β sheet), two different regions of the same polypeptide chain (or two separate polypeptide chains) are held together by hydrogen bonds that create an extended, sheetlike structure (see Figure 1-21, *right*). The protein regions that form β sheets can interact with each other in two different ways. If the two interacting regions run in the same N-terminus → C-terminus direction, the structure is called a *parallel β sheet;* when the two strands run in opposite N-terminus → C-terminus directions, the structure is called an *antiparallel β sheet.* Because the carbon atoms that comprise the backbone of the polypeptide chain are successively located a little above and a little below the plane of the β sheet, such structures are sometimes called *pleated β sheets.*

In order to help visualize the location of α helices and β sheets within a protein molecule, a set of conventions has recently been established for drawing the structure of protein molecules. These conventions distinguish between three regions of a polypeptide chain: (1) α helices, which are drawn either as helices or as cylinders, (2) β sheets, which are drawn as flat arrows with the arrowhead pointing in the direction of the C-terminus, and (3) looped regions connecting the α helices and β sheets, which are drawn as narrow ribbons (Figure 1-22). Examination of such diagrams has led to the realization that certain combinations of α helix and β sheet tend to recur frequently in different kinds of proteins. These units of secondary structure, called *supersecondary structures* or **motifs,** consist of small segments of α helix and/or β sheet connected to one another by looped regions of varying length. Among the commonly encountered motifs are the *β-α-β* motif, the *hairpin-loop* motif, and the *helix-turn-helix* motif. When a particular type of motif recurs in several different proteins, it often performs the same function. For example, we will learn in Chapter 10 that the helix-

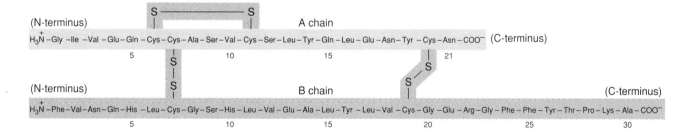

	A chain				B chain		
Position:	4	8	9	10	3	29	30
Bovine (shown above)	Glu	Ala	Ser	Val	Asn	Lys	Ala
Human	—	Thr	—	Ile	—	—	Thr
Sheep	—	—	Gly	—	—	—	—
Dog	—	Thr	—	Ile	—	—	—
Rat	Asp	Thr	—	Ile	Lys	—	Ser

Figure 1-20 *Amino Acid Sequence of Insulin   Differences in the primary structure of insulin from several different organisms are summarized in the lower box. Amino acid abbreviations are defined in Figure 1-18.*

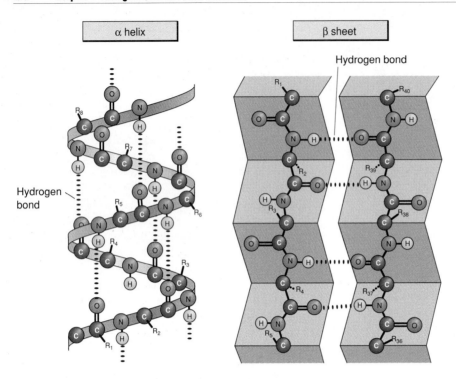

Figure 1-21 *The α-Helix and β-Sheet Forms of Secondary Structure* (Left) *An α helix is generated by hydrogen bonding between the —C=O and —NH groups of the peptide bonds. The amino acid side chains ($R_1 \rightarrow R_8$) project outward from the α helix.* (Right) *A β sheet is also generated by hydrogen bonding between the —C=O and —NH groups of the peptide bonds. However, the hydrogen bonds are formed between amino acids situated in different regions of the polypeptide chain. In this example, the two regions of polypeptide chain are running in opposite N-terminus → C-terminus directions, forming an antiparallel β sheet. The amino acid side chains ($R_1 \rightarrow R_5$ and $R_{36} \rightarrow R_{40}$) project either above or below the plane of the pleated β sheet.*

turn-helix motif is one of several secondary structure motifs that help to bind protein molecules to specific base sequences in DNA.

Proteins vary widely in the amount of α helix and β sheet they contain. Although regions of both kinds of secondary structure often occur in the same molecule, such is not always the case. One extreme situation is illustrated by silk *fibroin,* the protein that makes up the fibers spun by the silkworm. Fibroin molecules consist almost entirely of long regions of antiparallel β sheet, a structural organization that creates strong and virtually nonstretchable silk fibers. At the other extreme is the protein *α-keratin,* the predominant constituent of hair and wool. Molecules of α-keratin are constructed from long extended polypeptide chains that are almost entirely in the α-helical configuration, a structural organization that helps to make hair and wool more stretchable than silk. Fibroin and α-keratin are examples of *fibrous proteins,* which are defined as proteins that form extended fibers in which a particular type of secondary structure tends to predominate.

Tertiary Structure Creates Globular Proteins That Are Folded into Domains

Although fibrous proteins perform a number of important tasks, especially as structural materials, they consti-

tute only a tiny fraction of all proteins. Most proteins are not organized into extended fibers like the fibrous proteins, but instead exhibit a **tertiary structure** that is created by folding the polypeptide chain into a compact, globular shape. In such *globular proteins,* the polypeptide chain is organized into localized regions of compact folding known as **domains.** Small proteins are typically folded into a single domain, whereas larger proteins may contain anywhere from one to several dozen domains. In proteins that consist of more than a single domain, each domain may carry out a different function. For example, an enzyme may contain one domain that catalyzes a particular chemical reaction and another domain that binds some factor that is required for the reaction to occur (Figure 1-23). Moreover, a domain that carries out a particular function in one protein may also occur in other proteins, where it performs a similar function. At the structural level, three classes of domains can be distinguished on the basis of whether they are constructed from α helices, β sheets, or a mixture of both (Figure 1-24).

What causes a polypeptide chain to become folded into its proper tertiary structure? Unlike secondary structure, which depends only on hydrogen bonds between —C=O and —NH groups, a variety of bonds and interactions between amino acid side chains have been

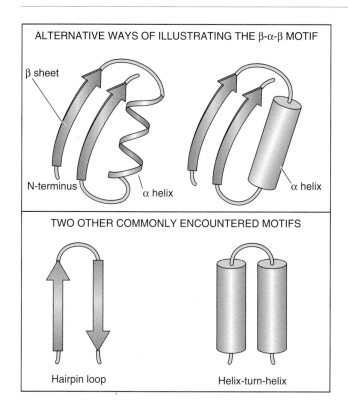

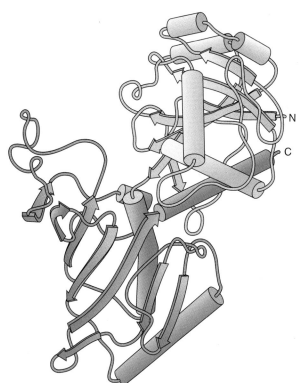

Figure 1-22 *Secondary Structure Motifs in Protein Molecules* (Top) *Two different ways of illustrating the β-α-β motif; the α-helical region is represented as a helix in one diagram and as a cylinder in the other.* (Bottom) *Two other commonly encountered protein motifs, the hairpin-loop motif and the helix-turn-helix motif. In each of the four diagrams, only a small portion of an entire protein molecule is shown.*

Figure 1-23 *An Example of a Protein Containing Two Functional Domains* *This model of the enzyme glyceralde-hyde-phosphate dehydrogenase shows a single polypeptide chain folded into two domains: one domain binds to the substance being metabolized, whereas the other domain binds to a chemical factor required for the reaction to occur. The two domains are indicated by different colored shading. The letters N and C refer to the N-terminus and C-terminus respectively.*

implicated in the formation of tertiary structure. Included in this category are disulfide bonds, electrostatic interactions, hydrogen bonds, van der Waals interactions, and the hydrophobic effect.

1. Disulfide bonds. The amino acid cysteine contains a sulfhydryl (—SH) group that can form a *disulfide bond* (—S-S—) with another cysteine. Unlike the other bonds involved in tertiary structure, the disulfide bond is a **covalent bond,** which means that it is a strong chemical bond based on the sharing of a pair of electrons between the two atoms being bonded. Disulfide bonds between cysteine groups located in different regions of the polypeptide chain produce crosslinks that stabilize the protein's three-dimensional folding. This type of covalent bond is the only one that is routinely involved in protein folding; the remaining bonds and interactions all involve noncovalent forces, which are considerably weaker than covalent bonds.

2. Electrostatic (ionic) bonds. Chemical groups carrying a positive charge are electrically attracted to groups containing a negative charge. This type of noncovalent bond, called an *electrostatic* or *ionic bond,* oc-

curs between basic amino acids (lysine, arginine, and histidine), which carry a positive charge, and acidic amino acids (glutamic and aspartic acids), which carry a negative charge.

3. Hydrogen bonds. Hydrogen bonds are noncovalent bonds that are formed between a hydrogen atom in a donor group such as —NH or —OH, and an oxygen or nitrogen atom in an acceptor group. Examples of amino acid side chains that can form hydrogen bonds with each other are tyrosine-histidine and serine-aspartic acid.

4. Van der Waals interactions. When two atoms are in close proximity, they become nonspecifically attracted to each other by a weak force known as the *van der Waals force.* This weak attraction is optimal when atoms are about 0.3–0.4 nm apart. If the two atoms get any closer, a mutual repulsion occurs.

5. Hydrophobic effect. Many amino acid side chains are hydrophobic; that is, they tend to repel water molecules. This property creates an inherently unstable situation in an unfolded protein molecule, where

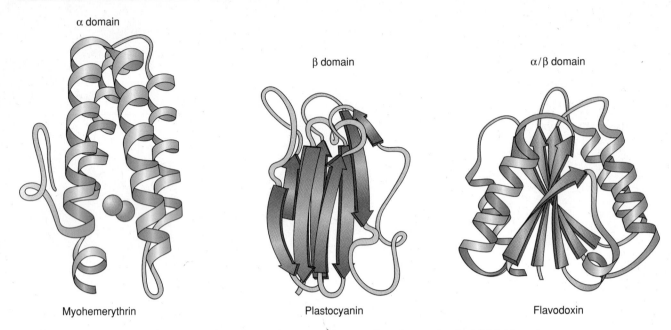

Figure 1-24 *The Three Main Types of Protein Domains, and Examples of Proteins in Which They Occur*

hydrophobic side chains are exposed to the aqueous environment. By folding the polypeptide chain into a compact globular structure, the hydrophobic side chains become buried in the interior of the protein molecule where water is largely absent. This mode of stabilizing protein folding is called the *hydrophobic effect.*

Quaternary Structure Creates Proteins That Consist of Multiple Subunits

Thus far we have seen how a single polypeptide chain can be twisted and folded to generate a protein with secondary and tertiary structure. Although some proteins are comprised of only a single polypeptide chain, a fairly large number consist of multiple polypeptide chains organized together into a **quaternary structure.** Such proteins are called *multisubunit proteins,* and their individual polypeptide chains are referred to as *subunits.* The forces that hold the subunits together are of the same general types as we encountered in tertiary structure: electrostatic bonds, hydrogen bonds, van der Waals forces, hydrophobic effects, and occasional disulfide bonds. In the case of quaternary structure, however, these interactions occur between amino acid side chains located in *different* polypeptide subunits and hence function to hold the subunits together. Proteins exhibiting quaternary structure contain anywhere from two subunits to dozens or even hundreds of subunits. The subunits may be identical polypeptide chains or a variety of different polypeptides.

Proper Conformation Is Essential for Protein Function

We have now seen how the overall shape of a protein molecule is generated by the bonds and interactions that create the protein's primary, secondary, tertiary, and in some cases, quaternary structure (Figure 1-25). The final three-dimensional shape generated by these forces is referred to as protein **conformation.** The importance of a protein's proper conformation can be demonstrated experimentally by treating proteins with agents that disrupt the bonds or interactions involved in maintaining secondary, tertiary, and/or quaternary structure. For example, a solution of protein molecules might be exposed to (1) urea, which disrupts hydrogen bonds, (2) salt or an altered pH, which disrupts electrostatic bonds, (3) mercaptoethanol, which disrupts disulfide bonds, (4) organic solvents, which disrupt hydrophobic interactions, or (5) moderate increases in temperature, which disrupt all of these weak bonds and interactions. Such treatments usually cause protein molecules to lose both their normal conformation and the ability to carry out their normal functions, even though the protein's primary structure has not been altered. This loss in function caused by disrupting a protein's conformation is referred to as **denaturation.**

The question of how proteins normally acquire their proper conformation was first explored in the late 1950s by Christian Anfinsen and his colleagues. This pioneering work focused on the enzyme *ribonuclease,* a protein that degrades RNA. Ribonuclease is an attractive model for study because its conformation is maintained by four disulfide bridges that can be experimentally disrupted, and it exhibits a biological activity that is easily measured. Anfinsen discovered that ribonuclease can be denatured by treating it with a combination of urea and mercaptoethanol, which disrupts hydrogen bonds and disulfide bonds, respectively. Such treatment causes ribonuclease to lose both its normal conformation and its ability to function. However, both of these proper-

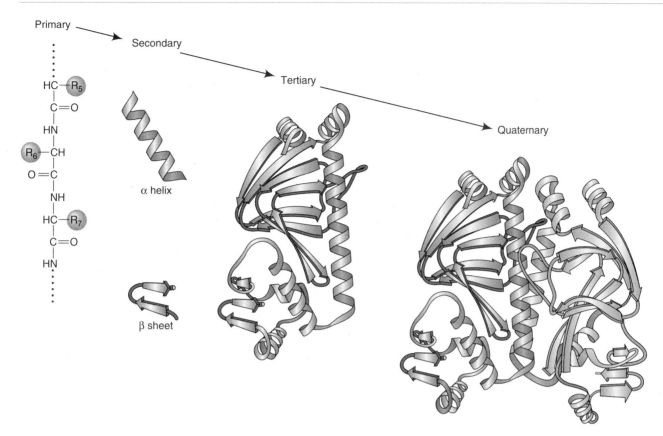

Figure 1-25 *The Four Levels of Protein Structure* *Primary structure refers to the linear sequence of amino acids in a polypeptide chain. Secondary structure involves the folding of the primary structure into α helices and β sheets. Tertiary structure refers to the further folding of the polypeptide chain into one or more compact structures known as domains. Quaternary structure involves the formation of a multisubunit protein from more than one polypeptide chain.*

ties reappear spontaneously when the urea and mercaptoethanol are removed (Figure 1-26). In other words, this denatured protein spontaneously regains its proper conformation and ability to function when the disrupting conditions are eliminated.

The discovery that some denatured proteins can spontaneously refold into their proper conformation indicates that all the information required for proper protein folding is inherent in the amino acid sequence of the polypeptide chain. In theory, this finding means that it should be possible to predict the final three-dimensional conformation of a polypeptide chain once its amino acid sequence is known. Although such a goal is yet to be accomplished, some progress has been made in predicting secondary structures. The most common approach to this problem is to examine the amino acid sequences of known stretches of α helix or β sheet to determine which kinds of amino acids are most commonly involved in each type of structure. Such studies have revealed that certain amino acids have a strong tendency to form α helices, others have a tendency to form β sheets, and a few (especially proline) tend to promote the formation of loops or turns that disrupt regions of secondary structure. Attempts to predict tertiary structure have met with much less success, most likely because it involves so many different kinds of bonds

between side chains that are far removed from one another in the primary sequence. Moreover, a group of proteins called *chaperones* have recently been discovered that appear to promote the proper folding of other kinds of proteins within living cells (page 503). The involvement of these additional components during the folding process may further complicate the picture of how proteins acquire their proper tertiary structure.

WHAT ARE VIRUSES?

An introductory chapter on cells and their molecules would not be complete without considering the nature and origin of viruses. These tiny parasites consist of molecules that cannot, by themselves, carry out most cell functions. Yet viruses are able to infect cells and reproduce using the cell's molecular machinery. Although the ability of viruses to cause disease is commonly recognized, we will see in later chapters that viruses are also important experimental tools for studying cell biochemistry and function, and they play an especially powerful role as experimental agents for introducing new genetic information into cells (page 101).

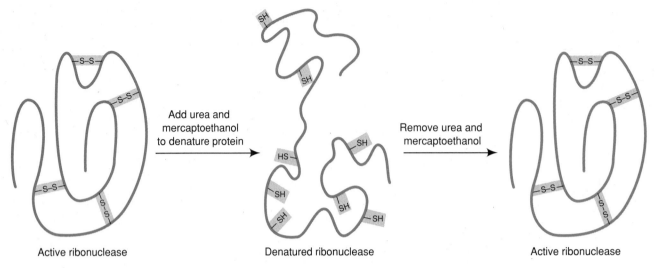

Active ribonuclease   Denatured ribonuclease   Active ribonuclease

Figure 1-26 *Evidence That a Protein Can Spontaneously Regain Its Proper Conformation after Being Denatured*
Ribonuclease that has been denatured by exposure to urea and mercaptoethanol refolds into its proper conformation when the urea and mercaptoethanol are removed. In intact cells, specialized proteins called chaperones often facilitate the process of protein folding.

Viruses Are Infectious Agents That Are Smaller Than Cells

The discovery of viruses in the late nineteenth century was originally based on their ability to cause disease. It was already known that bacteria cause many illnesses in animals and plants, but the agents responsible for certain diseases, such as smallpox and tobacco mosaic disease, would not grow or reproduce in the laboratory outside of living cells, nor could they be seen with the light microscope as could bacteria. In 1892 the Russian biologist Dmitri Iwanowsky showed that the infectious agent responsible for tobacco mosaic disease will pass through a porcelain filter containing pores that are too small to permit the passage of bacterial cells. The term **virus** was introduced to refer to such infectious agents, and the term **bacteriophage** was introduced later to refer to viruses that attack bacterial cells.

As experimentation progressed into the early twentieth century, it became apparent that viruses resemble complex chemicals more than living cells. In 1935 Wendell Stanley showed that tobacco mosaic virus can be crystallized, and that such crystals retain the ability to cause disease and reproduce when injected into tobacco plants. The ability to form crystals is typically a property of molecules, not cells. The development of electron microscopy in the early 1950s finally confirmed that viruses are quite different from living cells in appearance.

Viruses Consist of DNA or RNA Surrounded by a Protein Coat

In electron micrographs, viruses typically appear as tiny particles ranging from about 10 to 100 nm in size,

which is much smaller than cells (see Figure 1-5). The genetic information that determines a virus's properties is stored in a molecule of either DNA or RNA, which forms the core of the virus particle. Surrounding the nucleic acid core is a protein coat. The nucleic acid and protein are usually packaged together to form a particle with a symmetrical shape. Most virus particles are either *polyhedral* (solids with multiple plane faces) or *helical*, or some combination of the two (Figure 1-27). One of the most commonly encountered polyhedral shapes is a 20-sided polyhedron called an *icosahedron*, whose faces consist of equilateral triangles. The inherent stability and efficiency of such a structure was dramatically demonstrated by the architect and philosopher Buckminster Fuller, who applied this architectural organization to the construction of his famous geodesic domes. Helical viruses also exhibit a regular symmetry, although in this case the protein subunits surrounding the nucleic acid core form a cylinder composed of protein subunits arranged in a spiral configuration.

Some viruses have a more elaborate structure than these simple polyhedral or helical configurations. In the *enveloped* viruses, a polyhedral or helical protein coat is enclosed by a membranous envelope consisting of a mixture of protein, lipid, and/or carbohydrate. Unlike the membranes of cells, these envelopes are not capable of regulating the flow of materials into and out of the virus. Other viruses may exhibit a complex combination of polyhedral and helical symmetry. The *T-even bacteriophages,* for example, consist of a polyhedral head attached to a helical tail. The remarkable efficiency with which DNA is packed into the tiny head of such bacteriophages is dramatically illustrated by electron micrographs of ruptured virus particles, which

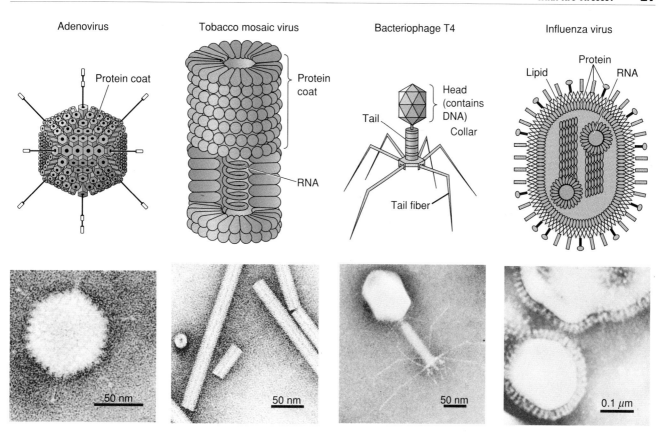

Adenovirus — Protein coat

Tobacco mosaic virus — Protein coat — RNA

Bacteriophage T4 — Head (contains DNA), Tail, Collar, Tail fiber

Influenza virus — Lipid, Protein, RNA

50 nm — 50 nm — 50 nm — 0.1 µm

Figure 1-27 *Diagrams and Micrographs Illustrating the Symmetrical Organization of Several Kinds of Viruses* *Micrographs courtesy of R. C. Williams (adenovirus, tobacco mosaic virus, and bacteriophage T4) and W. G. Laver (influenza virus).*

reveal the enormous amount of DNA present (Figure 1-28).

Viruses Reproduce inside Living Cells

Viruses have evolved to reproduce in specific cell types, whether it be a bacterium, a plant cell, or an animal cell. In some cases the interaction is so specific that only certain genetic strains of an organism are susceptible to infection by a particular virus. This specificity is governed by a number of factors, including viral recognition of the host cell surface and the ability of a virus to utilize intracellular components to aid in its own replication. Although the details of the infection process vary among viruses, five basic steps can be distinguished: adsorption, penetration, replication, maturation, and release.

1. Adsorption. Viral infections are typically initiated by the binding of a viral coat protein to a specific chemical receptor site on the surface of the host cell. Such receptor sites consist of molecules that are normal components of the plasma membrane or cell wall.

2. Penetration. After a virus attaches to the host cell surface, its nucleic acid enters the cell. Viruses that attack bacterial cells typically contain enzymes that digest a hole in the bacterial cell wall; their nucleic acid is then injected

through the hole. In contrast, animal viruses tend to be engulfed intact by an infolding of the plasma membrane.

3. Replication. After it has entered the cell, there are two possible fates for the viral nucleic acid. In the case of a **virulent virus,** the nucleic acid quickly directs the host cell to make more viral nucleic acid and protein molecules. With **lysogenic viruses,** the information from the viral nucleic acid becomes incorporated into the DNA of the host cell. This arrangement persists for an indefinite period of time, until certain stimuli trigger the release of viral nucleic acid and an ensuing virulent infection.

4. Maturation. During the next step, the newly synthesized viral nucleic acid and protein molecules are assembled into intact virus particles. For most viruses, this maturation process appears to proceed by a simple self-assembly process. It was first shown in the early 1950s by Heinz Fraenkel-Conrat that a mixture of purified RNA and protein isolated from tobacco mosaic virus will spontaneously assemble into reconstituted virus particles that can infect cells and reproduce normally. If purified RNA from one strain of virus is assembled with protein obtained from another strain of virus, the hybrid virus is also capable of infecting cells and reproducing, although it is the source of the RNA and not the protein that determines the type of virus that is reproduced by the infected cells.

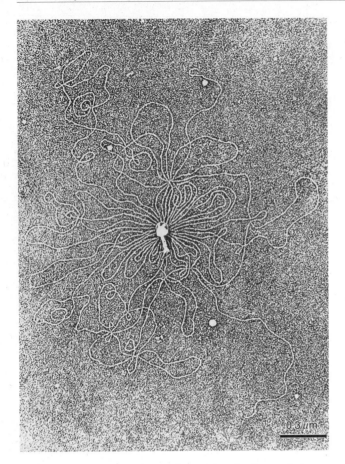

0.3 μm

Figure 1-28 *Release of DNA from Bacteriophage T2 Ruptured by Osmotic Shock* *The long tangled fiber represents a single DNA molecule that was packed into the tiny head of the virus. Courtesy of A. K. Kleinschmidt.*

5. Release. After new virus particles have been assembled, they are released from the infected cell into the extracellular environment and move on to infect other cells. Virus release is often accompanied by rupture of the plasma membrane and destruction of the host cell. This process of cell **lysis** does not always occur, however. With certain types of viruses, the newly formed virus particles are continuously released from the infected cell by budding off from the cell surface. Part of the plasma membrane may be taken along with the virus in the process, contributing to the viral envelope. With such viruses, the normal functions of the host cell continue during the period of infection. Because the cells are not destroyed during the process of virus release, they act as perpetual carriers of the infecting virus.

Viruses May Have Evolved from Ancient Cells

It should be clear from the foregoing description that viruses are quite different from living cells. The question therefore arises as to whether or not viruses are "alive." Such a question does not have an absolute an-

swer because it depends on how one defines life. But some important insights into the relationship between viruses and living cells have come from a consideration of the evolutionary origin of viruses.

Two main hypotheses have been proposed to explain the origin of viruses. One states that viruses are a precellular form of life that first occurred on the primordial earth prior to the evolution of cells. This hypothesis appears somewhat unlikely, for it is hard to picture how viruses could have evolved before living cells when they require the existence of cells for their replication. The other viewpoint considers viruses to be ancient derivatives of degenerate cells or cell fragments. The most extreme form of this hypothesis is the view that viruses are simply pieces of DNA or RNA that have escaped from cells and have sufficient genetic information to be able to replicate in other cells. This theory could explain why viruses tend to preferentially infect certain cell types, since such nucleic acid fragments might be expected to function only in cells that are similar to the one from which they were originally derived.

If viruses are in fact cellular derivatives, it is not surprising to find that they are capable of entering cells and taking over many cellular functions. But by themselves, virus particles lack most of the basic features of cells. Earlier in the chapter we defined four basic functions that are performed by all cells (page 5). Viruses carry out only one of these functions, namely the utilization of genetic information. Even this property depends upon living cells for its occurrence, since viral reproduction requires a host cell and its synthetic machinery. The complete and integrated performance of all four functions that are essential to life clearly requires more structural complexity and compartmentalization than occurs in virus particles.

SUMMARY OF PRINCIPAL POINTS

• All organisms are composed of cells that perform the same basic functions: they maintain a selective plasma membrane barrier that regulates the flow of materials into and out of the cell, they duplicate and utilize genetic information to guide the synthesis of cell components, they contain enzymes that catalyze metabolic reactions, and they exhibit motility.

• All cells are enclosed by a plasma membrane; in plant and bacterial cells, a rigid cell wall is present as well. In eukaryotic cells, intracellular membranes form a series of cytoplasmic compartments known as the endoplasmic reticulum, Golgi complex, lysosomes, peroxisomes, and vacuoles.

• Eukaryotic cells, which store their genetic information in a membrane-enclosed nucleus, are usually larger and have a more extensive internal membrane system than prokaryotic cells, which store their genetic material in a nucleoid that is not membrane-enclosed. In both cell types, protein synthesis occurs on ribosomes.

- In eukaryotic cells, mitochondria are involved in breaking down nutrients and converting their energy into a useful chemical form. The chloroplasts of plant cells carry out photosynthesis, which traps energy from sunlight. In prokaryotic cells, the plasma membrane carries out most of the reactions that occur in the mitochondria of eukaryotes; in photosynthetic prokaryotes, photosynthesis is carried out by intracellular membranes instead of chloroplasts.

- Motility in eukaryotic cells is made possible by microtubules and actin filaments, which are not present in prokaryotic cells. Prokaryotes and eukaryotes both utilize motile cellular appendages called flagella, but the flagella of prokaryotes are much simpler in structure and function than their eukaryotic counterparts.

- Cell size is limited by the ratio of a cell's surface area to its volume.

- The most abundant substance in cells is usually water, which functions as both a chemical reactant and as a solvent for ions and molecules.

- The major building blocks of the cell are sugars, fatty acids, nucleotides, and amino acids. These compounds are used in the construction of polysaccharides, lipids, nucleic acids, and proteins.

- Sugars are the building blocks of polysaccharides such as glycogen and starch, which store glucose, and cellulose, which contributes to the strength of the plant cell wall.

- Lipids are a heterogeneous group of hydrophobic molecules defined on the basis of their solubility in nonpolar organic solvents. Included in this group are fatty acids, triacylglycerols (triglycerides), phospholipids, glycolipids, steroids, and terpenes. Fatty acids function both as energy sources and as building blocks for more complex lipids, whereas phospholipids, glycolipids, and certain steroids are important constituents of biological membranes.

- Nucleotides are the building blocks of nucleic acids, which function in the storage, reproduction, and transmission of genetic information. DNA contains the sugar deoxyribose and the bases A, G, C, and T, whereas RNA contains the sugar ribose and the bases A, G, C, and U.

- Twenty different amino acids are employed in the construction of protein molecules. The primary structure of a protein refers to its linear sequence of amino acids, which is unique for each type of protein. Secondary structure consists of α helices and β sheets. Tertiary structure refers to the folding of the polypeptide chain into one or more compact structures known as domains. Quaternary structure involves the formation of a multisubunit protein from more than one polypeptide chain. If a protein's final three-dimensional conformation is disrupted by denaturing conditions, the protein usually loses its capacity to function.

- Viruses consist of a nucleic acid core surrounded by a protein coat. Viruses can reproduce only when they infect a living cell. The infection process involves five stages: adsorption, penetration, replication, maturation, and release.

SUGGESTED READINGS

Books

Branden, C., and J. Tooze (1991). *Introduction to Protein Structure,* Garland, New York.

Mathews, C. K., and K. E. van Holde (1990). *Biochemistry,* Benjamin/Cummings, Redwood City, CA, Ch. 4–9.

Articles

Albrecht-Buehler, G. (1990). In defense of "nonmolecular" cell biology, *Int. Rev. Cytol.* 120:191–241.

Anfinsen, C. B. (1973). Principles that govern the folding of protein chains, *Science* 181:223–230.

Angert, E. R., K. D. Clements, and N. R. Pace (1993). The largest bacterium, *Nature* 362:239–241.

Chothia, C. (1984). Principles that determine the structure of proteins, *Annu. Rev. Biochem.* 53:537–572.

Clegg, J. S. (1984). Properties and metabolism of the aqueous cytoplasm and its boundaries, *Amer. J. Physiol.* 246:R133–R151.

Horowitz, S. B., and P. L. Paine (1976). Cytoplasmic exclusion as a basis for asymmetric nucleocytoplasmic solute distributions, *Nature* 260:151–153.

Karplus, M., and J. A. McCammon (1986). The dynamics of proteins, *Sci. Amer.* 254 (April):42–51.

Richards, F. M. (1991). The protein folding problem, *Sci. Amer.* 264 (January):54–63.

Wiggins, P. M. (1990). Role of water in some biological processes, *Microbiol. Rev.* 54:432–449.

C h a p t e r   2

Energy and Enzymes

Chapter 1 provided a brief introduction to the internal organization of cells and the functions that cells perform. Before examining the properties of cellular organelles in more detail, we first need to address a profoundly important issue: How is it possible that the trillions of molecules that make up a living cell are synthesized and brought together in exactly the right way to create a properly functioning cell? At face value, this task seems to be extraordinarily daunting.

In the broadest terms, four conditions must be met in order to produce functioning cells: (1) Cells require *chemical building blocks* such as sugars, fatty acids, nucleotides, and amino acids, as well as water, inorganic salts, metal ions, oxygen, and carbon dioxide. Some of these materials are produced by cells, whereas others must be obtained from the external environment. (2) Cells need *energy* both to drive the chemical reactions involved in building cell components and to power the activities that these components carry out. (3) Most of the chemical reactions that take place within cells would normally occur too slowly to maintain life as we know it. Cells therefore utilize protein catalysts called *enzymes* to speed up chemical reactions by many orders of magnitude. (4) Finally, cells require a set of instructions or *information* that guides all cell activities. In other words, cells need to know what kinds of chemical reactions to carry out, what kinds of structures to create, and what kinds of functions to perform. The information that guides all of these activities is contained in the nucleic acids DNA and RNA.

How do cells fulfill these requirements for chemical building blocks, energy, enzymes, and information? Chemical building blocks, which were described in Chapter 1, are obtained by consuming nutrients from the environment. The second and third requirements, which involve the utilization of energy and the action of enzymes, will be described in the current chapter. Finally, the principles that guide the flow of information will be introduced in Chapter 3.

ENERGY FLOW IN LIVING CELLS

The notion that cells require energy makes intuitive sense to us because the word "energy" is part of our everyday language, and we know that it takes energy to do things. For example, running an automobile requires a source of energy such as gasoline, and running a refrigerator requires a source of energy such as electricity. Cells, too, perform a variety of tasks that require energy: They grow, divide, move, synthesize macromolecules, and transport materials. But what exactly is meant when we say that cells require a source of "energy"? In scientific contexts, **energy** is defined as the capacity to do work, or in other words, the ability to cause some kind of change to occur. Thus in order for cells to function, energy must flow into cells from their surroundings. Although energy occurs in many different forms, cells usually obtain energy either in the form of organic molecules that contain chemical energy or, in the case of photosynthetic organisms, as photons of light. After energy has been taken up by a cell, it can be transformed and utilized in a variety of different ways.

To understand the principles that guide the use of energy by living cells, biologists need to have a basic familiarity with the field of **thermodynamics,** which is devoted to the study of energy transformations. Investigations in this area have led to the discovery of several universal principles that apply to all energy conversions, whether they occur in living organisms or not. In the following sections, we will see how these thermodynamic principles apply to the use of energy by cells.

The First Law of Thermodynamics States That Energy Can Be neither Created nor Destroyed

The two thermodynamic principles that are most relevant for biologists are called the first and second laws of thermodynamics. The **first law of thermodynamics** states that the total amount of energy in the universe is constant. In other words, energy can be neither created nor destroyed. The only changes that can occur are those that involve transformations in either the kind of energy or its physical location. Thus the total amount of energy that is present in any particular location can change, and transformations can take place between various forms of energy such as heat, light, electricity, mechanical energy, and chemical energy. However, the total energy content of the universe always remains constant.

In thermodynamic terms, the particular location in which energy changes are being studied is usually referred to as the *system,* and the rest of the universe is called the *surroundings.* According to the first law of thermodynamics, the energy content of an individual system can change, but the total energy content of any system plus its surroundings always remains the same. Cells, like everything else in the universe, must obey the laws of thermodynamics. Since the first law states that energy can be neither created nor destroyed, the energy requirements for cellular activities can be met only by taking up energy from the surroundings and transforming it into forms that are useful to the cell. Thus the total energy content of a cell plus its surroundings (i.e., the rest of the universe) always remains constant.

The Second Law of Thermodynamics States That the Entropy of the Universe Is Always Increasing

Although the first law of thermodynamics tells us that energy is neither created nor destroyed when a biological process takes place, it tells us nothing about the probability that any such process will actually occur. For example, thousands of different chemical reactions take place within cells. As biologists, we would like to know the direction in which a particular reaction will tend to proceed under conditions that prevail in the cell, and whether energy will be consumed or released by each reaction. To answer questions about the direction in which reactions tend to proceed, we need to turn to the **second law of thermodynamics.** This law states that every event in the universe proceeds in the direction that causes the system plus its surroundings to exhibit a net increase in randomness, or **entropy.** In other words, the universe is constantly becoming more disordered and random. This propensity to increase the amount of entropy in the universe is the ultimate driving force for all events in the universe, including every biological process.

Although the second law of thermodynamics states that the entropy of a system plus its surroundings always increases, it is not necessarily true that the entropy of an individual system always increases. It may increase, decrease, or stay the same. However, if a system experiences a decrease in entropy (becomes less random), the entropy of the surroundings must increase by a sufficient amount to allow the total entropy of the system plus its surroundings to increase. This is exactly what happens during the development of living organisms. The formation of highly organized cells involves a large decrease in the entropy of the organism, but only at the expense of an even greater increase in the entropy of the rest of the universe.

Although the propensity to increase the total entropy of the universe is the driving force for all biological processes, this way of stating the second law of thermodynamics is of limited usefulness to biologists because it implies that we need to assess the entropy changes that occur in the entire universe each time we wish to determine whether a given chemical reaction will proceed or not. What is obviously required is a thermodynamic parameter that allows us to apply the second law of thermodynamics to an individual chemical reaction without requiring us to measure the entropy changes that occur in the entire universe. The thermodynamic parameter that allows us to do this is called **free energy,** and is symbolized with the letter *G* in honor of Josiah Willard Gibbs, the biochemist who for-

mulated this crucial concept. Free energy is an important principle for biological systems because it represents *the energy that can be harnessed to do useful work.*

Changes in Free Energy Determine the Direction in Which Reactions Proceed

For living organisms, where pressure and volume remain constant, the change in free energy that accompanies any biological process is determined by two parameters: the change in the total internal energy (ΔE) of the system, and the change in the entropy (ΔS) of the system. The change in free energy (ΔG) is related to these two parameters by the equation

$$\Delta G = \Delta E - T\Delta S \qquad (2\text{-}1)$$

where T is the temperature in Kelvins (degrees Celsius + 273). Using this equation, it has been shown mathematically that if the entropy of the universe is always increasing ($\Delta S_{universe}$ = positive), it follows that the free energy of a system is always decreasing (ΔG_{system} = negative). Thus another way of stating the second law of thermodynamics is to say that all reactions tend to proceed in the direction that causes a *decrease in the free energy of the system.*

This means that knowing the value of ΔG for any given chemical reaction tells us the direction in which the reaction will occur: Reactions proceed in the direction that causes the free energy of the system to decrease—that is, the direction in which ΔG is negative. Such reactions are called **exergonic** reactions, which means that they release free energy. Exergonic reactions are said to take place *spontaneously,* although the choice of the word "spontaneous" is an unfortunate one because it implies that the process occurs quickly. In fact, thermodynamic considerations provide no information about the rate at which reactions occur; they simply indicate the *direction* in which a reaction will proceed. For example, if the conversion of substance A to substance B has a negative ΔG (i.e., is exergonic), the reaction will progress in the direction of A → B rather than B → A. Although the conversion of A → B is therefore said to be *thermodynamically spontaneous,* the phrase *thermodynamically favorable* is actually more appropriate because thermodynamic considerations do not provide any information about how quickly the conversion of A → B will proceed.

Conversely, reactions in which ΔG is positive, called **endergonic** reactions, do not progress in the direction specified because it would involve an increase in the free energy of the system (which would violate the second law of thermodynamics). Fortunately, reactions that are endergonic under one set of conditions can become exergonic under another set of conditions.

As we will see later, living cells often use this tactic to convert endergonic reactions into exergonic reactions, thereby making it possible to carry out reactions that would not otherwise occur.

$\Delta G^{\circ\prime}$ Determines the Direction in Which Reactions Proceed under Standard Conditions

Since ΔG tells us the direction in which a reaction will proceed, it is important to know how to calculate the actual value of ΔG for any given chemical reaction. Equation 2-1 is not a practical approach for calculating ΔG because it requires knowing the change in entropy (ΔS), and entropy changes are difficult to measure. However, another approach has been devised for calculating ΔG that is much simpler. For the generalized chemical reaction

reactants $\rightarrow$ products

the change in free energy (ΔG) can be calculated from the equation

$$\Delta G' = \Delta G^{\circ\prime} + 2.303RT \log_{10} \frac{[products]}{[reactants]} \qquad (2\text{-}2)$$

where R is the gas constant, T is the temperature in Kelvins, [reactants] is the mathematical term obtained by multiplying together the initial molar concentrations of each of the reactants, [products] is the mathematical term obtained by multiplying together the initial molar concentrations of each of the products of the reaction, and $\Delta G^{\circ\prime}$ is the standard free-energy change. This **standard free-energy change ($\Delta G^{\circ\prime}$)** is a measure of the amount of free energy released during conversion of reactants to products under "standard conditions," which are defined as a temperature of 25°C (298 K), a pressure of 1 atmosphere, and all reactants and products maintained at concentrations of 1.0 M. The prime (') used in writing $\Delta G'$ and $\Delta G^{\circ\prime}$ indicates that a pH value of 7.0 is being specified. The relationship between the value of $\Delta G^{\circ\prime}$ and the direction in which a reaction will proceed under standard conditions is summarized in Table 2-1.

It is critical that the distinction between $\Delta G'$ and $\Delta G^{\circ\prime}$ be clearly understood. $\Delta G'$ is a measure of the actual change in free energy that occurs with a particular

mixture of reactants and products at any given concentration; the value of $\Delta G'$ thus varies, depending on the conditions involved. In contrast, $\Delta G^{\circ\prime}$ is a constant for any given reaction determined under standard conditions. How is the value of $\Delta G^{\circ\prime}$ actually calculated? By definition, $\Delta G^{\circ\prime}$ is determined under conditions of **equilibrium**—that is, after no further net change in the concentration of reactants or products occurs. Therefore no additional net change in free energy can occur or, in other words, $\Delta G' = 0$. Substituting $\Delta G' = 0$ into Equation 2-2, we obtain the equilibrium expression

$$0 = \Delta G^{\circ\prime} + 2.303RT \log_{10} \frac{[products]_{eq}}{[reactants]_{eq}}$$

where $[reactants]_{eq}$ and $[products]_{eq}$ are the molar concentrations of reactants and products at equilibrium. Rearranging terms we obtain

$$\Delta G^{\circ\prime} = -2.303RT \log_{10} \frac{[products]_{eq}}{[reactants]_{eq}} \qquad (2\text{-}3)$$

Because the **equilibrium constant (K_{eq})** for any chemical reaction is defined as

$$K_{eq} = \frac{[products]_{eq}}{[reactants]_{eq}} \qquad (2\text{-}4)$$

K_{eq} can be substituted into Equation 2-3 to obtain the relationship

$$\Delta G^{\circ\prime} = -2.303RT \log_{10} K'_{eq} \qquad (2\text{-}5)$$

(The parameter K'_{eq} is employed instead of K_{eq} to indicate that the equilibrium constant is being specified at pH = 7.0).

Equation 2-5 provides a relatively easy way of calculating $\Delta G^{\circ\prime}$ for any given chemical reaction. One simply measures the concentrations of reactants and products that are present after equilibrium has been achieved, uses this information to calculate K'_{eq} (Equation 2-4), and then substitutes this value of K'_{eq} into Equation 2-5 to obtain $\Delta G^{\circ\prime}$. To illustrate, let us calculate the value of $\Delta G^{\circ\prime}$ for the following chemical reaction that cells

Table 2-1 The Relationship between $\Delta G^{\circ\prime}$ and the Direction in Which a Chemical Reaction Will Proceed Under Standard Conditions

$\Delta G^{\circ\prime}$	Direction of Reaction (Under Standard Conditions)	Reaction Type	K'_{eq}	Concentration at Equilibrium
Negative	A $\rightarrow$ B	Exergonic	>1	B > A
Positive	A $\leftarrow$ B	Endergonic	<1	A > B
Zero	Reaction is already at equilibrium	Neither exergonic nor endergonic	$=1$	A = B

carry out as part of the metabolic pathway for degrading glucose:

$$
\begin{array}{ccc}
& & H \\
& & | \\
CH_2OH & & C=O \\
| & & | \\
C=O & \rightleftharpoons & H-C-OH \\
| & & | \\
CH_2OPO_3{}^{2-} & & CH_2OPO_3{}^{2-}
\end{array}
$$

Dihydroxyacetone   Glyceraldehyde
phosphate   3-phosphate
(DHAP)   (G3P)

For this chemical reaction, the ratio of G3P to DHAP is always found to be 0.0475 once equilibrium has been attained. In other words, $K'_{eq} = 0.0475$; substituting this value into Equation 2-5, we obtain

$$\Delta G^{\circ\prime} = -2.303RT \log_{10}(0.0475)$$

Inserting appropriate values for the gas constant R (0.00198 kcal/mol/K) and absolute temperature T (25°C = 298K), we can calculate

$$\Delta G^{\circ\prime} = -2.303 \times 0.00198 \times 298 \times \log_{10}(0.0475)$$
$$= +1.8 \text{ kcal/mol}$$

Because the calculated value of $\Delta G^{\circ\prime}$ turns out to be positive, it tells us that the conversion of DHAP to G3P is endergonic under standard conditions and hence will not proceed in the forward direction; the reaction will instead proceed in the opposite direction, and so G3P will be converted to DHAP. Since the forward reaction (DHAP → G3P) has a $\Delta G^{\circ\prime}$ of +1.8 kcal/mol, this reverse reaction (G3P → DHAP) must have a $\Delta G^{\circ\prime}$ of −1.8 kcal/mol. In other words, the reverse reaction is exergonic, liberating 1.8 kcal of free energy for every mole of G3P converted to DHAP. Thus we see that a reaction that is endergonic in one direction is exergonic in the opposite direction (and vice versa).

ΔG′ Determines the Direction in Which Chemical Reactions Actually Proceed inside Cells

Although determining the value of $\Delta G^{\circ\prime}$ provides information about the behavior of chemical reactions under standard conditions, chemical reactions in living cells do not typically start with 1.0 M concentrations of reactants and products, the standard conditions that apply to the calculation of $\Delta G^{\circ\prime}$. Under nonstandard conditions it is $\Delta G'$, not $\Delta G^{\circ\prime}$, that determines the direction in which a reaction will proceed and the amount of free energy released.

Let us therefore return to the conversion of DHAP to G3P discussed in the preceding section and determine what would happen under conditions that might

exist in the cell, such as a DHAP concentration of 10^{-4} M and a G3P concentration of 10^{-6} M. The actual change in free energy, $\Delta G'$, can be calculated for these particular conditions by substituting into Equation 2-2 as follows:

$$\Delta G' = \Delta G^{\circ\prime} + 2.303RT \log_{10} \frac{[G3P]}{[DHAP]}$$

$$= 1.8 \text{ kcal/mol} + 2.303RT \log_{10} \frac{10^{-6}}{10^{-4}}$$

$$= 1.8 \text{ kcal/mol} + (2.303 \times 0.00198 \times 298 \times -2)$$

$$= 1.8 \text{ kcal/mol} - 2.7 \text{ kcal/mol}$$

$$= -0.9 \text{ kcal/mol}$$

Thus, in spite of the positive value of $\Delta G^{\circ\prime}$ previously calculated for the conversion of DHAP to G3P under standard conditions, $\Delta G'$ turns out to be negative for this same chemical reaction when the DHAP concentration is 10^{-4} M and the G3P concentration is 10^{-6} M. Under these conditions $\Delta G' = -0.9$ kcal/mol, which means that the reaction will proceed spontaneously with a release of 0.9 kcal of free energy for every mole of DHAP converted to G3P (in spite of the fact that the reaction would not occur in this direction at all under standard conditions). Since $\Delta G'$ is calculated on the basis of real-life concentrations of reactants and products rather than arbitrary standard conditions, it is $\Delta G'$ rather than $\Delta G^{\circ\prime}$ that determines the direction in which chemical reactions actually proceed within cells.

ENZYMES AND CATALYSIS

Thousands of different chemical reactions take place within cells. These reactions underlie such diverse activities as the synthesis and breakdown of chemical building blocks and macromolecules, the conversion of chemical energy, the transmission of genetic information, the transport of materials across membranes, and motility. Thus every cell function introduced in Chapter 1 is ultimately dependent upon chemical reactions. It is therefore crucial to understand the factors that govern the rate at which these reactions proceed.

Although thermodynamic calculations reveal the direction in which a given reaction will tend to proceed, this information tells us nothing about how fast the reaction will occur. To illustrate the importance of this distinction, let us consider the behavior of the cellulose molecules found in the cell walls of plants. In the presence of oxygen, cellulose can be oxidized to CO_2 and H_2O. If one determines the equilibrium constant for this reaction and uses this information to calculate $\Delta G'$, it is found to be an exergonic reaction with a very negative value of $\Delta G'$. Thermodynamic considerations thus indicate that the oxidation of cellulose is an energetically favorable reaction that proceeds "spontaneously." In practice, however, the oxidation of cellulose does not occur at a perceptible rate under normal environmental conditions. If it did, the pages of this book (which consist largely of cellulose) would be in danger of breaking down to CO_2 and H_2O before you could finish reading the chapter!

The same principle applies to the thousands of chemical reactions that occur within cells. Under normal conditions most of these reactions would not, by themselves, proceed fast enough to sustain life, even though they may be thermodynamically favorable. Living organisms have therefore developed special molecules called **enzymes** to speed up the rate of chemical reactions. Cells manufacture several thousand different enzymes, each enhancing the rate of a different chemical reaction. Biochemists have been studying the properties of enzymes for more than a hundred years, telling us much about what enzymes are, how they work, and how they are regulated.

Enzymes Function as Biological Catalysts

The first clue to the existence of enzymes was provided in the early nineteenth century by the Swedish chemist Jon Berzelius, who discovered that an extract of potatoes is more effective than concentrated sulfuric acid in promoting the breakdown of starch. Berzelius concluded that potatoes contain substances that function as **catalysts,** which are defined as agents that speed up chemical reactions without being consumed in the process. With remarkable insight, Berzelius went on to predict that all materials found in living organisms are synthesized under the influence of such catalysts. Shortly thereafter, Louis Pasteur postulated that the catalytic effect is intimately associated with cell structure and so cannot be separated from living cells. If true, then biological catalysts could not be isolated and studied in purified form. It was therefore a great milestone when Hans and Eduard Büchner demonstrated in 1897 that an extract of yeast, from which all intact cells had been removed, was capable of catalyzing the breakdown of glucose. This was an especially significant observation because glucose degradation is a major metabolic pathway in all cells. This finding firmly established that biological catalysts, in the absence of any cell organization, are capable of enhancing the rates of metabolic reactions.

As more and more attention began to be paid to biological catalysts, it became apparent that a systematic nomenclature system was needed to facilitate communication. The term *enzyme* was adopted as a general designation for biological catalysts, and it was decided that individual enzymes would be named by adding the suffix -*ase* to the name of the **substrate** (the substance upon which the enzyme acts). For example, the enzyme catalyzing the degradation of urea was termed *urease* and enzymes catalyzing hydrolysis of nucleic acids were designated *nucleases.* It also became common practice to add the -*ase* suffix to the term designating the type of reaction being catalyzed; thus the enzyme catalyzing glucose oxidation is called *glucose oxidase,* the enzyme catalyzing glycogen synthesis is called *glycogen synthase,* and so forth. This basic approach to enzyme nomenclature is still in use today, although some older names introduced prior to the adoption of this system (e.g., trypsin, pepsin) have been retained.

Enzymes Are Almost Always Proteins

In the years following the Büchners' discovery that enzymes can catalyze chemical reactions outside living cells, a large number of other enzymes were identified by biochemists. But attempts to determine the chemical makeup of enzymes encountered many difficulties. An example of the problems experienced during this period is the classic story of Richard Willstätter's encounters with *peroxidase,* one of the first enzymes to be successfully purified. When Willstätter analyzed the chemical composition of his purified enzyme preparation, he found no lipid, no sugar, no nucleic acid, and no protein. These results led to the prevailing opinion in the early twentieth century that enzymes belong to none of the commonly recognized classes of cellular molecules.

In 1926, however, James Sumner reported the first crystallization of an enzyme, in this case urease. He found that his urease crystals consisted entirely of protein, even after repeated cycles of recrystallization. Since crystallization is generally recognized as the ultimate criterion of purity, these results suggested to Sumner that urease is a protein. Although this conclusion initially met with skepticism, similar experiments by John Northrop and his colleagues on other enzymes soon led to widespread acceptance of the idea that enzymes are made of protein. Willstätter, it turns out, had been misled by the fact that enzymes are active in extremely small quantities. Thus his peroxidase preparation exhibited measurable enzyme activity even though the protein content of the sample was too low to be detected by the analytical techniques then available.

In the years since enzymes were first crystallized by Sumner and Northrop, more than a thousand enzymes have been purified and identified as proteins. Although the number of different reactions catalyzed by these enzymes is very large, the group can be subdivided into a relatively small number of categories (Table 2-2). All cellular reactions are catalyzed by enzymes that fit into one of these basic categories.

The idea that all enzymes are made of protein was accepted for more than 50 years after the pioneering

Table 2-2   The Six Major Classes of Enzymes

Enzyme Class	Examples	Type of Reaction Catalyzed	Sample Reaction
Oxidoreductases	Dehydrogenases, oxidases, peroxidases, hydroxylases, reductases, oxygenases	Oxidation-reduction	$AH_2 + B \rightarrow A + BH_2$
Transferases	Transaminases, transmethylases	Transfer of a functional group from one molecule to another	$AX + B \rightarrow A + BX$
Hydrolases	Esterases, amidases, glycosidases, peptidases, phosphatases	Hydrolytic cleavage	$AB + H_2O \rightarrow AH + BOH$
Lyases	Aldolases, synthases, deaminases, hydrases, decarboxylases, nucleotide cyclases	Cleavage of C–C, C–O, C–N, etc. by elimination, leaving a double bond; or addition to double bonds	$A\text{–}COO^- \rightarrow A + CO_2$
Isomerases	Isomerases, racemases, epimerases, mutases	Intramolecular rearrangements	$\begin{array}{l} X\text{–}C\text{–OH} \rightarrow X\text{–}C\text{–O–}\boxed{P} \\ \quad\ \ \| \qquad\qquad\ \| \\ \quad C\text{–O–}\boxed{P} \qquad C\text{–OH} \end{array}$
Ligases	Ligases, synthetases	Joining of two molecules coupled to the cleavage of a high-energy bond	$X + Y + ATP \rightarrow XY + ADP + P_i$

discoveries of Sumner and Northrop. Then, in the early 1980s, Thomas Cech and Sidney Altman independently discovered that certain types of RNA molecules can function as catalysts that exhibit many of the properties of enzymes. These RNA enzymes, called **ribozymes,** will be described in later chapters when we discuss their role in the flow of genetic information (page 482).

Enzymes Are More Efficient, Specific, and Controllable Than Other Catalysts

Over the years, chemists have discovered that many kinds of organic and inorganic molecules can function as catalysts. It is therefore important to understand the similarities and differences between enzymes and other kinds of chemical catalysts. All catalysts, including enzymes, exhibit several common features:

1. Catalysts are substances that *increase the rate* of chemical reactions.

2. Catalysts are *not consumed* during the process of speeding up reaction rates.

3. Only *small quantities* of a catalyst are required.

4. Catalysts *do not alter the equilibrium* of the reaction being catalyzed; they simply decrease the amount of time required to achieve equilibrium. Since equilibrium conditions are not altered, the presence of a catalyst does not change the amount of free energy released ($\Delta G'$) or the direction in which a reaction will proceed.

In addition to the preceding attributes, which are shared by all catalysts, enzymes exhibit several unique features that distinguish them from the catalysts routinely encountered in organic or inorganic chemistry. These unique properties of enzymes relate to their efficiency, specificity, and regulation.

1. Enzyme Efficiency. Enzymes tend to be more efficient than other kinds of catalysts. To illustrate the point, let us briefly consider the reaction in which hydrogen peroxide is broken down to water and oxygen:

$$2H_2O_2 \rightleftharpoons 2H_2O + O_2$$

The $\Delta G'$ for this reaction is negative, thus indicating that the breakdown of H_2O_2 is thermodynamically favorable. However, in the absence of a catalyst the reaction does not proceed very quickly. One agent known to catalyze this reaction is the ferric ion (Fe^{3+}), which causes the breakdown of H_2O_2 to occur around 30,000 times faster than the uncatalyzed reaction. Although free Fe^{3+} is thus an effective catalyst for this reaction, it is not nearly as efficient as the enzyme *catalase,* an Fe^{3+}-containing protein made by cells to protect themselves from the toxic effects of H_2O_2. In the presence of catalase, the breakdown of H_2O_2 occurs 10^8 times faster than in the absence of a catalyst. This extraordinary efficiency is typical of enzyme-catalyzed reactions, which proceed up to 10^8 to 10^{11} times faster than uncatalyzed reactions.

2. Enzyme Specificity. The fact that a substance acts as a catalyst does not guarantee that the catalytic effect will be a selective one. The inorganic and organic catalysts typically encountered in the chemical laboratory tend to act upon a broad spectrum of different substrates, and a given catalyst may even be effective on different kinds of chemical reactions. One of the most striking properties that distinguishes enzymes from these other kinds of catalysts is that enzymes are extraordinarily selective, both in terms of the type of reaction a given enzyme will catalyze and the structures of the substances it will act upon. To illustrate this point, consider the enzyme *glucose oxidase.* This enzyme only catalyzes oxidation reactions, and is highly selective for the six-carbon sugar, α-glucose. The recognition of α-glucose is so precise that glucose oxidase catalyzes the oxidation of α-glucose, but not the other three-dimensional form of this sugar, β-glucose (see Figure 1-11).

3. Enzyme Regulation. A final property that distinguishes enzymes from other kinds of catalysts is that the catalytic activity of enzymes can be controlled. In response to the changing needs of the organism, the activity of many enzymes can be increased, decreased, or even turned off entirely. The mechanisms involved in this type of enzyme regulation will be discussed later in the chapter after we have described how enzymes function.

MECHANISM OF ENZYME ACTION

The unique properties of enzymes arise largely from the fact that they are proteins. Since protein molecules can be assembled from countless combinations of 20 different amino acids (page 21), enzymes exhibit an enormous structural and functional diversity. As we now examine the mechanism of enzyme action, we will see how this structural diversity allows each enzyme to bind to a specific substrate and convert it to an appropriate product.

Catalysts Speed Up Reactions by Lowering the Activation Energy ($\Delta G^{\ddagger}$)

To understand how enzymes work, it is first necessary to know what determines the rate at which reactions normally proceed. In addressing this issue, let us consider the hypothetical chemical reaction illustrated in Figure 2-1 (*red curve*). In this reaction, the initial free energy of the reactants is higher than the free energy of the products. Free energy is therefore released during conversion of reactants to products, indicating that the reaction is exergonic and hence thermodynamically favorable. However, in order for the reactants to be converted to products, they must first proceed through an intermediate chemical stage, called a **transition state,** whose free energy is higher than that of the initial reactants. In other words, energy must first be added to the reactants before they can be converted to products. The

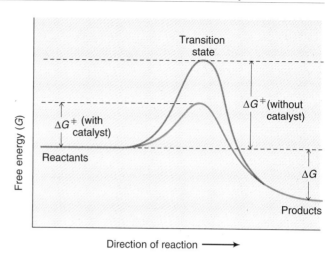

Figure 2-1 *Free Energy Diagram for a Chemical Reaction in the Presence or Absence of a Catalyst* *In the presence of a catalyst* (blue curve), *less energy is required to achieve the transition state that leads to the formation of products.*

term **activation energy** or $\Delta G^{\ddagger}$ is employed when referring to the amount of energy required for converting reactants from their initial state to this activated transition state.

The rate at which a chemical reaction proceeds is directly proportional to the number of molecules reaching the transition state. Hence any factor that increases the number of molecules having sufficient energy to attain the transition state will increase the rate of the reaction. The simplest way to increase the number of molecules with sufficient energy is to raise the temperature. As is illustrated in Figure 2-2 (*top graph*), a typical reaction mixture consists of a population of molecules of varying free-energy levels. At any given temperature, a certain fraction of the molecules will have sufficient energy to attain the transition state. Raising the temperature increases the overall thermal energy of the population as a whole, thereby increasing the relative number of molecules whose free energy is high enough to reach the transition state. Hence the reaction proceeds faster at higher temperatures.

Although raising the temperature is an effective way of increasing reaction rates, it is not very practical for living organisms. First of all, higher temperatures would disrupt many of the molecular and structural components normally required by living cells. Moreover, raising the temperature would indiscriminately increase the rates of all chemical reactions, and hence would not allow cells to selectively determine which kinds of reactions will occur. Cells have therefore turned to an alternative approach for increasing reaction rates—namely, the use of catalysts. A catalyst increases the rate of a chemical reaction by creating a new reaction pathway whose transition state has a lower activation energy ($\Delta G^{\ddagger}$) than that of the uncatalyzed reaction (see Figure 2-1, *blue curve*). Hence at

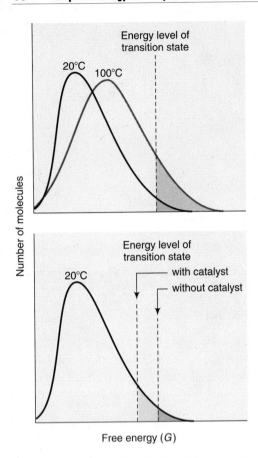

Figure 2-2 Energy Distribution Diagrams Illustrating How Temperature and Catalysts Influence Reaction Rates
Each curve shows that a reaction mixture under any given set of conditions consists of a population of molecules of varying free-energy levels. The shaded areas of each curve represent the number of molecules that have sufficient free energy to reach the transition state. (Top) Raising the temperature increases the number of molecules having sufficient free energy to attain the transition state. (Bottom) The presence of a catalyst creates a transition state with a lower free energy. Hence in the presence of a catalyst, more molecules possess sufficient free energy to attain the transition state at any given temperature.

any given temperature, the number of molecules possessing sufficient free energy to attain the transition state is higher for the catalyzed reaction than for the uncatalyzed reaction (see Figure 2-2, *bottom*).

All catalysts, including enzymes, act by lowering the activation energy. This lowering of the activation energy does not change the value of ΔG, which is determined solely by the difference in free energy between the initial reactants and final products. Since it is ΔG (rather than $\Delta G^{\ddagger}$) that determines the concentration of reactants and products that will be present at equilibrium, catalysts do not influence the final equilibrium conditions of chemical reactions; they only determine the rate at which reactions proceed toward equilibrium.

Enzymes Lower Activation Energy by Reducing the Energy and/or Entropy Changes Associated with the Transition State

The realization that enzymes act by lowering the activation energy ($\Delta G^{\ddagger}$) of chemical reactions raises the question of how enzymes accomplish this reduction in $\Delta G^{\ddagger}$. Equation 2-1, discussed earlier in the chapter, showed that changes in free energy are caused by changes in both internal energy (ΔE) and entropy (ΔS). Applying this same principle to activation energy yields the expression

$$\Delta G^{\ddagger} = \Delta E^{\ddagger} - T\Delta S^{\ddagger}$$

This equation tells us that enzymes can reduce $\Delta G^{\ddagger}$ by altering the energy change ($\Delta E^{\ddagger}$) and/or the entropy change ($\Delta S^{\ddagger}$) associated with attaining the transition state.

It is relatively easy to see how enzymes can alter the amount of energy ($\Delta E^{\ddagger}$) required to attain the transition state. Chemical reactants are often required to pass through highly strained and distorted transition states before the conversion to products will take place. Attaining such a distorted state usually requires energy. Enzymes can reduce this energy requirement by interacting with reactants to form alternative transition states that require less energy to create and yet still permit the ensuing reaction to occur. As we will see shortly, these alternative transition states typically involve the transfer of protons or electrons between enzyme and reactants.

Enzymes also influence the entropy (randomness) of reaction pathways. In most chemical reactions, the reacting molecules must assume a particular *orientation* relative to one another before the reaction can occur. In uncatalyzed reactions, which depend upon random collisions between molecules, the likelihood of the proper orientation occurring by chance during these haphazard collisions is quite small. By binding to reactants and holding them in the proper orientation, enzymes increase the probability that the reaction will occur.

An Enzyme-Substrate Complex Is Formed During Enzymatic Catalysis

The preceding discussion implies that enzymes transiently bind to the reactants (substrates) involved in the reaction being catalyzed. This idea that enzymes bind to their substrates was first proposed in 1913 by Leonor Michaelis and Maud Menten to explain an unusual aspect of enzyme-catalyzed reactions. In uncatalyzed reactions, the rate of product formation is directly proportional to the concentration of reactants because increasing their concentration increases the probability of collisions between the reacting molecules. For enzymatic reactions this same principle holds true at low substrate concentrations, where the rate of product formation is almost directly proportional to substrate concentration. As the substrate concentration is raised, however, the reaction velocity does not continue to increase proportionally, and a point is eventually reached where further increases in substrate concentration cause no further change in reaction rate. At this point, enzyme **saturation** has been achieved (Figure 2-3).

To explain the saturation effect, Michaelis and Menten proposed that an enzyme (E) binds directly to its substrate (S), forming a transient **enzyme-substrate complex** (ES). The enzyme-substrate complex in turn breaks down to regenerate free enzyme plus product (P):

$$E + S \rightleftharpoons ES \rightleftharpoons E + P \qquad (2\text{-}6)$$

According to this equation, the rate of product formation is directly related to the concentration of the enzyme-substrate complex. One can therefore envision a simple explanation for the saturation effect: Enzyme molecules are present in small amounts compared to the substrate, so at high concentrations of substrate all the enzyme molecules will be converted to the form of the enzyme-substrate complex. At this point, raising the substrate concentration any further has no effect because all the enzyme molecules are already binding to substrate and producing product as fast as they can.

In spite of the attractive features of the Michaelis-Menten theory, it was not until 25 years later that the existence of the postulated enzyme-substrate complex was verified experimentally. In the late 1930s David Keilin carried out an elegant set of investigations on the enzyme *peroxidase*. This enzyme, which catalyzes the degradation of hydrogen peroxide, is normally a reddish-brown color in solution. Keilin discovered that when peroxidase is mixed with hydrogen peroxide in the absence of the second substrate needed to complete the reaction, the color of the enzyme solution changes dramatically. Such a change in color indicates an alteration in the chemical components present. Since the absence of the second substrate prevented formation of the reaction product, Keilin concluded that the color change was due to the formation of a chemical complex between the enzyme peroxidase and its substrate hydrogen peroxide, the only two ingredients present.

In the years since Keilin's pioneering work, the existence of enzyme-substrate complexes has been confirmed by a variety of other techniques. In some cases enzyme-substrate complexes have even been purified and crystallized, permitting analysis of their three-dimensional structure. This information has provided solid support for the Michaelis-Menten notion that enzymes act by combining, however transiently, with their appropriate substrates.

Enzyme Specificity Is Explained by the Shape of the Active Site

Since enzymes act by binding to their substrates, the exquisite specificity of enzymatic catalysis requires that each enzyme be able to recognize its proper substrates. As we saw in Chapter 1, every protein molecule has a unique three-dimensional shape that is dictated by its amino acid sequence. The role played by the shape of enzyme molecules in substrate recognition has interested biologists and chemists for many years. In the late nineteenth century Emil Fischer first noted that enzymes can distinguish between closely related three-dimensional configurations of the same substrate, leading him to propose that each enzyme possesses a specific region, called an **active site,** whose three-dimensional shape is exactly complementary to the shape of the substrate molecule upon which it acts (Figure 2-4, *top*). According to this **lock-and-key theory,** substrate binding will not occur unless the shape of the substrate is precisely complementary to that of the active site.

In spite of its usefulness in explaining enzyme specificity, the lock-and-key model has turned out to be an oversimplification. More recent investigations suggest that enzymes are dynamic structures that undergo changes in conformation upon binding of substrate. According to the **induced-fit model** proposed by Daniel Koshland, binding of substrate to enzyme triggers alterations in the shape of the enzyme as well as the substrate (see Figure 2-4, *bottom*). The final proper fit between enzyme and substrate occurs only after these changes in enzyme and substrate conformation have

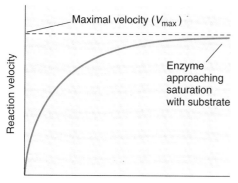

Figure 2-3 *Effect of Increasing Substrate Concentration on the Rate of an Enzyme-Catalyzed Reaction* *At high substrate concentrations the enzyme becomes saturated with substrate, and further increases in substrate concentration produce no further increase in reaction rate.*

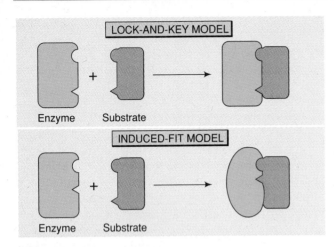

Figure 2-4 *The Lock-and-Key and Induced-Fit Models of Enzyme-Substrate Interaction* *The induced-fit model is a more recent elaboration upon the original lock-and-key idea. In the lock-and-key model, the shape of the substrate must be precisely complementary to that of the active site. In the induced-fit model, the interaction between enzyme and substrate causes them both to undergo some distortion. The distortion of the substrate in turn promotes the formation of the transition state.*

taken place. The significance of the induced-fit arrangement is that the change in the shape of the substrate induced by interaction with the enzyme helps to push the substrate toward its transition state configuration.

In many enzymatic reactions, the substrate is considerably smaller than the enzyme. For example the enzyme *catalase*, with a molecular weight of 250,000 daltons, catalyzes the breakdown of hydrogen peroxide, whose molecular weight is 34 daltons. It therefore follows that only a small portion of an enzyme's amino acid sequence is directly involved in the portion of the active site that binds to the substrate. A variety of techniques have been devised for mapping the arrangement of these particular amino acids. One such approach is to treat enzymes with molecules that form covalent bonds with specific amino acid side chains. For example, treating the enzyme ribonuclease with the alkylating agent iodoacetate causes two histidines in the ribonuclease molecule to become modified by the addition of an acetyl group. Modification of either of these two histidines (amino acids 12 and 119) destroys the catalytic activity of the enzyme, suggesting that they are both part of the active site. Such observations have made it clear that amino acids that are far removed from one another in terms of the linear amino acid sequence come together to form the active site (Figure 2-5). Hence any disruptions in protein conformation that alter the folding of the polypeptide chain can change the spatial relationship among amino acids forming the active site and lead to a loss in enzymatic activity.

The enzyme inhibitor *diisopropyl phosphofluoridate (DIPF)*, which inactivates enzymes by attaching to the hydroxyl group of the amino acid serine, has also

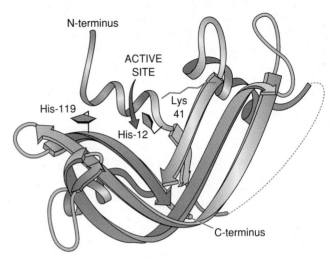

Figure 2-5 *The Active Site of the Enzyme Ribonuclease* *In this three-dimensional model of ribonuclease, regions of α helix are represented as spirals, and β sheets are shown as flat ribbons with arrows pointing toward the C-terminus. The three amino acids that play key roles in the active site (His-12, Lys-41, and His-119) are brought together by the proper folding of the polypeptide chain. In order to produce the crystals that were employed to generate this model of ribonuclease, the peptide bond that normally exists between amino acids 20 and 21 was broken* (dashed line).

been used to map active sites. DIPF selectively interacts with a class of enzymes that share the presence of an essential serine residue at the active site. By analyzing the amino acid sequences adjacent to the serine with which the DIPF has reacted, information about the arrangement of amino acids at the active site can be obtained. This approach has led to the discovery that similar amino acid sequences tend to recur in the active sites of different enzymes (Table 2-3).

The Catalytic Efficiency of the Active Site Is Based on a Combination of Factors

Once the substrate(s) has become bound to the active site, several phenomena contribute to the large acceler-

Table 2-3   Amino Acid Sequence of the Active Sites of Several Related Enzymes

Enzyme	Sequence of Active Site*
Trypsin	Gly-Asp-Ser-Gly-Gly-Pro
Chymotrypsin	Gly-Asp-Ser-Gly-Gly-Pro
Elastase	Gly-Asp-Ser-Gly-Gly-Pro
Thrombin	Asp-Ser-Gly
Alkaline phosphatase	Asp-Ser-Ala
Butyrylcholine esterase	Gly-Glu-Ser-Ala-Gly
Acetylcholine esterase	Glu-Ser-Ala
Aliesterase	Glu-Ser-Ala

*Asp and Glu are acidic amino acids, whereas Gly and Ala are hydrophobic amino acids.

ation in reaction rate that typically occurs in enzyme-catalyzed reactions. First, in reactions involving more than one substrate, the active site brings the reacting molecules close to one another and situates them in the proper orientation, thereby optimizing the chances of the reaction occurring. Second, some enzymes act by forming an unstable covalent intermediate between enzyme and substrate. Such covalent intermediates generate the final reaction product more readily than would the substrate by itself. Third, active sites often contain functional groups that are capable of donating and accepting protons or electrons. The transfer of protons or electrons between enzyme and substrates can increase the reaction rate by creating transitional states that promote the formation of products. Finally, the binding of substrate to an active site is thought to induce a strain or distortion in the structure of the substrate molecule that enhances its reactivity. The fact that enzymes and substrates undergo an induced change in conformation upon binding to one another is consistent with this notion, since the resulting conformational changes may distort the enzyme and/or substrate in such a way as to make the molecules more reactive.

The catalytic activity of most enzymes is based on a combination of the above four factors. To cite an example, let us briefly consider the hydrolysis of peptide bonds catalyzed by the protein-digesting enzyme, *chymotrypsin.* When a molecule of substrate binds to chymotrypsin, serine number 195, which is located at the active site of this enzyme, forms a covalent bond with the carbon atom of the peptide bond to be hydrolyzed (Figure 2-6). Formation of this covalent intermediate is facilitated by histidine number 57, which acts as a proton acceptor. Histidine 57 then donates this proton back to the nitrogen atom of the peptide bond being hydrolyzed, thereby facilitating cleavage of the peptide bond. In the final step of the reaction, histidine 57 accepts a proton from a water molecule and transfers it back to serine 195 as part of the process that releases the covalently bound peptide fragment from the enzyme. Although proton transfer and the formation of a covalent intermediate are therefore crucial contributors to the mechanism by which chymotrypsin cleaves peptide bonds, they are not the only contributors to the catalytic efficiency of this enzyme. Orientation and strain effects must also be postulated in order to account for the overall efficiency with which chymotrypsin cleaves peptide bonds.

ENZYME KINETICS AND REGULATION

We have now seen that enzymes provide reactions with a new transition state of lower free energy, thereby speeding up the rate at which equilibrium is attained. Therefore, in the presence of enzymes, reactants need less input of energy to form their products.

This phenomenon is of profound biological significance because it permits metabolic reactions to occur at the moderate temperatures that are compatible with living organisms.

The preceding considerations do not, however, tell us very much about the factors that determine how fast a given enzyme-catalyzed reaction will proceed under the conditions that prevail in a particular cell. Analysis of the factors involved in determining the rate at which an enzyme-catalyzed reaction will occur is referred to as the study of **enzyme kinetics.** In the following sections, we will discuss some of the major factors that influence enzyme kinetics in living cells.

V_{max} and K_m Provide Information about an Enzyme's Catalytic Efficiency and Substrate Affinity

The earliest attempts to analyze the kinetics of enzyme-catalyzed reactions were made by Michaelis and Menten, whose work led to the description of two fundamental kinetic parameters of enzymatic reactions. The first of these parameters, called V_{max}, is defined as the **maximum velocity** (rate) that a reaction can attain in the presence of a given amount of enzyme. In other words, V_{max} represents the rate of an enzymatic reaction under conditions in which the enzyme has become saturated with substrate. The value of V_{max} for any given enzymatic reaction varies with enzyme concentration, since the more enzyme molecules present, the faster the reaction will proceed at saturation. V_{max} is a reflection of how quickly individual enzyme molecules can catalyze the conversion of substrate to product. The term *turnover number* is employed to refer to the actual number of substrate molecules that one enzyme molecule can convert to product per unit time when the enzyme is saturated with substrate. The most efficient enzymes have very high turnover numbers, catalyzing the conversion of several hundred thousand molecules of substrate to product per second; on the other hand, less efficient enzymes may have turnover numbers that are less than one molecule per second (Table 2-4). Enzymes with higher turnover numbers will naturally exhibit higher values of V_{max} at any given enzyme concentration than will enzymes with lower turnover numbers.

The other important kinetic parameter of enzymatic reactions is the **Michaelis constant** or K_m, which is defined as the substrate concentration at which the reaction velocity is equal to one-half the maximal velocity (V_{max}). The substrate concentration at which a reaction reaches half its maximal velocity is always found to be the same for any given enzyme, regardless of the enzyme concentration (Figure 2-7, *top*); in other words K_m, unlike V_{max}, does not change when the enzyme concentration is varied. The value of K_m is a useful indicator of the relative affinity of an enzyme for its substrate. When the K_m for a given enzyme is high, it means that a

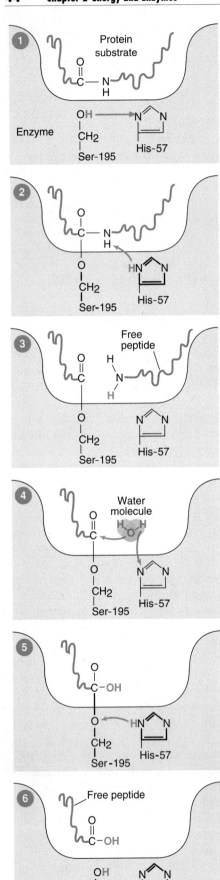

high substrate concentration is required before half the active sites become filled; in other words, the enzyme has a low affinity for the substrate and so the substrate concentration must be relatively high before the reaction will proceed at an appreciable rate. Conversely when the K_m is low, it means that a low concentration of substrate will cause the active sites to become filled; in this case, the enzyme has a high affinity for the substrate and so the reaction will proceed at relatively low substrate concentrations.

The values of K_m vary widely among enzymes, falling in the general range of 10^{-7} to 10^{-1} M. Since the Michaelis constant is a measure of the affinity of an enzyme for a substrate, enzymes that catalyze reactions involving more than one substrate have multiple K_m values, one for each substrate. Inside cells, the concentrations of most substrates fall within an order of magnitude of the K_m values for their respective enzymes. In this range the rate of an enzyme-catalyzed reaction is sensitive to changes in substrate concentration, indicating that enzymes have evolved with substrate binding affinities in the optimal range for biological efficiency and regulation. The evolution of enzyme molecules with such attributes has been an important step in permitting precise control of cellular reaction rates.

Values for V_{max} and K_m Can Be Determined Using the Michaelis-Menten Equation

Because K_m and V_{max} provide information about an enzyme's substrate affinity and catalytic efficiency, it is important to know how to determine the value of these parameters for individual enzymes. Approximate values for V_{max} and K_m can be estimated by examining a graph of reaction velocity plotted against substrate concentration (see Figure 2-7, *top*), but simpler and more accurate approaches are based on mathematical derivations of the Michaelis-Menten equation. The derivations of these equations are the province of biochemistry textbooks and so will not be described in detail here. For our purposes, it is sufficient to start with the fact that the original formulation of the enzyme-substrate com-

Figure 2-6 Catalytic Mechanism by Which the Enzyme Chymotrypsin Cleaves a Polypeptide Chain   *Serine 195 (Ser-195) and histidine 57 (His-57) are crucial constituents of the active site of chymotrypsin. During catalysis, Ser-195 forms a transient covalent bond with the substrate and His-57 serves as a proton donor and acceptor. The diagram illustrates the following six steps in the catalytic mechanism: (1) the substrate binds noncovalently to the active site and a proton is donated from Ser-195 to His-57, allowing a bond to form between Ser-195 and the substrate, (2) the proton from His-57 helps catalyze peptide-bond breakage in the substrate, (3) one of the peptide products is released, (4) a water molecule donates its proton to His-57 and its hydroxyl group to the remaining peptide fragment, (5) the proton from His-57 is transferred to Ser-195, leading to (6) release of the second peptide.*

Table 2-4 Turnover Numbers of Selected Enzymes

Enzyme	Turnover Number*
Carbonic anhydrase	600,000
3-Ketosteroid isomerase	285,000
Catalase	93,000
Acetylcholinesterase	25,000
Lactate dehydrogenase	1,000
Ribonuclease	800
β-Galactosidase	200
DNA polymerase I	15
Tryptophan synthetase	2
Lysozyme	0.5

*Turnover numbers are expressed as the number of substrate molecules converted to product per second.

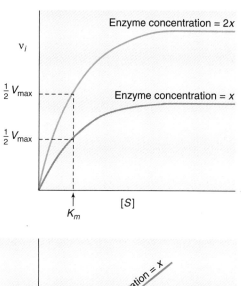

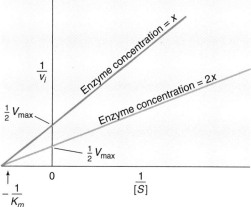

Figure 2-7 *Effect of Enzyme Concentration on the Rate of an Enzyme-Catalyzed Reaction* (Top) *In a plot of reaction velocity versus substrate concentration, the reaction is seen to proceed at half its maximal velocity when the substrate concentration is equal to the Michaelis constant (K_m). When the enzyme concentration is doubled, the maximal velocity (V_{max}) is also doubled, but K_m stays the same. (Bottom) In a Lineweaver-Burk plot of the same data, the y intercept corresponds to the reciprocal of V_{max} and the x intercept equals the negative reciprocal of K_m.*

plex (Equation 2-6) allows the mathematical derivation of the following relationship between V_{max} and K_m:

$$v_i = \frac{V_{max}\,[S]}{K_m + [S]} \tag{2-7}$$

In this equation, v_i stands for the *initial velocity,* which is the rate of product formation during the early stages of an enzyme reaction before product accumulation or other secondary events have caused the reaction to slow down, and [S] is the substrate concentration.

Two points are worth emphasizing about Equation 2-7, which is commonly referred to as the **Michaelis-Menten equation.** The first is that it accurately predicts the hyperbolic shape of the curves obtained when one plots experimental data for reaction velocity versus substrate concentration (see Figure 2-7, *top*). The second point relates to what happens when the substrate concentration in Equation 2-7 is set equal to K_m:

$$v_i = \frac{V_{max} K_m}{K_m + K_m}$$

$$= \frac{V_{max} K_m}{2K_m}$$

$$= \frac{V_{max}}{2}$$

The above derivation indicates that when the substrate concentration equals the Michaelis constant, the reaction velocity is equal to V_{max} divided by two. In other words, we have just provided the mathematical derivation of our original definition of the Michaelis constant, which was defined as the substrate concentration at which an enzymatic reaction proceeds at one-half maximal velocity.

In its original form, the Michaelis-Menten equation has one serious drawback: Graphing the equation yields a hyperbolic curve that cannot be used easily to determine actual values of V_{max} and K_m. If one tries to determine the value of V_{max} from such a hyperbolic plot (see Figure 2-7, *top*), it is difficult to know exactly when V_{max} has been attained because of the gradually changing slope of the curve. To overcome this difficulty, H. Lineweaver and D. Burk converted the Michaelis-Menten equation into a linear form by inverting Equation 2-7 as follows:

$$\frac{1}{v_i} = \frac{K_m + [S]}{V_{max}\,[S]}$$

$$= \frac{K_m}{V_{max}}\left[\frac{1}{[S]}\right] + \frac{1}{V_{max}} \tag{2-8}$$

When Equation 2-8 is graphed as $(1/v_i)$ versus $(1/[S])$, a straight line is obtained (see Figure 2-7, *bottom*). In such a graph, called a **Lineweaver-Burk plot** (or double-reciprocal plot), the intercept with the y axis corresponds to $1/V_{max}$ and the intercept with the x axis corresponds to $-1/K_m$. Figure 2-8 illustrates how the Lineweaver-Burk plot is employed in practice to determine V_{max} and K_m. As can be seen in this example, the power of the Lineweaver-Burk plot stems from the fact that it permits V_{max} and K_m to be determined from a relatively small amount of kinetic data.

Finding out the value of V_{max} and K_m provides crucial information about the potential rate at which an enzyme can function, but it is only the first step in assessing the rate at which the reaction will actually proceed in living cells. As we will see in the following sections, the rate of an enzyme-catalyzed reaction is influenced by a variety of other factors, such as environmental conditions and the presence or absence of various molecules that influence enzymatic activity.

Tube number	Substrate concentration (mM) [S]	Reaction velocity (nmol/min) v_i	$\dfrac{1}{[S]}$	$\dfrac{1}{v_i}$
❶	0.25	0.48	4.0	2.08
❷	0.50	0.63	2.0	1.58
❸	1.0	0.75	1.0	1.33
❹	2.5	0.85	0.4	1.18

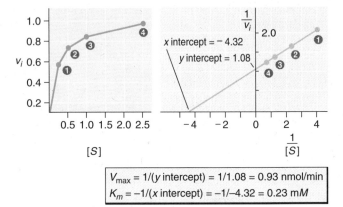

$V_{max} = 1/(y \text{ intercept}) = 1/1.08 = 0.93$ nmol/min
$K_m = -1/(x \text{ intercept}) = -1/-4.32 = 0.23$ mM

Figure 2-8 An Example Showing How K_m and V_{max} Are Determined Experimentally Using the Lineweaver-Burk Plot *In this experiment, four test tubes containing equal amounts of the enzyme acid phosphatase were incubated with varying amounts of substrate.* (Left graph) *When the initial rate of product formation* (v_i) *is plotted against substrate concentration* [S], *it is difficult to determine* K_m *and* V_{max} *because the reaction has not reached saturation.* (Right graph) *A Lineweaver-Burk plot of the same data yields a straight line, making it relatively easy to determine the values of* K_m *and* V_{max}.

Enzyme Activity Is Influenced by pH and Temperature

Enzymatic activity is sensitive to relatively small changes in temperature or pH, even though such changes do not alter the covalent structure of enzyme molecules. Changes in pH affect enzymatic activity by altering the ionization of the enzyme and/or substrate, which in turn influences enzyme conformation and substrate binding. Lowering the pH (increasing the proton concentration) causes protons to bind to various enzyme side chains and substrate groups, causing these molecules to become more positively charged; conversely, increasing the pH induces protons to dissociate, causing molecules to become more negatively charged. Such changes in electric charge can disrupt ionic bonds that are involved in maintaining an enzyme's proper conformation, and may even denature the enzyme completely. For this reason, extreme changes in pH usually lead to enzyme inactivation. Optimum enzyme activity is typically observed around pH 7, which is close to the pH prevailing within cells. However, enzymes functioning in environments that are more acidic or basic exhibit optimal activity at pH values that are closer to those of their natural surroundings. For example pepsin, the protein-digesting enzyme secreted into the highly acidic gastric juices of the stomach, functions optimally at a pH of about 2 (Figure 2-9, *left*).

Temperature, like pH, can influence enzyme activity by altering enzyme conformation. It was mentioned earlier in the chapter that raising the temperature enhances the rate of chemical reactions by increasing the frequency of collisions between reacting molecules. Up to a certain point, this principle holds for enzyme-catalyzed reactions as well. However, a temperature is eventually reached that begins to disrupt the weak bonds involved in maintaining the secondary, tertiary, and quaternary structure of enzyme molecules. At this stage, loss of proper conformation and eventual denaturation ensue. Thus the effect of temperature on enzyme-catalyzed reactions exhibits two distinct phases: an increase in reaction rate that accompanies moderate increases in temperature, and a subsequent decline in reaction rate that occurs as further increases in temperature lead to enzyme denaturation (see Figure 2-9, *right*). Mammalian enzymes usually exhibit maximum activity near the typical body temperature of 37°C, but significant variations from this norm are observed, especially in organisms where the prevailing temperature is significantly warmer or colder. For example, enzymes from bacteria that live in hot springs can function efficiently at temperatures exceeding 90°C, whereas enzymes from arctic microorganisms are active at temperatures near freezing.

Enzymes May Require Coenzymes, Prosthetic Groups, and Metal Ions for Optimal Activity

Some enzymes consist solely of protein and exhibit optimum catalytic activity in the absence of any other components. Many enzymes, however, require the presence of nonprotein enzyme constituents. If these nonprotein components are lacking or are present in insufficient amounts, such enzymes are unable to function at their optimal rates. The nonprotein components utilized by enzymes can be grouped into three broad categories: *coenzymes, prosthetic groups,* and *metal ions.*

Coenzymes are small organic molecules that bind *reversibly* to enzymes and participate directly in the catalytic process. Such compounds are usually derived from *vitamins,* which are essential organic molecules that an organism cannot synthesize and hence must obtain from its food sources. Table 2-5 summarizes the kinds of catalytic mechanisms in which the main coenzymes participate. Later in the chapter we will discuss the roles of the coenzymes *NAD* and *FAD,* whose oxidation and reduction are central to the process by which cells obtain energy from their surroundings and transform it into useful chemical forms.

Prosthetic groups are similar to coenzymes in function, but they are attached to enzymes more tightly. Like coenzymes, prosthetic groups are an essential component of the catalytic mechanism in the enzymes in which they occur. An example of this type of interaction occurs in *catalase,* a H_2O_2-degrading enzyme that is normally a reddish-brown color in solution. When treated with a mild acid, catalase dissociates into a colorless protein and a red, iron-containing prosthetic group known as a *heme group.* The protein portion of the enzyme, lacking its normal prosthetic group, is referred to as an **apoprotein.** By itself, the catalase apoprotein is incapable of catalyzing the breakdown of

H_2O_2 because the heme group plays an essential role in the catalytic mechanism.

The final type of nonprotein component utilized by enzymes consists of *metal ions* such as iron, copper, manganese, magnesium, calcium, and zinc. These metal ions either stabilize the proper conformation of the enzyme and/or substrate, or participate directly in the chemical reaction being catalyzed. Metal ions are usually held in place by bonds to amino acid side chains, although they may also be bound to prosthetic groups. In catalase, for example, the heme prosthetic group consists of an *iron atom* bound to an organic ring compound called *protoporphyrin.* The ability of this iron atom to donate and accept electrons plays a central role in the mechanism by which catalase degrades H_2O_2.

Enzyme Inhibitors Can Act Irreversibly or Reversibly

Many naturally occurring and synthetic chemicals are known to interfere with enzyme activity. Laboratory studies of these enzyme inhibitors have provided important insights into issues such as the mechanism of enzyme action, the nature of catalytic sites, and the role of enzyme conformation in enzyme activity. The study of enzyme inhibition has also had practical benefits, helping us to deal more effectively with toxins and poisons that act on human enzymes, as well as facilitating the design of specific enzyme inhibitors to be used as therapeutic drugs.

Based on differences in their mechanisms of action, enzyme inhibitors can be classified into three broad cat-

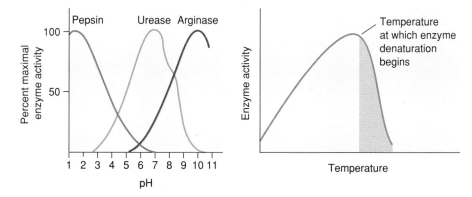

Figure 2-9 *The Effects of pH and Temperature on Enzyme Activity* (Left) *The effect of pH on the catalytic activity of three different enzymes. The curve for urease, which exhibits optimum activity around pH 7, is typical for most enzymes. However, enzymes functioning in environments that are more acidic or basic, such as pepsin and arginase, exhibit optimal activity at pH values closer to those of their natural surroundings.* (Right) *The effect of temperature on the rate of a typical enzyme-catalyzed reaction. Increasing the temperature causes an increase in reaction rate up to a certain point (usually around 37°C for mammalian enzymes), after which the reaction rate declines because of enzyme denaturation.*

Table 2-5 Major Coenzymes and the Vitamins from Which They Are Derived

Vitamin	Coenzyme	Function of Coenzyme
Nicotinic acid (niacin)	NAD (nicotinamide adenine dinucleotide)	Oxidation-reduction
	NADP (nicotinamide adenine dinucleotide phosphate)	Oxidation-reduction
Riboflavin (vitamin B_2)	FMN (flavin mononucleotide)	Oxidation-reduction
	FAD (flavin adenine dinucleotide)	Oxidation-reduction
Pantothenic acid	Coenzyme A	Acyl group transfer
Thiamine (vitamin B_1)	Thiamine pyrophosphate	Oxidative decarboxylation
Pyridoxine (vitamin B_6)	Pyridoxal phosphate	Amino group transfer
Biotin	Biocytin	CO_2 transfer
Folic acid	Tetrahydrofolate coenzymes	One-carbon transfer
Cobalamine (vitamin B_{12})	Cobamide coenzymes	Alkyl group transfer

egories: *irreversible inhibitors, competitive inhibitors,* and *noncompetitive inhibitors.*

Irreversible Inhibitors Are Toxic Substances That Bind Covalently to Enzymes

Substances that bind covalently to enzymes and permanently disrupt their catalytic activity are referred to as **irreversible inhibitors.** Agents in this category tend to be highly toxic and are cause for extreme concern when encountered in the environment. A classic example of an irreversible inhibitor is DIPF (page 42), a prominent component of insecticides and nerve gas. DIPF combines covalently with the hydroxyl group of the amino acid serine (Figure 2-10). Since serine hydroxyl groups play a role in the catalytic sites of many enzymes (see Table 2-3), this reaction often results in enzyme inactivation. Enzymes involved in the transmission of nerve impulses are quickly inactivated in this way, explaining the extremely toxic effects of DIPF and its derivatives on the nervous system.

In some cases, irreversible inhibitors are very selective in the enzymes they attack because the inhibitor closely resembles a particular substrate. This structural resemblance allows them to bind preferentially to the active site of a specific kind of enzyme. A covalent bond then forms between inhibitor and enzyme, permanently blocking the active site. Numerous synthetic inhibitors function in this way, as do some naturally occurring toxins. For example penicillin, the antibiotic manufactured by the fungus *Penicillium,* is a selective inhibitor of enzymes involved in the synthesis of bacterial cell walls (page 251). Because penicillin does not act upon other enzymes, humans can take penicillin to treat a bacterial infection and not be concerned about toxic effects on enzymes other than those involved in bacterial cell wall formation.

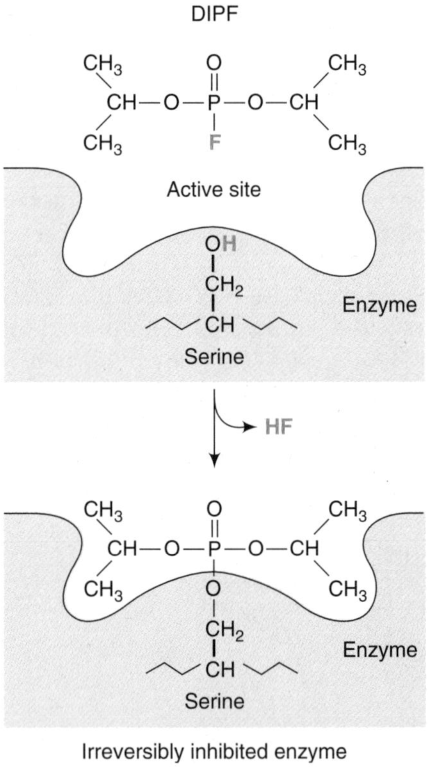

Figure 2-10 *Mechanism of Action of Diisopropyl Phosphofluoridate (DIPF), an Irreversible Enzyme Inhibitor* *DIPF causes irreversible enzyme inhibition by forming a covalent bond with the amino acid serine, which is part of the active site of many enzymes.*

Competitive Inhibitors Are Reversible Inhibitors That Bind to the Active Site, Altering K_m But Not V_{max}

In contrast to the irreversible inhibitors just described, a diverse group of inhibitors bind *noncovalently* and *reversibly* to enzyme molecules. Some of these reversible inhibitors are called **competitive inhibitors** because they resemble one of the substrates of the reaction being inhibited, and are therefore capable of competing with the substrate for binding to the enzyme's active site. A typical example of competitive inhibition involves the enzyme *succinate dehydrogenase,* which catalyzes the oxidation of succinate in mitochondria. Several substances that resemble succinate in structure can bind to the enzyme's active site, but unlike succinate, these substances are not oxidized by the enzyme (Figure 2-11). Succinate dehydrogenase molecules containing such competitive inhibitors bound to their active sites cannot bind to the normal substrate, resulting in a net decrease in reaction rate.

Because the binding of competitive inhibitors to their respective enzymes is reversible, an equilibrium is established between free and enzyme-bound inhibitor molecules. In other words, a cycle is established in which inhibitor molecules are continually binding to the enzyme's active site and then dissociating from it. Each time an inhibitor molecule dissociates from the enzyme, a moment occurs when the active site is free and therefore potentially available to bind to either another inhibitor molecule *or* a molecule of substrate. Hence a competition is established between inhibitor and substrate for the active site. If the substrate concentration is low, the active site is more likely to bind to the inhibitor and therefore most enzyme molecules will be inhibited; but if the substrate concentration is high, the active site is more likely to bind to the substrate than the inhibitor, and so fewer enzyme molecules will be inhibited (Figure 2-12). Hence the rate of the reaction can be increased by raising the concentration of substrate relative to inhibitor; at very high substrate concentrations the inhibition can even be overcome entirely, since the probability becomes very small that any free enzyme molecule will encounter a molecule of inhibitor before encountering a molecule of substrate.

Thus by definition, *inhibition by competitive inhibitors can be overcome by adding excess substrate.* This means that the maximal velocity (V_{max}) of an enzymatic reaction is unaffected by competitive inhibitors. But due to the binding of inhibitor molecules to the enzyme, it takes a higher substrate concentration to achieve V_{max} or one-half V_{max}; hence the K_m will be higher. An unaltered V_{max} and an increased K_m are therefore the hallmarks of competitive inhibition (Figure 2-13).

A striking example of how the principles of competitive inhibition can be exploited for medical purposes is provided by the use of sulfa drugs to fight bacterial infections. The structure of the sulfa drug *sulfanilamide* closely resembles that of *p*-aminobenzoic acid, a substance that bacteria utilize as a precursor for the synthesis of the essential compound, folic acid. By competitively inhibiting the bacterial enzyme responsible for metabolizing *p*-aminobenzoic acid, sulfanilamide blocks the production of folic acid and eventually leads to bacterial death. Humans are not adversely affected by sulfanilamide because we do not employ this enzymatic pathway for manufacturing folic acid.

Noncompetitive Inhibitors Are Reversible Inhibitors That Alter V_{max} by Binding Away from the Active Site

Another class of inhibitors, called **noncompetitive inhibitors,** also bind noncovalently and reversibly to enzyme molecules. But in contrast to the behavior of competitive inhibitors, the binding of noncompetitive inhibitors to enzymes cannot be reversed by adding excess substrate. The reason for this differing behavior is that noncompetitive inhibitors do not resemble an enzyme's substrate and hence do not compete with the substrate for binding to the active site. Noncompetitive inhibitors instead bind to regions of the enzyme away from the active site, triggering changes in enzyme conformation that interfere with an enzyme's catalytic efficiency (turnover number) rather than decreasing the proportion of enzyme molecules that are bound to substrate (Figure 2-14).

Because of this reduction in an enzyme's turnover number, the value of V_{max} is always decreased in the presence of a noncompetitive inhibitor, regardless of the substrate concentration. In the simplest type of noncompetitive inhibition, the inhibitor has no effect on the binding affinity of the substrate for the active site. The result is therefore a reduction in V_{max}, but no change in the value of K_m (Figure 2-15). In actuality, many noncompetitive inhibitors exert effects on substrate binding as well as catalytic efficiency. In such cases, K_m as well as V_{max} will be altered.

Since noncompetitive inhibitors do not resemble substrates and hence do not bind preferentially to a

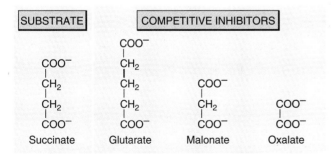

Figure 2-11 *Competitive Inhibitors of the Enzyme Succinate Dehydrogenase* *Glutarate, malonate, and oxalate structurally resemble the normal substrate succinate, and therefore compete for binding to the enzyme's active site.*

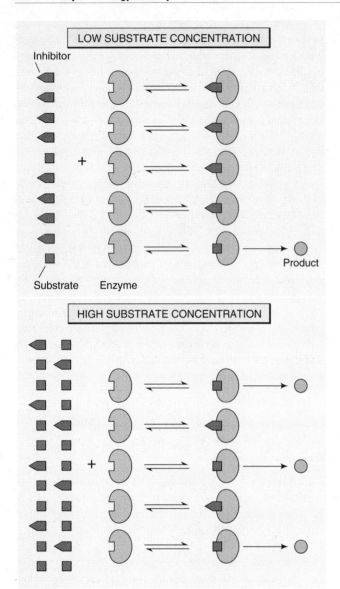

Figure 2-12 *Mechanism of Competitive Inhibition*
Inhibitor and substrate are both capable of binding to the active site, but only the substrate can be converted to product. (Top) At low substrate concentration, most enzyme molecules are bound to the inhibitor and therefore are not generating product. (Bottom) At higher substrate concentration, more enzyme molecules are bound to substrate and hence are generating product.

specific type of active site, a single noncompetitive inhibitor often acts on a broad spectrum of enzymes. For example, enzymes that require metal ions are subject to noncompetitive inhibition by any agent that binds reversibly to the required ion. Among the noncompetitive inhibitors that function in this way are *cyanide*, which binds to iron, and *EDTA* (ethylenediaminetetraacetate), which binds to magnesium. Because iron-containing proteins play roles in mitochondrial oxidation-reduction reactions, and magnesium ions are required for all reactions involving ATP, the toxic effects

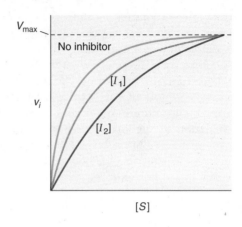

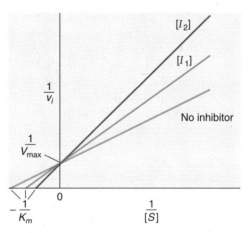

Figure 2-13 *Kinetics of Competitive Inhibition* (Top) *Plot of reaction velocity against substrate concentration at low [I₁] and high [I₂] concentrations of inhibitor. (Bottom) The same data plotted by the Lineweaver-Burk method. Note that competitive inhibition changes the value of K_m but not V_{max}.*

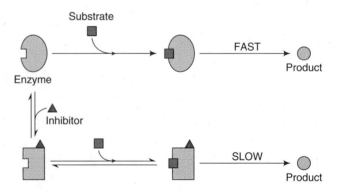

Figure 2-14 *Mechanism of Noncompetitive Inhibition* *A noncompetitive inhibitor binds to a different region of the enzyme than does the substrate. Binding of inhibitor to the enzyme triggers a conformational change that decreases the enzyme's catalytic efficiency, but does not affect its affinity for substrate.*

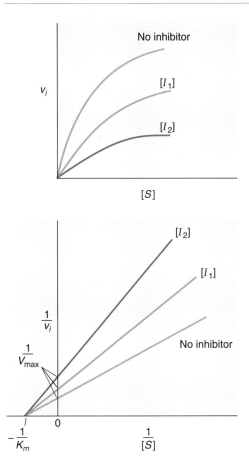

Figure 2-15 *Kinetics of Noncompetitive Inhibition* (Top) *Plot of reaction velocity against substrate concentration at low [I_1] and high [I_2] concentrations of inhibitor.* (Bottom) *The same data plotted by the Lineweaver-Burk method. Note that noncompetitive inhibition decreases the value of* V_{max} *but leaves* K_m *unchanged.*

of such inhibitors on cell function can be extremely widespread.

Allosteric Regulation Involves Reversible Changes in Protein Conformation Induced by Allosteric Activators and Inhibitors

One of the unique features that distinguishes enzymes from nonbiological catalysts is that enzymes can be regulated. The ability to control enzyme activity is crucial if cells are to maintain their metabolic pathways in proper balance, with each pathway generating neither too much nor too little of its particular product. This type of control is often mediated by molecules that are products of the metabolic pathways involved. When a metabolic product functions to inhibit one of the enzymes participating in the pathway by which the product is synthesized, the phenomenon is termed **feedback inhibition.** This mechanism prevents a product from being overproduced, for as more product is formed, the pathway generating the product becomes progressively more inhibited (Figure 2-16).

The molecular mechanism underlying feedback inhibition was not comprehended until 1963, when Jacques Monod, Jean-Pierre Changeux, and Francois Jacob first proposed a theory of enzyme control known as **allosteric regulation.** This theory was founded on the recognition that feedback inhibitors do not resemble the substrates of the enzymes they inhibit; it was therefore proposed that feedback inhibitors bind not to an enzyme's active site, but rather to a special regulatory region that was termed an **allosteric site.** Allosteric sites were envisioned as being highly selective, binding to specific kinds of regulatory molecules just as active sites bind to specific substrates. According to this model, the binding of an allosteric regulator to an allosteric site is a transient event that triggers a transient change in enzyme conformation. In the case of enzymes controlled by feedback inhibition, the conformational change causes a decrease in the enzyme's catalytic activity. Moreover, it was subsequently discovered that some enzymes are activated rather than inhibited by the binding of an appropriate allosteric regulator. In other words, some allosteric sites are designed to bind to *allosteric inhibitors* that trigger a decrease in enzyme activity, whereas others bind to *allosteric activators* that increase enzyme activity (Figure 2-17).

When this model of allosteric control was first proposed in 1963, there was little experimental evidence to support it. A few years later, however, John Gerhart and Howard Schachman discovered that the behavior of the enzyme *aspartate transcarbamoylase (ATCase)* closely conforms to the allosteric model. ATCase catalyzes the first step in the biosynthesis of the nucleotides cytidine triphosphate (CTP) and uridine triphosphate (UTP), and is subject to feedback inhibition by one of the end products of this pathway, CTP. When Gerhart and Schachman treated ATCase with the denaturing agent *p*-chloromercuribenzoate (PCMB), the enzyme maintained its catalytic activity but lost its sus-

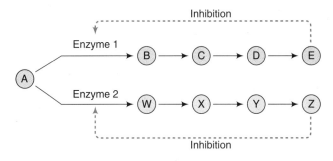

Figure 2-16 *The Principle of Feedback Inhibition* *The final product of each reaction pathway regulates its own formation by inhibiting an enzyme that catalyzes one of the early steps in the formation of that product.*

ceptibility to inhibition by CTP. Analysis of the denatured enzyme by centrifugation revealed the presence of two types of polypeptide subunits instead of the intact enzyme (Figure 2-18). One type of subunit possessed enzymatic activity but could not be inhibited by CTP. The other had no enzymatic activity but was capable of binding CTP. When the two types of subunits were recombined, normal enzyme activity susceptible to inhibition by CTP was regenerated. These results led to the conclusion that ATCase consists of two kinds of subunits, a **catalytic subunit** that carries out the reaction and a **regulatory subunit** that binds the allosteric inhibitor, CTP. The multisubunit makeup of this enzyme has since been confirmed by electron microscopy (Figure 2-19). It is now known that ATCase consists of six catalytic subunits and six regulatory subunits, and that in addition to allosteric inhibition by CTP, it is susceptible to allosteric activation by ATP (Figure 2-20).

In the years following this pioneering work with ATCase, a large number of other enzymes have been found to be subject to allosteric regulation. Like ATCase, these allosteric enzymes consist of multiple subunits, with catalytic and allosteric sites present on

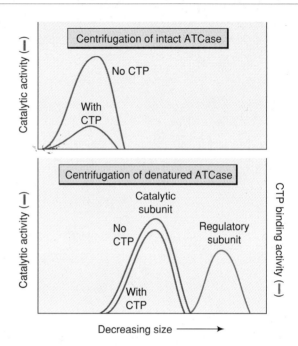

Figure 2-18 Experiment Demonstrating the Existence of Separate Catalytic and Regulatory Subunits in the Allosteric Enzyme Aspartate Transcarbamoylase (Top) *Purification of aspartate transcarbamoylase (ATCase) by centrifugation yields an intact enzyme exhibiting ATCase activity that is subject to inhibition by the allosteric regulator CTP. (Bottom) When ATCase is denatured by treatment with PCMB and again purified by centrifugation, two components are revealed: a catalytic subunit whose enzymatic activity is not influenced by CTP, and a smaller regulatory subunit that binds CTP but has no catalytic activity.*

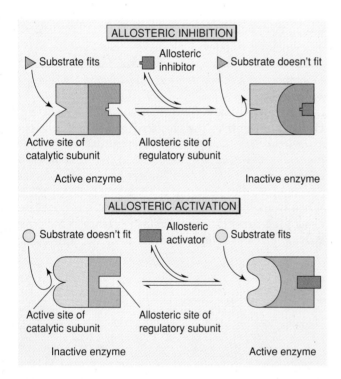

Figure 2-17 Allosteric Control of Enzyme Activity *The enzyme in this diagram consists of two subunits, a catalytic subunit that carries out the reaction and a regulatory subunit that binds to allosteric regulators. (Top) The reversible binding of an allosteric inhibitor to its allosteric site induces a change in enzyme conformation that decreases the enzyme's catalytic activity. (Bottom) The reversible binding of an allosteric activator to its allosteric site induces a change in enzyme conformation that increases the enzyme's catalytic activity.*

different subunits. Individual allosteric enzymes often have sites for several different kinds of allosteric inhibitors and/or activators, making allosteric control an extremely powerful means of regulating an enzyme's catalytic activity. Allosteric inhibitors and activators usually act by altering the value of K_m (see Figure 2-20), but effects on V_{max} are sometimes observed as well. In either case, the reaction kinetics do not tend to fit the typical inhibition kinetics observed with competitive or noncompetitive inhibitors. This differing behavior is caused by the presence of multiple catalytic and allosteric sites, which leads to the phenomenon of *cooperativity* described in the following section.

Allosteric Enzymes Exhibit Cooperative Interactions between Subunits

Allosteric enzymes and other proteins composed of multiple subunits do not usually exhibit typical Michaelis-Menten kinetics. To illustrate this point, let us consider the behavior of the allosteric enzyme ATCase. If one measures the rate of the reaction catalyzed by this enzyme as a function of increasing substrate con-

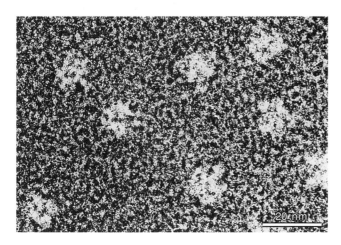

Figure 2-19 *Electron Micrograph of Negatively Stained Aspartate Transcarbamoylase Molecules* *The micrograph reveals that each enzyme molecule is composed of multiple subunits. Courtesy of R. C. Williams.*

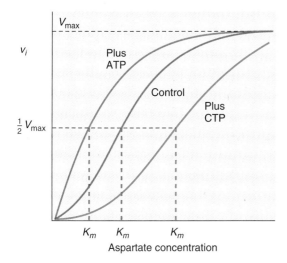

Figure 2-20 *Allosteric Regulation of Aspartate Transcarbamoylase* *The allosteric inhibitor CTP increases the K_m of the reaction, whereas the allosteric activator ATP decreases the K_m.*

centration, a sigmoidal (S-shaped) curve is obtained instead of the normal hyperbola (compare Figures 2-3 and 2-20). This sigmoidal shape means that at low substrate concentrations, small increases in the concentration of substrate cause the enzyme to become more efficient. This effect, known as **positive cooperativity,** occurs because the binding of the first substrate molecule to an active site on one catalytic subunit of ATCase induces a conformational change that causes active sites on other catalytic subunits to bind substrate with increased affinity. Conversely, some enzymes exhibit **negative cooperativity,** in which the binding of a substrate molecule to the active site changes the enzyme's conformation so that other catalytic sites bind the substrate with less

affinity. In the latter case the curve relating reaction velocity to substrate concentration resembles a hyperbola, but the shape of the curve is significantly distorted from the normal hyperbola observed with enzymes that do not exhibit cooperativity (Figure 2-21).

The existence of cooperativity allows cells to produce enzymes that are more sensitive or less sensitive to changes in substrate concentration. Positive cooperativity causes an enzyme's catalytic activity to increase faster than normal as the substrate concentration is increased, whereas negative cooperativity causes catalytic activity to increase more slowly as the substrate concentration is increased. Under the conditions that prevail in living cells, the need exists for both kinds of enzymes: enzymes that are extremely sensitive to changes in substrate concentration and enzymes whose sensitivity to substrate concentration is diminished. The existence of positive and negative cooperativity gives cells a way of making enzymes of both types. In addition to these effects on substrate binding, cooperativity can also influence the binding of allosteric regulators to allosteric sites. Such effects produce enzymes with greater or lesser sensitivity to fluctuations in the concentration of allosteric regulators.

Enzyme Activity Can Be Regulated by Covalent Modifications of Protein Structure

Allosteric regulation is only one of several ways that have evolved to regulate enzyme activity (Figure 2-22).

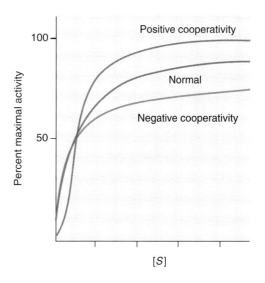

Figure 2-21 *Positive and Negative Cooperativity* *Enzymes exhibiting positive cooperativity are more sensitive to changes in substrate concentration and achieve maximal velocity at lower substrate concentrations than in the absence of cooperativity. Enzymes exhibiting negative cooperativity display the opposite characteristics; that is, they are less sensitive to changes in substrate concentration and achieve maximal velocity at higher substrate concentrations than in the absence of cooperativity.*

Another widely employed strategy involves the covalent modification of enzyme structure. Depending on the enzyme involved, covalent modification may either increase or decrease catalytic activity, and the process may be either reversible or irreversible. In many cases, covalent modification is required for an enzyme to exhibit any catalytic activity at all.

Covalent enzyme modifications can be divided into two general categories: (1) adding and removing chemical groups and (2) cleaving peptide bonds. The first category involves the addition (and removal) of chemical groups such as methyl, acetyl, ADP-ribosyl, and phosphate in reactions referred to as *methylation, acetylation, ADP-ribosylation,* and *phosphorylation.* Of these, **phosphorylation** (addition of phosphate groups) and **dephosphorylation** (removal of phosphate groups) are the most commonly encountered mechanisms for regulating enzyme activity. Among the numerous enzymes known to be regulated by phosphorylation and dephosphorylation is *glycogen phosphorylase,* the enzyme that catalyzes the breakdown of glycogen to glu-

cose 1-phosphate. Phosphorylase exists in two forms that differ in both structure and activity. One form, called *phosphorylase b,* is a dimer (two subunits) that is catalytically inactive. The two identical polypeptide chains comprising this dimer each contain a serine at position 14 that can be phosphorylated. When a cell containing phosphorylase b requires glucose, the two serines are phosphorylated by another enzyme called phosphorylase kinase; as a result, the phosphorylase b dimer is converted to a catalytically active tetramer (four subunits) called *phosphorylase a.* The opposite reaction, which involves the dephosphorylation of phosphorylase a, is employed to convert the enzyme back to its inactive form (Figure 2-23).

Reversible cycles of phosphorylation and dephosphorylation are also employed in the regulation of *pyruvate dehydrogenase,* an enzyme that catalyzes the conversion of pyruvate to acetyl-CoA. Pyruvate dehydrogenase is part of a larger *multienzyme complex* which consists of several enzymes that function in sequence. Such complexes are fairly common because they limit the distance each substrate molecule must diffuse as it is acted upon by subsequent enzymes in a metabolic pathway. Among the enzymes present in the pyruvate dehydrogenase complex are a protein kinase and a protein phosphatase, which catalyze the phosphorylation and dephosphorylation of pyruvate dehydrogenase, respectively. In contrast to the situation with glycogen phosphorylase, the phosphorylation of pyruvate dehydrogenase inhibits enzyme activity and dephosphorylation activates it.

The second category of covalent enzyme modification, which involves the cleavage of polypeptide chains, is an altogether different process from adding and removing chemical groups. Certain types of enzymes are synthesized in the form of inactive precursors known as **proenzymes** (or *zymogens*). Proenzymes are catalyti-

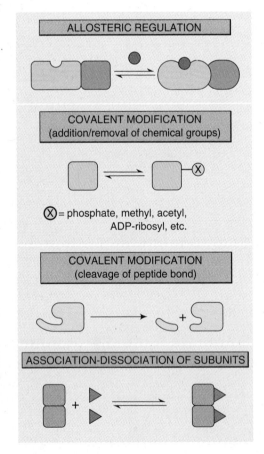

Figure 2-22 Mechanisms Employed by Cells to Regulate Enzyme Activity   *The binding of allosteric regulators, the covalent modification of protein structure, and the association and dissociation of protein subunits can be employed either to increase or decrease the catalytic activity of an enzyme molecule.*

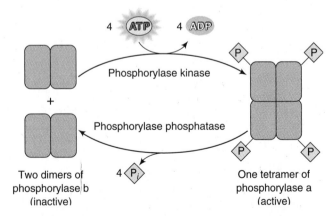

Figure 2-23 Contol of Glycogen Phosphorylase Activity by Phosphorylation and Dephosphorylation   *The phosphorylation of each subunit and the resulting joining of two dimers to form a tetramer results in enzyme activation.*

cally inactive because they contain extra amino acids. At an appropriate time and place, these extra sequences are removed by enzymes that cleave peptide bonds. This type of regulation is exemplified by the digestive enzymes pepsin, trypsin, and chymotrypsin, which are synthesized as inactive proenzymes called *pepsinogen, trypsinogen,* and *chymotrypsinogen,* respectively. These proenzymes are converted into their active forms after they have been secreted from the cell into the digestive tract. In the case of trypsinogen, activation involves the removal of a segment containing six amino acids from one end of the molecule in a reaction catalyzed by the enzyme *enterokinase.* Similar reactions involving the removal of small peptides mediate the activation of the other proenzymes. Enzymes manufactured as inactive proenzymes tend to be molecules that would have disastrous effects if they were active inside the cell. For example, the enzymes pepsin, trypsin, and chymotrypsin would degrade many essential cellular proteins if they were allowed to function while still residing within cells. Hence these enzymes are activated only after they are safely outside the cell.

Enzyme Activity Can Be Regulated by the Association and Dissociation of Subunits

Yet another approach for regulating enzyme activity involves adding and removing subunits from a multisubunit enzyme. In some cases, multiple copies of the same subunit interact with each other. For example, we have already seen how the catalytic activity of the enzyme phosphorylase is regulated through the association and dissociation of multiple copies of the same polypeptide chain (see Figure 2-23). But more commonly, the interacting subunits differ in structure. An interesting example of the latter phenomenon involves *lactate dehydrogenase (LDH),* the enzyme that catalyzes the conversion of pyruvate to lactate. Lactate dehydrogenase is constructed from two different kinds of polypeptide chains (designated H and M) that come together in various combinations to form enzyme molecules consisting of four subunits. A total of five different types of lactate dehydrogenase can therefore be created. Two consist of four identical subunits and are designated H_4 and M_4 respectively. The other three are composed of varying mixtures of the two types of subunits, and are designated H_3M, H_2M_2, and HM_3. These differing forms of the same enzyme are referred to as **isozymes.**

Kinetic analyses have revealed that even though the five LDH isozymes catalyze the same reaction, they differ in both K_m and V_{max}. The M_4 isozyme has a relatively low K_m and a high V_{max}, whereas the opposite is true for the H_4 isozyme. This difference means that the M_4 isozyme is more efficient at converting pyruvate to lactate than is the H_4 isozyme. It is interesting to note that the more efficient M_4 isozyme is enriched in skeletal muscle and embryonic tissue, where conditions often require the formation of lactate. In contrast, the less efficient H_4 isozyme is enriched in heart muscle, where lactate production is only appropriate under certain emergency conditions. The association of two different kinds of subunits therefore allows the same enzyme to be specifically adapted to the needs of differing cell types.

COUPLED REACTIONS AND THE ROLE OF ATP

Since enzyme activity can be regulated, it is evident that enzyme molecules are not just catalysts that automatically make reactions go faster. Changes in enzyme activity allow enzymes to serve as control agents that determine how much product each metabolic pathway will generate, thereby helping to keep the various metabolic activities of a cell in proper balance.

In addition to acting as catalysts and control agents, enzymes serve a third important function; they act as *coupling agents* that connect thermodynamically favorable (energy-releasing) processes to thermodynamically unfavorable (energy-requiring) processes. This aspect of enzyme function is crucially important because many of the reactions that occur within cells would be thermodynamically unfavorable if they were not coupled to energy-releasing, thermodynamically favorable reactions. As an example, let us consider the following hypothetical reactions:

$$A \rightarrow B \quad \Delta G' = +4 \text{ kcal/mol (endergonic)}$$
$$C \rightarrow D \quad \Delta G' = -7 \text{ kcal/mol (exergonic)}$$

Suppose that cells need to produce substance B. As written, the conversion of A →B is a thermodynamically unfavorable reaction and so will not proceed in the direction designated. However, if the conversion of A → B is coupled to the conversion of C →D, the following reaction is obtained:

$$A + C \longrightarrow B + D \quad \Delta G' = -3 \text{ kcal/mol}$$
$$\text{(exergonic)}$$

The value of $\Delta G' = -3$ kcal/mol for the above coupled reaction is obtained by adding the values of $\Delta G'$ for the two individual reactions (+4 kcal/mol *plus* −7 kcal/mol = −3 kcal/mol). Since the $\Delta G'$ for this coupled reaction is negative, it is thermodynamically favorable and will proceed in the direction written. Hence coupling the two reactions provides cells with a way of synthesizing substance B. The ability of enzymes to catalyze coupled reactions is crucial to living organisms, for it allows cells to carry out activities that otherwise would not be thermodynamically feasible.

ATP Plays a Central Role in Transferring Free Energy

Although any two chemical reactions could theoretically be coupled by an enzyme containing the appropriate catalytic sites, in practice the coupled reactions that occur in living organisms tend to share some features in common. One widely occurring pattern involves the use of *high-energy phosphorylated compounds* that release energy when their phosphate groups are removed (Table 2-6). Such compounds play a key role in transferring energy from thermodynamically favorable to thermodynamically unfavorable processes.

The most prominent example of such a high-energy compound is **adenosine triphosphate (ATP).** The structure of ATP and the reactions involved in the removal of its phosphate groups are summarized in Figure 2-24. This figure illustrates how the terminal phosphate group of ATP can be removed by a hydrolysis reaction that is exergonic, yielding **adenosine diphosphate (ADP)** and a free phosphate group (**P**$_i$). The terminal phosphate of ADP can also be removed in another exergonic hydrolysis reaction, yielding AMP and P$_i$. Each of these two hydrolysis reactions has a $\Delta G^{\circ\prime}$ of -7.3 kcal/mol; thus 7.3 kcal of usable energy is released per mole of ATP or ADP hydrolyzed under standard conditions. Under the nonstandard conditions that prevail within cells, the actual free-energy change ($\Delta G^{\prime}$) is thought to be closer to -11 kcal/mol (Figure 2-25).

The central role played by ATP in cellular energy flow was first perceived with remarkable foresight by Fritz Lipmann in 1940. On the basis of what was known about the energy released during the hydrolysis of ATP, Lipmann postulated the existence of an *ATP cycle* within cells. According to this model, the ultimate source of energy for energy-requiring cellular activities is the oxidation of foodstuffs (or absorption of light in the case of photosynthetic organisms). Lipmann postulated that the energy released during the oxidation of foodstuffs or during photosynthesis is employed to drive the formation of ATP from ADP and P$_i$. When the energy contained in the ATP mole-

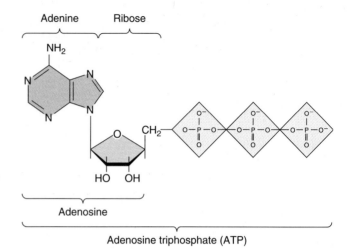

Figure 2-24 *Structure of ATP and Its Relationship to ADP and AMP* *The breakdown of ATP to ADP yields energy, as does the breakdown of ADP to AMP.*

cule is needed, ATP is broken down to ADP and P$_i$, releasing the free energy for use by the cell. In this way, a cycle is established in which ATP couples energy-yielding reactions with energy-requiring ones (Figure 2-26).

The Nucleotides GTP, UTP, and CTP Are Also Involved in Transferring Free Energy

ATP belongs to a group of high-energy nucleotides that differ from ATP only in the nitrogenous base they contain. Like ATP, these triphosphate compounds release free energy when their phosphate groups are removed, and they can therefore be used for transferring energy from thermodynamically favorable to thermodynamically unfavorable reactions. Included in this group of nucleotides is **guanosine triphosphate (GTP),** the most widely used energy-transferring triphosphate compound next to ATP. Among its many important functions, GTP transfers energy to key steps involved in protein synthesis (page 491), the budding and fusion of membrane vesicles (page 285), and the transmission of signals received by the cell surface (page 213). Many of

Table 2-6 $\Delta G^{\circ\prime}$ Values for the Hydrolysis of Selected Phosphate Compounds

Phosphate Compound		Hydrolysis Products	$\Delta G^{\circ\prime}$ (kcal/mol)
Phosphoenolpyruvate	→	Pyruvate + P$_i$	−14.8
Creatine phosphate	→	Creatine + P$_i$	−10.3
Pyrophosphate	→	P$_i$ + P$_i$	−8.0
ATP	→	ADP + P$_i$	−7.3
ATP	→	AMP + PP$_i$	−7.3
ADP	→	AMP + P$_i$	−7.3
Glucose 1-phosphate	→	Glucose + P$_i$	−5.0
Glucose 6-phosphate	→	Glucose + P$_i$	−3.3
Glycerol 3-phosphate	→	Glycerol + P$_i$	−2.2

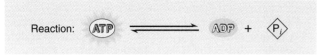

Reaction: $ATP \rightleftharpoons ADP + P_i$

Typical intracellular concentrations
of reactants and products:

$ATP = 10^{-2}M$ $ADP = 10^{-3}M$ $P_i = 10^{-2}M$

The free energy change under these conditions
can be determined using Equation 2-2:

$$\Delta G' = \Delta G^{\circ\prime} + 2.303RT \log_{10} \frac{[\text{products}]}{[\text{reactants}]}$$

$$= -7.3 \text{ kcal/mol} + 2.303RT \log_{10} \frac{[ADP][P_i]}{[ATP]}$$

$$= -7.3 \text{ kcal/mol} + 2.303RT \log_{10} \frac{10^{-3} \times 10^{-2}}{10^{-2}}$$

$$= -7.3 \text{ kcal/mol} + 2.303RT \log_{10} 10^{-3}$$

$$= -7.3 \text{ kcal/mol} + (2.303 \times 0.00198 \times 298 \times -3)$$

$$= -7.3 \text{ kcal/mol} - 4.1 \text{ kcal/mol}$$

$$\Delta G' = -11.4 \text{ kcal/mol}$$

Figure 2-25 *Determining the Actual Free-Energy Change (△G′) That Accompanies the Hydrolysis of ATP under Typical Conditions inside Cells* *Note that the value obtained, -11.4 kcal/mol, is significantly greater than the free-energy change for ATP hydrolysis under standard conditions (-7.3 kcal/mol).*

the actions of GTP are mediated by a special class of GTP-binding proteins known as *G proteins*.

Although their use is more restricted than ATP and GTP, the nucleotides *uridine triphosphate (UTP)* and *cytidine triphosphate (CTP)* also provide energy for thermodynamically unfavorable reactions. Among their roles, derivatives of UTP transfer energy to reactions involved in building up oligosaccharide and polysaccharide chains (see Figure 7-16) and derivatives of CTP provide energy for the synthesis of certain kinds of membrane lipids (see Figure 7-6).

Coenzymes Such as NAD⁺ and FAD Play a Central Role in Coupled Oxidation-Reduction Reactions

High-energy phosphate compounds are not the only agents employed by cells to couple thermodynamically unfavorable reactions to reactions that are favorable. Reactions that involve the transfer of electrons from one substance to another are often coupled by a group of coenzymes that are specialized for electron transfer. In such reactions, the substance that gives up electrons is said to undergo **oxidation** and the substance that gains

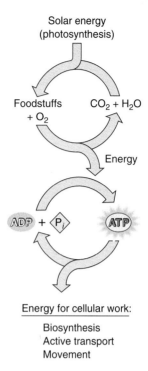

Solar energy
(photosynthesis)

Foodstuffs + O_2

$CO_2 + H_2O$

Energy

$ADP + P_i$ ATP

Energy for cellular work:

Biosynthesis
Active transport
Movement

Figure 2-26 *The ATP Cycle* *Energy released during the oxidation of foodstuffs is conserved by utilizing it to drive the formation of ATP. The energy stored in ATP is then used by cells to drive energy-requiring processes such as biosynthesis, active transport, and movement.*

electrons is said to undergo **reduction.** The coenzyme most commonly employed for coupling oxidation-reduction reactions is **nicotinamide adenine dinucleotide,** a derivative of the vitamin niacin. In its oxidized and reduced states this coenzyme is written as **NAD⁺** and **NADH,** respectively.

To illustrate the coupling role of NAD^+, let us consider a hypothetical reaction in which an organic molecule (XH_2) is oxidized by an enzyme that requires NAD^+ as a coenzyme. As we will learn in Chapter 8, biological oxidation reactions usually involve the removal of hydrogen atoms and their associated electrons. Thus when the molecule XH_2 is oxidized, it gives up its two hydrogen atoms along with two electrons. If oxidation is carried out by an enzyme utilizing NAD^+ as a coenzyme, the reaction can be written as follows:

$$\underset{\text{(oxidized)}}{NAD^+} \qquad \underset{\text{(reduced)}}{NADH + H^+}$$

$$XH_2 \xrightarrow{\quad \text{Enzyme A} \quad} X \qquad (2\text{-}9)$$

In this reaction, which is thermodynamically favorable, NAD^+ receives both of the electrons and one of the hydrogen atoms being given up by XH_2, thereby converting NAD^+ to its reduced form, NADH. The second hydrogen atom given up by XH_2 is released as a free proton (H^+).

The reduced NADH generated by the preceding reaction then dissociates from enzyme A and diffuses to enzyme B, where it drives the reduction of substance Y:

$$
\begin{array}{cc}
\text{(reduced)} & \text{(oxidized)} \\
\text{NADH} + \text{H}^+ & \text{NAD}^+
\end{array}
$$

$$Y \xrightarrow[\text{Enzyme B}]{} YH_2 \qquad (2\text{-}10)$$

By itself the conversion of Y to YH_2 would have been thermodynamically unfavorable, but the participation of NADH makes the reaction favorable.

If Equations 2-9 and 2-10 are now added together, we obtain the following:

$$XH_2 + Y \rightarrow X + YH_2$$

Note that NAD^+ and NADH do not appear in this summary equation because no net change has occurred in these coenzymes; the NAD^+ that had been reduced to NADH in Equation 2-9 is simply oxidized back to NAD^+ in Equation 2-10. The net effect is that NAD^+ and NADH have acted to couple the oxidation of substance X to the reduction of substance Y. This role is made possible by the fact that NAD^+ and NADH tend to be bound rather loosely to the enzymes with which they interact. These coenzymes can therefore function as mobile coupling agents that link oxidation-reduction reactions by diffusing from one enzyme to another.

Several other coenzymes play roles that are similar to those of NAD^+. A phosphorylated derivative of NAD^+ known as **nicotinamide adenine dinucleotide phosphate ($NADP^+$)** functions in much the same way as NAD^+. **Flavin adenine dinucleotide (FAD)** and **flavin mononucleotide (FMN),** which are derived from the vitamin riboflavin, have chemical structures that are rather different from NAD^+ and $NADP^+$ but play a comparable role in oxidation-reduction reactions. The behavior of these various compounds in oxidation-reduction reactions will be described more thoroughly in Chapter 8 in our discussion of the pathways involved in capturing energy from nutrients.

SUMMARY OF PRINCIPAL POINTS

• The use of energy by cells is governed by the laws of thermodynamics. The first law states that the total amount of energy in the universe is constant, which means that a cell's energy requirements can be met only by taking in energy from the surroundings and transforming it into useful forms. The direction in which chemical reactions proceed is governed by the second law of thermodynamics, which states that events proceed in the direction that causes the entropy of the universe to increase. All reactions proceed in the direction that causes the free energy (G) of the system to decrease. In thermodynamically favorable (exergonic) reactions, ΔG is negative and the products of the reaction predominate over reactants at equilibrium. In reactions that are thermodynamically unfavorable (endergonic), ΔG is positive and the reactants predominate over the products at equilibrium.

• Enzymes increase reaction rates without affecting the final equilibrium conditions and act in small quantities without being consumed. Enzymes can be distinguished from nonbiological catalysts by their efficiency, specificity, and ability to be regulated.

• Enzymes speed up reaction rates by lowering the activation energy of chemical reactions. This is accomplished by the formation of a transient complex between an enzyme and its substrate(s), which generates a transition state that requires less energy to create and which places substrates in the proper orientation for the reaction to occur.

• The ability of an enzyme to recognize its proper substrates is based upon the shape of the active site, which is complementary to that of the substrates.

• V_{max} is the maximum reaction velocity that can be attained with a given amount of enzyme, and hence is a measure of an enzyme's inherent catalytic efficiency. The Michaelis constant, or K_m, is the substrate concentration at which half maximal velocity is achieved, and is a measure of the binding affinity of an enzyme for its substrate.

• Changes in pH influence enzyme activity by altering the ionization of the substrate and/or enzyme, which in turn influence enzyme conformation and substrate recognition. Increasing the temperature causes the rate of enzymatic reactions to increase until enzyme molecules begin to denature, at which point the reaction velocity begins to fall. Many enzymes require nonprotein components such as coenzymes, prosthetic groups, and metal ions in order to be catalytically active.

• Irreversible inhibitors bind covalently to enzymes, permanently disrupting proper conformation. Competitive inhibitors resemble an enzyme's normal substrate, reversibly competing with the substrate for binding to the active site. Noncompetitive inhibitors bind reversibly to enzymes at locations other than the active site, causing alterations in conformation that decrease enzymatic activity.

• Allosteric inhibitors and activators bind reversibly to allosteric sites, altering enzyme conformation and catalytic activity. Allosteric enzymes often exhibit cooperativity, in which the binding of a substrate molecule to one enzyme subunit influences the binding of subsequent substrate molecules to other subunits.

• Enzyme activity can also be regulated by the cleavage of peptide bonds, the association and dissociation of polypeptide subunits, and covalent modifications such as methylation, acetylation, ADP-ribosylation, and phosphorylation.

• Besides functioning as catalysts and control agents, enzymes act as coupling agents that link thermodynamically favorable reactions to thermodynamically unfavorable ones. ATP plays a central role in such coupled reactions, trapping energy released by the oxidation of foodstuffs and using this energy to drive energy-requiring activities. In reactions involving oxidation-reduction, coenzymes like NAD^+ and FAD often couple the oxidation of one substance to the reduction of another.

SUGGESTED READINGS

Books

Becker, W. M., and D. W. Deamer (1991). *The World of the Cell,* 2nd Ed., Benjamin/Cummings, Redwood City, CA, Chs. 5 and 6.

Fersht, A. (1985). *Enzyme Structure and Mechanism,* Freeman, New York.

Lehninger, A. L., D. L. Nelson, and M. M. Cox (1993). *Principles of Biochemistry,* Worth, New York, Ch. 8.

Mathews, C. K., and K. E. van Holde (1990). *Biochemistry,* Benjamin/Cummings, Redwood City, CA, Chs. 3, 10, and 11.

Perutz, M. F. (1990). *Mechanisms of Cooperativity and Allosteric Regulation in Proteins,* Cambridge University Press, New York.

Articles

Gerhart, J. C., and H. K. Schachman (1965). Distinct subunits for the regulation and catalytic activity of aspartate transcarbamoylase, *Biochemistry* 4:1054-1062.

Kraut, J. (1988). How do enzymes work?, *Science* 242:533-540.

Lipmann, F. (1941). Metabolic regulation and utilization of phosphate bond energy, *Adv. Enzymol.* 1:99-162.

Monod, J., J.-P. Changeux, and F. Jacob (1963). Allosteric proteins and cellular control systems, *J. Mol. Biol.* 6:306-309.

Sumner, J. B. (1926). The isolation and crystallization of the enzyme urease, *J. Biol. Chem.* 69:435-441.

C h a p t e r 3

The Flow of Genetic Information

Virtually every activity that takes place in living cells is made possible in one way or another by inherited information that has been encoded in DNA molecules. Although our current understanding of the role played by DNA is taken largely for granted, achieving this understanding required almost a hundred years of experimentation and the independent contributions of hundreds of different investigators. In this chapter we will summarize the basic principles that govern DNA action and the major lines of experimentation that led to the formulation of these principles. This material is placed in the introductory section of the book because a basic understanding of genes and nucleic acids is essential background information for virtually every area of cell biology. In Chapters 10 through 12 we will return to the role played by DNA within the cell, examining in depth how many of the activities introduced in this chapter are carried out.

In the process of introducing the basic principles that govern the role played by DNA within cells, it will become apparent that scientific progress is not always straightforward. Biologists' views concerning DNA molecules have undergone many shifts over the years, and for long periods many individuals argued that DNA could not possibly store and transmit genetic information. One of the most important lessons to be learned from examining the history of our current ideas concerning DNA is that scientific data can be easily misinterpreted, and that once incorrect ideas have become established, they can be very difficult to dislodge. Scientists must therefore always be on guard against the view that a particular concept, no matter how widely accepted, is a "fact." Science is not a collection of facts, but an ongoing intellectual process involving hypotheses, experimentation, formulation of models, and modification of these hypotheses and models as new information becomes available. In this chapter we will see how well-established concepts can undergo a complete reversal as the result of unexpected new findings. It therefore follows that some of our current ideas, no matter how firmly entrenched, may be subject to similar upheaval in the future.

IDENTIFYING DNA AS THE GENETIC MATERIAL

Every time an organism reproduces, the instructions that guide the development of a new organism must be transmitted from parent to offspring. Prior to the late nineteenth century most biologists believed that transmission of these instructions involves a complete "blending" of the traits of the two parents, and that the characteristics of the individual parents therefore disappear in the process. Such a theory made experimentation on the chemical basis of inheritance virtually impossible, since it is difficult to carry out a systematic study of traits that are continually disappearing. This viewpoint changed dramatically, however, as a result of experiments carried out in the 1860s by an Austrian monk, Gregor Mendel. These experiments, in which Mendel quantified the inheritance patterns of certain traits in garden peas, are thoroughly covered in introductory biology textbooks and so need not be described here. These studies led Mendel to conclude that hereditary information is transmitted in the form of distinct entities, later termed **genes,** which do not blend and disappear but rather maintain their unique identities over many generations.

The Scientist Who Discovered DNA Concluded That It Could Not Be Involved in the Transmission of Hereditary Information

The idea that hereditary information is stored in discrete entities called genes raised a profound question: What chemical substance do genes contain that allows them to store and transmit inherited information? The chemical nature of genes was not investigated by Mendel himself, but was unknowingly identified a few years later by a Swiss physician named Johann Friedrich Miescher. Miescher was one of the first biologists to use the tools of biochemistry rather than microscopy to study the molecular composition of cells. Because Miescher was particularly interested in the chemical makeup of the nucleus, he devised a procedure for isolating and purifying nuclei from intact cells. The development of this technique, which involved incubating white blood cells with protein-digesting enzymes to destroy the cytoplasm, represented an important milestone because it was the first time any organelle had been successfully isolated from cells in pure form. Miescher then discovered that extracting these isolated nuclei with dilute sodium hydroxide removed a chemical substance enriched in phosphorus that had never been observed before. He named this substance *nuclein* because it had been extracted from the nucleus; when later analyses showed the molecule to be highly acidic, it was renamed *nucleic acid.*

Miescher initially believed that nucleic acids function in the transmission of hereditary information, a conclusion based largely on the discovery that nucleic acid accounts for the bulk of the mass of sperm cells. He soon rejected this idea, however, based on data which suggested that the amount of nucleic acid in a hen's egg is hundreds of times greater than in the sperm cell that fertilizes the egg. Since Miescher reasoned that both parental cells (sperm and egg) must contribute roughly equal amounts of hereditary information to the offspring, it seemed to him that nucleic acid could not be carrying the hereditary information. This erroneous conclusion arose from an unfortunate coincidence based on the fact that the property used by Miescher to identify nucleic acid was its high content of phosphorus. Unfortunately hen's eggs contain an unusual protein called phosvitin that also contains a large amount of phosphorus. Of the thousands of proteins that have been identified and characterized in the years since Miescher's studies, phosvitin is the only one containing so much phosphorus that it could be easily mistaken for a nucleic acid. The presence of phosvitin caused Miescher to inadvertently overestimate the nucleic acid content of the hen's egg, and hence to conclude that nucleic acid cannot be involved in the transmission of hereditary information.

The Idea That DNA Stores Genetic Information Was Widely Held by the Late Nineteenth Century

Although Miescher was led astray concerning the role of the substance he had discovered, other biologists were soon back on the right track. In the early 1880s a botanist named Eduard Zacharias reported that extracting nucleic acids from cells also causes the staining of the chromosomes to disappear. Zacharias concluded on this basis that chromosomes contain nucleic acids. It was known at the time that chromosomes are duplicated prior to cell division, and that each of the two duplicate sets of chromosomes is passed on to one of the newly forming cells. This observation suggested that chromosomes transmit hereditary information to the new cells formed at the time of cell division. When Zacharias discovered that chromosomes contain nucleic acids, it seemed logical to infer that nucleic acids are involved in the transfer of hereditary information. Within a few years this viewpoint came to be widely accepted, as is evidenced by the following statement written in the most prominent cell biology textbook of the era:

> It is the remarkable substance [nucleic acid] which is almost certainly identical with [chromosomes]. The most essential material handed on by the mother cell to its progeny is the [chromosomes], and this substance therefore has a special significance in inheritance. (From E. B. Wilson, *The Cell in Development and Inheritance,* 1896, pp. 332, 352.)

This statement makes it clear that by the year 1896, biologists had reached a conclusion that closely parallels our current understanding of the role played by DNA in transmitting hereditary information.

The Idea That DNA Stores Genetic Information Was Rejected Shortly after It Was First Proposed

The idea that genetic information is stored in DNA molecules was not popular for very long. During the next decade, some surprising results emerged from studies in which staining procedures were utilized to monitor the amount of DNA present in cells at different developmental stages. If DNA molecules contain a cell's inherited instructions and these instructions must be passed on intact to the next generation of cells during the process of cell division, then a cell would be expected to maintain a constant amount of DNA. However, when staining procedures were used to quantify the amount of DNA present in cells under different conditions, dramatic variations in DNA staining were observed. These unexpected results soon led to a repudiation of the idea that nucleic acids carry genetic information.

In the following decades attention turned away from the study of nucleic acids, and most biologists came to believe that genes are made of protein molecules. Fortunately, the chemist Phoebus Levene and a small number of other investigators retained an interest in nucleic acids and proceeded to investigate their structural organization. As a result of the combined efforts of these individuals, it was discovered by the mid-1930s that the fundamental building block of nucleic acids is the *nucleotide* (page 19) and that the nucleic acid originally discovered by Miescher is DNA, which contains the bases A, G, C, and T. Although this information was crucial in furthering our understanding of nucleic acids, the insensitivity of the methods employed by Levene had some unfortunate consequences. Levene's data suggested that DNA contains roughly equivalent amounts of the four bases A, G, C, and T, leading him to propose that DNA is a short *tetranucleotide* containing one each of the four bases. When improved isolation techniques later revealed that DNA molecules are much longer than four nucleotides, Levene modified his theory and claimed that DNA is a repetitious polymer containing the same tetranucleotide sequence repeated again and again (e.g., ... AGCTAGCTAGCT. . . .). Since it is obvious that such a repetitive pattern cannot store genetic information, Levene's *tetranucleotide theory* further reinforced the belief that DNA, in spite of its presence in chromosomes, does not contain hereditary information.

By this time it had become well established that chromosomes consist of protein as well as DNA. Since DNA did not seem to have the attributes expected of a molecule encoding genetic information, most biologists came to believe that genes must be made of protein. It was argued that proteins are constructed from 20 different amino acids that can be assembled into a vast number of different sequences, thereby generating the sequence diversity and complexity expected of a molecule that stores and transmits genetic information. The conclusion that genes must consist of protein rather than DNA seemed so logical and inescapable that it was taught for many years as an established fact!

Genetic Transformation Studies Revived the Idea That DNA Stores Hereditary Information

A great surprise was in store, however, for biologists who were studying protein molecules to determine how genetic information is stored and transmitted. The background for this surprise was provided in England during the late 1920s by Frederick Griffith, a microbiologist who was studying the organism that causes bacterial pneumonia. Griffith discovered that this bacterium, *Streptococcus (diplococcus) pneumoniae,* exists in two forms: an *S strain* that forms smooth shiny colonies when grown on agar plates and an *R strain* whose colonies are rough and dry. This difference in appearance is caused by the fact that S-strain cells produce a polysaccharide coat or *capsule* (page 251), which is absent in R-strain cells. When injected into mice, S-strain but not R-strain bacteria trigger a fatal pneumonia.

One of the most intriguing discoveries made by Griffith, however, was that pneumonia can also be induced by injecting animals with a mixture of live R-strain bacteria and dead S-strain organisms (Figure 3-1). This finding

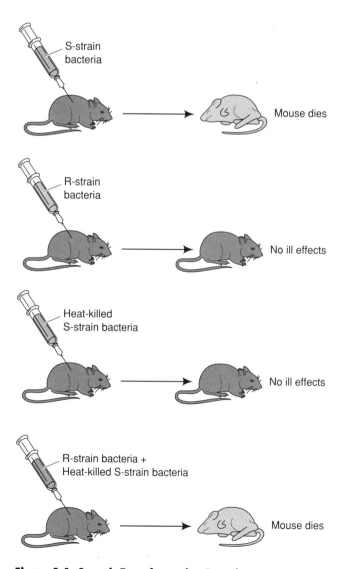

Figure 3-1 *Genetic Transformation Experiments on Streptococcus pneumoniae Performed by Griffith* *These experiments revealed that heat-killed S-strain bacteria are able to convert live R-strain cells into S-strain cells.*

was surprising because neither the live R-strain nor the dead S-strain bacteria cause pneumonia if injected alone. When Griffith autopsied the animals that had been injected with the mixture of live R-strain and dead S-strain bacteria, he found them teeming with live S-strain bacteria. Since the animals had not been injected with any live S-strain bacteria, he concluded that the dead S-strain bacteria had somehow converted the live R-strain cells into S-strain cells. This process was termed **genetic transformation** because it involved an alteration in the hereditary properties of the cells; that is, the newly formed S-strain bacteria transmitted the S-strain trait to their progeny, and thus an alteration in hereditary makeup had occurred.

Griffith's pioneering discovery opened the door for studies on the biochemical nature of the substance responsible for genetic transformation. The first important observation emerged a few years later, when James Alloway reported that soluble extracts of dead S-strain cells can transform R-strain bacteria into S-strain bacteria. During the next decade, Oswald Avery and his associates worked on identifying the component in the bacterial extracts that is responsible for inducing this genetic transformation. Their first important discovery was that treatment of bacterial extracts with nucleases but not proteases abolishes the ability to cause genetic transformation, suggesting that the transforming agent is a nucleic acid rather than a protein. After many years of effort, Avery succeeded in isolating the active nucleic acid molecule from bacterial extracts and identifying it as DNA. Microgram quantities of this DNA were found to be capable of transforming bacterial cells from R-strain into S-strain. In 1944, after a decade of painstaking work, Avery announced his inescapable conclusion to the world: The chemical substance responsible for genetic transformation is DNA!

The Idea That DNA Encodes Genetic Information Was Experimentally Verified by the Early 1950s

At first, Avery's conclusions fell on deaf ears. Since scientists had believed for 40 years that genes are made of protein and not DNA, a variety of arguments were quickly formulated to discredit Avery's work. It was claimed, among other things, that Avery's nucleases and proteases were impure, that his final DNA preparation was contaminated with protein, and that genetic transformation in bacteria is a special case with no relevance to inheritance in other organisms.

Within a few years, however, evidence supporting Avery's contention that DNA stores genetic information began to accumulate. In 1949, Alfred Mirsky performed direct chemical measurements on the amount of DNA present in different kinds of cells. His data revealed that the amount of DNA is usually the same in all cells of an organism other than sperm and egg cells, which contain half the normal amount of DNA. This finding of *DNA constancy* is exactly what would be expected of the chemical substance that stores genetic information. Even the presence of half the normal amount of DNA in sperm and eggs is easily explained, since these two cell types join together to form a new organism. The principle of DNA constancy was in direct opposition to reports early in the century which had suggested that DNA is present in variable amounts and therefore cannot be the genetic material (page 63). In retrospect, it is apparent that these earlier reports contained errors that arose from the use of staining methods rather than direct chemical measurements to determine DNA content. Staining methods are influenced by the amount of protein associated with DNA, and we now know that it was variations in nuclear protein content rather than DNA content that were responsible for the observed variability in DNA staining.

Shortly after Mirsky's observations concerning DNA constancy were published, Erwin Chargaff provided further support for the idea that DNA is the genetic material. Using sensitive techniques to measure the nucleotide base composition of DNA obtained from various organisms, Chargaff discovered that Levene's generalization about the tetranucleotide nature of DNA was incorrect. Instead of equal amounts of the four bases, Chargaff found that DNAs from different kinds of organisms are characterized by differing base compositions (Table 3-1). Hence rather than being constructed from a simple repeating sequence of four bases, DNA must have a more complex sequence structure that varies from organism to organism. This is exactly what would be expected of the chemical substance that stores genetic information. Moreover, Chargaff discovered that DNA molecules isolated from different cell types within the same organism have similar base compositions, which is also compatible with a genetic role because all cells of an individual organism would be expected to inherit the same genetic information.

The final evidence which convinced the remaining skeptics that DNA is the genetic material was obtained in 1952 by Alfred Hershey and Martha Chase, who studied the infection of bacterial cells by a virus called *bacteriophage T2*. During infection, this virus attaches to the bacterial cell surface and injects material into the cell (Figure 3-2). Shortly thereafter the bacterial cell bursts, releasing thousands of new virus particles. This scenario suggests that the material injected into the bacterial cell carries genetic information coding for the production of new virus particles. What is the chemical nature of this injected material? Only two possibilities exist because the T2 virus is constructed from only two kinds of molecules: DNA and protein. In trying to determine which of these two substances is the injected material, Hershey and Chase took advantage of the fact

Table 3-1   Base Composition of DNA from Various Sources

Source of DNA	A	T	G	C	A/T	G/C	(A + T)/(G + C)
Bovine thymus	28.4	28.4	21.1	22.1	1.00	0.95	1.31
Bovine liver	28.1	28.4	22.5	21.0	0.99	1.07	1.30
Bovine kidney	28.3	28.2	22.6	20.9	1.00	1.08	1.30
Bovine brain	28.0	28.1	22.3	21.6	1.00	1.03	1.28
Human liver	30.3	30.3	19.5	19.9	1.00	0.98	1.53
Locust	29.3	29.3	20.5	20.7	1.00	1.00	1.41
Sea urchin	32.8	32.1	17.7	17.3	1.02	1.02	1.85
Wheat germ	27.3	27.1	22.7	22.8	1.01	1.00	1.19
Marine crab	47.3	47.3	2.7	2.7	1.00	1.00	17.50
Aspergillus (mold)	25.0	24.9	25.1	25.0	1.00	1.00	1.00
Saccharomyces cerevisiae (yeast)	31.3	32.9	18.7	17.1	0.95	1.09	1.79
Clostridium (bacterium)	36.9	36.3	14.0	12.8	1.02	1.09	2.73

that the viral DNA contains phosphorus but no sulfur, whereas the viral protein contains sulfur but no phosphorus. They began by allowing the T2 virus to reproduce in the presence of radioactive phosphorus (^{32}P) and radioactive sulfur (^{35}S) to label the viral DNA and protein respectively. When this radioactive virus was then allowed to infect a new population of nonradioactive bacterial cells, the ^{32}P but not the ^{35}S was found to enter the bacteria (Figure 3-3). It was therefore concluded that DNA, not protein, carries the genetic information that guides the production of new virus particles.

Thus by the early 1950s, the role of DNA as the carrier of genetic information came to be widely accepted. Unfortunately Oswald Avery, the visionary most responsible for this complete turnabout in our views concerning the function of DNA, never received the credit he so richly deserved. The Nobel Prize Committee discussed Avery's work but decided he had not done enough. Perhaps Avery's modest and unassuming nature was responsible for this lack of recognition. After Avery died in 1955, Erwin Chargaff wrote in tribute: "He was a quiet man; and it would have honored the world more, had it honored him more."

As the scientific community gradually came to realize that DNA encodes genetic information, an entirely new set of questions began to emerge. The most crucial issues were the following: (1) How is DNA accurately replicated so that duplicate copies of the genetic information can be passed on from cell to cell during cell division, and from parent to offspring? (2) How does the information encoded within the DNA molecule determine the properties of cells and organisms? (3) How does information in DNA undergo the changes that are required in order for evolution to occur? We will provide some basic answers to each of these ques-

tions in the remainder of this chapter. Then in later chapters devoted to the nucleus (Chapter 10), ribosomes (Chapter 11), and cell division (Chapter 12) we will elaborate upon these issues in more detail, emphasizing their relationship to the organelles and structures involved in the storage, transmission, and utilization of genetic information.

DNA STRUCTURE AND REPLICATION

Every time a cell divides, it must duplicate its DNA molecules so that each of the two newly forming cells will contain a complete set of genetic instructions. In a human being, which contains about 10^{13} cells at birth, this DNA replication process must be successfully repeated trillions and trillions of times. In order to determine how the process of DNA replication is carried out at the molecular level, biologists first needed to unravel the three-dimensional structure of the DNA molecule.

DNA Is a Double Helix

One of the first insights into the structural organization of the DNA molecule emerged from the previously mentioned studies of Erwin Chargaff on DNA base composition. In the process of analyzing DNA molecules isolated from a wide variety of organisms, Chargaff discovered a remarkable pattern: The amount of adenine and thymine tended to be present in equal amounts, as did the amounts of guanine and cytosine. This A = T and G = C pattern, which came to be known as *Chargaff's rule,* appeared to hold regardless of the source of the DNA or its overall base composition (see Table 3-1).

Although Chargaff felt that this regularity must reflect some underlying property of the DNA molecule,

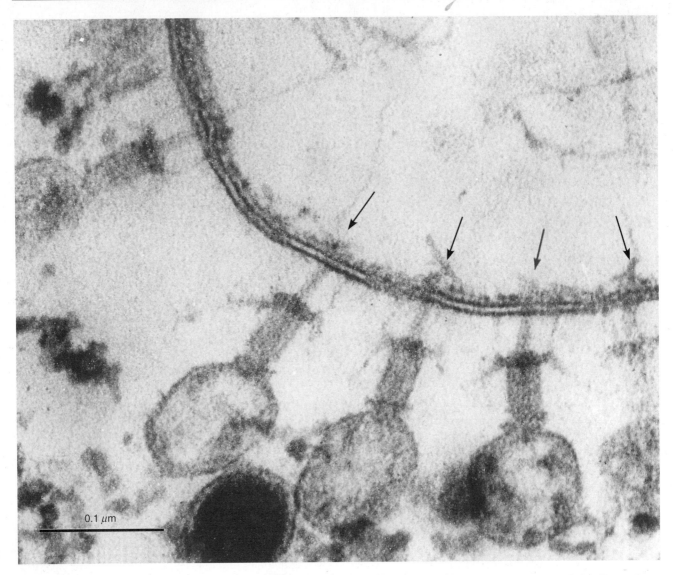

Figure 3-2 *Electron Micrograph Showing Bacteriophage Particles Attached to the Surface of a Bacterial Cell*   *The arrows point to material being injected from the attached virus particles into the cell. Courtesy of L. D. Simon and Photo Researchers Inc.*

its exact significance eluded him. The importance of the A-T and G-C relationships first became apparent in 1953, when James Watson and Francis Crick published a three-dimensional model of the structure of DNA that was to have far-reaching implications. This model was based on X-ray diffraction pictures of DNA taken by Rosalind Franklin and Maurice Wilkins, which suggested that DNA is coiled into a helically twisted chain. When Watson and Crick built molecular models of possible helical configurations, they discovered that two intertwined helical chains of DNA could be held together by hydrogen bonds between the nitrogenous bases of the two opposing DNA strands (Figure 3-4). In such a DNA **double helix,** the sugar-phosphate backbones of the two DNA strands are on the outside of the helix and the nitrogenous bases project toward the interior of the DNA molecule, where they form hydrogen bonds with

each other. As we learned in Chapter 1, nucleic acid chains exhibit an inherent directionality that allows the 5' end of the chain to be distinguished from the 3' end (page 19). In a DNA double helix, the two DNA chains run in opposite directions; that is, one chain runs in the $5' \rightarrow 3'$ direction while the other runs in the $3' \rightarrow 5'$ direction.

But the most remarkable feature of this model was the realization that the hydrogen bonds holding together the two strands of the helix only fit when they are formed between the base adenine in one chain and thymine in the other, or between the base guanine in one chain and cytosine in the other (Figure 3-5). This means that the base sequence of one chain determines the base sequence of the opposing chain; the two chains of the DNA double helix are therefore said to be **complementary** to each other. It was immediately

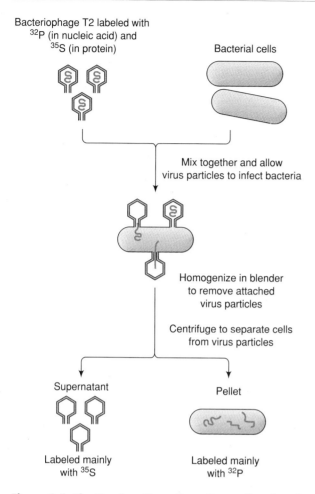

Bacteriophage T2 labeled with
^{32}P (in nucleic acid) and
^{35}S (in protein)

Bacterial cells

Mix together and allow
virus particles to infect bacteria

Homogenize in blender
to remove attached
virus particles

Centrifuge to separate cells
from virus particles

Supernatant

Pellet

Labeled mainly
with ^{35}S

Labeled mainly
with ^{32}P

Figure 3-3 *The Hershey-Chase Experiment Showing That Viral DNA Enters Bacterial Cells Infected by Bacteriophage T2* *The only radioisotope to enter the bacterial cells during viral infection was ^{32}P, which was contained in the viral DNA.*

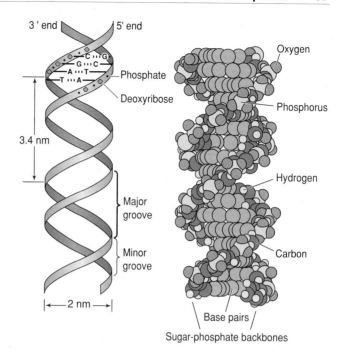

3' end   5' end

Phosphate

Deoxyribose

3.4 nm

Major
groove

Minor
groove

|← 2 nm →|

Oxygen

Phosphorus

Hydrogen

Carbon

Base pairs

Sugar-phosphate backbones

Figure 3-4 *The DNA Double Helix*   (Left) *A stylized representation of the DNA double helix showing the two DNA strands held together by hydrogen bonds between adenine (A) and thymine (T), and between guanine (G) and cytosine (C). Note that the two chains run in opposite directions; that is, one chain runs in the 5' → 3' direction while the other runs in the 3' → 5' direction.* (Right) *A space-filling molecular model of the double helix.*

obvious, of course, that such a model explained why Chargaff had observed that DNA molecules contain equal amounts of the bases A and T and equal amounts of the bases G and C.

DNA Is Replicated by a Semiconservative Mechanism Based on Complementary Base-Pairing

The most profound implication of the Watson-Crick DNA model was that it suggested a mechanism by which cells can duplicate their genetic information: The two strands of the DNA double helix simply separate from each other prior to cell division, and each strand then functions as a **template** that dictates the synthesis of a new complementary DNA strand using the base-pairing rules. In other words, the base A in the template strand causes the base T to be inserted in the newly forming strand, the base G causes the base C to be inserted, the base T causes the base A to be inserted, and the base C causes the base G to be inserted (Figure 3-6).

This process is called **semiconservative replication** because each newly formed DNA molecule consists of one old DNA strand that had served as the template, plus one new DNA strand.

A few years after this idea was first proposed, an elegant test of its validity was designed by Matthew Meselson and Franklin Stahl. In their studies two isotopic forms of nitrogen, ^{14}N and ^{15}N, were used to distinguish newly synthesized strands of DNA from old strands. Bacterial cells were first grown for several generations in a medium containing ^{15}N-labeled ammonium chloride to incorporate this *heavy* isotope of nitrogen into their DNA molecules. Cells containing ^{15}N-labeled DNA were then transferred to a growth medium containing the normal *light* isotope of nitrogen, ^{14}N. Any new strands of DNA synthesized after this transfer would therefore incorporate ^{14}N rather than ^{15}N. DNA labeled with ^{15}N is significantly more dense than DNA labeled with ^{14}N, so the old and new DNA strands can be distinguished from one another by a technique called *isodensity centrifugation* (page 135), which separates molecules based on differences in density.

Figure 3-7 summarizes the experiments designed by Meselson and Stahl to confirm the existence of semiconservative replication. In one set of experiments, cells containing ^{15}N-labeled DNA were allowed to divide *once* in

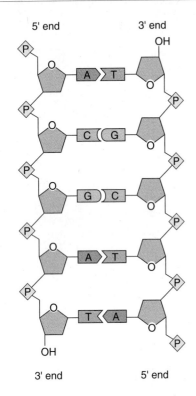

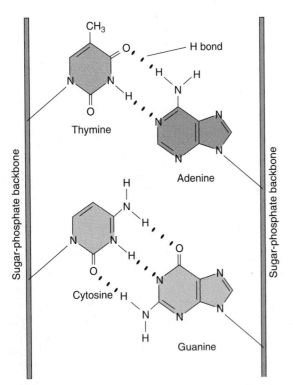

Figure 3-5 *Base-Pairing in DNA* (Top) *A diagrammatic representation of a short stretch of DNA, showing the relationship between the bases in the two chains.* (Bottom) *Chemical structures of the four bases showing the location of the hydrogen bonds that link adenine to thymine and guanine to cytosine.*

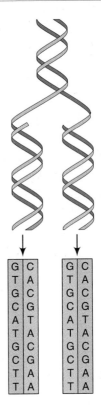

Figure 3-6 *The Semiconservative Model of DNA Replication* *During replication, the DNA double helix unwinds and each strand serves as a template for the formation of a new complementary strand.*

the presence of ^{14}N. When DNA isolated from the cells was analyzed by isodensity centrifugation, it exhibited a density halfway between the densities of ^{15}N-labeled DNA and ^{14}N-labeled DNA (see Figure 3-7, *graph B*). This result is expected if the two strands of the DNA double helix separate during replication and each of the old ^{15}N-containing strands serves as a template for the synthesis of a new ^{14}N-containing strand. Further confirmation for this interpretation was obtained by heating the DNA to separate the two strands of the double helix from each other. As expected, one strand exhibited the density of a ^{15}N-containing strand and the other exhibited the density of a ^{14}N-containing strand (see Figure 3-7, *graph E*).

In a second set of experiments, cells containing ^{15}N-labeled DNA were allowed to divide *twice* in the presence of ^{14}N. When the resulting DNA was analyzed by isodensity centrifugation, 50 percent of the DNA molecules exhibited the density of DNA molecules constructed from one ^{15}N-containing strand and one ^{14}N-containing strand, and the other 50 percent of the DNA molecules exhibited the density of DNA molecules constructed from two ^{14}N-containing strands (see Figure 3-7, *graph C*). As is illustrated in the DNA diagrams included in Figure 3-7, such results are consistent with the existence of semiconservative replication.

Isodensity centrifugation of DNA

Interpretation of data

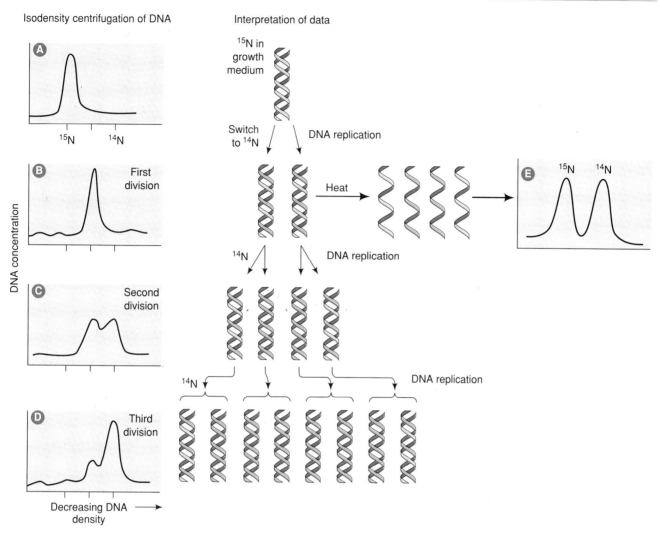

Figure 3-7 *The Meselson-Stahl Experiment Demonstrating That DNA Replication Is Semiconservative* *Bacterial cells growing in a medium containing heavy nitrogen (^{15}N) were switched to a medium containing light nitrogen (^{14}N). Isodensity centrifugation was then employed to separate newly synthesized DNA molecules based on their differing contents of ^{14}N and ^{15}N. Each graph on the left side of the diagram represents a new round of DNA replication. The changes in DNA density that occur after each cell division are compatible with the conclusion that newly synthesized DNA molecules consist of one old strand and one new strand, thus confirming the semiconservative model of DNA replication.*

DNA Is Replicated by DNA Polymerase

The Meselson-Stahl experiments provided strong verification for the idea that DNA is synthesized by a process in which each strand of the DNA double helix serves as a template for the synthesis of a new complementary strand. At first this replication process was thought to be inseparable from living cells, but in the mid-1950s Arthur Kornberg showed that an enzyme isolated from bacterial cells can carry out DNA synthesis in a test tube. This enzyme, which he named **DNA polymerase,** requires that at least a small amount of DNA be initially present to act as a template. Under such conditions, DNA polymerase catalyzes a nucleotide polymerization reaction that utilizes as substrates the triphosphate derivatives of the four bases found in DNA. These

deoxynucleoside triphosphates are called dATP, dTTP, dGTP, and dCTP (Figure 3-8). As each substrate molecule is incorporated into a growing DNA chain, its two terminal phosphate groups are released. Since deoxynucleoside triphosphates are high-energy compounds whose free energy of hydrolysis is comparable to that of ATP (page 56), the energy equivalent of one molecule of ATP is consumed for every nucleotide incorporated into DNA. This energy input drives what would otherwise be a thermodynamically unfavorable polymerization reaction.

A crucial feature of the DNA polymerase reaction became apparent when Kornberg compared the properties of the newly synthesized DNA molecules with the DNA that had been added to begin the reaction. Not only were the initial and final DNA molecules found to

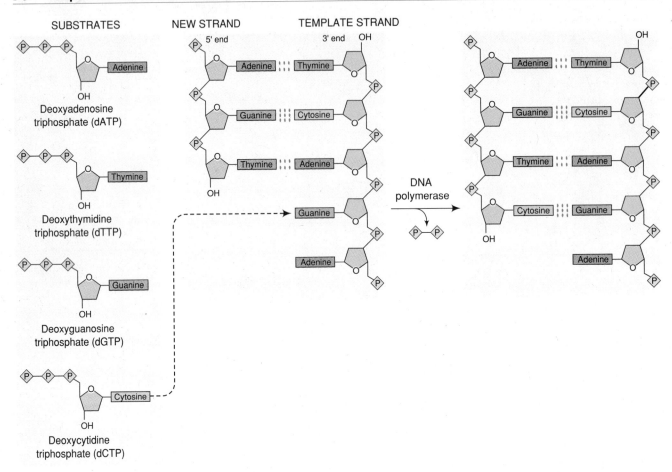

Figure 3-8 The DNA Polymerase Reaction *DNA polymerase catalyzes the addition of deoxynucleoside triphosphates to the 3' end of a DNA chain, using an existing DNA strand as a template. As each deoxynucleoside triphosphate is added to the growing chain, its two terminal phosphate groups are released. The base sequence of the newly forming strand is determined by complementary base-pairing with the template strand.*

resemble each other in base composition, but measurements of the frequencies with which the various bases appeared adjacent to one another indicated that the sequence of bases in the two DNAs was the same. Kornberg therefore concluded that the starting DNA was acting as a template whose base sequence was accurately copied by DNA polymerase. Definitive proof for this idea was finally obtained in 1967, when it was shown that DNA polymerase can utilize purified viral DNA as a template for the synthesis of new DNA molecules that are capable of infecting cells just like the natural virus. This milestone experiment was the first to demonstrate that an accurate and biologically active copy of a naturally occurring DNA molecule can be synthesized by DNA polymerase outside a living cell.

In spite of the importance of these discoveries, the enzyme isolated by Kornberg is not the principal enzyme responsible for DNA replication in intact cells. This fact became apparent in 1969 when Peter DeLucia and John Cairns and his associates reported that mutant strains of bacteria defective in the Kornberg enzyme can still replicate their DNA and divide in a normal man-

ner. With the Kornberg enzyme missing, it was possible to detect the presence of two other bacterial enzymes that synthesize DNA. These enzymes were named *DNA polymerases II and III* to distinguish them from the original Kornberg enzyme, now called *DNA polymerase I.* When the rates at which these three enzymes synthesize DNA in a test tube are compared, only DNA polymerase III is found to work fast enough to account for the rate of DNA replication in intact cells, which averages about 800 nucleotides per second in bacteria.

Such observations suggest that DNA polymerase III is responsible for DNA replication in intact bacterial cells, but the evidence would be more convincing if it could be shown that cells lacking DNA polymerase III are unable to replicate their DNA. But how can one grow and study cells that have lost the ability to carry out an essential function such as DNA replication? One powerful approach is the use of **temperature-sensitive mutants,** which are organisms that produce proteins that function properly at normal temperatures but become seriously impaired when the temperature is raised slightly. For example, DNA polymerase III mutants have been isolated in which this enzyme functions normally at 37°C, but loses its capacity to function when the temperature is raised to 42°C. Bacteria containing this temperature-sensitive form of DNA polymerase III grow normally at 37°C, but lose the ability to replicate their DNA when the temperature is elevated to 42°C. Such observations indicate that DNA polymerase III must play an essential role in the process of normal DNA replication. We will see shortly, however, that DNA polymerase III is not the only protein required for DNA replication; a variety of other proteins, including DNA polymerase I, also play important roles in the replication process.

Like bacteria, eukaryotic cells contain several different types of DNA polymerase. The four main eukaryotic enzymes are called DNA polymerases α, β, γ, and δ. Although these enzymes have not been as well studied as their bacterial counterparts, current evidence suggests that DNA polymerase α is the main enzyme responsible for cellular DNA replication. Studies employing an inhibitor of eukaryotic DNA replication called *aphidicolin* have been particular helpful in pointing to the importance of DNA polymerase α. In one such study, mutant cells were isolated which are resistant to the effects of aphidicolin on cellular DNA replication. When DNA polymerases were isolated from such cells, DNA polymerase α was found to have developed resistance to aphidicolin.

DNA Polymerases Catalyze DNA Synthesis in the 5' → 3' Direction

The discovery of DNA polymerase was merely the first step in unraveling the mechanism by which DNA is replicated. One of the earliest problems that arose in trying to understand DNA replication emerged from the discovery that DNA polymerases only catalyze the attachment of nucleotides to the 3' end of an existing polynucleotide chain; in other words, DNA polymerases only synthesize DNA chains in the 5' → 3' direction (see Figure 3-8). Yet the two strands of the DNA double helix run in opposite directions. So how does an enzyme that functions solely in the 5' → 3' direction manage to replicate a DNA molecule containing one chain running in the 5' → 3' direction and one chain running in the 3' → 5' direction?

An answer to this question was first proposed in 1968 by Reiji Okazaki, whose experiments suggested that DNA is synthesized as small pieces that are later joined together. In these studies, DNA was first isolated from bacterial cells that had been briefly exposed to a radioactive substrate that is incorporated into DNA. When this DNA was analyzed, most of the radioactivity was found to be associated with small fragments of DNA measuring about a thousand nucleotides in length (Figure 3-9, *left*). After a longer labeling periods, the

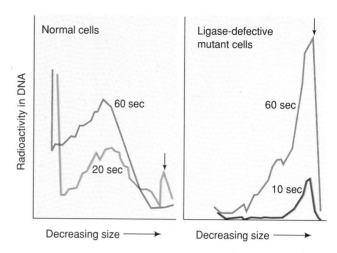

Figure 3-9 *Summary of Okazaki's Experiments on the Mechanism of DNA Replication in Bacterial Cells* Bacteria were incubated for brief periods with radioactive thymidine to label newly synthesized DNA. The DNA was then isolated, dissociated into its individual strands, and fractionated by centrifugation into molecules of differing size. (Left) In normal bacteria incubated with ³H-thymidine for 20 seconds, a significant amount of radioactivity is present in small DNA fragments (arrow). By 60 seconds the radioactivity present in small DNA fragments has all shifted to larger DNA molecules. (Right) In bacterial mutants deficient in the enzyme DNA ligase, radioactivity remains in small DNA fragments even after 60 seconds of incubation. It was therefore concluded that DNA ligase normally functions to join small DNA fragments together, forming longer DNA chains.

radioactivity was associated with larger DNA molecules. This finding suggested to Okazaki that the smaller DNA pieces, often called **Okazaki fragments,** are precursors of newly forming larger DNA molecules. It was later discovered that the conversion of Okazaki fragments into larger DNA molecules does not occur in mutant bacteria that are deficient in the enzyme **DNA ligase,** which joins DNA fragments together by catalyzing the ATP-dependent formation of a phosphodiester bond between the 3' end of one nucleotide chain and the 5' end of another (see Figure 3-9, *right*).

The preceding observations suggest a model of DNA replication that is consistent with the fact that DNA polymerase only synthesizes DNA in the 5' → 3' direction. According to this model, which is illustrated in Figure 3-10, the two strands of the DNA double helix unwind in the region where DNA replication is occurring, forming a Y-shaped structure called a **replication fork.** As the replication fork moves along the DNA double helix, DNA polymerase synthesizes two new DNA strands in slightly different ways. One of the two newly forming strands, called the **leading strand,** is synthesized as a continuous chain because it is growing in the 5' → 3' direction. Because of the opposite orientation of the two DNA strands, the other newly forming strand, called the **lagging strand,** must grow in the 3' → 5'

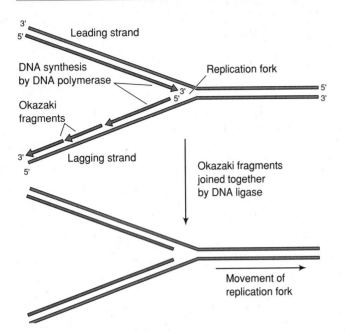

Figure 3-10 *Direction of DNA Synthesis at a Replication Fork* *DNA polymerase synthesizes a continuous strand of DNA in the 5' → 3' direction on the leading strand, and short Okazaki fragments of DNA in the 5' → 3' direction on the lagging strand. DNA ligase later joins the Okazaki fragments together.*

direction. But DNA polymerase cannot add nucleotides in the 3' → 5' direction, so the lagging strand is instead formed as a series of short, discontinuous Okazaki fragments that are synthesized in the 5' → 3' direction. These fragments are then joined together by DNA ligase to make a continuous new 3' → 5' DNA strand.

Given the complexity of this model, one may wonder why cells don't simply produce an enzyme that synthesizes DNA in the 3' → 5' direction. One possible answer is related to the need for error correction during DNA replication. About one out of every 10,000 nucleotides incorporated during DNA replication is incorrectly base-paired with the template DNA strand. Such mistakes are usually corrected by a **proofreading** mechanism, which utilizes the same DNA polymerase molecule that catalyzes DNA synthesis. Proofreading is

made possible by the fact that DNA polymerase exhibits a *3'-exonuclease* activity, which catalyzes the removal of improperly base-paired nucleotides from the 3' end of a polynucleotide chain (Figure 3-11). This proofreading capability improves the fidelity of DNA replication to the point where an average of only one error occurs for every billion base pairs replicated. If cells did contain an enzyme capable of synthesizing DNA in the 3' → 5' direction, proofreading would not work because a DNA chain growing in the 3' → 5' direction would contain a nucleotide triphosphate at its 5' end. If this 5' nucleotide were an incorrect base that needed to be removed during proofreading, its removal would eliminate the triphosphate group that provides the free energy that allows DNA polymerase to add nucleotides to a growing DNA chain. Hence no further elongation of the DNA chain could take place.

DNA Synthesis Is Initiated Using Short RNA Primers

Since DNA polymerase can only add nucleotides to the 3' end of an existing nucleotide chain, how is the replication of a DNA double helix started? For the leading strand, initiation only needs to occur once when a replication fork first forms; DNA polymerase can then add nucleotides to the chain continuously in the 5' → 3' direction. But the lagging strand is synthesized as a series of discontinuous Okazaki fragments, each of which must be initiated separately. Shortly after Okazaki fragments were first discovered, RNA was implicated in the initiation of DNA replication by the following observations: (1) Okazaki fragments often have short stretches of RNA at their 5' ends; (2) DNA polymerase can catalyze the addition of nucleotides to the 3' end of RNA chains as well as DNA chains; (3) cells contain an enzyme called **DNA primase** that synthesizes RNA fragments about ten bases long using DNA as a template, and (4) unlike DNA polymerase, which only adds nucleotides to the end of existing chains, DNA primase can initiate the synthesis of RNA chains from scratch by joining two nucleotides together.

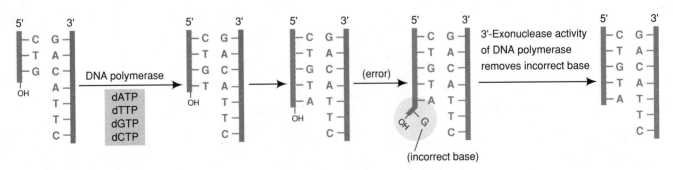

Figure 3-11 *Proofreading by 3'-Exonuclease* *If an incorrect base is inserted during DNA replication, the 3'-exonuclease activity which is part of the DNA polymerase molecule catalyzes its removal so that the correct base can be inserted.*

DNA primase = makes small RNA from DNA.

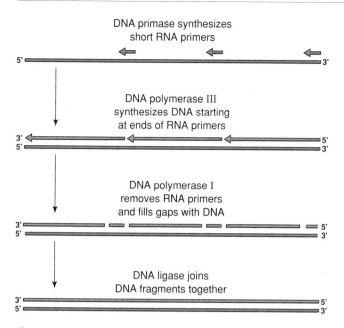

DNA primase synthesizes
short RNA primers

DNA polymerase III
synthesizes DNA starting
at ends of RNA primers

DNA polymerase I
removes RNA primers
and fills gaps with DNA

DNA ligase joins
DNA fragments together

Figure 3-12 *The Role of RNA Primers in DNA Synthesis*
The steps are shown for synthesis of the lagging strand,
which requires the formation of multiple RNA primers.

The preceding observations led to the conclusion that DNA synthesis is initiated by the formation of short **RNA primers** (Figure 3-12). After each RNA primer is formed, DNA polymerase catalyzes the addition of nucleotides to its 3' end. In the lagging strand, this process creates Okazaki fragments containing short stretches of RNA at their 5' ends. These RNA sequences are later removed and replaced by DNA. In bacterial cells, where the enzymes involved in replication have been the most extensively studied, DNA polymerase III carries out the continuous synthesis of DNA in the leading strand as well as the synthesis of Okazaki frag-

ments in the lagging strand, whereas DNA polymerase I is responsible for removing the RNA primers and filling in the resulting gaps.

Why do cells employ RNA primers that must later be removed rather than simply using a DNA primer in the first place? The answer may again be related to the process of error correction. We have already seen that DNA polymerase has a 3'-exonuclease activity that allows it to remove incorrect nucleotides from the 3' end of a DNA chain. In fact, DNA polymerase will only elongate an existing DNA chain if the nucleotide present at the 3' end is properly base-paired. An enzyme that *initiates* the synthesis of a new chain cannot carry out such a proofreading function because it is not adding a nucleotide to an existing base-paired end. As a result, enzymes that initiate nucleic acid synthesis are not very good at correcting errors. By using RNA rather than DNA to initiate DNA synthesis, any errors that do occur during initiation are restricted to RNA sequences that are later removed by DNA polymerase I.

Unwinding a DNA Double Helix Requires Helicases, Single-Stranded DNA Binding Proteins, and Topoisomerases

At a replication fork, the two strands of the DNA double helix must unwind so that each can serve as a template for DNA polymerase. At least three classes of proteins facilitate this unwinding process: *DNA helicases, single-stranded DNA binding proteins,* and *topoisomerases* (Figure 3-13).

1. DNA helicases are proteins that bind to single-stranded DNA and travel along it, driven by energy derived from the hydrolysis of ATP. When they encounter

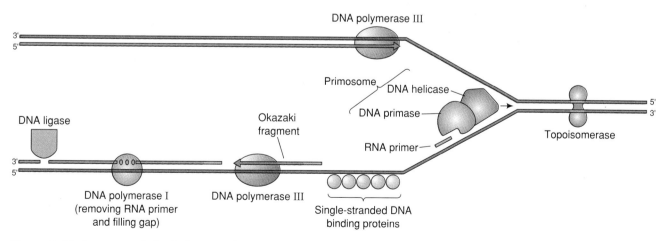

Figure 3-13 *Summary of the Major Proteins Involved in DNA Replication in Bacterial Cells* *The primosome moves along the lagging strand, unwinding the double helix and synthesizing primers as it proceeds. Single-stranded DNA binding proteins maintain the unwound DNA in a single-stranded state until DNA polymerase III can synthesize the next Okazaki fragment. DNA polymerase I removes the RNA primers and fills the resulting gaps, and DNA ligase joins the fragments. Topoisomerase introduces transient breaks in one of the two DNA strands ahead of the replication fork, thereby allowing the helix to unwind locally without having to rotate the entire DNA molecule.*

a region of double-stranded DNA, unwinding occurs. Some helicases travel in the 3' → 5' direction, while others move in the 5' → 3' direction. The helicase that functions at the replication fork forms a tight complex with DNA primase, producing a structure called a **primosome.** The primosome is associated with the lagging strand and moves along with the replication fork, unwinding the double helix and synthesizing primers as it proceeds.

2. **Single-stranded DNA binding proteins** (also called **helix-destabilizing proteins**) function to maintain DNA in a single-stranded state after the DNA double helix has been unwound by the action of helicases. These proteins are associated predominantly with the lagging strand, where they keep the single-stranded DNA unwound until DNA polymerase can begin to synthesize the next Okazaki fragment.

3. **Topoisomerases** are nucleases that facilitate DNA unwinding by introducing transient breaks in one or both strands of the DNA double helix. Figure 3-14 illustrates why such breaks are required. In a DNA molecule that is replicated at a rate of about 800 nucleotides per second, which is typical for bacterial cells, the DNA double helix ahead of the replication fork would need to rotate at a rate of about 80 turns per second. For long

DNA molecules, large amounts of energy would be required to drive this rapid rotation. A topoisomerase solves this problem by introducing a transient break or "nick" in one of the two DNA strands ahead of the replication fork, which allows the helix to unwind locally without having to rotate the entire DNA molecule. The transient nick is repaired by the topoisomerase as part of its mechanism of action.

The Process of DNA Replication Is Complicated by Chromosome Structure

We have now seen how DNA synthesis is carried out at the replication fork by a process that requires the participation of several classes of proteins (Table 3-2). Although the roles played by most of these components have been established in studies involving bacterial cells, the basic principles underlying DNA synthesis are similar in eukaryotes as well. However, the process of DNA replication in eukaryotic cells is complicated by the fact that their nuclear DNA molecules are bound to large amounts of protein and folded into complex structures known as *chromatin fibers,* which are in turn packaged into enormous *chromosomes.* The DNA molecule contained in a typical human chromosome, for example, would measure about 2 cm in length if completely extended. The relationship between DNA replication and chromosomal organization will be described in Chapter 12, when we discuss the organization of both prokaryotic and eukaryotic chromosomes and the mechanisms by which they are duplicated prior to cell division.

Retroviruses Employ Reverse Transcriptase to Synthesize DNA

Although DNA molecules are almost always synthesized using an existing DNA molecule as a template, an exception occurs in a class of RNA viruses known as **retroviruses.** Included in this group are certain cancer-causing viruses and the virus associated with AIDS.

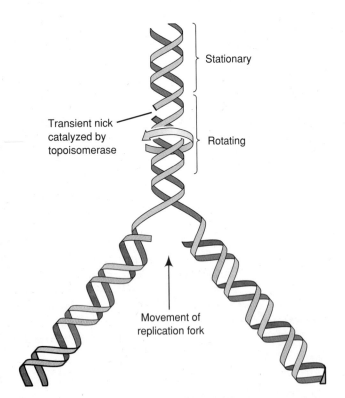

Figure 3-14 *Role of Topoisomerase in DNA Unwinding*
Topoisomerase catalyzes the reversible breaking or "nicking" of one of the two strands of the DNA double helix. This transient breakage permits the helix to unwind locally without the need to rotate the entire DNA molecule ahead of the replication fork.

Table 3-2 Some of the Proteins Involved in Bacterial DNA Replication	
Protein	**Main Function**
DNA polymerase III	Elongates DNA
DNA polymerase I	Removes primers and fills gaps
DNA ligase	Joins DNA fragments together
DNA primase	Synthesizes RNA primers
Helicases	Unwind DNA double helix
Single-stranded DNA binding proteins	Stabilize single-stranded DNA
Topoisomerases	Create transient breaks to facilitate DNA unwinding

Retroviruses store their genetic information in molecules of RNA rather than DNA. When a retrovirus infects a cell, it brings along a novel DNA-synthesizing enzyme called **reverse transcriptase.** Unlike DNA polymerase, which uses DNA as a template for directing DNA synthesis, reverse transcriptase utilizes RNA as a template for directing DNA synthesis (Figure 3-15). The net result is that the genetic information stored in the original viral RNA molecule is copied into a newly forming molecule of double-stranded DNA. This viral DNA then becomes integrated into the host cell DNA, where it can be maintained for an indefinite period of time.

In addition to its role in viral infection, reverse transcriptase is a powerful experimental tool because it allows DNA molecules to be synthesized in the laboratory that are complementary to any given molecule of RNA. As we will see later (page 100), such complementary DNA molecules or **cDNAs** can be used in a variety of ways to study the organization and behavior of individual genes.

DNA TRANSCRIPTION AND TRANSLATION

Now that we have seen how DNA is replicated, we are ready to address the second major question posed earlier in the chapter: How does the information encoded within DNA determine the properties of cells and organisms? Avery's experiments, of course, had led to the realization that DNA is the molecule from which genes are constructed. But what, exactly, does a gene do?

Experiments on *Neurospora* Led to the One Gene–One Enzyme Theory

The first major contribution to our understanding of what genes do was made by Sir Archibald Garrod, an English physician who studied inherited diseases in the early 1900s. Garrod was especially intrigued by *alkap-*

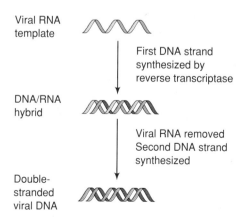

Figure 3-15 *Reverse Transcriptase Catalyzes the Synthesis of DNA Using RNA as a Template* *Although this viral enzyme normally makes DNA copies of viral RNA, it can be used experimentally to make DNA copies of other kinds of RNA as well.*

Viral RNA template

First DNA strand synthesized by reverse transcriptase

DNA/RNA hybrid

Viral RNA removed Second DNA strand synthesized

Double-stranded viral DNA

tonuria, a disease that causes the urine of afflicted individuals to blacken upon exposure to air. When Garrod analyzed urine obtained from such patients, he discovered that it contained large quantities of *homogentisic acid.* In normal individuals homogentisic acid is degraded to carbon dioxide and water, and hence it does not appear in the urine. Garrod concluded that patients with alkaptonuria must therefore lack the enzyme that ordinarily catalyzes the breakdown of homogentisic acid. Because alkaptonuria is an inherited disease, this conclusion implied that hereditary information controls the production of enzymes. Although many textbooks credit Garrod with being ahead of his time in seeing this relationship, his written reports published in 1909 suggest that Garrod failed to see the profound implications of this work.

Thirty years passed before the next opportunity arose for investigating the relationship between genes and enzymes. In the early 1940s, George Beadle and Edward Tatum embarked upon a study of the nutritional requirements of the common bread mold, *Neurospora crassa. Neurospora* is a relatively self-sufficient organism that can grow in a *minimal medium* containing only sugar, salts, and the vitamin biotin. From these few ingredients, *Neurospora's* metabolic pathways generate everything else the organism requires. To investigate the influence of genes on these metabolic pathways, Beadle and Tatum treated a *Neurospora* culture with X-rays to induce genetic changes or **mutations.** Such treatments produced mutant organisms that lost their ability to survive in the minimal medium, although they could be grown on a *complete medium* supplemented with amino acids and vitamins.

Such observations suggested that the *Neurospora* mutants had lost the ability to synthesize certain amino acids and vitamins and therefore required these substances in the growth medium. To determine exactly which nutrients were required, Beadle and Tatum transferred the mutant organisms to a variety of different growth media, each containing a single amino acid or vitamin added as a supplement to the minimal medium. This approach led to the discovery that one mutant strain would grow only in a medium supplemented with vitamin B_6, a second mutant would only grow in a medium supplemented with the amino acid arginine, and so forth (Figure 3-16). A large number of different mutants were eventually characterized, each impaired in its ability to synthesize a particular amino acid or vitamin.

Because the synthesis of amino acids and vitamins is accomplished through multistep metabolic pathways, Beadle and Tatum worked to identify the particular step in each pathway that had become defective. They approached this question by supplementing the minimal medium with metabolic precursors of a given amino acid or vitamin rather than with the amino acid or vitamin itself. By determining which precursors support the

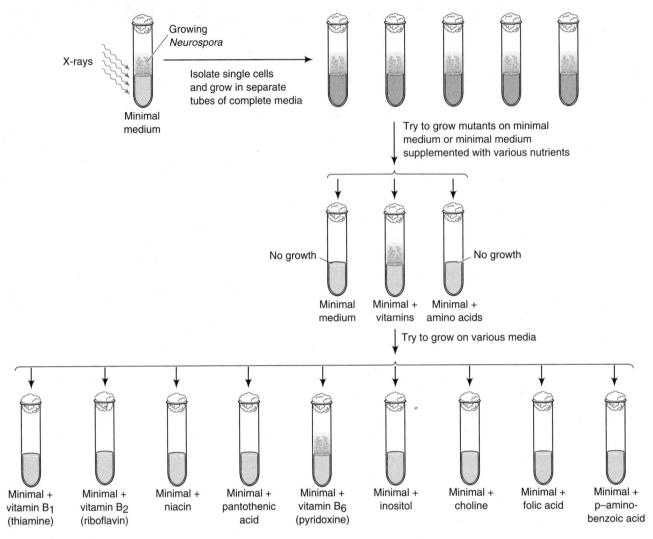

Figure 3-16 *Experimental Approach Utilized by Beadle and Tatum to Identify Metabolic Defects in* Neurospora *Mutants* This example illustrates the behavior of a mutant strain of Neurospora that has lost the ability to grow on the minimal medium, but is capable of growing on the complete medium or on a medium that is supplemented with vitamin B_6. It can therefore be concluded that this mutant has lost the ability to synthesize vitamin B_6.

growth of a particular mutant organism, the defective step in each metabolic pathway could be pinpointed (Table 3-3). These experiments revealed that each *Neurospora* mutant was deficient in a single metabolic reaction. Since metabolic reactions are catalyzed by enzymes, it was concluded that each mutant was defective in a particular enzyme. Moreover, genetic studies revealed that every metabolic defect corresponded to a mutation in a single gene. Beadle and Tatum therefore combined these observations into a unifying concept, called the *one gene–one enzyme theory,* which stated that the function of a gene is to control the production of an enzyme molecule.

The Base Sequence of a Gene Usually Codes for the Sequence of Amino Acids in a Polypeptide Chain

The theory that genes direct the production of enzyme molecules represented a major advance in our understanding of gene action, but it did not provide much in-

sight into the question of how genes accomplish this task. The first clue to the underlying mechanism emerged a few years later in the laboratory of Linus Pauling, who was studying the inherited disease *sickle-cell anemia.* The red blood cells of individuals suffering from sickle-cell anemia exhibit an abnormal shape, and thus they become trapped and damaged when passing through small blood vessels (Figure 3-17). In trying to determine the molecular defect responsible for this behavior, Pauling decided to analyze the properties of *hemoglobin,* the major protein of red cells. The properties of normal and sickle-cell hemoglobin were investigated using *electrophoresis,* a technique in which charged molecules are separated from one another by placing them in an electric field (page 144). During electrophoresis the sickle-cell hemoglobin was found to migrate at a different rate than normal hemoglobin, suggesting that the two proteins differ in electric charge. Since some amino acids have charged side chains, Pauling proposed that the dif-

Table 3-3   Method for Identifying the Metabolic Step That Is Defective in *Neurospora* Mutants That Have Lost the Ability to Synthesize Arginine Using the Pathway: A → B → C → Arginine

Type of Neurospora	Cell Growth on Minimal Medium Supplemented with:					Conclusion About Location of Defect
	A	B	C	Arginine	No Supplement	
Normal	Yes	Yes	Yes	Yes	Yes	No defect
Mutant 1	No	Yes	Yes	Yes	No	Enzyme catalyzing A → B
Mutant 2	No	No	Yes	Yes	No	Enzyme catalyzing B → C
Mutant 3	No	No	No	Yes	No	Enzyme catalyzing C → Arginine

Note: In this hypothetical experiment, mutants that require arginine for growth were distinguished from one another by comparing their abilities to grow on various intermediates involved in the pathway for synthesizing arginine. For example, mutant 1 can grow when substance B or C is provided, but not when substance A is provided. Hence mutant 1 must be defective in the enzyme that catalyzes the conversion of A → B.

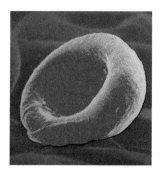 

Figure 3-17 *Scanning Electron Micrographs of Normal and Sickled Red Blood Cells* *The abnormal shape of the sickled red cells causes them to become trapped and damaged when passing through small blood vessels. Their aberrant shape is caused by the presence of a mutated form of hemoglobin. Courtesy of B. Longcore (left) and J. Lewin (right).*

ference between normal and sickle-cell hemoglobin could be explained by a difference in amino acid makeup. To test this hypothesis, it seemed necessary to determine the amino acid sequence of the hemoglobin molecule. At the time of Pauling's discovery in the early 1950s, the largest protein to have been sequenced was less than one-tenth the size of hemoglobin. Determining the complete amino acid sequence of hemoglobin would therefore have been a monumental undertaking.

Fortunately, an ingenious shortcut devised by Vernon Ingram soon made it possible to identify the amino acid abnormality in sickle-cell hemoglobin without determining its complete amino acid sequence. Ingram cleaved hemoglobin into smaller peptide fragments using the protease *trypsin,* and then separated these peptides from each other using a combination of electrophoresis and chromatography. This procedure separates peptides into a two-dimensional series of spots known as a *fingerprint* (Figure 3-18). When Ingram examined the fingerprints of normal and sickle-cell hemoglobin, he discovered that only one peptide differed between the two proteins. Analyzing the amino acid composition of this peptide revealed that a glutamic acid present in normal hemoglobin

had been replaced by a valine in sickle-cell hemoglobin. Since glutamic acid is negatively charged and valine is neutral, this substitution explains the difference in electrophoretic behavior between normal and sickle-cell hemoglobin originally observed by Pauling.

Ingram's studies, which were published in the mid-1950s, confirmed the idea that genes specify the amino acid sequence of protein molecules. We have just seen that in the gene coding for hemoglobin, a mutation that alters a single amino acid in a protein's sequence changes the protein sufficiently to induce a debilitating, and often fatal, disease. Later studies revealed the existence of other abnormal forms of hemoglobin, some of which involve mutations in a gene that differ from the one altered in sickle-cell anemia. Two different genes are able to influence the amino acid sequence of the same protein because hemoglobin is a multisubunit protein containing two different kinds of polypeptide chains. The amino acid sequences of these two types of chains, called the α and β subunits, are specified by two separate genes.

The preceding discoveries necessitated several refinements in the one gene–one enzyme theory of Beadle and Tatum. First, the fact that hemoglobin is not an enzyme indicates that genes encode the amino acid sequences of proteins in general, not just enzymes. In addition, the discovery that different genes code for the α and β chains of hemoglobin reveals that each gene encodes the amino acid sequence of a polypeptide chain, not necessarily a complete protein. This refined view came to be known as the *one gene–one polypeptide chain theory.* According to this theory, the nucleotide base sequence of a gene determines the order of amino acids in a polypeptide chain. In the mid-1960s this prediction was confirmed in the laboratory of Charles Yanofsky, where the locations of dozens of mutations in the bacterial gene coding for the enzyme tryptophan synthetase were determined. The positions of these mutations within the gene were found to correlate with the positions of the resulting amino acid alterations in the tryptophan synthetase polypeptide chain (Figure 3-19).

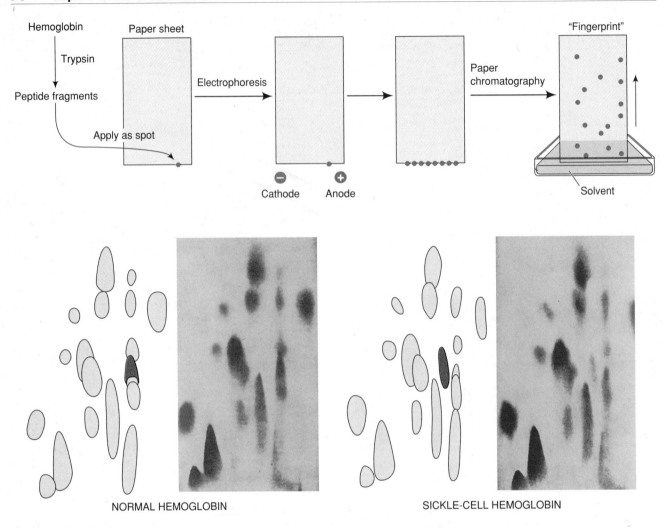

Figure 3-18 *Comparison of the Fingerprints of Normal and Sickle-Cell Hemoglobin* (Top) *Protein fingerprinting is carried out by degrading proteins into peptide fragments using a protease such as trypsin, and then separating the peptides using a combination of electrophoresis and chromatography.* (Bottom) *The photographs show the fingerprint patterns obtained with normal and sickle-cell hemoglobin. The colored spots in the drawings that accompany the photographs indicate the regions where the two patterns differ.*

The conclusion that the base sequence of a gene codes for the amino acid sequence of a polypeptide chain is now known to apply to virtually all genes, with the exception of a few kinds of genes that code for RNA molecules such as transfer RNA (page 88) and ribosomal RNA (page 479). A direct linear correspondence between the complete base sequence of a gene and the amino acid sequence of the polypeptide chain it encodes is typical for bacterial genes, but the organization of eukaryotic genes is often more complex. This additional complexity will be described later in the chapter, after we have seen how genetic information is transferred from DNA base sequences to the amino acid sequences of polypeptide chains.

Messenger RNA Carries Information from DNA to Newly Forming Polypeptide Chains

Once biologists realized that the nucleotide base sequence of a gene determines the amino acid sequence

of a polypeptide chain, the question arose as to how this transfer of information occurs. Several findings suggested that DNA cannot directly guide the assembly of amino acids into polypeptide chains. First, studies carried out on eukaryotic cells incubated with radioactive amino acids revealed that the incorporation of radioactivity into protein molecules occurs in the cytoplasm, whereas most of the cell's DNA is located in the nucleus. Moreover, cells whose nuclei have been artificially removed continue synthesizing proteins for considerable periods of time, in spite of the absence of nuclear DNA. Such observations suggest that some molecule must serve as an intermediary that carries information from the nuclear DNA to the site of protein synthesis in the cytoplasm.

Experiments carried out in the early 1940s in the laboratories of Jean Brachet and Torbjöern Caspersson provided some early clues as to the nature of this intermediary even before the genetic role of DNA had been

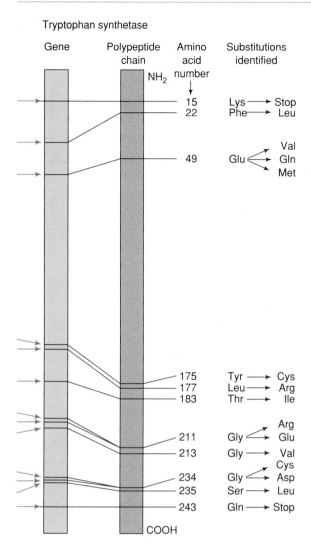

Tryptophan synthetase

Figure 3-19 *Relationship between Mutation Sites in a Gene and Amino Acid Substitutions in the Corresponding Polypeptide Chain* *In these data for the bacterial tryptophan synthetase gene, the linear order of mutation sites in the gene (colored arrows) corresponds to the linear sequence of amino acid alterations in the polypeptide chain. The mutation sites in the gene and polypeptide do not line up precisely because the gene mutation sites were located using genetic recombination frequencies, a technique that does not permit the exact localization of sites within a DNA base sequence.*

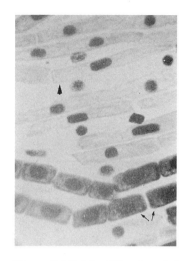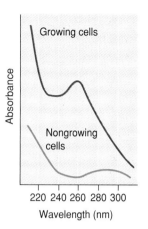

Figure 3-20 *Identification of Cytoplasmic RNA in Cells of the Growing Onion Root Tip* (Left) *Light microscopy of root cells stained for nucleic acid reveals darkly stained cytoplasm in cells located near the growing tip (arrows), whereas cells located in the nongrowing region of the root have unstained cytoplasm (arrowhead). (Right) Absorption spectra of the cytoplasm of root tip cells analyzed by ultraviolet light microscopy. The cytoplasm of the cells from the root tip exhibits an absorption peak at 260 nm, which is characteristic of nucleic acids. The curve for nongrowing cells shows less overall absorption and a peak at 280 nm, which is characteristic of proteins. Such data indicate that cytoplasmic nucleic acid is present in higher concentration in the growing cells. Micrograph courtesy of T. O. Caspersson.*

demonstrated. Both laboratories were attempting to measure the quantity of RNA present in the cytoplasm, either by staining cells with basic dyes or by measuring the amount of ultraviolet light they absorb (Figure 3-20). These studies revealed that cells that are active in protein synthesis tend to contain large amounts of RNA in their cytoplasm.

The significance of this observation did not become apparent until a decade later, when subcellular fractionation techniques (page 133) were first applied to the study of protein synthesis. In these pioneering experiments, animals were injected with radioactive amino acids to allow the most recently formed protein molecules to become

radioactive. Various subcellular components were then isolated and analyzed for the presence of radioactive protein. The bulk of the radioactivity was detected in the *microsomal* fraction, which consists of a mixture of ribosomes and membranes derived from the endoplasmic reticulum. When the microsomal fraction was separated into ribosomes and membranes, the newly synthesized proteins were found to be associated predominantly with the ribosomes (Figure 3-21). It was therefore concluded that protein synthesis occurs on ribosomes.

Since RNA accounts for roughly half the mass of the ribosome, the discovery that proteins are synthesized on ribosomes explained why Brachet and Caspersson had detected large quantities of RNA in cells that are active in protein synthesis. But the mechanism by which the genetic information guiding protein assembly is transmitted from the nuclear DNA to cytoplasmic ribosomes remained to be determined. The notion that RNA molecules might serve as intermediaries first surfaced in the late 1950s, primarily as the result of studies involving cells incubated with radioactive precursors to label newly synthesized RNA molecules. In cells briefly exposed to such precursors, the newly formed radioactive RNA was found predominantly in the nucleus. During subsequent incubation in a nonradioactive medium, the radioactivity moved

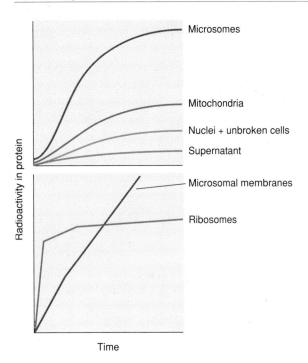

Figure 3-21 *Data from Two Experiments Designed to Determine Where Protein Synthesis Occurs within the Cell* (Top) *Rat liver exposed to radioactive amino acids to label newly synthesized protein was separated into various subcellular components. The newly synthesized radioactive protein appears first in the microsomal fraction.* (Bottom) *When the microsomal fraction is subdivided into ribosomes and membranes, newly synthesized proteins are found to be associated initially with the ribosomes. These results suggest that proteins are synthesized on ribosomes and subsequently shipped to other parts of the cell.*

to the cytoplasm (Figure 3-22). It was therefore concluded that RNA is synthesized in the nucleus and later travels to the cytoplasm, a behavior expected of a molecule that transfers information from the nuclear DNA to the cytoplasmic protein-synthesizing machinery.

At first it was thought that **ribosomal RNA (rRNA)**, which accounts for the bulk of the RNA in ribosomes, is the molecule that migrates to the cytoplasm to guide the synthesis of specific proteins. It was soon discovered, however, that rRNA isolated from different cells and organisms is similar in size and base composition, whereas the proteins synthesized by different ribosomes vary widely in size and amino acid sequence. Hence rRNA did not appear to be a likely candidate for the molecule that specifies the type of protein being synthesized.

A clever study carried out by Francois Jacob and Jacques Monod reinforced the view that rRNA is not the intermediary that carries protein-coding information from DNA to the ribosome. In this experiment, bacteria were grown in media containing abnormally high concentrations of radioactive phosphate. Upon incorporation into newly forming DNA molecules, the radioactivity caused the DNA to break apart. Shortly thereafter the synthesis of proteins stopped, even though the cell's rRNA molecules remained intact. It was therefore concluded that the molecules that guide protein synthesis are transient intermediaries that must be continually transmitted from DNA to the ribosomes. Ribosomal RNA, which is a stable component of the ribosome, clearly could not function in such a capacity.

As an alternative to the discredited idea that rRNA guides the process of protein synthesis, Jacob and

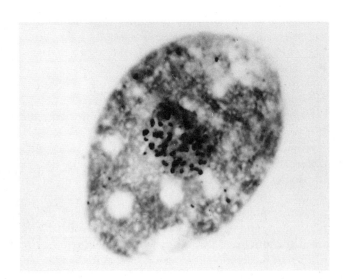

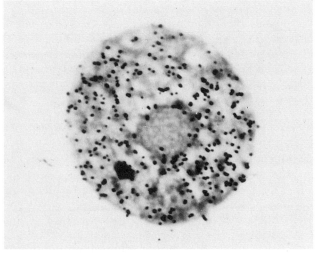

Figure 3-22 *Evidence That RNA Is Synthesized in the Nucleus and Then Migrates to the Cytoplasm* Tetrahymena *cells were incubated for 15 minutes in the presence of a radioactive RNA precursor to label newly synthesized RNA. The cells were then coated with a layer of photographic emulsion, which produces silver grains wherever the radioactive RNA is located. (Left) After 15 minutes of labeling, the silver grains indicating the presence of newly synthesized RNA are localized mainly over the nucleus. (Right) After an additional 88 minutes of incubation in a nonradioactive medium, the radioactive RNA has moved to the cytoplasm. Courtesy of D. M. Prescott.*

Monod proposed that the molecule that carries information from DNA to ribosomes is a less stable type of RNA, which they called **messenger RNA (mRNA).** According to their theory, each gene produces a unique mRNA that subsequently becomes associated with ribosomes to guide the synthesis of a particular polypeptide chain. Ribosomes were thus viewed as nonspecific protein-synthesizing machines that are in theory capable of producing any kind of protein, depending upon the particular mRNA to which they are attached.

As Jacob and Monod were quick to point out, there was already evidence for the existence of RNA molecules exhibiting some of the properties predicted for messenger RNA. A few years earlier, Elliot Volkin and Lazarus Astrachan had discovered that infection of bacterial cells with bacteriophage T2 leads to the production of a new type of RNA whose base composition resembles that of the bacteriophage DNA. To verify that such RNA molecules function as messengers, the following predictions of the messenger RNA model had to be confirmed: (1) that the base sequence of mRNA molecules is copied from DNA; (2) that mRNA becomes associated with ribosomes; and (3) that the base sequence of mRNA determines the linear order of amino acids in newly forming polypeptide chains. As we will see in the following sections, all three predictions were confirmed shortly after the existence of messenger RNA was first postulated.

The Base Sequence of Messenger RNA Is Copied from DNA by the Enzyme RNA Polymerase

The prediction that the base sequence of mRNA molecules is copied from DNA was confirmed in several ways. The first major breakthrough occurred when Samuel Weiss, Jerard Hurwitz, and Audrey Stevens independently isolated the enzyme responsible for catalyzing RNA synthesis. The enzyme, called **RNA polymerase,** catalyzes a nucleotide polymerization reaction that utilizes as substrates the triphosphate derivatives of the four bases found in RNA. These ribonucleoside triphosphates are ATP, GTP, CTP, and UTP (Figure 3-23). Although the product of the RNA polymerase reaction is a *single-stranded* molecule of RNA, the reaction resembles the synthesis of double-stranded DNA molecules by DNA polymerase in several ways: (1) The energy that drives the polymerization reaction is obtained by using high-energy triphosphate compounds as substrates for the reaction. (2) RNA synthesis requires a DNA template. (3) RNA polymerase employs the complementary base-pairing mechanism to copy base sequence information from one strand of the DNA double helix into a complementary chain of RNA. The same base-pairing rules apply as in DNA synthesis, with the single exception that the base uracil (U) substitutes in RNA for the base thymine (T) in DNA. This substitution is permitted because uracil and thymine can both form hydrogen bonds with adenine (Figure 3-24). Hence adenine pairs with

thymine in a complementary chain of DNA and with uracil in a complementary chain of RNA.

The most convincing support for the conclusion that the base sequences of RNA molecules are copied from DNA has emerged from experiments utilizing nucleic acid hybridization, a technique to be described in detail later in the chapter. In essence, nucleic acid hybridization permits one to determine the extent to which the base sequences of two nucleic acids are complementary to each other. By means of this approach, it has been shown that RNA molecules synthesized by purified preparations of RNA polymerase are complementary in sequence to the DNA molecules employed as templates. Similar results have been obtained in experiments carried out on RNA isolated from intact cells. RNA molecules synthesized after viral infection of bacteria, for example, have been shown to be complementary in sequence to viral, but not bacterial, DNA.

Because DNA contains two strands while RNA is usually single stranded, the question arises as to whether one or both of the DNA strands is copied into complementary molecules of RNA. Investigation of this issue has been facilitated by the existence of certain viruses, such as bacteriophage SP8, whose two DNA strands differ enough in base composition to permit them to be separated by *isodensity centrifugation*, a technique that separates molecules based on differences in density (page 135). Exploiting this approach, Julius Marmur and his colleagues separated the two strands of bacteriophage SP8 DNA from one another and tested their ability to form complementary hybrids with RNA molecules synthesized in virus-infected cells. Only one of the isolated DNA strands was found to be complementary to the newly synthesized RNA (Figure 3-25), indicating that RNA polymerase is capable of selectively copying one of the two DNA strands.

Messenger RNA Becomes Associated with Ribosomes

The second prediction made by the messenger RNA model is that after a DNA base sequence has been copied into a complementary molecule of mRNA, the RNA then becomes bound to ribosomes. This prediction was first supported by experiments in which bacteria that had been growing in a ^{15}N-containing medium were switched to ^{14}N-containing media and infected with bacteriophage T4. It was reasoned that any new ribosomes made by the cells after they had been infected by the virus would contain ^{14}N, and hence could be distinguished by isodensity centrifugation from the heavier ^{15}N-labeled ribosomes formed prior to infection. When the ribosomes from infected cells were analyzed by centrifugation, it was found that the old ^{15}N-containing ribosomes contained the bulk of the newly synthesized

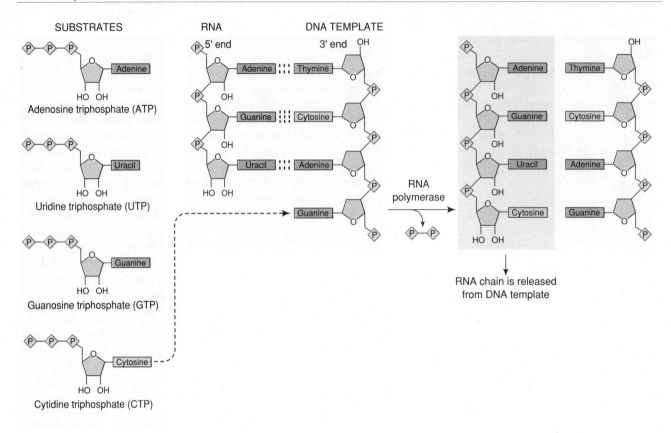

SUBSTRATES

Adenosine triphosphate (ATP)

Uridine triphosphate (UTP)

Guanosine triphosphate (GTP)

Cytidine triphosphate (CTP)

RNA

5' end

DNA TEMPLATE

3' end

RNA polymerase

RNA chain is released from DNA template

Figure 3-23 *The RNA Polymerase Reaction* *The enzyme RNA polymerase catalyzes the addition of ribonucleoside triphosphates to the 3' end of an RNA chain, using an existing DNA strand as a template. As each ribonucleoside triphosphate is added to the growing chain, its two terminal phosphate groups are released. The base sequence of the newly forming RNA molecule is determined by complementary base-pairing with the DNA template strand. In RNA, the base uracil (U) is incorporated in places where thymine (T) would have been incorporated during DNA synthesis (see Figure 3-8).*

viral mRNA (Figure 3-26). Newly forming viral proteins were also found on these same ribosomes. Such data are compatible with the idea that the viral mRNA was able to associate with preexisting bacterial ribosomes and program them to synthesize viral proteins.

The Base Sequence of Messenger RNA Determines the Amino Acid Sequence of a Polypeptide Chain

The third and final prediction made by the messenger RNA model is that the base sequence in an mRNA molecule determines the order in which amino acids are joined together during protein synthesis. This crucial prediction was initially verified in 1961 by Marshall Nirenberg, who pioneered the development of *cell-free systems* for studying the process of protein synthesis. In such cell-free systems, protein synthesis can be studied outside of living cells by mixing together isolated ribosomes, amino acids, an energy source, and a cellular extract containing soluble components of the cytoplasm.

Nirenberg observed that the rate of protein synthesis by cell-free systems could be stimulated by adding

RNA, raising the question of whether the added RNA molecules function as messengers that determine the amino acid sequences of the proteins being manufactured. To address this question, Nirenberg added synthetic RNA molecules of known base sequence to the cell-free protein-synthesizing system. One such RNA molecule was polyuridylic acid, or *poly U,* an RNA that contains the single base, uracil, repeated again and again. When poly U was added to the cell-free system, a marked stimulation was observed in the incorporation of one particular amino acid, phenylalanine, into polypeptide chains. Synthetic RNA molecules containing bases other than uracil did not stimulate phenylalanine incorporation, while poly U enhanced the incorporation of only phenylalanine (Table 3-4). Nirenberg concluded on this basis that poly U directs the synthesis of polypeptide chains that consist solely of phenylalanine. This observation represented a crucial milestone in the development of the mRNA concept, for it was the first time that it had been shown that the nucleotide base sequence of an RNA molecule influences the sequence of amino acids being incorporated during protein synthesis.

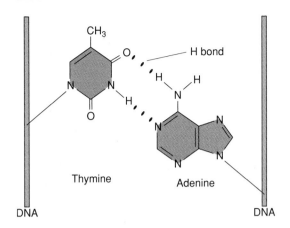

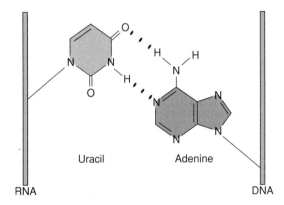

Figure 3-24 *Similarities in the Structures of Thymine and Uracil Allow Them Both to Form Hydrogen Bonds with Adenine* In DNA thymine pairs with adenine, whereas in RNA uracil pairs with adenine.

Genetic Information Encoded in DNA Is Expressed by a Two-Stage Process Involving Transcription and Translation

The evidence outlined in the preceding three sections led to the conclusion that the expression of genetic information encoded in DNA involves a two-stage process. During the first stage, termed **transcription,** information encoded in the base sequence of DNA directs the synthesis of complementary molecules of mRNA. Depending on the length of the polypeptide chain to be synthesized, mRNAs may be hundreds or thousands of bases long. In the second stage of the process, referred to as **translation,** the base sequence of mRNA molecules is used to guide the sequential incorporation of amino acids into polypeptide chains.

Although it is easy to phrase this two-stage model in relatively simple terms, the transcription-translation pathway is comprised of a large number of individual steps, many of which are quite complicated. In the remainder of this section, we will provide an overview of the main steps that are involved. Later in the book the topics of transcription and translation will be discussed in much greater detail, when we focus upon the struc-

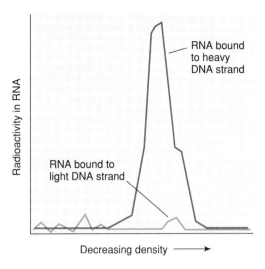

Figure 3-25 *Evidence That Only One Strand of the DNA Double Helix Serves as a Template for RNA Synthesis in Bacteriophage SP8* The two strands of the bacteriophage DNA were separated from each other by isodensity centrifugation. Each DNA strand was then individually mixed with radioactive bacteriophage RNA to determine which DNA strand would bind by complementary base-pairing to the RNA. Centrifugation of the resulting DNA-RNA hybrids showed that the radioactive RNA binds to only the heavy DNA strand.

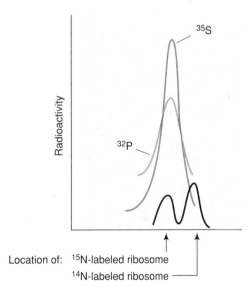

Figure 3-26 *Data from an Early Experiment Supporting the Existence of Messenger RNA* Bacteria that had been growing in ^{15}N-containing media were switched to a medium containing ^{14}N, and were then infected with bacteriophage T4 in the presence of radioactive precursors of RNA and protein synthesis. When bacterial homogenates were subsequently analyzed by isodensity centrifugation, the newly synthesized viral RNA (labeled with ^{32}P) and viral-specific proteins (labeled with ^{35}S) were found to be predominantly associated with the old ^{15}N-containing bacterial ribosomes.

Table 3-4 Data Illustrating the Effects of Some Synthetic RNAs on Cell-Free Protein Synthesis

Synthetic RNA Added	Radioactive Amino Acid Added	Amount of Radioactivity Incorporated into Polypeptides (counts/min)
None	Phenylalanine	68
Poly U	Phenylalanine	38,300
Poly U	Glycine, alanine, serine, aspartic acid, and glutamic acid	33
Poly A	Phenylalanine	50
Poly C	Phenylalanine	38

Note: Data from M. W. Nirenberg and J. H. Matthaei (1961), *Proc. Nat. Acad. Sci. USA* 47:1588–1602.

ture and function of the nucleus and ribosomes (Chapters 10 and 11).

RNA Splicing Often Accompanies the Production of Eukaryotic Messenger RNAs

The genes of bacterial cells are typically organized as continuous DNA stretches that can be directly copied by RNA polymerase into functional molecules of mRNA. For many years this model was thought to apply to eukaryotic cells as well, but evidence that appeared in the late 1970s indicated that most eukaryotic genes are interrupted by DNA sequences that do not code for any portion of the gene's polypeptide product (page 429). The DNA segments that code for the gene's polypeptide product are referred to as **exons,** whereas the noncoding sequences interspersed between the exons are called **introns** (Figure 3-27). During transcription of eukaryotic genes, RNA polymerase copies both exons and introns. The resulting RNA product then undergoes a **splicing** process that removes the noncoding sequences and produces a functional mRNA molecule.

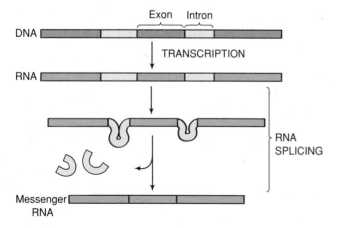

Figure 3-27 RNA Splicing in Eukaryotic Cells *Many eukaryotic genes contain noncoding sequences called introns that are transcribed into RNA, but must then be removed by a splicing reaction in order to generate mRNA.*

What is the advantage of synthesizing RNA molecules that contain extraneous sequences that need to be removed? At least two answers have been proposed to this question. First, the RNA molecules transcribed from certain eukaryotic genes can be spliced in several different ways, producing mRNAs that code for different polypeptides (page 465). This gives genes added flexibility by allowing them to code for more than one kind of polypeptide chain. A second advantage to the presence of exons and introns is that it may facilitate the evolution of new types of proteins. As we learned in Chapter 1, polypeptide chains are often organized into a series of discrete functional regions called *domains.* In genes that are organized into multiple exons and introns, the exons often code for different domains of the same polypeptide chain. During evolution, DNA molecules undergo random breakage and rejoining events that rearrange the genetic information. By separating exons with noncoding sequences, the presence of introns increases the chances that new functional proteins will be created by bringing together new combinations of exons, which in turn code for proteins exhibiting new combinations of domains.

The Genetic Code Is a Triplet Code

Now that we have established that the base sequences of mRNA molecules determine the amino acid sequences of newly synthesized proteins, the question arises as to the nature of the coding process. Nirenberg's discovery that poly U stimulates the synthesis of polypeptide chains containing only phenylalanine does not in itself define the coding relationship between the base U in RNA and the phenylalanine residues that make up the polypeptide chain. It is clear that a single base in RNA cannot code for an amino acid because 20 different amino acids must be encoded, and there are only four different kinds of bases in RNA (A, G, C, and U). If the code for each amino acid were a doublet involving two adjacent bases (i.e., UU = phenylalanine), there would still only be $4 \times 4 = 16$ possible combinations of bases.

To accommodate 20 different amino acids, the coding system must be based at the very least on nucleotide triplets (i.e., UUU = phenylalanine). In such a **triplet code,** a total of $4 \times 4 \times 4 = 64$ different combinations of four bases is possible. which is more than sufficient to encode 20 different amino acids.

Although this mathematical argument first led biologists to suspect the existence of a triplet code in the early 1950s, experimental evidence for such a code did not appear until 1961. In that year Francis Crick, Sidney Brenner, and their associates reported the results of a series of studies in which the dye acridine orange was utilized to induce mutations in bacteriophage T4. Acridine dyes insert between adjacent bases in DNA, distorting the double helix in such a way that extra bases tend to be incorporated during DNA replication. Treating virus-infected bacteria with acridine orange therefore induces the production of virus particles with varying numbers of extra bases inserted in their DNA. When one or two extra bases were inserted, the resulting virus particles were rendered incapable of carrying out subsequent infections. The presence of three extra bases, however, altered the properties of the virus only slightly.

On the basis of these observations, it was concluded that the insertion of extra bases alters the way in which the triplet code is read during translation. To illustrate this concept of a **frameshift mutation,** let us consider a hypothetical gene in which the triplet CAT is repeated over and over again:

$$\cdots \text{(CAT)(CAT)(CAT)(CAT)(CAT)(CAT)} \cdots$$

A mutation involving the insertion a single extra base (X) will alter the reading of all mRNA triplets beyond the point of insertion, creating a garbled sequence:

$$\cdots \text{(CAT)(XCA)(TCA)(TCA)(TCA)(TCA)(T} \cdots$$

The insertion of a second base (Y) shifts the triplets by one additional nucleotide, leaving the message still garbled:

$$\cdots \text{(CAT)(XYC)(ATC)(ATC)(ATC)(ATC)(AT} \cdots$$

If a third base (Z) is inserted, however, the triplets beyond the mutation point become properly aligned again:

$$\cdots \text{(CAT)(XYZ)(CAT)(CAT)(CAT)(CAT)(CAT)} \cdots$$

Even though this DNA sequence contains a small segment of misinformation in the region of the three inserted bases (XYZ), the vast majority of the sequence codes for the proper amino acid sequence. This phenomenon would only be expected if the genetic code were read in units of three bases and explains why bacteriophage mutants containing three insertions behave almost normally, whereas those with one or two do not.

The Coding Dictionary Was Established Using Synthetic RNA Polymers and RNA Triplets

Once the existence of a triplet code had been established, the next step was to identify the triplets that specify each of the 20 amino acids. Nirenberg's experiments with poly U indicated that the sequence UUU in mRNA must code for the amino acid phenylalanine. The coding properties of other homogeneous RNA polymers, such as poly A, poly C, and poly G, were also studied in cell-free protein-synthesizing systems to determine the amino acids encoded by the triplets AAA, CCC, and GGG. Determining the coding assignments of triplets that contain more than one type of base, however, required more complicated approaches.

One clever strategy devised by Nirenberg and his associate Philip Leder involved the use of individual triplets, rather than long RNA molecules, to identify specific codes. Since individual triplets code for single amino acids, not entire polypeptide chains, monitoring the incorporation of radioactive amino acids into polypeptides could not be employed as an assay for determining which amino acid is encoded by a given triplet. However, we will learn shortly that amino acids are carried to the ribosome by small RNA molecules called *transfer RNAs* or *tRNAs* (page 88); Nirenberg and Leder therefore reasoned that they could measure the ability of individual triplets to induce binding of these tRNA–amino acid complexes to the ribosome. Adding the synthetic triplet UUU, for example, was found to cause the tRNA containing phenylalanine to bind to the ribosome. Utilizing the same approach with other triplets, Nirenberg and Leder were able to identify the amino acids encoded by most of the 64 possible RNA triplets. The few cases that yielded ambiguous data were soon resolved by Har Gobind Khorana, who developed techniques for synthesizing short mRNAs of defined sequence. Examination of the amino acid composition of the polypeptides synthesized in the presence of these synthetic RNAs verified the assignments generated by the triplet binding assay and resolved the questionable cases.

Of the 64 Possible Codons in Messenger RNA, 61 Code for Amino Acids

As a result of the pioneering work carried out in the laboratories of Nirenberg and Khorana, the coding assignments for all 64 triplets were established by the mid-1960s. These coding assignments are summarized

Table 3-5   Table of the Genetic Code

First Base of mRNA Codon	Second Base of mRNA Codon				Third Base of mRNA Codon
	U	**C**	**A**	**G**	
U	UUU Phe UUC UUA Leu UUG	UCU UCC Ser UCA UCG	UAU Tyr UAC UAA Stop UAG Stop	UGU Cys UGC UGA Stop UGG Trp	U C A G
C	CUU CUC Leu CUA CUG	CCU CCC Pro CCA CCG	CAU His CAC CAA Gln CAG	CGU CGC Arg CGA CGG	U C A G
A	AUU AUC Ile AUA AUG Met	ACU ACC Thr ACA ACG	AAU Asn AAC AAA Lys AAG	AGU Ser AGC AGA Arg AGG	U C A G
G	GUU GUC Val GUA GUG	GCU GCC Ala GCA GCG	GAU Asp GAC GAA Glu GAG	GGU GGC Gly GGA GGG	U C A G

Note: This table summarizes the coding assignments of the 64 possible triplet sequences (codons) that occur in messenger RNA. Of the 64 codons, 61 code for amino acids and the remaining three (UAA, UAG, and UGA) function as stop codons. The codon AUG is employed in the initiation of protein synthesis.

in Table 3-5, which lists the amino acids specified by each of the 64 possible triplets occurring in mRNA. By convention, such messenger RNA triplets, or **codons,** are always written in the 5' → 3' direction. Examination of the genetic code assignments reveals several important features. Perhaps the most striking is that 61 of the 64 triplets code for amino acids, even though only 20 different amino acids are incorporated into proteins. Hence most amino acids are specified by more than one codon. When more than one codon exists for the same amino acid, the first two bases of each codon are almost always the same. For example, the four codons specifying the amino acid proline all begin with CC-. An analogous pattern holds for threonine (AC-), alanine (GC-), and glycine (GG-). In the case of amino acids specified by only two triplets, the first two bases are again the same. Such observations indicate that the first two bases of a triplet are more important than the third in specifying a particular amino acid.

Because most amino acids are encoded by more than one codon, the genetic code is said to be *degenerate.* In spite of the somewhat derogatory connotation of this word, degeneracy in no way implies malfunction. Each of the 61 codons that codes for an amino acid specifies one and only one amino acid, so

there is no ambiguity during protein synthesis as to which amino acid is to be incorporated for each codon. Degeneracy even serves a useful function in enhancing the adaptability of the coding system. If only 20 codons existed, one for each of the 20 amino acids, then any mutation in DNA that led to the formation of one of the other 44 triplets would stop the synthesis of the growing polypeptide chain at that point. Since incomplete polypeptide chains are almost always nonfunctional, the susceptibility of such a coding system to disruption would be very great. With a degenerate code, most mutations produce codon changes that simply alter the amino acid being specified. The resulting change in protein behavior caused by a single amino acid substitution is often quite small, and in some cases may even be advantageous. Moreover, mutations in the third base of a codon frequently do not change the amino acid at all.

Although the codon assignments summarized in Table 3-5 were originally derived from experiments involving synthetic nucleotides and cell-free protein-synthesizing systems, their validity has subsequently been confirmed by analyzing the amino acid sequences of mutant proteins. It was seen earlier in the chapter, for example, that sickle-cell hemoglobin differs from normal hemoglobin in a single amino acid position, where valine is substituted for glutamic acid. Examination of the genetic code assignments reveals that glutamic acid may be encoded by either GAA or GAG. Whichever triplet is used, a single base change could create a codon for valine. For example, GAA might have been changed to GUA, or GAG might have been changed to GUG. In either case, a glutamic acid codon would be converted into a valine codon. Many other mutant proteins have been examined in a similar fashion, and in nearly all cases the amino acid substitutions can be explained by a single base change in a triplet codon (Table 3-6).

The Genetic Code Contains Special Stop and Start Signals

Of the 64 triplets that can occur in mRNA, all but three code for an amino acid. The existence of these three noncoding triplets was first demonstrated in experiments in which Har Gobind Khorana synthesized artificial mRNAs containing repeated patterns of three bases such as the following:

AGUAGUAGUAGUAGU ⋯

Depending on where the ribosome begins reading such a message, one might expect the formation of three possible polypeptide products:

(AGU)(AGU)(AGU)(AGU)(AGU) ⋯ → polypeptide 1
A(GUA)(GUA)(GUA)(GUA)(GU ⋯ → polypeptide 2
AG(UAG)(UAG)(UAG)(UAG)(U ⋯ → polypeptide 3

However, when Khorana added this particular mRNA to a cell-free protein-synthesizing system, only two products were detected: a polypeptide consisting entirely of serine (the amino acid encoded by AGU), and a polypeptide consisting entirely of valine (the amino acid encoded by GUA). The lack of a third product suggested one of two possibilities. Either the third triplet, UAG, also codes for serine or valine, or it does not code for any amino acid at all.

To distinguish between these alternatives, Khorana synthesized an RNA molecule with the following repeating sequence:

AGAUAGAUAGAUAGAUAGAU ⋯

Examination of this pattern reveals that no matter where protein synthesis begins, the triplet UAG will soon be encountered. When this particular RNA is added to a cell-free protein synthesizing system, short chains of two or three amino acids are synthesized instead of the usual long polypeptide chains. The most reasonable explanation for this phenomenon is that every time the triplet UAG is encountered, it causes the growing peptide chain to be terminated:

(AGA) (UAG) (AUA) (GAU) (AGA) (UAG) (AUA) (GAU) (A
Arg Stop Ile   Asp   Arg   Stop Ile   Asp

A (GAU) (AGA) (UAG) (AUA) (GAU) (AGA) (UAG) (AUA)
Asp   Asp   Stop Ile   Asp   Arg   Stop Ile

AG (AUA) (GAU) (AGA) (UAG) (AUA) (GAU) (AGA) (UA
Ile   Asp   Arg   Stop Ile   Asp   Arg

The preceding data suggest that rather than coding for an amino acid, the triplet UAG functions as a punctuation mark that signals the end of the polypeptide chain. Two other triplets, UAA and UGA, have been found to play a similar role. Although these three **stop codons** are sometimes referred to as *nonsense* codons because they do not code for amino acids, it should be emphasized that they play an important biological function in signaling when the synthesis of a particular polypeptide chain is complete.

In addition to these stop codons, a special **start codon** also exists for beginning the synthesis of polypeptide chains. The main triplet utilized for this purpose is AUG, which is the codon for methionine. Methionine (or its derivative N-formylmethionine in

Table 3-6   Some Amino Acid Substitutions Observed in Mutant Proteins

Amino Acid Substitution	Corresponding Alteration in Triplet Code
Hemoglobin	
Glu → Val (sickle-cell anemia)	GAA → GUA or GAG → GUG
His → Tyr	CAU → UAU or CAC → UAC
Asn → Lys	AAU → AAA or AAC → AAG
Tobacco Mosaic Virus Coat Protein	
Glu → Gly	GAA → GGA or GAG → GGG
Ile → Val	AUU → GUU or AUC → GUC or AUA → GUA
Tryptophan Synthetase	
Tyr → Cys	UAU → UGU or UAC → UGC
Gly → Arg	GGA → CGA or GGG → CGG
Lys → Stop	AAA → UAA or AAG → UAG

prokaryotes) is therefore the first amino acid incorporated into polypeptide chains, although it is frequently removed in the process of generating the final protein product. Unlike stop codons, which function only at the end of polypeptide chains, AUG serves as both a start codon and as the codon that specifies the incorporation of methionine elsewhere in polypeptide chains. In prokaryotic cells, the codons GUG and UUG are also employed in rare instances as start codons. Although these two codons normally specify valine and leucine respectively, as start codons they act like AUG and specify N-formylmethionine as the first amino acid.

The Genetic Code Is Nearly Universal

In recent years, sophisticated new sequencing techniques have provided a wealth of information concerning the nucleotide base sequences of genes and the amino acid sequences of their corresponding protein products. In addition to providing further verification for the codon assignments summarized in Table 3-5, such observations have revealed that the same set of codons is employed by all organisms and cell types, with a few minor exceptions such as certain mitochondrial DNAs (page 657). The reason that the genetic code has remained virtually unchanged throughout evolution is not difficult to comprehend. Any mutation that alters the identity of the amino acid specified by a given triplet would cause amino acid substitutions to occur in every protein whose mRNA contains that particular triplet. Although an occasional mutation may be advantageous, the simultaneous occurrence of amino acid substitutions in virtually every protein would certainly be deleterious, and organisms exhibiting such widespread mutations would not be expected to survive.

Transfer RNAs Bring Amino Acids to the Ribosome During Protein Synthesis

Since the linear sequence of codons in mRNA ultimately determines the amino acid sequence of polypeptide chains, a mechanism must exist that enables codons to arrange amino acids in the proper order. The general nature of this mechanism was first proposed in 1957 by Francis Crick, before the triplet code or the existence of mRNA had even been demonstrated. With remarkable foresight Crick postulated that amino acids are not capable of directly recognizing nucleotide base sequences, and that "adaptor" molecules must therefore mediate the interaction between amino acids and base sequences. He further proposed that each of these hypothetical adaptor molecules contains two sites, one capable of binding a specific amino acid and the other capable of recognizing a nucleotide base sequence that codes for this amino acid.

In the year following Crick's proposal, the existence of adaptor molecules exhibiting these properties

was demonstrated by Mahlon Hoagland. While studying the process of protein synthesis in cell-free systems, Hoagland discovered that radioactive amino acids become attached to small molecules of RNA prior to becoming incorporated into polypeptide chains. These small RNA molecules were found in the soluble portion of the cytoplasm, which is called the *cytosol* (Figure 3-28). When the cytosol was incubated with radioactive amino acids and ATP, the amino acids became covalently bound to these small RNAs, even in the absence of ribosomes. Adding the resulting amino acid–RNA complexes to ribosomes led to the onset of protein synthesis, and the radioactive amino acids became incorporated into newly forming polypeptide chains. These observations led Hoagland to conclude that amino acids are initially bound to small RNA molecules, which then bring the amino acids to the ribosome for subsequent insertion into newly forming polypeptide chains.

The RNA discovered by Hoagland, called **transfer RNA (tRNA),** consists of a family of small RNA molecules that bind to different amino acids. Evidence for the existence of multiple tRNAs was initially obtained from experiments in which isolated transfer RNA preparations were incubated with a single radioactive amino acid until no additional radioactivity became bound to the RNA. When a second radioactive amino acid was then added, additional radioactivity bound to the RNA. Subsequent increases in bound radioactivity occurred as each of the 20 amino acids was added, suggesting that a separate population of tRNA molecules exists for every amino acid. The ultimate verification of this conclusion was obtained by separating transfer RNA into its individ-

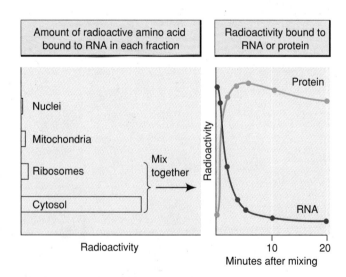

Figure 3-28 Data That Led to the Discovery of Transfer RNA (Left) *Radioactive amino acids were mixed with various subcellular components; only the cytosol fraction was found to bind these added amino acids to RNA in significant amounts.* (Right) *When the cytosol fraction containing radioactive amino acids bound to RNA was mixed with ribosomes, the labeled amino acids were incorporated into newly synthesized proteins.*

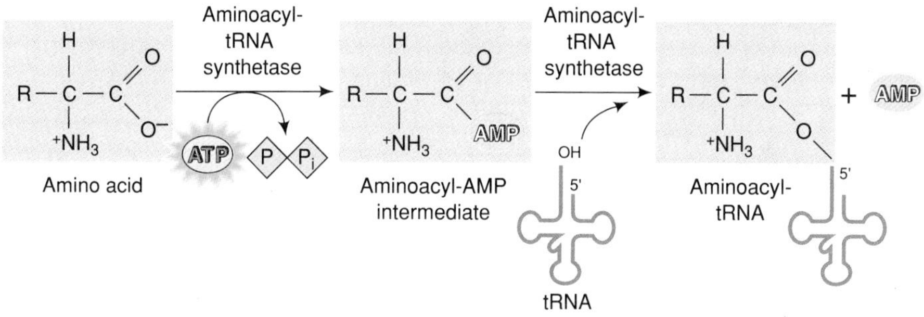

Figure 3-29 *The Aminoacyl-tRNA Synthetase Reaction* *In this reaction for joining an amino acid to its appropriate tRNA, the amino acid first reacts with ATP to form a high-energy aminoacyl-AMP intermediate. In the second step, the aminoacyl-AMP reacts with the appropriate tRNA to produce the final aminoacyl-tRNA.*

ual components using biochemical fractionation techniques. This approach led to the discovery that at least one tRNA exists for each amino acid, and that for many amino acids more than one tRNA occurs. Different tRNAs that bind to the same amino acid are referred to as *isoaccepting tRNAs,* and are designated by a combination of subscripts and superscripts. For example, two different tRNAs that both bind to the amino acid leucine would be designated as $tRNA_1^{leu}$ and $tRNA_2^{leu}$.

Each Transfer RNA Binds to a Specific Amino Acid and to Messenger RNA

The existence of separate tRNAs for each of the 20 amino acids raises the question of how the proper amino acid becomes joined to its corresponding tRNA. This selectivity is made possible by a family of enzymes, termed **aminoacyl-tRNA synthetases,** which catalyze the covalent attachment of an amino acid to the 3' end of a tRNA molecule (Figure 3-29). The final product of this reaction is called an **aminoacyl-tRNA.** At least 20 different aminoacyl-tRNA synthetases exist, each catalyzing the attachment of one type of amino acid to its corresponding tRNAs.

Once an aminoacyl-tRNA complex has been formed, the amino acid must be transferred to a newly forming polypeptide chain. In order to be able to insert the amino acid into the proper location within the growing polypeptide, the aminoacyl-tRNA complex must recognize the proper codon in mRNA. If tRNA molecules act as adaptors in the sense originally suggested by Crick, it follows that it is the tRNA molecule rather than its attached amino acid that recognizes the appropriate codon in mRNA. The first critical test of this idea was carried out by Francois Chapeville and Fritz Lipmann, who designed an elegant experiment involving the tRNA for the amino acid cysteine. It was known at the time that treatment of cysteine with a nickel catalyst converts it into the amino acid alanine. Chapeville and Lipmann therefore set out to determine what would happen if cysteine were converted to alanine by nickel treatment after it had been attached to its appropriate tRNA. The result, of course, is an alanine residue covalently at-

tached to a tRNA molecule that normally carries cysteine. When they added this abnormal aminoacyl-tRNA to a cell-free protein-synthesizing system, alanine was found to be inserted into polypeptide chains in locations normally occupied by cysteine. It was therefore concluded that codons in mRNA must recognize tRNA molecules rather than their attached amino acids.

The most straightforward mechanism that can be envisioned for this recognition process is that codons in mRNA recognize complementary triplet sequences in tRNA. To test this prediction, it is necessary to find out whether tRNA molecules contain the appropriate triplet base sequences.

Transfer RNAs Contain Anticodons That Recognize Codons in Messenger RNA

The first person to determine the base sequence of a tRNA molecule was Robert Holley, who reported the sequence of an alanine tRNA from yeast in 1965. In the ensuing years, more than a hundred other tRNAs have been sequenced. One of the most important generalizations to emerge from these data is that each tRNA molecule contains a triplet whose sequence is complementary to a codon specifying the amino acid carried by that tRNA. This triplet sequence is referred to as an **anticodon** because its identical location in all tRNAs suggests that it recognizes and binds to mRNA codons during the process of protein synthesis. As in all situations that involve complementary base-pairing between two nucleic acid sequences, the messenger and transfer RNA chains are oriented in opposite directions in regard to their 5' → 3' polarity. For example, the messenger RNA codon CCU and its corresponding anticodon in transfer RNA interact in the following manner:

Codon
→

mRNA   5' —— CUU —— 3'
tRNA   3' —— GAA —— 5'

←
Anticodon

Since base sequences are always written in the 5' → 3' direction, the tRNA anticodon that recognizes the codon CUU in mRNA is properly designated as AAG, not GAA.

Transfer RNA molecules exhibit several other characteristic features in addition to the presence of an anticodon. All tRNAs possess the sequence CCA at their 3' ends; the terminal A residue of this sequence is the site to which amino acids are covalently bound during the aminoacyl-tRNA synthetase reaction. A typical tRNA molecule contains about 75–90 bases, of which 10 to 15 percent are modified in some way. The principal modifications include methylation of bases and sugars, and creation of unusual bases such as *pseudouridine* (ψ), *inosine, dihydrouridine,* and *ribothymidine* (Figure 3-30). Each tRNA molecule contains complementary sequences that hydrogen bond with each other, generating a *stem-and-loop* structure in which the "stem" of each loop is held together by hydrogen bonds between the complementary bases (Figure 3-31, *left*). Transfer RNAs typically contain three or four stem-and-loop regions, giving diagrams of tRNA molecules the appearance of a *cloverleaf.* The anticodon always occurs in the second loop from the 5' end.

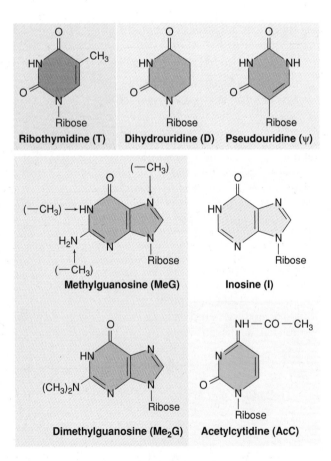

Figure 3-30 *Some of the Modified Bases Found in Transfer RNAs* *Arrows in methylguanosine point to the three places where methyl groups can be attached.*

X-ray crystallographic studies carried out in the laboratories of Alexander Rich and Aaron Klug have permitted the three-dimensional structure of the transfer RNA molecule to be elucidated. In addition to confirming the presence of the base-paired regions predicted by the cloverleaf model, their studies revealed that the three-dimensional shape of a tRNA molecule resembles a twisted letter L. The amino acid acceptor site and the anticodon loop are situated at the opposite ends of the L, whereas the other loops occur at the bend (see Figure 3-31, *right*).

The base sequence of each tRNA molecule plays an important role in allowing an aminoacyl-tRNA synthetase to recognize its appropriate tRNA. Changes in the base sequence of either the anticodon sequence or the acceptor end of a tRNA molecule have been shown to alter the amino acid to which a tRNA becomes bound. Thus bases located in at least two regions of the tRNA molecule are recognized by aminoacyl-tRNA synthetases when they pick out the tRNA that is to be bound to a particular amino acid.

DNA MUTATION, REPAIR, AND REARRANGEMENT

We have now seen that complementary base-pairing governs the replication of DNA, the transcription of DNA into mRNA, and the interactions between mRNA and tRNA that occur during the translation of mRNA into polypeptide chains. The final question to be addressed concerning the flow of genetic information concerns the mechanism by which DNA undergoes the alterations in base sequence that are a prerequisite for evolutionary change. Whereas the short-term survival of an individual organism requires that DNA replication be extremely accurate, evolution of a species depends on changes in the information content of DNA. A finely tuned balance must therefore be achieved to ensure that some, but not too many, changes in DNA base sequence continually arise.

Spontaneous Mutations Limit the Fidelity of DNA Replication

Changes in the base sequence of a DNA molecule are called **mutations.** During the normal process of DNA replication, a tiny number of mutations occur spontaneously. However, the spontaneous mutation rate is very low; the chance that a given base pair will be altered during DNA replication is only about one in a billion (or 10^{-9}). The main factors responsible for the scarcity of mutations are the precision of the complementary base-pairing mechanism that initially selects the proper base, the proofreading ability of DNA polymerase (page 72),

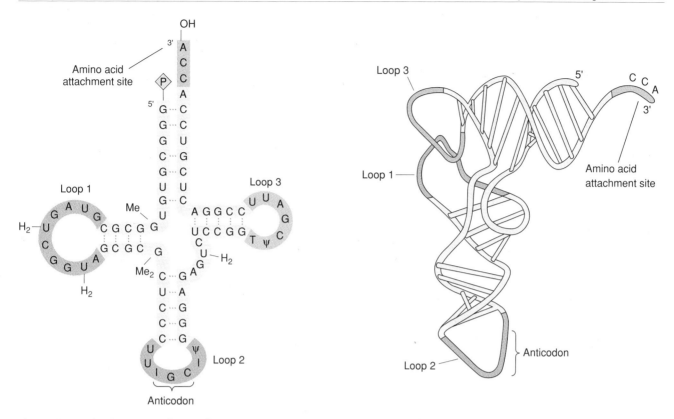

Figure 3-31 *The Structure of Transfer RNA*   (Left) *The base sequence of an alanine tRNA from yeast is shown in the two-dimensional cloverleaf configuration. All tRNAs are folded in a similar way, although some have an extra loop between loops 2 and 3. Abbreviations for the unusual bases are defined in the preceding figure.* (Right) *A three-dimensional model of yeast phenylalanine tRNA, whose overall shape resembles an upside-down letter L.*

and the existence of DNA repair mechanisms, which will be described later in the chapter.

Balanced against these forces are several factors that promote the occurrence of mutations. One such factor may be the tendency of DNA bases to exist in multiple structural forms, called **tautomers,** which are in equilibrium with one another. The base thymine, for example, exists in two forms that differ in the arrangement of the chemical groups involved in hydrogen bonding. In the predominant form of thymine, called the *keto* form, these groups are arranged in a way that allows them to form hydrogen bonds with the base adenine (Figure 3-32). In the less prevalent *enol* form of thymine, these same chemical groups are oriented in a way that permits them to form hydrogen bonds with the base guanine. Since a small amount of the enol form of thymine is always in equilibrium with the more prevalent keto form, a few thymine-guanine base pairs will always be generated during DNA replication, along with the expected thymine-adenine pairs.

Another factor thought to contribute to spontaneous mutations is any imbalance that might exist in the concentrations of the four deoxynucleoside triphosphates that are utilized as substrates for DNA replication. If one of the four is present in significantly higher concentration than the others, it is more likely to be selected as the next

base during any given step in the DNA polymerase reaction, even if it is not the proper base. Although the proofreading ability of DNA polymerase tends to remove such errors after they have occurred, the chance that some of these mistakes will go uncorrected increases when more of these errors are produced in the first place.

Additions and deletions of base sequences also occur spontaneously during DNA replication. Such errors are generated by the existence of short sequences in a DNA strand that are complementary to nearby sequences in the same strand. This arrangement fosters the formation of stem-and-loop structures that lead to both insertion and deletion mutations as DNA is replicated (Figure 3-33). DNA is also subject to spontaneous chemical changes called **depurination** and **deamination,** which can alter hundreds or thousands of bases in a typical mammalian cell per day. Depurination refers to the spontaneous breaking of the bond between deoxyribose and either adenine or guanine, which leaves no base attached to the DNA strand at that point. Deamination is the removal of a base's amino group, which can alter its base-pairing properties. For example, deamination of cytosine produces the base uracil. If the uracil is left uncorrected, it will base-pair with adenine during the next round of DNA replication instead of the guanine that had been originally paired with cytosine.

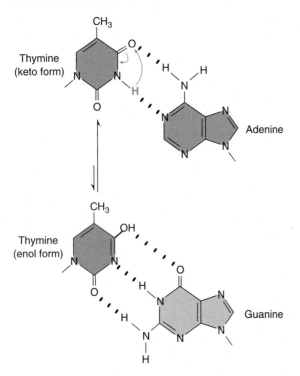

Figure 3-32 *Structures of the Two Tautomeric Forms of Thymine*   *The rare enol configuration of thymine* (bottom left) *forms hydrogen bonds with guanine instead of adenine.*

Mutations Are Also Caused by Environmental Chemicals and Radiation

Although spontaneous mutations can play an important role in generating the genetic changes that are required in order for evolution to occur, the immediate effects of most mutations are deleterious, and hence an excessive number of mutations is undesirable. For this reason, concern has been raised about recent increases in the number of mutation-causing agents, or **mutagens,** in our environment. Environmental mutagens fall into two major categories: chemicals and radiation.

Mutagenic chemicals are substances that induce DNA sequence alterations by one of three general mechanisms. (1) **Base analogs** are substances whose chemical structures resemble those of the bases that normally occur in DNA. For this reason, they may be incorporated in place of one of the usual bases during DNA replication. Base analogs are typically capable of forming abnormal base pairs, which leads to subsequent mutations. (2) **Base-modifying agents** interact with bases that are already present in DNA, changing their structures and hence their base-pairing properties. Among the substances that fit in this category are *deaminating agents* that act by removing amino groups, *hydroxylating agents* that act by adding hydroxyl groups, and *alkylating agents* that act by adding hydrocarbon groups. (3) **Intercalating agents** are chemicals that insert between adjacent bases of the DNA double helix, distorting the helical structure and increasing the chance that a base will be deleted or an extra base will be inserted during DNA replication. If this occurs in a DNA region that codes for a polypeptide chain, a frameshift mutation will result.

Mutations can also be induced by several types of radiation. The sun is a strong source of **ultraviolet radiation,** which causes bonds to form between adjacent pyrimidines located in the same or opposite strands of a DNA molecule. The most common target is adjacent thymine

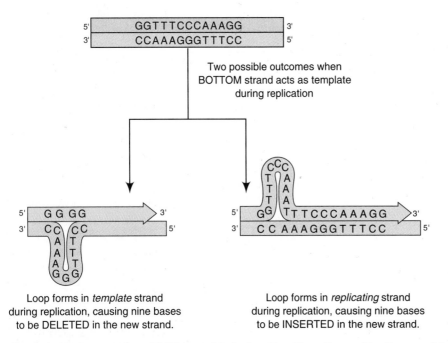

Loop forms in *template* strand during replication, causing nine bases to be DELETED in the new strand.

Loop forms in *replicating* strand during replication, causing nine bases to be INSERTED in the new strand.

Figure 3-33 *Spontaneous Addition and Deletion Mutations Created by Stem-and-Loop Structures That Form During DNA Replication*   *Such structures are caused by the presence of complementary regions within a given DNA strand.*

residues situated in the same DNA strand. Exposure to ultraviolet light causes adjacent thymines to become linked together, forming *thymine dimers* that interfere with proper DNA replication. X-rays and the gamma rays emitted by certain radioactive substances are examples of **ionizing radiation,** which acts by removing electrons from molecules. This process generates highly reactive intermediates that can cause various kinds of DNA damage.

Mutations Are Stabilized by Subsequent Rounds of DNA Replication

Once a DNA molecule containing a base sequence mutation undergoes subsequent rounds of replication, the mutation becomes very difficult, if not impossible, to repair. To illustrate this point, let us consider what happens to a DNA molecule that has been briefly exposed to the deaminating agent *nitrous acid.* This mutagen removes the amino group from the base cytosine, converting it to uracil (Figure 3-34). Since cytosine was originally paired with guanine, the immediate result is

the conversion of a C-G base pair to a U-G base pair. We will see in a moment that the presence of the abnormal base U in DNA can be recognized and corrected by enzymes that function in DNA repair. However, if a DNA molecule containing a U-G base pair undergoes replication prior to being repaired, the incorrect base (U) will serve as a template for its complement (A), yielding a new DNA double helix with a U-A base pair.

During the next round of DNA replication, the adenine of this U-A base pair will serve as a template for the incorporation of its normal complement, thymine. The final result is the presence of a T-A base pair where a C-G base pair had originally been located. Since DNA repair mechanisms do not recognize anything abnormal about a T-A base pair, this mutation will now be replicated indefinitely. The preceding type of scenario is applicable to many different kinds of mutations. For this reason, it is important that mutations be repaired swiftly before subsequent rounds of DNA replication have had a chance to occur.

Excision Repair Corrects Mutations That Involve Abnormal Bases

Fortunately, mechanisms exist for repairing DNA mutations before they have become stabilized by subsequent rounds of DNA replication. One type of DNA repair mechanism, called **excision repair,** is designed to re-

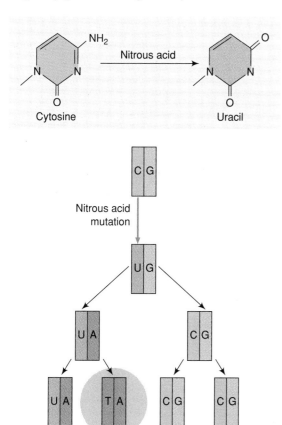

Figure 3-34 *DNA Mutation Induced by Nitrous Acid* (Top) *Nitrous acid causes bases to undergo deamination (loss of amino groups), producing new bases with differing base-pairing properties. Deamination of cytosine, for example, produces uracil, which forms hydrogen bonds with adenine instead of guanine. (Bottom) Schematic diagram illustrating how the conversion of cytosine to uracil in DNA can lead to the substitution of a T-A base pair for the original C-G base pair during two subsequent rounds of replication.*

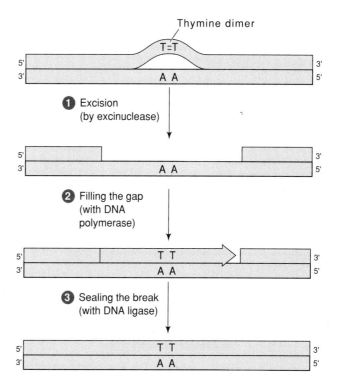

Figure 3-35 *The Steps Involved in Excision Repair* *This type of DNA repair removes abnormal bases from DNA and fixes the damaged region. The enzyme excinuclease is only one of several kinds of DNA repair nucleases that are involved in recognizing and excising abnormal bases during Step 1 of the repair sequence. In this particular example, the defect being repaired is a thymine dimer.*

move and replace abnormal DNA bases. Figure 3-35 illustrates how this three-step process operates. (1) In the first step, the damaged base is removed by enzymes that specifically recognize abnormal bases. One enzyme in this category is a special nuclease that removes thymine dimers by cutting the DNA strand on both sides of the dimer. Because of its role in excision repair, it is called an **excinuclease.** Another group of enzymes, called **DNA glycosylases,** recognize deaminated bases, alkylated bases, and bases with abnormal rings. DNA glycosylases cleave the bond that joins these aberrant bases to deoxyribose, leaving an empty spot in the DNA strand where the deoxyribose contains no attached base. Such empty spots, which also occur as a result of spontaneous depurination (page 91), are recognized by the enzyme **AP endonuclease,** which cuts the phosphodiester backbone adjacent to the altered site. (2) The gap created by the excision of a damaged DNA segment in the preceding step is filled in by **DNA polymerase,** which synthesizes a new complementary stretch of nucleotides. (3) Finally, the break in the repaired strand is sealed by the action of **DNA ligase.**

Mutant bacteria that are missing DNA polymerase I have been found to be impaired in their ability to repair damaged DNA segments, indicating that this particular form of DNA polymerase is the enzyme involved in excision repair. Cells defective in DNA repair have also been identified in humans suffering from an inherited disease known as *xeroderma pigmentosum.* Such individuals are extremely sensitive to sunlight and tend to develop multiple skin cancers, suggesting that DNA repair plays an important role in protecting people from the mutagenic effects of ultraviolet light.

The properties of the excision repair system provide some interesting insights into the question of why DNA molecules contain the base thymine in place of the uracil found in RNA. The most likely explanation involves the need to be able to repair DNA mutations involving cytosine. If the amino group normally present on cytosine is removed by exposure to a mutagen that acts as a deaminating agent, this action converts cytosine into the base uracil. This abnormal DNA base is then removed by a DNA glycosylase that specifically recognizes uracil. But if uracil were one of the four normal bases used in the construction of DNA (as it is in RNA), this uracil-specific DNA glycosylase could not distinguish normal uracils from uracils generated by the deamination of cytosine. Thus the presence of uracil as a normal base in DNA would make it impossible to repair mutations resulting from the deamination of cytosine.

Mismatch Repair Corrects Mutations That Involve Noncomplementary Base Pairs

Excision repair is a powerful mechanism for correcting damaged DNA sequences, but it applies solely to situations in which abnormal nucleotides are present. This type of error is not the only one that can occur in DNA. During the process of DNA replication, noncomplementary base pairs are occasionally formed that escape the proofreading mechanism discussed earlier (page 72). Because the bases involved in such a mismatched pair do not hydrogen bond properly with each other, their presence can be detected and corrected by a specially designed **mismatch repair** system.

In order to operate properly, this mismatch repair system must solve a problem that puzzled biologists for many years: How is the incorrect member of the base pair distinguished from the correct member? Unlike the situation in excision repair, neither of the two bases exhibits any chemical alteration that allows it to be recognized as an abnormal base. The base pair simply consists of two normal bases that are inappropriately paired with one another, such as the base A paired with C, or G paired with T. If the incorrect member of an A-C base pair were the base C, and the repair system instead removed the base A, the repair system would be creating a permanent mutation instead of correcting a mismatched base pair!

To solve this problem, the mismatch repair system must recognize which of the two DNA strands represents the newly synthesized strand from the last round of DNA replication, since the newly synthesized strand would be the one containing the incorrectly inserted base. In the early 1980s, it was shown that the detection system used by the bacterium *E. coli* for this purpose involves the addition of a methyl group to the base adenine (A) whenever it occurs in the sequence GATC in DNA. This process of **DNA methylation** does not occur until a short time after a new DNA strand has been synthesized; hence the old and new DNA strands can be distinguished from each other during the interim period by the fact that the new strand is not yet methylated. The mismatch repair system detects this difference and selectively repairs the unmethylated strand (Figure 3-36).

Most of the evidence for mismatch repair, as well as for proofreading by DNA polymerase, has come from studies on bacterial cells. Eukaryotic cells are thought to contain comparable mechanisms for correcting replication errors because the overall accuracy of DNA replication in higher organisms and bacteria appears to be comparable. Yet the detailed mechanisms involved in error correction may be somewhat different in higher organisms. For example, the methylation of GATC sequences is not a universal recognition signal for mismatch repair because DNA methylation does not occur in most lower eukaryotes, and DNA methylation in higher eukaryotes involves cytosine rather than adenine and has been implicated in gene regulation rather than DNA repair.

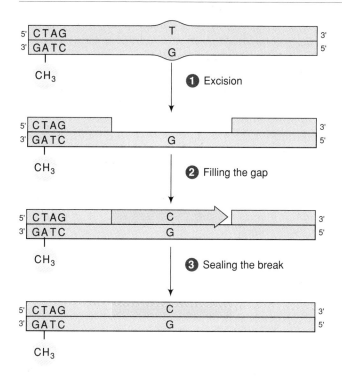

Figure 3-36 *The Steps Involved in Mismatch Repair* *This type of DNA repair fixes noncomplementary base pairs. The steps are similar to those in excision repair, but the process requires a mechanism for recognizing the newly synthesized DNA strand. In* E. coli, *this recognition signal is the presence of methyl groups on the parental DNA strand.*

Transposable Elements Promote the Rearrangement of DNA Base Sequences

So far our discussion of DNA sequence alterations has focused on localized mutations involving one or a few bases. In addition to such localized DNA changes, base sequence rearrangements involving much longer stretches of DNA can also occur. In the early 1940s, studies of inheritance patterns in corn led Barbara McClintock to postulate the existence of movable genetic elements that permit genes to migrate from one location to another. Although these ideas did not fit into the main body of genetic knowledge at the time, the reality of such mobile DNA sequences was later confirmed using modern DNA isolation and sequencing techniques. Such mobile DNA sequences, now referred to as **transposable elements,** have been identified in a variety of eukaryotic and prokaryotic organisms. Transposable elements can be subdivided into two general categories: **insertion sequences,** which are less than a few thousand bases in length and contain only genes involved in the movement of the transposable element itself, and **transposons,** which tend to be larger and may contain numerous genes in addition to those required for moving the sequence. In some cases, viral DNA may act as a transposon.

Transposable elements typically contain short *inverted repeat sequences* at both ends. This means that the same sequence is found at the two ends of the

transposable element but running in opposite directions. The movement of transposable elements is made possible by a protein encoded by a gene situated within the transposable element itself. This protein, called a **transposase,** catalyzes the removal of a transposable element from one region of a DNA molecule and its insertion somewhere else. One model for how a transposase might work is illustrated in Figure 3-37. According to this model, the transposase binds to the two repeated sequences that are located at opposite ends of the transposable element. The DNA is then cleaved at the ends of these repeated sequences, excising the transposable element in the form of a circular molecule that is subsequently inserted elsewhere in the DNA. Alternatively, in

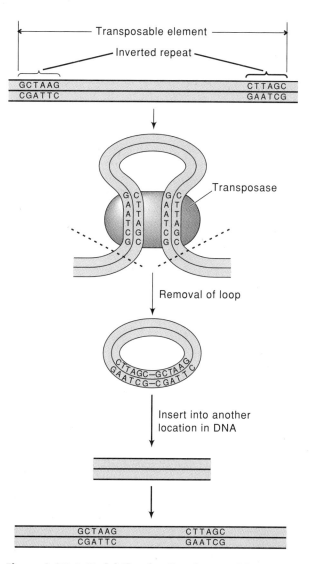

Figure 3-37 *A Model Showing How Inverted Repeat Sequences Might Facilitate the Movement of Transposable Elements* *In this model the transposase binds to the repeated sequences located at opposite ends of the transposable element. The resulting loop is cleaved at the ends of the repeated sequences, excising the transposable element as a circular DNA molecule that can then be inserted elsewhere in the DNA.*

some instances the original transposable element is replicated rather than excised; in this case the duplicate copy is inserted at the new DNA location, leaving the original copy at its initial location. Some transposable elements are selectively inserted at particular DNA locations, while others move about more randomly.

One of the most interesting properties exhibited by transposable elements is that in the process of moving from site to site, they influence the expression and organization of neighboring genes. Since transposable elements can migrate with considerable frequency, their behavior has the potential for causing rapid and dramatic genetic alterations. This phenomenon is now believed to be a major contributing factor to evolutionary change.

RECOMBINANT DNA AND GENE CLONING

The intent of this chapter has been to describe the fundamental principles that govern the flow of genetic information within cells. Much of our current understanding in this area has been made possible by the development of sophisticated techniques for isolating and analyzing the properties of nucleic acid molecules. In the concluding sections of this chapter we will describe the techniques that have been most useful in helping us study gene organization and function, and will then examine some of the practical applications of this technology.

Nucleic Acid Hybridization Techniques Are Used to Determine Whether Two Nucleic Acids Contain Similar Base Sequences

In 1960, Julius Marmur and Paul Doty discovered that the two strands of a DNA double helix can be separated from each other by raising the temperature, and that the two separated strands then reassociate back into a double helix when the temperature is lowered. This pioneering discovery spurred the development of a family of procedures that identify nucleic acids based on the ability of single-stranded chains with complementary base sequences to bind to or *hybridize* with each other. **Nucleic acid hybridization** techniques can be applied to DNA-DNA, DNA-RNA, and even RNA-RNA interactions. Because hybridization procedures have had a profound impact on our ability to study gene organization and expression, we will briefly describe how these techniques are carried out.

In hybridization experiments involving double-stranded DNA molecules, the DNA must first be dissociated into its individual strands. This is usually accomplished by exposing DNA either to high temperatures or to alkaline pH, both of which disrupt the hydro-

gen bonds holding the two strands together. DNA molecules that have been dissociated into two separate chains are said to be *denatured*. **Denaturation** is accompanied by several changes in the physical properties of a DNA solution, including a decrease in viscosity, an increase in density, and an increase in the ability to absorb ultraviolet light in the 260-nm region of the spectrum. Of the three properties, the increase in absorption at 260 nm is easiest to use when monitoring the process of DNA denaturation. If the absorbance at 260 nm is measured as a DNA solution is heated, a temperature is eventually reached at which the absorbance begins to rise as the two DNA strands start to separate; the absorbance then continues to increase until denaturation is complete (Figure 3-38).

The temperature range over which denaturation occurs is typically quite narrow, just as in the melting of a simple crystal. Thermal denaturation is therefore referred to as DNA *melting*, and the **melting temperature (T_m)** is defined as the temperature at which the transition from double-stranded to single-stranded DNA is halfway complete. The melting temperature of a DNA sample is strongly influenced by its base composition. DNA molecules with higher G-C contents melt at higher temperatures because the G-C base pair is held together by three hydrogen bonds, whereas the A-T pair is only held together by two (see Figure 3-5). Therefore more energy is required to disrupt G-C base pairs during DNA melting.

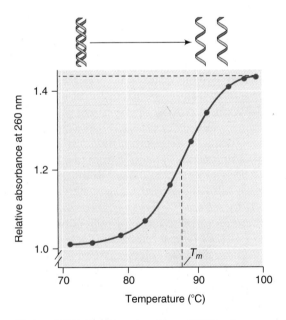

Figure 3-38 *The Denaturation of DNA* *Denaturation was monitored by measuring the ability of a DNA solution to absorb ultraviolet light at 260 nm. As the temperature is increased, the absorption at 260 nm increases as the double-stranded DNA dissociates into single strands. The temperature at which the increase in absorption is halfway completed is referred to as the melting temperature (T_m).*

After a DNA sample has been dissociated into single strands, its ability to hybridize to other single-stranded DNA or RNA chains can be tested. Hybridization is carried out either in solution or with one of the nucleic acids immobilized on an inert support such as nitrocellulose filter paper. It has been empirically determined that a complementary region as short as a dozen bases is sufficient to cause two nucleic acid strands to hybridize to each other. Since nucleic acid molecules are usually much longer than this, a small region of complementarity can hold together two strands whose sequences are largely noncomplementary. For this reason, nucleic acid hybridization studies often include treatment of the hybrids with an enzyme like *S1 nuclease,* which selectively degrades single-stranded nucleic acids. Only base sequences that have become hybridized survive treatment with such nucleases (Figure 3-39).

Base sequences do not need to be perfectly complementary in order to hybridize. By altering the temperature, salt concentration, and pH employed during hybridization, one can establish conditions in which pairing will occur between nucleotide sequences that contain numerous mismatched bases. Under such conditions of *reduced stringency,* it is possible to detect gene sequences that are related to one another but not identical. This approach is useful for identifying families of related genes, both within a given type of organism and among different kinds of organisms.

Southern and Northern Blotting Allow Hybridization to Be Carried Out with Nucleic Acids Separated by Electrophoresis

Hybridization techniques are often used to detect the presence of DNA or RNA molecules that contain a particular kind of base sequence. In such cases the DNA or RNA is hybridized to a purified, single-stranded radioactive DNA fragment called a **DNA probe** whose sequence is complementary to the base sequence of interest. DNA probes range in size from about a dozen to several thousand bases in length. As we will see shortly, DNA probes can be produced both by gene cloning techniques and by direct chemical synthesis of nucleotide chains of defined sequence.

DNA probes are most commonly used to identify nucleic acid molecules that have been fractionated by *gel electrophoresis,* a technique that separates nucleic acids based on differences in size (page 145). In the **Southern blotting** procedure, first described by Edwin Southern, DNA molecules that have been fractionated by gel electrophoresis are transferred to nitrocellulose filter paper by "blotting" the gel with the paper. The nitrocellulose filter is then incubated with a radioactive DNA probe to permit the probe to hybridize to any

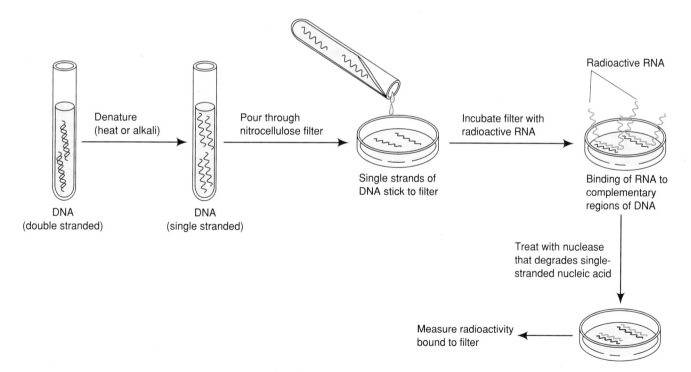

Figure 3-39 Nucleic Acid Hybridization *In this example, which involves DNA-RNA hybridization, the DNA is dissociated into single strands, bound to a nitrocellulose filter, and incubated with radioactive RNA. A nuclease that degrades single-stranded nucleic acids is then used to destroy any unhybridized RNA, and the radioactivity bound to the filter is measured. This particular approach represents only one of many different ways in which hybridization can be carried out.*

complementary DNA sequences, and the position of the bound radioactivity is determined by laying the filter paper against X-ray film. Developing the film yields an *autoradiogram* in which dark bands appear wherever the radioactive probe has hybridized to DNA molecules present on the nitrocellulose paper. This procedure, called **autoradiography,** allows one to identify those DNA molecules that are complementary to the radioactive nucleic acid employed as probe (Figure 3-40).

An analogous procedure, termed **Northern blotting,** utilizes a similar approach to analyze electrophoretically separated samples of RNA (rather than DNA) for the presence of molecules that are complementary to a radioactive DNA probe. The ability to detect RNA molecules containing sequences that are complementary to a specific DNA probe is extremely helpful when one is trying to identify specific kinds of messenger RNAs or molecules that are precursors to messenger RNA.

In Situ and Colony Hybridization Allow Nucleic Acids to Be Identified in Cells

In addition to their usefulness in characterizing isolated nucleic acid molecules, hybridization techniques can also be applied to situations in which nucleic acids reside in their original location within cells. As an example, this ***in situ* hybridization** approach has been employed to determine the subcellular localization of the DNA sequences that code for rRNA. To accomplish such an objective, thin tissue slices are incubated at high pH to denature the DNA, and radioactive rRNA is then added. After incubating under suitable hybridization conditions, the unhybridized radioactivity is washed away and the tissue is coated with a layer of photographic emulsion to permit localization of the radioactivity. This latter technique, called *microscopic autoradiography* (page 132), generates silver grains over the radioactive site that can be observed when the specimen is examined with a microscope. As we will see later, such experiments have revealed that the DNA coding for rRNA is localized in the nucleolus (see Figure 11-13). By varying the type of radioactive nucleic acid utilized as a probe, this same general approach can be used for determining the subcellular localization of other kinds of DNA and RNA.

The principle of nucleic acid hybridization has also been utilized to determine whether or not a given population of cells contains a particular kind of nucleic acid. In one version of this approach, termed **colony hybridization,** an agar plate containing bacterial colonies is blotted with a piece of nitrocellulose filter paper to make a replica of the pattern of colonies on the original plate. The cells on the filter paper are then broken open and hybridized with a radioactive DNA probe. Examining the distribution of hybridized radioac-

tive DNA on the filter paper by autoradiography allows one to determine which colonies contain nucleic acids that are complementary to the radioactive probe. Colony hybridization is widely used in conjunction with recombinant DNA cloning techniques, which will be discussed shortly.

The Discovery of Restriction Endonucleases Paved the Way for Recombinant DNA Technology

The nuclear DNA of a typical eukaryotic cell contains more than a billion base pairs, while an average gene measures only a few hundred to several thousand base pairs in length. It is therefore obvious that the organization and behavior of an individual gene can be studied only after it has been isolated from the large mass of DNA with which it is normally associated. Techniques for obtaining individual gene sequences in large quantity have developed rapidly in recent years as a result of several important discoveries.

The first of these breakthroughs emerged from an unexpected source. In the early 1950s, Salvador Luria and Giuseppe Bertani discovered that bacteria have a primitive defense system that provides protection against attack by certain viruses. In the following decades, work in the laboratories of Werner Arber, Hamilton Smith, and Daniel Nathans revealed that the protective mechanism involves the production of bacterial enzymes called **restriction endonucleases** (or simply **restriction enzymes**), which *restrict* viral growth by attacking and digesting the DNA of invading viruses. The DNA of the bacterium is protected from degradation because bacterial DNA is methylated at or near the sites attacked by restriction enzymes. The methyl groups block the action of restriction enzymes, so the bacterial DNA remains intact while the viral DNA is destroyed.

Restriction enzymes are called **endonucleases** because they cut DNA internally within the polynucleotide chain; in contrast, **exonucleases** digest nucleic acids by removing nucleotides from one end of the chain or the other. Restriction endonucleases are extraordinarily specific in the DNA sites they attack; each enzyme cleaves DNA only in places where it encounters a specific recognition sequence four to eight bases long called a **restriction site.** More than 200 different kinds of restriction enzymes have been identified, each specific for a different restriction site. Individual restriction enzymes are named using a three-letter abbreviation for the genus and species of the bacterium from which they were obtained, followed by a single letter (if applicable) referring to the bacterial strain and a Roman numeral indicating the order in which the enzyme was discovered. Hence *Hae*III refers to the third restriction enzyme to be isolated from the bacterium *Haemophilus aegyptius,*

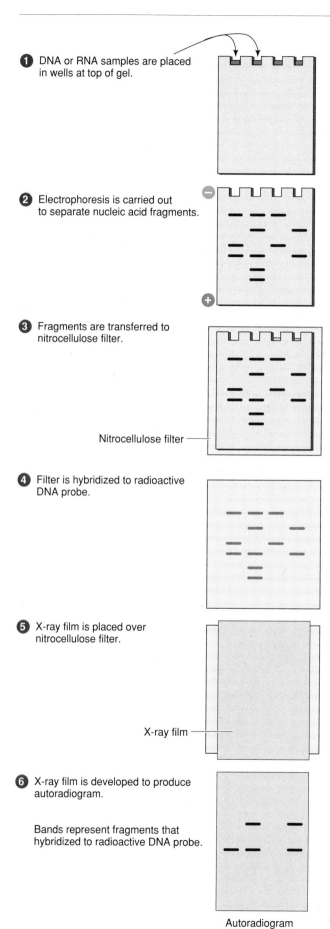

1 DNA or RNA samples are placed in wells at top of gel.

2 Electrophoresis is carried out to separate nucleic acid fragments.

3 Fragments are transferred to nitrocellulose filter.

Nitrocellulose filter

4 Filter is hybridized to radioactive DNA probe.

5 X-ray film is placed over nitrocellulose filter.

X-ray film

6 X-ray film is developed to produce autoradiogram.

Bands represent fragments that hybridized to radioactive DNA probe.

Autoradiogram

*Eco*RI refers to the first restriction enzyme isolated from the R strain of *Escherichia coli,* and so forth.

Most restriction sites are **palindromes,** which means that the base sequence read in the 5' → 3' direction of each strand is the same. For example, the following sequence is a palindrome:

5' — AAGCTT — 3'
3' — TTCGAA — 5'

Restriction enzymes can cut such sites in one of two different ways (Figure 3-41). Some restriction enzymes cleave both strands at the same point, generating fragments with flush or *blunt ends.* Other enzymes cut the two DNA strands in a staggered fashion, producing fragments with short protruding regions of single-stranded DNA. Since the single-stranded regions of such fragments are complementary to each other, they can reassociate by complementary base-pairing. These single-stranded ends are therefore called "sticky" or *cohesive ends.*

The ability of restriction enzymes to cleave DNA at precise locations makes them an invaluable tool for cutting large DNA molecules into reproducible sets of smaller fragments. If the sizes of these fragments are determined after DNA has been digested with various restriction enzymes, it is possible to construct a **restriction map** which identifies the locations of the restriction sites relative to one another (Figure 3-42). In such maps the distances between restriction sites are expressed in units called **kilobases (kb),** which correspond to thousands of base pairs.

Gene Cloning Techniques Permit Individual DNA Sequences to Be Produced in Large Quantities

The discovery of restriction endonucleases was a landmark event in the history of cell biology because it provided biologists with a set of tools that function like molecular scissors, allowing them to cut DNA in a precise and reproducible fashion. This capability represented a crucial first step on the road to isolating individual genes, but a second problem still had to be overcome: How can individual DNA sequences be purified in quantities that are large enough to permit them to be thoroughly characterized and studied? Pioneering experiments carried out in the early 1970s in the laboratories of Herbert Boyer, Stanley Cohen, and Paul Berg provided an answer to this question that has

Figure 3-40 *Southern and Northern Blotting* *In Southern blotting, DNA fragments that have been separated by electrophoresis are identified by hybridization to a radioactive DNA probe of known sequence. In Northern blotting, a similar approach is used to identify RNA molecules that have been separated by electrophoresis.*

revolutionized modern biology. They showed that DNA fragments derived from different sources can be joined together to form **recombinant DNA** molecules that will replicate in bacterial cells. Bacteria containing recombinant DNA molecules of interest are then identified and grown up in large quantity.

This technique, called **DNA cloning,** has permitted hundreds of different genes to be isolated for detailed investigation. Although the specific details of DNA cloning experiments vary, the following five steps are typically involved: (1) isolation of DNA for cloning, (2) insertion of DNA into a cloning vector, (3) replication of the recombinant cloning vector in a host cell, (4) selection of cells containing the cloning vector, and (5) identification of recombinant clones containing the DNA of interest. Each of these steps will now be briefly described.

Step 1: Isolation of DNA for Cloning.   DNA molecules containing gene sequences to be cloned can be obtained in two different ways. One approach is to isolate the total DNA from a cell population and break it into smaller fragments by physical shearing or treatment with restriction enzymes. The resulting DNA fragments collectively contain all of the organism's genes. If such fragments are cloned, the resulting group of clones is called a **genomic library** because it contains all of the organism's gene sequences. One drawback to using restriction enzymes to generate DNA fragments for a genomic library is that some genes will contain restriction

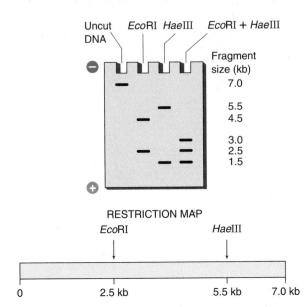

RESTRICTION MAP

Figure 3-42 *Restriction Mapping*   *In this hypothetical example, the location of restriction sites for EcoRI and HaeIII is determined in a DNA fragment 7.0 kb long. Treatment with EcoRI cleaves the DNA into two fragments measuring 2.5 kb and 4.5 kb, indicating that DNA has been cleaved at a single point located 2.5 kb from one end. Treatment with HaeIII cleaves the DNA into two fragments measuring 1.5 kb and 5.5 kb, indicating that DNA has been cleaved at a single point located 1.5 kb from one end. Simultaneous treatment with EcoRI and HaeIII cleaves the DNA into three fragments measuring 3.0 kb, 2.5 kb, and 1.5 kb. Since the 3.0-kb fragment is not produced by either EcoRI or HaeIII treatment alone, it must represent the DNA located between the sites of EcoRI and HaeIII cleavage. The restriction map illustrated at the bottom of the figure is the only map consistent with all of these data.*

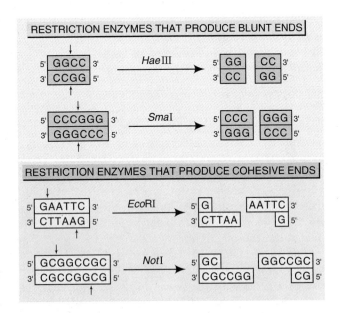

Figure 3-41 *The Two Classes of Restriction Enzymes*   *HaeIII and SmaI are examples of restriction enzymes that cut both DNA strands in the same location, generating fragments with blunt ends. EcoRI and NotI are examples of enzymes that cut DNA in a staggered fashion, generating fragments with cohesive ends. The restriction sites illustrated in these examples are all palindromes.*

sites for the enzyme being utilized to cut the DNA, and hence will be cleaved into two or more pieces. This problem can be circumvented by carrying out a *partial DNA digestion* in which a limited amount of restriction enzyme is used for a brief period of time. Under such conditions, some restriction sites remain uncut and the probability improves that at least one intact copy of each gene will be present in the genomic library.

The alternate DNA source for cloning experiments is DNA that has been generated by copying mRNA with the enzyme *reverse transcriptase* (page 75). This enzymatic reaction produces a population of DNA molecules referred to as **complementary DNA (cDNA)** because they are complementary in sequence to the mRNA employed as template. If the entire mRNA population of a cell is isolated and copied into cDNA for cloning, the resulting group of clones is called a **cDNA library.** The advantage of a cDNA library is that it only contains gene sequences that are expressed in the form of mRNA in a particular cell type, rather than reflecting the entire DNA content of the cell.

Step 2: Insertion of DNA into a Cloning Vector.
In order to clone DNA fragments of either of the above types, it is first necessary to link them to a second kind of DNA, termed a **cloning vector,** that is capable of rapid replication in an appropriate host cell (usually a bacterium or yeast, although certain cloning vectors can be incorporated directly into mammalian cells as well). Several different kinds of DNA molecules are commonly employed as cloning vectors. (1) **Plasmids** are small circular DNA molecules that are useful as cloning vehicles because they can replicate in the cytoplasm of bacterial cells independently of the chromosomal DNA (Figure 3-43). Plasmids typically carry genes for antibiotic resistance, whose presence can be used to select for bacterial cells containing the plasmid. (2) **Viral cloning vectors** are most often derived from DNA of the bacterial virus, *bacteriophage lambda* (λ). DNA fragments about 15 kb in length can be inserted into these viral cloning vectors, which are then used to construct bacteriophage particles that can infect bacterial cells. (3) **Cosmids** are DNA molecules that share some of the features of both plasmids and lambda cloning vectors, but are used for cloning larger pieces of DNA. DNA fragments about 40–45 kb in length are typically inserted into cosmids, which are then packaged into bacteriophage particles. Although such virus particles do not contain the genes required for viral replication, they can infect bacterial cells and release the cosmid DNA inside the cell. The released cosmid DNA then replicates like a plasmid. (4) **Shuttle vectors** are DNA molecules that are capable of replicating in two or more cell types. Some of the most widely used shuttle vectors are DNA molecules that are capable of replicating in the cytoplasm of both bacteria and yeast.

How are DNA fragments inserted into such vectors for cloning? One approach exploits the fact that cloning vectors typically contain restriction sites for enzymes that generate cohesive ends, such as *Eco*RI, *Pst*I, *Bam*HI, and *Hin*dIII. In the commonly used plasmid called *pUC19,* single restriction sites for each of these enzymes are clustered in a small region of the plasmid. When pUC19 is cleaved by one of these enzymes, the circular plasmid is converted into a linear DNA molecule containing cohesive ends (Figure 3-44). During cloning experiments the DNA to be cloned is cleaved with the same restriction enzyme, generating DNA fragments with the same kinds of cohesive ends as the cut plasmid. The plasmid and DNA fragments are then mixed together so that the cohesive ends of the plasmid and the cohesive ends of the DNA fragments can hybridize to one another. The result is the formation of some recombinant plasmid molecules containing an inserted DNA fragment. In a similar fashion, restriction enzymes that generate cohesive ends can be used to insert DNA fragments into viral cloning vectors.

In some cases, the preceding approach is not applicable because the DNA being cloned lacks an appro-

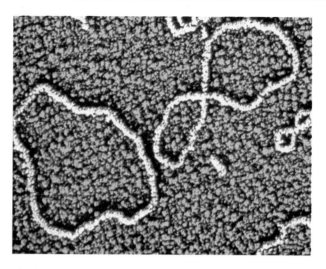

Figure 3-43 *Electron Micrograph of Isolated Plasmid Molecules* *Each circle represents a single molecule of plasmid DNA. Courtesy of SPL/SS/Photo Researchers.*

priate restriction site. Two alternative techniques have therefore been developed for introducing DNA into cloning vectors (Figure 3-45). In one of these approaches short synthetic DNA molecules, called **linkers,** are covalently joined to the ends of the DNA molecule being cloned. Many kinds of linkers are available, each containing the cleavage site for a different restriction enzyme. After a linker containing a particular restriction site has been joined to a DNA molecule, the DNA is treated with the appropriate restriction enzyme to generate cohesive ends. The other approach for inserting DNA into a cloning vector employs the enzyme *terminal transferase* to add complementary nucleotide "tails" to the 3' ends of the two DNA molecules being joined. For example, this enzyme can be employed to add G residues to the 3' ends of a cut plasmid and C residues to the 3' ends of the DNA to be cloned. Hybridization of the plasmid to the DNA being cloned is then mediated by complementary base-pairing between the G and C residues.

Step 3: Replication of the Recombinant Cloning Vector in Bacteria. Once DNA sequences have been inserted into a cloning vector, the resulting recombinant DNA molecules are then produced in large quantity by introducing them into an appropriate host cell. The bacterium *Escherichia coli* is commonly employed for this purpose. Under favorable conditions *E. coli* divides every 22 minutes, giving rise to a billion cells in less than 11 hours. Moreover, each of these cells can produce hundreds of copies of a cloning vector, so a single recombinant DNA molecule introduced into one bacterial cell will be amplified several hundred billionfold in less than half a day. In some cases bacterial cells also transcribe recombinant DNA molecules into mRNA, which is in turn translated into polypeptide chains. Hence cloning can be employed not just to produce

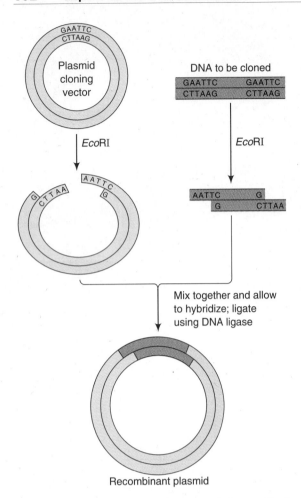

Figure 3-44 *Inserting a DNA Fragment into a Cloning Vector Using a Restriction Enzyme That Generates Cohesive Ends* *In this example, a plasmid and the DNA being cloned are both cleaved with the restriction enzyme EcoRI. The plasmid and DNA fragments are then mixed together so that the cohesive ends of the plasmid and the cohesive ends of the DNA fragments can hybridize to one another. DNA ligase is then used to join the DNA fragment to the plasmid.*

large quantities of an individual gene's DNA sequence, but large amounts of the gene's protein product as well.

Cloning vectors containing recombinant DNA are introduced into cells in one of two ways. If the cloning vector is a virus, it is allowed to infect an appropriate cell population. Plasmid vectors, on the other hand, are simply introduced into the medium surrounding the target cells. Both prokaryotic and eukaryotic cells take up plasmid DNA that has been placed in the external medium, although the efficiency of this process is relatively low and special treatments are usually necessary to enhance its rate. The addition of calcium ions, for example, markedly increases the efficiency with which cells take up DNA from their external environment.

Step 4: Selection for Bacteria Containing the Cloning Vector. The next step in the cloning process

is to select for those cells that have successfully taken up the cloning vector. In experiments involving plasmid vectors, antibiotics are often employed as a selection tool because plasmids typically contain genes that make cells resistant to specific antibiotics. For example, many of the plasmid vectors employed in cloning experiments contain a gene that codes for resistance to *ampicillin* (a type of penicillin). Bacteria employed in such experiments are therefore grown in the presence of ampicillin so that only cells that have taken up the plasmid will survive and grow.

A different approach is employed with lambda cloning vectors, which are derived from a bacteriophage lambda DNA molecule that is only about 70 percent as long as normal lambda DNA. As a result, it is too small to be packaged into a functional virus particle. But if a fragment of DNA is inserted into the middle of the cloning vector, this produces a larger piece of DNA that is again capable of being assembled into a functional virus (Figure 3-46). Hence when lambda cloning vectors are employed, the only bacteriophage particles that can successfully infect bacterial cells are those that contain an inserted recombinant DNA sequence.

Step 5: Identification of Recombinant Clones Containing the DNA of Interest. The preceding steps typically generate large numbers of bacteria producing many different kinds of recombinant DNA. How does one go about identifying an individual population of cells, called a **clone,** that has arisen from a single cell that has incorporated a cloning vector containing a specific DNA sequence of interest? The most common procedure for identifying such clones involves nucleic acid hybridization with a radioactive DNA probe. The mechanism for producing these probes varies, depending on what is known about the gene of interest. In some situations a DNA fragment related to the gene being pursued may already be available. Information may also be available about the amino acid sequence of the protein encoded by the gene. In the latter case the genetic code is utilized to predict the nucleotide base sequences that might code for amino acid sequences contained in the protein. Short DNA fragments containing these sequences are then chemically synthesized for use as radioactive probes.

The way in which nucleic acid hybridization is carried out depends upon the type of cloning vector employed. With plasmids and other nonviral cloning vectors, the colony hybridization technique is employed (page 98). In this procedure, a dilute preparation of bacteria is first spread over a solid culture medium so that each individual cell can give rise to a separate colony of cells. The culture dish is then blotted with a nitrocellulose filter to make a replica copy of the colonies in the dish, and a radioactive DNA probe is hybridized to the filter to determine the location of colonies containing

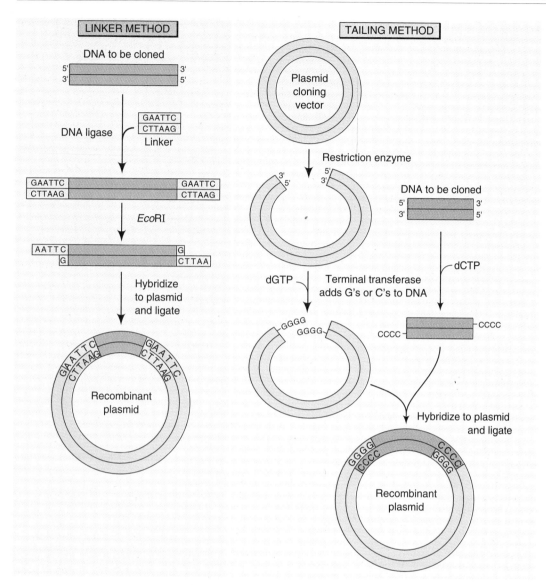

Figure 3-45 Two Methods for Inserting DNA Fragments into Cloning Vectors When the Fragments Lack an Appropriate Restriction Site (Left) *In the example illustrating the linker method, a synthetic DNA linker containing an EcoRI restriction site is covalently joined to the ends of the DNA molecule to be cloned. The DNA is treated with EcoRI to generate cohesive ends and is then cloned in a vector containing an EcoRI site.* (Right) *In the example illustrating the tailing method, terminal transferase is employed to add a tail of G residues to the plasmid cloning vector and a tail of C residues to the DNA being cloned. Complementary base-pairing between the C and G residues then allows the DNA being cloned to be hybridized to the plasmid DNA.*

DNA sequences of interest (Figure 3-47). Recent advances permit hundreds of thousands of colonies to be screened simultaneously in this fashion.

When viral cloning vectors are employed, a slightly different procedure is used. In this case, a population of bacteriophage particles containing recombinant DNA is mixed with bacterial cells, and the mixture is placed on a continuous growth medium under conditions that produce a continuous "lawn" of bacteria across the plate. Each time a bacteriophage infects one of these cells, the virus is replicated and eventually causes the cell to rupture and die. The released virus particles then infect neighboring cells, repeating the same process again. This cycle eventually produces a clear zone of dead bacteria called a **plaque**, which contains large numbers of replicated bacteriophage particles derived by replication from a single type of recombinant virus. To determine the type of recombinant DNA contained in any given plaque, the culture dish is simply blotted with a piece of nitrocellulose filter to make a replica, and the nitrocellulose filter is then hybridized with a radioactive DNA probe.

In some situations the ultimate goal of a recombinant DNA experiment is to generate clones that produce large

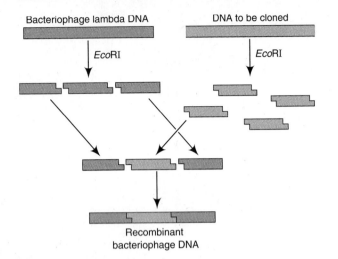

Bacteriophage lambda DNA

DNA to be cloned

EcoRI

EcoRI

Recombinant
bacteriophage DNA

Figure 3-46 *Using Bacteriophage Lambda as a Cloning Vector*   *The middle segment of the lambda DNA molecule is removed following EcoRI cleavage and is then replaced by the DNA fragment to be cloned. The presence of the inserted DNA fragment is required to make the lambda DNA molecule large enough to be packaged into a functional virus.*

amounts of the protein product of a given gene rather than to isolate the gene itself. Such cloning experiments often employ specially modified cloning vectors called **expression vectors,** which contain DNA sequences that promote the transcription of the cloned gene. To identify clones producing a particular protein product, nitrocellulose filter replicas of a group of bacterial colonies are incubated with radioactive antibodies that bind to the protein of interest. In such experiments, steps must usually be taken to increase the efficiency with which the cloned gene is transcribed and translated into a protein product.

This is especially true when eukaryotic genes are cloned in bacteria, for the transcription and translation mechanisms of bacterial cells are slightly different from those of higher organisms. In recent years, significant advances have been made in developing cloning vectors that optimize the efficiency with which eukaryotic genes are expressed as functional protein products in bacterial cells. In addition, cloning vectors have been developed that replicate in yeast, a eukaryotic microorganism whose transcription and translation mechanisms more closely resemble those of higher organisms.

Rapid Procedures Exist for Determining the Base Sequence of DNA Molecules

Shortly after recombinant DNA techniques were first introduced, their usefulness was significantly enhanced by the development of two rapid procedures for determining the linear sequence of bases in DNA. The *chemical cleavage method,* developed by Allan Maxam and Walter Gilbert, utilizes four different chemical reagents to cleave single-stranded DNA chains adjacent to particular bases. These

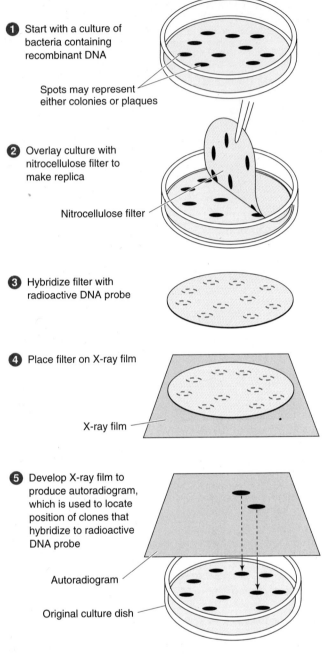

1 Start with a culture of bacteria containing recombinant DNA

Spots may represent either colonies or plaques

2 Overlay culture with nitrocellulose filter to make replica

Nitrocellulose filter

3 Hybridize filter with radioactive DNA probe

4 Place filter on X-ray film

X-ray film

5 Develop X-ray film to produce autoradiogram, which is used to locate position of clones that hybridize to radioactive DNA probe

Autoradiogram

Original culture dish

Figure 3-47 *Method for Detecting Bacterial Colonies or Plaques Containing a Particular Cloned Gene*   *In this procedure, cells containing a particular kind of cloned DNA fragment are detected by hybridizing with a radioactive DNA probe containing the sequence of interest.*

reactions cleave DNA adjacent to either (1) guanine, (2) cytosine, (3) adenine and guanine, or (4) cytosine and thymine. When such reactions are carried out on a single-stranded DNA molecule that has first been labeled at its 5'-end with ^{32}P, radioactive fragments are produced whose relative sizes are determined by the location of the base for which the cleavage reaction is specific. For example, if a DNA molecule contains guanine as its second, fifth, and tenth bases, the guanine-specific cleavage reaction will generate radioactive fragments 1, 4, and 9 bases in length. Analysis of the pattern of radioactive fragments by gel electrophoresis allows one to elucidate the sequential arrangement of bases in the original DNA molecule (Figure 3-48).

An alternative sequencing approach is the *chain termination method* of Frederick Sanger, which utilizes dideoxynucleotides (nucleotides lacking a 3' hydroxyl group) to interfere with the normal enzymatic synthesis of DNA. When a dideoxynucleotide is incorporated into a growing DNA chain instead of the normal deoxynucleotide, further elongation of the DNA chain stops because the absence of the 3' hydroxyl group makes it impossible to form a bond with the next nucleotide. In the Sanger procedure, DNA synthesis is carried out in the presence of the four normal deoxynucleotide substrates plus low concentrations of one of the four dideoxynucleotides. Every time a dideoxynucleotide is incorporated, the synthesis of the DNA chain will be prematurely terminated. Hence a series of incomplete DNA fragments of varying sizes is produced whose sizes provide information concerning the sequential arrangement of bases. For example, if DNA synthesis in the presence of dideoxy-CTP (ddCTP) yields DNA fragments that are 4, 9, and 12 bases long, then the DNA strand being synthesized must contain cytosine as its fourth, ninth, and twelfth base (Figure 3-49).

The relative ease with which DNA sequences can now be determined has provided a new way of elucidating the amino acid sequences of protein molecules. Until relatively recently, the most straightforward way of determining the amino acid sequence of protein molecules involved the use of proteases to cleave proteins into fragments, followed by chemical reactions that sequentially remove amino acids from the ends of these fragments. Although this approach has been automated and successfully applied to a large number of proteins, it is not nearly as efficient or rapid as DNA sequencing techniques, especially for large proteins. For this reason, an increasingly common alternative is to clone and sequence the gene coding for a protein, and to utilize the genetic code to predict the amino acid sequence encoded by the gene. Although there are six possible reading frames for any DNA sequence (three in each direction), the one that codes for the protein product can be recognized because it is the only one to proceed for long stretches without interruption by stop codons. In many cases, this approach has allowed the amino acid sequence of a protein molecule to be predicted before the protein itself has been isolated and purified.

DNA Molecules of Defined Sequence Can Be Synthesized by Chemical Procedures

In addition to the important contributions of DNA sequencing techniques, the power of recombinant DNA technology has also been enhanced by the development of rapid procedures for the synthesis of DNA fragments of defined sequence. Synthetic procedures for linking together nucleotide bases in a defined sequence were initially pioneered by Har Gobind Khorana, who was the first person to create an entirely synthetic gene in this fashion. Machines that automate the procedure for joining bases in a defined sequence are now available, and many gene fragments and small DNA pieces needed for recombinant DNA experiments are routinely made in this way. DNA sequences generated by chemical synthesis can be incorporated directly into recombinant DNA cloning vectors, opening the possibility for creating new genes and modifying existing ones in predetermined ways.

The Polymerase Chain Reaction Allows Individual Gene Sequences to Be Produced in Large Quantities without Cloning

In the mid-1980s, a simple and rapid procedure was developed by Kary Mullis that can produce an unlimited number of copies of an individual gene sequence without the need for gene cloning. This procedure, called the **polymerase chain reaction (PCR),** is capable of copying DNA sequences that are initially present in extremely small amounts. The only requirement is that part of the base sequence of the gene being copied must be known. Based on this information, short DNA primers are chemically synthesized that are complementary to sequences located at the two ends of the gene. The way in which the primers are then employed is summarized in Figure 3-50. During the first step, the DNA to be copied is denatured into single strands by heating, the primers are added, and the mixture is cooled to allow the primers to hybridize to the target DNA sequences. DNA polymerase is then added to catalyze the synthesis of complementary DNA strands using the two primers as starting points. After allowing the enzyme a minute or two to copy the gene, the solution is again heated to denature the DNA and then cooled to allow more primer molecules to bind. Each time this cycle is repeated, the number of DNA molecules doubles. Within a few hours, this doubling process generates several hundred billion copies of the original DNA sequence.

The speed and sensitivity of the polymerase chain reaction allows it to be used in some unique situations where gene cloning would not be practical. For example, tiny amounts of DNA isolated from ancient fossils can be amplified using the polymerase chain reaction, thereby allowing the study of genes that are thousands of years old. In a similar fashion, DNA isolated from trace amounts of blood found at a crime scene can be copied by the polymerase chain reaction to produce enough DNA for possible genetic identification.

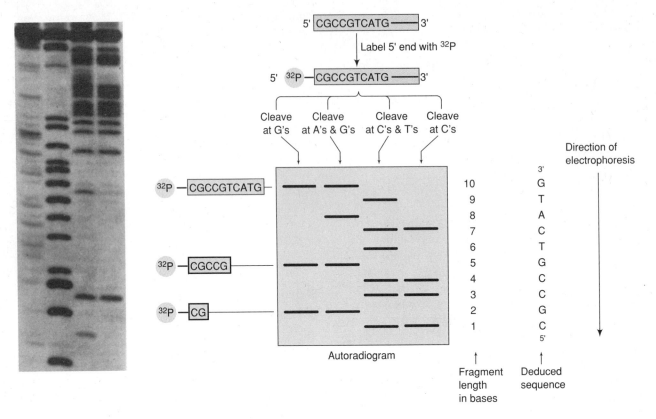

Figure 3-48 *The Maxam-Gilbert Chemical Cleavage Method for Sequencing DNA* *In this procedure, four different chemical reagents are employed to cleave single-stranded DNA chains adjacent to particular bases. The radioactive fragments are then separated on the basis of size by gel electrophoresis (page 145), which allows the sequential arrangement of bases in the original DNA molecule to be determined. The autoradiogram on the left shows what such gels look like. The diagram illustrates the banding pattern that would be observed for a DNA molecule whose sequence is CGCCGTCATG read in the 5' → 3' direction. To help show what such gel patterns mean, the sequences of the three fragments produced by the reagent that cleaves adjacent to G residues are provided at the left of the diagram.*

PRACTICAL APPLICATIONS OF DNA TECHNOLOGY

As we will see in the remainder of this book, recombinant DNA technology has had an enormous impact on the field of cell biology, leading to many new insights into the organization, regulation, and functioning of genes and their protein products. Many of these discoveries would have been virtually inconceivable without this powerful new way of isolating and characterizing DNA sequences. The rapid advances that have occurred in our ability to isolate and manipulate gene sequences has also opened up a new field, called *genetic engineering*, which refers to the application of recombinant DNA technology to practical problems. In concluding this chapter, we will briefly examine some of the areas in which the practical benefits of recombinant DNA technology are beginning to be seen.

Genetic Engineering Has Facilitated the Production of Useful Proteins

One of the first practical benefits to be derived from recombinant DNA technology was the ability to clone genes coding for useful proteins that are difficult to obtain by other means. Included in this category are human proteins such as *insulin* needed for the treatment of diabetes, *blood clotting factor* used in the treatment of hemophilia, *growth hormone* needed for the treatment of pituitary dwarfism, *tissue plasminogen activator (TPA)* employed for dissolving blood clots in heart attack patients, and *interferon, interleukin,* and *tumor necrosis factor,* which are used in the treatment of certain kinds of cancer. Traditional methods for isolating and purifying these proteins from natural sources are quite cumbersome and yield only tiny amounts of material; hence prior to the advent of recombinant DNA technology, inadequate supplies of such substances were available and their cost was extremely high. By cloning the appropriate genes in bacteria and yeast, however, it has been possible to create cell lines that produce these proteins in large quantities.

Genes have also been introduced into bacteria to create bacterial strains with important practical applications. Included in this category are bacteria that accelerate the degradation of oil spills, as well as bacteria that destroy dangerous chemicals present in toxic wastes.

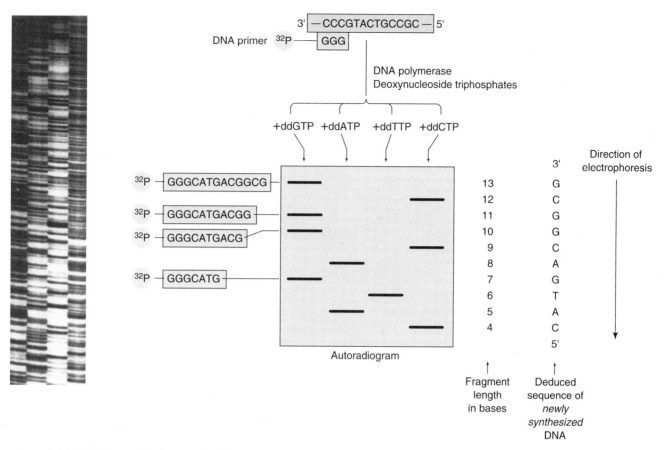

Figure 3-49 *The Sanger Chain Termination Method for Sequencing DNA* *The dideoxynucleotides ddGTP, ddATP, ddTTP, and ddCTP are employed to block chain elongation at various points. A short primer must be provided in order to initiate synthesis of the DNA strand being sequenced. The autoradiogram on the left shows what such gels look like. The diagram illustrates the banding pattern that would be observed for a DNA molecule who sequence is CATGACGGCG read in the 5′ → 3′ direction (not including the primer). To help show what such gel patterns mean, the sequences of the four fragments produced in the presence of ddGTP are provided at the left of the diagram.*

Genetic Engineering of Plants Is Helping to Increase Crop Yields

Recombinant DNA technology is also being used to create plants with new genes for improved disease resistance, photosynthetic efficiency, nitrogen-fixing abilities, ability to grow under adverse weather conditions, and nutritional value. The most widely used vector for introducing cloned genes into plant cells is the **Ti plasmid,** a naturally occurring DNA molecule that can be introduced into a broad spectrum of plants by the bacterium *Agrobacterium tumefaciens*. In nature, infection of plant cells by this bacterium leads to an insertion of a small part of the Ti plasmid into the plant cell DNA; expression of a portion of this inserted DNA then triggers the formation of an uncontrolled growth known as a *crown gall tumor* (page 802). However, the DNA sequences involved in triggering tumor formation can be removed from the Ti plasmid without preventing the transfer of DNA from the plasmid to the host cell chromosome. By inserting genes of interest into such modified plasmids, vectors can be created that are capable of transferring foreign genes into plant cells.

Progress in using this approach has been rapid in the past few years, and genetically engineered tomato, soybean, cotton, rice, corn, sugarbeet, and alfalfa crops are expected to enter the marketplace by the year 2000. Thus far the greatest success has been obtained in producing crops with increased resistance to insects, herbicides (weed killers), viral diseases, and spoilage. For example, significant resistance to a broad spectrum of plant viruses has been achieved by introducing viral coat protein genes into tomato and potato plants (Figure 3-51). These genetically engineered plants produce small amounts of viral coat protein, which in turn interferes with the ability of an infecting virus to shed its coat and replicate.

DNA Fingerprinting Is a New Way of Identifying Individuals for Legal Purposes

Recombinant DNA technology has recently led to a new way of identifying people that soon may be used more widely than fingerprints. The theory underlying the use of fingerprinting in criminal investigations is that no two people have exactly the same fingerprint patterns.

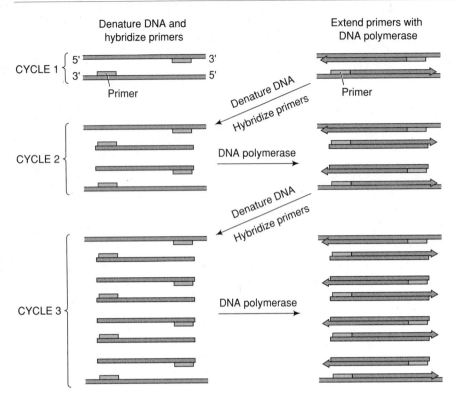

Denature DNA and hybridize primers

Extend primers with DNA polymerase

CYCLE 1

5'   3'
3'   5'
Primer

Primer

Denature DNA Hybridize primers

CYCLE 2

DNA polymerase

Denature DNA Hybridize primers

CYCLE 3

DNA polymerase

Figure 3-50 *The Polymerase Chain Reaction*   *The DNA being copied is denatured into single strands by heating and is then hybridized to DNA primers that are complementary to sequences located at opposite ends of the gene sequence. Next, DNA polymerase is added to catalyze the synthesis of complementary DNA strands using the two primers as starting points. After allowing the enzyme a minute or two to copy the gene, the solution is again heated to denature the DNA and the cycle is repeated. Within a few hours, this doubling process generates several hundred billion copies of the original DNA sequence.*

It is also true, however, that no two people other than identical twins have the same exact set of DNA base sequences. Although the differences in DNA base sequence between any two individuals are very small, a recently developed technique called **DNA fingerprinting** is capable of detecting these differences. DNA fingerprinting was made possible by the discovery that human DNA contains short repeated sequences, called **variable number of tandem repeats** or **VNTRs,** which vary appreciably from individual to individual. Hence the pattern of VNTRs in a person's DNA can be used to identify that individual uniquely.

In DNA fingerprinting, DNA is isolated from a small sample of blood, semen, or other bodily fluid or tissue. After amplification by the polymerase chain reaction, the isolated DNA is cleaved with a restriction enzyme to produce fragments that are separated by electrophoresis, and a Southern blot is then performed using a radioactive DNA probe specific for VNTRs. The resulting pattern of fragments serves as a "fingerprint" that identifies the individual from which the DNA was obtained (Figure 3-52). Although a remote chance exists that an individual could be erroneously identified by this procedure, DNA fingerprinting is becoming a common tool in criminal investigations.

Recombinant DNA Technology Is Being Used in the Diagnosis and Treatment of Genetic Diseases

A large number of human diseases arise from the inheritance of faulty genes, such as the genes that cause sickle-cell anemia, cystic fibrosis, Huntington's disease, and diabetes. In many of these cases, it is now possible to detect the faulty gene given a small sample of DNA from an affected individual. The method used for detecting these genetic abnormalities is based on the fact that gene mutations often cause changes in the recognition sites for restriction enzymes. Hence the restriction map for a DNA sample obtained from normal individuals will not look the same as the restriction map for an individual carrying a mutated gene. These differing restriction maps are referred to as **restriction fragment length polymorphisms, or RFLPs.** The power of the RFLP approach is that it can detect individuals who carry a defective gene but may not yet exhibit symptoms of the disease.

As an example of how this approach works, let us consider an RFLP associated with the mutation that causes sickle-cell anemia. The mutation responsible for sickle-cell anemia creates an extra recognition site for the restriction enzyme *Dde*I (Figure 3-53). To detect

Normal
tomato plant

Genetically
engineered
tomato
plant

Figure 3-51 *Resistance to Viral Disease in Genetically Engineered Tomato Plants* *The coat protein gene of the TMV virus was introduced into tomato plants by infection with* Agrobacterium tumefaciens *containing the TMV coat protein gene in a Ti plasmid. These pictures show the growth of these tomato plants in a greenhouse* (top) *and in a field* (bottom) *after deliberate infection with the TMV virus. Note that the genetically engineered plants (located on the right) are more resistant to the damaging effects of viral infection than are normal tomato plants (located on the left). Courtesy of R. T. Fraley.*

this abnormality, a DNA sample obtained from an individual suspected of carrying the sickle-cell gene is cut with the restriction enzyme *Dde*I and the resulting fragments are separated by electrophoresis. A Southern blot is then performed using a radioactive cDNA probe specific for the gene encoding the β subunit of hemoglobin. This type of experiment reveals that a 376 base-pair fragment present in normal DNA is split in sickle-cell DNA into fragments of 201 and 175 base pairs respectively. In an individual carrying both the sickle-cell gene and the normal gene, all three fragments are observed.

Besides facilitating the diagnosis of genetic diseases, recombinant DNA techniques may soon make it possible to treat such conditions by transplanting normal genes into individuals who possess disease-causing genes. One of the first successful attempts at gene transplantation in animals was carried out by Richard

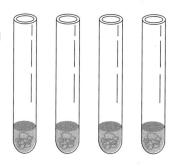

1 Isolate DNA from blood or tissue samples obtained from crime scene and three suspects.

2 Digest total DNA with a restriction enzyme and separate fragments by electrophoresis.

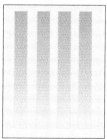

3 Prepare Southern blot and hybridize with radioactive DNA probe specific for VNTRs.

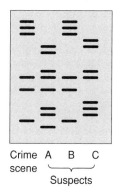

4 Prepare autoradiogram and compare banding patterns; these data implicate suspect B.

Crime scene   A   B   C
Suspects

Figure 3-52 *DNA Fingerprinting* *This hypothetical example shows how DNA fingerprinting can be used in criminal investigations to link a suspect to a crime scene. DNA fingerprinting is based on the discovery that human DNA contains short repeated sequences (VNTRs) which vary from individual to individual. Hence the pattern of VNTRs in a person's DNA uniquely identifies that individual.*

Palmitter and his associates, who injected a plasmid containing the gene for *human growth hormone* into fertilized mouse eggs. The eggs were then reimplanted into mice and allowed to develop normally. In order to increase the likelihood that the human gene would be expressed in mouse cells, the injected plasmid contained a mouse DNA sequence that normally promotes

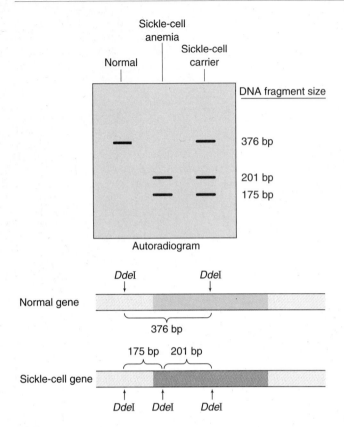

Figure 3-53 *The Use of Restriction Fragment Length Polymorphism (RFLP) in the Diagnosis of a Genetic Disease* *In this example, human DNA was cleaved with the restriction enzyme* DdeI *and subjected to electrophoresis. A Southern blot was then performed using a radioactive cDNA probe specific for the β subunit of hemoglobin. The resulting autoradiogram shows an RFLP associated with sickle-cell anemia.*

Figure 3-54 *A Transgenic Mouse* *The mouse on the right developed from an egg into which a plasmid containing the human growth hormone gene had been injected. The resulting production of human growth hormone in the cells of this mouse caused the mouse to grow to twice the size of its normal littermate on the left. Courtesy of R. L. Brinster and R. E. Hammer, School of Veterinary Medicine, University of Pennsylvania.*

the transcription of mouse genes. By inserting this sequence adjacent to the human growth hormone gene, it was hoped that it would facilitate transcription of the human gene. When the mice that developed from these injected eggs were born and allowed to develop into mature adults, many were found to grow significantly larger than normal (Figure 3-54). Blood samples obtained from these affected animals revealed that they were producing human growth hormone. These experiments represented one of the first successful creations of a **transgenic** animal, which is an animal that carries a functional gene transplanted from another organism.

Given the success of the preceding experiments, the question arises as to whether gene transplantation techniques can eventually be used in humans to replace defective genes. Humans certainly suffer from many genetic diseases that could theoretically be cured by replacing the defective gene with a normal copy. Potential candidates for such treatment include cystic fibrosis, hemophilia, hemoglobin disorders, hyperchol-

esterolemia, muscular dystrophy, lysosomal storage diseases, and an immune disorder called ADA deficiency. Procedures have already been worked out for removing cells from sick patients, introducing cloned genes into these cells using altered viruses designed for this purpose, and then reinjecting the genetically altered cells back into the individual.

The first patient to be treated this way in the United States was a 4-year-old girl suffering from *ADA deficiency*, an inherited deficit in the enzyme *adenosine deaminase (ADA)*. The lack of adenosine deaminase causes a potentially lethal immune disorder in which individuals are unable to fight infections because they do not produce enough white blood cells called *T lymphocytes*. Prior to treatment, the girl was suffering from frequent infections and was generally lethargic. In 1990 an experiment was initiated in which some T lymphocytes were removed from the patient, a cloned gene for ADA was introduced into the cells, and the genetically altered cells were then injected back into her bloodstream. After a year of treatment her immune function had improved dramatically and her energy returned. A second patient treated in a similar fashion also responded positively. Although experiments along these lines are still in their early stages, and major ethical and legal questions still need to be addressed about the widespread use of such approaches, these preliminary results are extremely encouraging. The idea of using normal genes to treat genetic diseases appears to be a realistic prospect that has the potential of revolutionizing the practice of modern medicine.

SUMMARY OF PRINCIPAL POINTS

• DNA was discovered in the 1860s, but almost a hundred years passed before it was established that DNA stores genetic information. The first breakthrough was Avery's discovery that bacteria can be transformed from one genetic type to another by purified DNA. Then it was shown that an organism's DNA content is constant from cell to cell, that DNA base composition differs among different types of organisms, and that viral DNA is injected into bacterial cells during viral infections.

• In 1953 Watson and Crick proposed that DNA is a double helix held together by hydrogen bonds between the base A in one strand and T in the other, and between G in one strand and C in the other. The Meselson-Stahl experiments revealed that DNA is replicated by a semiconservative mechanism in which the two strands of the double helix first separate, and then serve as templates upon which new strands are copied by complementary base-pairing.

• DNA replication is catalyzed by the enzyme DNA polymerase, which adds nucleotides to DNA chains in the $5' \rightarrow 3'$ direction. The synthesis of one of the two DNA strands is discontinuous, generating small DNA fragments that are later joined by the enzyme DNA ligase. DNA replication is initiated by the enzyme DNA primase, which catalyzes the formation of RNA primers that are later removed and replaced by DNA. During DNA replication, the DNA double helix is unwound through the action of helicases, single-stranded DNA binding proteins, and topoisomerases.

• Some RNA viruses contain an enzyme called reverse transcriptase, which permits them to use RNA as a template for DNA synthesis.

• The Beadle-Tatum experiments involving X-ray induced mutations in *Neurospora* led them to conclude that genes control the production of enzymes. Later studies broadened this conclusion to encompass the idea that genes code for polypeptide chains.

• The information encoded in the base sequence of DNA is transcribed by RNA polymerase into messenger RNA (mRNA), which binds to ribosomes and is translated to form a polypeptide chain. Information in mRNA is encoded in units of three bases called codons. Most codons specify an amino acid, but a few function as stop signals that mark the end of a polypeptide chain. Amino acids are brought to the ribosome by transfer RNA (tRNA) molecules. Each type of tRNA contains a bound amino acid and an anticodon that binds to the mRNA codon calling for that particular amino acid.

• Mutations in DNA arise spontaneously as well as through the action of environmental mutagens. Several mechanisms can repair mutations before they become stabilized by subsequent rounds of DNA replication. The 3'-exonuclease activity of DNA polymerase allows it to proofread as it replicates, removing incorrect nucleotides that were mistakenly incorporated. Mismatch repair corrects noncomplementary base pairs that escape the proofreading process, whereas excision repair corrects mutations involving abnormal bases.

• Transposable elements promote the movement of DNA segments from one location to another, leading to changes in the expression and organization of neighboring genes.

• Nucleic acid hybridization techniques are employed to determine whether two nucleic acids contain similar base sequences. Examples include Southern and Northern blotting, *in situ* hybridization, and colony hybridization.

• Recombinant DNA techniques permit individual genes to be isolated from large masses of DNA and produced in virtually unlimited quantities. Such technology utilizes restriction endonucleases, which cleave DNA molecules at specific sequences four to eight bases long. As an alternative to gene cloning, the polymerase chain reaction allows an unlimited number of copies of an individual gene sequence to be produced from tiny amounts of DNA if part of the base sequence of the gene being copied is known.

• Recombinant DNA technology has a growing number of practical benefits, such as producing useful proteins from genetically engineered bacteria, increasing crop yields through the use of genetically engineered plants, using DNA fingerprinting to identify individuals for legal purposes, and employing recombinant DNA techniques in the diagnosis and treatment of human genetic diseases.

SUGGESTED READINGS

Books

Adams, R. L. P. (1991). *DNA Replication,* IRL Press, Oxford, England.

Glick, B. R., and J. J. Pasternak (1994). *Molecular Biotechnology: Principles and Applications of Recombinant DNA,* ASM Press, Herndon, VA.

Kornberg, A., and T. A. Baker (1991). *DNA Replication,* 2nd Ed., Freeman, New York.

Miklos, D. A., and G. A. Freyer (1990). *DNA Science,* Carolina Biological Supply, Burlington, NC.

Portugal, F. H., and J. S. Cohen (1977). *The Century of DNA: A History of the Discovery of the Structure and Function of the Genetic Substance,* MIT Press, Cambridge, MA.

Russell, P. J. (1992). *Genetics,* 3rd Ed., HarperCollins, New York.

Watson, J. D., and J. Tooze (1981). *The DNA Story,* Freeman, San Francisco.

Watson, J. D., M. Gilman, J. Witkowski, and M. Zoller (1992). *Recombinant DNA,* 2nd Ed., Freeman, New York.

Wilson, E. B. (1896). *The Cell in Development and Inheritance,* Macmillan, New York.

Articles

Chargaff, E. (1971). Preface to a grammar of biology, *Science* 172:637–642.

Chilton, M.-D. (1983). A vector for introducing new genes into plants, *Sci. Amer.* 248 (June):51–59.

Cohen, S. N., A. C. Y. Chang, H. W. Boyer, and R. B. Helling (1973). Construction of biologically functional bacterial plasmids *in vitro, Proc. Natl. Acad. Sci. USA* 70:3240–3244.

Cohen, S. N., and J. A. Shapiro (1980). Transposable genetic elements, *Sci. Amer.* 242 (February):40-49.

Darnell, J. E., Jr. (1985). RNA, *Sci. Amer.* 253 (October):68-78.

Dickerson, R. E. (1983). The DNA helix and how it is read, *Sci. Amer.* 249 (December):94-111.

De Lucia, P., and J. Cairns (1969). Isolation of an *E. coli* strain with a mutation affecting DNA polymerase, *Nature* 224:1164-1166.

Echols, H., and M. F. Goodman (1991). Fidelity mechanisms in DNA replication, *Annu. Rev. Biochem.* 60:477-511.

Felsenfeld, G. (1985). DNA, *Sci. Amer.* 253 (October):58-67.

Gasser, C. S., and R. T. Fraley (1992). Transgenic crops, *Sci. Amer.* 266 (June):62-69.

Hou, Y.-M., and P. Schimmel (1988). A simple structural feature is a major determinant of the identity of a transfer RNA, *Nature* 333:140-145.

Joyce, C. M., and T. A. Steitz (1994). Function and structure relationships in DNA polymerases, *Annu. Rev. Biochem.* 63: 777-822.

Lander, E. S. (1989). DNA fingerprinting on trial, *Nature* 339:501-505.

Marmur, J. (1994). DNA strand separation, renaturation and hybridization, *Trends Biochem. Sci.* 19: 343-346.

Miller, A. D. (1992). Human gene therapy comes of age, *Nature* 357:455-460.

Modrich, P. (1987). DNA mismatch correction, *Annu. Rev. Biochem.* 56:435-466.

Morgan, R. A., and W. F. Anderson (1993). Human gene therapy, *Annu. Rev. Biochem.* 62:191-217.

Mullis, K. B. (1990). The unusual origin of the polymerase chain reaction, *Sci. Amer.* 262 (April):56-65.

Radman, M., and R. Wagner (1988). The high fidelity of DNA duplication, *Sci. Amer.* 259 (August):40-46.

Saks, M. E., J. R. Sampson, and J. N. Abelson (1994). The transfer RNA identity problem: A search for rules, *Science,* 263: 191-197.

Stillman, B. (1989). Initiation of eukaryotic DNA replication in vitro, *Annu. Rev. Cell Biol.* 5:197-246.

Verma, I. M. (1990). Gene therapy, *Sci. Amer.* 263 (November):68-84.

Chapter 4

Experimental Approaches for Studying Cells

The beginning chapters of this book have provided background information that is required before we can examine the contributions each organelle makes to the life of the cell. A final background topic that also needs to be covered concerns the experimental techniques used by biologists to investigate cell structure and function. Discussing methodology early in the text is especially important because the remainder of the book frequently emphasizes the analysis of experimental data. This experimental focus has been chosen because our concepts about the cell are in a state of continual flux, changing in either subtle or significant ways as each major new set of experiments is carried out. To try to teach nothing but a set of "facts" about the cell would therefore be a disservice, for most "facts" will sooner or later be superseded by newer information. What is more significant for the student to understand is the way in which our ideas about the cell have developed. In practice, the science of cell biology is the process of asking questions, carrying out experiments designed to investigate these questions, drawing conclusions and formulating new questions based on the results, and repeating this cycle again and again. Our conclusions are always tentative, subject to further investigation and refinement. Unfortunately, few nonscientists understand what the scientist means by the word "truth." As Lynn White, Jr., author of *Machina ex Deo*, so aptly put it, scientific truth ". . . is not a citadel of certainty to be defended against error: it is a shady spot where one eats lunch before tramping on."

The experimental approaches employed by cell biologists encompass a broad spectrum of microscopic and biochemical techniques for examining the properties of cells, organelles, and macromolecules. Many of the techniques involved in investigating the properties of nucleic acids have already been described in Chapter 3. In this chapter we will provide an overview of some of the other important techniques that are used in the numerous experiments mentioned elsewhere in the book. The advantage of collecting these techniques into a single chapter near the beginning of the text, rather than spreading them throughout the book, is that it provides a more readily accessible reference source for the reader.

GROWING CELLS IN THE LABORATORY

Ideally cells should be studied under conditions that are as close to natural as possible. Experiments carried out while cells are still in their native environment within intact animals or plants are said to be carried out **in vivo** ("in life"). Unfortunately, the types of experiments that can be carried out *in vivo*, especially in the case of complex multicellular organisms, are quite limited. It is possible to take cells from living organisms for microscopic analysis or to administer radioisotopes to intact organisms for monitoring metabolic activities. But many variables cannot be controlled *in vivo* and the kinds of experimental manipulations that can be carried out are severely restricted. For this reason, biologists often study cells that are grown under artificial laboratory conditions. Such experiments involving *cell cultures* are said to be carried out **in vitro** ("in glass").

The main advantage to using cell culture techniques is that they allow cells to be studied under carefully controlled conditions. Cellular responses to various external agents, such as hormones or metabolic inhibitors, can be gauged without the complicating effects of secondary factors present in intact organisms. Cultured cells are also excellent models for studying the regulation of cell growth and division, and certain cells even retain differentiated properties characteristic of their native state, permitting detailed analysis of specialized cell functions.

Animal, Plant, and Bacterial Cells Can Be Grown in the Laboratory

The science of eukaryotic cell culture dates back to the turn of the century, when pioneers such as Ross Harrison and Alexis Carrel first showed that tissues and cells can survive outside the body in a culture medium containing blood serum or plasma. Carrel's discovery that extracts of chicken embryos stimulate cells to grow and divide led to the establishment of permanent cell lines and focused attention on cell culture techniques as a powerful approach for studying the regulation of cell growth, division, and differentiation. Many cell culturing procedures have subsequently been developed, and hundreds of different cell types are now routinely grown in the laboratory.

The initial step in creating a cell culture is to isolate the cell population that one wishes to grow. Animal tissues are typically dispersed into individual cells by mechanically agitating or disrupting the tissue in a medium containing two ingredients: (1) proteases such as trypsin and collagenase, which break down extracellular materials that tend to hold cells together, and (2) Ca^{2+}-binding agents such as *EDTA,* which remove calcium ions required for adhesion between cells. Cell suspensions produced by such treatments usually consist of a heterogeneous mixture of cell types. If the goal is to culture a particular kind of cell from the suspension, then some type of cell separation procedure must be employed. One of the most powerful techniques for separating cells utilizes antibodies that selectively bind to the surface of a particular cell type. Such antibodies are linked to a fluorescent dye and then added to an appropriate cell suspension, where they selectively bind to their target cells. The cells containing bound fluorescent antibody are then passed through a machine called a *fluorescence-activated cell sorter,* which separates fluorescent cells from the rest of the cells in the population.

Once cells have been isolated, they can be grown either *suspended* in a liquid medium or *attached* to a solid surface. Culturing in liquid suspension allows a greater mass of cells to be produced, but most cells derived from animal tissues do not grow well in suspension. In such cases, cells are placed in a test tube or flask and covered with a thin layer of nutrient medium. The cells attach to the glass or plastic surface and proliferate until they form a confluent *monolayer* of cells covering the entire surface of the container (Figure 4-1). At this stage, some of the cells are removed and transferred to another container to keep them growing and dividing.

Unlike multicellular plants and animals, bacteria are single-celled organisms that can be placed directly into culture without the need for disrupting cell-cell connections. They can be easily grown either in suspension culture or attached to the surface of a solid medium such as agar. *Agar* is a carbohydrate material derived from seaweed that is dissolved in a nutrient medium to make a semisolid gel on which bacteria can grow. Bacteria are grown on agar by streaking a sample of cells across the gel surface. From this initial inoculation, individual colonies that have grown from single cells can be selected for further analysis (Figure 4-2).

The composition of the nutrient medium has a critical influence on the properties of cells grown in culture. Many bacteria require only a source of carbon such as glucose, and a mixture of inorganic salts that provide essential elements, including N, P, S, Mg, Na, and K. Trace elements required for cell growth are usually present in sufficient quantities as impurities in the other ingredients. The nutritional requirements of eukaryotic cells are more complex than those of bacteria. For this reason early culture work utilized natural biological substances, such as serum or plasma, as nutrient sources. Harry Eagle spent many years in defining the nutritional needs of mammalian cells and formulating media to optimize their growth. This work eventually led to the formulation of chemically defined media composed of known mixtures of salts, amino acids, vitamins, and carbohydrates.

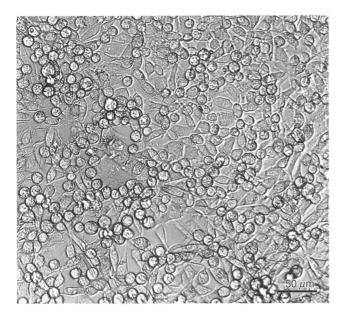

Figure 4-1 *Fibroblasts Growing as a Monolayer in Cell Culture* *Courtesy of K. K. Sanford.*

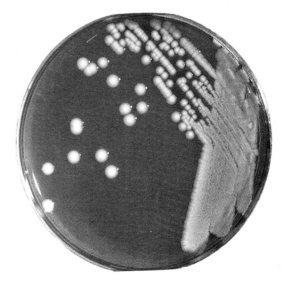

Figure 4-2 *Bacterial Colonies Growing on an Agar Plate* *Each round white colony represents a clone of bacterial cells derived from the division of a single cell that initially landed in that location.*

In order to grow in such chemically defined media, most eukaryotic cells require the addition of one or more **growth factors** that are normally produced by tissues to regulate cell growth and division. For example, some animal cells produce a protein called *nerve growth factor* that is required for the survival of certain kinds of nerve cells in culture. In plant cell cultures, the effects of growth factors can be especially striking. Plants manufacture a series of *plant growth substances,* such as gibberellins, auxin, and cytokinins, which regulate cell growth and division in various parts of the plant. When dividing cells removed from an intact plant are placed in culture, they develop into an undifferentiated mass of cells known as *callus tissue.* Addition of the proper combination of plant growth substances can induce the cell mass to develop into leaves, shoots, roots, and even a complete plant (Figure 4-3).

Primary Cell Cultures and Transformed Cell Lines Differ in Their Properties

Cells such as bacteria and single-celled eukaryotes usually adapt to growth in culture with relative ease because they normally exist as single cells in nature. But many of the cells obtained from multicellular organisms are not accustomed to growing as individual cells, and so do not adjust well to culture. Moreover, those cells that do successfully adapt to growing in culture often lose their capacity to divide after a certain period of time. This period varies among cell types, averaging about 50 cell divisions for human cells. Such cell populations that are only capable of dividing a limited number of times in culture are referred to as **primary cultures.**

In contrast, some cells obtained from multicellular organisms have an unlimited capacity to divide in culture and hence can be used to produce permanently dividing *cell lines.* These **transformed cells** are obtained either from tumor tissue or from primary cell cultures that have been altered by exposure to viruses, chemical mutagens, or radiation. When transformed cells are used to establish permanent cell lines, it is desirable to work with a homogeneous cell population that has been derived from a single parental cell. This goal is accomplished by isolating single cells from a cultured cell population and allowing each individual cell to grow by successive divisions into a uniform cell population or **clone** (Figure 4-4).

The rapid growth and unlimited life spans of transformed cell lines make them an ideal source of material for experimentation, but it is important to realize that transformed cells exhibit many aberrant properties that distinguish them from the normal cells of an intact organism or the cells of a primary culture. For example, transformed cells usually possess an abnormal number of chromosomes, lack many of the specialized properties associated with normal cells, grow without attaching to a surface, and produce tumors

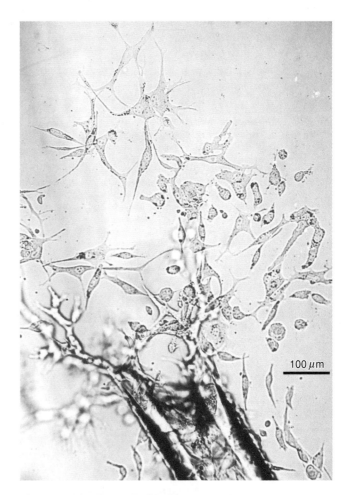

Figure 4-4 *A Clone of Fibroblasts* *A single fibroblast cell was isolated in a glass capillary tube, which is seen at the bottom of this photograph. The cells that have arisen by division from this original parental cell are migrating out through the end of the tube into the culture medium. Courtesy of K. K. Sanford.*

Figure 4-3 *Plant Cell Cultures* *When isolated plant cells are cultured in the presence of plant growth substances, they begin to form leaves, roots, and shoots. Courtesy of Sigma Chemical Company.*

when injected back into an appropriate host organism. One of the most thoroughly studied transformed cell lines was established in 1951 from a piece of uterine cancer tissue removed from a woman named Henrietta Lacks. Although she died a year later, her tumor cells have been growing in culture ever since. These cells, called *HeLa* cells in her honor, are now used in hundreds of laboratories and have contributed to the solution of a large number of important biological and medical problems.

Although the vigorous way in which HeLa cells grow in culture has contributed to their usefulness as a scientific tool, these same properties also pose some serious problems. If another cell line is accidentally contaminated by even a single HeLa cell, that culture may eventually become overgrown with HeLa cells. Because most cell lines look alike, this overgrowth may occur without any obvious visual changes. However, careful examination of the number and appearance of the chromosomes usually allows the real identity of a cell culture to be determined. Using this approach, Walter Nelson-Rees discovered in the 1970s that almost a hundred different human cell lines had actually been taken over by HeLa cells. Some of these contaminated cell cultures came to him from locations as far away as Russia. By the time these investigations were complete, it had become apparent that roughly one-third of the most popular human cell lines used in biomedical research had been destroyed and overtaken by HeLa cells. This story emphasizes the importance of continually making sure that the real identity of a cell line being maintained in culture is known.

Hybrid Cells Can Be Created by the Technique of Cell Fusion

The ability to study the behavior of cultured cells has been significantly aided by the development of a technique called **cell fusion.** In 1960, a research team in Paris headed by Georges Barski first discovered that cells of two different types can spontaneously fuse to create a new kind of cell that contains genetic information derived from the two original cells. Barski had been studying two mouse cell lines that could be distinguished from each other by differences in their chromosomal makeup. When cells of both types were grown together in the same culture, a new kind of cell eventually appeared that contained chromosomes derived from each of the original cell types.

Shortly thereafter Henry Harris discovered that cell fusion can be artificially induced by treating cells with *Sendai virus* that has been inactivated by exposure to ultraviolet light. The interaction of this virus with plasma membranes enhances their ability to fuse with each other. Fusion of the plasma membranes of two cells creates a *heterokaryon* in which the nuclei of the

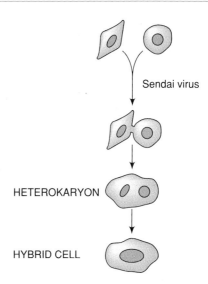

Figure 4-5 *The Cell Fusion Technique* *In addition to using Sendai virus, exposure to other inactivated viruses or polyethylene glycol can also induce cells to fuse.*

two original cells share the same cytoplasm (Figure 4-5). If heterokaryons are permitted to undergo cell division, the nuclear envelopes break down and a single nucleus is formed containing chromosomes derived from both of the original cells. The resulting **hybrid cells** can be used to establish permanent cell lines.

The technique of cell fusion has played an important role in cell biology because the mechanisms underlying various cell functions can be investigated by fusing cells with and without the function in question. For example, the basis of the uncontrolled growth exhibited by cancer cells has been investigated in cell hybrids created by fusing normal cells with cancer cells (page 787). Cell fusion has also been used to demonstrate that proteins move around in the plasma membrane (page 169), and to assign human genes to particular chromosomes. The latter process takes advantage of the fact that hybrids formed between human cells and mouse cells gradually tend to lose human chromosomes. Human genes can therefore be assigned to individual chromosomes by correlating the presence or absence of various genetic traits with the presence or absence of particular human chromosomes in human-mouse hybrid cell lines.

VIEWING CELLS WITH A MICROSCOPE

Our current appreciation of how cells carry out their various activities has been made possible by a combination of microscopic and biochemical tools. It was the microscope, of course, that first opened our eyes to the existence of cells. The use of curved surfaces for magnifying objects was reported as early as A.D. 127, although it was not until the year 1235 that this principle led to the invention of eyeglasses by the Englishman Roger Bacon. In 1590 Jans and Zacharias Janssen combined two convex lenses within a tube to construct the forerunner of the light microscope, an instrument that was soon improved by Antonie van Leeuwenhoek to the point where objects could be magnified almost 300-fold. As we saw in Chapter 1, it was the development of such microscopes that led to the discovery of cells and the formulation of the cell theory.

The Light Microscope Can Visualize Objects as Small as 200 Nanometers in Diameter

Typical prokaryotic and eukaryotic cells fall within the size range of 1 to 100 μm in diameter. Because the unaided human eye cannot resolve objects smaller than 100 μm in diameter, microscopes are needed for detecting the existence of cells and visualizing their internal architecture. In a light microscope, the material being examined is placed on a glass slide and viewed with visible light. After the light waves have passed through the specimen, the image is magnified by passing the light through two lenses: an *objective lens* and an *ocular lens* or *eyepiece* (Figure 4-6).

Although these lenses magnify the image of the object being observed, magnification is not the critical issue. One can magnify a photograph, for example, to any size that is desired, but eventually nothing more is gained because the image simply becomes blurred. The crucial issue is not magnification but **resolving power,** a term that refers to the ability to distinguish adjacent objects as separate entities. The resolving power of any optical system can be calculated from the equation

$$r = \frac{0.61\lambda}{n \sin \alpha} \qquad (4\text{-}1)$$

where the resolving power r is defined as the minimum distance that can separate two points that still remain recognizable as separate points, λ is the wavelength of light (or any other type of radiation) used to illuminate the object, n is the refractive index of the medium in which the object is placed, and $\sin \alpha$ is the sine of half the angle between the specimen and the objective lens. The term $n \sin \alpha$ is often called the *numerical aperture.*

It is apparent from this equation that only a small number of variables influence the resolving power of a microscope. The refractive index n can be increased by immersing the sample in oil ($n = 1.5$) rather than air ($n = 1.0$), and moving the lens closer to the specimen to increase α. The upper theoretical limit of α is 90°, meaning that the value of $\sin \alpha$ cannot exceed 1.

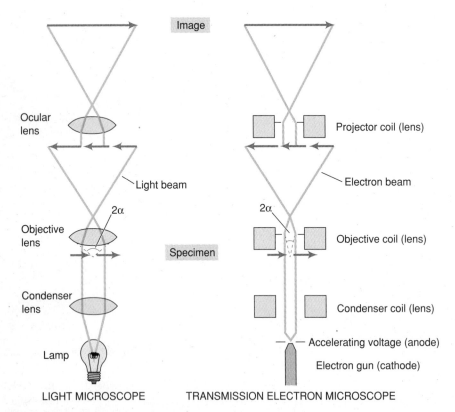

Figure 4-6 *Comparison of Image Formation in the Light and Transmission Electron Microscopes* *The specimen being viewed is represented by red arrows. The aperture angle (2α) in the electron microscope is usually no more than a few tenths of a degree, but is drawn larger to make it visible. To facilitate this side-by-side comparison, the electron microscope is drawn upside down.*

Hence the maximum numerical aperture of an optical system employing an oil immersion lens will be $1.5 \times 1 = 1.5$. Under these conditions, a microscope illuminated by visible light with an average wavelength of about 500 nm will have a resolving power of $(0.61 \times 500)/1.5 \approx 200$ nm. This means that objects that are closer to one another or smaller than 200 nm cannot be distinguished as separate objects by light microscopy. A resolving power of 200 nm is adequate for visualizing larger organelles, such as the nucleus, mitochondria, and chloroplasts; however, one cannot see smaller structures like ribosomes, membranes, and individual cytoskeletal filaments or chromatin fibers.

In order to take maximum advantage of the resolving power of the light microscope, samples must be prepared in a way that produces *contrast*—that is, differences in the darkness or color of the structures being examined. Without special treatment to increase contrast, little discernible structure will be visible because most cell structures appear either opaque or transparent. Contrast is usually enhanced by applying specific **stains** that color or otherwise alter the light-transmitting properties of cell constituents. To prepare cells for staining, tissues are first treated with **fixatives** that kill the cells while preserving their structural appearance. The most widely employed fixatives are acids and aldehydes such as acetic acid, picric acid, formaldehyde, and glutaraldehyde. The fixed tissue is then sliced into sections that are thin enough to be transparent under the microscope. To provide the physical support needed for cutting thin sections, tissues are either embedded in paraffin wax or quick-frozen. Sections are cut with a special instrument called a *microtome* and are then spread on glass and stained. Although a fairly reliable picture of cell structure can be obtained from the examination of tissues that have been sectioned and stained, distortions are sometimes introduced by the fixation, sectioning, and staining procedures.

Specialized Kinds of Light Microscopy Permit the Visualization of Living Cells

Fixation, sectioning, and staining are useful procedures for visualizing a cell's internal architecture, but little can be learned about the dynamic aspects of cell behavior by examining cells that have been killed and sliced. Therefore techniques have also been developed for using light microscopy to observe cells that are intact and, in many cases, still living. However, the thickness and opacity of intact cells make it difficult to obtain adequate resolution and contrast. In a standard **bright-field light microscope,** which uses visible light that has been transmitted through a specimen, relatively little structural detail can be seen when examining living cells (Figure 4-7, *top*). To obtain better contrast and resolution, several modifications have been introduced to this basic light microscope.

Phase-Contrast Microscopy Detects Differences in Refractive Index and Thickness

Phase-contrast microscopy is a type of light microscopy that improves contrast by exploiting differences in the thickness and refractive index of various regions of the cells being examined. The *refractive index* of a material is a measure of the velocity with which light waves pass through it. A beam of light passing through an object of high refractive index will be slowed down during its passage, resulting in a change in *phase* relative to light waves that have not passed through the object. (Light waves are said to be traveling "in phase" when the crests and troughs of the waves match each other.) Many cellular structures have high refractive indices and thus alter the phase of light waves, but the human eye cannot detect such phase changes directly. The phase-contrast microscope overcomes this problem by converting phase differences into alterations in brightness.

This conversion is accomplished by taking waves that have been refracted and using them to *interfere* with unrefracted waves. **Interference** refers to the process by which two or more light waves combine to reinforce or cancel one another, producing a wave equal to the sum of the two combining waves. Refracted and unrefracted waves can be easily separated from each other because light waves passing obliquely into an area of changing refractive index not only have their phase altered, but also have the angle of their paths deflected. Therefore in a phase-contrast microscope, cells are illuminated by a cone of light whose rays pass obliquely through the specimen, causing the refracted waves to deviate away from the unrefracted rays. When the unrefracted and refracted rays are focused back together, they undergo interference.

However, for most biological materials the phase difference between the unrefracted and refracted waves is only about one-fourth wavelength, which produces minimal interference. This small effect would be undetectable to the eye in an ordinary light microscope. But in a phase-contrast microscope, a *phase plate* of refracting material designed to retard the refracted waves by an additional one-fourth wavelength is placed in the path of the refracted waves (Figure 4-8). When the refracted and unrefracted waves are focused back together, the additional one-fourth wavelength delay increases the interference between the two waves to such a degree that a significant decrease in the amplitude of the resulting wave occurs. Since changes in amplitude are detected by the human eye as changes in brightness, the reduced amplitude appears as a darkened area in the specimen. In this way, the phase-contrast microscope converts differences in refractive index and thickness to contrasting degrees of brightness (see Figure 4-7).

Bright-field light microscopy

Phase-contrast light microscopy

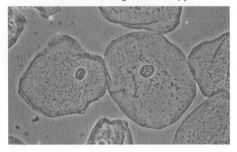

Nomarski interference light microscopy

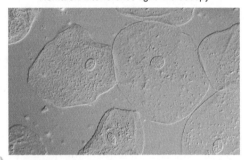

Dark-field light microscopy

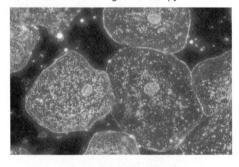

Bright-field light microscopy
(cells fixed and stained)

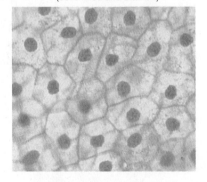

Polarization Microscopy Uses Polarized Light to Detect Structures That Are Birefringent

One of the oldest methods for improving contrast with the light microscope involves the use of *polarized light,* a term that refers to light whose waves travel in a single plane. In **polarization microscopy,** light is first passed through a polarizing filter that only allows waves traveling in a single plane to pass. The resulting polarized light is transmitted through the specimen and then through a second polarizing filter whose transmission plane is oriented at right angles to the first filter. Because the second filter is oriented in the wrong plane, none of the light would normally be expected to pass through the filter and so no image would be seen with the microscope. However, some materials exhibit a property called *birefringence,* which allows them to rotate polarized light. Birefringent structures present in cells can rotate polarized light in such a way that it passes through the second filter and hence creates an image. Birefringence is most commonly exhibited by highly ordered structures, such as cytoskeletal filaments built from a repetitive pattern of chemical building blocks.

Interference Microscopy Detects Tiny Differences in Refractive Index

Interference microscopy resembles phase-contrast microscopy in principle, but is more sensitive because it employs special mirrors to split the light beam into two separate rays. One beam passes through the specimen while the second beam, called the *control beam,* is routed through a comparable medium lacking the specimen. When the two beams are recombined, any changes that have occurred in the phase of the specimen beam cause it to interfere with the control beam. Complete separation of the two beams enhances sensitivity to small differences in refractive index. The magnitude of the phase changes can even be quantified and related to the dry weight of cellular components. A special type of interference microscope, known as the **differential** or **Nomarski interference microscope,** enhances contrast by recombining the split beam in polarized light. The remarkable enhancement in resolution obtained in this way produces a striking three-dimensional image of the specimen being examined (see Figure 4-7).

Figure 4-7 The Same Cell Type Viewed by Several Different Kinds of Microscopy *All five photographs are of epithelial cells.* (Top to bottom) *The first four photographs show the same group of living epithelial cells viewed by bright-field light microscopy, phase-contrast light microscopy, Nomarski interference light microscopy, and dark-field light microscopy. The fifth photograph shows epithelial cells that have been fixed and stained prior to examination by bright-field light microscopy. Courtesy of J. Solliday/Biological Photo Service* (top, center, bottom center), *M.I. Walker/Photo Researchers* (top center), *and Ray Simons Photoresearchers* (bottom).

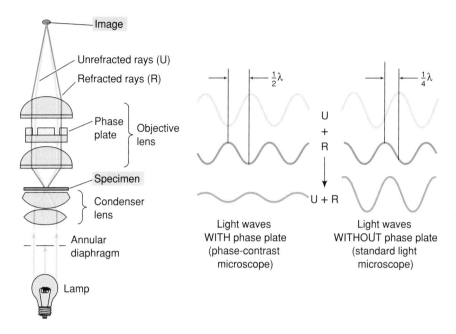

Light waves
WITH phase plate
(phase-contrast
microscope)

Light waves
WITHOUT phase plate
(standard light
microscope)

Figure 4-8 *The Phase-Contrast Microscope* *In this instrument, light rays refracted by the specimen are delayed an extra one-fourth wavelength by the phase plate. The wave diagrams show how this extra delay increases the interference produced when the refracted (R) and unrefracted (U) rays are focused back together (U + R).*

Dark-Field Microscopy Detects Scattered Light Deflected from a Specimen

Yet another way of increasing contrast with living material is the use of **dark-field microscopy,** which employs a special condenser to illuminate the specimen at an angle that prevents direct light from entering the lens. Under these conditions, the only light reaching the observer is that which has been scattered as a result of interacting with the specimen. Hence in dark-field microscopy the background appears dark and cell structures are seen as areas whose brightness is directly proportional to the amount of light scattered (see Figure 4-7).

Fluorescence Microscopy Detects the Location of Specific Molecules

Fluorescence microscopy allows fluorescent molecules to be visualized in either intact living cells or cells that have been treated with fixatives. A *fluorescent* molecule is a substance that absorbs light energy at one wavelength and emits it at a longer wavelength. In fluorescence microscopy, cells are usually stained with a fluorescent molecule that is designed to interact with a specific cellular component (see Figure 13-28). The cells are then are illuminated with light of a wavelength that can be absorbed by the fluorescent molecules. The light emitted by the specimen is passed through a colored filter so that only light of the wavelength emitted by the fluorescent molecules is seen by the observer. In

other words, if a cell has been exposed to a molecule that emits a green fluorescence, then a green filter is placed between the objective lens and the eyepiece so that the observer only sees light of the wavelength emitted by the fluorescent molecules.

The Confocal Scanning Microscope Eliminates Blurring by Focusing the Illuminating Beam on a Single Plane within the Image

When intact cells are viewed, the resolving power of fluorescence microscopy is limited by the fact that fluorescence is emitted throughout the entire depth of the specimen, but the viewer can only focus the objective lens on a single plane at any given time. As a result, light emitted from regions of the specimen above and below the focal plane cause a blurring of the visible image (Figure 4-9, *left*). Because of this limitation, fluorescence microscopy has traditionally been used on cells that are well flattened, such as cultured cells attached to the surface of a culture dish. However, a new instrument called the **confocal scanning microscope** overcomes this problem by employing a laser beam to illuminate a single plane of the specimen at a time. Because it eliminates the blur normally caused by fluorescence emitted from adjacent focal planes, confocal scanning microscopy gives much better resolution than traditional fluorescence microscopy (see Figure 4-9, *right*). Moreover, it is possible to use the laser beam to sequentially illuminate different focal planes, generating a series of images that can be combined to construct a three-dimensional representation of the cell.

The Electron Microscope Can Visualize Objects as Small as 0.2 Nanometers in Diameter

In spite of the many modifications that have been introduced to increase the usefulness of the light microscope, one cannot overcome its inherently limited resolving power. According to Equation 4-1, the best resolution that can be obtained when using visible light to illuminate a specimen is approximately 200 nanometers (0.2 μm). The only way to improve the resolving power beyond this point is to decrease the wavelength of the illumination source. This is the rationale that fostered the development of the **electron microscope,** an instrument which uses electrons instead of visible light to illuminate specimens. Because the effective wavelength of a typical electron beam is 100,000 times shorter than the wavelength of light, an enormous enhancement in resolving power is theoretically possible with an electron microscope.

The discovery that electrons have wavelike properties and can be focused on objects using an electromagnetic field was the major impetus to the construction of such a microscope. By the late 1930s a system of magnetic coils capable of focusing an electron beam had been designed, and it was shown that the effective magnification of images created by illuminating objects with such a beam can be regulated by adjusting the current flowing through the coils. This finding made possible the development of the first electron microscope, an instrument that forms an image with a focused electron beam in much the same way as a light microscope uses a focused light beam (see Figure 4-6).

Although the development of the electron microscope has led to an enormous improvement in resolving power, practical limitations in the design of the magnetic coils used to focus the electron beam have prevented the electron microscope from achieving its full theoretical potential. The main problem is that magnetic coils produce considerable distortion when the angle of illumination (2α) is more than a few tenths of a degree. This tiny angle is orders of magnitude less than that of a good glass lens, giving the electron microscope a numerical aperture that is considerably smaller than that of the light microscope. The practical resolving power for the best electron microscope is therefore only about 0.2 nm, far from the theoretical limit of 0.002 nm. Of course 0.2 nm is still about a thousand times better than the capabilities of a typical light microscope. The development of the electron microscope and its first application to biological materials in the late 1940s therefore ushered in a new era in cell biology, opening our eyes to an exquisite subcellular architecture never before seen.

The Transmission Electron Microscope Forms an Image from Electrons That Pass through the Specimen

The most commonly used type of electron microscope is called the **transmission electron microscope (TEM)** because it forms an image from electrons that are transmitted through the specimen being examined. Since electrons are not visible to the human eye, they are detected in the TEM by allowing the electron beam to strike a fluorescent screen or photographic plate. Some molecules and structures in the specimen scatter electrons, preventing the electrons from passing straight through to the photographic plate. Such areas of the specimen appear as dark regions in the resulting photographic image and are referred to as being *electron-opaque.* Conversely, regions containing molecules with less electron-scattering ability appear as lighter areas in the photographic image. The final image thus depends on differences in the electron-scattering abilities of the various structures that make up the cell.

The electron-scattering ability of a structure depends on the atomic number of the chemical elements it contains. The higher the atomic number, the greater the ability of an element to scatter electrons. The most frequently encountered elements in biological molecules (C, H, O, N, P, S) have relatively low atomic numbers and scatter electrons poorly, resulting in an image with little contrast. In order to improve the visibility of cellular structures, tissues must therefore be stained with substances that are more effective at scattering electrons. Metals of high atomic number, such as osmium, uranium, and lead, are most commonly used for this purpose. These *electron-opaque stains* form complexes with biological molecules and thereby enhance their visibility.

In spite of the enhanced resolution made possible by the electron microscope, this instrument is not without its inherent limitations. An electron beam is too weak to pass an appreciable distance through air, so the internal chamber of an electron microscope must be kept under vacuum. The poor penetrating power of an electron beam also limits the thickness of the specimen to no more than a few hundred nanometers when using a conventional electron microscope. Such restrictions create technical difficulties in preparing biological material for observation and make examination of living cells virtually impossible.

Specimens Are Usually Sectioned and Stained Prior to Examination with the Transmission Electron Microscope

Procedures for preparing tissue samples for transmission electron microscopy were first developed in the early 1950s by Keith Porter, George Palade, Fritiof

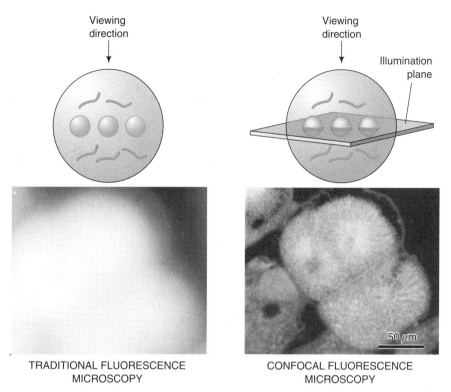

Viewing
direction

Viewing
direction

Illumination
plane

TRADITIONAL FLUORESCENCE
MICROSCOPY

CONFOCAL FLUORESCENCE
MICROSCOPY

50 μm

Figure 4-9 *Comparison of Traditional Fluorescence Microscopy with Confocal Fluorescence Microscopy* (Top left) *In traditional fluorescence microscopy the entire specimen is illuminated, so fluorescent material located above and below the plane of focus tends to blur the image of the three spherical objects.* (Top right) *In confocal fluorescence microscopy, light is focused on a single plane at a time, so only the three spherical objects fluoresce.* (Bottom) *The two bottom photographs show cells of a sea urchin embryo stained with fluorescent antibodies that bind to tubulin. The image is blurred when the specimen is examined by traditional fluorescence microscopy, but with confocal fluorescence microscopy the image is much sharper because the fluorescence is limited to molecules located in the plane of focus. Micrographs courtesy of R. G. Summers and D. W. Piston.*

Sjøstrand, and Humberto Fernández-Morán. In the techniques pioneered by these investigators, tissues are first treated with oxidized metallic compounds such as osmium tetroxide, potassium permanganate, and phosphotungstic acid. These substances function both as fixatives that preserve cell structure, and as electron-opaque stains. Formaldehyde and glutaraldehyde cause less protein denaturation and are therefore employed as fixatives when gentler treatment of the specimen is required. After fixation, tissues are embedded in a hard plastic resin and sliced with an *ultramicrotome* into sections measuring no more than 50–100 nm in thickness. Examination of these **thin sections** with the electron microscope provided biologists with their first detailed views of the cell interior, and led to the discovery of components that were too small to be seen with the light microscope, such as ribosomes and membranes (Figure 4-10).

Viewing thin sections with a transmission electron microscope usually provides a flat, two-dimensional view of the cell interior. However, a limited amount of

three-dimensional information can be obtained by photographing the same specimen at two slightly different angles, and then either viewing the two photographs through a stereoscope or crossing your eyes to view the two photographs as a single image (Figure 4-11). The depth of the three-dimensional effect produced by this technique of **stereo electron microscopy** depends upon the thickness of the tissue section, which usually does not exceed 100 nm because of the limited penetrating power of the electron beam. To examine thicker sections, one must employ a special *high-voltage electron microscope* whose electron beam is roughly ten times more powerful than the electron beams of conventional electron microscopes.

Although cutting tissues into thin sections is the most common method for preparing material for transmission electron microscopy, other techniques have been developed for particular purposes. For example, the shape and surface appearance of small particles or organelles can be examined without the need for cutting them into sections. In the **negative staining** technique, the sample is simply suspended in a solution containing a heavy metal that does not stain the specimen, allowing it to be visualized in relief against the darkly stained background (Figure 4-12, *left*). In the closely related technique of **positive staining,** a specimen is first treated with an electron-opaque stain and the stain not bound to the specimen is then washed away, leaving the stained sample visible against an unstained background (see Figure 4-12, *middle*).

Isolated particles or macromolecules can also be visualized by placing them in an evacuated chamber and spraying heavy metal vapor across their surfaces. This **shadow casting** process causes metal to be deposited on one side of the specimen, creating a "shadow" and a resulting three-dimensional appearance (see Figure 4-12, *right*). A related procedure has been developed by Albrecht Kleinschmidt for visualizing purified macromolecules such as DNA and RNA. In this technique a solution of DNA and/or RNA is spread on an air-water interface, creating a molecular monolayer that is collected on a thin film and visualized by uniformly depositing heavy metal on all sides (see Figure 1-28).

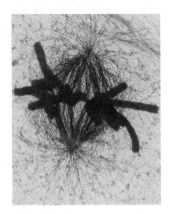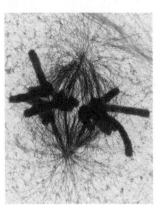

2 μm

Figure 4-11 *Stereo Electron Microscopy* *Two high-voltage electron micrographs of a mammalian cell in mitosis photographed by tilting the specimen several degrees to the left or to the right in the electron beam. To see the three-dimensional effect, look at the two micrographs and allow your eyes to cross until you see the two photos as a single image. Special glasses designed for stereo viewing can also be used to visualize three dimensions. Courtesy of P. Favard, N. Carosso, and R. McIntosh.*

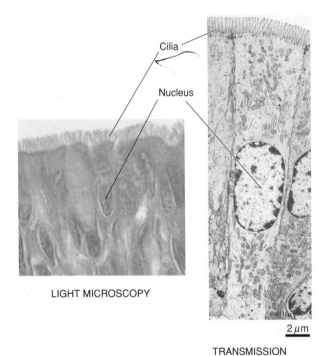

Cilia

Nucleus

LIGHT MICROSCOPY

2 μm

TRANSMISSION ELECTRON MICROSCOPY

Figure 4-10 *A Comparison of the Resolution Obtained by Light Microscopy Versus Transmission Electron Microscopy* *The epithelial cells shown in these two photographs were fixed and sectioned using techniques appropriate for examination by either light or electron microscopy. The enhanced resolution afforded by electron microscopy reveals the presence of cytoplasmic organelles that are not clearly visible with light microscopy. Courtesy of M. Abbey/Photoresearchers (left) and H. Cheng and C. P. Leblond (right).*

Freeze-Fracturing and Freeze-Etching Are Useful Techniques for Examining the Interior Organization of Cell Membranes

Freeze-fracturing is an approach to sample preparation that is fundamentally different from the methods described thus far. Instead of cutting uniform slices through a tissue sample, specimens are rapidly frozen at the temperature of liquid nitrogen or liquid helium, placed in a vacuum, and struck with a sharp knife edge. Samples frozen at such low temperatures are too hard to be cut, so they instead fracture along lines of natural weakness. Most of these weak areas tend to be associated with cellular membranes, which are often split down the middle by the fracturing process. A replica of the fractured specimen is made by shadowing its surface with a heavy metal, such as platinum, and then backing it with a carbon film. After dissolving the tissue in strong acid, the metal replica is viewed with the electron microscope.

Where a fracture plane runs through the interior of a membrane, two fracture faces called the **P face** (protoplasmic face) and **E face** (exoplasmic face) are revealed. The P face is derived from the half of the membrane that lies adjacent to the cytoplasm, and the E face is derived from the opposite side of the membrane (Figure 4-13). Both fracture faces typically appear as smooth surfaces interrupted by randomly dispersed protein particles, although in some cases these particles may be organized into regular arrays (Figure 4-14). The particles observed in membrane fracture faces represent proteins that are integral components of membrane structure.

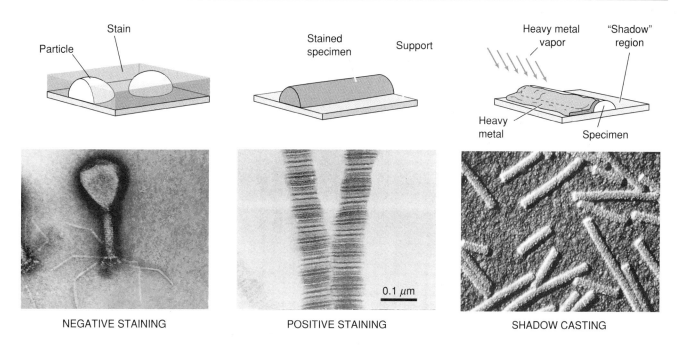

NEGATIVE STAINING   POSITIVE STAINING   SHADOW CASTING

Figure 4-12 *Negative Staining, Positive Staining, and Shadow Casting Techniques for Electron Microscopy* (Left) *In negative staining, the specimen is suspended in an electron-opaque stain that does not stain the material being examined, allowing it to be visualized in relief against a stained background. The electron micrograph shows a negatively stained preparation of a bacteriophage.* (Middle) *In positive staining, the specimen is reacted with an electron-opaque stain and the unreacted stain is then removed. The electron micrograph shows positively stained collagen fibers, which exhibit a repetitive pattern of cross striations.* (Right) *In shadow casting, heavy metal vapor is sprayed at an angle across the specimen, causing an accumulation of metal on one side and a "shadow" region lacking metal on the other side. The electron micrograph shows a shadow-casted preparation of tobacco mosaic virus. Micrographs courtesy of Dr. Michael F. Moody (left), J. Woodhead–Galloway (middle), and Omikron/Photoresearchers (right).*

In a closely related technique called **freeze-etching,** a frozen and fractured tissue sample is briefly exposed to a vacuum prior to forming a replica of the fractured surface. This process results in sublimation of water from the specimen's surfaces, which produces an "etching" effect—that is, an accentuation of surface detail. Brief etching enhances the view of outer membrane surfaces, which are covered with ice and difficult to see in a typical freeze-fracture specimen. Etching that is more prolonged exposes structures that are located deep within the cell interior. This process of **deep-etching** can provide striking views of the three-dimensional arrangement of intracellular components (Figure 4-15).

The Scanning Electron Microscope Reveals the Surface Architecture of Cells and Organelles

All of the electron microscopic techniques described thus far utilize the transmission electron microscope, which forms images from electrons that have been transmitted through a specimen. A second type of instrument, called the **scanning electron microscope (SEM),** produces images from electrons that are de-

flected off a specimen's outer surfaces. The SEM employs a fine beam of electrons that is moved back and forth across the specimen in much the same way as a moving electron beam is used to form a television image (Figure 4-16). As the beam traverses the surface of the object, the electrons are deflected from the sample and secondary electrons are induced and emitted. These deflected and emitted electrons are detected by a photomultiplier tube and used to form a three-dimensional image of the object's surface features (Figure 4-17).

The preparation of samples for scanning electron microscopy usually begins with fixation or freeze-drying to preserve the specimen's shape, though simple air drying is suitable if shape changes can be tolerated. The sample is then shadowed with a heavy metal and viewed directly with the microscope. Since the scanning microscope utilizes deflected electrons rather than electrons transmitted through the sample, the image depicts only surface features. Neither light microscopy nor transmission electron microscopy are particularly well suited for this purpose, so the development of the scanning microscope has been a major help in unraveling the three-dimensional surface architecture of cells and organelles.

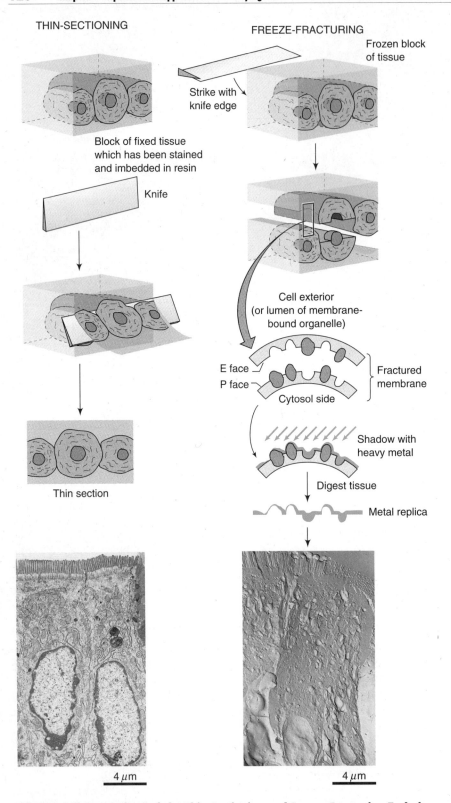

THIN-SECTIONING

Block of fixed tissue
which has been stained
and imbedded in resin

Knife

Thin section

FREEZE-FRACTURING

Frozen block
of tissue

Strike with
knife edge

Cell exterior
(or lumen of membrane-
bound organelle)

E face

P face

Fractured
membrane

Cytosol side

Shadow with
heavy metal

Digest tissue

Metal replica

4 μm

4 μm

Figure 4-13 *Comparison of the Thin-Sectioning and Freeze-Fracturing Techniques*
(Left) *In thin-sectioning procedures, a sharp knife is used to cut thin slices through the specimen.
The individual slices are then stained and examined by transmission electron microscopy.*
(Right) *In freeze-fracture procedures, a frozen sample is fractured by striking it with a sharp
knife edge. The fractured surface is shadowed with a heavy metal, the tissue is dissolved in
strong acid, and the heavy metal replica is viewed with the transmission electron microscope.
The fracture surfaces which are produced by splitting a membrane through the middle are called
the P face and the E face. Micrographs courtesy of R. L. Roberts, R. G. Kessel, and H.-N. Tung.*

Scanning Probe Microscopes Reveal the Surface Features of Individual Molecules

The main shortcoming of the scanning electron microscope is that its resolving power is currently in the range of 2–10 nm, which is considerably less than that of the transmission electron microscope. Because of this limi-

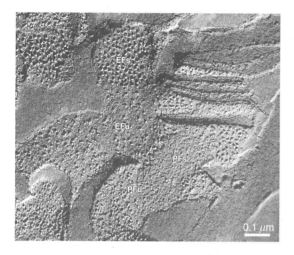

Figure 4-14 *Freeze-Fracture Electron Microscopy*
(Top) *Freeze-fracture electron micrograph of thylakoid membranes in wheat chloroplasts. The labels EF and PF stand for E faces and P faces, respectively. The letters s or u after the EF and PF designations indicate whether the thylakoid membranes are "stacked" upon one another, or "unstacked." The membrane particles represent protein complexes that are involved in the process of photosynthesis. Note that the distribution of membrane particles varies among the different types of fracture faces. (Bottom) Freeze-fracture electron micrograph showing the P face of the plasma membrane of a green alga. The highly ordered array of membrane particles represent enzyme complexes involved in cell wall formation. Courtesy of K. D. Allen (top) and T. H. Giddings, Jr. (bottom).*

tation, scientists have had to devise new ways to examine surface details at the molecular level. In 1981 Gerd Binnig and Heinrich Rohrer invented an instrument called the *scanning tunneling microscope (STM),* which can explore the surface structure of a specimen at the atomic level. The STM consists of a tiny probe that is brought to within 1 nm of the specimen's surface. A voltage is applied to the tip of the probe as it scans across the specimen, causing electrons to leak or "tunnel" across the gap between the probe and the sample. As the probe scans the sample, the tip of the probe is automatically moved up and down to keep the tunneling current constant. A computer measures this movement and uses the information to generate a map of the sample's surface. This approach is so sensitive that it allows the two strands of a DNA double helix to be clearly visualized (Figure 4-18).

In spite of the enormous power of the STM, it suffers from two limitations: the specimen must be an electrical conductor, and it only provides information about the electrons associated with the specimen's surface. Therefore in the past few years, researchers have begun to develop a variety of other **scanning probe microscopes** that scan a sample just like the STM, but measure different kinds of interactions between the tip and the sample surface. For example, in the *atomic force microscope,* the scanning tip is pushed right up against the surface of the sample. As it scans along the surface, it moves up and down as it runs into the microscopic hills and valleys formed by the atoms present at the sample's surface. A variety of other scanning probe microscopes have also been designed to detect such

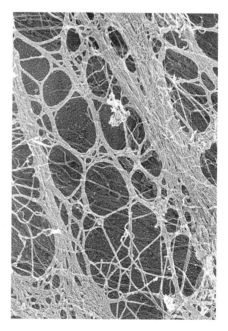

Figure 4-15 *Deep-Etch Electron Microscopy* *This deep-etch electron micrograph of a fibroblast cell shows a network of cytoskeletal filaments within the cytoplasm. Courtesy of D. W. Fawcett, J. Heuser, and Photo Researchers, Inc.*

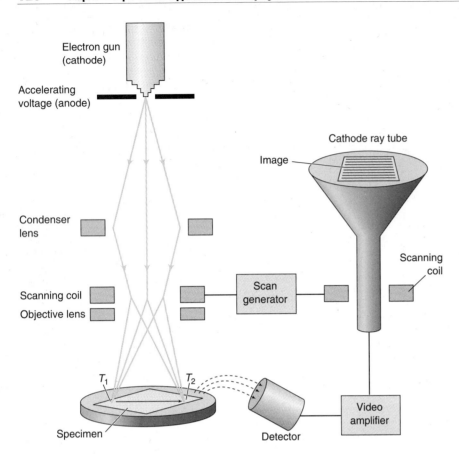

Figure 4-16 *Image Formation in the Scanning Electron Microscope* *The electron beam is moved back and forth across the specimen by a scanning coil, illuminating different points in the sample at different times (T_1, T_2). The scan generator synchronizes the movement of this beam with the beam in a cathode-ray tube (television tube). Electrons scattered or emitted from the specimen are picked up by a detector that modulates the beam in the cathode-ray tube, thereby forming an image of the specimen.*

properties as friction, magnetic force, electrostatic force, van der Waals forces, heat, and sound.

USING MICROSCOPY TO LOCALIZE MOLECULES INSIDE CELLS

So far our discussion of light and electron microscopy has focused mainly on the use of these techniques to view the structural organization of cells. By employing appropriate **cytochemical procedures,** microscopy can also be used to localize specific kinds of chemical substances and metabolic activities within the cell. In the following sections, we will briefly describe some of the more important cytochemical methods used by cell biologists.

Staining Reactions Can Localize Specific Kinds of Molecules within Cells

Staining reactions have been widely used in conjunction with light microscopy to provide information about the subcellular localization of particular kinds of molecules. For example, the large number of acidic phosphate groups in DNA and RNA permits these molecules to be easily detected with basic dyes, some of which can even distinguish between DNA and RNA. *Methyl green,* for example, selectively stains DNA green, whereas *pyronin G* colors both DNA and RNA red. Nucleic acids can also be stained by reagents that bind to nitrogenous bases. A typical example is the dye *acridine orange,* which imparts a yellow-green fluorescence to DNA and a reddish fluorescence to RNA. This example illustrates that the same dye can stain structures different colors, depending on the particular molecule with which it interacts.

One of the first cytochemical reactions for nucleic acids to be devised was the **Feulgen reaction** for staining DNA. In this procedure, tissue sections are first treated with dilute hydrochloric acid to remove purine residues from DNA, liberate free aldehyde groups on the adjacent deoxyribose moieties, and degrade any RNA that may be present. The free aldehyde groups on the

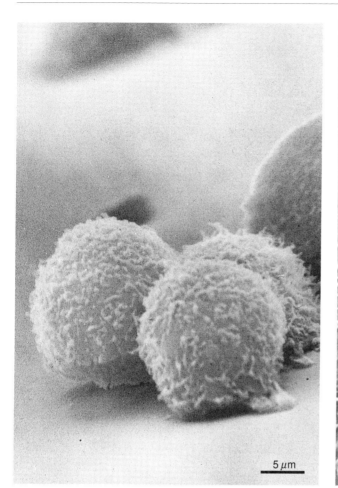
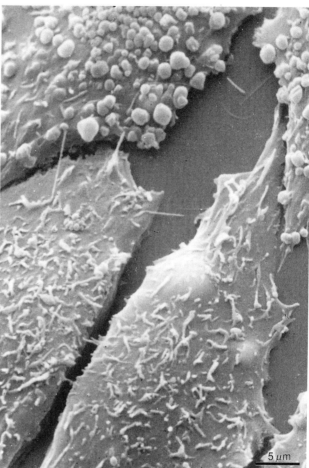

Figure 4-17 *Three-Dimensional View of Cell Shape and Cell Surface Features Made Possible by Scanning Electron Microscopy* (Left) *Scanning electron micrograph of cultured Chinese hamster ovary cells that are just settling down on the surface of the culture vessel. The cells are spherical and contain small projections (microvilli) emerging from the cell surface.* (Right) *After several hours, the cells have become flattened and are covered with small rounded blebs (protrusions) in addition to microvilli. Courtesy of K. R. Porter.*

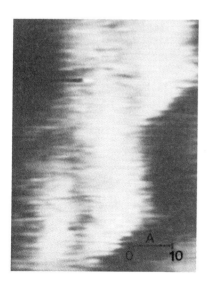

Figure 4-18 *A DNA Double Helix Visualized by Scanning Tunneling Microscopy* *Courtesy of P. Arscott.*

depurinated DNA are then reacted with the *Schiff reagent,* yielding a vivid red color. Cells treated in this way exhibit intense nuclear staining because the nucleus contains the bulk of the cell's DNA (Figure 4-19). The Feulgen reaction is so selective that the staining intensity can be quantified with a light meter attached to the microscope, and the resulting data can be employed to determine the amount of DNA present in the nucleus.

In addition to these specific reactions for nucleic acids, other dyes can be used to reveal the presence of proteins, polysaccharides, and lipids. Stains for protein molecules usually bind to amino acid side chains, especially the sulfhydryl group of cysteine residues. Polysaccharides are stained using the *periodic acid–Schiff* technique, which employs periodic acid to oxidize the hydroxyl groups present in sugar molecules to aldehyde groups. The aldehyde groups are then treated with the Schiff reagent to produce a red color. Cytochemical identification of lipids is based on the use of hydrophobic dyes, such as *Sudan red* and *Sudan black,* whose solubility in lipid causes them to accumulate within intracellular fat droplets.

In contrast to the specificity of the stains employed in light microscopy, the heavy metal stains normally employed in electron microscopy are not very selective. However, salts of uranium, iridium, and bismuth exhibit a preference for interacting with nucleic acids, whereas phosphotungstic acid and osmium are more effective for staining proteins. Potassium permanganate, which acts as a fixative as well as a general stain, provides special contrast for cellular membranes, and osmium reacts efficiently with lipids.

Antibodies Are Powerful Tools for Localizing Proteins and Other Antigens inside Cells

The ability to stain particular substances selectively can be enormously enhanced by exploiting the specificity of **antibody** molecules, which are naturally occurring proteins whose differing three-dimensional shapes allow each antibody to specifically recognize and bind to a particular type of molecule. Antibodies are normally made by vertebrates in response to infection by an invading organism, but they can be produced in the laboratory by injecting an appropriate foreign substance into an experimental animal such as a rabbit or a mouse. Millions of different antibodies can be produced, each capable of binding to the specific foreign substance or **antigen** that triggered its formation. The antibodies utilized in biological research are most commonly directed against protein molecules, but they can be produced against other kinds of antigens as well. In recent years a highly selective approach, called the *monoclonal antibody technique,* has been devised for obtaining large quantities of identical antibody molecules all directed against the same antigen. This procedure, which has greatly facilitated the utilization of antibodies in biological and medical research, is described in detail later in the chapter (page 142).

Because antibodies are not easily seen with either the light or electron microscope, they must be linked to some visible substance before they can be employed in conjunction with microscopy. When light microscopy is used, antibodies are usually linked to a fluorescent dye such as *fluorescein,* which emits a green fluorescence, or *rhodamine,* which emits a red fluorescence. An antibody that has been linked to one of these fluorescent dyes is simply applied to a cell or tissue specimen, and its location is then determined by fluorescence microscopy. For example, if cells are stained with a fluorescent antibody directed against the protein tubulin, only the microtubules, which contain tubulin, become fluorescent (Figure 4-20).

Antibodies are also employed to localize substances with the electron microscope, although the

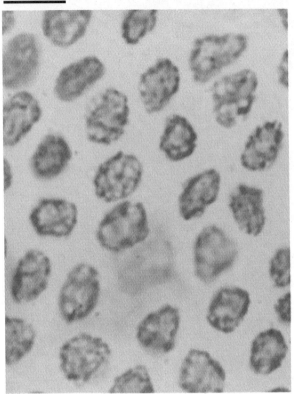

20 μm

Figure 4-19 *The Feulgen Reaction for DNA* *In this light micrograph of rat uterine epithelial cells stained by the Feulgen method, the nuclei stain bright red because of their high DNA content. Courtesy of L. Lavia.*

antibodies must be visualized with substances that are electron-opaque rather than fluorescent. One common procedure utilizes the enzyme *horseradish peroxidase,* which is linked directly to the antibody molecule. After the antibody has become bound to the cell, diaminobenzidine is added; peroxidase catalyzes the conversion of this substance to a brown, electron-opaque product that can be seen by both electron and light microscopy (Figure 4-21). Antibodies can also be made visible for electron microscopy by linking them directly to an electron-opaque substance such as the iron-containing protein *ferritin,* or particles of *gold* (Figure 4-22).

As an alternative to using a single-step antibody procedure for localizing substances within cells, a two-step indirect approach can be employed to increase the sensitivity of the staining process. For example, when localizing fluorescent antibodies by light microscopy, the specimen can be reacted first with a nonfluorescent *primary antibody* directed against the substance one wishes to localize. After removing the excess unbound primary antibody, the sample is then reacted with a fluorescent *secondary antibody* directed against the primary antibody. In other words, if the primary antibody had been made in rabbits, the secondary antibody might be a goat antibody directed against rabbit antibody molecules. This approach yields a more intense staining reaction because each molecule of primary antibody is bound by several molecules of the fluorescent secondary antibody.

A modification of the preceding approach has recently been introduced that takes advantage of the strong binding that occurs between the vitamin *biotin* and a biotin-binding protein called *avidin.* The primary antibody is chemically linked to biotin, and the location of the primary antibody is then determined by staining the specimen with avidin, which in turn is reacted with

biotin attached to a visible marker. In this way, each molecule of primary antibody is visualized by virtue of its attachment to an extensive network of biotin and avidin molecules.

Cytochemical Procedures Can Localize the Activity of Specific Enzymes within a Cell

One limitation of using antibodies to localize molecules within the cell is that the molecule in question must first be isolated and injected into animals to trigger the formation of antibodies. As an alternative, many enzymes can be localized within cells by techniques that assay for the enzyme's metabolic activity without the need for removing and purifying the enzyme. The basic principle involved in this approach is quite simple. A tissue section is first incubated with a substrate of the enzyme one wishes to localize. The reaction product is then detected by converting it to a substance that is either colored or electron-opaque. Because harsh fixatives

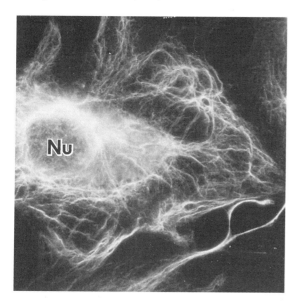

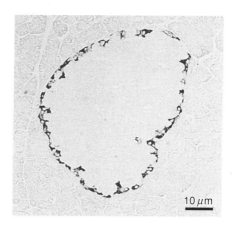

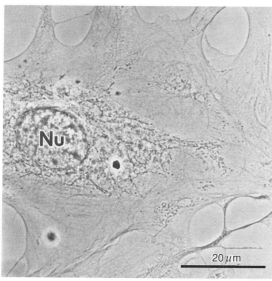

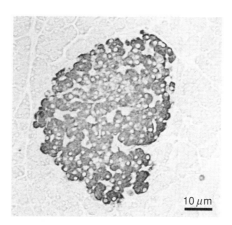

Figure 4-20 *The Use of Fluorescent Antibodies in Light Microscopy* (Top) *Fluorescence microscopy of a cell stained with a fluorescent antibody directed against the microtubule protein, tubulin. (Bottom) Phase contrast micrograph of the same cell. The nucleus is labeled Nu in each case. Courtesy of S. H. Blose.*

Figure 4-21 *The Use of Peroxidase-Labeled Antibodies in Light Microscopy* *Tissue obtained from the islets of Langerhans in the pancreas was incubated with peroxidase-labeled antibodies specific for the hormone insulin* (top) *or glucagon* (bottom). *Use of this technique shows that the two hormones are localized in different cell types. Courtesy of S. L. Erlandsen.*

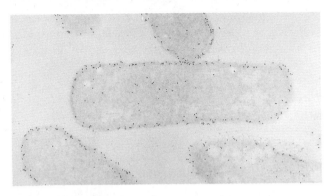

Figure 4-22 *The Use of Gold-Labeled Antibodies in Electron Microscopy* *Cells of the bacterium E. coli were stained with gold-labeled antibodies directed against a plasma membrane protein. The small dark granules distributed around the periphery of the cell are the gold-labeled antibody molecules. Courtesy of J. R. Maddock.*

inactivate enzymes, sections must be cut either from unfixed frozen tissue or from tissue treated with mild fixatives, such as cold formaldehyde or glutaraldehyde.

The preceding approach is nicely illustrated by the *Gomori technique* for localizing acid phosphatase, an enzyme that hydrolyzes phosphate ester bonds present in a wide variety of substrates. In this procedure, tissue sections are incubated at low pH in a medium containing a phosphate ester, such as β-glycerophosphate, and lead nitrate. Acid phosphatase catalyzes hydrolysis of the β-glycerophosphate, releasing inorganic phosphate. The released inorganic phosphate in turn reacts with lead nitrate to form an insoluble precipitate of lead phosphate. This precipitate, which marks the site where acid phosphatase is located within the cell, is electron-opaque and therefore visible in the electron microscope (Figure

4-23). In adapting this technique for light microscopy, the lead phosphate is treated with ammonium sulfide to produce lead sulfide, a readily visible brown pigment.

Similar principles have been used in the microscopic localization of many other enzymes, including esterases, oxidoreductases, glycosidases, and proteases. The information obtained in this way has added much to our understanding of where biochemical activities occur within cells.

Microscopic Autoradiography Is Employed to Locate Radioactive Molecules within a Cell

An alternative approach for localizing molecules and enzymatic activities inside cells utilizes radioactive substances and is called **microscopic autoradiography**. In this procedure, radioactive compounds are incubated with tissue sections or administered to intact cells or organisms. After sufficient time has elapsed for the radioactive substance to become incorporated into newly forming intracellular molecules and structures, the remaining unincorporated radioactivity is washed away and the specimen is sectioned and coated with a layer of photographic emulsion. Radiation given off by radioactive molecules located within the cell interacts with silver bromide crystals present in the photographic emulsion; when the emulsion is later developed, *silver grains* appear wherever radiation had bombarded the emulsion. The location of these silver grains, which are readily visible with both the light and electron microscope, can be used to pinpoint the regions of the cell containing the radioactivity (Figure 4-24).

In practice, substances must emit relatively weak forms of radiation to be useful for microscopic autoradi-

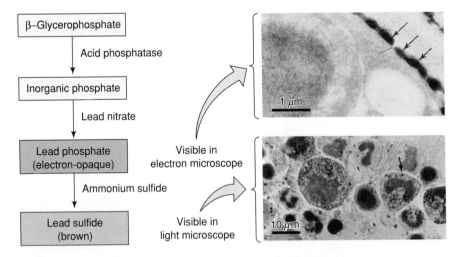

Figure 4-23 *The Gomori Procedure for Localizing Acid Phosphatase* *In the top electron micrograph, the dark material (arrows) represents lead phosphate produced by the action of acid phosphatase present in the cytoplasm. In the bottom light micrograph, the dark granules (arrow) represent sites where acid phosphatase has been converted to lead sulfide. Micrographs courtesy of J.-C. Roland (top) and D. F. Bainton (bottom).*

ography because stronger radiation penetrates the emulsion too far to permit accurate localization. For this reason the most widely used radioisotope in autoradiography is tritium (^{3}H), an atom whose low-energy radiation permits a resolution of about 1 μm with the light microscope and close to 0.1 μm with the electron microscope. Since hydrogen is ubiquitous in biological molecules, a wide range of ^{3}H-labeled compounds are potentially available for use in autoradiography. For example, ^{3}H-amino acids are employed for locating newly synthesized proteins, ^{3}H-thymidine is used for monitoring DNA synthesis, ^{3}H-uridine is employed for localizing newly made RNA molecules, and ^{3}H-glucose is used to study the synthesis of polysaccharides.

CENTRIFUGATION AND SUBCELLULAR FRACTIONATION

Although microscopy in its various forms has provided biologists with a detailed picture of cell organization, the amount of functional information that can be obtained in this way is quite limited. By far the most powerful tool for investigating the functional organization of cells is the isolation of individual organelles and structural components by **subcellular fractionation.** Once a cell's component parts have been isolated, they can be subjected to experimental manipulations and biochemical analyses that would be inconceivable with the microscope. Subcellular fractionation is one of the most widely used techniques in cell biology and is responsible for much of our current understanding of how cells carry out their various functions. Because almost all subcellular fractionation procedures involve the isolation of individual cellular components by centrifugation, we will begin this section with a discussion of the principles that govern the behavior of particles in a centrifugal field.

Differences in Size and Density Are the Main Factors That Govern the Behavior of Particles During Centrifugation

The centrifuge is the most important tool available to cell biologists for the separation and analysis of subcellular organelles and their constituent macromolecules. Centrifugal force is generated by placing test tubes containing samples to be fractionated in a holder called a *rotor,* and then spinning the rotor at high speeds in a *centrifuge.* The **ultracentrifuge,** developed by the Swedish chemist Theodor Svedberg between 1920 and 1940, is a special high-speed instrument capable of spinning rotors at speeds of 100,000 revolutions per minute or more. The large forces generated by these high speeds make it possible to isolate not just organelles, but macromolecules such as proteins and nucleic acids as well. The ultracentrifuge is a central instrument in

every cell biology laboratory, without which we would lack much of our current understanding of the functions carried out by cellular organelles.

Several different kinds of centrifugation can be used for isolating subcellular components. In order to understand how they work, it is first necessary to know something about the principles that govern the movement of materials in a centrifugal field. The rate of movement of any particle subjected to a centrifugal force is given by the *Stokes formula:*

$$\frac{dx}{dt} = \frac{2r^2(\rho_P - \rho_M)}{9\eta} \cdot g \qquad (4\text{-}2)$$

where *dx/dt is* the velocity with which the particle moves toward the bottom of the tube, *r* is the particle radius, ρ_P and ρ_M are the respective densities of the particle and the suspending medium, η is the viscosity of the suspending medium, and *g* is the centrifugal force exerted on the particle. This formula tells us that when the density of a particle exceeds that of the medium, its velocity will be positive and the particle will migrate in the direction of the force field (toward the bottom of the tube). If the particle is less dense than the medium, its velocity will be negative, meaning that it will move toward the top of the tube. Finally, if the density of the particle is exactly equal to the density of the suspending medium, the particle will not move because the term $\rho_P - \rho_M = 0$.

Of the various terms included in Equation 4-2, two are relatively easy to measure: (1) The particle velocity *(dx/dt)* can be determined by measuring the distance a particle has migrated in the centrifuge tube as a function of time. Special analytical ultracentrifuges have even been designed to make such measurements automatically. (2) The centrifugal force (*g*) can be calculated from the following formula:

$$g = \frac{4\pi^2 \, (\text{rpm})^2}{3600} \cdot x \qquad (4\text{-}3)$$

where rpm is the rotor speed in revolutions per minute and *x* is the distance between the central axis of the centrifuge and the point at which the centrifugal force is being measured in the rotor.

The other four variables in Equation 4-2 (*r*, ρ_P, ρ_M, and η) are not as easy to measure experimentally. Instead they are grouped together to yield a **sedimentation coefficient** (*s*), which is defined as follows:

$$s = \frac{2r^2(\rho_P - \rho_M)}{9\eta} \cdot x \qquad (4\text{-}4)$$

This expression for the sedimentation coefficient *s* can be substituted in Equation 4-2 to yield

$$\frac{dx}{dt} = sg$$

which rearranges to

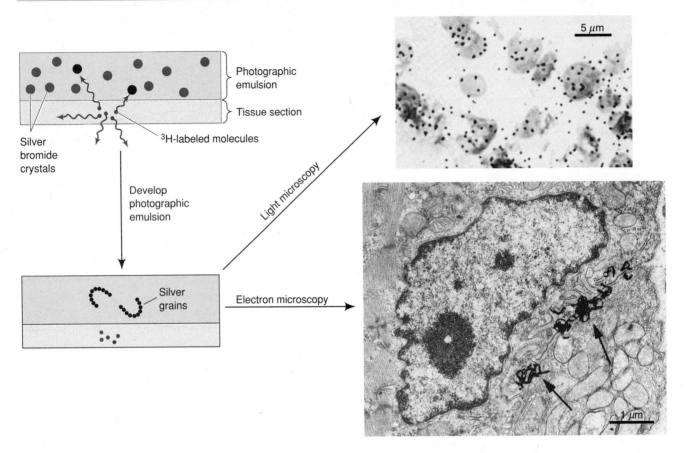

Figure 4-24 *The Technique of Microscopic Autoradiography* *In cells containing radioactive molecules, the location of the radioactivity can be determined using a photographic emulsion. The location of the radioactivity within the cell is marked by the formation of silver grains that are visible by both light and electron microscopy. (Top right) Light microscopic autoradiograph of toad bladder cells exposed to radioactive aldosterone, a steroid hormone that acts on the nucleus of these cells. Note that the small dark silver grains are localized predominantly over the nuclei. (Bottom right) Electron microscopic autoradiograph of the junction between a nerve ending and a muscle cell. The tissue was incubated with a radioactive toxin that binds to the receptor for the neurotransmitter, acetylcholine. The location of the receptor is indicated by the presence of silver grains (arrows). Micrographs courtesy of I. S. Edelman (top) and M. M. Salpeter (bottom).*

$$s = \frac{dx/dt}{g} \qquad (4\text{-}5)$$

Because the values of dx/dt and g are easy to measure experimentally, this equation can be used to calculate the value of s for any particle being centrifuged.

What is the usefulness of knowing the value of s for a given molecule or particle? Equation 4-4 showed that the value of s is determined by four factors: r, ρ_P, ρ_M, and η. However, when comparing the value of s for different particles, the last two factors usually remain the same. In other words, the sedimentation rates of different particles are compared in a medium of the same density (ρ_M) and viscosity (η). If these two factors are held constant, any differences observed in the value of s for different particles must be due to differences in their size (r) and density (ρ_P). For most biological particles and molecules, differences in size (r) are much larger than differences in density (ρ_P). Therefore the impor-

tance of knowing the value of s for different materials is that it provides a rough estimate of the value of r, which is a measure of a particle's size and shape.

For most biological particles and molecules, the value of s lies in the range of 1×10^{-13} to 200×10^{-13} sec. In order to avoid repetition of the term "10^{-13} sec," a sedimentation coefficient of 10^{-13} sec has been defined as a **Svedberg unit (S)**. Thus a particle with a sedimentation coefficient of 18×10^{-13} sec is said to have a sedimentation coefficient of 18S. The convenience of this nomenclature is that the sedimentation coefficients of biological molecules can be expressed as simple integers. Because the sedimentation coefficient is an estimate of a particle's size and shape, it is not surprising to find that the S values obtained for various particles and molecules correlate roughly with their molecular weights (Table 4-1). The relationship between the two is not directly proportional, however, which means that S values cannot be added to one another. For example, mammalian ribosomes are comprised of a 40S subunit joined to a larger 60S subunit, but together the two subunits generate a ribosome that is 80S in size, not 100S.

If one wants to determine the molecular weight of a molecule or particle from its sedimentation coefficient, it is necessary to know the diffusion constant (*D*) as well. Molecular weight can then be calculated using the *Svedberg equation:*

$$M = \frac{RTs}{D(1 - \bar{v}\rho_M)} \qquad (4\text{-}6)$$

where *M* is the molecular weight, *R* the gas constant, *T* the absolute temperature, *s* the sedimentation coefficient, *D* the diffusion constant, $\bar{v}$ the partial specific volume of the particle (the volume of fluid displaced per gram of particle), and ρ_M the density of the medium.

The main use of centrifugation in cell biology is not, however, for calculating molecular weights, but for isolating organelles and macromolecules. In the following sections, we will describe the types of centrifugation that are employed for such purposes.

Velocity Centrifugation Separates Organelles and Molecules Based Mainly on Differences in Size

The most common type of centrifugation used for isolating cellular components is **velocity centrifugation.** During velocity centrifugation, the density of the components being isolated is greater than the density of the medium in which they are being centrifuged. Under such conditions, materials move toward the bottom of the tube. The rate at which different cellular components move varies with their size, with larger particles sedimenting more rapidly than smaller particles (see Equation 4-2).

Velocity centrifugation exploits this difference in migration velocity in one of two different ways. One approach, called **differential centrifugation,** utilizes a successive series of centrifugations at increasing speeds. The initial centrifugation step is fast enough to

sediment the largest particles into a pellet at the bottom of the tube, leaving the smaller particles suspended in the remaining solution, or **supernatant.** The supernatant is removed and centrifuged again at higher speeds to sediment the smaller particles (Figure 4-25). As will be seen in a moment, this approach is widely employed in subcellular fractionation experiments, but its usefulness is restricted to the separation of particles that exhibit relatively large differences in size.

When separating particles exhibiting smaller differences in size, a second type of velocity centrifugation is employed. In this technique, termed **moving-zone centrifugation,** the sample is applied as a narrow layer at the top of the solution contained within a centrifuge tube. Centrifugation is then carried out long enough for the particles to move part way down the tube. Because the migration velocity of different sized particles will vary, the particles separate from one another as they move down the tube. Centrifugation is stopped *before the particles reach the bottom of the tube* and the various components are recovered by puncturing the tube and collecting sequential fractions (Figure 4-26). To minimize diffusion of the particles as they move toward the bottom of the tube, the solution in the tube is modified by including a shallow gradient of sucrose or glycerol which is denser at the bottom of the tube than at the top. But it is important to emphasize that even when such gradients are employed, the density of the particles is greater than the density of the separating medium everywhere in the tube.

Isodensity Centrifugation Separates Organelles and Molecules Based on Differences in Density

In contrast to velocity centrifugation, which separates components based primarily on differences in size, **isodensity centrifugation** separates on the basis of *differences in density.* We already noted in our

Table 4-1   Relationship between Molecular Weight and Sedimentation Coefficient for Selected Molecules and Particles

	Molecular Weight (daltons)	Sedimentation Coefficient
Myoglobin	16,900	2.0 S
Lysozyme	17,200	2.2 S
Transfer RNA	25,000	4.0 S
Hemoglobin	64,500	4.5 S
Catalase	248,000	11.3 S
Urease	483,000	19 S
Small ribosomal RNA (bacterial)	550,000	16 S
Ribosome (mammalian)	4,500,000	80 S
Tobacco mosaic virus	40,600,000	198 S

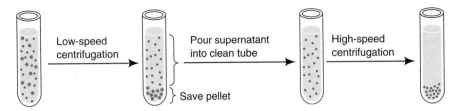

Figure 4-25 *The Principle of Differential Centrifugation* *The sample initially contains a mixture of small and large particles. After an initial low-speed centrifugation to sediment the larger particles, the supernatant is removed and centrifuged at higher speed to sediment the smaller particles.*

previous examination of Equation 4-2 that no movement occurs when the density of a particle (ρ_P) equals the density of the medium (ρ_M). Isodensity centrifugation exploits this fact by creating a gradient whose *density overlaps the density of the components being separated.* In such a gradient, particles migrate until they reach the region in which their density is equal to that of the suspending medium. At this point, they reach equilibrium and can no longer move in either direction. After equilibrium has been achieved, centrifugation is stopped and fractions are collected (Figure 4-27, *top*). Because a particle's density is the only factor that influences its final equilibrium position in such a gradient, this method separates components solely on the basis of differences in density. In practice, it is possible to separate components whose densities differ by less than 1 percent.

The gradients used in isodensity centrifugation can be generated from several different kinds of materials. Sucrose or glycerol, which have a maximum density of about 1.3 g/cm³, are usually employed for isolating membrane-bound organelles such as the Golgi complex, endoplasmic reticulum, lysosomes, and mitochondria. It is important to understand that the function of such sucrose or glycerol gradients differs in principle from the moving-zone gradients mentioned in the preceding section. In moving-zone centrifugation, the only purpose of the gradient is to minimize diffusion; even at the bottom of the tube, the density of the particles is greater than that of the medium and the particles continue to move. In contrast, isodensity centrifugation employs gradients that are dense enough to cause particles to stop moving when they reach the region of the gradient where the density of the medium equals the density of the particle.

Fractionation of materials whose densities are greater than 1.3 g/cm³, such as DNA and RNA, requires gradients formed from substances that are denser than sucrose or glycerol. Heavy metal salts like *cesium chloride* and *cesium sulfate* are typically used for this purpose. The density of such salts is so great that gradients can be formed by the process of centrifugation itself. A solution of a heavy metal salt like cesium chloride is simply centrifuged at high gravitational forces until the centrifugal field causes the ce-

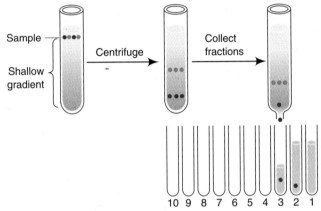

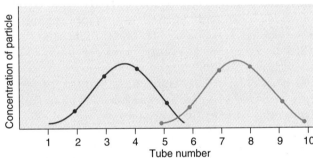

Figure 4-26 *The Principle of Moving-Zone Centrifugation* *A sample containing two types of particles of slightly different size is layered on the surface of a shallow density gradient of sucrose or glycerol. After centrifuging for an appropriate period of time, the process is stopped and the bottom of the tube is punctured to permit collection of sequential fractions. Each fraction is then assayed for the presence of the two particles.*

sium salt to become more concentrated at the bottom of the tube and less concentrated toward the top. Since such gradients form spontaneously during centrifugation, the material to be fractionated can be mixed directly with the cesium chloride solution and the entire mixture centrifuged. As the density gradient begins to form during centrifugation, particles or molecules dispersed throughout the tube either sediment toward the bottom or move upward until they reach the point at which the density of the medium equals

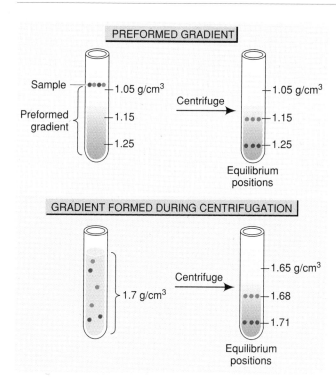

PREFORMED GRADIENT

Sample — 1.05 g/cm³ ···· → 1.05 g/cm³

Preformed gradient — 1.15 Centrifuge → ···· 1.15

— 1.25 → ···· 1.25

Equilibrium positions

GRADIENT FORMED DURING CENTRIFUGATION

1.7 g/cm³ Centrifuge → 1.65 g/cm³

···· 1.68

1.71

Equilibrium positions

Figure 4-27 The Principle of Isodensity Centrifugation
(Top) *Density gradients made of sucrose or glycerol are formed in a centrifuge tube and the sample to be separated is then layered on top. During centrifugation, sample components move downward until they reach the region of the gradient corresponding to their own density, at which point they stop. In this example, the two components being separated have densities of 1.15 g/cm³ and 1.25 g/cm³ respectively.* (Bottom) *Gradients made of heavy metal salts such as cesium chloride are generated directly by centrifugal force. The components being separated move either upward or downward until each reaches the region corresponding to its own density. In this example, the two components being separated have densities of 1.68 g/cm³ and 1.71 g/cm³ respectively.*

their own density (see Figure 4-27, *bottom*). Isodensity centrifugation in cesium salts is commonly used to separate DNA molecules that differ in base composition; the G-C base pair is more compact than the A-T base pair, increasing the density of DNA molecules with higher G-C contents.

Isolating Organelles Requires That Cells First Be Broken Open in an Appropriate Medium

We have now seen that centrifugation can be used in two different ways to isolate cellular components: Organelles or molecules that differ in size are separated from each other by velocity centrifugation (either differential or moving zone), whereas components that differ in density are separated by isodensity centrifugation. Because virtually all of a cell's major constituents differ from one another in either size and/or density (Figure 4-28),

these two approaches make centrifugation an extremely powerful tool for separating, isolating, and purifying cell organelles.

Although the first attempts at isolating organelles were reported more than a century ago, only recently have comprehensive schemes been developed for fractionating cells into their constituent organelles. The invention of the ultracentrifuge played a key role in these developments, but it is not the only important factor. Cells must first be broken open to release their organelles into solution, and the medium used for this purpose is of critical importance. Pioneering studies carried out by Albert Claude in the early 1940s revealed that disrupting cells in buffered salt solutions permits the subsequent isolation of several discrete organelle fractions by centrifugation. But salt solutions cause considerable aggregation of cell components, with an accompanying loss in their structural integrity. In 1948 George Hogeboom, Walter Schneider, and George Palade reported that sucrose is a better medium for cell breakage and fractionation because it produces minimal aggregation and structural alterations. This technically simple change revolutionized modern cell biology. Virtually every important subcellular fractionation experiment performed since that time has employed a sucrose-containing medium for the isolation of subcellular organelles.

The use of sucrose is not entirely without its drawbacks, however. Many biological substances are water soluble and therefore tend to leak out of organelles during isolation. To counteract this problem nonpolar solvents, like carbon tetrachloride and benzene, have also been employed in cell fractionation studies. Balancing the advantage of this approach in minimizing the loss of water-soluble molecules is the disadvantage that nonpolar solvents extract lipids and denature many proteins. Although nonpolar solvents are therefore useful for isolating organelles that retain all their water-soluble constituents, the inherent disadvantages of this approach severely limit its general usefulness.

Another factor critical to the success of subcellular fractionation studies is choosing a reliable means for breaking cells open. Ideally the plasma membrane (and cell wall in plants) should be disrupted under conditions that cause minimal damage to the rest of the cell, releasing organelles into solution with their structural features and functional integrity intact. Unfortunately this goal is difficult to achieve in practice, so the method chosen for breaking cells tends to involve some compromises based on optimizing conditions for the particular organelle in which one is most interested. Organelles such as mitochondria, chloroplasts, and nuclei are relatively easy to release intact, whereas membrane systems like the endoplasmic reticulum, Golgi complex, and plasma membrane tend to break into smaller fragments.

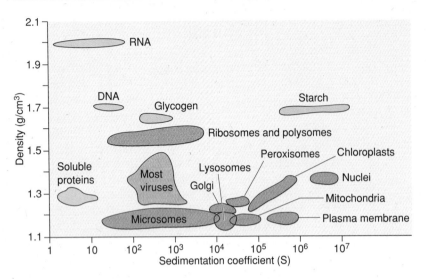

Figure 4-28 *Sedimentation Coefficients and Densities of Various Organelles,* ***Macromolecules, and Viruses*** *Because each component exhibits a relatively unique combination of size and density, velocity and isodensity centrifugation can be employed to separate most of these materials from one another.*

The most commonly used procedures for disrupting cells are based on homogenizing a cell or tissue suspension in a tube containing a tightly fitted glass or Teflon pestle. Rotating the pestle and/or moving it up and down shears cells and releases their organelles into the suspending medium, producing a solution called a **homogenate.** Other procedures for producing homogenates include blending tissue in a high-speed mixer, freezing and thawing, grinding with a mortar and pestle, forcing cells through a narrow orifice at high velocity, subjecting cells to ultrasonic vibrations, and osmotic rupturing. The choice concerning the most appropriate technique to use is generally based on the specific objectives of the experiment being performed.

Homogenates Are Separated by Differential Centrifugation into Nuclear, Mitochondrial, Microsomal, and Cytosol Fractions

Once cells have been disrupted, differential centrifugation can be employed to separate the resulting homogenate into fractions that contain different organelles (Figure 4-29). The first step involves centrifugation at 500–1000g (500 to 1000 times the force of gravity), which sediments the cell's largest components. This fraction, referred to as the **nuclear fraction,** is enriched in nuclei but also contains unbroken cells, tissue debris, and perhaps some chloroplasts in the case of plant tissue. The supernatant from the first centrifugation step is removed and centrifuged at about 10,000g, which sediments organelles of intermediate size such as mitochondria, lysosomes, and peroxisomes. This material is called the **mitochondrial fraction,** even though other organelles are present as well. Finally, the remaining

supernatant is centrifuged at high speed (~100,000g) to sediment the smallest cellular components. This step produces the **microsomal fraction,** which consists predominantly of fragments of smooth and rough endoplasmic reticulum as well as free ribosomes. The final supernatant contains the soluble molecules of the cytoplasm and is therefore called the **cytosol fraction.**

None of the fractions generated by the preceding scheme are completely pure. Therefore the fractions obtained by differential centrifugation are usually subjected to other kinds of centrifugation for further purification. For example, isodensity centrifugation is used for purifying mitochondria, lysosomes, and peroxisomes because these organelles differ from one another in density. On the other hand, moving-zone centrifugation is more useful when purifying ribosomes, and differential centrifugation in dense sucrose (~2.0 *M*) is employed to purify nuclei.

The purity of organelle preparations generated by the preceding techniques is often assessed by examining them with the electron microscope. Analyzing for specific chemical components is also useful in assessing purity because certain macromolecules and enzymes are preferentially localized in particular organelles (Table 4-2). Figure 4-30 illustrates how testing for the presence of such chemical markers can be employed to confirm the identity of organelle fractions generated when cells are homogenized and fractionated by differential centrifugation.

In spite of the enormous power of the subcellular fractionation approach, the possibility of artifacts should always be kept in mind. For example, the process of homogenization may cause water-soluble molecules to leak out of organelles and into the cytosol. Conversely, substances that are normally located in the cytosol may stick to cytoplasmic organelles under the

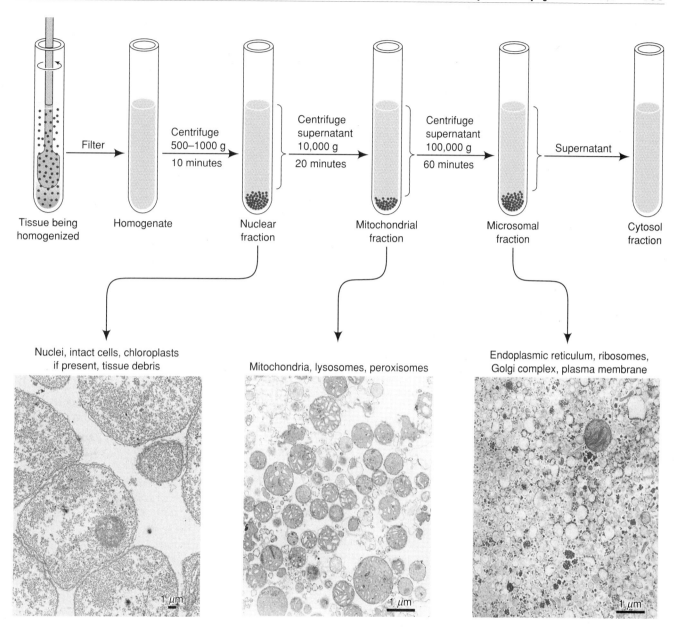

Figure 4-29 *Subcellular Fractionation by Differential Centrifugation* *The four fractions typically produced by differential centrifugation are called the nuclear, mitochondrial, microsomal, and cytosol fractions. In spite of these names, none of the fractions obtained by differential centrifugation consists of a single completely purified component. For this reason, fractions obtained by differential centrifugation are often subjected to other kinds of centrifugation for further purification. Micrographs courtesy of D. J. Morré (left) and P. Baudhuin (middle, right).*

artificial conditions employed for subcellular fractionation. Aggregation of organelles may also influence the fraction in which they appear. Membranes of the endoplasmic reticulum, for example, typically sediment as part of the microsomal fraction. But if the pH is too low, these membranes may stick together and form large masses that sediment as part of the nuclear fraction.

In spite of these problems, subcellular fractionation is an extraordinarily powerful tool that has provided us with a wealth of information about how cells are organized and how they carry out their various functions. As each of the cell's organelles is discussed in detail in later

chapters, it will become increasingly apparent that much of our current understanding of cell organization and function depends in one way or another on data obtained from subcellular fractionation experiments.

TECHNIQUES FOR STUDYING MACROMOLECULES

In the course of studying the properties of individual organelles, it is often necessary to isolate and characterize the molecules from which they are constructed. Much

Table 4-2   Chemical Markers Used in the Identification of Organelle Preparations Derived from Subcellular Fractionation

Organelle	Marker
Nuclei	DNA, histones
Mitochondria	Succinate dehydrogenase, cytochrome oxidase, monoamine oxidase
Chloroplasts	Ribulose-bisphosphate carboxylase, photosystems I and II
Lysosomes	Acid phosphatase, aryl sulphatase, acid deoxyribonuclease
Peroxisomes	Catalase, urate oxidase, D-amino acid oxidase
Endoplasmic reticulum	Glucose 6-phosphatase (liver), cytochrome b_5, cytochrome P-450
Ribosomes	RNA
Golgi complex	Glycosyl transferases, thiamine pyrophosphatase, nucleoside diphosphatase,
Plasma membrane	Adenylyl cyclase, Na^+-K^+ ATPase
Cytosol	Enzymes of glycolysis (e.g., phosphoglucomutase)

Spectrophotometry Can Be Used to Detect Proteins and Nucleic Acids

Most biological compounds either absorb light in the ultraviolet region of the electromagnetic spectrum or react with certain chemical reagents to form colored compounds that absorb visible light. The exact amount of light absorbed by a solution can be quantified in an instrument called a *spectrophotometer* and the resulting **absorbance** value used to calculate the concentration of the absorbing material. Altering the wavelength of the illuminating light beam allows the presence of compounds with differing light-absorbing properties to be detected. One can also measure absorption at various wavelengths, generating an **absorption spectrum** that can be used to identify the type of molecule present.

Both proteins and nucleic acids absorb ultraviolet light. Although the absorption spectra of these two substances overlap, the peak absorption of proteins occurs at 280 nm and the peak of nucleic acids is at 260 nm (Figure 4-31). It is therefore common practice to measure absorption at either 260 or 280 nm when analyzing for nucleic acids or proteins, respectively. Caution is needed when using this approach, however, because nucleic acids absorb more strongly than proteins on a per-weight basis. This means that tiny amounts of contaminating nucleic acid can easily overshadow the presence of protein molecules. The difference in absorption intensity between proteins and nucleic acids occurs because *unsaturated rings* (carbon-containing rings with double bonds) are responsible for the absorption peaks at 260 and 280 nm. Every base present in a nucleic acid molecule contains one or two such rings, whereas only a few amino acids have unsaturated rings (phenylalanine, tyrosine, and tryptophan). In addition to the absorption peaks at 260 and 280 nm generated by unsaturated rings, the peptide bonds found in protein molecules ab-

of our understanding of the roles played by nucleic acids and proteins, for example, has depended on the development of procedures for purifying and characterizing these macromolecules. In this section we will briefly review some of the more widely used methods for studying biological molecules.

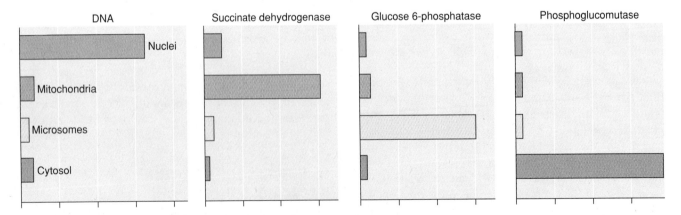

Figure 4-30 *Using Chemical Markers to Identify Subcellular Fractions*   *In this set of experiments, a rat liver homogenate was fractionated by differential centrifugation according to the scheme outlined in the preceding figure. The individual fractions were then analyzed for the presence of DNA (to identify nuclei), succinate dehydrogenase (to identify mitochondria), glucose 6-phosphatase (to identify endoplasmic reticulum membranes), and phosphoglucomutase (to identify the cytosol).*

sorb ultraviolet light strongly at 220–230 nm. Measuring the absorption at 220–230 nm is therefore a sensitive way of detecting the presence of protein, although care must be taken in interpreting such data because many compounds absorb light at this wavelength.

Using spectrophotometry to measure ultraviolet absorption is a quick and simple way of detecting proteins or nucleic acids in solution, particularly if the sample is relatively pure. The specificity of the spectrophotometric approach can also be increased by treating the sample with chemical reagents that yield characteristic colored compounds in the presence of specific types of macromolecules. Spectrophotometry is then employed to measure the amount of color produced.

Radioactive Isotopes Are a Sensitive Way of Detecting Tiny Amounts of Material

The use of chemical isotopes for detecting the presence of biological molecules has had a tremendous impact on cell biology in recent years. **Isotopes** are defined as different forms of the same chemical element that have the same number of electrons (and hence similar chemical properties) but differing atomic weights. In biology the most widely used isotopes are **radioactive;** that is, they spontaneously disintegrate into other kinds of atoms, emitting radiation in the process. Because extremely sensitive techniques are available for detecting radiation, radioactive isotopes have come to be widely used as "tracers" that permit the behavior of small numbers of molecules to

be monitored (Table 4-3). The radioactive isotopes ^3H, ^{14}C, and ^{32}P are among the most commonly used isotopes in cell and molecular biology because these three elements occur so widely as constituents of cellular molecules.

Two general approaches are employed for detecting radioactivity that has become incorporated into biological molecules: *autoradiography,* which utilizes a photographic film to determine the location of emitted radiation (pages 98 and 132), and instruments called *radiation detectors,* which are designed to detect and quantify individual radioactive disintegrations. Radioactivity measured by such instruments is usually expressed in terms of *counts per minute (cpm),* which refers to the number of disintegrations recorded by the instrument. Radiation detectors are sensitive enough to measure radiation levels as low as 5–10 cpm, which may correspond to as little as 10^{-13} or 10^{-14} grams of the radioactive molecule being detected. Radioactivity can therefore be used to detect substances that occur in amounts that are far too small to be detected by ordinary chemical means.

The radioactive isotopes employed in biology generally emit relatively weak forms of radiation. The most efficient method for detecting such weak radiation is to place the radioactive material into vials containing a special solution that gives off light in response to radiation. The flashes of light are then detected in an instru-

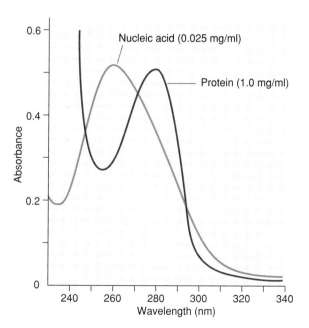

Figure 4-31 Ultraviolet Absorption Spectra of Proteins and Nucleic Acids *Note that the protein concentration must be 40 times higher than the nucleic acid concentration to achieve absorbance of comparable magnitude. For this reason, tiny amounts of contaminating nucleic acid can easily overshadow the presence of protein molecules.*

Table 4-3 Some Commonly Used Isotopes		
Predominant Natural Isotope	**Useful Isotopic Tracers**	**Half-Life**
	Radioactive	
^1H	^3H (tritium)	12.3 years
^{127}I	^{125}I	59.7 days
^{12}C	^{14}C	5,570 years
^{32}S	^{35}S	87.1 days
^{40}Ca	^{45}Ca	164 days
^{127}I	^{131}I	8.1 days
^{35}Cl	^{36}Cl	310,000 years
^{23}Na	^{24}Na	15 hours
^{31}P	^{32}P	14.3 days
^{39}K	^{42}K	12.5 hours
	Nonradioactive	
^1H	^2H (deuterium)	
^{14}N	^{15}N	
^{16}O	^{18}O	

Note: The radioactive isotopes are arranged in order of increasing energy of the emitted radiation. The term *half-life* refers to the time required for half of the atoms in a sample to undergo radioactive disintegration.

ment called a *liquid scintillation counter.* In liquid scintillation counting, the intensity of the light flashes varies with the energy of the radioactive emission. Thus ^{32}P, which emits a stronger form of radiation than ^{14}C or ^{3}H, causes brighter flashes of light. A scintillation counter is able to detect these differences and can therefore be used to distinguish between ^{32}P, ^{14}C, and ^{3}H. In this way the presence of two or more radioactive isotopes can be measured simultaneously.

By choosing the appropriate radioactive isotope, one can obtain information about specific metabolic processes. For example, if cells are incubated with inorganic phosphate labeled with the radioactive isotope ^{32}P, any cellular molecule that contains phosphate will incorporate radioactivity as it is being synthesized. If one wishes to detect only the synthesis of DNA, cells would instead be incubated with ^{3}H- or ^{14}C-thymidine because thymidine is not incorporated into any other macromolecule. In a similar fashion, radioactive uridine is used to monitor the synthesis of RNA, radioactive amino acids are utilized to monitor protein synthesis, and radioactive sugars are employed for studies of carbohydrate formation.

In addition to their usefulness in studying the synthesis of specific molecules, radioactive isotopes are employed in **pulse-chase experiments** to determine what happens to molecules after they have been synthesized. In this type of experiment, cells are first incubated with a particular radioactive compound for a brief period of time called the "pulse"; the radioactive compound is then removed and incubation is continued (the "chase"). During the chase period no additional radioactivity is incorporated by the cell, so the fate of the radioactive molecules labeled during the pulse period can be followed. This approach has been employed to determine the sequence of steps involved in metabolic pathways, and has also led to the discovery that many macromolecules are not permanently retained by cells once they have been synthesized, but are instead continually being broken down and synthesized again.

Although most of the isotopes employed in biological experimentation are radioactive, nonradioactive isotopes such as ^{15}N, ^{18}O, and ^{2}H (deuterium) also have important biological applications. Because they are not radioactive, these isotopes can only be detected because their density differs from that of the normally prevalent forms of their respective elements. The use of nonradioactive isotopes is practical only when the element involved accounts for a substantial portion of the mass of a particular biological molecule. For example, the fact that DNA and RNA contain about 15 percent nitrogen by weight means that the density of these molecules is altered appreciably when ^{15}N is substituted for the normally prevalent isotope ^{14}N. Hence DNA synthesized in the presence of ^{15}N can be easily separated from normal DNA by isodensity centrifugation, an ap-

proach that has been used extensively in the study of DNA replication (page 67).

Antibodies Are Sensitive Tools for Detecting Specific Proteins

Earlier in the chapter we discussed the use of microscopic techniques that employ antibodies to localize specific proteins and other antigens inside cells (page 130). Antibodies are also sensitive tools for detecting specific proteins after they have been extracted from cells. Moreover, antibodies are so selective that they are capable of distinguishing between proteins that differ from one another by only a single amino acid.

One commonly employed method for using antibodies to monitor the presence of specific proteins is called the **enzyme-linked immunosorbent assay (ELISA).** In this procedure, the protein or other antigen whose concentration is being measured is bound to a plastic support medium (Figure 4-32). A drop of solution containing an antibody directed against this antigen is then added to generate an antigen-antibody complex. After the unreacted antibody is washed away, a second type of antibody that binds specifically to the first antibody is added. Before being added, this secondary antibody is linked to an enzyme that catalyzes the formation of a colored product when an appropriate substrate is added. The binding of the secondary antibody to the initial antigen-antibody complex can therefore be quantified by measuring the amount of colored product that is formed when substrate is introduced. Because each enzyme molecule can catalyze the formation of thousands of molecules of colored product, such measurements provide an extremely sensitive way of detecting tiny amounts of antigen-antibody complex. Under appropriate conditions, this type of assay can detect the presence of less than a nanogram of the original protein antigen.

The Monoclonal Antibody Technique Generates Large Quantities of Identical Antibody Molecules

For many years, the use of antibodies for detecting the presence of specific proteins was hampered by the lack of a reproducible method for producing large quantities of identical antibody molecules all directed against the same antigen. This problem existed because each animal injected with an antigen generates a heterogeneous collection of antibodies. Then in 1975, Georges Köhler and César Milstein reported the development of a novel approach to this problem called the *monoclonal antibody technique.* This procedure is designed to produce permanent cell lines that grow well in culture and synthesize a single type of antibody. To generate cells ex-

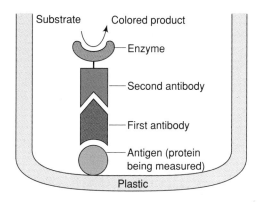

Figure 4-32 *The Principle Underlying the Enzyme-Linked Immunosorbent Assay (ELISA)* *In this technique, the antigen whose concentration is being measured is bound to a plastic support medium and reacted with an antibody directed against the antigen. A second antibody directed against the first antibody is then added. The second antibody contains a bound enzyme that catalyzes the formation of a colored product when an appropriate substrate is added.*

hibiting these two properties, hybrid cells are formed by using inactivated Sendai virus to fuse together two populations of cells: a transformed mouse tumor cell line that grows well in culture, and a population of antibody-producing spleen cells called *B lymphocytes* that have been isolated from a mouse injected with the antigen of interest (Figure 4-33). The reason for fusing the two cell types is that each cell exhibits one of the properties desired: the tumor cells grow well in culture, while the B lymphocytes each produce a single type of antibody.

After cell fusion has occurred, the cells are grown in a selective medium that permits only the hybrid cells to survive. This selection is made possible by the fact that the tumor cells employed in the cell fusion step lack *hypoxanthine-guanine phosphoribosyltransferase (HGPRT)*, an enzyme involved in the synthesis of purines and pyrimidines. In the presence of a special type of growth medium called the *HAT medium,* cells lacking HGPRT die while other cells grow normally. Hence when a culture containing the fused cells is grown in the HAT medium, the original tumor cells die out because of their lack of HGPRT; in addition, the original B lymphocytes also die because they have a limited life span in culture. However, hybrid cells formed by the fusion of the tumor cells and B lymphocytes survive and proliferate because they exhibit both normal HGPRT (a trait acquired from the B lymphocyte) and an unlimited capacity to divide (a trait acquired from the tumor cell).

The process of cell fusion generates many kinds of hybrid cells, each derived from a different B lymphocyte and thus producing a different antibody (page 722). To obtain a pure cell line producing a single

type of antibody, individual hybrid cells are isolated and allowed to proliferate into separate cell populations called **hybridomas.** Because each hybridoma is derived from a single cell, it makes a single type of antibody called a **monoclonal antibody.** To determine which hybridomas make monoclonal antibodies directed against the antigen of interest, an ELISA assay is performed in which the antigen in question is bound to the surface of a plastic culture dish containing multiple compartments. Samples of culture medium containing monoclonal antibody secreted by various hybridomas are then added to the various compartments. Those samples that generate a colored reaction product in the ELISA assay indicate the presence of a hybridoma that produces antibody directed against the original antigen. Such hybridomas can be maintained in culture indefinitely and thereby provide an inexhaustible source of monoclonal antibodies directed against the antigen of interest.

Dialysis and Precipitation Are Used to Separate Large Molecules from Small Molecules

We learned in Chapter 1 that biological molecules can be divided into two general categories: large macromolecules and smaller building blocks. In experiments designed to investigate the properties of cellular molecules, it is often necessary to separate these two categories of substances from one another. One way is by **dialysis,** a process that utilizes artificial membranes containing pores that are small enough to permit the passage of water and small molecules, but not large molecules. During dialysis, a solution containing a mixture of small and large molecules is placed in a membrane bag surrounded by a large volume of solution. Small molecules pass through the pores in the membrane until their concentrations outside and inside the bag become equal. In contrast, large molecules are retained inside the bag (Figure 4-34). Dialysis is usually used to remove and change the salts, buffers, and other small molecules that are present in solutions of macromolecules. Membranes with pores that permit the passage of substances with molecular weights less than 10,000–15,000 daltons are most often utilized, but membranes with cutoffs ranging from as small as a few thousand to greater than 50,000 daltons are available for appropriate applications.

Small and large molecules can also be separated from one another by precipitating with *trichloroacetic acid (TCA)* or *perchloric acid (PCA)*. The anions of TCA and PCA form insoluble salts with proteins and nucleic acids, causing them to precipitate out of solution. Any lipid present in the precipitate can be extracted with a nonpolar solvent, and the precipitated

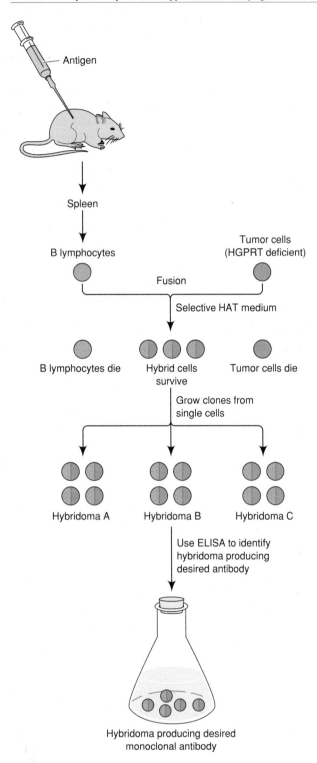

Figure 4-33 *The Production of Monoclonal Antibodies*
B lymphocytes and tumor cells are fused together to create hybrid cells, each of which makes a single type of antibody. Individual hybrid cells are isolated and grown into immortal cell lines (hybridomas) that can be maintained in culture indefinitely.

nucleic acids can be resolubilized by mild heating. These procedures permit a crude fractionation of the major classes of macromolecules and permit their separation from the smaller molecules found in cells. Acid precipitation denatures most macromolecules, so this approach tends to be used in experiments where the rates of synthesis of macromolecules are of interest rather than in experiments focusing on the purification of macromolecules.

Electrophoresis Separates Macromolecules Based on Differences in Size and Electric Charge

Cells contain thousands of different macromolecules that must be separated from one another before the properties of individual components can be investigated. One of the most common approaches for separating individual molecules from each other involves **electrophoresis,** a group of related techniques that utilize an electric field to separate electrically charged molecules. During electrophoresis, positively charged molecules move toward the negative pole or *cathode* and are hence called **cations,** whereas negatively charged molecules migrate toward the positive pole or *anode* and are therefore called **anions.** The rate at which any given molecule moves during electrophoresis depends upon its charge as well as its size.

Electrophoresis can be carried out using a variety of support media, such as paper, cellulose acetate, starch, polyacrylamide or agarose (a polysaccharide obtained from seaweed). Of these media, gels made of polyacrylamide or agarose provide the best resolution and are most commonly employed for the electrophoresis of nucleic acids and proteins. The sample being analyzed is placed on top of a gel and voltage is applied to induce migration of charged molecules. Depending on their charge and size, molecules move through the gel at dif-

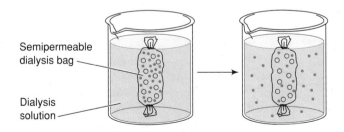

Figure 4-34 *The Technique of Dialysis* *The pores in the dialysis bag permit small molecules but not large molecules to diffuse across the membrane. By replacing the dialysis solution several times in succession, the concentration of the small molecules within the dialysis bag can be lowered to trace amounts.*

fering rates. The resulting bands of material are then visualized by appropriate staining or autoradiographic techniques. Because the procedures involved in the electrophoresis of nucleic acids and of proteins differ somewhat, these two types of electrophoresis will be described separately.

Electrophoresis of Nucleic Acids

Electrophoresis of DNA and RNA is carried out in gels made of polyacrylamide or agarose, both of which contain pores through which the moving nucleic acid molecules pass during electrophoresis. Since all nucleic acids are constructed from nucleotides that contain negatively charged phosphate groups, differences in electric charge are not important in determining the behavior of nucleic acids during electrophoresis. However, nucleic acids do differ in size, and this size difference influences the way in which they move through the pores of polyacrylamide or agarose gels. Larger molecules are hindered more as they pass through the tiny pores in the gel, so larger nucleic acids move more slowly during electrophoresis than do smaller nucleic acids. The net result is that gel electrophoresis separates nucleic acids based on differences in size, with the smallest nucleic acids moving most rapidly toward the bottom of the gel. The main difference between the behavior of agarose and polyacrylamide gels is that agarose is more porous; hence agarose gels can be used to separate larger nucleic acids than polyacrylamide gels.

Polyacrylamide gel electrophoresis is generally used for separating DNA molecules and fragments ranging up to about 500 base pairs in size, whereas agarose gels can handle molecules up to about 10,000 base pairs in length (Figure 4-35). Beyond this size, larger DNA molecules tend to travel through gels at a constant speed that is independent of their size. This problem in separating large DNA molecules can be overcome, however, by a special technique called **pulsed-field gel electrophoresis.** In this type of electrophoresis the direction of the electric field is repeatedly altered, forcing the migrating DNA molecules to continually reorient themselves from one direction to another. The time occupied by each reorientation event is longer for large DNA molecules than for smaller DNA molecules; therefore large DNA molecules move more slowly than smaller ones. This novel approach allows DNA molecules containing hundreds of thousands or even millions of base pairs to be separated from one another by gel electrophoresis.

Nucleic acids that have been fractionated by gel electrophoresis can be visualized in several different ways. If the nucleic acid is radioactive, autoradiography is carried out by placing X-ray film over the gel to local-

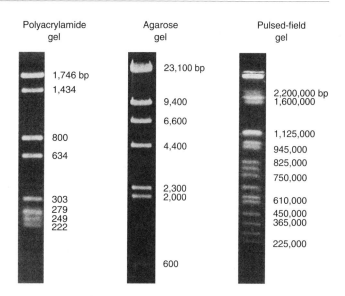

Figure 4-35 *Using Gel Electrophoresis to Separate DNA Molecules Based on Differences in Size* *The direction of electrophoresis in all three photographs is from top to bottom.* (Left) *Polyacrylamide gel electrophoresis is used for separating small DNA fragments up to about 500 base pairs (bp) in length.* (Middle) *Agarose gel electrophoresis is best suited for separating DNA molecules in the range of 500 to 10,000 base pairs in length.* (Right) *Pulsed-field agarose gel electrophoresis utilizes an electric field whose direction is repeatedly altered, forcing the migrating DNA molecules to constantly reorient themselves. This approach allows DNA molecules containing hundreds of thousands or even millions of base pairs to be separated from one another.*

ize the radioactive bands. Gels containing DNA also can be directly stained with *ethidium bromide,* which fluoresces intensely when illuminated with ultraviolet light. Finally, Southern or Northern blotting can be employed to locate DNA or RNA molecules containing specific base sequences (page 97).

Electrophoresis of Proteins

Unlike nucleic acids, the electric charge of protein molecules varies significantly for different proteins and electrophoresis is therefore influenced by differences in charge as well as size. However, a widely used technique called **SDS polyacrylamide gel electrophoresis** allows proteins to be separated solely on the basis of size, just as in the case of nucleic acids. In this technique proteins are first treated with the detergent *sodium dodecyl sulfate (SDS),* an anion that coats the surfaces of protein molecules. This coating disrupts the weak bonds involved in maintaining protein conformation and, in the case of multisubunit proteins, dissociates the molecule into its individual polypeptide chains. The large negative charge contributed by the bound SDS molecules masks any inherent charge of the protein chains themselves.

Since all protein molecules are equally coated with negatively charged SDS molecules, protein size is the only variable that influences electrophoretic mobility. As in the case with nucleic acids, smaller proteins migrate more rapidly through the pores of a polyacrylamide gel during electrophoresis than do larger proteins (Figure 4-36). Therefore SDS polyacryl-amide gel electrophoresis separates proteins based on differences in size just as gel electrophoresis separates nucleic acids based on differences in size.

Protein mixtures that have been fractionated by gel electrophoresis are visualized either by staining with a silver stain or a dye such as Coomassie blue, or by autoradiography if the proteins are radioactive. In addition, a technique called **Western blotting** can be used to detect the presence of proteins that react with a particular kind of antibody. As in the case of the Northern and Southern blotting procedures used with nucleic acids, a replica of the electrophoretic gel is made by transferring the separated proteins to a sheet of nitrocellulose paper. The nitrocellulose sheet is then treated with a radioactive antibody to locate the presence of proteins that react with the antibody.

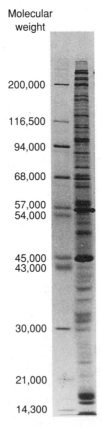

Molecular
weight

200,000 —

116,500 —

94,000 —

68,000 —

57,000 —
54,000 —

45,000 —
43,000 —

30,000 —

21,000 —

14,300 —

Figure 4-36 *Separation of Proteins by SDS Polyacryl-amide Gel Electrophoresis*   (Left) *Marker proteins with molecular weights indicated.* (Right) *Cell extract fractionated by this procedure showing a diversity of polypeptides. Courtesy of S. H. Blose.*

The major limitation of SDS polyacrylamide gel electrophoresis is that it can only separate several dozen polypeptides at a time, and of course cannot separate two polypeptide chains of similar molecular weight. In 1975 Patrick O'Farrell designed a procedure for carrying out *two-dimensional gel electrophoresis* that allows more than a thousand proteins to be separated simultaneously. The first step in this two-dimensional separation procedure involves **isoelectric focusing,** a technique that exploits the fact that a molecule's electric charge varies with pH. As the pH is increased, molecules tend to lose positive charge and/or increase negative charge, whereas a decrease in pH results in the opposite effect. By varying the pH, a point is reached where the negative and positive charges in a molecule exactly balance one another. At this pH, called the *isoelectric point,* the molecule's net charge is zero. Proteins differ from one another in their isoelectric points because they contain different numbers of charged amino acid side chains.

Isoelectric focusing permits proteins to be separated from one another based on differences in their isoelectric points. In this procedure, proteins are subjected to electrophoresis in a polyacrylamide gel in which a gradient of pH is maintained. Each protein migrates to the point in the pH gradient that corresponds to its isoelectric point; at that point, the net charge of the protein chain becomes zero and it will no longer move (Figure 4-37). Each protein eventually becomes localized in a sharp band corresponding to the position of its isoelectric point in the pH gradient.

After isoelectric focusing has been completed, SDS is added and SDS polyacrylamide gel electrophoresis is carried out at right angles to the direction in which the isoelectric focusing was originally performed. The result is a two-dimensional separation procedure in which a thousand or more polypeptides can be separated in the same sample (Figure 4-38).

Chromatography Separates Molecules Based on Differences in Their Affinities for Two Phases

Besides electrophoresis, the other main technique for separating biological molecules is chromatography. The term **chromatography** encompasses a broad spectrum of procedures that separate molecules based on differences in the way they become distributed between two phases. In such techniques, molecules dissolved in a first phase consisting of liquid or gas are passed through a bed of material that constitutes the second phase. Depending on their relative affinities for the two phases, molecules move at different rates and so become physically separated from one another. The following kinds of chromatography are commonly encountered in biological research:

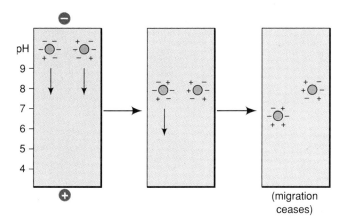

Figure 4-37 *Separation of Proteins by Isoelectric Focusing* *A stable pH gradient is created in the gel before subjecting the proteins to electrophoresis. Each protein migrates to the position where its net charge becomes zero.*

Isoelectric focusing pH gradient

Figure 4-38 *Separation of All the Proteins in a Bacterial Cell by Two-Dimensional Electrophoresis* *Isoelectric focusing was first carried out in the horizontal direction, followed by SDS polyacrylamide gel electrophoresis in the vertical direction. Each spot represents a different polypeptide. Courtesy of P. O'Farrell.*

1. Ion-exchange chromatography. Biological molecules are often separated and purified by chromatographic procedures carried out in cylindrical columns. In this approach, called *column chromatography,* a solution containing molecules to be separated is passed through a column filled with a matrix material that is capable of binding to the molecules as they pass through. Interactions with the matrix slow down the passage of different molecules to varying extents, allowing them to be collected as separate fractions as they emerge from the bottom of the column.

Many kinds of matrix materials are employed in column chromatography. In ion-exchange chromatogra-

phy, which is used to separate molecules that are electrically charged, the matrix contains either positively or negatively charged side chains. For example, if the molecules being separated are positively charged, a matrix with negatively charged side chains is employed. As they pass through the column, positively charged molecules bind to the negatively charged matrix. These bound molecules are then removed by changing the pH and/or ionic strength of the solution passing through the column, either in discrete steps or in a continuous gradient. Molecules of differing charge emerge from the column at different rates, and can therefore be collected as separate fractions (Figure 4-39, *left*).

2. Adsorption chromatography. In this related type of column chromatography, the molecules being separated are adsorbed to the column matrix by binding forces other than electric charge. An example of such a matrix is *hydroxylapatite,* a hydrated form of calcium phosphate that is frequently employed for fractionating proteins and nucleic acids. Hydroxylapatite is especially useful in fractionating DNA because double-stranded DNA binds to such columns more tightly than does single-stranded DNA.

3. Gel filtration chromatography. This kind of column chromatography separates molecules on the basis of size rather than their ability to bind to the matrix. The column matrix in gel filtration chromatography consists of small beads of polymer or glass containing numerous microscopic pores. Molecules too large to enter the pores flow around the beads and therefore pass through the column more quickly than do small molecules, which pass into the pores of the beads and are temporarily retarded (see Figure 4-39, *right*). A molecule's rate of passage through such a column is therefore a direct function of its size. Beads in a variety of pore sizes are available for the separation of molecules with widely differing molecular weights.

4. Affinity chromatography. This procedure is one of the most selective types of column chromatography available because it utilizes a matrix containing chemical groups that specifically bind to the molecule that one wishes to purify. Because of its selectivity, affinity chromatography can be used to purify individual proteins from a complex mixture of molecules. For example, the substrate for a specific enzymatic reaction could be covalently attached to a column matrix and used to purify its corresponding enzyme. If a crude mixture of enzymes is passed through such a column, any enzyme molecule that acts upon this substrate will be selectively bound to the column. In a related approach, called **immunoaffinity chromatography,** an antibody directed against a specific protein is coupled to the column matrix. When a mixture of proteins is

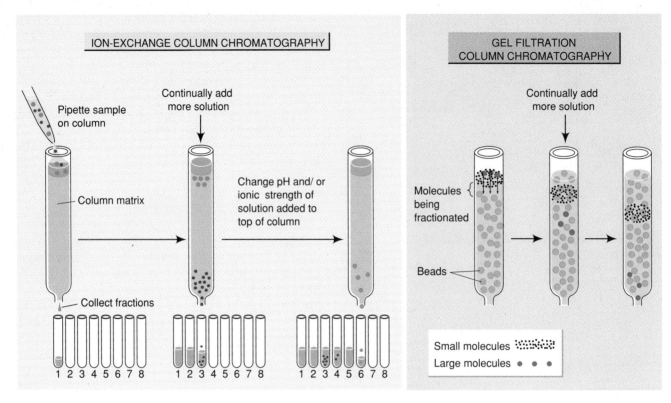

Figure 4-39 *Separation of Molecules by Two Types of Column Chromatography* (Left) *In ion-exchange chromatography, separation is based on differences in electrostatic interaction between the charged column matrix and the molecules being fractionated. When the sample enters the column, molecules with charge opposite to that of the column matrix stick to the matrix, but other molecules pass through. A subsequent change in the ionic strength or pH of the buffer causes the bound molecules to be released from the matrix. (Right) In gel filtration chromatography, separation is based on size. Larger molecules, which cannot pass into the beads of the column matrix, move around them and through the column more quickly than smaller molecules, which are momentarily retained in the beads during passage.*

passed through such a column, the only protein retained is the protein that binds to the antibody. This protein can then be removed by altering the pH, which weakens the binding between antibody and protein.

5. High-performance liquid chromatography (HPLC). Most types of column chromatography are slow procedures because of the time it takes for fluids to pass through the column matrix. In addition to its inconvenience, this delay hampers resolution by giving the sample time to spread out as it moves through the column. In *high-performance liquid chromatography,* high-pressure pumps are used to force fluid through a column at rapid rates. This procedure requires strong metal columns and special matrix materials consisting of tiny, noncompressible spheres. With such a system, separations that would have taken hours can be accomplished in minutes, and with better resolution.

6. Paper and thin-layer chromatography. Although the preceding examples all involve a matrix phase packed into a column, the solid phase used in chromatographic techniques can also be spread out as a sheet. In *paper chromatography,* for example, the

sample is applied as a small spot to a piece of filter paper and the solvent is then allowed to flow across the paper by capillary action. Molecules in the sample will migrate at differing rates, depending on their relative affinities for the solvent and the paper. The separated compounds are then visualized by spraying the sheet with an appropriate staining reagent or by autoradiography. In the closely related technique of *thin-layer chromatography (TLC),* the solid phase is spread out as a thin layer on a rigid support of glass or plastic (Figure 4-40). The thinner the layer, the better the resolution of separated components. By varying the chemical nature of the solid phase, it is possible to separate molecules by the principles of ion-exchange, adsorption, or gel filtration chromatography. Paper and thin-layer chromatography are most commonly used in the fractionation of relatively small molecules because macromolecules do not migrate sufficiently to produce good separation.

7. Gas-liquid chromatography (GLC). In addition to using chromatographic techniques for separating molecules in liquid solution, chromatography has also been adapted for the separation of molecules in a

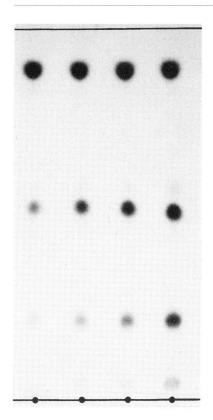

Figure 4-40 *Separation of Lipids by Thin-Layer Chromatography*   *Samples are applied as separate spots at the bottom and the lipids are separated from one another as the solvent moves upward. The leading edge of the solvent migration is indicated by the line at the top. Courtesy of R. Douce.*

gaseous phase. This type of chromatography is limited to molecules that are readily volatilized, such as fatty acids, steroids, and other lipids. The basic principle is similar to liquid chromatography in that the molecules being separated flow across a bed of material for which they have differing affinities. However, in gas-liquid chromatography the matrix material is contained in a long coil through which the gas phase containing the sample is passed. The matrix material itself is inert, but it is coated with a heavy liquid with which the molecules being separated can interact. The sample molecules will therefore pass through the coil at different rates, depending on their relative affinities for the heavy liquid phase.

X-Ray Diffraction Reveals the Three-Dimensional Structures of Macromolecules

Earlier in the chapter, we saw how microscopic techniques are used to examine the structural organization of cells. When investigating the properties of cellular macromolecules, it is likewise important to understand the three-dimensional structure of the molecule being examined. Some information about the structural organization of certain macromolecules has been provided by transmission electron microscopy and scanning probe microscopy, but the most powerful tool for elucidating the three-dimensional structure of macromolecules is **X-ray diffraction.** Although this technique is quite complex and does not yield a direct visual image the way microscopy does, it has produced a wealth of information about the three-dimensional structures of a variety of individual proteins and nucleic acids.

When X-ray diffraction is used to investigate molecular structure, a narrow beam of X-rays is transmitted through the specimen to be examined. In various regions of the molecule, the beam encounters atoms that are opaque to X-rays. When a wave passes by an opaque object its path is altered—a phenomenon known as *diffraction.* Since the wavelength of X-rays is very short (~0.1 nm), all atoms larger than hydrogen act as separate diffraction edges. By placing a photographic plate on the opposite side of the specimen, the pattern of X-ray diffraction can be recorded and used to determine the position of individual atoms within the molecule.

Unless the specimen exhibits considerable molecular order, such diffraction patterns are virtually impossible to interpret. For example, diffraction patterns produced by randomly distributed molecules in dilute solution are unintelligible because of interference between individual molecules. On the other hand, crystals or fibers with repeating structures generate good diffraction patterns because the patterns generated by individual atoms reinforce one another. Since every atom other than hydrogen contributes to the diffraction pattern, X-ray analysis of macromolecules containing thousands of atoms is a singularly complex task. The situation can be simplified somewhat by analyzing diffraction produced by arrays of atoms rather than by each atom itself. This type of analysis is carried out by rotating the specimen in the X-ray beam. If the sample is highly ordered, a distinct pattern of spots is obtained (Figure 4-41, *top and middle*). Spreading of the spots occurs as disorder increases, until a concentric pattern of spots is eventually obtained (see Figure 4-41, *bottom*). The latter pattern represents an analysis of randomly oriented crystals.

Determining the three-dimensional structure of macromolecules from X-ray diffraction patterns requires complicated mathematical analysis that is beyond the scope of this text. One of the key developments in the application of this type of analysis to protein structure has been the introduction of heavy-metal atoms into specific regions of the molecule being examined. The resulting alteration in the diffraction pattern permits one to identify the particular region of the molecule to which the metal is bound. This information in turn provides a point of departure from which the rest of the structure can be deduced. Pioneers such as Linus Paul-

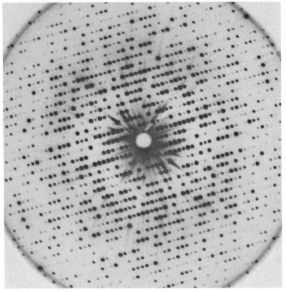

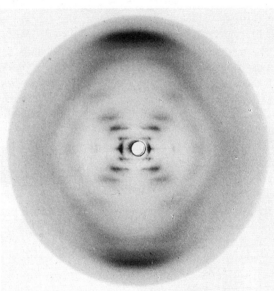

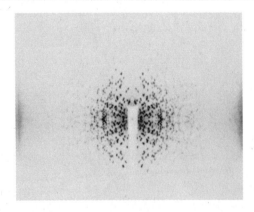

Figure 4-41 *Examples of X-Ray Diffraction Patterns*
From top to bottom the three patterns represent a protein (myoglobin), a nucleic acid (DNA), and a randomly oriented set of crystals. Courtesy of J. C. Kendrew (top), *M. H. F. Wilkins* (middle), *and P. Luger* (bottom).

ing, John Kendrew, Max Perutz, and Maurice Wilkins have played crucial roles in applying such analyses to the structure of biological macromolecules.

X-ray diffraction has been especially important in determining the three-dimensional structure of DNA, which in turn has had a dramatic impact on our understanding of how DNA stores and transmits genetic information (page 66). X-ray diffraction also has been used to unravel the three-dimensional structure of many important proteins, yielding crucial information about protein specificity and the mechanism of enzyme action. Finally, this approach has been useful in analyzing the structure of membranes and muscle fibers, and promises to continue to be an indispensable tool for studying the structure and interactions of biological macromolecules.

SUMMARY OF PRINCIPAL POINTS

• Cells used for experimental purposes are obtained either directly from intact organisms or from cell cultures, where conditions can be carefully controlled and a wide variety of manipulations are possible. The technique of cell fusion permits cell functions to be studied by artificially fusing cultured cells of differing types together.

• The light microscope was responsible for the discovery of cells and the formulation of the cell theory, but its limited resolving power (~200 nm) restricts its usefulness for visualizing structural details. Modifications of conventional light microscopy such as phase-contrast microscopy and differential interference microscopy allow the behavior of living cells to be observed, while fluorescence microscopy and confocal scanning microscopy permit the location of various kinds of molecules to be investigated.

• In transmission electron microscopy, the use of electrons rather than light as an illumination source improves the resolving power from 200 nm to about 0.2 nm. The ability of electron microscopy to reveal the details of subcellular architecture is enhanced by a diverse set of procedures for specimen preparation, such as thin sectioning, negative staining, positive staining, shadow casting, and freeze-etching. Scanning electron microscopy provides a three-dimensional view of the cell surface, and various forms of scanning probe microscopy allow the surface features of individual molecules to be visualized.

• By employing cytochemical staining techniques, microscopy can be used to localize specific molecules and metabolic activities within cells. Stains exhibiting broad specificity are employed to localize proteins, nucleic acids, polysaccharides, and lipids. Staining with antibodies permits specific proteins to be located, the use of substrates that generate visible products allows individual enzymes to be localized, and microscopic autoradiography permits radioactive molecules to be located.

• Two types of centrifugation are commonly employed in cell biology, velocity centrifugation and isodensity centrifugation. Velocity centrifugation separates components based on differences in size, and can be carried out on a sample suspended in solution (differential centrifugation) or on a sample applied to the top of a tube as a discrete zone (moving-zone centrifugation). Isodensity centrifugation separates components based on differences in density.

• Subcellular fractionation techniques allow individual organelles to be isolated from cells for further study. During traditional subcellular fractionation experiments, cells are homogenized and then separated by differential centrifugation into nuclear, mitochondrial, microsomal, and cytosol fractions.

• Spectrophotometry, radioactive isotopes, and antibodies are used to detect the presence of specific kinds of molecules, whereas dialysis and precipitation are important tools for separating large molecules from small molecules. Electrophoresis separates molecules based on differences in size and electric charge, and chromatography permits molecules to be separated on the basis of their affinities for various solvents and matrix materials. Finally, X-ray diffraction is a powerful approach for determining the three-dimensional structure of macromolecules, especially proteins and nucleic acids.

SUGGESTED READINGS

Books

Andrews, A. T. (1986). *Electrophoresis,* 2nd Ed., Clarendon Press, Oxford, England.

Bozzola, J. J., and L. D. Russell (1992). *Electron Microscopy: Principles and Techniques for Biologists,* Jones and Bartlett, Boston.

Flegler, S. L., J. W. Heckman, and K. L. Klomparens (1993). *Scanning and Transmission Electron Microscopy: An Introduction,* Freeman, New York.

Ford, T. C., and J. M. Graham (1991). *An Introduction to Centrifugation,* Bios Scientific Publishers, Oxford, England.

Freifelder, D. (1982). *Physical Biochemistry. Application to Biochemistry and Molecular Biology,* 2nd Ed., Freeman, San Francisco.

Freshney, R. I. (1993). *Culture of Animal Cells: A Manual of Basic Technique,* 3rd Ed., Wiley-Liss, New York.

Scopes, R. (1982). *Protein Purification: Principles and Practice,* Springer-Verlag, New York.

Slayter, E. M., and H. S. Slayter (1992). *Light and Electron Microscopy,* Cambridge Univ. Press, New York.

White, L., Jr. (1968). *Machina ex Deo: Essays in the Dynamism of Western Culture,* MIT Press, Cambridge, MA.

Articles

Arscott, P. G., G. Lee, V. A. Bloomfield, and D. F. Evans (1989). Scanning tunnelling microscopy of Z-DNA, *Nature* 339:484–486.

Clark, S. M., E. Lai, B. W. Birren, and L. Hood (1988). A novel instrument for separating large DNA molecules with pulsed homogeneous electric fields, *Science* 241:1203–1205.

de Duve, C., and H. Beaufay (1981). A short history of tissue fractionation, *J. Cell Biol.* 91:293s–299s.

Herman, E. M. (1988). Immunocytochemical localization of macromolecules with the electron microscope, *Annu. Rev. Plant Physiol. Plant Mol. Biol.* 39:139–155.

Howell, K. E., E. Devaney, and J. Gruenberg (1989). Subcellular fractionation of tissue culture cells, *Trends Biochem. Sci.* 14:44–47.

Lichtman, J. W. (1994). Confocal microscopy, *Sci. Amer.* 271(August): 40–45.

O'Farrell, P. H. (1975). High-resolution two-dimensional electrophoresis of proteins, *J. Biol. Chem.* 250:4007–4021.

Milstein, C. (1980). Monoclonal antibodies, *Sci. Amer.* 243(October):66–74.

Nelson-Rees, W. A., D. W. Daniels, and R. R. Flandermeyer (1981). Cross-contamination of cells in culture, *Science* 212:446–452.

Pease, D. C., and K. R. Porter (1981). Electron microscopy and ultramicrotomy, *J. Cell Biol.* 91:287s–292s.

Roberts, C. J., P. M. Williams, M. C. Davies, D. E. Jackson, and S. J. B. Kendler (1994). Atomic force microscopy and scanning tunnelling microscopy: refining techniques for studying biomolecules, *Trends Biotechnology* 12:127–132.

Part 2

*Cell
Organization*

C h a p t e r 5

Membranes and Membrane Transport

At an early point during the evolutionary origin of cells a membrane barrier must have evolved, leading to the separation of the cell interior from the potentially disruptive outside world. During the subsequent evolution of eukaryotic cells, this external membrane gave rise to a series of intracellular membranes that subdivided the cytoplasm into multiple compartments. In present-day eukaryotes, these internal membranes have become so vital that virtually every important cellular activity involves events that occur within, on, attached to, or through membranes. Knowing how membranes are constructed and how they function is therefore crucial to our basic understanding of cell function. In this chapter, we will examine the molecular architecture of membranes and the way in which they control the transport of materials from one side of a membrane to the other. The following four chapters will then describe the specialized functions of the plasma membrane, cytoplasmic membranes, and mitochondrial and chloroplast membranes.

MODELS OF MEMBRANE ARCHITECTURE

For more than a hundred years, scientists have been trying to unravel the molecular organization of biological membranes. Because cells contain many different kinds of membranes, finding the common features that underlie the construction of all membranes has been an exceptionally challenging task. This intense research effort has, however, finally produced a model of membrane architecture that is believed to be universally applicable. In the following sections we will describe the basic elements of this model and the experimental observations that led to its formulation.

Cellular Membranes Contain a Lipid Bilayer

Because membranes are too thin to be resolved by light microscopy, no one had ever seen a cellular membrane prior to the early 1950s, when electron microscopy was first applied to the study of cell structure. Yet indirect evidence led biologists to postulate the existence of membranes long before they actually could be seen. In the late nineteenth century, C. Nageli, W. Pfeffer, and Charles Overton discovered that different substances enter and leave cells at significantly different rates, leading these scientists to suggest that cells are surrounded by a **plasma membrane** that regulates the passage of materials into and out of cells. Overton subsequently reported that the rate at which different molecules pass into cells is directly related to their solubility in lipid; the more soluble a molecule is in lipid, the more readily it enters cells. This correlation led Overton to propose that the plasma membrane is constructed of lipid molecules.

Insight into the question of how lipid molecules might be arranged in cellular membranes was provided a few years later by Irving Langmuir, who discovered that purified phospholipids spread on a water surface spontaneously generate a film one molecule thick. In interpreting these results, Langmuir took into account the fact that phospholipids contain both **hydrophilic (polar)** regions that are soluble in water, as well as **hydrophobic (nonpolar)** regions that are insoluble in water. He theorized that when phospholipids are placed on a water surface, the molecule's hydrophilic end (called the *head* group) becomes aligned next to the water surface, while the molecule's hydrophobic end (called the *tail*) is repelled by the water and protrudes out toward the air (Figure 5-1, *top*).

Langmuir's methods were soon adopted by E. Gorter and F. Grendel, who studied the behavior of lipid molecules extracted from red blood cells. When spread on water, the lipid formed a monomolecular film that covered an area roughly twice as large as the surface area of the red cells from which the lipids had been extracted. In other words, red cells contain

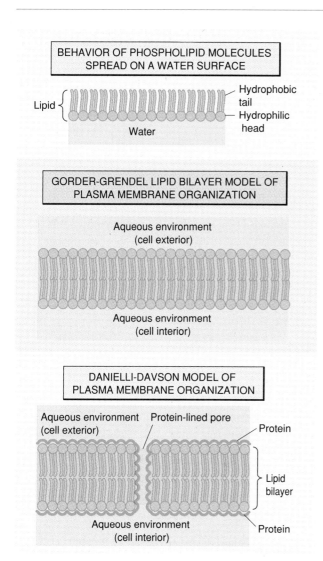

Figure 5-1 *Historical Development of the Idea That Cellular Membranes Contain a Lipid Bilayer* (Top) *In Langmuir's experiments, purified phospholipids spread on a water surface were found to produce a film one molecule thick. The hydrophilic head group of each lipid molecule protrudes into the water surface, whereas the hydrophobic tails are repelled by the water and stick out into the air. (Middle) Gorter and Grendel discovered that lipids extracted from red blood cells and spread on water form a film that covers an area roughly twice as large as the surface area of the cells from which the lipids were extracted. They therefore proposed that cell membranes consist of a lipid bilayer. In this arrangement, the hydrophilic head groups are in contact with aqueous environments, while the hydrophobic tails are not. (Bottom) In the model of membrane organization proposed by Davson and Danielli, a lipid bilayer is covered on each surface by a layer of protein molecules. Protein-lined pores were included to accommodate the passage of water-soluble molecules through the hydrophobic core of the lipid bilayer.*

enough lipid to create a surface membrane two molecules thick. Gorter and Grendel proposed that such a **lipid bilayer** would be most stable if the hydrophilic head groups were exposed to the aqueous environments at the two membrane surfaces, and the hydrophobic tails were sequestered away from water in the membrane interior (see Figure 5-1, *middle*). Although it is clear in retrospect that these experiments had several shortcomings, the conclusion that membranes contain a bilayer of lipid molecules has turned out to be basically sound.

The Davson-Danielli Model Was the First Detailed Representation of Membrane Organization

Shortly after Gorter and Grendel proposed the lipid bilayer model in 1925, its shortcomings were revealed by measurements of membrane surface tension reported by E. Harvey and James Danielli. These new data indicated that the surface tension of cellular membranes is significantly lower than that of pure lipid films; hence membranes cannot consist of lipid alone. Subsequent studies showed that the addition of protein to pure lipid droplets causes the surface tension of the droplets to decrease to values more typical of cellular membranes. This finding suggested that cellular membranes contain protein molecules as well as lipids.

Additional evidence for the existence of membrane proteins derived from studies of membrane permeability. Certain sugars, ions, and other hydrophilic molecules pass into cells more readily than can be explained by the permeability of pure lipid bilayers to water-soluble molecules. Such considerations led Danielli and Hugh Davson to propose a membrane model in which a bilayer of lipid molecules, oriented with their hydrophilic head groups toward the membrane surfaces, is covered on both sides by layers of protein. To account for the permeability of membranes to water-soluble molecules, the model was eventually modified to include intermittent protein-lined pores (see Figure 5-1, *bottom*). This *Davson-Danielli model* of membrane organization, first proposed in the early 1930s, dominated the thinking of cell biologists for many decades.

Membranes Exhibit a Trilaminar Appearance in Electron Micrographs

During the early 1950s, the use of electron microscopy permitted cellular membranes to be seen for the first time. In addition to confirming the presence of a plasma membrane at the surface of all cells, electron microscopy revealed the existence of other kinds of membranes, including the endoplasmic reticulum and its derivatives, the membranes of the nuclear envelope, and mitochondrial and chloroplast membranes. At high

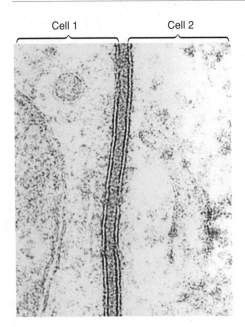

Figure 5-2 *Trilaminar Appearance of Cellular Membranes* *This thin-section electron micrograph shows two cells lying next to each other with their plasma membranes separated by a small intercellular space. Each membrane appears as two dark lines separated by a lightly stained central zone, an appearance known as the trilaminar staining pattern. Courtesy of D. W. Fawcett.*

magnification, membranes typically measured 7–8 nm in thickness and exhibited the same staining appearance: two dark lines separated by a lightly stained central zone (Figure 5-2). The fact that this same three-layered or *trilaminar* staining pattern is exhibited by a broad spectrum of membrane types led J. David Robertson to first postulate that all cellular membranes share a common underlying structure.

One of the first membranes to have its structure investigated was **myelin,** a membranous sheath that surrounds the axons of certain kinds of nerve cells. Myelin is derived from the plasma membrane of specialized cells called Schwann cells. During development of the axon, the plasma membrane of the Schwann cell wraps around the axon several times (Figure 5-3). This process covers the axon with a stack of tightly packed membranes whose repeating, quasi-crystalline structure makes it amenable to analysis by X-ray diffraction. When X-ray analyses were carried out in the laboratories of F. O. Schmitt and J. B. Finean, their data revealed that myelin consists of alternating layers of lipid and protein. One possible interpretation of such a pattern is that each membrane within the stack consists of an interior region of lipid covered on both sides by protein (a view that was consistent with the Davson-Danielli model).

Direct support for this interpretation was soon provided by Robertson, who investigated the properties of the electron-opaque stains used by electron microscopists to generate the trilaminar staining pattern. Robertson's studies revealed that such stains bind both to phospholipid head groups and to protein molecules. Applying this information to the appearance of membranes in electron micrographs, Robertson proposed that the lightly stained interior zone of the trilaminar staining pattern contains the hydrophobic lipid tails, whereas the two darkly stained outer regions contain the phospholipid head groups and associated protein. This conclusion appeared to provide strong support for the Davson-Danielli view that membranes consist of a lipid bilayer covered on both surfaces with protein.

The Davson-Danielli Model Failed to Explain Many Aspects of Membrane Behavior

In spite of the growing support for the Danielli-Davson model, data began to emerge during the 1960s that

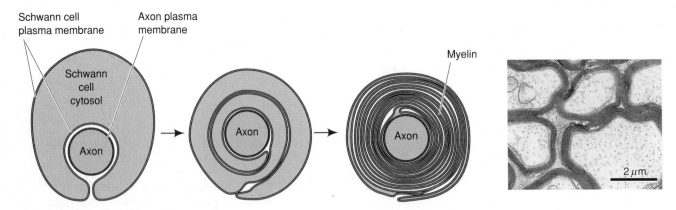

Figure 5-3 *Origin of Myelin Membranes* (Left diagram) *During the early stages of myelin formation, nonmyelinated axons become enveloped by Schwann cells.* (Middle diagram) *The Schwann cell plasma membrane begins to wrap around the axon.* (Right diagram) *The multiple layers of Schwann cell membrane become closely stacked on each other, eliminating all cytoplasmic material from the spaces between the membranes. The final stack of closely packed membranes is called myelin. The electron micrograph is a cross section through several axons surrounded by mature stacks of myelin membranes. Micrograph courtesy of C. S. Raines.*

were difficult to reconcile with this view of membrane architecture. Among the more significant problems to arise were the following:

1. In the Davson-Danielli model, both surfaces of the lipid bilayer are covered by a layer of protein. Membranes typically measure 7–8 nm in thickness, with about 5 nm occupied by the lipid bilayer. This leaves only 1–2 nm of space on each surface of the bilayer for the membrane protein. Only a thin monolayer of protein dominated by extended regions of β sheet (page 23) would fit in such a small space. Yet physical measurements using infrared spectroscopy and circular dichroism (differential absorption of right-hand and left-hand circularly polarized light) revealed that most membrane proteins are globular proteins containing extensive regions of α helix as well as β sheets. Such globular proteins would be too large to form a thin monolayer at the membrane surface, and must therefore protrude into the membrane interior where the lipid bilayer is located. Direct support for this conclusion emerged from freeze-fracture electron microscopy, which revealed that membranes often contain large numbers of protein particles embedded within the membrane interior (Figure 5-4).

2. According to the Davson-Danielli model, membranes are held together by electrostatic (ionic) bonds between the hydrophilic head groups of lipid molecules and the ionic side chains of protein molecules. It therefore follows that increasing the ionic strength should disrupt membrane structure, releasing proteins whose ionic side chains make them readily soluble in water. These premises are not supported by the available experimental data. Most membrane proteins are not extracted by increasing the ionic strength and, once isolated by more drastic extraction techniques (organic solvents or detergents), they are found to be relatively insoluble in water. Such observations suggest that many membrane proteins are hydrophobic molecules that are bound to the lipid bilayer by interactions with the hydrophobic tails.

3. The accuracy of calculations indicating that membranes contain exactly enough lipid for a complete layer two molecules thick is questionable. The data do not rule out the possibility that the amount of lipid present is less than that required for a complete bilayer, and that the bilayer is therefore interrupted by protein molecules.

4. According to the Davson-Danielli model, the hydrophilic lipid head groups are covered by a layer of protein. Yet depending on the cell type, up to 75 percent of the membrane lipid can be degraded by exposing membranes to enzymes called *phospholipases,* which selectively remove phospholipid head groups. This susceptibility to enzymatic digestion suggests that

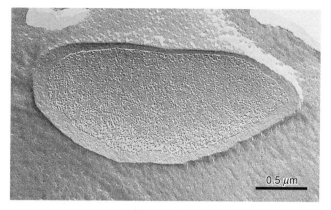

Figure 5-4 Membrane Proteins Visualized by Freeze-Fracture Electron Microscopy *Membrane proteins appear as discrete particles embedded within the lipid bilayer. The two micrographs show membrane particles in the red cell plasma membrane (top) and chloroplast membranes (bottom). Courtesy of D. Branton (top) and R. B. Park (bottom).*

many of the phospholipid head groups are exposed at the membrane surface rather than covered by a layer of protein.

5. Optimum thermodynamic stability requires that hydrophobic regions of membrane lipids and proteins be sequestered from contact with water; hydrophilic regions of membrane lipids and proteins, on the other hand, should be exposed to the aqueous environment. The Davson-Danielli model does not meet these criteria for thermodynamic stability because the hydrophilic lipid head groups are sequestered from aqueous contact (by a covering layer of membrane proteins), whereas the membrane proteins, which contain extensive hydrophobic regions (see item 2 above), are exposed to the aqueous environment at the membrane surface.

6. Data to be described later in the chapter indicate that membranes are fluid structures in which both the lipid and the protein molecules have significant freedom of movement. Movement of membrane proteins and lipids is not easy to envision in the Davson-Danielli model, which postulates that the membrane proteins form a continuous sheet linked by ionic bonds to the underlying lipid bilayer.

Table 5-1   Protein and Lipid Composition of Some Typical Membranes

Membrane	Protein/Lipid Ratio	Percent of Total Lipid		
		Phospholipid	**Glycolipid**	**Cholesterol**
Myelin	0.25	45	25	25
Plant plasma membrane (rye leaf)	0.7	32	16	33*
Red cell plasma membrane (human)	1.1	65	3	25
Endoplasmic reticulum	1.5	90	—	8
Chloroplast thylakoid membrane	2.3	10	60	—
Bacterial plasma membrane (*E. coli*)	2.5	90	—	—
Mitochondrial inner membrane	3.2	90	—	5

*The value marked with an asterisk represents the cholesterol derivatives sitosterol, campesterol, and stigmasterol, which occur in plant plasma membranes in place of cholesterol. The membranes in this table are arranged in order of increasing protein content. Blank entries indicate either the absence of the component in question or values that are too low to be experimentally reliable.

7. Different kinds of membranes vary significantly in their chemical makeup (Table 5-1). Although most membranes have roughly equal amounts of protein and lipid, the protein/lipid ratio ranges from as high as 3.2 in some mitochondrial membranes to as low as 0.25 in myelin. Such enormous variations in protein content are difficult to reconcile with the uniform protein layers inherent to the Davson-Danielli model.

The Fluid Mosaic Model Is a More Accurate Representation of Membrane Architecture

The preceding problems stimulated considerable interest in the design of new models of membrane organization, culminating in 1972 with the **fluid mosaic model** proposed by S. Jonathan Singer and Garth Nicolson (Figure 5-5). The fluid mosaic model, which now dominates

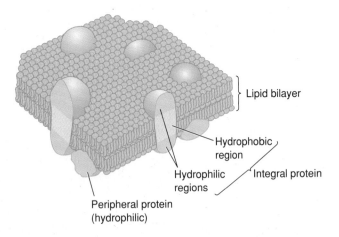

Figure 5-5   The Fluid Mosaic Model of Membrane Organization   *In this model, originally proposed in 1972 by Singer and Nicolson, protein molecules are dispersed in a fluid lipid bilayer. The hydrophobic regions of membrane proteins are embedded within the lipid bilayer, and their hydrophilic regions are exposed at the membrane's surfaces. Lipid-anchored proteins were not part of the initial fluid mosaic model, but have been added to it more recently (see Figure 5-17).*

our view of membrane organization, is based on three guiding principles. (1) Membrane lipids are arranged in the form of a bilayer, but the bilayer is interrupted by embedded proteins. The amount of embedded protein can vary significantly, depending on the type of membrane. (2) The lipid bilayer is fluid, thereby permitting lateral movement of proteins and lipids within the plane of the membrane. (3) Proteins are bound to membranes in several different ways. *Integral proteins* are embedded within the lipid bilayer, held by hydrophobic forces involving hydrophobic regions of the protein and hydrophobic fatty acid chains of the lipid bilayer. *Peripheral proteins* reside outside the lipid bilayer, bound to the membrane's outer and inner surfaces by ionic bonds that link the protein to phospholipid head groups and/or to other membrane proteins. And finally, *lipid-anchored proteins* also reside outside the lipid bilayer, but they are covalently bound to lipid molecules in the bilayer. (Lipid-anchored proteins were not part of the initial fluid mosaic model, but have been added to it more recently.)

The major strength of the fluid mosaic model is that it provides answers for most of the criticisms of the Davson-Danielli model. For example, the presence of proteins embedded in the lipid bilayer of the fluid mosaic model is consistent with the protein particles that have been observed within membranes in freeze-fracture micrographs. The idea that integral proteins are bound to the membrane by hydrophobic forces is compatible with the observation that many membrane proteins are not readily extracted by increasing the ionic strength, and hence cannot be attached to the membrane by ionic bonds. The exposure of lipid head groups at the membrane surface is compatible with their susceptibility to phospholipase digestion. And finally, the fluid mosaic model is thermodynamically favorable because the hydrophilic regions of the membrane lipids and proteins are exposed to the aqueous environment, whereas their hydrophobic regions are buried in the membrane interior away from contact with water (see Figure 5-5). Even the trilaminar staining pattern observed in electron micrographs (see Figure 5-2) is consistent with the fluid mosaic model if one postulates that the electron-opaque stains are binding to the hydrophilic regions of membrane proteins and lipids. Although the fluid mosaic model has undergone some small revisions since it was first proposed and will no doubt continue to be refined as new information becomes available, the principles underlying this model are now widely accepted as the basic starting point for understanding membrane organization and function.

MEMBRANE LIPIDS

Cells contain many different kinds of membranes, each characterized by a unique chemical composition and function. But in spite of this diversity, all membranes possess a similar structural backbone consisting of a bilayer of lipid molecules. Investigating the properties of lipid bilayers is therefore crucial if we are to understand how membranes carry out their diverse functions.

Phospholipids, Glycolipids, and Steroids Are the Predominant Membrane Lipids

Although the lipid bilayer is a universal feature of all cellular membranes, the types of lipids utilized in membrane construction can vary significantly. **Phospholipids** are the most prevalent class of membrane lipid, but many kinds of phospholipids are employed and their relative proportions differ among membranes (Figure 5-6). Sphingomyelin, for example, is one of the main phospholipids of animal plasma membranes, but it is scarce or absent in mitochondrial, chloroplast, and

plant and bacterial plasma membranes. In addition to phospholipids, the other classes of lipids commonly encountered in membranes are **glycolipids** and **steroids.** Variations in membrane glycolipid and steroid content are as common as variations in phospholipid makeup. Glycolipids, for example, are abundant in myelin and chloroplast membranes, but present in only trace amounts in bacterial plasma membranes. Cholesterol is the principal steroid found in animal cell membranes, whereas cholesterol derivatives such as sitosterol, campesterol, and stigmasterol are the predominant steroids of plant plasma membranes.

The fatty acid chains that comprise the hydrophobic tails of membrane phospholipids and glycolipids also vary significantly among membranes. Less than 10 percent of the fatty acid chains in myelin membrane lipids are unsaturated (contain double bonds), but more than 50 percent of the chains in mitochondrial and chloroplast membrane lipids are unsaturated. Chloroplast membranes are rich in branched-chain fatty acids, whereas bacterial membranes often contain unusual fatty acids with branched chains or cyclopropane rings. Changes in dietary intake or physiological conditions can induce significant alterations in the spectrum of fatty acids used in the construction of membrane lipids, indicating that their fatty acid makeup is not rigidly specified.

Membrane Lipids Spontaneously Form Bilayers Because They Are Amphipathic

In spite of the chemical differences observed among the various types of membrane lipids, they all share an important property in common—*the ability to form bilayers.* This shared property reflects an underlying

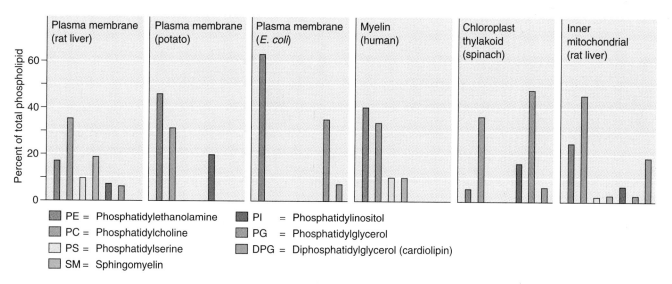

Figure 5-6 Phospholipid Composition of Several Types of Membranes *Analysis is generally carried out by using organic solvents to extract the lipids from purified membrane preparations and then fractionating the isolated lipids by thin-layer chromatography (page 148).*

structural resemblance among the different lipids. The resemblance is most striking for phospholipids and glycolipids, both of which consist of a hydrophilic head group and two hydrophobic fatty acid tails (Figure 5-7). Steroids such as cholesterol have a somewhat smaller hydrophilic head group and a single hydrophobic tail instead of two, but the overall separation of these molecules into distinct hydrophilic and hydrophobic regions is still apparent. Substances that exhibit both hydrophilic and hydrophobic properties in the same molecule are said to be **amphipathic.**

It is the amphipathic nature of membrane lipids that promotes the formation of bilayers. When placed in an aqueous environment, amphipathic molecules tend to organize in a way that minimizes contact between their hydrophobic regions and the surrounding water molecules. This phenomenon can be demonstrated experimentally by either (1) placing a drop containing amphipathic lipids in a small hole separating two aqueous compartments or (2) exposing an amphipathic lipid-water suspension to ultrasonic vibrations. The first procedure generates planar lipid bilayers, whereas the second produces enclosed vesicles called **liposomes** (Figure 5-8). In both cases the lipid molecules spontaneously arrange into bilayers in which the hydrophilic head groups are exposed to the surrounding water molecules while the hydrophobic tails are buried within the bilayer to shield them from contact with water.

Among the properties of artificial lipid bilayers that have been studied are thickness, electrical resistance, permeability to water and small molecules, temperature-dependent changes in state, electron spin resonance spectra, and X-ray diffraction patterns. In most cases the data obtained are similar to those derived from studies of natural membranes, supporting the theory that the lipid molecules in cellular membranes are arranged in the form of a bilayer.

The Lipid Bilayer Is Fluid

The first indication that the lipid bilayer may be fluid rather than rigid came from studies employing the technique of *electron spin resonance (ESR)* spectroscopy to monitor the freedom of movement of membrane lipids. In this technique a chemical group containing an unpaired electron, usually a nitroxide group, is attached to the fatty acid tail of a phospholipid molecule. The term *spin-label* is generally employed to refer to such substances. The presence of the unpaired electron in the spin-label causes it to absorb and emit energy when exposed to an external magnetic field of appropriate intensity. If one plots the amount of energy absorbed as a function of the intensity of the magnetic field, an ESR spectrum is obtained. The shape of such a spectrum is influenced by the mobility of the molecule containing the unpaired electron. Therefore one can introduce

phospholipids containing a nitroxide spin-label into cellular membranes and use ESR spectroscopy to monitor the mobility of these molecules within the bilayer.

Such experiments, pioneered in the laboratories of Harden McConnell and O. Hayes Griffith in the late 1960s, provided the first direct evidence for the fluidity of the lipid bilayer. In these studies, spin-labels inserted into cellular membranes were found to exhibit an ESR spectrum whose shape is intermediate between that of a spectrum produced by a rigidly fixed molecule and that produced by a completely mobile molecule (Figure 5-9). This intermediate pattern indicates that the lipid molecules in the bilayer are neither fixed in position, as in a crystal, nor completely free to move, as in a liquid. In this partially fluid state the individual lipid molecules are capable of **lateral diffusion** within the bilayer, but they always retain the same orientation, with their hydrophilic head groups pointed toward the membrane surface and their hydrophobic tails projecting toward the membrane interior (Figure 5-10). Lateral diffusion of lipid molecules is so rapid that individual lipid molecules can diffuse from one end of a cell to the other in a matter of a few seconds.

In contrast to the speed of lateral migration, the movement of lipid molecules from one side or **leaflet** of the bilayer to the other is normally quite slow. This type of movement, in which membrane lipids cross from one leaflet to the other, is called **transverse diffusion** or **flip-flop.** In an elegant experiment designed to investigate the rate at which flip-flop occurs, Roger Kornberg and Harden McConnell generated artificial phospholipid vesicles containing a nitroxide spin-label, and then incubated these vesicles with the reducing agent ascorbate, to which the vesicles are impermeable. Ascorbate converts the nitroxide group to an amine, thereby abolishing the characteristic ESR spectrum of the spin-label. Within a few minutes of ascorbate treatment the magnitude of the ESR signal decreased by half, indicating that all the spin-label groups exposed on the exterior of the vesicle had been degraded. In contrast, the spin-label groups associated with lipid molecules in the inner leaflet initially remained intact because the vesicles were impermeable to ascorbate. But these remaining spin-label groups gradually disappeared over a period of many hours; this gradual loss was caused by the occasional flip-flop of molecules from the inner leaflet of the bilayer to the outer half, where they were degraded by the ascorbate present in the external medium. From the rate of decay of the ESR signal, it was calculated that flip-flop occurs once every several hours for a typical membrane lipid. Thus the movement of lipid molecules from one side of a pure lipid bilayer to the other is extremely slow. In a later chapter we will see that this process can be speeded up by special membrane proteins known as *phospholipid translocators* (page 261).

In addition to electron spin resonance spectroscopy, the technique of *differential scanning*

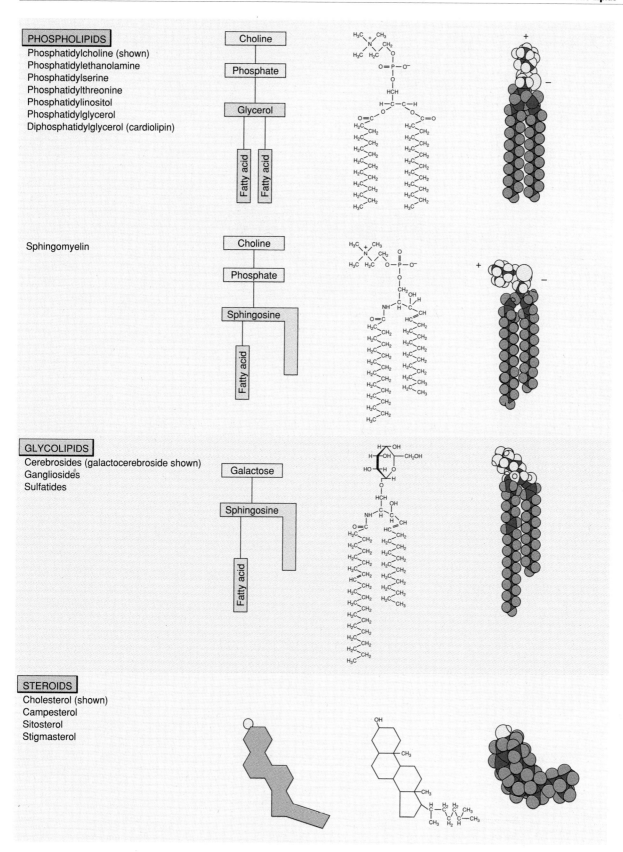

PHOSPHOLIPIDS
Phosphatidylcholine (shown)
Phosphatidylethanolamine
Phosphatidylserine
Phosphatidylthreonine
Phosphatidylinositol
Phosphatidylglycerol
Diphosphatidylglycerol (cardiolipin)

Sphingomyelin

GLYCOLIPIDS
Cerebrosides (galactocerebroside shown)
Gangliosides
Sulfatides

STEROIDS
Cholesterol (shown)
Campesterol
Sitosterol
Stigmasterol

Figure 5-7 Structures of the Three Major Classes of Membrane Lipids: Phospholipids, Glycolipids, and Steroids
Phospholipids contain phosphate groups covalently attached to a lipid backbone, glycolipids contain carbohydrate groups attached to a lipid backbone, and steroids are derivatives of a four-membered ring compound (page16). In spite of their structural differences, all three types of lipid are amphipathic molecules containing hydrophilic head groups and hydrophobic tails.

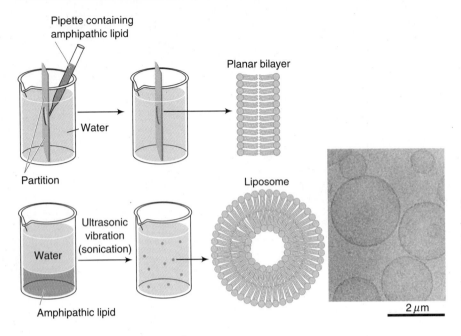

Figure 5-8 *Procedures for Creating Artificial Lipid Bilayers* *Depending on the procedure employed, artificial bilayers take the form of either planar membranes or enclosed vesicles (liposomes). Electron micrograph of liposomes courtesy of P. M. Frederik.*

2 μm

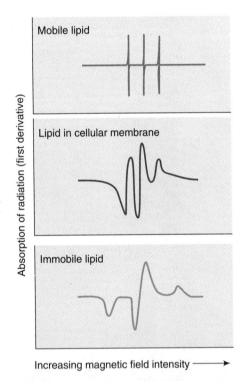

Figure 5-9 *Demonstration of the Mobility of Membrane Lipids Using Electron Spin Resonance Spectroscopy* (Top) *A spin-labeled phospholipid in solution at room temperature yields an ESR spectrum typical of a completely mobile molecule. (Bottom) When the temperature of a solution containing the same spin-labeled phospholipid is lowered to −65°C, the ESR spectrum changes to that of an immobilized molecule. (Middle) The ESR spectrum of the same spin-labeled phospholipid when inserted into a cellular membrane at room temperature is intermediate between the above two extremes. Such a pattern indicates that membrane lipids are in a partially fluid state.*

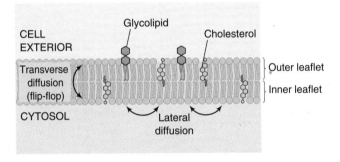

Figure 5-10 *The Difference between Lateral Diffusion and Transverse Diffusion (Flip-Flop) of Membrane Lipids* *Membrane fluidity allows rapid lateral diffusion of membrane lipids, but not flip-flop between the two halves of the bilayer. This diagram also illustrates that membrane lipids are distributed asymmetrically, giving the inner and outer halves of the bilayer different lipid compositions.*

calorimetry has also made important contributions to our understanding of membrane fluidity. This procedure monitors the uptake of heat that is known to occur during the transition between different physical states (e.g., from solid to liquid or liquid to gas). At low temperatures lipid bilayers are immobilized in a gel state, but raising the temperature causes them to "melt" into a more fluid state. The *transition temperature* (T_m) at which the gel-to-fluid melting occurs can be determined by placing membranes into a sealed chamber (calorimeter) in which the uptake of heat is measured as the temperature is increased. The point of maximum heat absorption corresponds to the transition temperature (Figure 5-11). For most membranes the transition temperature is lower than the normal temperature at which

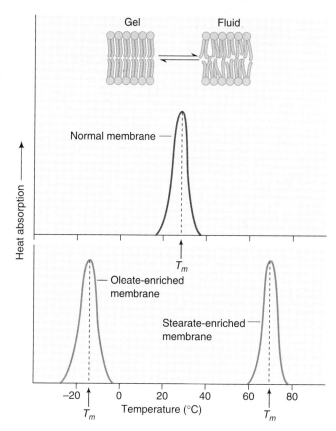

Figure 5-11 *Using Differential Scanning Calorimetry to Monitor the Transition of Membranes from the Gel State to the Fluid State* (Top) *As the temperature is increased, membranes absorb heat at the gel-to-fluid transition point. (Bottom) Membranes from cells grown in media enriched in oleate, an unsaturated fatty acid, are more fluid than normal membranes and therefore have a lower transition temperature. Membranes from cells grown in media enriched in stearate, a saturated fatty acid, are less fluid than normal membranes and therefore have a higher transition temperature.*

the membrane exists; hence the bilayer tends to remain in a fluid state.

Membrane Fluidity Can Be Altered by Changes in Lipid Composition

Transition temperatures vary significantly for different kinds of membranes, indicating that membranes are not all equally fluid. Three aspects of a membrane's lipid makeup are especially important in determining fluidity: fatty acid chain length, fatty acid saturation, and steroid content. (1) Long-chain fatty acids have higher transition temperatures than shorter-chain fatty acids, which means that membranes enriched in long-chain fatty acids tend to exhibit decreased fluidity. (2) Unsaturated fatty acids have lower transition temperatures than saturated fatty acids, giving membranes enriched in unsaturated fatty acids an increased fluidity. (3) Steroids such as cholesterol, sitosterol, and stigmasterol contain in-

flexible rings that make the bilayer less fluid. Steroid molecules also exert an opposing effect; they prevent fatty acid chains from forming tightly packed arrays, thus *counteracting* the decrease in fluidity that normally occurs as the temperature is lowered. As a result of these two opposing effects, steroid molecules tend to minimize both decreases and increases in membrane fluidity that might otherwise occur in response to temperature fluctuations.

The effects of lipid composition on membrane fluidity can be quite dramatic. Membranes enriched in lipids containing the unsaturated fatty acid oleate, for example, have transition temperatures that are as much as 80°C lower than membranes enriched in lipids containing the saturated fatty acid stearate (see Figure 5-11). The ability to regulate membrane fluidity by changing the lipid composition is important to organisms that do not maintain a constant temperature, such as microorganisms, plants, and cold-blooded animals. When organisms of this type are exposed to low temperatures, membrane fluidity would be expected to decrease. The organisms adapt, however, by increasing the degree of lipid unsaturation or decreasing the average length of the fatty acid chains, thereby maintaining membrane fluidity.

The existence of mechanisms for regulating membrane fluidity suggests that a certain degree of fluidity is required for normal membrane function. Strong evidence for this assertion has emerged from studies on *Acholeplasma laidlawii*, a small bacteriumlike organism that uses whatever fatty acids are available in the environment for constructing its plasma membrane. When grown in a medium enriched with saturated fatty acids, *Acholeplasma* is forced to construct membranes with an abnormally high content of saturated fatty acids. This abnormal lipid composition causes the transition temperature of the plasma membrane to increase to almost 60°C, immobilizing the lipid bilayer in the gel state. As a result the cells tend to swell and rupture, indicating that normal membrane permeability and transport depend upon the fluid state of the lipid bilayer. These observations are potentially applicable to other organisms as well. In animals the fatty acid composition of cellular membranes, and hence membrane fluidity, is influenced by diet. Hence an organism's dietary habits may directly affect the properties of its cellular membranes.

Lipids Are Arranged Asymmetrically across the Bilayer

Since membranes are constructed from a mixture of lipid types, the question arises as to whether the various kinds of lipids are randomly distributed across both leaflets of the bilayer. This question was first addressed in studies involving mammalian red blood cells (*erythrocytes*), which lack intracellular organelles and membranes and are therefore ideal sources of purified plasma membranes. One of the easiest ways to isolate the plasma membrane from red blood cells is to decrease the ionic strength of the surrounding medium, thereby causing water to enter the cells by osmosis (page 176). As a consequence the cells swell and the plasma membrane ruptures, allowing the intracellular contents to leak out. The remaining empty plasma membrane sac is called a red cell *ghost* (Figure 5-12).

A clever trick for analyzing the arrangement of lipids in these membranes has emerged from the discovery that red cell ghosts can be converted into either right-side-out or inside-out membrane vesicles by altering the extracellular Mg^{2+} concentration. The spatial arrangement of membrane lipids can then be investigated by exposing the vesicles to reagents that cannot cross the membrane. One popular approach involves the use of phospholipid-hydrolyzing enzymes called *phospholipases*. Because they are too large to pass through the membrane, the phospholipase molecules can only attack lipids that are exposed on the outer surface of the vesicle. One can therefore investigate lipid

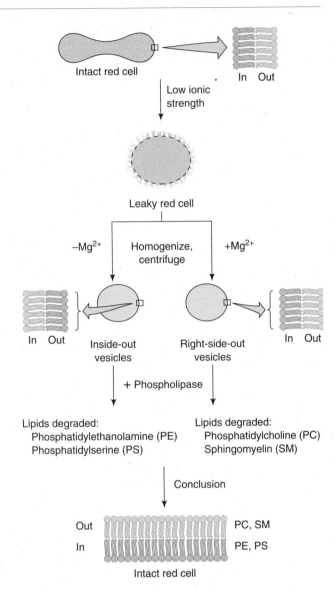

Figure 5-13 *Studying Lipid Distribution in Right-Side-Out and Inside-Out Red Cell Membrane Vesicles* *Exposing red blood cells to low ionic strength causes the cells to become leaky. Homogenization in the presence or absence of Mg^{2+} converts these leaky ghosts to right-side-out or inside-out vesicles. Subsequent digestion with phospholipase establishes which phospholipids are situated in the inner and outer halves of the bilayer.*

distribution in the two halves of the lipid bilayer by determining which lipids are degraded when phospholipases are added to right-side-out vesicles and which ones are degraded when phospholipases are added to inside-out vesicles (Figure 5-13).

Experiments of this type have revealed that phosphatidylcholine and sphingomyelin are located mainly in the outer leaflet of the plasma membrane, whereas phosphatidylethanolamine and phosphatidylserine occur predominantly in the inner leaflet. Because phosphatidylserine carries a net negative charge, its prevalence in the inner half of the bilayer creates a difference in charge between the two sides of the bilayer.

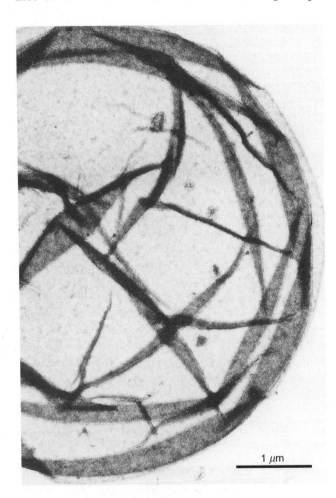

Figure 5-12 *Red Blood Cell Ghost* *Empty plasma membrane sacs or "ghosts" are prepared by exposing red blood cells to hypotonic (low ionic strength) solutions, which causes cells to rupture and their intracellular contents to leak out. Courtesy of T. Steck.*

Glycolipids are restricted to the outer half of the bilayer, with their carbohydrate groups protruding away from the cell surface. Lipids in the outer leaflet also tend to have fatty acid chains that are less saturated (have fewer double bonds) than lipids in the inner leaflet, making the outer half of the plasma membrane slightly less fluid than the inner half.

An asymmetrical arrangement of membrane lipids has also been reported for several types of membranes other than the plasma membrane (Figure 5-14). Although the functional significance of lipid asymmetry is not completely understood, in some cases a direct relationship to a particular membrane property has been noted. For example, glycolipids are restricted to the outer leaflet of the plasma membrane because their carbohydrate groups are involved in signaling events that occur at the outer cell surface. Phosphatidylserine and phosphatidylinositol, on the other hand, are more highly concentrated in the inner leaflet of the plasma membrane, where they are involved in transmitting certain kinds of signals from the plasma membrane to the cell interior (page 218).

MEMBRANE PROTEINS

While the lipid bilayer imparts an underlying structural similarity to all cellular membranes, proteins are responsible for most of the functional traits that distinguish dif-ferent kinds of membranes from one another. Hundreds of different membrane proteins have been identified to date, most of which can be grouped into one of four basic categories depending on whether they function as: (1) transport proteins, (2) receptors for extracellular signals, (3) attachment sites for intracellular or extracellular components, or (4) catalysts for metabolic reactions (Figure 5-15). Each of these categories includes dozens of different proteins with wide-ranging effects on cellular activities.

Radioactive Labeling Procedures Permit the Orientation of Membrane Proteins to Be Studied

The spatial arrangement of membrane proteins is often investigated using purified membrane preparations that have been obtained by subcellular fractionation procedures. After membranes have been isolated and purified, their protein composition is determined by extracting with the detergent *sodium dodecyl sulfate (SDS)* and fractionating the extracted polypeptides by one- or two-dimensional SDS polyacrylamide gel electrophoresis (page 145). To study the properties of individual membrane proteins, they must first be removed from the membrane and purified. Proteins bound to membrane surfaces by ionic bonds are usually isolated by extracting membranes with solutions of high ionic strength. In contrast, proteins embedded in the phospholipid bilayer must be extracted with detergent solutions. Detergents are amphipathic molecules whose

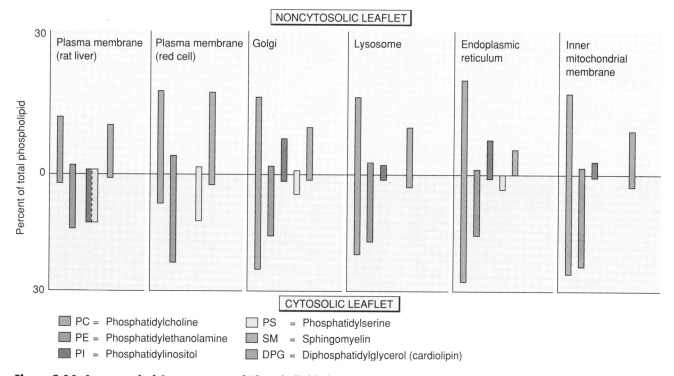

Figure 5-14 Asymmetrical Arrangement of Phospholipids in Several Types of Membranes *For each membrane the bottom set of bars represents the phospholipid makeup of the half of the bilayer that faces the cytosol, and the top set of bars represents the phospholipid makeup of the noncytosolic (exoplasmic) side of the bilayer.*

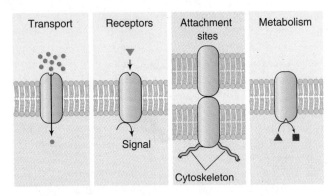

Figure 5-15 *Major Functions of Membrane Proteins*   *In their most prominent roles, membrane proteins function as solute transporters, receptors for external signals, attachment sites for intracellular and extracellular structures, and enzymes that catalyze metabolic reactions.*

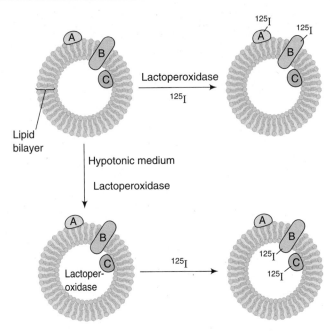

Figure 5-16 *A Method for Labeling Proteins Exposed at Either the Outer or Inner Surface of a Membrane Vesicle*   (Top) *Lactoperoxidase located outside the membrane vesicle is employed to catalyze the labeling of membrane proteins exposed at the outer membrane surface. In this example, proteins A and B become radioactively labeled under these conditions. (Bottom) Membrane vesicles are incubated in a hypotonic medium to make them permeable to lactoperoxidase, which then catalyzes the labeling of membrane proteins exposed at the inner membrane surface. In this example, proteins B and C become radioactively labeled under these conditions. The labeling patterns indicate that protein A is exposed at the outer membrane surface, protein B is exposed at both the inner and outer membrane surfaces, and protein C is exposed at the inner membrane surface.*

hydrophobic regions interact with the hydrophobic segments of protein molecules, and whose hydrophilic regions permit the detergent-protein complex to be solubilized in water.

To determine how proteins are oriented in membranes, radioactive labeling procedures have been devised that distinguish between proteins exposed on the inner and outer surfaces of membrane vesicles. A typical example is the use of the enzyme *lactoperoxidase* to catalyze the labeling of membrane proteins with radioactive iodine. Because lactoperoxidase is too large to pass through membranes, only proteins exposed on the outer surface of intact membrane vesicles are labeled. To label proteins exposed on the inner membrane surface, vesicles are first exposed to a hypotonic (low ionic strength) solution to make them more permeable; the lactoperoxidase can then enter the vesicles and catalyze the labeling of proteins exposed on the vesicle's inner surface. In this way, it is possible to determine whether a given membrane protein is exposed on the inner membrane surface, the outer membrane surface, or both (Figure 5-16). Similar experiments are also carried out using the enzyme *galactose oxidase* to catalyze labeling of membrane-associated carbohydrate groups with ^{3}H-borohydride. This allows the orientation of carbohydrate chains attached to membrane glycoproteins to be investigated.

Membranes Contain Integral, Peripheral, and Lipid-Anchored Proteins

The fluid mosaic model as originally formulated envisioned two main classes of membrane proteins. The first class consists of amphipathic molecules, called **integral proteins,** whose hydrophobic regions are embedded within the lipid bilayer and hydrophilic regions are exposed at the membrane's surfaces (Figure 5-17). Proteins of this type account for most membrane-

associated receptors and transporters. Integral proteins are attached to membranes by their hydrophobic amino acids, which bind to the hydrophobic fatty acid chains of the lipid bilayer. Hence they can only be removed from membranes by treatments that disrupt hydrophobic interactions, such as extraction with detergents or organic solvents. Almost all integral proteins span the lipid bilayer, passing from one side of the membrane to the other. Most of these *transmembrane* proteins are anchored to the bilayer by one or more α-helical stretches, each consisting of 15–25 mainly hydrophobic amino acids. For membrane proteins whose amino acid sequences are known, the tentative position of the transmembrane α-helical stretches can be determined by a computer program that examines the amino acid sequence of the protein and identifies the location of hydrophobic amino acid clusters. This kind of analysis generates a map, called a **hydropathy plot,** that predicts how many membrane-spanning regions are contained within a given membrane protein (Figure 5-18).

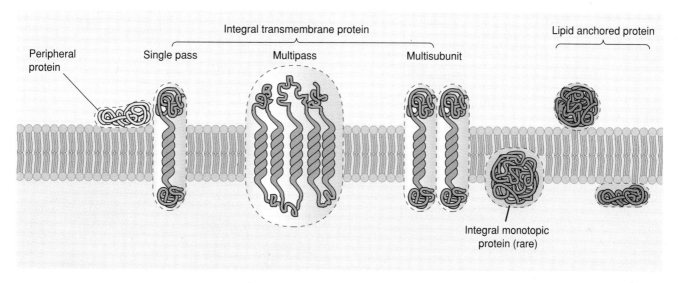

Figure 5-17 ***The Three Main Classes of Membrane Proteins*** *Peripheral proteins are attached to the membrane by ionic bonds that link them to other membrane proteins or to phospholipid head groups. Integral proteins contain hydrophobic regions that are embedded within the lipid bilayer. Most integral proteins are transmembrane proteins that span the lipid bilayer, but a few integral proteins may be embedded on only one side of the bilayer (integral monotopic proteins). Lipid-anchored proteins are attached to the membrane by covalently bound lipid molecules that are embedded in the lipid bilayer.*

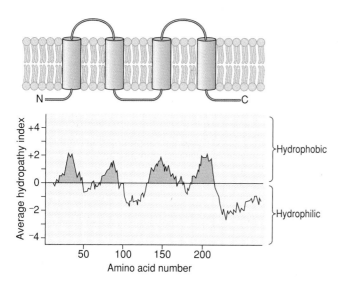

Figure 5-18 ***Hydropathy Plot for a Typical Transmembrane Protein*** *The hydropathy index, which is plotted on the vertical axis, represents the calculated water solubility of successive segments of the protein chain based on amino acid composition. This particular molecule, a plasma membrane protein known as connexin, exhibits four distinct hydrophobic regions. These four regions correspond to four α-helical segments (illustrated as cylinders) that span the plasma membrane.*

The second class of proteins envisioned by the fluid mosaic model consists of hydrophilic molecules whose polypeptide chains are situated entirely outside the lipid bilayer. Such molecules, termed **peripheral proteins,** are bound mainly by ionic bonds to other membrane proteins or to phospholipid head groups. Because the ionic forces holding these proteins to membranes are

relatively weak, peripheral proteins are easily removed by extracting membranes with solutions of increased ionic strength.

In addition to integral and peripheral proteins, membranes are now known to contain a third class of proteins that are strictly speaking neither integral nor peripheral, but have some of the characteristics of both. The polypeptide chains of this third type of membrane protein are situated outside the bilayer, but are covalently bound to lipid molecules in the bilayer. Such **lipid-anchored proteins** can protrude from either side of a membrane. Those associated with the cytoplasmic surface of the plasma membrane exhibit a particularly interesting property; that is, they exist in the form of soluble cytoplasmic proteins as well as in a membrane-attached form. These proteins are attached to the plasma membrane by the addition of a lipid sidechain that subsequently becomes inserted into the plasma membrane bilayer (page 304).

Some Membrane Proteins Are Free to Move within the Lipid Bilayer

Earlier in the chapter we established that lipid molecules can diffuse laterally within the plane of a membrane. Are membrane proteins also free to move within the membrane? In a classic experiment designed to address this question, David Frye and Michael Edidin investigated the behavior of membrane proteins in human and mouse cells that had been artificially fused (Figure 5-19). The presence of mouse membrane proteins was detected using mouse-specific antibodies linked to a green fluorescent dye called *fluorescein,* while human membrane proteins were located with human-specific antibodies linked to a red fluorescent dye *rhodamine.* The plasma membrane of the newly fused cell hybrids was found to exhibit distinct regions fluorescing either green or red, indicating the presence of separate regions of plasma membrane containing either mouse or human proteins. Within an hour, however, the separate regions of green and red fluorescence were completely intermixed, indicating that the mouse and human membrane proteins had diffused through the lipid bilayer and intermingled. If the fluidity of the membrane was depressed by lowering the temperature below the transition temperature of the lipid bilayer, the intermixing could be prevented.

Additional evidence for the mobility of membrane proteins emerged from studies in which white blood cells were incubated with fluorescent antibodies that bind to specific membrane proteins; fluorescence microscopy was then employed to determine the location of the fluorescent antibodies (and hence the membrane proteins to which they had bound). Initially each cell exhibited a diffuse pattern of fluorescence, indicating that the fluorescent antibodies were binding to membrane protein molecules distributed over the entire surface of the cell. Within an hour, however, the fluorescence had become concentrated at one end of the cell, an indication that the membrane proteins had moved together to form a single large cluster (Figure 5-20). This migration of material to one end a cell—a phenomenon called **capping**—demonstrates that membrane proteins are capable of lateral movement through the lipid bilayer.

If proteins are completely free to diffuse within the plane of the membrane, then they should eventually become randomly distributed. Support for the idea that at least some membrane proteins behave in this way has emerged from freeze-fracture microscopy, which

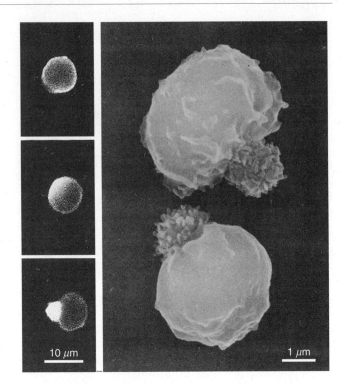

Figure 5-20 *Evidence for the Mobility of Membrane Proteins Provided by the Phenomenon of Lymphocyte Capping* *Lymphocytes were treated with fluorescein-labeled antibodies that bind to plasma membrane proteins. (Top left) Initially the fluorescence is distributed over the entire cell surface. (Middle left) Somewhat later a patchy distribution of fluorescence develops. (Bottom left) Within an hour the cell-surface fluorescence has become concentrated in a single region referred to as a "cap," indicating that the membrane proteins have moved together to form a single cluster. Capping occurs in the presence of antibody molecules because antibodies are capable of crosslinking proteins into large aggregates, which then become anchored at one end of the cell. (Right) Scanning electron micrograph of capped cells, Courtesy of I. Yahara.*

directly visualizes proteins embedded within the lipid bilayer (page 124). When plasma membranes are examined in freeze-fracture micrographs, their embedded protein particles often tend to be randomly distributed. Such evidence for protein mobility is not restricted to the plasma membrane. It has also been found, for example, that the protein particles of the inner mitochondrial membranes are randomly arranged. If isolated mitochondrial membrane vesicles are exposed to an electric potential, the protein particles, which bear a net negative charge, all move to one end of the vesicle (Figure 5-21). Removing the electric potential causes the particles to become randomly distributed again, indicating that these proteins are free to move within the lipid bilayer.

The Mobility of Membrane Proteins Is Variable

Although many membrane proteins are now known to diffuse through the lipid bilayer, their rate of move-

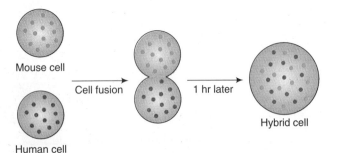

Figure 5-19 *Evidence for the Mobility of Membrane Proteins Obtained by Fusing Mouse and Human Cells* *Mouse membrane proteins were detected using mouse-specific antibodies linked to a green dye, whereas human membrane proteins were located with human-specific antibodies linked to a red dye. Within an hour after fusing mouse and human cells, the human and mouse membrane proteins had diffused through the lipid bilayer and become intermingled.*

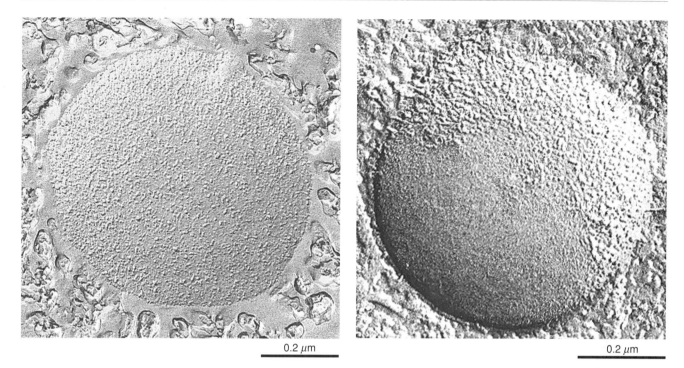

0.2 μm

0.2 μm

Figure 5-21 *Evidence for the Mobility of Membrane Proteins Obtained from Studying the Behavior of Protein Particles in Inner Mitochondrial Membrane Vesicles* (Left) *Freeze-fracture micrograph showing the random distribution of membrane protein particles.* (Right) *Exposing the membrane vesicle to an electric field causes the membrane particles to migrate to one end of the vesicle (upper right side of micrograph). Courtesy of A. E. Sowers.*

ment is variable. One widely used approach for quantifying the rate at which membrane proteins (and lipids) diffuse is *fluorescence photobleaching recovery.* In this technique, fluorescent chemical groups are first attached to membrane proteins or lipids; a narrow beam of laser light is then focused upon the membrane, which bleaches (removes the fluorescence from) a small spot on the membrane. The subsequent diffusion of unbleached molecules from adjacent parts of the membrane into the bleached region causes the fluorescence of the bleached area to gradually recover (Figure 5-22). The rate at which recovery occurs can be used to calculate the diffusion rate of the fluorescent lipid or protein molecules.

Measurements of this type have revealed that membrane lipids typically diffuse at rates approaching several micrometers per second, while membrane proteins are more variable in their diffusion rates. A few membrane proteins diffuse almost as rapidly as lipids, but most move more slowly than would be expected if they were completely free to diffuse within the lipid bilayer. Moreover, the diffusion of many membrane proteins is restricted to a limited membrane area. Such restrictions give membranes a "patchy" organization; that is, membranes consist of a series of separate *membrane domains* exhibiting different protein compositions and functions. In many cases, the reason for restricting a protein to a particular membrane domain is clear. For example, muscle cells possess membrane proteins called *acetylcholine receptors* (page 208) that are de-

signed to receive signals from adjacent nerve cells; acetylcholine receptors are selectively located where the plasma membrane of a muscle cell makes close contact with a stimulating nerve cell. Another illustration of restricted protein mobility occurs in cells lining the intestinal tract, where membrane proteins involved in solute transport are preferentially located on the side of the cell where the corresponding type of transport is required. Such examples make it clear that mechanisms for restricting the mobility and distribution of specific membrane proteins must exist. Among the mechanisms known to be utilized for restricting protein mobility are (1) aggregation of membrane proteins with each other; (2) anchoring membrane proteins to elements of the cytoskeleton and/or extracellular structures, including other cells; and (3) formation of protein barriers such as *tight junctions* (page 236), which create obstacles to the diffusion of other membrane proteins.

Membrane Proteins Are Oriented Asymmetrically— The Red Cell Membrane as a Model

The mammalian red blood cell is one of the first cell types in which the spatial arrangement of proteins within a biological membrane was investigated. As we learned earlier in the chapter, the red cell is ideally suited for membrane studies because the absence of a nucleus and cytoplasmic organelles makes its plasma membrane relatively easy to isolate. After the plasma membrane has been isolated, its lipid components can

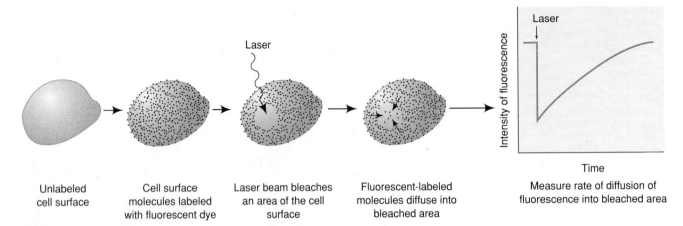

Figure 5-22 *Measuring Membrane Fluidity Using Fluorescence Photobleaching Recovery* *Membrane proteins or lipids are first labeled with a fluorescent compound, and the fluorescence in a local area is then bleached by irradiation with a laser beam. Membrane fluidity is measured by determining the rate at which the diffusion of fluorescent molecules from surrounding regions into the bleached area causes fluorescence to reappear in the laser-bleached spot.*

be removed by extracting isolated membranes with organic solvents, leaving the protein behind. When the proteins are fractionated by SDS-polyacrylamide gel electrophoresis and stained with a protein-specific dye, a relatively small number of components are seen. Prominent among them are *spectrin, ankyrin, band 3 protein, glycophorin, band 4.1 protein,* and *actin* (Figure 5-23). Several of the proteins can also be stained with a carbohydrate-specific dye called the *periodic acid–Schiff (PAS) reagent,* indicating that they are **glycoproteins** (proteins containing covalently bound carbohydrate groups). The most intense PAS-staining reaction is observed with glycophorin, which is two-thirds carbohydrate by weight.

Identifying the major protein components of the red cell membrane is only the first step in determining how this membrane is organized. To investigate the orientation of these proteins, isolated membranes have been extracted with solutions of increasing ionic strength. Spectrin, ankyrin, band 4.1, and actin are readily extracted in this way, indicating that they are peripheral proteins. The remaining molecules, mainly glycophorin and band 3, are integral proteins embedded within the lipid bilayer and therefore require harsher extraction techniques, such as treatment with detergents. The properties of these two classes of red cell membrane proteins are examined in the following sections.

The Integral Proteins of the Red Cell Membrane: Glycophorin and Band 3 Protein

The orientation of glycophorin and band 3 protein in the red cell membrane has been investigated by radioactive labeling of right-side-out and inside-out membrane vesicles (page 166). Glycophorin and band 3 can be radioactively labeled in both types of vesicles, indicating that the two proteins are exposed at the cytoplasmic as

well as the exterior surface of the plasma membrane. Such findings raise two possibilities: either glycophorin and band 3 molecules are present on both sides of the membrane, or each individual glycophorin and band 3 molecule is a transmembrane protein that spans the membrane from one side to the other. To distinguish between these alternatives, red cell membranes were radioactively labeled from either the inner or outer mem-

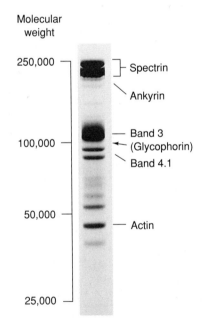

Figure 5-23 *Red Cell Membrane Proteins Separated by SDS Polyacrylamide Gel Electrophoresis* *The arrow marks the position of glycophorin, whose high carbohydrate content inhibits staining and decreases its electrophoretic mobility. This decreased mobility makes glycophorin appear to have a higher molecular weight than its actual value of 30,000 daltons. Some proteins whose functions were unknown when such electrophoretic separations were first performed have been named using numbers (e.g., band 3, band 4.1). Courtesy of T. Steck.*

brane surface. The radioactive glycophorin and band 3 molecules generated by the treatment were then digested with a protease to cleave the proteins into peptide fragments. When the red cell was labeled from the outside, one set of fragments from glycophorin and band 3 protein were found to be radioactive, whereas labeling from the inside produced a different set of radioactive fragments. This means that different regions of each glycophorin and band 3 molecule are exposed on opposite sides of the membrane, suggesting that glycophorin and band 3 are transmembrane proteins that span the membrane asymmetrically and hence expose unique portions of the polypeptide chain at each membrane surface.

Because glycophorin and band 3 protein are both glycoproteins, studies have been carried out to determine on which side of the membrane their attached carbohydrate groups are located. When right-side-out vesicles are treated with proteases, the attached carbohydrate groups are released as glycopeptides; comparable treatment of inside-out vesicles does not release any carbohydrate. Likewise, radioactive labeling of carbohydrate groups using galactose oxidase and ^{3}H-borohydride (page 168) has shown that carbohydrate groups can be labeled only from the outside of right-side-out vesicles. Both observations indicate that the carbohydrate groups of glycophorin and band 3 protein face the cell exterior. This asymmetric arrangement typifies the orientation of membrane-associated carbohydrate groups, which are located on the noncytoplasmic (*exoplasmic*) surface of cellular membranes. This means that the carbohydrate groups of the plasma membrane face the cell exterior, whereas the carbohydrate groups of intracellular membranes (like mitochondrial or ER

membranes) face the lumen (interior) of the organelle or compartment.

Much is now known about the orientation of glycophorin and band 3 within the plasma membrane (Figure 5-24). Glycophorin is a single polypeptide 131 amino acids long that contains 16 covalently attached carbohydrate chains. The polypeptide chain can be subdivided into three distinct regions. The N-terminal region, which is the site of the attached carbohydrate chains, protrudes from the cell surface. The middle portion of the chain consists of a single hydrophobic α-helical segment that passes from one side of the lipid bilayer to the other. And finally, the C-terminal region is enriched in hydrophilic amino acids and is exposed at the inner surface of the membrane. Glycophorin is thus arranged according to the principles of the fluid mosaic model; its hydrophobic region is buried in the membrane interior, while its hydrophilic ends are associated with the aqueous environments outside and inside the cell.

In contrast to glycophorin, which consists of a single polypeptide chain, band 3 protein is a dimer consisting of two identical chains. Both the N-terminus and C-terminus of each polypeptide are situated at the inner surface of the membrane, while the middle region of the molecule consists of about a dozen hydrophobic α-helical segments that pass back and forth across the lipid bilayer. A single hydrophilic carbohydrate chain extends outward from the cell surface. Thus each subunit of the band 3 protein is organized like glycophorin in that its hydrophobic regions are buried in the membrane interior; however, unlike glycophorin, the N-terminus and C-terminus of band 3 are located on the same side of the membrane and multiple hydrophobic regions (rather than one) traverse the membrane.

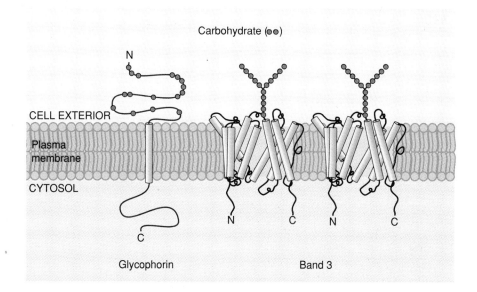

Figure 5-24 *Orientation of Glycophorin and Band 3 Protein in the Red Cell Membrane*
The cylindrical shapes represent regions of α helix. Glycophorin is a single-pass transmembrane protein, whereas band 3 is a multipass transmembrane protein consisting of two subunits.

The Peripheral Proteins of the Red Cell Membrane: Spectrin, Ankyrin, Band 4.1, and Actin

The peripheral proteins of the red cell membrane can be radioactively labeled in inside-out but not right-side-out vesicles. Thus, unlike glycophorin and band 3 protein, which are exposed at both the outer and inner surfaces of the plasma membrane, the peripheral proteins of the red cell membrane are associated exclusively with the inner membrane surface. In electron micrographs, these proteins form a meshwork associated with the inner surface of the plasma membrane. The main constituent of the meshwork is a long thin molecule called **spectrin;** although purified spectrin does not form an interconnected network by itself, mixing spectrin with actin and band 4.1 protein results in the formation of a meshwork that closely resembles the material associated with the cytoplasmic surface of the red cell membrane. The spectrin network is bound to the plasma membrane by two proteins: the **band 4.1 protein,** which links spectrin to glycophorin, and **ankyrin,** which links spectrin to band 3 (Figure 5-25). Spectrin and its associated proteins form a skeletal framework that supports the plasma membrane and helps the red blood cell maintain its distinctive biconcave shape (see Figure 3-17, *left*).

Many types of evidence indicate that the spectrin meshwork functions as a cytoskeleton that stabilizes the plasma membrane and restricts the mobility of the membrane's integral proteins. It has been shown, for example, that removing spectrin from red cell membranes increases the lateral mobility of the membrane's integral proteins and makes the membrane more fragile. Likewise, fluorescence photobleaching recovery measurements have revealed that integral membrane proteins diffuse 50 times more rapidly in spectrin-deficient mutant red cells than in normal red cells. Although spectrin is unique to red blood cells, similar cytoskeletal proteins appear to play comparable roles in other cell types.

Membrane Proteins Can Consist of Different Kinds of Subunits

In examining the red cell membrane, we encountered examples of integral membrane proteins that consist of either one or two copies of the same polypeptide chain. Membrane proteins can also be constructed from different kinds of polypeptides that interact with one another to form functional complexes. The first such multisubunit membrane protein complex to have its molecular structure elucidated was the *photosynthetic reaction center complex* of the purple photosynthetic bacterium, *Rhodopseudomonas viridis.* This protein complex, whose role in trapping light energy for photo-

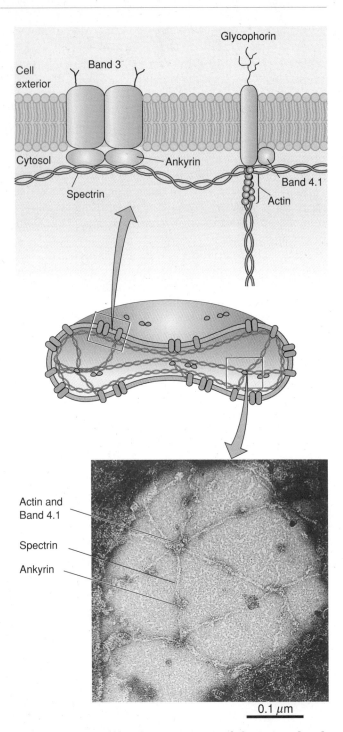

Figure 5-25 Postulated Arrangement of the Integral and Peripheral Proteins of the Red Cell Plasma Membrane
The peripheral proteins spectrin, actin, and band 4.1 form a cytoskeletal meshwork bound to the cytoplasmic surface of the membrane. Micrograph courtesy of D. Branton.

synthesis is described in Chapter 9, consists of four separate polypeptide chains. The three-dimensional arrangement of the four polypeptides has been determined by extracting the protein complex from the plasma membrane with detergents, crystallizing the isolated protein,

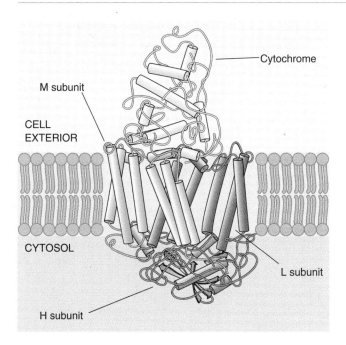

Figure 5-26 *Structure of a Bacterial Photosynthetic Reaction Center* *The cylindrical shapes represent regions of α helix. Each of the four subunits is indicated by a different color.*

and analyzing the structure of the crystallized complex by X-ray diffraction (Figure 5-26).

Such studies have revealed that three of the polypeptide chains, called the *L, M,* and *H* subunits, are integral membrane proteins containing α-helical segments that are buried within the lipid bilayer. The L and M subunits are each folded into five α helices that pass back and forth across the lipid bilayer, while the H subunit contains only a single α-helical segment buried within the bilayer; the remainder of the H subunit is folded into a globular hydrophilic domain that is exposed on the cytoplasmic surface of the membrane. The fourth subunit, called a *cytochrome,* is a peripheral protein that is bound to the exterior surface of the membrane.

As predicted by the fluid mosaic model, the transmembrane regions of the L, M, and H subunits are enriched in hydrophobic amino acids that anchor the protein to the surrounding hydrophobic lipid tails. However, several hydrophilic amino acids are buried in the interior of the protein complex itself, where they function in binding metal ions associated with the reaction center (page 396). This situation, with hydrophobic regions located on the exterior of the protein molecule and hydrophilic regions buried in the interior, is the opposite of what is observed with water-soluble proteins, where the exterior surface of the molecule is hydrophilic and the interior is more hydrophobic.

TRANSPORT ACROSS CELLULAR MEMBRANES

The existence of membranes is profoundly important to cells because it permits them to create compartments in which differing environmental conditions can be maintained. The plasma membrane, for example, defines the outer boundary of the cell and permits conditions to be established within cells that are quite different from those prevailing in the external environment. Other types of membranes subdivide cells into multiple internal compartments, creating organelles such as mitochondria, chloroplasts, nuclei, and the endoplasmic reticulum. Membranes allow the local conditions prevailing within each compartment to be adjusted independently. The ability to create such distinctive local environments requires that membranes be capable of transporting materials into and out of the compartments whose boundaries they define. In the remainder of the chapter we will discuss how this transport is accomplished.

Diffusion Is the Net Movement of a Substance Down Its Concentration Gradient

The most straightforward mechanism for moving materials from one side of a membrane to the other side is **simple diffusion.** For uncharged molecules, simple diffusion refers to the unaided net movement of a substance from a region where its concentration is higher to a region where its concentration is lower; in other words, the substance is moving down its concentration gradient. Such movement is generated by the random motion of individual molecules driven by their inherent thermal energies. No matter how a population of molecules is initially distributed, diffusion will tend to create a random solution in which the concentration everywhere is the same. To illustrate this point, let us consider a two-chambered apparatus containing different concentrations of the same dissolved substance, or **solute,** separated by a membrane through which the solute can freely pass (Figure 5-27, *top*). Random movements of individual solute molecules back and forth through the membrane will eventually cause the concentrations in the two chambers to become equal. Hence the random motion of individual molecules leads to a net movement of solute from the chamber where its initial concentration was higher to the chamber where its initial concentration was lower. After the concentration has been equalized on both sides of the membrane, random back-and-forth movement of individual molecules continues, but no further net change in concentrations occurs. At this point, **equilibrium** is said to be achieved.

Although diffusion generally occurs from regions of higher to regions of lower solute concentration, in strict thermodynamic terms it is differences in free energy rather than concentration that determine the net direction of diffusion. As we saw in Chapter 2, the laws of

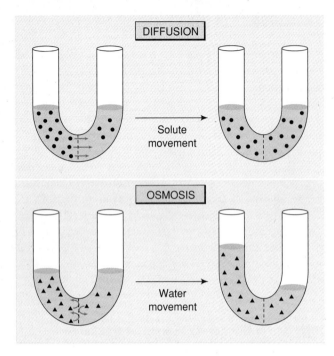

Figure 5-27 Comparison of Simple Diffusion and Osmosis (Top) *In simple diffusion, the membrane is permeable to the dissolved solute; the solute therefore diffuses from the chamber where its concentration is higher to the chamber where its concentration is lower. Equilibrium is achieved when the solute concentration is the same in both chambers.* (Bottom) *In osmosis, the membrane is not permeable to the dissolved solute; instead, water diffuses from the chamber where the solute concentration is lower (water concentration is higher) to the chamber where the solute concentration is higher (water concentration is lower). At equilibrium, the pressure that builds up in the chamber containing the excess water counterbalances the tendency of water to continue diffusing into that chamber.*

thermodynamics dictate that all events proceed in the direction of decreasing free energy. Although the free energy of any solution is directly related to the concentration of its constituent molecules, factors such as heat, pressure, and entropy also contribute to the total free energy. To illustrate this point, let us consider an example in which two compartments, A and B, are separated by a membrane permeable to substance X. If the concentration of substance X is slightly lower in compartment A than in compartment B, but the heat or pressure is higher in compartment A, then the free energy added by the heat or pressure may cause diffusion to occur in the direction A → B rather than B → A. Thus properly speaking, diffusion proceeds from areas of higher to lower free energy rather than from areas of higher to lower concentration. At thermodynamic equilibrium, no further net movement occurs because the free energy of the system is at a minimum.

Osmosis Is Caused by the Diffusion of Water Molecules through a Semipermeable Membrane

The preceding principle applies to the behavior of water molecules as well as to solutes. When any solute is dissolved in water, the solute molecules disrupt the ordered three-dimensional interactions that normally occur between individual molecules of water, thereby decreasing the concentration and free energy of the water molecules. Water, like other substances, tends to diffuse from areas where its free energy is higher to areas where its free energy is lower; therefore *water diffuses from areas of low solute concentration (high water concentration) to areas of high solute concentration (low water concentration).* This principle can be verified by placing solutions of differing solute concentrations in two compartments separated by a *semipermeable membrane* that is permeable to water but not to the dissolved solute (see Figure 5-27, *bottom*). As expected, net movement of water across the membrane occurs from the region of low solute concentration (high water concentration) to the region of high solute concentration (low water concentration). Such movement of water through a semipermeable membrane is called **osmosis.**

Osmosis provides an explanation for a well-known biological phenomenon: cells tend to shrink or swell when the solute concentration of the extracellular medium is changed. Mammalian cells, for example, lose water and shrink when placed in sucrose solutions more concentrated than 0.25 *M;* conversely, they take up water and swell when placed in sucrose solutions less concentrated than 0.25 *M* (Figure 5-28). These movements of water take place because cells are impermeable to sucrose. As a result, when cells are placed in a concentrated solution of sucrose, water moves from the area of low solute concentration (high water concentration) inside the cell to the area of high solute concentration (low water concentration) outside the cell. Conversely when cells are placed in a dilute sucrose solution, water again moves from the area of lower solute concentration (this time outside the cell) to the area of higher solute concentration (this time inside the cell). Solutions whose solute concentration is higher than that inside cells and which therefore cause water molecules to diffuse out of the cell are called **hypertonic** solutions; those whose solute concentration is lower than that inside cells and which therefore cause water to diffuse into the cell are referred to as **hypotonic** solutions. A sucrose solution that causes water to flow neither into nor out of the cell must have the same solute concentration as the cell interior; such solutions are said to be **isotonic.**

Although sucrose solutions are isotonic for mammalian cells at a concentration of 0.25 *M,* this value is not

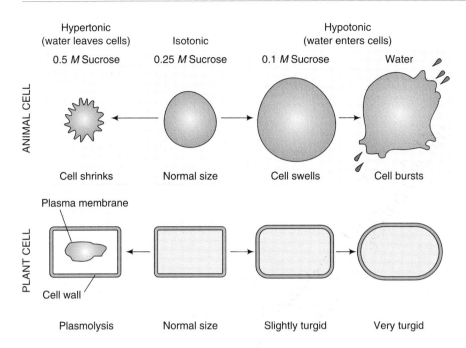

Figure 5-28 *Behavior of Animal and Plant Cells Placed in Sucrose Solutions of Differing Concentrations* (Top) *When an animal cell is placed in a hypertonic solution, water leaves the cell and the cell shrinks. In hypotonic solutions, water enters the cell and it swells and bursts. (Bottom) When a plant cell is placed in a hypertonic solution, the cytoplasm shrinks away from the cell wall (a phenomenon called plasmolysis). In hypotonic solutions, the plant cell wall prevents extensive swelling; however, the intracellular pressure is increased by the entering water and the cell therefore becomes turgid (rigid).*

the same for all solutes. The isotonic concentration of NaCl solutions, for example, is roughly 0.14 M. This value is lower than that for sucrose because NaCl dissociates into two separate ions, Na^+ and Cl^-, when dissolved in water. Sucrose, on the other hand, remains as a single intact molecule in solution. Thus at the same molar concentration, NaCl yields twice as many dissolved particles as sucrose, and so only about half as much NaCl is needed to achieve the same effective concentration of dissolved particles. The difference between 0.14 M and 0.25 M is not exactly twofold because the osmotic effect of ions such as Na^+ and Cl^- differs slightly from that of uncharged molecules such as sucrose.

Simple Diffusion of Solute Molecules Is Influenced by Their Lipid Solubility, Size, and Electric Charge

The discovery that cells shrink or swell when placed in hypertonic or hypotonic solutions has had an important practical benefit in that it provided the basis for the first experimental procedure designed to measure membrane permeability. This procedure, developed in the late nineteenth century by Charles Overton, involves two steps; cells are first placed in concentrated solutions of various substances, and are then examined

under the microscope to determine whether a change in cell volume has occurred. Plant cells are especially well suited for this purpose because loss of water causes the cytoplasm to shrink and pull away from the cell wall. This process, known as **plasmolysis,** can be readily observed under the microscope. Overton noted that extensive plasmolysis occurs when plant cells are suspended in concentrated solutions of sucrose. It does not occur, however, when cells are placed in concentrated solutions of alcohols such as ethanol. He therefore concluded that ethanol, unlike sucrose, diffuses rapidly through the plasma membrane. This causes the intracellular and extracellular concentrations of ethanol to equalize, leaving no difference in solute concentration to drive the flow of water molecules out of the cell.

When experiments of this type were repeated with various substances, it became apparent that some solutes resemble ethanol in being able to diffuse quickly through the plasma membrane, others resemble sucrose in not being able to penetrate the membrane at all, and many fall somewhere in between. To quantify the penetration rates of these substances, Overton introduced a refinement to the plasmolysis assay based upon the following rationale. If a molecule diffuses through the plasma membrane at a relatively slow rate, cells exposed to a concentrated solution of this substance will initially undergo plasmolysis because of the high extra-cellular concentration of solute. But as the solute gradually diffuses into the cell, the difference in solute concentration between the extracellular and intracellular compartments gradually disappears and plasmolysis will therefore be reversed. By measuring the amount of time required for the reversal of plasmolysis, the relative permeability of the plasma membrane to different substances can be estimated.

The results of such studies led Overton to conclude that cells are surrounded by a membrane through which some solutes diffuse more readily than others. While investigating the factors that might be responsible for this *selective permeability,* Overton discovered that solubility in lipid strongly influences a solute's ability to diffuse through membranes. The solubility of a substance in lipid can be measured experimentally by shaking the solute under study in an oil/water mixture, allowing the two phases to separate, and measuring the final concentration of the solute in the oil and water phases. The ratio of these two values, termed the oil/water *partition coefficient,* is a measure of the lipid solubility of the solute. Overton discovered that the higher the oil/water partition coefficient, the more readily a substance will diffuse through the plasma membrane. As mentioned earlier in the chapter, the discovery that a molecule's solubility in lipid influences its ability to enter into cells first led to the idea that cells are surrounded by a lipid-containing membrane (page 156).

Although subsequent studies have confirmed Overton's conclusion that lipid solubility influences the ability of solute molecules to diffuse through cellular membranes, other factors also play a role. Investigations of cell permeability carried out in the laboratory of R. Collander in the early 1930s led to the realization that molecular size is also important. He discovered, for example, that small hydrophilic molecules such as water and methanol enter into cells more readily than would be expected on the basis of their low lipid solubility (Figure 5-29). Certain large molecules, on the other hand, diffuse into cells more slowly than would be predicted on the basis of their lipid solubility. Such observations suggest that the plasma membrane acts as a molecular sieve, differentially permitting the passage of smaller molecules.

The discovery that small molecules can pass through membranes in spite of being insoluble in lipid suggests that membranes contain tiny channels through which small hydrophilic substances can pass. Calculations based on the observed permeability of cells to various small hydrophilic molecules indicate that such channels are no more than 0.5–1.0 nm in diameter. The major substance of physiological importance small enough to traverse these postulated channels is water, which passes into and out of cells more readily than

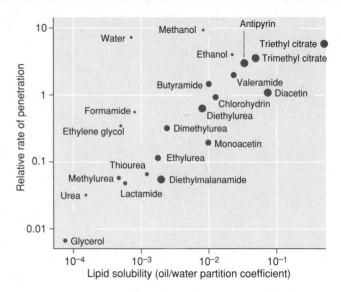

Figure 5-29 *Penetration of Various Solutes into Cells as a Function of Their Lipid Solubility* *The diameter of each point is roughly proportional to the size of the molecule it represents. Note that in spite of the general correlation between penetration rate and lipid solubility, significant deviations occur. Those molecules penetrating faster than expected* (above the shaded area) *are relatively small, whereas those penetrating more slowly than expected* (below the shaded area) *are relatively large.*

would be expected on the basis of its negligible lipid solubility. Small ions can also pass through such channels, although the widely differing rates at which ions pass through membranes indicates that more selective transport mechanisms are usually involved.

Charge is another factor that influences the ability of solute molecules to diffuse through the plasma membrane. Strongly charged substances typically diffuse through membranes less readily than weakly charged or uncharged substances. This behavior results from a combination of the lipid solubility and size effects mentioned above; charged molecules tend to attract a shell of water molecules around themselves, making them insoluble in lipid and increasing their effective size. Both factors decrease the ability of charged molecules to diffuse through the plasma membrane.

Membrane Transport Proteins Have Evolved to Aid the Transport Process

Simple diffusion gives membranes a rudimentary selectivity; that is, lipid-soluble substances and small molecules, such as water and dissolved gases, diffuse through membranes readily, whereas large and/or electrically charged substances pass through slowly or not at all. This limited selectivity is not sufficient, however, for meeting all of the cell's needs. For many kinds of molecules, unaided diffusion is simply too slow. This is especially true for ions, sugars, amino acids, and other hydrophilic substances that are insoluble in lipid and therefore have difficulty diffusing through membranes. Because simple diffusion is inadequate for meeting the need for such substances, a special class of membrane proteins has evolved for the specific purpose of transporting ions and molecules through membranes. These membrane **transport proteins** move substances across membranes by a process that differs from simple diffusion in the following ways:

1. Transport of solutes by membrane transport proteins is typically several orders of magnitude faster than simple diffusion.

2. Whereas the rate of simple diffusion is directly proportional to solute concentration, the rate of solute movement by membrane transport proteins approaches a maximum value as the solute concentration is increased (Figure 5-30). The curve obtained when solute transport rate is plotted as a function of solute concentration has the same hyperbolic shape as the curve generated when the rate of an enzyme-catalyzed reaction is plotted as a function of substrate concentration (compare Figure 5-30 with Figure 2-3). This similarity means that the rate of solute transport by membrane transport proteins can be described mathematically using the Michaelis-Menten equation for enzyme kinetics (Equation 2-7):

$$v_i = \frac{V_{\max}[S]}{K_m + [S]}$$

In the case of solute transport, v_i stands for the rate of solute diffusion, $[S]$ is the solute concentration, $V_{\max}$ is the maximal rate of solute transport observed at saturation, and K_m is the solute concentration at which the rate of transport is half-maximal. The saturation kinetics exhibited by membrane transport indicate that solute molecules are interacting with a limited number of membrane transport proteins (just as enzymatic catalysis involves interaction of substrate with a limited number of enzyme molecules). At saturation all of a membrane's transport proteins are occupied with solute, so additional increases in solute concentration cannot increase the transport rate any further.

3. With simple diffusion, structurally related molecules tend to diffuse through membranes at comparable rates; membrane transport proteins, on the other hand, are highly selective in distinguishing between closely related molecules. The transport protein that speeds up the movement of glucose, for example, does not act upon closely related sugars. The solute specificity of membrane transport proteins is comparable to the

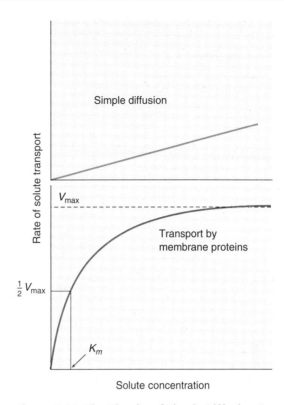

Figure 5-30 *The Kinetics of Simple Diffusion Compared with the Kinetics of Solute Transport Mediated by Membrane Proteins* *The rate of simple diffusion increases in direct proportion to the solute concentration* (top), *whereas transport involving membrane proteins exhibits saturation kinetics* (bottom).

substrate specificity of enzymes, suggesting that transport proteins possess solute-binding sites that are analogous to the active sites of enzyme molecules.

4. Unlike simple diffusion, membrane transport is subject to various kinds of inhibition. For example, transport mediated by membrane proteins can usually be inhibited by molecules that resemble the solute in chemical structure. This phenomenon, which is reminiscent of the action of competitive inhibitors on enzyme-catalyzed reactions (page 48), is caused by competition between inhibitor and solute for the solute-binding site on the transport protein. Membrane transport proteins are also inhibited by agents that denature proteins, just as enzymes can be inhibited by denaturing agents.

The preceding characteristics indicate that membrane transport proteins interact with solutes in ways that resemble the interaction of an enzyme with its substrate. The principal difference is in the final outcome: Enzymes catalyze the conversion of substrate to product, whereas transport proteins catalyze the movement of substances across membranes.

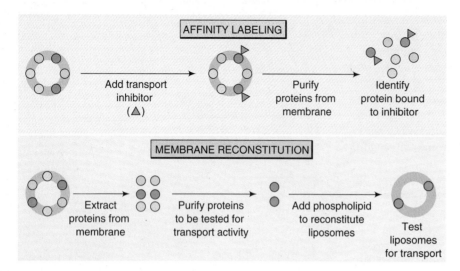

Figure 5-31 *Two Approaches for Identifying Membrane Transport Proteins*
(Top) *Affinity labeling employs radioactive molecules that bind to a specific type of transport protein. Inhibitors of specific transport systems are often used for this purpose.*(Bottom) *In an alternative approach, membranes are reconstituted from a mixture of phospholipids and a purified protein that has been implicated in a particular type of transport. The transport properties of the reconstituted membrane are then evaluated.*

Membrane Transport Proteins Are Identified by Affinity Labeling and Membrane Reconstitution Techniques

The discovery that proteins can move substances across cellular membranes raises the question of how specific transport proteins are identified. Figure 5-31 illustrates two ways of identifying the membrane protein that is involved in transporting a particular substance. One of the methods, called *affinity labeling,* utilizes radioactive molecules that are expected to bind to a specific transport protein. For example, *cytochalasin B* is known to be a potent inhibitor of glucose transport; hence membranes that have been exposed to radioactive cytochalasin B would be expected to contain radioactivity bound to the protein responsible for transporting glucose.

A more definitive method for identifying membrane transport proteins involves the *reconstitution* of artificial membranes from individual purified components. In this approach, proteins are extracted from membranes with detergent solutions and separated into different protein components; the purified proteins are then mixed together with phospholipids under conditions known to promote the formation of liposomes. These reconstituted liposomes can then be tested for their ability to carry out specific types of transport.

Dozens of different transport proteins have been identified using the preceding methods. Some of these proteins, called **uniports,** move a single substance from one side of a membrane to the opposite side. Others couple the movement of two substances to one another. When two substances are moved in the same direction, the transport protein is called a **symport;** if

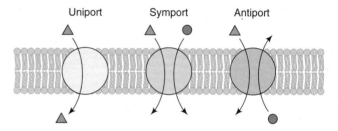

Figure 5-32 *A Comparison between Uniports, Symports, and Antiports*   *A uniport is a membrane transport protein that moves a single substance across a membrane; a symport moves two substances in the same direction; and an antiport moves two substances in opposite directions.*

movement occurs in opposite directions, the protein is termed an **antiport** (Figure 5-32).

Transport proteins can also be classified on the basis of their energy requirements. If solute transport requires an input of energy, the membrane protein is performing **active transport;** when energy is not required, the process is called **facilitated diffusion.** To determine whether active transport or facilitated diffusion is involved in the transport of a particular solute, it is therefore necessary to calculate the change in free energy that accompanies the solute's movement across the membrane.

How to Determine the Energy Requirements for the Transport of Uncharged Molecules

How do we determine the energy change that occurs when a substance is moved from one side of a membrane to the other? In Chapter 2, we learned that the change in free energy ($\Delta G'$) that accompanies any chemical reaction is given by the equation

$$\Delta G' = \Delta G^{\circ\prime} + 2.303RT \log_{10} \frac{[\text{products}]}{[\text{reactants}]}$$

where R is the gas constant (1.987 cal/mol-K) and T is the absolute temperature. This equation can be applied to the transport of a solute *into* a membrane-bounded compartment by rewriting it in the form

$$\Delta G' = \Delta G^{\circ\prime} + 2.303RT \log_{10} \frac{[C]_{\text{in}}}{[C]_{\text{out}}} \qquad (5\text{-}1)$$

where $[C]_{\text{out}}$ is the initial concentration of solute outside the compartment and $[C]_{\text{in}}$ is the initial concentration of solute inside the compartment. By definition, the term $\Delta G^{\circ\prime}$ in the above equation refers to the standard free-energy change that occurs when $[C]_{\text{in}}$ and $[C]_{\text{out}}$ are both equal to 1.0 M and the system then proceeds to equilibrium. But at equilibrium, the solute concentrations on opposite sides of the membrane will be equal, so no change in free energy would occur in a system that had already started with equal concentrations of solute (1.0 M) on opposite sides of the membrane. Hence for solute transport, the term $\Delta G^{\circ\prime}$ is always equal to zero. Equation 5-1 therefore simplifies to:

$$\Delta G' = 2.303RT \log_{10} \frac{[C]_{\text{in}}}{[C]_{\text{out}}} \qquad (5\text{-}2)$$

According to this equation, the calculated value of $\Delta G'$ will be negative and hence energy will be released when $[C]_{\text{out}}$ exceeds $[C]_{\text{in}}$—that is, when the solute entering the compartment is present in higher concentration outside the compartment than inside. This means that movement of a solute down its concentration gradient is thermodynamically spontaneous, and so proceeds without an input of energy. However, when $[C]_{\text{in}}$ is greater than $[C]_{\text{out}}$, the calculated value of $\Delta G'$ will be positive and hence energy is required for solute uptake. In other words, active transport is required when the solute entering the compartment is present in lower concentration outside the compartment than inside.

To illustrate the usefulness of Equation 5-2, let us consider what happens during the uptake of the sugar maltose by bacterial cells. Maltose can be accumulated by bacteria to concentrations that are more than 100,000-fold higher inside the cell than outside. This means that bacteria growing in an environment where the external maltose concentration is only 0.000001 mM can transport this sugar into the cell to concentrations that reach 0.1 mM. The energy requirement for transporting maltose into the cell at 25°C against such a steep concentration gradient can be calculated using Equation 5-2:

$$\Delta G' = 2.303RT \log_{10} \frac{[0.1]}{[0.000001]}$$
$$= 2.303(1.987)(273 + 25)\log_{10}(100,000)$$
$$= 6818 \text{ cal/mol} = 6.818 \text{ kcal/mol}$$

Because this calculated value of $\Delta G'$ is positive, it means that roughly 6.8 kcal of energy are required per mole of maltose transported into the cell. For comparison, the hydrolysis of ATP releases about 7 kcal/mol of free energy under standard conditions. Hence the energy required to transport maltose against a 100,000-fold concentration gradient can be met by coupling its uptake to the hydrolysis of ATP.

How to Determine the Energy Requirements for Ion Transport

When the substance being transported across a membrane is electrically charged, the energy requirement for solute transport cannot be determined from Equation 5-2 alone. Under such conditions, we must consider the electrical properties of the membrane. In Chapter 6 we will learn that a voltage, or **membrane potential,** commonly exists across membranes that makes one side of the membrane negative and the other side positive. To determine how much energy is required to transport a charged ion across such a membrane, the magnitude of the membrane potential must be taken into account along with the magnitude of the concentration gradient. This is accomplished by modifying Equation 5-2 to include the contribution of the membrane potential:

$$\Delta G' = 2.303RT \log_{10} \frac{[C]_{\text{out}}}{[C]_{\text{in}}} + zF\Delta\psi \qquad (5\text{-}3)$$

where z is the charge on the molecule, F is the faraday (23,062 cal/mol-V), and $\Delta\psi$ is the membrane potential in volts. This equation tells us that during ion transport, it is the combination of the concentration gradient ($[C]_{\text{out}}/[C]_{\text{in}}$) and the electric charge gradient ($zF\Delta\psi$) that determines how much energy is required or released. This combination of the concentration and charge gradients is referred to as the **electrochemical gradient.**

One cannot always rely on intuition to recognize whether a charged solute requires energy for its transport. To illustrate this point, let us consider what happens when nerve cells with an intracellular Cl^- concentration of 50 mM are placed in a solution containing 100 mM Cl^-. Since the Cl^- concentration is twice as high outside the cell as inside, the chloride ions might be expected to passively diffuse into the cell without the need for active transport. However, this simple view ignores the membrane potential of roughly −60 millivolts (−0.06 volts) that exists across the plasma membrane of nerve cells. By convention, the minus sign indicates that the inside of the cell is negative relative to the outside. This negatively charged environment inside the cell tends to repel the movement of the negatively charged Cl^- ions into the cell. Hence the membrane potential, which acts to *inhibit* the movement of Cl^- into the cell, counteracts the effect of the concentration gradient, which tends to *promote* the movement of Cl^- into the cell. To quantify the relative magnitude of these opposing forces, we can insert the values of both the concentration gradient and the membrane potential into Equation 5-3:

$$\Delta G' = 2.303RT \log_{10} \frac{[50]}{[100]} + zF(-0.06)$$
$$= 2.303(1.987)(273 + 25)\log_{10}(0.5)$$
$$+ (-1)(23,062)(-0.06)$$
$$= 973 \text{ cal/mol} = 0.973 \text{ kcal/mol}$$

The fact that $\Delta G'$ turns out to be positive rather than negative means that even though the Cl^- concentration is twice as high outside the cell as inside, energy is still required to drive the movement of Cl^- into the cell. This unexpected result arises because of the presence of the membrane potential.

The equations that we have been using to determine the energy requirements for solute transport apply to the movement of substances *into* a membrane-enclosed compartment. Any process that consumes energy in one direction will release the same amount of energy when it occurs in the opposite direction. Hence if the transport of a solute such as maltose *into* the cell is an energy-requiring reaction with a $\Delta G'$ of 6.8 kcal, then the movement of maltose *out of* the cell under the same conditions will be an energy-releasing process with a $\Delta G'$ of -6.8 kcal/mol.

Facilitated Diffusion Is the Assisted Movement of a Substance Down Its Electrochemical Gradient

When the calculated value of $\Delta G'$ for the movement of a substance across a membrane is negative, it signifies that energy is released rather than consumed. This means that the substance will tend to pass across the membrane spontaneously because it is moving "down" its electrochemical gradient—that is, from a region of higher free energy to a region of lower free energy. This natural tendency of substances to move down their electrochemical gradient is the driving force that propels both simple diffusion (which we discussed earlier) and facilitated diffusion. The main difference between the two events is that facilitated diffusion utilizes membrane components that assist the diffusion process, conveying substances across membranes faster than would be possible by diffusion alone.

Considerable insight into the mechanism of facilitated diffusion has come from the study of **ionophores,** which are small molecules manufactured by certain microorganisms for the purpose of increasing the diffusion rate of specific ions across membranes. One of the most thoroughly investigated ionophores is *valinomycin,* a small cyclic molecule constructed from 12 alternating hydroxy acids and amino acids. When valinomycin is added to liposomes, the rate of K^+ diffusion through the lipid bilayer increases almost 100,000-fold. In contrast, permeability to Na^+ increases only slightly, indicating that valinomycin shares two important features with membrane transport proteins; that is, it accelerates the rate of solute movement across membranes and it exhibits solute specificity. Valinomycin is a doughnut-shaped molecule containing a central cavity in which a potassium ion fits snugly, bonded to six oxygen atoms. Smaller ions such as sodium and lithium do not fit into the site as tightly and so are transported less efficiently. Valinomycin is particularly well suited for transporting ions across a hydrophobic lipid bilayer because the charge of the bound potassium ion is masked in the molecule's central cavity, leaving the hydrophobic groups of the valinomycin molecule exposed on the molecule's exterior surface.

One can envision at least two mechanisms by which ionophores like valinomycin might increase the rate of ion movement across membranes. First, an ionophore might act as a *mobile carrier* that binds to an ion at one membrane surface, diffuses across the lipid bilayer, and then releases the bound ion on the other side of the membrane. Alternatively, ionophores might form transmembrane *channels* that allow ions to enter either end of the channel, diffuse through it, and exit at the opposite end (Figure 5-33). A simple test has been designed to distinguish between the two possibilities. The test is based on the assumption that the rate of ion transport by a mobile carrier should be dramatically affected by major changes in membrane fluidity because the carrier must diffuse through the fluid lipid bilayer; ion transport through a fixed transmembrane channel, on the other hand, should be relatively unaffected by membrane fluidity because the channel itself does not need to move (Figure 5-34).

When the preceding test is applied to a lipid bilayer containing valinomycin, the rate of ion transport is found to increase dramatically when the gel-to-fluid transition for the bilayer is triggered by raising the temperature. This strong connection between membrane fluidity and ion transport suggests that valinomycin acts as a mobile carrier, diffusing back and forth across the lipid bilayer. But not all ionophores act in the same way. *Gramicidin,* an ionophore consisting of a linear chain of 15 amino acids, transports ions at a rate that does not change dramatically when lipid bilayers undergo the gel-to-fluid transition. This finding suggests that grami-

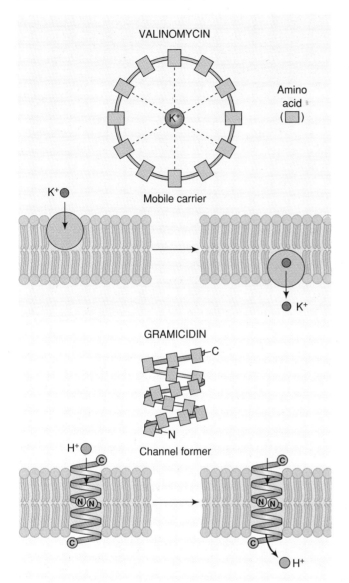

VALINOMYCIN

Amino acid (☐)

K⁺

Mobile carrier

K⁺

GRAMICIDIN

C

N

H⁺

Channel former

H⁺

Figure 5-33 Structure and Mechanism of Action of the Ionophores Valinomycin and Gramicidin (Top) *Valinomycin functions as a mobile carrier, diffusing back and forth through the lipid bilayer.* (Bottom) *Gramicidin functions as a channel former, creating a fixed transmembrane channel through which ions can diffuse. The gramicidin channel consists of two helical gramicidin molecules joined end to end.*

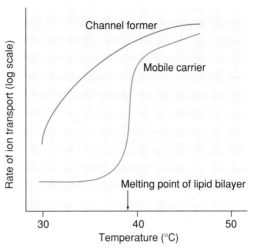

Figure 5-34 Distinguishing between Ionophores That Act as Either Mobile Carriers or Channel Formers *The two classes of ionophores can be distinguished by the way they respond to changing temperature. Because a mobile carrier diffuses back and forth across the lipid bilayer, the rate at which it transports ions increases dramatically when the temperature is raised to the melting point for the lipid bilayer (the temperature at which the bilayer undergoes the gel-to-fluid transition). Ion transport by a channel-forming ionophore does not undergo a sudden increase in rate at the melting temperature because the transport mechanism does not require the ionophore to diffuse across the bilayer.*

The Glucose Transporter Mediates the Facilitated Diffusion of Glucose

The discovery that ionophores speed up diffusion either by forming channels or by functioning as mobile carriers raises the question of whether membrane transport proteins act in a similar manner. In addressing this issue, we will first consider the facilitated diffusion of glucose in red blood cells, which occurs at a rate that is at least a hundred times faster than simple diffusion. Glucose uptake by red blood cells exhibits saturation kinetics, susceptibility to competitive and noncompetitive inhibition, and solute specificity, all of which are classic features of a process mediated by a membrane transport protein. The solute specificity of the glucose transport system is flexible enough to permit uptake of a few related sugars, such as galactose and mannose, but most sugars are excluded (Figure 5-35).

Affinity labeling and membrane reconstitution experiments have led to the conclusion that the facilitated diffusion of glucose is carried out by a membrane glycoprotein of 55,000 molecular weight called the *glucose transporter.* The single carbohydrate chain of this protein can be labeled by incubating intact red cells or ghosts with ³H-borohydride and galactose oxidase, indicating that the glucose transporter is exposed at the outer membrane surface. The discovery that treating

cidin does not diffuse across the lipid bilayer, but instead forms a relatively fixed channel through which ions can pass. Studies on the structure of gramicidin have confirmed the existence of such a channel, which is constructed from two helically twisted gramicidin molecules that form a passageway through which ions travel. Like valinomycin, gramicidin is selective in the ions it transports; the selectivity of the gramicidin channel follows the order $H^+ > NH_4^+ > K^+ > Na^+ > Li^+$.

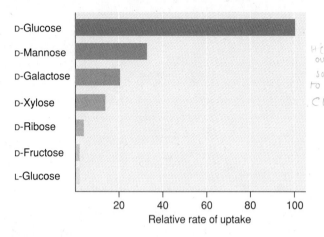

Figure 5-35 Solute Specificity of the Glucose Transporter in Human Red Blood Cells *The selectivity of the glucose transporter is flexible enough to permit uptake of a few related sugars, such as galactose and mannose, but glucose is transported most efficiently.*

inside-out vesicles with the protease trypsin abolishes glucose transport indicates that the glucose transporter is also exposed at the inner membrane surface. Taken together, these observations indicate that the glucose transporter is a transmembrane protein. Amino acid sequence analysis has revealed that the transmembrane regions consist of twelve α helices that pass back and forth across the lipid bilayer.

How does this protein speed up the diffusion of glucose molecules? The glucose transporter contains a single glucose binding site that exists in two mutually exclusive conformations, one facing the cell exterior and the other facing the cytoplasm. This suggests that the mechanism of glucose transport involves changes in the shape of the transport protein that cause its glucose binding site to alternately face one side of the membrane and then the other (Figure 5-36, *top*). Transport proteins that function through such *alternating conformational changes* are referred to as **carrier proteins.** In contrast to mobile carriers such as valinomycin, carrier proteins do not actually diffuse back and forth across the lipid bilayer; rather it is the changing three-dimensional shape of carrier proteins that propels the movement of solutes across membranes.

Band 3 Protein Is an Anion Exchanger That Mediates the Facilitated Diffusion of Bicarbonate and Chloride Ions

Another well-studied example of facilitated diffusion involves the transport of bicarbonate and chloride ions by red blood cells. These two anions plays a central role in the process by which CO_2 produced in metabolically active tissues is delivered to the lung. CO_2 diffuses from tissues into red blood cells, where the enzyme *carbonic anhydrase* converts it to HCO_3^- (bicarbonate) and H^+.

As the amount of HCO_3^- in the red cell rises, it diffuses out of the cell. To prevent a net charge imbalance, the expulsion of each negatively charged bicarbonate ion is accompanied by the uptake of one negatively charged chloride ion. In the lungs, this entire process is reversed; Cl^- is transported out of the red cell accompanied by the uptake of HCO_3^-, which is then converted to CO_2 (Figure 5-37). The net result is the movement of CO_2 (in the form of HCO_3^-) from tissues to the lung, where the CO_2 is exhaled from the body.

Several kinds of evidence indicate that the **band 3 protein** mediates the facilitated diffusion of bicarbonate and chloride anions across the red cell membrane. For example, affinity labeling experiments (page 180) have revealed that exposing red cells to radioactive inhibitors of anion transport causes the band 3 protein to become labeled. In addition, liposomes acquire the capacity to transport HCO_3^- and Cl^- when the vesicles are reconstituted with purified band 3 protein. The ability of band 3 protein to transport HCO_3^- depends upon the presence of Cl^- on the opposite side of the membrane; each time a bicarbonate ion is transported in one direction, a chloride ion is transported in the opposite direction. Thus band 3 protein is an antiport (page 180) that functions as an *anion exchanger,* coupling the transport of HCO_3^- in one direction with the transport of Cl^- in the opposite direction.

The coupled transport of HCO_3^- and Cl^- by band 3 protein is thought to be generated by alternating conformational changes similar to the ones described for glucose transport (see Figure 5-36, *bottom*). The major difference is that during anion transport by band 3 protein, the solute binding site interacts with different ions on opposite sides of the membrane. In tissues where the CO_2 concentration is high, the binding site of band 3 protein binds to HCO_3^- at the interior membrane surface and Cl^- at the exterior surface. In the lung, where CO_2 levels are lower because CO_2 is being exhaled from the body, the reciprocal process occurs; HCO_3^- binds at the exterior membrane surface and Cl^- at the interior surface.

Ion Channels Mediate the Facilitated Diffusion of Small Ions

The use of alternating conformational changes, as exemplified by the glucose transporter and the band 3 anion exchanger, is one of two basic ways in which membrane proteins can facilitate diffusion. The other mechanism involves the formation of *protein channels* through which water-soluble molecules and ions can pass. Some protein channels are relatively large and nonspecific, such as those occurring in the outer membranes of bacteria, mitochondria, and chloroplasts. Other protein channels are smaller and more selective. Because they are usually involved in transporting ions, these smaller channels are referred to as **ion channels.**

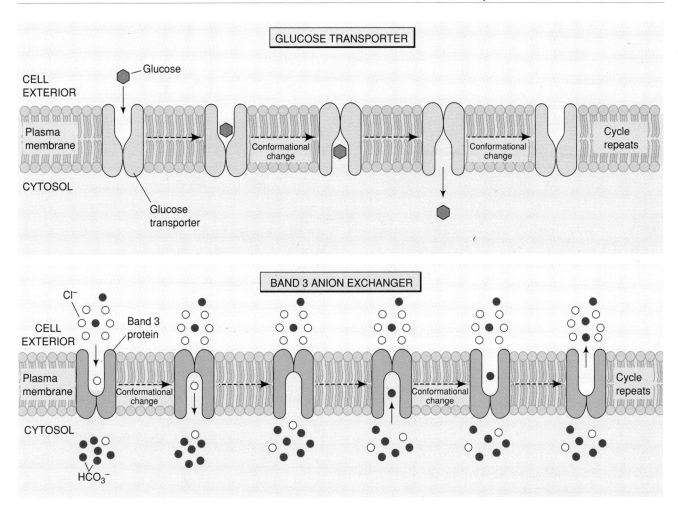

Figure 5-36 *Facilitated Diffusion by Transport Proteins That Undergo Alternating Conformational Changes*   (Top) *A model for facilitated diffusion by the red cell glucose transporter. The glucose transporter is thought to undergo changes in conformation that cause its glucose binding site alternately to face one side of the membrane and then the other.* (Bottom) *A model for bicarbonate-chloride exchange by the band 3 anion exchanger of the red cell membrane. According to this model, the anion binding site accepts either bicarbonate or chloride ions, depending on which is present in higher concentration. The diagram illustrates the predominant direction of transport in body tissues where CO_2 levels are high. In the lung, where CO_2 levels are low, bicarbonate and chloride move in opposite directions from what is shown here.*

Because diffusion through ion channels does not require complex conformational changes, it is significantly faster than transport mediated by protein carriers. This does not mean, however, that ion channels are simply open holes in the membrane. Two features distinguish ion channels from such simple openings. First, they are capable of discriminating between different ions; thus separate channels exist for transporting important cellular ions such as Na^+, K^+, Ca^{2+}, and Cl^-. In addition, ion channels have *gates* that can be opened or closed in response to appropriate stimuli. We will see in later chapters how the opening of these gates is controlled by signaling molecules (pages 210 and 768), as well as by electrical or mechanical stimulation of the cell surface (pages 201 and 592).

The ability of ion channels to regulate the diffusion of specific ions across cellular membranes plays a prominent role in many types of cellular communication. For example, we will learn in Chapter 6 that changes in the permeability of Na^+ and K^+ channels play a central role in the transmission of electrical signals by nerve cells. Many toxins, drugs, and anesthetics that act on the nervous system do so by interfering with the activity of ion channels; for example, a few micrograms of *tetrodotoxin,* a poison produced by the Japanese puffer fish, induces paralysis and death in humans by blocking Na^+ channels. Another illustration of the widespread importance of ion channels occurs in *cystic fibrosis,* an inherited disease that is characterized by abnormally viscous secretions in the lungs, pancreas, and intestinal tract. Cystic fibrosis results from a defect in a Cl^- channel found in epithelial cells. The diminished transport of Cl^- across the surface of glandular epithelial cells leads to decreased fluid secretion; as a result,

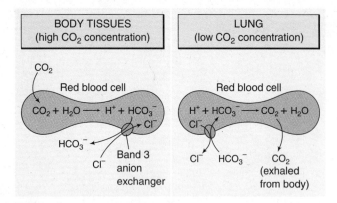

Figure 5-37 *Role Played by the Band 3 Anion Exchanger in Transporting CO₂ from Body Tissues to the Lung* CO_2 *produced in body tissues diffuses into red blood cells, where the enzyme carbonic anhydrase converts it to HCO_3^- and H^+. The band 3 anion exchanger in the red cell membrane then facilitates the diffusion of HCO_3^- out of the cell coupled to the uptake of Cl^-. In the lungs the process is reversed; Cl^- is transported out of the red cell accompanied by the uptake of HCO_3^-, which is converted to CO_2 that is subsequently exhaled from the body.*

the mucus that coats the respiratory airways, pancreatic duct, and intestinal tract is dehydrated and tends to obstruct the passageways.

Active Transport Involves Membrane Carriers That Move Substances against an Electrochemical Gradient

Although facilitated diffusion is an important mechanism for speeding up the movement of substances across cellular membranes, it only involves transport of molecules down an electrochemical gradient. What happens when a substance needs to be transported *against* an electrochemical gradient? Such situations require active transport, a process that differs from facilitated diffusion in one crucial aspect: Active transport utilizes membrane transport proteins that can move substances against ("up") an electrochemical gradient (Table 5-2).

Since transporting a substance against an electrochemical gradient is an energy-requiring activity, active transport must be coupled to a process that provides the necessary energy. The energy-releasing events employed for this purpose include the hydrolysis of high-energy molecules such as ATP, the absorption of photons of light, the transfer of electrons (oxidation-reduction reactions), and the movement of other substances *down* their respective electrochemical gradients. In the following sections, we will see how energy derived from these various sources is used to drive active transport.

ATP Can Provide Energy for Active Transport—The Na⁺-K⁺ Pump as a Model

In Chapter 2 we learned that ATP molecules store energy that can be used by cells to drive energy-requiring activities (see Figure 2-26). The active transport of sodium and potassium ions is a classic example of an energy-requiring transport system powered by the hydrolysis of ATP. In most cells the concentration of K^+ is higher and the concentration of Na^+ is lower than the concentration of these same ions in the extracellular fluid (Figure 5-38). Maintaining this unequal distribution of Na^+ and K^+ is important because K^+ is needed inside cells for the activity of certain enzymes, and the Na^+ gradient is required for the transmission of nerve impulses (page 199) and for the active transport of sugars and amino acids (page 190).

The first indication that the transport of Na^+ and K^+ might be related was provided in 1955 by Alan Hodgkin and Richard Keynes, who reported that transport of Na^+ out of squid nerve cells requires the presence of K^+ in the external medium. Other investigators soon observed the same phenomenon in red blood cells. In the latter studies, the rate at which Na^+ is transported out of red cells was slowed down by lowering the temperature, thereby causing Na^+ to accumulate inside the cell. Returning the temperature to normal caused the accumulated Na^+ to be pumped out of the cell, but this Na^+ expulsion occurred only when K^+ was present in the external medium. The transport of K^+ from the external medium into the cells

Table 5-2   Comparison of Simple Diffusion, Facilitated Diffusion, and Active Transport

Property	Simple Diffusion	Facilitated Diffusion	Active Transport
Rate of solute movement	Slow	Fast	Fast
Saturation kinetics	No	Yes	Yes
Solute specificity	Low	High	High
Direction relative to electrochemical gradient	Down	Down	Up
Inhibited by competitive and noncompetitive inhibitors	No	Yes	Yes
Energy requirement	No	No	Yes
Type of membrane transport protein	None	Carriers and channels	Carriers (pumps)

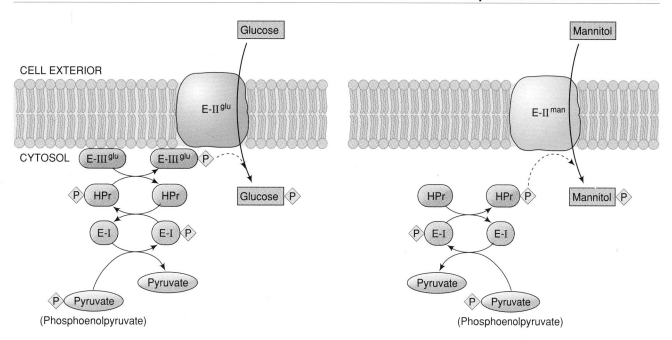

Figure 5-41 *Proposed Mechanism of the Bacterial Phosphotransferase System for Sugar Transport*　*The energy driving this type of sugar transport is provided by the high-energy phosphate compound, phosphoenolpyruvate, whose phosphate group is transferred first to enzyme I and then to HPr. From HPr the phosphate group is transferred to enzyme II either directly (as in mannitol transport, illustrated on the right), or via enzyme III (as in glucose transport, illustrated on the left). Finally, the phosphorylated enzyme II molecule donates the phosphate group to a sugar molecule as it transfers the sugar across the membrane.*

glucose phosphate in the cytosol. Although differing forms of enzymes II and III catalyze a similar process for other sugars, enzyme III is not always required. For example, during transport of the sugar mannitol (see Figure 5-41, *right*), HPr transfers its phosphate group directly to a mannitol-specific form of enzyme II (E-II^man), which then phosphorylates mannitol as it transfers the mannitol across the membrane.

A unique feature of the phosphotransferase system is that the sugar molecules being transported are chemically altered by the addition of a phosphate group as they pass through the plasma membrane. The newly acquired phosphate group, with its negative charge, traps the transported sugar molecule inside the cell because the plasma membrane is relatively impermeable to charged substances. In this way, high concentrations of sugar-phosphate can be accumulated within the cell under conditions where the external concentration of free sugar is relatively low.

Light Can Provide Energy for Active Transport—Bacteriorhodopsin as a Model

As an alternative to energy-rich compounds like ATP and phosphoenolpyruvate, some active transport systems obtain energy directly from photons of light. A classic example of such a system occurs in the oxygen-requiring bacterium, *Halobacterium halobium.* When this organism is grown under oxygen-deficient conditions, membrane patches containing a purple pigment

appear within the plasma membrane. In freeze-fracture micrographs these **purple membrane** patches are seen to consist of a distinctive arrangement of hexagonally packed particles, quite unlike anything observed in the rest of the membrane. Such an orderly arrangement implies that the mobility of membrane proteins is restricted in the regions of the purple patches; otherwise the purple color would be expected to diffuse throughout the entire plasma membrane. Aggregation of the membrane proteins making up the purple patches, rather than attachment to an underlying cytoskeletal network, appears to be responsible for the restriction in mobility.

Purple patches can be isolated for biochemical study by lowering the ionic strength of the medium to disrupt the plasma membrane. Under these conditions, the purple patches retain their structural integrity and can be isolated free of other membrane contaminants by centrifugation. Over two-thirds of the mass of isolated purple membranes is accounted for by a single protein, **bacteriorhodopsin.** Associated with this protein is the prosthetic group **retinal,** a derivative of vitamin A that gives the protein its purple color. The bacteriorhodopsin molecule contains seven regions of α helix connected to one another by short nonhelical regions. Each helical segment measures 3–4 nm in length, which fits across the width of the lipid bilayer (Figure 5-42). The retinal group, located in the middle of the bacteriorhodopsin molecule, is responsible for absorbing light. Absorption of light by retinal causes a confor-

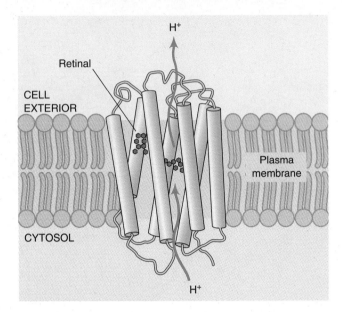

Figure 5-42 *Structure of Bacteriorhodopsin* *The cylindrical shapes represent regions of α helix. Absorption of a photon of light by the retinal group induces a conformational change that drives the transport of protons (H⁺) through the central channel of the protein.*

mational change in bacteriorhodopsin that triggers the transport of two protons (H⁺) out of the cell. The net result is the creation of an electrochemical proton gradient in which the proton concentration is higher outside the cell than inside. In Chapters 8 and 9 we will see how such proton gradients are used by cells to drive the production of ATP. Thus when *Halobacterium halobium* creates patches of purple membrane under oxygen-deficient conditions, the cells acquire a way of making ATP that does not depend upon the presence of oxygen.

Electron Transfer Reactions Can Provide Energy for Active Transport

Another way of obtaining energy for active transport is to couple solute transport to an oxidation-reduction reaction. By definition, oxidation-reduction reactions involve the transfer of electrons from one substance to another accompanied by the release of free energy. Electron transfer reactions play a crucial role in the mechanism by which cells capture energy from food and from sunlight. Both processes involve electron transfer reactions that release energy that is used to drive the active transport of protons (H⁺) across a membrane, creating an electrochemical proton gradient. Depending on the cell type and energy source, the membrane may be the mitochondrial inner membrane, the chloroplast thylakoid membrane, or the plasma membrane. In each case, energy stored in the proton gradient can be used to drive the formation of ATP. The details of these events will be described in Chapters 8

and 9, when we examine how cells capture energy derived from food and sunlight.

Ion Gradients Can Provide Energy for Active Transport—Na⁺-Linked Transport of Sugars and Amino Acids as a Model

We have now seen that active transport can be powered by energy derived from electron transfer reactions, or photons of light, or energy-rich compounds such as ATP and phosphoenolpyruvate. The driving force for active transport can also be provided by the movement of an ion down its electrochemical gradient. During the early 1950s, Halvor Christensen and his colleagues were the first to show that the rate of amino acid uptake by animal cells is directly related to the extracellular concentration of sodium ions. Shortly thereafter the transport of certain sugar molecules was also found to depend on the presence of sodium ions (Figure 5-43). The explanation for this relationship soon became clear; the active uptake of sugars and amino acids into the cell is linked to the simultaneous uptake of Na⁺, which is moving down its electrochemical gradient and hence drives the transport process.

Studies employing competitive inhibitors to investigate the specificity of Na⁺-dependent transport have revealed that different inhibitors depress the transport of differing sets of sugars or amino acids. This finding indicates that several carrier proteins are involved in Na⁺-dependent transport, each specific for a small group of related amino acids or sugars. Each carrier molecule contains two binding sites, one for Na⁺ and the other

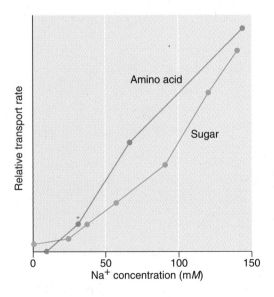

Figure 5-43 *Effect of External Na⁺ Concentration on the Rate of Amino Acid and Sugar Uptake into Cells* *The data represent the transport rate of the amino acid glycine into red blood cells, and the sugar 7-deoxy-d-glucoheptose into intestinal lining cells. Both types of transport are stimulated by Na⁺ in the extracellular medium.*

for the particular sugar or amino acid being transported. Only when both sites are occupied does transport occur. In other words, these transport proteins function as symports (page 180), moving both sugar and Na$^+$ in the same direction. Although Na$^+$-dependent symports are theoretically capable of moving both solutes either into or out of the cell, the high concentration of Na$^+$ outside the cell relative to that inside the cell preferentially drives the process inward. Hence the sugar or amino acid being transported by Na$^+$-dependent symports is preferentially moved into the cell, even though it is usually moving against its own concentration gradient.

The Na$^+$ gradient that provides the driving force for Na$^+$-dependent transport is created by the Na$^+$-K$^+$ pump, which actively expels Na$^+$ from cells. The interdependence between the Na$^+$-K$^+$ pump and the Na$^+$-dependent transport of amino acids and sugars is well illustrated in the cells that line the intestinal tract of higher organisms. These cells, which are specialized for absorbing nutrients from the intestinal tract and transferring them to the bloodstream, exhibit a marked *polarity;* that is, the plasma membrane exhibits morphologically and functionally distinct features at opposite ends of the cell (Figure 5-44). At the *apical* surface of the cell, which faces the intestinal lumen (cavity), the plasma membrane contains an Na$^+$-dependent glucose symport. Because the Na$^+$ concentration is higher outside the cell than inside, this symport couples the movement of Na$^+$ down its electrochemical gradient to the transport of glucose into the cell against its concentration gradient. At the opposite, or *basolateral,* surface of the cell, the plasma membrane contains a facilitated diffusion transporter that speeds the passage of glucose out of the cell and into the bloodstream. The basolateral surface of the cell also contains the Na$^+$-K$^+$ pump, which actively expels Na$^+$ from the cell interior. Without the Na$^+$-K$^+$ pump, the Na$^+$ concentration inside the cell would gradually rise due to the activity of the Na$^+$-dependent glucose symport; eventually the Na$^+$ gradient would disappear, and so the Na$^+$-dependent glucose symport would no longer be able to transport glucose from the intestinal lumen into the cell. Hence the movement of glucose from the intestinal tract into the bloodstream ultimately depends upon interactions among several different transport systems located in different regions of the cell.

Proton Gradients Can Provide Energy for Active Transport—H$^+$-Linked Transport of Lactose as a Model

The Na$^+$ gradient is not the only ion gradient capable of driving active transport. In bacteria and plants, gradients of H$^+$ rather than Na$^+$ are coupled to the uptake of sugars and amino acids. One of the first proton-linked transport systems to be described was the *lactose car-*

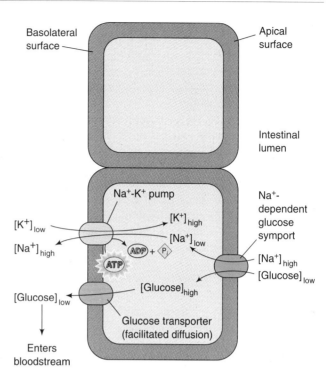

Figure 5-44 *Arrangement of Membrane Transport Proteins Involved in Transporting Glucose from the Intestinal Tract into the Bloodstream* *Glucose is taken up from the intestinal lumen (cavity) by an Na$^+$-dependent glucose symport located on the side of the cell facing the lumen. Because the Na$^+$ concentration is higher outside the cell than inside, this transport system selectively moves glucose into the cell against its own concentration gradient. An Na$^+$-K$^+$ pump located on the opposite side of the cell prevents the internal Na$^+$ concentration from rising by pumping Na$^+$ out of the cell. This same side of the cell also contains a facilitated diffusion system that speeds the passage of glucose out of the cell and into the bloodstream.*

rier of the bacterium *Escherichia coli,* discovered in the early 1950s by Georges Cohen, Howard Rickenberg, and Jacques Monod. Their studies revealed that bacteria grown in the absence of lactose are incapable of actively transporting this sugar, but they quickly develop the ability to do so when lactose is added to the growth medium. If inhibitors of protein synthesis are added along with the lactose, however, the ability to transport lactose is not acquired. It was therefore concluded that the presence of lactose causes cells to synthesize a new membrane protein that is involved in the transport of lactose.

In the early 1970s a major breakthrough occurred in our understanding of how the lactose carrier operates. Ian West and Peter Mitchell discovered that lactose uptake by bacterial cells is accompanied by an increase in the pH (i.e., a decrease in proton concentration) of the external medium. This discovery led them to propose that protons are transported along with lactose by the lactose carrier. Howard Kaback and his associates tested the idea by exposing isolated bacterial membrane vesicles to solutions in which the proton concentration had been artificially elevated. Such treatment stimulated the active transport of lactose into the vesicles, supporting the conclusion that movement of protons down their free-energy gradient drives the accompanying transport of lactose.

How is this transport accomplished? As in the case of other transport proteins, alternating conformational states appear to be involved. The lactose carrier is a hydrophobic transmembrane protein that contains a single lactose binding site and a single proton binding site. In one conformation the two sites face the cell exterior; in the other conformation they face the cell interior. Transport is initiated by the binding of lactose and a proton to their respective binding sites when the sites face the cell exterior. This binding induces a conformational change that reorients the lactose carrier so that its binding sites face the cell interior, allowing the lactose and proton to be released within the cell. The release of lactose and the proton triggers a conformational change that reorients the binding sites, causing them again to face the outside of the cell.

Membranes Also Move Materials by the Budding and Fusion of Membrane Vesicles

In the preceding sections we have seen that simple diffusion, facilitated diffusion, and active transport all involve the passage of substances across the lipid bilayer. Membranes also transfer materials in a fundamentally different way: the budding and fusion of membrane vesicles. For example, we will learn in Chapter 7 that during the process of *endocytosis,* small regions of plasma membrane fold inward and bud off as tiny vesicles that bring materials dissolved in the extracellular fluid into the cell. Transport in the opposite direction occurs during *exocytosis,* when cytoplasmic membrane vesicles fuse with the plasma membrane and discharge their contents into the extracellular space. In addition to moving substances into and out of cells, budding and fusion of membrane vesicles is also employed in transporting substances from one membrane-enclosed compartment to another. Because movements involving the budding and fusion of membrane vesicles involve interactions among different kinds of membranes, we will delay a detailed description of these events until the relevant membrane systems are described in Chapter 7.

SUMMARY OF PRINCIPAL POINTS

• Cellular membranes are constructed from lipid, protein, and small amounts of covalently bound carbohydrate.

• Membrane phospholipids form a bilayer in which their hydrophobic tails are buried in the membrane interior and their hydrophilic head groups are exposed at the membrane surfaces. Phospholipids, glycolipids, and steroids are employed in varying proportions in the construction of the lipid bilayer, which is a fluid structure that permits lateral diffusion of membrane proteins as well as lipids.

• Membrane proteins are divided into three classes: integral proteins that are embedded within the lipid bilayer, peripheral proteins that are bound to the membrane surface by ionic bonds, and lipid-anchored proteins that are covalently attached to lipid molecules embedded in the bilayer.

• Membrane proteins and lipids are oriented asymmetrically in cellular membranes, with the carbohydrate groups of glycoproteins and glycolipids usually restricted to a membrane's noncytosolic surface.

• The red cell membrane contains several predominant transmembrane proteins (glycophorin and band 3) and peripheral proteins (spectrin, ankyrin, band 4.1, and actin). The latter proteins form a cytoskeletal meshwork that stabilizes the membrane and restricts the mobility of its integral proteins.

• Substances pass across cellular membranes by simple diffusion, facilitated diffusion, and active transport. Simple diffusion refers to the unaided net movement of substances from regions of higher to lower concentration. Simple diffusion drives the net movement of dissolved solutes as well as water molecules; in the latter case the process is termed osmosis. Substances that are lipid soluble, small, and uncharged tend to diffuse through membranes most readily.

• Facilitated diffusion and active transport are carried out by membrane transport proteins; these processes are faster and more selective than simple diffusion, exhibit Michaelis-Menten saturation kinetics, and are susceptible to competitive and noncompetitive inhibitors.

• Facilitated diffusion refers to the assisted movement of a substance down its electrochemical gradient. Microorganisms produce small molecules called ionophores that facilitate the diffusion of specific ions across membranes. Some ionophores act as mobile carriers that diffuse back and forth across the lipid bilayer, whereas others form relatively fixed channels through which ions can diffuse.

• Membrane transport proteins that carry out facilitated diffusion are divided into carrier proteins and channel proteins. Carrier proteins undergo alternating conformational changes that cause their solute binding sites to alternately face one side of the membrane and then the other. Channel proteins form small channels through which specific ions can pass; the permeability of these ion channels is regulated by gates that open and close in response to various signals.

• Active transport is carried out by membrane carrier proteins that are capable of moving substances against an electrochemical gradient. The energy required for active transport can be provided by the hydrolysis of energy-rich compounds such as ATP and phosphoenolpyruvate, by photons of light, by electron transfer reactions, or by the movement of other substances down an electrochemical gradient.

SUGGESTED READINGS

Books

Evans, W. H., and J. M. Graham (1989). *Membrane Structure and Function,* IRL Press, Oxford, England.

Gennis, R. D. (1989). *Biomembranes: Molecular Structure and Function,* Springer-Verlag, New York.

Jain, M. K. (1988). *Introduction to Biological Membranes,* 2nd Ed., Wiley, New York.

Larsson, C., and I. M. Møller, eds. (1990). *The Plant Plasma Membrane,* Springer-Verlag, New York.

Petty, H. R. (1993). *Molecular Biology of Membranes: Structure and Function,* Plenum, New York.

Robertson, R. N. (1983). *The Lively Membranes,* Cambridge University Press, New York.

Stein, W. D. (1990). *Channels, Carriers, and Pumps: An Introduction to Membrane Transport,* Academic Press, San Diego.

Articles

Bloch, R. J., and D. W. Pumplin (1992). A model of spectrin as a concertina in the erythrocyte membrane skeleton, *Trends Cell Biol.* 2:186–189.

Bretscher, M. S. (1985). The molecules of the cell membrane, *Sci. Amer.* 253 (October):100–108.

Cross, G. A. M. (1990). Glycolipid anchoring of plasma membrane proteins, *Annu. Rev. Cell Biol.* 6:1–39.

Edidin, M. (1992). Patches, posts, and fences: proteins and plasma membrane domains, *Trends Cell Biol.* 2:376–380.

Frye, L. D., and M. Edidin (1970). The rapid intermixing of cell-surface antigens after formation of mouse-human heterokaryons, *J. Cell Sci.* 7:319–335.

Gould, G. W., and G. I. Bell (1990). Facilitative glucose transporters: an expanding family, *Trends Biochem. Sci.,* 15:18–23.

Hedrich, R., and J. I. Schroeder (1989). The physiology of ion channels and electrogenic pumps in higher plants, *Annu. Rev. Plant Physiol.,* 40:539–569.

Hilgemann, D. W. (1994). Channel-like function of the Na, K pump probed at microsecond resolution in giant membrane patches, *Science* 263:1429–1432.

Luna, E. Z., and A. L. Hitt (1992). Cytoskeleton-plasma membrane interactions, *Science* 258:955–964.

Nikaido, H., and M. H. Saier, Jr. (1992). Transport proteins in bacteria: common themes in their design, *Science* 258:936–942.

Robertson, J. D. (1981). Membrane structure, *J. Cell Biol.* 91:189s–204s.

Singer, S. J. (1990). The structure and insertion of integral proteins in membranes, *Annu. Rev. Cell Biol.* 6:247–296.

Singer, S. J., and G. L. Nicolson (1972). The fluid mosaic model of the structure of cell membranes, *Science* 175:720–731.

Stoeckenius, W. (1976). The purple membrane of salt-loving bacteria, *Sci. Amer.* 234 (June):38–46.

Sussman, M. R. (1994). Molecular analysis of proteins in the plant plasma membrane, *Annu. Rev. Plant Physiol. Plant Molec. Biol.* 45:211–234.

Unwin, N., and R. Henderson (1984). The structure of proteins in biological membranes, *Sci. Amer.* 250 (February):78–94.

Welsh, M. J., and A. E. Smith (1993). Molecular mechanisms of CFTR chloride channel dysfunction in cystic fibrosis, *Cell* 73:1251–1254.

Zachowski, A. (1993). Phospholipids in animal eukaryotic membranes: transverse asymmetry and movement, *Biochem. J.* 294:1–14.

C h a p t e r   6

The Cell Surface and Cellular Communication

The cell surface is a region of unique importance because it is the only place where a cell makes direct physical contact with its surrounding environment. In multicellular organisms the surface of most cells is in contact with neighboring cells or an extracellular matrix, whereas in unicellular organisms it more commonly faces an aqueous environment. In either case, the cell surface represents the primary link to the external world and is intimately involved in the communication between cells and their surroundings. Now that the general properties of membranes have been introduced in the preceding chapter, we will discuss some of the roles played by the plasma membrane and its associated cell surface components.

THE MEMBRANE POTENTIAL

A plasma membrane is located at or near the external surface of every living cell. Because it is selectively permeable and capable of solute-specific transport, the presence of a plasma membrane allows the concentrations of various substances inside the cell to differ from those prevailing in the extracellular environment. Many of the solutes that become unequally distributed inside and outside the cell carry a net charge, which can in turn generate a voltage across the plasma membrane. Such a voltage, or **membrane potential,** exists at the surface of almost all cells. Although the magnitude of the membrane potential varies among cell types, it generally falls in the range of −20 to −300 millivolts (the minus sign indicates that the inside of the cell is negatively charged relative to the outside). Because the electrical potential across the plasma membrane influences the movement of substances into and out of the cell, and plays a prominent role in certain types of cellular communication, it is important to understand the factors that determine its magnitude and polarity.

The Donnan Equilibrium and Nernst Equation Describe How the Passive Distribution of Ions Influences the Membrane Potential

In 1927 an English chemist, Frederick Donnan, first described how an unequal distribution of diffusible ions across a membrane can be generated by purely passive forces. His model was based on the fact that cells are enclosed by a membrane that is permeable to small ions, but impermeable to intracellular organic molecules such as amino acids, proteins and nucleic acids. These organic molecules typically carry a net negative charge and so will be referred to as organic anions (A^-). In order to maintain the electrical neutrality of the cell, these negatively charged anions must be balanced by positively charged cations; in cells, the predominant cation is K^+.

Figure 6-1 illustrates what would happen if a hypothetical cell containing negatively charged organic anions (A^-) and positively charged potassium cations (K^+) is placed in a solution of KCl. Because Cl^- is present in higher concentration outside the cell, it begins to diffuse into the cell. The negative charge of the entering chloride ions quickly establishes a charge imbalance across the plasma membrane in which the inside of the cell is

negatively charged with respect to the outside. This negative membrane potential creates a driving force that accelerates the diffusion of positively charged potassium ions into the cell, but at the same time slows the inward diffusion of the negatively charged chloride ions. Eventually the membrane potential stabilizes at a value that permits K^+ and Cl^- to diffuse into and out of the cell at equal rates. Donnan showed that when this final equilibrium is achieved, the internal and external concentrations of K^+ and Cl^- are related by the equation:

$$\frac{[K^+]_{in}}{[K^+]_{out}} = \frac{[Cl^-]_{out}}{[Cl^-]_{in}} \tag{6-1}$$

Although Equation 6-1 describes the conditions prevailing at equilibrium, K^+ and Cl^- are still unequally distributed across the membrane; the K^+ concentration is higher inside the cell than outside, and the Cl^- concentration is higher outside the cell than inside. Since the membrane is permeable to both K^+ and Cl^-, what prevents these ions from diffusing down their respective concentration gradients and becoming equally distributed on both sides of the membrane? In other words, how are concentration gradients of K^+ and Cl^- maintained in the absence of an energy-requiring transport process? The answer is to be found in the balance between two opposing forces, one chemical and the other electrical. The chemical concentration gradient tends to drive K^+ out of the cell, while the electrical membrane potential tends to drive K^+ into the cell. Conversely, the chemical concentration gradient tends to drive Cl^- into the cell, while the electrical membrane potential drives Cl^- out of the cell. In each case, the chemical force driving diffusion in one direction is exactly counteracted by the electrical force driving diffusion in the opposite direction. When these opposing forces are balanced for any given ion, no net diffusion occurs in either direction and the ion is said to be in **electrochemical equilibrium.**

How does one determine the magnitude and polarity of the membrane potential that is required to produce electrochemical equilibrium for any given ion? In Chapter 5 (page 181), we saw that the change in free energy that accompanies the movement of an electrically charged substance across a membrane is defined by the expression:

$$\Delta G' = 2.303RT \log_{10} \frac{[C]_{out}}{[C]_{in}} + zF\,\Delta\psi$$

When electrochemical equilibrium has been achieved, no further change in free energy occurs and so $\Delta G'$ equals zero. The above equation therefore reduces to the following:

$$0 = 2.303RT \log_{10} \frac{[C]_{out}}{[C]_{in}} + zF\,\Delta\psi$$

Rearranging and solving for the membrane potential yields the equation:

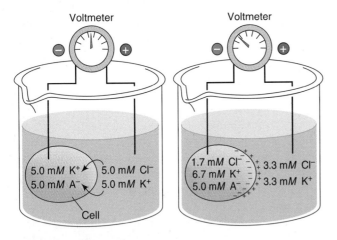

Figure 6-1 The Donnan Equilibrium (Left) *In this hypothetical example, a cell is enclosed by a membrane that is permeable to K^+ and Cl^-, but not A^-. Initially the concentrations of K^+ and Cl^- outside the cell, and K^+ and A^- inside the cell, are equal to 5.0 mM. (Right) At equilibrium the movement of K^+ and Cl^- into the cell has established a membrane potential that counterbalances the concentration gradients for each ion. Thus the concentration gradient for K^+, which tends to drive K^+ out of the cell, is balanced by the membrane potential, which drives K^+ into the cell. Likewise, the concentration gradient for Cl^-, which tends to drive Cl^- into the cell, is balanced by the membrane potential, which drives Cl^- out of the cell.*

$$\Delta \psi = \frac{2.303RT}{zF} \log_{10} \frac{[C]_{out}}{[C]_{in}} \quad (6\text{-}2)$$

Since the above expression, called the **Nernst equation,** describes electrochemical equilibrium, the value of $\Delta\psi$ must represent the magnitude and polarity of the membrane potential that will exactly counterbalance the tendency of a given ion to diffuse down its concentration gradient. This equilibrium value is referred to as the **equilibrium potential** or **Nernst potential.** Use of the Nernst equation to determine the equilibrium potential can be simplified by inserting values for R, T, z, and F, and calculating the numerical value of the expression $2.303RT/zF$. For an ion carrying a single positive charge at room temperature, such a calculation reveals that the expression $2.303RT/zF$ is equal to about 58 millivolts (mV). The Nernst equation therefore simplifies to

$$\Delta \psi = 58 \log_{10} \frac{[C^+]_{out}}{[C^+]_{in}} \quad (6\text{-}3)$$

where $\Delta\psi$ is expressed in millivolts. For negatively charged ions, the concentration gradient will be oriented in the opposite direction relative to the membrane potential. Hence for an ion carrying a single negative charge, the comparable equation would be

$$\Delta \psi = 58 \log_{10} \frac{[C^-]_{in}}{[C^-]_{out}} \quad (6\text{-}4)$$

To illustrate the use of the above equations, let us consider an example in which the K^+ concentration is 100 mM inside the cell and 10 mM outside the cell. Substituting these values in Equation 6-3, we obtain the following expression:

$$\Delta \psi = 58 \log_{10} \frac{10}{100} = 58 \log_{10}(0.1) = -58 \text{ mV}$$

Hence a membrane potential of -58 mV is required to balance the tendency of K^+ to diffuse out of the cell given such a concentration gradient. Since the minus sign means that the inside of the cell is negative relative to the outside, it is this negatively charged interior that counteracts the tendency of the positively charged potassium ions to diffuse out of the cell. The existence of the negative membrane potential does not mean, however, that the inside of the cell contains a large excess of negative charges. The separation of electric

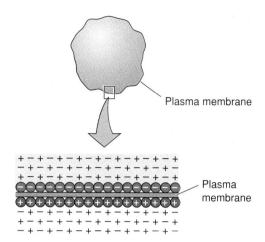

Figure 6-2 *Charge Distribution Inside and Outside a Cell* *The net excess of positive charges associated with the outer surface of the plasma membrane and negative charges associated with the inner surface of the plasma membrane represents only a tiny fraction of the total charges present inside and outside the cell.*

charges that creates the membrane potential occurs only in the immediate vicinity of the plasma membrane, and so involves only a tiny fraction of the total number of positive and negative charges that exist in the cell. The bulk of the cytoplasm and extracellular fluid therefore remains electrically neutral (Figure 6-2).

Active Transport Influences the Membrane Potential

The Donnan equilibrium provides a model for explaining how passive forces can generate an unequal distribution of ions across membranes. But to what extent are such passive forces actually responsible for the ionic distributions observed in living cells? Let us address this question for K^+, Na^+, and Cl^-, which are the most abundant diffusible ions in cells and are therefore crucial in determining the magnitude and polarity of the membrane potential. K^+ is present in higher concentration inside cells than outside, whereas the opposite is true for Na^+ and Cl^- (Table 6-1). If passive forces were entirely responsible for these unequal distributions, then K^+, Na^+, and Cl^- would each be in electrochemical equilibrium across the plasma membrane. In other

Table 6-1 Ionic Concentrations and Equilibrium Potentials in Three Cell Types

| | Frog Muscle Cell | | | Mammalian Nerve Cell | | | Squid Nerve Cell | | |
	Na^+	K^+	Cl^-	Na^+	K^+	Cl^-	Na^+	K^+	Cl^-
Concentration inside cells (mM)	13	140	3	10	140	10	50	400	52
Concentration outside cells (mM)	110	2.5	90	145	5	110	440	20	560
Equilibrium potential (mV) (calculated from Nernst equation)	+54	−101	−86	+67	−84	−60	+55	−75	−60

words, the equilibrium potential for each ion calculated by the Nernst equation would be equal to the observed membrane potential. When such calculations are carried out, however, only Cl$^-$ appears to be in electrochemical equilibrium. The equilibrium potential for Cl$^-$ calculated from the Nernst equation is typically quite close to the observed membrane potential. One can therefore conclude that chloride ions are largely in electrochemical equilibrium (passively distributed) across the plasma membrane.

In contrast, the equilibrium potential for K$^+$ calculated from the Nernst equation is usually more negative than the actual membrane potential. For example, the equilibrium membrane potential calculated from the internal and external distribution of K$^+$ in squid nerve cells is −75 mV; however, if an electrode is inserted into the cell to determine the actual magnitude of the membrane potential, a value of −60 mV is obtained. This means that the intracellular concentration of K$^+$ is somewhat higher than that expected for electrochemical equilibrium. The slightly higher-than-expected internal K$^+$ concentration is maintained by the Na$^+$-K$^+$ pump (page 186), which actively transports K$^+$ into the cell.

Na$^+$ is even further from electrochemical equilibrium than is K$^+$. The equilibrium potential for Na$^+$ in squid nerve cells calculated from the Nernst equation is +55 mV, a value that is far from the actual membrane potential of −60 mV. This discrepancy in polarity occurs because Na$^+$ is present in higher concentration outside the cell than inside, whereas at electrochemical equilibrium it would be expected to be present in higher concentration inside the cell than outside. The unexpected distribution of Na$^+$ is also maintained by the Na$^+$-K$^+$ pump, which actively transports Na$^+$ out of the cell.

The Goldman Equation Takes into Account the Contributions of Multiple Ions to the Membrane Potential

We have now seen that both passive and active forces contribute to the unequal distribution of diffusible ions across the plasma membrane. These unequally distributed ions in turn generate an electrical potential across the membrane. But how is the magnitude and polarity of the potential determined when multiple ions are involved? In squid nerve cells, for example, the Nernst equation predicts an equilibrium potential of −75 mV for K$^+$, +55 mV for Na$^+$, and −60 mV for Cl$^-$. How is the actual membrane potential determined in such situations?

The answer is that each of these ions contributes to the membrane potential. The magnitude of each ion's contribution is determined by the steepness of its concentration gradient as well as the relative permeability of the membrane to that ion. To help explain why relative permeability is so important, let us consider a hypothetical cell containing a plasma membrane that is permeable only to K$^+$. Since the intracellular concentra-

tion of K$^+$ is higher than that outside the cell, the positively charged potassium ions would tend to diffuse out of the cell, creating a charge imbalance in which the *cell exterior is more positively charged* than the cell interior. This would not occur, however, if the hypothetical membrane were permeable only to Na$^+$; since the extracellular concentration of Na$^+$ is higher than that inside the cell, the positively charged sodium ions would tend to diffuse into the cell and create a charge imbalance in which the *cell interior is more positively charged* than the cell exterior. Thus the relative permeability of the membrane to Na$^+$ and K$^+$ strongly influences the magnitude and polarity of the final membrane potential.

To account for this influence of membrane permeability, the Nernst equation has been modified to take into consideration the contribution of multiple ions of differing permeabilities:

$$\Delta\psi = 58 \log_{10} \frac{P_K[\text{K}^+]_{\text{out}} + P_{Na}[\text{Na}^+]_{\text{out}} + P_{Cl}[\text{Cl}^-]_{\text{in}}}{P_K[\text{K}^+]_{\text{in}} + P_{Na}[\text{Na}^+]_{\text{in}} + P_{Cl}[\text{Cl}^-]_{\text{out}}} \quad (6\text{-}5)$$

In this expression, called the **Goldman equation,** P_K, P_{Na}, and P_{Cl} are permeability coefficients that refer to the relative permeabilities of the membrane to K$^+$, Na$^+$, and Cl$^-$, respectively. In theory all ions that are unequally distributed across the plasma membrane could be included in this equation, but in practice only K$^+$, Na$^+$, and Cl$^-$ have permeability coefficients that are high enough to make significant contributions to the final membrane potential.

The Goldman equation appears to fit the available data quite well. In squid nerve cells, for example, the relative permeabilities for K$^+$, Na$^+$, and Cl$^-$ are $P_K = 1.0$, $P_{Na} = 0.04$, and $P_{Cl} = 0.45$. Inserting these values into Equation 6-5, along with the intracellular and extracellular ion concentrations given in Table 6-1, we obtain

$$\Delta\psi = 58 \log_{10} \frac{(1.0)(20) + (0.04)(440) + (0.45)(52)}{(1.0)(400) + (0.04)(50) + (0.45)(560)}$$

$$= 58 \log_{10} (61/654) = -60 \text{ mV} \quad (6\text{-}6)$$

This predicted membrane potential of −60 mV agrees closely with values for the membrane potential of squid nerve cells that have been measured experimentally by inserting electrodes directly into the cytoplasm.

Changes in Membrane Potential Act as a Signaling Device in Nerve and Muscle Cells

The Goldman equation tells us that the value of the membrane potential is determined by a membrane's permeability to different ions. It logically follows, therefore, that changes in membrane permeability will alter the membrane potential. This principle lies at the heart of the electrical signaling mechanism that allows nerve and muscle cells to communicate with one another.

The idea that electrical signaling is employed as a cellular communication mechanism was first suggested 200 years ago by Luigi Galvani after he observed that an isolated muscle will contract if its nerves are electrically stimulated. To investigate the nature of this signaling process, Hermann von Helmholtz measured the time required for a muscle to contract after the nerve leading to the muscle had been electrically stimulated at varying points along its length. These data revealed that the electrical signal passes down the nerve at a velocity of about 40 meters per second, which is orders of magnitude slower than the speed of an electric current passing through a wire. Hence the electrical signals carried by nerve cells differ in a fundamental way from a conventional electric current.

Our understanding of how this electrical signaling works has been greatly aided by studies involving the giant nerve cells of squid. Nerve cells possess long cytoplasmic outgrowths, called **axons,** that play a key role in transmitting electrical signals from one nerve cell to another. The axons of squid nerve cells are uniquely suited for experimental study because they measure up to a millimeter in diameter, which is hundreds of times larger than the axons of vertebrate nerve cells. It is relatively easy, therefore, to insert electrodes into squid axons and measure the electrical potential across the plasma membrane.

In the late 1930s two teams of scientists, Alan Hodgkin and Andrew Huxley in England and Howard Curtis and Kenneth Cole in the United States, measured the changes that occur in the membrane potential of the squid axon when the cell is electrically stimulated. They discovered that electrically stimulating the cell causes a signal to pass down the axon that involves a rapid series of changes in membrane potential. The first change to be observed when the signal reaches a given point on the axon is membrane **depolarization**—that is, a change in membrane potential to a less negative value. During the depolarization phase, the membrane potential rapidly shifts from its normal resting value of −60 mV to a positive value of about +40 mV. In other words, the inside of the cell becomes positively charged relative to the outside. This reversed polarity lasts for only about a thousandth of a second. The membrane potential then reverts back to a **hyperpolarized** state that is slightly more negative than the initial resting potential, and finally returns to its normal resting value of −60 mV. This sequence of changes in the membrane potential is collectively referred to as an **action potential** (Figure 6-3, *top*).

The Action Potential Is Produced by Changes in Membrane Permeability to Na⁺ and K⁺

What causes the changes in membrane potential that constitute an action potential? Hodgkin and Huxley, in collaboration with Bernard Katz, were the first to pro-

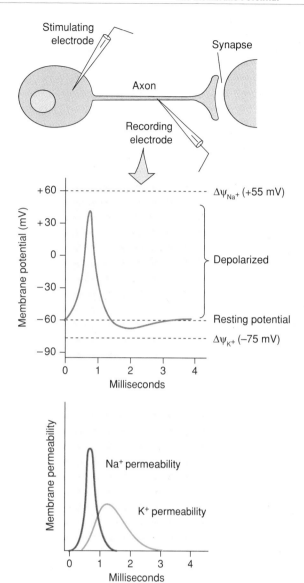

Figure 6-3 *The Action Potential* (Top diagram and graph) *A nerve cell is stimulated with an electrode while a second electrode measures the change in membrane potential that occurs as the resulting action potential passes along the axon. During the first stage of an action potential the membrane depolarizes, shifting the membrane potential from −60 mV to about +40 mV; the membrane potential then reverts back to a hyperpolarized state that is slightly more negative than the initial resting value, and finally returns to a value of −60 mV. (Bottom graph) An action potential is caused by two changes in membrane permeability acting in sequence: a transient increase in permeability to Na⁺, followed by a transient increase in permeability to K⁺.*

pose that the initial reversal in membrane polarity is caused by a transient increase in the permeability of the plasma membrane to Na⁺. Because the Na⁺ concentration is higher outside the cell than inside, an increase in permeability would cause the positively charged sodium ions to diffuse into the cell and hence increase the positive charge inside the cell. This prediction can be quantified using the Goldman equation. To illustrate, let us repeat our earlier calculation in which we used the Goldman equation to predict the membrane potential of the squid axon (Equation 6-6); however, this time we will increase the permeability to Na^+ 400-fold from $P_{Na} = 0.04$ to $P_{Na} = 16$. Under such conditions, we obtain the following:

$$\Delta\psi = 58 \log_{10} \frac{(1.0)(20) + (16)(440) + (0.45)(52)}{(1.0)(400) + (16)(50) + (0.45)(560)}$$

$$= 58 \log_{10} (7083/1452) = +40 \text{ mV} \qquad (6\text{-}7)$$

Thus increasing the membrane permeability to sodium ions can indeed convert the predicted membrane potential from −60 mV to the value of about +40 mV observed at the peak of the action potential. Because the permeability to Na^+ is so high relative to the permeability to K^+, the membrane potential is in fact approaching the equilibrium potential for Na^+ (+55 mV) predicted by the Nernst equation.

If an increase in permeability to Na^+ is the correct explanation for the initial change in membrane potential that occurs during an action potential, then the occurrence of an action potential should require the presence of sodium ions outside the cell. This prediction has been confirmed experimentally; action potentials have been shown to decrease in magnitude as the external Na^+ concentration is lowered (Figure 6-4) and to disappear entirely in the absence of external Na^+.

If the change in membrane potential from −60 mV to +40 mV is caused by an increased permeability to Na^+, how is the potential restored to its initial value of −60 mV? To address this question, Hodgkin and Huxley developed a procedure called the *voltage clamp technique*. In this approach, two electrodes are inserted

into the cell; one monitors the membrane potential and the other is used to pass current across the membrane, thereby maintaining or "clamping" the membrane potential at a fixed value. This procedure allows the movement of Na^+ and K^+ to be monitored under conditions where the membrane potential is artificially held at a constant value. Hodgkin and Huxley found that if they depolarized an axon with an electric current and subsequently held the membrane potential at zero using a voltage clamp, a small electric current first flows into the axon and then out of the axon. If the experiment is repeated in a medium lacking Na^+, the inward flow of current is abolished, but the outward flow still occurs. They concluded, therefore, that the inward current is caused by an influx of sodium ions. Since the outward flow of current is not dependent upon Na^+, Hodgkin and Huxley proposed that it involved the movement of potassium ions out of the cell. To test this theory, a radioactive isotope of the potassium ion ($^{42}K^+$) was introduced into the cells. As predicted, radioactivity was found to flow out of the cell during an action potential.

From these experiments, Hodgkin and Huxley concluded that an action potential is produced by two changes in membrane permeability acting in sequence: a transient increase in permeability to Na^+ followed by a transient increase in permeability to K^+ (see Figure 6-3, *bottom*). Because the external Na^+ concentration is higher than that inside the cell, the initial increase in permeability to Na^+ allows this ion to diffuse into the cell. Although the amount of Na^+ entering the cell is extremely small (less than 0.001 percent of the total number of Na^+ ions is involved in this movement), it is sufficient to cause the inner surface of the plasma membrane to become positively charged relative to the exterior. The subsequent increase in membrane permeability to K^+ facilitates the diffusion of K^+ out of the cell, which occurs because the concentration of K^+ is higher inside the cell than outside. The total amount of K^+ diffusing out of the cell is also very small, but the exit of positively charged ions is sufficient to restore the membrane potential to its original negative value.

This model suggests that the action potential can be entirely explained by changes in membrane permeability to Na^+ and K^+. It has been shown, in fact, that the cytoplasm can be squeezed out of the squid axon without destroying the ability of the axon to transmit nerve impulses. The only requirement is that the normal concentration gradients of Na^+ and K^+ be maintained by injecting the inside of the axon with a solution that has a high concentration of K^+ and a low concentration of Na^+. In addition to confirming the essential roles of Na^+ and K^+ in generating the action potential, such findings suggest that the nerve impulse is a purely membrane-derived phenomenon; that is, it does not require any component of the cytoplasm other than the proper internal concentrations of Na^+ and K^+.

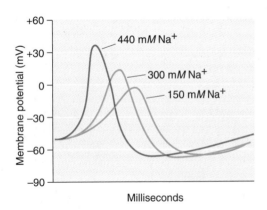

Figure 6-4 Influence of Extracellular Na+ Concentration on the Magnitude of the Action Potential *Decreasing the external Na+ concentration in seawater causes a reduction in the magnitude of the action potential measured in squid nerve cells.*

Voltage-Gated Ion Channels Control the Movement of Na⁺ and K⁺ During an Action Potential

If changes in membrane permeability to Na$^+$ and K$^+$ are responsible for generating the action potential, then what mechanism underlies these changes in permeability? In Chapter 5 we learned that the diffusion of ions through biological membranes is facilitated by a special class of membrane proteins known as *ion channels* (page 184). The behavior of ion channels during an action potential has been studied using the elegant **patch clamp technique** developed by Erwin Neher and Bert Sakmann. In this procedure, a finely polished micropipette tip is placed on the surface of a cell and suction is applied to draw a small portion of the plasma membrane into the pipette tip, forming a tight seal (Figure 6-5). The movement of ions through ion channels present in this small patch of membrane can then be detected by measuring the flow of current into and out of the pipette. By choosing a membrane possessing relatively few channels per unit area, it is possible to study a small patch of membrane containing only a single ion channel.

Such studies have revealed that separate ion channels exist for Na$^+$ and K$^+$. In nerve and muscle cells, the opening of both types of channels is controlled by changes in the membrane potential; hence they are referred to as **voltage-gated channels.** At the normal resting potential of -60 mV, most of the voltage-gated channels are closed. But when the membrane potential becomes less negative, the probability increases that any given channel will briefly assume the open configuration and allow ions to pass through. Hence when a nerve cell is subjected to an electrical stimulus that causes the membrane potential to become less negative, this small depolarization causes a few of the sodium channels to open. The resulting influx of positively charged sodium ions makes the membrane potential even less negative, which in turn causes additional sodium channels to open. This depolarizes the membrane even further, leading to the opening of more sodium channels. The original depolarization is thus amplified until the membrane potential approaches the equilibrium potential for Na$^+$.

The voltage-gated K$^+$ channels are slower to open in response to the changing membrane potential, so an increase in permeability to K$^+$ occurs after the increase in permeability to Na$^+$. As the K$^+$ channels open, the diffusion of positively charged potassium ions out of the cell causes the membrane potential to return to its initial negative value. The total number of Na$^+$ and K$^+$ ions that pass into and out of the cell during an action potential is quite small, typically less than a millionth of the total number of ions present. Hence no significant change in the intracellular and extracellular concentrations of Na$^+$ and K$^+$ occurs as a result of the opening of the voltage-gated ion channels during an action potential. Voltage-gated Na$^+$ and K$^+$ channels are typically restricted to the plasma mem-

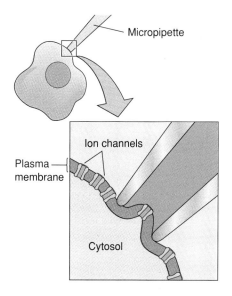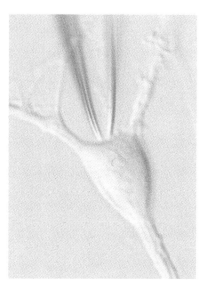

Figure 6-5 The Patch Clamp Technique (Left) *In this procedure, the movement of ions through individual ion channels is detected by monitoring the flow of current into and out of a tiny glass pipette applied to the surface of a cell.* (Right) *Electron micrograph showing a pipette tip sealed against the surface of a neuron. The pipette, measuring only 1/25,000 the diameter of a human hair, can record the opening and closing of individual ion channels. Micrograph courtesy of B. Sakmann.*

branes of nerve and muscle cells, explaining why they, but not other cells, exhibit action potentials.

Inhibitors of Ion Transport Can Block the Na⁺ and K⁺ Channels

Many drugs that act on the brain and nervous system do so by interfering with the function of ion channels. For example, *tetrodotoxin* and *saxitoxin* are toxic substances that selectively block Na⁺ channels. Tetrodotoxin is a potent poison produced by the Japanese puffer fish that can cause paralysis and death if consumed by humans in quantities as small as a few micrograms. Saxitoxin is produced by a marine dinoflagellate that periodically accumulates in great numbers in coastal seawater, causing the water to acquire a reddish color. These "red tides" are extremely dangerous because shellfish that feed on the dinoflagellates accumulate saxitoxin and then pass it on to humans who eat the shellfish. A small clam, for example, can contain enough saxitoxin to kill several dozen people. Tetrodotoxin and saxitoxin both exert their toxic effects by binding to the outer surface of the Na⁺ channel in a way that blocks the movement of Na⁺ through the channel and into the cell. This binding is so selective that measuring the amount of tetrodotoxin or saxitoxin bound to nerve cell membranes can be used to estimate how many Na⁺ channels are present. Such studies have revealed that the number of Na⁺ channels varies from a few dozen to more than 10,000 channels per square micrometer of membrane surface in different kinds of nerve cells.

Nerve cell function can also be disrupted by inhibitors that interfere selectively with the operation of K⁺ channels. The *tetraethylammonium ion,* for example, blocks the outward movement of K⁺ through the K⁺ channels without disrupting the inward flow of Na⁺ through the Na⁺ channels. Such specificity confirms the existence of separate channels for the diffusion of sodium and potassium ions.

The Permeability of Voltage-Gated Ion Channels Is Controlled by Sequential Conformational Changes

The ability of transport inhibitors to distinguish between Na⁺ and K⁺ channels has simplified the task of isolating and purifying the proteins that make up these ion channels. For example, the protein that forms the Na⁺ channel has been identified by measuring the ability of isolated membrane proteins to bind to radioactive tetrodotoxin or saxitoxin. The protein identified in this manner consists of four homologous domains that form the walls of a transmembrane channel through which sodium ions can pass. The four homologous regions each possess six α-helical segments; one of these segments exhibits a repeating pattern of positively charged

amino acids whose three-dimensional configuration is thought to be influenced by the membrane potential. This arrangement may explain how changes in membrane potential control the conformation of the Na⁺ channel and hence its permeability to Na⁺.

Current evidence suggests that changes in membrane potential cause the Na⁺ channel to adopt three distinct conformations known as "closed," "open," and "inactivated" (Figure 6-6). In unstimulated cells, the closed conformation is thermodynamically the most stable, making the channel impermeable to Na⁺. When the membrane becomes slightly depolarized, the open conformation is favored over the closed conformation, allowing Na⁺ to enter the cell. But the Na⁺ channel remains open for only a brief period before it assumes the inactivated conformation, which is even more stable than the open conformation in depolarized membranes. After the membrane is repolarized by opening of the K⁺ channels, the Na⁺ channels again assume the closed conformation typical of the resting cell.

Thus we have seen that the behavior of voltage-gated ion channels lies at the heart of the transient changes in membrane potential that comprise an action potential. In Chapter 17 we will discuss action potentials again, showing how they are used to transmit signals between nerve cells, and from nerve cells to muscle cells.

AN OVERVIEW OF SIGNALING MOLECULES AND THEIR RECEPTORS

The membrane potential is only one of several properties of the plasma membrane that plays a role in cellular communication. Because of its location at the cell surface, it is natural for the plasma membrane to be involved in transmitting signals from outside the cell to

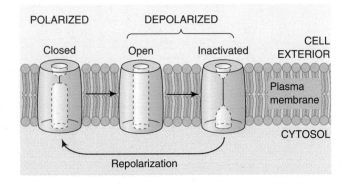

Figure 6-6 *Changes in Conformation of the Na⁺ Channel During an Action Potential* *When a membrane becomes depolarized, the Na⁺ channel opens for only a brief period before it assumes the inactivated conformation. This ensures that the influx of Na⁺ during an action potential is a transient event.*

the cell interior. This function is made possible by the presence of a special class of membrane proteins, termed **receptors,** that interact with signaling molecules produced by other cells. The binding of a signaling molecule to its receptor triggers a change in receptor conformation that is comparable to the change in enzyme conformation triggered by the binding of an allosteric regulator (page 51). The altered receptor in turn initiates a sequence of events that leads to various changes in cell behavior.

Hundreds of different molecules transmit signals between cells by binding to cell surface receptors. Based on differences in how far they travel, this chemically diverse group of signaling molecules is divided into three main categories: *hormones, neurotransmitters,* and *local mediators* (Figure 6-7).

Hormones Are Transported by the Circulatory System to Target Cells Located throughout the Body

Hormones are signaling molecules that are secreted directly into the circulatory system by specialized **endocrine** cells found in glands like the adrenal, testis, ovary, pancreas, thyroid, parathyroid, and pituitary. The circulatory system then carries the secreted hormones to target cells located throughout the body. The first experimental evidence for the existence of circulating hormones in animals was provided in 1848 by the German physiologist Arnold Berthold, who was exploring the effects of castration on roosters. He discovered that

removing the testes hinders the normal development of secondary sexual characteristics, such as wattles and combs, and behavior patterns, including normal crowing, belligerence toward other males, and attraction to hens. Transplanting a single testicle back into a castrated animal was found to restore normal physical and behavioral development. Examination of the implanted testicle revealed that it had developed an extensive blood supply, suggesting to Berthold that a substance produced by the testicle had been carried by the circulatory system, influencing those cells involved in sexual development. Although these studies provided the impetus for subsequent studies of blood-borne substances and their role in regulating physiological events, nearly 90 years passed before the substance produced by the testes was identified as testosterone.

In the early 1900s two Canadian physiologists, William Bayliss and Ernest Starling, provided direct evidence for the idea that circulating chemical messengers can trigger responses in specific target tissues. These investigators studied the factors involved in stimulating the release of pancreatic juice when the small intestine is exposed to acid released from the stomach. Injection of acid directly into the lumen of the small intestine was found to increase the flow of pancreatic juice. However, the increased flow of pancreatic secretions did not occur if the blood supply linking the intestine and the pancreas was disrupted. This result was the first unequivocal demonstration that chemical messengers released from one organ (in this case the intestine) are transported by the circulation to a target organ (the pancreas) where a specific physiological effect is elicited (increase in the flow of pancreatic juice). In 1905 Starling introduced the term *hormone* (derived from the Greek word *horman,* meaning "to arouse to activity") to refer to molecules that are transported by the bloodstream so that they can regulate events occurring in target cells located long distances from the cell producing the hormone.

In the years since these pioneering discoveries, dozens of different vertebrate hormones have been identified and purified. Based on differences in chemical structure, they are grouped into three main categories (Table 6-2). (1) *Protein* and *peptide hormones,* which represent 80 percent or more of all the vertebrate hormones, are composed of amino acid chains of varying lengths, from as few as three to as many as several hundred. Hormones in this category bind to plasma membrane receptors. (2) *Steroid hormones* are synthesized by enzymes of the smooth endoplasmic reticulum utilizing cholesterol as the parent molecule. Because they are not water soluble, steroid hormones are usually bound to proteins in the blood for efficient transport to their target cells. Unlike protein and peptide hormones, steroid hormones pass through the plasma membrane of their target cells and bind directly to intracellular receptors. (3) *Amino acid derivatives* such as epinephrine and thyrox-

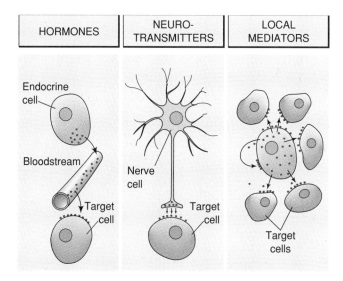

HORMONES	NEURO-TRANSMITTERS	LOCAL MEDIATORS

Figure 6-7 Cell-to-Cell Signaling by Hormones, Neurotransmitters, and Local Mediators *The main distinction between these three classes of signaling molecules involves the distance the molecule travels before encountering its target cell. Although most signaling molecules act on the cell surface, certain hormones and local mediators enter their target cells.*

Table 6-2   Examples of Hormones

Source and Hormone	Structure	Major Site of Action
Protein and Peptide Hormones		
Anterior pituitary		
Adrenocorticotropin (ACTH)	39 amino acids	Adrenal cortex
Somatotropin (growth hormone)	191 amino acids	Liver
Prolactin	199 amino acids	Mammary gland
Thyrotropin (TSH)	Subunits of 92 and 112 amino acids plus carbohydrate	Thyroid
Follicle-stimulating hormone (FSH)	Subunits of 92 and 118 amino acids plus carbohydrate	Ovary, testis
Luteinizing hormone (LH)	Subunits of 92 and 115 amino acids plus carbohydrate	Ovary, testis
Hypothalamus		
Thyrotropin releasing hormone (TRH)	3 amino acids	Anterior pituitary
LH-releasing hormone	10 amino acids	Anterior pituitary
Somatostatin	14 amino acids	Anterior pituitary
Posterior pituitary		
Oxytocin	9 amino acids	Uterus, mammary gland
Vasopressin (antidiuretic hormone)	9 amino acids	Kidney
Pancreas		
Insulin	Subunits of 21 and 30 amino acids	Fat tissue, liver, muscle
Glucagon	29 amino acids	Fat tissue, liver, muscle
Parathyroid		
Parathormone	84 amino acids	Bone, kidney, intestine
Steroid Hormones		
Ovary, placenta		
Estradiol		Uterus, vagina, breast, bone, brain
Testis		
Testosterone		Prostate, bone, brain, seminal vesicle
Ovary (corpus luteum), placenta		
Progesterone		Uterus
Adrenal cortex		
Cortisol		Many organs
Amino Acid Derivatives		
Adrenal medulla		
Epinephrine		Liver, heart, muscle, blood vessels, fat tissue
Thyroid		
Thyroxine		Many organs

ine comprise the third structurally distinct class of vertebrate hormones. These small molecules are derivatives of the amino acid tyrosine. Epinephrine is synthesized in the adrenal gland as well as in certain cells of the nervous system, and therefore can be classified as a hormone as well as a neurotransmitter. Epinephrine and its derivatives interact with plasma membrane receptors, whereas thyroxines pass through the plasma membrane and bind to intracellular receptors.

Neurotransmitters Are Released into Synapses by Nerve Cells

Neurotransmitters are signaling molecules that are released from nerve cells at specialized regions called **synapses,** where they diffuse only a tiny distance before encountering an adjacent target cell. For example, when an action potential reaches the end of an axon, the electrical signal must be transmitted across the synapse in order to trigger a response in the neighboring target cell. In most cases the signal is carried by chemical neurotransmitters that are released from the end of the axon, diffuse across the synapse, and trigger a response by binding to the surface of the adjoining cell. The earliest evidence for the existence of neurotransmitters was provided by an ingenious experiment carried out by Otto Loewi in 1921. Loewi knew that a heart removed from a frog will continue to beat if it is placed in a container filled with saline solution. Moreover, if one of the nerves leading to the heart is electrically stimulated, the heart will slow down. To investigate this effect Loewi allowed the solution bathing a beating heart to flow into a second flask containing another beating heart. Under such conditions stimulating the nerve of the first heart caused a slowing of the second heart as well, suggesting that the stimulated nerve releases a chemical substance into the saline solution that is capable of reducing the heart rate. The released substance was subsequently identified as *acetylcholine* by Sir Henry Dale.

It soon became apparent that acetylcholine is not the only substance utilized by nerve cells as a neurotransmitter. Among the chemicals that function in this capacity are the amines *acetylcholine, norepinephrine, dopamine, serotonin,* and *histamine;* the amino acids *glutamate, γ-aminobutyric acid (GABA),* and *glycine;* the peptides *enkephalin* and *β-endorphin;* and the nucleotide *ATP* (see Table 17-2).

Local Mediators Act on Neighboring Cells

Local mediators are signaling molecules that are produced by a variety of cell types and secreted into the extracellular fluid, where they act on neighboring cells located within a few millimeters of the cells that se-

creted the signaling molecule. In some cases, they also act upon the cell from which they were initially released. Regulation by local mediators is sometimes referred to as **paracrine** control (derived from the Greek *para-*, meaning "beside") to distinguish it from endocrine control by hormones circulating in the bloodstream. Protein *growth factors* (page 526), which regulate the growth and division of cells in multicellular organisms, are a prominent class of signaling molecules that act mainly on neighboring cells. Signaling proteins called *lymphokines* (page 730), which control the development and behavior of cells of the immune system, also exert their effects locally.

In addition to proteins and peptides, derivatives of amino acids and fatty acids can likewise function as local mediators (Table 6-3). A prominent group of substances in the latter category are the *prostaglandins,* named because they were originally detected in the seminal fluid secreted by the prostate gland. Prostaglandins are lipid molecules synthesized from fatty acids (principally arachidonic acid) within cellular membranes. Prostaglandins can exert control over activities occurring in nearby cells as well as in the cells where they were synthesized (**autocrine** stimulation). They are composed of a 20-carbon fatty acid skeleton that is folded into a loop by incorporating a ring at the bend in the molecule. Many types of prostaglandins have been identified, and most cells can produce at least one kind of prostaglandin. Although prostaglandins commonly induce contraction or relaxation of smooth muscle, their effects on target cells are numerous and diverse, ranging from inducing platelet aggregation, to stimulating contraction of smooth muscle in the uterus and bronchi of the lung, to causing inflammation. Some anti-inflammatory drugs, such as aspirin, function by inhibiting the production of prostaglandins.

Hormones, Neurotransmitters, and Local Mediators Transmit Signals by Binding to Receptors

Now that we have distinguished between hormones, neurotransmitters, and local mediators, it is important to emphasize that these are interrelated groups of signaling molecules with overlapping functions. Nerve cells can trigger the release of hormones and local mediators, hormones can influence nerve cells or trigger the release of local mediators and other hormones, and local mediators can modify the response of target tissues to the actions of hormones. In some cases, the same molecule can function in more than one category. For example, epinephrine, norepinephrine, and small peptides such as somatostatin act as both hormones and neurotransmitters.

Table 6-3 Examples of Local Mediators

Local Mediator	Structure	Major Site of Action
Proteins and Peptides		
Platelet-derived growth factor (PDGF)	Two subunits of 14,000 and 17,000 daltons	Fibroblasts
Nerve growth factor (NGF)	Two identical subunits of 118 amino acids each	Nerve cells
Epidermal growth factor (EGF)	53 amino acids	Epidermal and other cells
Lymphokines	Proteins of varying sizes	Lymphocytes
Amino Acid Derivatives		
Histamine	(structure)	Blood vessels
Fatty Acid Derivatives		
Prostaglandins (PGA$_1$)	(structure)	Smooth muscle, other widespread effects
Retinoic acid	(structure)	Epithelial tissues

Hormones, neurotransmitters, and local mediators also employ similar mechanisms when triggering responses in their target cells. In each case, the initial event is the binding of the signaling molecule to a specific receptor protein. Most signaling molecules are too hydrophilic or too large to pass through the lipid bilayer of the target cell's plasma membrane. Such molecules therefore act by binding to receptor proteins that are part of the plasma membrane. This binding in turn can trigger a variety of different changes inside the target cell. The term **signal transduction** is employed when referring to the process by which the binding of a signaling molecule to the outer cell surface triggers changes inside the target cell. Since this chapter is devoted to functions of the cell surface, we will restrict the remainder of our discussion to signaling events that involve such cell surface receptors. (Signaling molecules that pass through the plasma membrane and interact directly with receptors located inside the target cell will be discussed in Chapter 10. Steroid hormones and thyroxine, both of which act primarily upon the cell nucleus, are the main examples in this category.)

Receptors Are Identified by Testing Their Ability to Bind to Radioactive Signaling Molecules

Plasma membrane receptors for a wide variety of signaling molecules have been isolated and purified in recent years. One widely used approach for identifying recep-

tors involves exposing cells or isolated membranes to signaling molecules that have been radioactively labeled. For example, receptors for insulin or acetylcholine can be detected by incubating cells with radioactive insulin or acetylcholine. After incubation, the amount of radioactivity bound to the plasma membrane or to individual membrane proteins is measured. Demonstrating that a radioactive signaling molecule binds to a particular membrane protein does not, however, constitute proof that a true receptor has been identified. Before it can be stated with confidence that the observed binding represents an interaction between a signaling molecule and its receptor, it must be shown that the binding reaction exhibits the following five properties: *specificity, high affinity, saturation kinetics, reversibility,* and an appropriate *physiological response.*

1. Specificity. The binding of a signaling molecule to its receptor is a highly specific interaction that resembles the recognition between an enzyme and its substrate. (When discussing the interaction between a signaling molecule and its receptor, the term **ligand** is commonly employed to refer to the signaling molecule; *ligand* is a generic term that refers to any smaller molecule that binds to a larger molecule.) To assess the specificity of the binding between a signaling ligand and its receptor, one measures the ability of nonradioactive substances to compete with the radioactive signaling ligand for the receptor binding site. If binding to a true receptor is being observed, competition by substances unrelated to the signaling molecule should not occur. For example, the binding of radioactive insulin to its receptor is not inhibited by nonradioactive ligands such as glucagon and ACTH, but is inhibited by nonradioactive insulin or insulin derivatives (Figure 6-8). Hence the radioactive insulin must be binding to a receptor that is specific for insulin.

2. High affinity binding. Information concerning the affinity of a ligand for its receptor is obtained by analyzing the kinetics of the binding reaction. When the amount of ligand that binds to a receptor is graphed as a function of the total ligand concentration, a hyperbolic curve is obtained that is similar in shape to the curve relating enzyme activity to substrate concentration (Figure 6-9, *top*). The ligand concentration at which binding to receptor is half-maximal is called the **dissociation constant (K_d).** This dissociation constant (analogous to the Michaelis constant of enzymatic reactions) is inversely related to the affinity of the receptor for its ligand. A high value of K_d indicates that a large concentration of ligand is required before a significant amount of ligand will bind to the receptor; in other words, the receptor has a relatively low affinity for the ligand. Conversely, a low value of K_d means that the ligand binds to the receptor at low ligand concentration, indicating that the receptor has a high affinity for the ligand.

Interactions between signaling ligands and their receptors typically exhibit low K_d values (high affinity

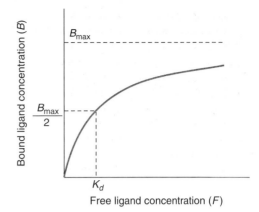

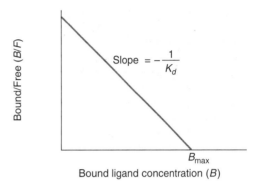

Figure 6-9 *Kinetic Analysis of the Binding of a Signaling Ligand to Its Receptor* (Top) *A standard Michaelis-Menten graph relating the concentration of bound ligand (B) to the free ligand concentration (F). Note that the receptor becomes saturated with bound ligand at high concentrations of free ligand. (Bottom) A Scatchard plot of the same data in which the ratio of bound to free ligand concentration (B/F) is plotted as a function of the free ligand concentration (F).*

binding). Precise measurement of K_d is therefore important in characterizing receptor preparations. Suppose, for example, one wished to identify the membrane receptor for insulin, a hormone whose normal concentration in the bloodstream is about 10^{-9} M. If an isolated membrane component were found to bind insulin with a K_d of 10^{-6} M, it could not be the true receptor for insulin because it does not have sufficient affinity to bind to insulin at its normally prevailing concentration of 10^{-9} M. Most isolated hormone receptors exhibit K_d values in the range of 10^{-8} to 10^{-11} M, corresponding well with the low concentrations at which most hormones act.

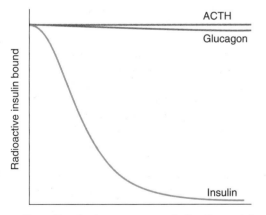

Figure 6-8 *Experiment Demonstrating Specificity in the Binding of a Hormone to Its Receptor* *In this example, radioactive insulin was incubated with isolated plasma membranes in the presence of varying concentrations of either nonradioactive insulin, glucagon, or ACTH. Note that hormones unrelated to insulin do not compete with the binding of radioactive insulin to its plasma membrane receptor.*

3. Saturation kinetics. Because cells contain a limited number of receptor molecules, increasing the concentration of signaling ligand should eventually lead to occupation of all receptor binding sites. At this point, the receptor is said to be *saturated* because further increases in ligand concentration do not result in additional binding of ligand to receptor. The number of binding sites present in an isolated receptor preparation can be determined by measuring the total number of ligand molecules bound to the receptor after saturation has been achieved. Because it is difficult to know when complete saturation has been attained, binding data are often analyzed by a transformation of the Michaelis-Menten equation termed a *Scatchard plot.* This type of analysis converts the data into a linear form by graphing the concentration of ligand bound to receptor against the ratio of bound to free ligand (see Figure 6-9, *bottom*). In such a plot the x intercept corresponds to the total number of binding sites, or B_{max}, while the slope of the line corresponds to the negative reciprocal of K_d (that is, $-1/K_d$). The actual number of receptors detected per cell for a given signaling molecule usually ranges from less than a dozen to several hundred thousand, depending on the type of receptor involved.

4. Reversibility. The binding of a signaling ligand to its receptor is usually a rapidly reversible interaction. This characteristic helps to prevent a receptor from becoming permanently activated by a signaling ligand.

5. Physiological response. The binding of a signaling ligand to its receptor should trigger an appropriate physiological response whose magnitude is directly related to the number of receptor molecules containing bound ligand. For example, the hormone insulin stimulates the transport of glucose into target cells. Thus during the binding of insulin to its plasma membrane receptor in intact cells, one should observe an increase in glucose transport that is directly related to the number of receptor molecules containing bound insulin.

Most Receptors for External Signaling Molecules Are Located in the Plasma Membrane

The receptors for a wide variety of hormones, neurotransmitters, and local mediators have been identified using the preceding methods. Several lines of evidence support the conclusion that most of these receptors (with the notable exception of those for steroid hormones and thyroxine) are localized within the plasma membrane of their target cells. First, subcellular fractionation studies have revealed that the receptors for signaling molecules are usually concentrated in isolated plasma membrane preparations. Second, hormones like ACTH, insulin, and glucagon have been shown to retain the ability to stimulate their respective target cells even when they are linked to inert beads that prevent the hormones from entering into cells. And finally, receptors that retain the ability to bind to their signaling ligands have been directly isolated by extracting isolated plasma membranes with detergent solutions.

The plasma membrane receptors identified in this way have turned out to be integral membrane proteins,

some of which contain covalently bound carbohydrate or lipid groups. To investigate the orientation of these receptors within the membrane, experiments have been carried out employing right-side-out and inside-out membrane vesicles. For example, it has been shown that right-side-out plasma membrane vesicles generated from fat cell membranes can bind to hormones like insulin, but inside-out vesicles from the same source do not. Studies of this type support the conclusion that membrane receptors are asymmetrically oriented across the membrane, exposing their binding sites for signaling molecules only on the exterior surface of the membrane.

Cells differ in the kinds of plasma membrane receptors they contain, permitting each cell type to respond selectively to a particular set of signaling molecules. Such an arrangement allows highly selective networks of cell-cell communication to be established. Since a large number of different receptors are involved in these signaling networks, it is beyond the scope of a cell biology textbook to discuss how each individual receptor operates. However, similarities that have been discovered in the basic mechanisms of receptor action have made it possible to group receptors into three main categories: *ion channel receptors, receptors linked to G proteins,* and *catalytic receptors* (Figure 6-10). In the following sections we will examine the basic properties of these three main groups of receptors.

ION CHANNEL RECEPTORS

The Nicotinic Acetylcholine Receptor Has Been Isolated and Purified

Plasma membrane receptors that function as ion channels are referred to as **ion channel receptors.** Receptors for the neurotransmitter **acetylcholine** are among the best understood receptors of this type. Among its many functions, acetylcholine triggers the contraction of muscle cells, slows the heartbeat, and transmits signals from one nerve cell to another. Acetylcholine is able to exert these diverse effects because it binds to several kinds of ion channel receptors, each triggering a different response. Among these receptors is the **nicotinic acetylcholine receptor,** which resides in the plasma membrane of vertebrate skeletal muscle cells and the electric organs of certain species of fish. To understand how the nicotinic acetylcholine receptor operates, let us briefly consider what happens when a nerve cell stimulates a muscle cell to contract.

As we saw earlier in the chapter, electrical signals pass along the axons of nerve cells in the form of an action potential. When an action potential reaches the end of an axon, it stimulates the release of neurotransmitter from membrane vesicles into the synaptic space

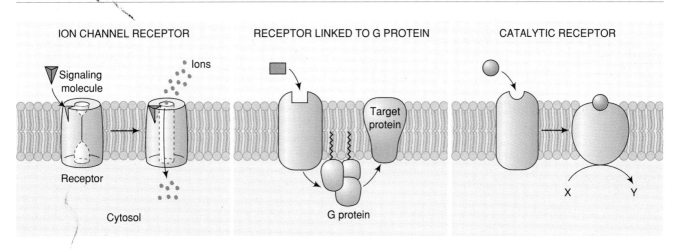

Figure 6-10 *Three Classes of Plasma Membrane Receptors* *(Left) An ion channel receptor is a membrane protein that functions both as a receptor for a signaling ligand and as an ion channel. The permeability of the channel is controlled by binding of the signaling ligand. (Middle) A G-protein linked receptor is a membrane protein that activates a G protein when a signaling ligand binds to the receptor. Activated G proteins in turn interact with other membrane proteins, which may be enzymes or ion channels. (Right) A catalytic receptor is a membrane protein that functions both as a receptor for a signaling ligand and as an enzyme. The receptor's enzymatic activity is controlled by binding of the signaling ligand.*

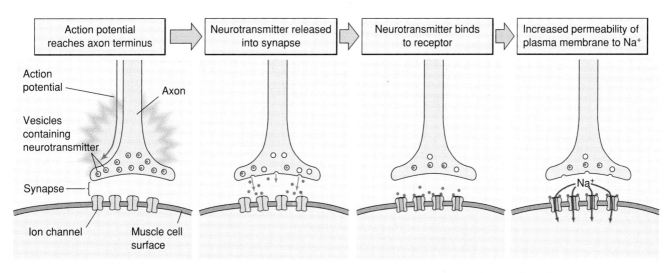

Figure 6-11 *Summary of the Events Occurring at a Synapse between a Nerve Axon and a Muscle Cell* *When an action potential reaches the end of an axon, it stimulates the release of neurotransmitters that diffuse across the synaptic space and bind to membrane receptors of the adjacent muscle cell, triggering membrane depolarization.*

(Figure 6-11). The released neurotransmitter diffuses across the synapse and binds to plasma membrane receptors of the adjacent muscle cell. This binding triggers a rapid but brief increase in the permeability of the muscle cell to small cations, mainly Na^+ and K^+. Since the electrochemical gradient is steeper for Na^+ than for K^+, the major effect is an increased flow of Na^+ into the target cell, which depolarizes the plasma membrane. Depolarization elicits a specific physiological response in the target cell, in this case muscle contraction.

The preceding events are described in greater depth in the chapter devoted to nerve cell function

(page 752). Here we are concerned primarily with the question of how the binding of acetylcholine to its receptor causes an increase in the permeability of the target cell plasma membrane to cations. To answer this question, it is necessary to study the behavior of purified receptor molecules. Purifying the acetylcholine receptor has been made possible by the discovery that certain snake-venom toxins, such as the small protein *α-bungarotoxin*, bind selectively and almost irreversibly to the acetylcholine receptor. These toxins prevent acetylcholine from interacting with its receptor, leading to respiratory paralysis and death. Radioactive α-bungaro-

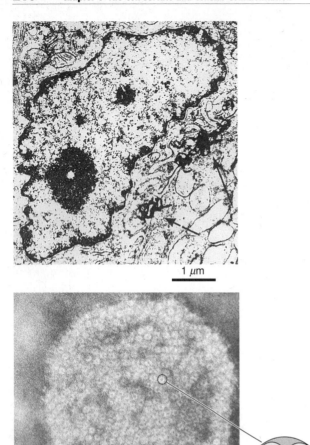

1 µm

0.05 µm

Figure 6-12 *Electron Micrographs of the Acetylcholine Receptor* (Top) *Autoradiograph of the junction between nerve and muscle cells that had been incubated with radioactive α-bungarotoxin, which binds to nicotinic acetylcholine receptors. The silver grains (arrows) are localized over the plasma membrane of the muscle cell, indicating the presence of acetylcholine receptors. (Bottom) Electron micrograph of a negatively stained plasma membrane isolated from electric organ cells of the ray fish, Torpedo. The circular structures are individual acetylcholine receptors embedded within the membrane. The diagram illustrates how each receptor is constructed from five subunits. Micrographs courtesy of M. M. Salpeter (top) and F. Hucho (bottom).*

toxin has been employed experimentally to locate acetylcholine receptors in microscopic tissue sections (Figure 6-12, *top*), and to test for the presence of acetylcholine receptors in purified protein preparations.

A convenient source of acetylcholine receptors for protein purification studies is the electric organ of the electric ray fish, *Torpedo*. This organ develops from the same embryonic tissue as muscle but does not accumu-

late the contractile proteins involved in muscle contraction. Instead its large flat cells, called *electroplaques,* become infiltrated on one side by neurons that produce acetylcholine. Stimulating these nerves causes the membrane potential on the synaptic side of each electroplaque to shift from a resting value of −90 mV to +60 mV; since the opposite side of the electroplaque remains at −90 mV, a potential difference of 150 mV is established across the two outer surfaces. In the electric organ, electroplaques are arranged in a linear series, which makes their individual voltages additive. A group of 5000 electroplaques therefore generates a total discharge of 5000×150 mV = 750 volts, allowing the fish to shock and incapacitate its prey.

The acetylcholine receptor has been purified by extracting electroplaque membranes with detergent solutions followed by affinity chromatography on columns containing snake toxins that bind specifically to the receptor. In electron micrographs, the purified receptor appears as a doughnut-shaped structure consisting of five subunits surrounding a central cavity; similar structures are observed in electron micrographs of intact plasma membranes containing acetylcholine receptors (Figure 6-12, *bottom*). Electrophoresis of the protein under denaturing conditions reveals the presence of four types of polypeptide chains (α, β, γ, and δ) in the ratio of 2:1:1:1. The smallest subunit (α) contains the binding site for acetylcholine.

The Nicotinic Acetylcholine Receptor Acts as a Neurotransmitter-Gated Ion Channel

Once the acetylcholine receptor had been successfully purified, it became possible to address a fundamental question: Does the binding of acetylcholine to its receptor alter the behavior of ion channels that are independent of the receptor molecule, or is the acetylcholine receptor itself an ion channel? The most direct evidence emerged from studies in which liposomes were reconstituted in the presence of the purified acetylcholine receptor. When acetylcholine was added to the liposomes, permeability to Na^+ increased. Since no other membrane components were present, it could be concluded that the acetylcholine receptor functions as a Na^+ channel.

These observations suggest that the acetylcholine receptor is a **neurotransmitter-gated ion channel**—that is, an ion channel whose opening is regulated by binding of the neurotransmitter acetylcholine. Amino acid sequence analysis has revealed that each of the receptor's five subunits contains four or five transmembrane α helices; one of the helices in each subunit is more hydrophilic than the rest and thus is thought to line the central hydrophilic channel through which the ions pass. Rings of negatively charged amino acids located at both ends of the channel appear to play a role

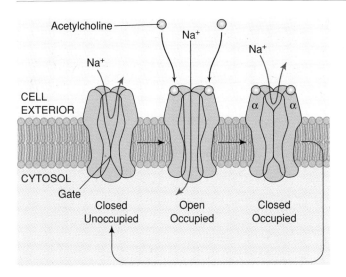

Figure 6-13 *Changes in Conformation of the Nicotinic Acetylcholine Receptor upon Binding of Acetylcholine* *Binding of acetylcholine to the nicotinic receptor causes the receptor's ion channel to open, allowing Na⁺ to diffuse through the channel. The open configuration of the channel is unstable and is superseded by a closed conformation that is impermeable to Na⁺, even though the receptor's binding site is still occupied by acetylcholine. Acetylcholine is then released and the receptor returns to its original conformation.*

in excluding negatively charged ions from the channel. Figure 6-13 illustrates how the permeability of the ion channel is regulated by acetylcholine. According to this model, the binding of acetylcholine to the receptor's α subunits induces a conformational change that causes the channel to open. The open form of the channel is unstable, however, being quickly superseded by a closed conformation whose free energy is lower. The acetylcholine molecules then dissociate from the receptor and are hydrolyzed by the enzyme *acetylcholinesterase.* After release of the bound acetylcholine, the receptor returns to its original conformation.

RECEPTORS LINKED TO G PROTEINS

The largest and most diverse group of plasma membrane receptors are those whose effects are mediated by **G proteins,** a family of related proteins that bind the nucleotide GTP. Many signaling molecules interact with receptors that are linked to G proteins, and G proteins in turn exert several kinds of effects on cells. The realization that G proteins play a prominent role in signal transduction first emerged from studies involving hormones whose actions are mediated by the nucleotide *cyclic AMP.*

Cyclic AMP Functions as a Second Messenger

The discovery of cyclic AMP by Earl Sutherland and his associates in 1957 was a crucial milestone in unraveling the mechanism by which external signaling molecules exert their effects. In the process of investigating how hormones induce the breakdown of glycogen in dog liver, Sutherland discovered that incubating liver homogenates with either epinephrine or glucagon stimulates the activity of *glycogen phosphorylase,* an enzyme that catalyzes the conversion of glycogen to glucose 1-phosphate. However, if membranes present in the homogenate were first removed by centrifugation, glycogen phosphorylase present in the supernatant could no longer be activated by epinephrine or glucagon. Addition of these hormones directly to isolated membranes triggered the formation of an unidentified substance that, when added to the membrane-free supernatant, could activate glycogen phosphorylase.

Sutherland isolated tiny amounts of the activating substance and tentatively identified it as a nucleotide. By coincidence David Lipkin had just discovered an unusual new nucleotide that can be synthesized in the laboratory by treating ATP with barium hydroxide. This substance turned out to be the same as the nucleotide identified by Sutherland, so Lipkin's procedure for synthesizing the compound in large quantities facilitated its identification as a cyclic derivative of AMP called *3′,5′-cyclic adenosine monophosphate,* or **cyclic AMP.** It also provided Sutherland with an abundant supply of this molecule for subsequent study.

Sutherland and his associates soon developed four lines of evidence supporting the idea that cyclic AMP mediates the action of glucagon and epinephrine on liver cells. First, they showed that adding either of the two hormones to liver elevates intracellular levels of cyclic AMP. Second, they found that adding the hormones to isolated membranes stimulates **adenylyl cyclase** (formerly called *adenylate cyclase*), a plasma-membrane-bound enzyme that catalyzes the formation of cyclic AMP from ATP (Figure 6-14). Third, they showed that the ability of glucagon and epinephrine to activate glycogen breakdown in liver cells can be mimicked by exposing cells to purified cyclic AMP in the absence of hormone. Finally, they established that the ability of the two hormones to stimulate glycogen breakdown can be amplified by drugs such as caffeine and theophylline, which depress the activity of the enzyme **phosphodiesterase.** Phosphodiesterase normally breaks down cyclic AMP into AMP, so inhibiting this enzyme with caffeine or theophylline enhances hormone action by maintaining abnormally high intracellular levels of cyclic AMP.

These four properties have been adopted as essential criteria for determining whether the effects of hormones other than glucagon and epinephrine are

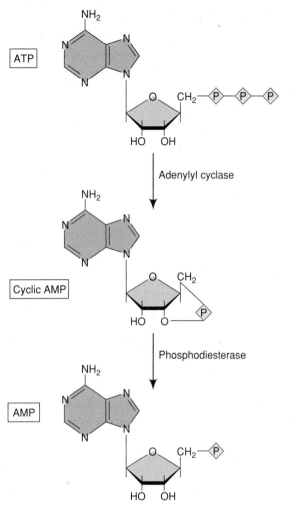

Figure 6-14 *The Formation and Breakdown of Cyclic AMP* *Adenylyl cyclase, an enzyme located in the plasma membrane, catalyzes the formation of cyclic AMP from ATP. Cyclic AMP is broken down to AMP by the cytoplasmic enzyme, phosphodiesterase.*

Table 6-4   Some Hormonal Responses Associated with Increased Cyclic AMP Levels

Hormone	Tissue	Response
ACTH	Adrenal	Hydrocortisone formation
TSH	Thyroid	Thyroxine formation
LH	Ovary	Progesterone formation
Vasopressin	Kidney	Water resorption
Glucagon	Adipose tissue	Lipid breakdown
Glucagon	Liver	Glycogen breakdown
Parathormone	Kidney	Phosphate excretion
Parathormone	Bone	Bone resorption
Secretin	Intestine	Pancreatic enzyme release
Epinephrine	Cardiac muscle	Increased contractility
Epinephrine	Liver	Glycogen breakdown
Epinephrine	Erythrocyte	Increased Na^+ permeability

would be that the hormone receptor itself catalyzes the formation of cyclic AMP from ATP, and that binding of a hormone to its receptor simply stimulates this enzymatic activity. The first convincing evidence that such is not the case was provided by Joseph Orly and Michael Schramm. In their experiments, two types of cells were artificially fused: erythrocytes containing epinephrine receptors but lacking adenylyl cyclase, and tumor cells possessing adenylyl cyclase but lacking epinephrine receptors. A few minutes after cell fusion had taken place, adenylyl cyclase could be stimulated by epinephrine, indicating that epinephrine receptors derived from one cell can establish communication with adenylyl cyclase molecules derived from another cell. It was therefore concluded that the epinephrine receptor and adenylyl cyclase are separate molecules (Figure 6-16).

Such an arrangement raises the possibility that receptors for different signaling molecules might interact with the same adenylyl cyclase molecule. Support for this notion is found in the behavior of fat cells, where adenylyl cyclase can be stimulated by at least six different hormones (ACTH, LH, TSH, epinephrine, glucagon, and secretin). If each of these hormones activates a separate adenylyl cyclase, then the effect of each hormone should be additive, a prediction that is not borne out by the experimental data. Instead, once adenylyl cyclase has been maximally stimulated by one hormone, addition of another hormone has no further effect, indicating that the same adenylyl cyclase is being activated by each of the six hormones.

If hormone receptors and adenylyl cyclase are separate molecules, how does the binding of a hormone to its receptor activate adenylyl cyclase? The first hint that a third component might be involved emerged from the finding that GTP enhances the ability of hormones to stimulate adenylyl cyclase. Identifying the component

mediated by cyclic AMP. In this way, cyclic AMP has been implicated in a remarkably diverse spectrum of hormonally induced responses (Table 6-4). Its widespread involvement in hormone action has led to the designation of cyclic AMP as a **second messenger,** since it transmits to the inside of cells signals originally brought to the outer surface by one of a variety of extracellular signaling molecules or "first messengers" (Figure 6-15).

G Proteins Transmit Signals between Plasma Membrane Receptors and Adenylyl Cyclase

There are several possible ways of explaining how the binding of a hormone to its cell surface receptor causes an increase in cyclic AMP. The simplest explanation

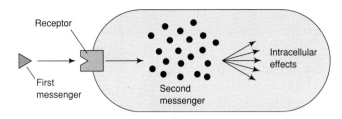

Figure 6-15 *The Second Messenger Concept   Various hormones, neurotransmitters, and local mediators function as first messengers that bind to the plasma membrane and trigger the formation of other signaling molecules, called second messengers, which function inside the cell.*

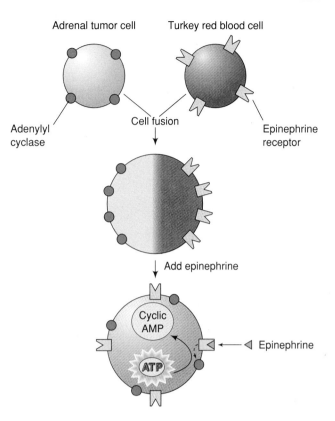

Figure 6-16 *Experiment Showing That the Epinephrine Receptor and Adenylyl Cyclase Are Separate Molecules   Erythrocytes containing epinephrine receptors but lacking adenylyl cyclase were fused with tumor cells containing adenylyl cyclase but lacking epinephrine receptors. Shortly after fusion, adenylyl cyclase can be stimulated by epinephrine, indicating that receptors derived from one cell can establish communication with adenylyl cyclase derived from another cell.*

with which GTP interacts was facilitated by the discovery of a mutant tumor cell line, designated cyc^-, which has normal epinephrine receptors and adenylyl cyclase, and yet fails to respond to epinephrine by increased production of cyclic AMP. However, epinephrine stimulation of cyclic AMP synthesis can be restored by providing the cyc^- cells with a GTP-binding membrane protein (G protein) isolated from normal cells. This particular G protein is called G_s because of its role in *stimulating* adenylyl cyclase.

Our current understanding of the role played by the G_s protein is illustrated in Figure 6-17. The G_s protein consists of three subunits (α, β, and γ) whose interactions are crucial to the regulation of adenylyl cyclase. In the absence of a stimulating hormone, GDP is bound to the α subunit of the G_s protein and the α, β, and γ subunits remain associated with one another. In this configuration, the G_s protein cannot activate adenylyl cyclase. But when a hormone or other appropriate signaling molecule binds to its plasma membrane receptor, the receptor binds to G_s; this event in turn triggers the dissociation of GDP from the G_s protein and its replacement by GTP. In the presence of bound GTP, the α subunit dissociates from the G_s protein and activates adenylyl cyclase. When the bound GTP is subsequently hydrolyzed to GDP, the α subunit reassociates with the β and γ subunits, and adenylyl cyclase is no longer stimulated.

Some G Proteins Inhibit Rather Than Stimulate Adenylyl Cyclase

Although many signaling molecules increase cyclic AMP levels in their target cells, some signals decrease cyclic AMP instead. The same molecule can even be stimulatory in one cell type and inhibitory in another. For example, epinephrine binds to several kinds of receptors referred to generically as **adrenergic receptors.** When epinephrine binds to β_1-*adrenergic receptors* present in heart muscle cells, it causes an increase in cyclic AMP formation and a resulting increase in heart rate and contractile force. In contrast, when epinephrine binds to

α_2-*adrenergic receptors* present in vascular smooth muscle, it causes a decrease in cyclic AMP formation and muscle contraction.

How can the binding of epinephrine to its receptor stimulate cyclic AMP formation in one case and inhibit it in another? The answer lies in the type of G protein involved. Cells in which epinephrine stimulates cyclic AMP formation contain the G_s protein, which mediates stimulation of adenylyl cyclase. In its place, cells in which adenylyl cyclase is inhibited by epinephrine contain an inhibitory G protein called G_i. As in the case of G_s, binding of epinephrine to its membrane receptor causes GTP to bind to the G_i protein, followed by dissociation of the α subunit. The released α subunit, as well as the β and γ subunits, contribute to the inhibition of adenylyl cyclase; the α subunit of G_i inhibits adenylyl cyclase directly, whereas the β and γ subunits combine with the α subunit of the stimulatory G_s protein, preventing the activation of adenylyl cyclase by G_s.

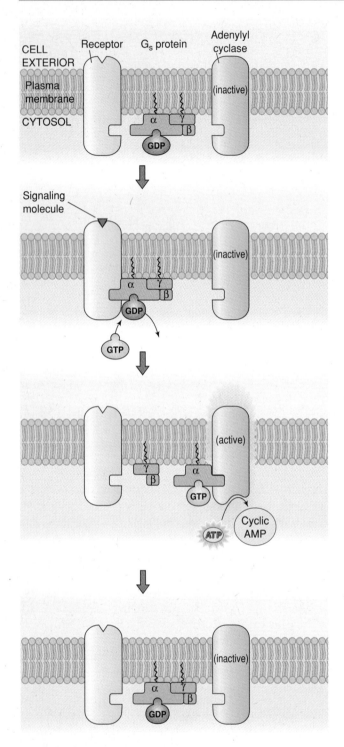

CELL EXTERIOR

Plasma membrane

CYTOSOL

Receptor   Gₛ protein   Adenylyl cyclase

α   γ   β

GDP

(inactive)

Signaling molecule

α   γ   β

GDP

GTP

(inactive)

(active)

γ   β

α

GTP

Cyclic AMP

ATP

(inactive)

α   γ   β

GDP

Figure 6-17 *Mechanism by Which the Gₛ Protein Mediates the Interaction between a Receptor and Adenylyl Cyclase* *The binding of certain signaling molecules to their plasma membrane receptors causes the receptor to bind to Gₛ (a lipid-anchored plasma membrane protein). This interaction triggers the dissociation of GDP from the Gₛ protein and its replacement by GTP. In the presence of bound GTP, the α subunit dissociates from the Gₛ protein and activates adenylyl cyclase. When GTP is subsequently hydrolyzed, the α subunit reassociates with β and γ subunits and the ability to stimulate adenylyl cyclase is lost.*

G Proteins Are Disrupted by Bacterial Toxins That Cause Cholera and Whooping Cough

Unexpected support for the role played by G proteins in regulating adenylyl cyclase has come from studies of two bacterial diseases, cholera and whooping cough. Humans suffering from cholera develop a debilitating form of diarrhea that is caused by a protein toxin produced by the bacterium *Vibrio cholerae*. When the toxin is taken up by cells lining the intestinal tract, it catalyzes the transfer of an ADP-ribose group from NAD^+ to the α subunit of Gₛ (Figure 6-18). This process of **ADP ribosylation** inhibits the hydrolysis of GTP by Gₛ, and hence locks Gₛ in its active GTP-bound state. Under such conditions, adenylyl cyclase remains permanently activated. The uncontrolled production of cyclic AMP by adenylyl cyclase causes an increased secretion of Na^+ and water into the intestinal tract, producing a profuse diarrhea.

Whooping cough, caused by the bacterium *Bordetella pertussis,* is another disease that illustrates the importance of G proteins in regulating adenylyl cyclase. Like cholera toxin, the toxin produced by *Bordetella pertussis* catalyzes ADP ribosylation of a G protein, although the target is Gᵢ rather than Gₛ. ADP ribosylation of Gᵢ prevents the protein from inhibiting adenylyl cyclase, again leading to an increase in cyclic AMP. Because whooping cough is an airborne infection, most of the infected cells line the respiratory passages. The increased production of cyclic AMP in these cells triggers a massive secretion of fluid into the lungs, causing severe coughing followed by a "whooping" sound when air is inhaled.

Cyclic AMP Activates Protein Kinase A

We have now seen how the binding of certain hormones or other signaling molecules to plasma membrane receptors causes the Gₛ protein to activate adenylyl cyclase, thereby stimulating the production of cyclic AMP. How does this increase in cyclic AMP affect the rest of the cell? A crucial milestone in our understanding of this signaling pathway occurred in the late 1960s, when Edwin Krebs and Donal Walsh reported that cyclic AMP binds to and activates an enzyme called *cyclic AMP-dependent protein kinase,* or **protein kinase A. Protein kinases** are a family of enzymes that catalyze the transfer of the terminal phosphate group from ATP to an amino acid in a protein molecule. This process of *protein phosphorylation* regulates the activity of the protein that is phosphorylated. Protein kinase A, which is activated by cyclic AMP, phosphorylates the amino acids *serine* and (to a lesser extent) *threonine* in a selected set of proteins. The enzyme consists of two regulatory subunits and two catalytic subunits. When the four subunits are bound together as a tetramer, pro-

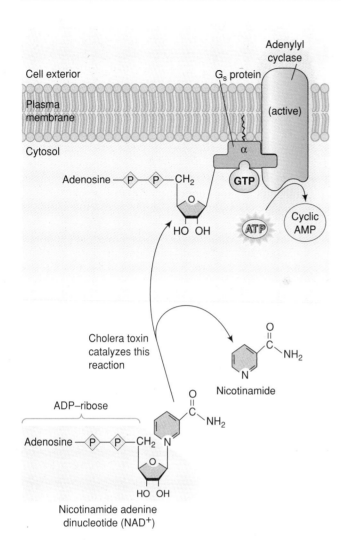

Figure 6-18 *Mechanism of Action of Cholera Toxin* *A fragment of the cholera toxin molecule catalyzes the transfer of an ADP-ribose group from NAD^+ to the α subunit of the G_s protein. This process of ADP-ribosylation keeps the G_s protein locked in an active state, causing cyclic AMP to be produced continuously.*

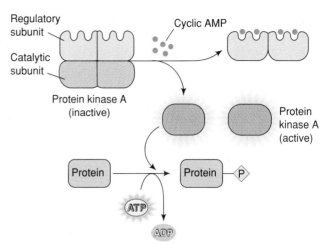

Figure 6-19 *Activation of Protein Kinase A by Cyclic AMP* *Cyclic AMP causes the regulatory subunits of protein kinase A to dissociate from the catalytic subunits; the free catalytic subunits in turn catalyze protein phosphorylation.*

Table 6-5 Examples of Proteins Phosphorylated by Protein Kinase A

Phosphorylase kinase

Glycogen synthase

Protein phosphatase-1

Protein phosphatase inhibitor-1

Hormone-sensitive lipase

Pyruvate kinase

6-Phosphofructo-2-kinase

Tyrosine hydroxylase

Phenylalanine hydroxylase

Troponin I

β_2-Adrenergic receptor

Tubulin

Cyclic AMP response element binding protein (CREB)

tein kinase A is inactive. Binding of cyclic AMP to the regulatory subunits causes them to dissociate, releasing two catalytic subunits that can then catalyze protein phosphorylation (Figure 6-19).

The widespread occurrence of protein kinase A in eukaryotic cells and its ability to catalyze the phosphorylation of a variety of proteins (Table 6-5) has led to the idea that most, if not all, of the biological effects of cyclic AMP are mediated by changes in protein phosphorylation. Support for this idea has been provided by the investigations of Philip Coffino and his associates on mouse lymphoid cells. Cyclic AMP triggers a variety of responses in this cell type, including inhibition of growth, induction of the enzyme phosphodiesterase, and cell rupture. Mutants resistant to the growth-inhibit-

ing effects of cyclic AMP were isolated by growing cells in the presence of high concentrations of cyclic AMP. Of several hundred mutants examined, all were found to be defective in protein kinase A activity. In addition, all three biological effects of cyclic AMP (growth inhibition, phosphodiesterase induction, and cell rupture) were altered in every mutant, supporting the conclusion that protein kinase A mediates each of these actions of cyclic AMP.

Additional support for the involvement of protein kinase A in mediating the biological effects of cyclic AMP has come from the discovery that a close correlation exists between the ability of a given dose of hormone to activate protein kinase A and the degree to which the final target tissue responds to that hormone.

For example, the effectiveness of varying concentrations of *luteinizing hormone (LH)* in stimulating testis cells to make testosterone or ovarian cells to make progesterone is closely related to the degree to which protein kinase A is activated by that particular hormone concentration (Figure 6-20).

Different Proteins Are Phosphorylated in Differing Target Cells by Protein Kinase A

If cyclic AMP activates the same enzyme (protein kinase A) in different cell types, how can each target cell respond in its own unique way to signaling molecules that act by stimulating cyclic AMP formation? This paradox is nicely illustrated by the behavior of the hormones glucagon, luteinizing hormone, and epinephrine. Glucagon stimulates glycogen breakdown in liver but lipid breakdown in fat cells. Luteinizing hormone enhances steroid synthesis in ovary and testis, while epinephrine increases sodium permeability in erythrocytes. But in spite of the differing outcomes, each of these hormones acts by stimulating the production of cyclic AMP, which in turn activates protein kinase A. The differing outcomes are made possible by the fact that each cell type contains its own unique set of target proteins that can be phosphorylated by protein kinase A.

Liver and muscle were among the first tissues in which the target proteins for protein kinase A were clearly identified. In these two tissues, hormones that elevate cyclic AMP levels and hence protein kinase A activity are known to trigger the breakdown of glycogen. Figure 6-21 shows how the activation of protein kinase A promotes glycogen breakdown by regulating the activity of several key enzymes:

1. The main enzyme phosphorylated by protein kinase A in liver and muscle is another protein kinase called **phosphorylase kinase** (①), which becomes activated upon phosphorylation. Activated phosphorylase kinase catalyzes the phosphorylation and resulting activation of **glycogen phosphorylase** (②), the enzyme responsible for catalyzing glycogen breakdown.

2. In addition to stimulating glycogen breakdown by activating phosphorylase kinase, protein kinase A depresses the rate of glycogen synthesis by phosphorylating **glycogen synthase** (③), the enzyme responsible for glycogen formation. In contrast to most enzymes phosphorylated by protein kinase A, glycogen synthase is inhibited rather than stimulated by phosphorylation.

3. A third point of control by protein kinase A involves **protein phosphatase**-1 or **PP1** (④), an enzyme that catalyzes protein *dephosphorylation* (removal of phosphate groups from protein molecules). Among the proteins dephosphorylated by PP1 are the three enzymes of glycogen metabolism already mentioned: phosphorylase

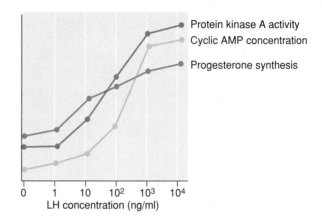

Figure 6-20 *Effects of Varying Concentrations of Luteinizing Hormone (LH) on Cyclic AMP Levels, Protein Kinase A Activity, and Progesterone Synthesis in Ovarian Cells* *The ability of varying doses of LH to stimulate progesterone synthesis is closely correlated with the ability of LH to simulate protein kinase A activity and cyclic AMP formation.*

kinase (①), glycogen phosphorylase (②), and glycogen synthase (③). Since the dephosphorylation of these three enzymes by PP1 would promote glycogen synthesis rather than breakdown, protein kinase A catalyzes reactions that inhibit the action of PP1. First, protein kinase A phosphorylates PP1 (④), causing it to lose catalytic activity. And second, protein kinase A catalyzes the phosphorylation of **protein phosphatase inhibitor-1** (⑤), a protein that inactivates PP1. This phosphorylation of protein phosphatase inhibitor-1 increases its ability to inactivate PP1.

Because each tissue contains its own unique set of proteins that can be phosphorylated by protein kinase A, cyclic AMP is able to trigger many other responses in addition to its effects on glycogen breakdown. For example, in fat cells protein kinase A catalyzes the phosphorylation and resulting activation of *hormone-sensitive lipase*. This enzyme catalyzes the breakdown of neutral fats to free fatty acids, which can then be used as an energy source. Hormone-sensitive lipase also catalyzes the release of free cholesterol in tissues that produce steroid hormones, such as the adrenal cortex; when adrenal cells are stimulated by adrenocorticotropin (ACTH) to produce steroid hormones, the activation of hormone-sensitive lipase by protein kinase A generates the cholesterol needed for steroid hormone synthesis. Yet another effect is observed when epinephrine binds to β_1-adrenergic receptors in heart muscle. In this case, protein kinase A helps to increase the rate of muscle contraction by catalyzing the phosphorylation of the contractile protein, troponin-I (page 576). Such examples illustrate how the same second messenger system is able to induce different effects in different target tissues (Figure 6-22).

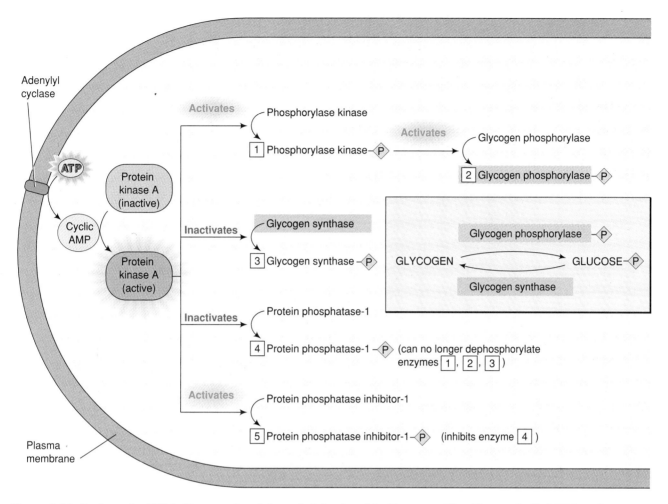

Figure 6-21 *Pathway by Which Glycogen Breakdown Is Stimulated by Hormones That Elevate Cyclic AMP Levels* *Cyclic AMP activates protein kinase A, which catalyzes the phosphorylation of phosphorylase kinase (becomes activated), glycogen synthase (becomes inactivated), protein phosphatase-1 (becomes inactivated), and protein phosphatase inhibitor-1 (becomes activated). These changes in enzyme activity lead to an enhancement of glycogen breakdown catalyzed by glycogen phosphorylase, and an inhibition of glycogen synthesis catalyzed by glycogen synthase.*

Advantages of the Cyclic AMP System Include Amplification and Flexibility

At first glance it might appear as if the cyclic AMP-mediated pathway of signal transduction is unnecessarily complex; the binding of a signaling molecule to its receptor (1) activates a protein (G_s) that (2) activates an enzyme (adenylyl cyclase) that (3) catalyzes the formation of a compound (cyclic AMP) that (4) activates another enzyme (protein kinase A) that (5) catalyzes the phosphorylation of other target proteins. Instead of this complex cascade, why doesn't the signaling molecule simply interact directly with the final target protein(s)? One possible answer is based on the phenomenon of signal amplification. Amplification of the incoming signal occurs because each step in the pathway is catalytic; that is, each activated receptor can activate many molecules of G_s, each of which can activate a molecule of adenylyl cyclase. Each adenylyl cyclase molecule in turn produces hundreds or thousands of mole-cules of cyclic AMP per second, each of which can activate a molecule of protein kinase. Every activated protein kinase molecule in turn catalyzes the phosphorylation of hundreds or thousands of molecules of phosphorylase kinase, glycogen synthase, hormone-sensitive lipase, or whatever enzyme is being acted upon. These enzymes may in turn convert hundreds or thousands of substrate molecules into products. In this way a single hormone molecule acting on the cell surface is ultimately capable of triggering a response involving millions of molecules.

A second advantage to a multistep cascade is the added flexibility that is gained when multiple components are included in a regulatory pathway. Each component can serve as a separate target for control by alternate mechanisms, creating the potential for a complex interplay between regulatory networks. For example, in addition to being activated by protein kinase A, phosphorylase kinase is also stimulated by calcium ions.

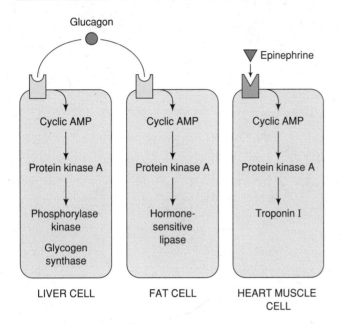

Figure 6-22 _Mechanism by Which Hormones Using the Same Second Messenger (Cyclic AMP) Elicit Different Responses in Different Cell Types_ *Protein kinase A activity is elevated in liver and fat cells stimulated by glucagon, and in heart muscle cells stimulated by epinephrine. Yet the effects of the elevated protein kinase A activity are different in the three cell types because each cell contains different proteins that are phosphorylated by protein kinase A.*

Hence any stimulus altering the intracellular concentration of calcium ions will influence glycogen breakdown by a mechanism that is independent of changes in cyclic AMP levels.

G Proteins Can Alter the Permeability of Ion Channels

Although many plasma membrane receptors utilize G proteins to send stimulatory or inhibitory signals to adenylyl cyclase, this is not the only way G proteins operate. Some G proteins directly control the permeability of ion channels. A prominent example involves the regulation of heart muscle contraction by the neurotransmitter **acetylcholine.** The plasma membrane of heart muscle cells contains **muscarinic acetylcholine receptors** that function differently from the nicotinic acetylcholine receptors of skeletal muscle described earlier (page 208). Binding of acetylcholine to muscarinic receptors slows the heartbeat by causing K^+ channels to open. Because the K^+ concentration is higher inside the cell than outside, K^+ diffuses out of the cell and the plasma membrane becomes hyperpolarized (i.e., the membrane potential is more negative than normal).

To investigate how these K^+ channels are opened, the patch clamp technique (page 201) has been employed to monitor the activity of K^+ channels in heart membrane fragments. Under conditions that limit the formation of second messengers such as cyclic AMP, the addition of purified G proteins to membrane fragments has been found to trigger the opening of individual K^+ channels. The K^+ channels can also be opened by purified α subunits of the G protein, but not the β and γ subunits. These findings suggest that binding of acetylcholine to muscarinic receptors activates a G protein whose α subunit then causes K^+ channels to open. At first a special type of G protein was thought to be involved, but subsequent studies have revealed that purified α subunits derived from several different G_i proteins are all capable of opening K^+ channels. Hence G_i proteins appear to perform more than one function, opening K^+ channels as well as inhibiting adenylyl cyclase.

G Proteins Are Involved in the Phosphoinositide Signaling Pathway

The discovery of cyclic AMP by Sutherland described earlier in the chapter (page 211) might never have occurred had he used rat liver instead of dog liver in his experiments. In dog liver, epinephrine binds to *β-adrenergic* receptors and triggers the activation of adenylyl cyclase. But rat liver has few β-adrenergic receptors; epinephrine instead binds to *α_1-adrenergic receptors* that do not activate adenylyl cyclase. Yet epinephrine still stimulates the breakdown of glycogen, using a second messenger pathway that mobilizes the release of calcium ions rather than promoting the formation of cyclic AMP.

This alternative second messenger system, called the **phosphoinositide signaling pathway,** involves derivatives of phosphatidylinositol, a phospholipid that occurs mainly in the inner leaflet of the plasma membrane. A small fraction of this phospholipid is normally phosphorylated, generating derivatives known as *phosphoinositides;* the most prominent among them is *phosphatidylinositol bisphosphate (PIP_2)*. The binding of certain extracellular signaling molecules to their plasma membrane receptors activates the enzyme **phospholipase C,** which cleaves PIP_2 into two breakdown products: **inositol trisphosphate (IP_3)** and **diacylglycerol (DAG).** As in the case of adenylyl cyclase, the coupling of membrane receptors to the activation of phospholipase C is mediated by a special class of G proteins, in this case called *G_p proteins* because they activate phospholipase (Table 6-6).

The two products generated by the action of phospholipase C, namely IP_3 and DAG, are both second messengers in their own right (Figure 6-23). In the following sections, we will describe how these two second messengers act.

Table 6-6	Targets of Plasma Membrane G Proteins	
Target	**G Protein**	**Action**
Adenylyl cyclase	G_s	Stimulates enzyme
	G_i	Inhibits enzyme
K^+ channel	G_i	Opens ion channel
Phospholipase C	G_p	Stimulates enzyme
Cyclic GMP phosphodiesterase*	G_t (transducin)	Stimulates enzyme

*The role of the G_t protein in regulating phosphodiesterase is discussed in Chapter 17.

Table 6-7	Some Signaling Molecules That Activate Phospholipase C	
Molecule	**Target Cell**	**Response**
Epinephrine	Liver (α_1-receptor)	Glycogen breakdown
Vasopressin	Liver	Glycogen breakdown
PDGF	Fibroblasts	Cell proliferation
Acetylcholine	Smooth muscle (muscarinic receptor)	Contraction

Inositol Trisphosphate (IP$_3$) Is a Second Messenger That Mobilizes Calcium Ions

The binding of a variety of hormones, neurotransmitters, and local mediators to their respective plasma membrane receptors is now known to induce the activation of phospholipase C (Table 6-7). These same signaling molecules also trigger a transient increase in the concentration of free Ca^{2+} in the cytosol. Since the Ca^{2+} concentration outside the cell (10^{-3} M) is typically much higher than that inside the cell (10^{-7} M), the transient increase in cytosolic Ca^{2+} might in theory be caused by a temporary increase in plasma membrane permeability to Ca^{2+}. However, current evidence suggests that this is not usually the correct explanation; instead, the IP$_3$ generated by phospholipase C serves as a

second messenger that diffuses from the plasma membrane, where it is produced, to the endoplasmic reticulum (ER), where it triggers the release of stored Ca^{2+} into the cytosol (Figure 6-24). Evidence for this scenario has come from cell fractionation experiments in which the addition of IP$_3$ to purified ER membranes has been shown to increase their permeability to Ca^{2+}. In intact cells the Ca^{2+} concentration is higher within the lumen of the ER than in the cytosol; hence the binding of IP$_3$ to ER membranes causes Ca^{2+} to diffuse from the ER lumen into the cytosol.

How does a transient increase in cytosolic Ca^{2+} concentration influence the rest of the cell? Calcium ions usually act by binding to calcium-binding proteins; prominent among them is **calmodulin,** which consists

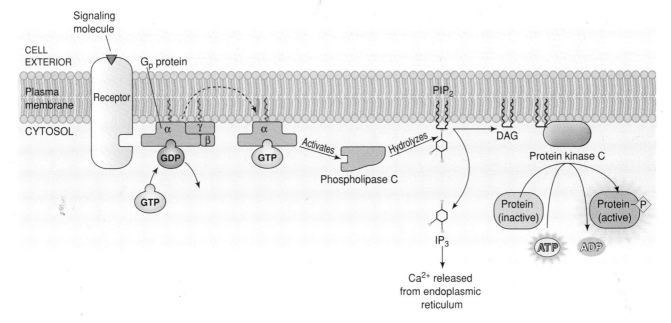

Figure 6-23 *The Phosphoinositide Pathway for Producing the Second Messengers Inositol Trisphosphate (IP$_3$) and Diacylglycerol (DAG)* *In this pathway, binding of a signaling molecule to its receptor activates a G_p protein, which in turn activates phospholipase C. Phospholipase C, which normally resides in the cytosol, is recruited to the inner surface of the plasma membrane, where it catalyzes the formation of DAG and IP$_3$. DAG functions by activating protein kinase C, while IP$_3$ mobilizes the release of Ca^{2+} from the endoplasmic reticulum.*

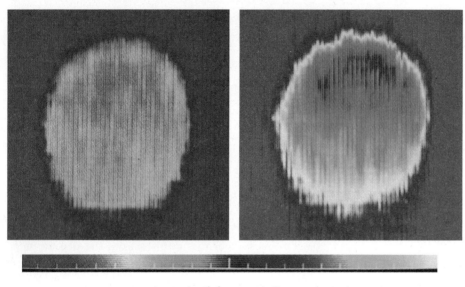

Increasing Ca²⁺ concentration ⟶

Figure 6-24 *Increase in Free Cytosolic Ca²⁺ Concentration Triggered by a Hormone That Stimulates the Formation of Inositol Trisphosphate* *Bovine adrenal chromaffin cells were stained with fura-2, a dye that fluoresces when bound to Ca²⁺. The color of the fluorescence changes with alterations in Ca²⁺ concentration. (Left) Fluorescence micrograph of an unstimulated adrenal cell. The yellow fluorescence indicates a relatively low concentration of free Ca²⁺. (Right) Fluorescence in an adrenal cell simulated by angiotensin, a hormone that triggers the formation of inositol trisphosphate. The green and blue fluorescence indicates an increased concentration of free Ca²⁺. Courtesy of R. D. Burgoyne.*

of a single polypeptide chain possessing four Ca²⁺-binding sites. Calmodulin is ubiquitous in eukaryotic cells, participating in a variety of regulatory activities. The binding of Ca²⁺ to calmodulin causes a change in protein conformation that allows calmodulin to reversibly bind to and activate other target proteins (Figure 6-25), although in at least one case, calmodulin is a permanent component of the regulated protein. Among the proteins regulated by calmodulin are enzymes involved in cyclic AMP metabolism (adenylyl cyclase and phosphodiesterase), smooth muscle contraction (myosin light chain kinase), Ca²⁺ transport (the plasma membrane Ca²⁺-dependent ATPase), glycogen breakdown (phosphorylase kinase), protein dephosphorylation (calcineurin), and protein phosphorylation (calmodulin-dependent multiprotein kinase). **Calmodulin-dependent multiprotein kinase** is an especially interesting enzyme because it catalyzes the phosphorylation of many different proteins. It therefore may play a pivotal role in calcium signaling pathways, comparable to that played by protein kinase A in cyclic AMP signaling pathways.

Examination of the list of proteins regulated by the Ca²⁺-calmodulin complex reveals that the Ca²⁺ and cyclic AMP signaling pathways interact with one another at several levels. For example, Ca²⁺-calmodulin alters cyclic AMP levels by stimulating adenylyl cyclase and phospho-

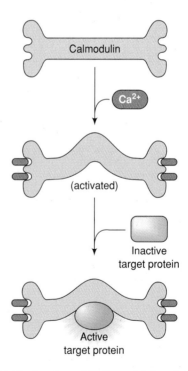

Figure 6-25 *Mechanism of Action of Calmodulin* *The binding of Ca²⁺ to calmodulin causes it to assume a new conformation in which it can bind to and activate other target proteins*

diesterase, the enzymes that catalyze the formation and breakdown of cyclic AMP respectively. Moreover, in some cases Ca^{2+} and cyclic AMP regulate the same target enzyme. An especially interesting example is phosphorylase kinase, an enzyme whose role in regulating glycogen breakdown was discussed on page 216. Phosphorylase kinase is a tetramer (four subunits) consisting of one catalytic subunit and three regulatory subunits. Two of the regulatory subunits are phosphorylated by protein kinase A, and are thus responsible for the fact that phosphorylase kinase can be activated by cyclic AMP. The other regulatory subunit allows phosphorylase kinase to be activated by Ca^{2+}. This Ca^{2+}-regulated subunit is none other than calmodulin itself. Thus cyclic AMP and Ca^{2+} promote glycogen breakdown by activating the same target enzyme, phosphorylase kinase.

Diacylglycerol (DAG) Is a Second Messenger That Activates Protein Kinase C

The other second messenger produced in conjunction with IP_3 is DAG (see Figure 6-23). The DAG molecule exerts its effects on the cell by stimulating yet another protein phosphorylating enzyme, **protein kinase C,** which requires Ca^{2+} and the membrane phospholipid phosphatidylserine for optimal activity. In the absence of DAG, protein kinase C resides in the cytosol in an inactive form. However, when DAG is generated by phospholipase C, the DAG molecule activates protein kinase C by causing it to bind to the inner surface of the plasma membrane. The activated protein kinase C then catalyzes the phosphorylation of assorted target proteins, triggering a variety of intracellular responses.

Although the target proteins phosphorylated by protein kinase C are not as well characterized as those phosphorylated by protein kinase A, much is known about the cellular effects that are triggered by the action of protein kinase C. Much of this information has been obtained from studies involving **phorbol esters,** a group of plant substances that bind to the plasma membrane and mimic the ability of DAG to stimulate protein kinase C. Phorbol esters trigger a variety of cellular responses, suggesting that protein kinase C acts upon many target proteins. One of the most prominent effects of phorbol esters is their ability to stimulate cells to grow and divide, which indicates that protein kinase C may be involved in the control of cell proliferation. We will see in Chapter 18 that phorbol esters promote the development of cancer in cells that have been previously exposed to cancer-causing agents, although phorbol esters alone do not cause cancer.

In addition to their ability to stimulate cell proliferation, phorbol esters and other agents that activate protein kinase C stimulate ion transport by the plasma membrane. One of the best studied examples involves the *Na+-H+ antiporter,* a membrane transport protein

that regulates intracellular pH by promoting the transport of Na^+ and H^+ in opposite directions across the plasma membrane. The Na^+-H^+ antiporter is stimulated in cells that have been exposed to either phorbol esters or growth factors that, like phorbol esters, trigger the activation of protein kinase C. Activation of the Na^+-H^+ antiporter triggered by protein kinase C causes the antiporter to pump protons (H^+) out of the cell, leading to an increase of about 0.2 pH units in the intracellular pH. The discovery that mutant cells lacking the Na^+-H^+ antiporter cannot be induced to proliferate by agents that stimulate protein kinase C suggests that this small increase in intracellular pH may play an important role in triggering cell proliferation.

Besides the effects on cell proliferation and membrane transport, activation of protein kinase C also stimulates cell secretion and enhances transcription of specific genes. At least some of the genes whose transcription is stimulated are regulated by the DNA-binding proteins *Jun* and *AP2.* Although activation of gene transcription by these DNA-binding proteins is enhanced in cells where protein kinase C has been stimulated, it is not clear whether Jun and AP2 are activated directly by protein kinase C or by some other mechanism.

CATALYTIC RECEPTORS

We have now discussed the properties of the first two classes of plasma membrane receptors defined earlier in the chapter—namely, ion channel receptors and receptors linked to G proteins (see Figure 6-10). The third group of receptors are known as **catalytic receptors** because the receptor protein functions as an enzyme whose activity transmits the incoming signal to the rest of the cell. In the following sections we will discuss the two best understood examples of catalytic receptors—that is, protein-tyrosine kinases and guanylyl cyclase.

Many Growth Factors Interact with Receptors That Function as Protein-Tyrosine Kinases

Growth factors such as platelet-derived growth factor (PDGF) and epidermal growth factor (EGF) bind to transmembrane receptors that exhibit protein kinase activity at the cytoplasmic surface of the plasma membrane. This protein phosphorylating activity differs in an important respect from that exhibited by the protein kinases discussed earlier in the chapter, which are called **protein-serine/threonine kinases** because they catalyze phosphorylation of the amino acids serine and threonine in protein molecules (page 214). Receptor protein kinases instead catalyze phosphorylation of the amino acid tyrosine, and are therefore called **protein-tyrosine kinases.** When a signaling ligand binds to a receptor exhibiting protein-tyrosine kinase activity, the receptor phosphorylates itself on several tyrosine residues. The tyrosines phosphorylated by this process of *autophosphorylation* are located in the receptor's cytoplasmic domain—that is, the region of the receptor exposed at the cytoplasmic surface of the plasma membrane (Figure 6-26, *top*). The phosphorylated tyrosines serve as selective binding sites that interact with specific cytoplasmic molecules, activating a cascade of events that trigger the appropriate cellular responses to the signaling growth factor (page 527).

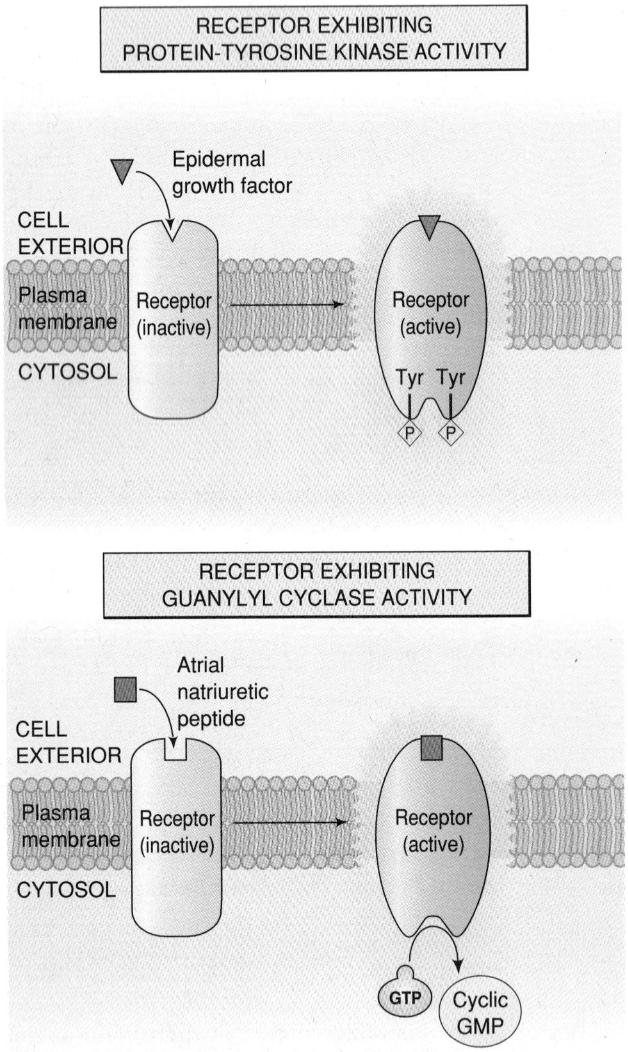

Figure 6-26 *Two Types of Catalytic Receptors* (Top) *In protein-tyrosine kinase receptors, the binding of a signaling molecule to its receptor stimulates the autophosphorylation of tyrosine residues in the receptor's cytoplasmic domain. The phosphorylated tyrosines then function as binding sites for specific cytoplasmic molecules, triggering an appropriate cellular response.* (Bottom) *In guanylyl cyclase receptors, the binding of a signaling molecule to its receptor stimulates the ability of the receptor to catalyze the formation of the second messenger, cyclic GMP.*

Receptors Exhibiting Guanylyl Cyclase Activity Produce the Second Messenger Cyclic GMP

Shortly after the discovery of cyclic AMP in the late 1950s, researchers began to look for other cyclic nucleotides that might also serve as second messengers. This search soon led to the identification of *3′,5′-cyclic guanosine monophosphate,* or **cyclic GMP,** a cyclic nucleotide synthesized from GTP by a mechanism comparable to the synthesis of cyclic AMP from ATP (see Figure 6-14). All the enzymes comparable to those involved in the cyclic AMP second messenger pathway were soon detected: **guanylyl cyclase** for the synthesis of cyclic GMP, *cyclic GMP phosphodiesterase* for the breakdown of cyclic GMP, and even a *cyclic GMP-dependent protein kinase* (**protein kinase G**) that is activated by cyclic GMP. It therefore seemed likely that cyclic GMP, like cyclic AMP, functions as a second messenger.

However, a frustrating period of many years ensued in which no extracellular signaling molecule was found to alter cyclic GMP levels, as would be expected if cyclic GMP were a second messenger. Finally, in the late 1980s it was discovered that cyclic GMP levels can be regulated by a small peptide hormone known as **atrial natriuretic peptide (ANP).** Released by the heart in response to high blood pressure, ANP lowers blood pressure by relaxing smooth muscle of blood vessels and by promoting excretion of Na^+ and water by the kidney. The ANP receptor is a transmembrane plasma membrane protein that possesses an ANP-binding site facing the cell exterior and a site exhibiting guanylyl cyclase activity facing the cell interior (see Figure 6-26, *bottom*).

Although binding of ANP to its plasma membrane receptor stimulates the receptor's guanylyl cyclase activity and hence elevates cyclic GMP levels, the mechanism by which cyclic GMP triggers enhanced excretion of Na^+ and water, or relaxes smooth muscle, has not been elucidated. Cyclic GMP is known to activate protein kinase G, which in turn phosphorylates a variety of proteins, but the functional significance of these effects is largely unknown. One situation in which the role played by cyclic GMP has been clearly established involves the effects of light on the retina. In this case, cyclic GMP influences the membrane potential of photoreceptor cells by controlling the permeability of cyclic GMP-gated ion channels (page 768).

Receptors Are Grouped into Superfamilies

As more and more receptors have been identified and characterized in recent years, it has become apparent that receptors specific for different signaling molecules often exhibit similarities in amino acid sequence and three-dimensional structure. Thus the rapidly growing number of different receptors can be reduced to a

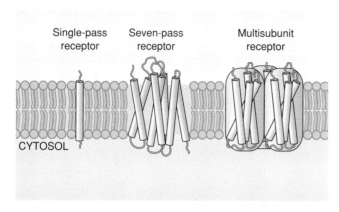

Single-pass receptor   Seven-pass receptor   Multisubunit receptor

CYTOSOL

Figure 6-27 *The Organization of Single-Pass, Seven-Pass, and Multisubunit Receptor Proteins within the Plasma Membrane* *The cylinders represent regions of α helix in each polypeptide chain. Single-pass receptors include catalytic receptors such as the protein-tyrosine kinase and guanylyl cyclase receptors, seven-pass receptors include the muscarinic acetylcholine receptor and the epinephrine receptor, and multisubunit receptors include the nicotinic acetylcholine receptor and receptors for the neurotransmitters γ-aminobutyric acid and glycine.*

smaller number of receptor classes or *superfamilies.* The importance of this discovery is that knowledge gained about any member of a given superfamily may help to elucidate the behavior of other receptors in the same family.

The catalytic receptors that we have just discussed, namely the protein-tyrosine kinase and guanylyl cyclase receptors, are members of a superfamily called *single-pass receptors* because the polypeptide chain contains a single transmembrane α helix (Figure 6-27). Two additional members of this group have recently been discovered; one is a *protein-tyrosine phosphatase* that catalyzes the removal of phosphate groups from tyrosine residues in proteins, and the other is a *protein serine/threonine kinase* that catalyzes the phosphorylation of serine and threonine residues in proteins.

A second superfamily consists of membrane proteins called *seven-pass receptors* because they contain seven transmembrane α helices. Included in this category are receptors for epinephrine, dopamine, serotonin, thyrotropin, luteinizing hormone, and acetylcholine (muscarinic receptor). The seven-pass structure is also exhibited by the visual pigment rhodopsin (page 768), even though this membrane protein is activated by the absorption of light rather than the binding of an extracellular signaling molecule. As might be expected from their structural similarities, seven-pass receptors share a common mode of action: They are all functionally coupled to G proteins.

A third superfamily is comprised of *multisubunit receptors* whose individual subunits resemble those that

make up the nicotinic acetylcholine receptor (page 210). Receptors for the neurotransmitters γ-amino butyric acid and glycine are two examples of molecules whose amino acid sequences resemble those of the nicotinic acetylcholine receptor. Such receptors are thought to belong to a family of related molecules that evolved from a common ancestral neurotransmitter receptor.

Signaling Responses Are Terminated by Reducing the Concentration of Signaling Molecules or Active Receptors

After the binding of an extracellular signaling molecule to its receptor has triggered an appropriate set of responses in a target cell, how are the responses terminated? Although cells attack this problem in many different ways, the mechanisms can be grouped into three general categories.

1. Removal of signaling molecules. The most direct way of terminating or limiting a cell's response to an extracellular signal is to remove the molecules that are involved in transmitting the signal. For example, the neurotransmitter acetylcholine is hydrolyzed by the enzyme *acetylcholinesterase,* which is present both in the synaptic space and bound to the muscle plasma membrane. Second messengers and their targets are also prime candidates for inactivation. Cyclic AMP and cyclic GMP are hydrolyzed by phosphodiesterases, and proteins phosphorylated by protein kinase A are dephosphorylated by protein phosphatases. In the phosphoinositide signaling pathway, the second messenger IP_3 is degraded by phosphatases while DAG is either hydrolyzed or phosphorylated. The hydrolysis of DAG is an especially interesting reaction because one of its products is *arachidonic acid,* a precursor for the synthesis of prostaglandins. Hence terminating the action of DAG can in turn activate an alternative signaling pathway involving prostaglandins.

Signaling events mediated by increased levels of Ca^{2+} in the cytosol are terminated in a different way because the calcium ion cannot be enzymatically altered. The plasma membrane of all cells contains an ATP-dependent **calcium pump** that maintains a low intracellular Ca^{2+} concentration by actively pumping calcium ions out of the cell. The activity of the Ca^{2+} pump is controlled in part by Ca^{2+}-calmodulin, which stimulates the pump. Hence the increase in cytosolic Ca^{2+} that occurs in response to the second messenger IP_3 is a transient effect because the Ca^{2+} (in conjunction with calmodulin) increases the activity of the Ca^{2+} pump that expels Ca^{2+} from the cell.

2. Receptor desensitization. When cells are exposed to an extracellular signaling molecule, the receptor for that signaling molecule often becomes rapidly inactivated. If the only receptors affected are those that bind to the initiating signal molecule, the process is called *homologous desensitization*. Regulation of this type is commonly mediated by phosphorylation of the receptor molecule. For example, the β_2-adrenergic receptor is subject to phosphorylation and inactivation by a protein kinase known as *β-adrenergic receptor kinase*. Phosphorylation occurs only when epinephrine is bound to the receptor, and hence serves to terminate the action of a receptor that has already functioned.

The β_2-adrenergic receptor is also phosphorylated by the cyclic AMP-dependent enzyme, protein kinase A. Since epinephrine is only one of many signaling molecules that increases the intracellular concentration of cyclic AMP, the β_2-adrenergic receptor can become desensitized in response to any signal that increases cyclic AMP levels. This phenomenon is called *heterologous desensitization* because the receptor is being desensitized through the action of a signaling molecule that acts on a different type of receptor. Direct inactivation of the receptor itself is not the only possible mechanism for heterologous desensitization. In some cases the action of a signaling molecule on a target cell causes G proteins to undergo alterations that inhibit their ability to transmit signals. Since more than one type of receptor can utilize the same pool of G proteins, such an inactivation would cause the cell to become unresponsive to all signaling molecules that employ the same type of G protein.

3. Receptor down-regulation.   In addition to the relatively rapid process of receptor desensitization, a slower inactivation response also occurs after the binding of a signaling ligand to its cell surface receptor. This slower response, called *down-regulation,* leads to a reduction in the number of receptor molecules present within the plasma membrane. During the process, cell surface receptors containing bound signaling ligands are gradually removed from the plasma membrane by invagination and pinching off of membrane vesicles. These membrane vesicles then migrate to the cell interior, where the signaling molecule and (in some cases) the receptor are degraded. This pathway for internalizing membrane receptors will be described in detail in Chapter 7, when we discuss *receptor-mediated endocytosis* (page 295).

THE EXTRACELLULAR MATRIX

Since this chapter is devoted to the cell surface, it is appropriate to discuss those structures outside the cell that interact with the outer cell surface. In unicellular organisms, such as algae growing in a pond or bacteria residing in the digestive tract, the outside of the cell usually faces a relatively simple fluid environment. The blood cells of multicellular animals are likewise suspended in a fluid medium. But most of the cells of multicellular organisms are grouped together into solid tissue masses. In some cases, such as the *epithelial* tissues that cover body and organ surfaces, the majority of the cells make direct contact with other cells. In most tissues, however, the cells secrete a group of macromolecules that together generate an interconnected molecular network called the **extracellular matrix.** The main cell type involved in producing and secreting the molecular components that comprise the extracellular matrix is the **fibroblast,** but in certain specialized tissues other related cell types are involved.

The characteristics of the extracellular matrix differ dramatically among tissues (Figure 6-28). Bone, for example, is constructed largely from a rigid extracellular matrix that contains a tiny number of interspersed cells. Hence the characteristics of bone are dictated to a large extent by the properties of the matrix. Cartilage is another tissue constructed almost entirely of matrix materials, although in this case the matrix is more flexible than in bone. In contrast to bone and cartilage, the connective tissue surrounding glands and blood vessels has a relatively gelatinous extracellular matrix containing numerous interspersed cells.

The preceding examples illustrate the crucial role played by the extracellular matrix in determining the shape and mechanical properties of organs and tissues. The matrix does more, however, than just provide structural support. It also influences properties such as cell shape and motility, growth and division, and the development of specialized cellular characteristics. These diverse effects are made possible by the various kinds of molecules found in the matrix. Although these molecules vary somewhat between tissues, the extracellular matrix of animal cells is almost always comprised of the same three classes of molecules: (1) glycosaminoglycans and proteoglycans, which form a hydrated gelatinous substance in which the other matrix components are embedded; (2) structural proteins, such as collagens and elastin, which impart strength and flexibility to the matrix; and (3) adhesive proteins, such as fibronectin and laminin, which promote the attachment of cells to the matrix. The properties of these three main constituents of the extracellular matrix are discussed in the following sections.

The Ground Substance of the Extracellular Matrix Is Formed by Glycosaminoglycans and Proteoglycans

The physical properties of the extracellular matrix are strongly influenced by the presence of a specialized group of polysaccharides called **glycosaminoglycans.**

Fibroblast

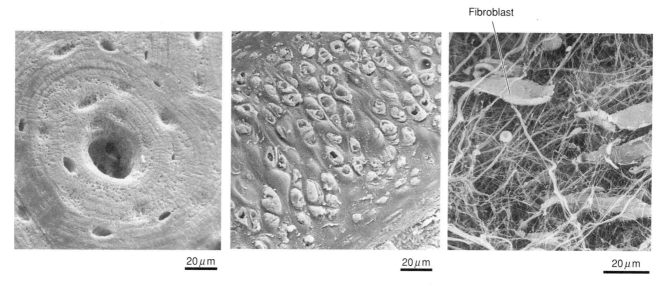

20 μm   20 μm   20 μm

Figure 6-28 *Scanning Electron Micrographs Illustrating the Differing Appearance of the Extracellular Matrix in Bone, Cartilage, and Connective Tissue* (Left) *In bone tissue, a hard calcified extracellular matrix is laid down in concentric rings around central canals. The small elliptical depressions represent regions in which the bone cells are found. (Middle) In cartilage the cells are embedded in a flexible matrix that contains large amounts of proteoglycan. (Right) In the connective tissue found under the skin, fibroblasts are surrounded by an extracellular matrix that contains large numbers of collagen fibers. Courtesy of Professors Richard Kessel and Randy Kardon, University of Iowa.*

These long polysaccharide chains are constructed from a repeating disaccharide unit containing one amino sugar and at least one negatively charged sulfate or carboxyl group. Among the more commonly encountered glycosaminoglycans are *chondroitin sulfate, keratan sulfate, heparin, heparan sulfate, hyaluronate (hyaluronic acid),* and *dermatan sulfate.* The disaccharide units employed in the construction of some of these glycosaminoglycans are illustrated in Figure 6-29. Because glycosaminoglycans are hydrophilic molecules containing large numbers of negative charges, they attract both positively charged cations and water molecules. The result is the creation of a hydrated, gelatinous material that forms the so-called "ground substance" of the extracellular matrix.

Most glycosaminoglycans are covalently bound to protein molecules to form protein-carbohydrate complexes known as **proteoglycans.** Each proteoglycan molecule contains multiple glycosaminoglycan chains attached to a single *core protein,* producing a complex in which carbohydrate accounts for as much as 95 percent of the total mass. By employing different core proteins and glycosaminoglycan chains of varying types and lengths, a diverse family of proteoglycans can be created. One of the most elaborate proteoglycans occurs in the extracellular matrix of cartilage tissue. This enormous molecular complex, which consists of dozens of proteoglycan molecules attached to a long backbone of hyaluronate, has a molecular weight of several million and can exceed several micrometers in total length (Fig-

ure 6-30). The unique resiliency and pliability of cartilage is based in large part on the properties of this proteoglycan complex.

Although most of the glycosaminoglycans found in the extracellular matrix exist only as components of proteoglycans and not as free glycosaminoglycans, **hyaluronate** is a unique exception. In addition to its role in proteoglycans, hyaluronate occurs as a free glycosaminoglycan chain comprised of hundreds or thousands of repeating disaccharide units. Hyaluronate is especially prevalent in the extracellular matrix of tissues where cells are actively proliferating or migrating. It is also found on the surface of migrating cells, but is removed when migration ceases and cell-cell contacts are established. Such observations suggest that hyaluronate facilitates cell migration, perhaps by attracting a shell of water around the outer surfaces of cells and making the surrounding matrix more watery. The hyaluronate molecule also has lubricating properties and is therefore located in places where friction must be reduced, such as the joints between moving bones.

Because glycosaminoglycans and proteoglycans tend to form hydrated gels, their presence exerts a profound impact on the physical characteristics of the extracellular matrix. Cartilage, for example, is a relatively pliable and resilient tissue because of its high content of proteoglycans. In many tissues, however, strength and/or rigidity of the matrix is as important, or more important, than pliability and resiliency. In such cases the presence of matrix components such as collagen, which imparts strength, is crucial.

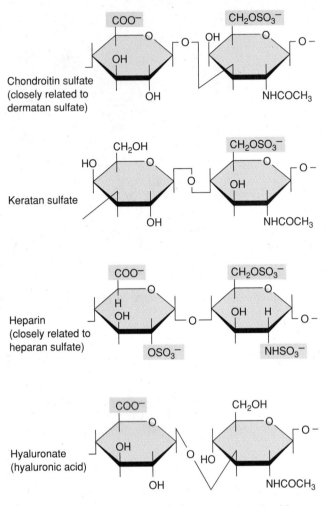

Figure 6-29 *Structures of the Repeating Disaccharide Units Employed in the Construction of Some Common Glycosaminoglycans* *The disaccharide units all have at least one negatively charged sulfate or carboxyl group.*

Collagen Is Primarily Responsible for the Strength of the Extracellular Matrix

The most prevalent component of the extracellular matrix is a family of related proteins called **collagens.** Collagen is the most abundant protein found in vertebrates, accounting for as much as 30 percent or more of the total body protein. Two defining features are shared by all collagens. First, collagen molecules are constructed from three intertwined polypeptide chains, termed α chains, wound together to form a rigid *triple helix.* In some collagens the triple helix is interrupted by nonhelical regions that allow the molecule to bend, but the triple helix is still crucial to the molecule's function. The second defining feature of the collagens is an unusual amino acid composition; as many as 25 percent of the amino acids are glycine and another 25 percent are the unusual amino acids hydroxyproline and hydroxylysine, which rarely occur in proteins other than collagen.

This unusual amino acid composition is an important chemical feature of the collagens because the three chains of the triple helix are held together by hydrogen bonds involving glycine and hydroxyproline. The significance of hydroxyproline is dramatically illustrated by the disease *scurvy,* which is caused by an inadequate dietary intake of ascorbic acid (vitamin C). Ascorbic acid is a reducing agent responsible for maintaining the activity of prolyl hydroxylase, the enzyme that catalyzes proline hydroxylation. In the absence of ascorbic acid, collagen is inadequately hydroxylated and therefore does not form a stable triple helix. This defect causes individuals with scurvy to suffer from a variety of disorders, including extensive bruising, hemorrhages, and breakdown of supporting tissues.

At least 14 different kinds of collagen are known to exist, and the list still appears to be growing. These varying forms of collagen, referred to as collagen types I through XIV, are constructed from different combinations of more than 20 types of α chains encoded by different genes. The collagen molecules constructed from these α chains can be divided into two main groups based on whether they form *banded fibrils* or *unbanded filamentous networks* (Figure 6-31). Banded collagen fibrils are so named because they exhibit a characteristic pattern of dark crossbands every 67 nm. Collagen types I, II, III, and V are the most prevalent collagens found in banded fibrils. These collagens create banded fibrils by lining up end-to-end and next to each other to generate structures that measure up to several hundred nanometers in diameter and several micrometers in length. Banded fibrils generally contain more than one type of collagen molecule. The banded collagen fibrils present in skin tissue, for instance, are constructed from a mixture of collagen types I and III, and the collagen fibrils present in the cornea contain collagen types I and V.

Collagen fibrils can aggregate together to produce even larger structures, known as **collagen fibers,** which measure several micrometers in diameter and are therefore visible with the light microscope. One of the most striking properties of these collagen fibers is their enormous physical strength. It has been reported, for example, that it takes a load of more than 20 pounds to break a collagen fiber measuring only a millimeter in diameter. Thus banded collagen is largely responsible for the mechanical strength of protective and supporting tissues such as skin, bone, tendon, and cartilage.

Among the collagens that do not form banded fibrils, types IV and VI are the best characterized. Collagens in this category tend to generate unbanded filamentous networks. For example, type IV collagen forms a meshwork of filaments localized in the basal lamina, a specialized matrix structure described later in the chapter (page 231). Type VI collagen forms a highly branched filamentous network located around nerves,

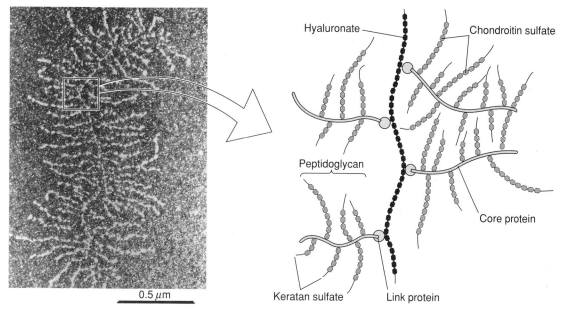

Figure 6-30 **The Cartilage Proteoglycan Complex** *The electron micrograph shows a proteoglycan complex isolated from bovine cartilage. The accompanying diagram illustrates the arrangement of the hyaluronate backbone and proteoglycan molecules that make up this proteoglycan complex.*

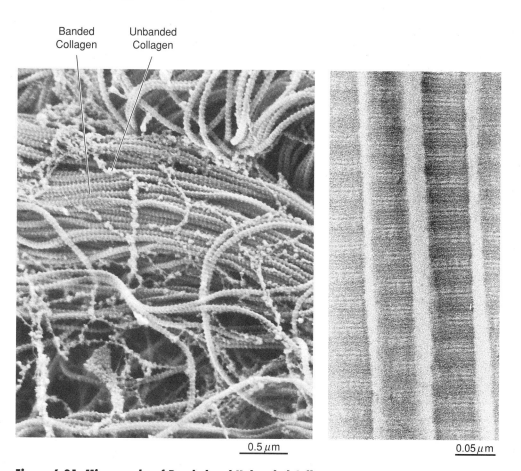

Figure 6-31 **Micrographs of Banded and Unbanded Collagen** (Left) *Scanning electron micrograph of collagen network in human skin, illustrating both banded and unbanded collagen fibrils. The banded fibrils have been stained with collagen-specific antibodies to make the banding pattern more visible. The unbanded collagen is stained with an antibody linked to gold particles.* (Right) *Transmission electron micrograph showing the banding pattern of banded collagen fibrils in greater detail. Courtesy of D. R. Keene (left) and N. Simionescu (right).*

blood vessels, and banded collagen fibers, where it may help to anchor these structures to the surrounding extracellular matrix.

Collagen Is Produced from a Precursor Molecule Called Procollagen

Since collagen fibrils are highly organized structures involving thousands of individual molecules, the question arises as to how such an elaborate structure is generated. If purified α chains are simply mixed together in a test tube, neither the triple-helical structure nor mature collagen fibrils are produced. Therefore cells must utilize some mechanism other than self-assembly of α chains to generate these higher-order structures. To investigate the mechanisms involved, fibroblasts have been incubated with radioactive amino acids to label newly forming proteins. Under these conditions, the first type of collagen to become radioactive is a larger precursor chain that contains extra amino acids at both ends. Interactions involving the extra segments are required during the initial formation of the triple-helical structure, explaining why mature α chains cannot self-assemble to form collagen.

Figure 6-32 illustrates our current view of the steps involved in generating mature collagen fibrils from

these precursors. The precursor chains are first assembled together in the lumen of the endoplasmic reticulum to form a triple-helical molecule called **procollagen.** The extra amino acids present at the two ends of the procollagen molecule block the ensuing formation of collagen fibrils while the procollagen remains within the cell. After procollagen is secreted from the cell into the extracellular space, it is converted to collagen by specific enzymes (procollagen peptidases) that remove the extra amino acids from the two ends of the molecule. The resulting collagen molecules can then spontaneously associate and polymerize to form mature collagen fibrils. In banded collagen fibrils, the 67-nm repeat distance between striations corresponds to roughly one-quarter of the length of an individual collagen molecule. As is shown in Figure 6-32, such a repeat pattern is created by packing together rows of collagen molecules in which each row is displaced by one-fourth the length of a single molecule.

The stability of the collagen fibril is reinforced by chemical bonds involving lysine and hydroxylysine residues. These bonds form crosslinks both within and between the individual collagen molecules that make up a collagen fibril. In addition, specialized types of collagen are often associated with the surface of collagen fibrils. These surface-associated collagens, which include

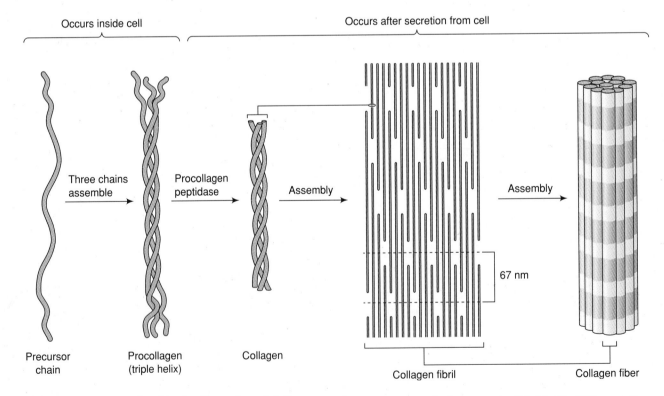

Figure 6-32 Steps Involved in the Formation of Collagen *Collagen precursor chains are assembled in the ER lumen to form triple-helical procollagen molecules. After secretion from the cell, procollagen is converted to collagen in a peptide cleavage reaction catalyzed by procollagen peptidase. The collagen molecules then bind to each other and polymerize into mature collagen fibrils. In banded collagen, the 67-nm repeat distance is created by packing together rows of collagen molecules in which each row is displaced by one-fourth the length of a single molecule.*

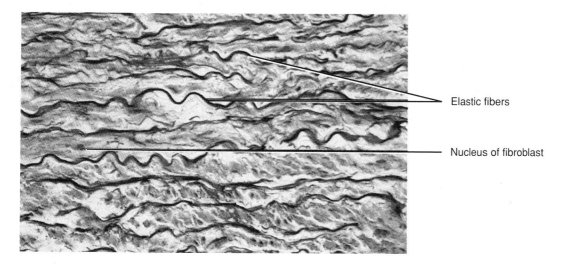

Figure 6-33 *Light Micrograph Showing Elastic Fibers   In this longitudinal section through an aorta, numerous elastic fibers are visible in the extracellular matrix. The elastic fibers allow the aorta to dilate and constrict as the heart pumps blood through it. Biophoto Associates/SS/Photo Researchers.*

collagen types IX and XII, have a triple-helical structure that is interrupted to allow the molecules to bend and hence serve as bridges between collagen fibrils, or between collagen fibrils and other matrix components.

Elastin Imparts Elasticity and Flexibility to the Extracellular Matrix

While collagen gives strength and toughness to the extracellular matrix, some tissues require a matrix that is also flexible and elastic. For example, lung tissue must expand and contract as an organism inhales and exhales. Likewise arteries, especially those close to the heart, dilate and constrict as the heart pumps blood through them. The elasticity of such tissues is imparted by stretchable *elastic fibers* located in the extracellular matrix (Figure 6-33). Like a rubber band, elastic fibers can be stretched to several times their normal length and will snap back to their original size when the tension is released. The relative degree of elasticity of any given tissue is controlled by varying the number of nonstretchable collagen fibrils that are interspersed among the elastic fibers.

The principal constituent of elastic fibers is the protein **elastin.** Like collagen, elastin is rich in the amino acids glycine and proline; however, in contrast to collagen the proline is not extensively hydroxylated and no hydroxylysine is present. Elastin molecules are crosslinked to one another by covalent bonds involving the amino acid lysine. The elasticity of the resulting crosslinked elastin network is based upon the ability of individual elastin molecules to adopt a variety of alter-

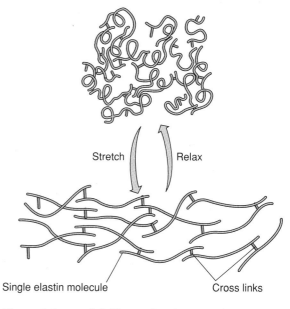

Figure 6-34 *Model Illustrating How Elastin Fibers Stretch and Recoil   Each elastin molecule in the crosslinked network can assume either an extended or compact conformation. Therefore the fiber can stretch when tension is exerted on it and will recoil when the tension is released.*

nate conformations. As shown in Figure 6-34, exerting tension on the network causes the individual molecules to adopt extended conformations that permit the overall network to stretch. When the tension is released, the individual molecules return to their normal, less extended conformations, and the crosslinks between molecules cause the network to recoil to its original shape.

Fibronectin Binds Cells to the Matrix and Guides Cellular Migration

Several glycoproteins present in the extracellular matrix act to bind cells to the matrix. The most prevalent of these adhesive glycoproteins is **fibronectin,** a large molecule containing about 5 percent carbohydrate and composed of two subunits with a combined molecular weight of about 500,000 daltons. Fibronectin occurs in a soluble form in blood and other body fluids, as insoluble fibrils in the extracellular matrix, and in an intermediate form that is loosely associated with cell surfaces. The various forms of fibronectin are all encoded by the same gene, but the RNA produced by this gene is spliced in various ways to generate messenger RNAs coding for a family of related proteins. Each fibronectin molecule is organized into a series of distinct globular domains whose individual properties can be studied by using proteases to cleave the protein into fragments containing single domains. Such studies have revealed that the domains carry out different functions—one domain binds the fibronectin molecule to collagen, a second domain binds to heparin, a third binds to the blood-clotting protein fibrin, and a fourth domain binds to cell surfaces (Figure 6-35). By using proteases to break the cell-binding domain into small fragments, it was discovered that the amino acid sequence *Arg-Gly-Asp* is responsible for binding fibronectin to cell surfaces. This same Arg-Gly-Asp sequence (also referred to as the *RGD sequence* in an alternative abbreviation system) has been detected in a variety of other adhesive proteins present in the extracellular matrix and in blood.

Since fibronectin binds to the cell surface as well as to matrix components like collagen and heparin, the fibronectin molecule can function as a bridging molecule that attaches cells to the extracellular matrix. This anchoring role has been demonstrated experimentally by placing cells in a culture flask whose surface has been coated with fibronectin. Under such conditions, the cells attach to the surface of the flask more efficiently than they do in the absence of fibronectin. After attaching, the cells flatten out and components of the intracellular cytoskeleton become aligned with the fibronectin molecules located outside the cells (Figure 6-36). Hence the interaction of cells with matrix components appears to influence overall cell shape through effects on cytoskeletal organization.

Another cellular property that can be influenced by fibronectin is motility. A prominent example occurs during early embryonic development, when specific groups of cells migrate from one region of the embryo to another. The pathways followed by migrating cells are rich in fibronectin, suggesting that the moving cells are guided by binding to fibronectin molecules along the way. This idea has received support from experiments in which developing amphibian embryos were injected with antibodies directed against fibronectin to block the ability of cells to bind to fibronectin. Such treatment disrupts normal cell migration, leading to the development of abnormal embryos. The ability of fibronectin to promote cell-matrix adhesion and guide cell migration is important in many other situations as well. For example, during blood clotting, fibronectin attaches platelets to fibrin in the newly forming blot clots. Fibronectin also guides the migration of cells of the immune system into wounded areas, thereby promoting wound healing.

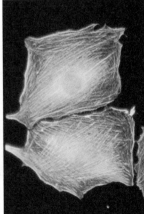

Figure 6-36 *Alignment of the Extracellular Matrix and the Intracellular Cytoskeleton* *These two fluorescence micrographs show the same cultured cells stained with fluorescent antibodies specific for either fibronectin (left) or actin (right). Note that the extracellular fibronectin matrix and intracellular actin filaments (part of the cytoskeleton) are aligned in a similar pattern. Courtesy of R. O. Hynes.*

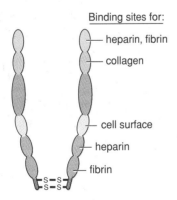

Binding sites for:
— heparin, fibrin
— collagen
— cell surface
— heparin
— fibrin

Figure 6-35 *Organization of the Fibronectin Molecule*
Fibronectin is composed of two polypeptide chains that are organized into a series of distinct globular domains specialized for different purposes.

Laminin Binds Cells to the Basal Lamina

Another adhesive glycoprotein found in the extracellular matrix is **laminin.** In contrast to fibronectin, which is dispersed widely throughout supporting tissues and body fluids, laminin is preferentially located in a specialized structure known as the **basal lamina.** The basal lamina is a thin sheet of matrix material that separates epithelial cell layers from underlying supporting tissues (Figure 6-37); in addition, basal laminae often surround nerve, muscle, and fat cells. This thin sheet of matrix material, typically measuring about 50 nm in thickness, acts as a structural support that maintains tissue organization, and as a permeability barrier that regulates the movement of both molecules and cells. In the kidney, for example, the basal lamina serves as a filter that allows small molecules but not blood proteins to be excreted into the urine. The basal lamina beneath epithelial cell layers prevents the passage of underlying connective tissue cells into the epithelium but permits the migration of white blood cells needed to fight infection. The influence of the basal lamina on cell migration is of special interest because it has recently been discov-

ered that the cell surface of some cancer cells is enriched in binding sites for laminin, suggesting that the resulting increase in the ability of cancer cells to bind to the basal lamina may facilitate their movement through this structure and hence allow them to migrate from one region of the body to another.

In addition to laminin, the main components of the basal lamina are the glycoprotein *entactin,* type IV collagen, and the proteoglycan heparin sulfate. Type IV collagen, one of the unbanded types of collagen, is arranged as a filamentous meshwork that forms the central layer of the basal lamina. Laminin molecules bind this collagen-containing core to the overlying layer of cells. Laminin is assembled from three polypeptide chains, generating a large, cross-shaped molecule with a molecular weight of about 820,000 daltons. As in the case of fibronectin, the laminin molecule is organized into several distinct domains (Figure 6-38). One domain binds to type IV collagen, a second domain binds to heparin, and yet another domain binds to cell surface receptors. These separate binding sites allow laminin to act as a bridging molecule that attaches cells to the basal lamina. Several forms of the laminin molecule have been detected in the basal laminae of different tissues, where they are thought to play distinct functional roles; included among these functions are the ability to influence patterns of cell growth, differentiation, and motility.

Integrin Receptors Bind to Fibronectin, Laminin, and Other Matrix Constituents

We have now seen that several components of the extracellular matrix are capable of binding to cells. The interaction between matrix components and cell surfaces is made possible by a family of cell surface receptors that bind specifically to molecules present in the matrix. The first such receptor to be isolated and well characterized was a *fibronectin receptor.* Purification of this molecule was facilitated by the discovery that the amino acid sequence Arg-Gly-Asp is responsible for binding fibronectin to cell surfaces (page 230). This information made it possible to purify the cell surface fibronectin receptor by extracting proteins from the plasma membrane, binding them to a fibronectin affinity chromatography column, and displacing the fibronectin receptor from the column by adding a small synthetic peptide containing the Arg-Gly-Asp sequence.

The fibronectin receptor is a transmembrane protein constructed of two subunits designated α and β. It contains two types of binding sites: an external site that binds to fibronectin, and a site exposed at the inner membrane surface that binds to the protein **talin.** Talin

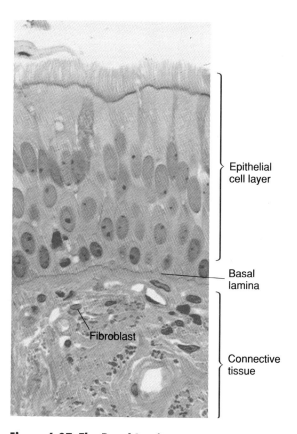

Figure 6-37 *The Basal Lamina* *The basal lamina is a thin sheet of matrix material that separates an epithelial cell layer from underlying tissues. Courtesy of G. W. Willis, M.D. and Biological Photo Service.*

Epithelial cell layer

Basal lamina

Fibroblast

Connective tissue

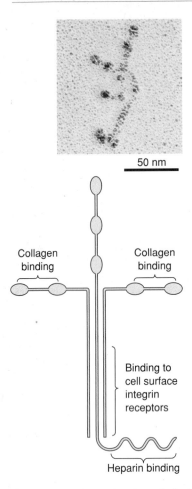

Figure 6-38 *Organization of the Laminin Molecule*
(Top) *Electron micrograph of isolated laminin molecule shadowed with platinum, showing its cross-shaped structure.* (Bottom) *Location of functional domains within the laminin molecule. Micrograph courtesy of J. Engel.*

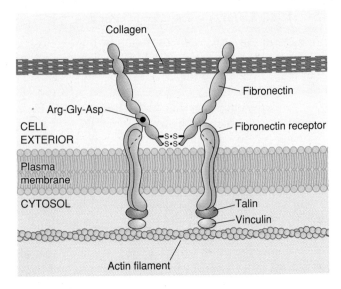

Figure 6-39 *The Fibronectin Receptor* *By binding to the cytoskeleton inside the cell and to fibronectin outside the cell, the fibronectin receptor integrates the organization of the cytoskeleton with that of the extracellular matrix. The fibronectin receptor is a member of the integrin family of cell surface receptors.*

in turn is linked to actin filaments of the cytoskeleton via the protein **vinculin** (Figure 6-39). Because of this arrangement, the binding of fibronectin to the outer cell surface can trigger a change in the organization of the cytoskeleton, which in turn influences cell shape and motility.

Cell surface receptors for several other matrix constituents, including laminin and certain collagens, resemble the fibronectin receptor in structure. This interrelated group of receptors, including the fibronectin receptor, is referred to as the **integrin** family of receptors because they *integrate* the organization of the cytoskeleton with that of the extracellular matrix. This integration of extracellular and intracellular structure is not a one-way street. The organization of the extracellular matrix influences the structure of the cytoskeleton, and the structure of the cytoskeleton influences the behavior of the matrix.

The Glycocalyx Is a Carbohydrate-Rich Zone Located at the Periphery of Many Animal Cells

The boundary between the extracellular matrix and the cell surface is a region that is enriched in carbohydrates. In plants, some fungi, and bacteria, this carbohydrate-rich zone is organized into a rigid cell wall that provides protection and structural support for the cell. Although animal cells do not have a rigid enclosure like a cell wall, a carbohydrate-rich zone called a **glycocalyx** is often present external to the plasma membrane (Figure 6-40). The glycocalyx has been implicated in functions such as cell recognition and adhesion, protection of the cell surface, and the creation of permeability barriers.

The glycocalyx is constructed from two kinds of components. The "attached glycocalyx" consists of inherent constituents of the cell surface that cannot be removed by mechanical means without simultaneously removing a portion of the plasma membrane itself. The carbohydrate chains attached to plasma membrane glycoproteins and glycolipids are major components of this layer. In contrast, the "unattached glycocalyx" consists of material located external to the plasma membrane that can be readily removed without affecting the viability of the cell or disrupting the plasma membrane. Included in this category are specialized structures such as the embryonic membranes surrounding most animal eggs and the outer coat of amoebas. The unattached glycoca-

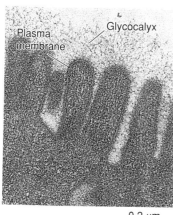

Figure 6-40 *The Glycocalyx   In this electron micrograph of a surface epithelial cell from rat colon, the glycocalyx appears as a thick carbohydrate-rich layer coating the cell surface. Courtesy of J. E. Michaels.*

lyx usually consists of glycoproteins and proteoglycans that are also constituents of the extracellular matrix. For this reason, it is often difficult to define exactly where the cell surface ends and the extracellular matrix begins.

CELL-CELL RECOGNITION AND ADHESION

In multicellular organisms, individual cells become associated in precise patterns to form tissues, organs, and organ systems. Such ordered interactions require that individual cells be able to recognize, adhere, and directly communicate with each other. The ability of cells to recognize one another was first demonstrated experimentally in 1907 by H. V. Wilson, who investigated the behavior of cells obtained from two differently colored species of marine sponge. Wilson isolated individual cells from a yellow-pigmented sponge and from a red-pigmented sponge, and then mixed the two cell populations together. Under these conditions the cells selectively bound together into clumps composed exclusively of either red or yellow cells, but not a mixture of the two. These results were one of the first indications that cells are capable of recognizing and selectively binding to other cells of the same type.

In the many years that have passed since these pioneering observations, examples of selective cell-cell recognition and adhesion have been described in a variety of organisms and cell types. From such studies a picture of the cell surface molecules involved in these processes is beginning to emerge.

N-CAMs and Cadherins Are Plasma Membrane Glycoproteins That Mediate Cell-Cell Adhesion

One way to identify molecules that mediate cell-cell adhesion is to develop antibodies against specific types of cells or cell membranes and then test the ability of the antibodies to block adhesion between these cells. When an antibody is detected that specifically perturbs cell-cell adhesion, the protein to which this antibody binds is a good candidate for a molecule involved in the adhesion process (Figure 6-41). This approach was first used successfully in the late 1970s by Gerald Edelman and his colleagues, who employed it to identify a membrane glycoprotein from nerve tissue called the *neural cell adhesion molecule* or **N-CAM.** When embryonic cells are exposed to antibodies directed against N-CAM, the cells can no longer bind to one another and the orderly formation of neural tissues is disrupted.

N-CAM is a plasma membrane glycoprotein that exists in at least three forms, all encoded by the same gene; two are transmembrane proteins, and the third is covalently attached to the outer membrane surface. In each case the bulk of the N-CAM molecule protrudes from the outer cell surface. The protruding region, which is similar in structure in all three types of N-CAM, contains the binding sites involved in cell-cell adhesion. In order to investigate the binding mechanism, experiments have been carried out in which purified N-CAM is incorporated into artificial phospholipid vesicles. Under these conditions, the membrane vesicles begin to adhere to one another; adhesion does not occur in the presence of antibodies directed against N-CAM, nor does it occur in vesicles that do not contain N-CAM. Such experiments have led to the conclusion that cell adhesion is mediated by the binding of N-CAMs located on one cell to N-CAMs located on another cell.

The use of antibodies that block cell adhesion has also led to the discovery of a group of cell adhesion molecules called **cadherins,** which can be distinguished from N-CAMs on the basis of their requirement for Ca^{2+}. It has been known for many years that removing Ca^{2+} from the extracellular medium causes cells to dissociate from one another. This calcium dependence arises at least in part because Ca^{2+} induces a conformational change in cadherins that allows them to mediate cell-cell adhesion; hence in the absence of extracellular Ca^{2+}, adhesion between cells is weakened. The best characterized members of the cadherin family are integral membrane glycoproteins known as *E-cadherin* (prevalent in epithelial tissue), *N-cadherin* (prevalent in nervous tissue), and *P-cadherin* (prevalent in placenta). Each type of cadherin occurs in a different spectrum of cell types, and the timing of their production during embryonic development correlates with the coming together of specific cell types to form tissues and organs.

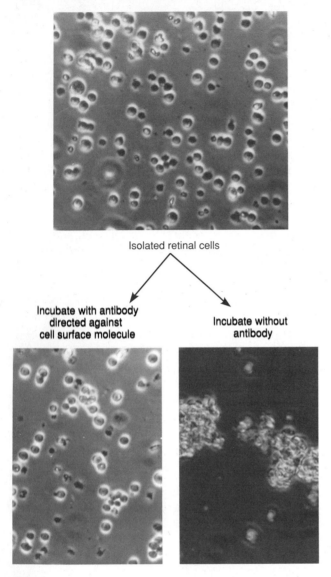

Isolated retinal cells

Incubate with antibody
directed against
cell surface molecule

Incubate without
antibody

Figure 6-41 *Using Antibodies to Detect Cell Adhesion Molecules* *Antibodies are first generated by injecting rabbits with foreign cells, in this case from chick retinal tissue. Isolated retinal cells are then incubated with the resulting antibodies to determine whether any of the antibodies block cell-cell adhesion. The antibodies isolated in this particular experiment prevent incubated retinal cells from adhering to each other* (bottom left). *The protein to which such an antibody binds is a good candidate for a molecule involved in the adhesion process. Courtesy of G. M. Edelman.*

The role played by cadherins in cell-cell adhesion has been investigated in cultured fibroblasts called *L cells,* which bind poorly to one another and contain little cadherin. When purified genes coding for E-cadherin or P-cadherin are introduced into L cells, the cells begin to produce cadherins and to bind more tightly to one another. Moreover, L cells that produce

E-cadherin bind preferentially to other cells producing E-cadherin; similarly, cells expressing P-cadherin selectively bind to other cells expressing P-cadherin (Figure 6-42). Such observations suggest that cadherins, like N-CAM, mediate cell-cell adhesion by a mechanism in which cadherin molecules present on one cell bind to cadherin molecules of the same type present on another cell.

Carbohydrate Groups Participate in Cell-Cell Recognition and Adhesion

Since N-CAMs and cadherins are both glycoproteins, the question arises as to the role played by their carbohydrate groups in cell adhesion. A variety of experimental observations suggest that the carbohydrate groups influence both the strength and specificity of cell-cell interactions. N-CAM molecules, for example, contain long repeating chains of the negatively charged carbohydrate *sialic acid.* The amount of sialic acid bound to N-CAM changes significantly during development, suggesting a possible role in regulating cellular adhesion. Support for this idea has emerged from the finding that membrane vesicles containing N-CAMs with small amounts of sialic acid bind together more tightly than membrane vesicles containing N-CAMs with large amounts of sialic acid.

Other kinds of carbohydrate groups have also been implicated in cell-cell adhesion. For example, the sugar galactose is thought to be involved in cell adhesion because removing galactose residues from the outer cell surface diminishes the ability of cells to adhere to each other. In a related set of experiments, isolated cells have been shown to bind to synthetic beads coated with galactose, but not to beads coated with glucose, thereby indicating the selectivity of the carbohydrate recognition process. A role for carbohydrate groups in cell adhesion is also suggested by the fact that many animal and plant cells secrete carbohydrate-binding proteins called **lectins,** which promote cell-cell adhesion by binding to a specific sugar or sequence of sugars exposed at the outer cell surface. Because a lectin molecule usually has more than one carbohydrate-binding site, it can bind to carbohydrate groups on two different cells, thereby linking the cells together (Figure 6-43).

Although the phenomena of cell-cell recognition and adhesion are often discussed together, recognition events are also important in contexts that do not lead to cell-cell adhesion. A well-known example involves the glycophorin molecules of the red cell membrane (page 172). One of the carbohydrate chains attached to glycophorin is responsible for determining the standard human blood types (A, B, AB, and O). Individuals with blood type A have N-acetylgalactosamine residues at the

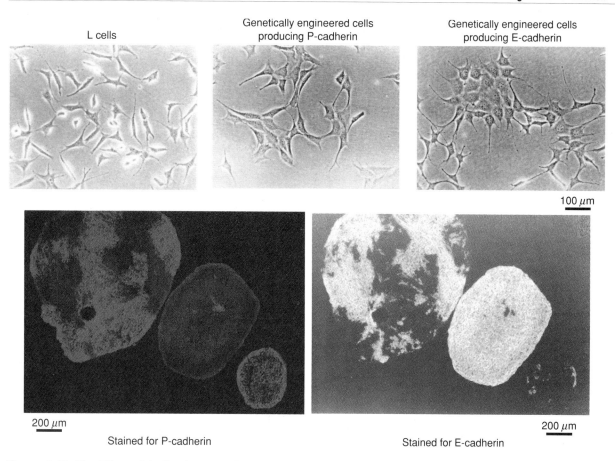

Figure 6-42 *The Effect of Cadherins on Cell Adhesion* (Top left) *A light micrograph of cultured L cells shows that these cells do not normally adhere to one another.* (Top middle and right) *When DNA sequences coding for E-cadherin or P-cadherin are introduced into these cells, cadherin is produced and cell-cell adhesion occurs.* (Bottom) *Fluorescence micrographs of cell aggregates that form when cells producing P-cadherin are mixed with cells producing E-cadherin. The same field is stained with either a red fluorescent dye specific for P-cadherin or a green fluorescent dye specific for E-cadherin. The staining pattern shows that cells producing P-cadherin are located in different areas from the cells producing E-cadherin. This indicates that cells making E-cadherin bind preferentially to other cells producing E-cadherin, and similarly, cells making P-cadherin bind preferentially to other cells producing P-cadherin. Courtesy of M. Takeichi.*

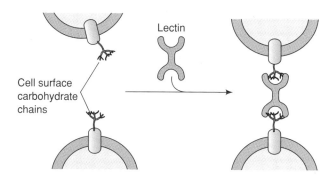

Figure 6-43 *Mechanism of Action of Lectins* *Lectins are proteins with multiple carbohydrate binding sites. This allows them to serve as molecular bridges that join cells together by binding to cell surface carbohydrate groups.*

ends of this branched carbohydrate chain, whereas individuals with blood type B have galactose instead. In blood type AB both N-acetylgalactosamine and galactose are present, and in type O blood cells these terminal sugars are missing entirely.

As is shown in Table 6-8, transfusions between individuals with differing blood types are hindered by the fact that people with blood types A, B, or O have antibodies in their bloodstream that recognize and bind to the terminal sugars in glycophorin. When an individual has antibodies against a particular type of glycophorin, they cannot accept a transfusion of blood cells containing such glycophorin molecules because the antibodies will bind to the glycophorin and cause the blood cells to coagulate. The following types of antibodies are asso-

Table 6-8   Molecular Basis of the ABO Blood Group Compatibilities

Blood Type of Individual	Sugar at Terminal End of Glycophorin Carbohydrate Chains		Serum Antibodies Directed Against Glycophorin Chains Terminating in:	
	N-Acetylgalactosamine	Galactose	N-Acetylgalactosamine	Galactose
A	Yes	No	No	Yes
B	No	Yes	Yes	No
AB	Yes	Yes	No	No
O	No	No	Yes	Yes

ciated with the standard blood types: (1) Individuals with type A blood have antibodies against glycophorin chains ending in galactose, which occur in type B and type AB blood. Hence they cannot be transfused with type B or type AB blood. (2) Individuals with type B blood have antibodies against glycophorin chains ending in N-acetylgalactosamine, which occur in type A and type AB blood. Hence they cannot be transfused with type A or type AB blood. (3) Individuals with blood type O have antibodies against glycophorin chains ending in both galactose (which occur in type B and type AB blood), and N-acetylgalactosamine (which occur in type A and type AB blood). Hence they cannot be transfused with type B, type AB, or type A blood.

Blood group compatibility is not the only recognition event in which the carbohydrate chains of glycophorin play a role. Like N-CAM, glycophorin contains large amounts of sialic acid. In the case of glycophorin, these sialic acid groups are selectively localized at the ends of oligosaccharide chains constructed from other types of carbohydrate. These terminal sialic acid groups have been implicated in the mechanism by which aging red cells are recognized and targeted for destruction. Red blood cells have an average life span of 3–4 months, after which they are destroyed in the spleen. The role of sialic acid in marking older cells for destruction has been studied by treating isolated red blood cells with the enzyme *sialidase (neuraminidase),* which catalyzes the removal of sialic acid groups. When cells treated in this manner are injected back into the animals from which they were originally obtained, they are rapidly destroyed by the host's spleen cells. In contrast, normal red cells that have not had their sialic acid groups removed survive for many weeks after being injected. Because removal of sialic acid exposes underlying galactose residues in the carbohydrate chains of glycophorin, it has been proposed that recognition of these newly exposed galactose residues by spleen cells causes the red cells to be targeted for destruction. Since the sialic acid content of the red cell membrane decreases as a cell becomes older, a model can be envisioned in which the loss of sialic acid groups eventually exposes galactose residues that permit an

aging red cell to be recognized and destroyed by the spleen.

CELL JUNCTIONS

Although cell adhesion molecules play an important role in allowing cells to recognize and adhere to one another, the establishment of long-term connections between cells usually requires complex structures known as **cell junctions.** Animal cells produce three types of junctions called *tight junctions, plaque-bearing junctions,* and *gap junctions* (Figure 6-44). Each type of junction is designed to perform a particular function: sealing (tight junctions), adhesion (plaque-bearing junctions), or communication (gap junctions). Because cell junctions involve an intimate association between the plasma membranes of adjacent cells, the presence of the cell wall around plant cells prevents them from forming such structures. However, we will see later in the chapter that the cell wall and its associated plasmodesmata carry out comparable functions for plants.

Tight Junctions Create a Permeability Barrier across a Layer of Cells

In cell layers that line the inner or outer surfaces of organs or body cavities, the cells are often joined together by physical connections called **tight junctions.** The properties of tight junctions have been investigated experimentally by incubating tissues in the presence of electron-opaque tracer molecules, and then using electron microscopy to observe the movement of the tracer through the extracellular space. Under such conditions, tracer molecules diffuse into the narrow spaces between adjacent cells until they encounter a tight junction, which blocks further movement (Figure 6-45). In other words, tight junctions prevent substances in the extracellular fluid from passing from one side of a cell layer to the other. The ability of tight junctions to restrict the movement of molecules across cell layers is especially important in organs like the bladder,

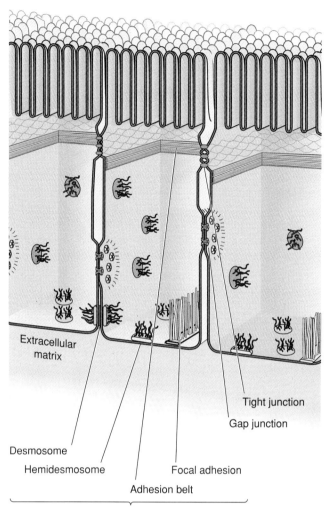

Tight junction

Gap junction

Desmosome

Hemidesmosome

Focal adhesion

Adhesion belt

Plaque-bearing junctions

Extracellular matrix

Figure 6-44 *Major Types of Cell Junctions in Animal Cells* *Tight junctions create an impermeable seal, gap junctions mediate cell-cell communication, and plaque-bearing junctions are specialized for cell-cell and cell-matrix adhesion. Adhesion belts, desmosomes, hemidesmosomes, and focal adhesions are all examples of plaque-bearing junctions.*

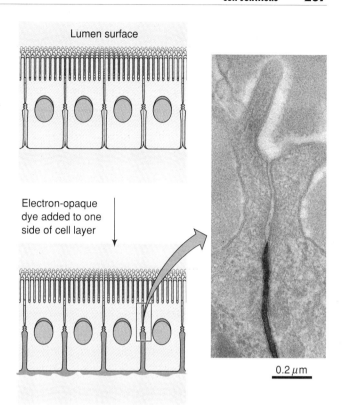

Lumen surface

Electron-opaque dye added to one side of cell layer

0.2 μm

Figure 6-45 *Experimental Evidence Showing That Tight Junctions Create a Permeability Barrier* *When an electron-opaque tracer is added to the extracellular space on one side of an epithelial cell layer, it penetrates into the space between adjacent cells only up to the point where a tight junction is present. Micrograph courtesy of D. S. Friend.*

where seepage of urine back into the body tissues must be prevented, and the intestinal tract, where the passage of ingested materials into body fluids must be carefully regulated. The only way for molecules to move from one side of these cell layers to the other is by passing through the cells themselves, a process that is precisely controlled by plasma membrane transport proteins.

In addition to creating a permeability barrier, tight junctions help to keep the plasma membrane of epithelial cells *polarized*—that is, organized into discrete functional domains at opposite ends of the cell. For example, in the epithelial cells lining the intestinal tract (see Figure 5-44), the plasma membrane facing the in-

testinal lumen (the *apical* cell surface) contains different transport proteins than does the plasma membrane located on the opposite side of the cell (the *basolateral* surface). Since the lipid bilayer is fluid, what prevents the transport proteins from becoming randomly distributed throughout the membrane? The role played by tight junctions in preventing the indiscriminate mixing of membrane proteins has been investigated by exposing epithelial cell layers to solutions lacking Ca^{2+}, a treatment that disrupts tight junctions. Following the removal of Ca^{2+}, membrane proteins that had previously been localized at one end of the cell become randomly distributed throughout the plasma membrane. Hence tight junctions appear to hinder the diffusion of plasma membrane proteins from one side of the junction to the other, thereby maintaining membrane polarity.

In thin-section electron micrographs, tight junctions appear as regions in which the space between adjacent cells has been completely obliterated; freeze-fracture electron micrographs, on the other hand, reveal an interconnected meshwork of raised ridges (Figure 6-46). At higher magnification, the ridges appear to be constructed from a series of repeating particles several nanometers in diameter. Based on such information, it has been postulated that a ridge of particles from the plasma membrane of one cell is joined to a ridge of particles in an adjacent membrane, fusing the two cells together. The ability of tight junctions to block the movement of small molecules through the intercellular space is directly related to the number of ridges present. In cell layers where the tight junctions are composed of a relatively small number of ridges, the seal is somewhat leaky to small molecules. When more ridges are present, the seal is effective enough to prevent the passage of substances as small as ions.

Plaque-Bearing Junctions Stabilize Cells against Mechanical Stress

Many tissues and organs are exposed to mechanical forces that subject the constituent cells to considerable stretching and distortion. To maintain proper tissue organization, such cells must be held together by junctions that exhibit significant mechanical strength. Junctions in this category share an important feature in common; that is, they are all connected to the cytoskeleton through dense fibrous structures, called *plaques,* which are associated with the inner surface of the plasma membrane. By connecting the cytoskeleton of one cell to that of its neighbors, such **plaque-bearing junctions** establish an interconnected cytoskeletal network that helps to maintain tissue integrity. Plaque-bearing junctions can be divided into two categories based on whether they are attached to the cytoskeleton through intermediate filaments (page 619) or actin filaments (page 582).

Plaque-Bearing Junctions Associated with Intermediate Filaments Are Called Desmosomes

The **desmosome** is the most commonly encountered type of cell junction that attaches to the cytoskeleton through intermediate filaments. Desmosomes are found

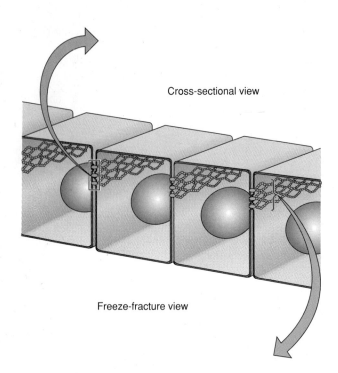

Cross-sectional view

Freeze-fracture view

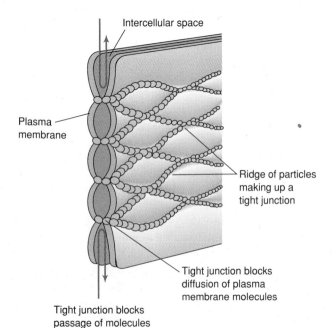

Intercellular space

Plasma membrane

Ridge of particles making up a tight junction

Tight junction blocks diffusion of plasma membrane molecules

Tight junction blocks passage of molecules in extracellular fluid

Figure 6-46 *Structural Organization of Tight Junctions* (Top) *Appearance of tight junctions in thin-section and freeze-fracture electron micrographs. The small arrows in the thin-section electron micrograph indicate points of membrane fusion that occur in the region of a tight junction. The freeze-fracture micrograph provides a view of the tight junction through the plane of the plasma membrane. The interconnected network of ridges (arrow) consists of rows of membrane protein particles 3–4 nm in diameter. (Bottom) The schematic model shows how tight junctions are constructed from rows of protein particles located in the plasma membranes of adjacent cells. The particles of neighboring cells bind tightly to each other, fusing the cells together and creating a barrier that prevents the passage of extracellular molecules through the narrow spaces between cells. Tight junction particles also block plasma membrane proteins from diffusing laterally through the lipid bilayer. Thin-section micrograph courtesy of R. S. Decker, freeze fracture micrograph courtesy of L. A. Staehelin.*

in most of the cell layers that cover or line the organs of multicellular animals, and are especially frequent in tissues where considerable mechanical stress is encountered, such as the skin and intestines. A typical desmosome occupies a circular area of plasma membrane measuring a few hundred nanometers in diameter. In the region where a desmosome joins two cells together, dense plaques of fibrous material can be seen in the cytoplasm just beneath the plasma membranes of the two adjoining cells (Figure 6-47). Small fibers measuring about 10 nm in diameter radiate from these plaques into the underlying cytoplasm. Cytoskeletal fibers of this size are collectively known as **intermediate filaments.** The protein makeup of intermediate filaments varies significantly among different cell types (page 619); intermediate filaments associated with the desmosomes of epithelial cells are constructed from the protein keratin, and are referred to as either *tonofilaments* or *keratin filaments.*

In the region where two cells are connected by a desmosome, the plasma membranes of the adjoining cells lie parallel to one another separated by a space of about 30 nm. This space is traversed by thin filaments that react with cytochemical stains for carbohydrates and that can be broken down by brief treatment with proteases. Breakdown of these filaments causes cells that had been held together by desmosomes to separate from each other. Taken together, the preceding

observations suggest that the thin filaments are membrane glycoproteins that link the two cells together in the region of the desmosome. The study of isolated plasma membranes enriched in desmosomes has led to the identification of several proteins associated with the desmosome. Among these components, the proteins *plakoglobin* and *desmoplakin* have been localized to the plaque, while *desmoglein* and *desmocollin* are membrane-associated glycoproteins that are thought to interact with the plaque at the inner membrane surface and to mediate cell-cell adhesion at the outer membrane surface (see Figure 6-47).

Besides joining cells to one another, junctions attached to intermediate filaments are also involved in connecting epithelial cells to the basal lamina. In such cases, the plaque and associated intermediate filaments occur in a single cell rather than two adjoining cells (Figure 6-48). This type of junction is therefore called a **hemidesmosome** ("half-desmosome").

Plaque-Bearing Junctions Associated with Actin Filaments Are Called Adherens Junctions

Cell junctions that are connected to the cytoskeleton through actin filaments are termed **adherens junctions.** In epithelial cells, adherens junctions typically

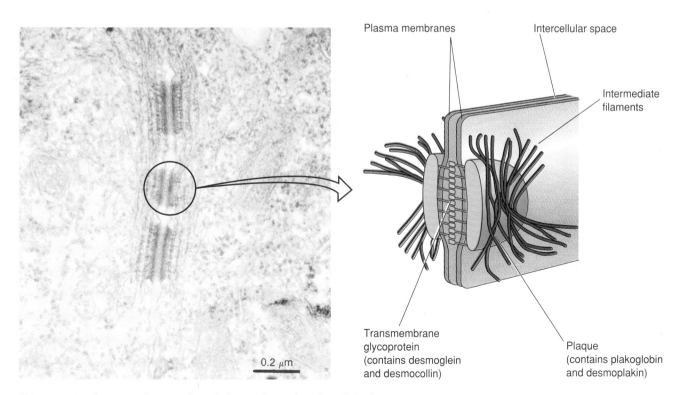

Figure 6-47 Electron Micrograph and Three-Dimensional Model of Desmosomes *In the electron micrograph, three desmosomes exhibiting dense plaques in the cytoplasm directly beneath the plasma membrane are clearly visible. The plaques are anchored to intermediate filaments that extend into the cytoplasm. Micrograph courtesy of R. S. Decker.*

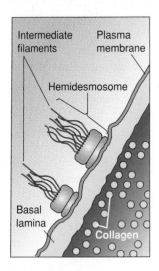
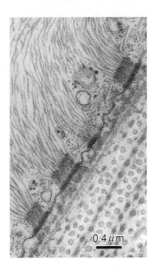

Figure 6-48 Diagram and Electron Micrograph Showing Several Hemidesmosomes *As in a desmosome, each hemidesmosome contains a plaque that is anchored to intermediate filaments extending into the cytoplasm. The external surface of the hemidesmosome abuts directly on the basal lamina. Micrograph courtesy of D. E. Kelly.*

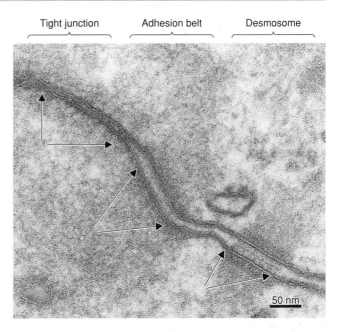

Figure 6-49 Electron Micrograph Showing Two Epithelial Cells Joined by a Tight Junction, an Adhesion Belt, and a Desmosome *In the region of the tight junction, the space between the two cells is obliterated. The intercellular space at the adhesion belt (a type of adherens junction) is about 25 nm, and expands to 30 nm at the desmosome. Adhesion belts and desmosomes both link the plasma membrane to elements of the cytoskeleton. Courtesy of R. L. Roberts and R. G. Kessel.*

form extensive zones called **adhesion belts** that completely encompass entire cells. Smaller adherens junctions also occur, primarily in nonepithelial cells. The electron microscopic appearance of an adherens junction differs in several ways from that of a desmosome (Figure 6-49). The space between the adjacent membranes is reduced from 30 nm to 20–25 nm, the underlying plaques are not as prominent, and the bundles of fibers that anchor the junction to the cytoskeleton consist of 6-nm diameter actin filaments rather than 10-nm diameter intermediate filaments. Since actin filaments are involved in contractile movements (see Chapter 13), adherens junctions may help to mediate coordinated movements and shape changes among adjacent cells.

Adherens junctions are constructed from a set of components that differ from those found in desmosomes (Figure 6-50). The plaques of adherens junctions contain the protein *vinculin* (page 232), a molecule that has been implicated in binding the adherens junction to actin filaments. The plaque is attached to the overlying plasma membrane by interactions involving transmembrane glycoproteins, which in turn mediate adhesion between the two adjacent cells. At least some of the transmembrane glycoproteins that hold adherens junctions together are members of the cadherin family, suggesting that cell-cell adhesion initiated by cadherins can ultimately lead to the formation of adherens junctions. Because cadherins are Ca^{2+}-dependent, removing Ca^{2+} from the extracellular medium causes adherens junctions to split apart.

In addition to their role in cell-cell adhesion, adherens junctions can also attach cells to the extracellular matrix. An adherens junctions of this type is called a

focal adhesion (or *adhesion plaque*). Focal adhesions bind to extracellular matrix components such as fibronectin rather than to the surface of another cell. Attachment of a focal adhesion to the extracellular matrix is mediated by the fibronectin receptor, a transmembrane glycoprotein that binds to fibronectin outside the cell and to a plaque associated with actin filaments inside the cell. The main components of the plaque are vinculin and an additional protein, *talin* (page 231), which is not present in the plaques of other types of adherens junctions.

Gap Junctions Permit Small Molecules to Move from One Cell to Another

Earlier in the chapter, we discussed the role of extracellular signaling molecules in cell-cell communication. In addition to this extracellular signaling pathway, some cells employ an alternative communication system that involves the direct movement of molecules from one cell to another. This type of communication is made possible by **gap junctions,** which permit small molecules to move directly from cell to cell without passing through the extracellular space. The idea that such channels permit the direct movement of molecules between adjoining cells owes much to the pioneering work of Werner Loewenstein and his colleagues, who employed microelectrodes to monitor the flow of electric current in insect salivary glands. When a small voltage was applied to electrodes placed

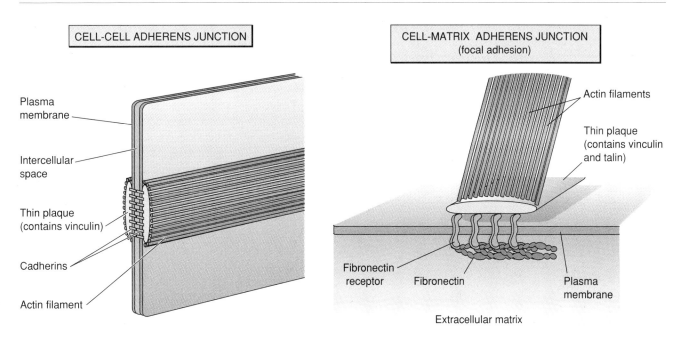

CELL-CELL ADHERENS JUNCTION

Plasma membrane

Intercellular space

Thin plaque (contains vinculin)

Cadherins

Actin filament

CELL-MATRIX ADHERENS JUNCTION (focal adhesion)

Actin filaments

Thin plaque (contains vinculin and talin)

Fibronectin receptor

Fibronectin

Plasma membrane

Extracellular matrix

Figure 6-50 *Models of the Two Major Types of Adherens Junctions* (Left) *Cell-cell adherens junctions are utilized for promoting long-term adhesion between cells. The plaques of this type of junction are bound to cytoplasmic actin filaments by the protein vinculin.* (Right) *Cell-matrix adherens junctions (focal adhesions) attach cells to fibronectin in the extracellular matrix. The plaques are attached to cytoplasmic actin filaments by the proteins vinculin and talin.*

in neighboring cells, the current that flowed between the cells was found to be several orders of magnitude higher than the current that could be measured by placing one of the electrodes in the external medium. In other words, current flowed readily from cell to cell, but not from cells to the extracellular fluid. Since current in living cells is carried by small ions, it was concluded that cells contain channels that permit ions and small molecules to pass directly from the cytoplasm of one cell to the cytoplasm of an adjoining cell without first appearing in the extracellular space.

To determine the size of these channels, cells were injected with fluorescent molecules of differing molecular weights. By using an ultraviolet microscope to monitor the movement of fluorescent molecules from cell to cell (Figure 6-51), it was determined that small molecules up to about 1000 daltons in molecular weight can move directly from cell to cell. Although initially some confusion existed as to which type of intercellular junction is responsible for this phenomenon, it eventually became clear that the gap junction is the only structure whose presence always correlates with the existence of direct molecular communication between cells.

In thin-section electron micrographs, gap junctions appear as regions in which the plasma membranes of two adjacent cells are aligned in parallel and separated by a tiny gap of about 3 nm. When the membrane surfaces in this region are examined by negative staining, they are found to be covered by hundreds of cylindrical structures termed **connexons**

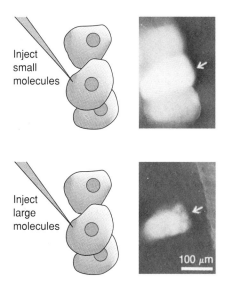

Inject small molecules

Inject large molecules

100 μm

Figure 6-51 *Permeability of Gap Junctions Studied by Injecting Fluorescent Molecules of Different Sizes* *In this experiment, fluorescent molecules were injected into the cells marked with arrows and movement of these molecules was then monitored by fluorescence light microscopy.* (Top) *Small fluorescent molecules (1158 daltons) are seen to migrate into adjacent cells.* (Bottom) *Larger fluorescent molecules (1926 daltons) do not pass into neighboring cells. Courtesy of W. R. Loewenstein.*

(Figure 6-52). Protein purification studies on isolated gap junctions have revealed that connexons are constructed from a single type of transmembrane protein called **connexin;** each connexon is formed by a cluster of six connexin molecules surrounding a central aqueous channel. Connexons located in the plasma membrane of one cell attach to connexons in the plasma membrane of the adjoining cell, forming channels across the 3-nm intercellular gap through which water-soluble molecules can pass.

An elegant study carried out by Loewenstein and his colleagues has revealed that gap junctions can be opened and closed. In this experiment, the effects of calcium ions on cell-cell communication were investigated by monitoring the migration of fluorescent molecules between cells microinjected with solutions containing varying concentrations of Ca^{2+}. The results showed that when the Ca^{2+} concentration is increased in an individual cell, gap junctions close and the movement of fluorescent molecules into that cell is inhibited (Figure 6-53). The closure of gap junctions is not an all-or-none event, for varying degrees of permeability are observed when the Ca^{2+} concentration is gradually increased from 10^{-7} M (maximum permeability) to 10^{-5} M (minimum permeability). Subsequent investigations have revealed that the permeability of gap junctions can also be altered by changes in membrane potential, intracellular pH, and cyclic AMP levels.

Gap junctions occur in almost every cell type found in vertebrates and invertebrates. They are especially abundant in tissues where extremely rapid communication between cells is required, such as nerve and muscle. In heart tissue they facilitate the flow of electric current that causes the heart to beat, whereas in the brain they are concentrated in the cerebellum, which is involved in coordinating rapid muscular activities. Such observations suggest that gap junctions are employed to speed up communication in situations where chemical transmission across a synapse is too slow.

In addition to their role in electrical coupling, gap junctions may have developmental and metabolic functions. A developmental role is suggested by the discovery that toad or mouse embryos develop abnormally if gap junctions are blocked experimentally using antibodies directed against connexin. It has also been suggested that gap junctions permit metabolic cooperation be-

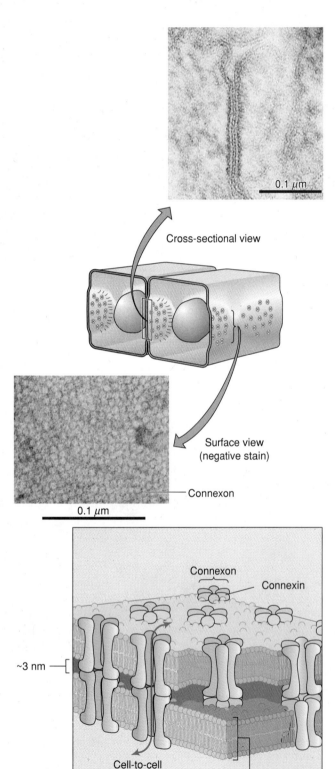

Cross-sectional view

Surface view
(negative stain)

Connexon

0.1 μm

Connexon

Connexin

~3 nm

Cell-to-cell channel

Plasma membrane

Figure 6-52 *Structural Organization of Gap Junctions*
(Top) Electron micrographs showing gap junctions in cross-sectional view and in surface view visualized by negative staining. The negatively stained specimen reveals the numerous circular pores that make up a gap junction. (Bottom) The schematic model shows how two plasma membranes are joined together by numerous cylindrical pores (connexons) in the region of a gap junction. Each connexon is formed by a cluster of six connexin molecules surrounding a central channel. When a connexon from one membrane attaches to a connexon from the adjoining membrane, an aqueous channel is created that permits the diffusion of water-soluble molecules between the two cells. Thin-section micrograph courtesy of R. S. Decker, negatively stained micrograph courtesy of E. L. Benedetti.

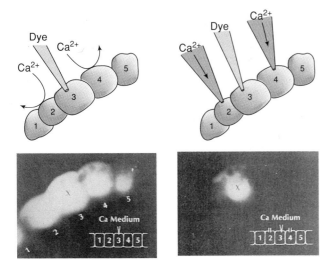

Figure 6-53 *Effects of Ca²⁺ on the Permeability of Gap Junctions* (Left) *A fluorescent dye injected into one cell (number 3) is seen to diffuse into adjacent cells. Although the experiment was carried out in a Ca²⁺-containing medium, the cells are impermeable to Ca²⁺. (Right) In the same Ca²⁺-containing medium, holes are punched in cells 2 and 4 to allow the Ca²⁺-containing medium to enter, thereby raising the intracellular Ca²⁺ concentration. Under these conditions, fluorescent dye injected into cell 3 does not diffuse into the adjacent cells, indicating that a high intracellular Ca²⁺ concentration blocks the movement of molecules through gap junctions. Courtesy of W. R. Loewenstein.*

tween neighboring cells. If one cell develops a defect that prevents it from carrying out a particular metabolic step, the substrate could pass into an adjacent cell for metabolic processing and then return to the original cell. Although the extent of this kind of metabolic cooperation in intact organisms is uncertain, studies involving isolated cells with differing metabolic capacities have shown that metabolic cooperation through gap junctions does occur in culture.

THE PLANT CELL SURFACE

Although the basic principles that have been described in this chapter are applicable to a wide variety of cell types, most of the examples have focused on the behavior of animal cells. The surfaces of plant and bacterial cells exhibit many of these same properties, but they also exhibit a few unique features that are not shared by the cells of animals. In the remainder of the chapter we will discuss some of these unique features of plant and bacterial cells.

Plant Cell Walls Provide a Supporting Framework for Intact Plants

Plants differ from animals in that they have no bones or related skeletal structures, and yet plants still ex-

hibit remarkable strength. It has been estimated, for example, that one large tree could support the combined weight of more than a thousand elephants. This enormous strength is imparted to plants not by bones, but by plant **cell walls**. All plant cells, with the exception of sperm and some eggs, are surrounded by a relatively rigid cell wall (Figure 6-54). In addition to providing mechanical support and strength for the plant as a whole, the cell wall protects individual cells from osmotic rupture and mechanical injury. The rigid cell wall also plays a central role in determining the characteristic shapes of plant cells; when the wall is removed, plant cells lose their normal shapes and become spherical. The presence of the cell wall is one of the major factors that causes the properties of plants to differ so much from those of animals; the rigidity of the wall makes cell movements virtually impossible, preventing plants from developing the kind of neuromuscular system that defines many of the unique properties of animals.

Although cell walls were once viewed as relatively inert secretions of the cells they surround, more recent studies have revealed the wall to be a dynamic structure that carries out many activities. For example, enzymes associated with the cell wall can degrade extracellular nutrients, thereby generating smaller compounds that are capable of passing through the plasma membrane and into the cell. The cell wall also acts as a permeability barrier and plays a role in certain types of secretory and metabolic events.

The Plant Cell Wall Is Constructed from Cellulose, Hemicellulose, Pectin, Lignin, and Glycoproteins

Cell walls are constructed from a complex group of macromolecules that are secreted by and surround the cells which they encase. In this sense the cell wall is fundamentally analogous to the extracellular matrix of animal cells. But instead of employing protein molecules like those found in the matrix of animal tissues, plant cell walls are constructed largely from polysaccharides. The predominant polysaccharide of the cell wall is **cellulose,** a molecule that imparts strength and rigidity to the cell wall just as collagen imparts similar characteristics to the extracellular matrix. As we saw in Chapter 1, cellulose is a repeating polymer constructed from thousands of glucose molecules linked together by $\beta(1 \rightarrow 4)$ bonds (see Figure 1-13). In cell walls, individual cellulose molecules are bundled together in groups of 50 or 60 cellulose molecules that line up in parallel and adhere to one another, forming long, rigid *microfibrils* measuring 5–15 nm in diameter and several micrometers in length. Cellulose microfibrils are often twisted together in a ropelike fashion to generate even larger structures, called *macrofibrils,* which are as strong as an equivalent-sized piece of steel.

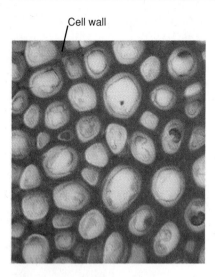

Cell wall

Figure 6-54 Light Micrograph of Woody Plant Tissue Showing Thickened Cell Walls *This tissue type, called sclerenchyma, is specialized for its enormous mechanical strength. The cells normally die at maturity, leaving behind their massive cell walls (red area) to provide structural support for the plant.*

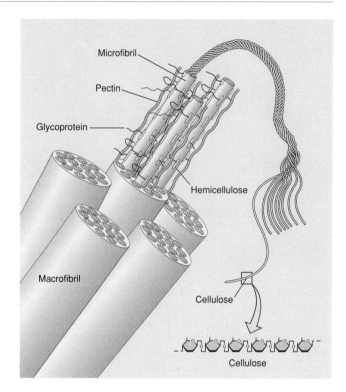

Figure 6-55 Relationship between Cellulose and Several Other Components of the Plant Cell Wall *Cellulose, hemicellulose, and glycoproteins are linked together to form a rigid interconnected network. Pectin molecules form a matrix in which the cellulose microfibrils are embedded. Lignins (not shown) occur mainly in woody tissues, where they are localized between the cellulose fibrils.*

Although cellulose is a crucial component, the cell wall is more than just a collection of cellulose fibrils. Among the other components that contribute to the structural integrity of the cell wall are several classes of macromolecules that interact with cellulose to form a complex, interconnected network (Figure 6-55). These macromolecules fall into four major categories: hemicelluloses, pectins, lignins, and glycoproteins.

1. Hemicelluloses are a heterogeneous group of carbohydrate polymers constructed from various five- and six-carbon sugars, including xylose, arabinose, mannose, and galactose. The hemicellulose xylan, which utilizes the pentose xylose as its main building block, accounts for as much as 50 percent of the cell wall in woody tissues. Hemicellulose molecules bind to the surface of cellulose microfibrils and to each other, creating a coating that helps to bond these fibrils together into a rigid interconnected network of cellulose and hemicellulose.

2. Pectins are polymers of the carbohydrate galacturonic acid and its derivatives. Like the glycosaminoglycans of animal cells, they readily form hydrated gels. This property of pectin molecules is responsible for the gelation that occurs during the process of making fruit jams and jellies. Pectin molecules are involved in binding adjacent cell walls together, and in forming the matrix in which cellulose microfibrils are embedded. Recent studies have shown that tomato cells grown in the presence of an inhibitor of cellulose synthesis produce intact cell walls that consist almost entirely of a crosslinked network of

pectin molecules. Hence cell walls appear to contain a pectin network that is largely independent of the cellulose-hemicellulose network.

3. Lignins are a group of polymerized aromatic alcohols that occur mainly in woody tissues, where they form a crosslinked network that contributes to the hardening of the cell wall. Lignin molecules are localized predominantly between the cellulose fibrils, where they function to resist compression forces. Lignin accounts for as much as 25 percent of the dry weight of woody plants, making it the second most abundant organic compound on earth (cellulose is the most abundant).

4. Glycoproteins are also important constituents of the plant cell wall, accounting for as much as 10 percent of the total mass. Prominent among the wall glycoproteins are a group of related glycoproteins called **extensins.** These molecules resemble collagens, the extracellular proteins of animal supporting tissues, in their high content of the unusual amino acid hydroxyproline. Extensins and related glycoproteins are thought to form crosslinked networks with each other as well as with cellulose microfibrils, generating a reinforced protein-polysaccharide complex.

In addition to the preceding components, a small percent of the total mass of the cell wall is accounted for by lipids, including waxes and other complex polymers. These hydrophobic molecules tend to be found on the external surface of the wall, where they form a *cuticle* layer that protects the cell against injury and desiccation. Minerals such as calcium and potassium also occur in plant cell walls, predominantly in the form of inorganic salts.

The Plant Cell Wall Is Synthesized in Several Discrete Stages

One of the difficulties encountered in trying to describe the organization of the cell wall occurs because the walls of different cell types differ in chemical composition, structural organization, and thickness. In certain plants, such as cotton, the cell wall is constructed almost entirely of cellulose fibrils, whereas in woody plants a typical cell wall contains only about 50 percent cellulose and in some plant seedlings the wall contains less than 1 percent cellulose. Likewise the cell wall may vary from as little as a hundred to more than several thousand nanometers in thickness. This diversity in cell wall structure and composition can be understood at least in part by examining the way in which cell walls are synthesized.

The plant cell wall is secreted from the cell in a stepwise fashion, creating a series of layers in which the first layer to be synthesized ends up farthest away from the plasma membrane (Figure 6-56). The first component to be formed, called the **middle lamella,** consists primarily of pectins. The middle lamella is shared by neighboring cell walls and functions to hold adjacent cells together. The next zone to be produced, called the **primary cell wall,** forms during the period when cell growth is still occurring. Primary walls measure about 100–200 nm in thickness, which is not much thicker than the basal lamina of animal cells. The primary wall consists of a loosely organized network of cellulose microfibrils associated with hemicelluloses, pectins, and glycoproteins (Figure 6-57, *top left*). Pectins are especially important in imparting flexibility to the primary cell wall, which makes it possible for the wall to expand during cell growth. The cellulose microfibrils are generated by cellulose-synthesizing enzyme complexes localized within the plasma membrane. These enzyme complexes, called **rosettes,** add glucose molecules to the growing microfibrils. Because the microfibrils are anchored to other wall components, the rosettes must move in the plane of the membrane as the cellulose microfibrils are lengthened (see Figure 6-57, *bottom left*).

The loosely textured organization of the primary cell wall creates a relatively thin flexible structure that is capable of expansion during cell growth. In some plant cells, development of the cell wall does not pro-

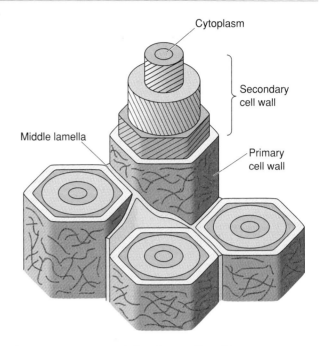

Figure 6-56 *Layers of the Plant Cell Wall* *Some plant cells only have a primary wall, while others have a primary wall as well as an extensive secondary wall.*

ceed beyond this point. However, cells that have stopped growing often add a thicker and more rigid set of layers to the wall collectively referred to as the **secondary cell wall.** The components of the multilayered secondary wall are superimposed on the *inner* surface of the primary wall after cell growth has ceased. Cellulose and lignins predominate in the secondary wall, and pectin is largely absent, making this structure significantly stronger, harder, and more rigid than the primary wall. Each layer of the secondary wall consists of densely packed bundles of cellulose microfibrils arranged in parallel and oriented so that they lie at an angle to the microfibrils of adjacent layers (see Figure 6-57, *right*). This organization imparts great mechanical strength and rigidity to the secondary cell wall and is ultimately responsible for the characteristic strength of woody tissues. Differences in the thickness and organization of the secondary wall are primarily responsible for the variations in strength and thickness of the cell wall observed between cell types.

The orderly arrangement of cellulose microfibrils in secondary walls appears to arise from an interaction between cellulose-synthesizing rosettes and microtubules that lie directly beneath the plasma membrane. During the formation of each layer of the secondary wall, the microtubules located beneath the plasma membrane are oriented in the same direction as the newly forming cellulose microfibrils. At this stage the cellulose-synthesizing rosettes are aggregated into large arrays containing dozens of rosettes (Figure 6-58). The underlying microtubules are thought to guide the movement of these

PRIMARY CELL WALL

SECONDARY CELL WALL

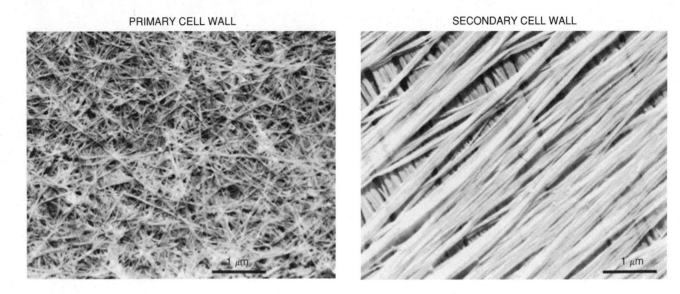

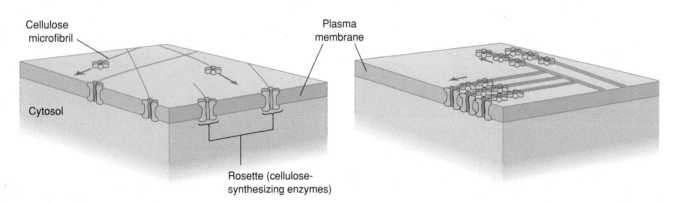

Figure 6-57 Cellulose Fibrils of Primary and Secondary Plant Cell Walls   (Top left) *Electron micrograph of a primary cell wall showing the loosely organized cellulose microfibrils.* (Top right) *Electron micrograph of a secondary wall showing densely packed cellulose macrofibrils oriented in parallel.* (Bottom) *Diagrammatic representation showing how cellulose fibrils are synthesized by rosettes. Each rosette is a cluster of cellulose-synthesizing enzymes embedded in the plasma membrane. As a rosette synthesizes a bundle of cellulose molecules (cellulose microfibril), the rosette moves through the plasma membrane in the direction indicated by the arrows. In primary walls, the cellulose microfibrils are synthesized as a loosely organized network. In secondary walls, rosettes form dense aggregates that synthesize large numbers of microfibrils in parallel, generating cellulose macrofibrils. Micrographs courtesy of K. Mühlenthaler.*

rosette aggregates, which synthesize large parallel bundles of cellulose microfibrils as they move.

Plasmodesmata Permit Direct Cell-Cell Communication through the Plant Cell Wall

Although the cell wall provides a thick encasement for plant cells, this barrier does not seal the cell completely from its surroundings. Plant cell walls usually contain small openings, or **plasmodesmata,** through which adjacent cells maintain direct contact with one another. In

electron micrographs, plasmodesmata appear as narrow channels in the cell wall that are lined by plasma membrane and often traversed by a tubule of endoplasmic reticulum (Figure 6-59). Thus the plasma membrane, cytoplasm, and endoplasmic reticulum are all in continuity from one cell to the next. Plasmodesmata tend to be concentrated in special areas of the cell wall, called *pit fields,* where the primary wall is thinner than normal and the secondary wall is absent.

To investigate the possible role played by the plasmodesmata in cell-cell communication, microelectrodes have been employed to monitor the flow of electric

0.1 μm

Figure 6-58 Cellulose-Synthesizing Rosettes *This freeze-fracture electron micrograph of the plasma membrane of a green alga shows a large group of rosettes that are moving together as a unit, synthesizing dozens of cellulose microfibrils arranged in parallel. Courtesy of T. H. Giddings, Jr.*

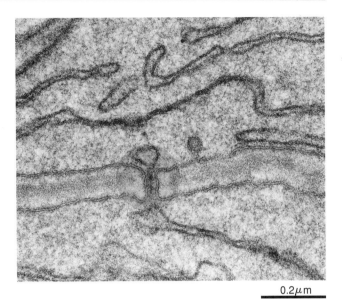
0.2 μm

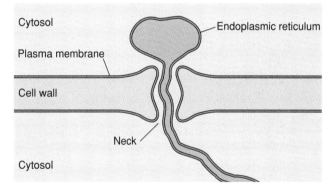

Figure 6-59 Electron Micrograph of a Plasmodesma Connecting Two Corn Plant Cells *The accompanying diagram illustrates how the plasma membrane, cytosol, and endoplasmic reticulum of the two adjacent cells are in continuity through the plasmodesma. Micrograph courtesy of H. H. Mollenhauer.*

current from cell to cell in plant tissues. As in the case of gap junctions between animal cells, such studies have revealed that electrical current passes between plant cells linked by plasmodesmata more readily than it does between cells that are not linked by plasmodesmata. Moreover, the magnitude of the current flow is directly related to the number of plasmodesmata present. This suggests that plasmodesmata play a role in cell-cell communication comparable to that played by gap junctions in animal cells, providing channels that permit ions and small molecules to pass directly between adjacent cells. Like gap junctions, the permeability of plasmodesmata is subject to regulation by the internal concentration of calcium ions. Although the mechanism of this regulation is not well understood, there is a growing consensus that the constricted *neck* regions that occur at the two opposite ends of the channel are involved.

Chemical and Electrical Signaling Occur in Plants as Well as Animals

The presence of plasmodesmata facilitates communication between neighboring plant cells, but communication over longer distances raises some unique problems for plants. The thickness and extensive crosslinking of the cell wall cause it to slow down the passage of molecules larger

than several thousand daltons in molecular weight, so extracellular signaling molecules must be relatively small in order to gain easy access to the cell surface. It is therefore not surprising that the best characterized group of plant signaling molecules, namely the *plant growth substances* (page 527), have turned out to be relatively small molecules. Although preliminary evidence indicates that some of these signaling molecules may interact with cell surface receptors, the mechanisms involved have not been as well established as those for animal signaling molecules.

The role of electrical signaling in plants is also poorly understood, although it has been known for many years that changes in membrane potential resembling action potentials can be induced in ordinary plants by localized wounding or elevated temperature. As in animal cells, action potentials in plant cells involve a transient depolarization of the plasma membrane, although changes in permeability to Ca^{2+} and Cl^- (rather than Na^+ and K^+) underlie the action potentials of at least some plants. Recent studies involving there response of tomato plants to wounding suggest that action potentials transmitted from cell to cell through plasmodesmata may be a long-distance signaling device. These investigations revealed that localized heating of tomato leaves is followed by the propagation of action potentials away from the initial wounding site. As part of a defensive response, the tissues through which the action potentials have spread then begin to produce proteinase inhibitors. If the plant is treated with chemical compounds known to block the propagation of action potentials, the production of proteinase inhibitors is also prevented. Such observations suggest that changes in membrane potential may be long-range signaling devices used by plants to trigger biochemical responses.

THE BACTERIAL CELL SURFACE

The surface of prokaryotic cells is even more complex than that of animal or plant cells. In bacteria, where the organization of the cell surface has been extensively studied, the cell is enclosed by several discrete structures collectively referred to as the **cell envelope.** This envelope is comprised of the plasma membrane and several layers external to it, including a relatively rigid *cell wall.* The wall is a strong protective structure that makes bacteria exceptionally resistant to damage by mechanical and chemical forces. For example, most bacteria exist under conditions in which the concentration of

dissolved solutes in the external environment is low compared to that within the cell cytoplasm. Water therefore tends to flow into the cell by the process of osmosis (page 176), generating a massive intracellular pressure that would cause cells to burst if a strong cell wall were not present.

The strength and rigidity of the bacterial cell wall also makes it an important determinant of cell shape. Bacteria occur in a variety of shapes, but most can be placed into one of three basic categories: spherical bacteria called **cocci,** rod-shaped bacteria termed **bacilli,** and spirally shaped bacteria known as **spirilla** (Figure 6-60). When the cell wall is removed from a bacterial cell with one of these unique shapes, the isolated cell wall retains the shape characteristic of the original cell (Figure 6-61).

Bacterial cell walls are divided into two general categories based on their reaction to a staining procedure developed in the late nineteenth century by the Danish physician, Hans Christian Gram. In this technique, bacteria are stained with a basic dye such as crystal violet and are then extracted with alcohol. Cells that retain the purple dye after the alcohol extraction step are referred to as **Gram-positive,** while those that release the dye are referred to as **Gram-negative.** This difference in staining behavior reflects an underlying difference in the organization of the cell envelope in Gram-negative and Gram-positive bacteria.

Gram-Positive Bacteria Have a Thick Cell Wall Made of Murein and Teichoic Acids

The envelope of Gram-positive bacteria is characterized by the presence of a thick **cell wall** located

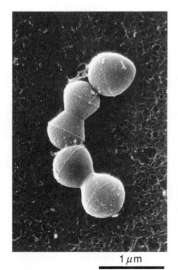
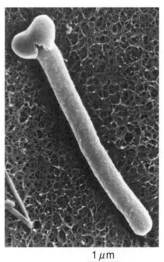
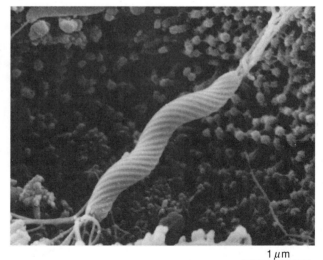

1 µm   1 µm   1 µm

Figure 6-60 *Scanning Electron Micrographs Showing the Principal Bacterial Cell Shapes* (Left) *Several cocci.* (Middle) *A bacillus (emerging from a spore coat).* (Right) *A spirillum. Courtesy of K. Amako* (left, middle) *and S. E. Erlandsen* (right).

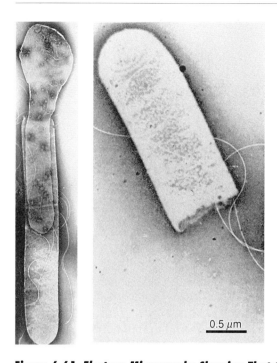

Figure 6-61 *Electron Micrographs Showing That the Bacterial Cell Wall Retains its Shape Upon Isolation* (Left) *A bacterial cell is partially released from the constraining cell wall, which retains its elongated shape.* (Right) *A portion of an isolated bacterial cell wall. Courtesy of D. Abram.*

immediately outside the plasma membrane (Figure 6-62). Chemical analyses carried out on the isolated walls of Gram-positive bacteria have revealed that the predominant molecular component is a polysaccharide-peptide complex termed **murein** (or **peptidoglycan**). The polysaccharide portion of murein consists of alternating residues of N-acetylglucosamine and N-acetylmuramic acid (a lactic acid derivative of glucosamine), with short peptide bridges connecting the N-acetylmuramic acid groups of adjacent chains to each other (Figure 6-63). These peptide bridges contain only a few kinds of amino acids, among which are several D-amino acids (page 20) and the unusual amino acid *diaminopimelic acid* (a close relative of lysine). Crosslinking of the polysaccharide chains by peptide bridges produces an intertwined three-dimensional network that imparts great strength to the cell wall.

Evidence that the murein backbone of the cell wall is responsible for determining bacterial cell shape has emerged from studies involving *lysozyme,* an enzyme that hydrolyzes the polysaccharide chains of murein. Bacteria treated with lysozyme lose their cell walls, creating wall-less cells called **protoplasts.** Bacterial cells that have had their murein walls destroyed by treatment with lysozyme become spheri-

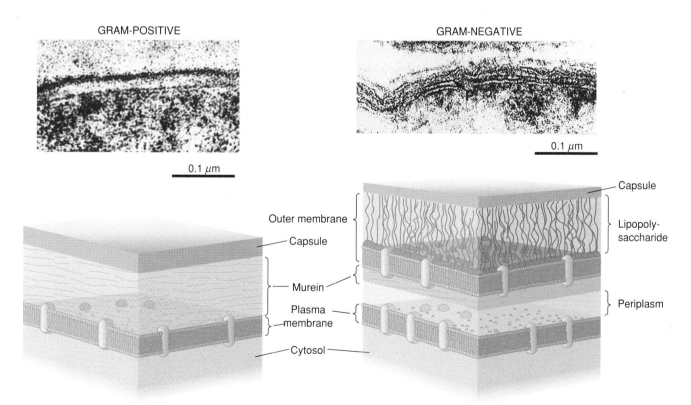

Figure 6-62 *Comparison of the Structure of Gram-Positive and Gram-Negative Bacterial Envelopes* *The envelopes of the two classes of bacteria differ in two main ways. First, the murein layer is thinner and less extensively crosslinked in Gram-negative bacteria. Second, Gram-negative bacteria possess an outer membrane that is located external to the murein wall and plasma membrane. Micrographs courtesy of J. W. Costerton.*

cal, regardless of what their original shape had been. Lysozyme is a normal constituent of tears, saliva, mucus, and egg whites, where it serves a protective function by destroying the cell walls of many invading bacteria.

Although murein typically accounts for the bulk of the cell wall in Gram-positive bacteria, other molecules are usually woven into its framework. Prominent among these are **teichoic acids,** which consist of glycerol or ribitol derivatives joined together to form long, negatively charged polymers. Such molecules, which are often covalently linked to the murein skeleton, have been implicated in the regulation of certain enzymes and in the binding of cations required for plasma membrane stability and function.

Gram-Negative Bacteria Have a Thin Murein Wall plus an Outer Membrane

The cell envelope of Gram-negative bacteria differs in two major ways from that of Gram-positive cells (see Figure 6-62). First, the murein layer is thinner and less extensively crosslinked, making it weaker than the murein wall of Gram-positive

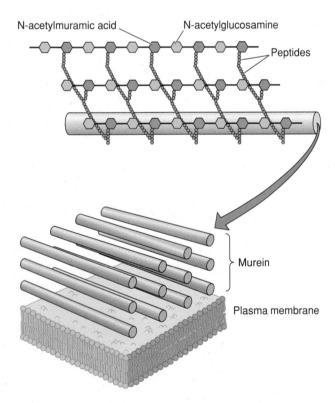

Figure 6-63 *Structure of Murein in the Bacterial Cell Wall*   *In the murein molecule, polysaccharide chains consisting of alternating residues of N-acetylmuramic acid and N-acetylglucosamine are connected by peptide bridges.*

cells. This difference underlies the differing behavior of Gram-positive and Gram-negative bacteria in the Gram-staining procedure. The thinner murein wall of Gram-negative bacteria permits removal of the Gram stain during the alcohol extraction step, while the thicker and stronger wall of Gram-positive bacteria prevents the dye from being extracted.

The second distinguishing feature of the envelope of Gram-negative bacteria is the presence of a unique structure called the **outer membrane.** This membrane, which is located external to the murein wall, has an unusual chemical makeup. The inner leaflet of its lipid bilayer contains typical membrane phospholipids, but the outer leaflet consists almost entirely of **lipopolysaccharides** rather than phospholipids. The lipopolysaccharide molecules consist of complex chains of sugar molecules, with fatty acid residues covalently attached at one end. Because sugar chains are hydrophilic and fatty acids are hydrophobic, lipopolysaccharides are amphipathic molecules that are inserted into the lipid bilayer with their fatty acids buried in the membrane interior and their sugar chains exposed at the outer membrane surface. This arrangement has a strong influence on the permeability properties of the outer membrane. Instead of being highly permeable to hydrophobic molecules, as is typical of cellular membranes, the outer membrane repels hydrophobic molecules because its surface is covered with hydrophilic sugar chains.

Lipopolysaccharides are of special medical interest because many disease symptoms associated with infection by Gram-negative bacteria are no more than host inflammatory responses to the presence of lipopolysaccharide. Cell wall lipopolysaccharides that trigger such responses are termed *endotoxins* to distinguish them from toxins that are secreted by bacteria. The ironic aspect of endotoxin behavior is that these substances are not inherently toxic; that is, they do not directly interfere with any normal activities occurring in the infected host. Yet when cells of the infected organism detect the presence of lipopolysaccharide endotoxins, they undergo what might aptly be termed a panic reaction, triggering a set of inflammatory responses that produce more damage than that directly caused by the bacteria.

In addition to lipids and lipopolysaccharides, the outer membrane contains several dozen kinds of protein molecules. The functions of these proteins have been investigated by studying the properties of mutant strains of bacteria in which particular proteins are defective or missing. Such studies have revealed that some outer membrane proteins are structural proteins that help to maintain membrane integrity, while others function as cell surface receptors, anchoring sites for external structures, or transporters for the uptake and export of specific molecules. Especially important in the latter regard are proteins called **porins,** which function as hydrophilic membrane pores for the passive diffusion of water soluble molecules smaller than 600–700 daltons.

Synthesis of the Bacterial Cell Wall Is Blocked by Penicillin

The mechanism by which bacterial cell walls are synthesized presents an interesting spatial problem because the bacterial wall, like the plant cell wall, is located outside the plasma membrane and so is separated from the cell interior by a permeability barrier. Current evidence suggests that the synthesis of the sugars and peptides involved in murein formation occurs in the cytoplasm, but these building blocks are joined together to form long chains by enzymes present in the plasma membrane. The resulting chains are then crosslinked into a murein network by a *transpeptidation* reaction that is catalyzed by an enzyme located at the outer surface of the plasma membrane (Figure 6-64). Thus as in plants, the bacterial plasma membrane plays a key role in the formation of the cell wall.

The synthesis of the murein wall is of special interest because the enzyme that catalyzes the transpeptidation reaction is inhibited by the antibacterial drug **penicillin.** Since no cells in nature other than bacteria have a murein wall, the effects of penicillin are directed solely against bacteria. Inhibition of the transpeptidation reaction by penicillin hinders cell wall formation and results in the production of fragile cells that tend to burst and die.

The Periplasmic Space Is a Unique Compartment outside the Plasma Membrane of Gram-Negative Bacteria

In Gram-negative bacteria, the existence of both a plasma membrane and an outer membrane creates a unique compartment between the two membranes referred to as the **periplasmic space.** The volume of this compartment often represents 25 percent or more of the total cell volume, so its presence is far from trivial. Proteins that normally reside in the periplasmic space have been identified by disrupting the murein wall and outer membrane, thereby releasing periplasmic proteins into the external medium. Among the periplasmic proteins that have been characterized are: (1) detoxifying enzymes for the inactivation of toxic substances; (2) binding proteins for specific sugars, amino acids, vitamins, and ions; and (3) hydrolytic enzymes such as phosphatases, nucleases, and proteases, which promote the breakdown of large molecules that cannot other-

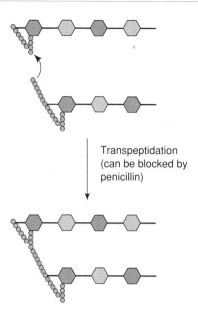

Figure 6-64 *The Transpeptidation Reaction* *This reaction, which is catalyzed by an enzyme located at the outer surface of the plasma membrane, creates the crosslinked murein network of the bacterial cell wall. Penicillin is a competitive inhibitor of the transpeptidation reaction.*

wise cross the plasma membrane. Thus periplasmic proteins both protect the cell from toxic compounds and facilitate the uptake of nutrients.

Gram-positive bacteria lack a well-defined periplasmic space because they have no outer membrane to serve as an exterior boundary. Instead, many of the proteins found in the periplasmic space of Gram-negative cells are either loosely bound to the cell surface or are secreted into the medium surrounding Gram-positive cells. Hence a unique group of degradative and detoxifying enzymes are associated with the cell surface of both Gram-positive and Gram-negative bacteria, acting upon molecules that are too large to pass through the plasma membrane and into the cell. These degradative and detoxifying functions are similar to those performed by lysosomal enzymes in eukaryotes (page 287), suggesting that lysosomes may owe their evolutionary origins to infoldings of the plasma membrane containing trapped periplasmic enzymes.

Capsules Are Often Produced by Gram-Positive and Gram-Negative Bacteria

Under appropriate environmental conditions, many Gram-negative and Gram-positive bacteria secrete a transparent zone of gelatinous material referred to as a **capsule** (see Figure 6-62). The capsule is usually tightly bound to the outermost surface of the cell envelope; when it is more loosely associated with the cell surface,

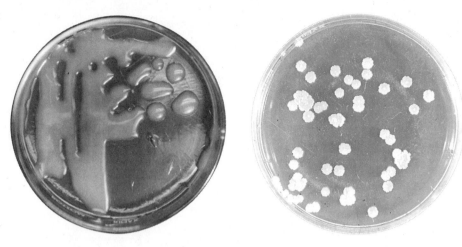

Figure 6-65 *Difference in Colony Morphology between Encapsulated and Unencapsulated Bacteria*   (Left) *Colonies of capsule-forming bacteria have a moist, glistening appearance. (Right) Colonies of bacteria that do not form capsules have a rough, coarse appearance. Courtesy of R. E. Levin.*

the term *slime layer* is sometimes employed. The gelatinous nature of the capsule layer causes bacteria with capsules to form moist, glistening colonies on agar, whereas those lacking a capsule typically form colonies with a rough, coarse appearance (Figure 6-65). The presence of a capsule can also be detected by suspending bacteria in India ink, which allows the capsule to be visualized as a clear zone surrounding the cell (Figure 6-66).

Capsules are typically comprised of polysaccharides containing the sugars glucose, galactose, mannose, ribose, fucose, and their derivatives. In some bacteria the capsule polysaccharide is constructed from a single type of sugar, whereas in others more than one type of sugar is employed. Capsule layers may also contain unusual polypeptides constructed from one or two kinds of amino acids repeated multiple times. In these polypeptides, the amino acids are in the unusual D-configuration rather than the L-configuration that is characteristic of the amino acids found in virtually all proteins (see Figure 1-17).

Cells that have had their capsules removed suffer no loss in viability, indicating that the capsule layer is not essential for survival. The presence of a capsule does, however, help to protect bacteria under adverse environmental conditions and influences their ability to cause disease. For example, *Streptococcus pneumoniae*, the organism responsible for bacterial pneumonia, can cause illness only when a capsule layer is present. The apparent explanation for this phenomenon is that immune cells responsible for engulfing and destroying invading bacteria are ineffective against the encapsulated forms of *Streptococcus*. Therefore bacteria that must travel through the bloodstream to cause illness are often encapsulated. Examples of other encapsulated bacteria responsible for serious disease are *Bacillus anthracis* (anthrax) and *Clostridium perfringens* (gas gangrene).

Figure 6-66 *Detecting the Presence of Bacterial Capsules with India Ink*   (Top) *In capsule-forming bacteria, the capsule appears as a clear zone surrounding cells that have been suspended in India ink. (Bottom) A comparable clear zone is absent in preparations of bacteria that do not make a capsule. Courtesy of R. N. Goodman.*

SUMMARY OF PRINCIPAL POINTS

• The permeability and transport properties of the plasma membrane cause the concentration of charged solutes to differ inside and outside the cell, generating a membrane potential. The magnitude and polarity of the potential is determined by three factors: passive ion distributions dictated by the Donnan equilibrium, active transport of charged ions, and the relative permeability of the membrane to different ions. The membrane potential typically falls in the range of -20 to -300 mV, with the minus sign indicating that the inside of the cell is negatively charged relative to the outside.

• Action potentials are caused by a transient increase in membrane permeability to Na^+ that allows a tiny amount of Na^+ to diffuse into the cell, followed by a transient increase in permeability to K^+ that allows a tiny amount of K^+ to diffuse out of the cell. These changes in ion permeability are caused by the opening and closing of voltage-gated ion channels for Na^+ and K^+.

• Most hormones, neurotransmitters, and local mediators transmit signals to their target cells by binding to plasma membrane receptors. Plasma membrane receptors are grouped into three functional classes: ion channel receptors, receptors linked to G proteins, and catalytic receptors.

• The nicotinic acetylcholine receptor is an example of a neurotransmitter-gated ion channel. Binding of acetylcholine to this receptor opens the channel to cations, leading to an increased flow of Na^+ into the cell and a resulting depolarization of the plasma membrane.

• Many cell surface receptors are linked to G proteins. Some receptors in this category activate G proteins that either stimulate or inhibit adenylyl cyclase. Cyclic AMP generated by adenylyl cyclase functions as a second messenger that activates protein kinase A, which in turn catalyzes the phosphorylation of a broad spectrum of proteins. Because each tissue contains a unique set of proteins subject to phosphorylation by protein kinase A, cyclic AMP triggers different responses in different tissues.

• Other receptors employ G proteins to activate phospholipase C, which catalyzes the formation of the second messengers inositol trisphosphate (IP_3) and diacylglycerol (DAG). DAG functions by activating protein kinase C, while IP_3 mobilizes the release of Ca^{2+} from the endoplasmic reticulum. The released Ca^{2+} exerts its effects by associating with calcium-binding proteins such as calmodulin. Yet another type of G-protein linked receptor is the muscarinic acetylcholine receptor, which activates a G protein that opens ion channels.

• Catalytic receptors are receptors that function as enzymes. Included in this category are growth factor receptors that act as protein-tyrosine kinases, and the atrial natriuretic peptide receptor, which exhibits guanylyl cyclase activity that catalyzes the formation of the second messenger, cyclic GMP.

• Cellular responses elicited by external signaling molecules are terminated by removing or degrading the signaling molecule, by desensitizing the receptor through chemical modification, or by reducing the number of receptors (down-regulation).

• The extracellular matrix of animal tissues consists of three classes of molecules: (1) glycosaminoglycans and proteoglycans, which form a hydrated gelatinous ground substance that contributes pliability and resiliency to the matrix, (2) structural proteins such as collagen, which imparts strength to the matrix, and elastin, which makes the matrix elastic and flexible, and (3) adhesive proteins such as fibronectin, which promotes attachment of cells to the matrix and guides cellular migration, and laminin, which forms part of

the basal lamina and attaches it to the overlying layer of cells. Adhesive matrix proteins bind to cell surface integrin receptors that integrate the organization of the intracellular cytoskeleton with that of the extracellular matrix.

• Two groups of cell surface glycoproteins have been implicated in cell-cell adhesion: N-CAM glycoproteins, which do not require Ca^{2+} for adhesion, and cadherins, which are Ca^{2+}-dependent. Members of both groups mediate cell-cell adhesion by a binding mechanism in which an N-CAM or cadherin molecule on one cell binds to an N-CAM or cadherin molecule of the same type on another cell.

• Long-term stable connections between animal cells usually involve the formation of tight junctions, plaque-bearing junctions, and gap junctions. Tight junctions seal cells together, forming a permeability barrier that prevents molecules from passing from one side of a cell layer to the other; tight junctions also maintain the polarity of epithelial cells by restricting the lateral diffusion of membrane proteins. Plaque-bearing junctions connect to the cytoskeleton and stabilize cells against mechanical stress. They are divided into two categories: desmosomes, which are attached to intermediate filaments, and adherens junctions, which are attached to actin filaments. Gap junctions permit small molecules to move directly from cell to cell, thereby permitting rapid electrical and metabolic communication between neighboring cells.

• The plant cell surface is covered by a cell wall that provides enormous strength and mechanical support for the plant as a whole. The plant cell wall is constructed from varying combinations of cellulose, hemicellulose, pectin, lignin, and glycoproteins. The wall is synthesized in a stepwise fashion, starting with a thin zone called the middle lamella, followed by the relatively thin and flexible primary wall, and finally (in some cases) the much thicker and stronger secondary wall. Channels in the cell wall known as plasmodesmata permit direct cytoplasmic communication between adjacent cells.

• The surface of bacterial cells is covered by a multilayered cell envelope that includes a strong murein wall. This wall makes bacteria resistant to damage by mechanical, chemical, and osmotic forces, and is a major factor in determining cell shape. Gram-positive bacteria have a thick wall made of murein and teichoic acids, whereas Gram-negative bacteria have a thinner murein wall covered by an outer membrane enriched in lipopolysaccharides. The presence of both a plasma membrane and an outer membrane in Gram-negative bacteria creates a unique compartment between these two membranes called the periplasmic space. Proteins localized in the periplasmic space help to protect the cell from toxic compounds and to facilitate the uptake of nutrients. Both Gram-positive and Gram-negative bacteria secrete capsules that help to protect them from adverse environmental conditions.

SUGGESTED READINGS

Books

Hardie, D. G. (1991). *Biochemical Messengers,* Chapman & Hall, London.

Hay, E. D., ed. (1991). *Cell Biology of Extracellular Matrix,* 2nd Ed., Plenum, New York.

Hille, B. (1991). *Ionic Channels of Excitable Membranes,* 2nd Ed., Sinauer, Sunderland, MA.

Kreis, T., and R. Vale, eds. (1993). *Guidebook to the Extracellular Matrix and Adhesion Proteins,* Oxford University Press, New York.

Morgan, N. G. (1989). *Cell Signaling,* Guilford Press, New York.

Articles

Berridge, M. J. (1993). Inositol trisphosphate and calcium signalling, *Nature* 361:315–325.

Boyer, B., and J. P. Thiery (1989). Epithelial cell adhesion mechanisms, *J. Membrane Biol.* 112:97–108.

Burgeson, R. E. (1988). New collagens, new concepts, *Annu. Rev. Cell Biol.* 4:551–577.

Casey, P. J., and A. G. Gilman (1988). G protein involvement in receptor-effector coupling, *J. Biol. Chem.* 263:2577–2580.

Changeux, J. P. (1993). Chemical signaling in the brain, *Sci. Amer.* 269 (November): 58–62.

Clapham, D. E., and E. J. Neer (1993). New roles for G-protein βγ-dimers in transmembrane signalling, *Nature* 365:403–406.

Fantl, W. J., D. E. Johnson, and L. T. Williams (1993). Signalling by receptor tyrosine kinases, *Annu. Rev. Biochem.* 62:453–481.

Geiger, B., and O. Ayalon (1992). Cadherins, *Annu. Rev. Cell Biol.* 8:307–332.

Gilman, A. G. (1987). G proteins: Transducers of receptor-generated signals, *Annu. Rev. Biochem.* 56:615–649.

Hunter, T. (1987). A thousand and one protein kinases, *Cell* 50:823–829.

Hynes, R. O. (1986). Fibronectins, *Sci. Amer.* 254 (June):42–51.

Hynes, R. O. (1992). Integrins: versatility, modulation, and signaling in cell adhesion, *Cell* 69:11–25.

Imoto, K., et al. (1988). Rings of negatively charged amino acids determine the acetylcholine receptor conductance, *Nature* 335:645–648.

Iñiguez-Lluhi, J., C. Kleuss, and A. G. Gilman (1993). The importance of G-protein βγ subunits, *Trends Cell Biol.* 3:230–236.

Jakobs, K. H., U. Gehring, B. Gaugler, T. Pfeuffer, and G. Schulz (1983). Occurrence of an inhibitory guanine nucleotide-binding component of the adenylate cyclase system in cyc⁻ variants of S49 lymphoma cells, *Eur. J. Biochem.* 130:605–611.

Kühn, K., and J. Eble (1994). The structural bases of integrin-ligand interactions, *Trends Cell Biol.* 4:256–261.

Lindner, M. E., and A. G. Gilman (1992). G proteins, *Sci. Amer.* 267 (July):56–65.

Lucas, W. J., and S. Wolf (1993). Plasmodesmata: the intercellular organelles of green plants, *Trends Cell Biol.* 3:308–315.

Mercurio, A. M. (1990). Laminin: multiple forms, multiple receptors, *Curr. Opinion Cell Biol.* 2:845–849.

Neher, E., and B. Sakmann (1992). The patch clamp technique, *Sci. Amer.* 266 (March):44–51.

Nose, A., A. Nagafuchi, and M. Takeichi (1988). Expressed recombinant cadherins mediate cell sorting in model systems, *Cell* 54:993–1001.

Roberts, K. (1990). Structures at the plant cell surface, *Curr. Opinion Cell Biol.* 2:920–928.

Ruoslahti, E. (1988). Fibronectin and its receptors, *Annu. Rev. Biochem.* 57:375–413.

Ruoslahti, E., and M. D. Pierschbacher (1987). New perspectives in cell adhesion: RGD and integrins, *Science* 238:491–497.

Rutishauser, U., S. Hoffman, and G. M. Edelman (1982). Binding properties of a cell adhesion molecule from neural tissue, *Proc. Natl. Acad. Sci. USA* 79:685–689.

Schulster, D., J. Orly, G. Seidel, and M. Schramm (1978). Intracellular cyclic AMP production enhanced by a hormone receptor transferred from a different cell: β-adrenergic responses in cultured cells conferred by fusion with turkey erythrocytes, *J. Biol. Chem.* 253:1201–1206.

Schwartz, M. A., K. Owaribe, J. Kartenbeck, and W. W. Franke (1990). Desmosomes and hemidesmosomes: constitutive molecular components, *Annu. Rev. Cell Biol.* 6:461–491.

Sharon, N., and H. Lis (1993). Carbohydrates in cell recognition, *Sci. Amer.* 268 (January):82–89.

Shockman, G. D., and J. F. Barrett (1983). Structure, assembly, and function of cell walls of gram-positive bacteria, *Annu. Rev. Microbiol.* 37:501–527.

Singer, S. J. (1992). Intercellular communication and cell-cell adhesion, *Science* 255:1671–1677.

Strader, C. D., T. M. Fong, M. R. Tota, D. Underwood, and R. A. F. Dixon (1994). Structure and function of G protein-coupled receptors, *Annu. Rev. Biochem.* 63:101–132.

Wildon, D. C., et al. (1992). Electrical signalling and systemic proteinase inhibitor induction in the wounded plant, *Nature* 360:62–65.

Wyatt, S. E., and N. C. Carpita (1993). The plant cytoskeleton-cell-wall continuum, *Trends Cell Biol.* 3:413–417.

Yatani, A., et al. (1988). The G protein-gated atrial K⁺ channel is stimulated by three distinct Gᵢ α-subunits, *Nature* 336:680–682.

C h a p t e r 7

Cytoplasmic Membranes and Intracellular Traffic

One of the most striking differences between the appearance of prokaryotic and eukaryotic cells is the presence in eukaryotes of an elaborate series of intracellular membranes that divide the cell into multiple compartments. In this chapter we focus on three groups of interrelated membranes: the endoplasmic reticulum, the Golgi complex, and lysosomes. The other internal membranes of eukaryotes, such as those enveloping chloroplasts, mitochondria, peroxisomes, and the nucleus, are sufficiently independent to warrant their discussion in

other chapters later in the text (Chapters 8, 9, and 10).

Our current understanding of the relationships between the endoplasmic reticulum, Golgi complex, and lysosomes is a tribute to the power of the side-by-side use of subcellular fractionation and electron microscopy. The awarding of the Nobel Prize in 1974 to Albert Claude, George Palade, and Christian de Duve recognized the contributions of three pioneers who were especially instrumental in fusing the study of subcellular biochemistry with subcellular morphology. Claude devised the first procedures for systematically isolating cell organelles by the process of subcellular fractionation. Palade applied subcellular fractionation and electron microscopic analysis to the study of the endoplasmic reticulum and Golgi complex, allowing him to uncover the significance of these membrane systems for the synthesis, transport, storage, and secretion of proteins. And finally, de Duve's subcellular fractionation studies led him to predict the existence of lysosomes before these structures had been identified by electron microscopy, a remarkable testimonial to the power of subcellular fractionation. In this chapter we will examine the experiments of these and other investigators who have explored the functions performed by the endoplasmic reticulum and related cytoplasmic membranes.

THE ENDOPLASMIC RETICULUM

In the late nineteenth century, light microscopists first noted that the cytoplasm of eukaryotic cells often contains regions that stain intensely with basic dyes. These cytoplasmic areas, referred to as *ergastoplasm*, were found to be especially prominent in cells involved in secretion. However, the limited resolving power of the light microscope caused people to question the reality of the ergastoplasm as a legitimate component until electron microscopic techniques for examining cell structure were introduced in the early 1940s.

In 1945 Keith Porter, Albert Claude, and Ernest Fullam discovered that cells grown in tissue culture become spread thin enough to be examined by electron microscopy without the need for sectioning. This realization permitted the first electron microscopic pictures of eukaryotic cells to be produced. The micrographs revealed the presence of an interconnected network of strands extending throughout the cytoplasm that came to be called the **endoplasmic reticulum** or **ER** (Figure 7-1). When thin-sectioning techniques were developed a few years later, it became clear that this interconnected network is comprised of a heterogeneous collection of membranous channels, vesicles, and sacs (Figure 7-2). Of course, thin-section electron micrographs only allow one to see thin slices

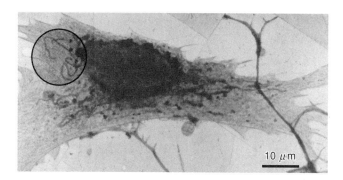

Figure 7-1 *Electron Microscopic Appearance of the Endoplasmic Reticulum in an Intact, Unsectioned Cell* *A series of individual micrographs of different regions of a fibroblast were spliced together to create this image of an entire cell. The color highlighting overlays an area enriched in endoplasmic reticulum. Courtesy of K. R. Porter.*

through the various components of the endoplasmic reticulum. As a result, the components of the endoplasmic reticulum look like separate membrane-enclosed spaces rather than interconnected channels and vesicles. It is therefore fortunate that the endoplasmic reticulum was first observed in intact cells, for it would have been difficult to comprehend from thin-sectioned material alone that this system represents a single, interconnected network of membrane-enclosed compartments (Figure 7-3).

The Endoplasmic Reticulum Consists of Two Components: The Smooth ER and the Rough ER

The relationship between the endoplasmic reticulum and the ergastoplasm of classical light microscopy first became apparent when George Palade discovered that

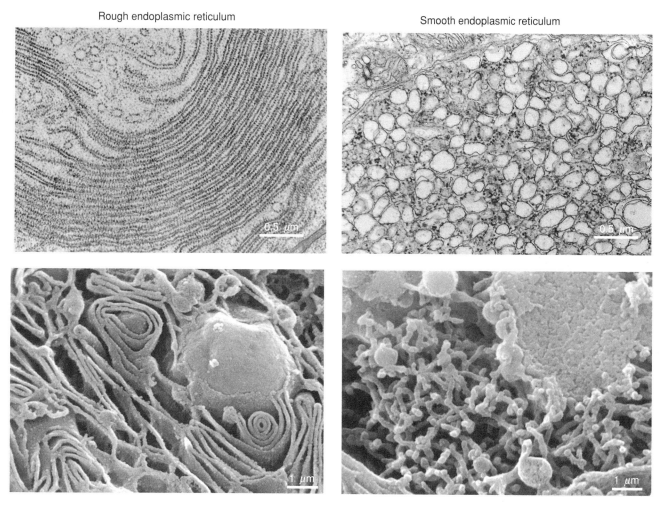

Figure 7-2 *The Endoplasmic Reticulum in Thin-Section and Scanning Electron Micrographs* (Top) *Thin-section electron micrographs of the endoplasmic reticulum.* (Bottom) *Scanning electron micrographs of the endoplasmic reticulum. In both sets of micrographs, the one on the left shows endoplasmic reticulum containing bound ribosomes (rough ER) and the one on the right shows endoplasmic reticulum lacking ribosomes (smooth ER). Courtesy of D. W. Fawcett (top left), M. Bielinska (top right), and K. Tanaka (bottom).*

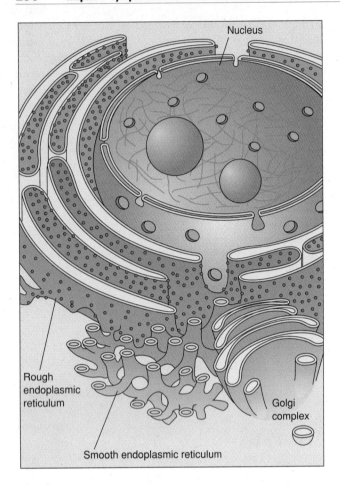

Figure 7-3 *Three-Dimensional Representation of the Endoplasmic Reticulum and Golgi Complex in an Intact Cell* *Although the endoplasmic reticulum look like a series of separate, membrane-enclosed spaces in thin-section electron micrographs, it actually represents an interconnected network of membrane-enclosed compartments.*

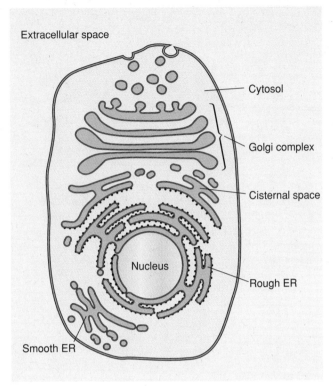

Figure 7-4 *Diagram Illustrating How the Endoplasmic Reticulum Divides the Cytoplasm into Two Compartments, the Cisternal Space and the Cytosol* *Functional continuity between the cisternal spaces of the ER, Golgi complex, and cytoplasmic vesicles is maintained by membrane vesicles that bud off from one membrane system and fuse with another.*

the endoplasmic reticulum exists in two forms: a "granular" or **rough ER** containing attached ribosomes, and an "agranular" or **smooth ER** lacking attached ribosomes (see Figure 7-2). Ribosomes contain large amounts of RNA, an acidic molecule exhibiting a strong affinity for basic dyes. Therefore the cytoplasm of cells containing large amounts of rough ER stains intensely when exposed to basic dyes.

The rough and smooth ER differ morphologically in ways other than the presence or absence of attached ribosomes. Rough ER is usually composed of large, flattened membrane sacs, whereas smooth ER more typically consists of an interconnected series of membrane tubules. The relative abundance of smooth and rough ER varies among cell types, as well as with differing metabolic and developmental states. In general, rough ER predominates in cells that are actively synthesizing protein for export, while an extensive smooth ER is associated with cells that are involved in the metabolism of lipids, drugs, and toxic substances.

It has been estimated that in rat liver cells, the endoplasmic reticulum accounts for about 20 percent of the total protein, 50 percent of the total lipid, and 60 percent of the total RNA. The endoplasmic reticulum can thus make a substantial contribution to a cell's overall mass.

The endoplasmic reticulum divides the cytoplasm into two compartments referred to as the **cytosol** and the **cisternal space** or **ER lumen** (Figure 7-4). The cytosol contains metabolic enzymes, ribosomes, transfer RNAs, and other components required for protein synthesis. Ribosomes bound to the ER are attached to the side of the ER membranes that faces the cytosol. The ER lumen is indirectly connected to the lumens of the Golgi complex and lysosomes, as well as to the outside of the cell. Instead of being maintained by permanent membrane connections, this functional continuity is established by shuttling membrane vesicles that bud off from one membrane system, travel some distance, and then fuse with another membrane system.

By subdividing the cytoplasm, the endoplasmic reticulum creates a series of membrane-bound compartments that can be adapted for different purposes. The conditions prevailing within these compartments are controlled by both the selective permeability of the ER membrane and its ability to carry out active transport. For example, smooth ER membranes are highly effective at establishing gradients of Ca^{2+}, which play important roles in cell signaling, cell secretion, nerve excitation, and muscle contraction. Another advantage to subdividing the cytoplasm into multiple compartments is that it creates a series of membrane-bound channels that can be used for moving materials through the cell. As will be discussed shortly, this routing function is especially important for conveying newly made proteins to their appropriate intracellular and extracellular destinations. Other materials are also routed via the endoplasmic reticulum. For example, triacylglycerols taken up by cells from the external environment are hydrolyzed in the cytosol to release free fatty acids, which then pass into the lumen of the ER for distribution to the rest of the cell via the channels of both the smooth and rough ER.

Our current understanding of the functions carried out by the endoplasmic reticulum is based to a large extent on studies of membranes isolated from cell homogenates. Cell homogenization procedures tend to disrupt the continuity of the ER, creating membrane fragments that spontaneously reseal into tiny vesicles called **microsomes.** Microsomes can be collected by high-speed centrifugation and further separated into fractions corresponding to smooth and rough ER

(Figure 7-5). Although microsomes are artificially generated derivatives of a larger interconnected membrane system, they have nonetheless been quite useful for studying the properties of the endoplasmic reticulum. Biochemical analyses of isolated microsomal vesicles have revealed the presence of enzymes involved in carbohydrate metabolism, lipid metabolism, the detoxification of drugs and other toxic chemicals, and the processing of protein molecules (Table 7-1). As we will see in the following sections, these four groups of enzymes reflect the major functions of the endoplasmic reticulum.

The Smooth ER Is Involved in Releasing Free Glucose from Glycogen

The enzyme *glucose 6-phosphatase,* which catalyzes the hydrolysis of glucose 6-phosphate to glucose and inorganic phosphate, is often employed as a marker for identifying microsome fractions because it is localized almost exclusively in membranes of the endoplasmic reticulum. This enzyme plays an especially important role in the liver, the organ responsible for maintaining a relatively constant level of glucose in the bloodstream. The liver stores glucose in the form of glycogen granules, which can be broken down when the body needs glucose. As we saw in the preceding chapter, the breakdown of glycogen by the liver is controlled by hormones that trigger an increase in cyclic AMP levels;

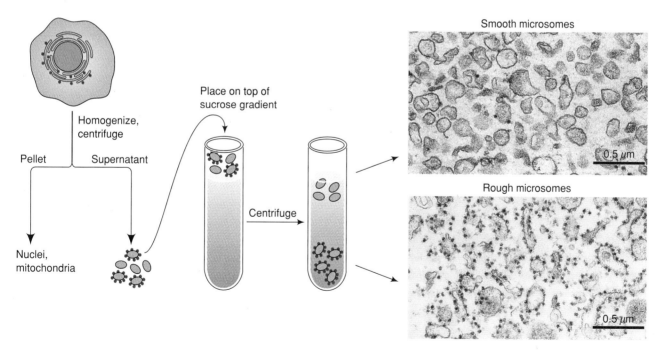

Figure 7-5 *Isolation of Microsomal Fractions Corresponding to Smooth and Rough ER* *When cells are homogenized, the continuity of the ER is disrupted. The resulting membrane fragments spontaneously reseal into tiny vesicles called microsomes, which can be separated by centrifugation into fractions corresponding to smooth and rough ER. The micrographs of the two fractions show that ribosomes are attached to rough microsomal membranes, but not to smooth microsomal membranes. Micrographs courtesy of C. de Duve.*

Table 7-1 Some Enzymes Detected in Isolated Microsomal Membranes

Category	Enzyme	Active Site Faces
Carbohydrate metabolism	Glucose 6-phosphatase	Lumen
	β-Glucuronidase	Lumen
	Glucuronyl transferase	Lumen
	Glycosyl transferases	Cytosol
Lipid metabolism	Fatty acid CoA ligase	Cytosol
	Phosphatidic acid phosphatase	Cytosol
	Cholesterol hydroxylase	Cytosol
	Choline phosphotransferase	Cytosol
	Phospholipid translocators (flippases)	Cytosol and lumen
Detoxification of drugs and related oxidases	Cytochrome P-450	Cytosol
	NADPH-cytochrome P-450 reductase	Cytosol
	Cytochrome b_5	Cytosol
	NADH-cytochrome b_5 reductase	Cytosol
	Acetanilide-hydrolyzing esterase	Lumen
Protein processing	Signal peptidase	Lumen
	Protein disulfide isomerase	Lumen

cyclic AMP activates protein kinase A, which in turn stimulates the enzymatic pathway for the conversion of glycogen to glucose 1-phosphate. However glucose 1-phosphate, like other phosphorylated sugars, cannot diffuse through membranes and exit from the cell. Instead, glucose 1-phosphate is converted to glucose 6-phosphate, which is then hydrolyzed to free glucose and inorganic phosphate by glucose 6-phosphatase associated with the smooth endoplasmic reticulum. The free glucose is then transported out of the cell and into the bloodstream, which distributes it to the rest of the body.

The Smooth ER Is Involved in Synthesizing Triacylglycerols and Steroids

In addition to its role in carbohydrate metabolism, the smooth ER is responsible for synthesizing several classes of lipids. For example, triacylglycerols are synthesized by the smooth ER and stored in the ER lumen, producing fat droplets. Animal cells specialized for storing fat are called *fat cells* or *adipocytes;* in such cells, fat droplets occupy almost the entire volume of the cytoplasm. Triacylglycerols stored in fat droplets can be broken down to glycerol and free fatty acids; oxidation of the fatty acids in turn yields large amounts of ATP. Thus a primary function of fat cells is to serve as a reservoir of stored energy. Fat cells also provide thermal insulation for organisms living in cold environments.

The smooth ER is also involved in the synthesis of cholesterol and in the subsequent oxidations, reductions, and hydroxylations that convert cholesterol to various steroid hormones. For this reason, the smooth ER is especially well developed in tissues that produce steroid hormones, such as the adrenal cortex and the interstitial region of the testis. Cholesterol and steroid hormones are synthesized by a multistage pathway that involves enzymes present in the cytosol as well as the smooth ER. The initial step in this pathway is the formation of a cholesterol precursor called *mevalonate,* whose synthesis is catalyzed by *HMG-CoA reductase* present in the smooth ER. Although the details of the subsequent pathway are beyond the scope of this text, the next several steps in cholesterol formation occur in the cytosol, and the pathway then reverts back to the endoplasmic reticulum for completion.

The Smooth ER Synthesizes the Phospholipids Needed for Cellular Membranes

Perhaps the most important group of lipids to be synthesized in the smooth ER consists of the phospholipids required for the construction of cellular membranes. Most of the phospholipid molecules destined to become incorporated into cellular membranes are initially synthesized as part of the lipid bilayer of the smooth ER (or the plasma membrane in the case of bacterial cells). Figure 7-6 illustrates the basic steps involved in phospholipid formation using phosphatidylcholine as an example. From the beginning of the process, the newly forming phospholipid molecules are embedded in the lipid bilayer of the ER on the side facing the cytosol. The substrates required for phospholipid synthesis tend to be constituents of the cytosol, so the enzymes involved in this pathway are either soluble components of the cytosol or membrane proteins whose catalytic sites face the cytosol.

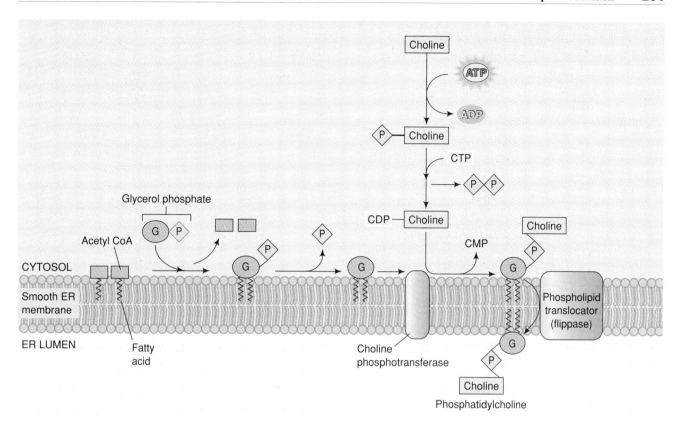

Figure 7-6 *Phospholipid Synthesis within the ER Membrane* *This example summarizes the synthesis of the membrane lipid, phosphatidylcholine. During the first step, fatty acids located in the ER membrane are joined to glycerol phosphate, whose phosphate group is then removed. Choline is then transferred to the newly forming phospholipid by the membrane-bound enzyme, choline phosphotransferase. The resulting phosphatidylcholine molecule is translocated by a phospholipid translocator (flippase) from the cytosolic side of the lipid bilayer to the lumenal side. The fact that other phospholipids are not translocated with such efficiency leads to an asymmetric distribution of different kinds of phospholipids across the bilayer.*

Since phospholipid molecules are only synthesized on one side of the lipid bilayer, how does the other side of the bilayer grow? As we learned in Chapter 5, flip-flop of membrane lipids from one leaflet of a bilayer to the other is an infrequent event in artificial phospholipid bilayers. The same limitation applies to most biological membranes, but the ER is different. The smooth ER contains special proteins called **phospholipid translocators** or **flippases** that catalyze the flip-flop of membrane lipids from one half of the bilayer to the other. For example, one of these proteins catalyzes the movement of phosphatidylcholine, but not other phospholipids, from the cytoplasmic side of the ER membrane to the side facing the lumen. This ability to selectively move lipid molecules from one side of the bilayer to the other contributes to the asymmetric arrangement of phospholipids across the membrane (page 165).

As newly forming phospholipid molecules are synthesized and retained within the membrane bilayer of the smooth ER, the lipid bilayer expands in size. In other words, *new membrane is formed by the expansion of previously existing membrane.* Besides expand-

ing the ER membrane, newly formed phospholipids also become incorporated into other cellular membranes. The details of the role played by the ER in the formation of other membranes will be discussed later in the chapter, after we have introduced some of the additional components that are involved.

The Smooth ER Oxidizes Foreign Substances Using Cytochrome P-450

A unique function of the smooth ER is its role in the detoxification of foreign substances such as drugs, pesticides, toxins, and pollutants. Most of the detoxification reactions involve oxidation, although some utilize reduction, hydrolysis, and conjugation as well. In nearly all cases, the net result is the conversion of a hydrophobic substance into a water-soluble one. This conversion is important because hydrophobic compounds tend to accumulate in body fats, whereas water-soluble materials are more easily excreted.

The first indication that cells contain a special enzyme system for oxidizing drugs was provided in 1953 by Gerald Mueller and James Miller, who reported that cancer-causing aminoazo dyes are oxidized by liver homogenates in a reaction requiring molecular oxygen and NADPH. Soon thereafter Bernard Brodie and his associates found that many other drugs and chemicals are oxidized by isolated liver microsomes under similar conditions. The oxidation of all these compounds involves a similar mechanism in which one molecule of oxygen is consumed per molecule of substrate oxidized, accompanied by the conversion of NADPH to $NADP^+$. Because one atom of oxygen appears in the product while the other appears in water, enzymes catalyzing this type of reaction are referred to as **mixed-function oxidases.**

The mixed-function oxidase system of microsomes consists of several elements. The central component is **cytochrome P-450,** a group of iron-containing integral membrane proteins of the smooth ER whose name is based on their ability to absorb light at a wavelength of 450 nm. Cytochrome P-450 occurs in many kinds of eukaryotic and prokaryotic cells, but it is especially prominent in the smooth ER of liver cells, where it accounts for nearly 20 percent of the ER protein and 2 to 3 percent of the total cellular protein. Cytochrome P-450 is involved in the hydroxylation of a variety of toxic substances, as well as steroids and fatty acids. As is shown in Figure 7-7, hydroxylation by cytochrome P-450 involves four basic steps: (1) binding of an oxidizable substrate to cytochrome P-450, (2) reduction of the iron atom in cytochrome P-450 by NADPH, (3) binding of oxygen to cytochrome P-450, and (4) oxidation of the substrate with one atom of the bound oxygen while the other oxygen atom is used to form water (the iron atom becoming reoxidized in the process). When the compound being hydroxylated is a toxic substance, hydroxylation usually increases its solubility and therefore promotes its excretion.

In addition to their role in detoxifying foreign substances, mixed-function oxidases participate in normal metabolic pathways for the oxidation of steroids and fatty acids. The utilization of a single enzyme system for catalyzing the oxidation of many diverse substrates means that alterations in this system will have widespread effects on cellular metabolism. To cite a case with practical implications, ingestion of sleeping pills containing phenobarbital triggers a proliferation of the smooth ER and an accompanying increase in mixed-function oxidase activity in liver cells (Figure 7-8). The resulting increase in the ability to degrade phenobarbital explains why habitual users of sleeping pills require higher and higher doses to achieve the same effect. Moreover, as a result of the broad specificity of mixed-function oxidases, the liver acquires not just an enhanced ability to degrade phenobarbital but an increased capacity to degrade other substances as well. Hence the chronic use of barbiturates enhances the rate at which the liver destroys other drugs, including therapeutically useful agents such as antibiotics, steroids, anticoagulants, and narcotics. The ability to stimulate the formation of drug-metabolizing enzymes is not restricted to phenobarbital; it is shared by hundreds of other chemical substances as well.

The proliferation of endoplasmic reticulum that occurs after the ingestion of foreign substances does not involve an equivalent increase in all smooth ER enzymes. Phenobarbital, for example, triggers an increase in cytochrome P-450 but not glucose 6- phosphatase (Figure 7-9, *left*). The selectivity of enzyme induction is further demonstrated by comparing the properties of the mixed-function oxidase systems induced by different drugs. To cite an example, the cancer-causing agent *3-methylcholanthrene* stimulates the formation of a mixed-function oxidase system whose cytochrome absorbs light maximally at 448 nm rather than 450 nm (see Figure 7-9, *right*). This cytochrome, called **cytochrome P-448,** metabolizes different substrates

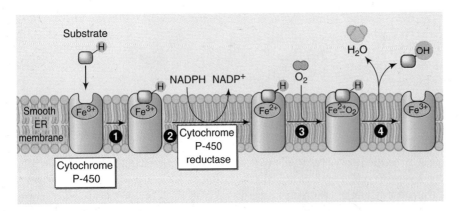

Figure 7-7 Mechanism by Which Substrates Are Hydroxylated by a Mixed-Function Oxidase *The hydroxylation reaction sequence involves four steps: (1) binding of a substrate to cytochrome P-450, (2) reduction of the iron atom in cytochrome P-450 by NADPH, (3) binding of oxygen to cytochrome P-450, and (4) oxidation of the substrate with one atom of oxygen while the other oxygen atom is used to form water.*

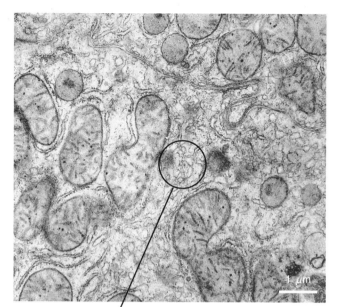

Smooth endoplasmic reticulum

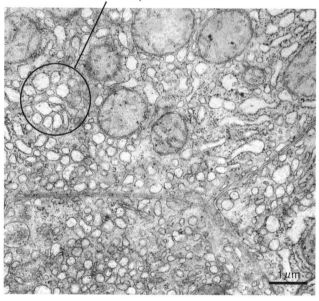

Figure 7-8 Proliferation of Smooth ER Caused by the Ingestion of Phenobarbital in Rats (Top) *Electron micrograph of liver cell of control rat.* (Bottom) *Liver cell of phenobarbital-treated rat, showing the increase in smooth endoplasmic reticulum. Courtesy of S. Orrenius.*

from those handled by cytochrome P-450. Because the mixed-function oxidase system employing cytochrome P-448 is most efficient in metabolizing polycyclic hydrocarbons, it is referred to as *aryl hydrocarbon hydroxylase.*

Although the oxidation of polycyclic hydrocarbons by aryl hydrocarbon hydroxylase might be assumed to be a detoxification process, in some cases the products generated are more toxic than the original chemicals. For example, mice with high levels of aryl hydrocarbon hydroxylase exhibit a greater incidence of spontaneous cancer than normal mice, and mice treated with inhibitors of this enzyme acquire fewer tumors than nor-

mal. Taken together, such observations suggest that aryl hydrocarbon hydroxylase may convert foreign substances into chemicals that cause cancer. It is interesting to note in this context that cigarette smoke is a potent inducer of aryl hydrocarbon hydroxylase.

The realization that cytochromes P-450 and P-448 differ in substrate specificity has spurred the search for other microsomal cytochromes. This effort has led to the discovery that cytochrome P-450 is not a single protein but a family of closely related molecules, each induced by and active upon a different class of substrates. In addition to the cytochrome P-450 family, another iron-containing protein localized in the endoplasmic reticulum is *cytochrome b_5.* In contrast to the other microsomal cytochromes, cytochrome b_5 employs NADH rather than NADPH as a coenzyme. Cytochrome b_5 is involved in the desaturation of fatty acids (which occurs during the formation of certain membrane lipids), and in NADH-dependent microsomal hydroxylation reactions.

The Rough ER Is Involved in Protein Targeting

The primary role of the rough ER is quite different from that of the smooth ER. Because the rough ER contains attached ribosomes, its main functions are directly related to activities taking place on these ribosomes. As we learned in Chapter 3, the information specifying a protein's amino acid sequence is encoded in messenger RNA (mRNA), which binds to ribosomes and directs the specific order in which amino acids are added during the process of protein synthesis. In eukaryotic cells, cytoplasmic ribosomes exist in two forms: attached to membranes of the rough ER, and free in the cytosol. (In Chapter 13 we will learn that many of the cell's "free" ribosomes may in fact be bound to the cytoskeleton, but we will continue to use the term "free ribosomes" as a general designation for all ribosomes that are not membrane-bound.) Shortly after the discovery of membrane-bound ribosomes, George Palade postulated that polypeptides synthesized on these attached ribosomes selectively pass into the lumen of the endoplasmic reticulum and are exported from the cell during the process of cell secretion, while proteins synthesized on free cytoplasmic ribosomes are retained for use within the cell.

The hypothesis that different classes of proteins are synthesized on free and membrane-attached ribosomes has subsequently been verified by examining the kinds of proteins synthesized on each type of ribosome (Table 7-2). Such data have revealed, however, that the situation is more complex than originally envisioned by Palade because the rough ER synthesizes several groups of proteins that differ in their mode of synthesis. *Secretory proteins* and *lumenal proteins* are synthesized on membrane-attached ribosomes and released into the lumen of the ER after completion. Secretory proteins are destined to be secreted from the cell, whereas lume-

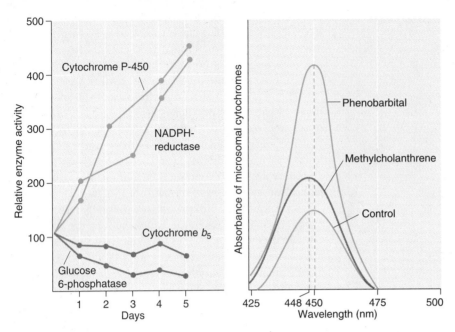

Figure 7-9 *Data Showing That Different Microsomal Enzymes Are Induced by Exposure to Different Drugs* (Left) *Injecting rats daily with phenobarbital causes an increase in cytochrome P-450 and NADPH-dependent cytochrome P-450 reductase, but not glucose 6-phosphatase or cytochrome b_5.* (Right) *Comparison of the absorption spectrum of microsomal cytochromes after injecting rats with phenobarbital or 3-methylcholanthrene. After treatment with 3-methylcholanthrene, the absorption maximum shifts from 450 to 448 nm, indicating the formation of a new cytochrome that metabolizes different substrates.*

nal proteins remain in the lumen of the ER, Golgi complex, lysosomes, or related cytoplasmic vesicles. In contrast, *integral membrane proteins* are also synthesized on membrane-attached ribosomes, but are retained within the lipid bilayer of the ER membrane after protein synthesis has been completed. Most of the cell's integral membrane proteins, including those of the plasma membrane, are formed in this way.

How is it determined whether a newly forming protein will be synthesized on free or membrane-attached ribosomes? In recent years it has become apparent that protein molecules contain amino acid sequences called **targeting signals** that guide proteins to their correct destinations. Targeting signals determine whether a newly forming polypeptide chain is synthesized on membrane-bound or free ribosomes, and guide the completed protein to its proper destination inside or outside the cell. Proteins synthesized on ribosomes attached to the rough ER are targeted for the ER, Golgi complex, lysosomes, plasma membrane, and secretion from the cell, whereas proteins synthesized on free ribosomes are released into the cytosol and can be targeted for incorporation into organelles such as mitochondria, chloroplasts, peroxisomes, and the nucleus (Figure 7-10). In this chapter we will discuss the targeting of proteins synthesized on ribosomes attached to the rough ER; the targeting of proteins synthesized on free ribosomes will be described in Chapter 11, which is devoted to the topic of protein synthesis.

Signal Sequences Target Proteins to the Endoplasmic Reticulum

The idea that the endoplasmic reticulum is involved in transporting newly synthesized proteins to their proper destination first emerged from studies carried out in the 1960s by Colvin Redman and David Sabatini, who studied protein synthesis in isolated vesicles of rough ER. After briefly incubating the vesicles in the presence of radioactive amino acids and the other components required for protein synthesis, the antibiotic *puromycin* was added to halt the process. In addition to blocking further protein synthesis, puromycin causes the partially completed polypeptide chains to be released from the ribosomes. At this point the rough ER vesicles were collected by centrifugation and treated with detergent to release the material contained within the lumen. The data revealed that most of the newly released radioactive protein appeared in the material isolated from the ER lumen (Figure 7-11). Such results suggest that newly forming polypeptide chains begin to pass through the ER membrane and into the lumen of the ER as they are being synthesized, and puromycin simply causes the chains to be prematurely released into the vesicle lumen. Thus, instead of being released into the cytosol, proteins synthesized on membrane-bound ribosomes are transported into the ER as they are being made, allowing them to be routed to their proper destinations.

Table 7-2 Examples of Proteins Synthesized by the Two Main Classes of Eukaryotic Cytoplasmic Ribosomes

Proteins Synthesized by Ribosomes Bound to the ER

Secretory proteins:

Peptide hormones

Growth factors

Digestive enzymes

Serum proteins

Extracellular matrix proteins

Other proteins released into the ER lumen:

Rough ER enzymes

Golgi complex enzymes

Lysosomal enzymes

Integral membrane proteins:

Rough ER membrane glycoproteins

Golgi membrane glycoproteins

Lysosomal membrane glycoproteins

Plasma membrane glycoproteins

Nuclear membrane glycoproteins

Glycolipid-anchored plasma membrane proteins

Peripheral proteins of the plasma membrane

(exterior membrane face)

Proteins Synthesized by Free Cytoplasmic Ribosomes

Soluble cytosol proteins

Lipid-anchored membrane proteins

(cytoplasmic membrane face)

Peripheral proteins of the plasma membrane

(cytoplasmic membrane face)

Mitochondrial proteins encoded by nuclear genes

Chloroplast proteins encoded by nuclear genes

Peroxisomal proteins

Nuclear proteins

Since the preceding studies indicate that certain proteins pass directly into the lumen of the ER as they are being synthesized, the question arises as to how the cell selects which proteins are to be handled in this way. An answer was first proposed in 1971 by Günter Blobel and David Sabatini, who suggested that special **signal sequences** are located near the N-terminal end of polypeptide chains that are destined to be synthesized on membrane-bound ribosomes and transported into the ER. They postulated that protein synthesis is initiated on free ribosomes, but the presence of a signal sequence, which is formed early during the process of protein synthesis, causes the newly forming chain and its associated ribosome to bind to the ER.

Evidence for the existence of such signal sequences was obtained shortly thereafter by César Milstein and his associates, who were studying the synthesis of the small polypeptide subunit, or *light chain*, of immunoglobulin G. In cell-free systems containing purified ribosomes and the components required for protein synthesis, the mRNA coding for the immunoglobulin light chain directs the synthesis of a polypeptide product that is 20 amino acids longer at its N-terminal end than the authentic light chain itself. However, adding ER membranes to this system leads to the production of an immunoglobulin light chain of the correct size. Such findings suggest that ER membranes are capable of cleaving the extra 20 amino acids from the N-terminus of the newly forming immunoglobulin light chain. Subsequent studies have revealed that other proteins destined for the ER also contain an extra amino acid sequence that is required for targeting the protein to the ER. Proteins containing such signal sequences at their N-terminus are often referred to as *pre*-proteins (e.g., prelysozyme, preproinsulin, pretrypsinogen, and so forth).

The amino acid composition of ER-specific signal sequences is surprisingly variable, but several unifying features tend to be present. These N-terminal signal sequences typically consist of three domains: a positively charged terminal region, a central hydrophobic region, and a more polar region adjoining the site where cleavage from the mature protein will take place (Figure 7-12).

Proteins Synthesized on Membrane-Bound Ribosomes Pass through Protein-Translocating Channels in the ER Membrane

How does the presence of a signal sequence at the beginning of a newly forming polypeptide chain cause it to become associated with membranes of the ER, and what role do components of the ER play after this association has taken place? Figure 7-13 summarizes the six steps that are involved in this interaction:

1. *Initiation of protein synthesis.* The synthesis of proteins targeted for the ER is initiated by the binding of mRNA molecules to free ribosomes present in the cytosol. Prior to mRNA binding, ribosomes are all functionally equivalent; it is the mRNA that determines what type of protein any given ribosome will make, and where this protein will eventually reside.

2. *Binding of SRP to the signal sequence.* If the mRNA that binds to a ribosome codes for a protein which contains a signal sequence at its N-terminus, then protein synthesis proceeds until the signal sequence has been formed and begins to emerge from the ribosome. At this stage protein synthesis is halted by a specialized protein-RNA complex, called the **signal recognition particle (SRP),** which binds to the emerging signal sequence. SRP consists of six polypeptide subunits associated with a 7S RNA.

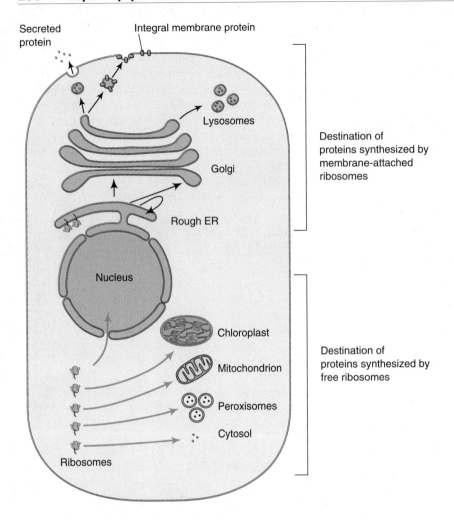

Figure 7-10 Protein Targeting Signals *Each class of cellular protein has a signal within its amino acid sequence that targets the protein to its correct destination. Proteins synthesized on ribosomes attached to the rough ER are targeted for the ER, Golgi complex, lysosomes, plasma membrane, and secretion from the cell. Proteins synthesized on free ribosomes are released into the cytosol and can be targeted for incorporation into mitochondria, chloroplasts, peroxisomes, and the nucleus.*

3. *Binding of the SRP-ribosome complex to the ER.* In addition to recognizing and binding to the signal sequence in the newly forming polypeptide chain, SRP also has an affinity for ER membranes. Therefore after it has bound to the signal sequence, SRP attaches the ribosome to the ER by binding to an ER membrane protein known as the *SRP receptor.*

4. *Release of SRP and binding of the ribosome to a protein-translocating membrane channel.* Binding of the SRP-ribosome complex to the SRP receptor is followed by release of SRP from the signal sequence in a reaction that is associated with the hydrolysis of GTP. As SRP is released from the signal sequence, the ribosome and its associated polypeptide chain become associated with a **protein-translocating channel** in the ER membrane. Several membrane proteins have been implicated in the functioning of this protein-translocat-

ing channel (*ribophorins, SSR-complex, sec61, sec62, and sec63*), but their precise roles are yet to be determined.

5. *Resumption of protein synthesis and cleavage of the signal sequence.* After the polypeptide chain has been inserted into the protein-conducting channel, protein synthesis resumes and the elongating polypeptide chain emerges into the lumen of the ER. This process exposes the cleavage site located next to the signal sequence, allowing the signal sequence to be removed by a membrane-bound enzyme called **signal peptidase.**

6. *Completion of the polypeptide chain and closing of the protein-translocating channel.* After the signal sequence has been removed, polypeptide synthesis continues. As we will see shortly, certain kinds of protein modification reactions occur while protein synthesis is

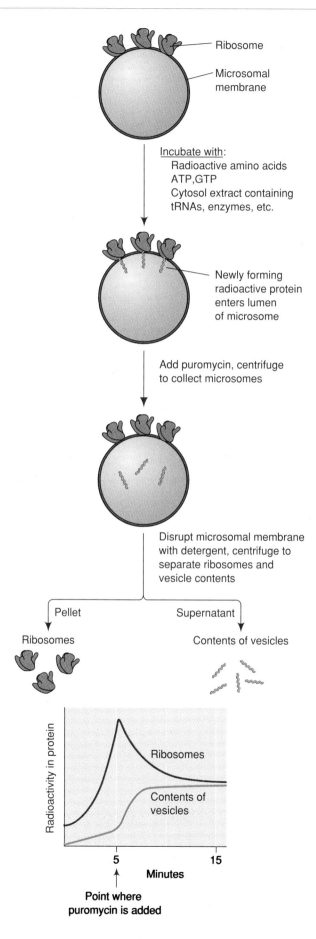

Figure 7-11 *Evidence That Proteins Synthesized on Membrane-Attached Ribosomes Pass Directly into the Lumen of the ER* Isolated rough ER vesicles (microsomes) were incubated with radioactive amino acids to label newly made polypeptide chains. Next, the addition of puromycin stops further protein synthesis and releases the newly forming polypeptide chains from the ribosomes. The microsomal vesicles were then collected by centrifugation and treated with detergent to separate the ribosomes from the internal contents of the vesicles. The graph shows that after the addition of puromycin, radioactivity is lost from the ribosomes and appears inside the vesicles. This suggests that the newly forming polypeptide chains are inserted through the ER membrane as they are being synthesized, and puromycin causes the chains to be prematurely released into the vesicle lumen.

still taking place. When the polypeptide chain is finally completed, it is either released into the lumen of the ER or, in the case of integral membrane proteins, retained within the lipid bilayer of the ER membrane. When the polypeptide chain exits from the protein-translocating channel, the ribosome dissociates from the ER membrane and the protein-translocating channel closes.

The major elements of the foregoing model have been verified by experiments in which various types of mRNAs, ribosomes, and membranes have been mixed together in different combinations. If, for example, mRNAs coding for proteins destined for secretion from the cell are mixed with free ribosomes and ER membranes stripped of their ribosomes, the newly synthesized proteins become associated with the ER membrane vesicles. When mRNAs coding for proteins that are not destined for the ER are employed in the same system, the newly synthesized proteins do not become associated with the ER membranes. Hence the information that specifies which proteins are to be synthesized in association with the ER must be contained in the mRNAs that code for these proteins. Studies involving genes with mutations in the bases coding for the signal sequence have verified the importance of this region in targeting the polypeptide chain to the endoplasmic reticulum.

The best evidence for the presence of protein-translocating channels in the endoplasmic reticulum has been obtained from studies using electrodes to measure the flow of current through isolated rough ER membranes. In these studies, puromycin was employed to release the polypeptide chains associated with membrane-attached ribosomes (Figure 7-14). Under such conditions, an increase in the flow of electric current through the membrane was measured. Presumably this increase in current flow occurred because the puromycin had cleared the transmembrane channels of their polypeptide chains, thereby allowing ions to pass freely through the channels. Current flow decreased

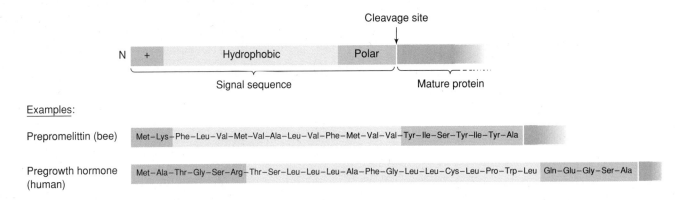

Figure 7-12 *General Features of the Signal Sequence That Targets Proteins to the ER* *The signal sequence typically consists of a positively charged terminal region, a central hydrophobic region, and a polar region adjoining the site where cleavage from the mature protein will take place.*

back to its original value when the ribosomes were subsequently removed from the membrane, indicating that the channel had closed. Such observations strongly support the idea that ER membranes contain protein-translocating channels that are open in the presence of attached ribosomes and closed in the absence of ribosomes.

Integral Membrane Proteins Contain Stop-Transfer Sequences as Well as Signal Sequences

Among the various classes of proteins synthesized on ribosomes attached to the rough ER, integral membrane proteins are unique in that they are retained in the membrane bilayer rather than being released into the lumen. What prevents these membrane proteins from being transferred into the lumen, and how do they become properly oriented within the membrane bilayer? In addressing these questions, it should be recalled that most integral proteins are anchored to the membrane bilayer by one or more transmembrane α-helical stretches of 15–25 mainly hydrophobic amino acids (page 168). Each of the hydrophobic stretches must be properly inserted into the membrane if a given integral protein is to be correctly oriented.

Let us first consider how the proper orientation is established for integral membrane proteins containing a single transmembrane α helix. Like other proteins synthesized on membrane-attached ribosomes, such *single-pass* integral membrane proteins have a signal sequence at their N-terminal end. The synthesis of single-pass integral membrane proteins therefore follows the steps outlined in Figure 7-13 up to the point where the newly forming polypeptide chain is emerging from the protein-translocating channel associated with the ER. However, at some point prior to completion of the polypeptide chain the hydrophobic stretch of 15–25 amino acids destined to form the transmembrane α helix is synthesized. This stretch of amino acids functions as a **stop-transfer sequence,** which halts the translocation

of the polypeptide chain through the membrane channel. At this stage the chain is thought to move laterally out of the channel and into the lipid bilayer of the surrounding membrane, where it becomes a permanent transmembrane segment (Figure 7-15, *left*). The transmembrane segment then anchors the protein in the membrane as the ribosome completes the process of protein synthesis. Support for this model has come from studies showing that mutant proteins lacking the hydrophobic stop-transfer sequence end up in the lumen and are secreted, rather than being retained in the membrane as single-pass integral proteins.

The preceding scenario creates a single-pass integral membrane protein in which the N-terminal end faces the lumen and the C-terminal end faces the cytosol. Single-pass proteins with the reverse orientation are created in a somewhat different manner (see Figure 7-15, *middle*). In this case the signal sequence is not cleaved from the growing polypeptide chain. Instead, its hydrophobic region is embedded in the membrane and forms the single transmembrane segment that anchors the protein to the membrane. Since the signal sequence creates the single transmembrane segment found in such proteins, a stop-transfer sequence is not required.

Membrane proteins with multiple membrane-spanning regions are formed in the same basic way as single-pass proteins, but an alternating pattern of signal sequences and stop-transfer sequences is involved. Figure 7-15 (*right*) illustrates how such an alternating pattern could create an integral protein with multiple hydrophobic segments passing back and forth across the membrane.

Insertion or Translocation into the ER Can Also Occur after a Protein Has Been Synthesized

For most proteins that are translocated across or integrated into ER membranes, translocation or integration

1 Protein synthesis initiated on unbound ribosomes

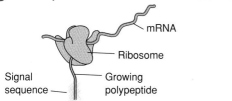

mRNA

Ribosome

Signal
sequence

Growing
polypeptide

2 SRP binds to signal sequence and halts protein synthesis

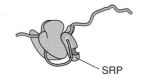

SRP

3 SRP-ribosome complex binds to ER membrane

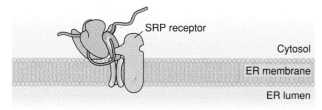

SRP receptor

Cytosol

ER membrane

ER lumen

4 Ribosome associates with protein-translocating channel

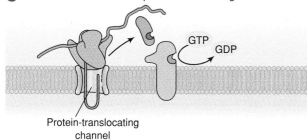

GTP GDP

Protein-translocating
channel

5 Protein synthesis resumes

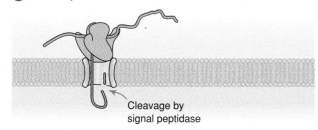

Cleavage by
signal peptidase

6 Protein synthesis is completed

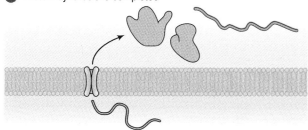

occurs while the protein is being synthesized. However, in several instances these events take place after protein synthesis has been completed and involve a process that requires neither SRP nor ribosomes. In such cases, insertion or translocation of polypeptide chains into the ER has been found to depend on an energy source such as ATP, as well as the presence of a class of protein molecules called **chaperones** that stabilize the unfolded state of the polypeptide chain being inserted or translocated. Chaperones are crucial components of this process because folded protein molecules cannot be inserted or translocated across membranes. Further research is needed to determine whether this transport pathway utilizes any of the same membrane components that are involved in the ER protein-translocating channel discussed earlier.

Newly Made Proteins Are Modified in the ER by N-Linked Glycosylation, Hydroxylation, and Linkage to Membrane-Bound Glycolipid

Proteins synthesized on ribosomes attached to ER membranes are subject to several kinds of modification reactions as they are being formed. Three of the more important modifications are briefly described below.

1. *N-linked glycosylation.* Most proteins made in the rough ER are **glycoproteins**—that is, proteins with covalently bound carbohydrate groups. During the formation of glycoproteins, sugar groups are added to the growing polypeptide chain before the process of protein synthesis has been completed. Because the sugar molecules are attached to the free NH_2 group of the amino acid asparagine, this process is referred to as **N-linked glycosylation.** The first step in N-linked glycosylation is the transfer of a 14-sugar *core oligosaccharide* to the amino acid asparagine in the newly forming polypeptide chain. The asparagines that serve as acceptor sites for these oligosaccharide chains are part of the amino acid sequence *Asn-X-Ser/Thr* (where X is any amino acid but proline, and Ser/Thr signifies that the third amino acid can be serine or threonine).

Figure 7-13 *Mechanism by Which Signal Sequences Target Newly Synthesized Proteins to the ER* *Protein synthesis is initiated on free cytoplasmic ribosomes. If the N-terminus of the newly forming polypeptide contains a signal sequence, it is bound by SRP, which in turn attaches the ribosome to the ER membrane. SRP is then released and protein synthesis resumes, allowing the elongating polypeptide chain to be translocated through the ER membrane as it is being synthesized. Integral membrane proteins are retained within the lipid bilayer after being synthesized, while other proteins are released into the ER lumen. Hence only Steps 1 through 4 (and sometimes Step 5) apply to the synthesis of integral membrane proteins.*

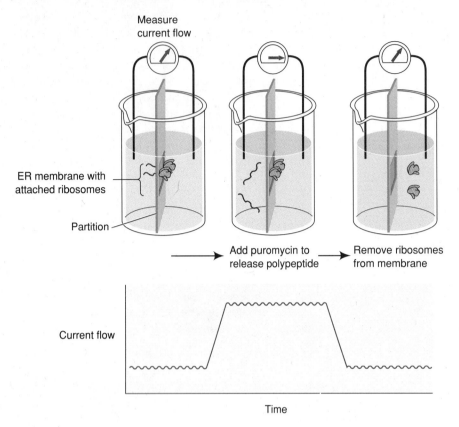

Figure 7-14 *Evidence for the Existence of Protein-Translocating Channels in Rough ER Membranes* *In these studies, electrodes were used to measure the flow of current through ER membranes containing attached ribosomes. When puromycin is added to trigger the release of the polypeptide chains, an increase in the flow of electric current is observed because the protein-translocating channels in the membrane remain open. When the ribosomes are removed from the membrane, the channels close and the flow of current returns to its original value.*

The core oligosaccharide always consists of the same combination of N-acetylglucosamine, mannose, and glucose molecules linked together in the same way (Figure 7-16). This 14-sugar oligosaccharide core is assembled from nucleoside-diphosphate sugar derivatives, which are sequentially added to a molecule of **dolichol phosphate** embedded within the ER membrane. Dolichol phosphate is a long-chain alcohol whose massive hydrophobic tail, 80–100 carbon atoms in length, anchors it firmly within the membrane bilayer. After the core oligosaccharide has been assembled on dolichol phosphate, it is transferred intact to the newly forming polypeptide as the growing chain emerges into the ER lumen. Several of the 14 sugars present in the core oligosaccharide are then removed before the glycosylated protein exits the ER.

2. *Hydroxylation.* In addition to N-linked glycosylation, proteins are also modified in the ER by hydroxylation of the amino acids proline and lysine. In contrast to the widespread occurrence of glycosylation, relatively few proteins are modified by hydroxylation. However, hydroxylation is a prominent event in cells that make collagen, an extracellular matrix protein that character-

istically contains large amounts of hydroxyproline and hydroxylysine (page 226).

3. *Attachment to membrane-bound glycolipid.* In Chapter 5 we learned that membranes contain a class of proteins that are anchored to the bilayer by covalent attachment to lipid molecules (page 169). One mechanism for creating such *lipid-anchored proteins* involves enzymes that catalyze the attachment of newly formed proteins to membrane-bound glycolipid in the ER (Figure 7-17). The lipid-anchored proteins generated by this process are usually destined for the exterior surface of the plasma membrane.

Newly Formed Proteins Acquire Their Proper Conformation in the ER

Before newly synthesized proteins can perform their normal biological functions, they must become folded into the appropriate three-dimensional conformation. Activities occurring in the ER lumen facilitate this process in several ways. One involves formation of the disulfide bonds that are required for the proper folding

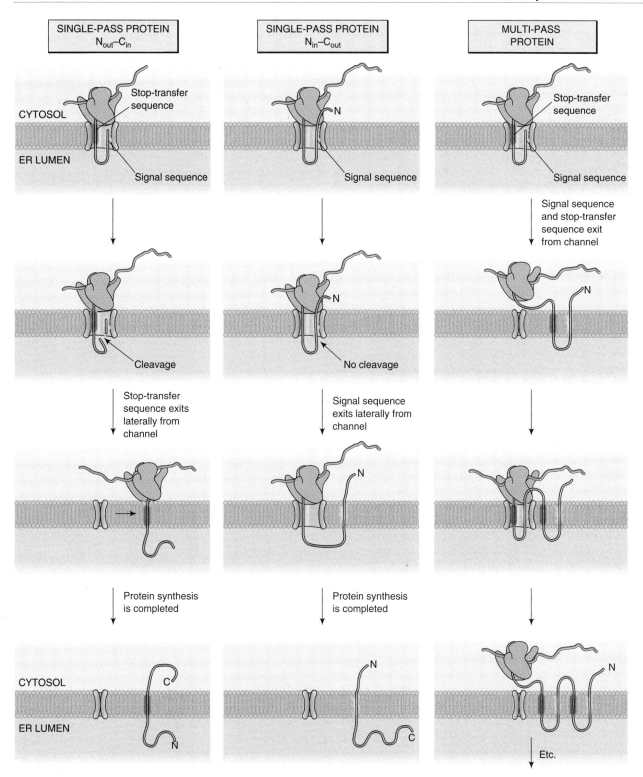

Figure 7-15 *Model Showing How Newly Forming Integral Membrane Proteins May Become Properly Oriented within the Membrane* (Left, middle) *Mechanism by which single-pass integral membrane proteins become properly oriented within the membrane. The term "N_{in}-C_{out}" refers to a protein whose N-terminus resides on the side of the membrane where protein synthesis was initiated and whose C-terminus resides on the opposite side of the membrane. N_{out}-C_{in} refers to the opposite orientation. Note that the signal peptide is not cleaved during the formation of proteins with the N_{in}-C_{out} orientation. (Right) Mechanism by which multiple-pass integral membrane proteins may become properly oriented within the membrane. In proteins of this type, an alternating pattern of signal sequences and stop-transfer sequences creates a polypeptide chain that passes back and forth across the bilayer.*

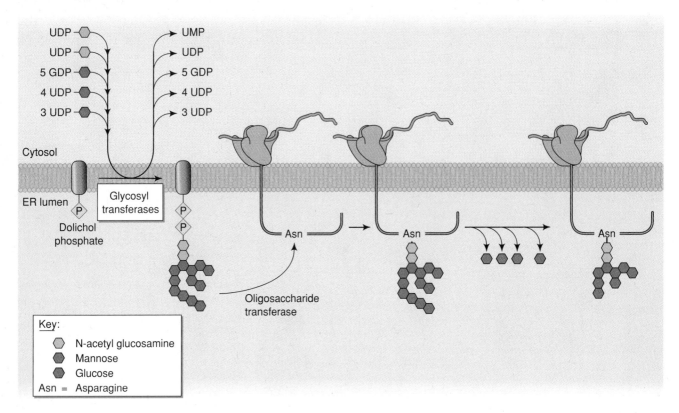

Figure 7-16 N-Linked Glycosylation of Newly Forming Proteins in the ER *A 14-sugar core oligosaccharide is assembled on dolichol phosphate embedded within the ER membrane. The 14-sugar oligosaccharide is then transferred intact to the amino acid asparagine in a newly forming polypeptide chain. Several of the sugars present in the core oligosaccharide are removed before the glycosylated protein exits from the ER.*

of most protein molecules (page 25). The lumenal surface of the ER membrane contains the enzyme **protein disulfide isomerase,** which catalyzes the formation and breakage of disulfide bonds between cysteine residues located in different parts of a polypeptide chain. This process often begins before the synthesis of a newly forming polypeptide has been completed, allowing various disulfide bond combinations to be tested until the most stable arrangement is found.

The development of proper conformation is also facilitated by several proteins which function as molecular chaperones that stabilize newly formed polypeptide chains during the folding process. One such molecule is *BiP,* a "binding-protein" present in the ER lumen that is thought to stabilize unfolded polypeptide chains by binding to their hydrophobic regions. This interaction is crucial because the hydrophobic regions of unfolded chains would otherwise be exposed to the surrounding aqueous environment, an inherently unstable arrangement. In addition to stabilizing polypeptides during the folding process, BiP binds to abnormally folded proteins and prevents them from leaving the ER.

In the case of proteins composed of multiple subunits, proper conformation also requires that the individual subunits become bound to one another. The joining together of independent subunits occurs in the lumen of the ER. Proteins consisting of more than one polypeptide chain can also be created by cleavage reactions that convert single polypeptide chains into proteins consisting of multiple chains. Although cleavage may take place while a polypeptide is still localized within the ER lumen, such reactions usually occur after the molecule has left the ER.

THE GOLGI COMPLEX

Most of the proteins synthesized on ribosomes attached to the endoplasmic reticulum are ultimately transported by this system of membrane channels to other regions of the cell. The first stop in this transport pathway is a specialized set of membranes collectively known as the **Golgi complex** (sometimes called *dictyosomes* in plant cells).

The Golgi Complex Consists of a Stack of Flattened Membrane Cisternae and Associated Vesicles

The Golgi complex was discovered in 1898 by the Italian cytologist Camillo Golgi, who stained nerve cells with osmium tetroxide and found that the stain became

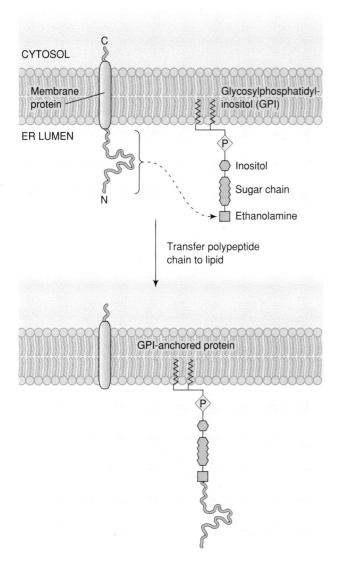

Figure 7-17 *Formation of Glycolipid-Anchored Membrane Proteins* *Enzymes localized in the ER catalyze the transfer of newly formed polypeptide chains to the membrane glycolipid, glycosylphosphatidylinositol (GPI).*

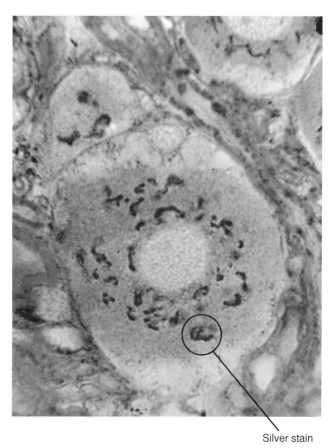

Silver stain

Figure 7-18 *Light Micrograph of a Silver-Stained Golgi Complex in Rat Nerve Cells* *Courtesy of R. G. Kessel.*

deposited in a threadlike network adjacent to the nucleus. Similar results were obtained when cells were stained with silver nitrate and other heavy-metal salts (Figure 7-18). Although staining of this structure was observed in a variety of cell types, the method failed to give consistent results and marked variations in staining pattern were evident. Since living cells viewed by phase microscopy exhibit no structure resembling the material stained by heavy metals, many cytologists viewed the region as an artifact of the staining process. In fact, the well-known Spanish microscopist Santiago Ramón y Cajal had observed the silver-stained material several years prior to Golgi, but had not reported the finding because, as he was to write later, "the confounded reaction never appeared again!"

The controversy surrounding the existence of the

Golgi complex was not settled until the advent of electron microscopy in the 1950s. Wherever light microscopists had seen heavy-metal staining, electron microscopy revealed the presence of a distinctive set of flattened membrane vesicles called **cisternae** (Figure 7-19). The reality of the Golgi complex as a legitimate cellular structure was therefore confirmed. Although the appearance of the Golgi complex varies significantly between cell types, it can almost always be recognized by one constant morphological feature: the presence of a stack of flattened membrane cisternae lying adjacent to one another (see Figure 7-3). The number of stacked cisternae present in the Golgi complex varies from a few to 20 or more. The membranes of the cisternae are often curved, giving a cuplike shape to the organelle. In addition to the stack of flattened cisternae, Golgi complexes have a variable number of small vesicles and membrane channels associated with them. The small vesicles are typically clustered around the cisternae, where they play a role in transporting materials between the Golgi complex and other cellular compartments.

The cisternae located at opposite ends of the Golgi complex differ from each other in size, shape, content, number of associated vesicles, and enzymatic activity. By convention, the end of the stack receiving newly

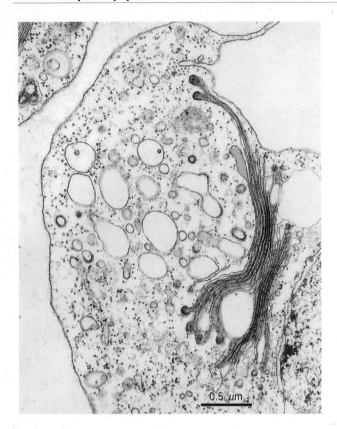

Figure 7-19 *Transmission Electron Micrograph of the Golgi Complex* *In this micrograph of a golden-brown alga, the prominent Golgi complex* (color) *appears as a series of flattened membrane cisternae lying next to the nuclear envelope. Courtesy of M. J. Wynne.*

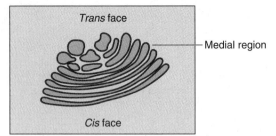

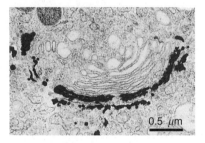

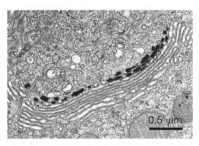

Figure 7-20 *Polarity of the Golgi Complex* (Top) *Diagram summarizing the naming systems used to refer to the opposite sides of the Golgi complex.* (Middle) *Transmission electron micrograph of mouse epididymis stained with osmium, which is deposited mainly in the* cis *cisternae of the Golgi complex.* (Bottom) *The same tissue stained cytochemically for the enzyme thiamine pyrophosphatase, which is localized in the* trans *cisternae of the Golgi complex. Courtesy of D. S. Friend.*

synthesized proteins from the ER is termed the *cis* **face,** the opposite end is termed the *trans* **face,** and the area in between is called the *medial region* (Figure 7-20). In cells where the Golgi membranes are curved, the *cis* face is easily identified by its location on the convex, outer surface of the complex. Networks of membrane channels and vesicles lie adjacent to both the *cis* and *trans* faces of the Golgi complex; these two regions are called the *cis* **Golgi network** and the *trans* **Golgi network,** respectively.

The enzymatic composition of Golgi membranes has been studied by both cytochemistry and subcellular fractionation. The most reliable cytochemical marker for the Golgi complex is the enzyme *thiamine pyrophosphatase* (see Figure 7-20). Despite the fact that the physiological function of this enzyme is unknown, it is an especially useful marker for Golgi membranes because it rarely occurs in other organelles. *Nucleoside diphosphatase* is another fairly consistent component of Golgi membranes, although it may be present in the endoplasmic reticulum as well.

Golgi membranes can be isolated and purified by a combination of moving-zone and isodensity centrifugation, using enzymatic assays as well as direct electron

microscope observation to confirm the identity of fractions containing Golgi membranes (Figure 7-21). Biochemical analyses of isolated Golgi membranes have revealed the presence of several enzymes whose presence had not been suspected from cytochemical tests. Most prominent among these are *glycosyl transferases,* which catalyze the attachment of sugar residues to protein chains, and *glucan synthetases,* which catalyze polysaccharide biosynthesis. As we will see shortly, the presence of these enzymes reflects the role played by the Golgi complex in carbohydrate metabolism.

Proteins Synthesized in the Rough ER Are Routed through the Golgi Complex

The first clues to the function of the Golgi complex were provided by light microscopists, who observed that vacuoles being secreted from living cells originate in the Golgi region, and that cells with intense secre-

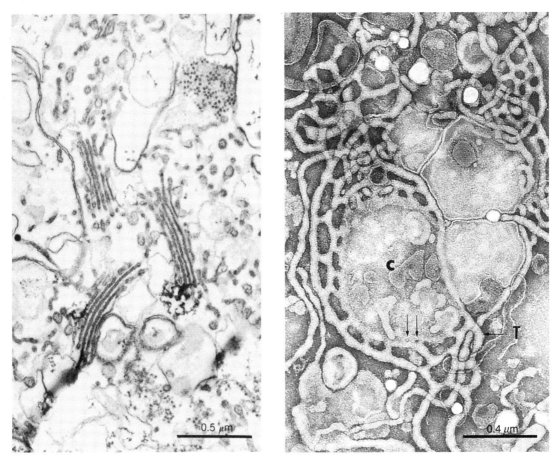

Figure 7-21 *Isolated Golgi Complexes from Plant and Animal Cells* (Left) *Thin-section electron micrograph of Golgi complexes isolated from onion tissue.* (Right) *Negatively stained Golgi membranes isolated from rat liver cells. These complexes are composed of a series of platelike structures containing a central discoid cisterna (C) surrounded by a network of branching tubules (T). Depressions at the periphery of the cisternae (arrows) represent the site of fusion of tubules with the cisternal membranes. Courtesy of H. H. Mollenhauer (left) and J. M. Sturgess (right).*

tory activity have a prominent Golgi complex. Both observations suggested that the Golgi complex plays a role in cell secretion. Further support for this view was later provided by electron microscopy, which revealed that vesicles associated with the *trans* face of the Golgi complex contain granular material similar to that present in vesicles destined for secretion. Electron microscopy alone, however, gives a static picture of a cell at any given instant, so it cannot be used to follow the dynamic flow of materials between compartments. Proof that substances destined for secretion from the cell pass through the Golgi complex therefore required the use of radioactive substrates to trace the movement of macromolecules through the various cell compartments.

In a classic study of this type, George Palade and his collaborators Lucien Caro and James Jamieson investigated the secretion of digestive enzymes by exocrine cells of the pancreas. Intact guinea pigs or isolated slices of pancreas from these animals were briefly exposed to the radioactive amino acid ^{3}H-leucine to label newly synthesized proteins. Electron microscopic autoradiographs revealed that in the first few minutes after exposure to ^{3}H-leucine, radioactive protein is associated with the rough endoplasmic reticulum. Within 10 minutes, however, it can be observed near the Golgi complex, after 30 minutes it is present in *trans* Golgi vesicles, and after an hour, most of the radioactivity has moved to large secretory vesicles (also called *zymogen granules*) that store protein destined for secretion from the cell (Figure 7-22). Shortly thereafter, the radioactive protein is released from the cell. By counting the number of silver grains localized over various regions of the cell at successive time intervals after labeling, one can view graphically the passage of newly synthesized protein through the various cellular compartments. The data obtained support the conclusion that secretory proteins are synthesized in the rough ER, transported through the Golgi complex into secretory vesicles, and finally secreted from the cell (Figure 7-23, *left*).

This conclusion, based on the use of autoradiography, was soon reinforced by data obtained from subcellular

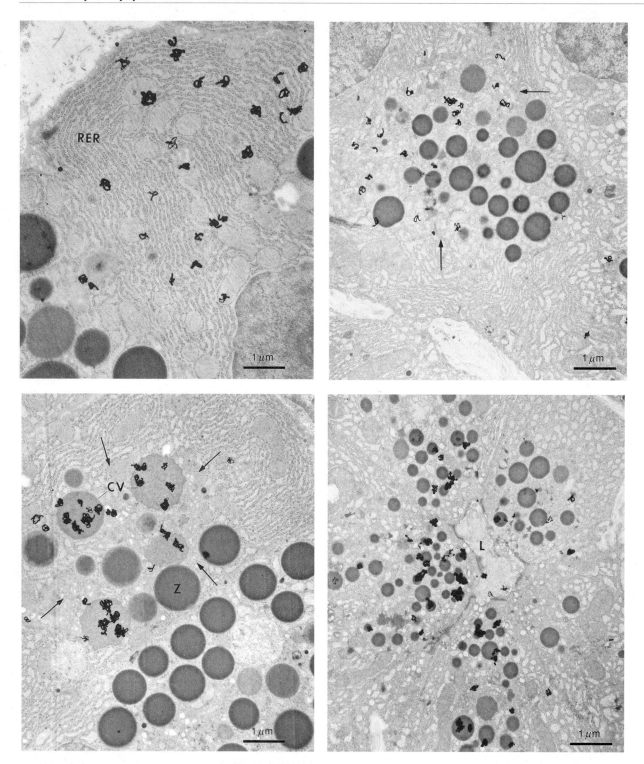

Figure 7-22 Demonstration of the Pathway Followed by Secretory Proteins Using Electron Microscopic Autoradiography
Pancreatic exocrine cells were incubated with ³H-leucine for 3 minutes to label newly synthesized proteins. The cells were then incubated for various periods in the absence of radioisotope, and the fate of the radioactive protein was monitored by electron microscopic autoradiography. (Top left) After 3 minutes of incubation the silver grains, representing newly synthesized proteins, are localized almost exclusively over the rough endoplasmic reticulum. (Top right) After 7 minutes the majority of the newly synthesized protein has moved to the periphery of the Golgi complex. (Bottom left) After 37 minutes the silver grains are concentrated over the condensing vacuoles (CV). Arrows indicate the periphery of the Golgi complex. Zymogen granules (Z) are still unlabeled at this point. (Bottom right) After 117 minutes the radioactivity is mainly localized over zymogen granules; some radioactive protein has even been secreted from the cell into the lumen (L) of the gland. Courtesy of J. D. Jamieson.

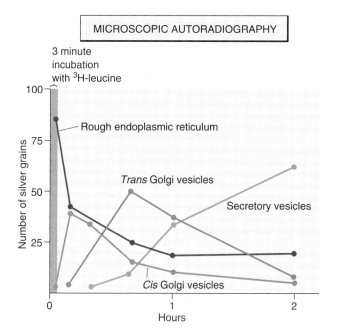

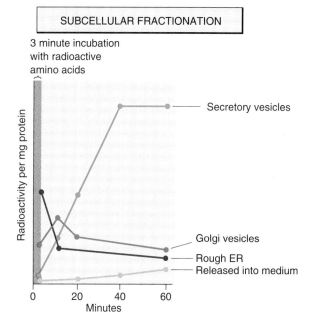

Figure 7-23 *Quantitative Data Obtained by Microscopic Autoradiography and Subcellular Fractionation Showing the Pathway Followed by Proteins Destined for Secretion* (Left) *Pancreatic exocrine cells were incubated for 3 minutes with ^{3}H-leucine to label newly synthesized protein, and the radioactive amino acid was then removed from the medium. After subsequent incubation for varying periods of time, cells were prepared for electron microscopic autoradiography and the number of silver grains counted over different regions of the cell. The data reveal a general pattern in which newly made proteins pass from the rough endoplasmic reticulum to the Golgi complex and finally to secretory vesicles.* (Right) *Isolated guinea pig pancreas was incubated with radioactive amino acids for 3 minutes to label newly made protein molecules. The radioactive isotope was then removed, incubation was continued for varying time periods, and the cells were homogenized and analyzed by subcellular fractionation. The data reveal that newly synthesized protein appears first in the rough ER, followed by vesicles derived from the Golgi complex, and finally in secretory vesicles. Secretion of radioactive protein from the cells occurs only after the secretory vesicles have become radioactive.*

fractionation experiments. In studies of this type, cells that have been briefly exposed to radioactive amino acids to label newly synthesized proteins are homogenized and fractionated into their subcellular components. Data obtained from such experiments revealed that newly synthesized proteins appear first in the rough ER, then in the Golgi complex, and finally in large secretory vesicles (see Figure 7-23, *right*). In other words, the same pattern that had been observed using electron microscopic autoradiography is seen in subcellular fractionation studies as well.

Additional evidence for this pattern has also been obtained from studies of yeast strains called *sec* mutants, which lose the ability to secrete proteins when grown at high temperatures. Under such conditions, proteins normally destined for secretion are accumulated within the cell. In one class of *sec* mutants, secretory proteins accumulate in the rough ER. In a second class of mutants, secretory proteins pass through the rough ER but become accumulated in the Golgi complex. And in a third class of mutants, secretory proteins pass through the rough ER and Golgi complex, but become accumulated in secretory vesicles without being secreted. Such findings provide further support for the conclusion that secretory proteins pass from the rough ER to the Golgi complex to

secretory vesicles to the cell exterior. Recent studies have revealed that many of the *sec* genes code for G proteins; studies designed to investigate the functions of these molecules should help to elucidate how proteins are transported through the ER, Golgi complex, and secretory vesicles prior to being secreted from the cell.

Proteins destined for secretion from the cell are not the only proteins to pass from the ER to the Golgi complex. Other proteins synthesized in the ER, such as those destined for incorporation into lysosomes and integral proteins destined for incorporation into the plasma membrane, also pass through the Golgi complex on the way to their final destinations.

Vesicles Transport Materials between the Endoplasmic Reticulum and the Golgi Complex

Proteins and lipids that have been synthesized in the ER are ultimately transported to a variety of other cellular compartments. In general, these molecules are transported by vesicles that bud off from one membrane compartment and are then specifically targeted to, and fuse with, another membrane compartment. During the first step in this process, materials that have been synthesized in the rough ER are transported to the Golgi complex by small *transitional vesicles* that bud from the ER and fuse with components of the *cis* Golgi network (Figure 7-24). In this way, lumenal proteins as well as integral membrane proteins are transferred from the ER to the Golgi complex.

Recent studies involving the drug **brefelden A** suggest that movement of materials between the ER and Golgi complex is a two-way process. In cells treated with brefelden A, the Golgi complex is disrupted and enzymes normally associated with the Golgi return back to the ER. It has been postulated that this type of reverse transport also occurs in normal cells, but is magni-

fied in cells treated with brefelden A because disruption of the Golgi stack prevents the normal forward transport of materials through the Golgi (Figure 7-25). Brefelden A disrupts Golgi membranes by binding to β-COP, a protein that normally coats the Golgi cisternae and associated vesicles.

The existence of reverse transport from the Golgi to the ER is also suggested by the discovery that the Golgi complex normally receives some proteins whose final destination is the ER. Such proteins contain a signal sequence at their carboxy-terminal end consisting of the amino acids Lys-Arg-Glu-Leu (KDEL in single-letter code) or a closely related sequence. To return these proteins to the ER, the Golgi complex is thought to em-

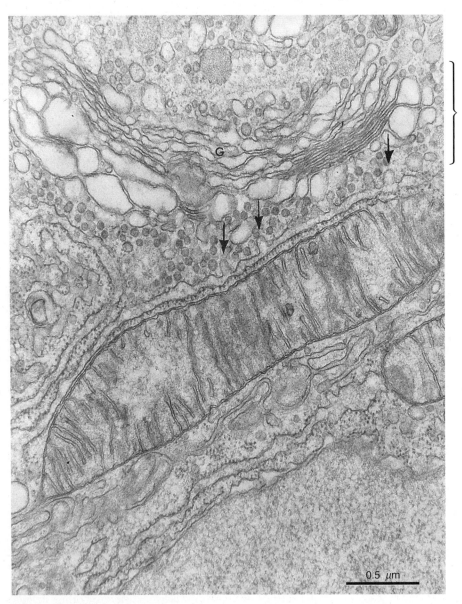

Golgi complex

0.5 μm

Figure 7-24 Electron Micrograph Showing the Transitional Vesicles That Transport Material from the Endoplasmic Reticulum to the Golgi Complex *This transmission electron micrograph of a mouse intestinal epithelial cell shows vesicles* (color) *budding from the endoplasmic reticulum* (arrows) *and moving to the* cis *face of the Golgi complex. Courtesy of D. S. Friend.*

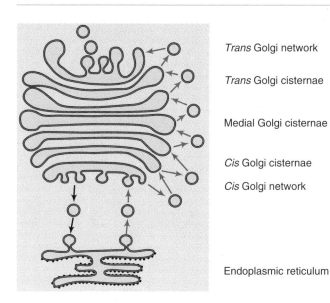

Trans Golgi network

Trans Golgi cisternae

Medial Golgi cisternae

Cis Golgi cisternae

Cis Golgi network

Endoplasmic reticulum

Figure 7-25 *Patterns of Vesicle Traffic in the Golgi Complex* *The forward movement of materials through the Golgi cisternae* (red arrows) *is inhibited in cells treated with the drug befrelden A, which disrupts the organization of the Golgi cisternae. Under such conditions the reverse flow of materials from the* cis *Golgi network back to the ER predominates* (black arrows).

ploy a receptor protein that binds to the KDEL sequence and delivers the protein back to the ER.

Proteins Are Glycosylated as They Pass through the Golgi Complex

Those proteins that are not returned to the ER from the *cis* Golgi network are permitted to travel onwards, passing through the *cis, medial,* and *trans* Golgi cisternae. This forward movement is accomplished by small vesicles that bud off from one cisterna and fuse with the membrane of the next (see Figure 7-25). Although the proteins contained in these vesicles have already been subjected to N-linked glycosylation in the rough ER, further glycosylation takes place as they pass through the Golgi complex.

The idea that the Golgi complex plays an important role in protein glycosylation was first clearly established in studies carried out by Marian Neutra and Charles Leblond on mucus-secreting cells of the intestine. The predominant constituent of intestinal mucous is a glycoprotein whose carbohydrate content accounts for the viscosity of this secretion. To determine the subcellular site where attachment of the carbohydrate occurs, rats were injected with ^{3}H-glucose and the fate of the radioactive sugar was monitored. In the first few minutes, electron microscopic autoradiographs revealed that most of the radioactive glucose was present in the Golgi region (Figure 7-26). Within 40 minutes it had migrated to mucous granules located adjacent to the *trans* face of the Golgi complex; several hours later these radioactive mucous granules had moved

to the cell periphery and begun to secrete their radioactive contents into the extracellular fluid.

The preceding observations provided the first indication that sugar chains are added to secretory proteins in the Golgi complex. Further support for this conclusion was provided later by the discovery that isolated Golgi membranes contain *glycosyl transferases,* the enzymes that catalyze the attachment of sugar residues to protein chains. These enzymes are integral membrane proteins whose active sites face the lumen of the Golgi cisternae. The relationship between the protein glycosylation reactions that occur in the Golgi complex and those taking place in the ER has been clarified by studies carried out on the synthesis of glycoproteins whose carbohydrate sequences are known. Such studies have revealed that the Golgi complex is involved in two types of glycosylation: (1) The N-linked oligosaccharide chains that had initially been joined to newly forming proteins in the rough ER are further modified in the Golgi complex. New sugars are added and others are removed as the proteins move through the Golgi complex, creating a variety of N-linked oligosaccharide structures. (2) In addition, the Golgi complex contains glycosyl transferases that attach sugar chains to the hydroxyl groups of the amino acids serine, threonine, and hydroxylysine. Because the sugar molecules become attached to the oxygen atom of the hydroxyl group, this process is referred to as **O-linked glycosylation.** The length and composition of O-linked oligosaccharide chains is quite variable. The most extensive example of O-linked glycosylation occurs in proteoglycans (page 225), where glycosaminoglycan chains such as chondroitin sulfate and keratan sulfate are attached to multiple serine hydroxyl groups of a core protein. The net result is a glycosylated protein in which carbohydrate accounts for as much as 95 percent of the total mass.

The Golgi Complex Is Also Involved in Synthesizing Polysaccharides

In addition to catalyzing the addition of sugar groups to protein molecules, the Golgi complex is also involved in linking sugar molecules together to form polysaccharides. The main polysaccharide synthesized in the Golgi complex of animal cells is hyaluronate, a unique glycosaminoglycan because it is secreted into the extracellular matrix without being covalently linked to protein. In plant cells, several polysaccharide components of the cell wall are synthesized in the Golgi complex prior to being secreted into the extracellular space. Included in this category are the hemicelluloses and pectins, but not cellulose. Cellulose is synthesized by enzyme complexes located within the plasma membrane (page 245), although the Golgi complex is responsible for packaging these enzyme complexes into membrane vesicles that migrate to the cell surface and fuse with the plasma membrane.

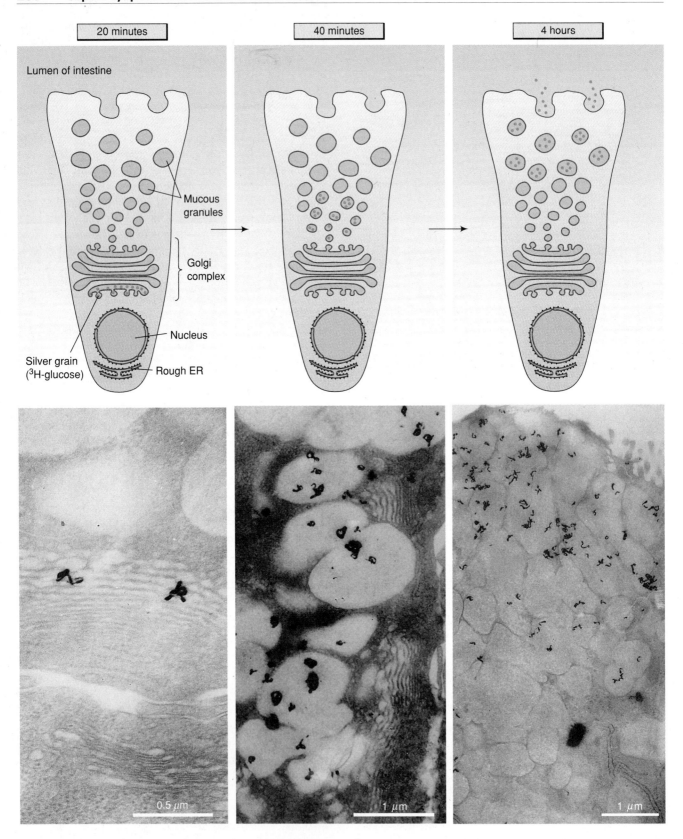

Figure 7-26 *The Localization of Radioactive Glucose in Rat Intestinal Cells at Varying Times after Injection of ³H-Glucose*
(Left) *In an electron microscopic autoradiograph prepared 20 minutes after injection of ³H-glucose, radioactivity is localized in the Golgi complex. (Middle) Forty minutes after injection, radioactivity is localized in mucous granules adjacent to the* trans *face of the Golgi complex. (Right) Four hours after injection, the radioactive mucous granules have moved to the cell periphery and are secreting their radioactive contents into the extracellular fluid. Courtesy of C. P. LeBlond*

Proteins Are Sorted by the Golgi Complex Using Chemical Signals and Bulk Flow

The proteins and polysaccharides processed in the Golgi complex do not all share the same destination. Some are eventually secreted from the cell, whereas others are destined for intracellular sites such as the plasma membrane or lysosomes. How are these molecules sorted from one another so that each can be routed to its proper destination? In several cases, the molecular signals that guide the sorting process have been identified. For example, the KDEL sequence mentioned earlier in the chapter signals that a protein is to be returned to the ER (page 278). We will see shortly that proteins destined for lysosomes are recognized by another type of chemical signal, the sugar mannose 6-phosphate.

In the absence of a specific targeting signal, materials passing through the Golgi complex are transported into secretory vesicles that fuse with the plasma membrane. Molecules present in the lumen of such vesicles are expelled from the cell, while the membrane proteins and lipids become incorporated into the plasma membrane (Figure 7-27). The existence of this so-called *bulk flow* pathway for moving

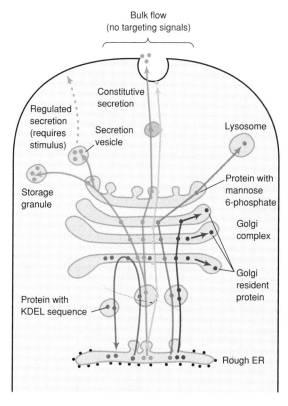

Figure 7-27 Protein Sorting in the Golgi Complex *Molecules that lack targeting signals are continuously carried by bulk flow to the cell surface. Other proteins are targeted to various cellular locations by the presence of specific chemical signals. The two best characterized signals are KDEL, which targets proteins to the ER, and mannose 6-phosphate, which targets proteins to lysosomes.*

materials to the cell surface has been elegantly demonstrated by experiments in which cells were incubated with small peptides containing the acceptor sequence for N-linked glycosylation, *Asn-X-Ser/Thr.* Such peptides diffuse into the lumen of the ER, where they are glycosylated and transported to the Golgi complex; after passing through the Golgi complex, they are rapidly secreted from the cell.

The preceding observations indicate that in the absence of a specific targeting signal, molecules present in the ER are automatically transported via the Golgi complex to the cell surface. Although glycosylated peptides were involved in these particular studies, even glycosylation seems to be unnecessary for this process of bulk flow. Studies utilizing the inhibitor *tunicamycin,* which blocks N-linked glycosylation, have shown that at least some proteins made in the ER are still secreted even though glycosylation has not taken place. Thus no particular chemical signal appears to be required for transporting materials from the ER to the Golgi to the cell surface.

The Golgi Complex Produces Various Kinds of Intracellular Granules

Most of the materials that pass through the Golgi complex have one of three possible destinations: secretion from the cell, incorporation into lysosomes, or incorporation into the plasma membrane. In the remaining sections of this chapter, each of these three outcomes will be examined in detail. Before we proceed, however, it should be emphasized that these destinations are not the only ones for materials processed by the Golgi complex. Under appropriate circumstances, the Golgi can also produce various kinds of cytoplasmic storage vesicles containing materials such as enzymes, pigments, and yolk.

One of the more striking examples of this phenomenon occurs in *neutrophils,* the white blood cells that ingest and destroy invading bacteria. From the extensive studies of Dorothy Bainton and Marilyn Farquhar, we know that mature neutrophils contain two types of cytoplasmic granules, called *azurophil granules* and *specific granules,* which can be distinguished from each other on the basis of differences in size, density, staining characteristics, and mode of origin (Figure 7-28, *bottom*). Azurophil granules, so named because of their affinity for the basic dye methylene azure, are large, dense granules about 800 nm in diameter that form relatively early in neutrophil development. In contrast, specific granules are less dense, measure about 500 nm in diameter, and form later in development. Cytochemical tests for individual enzymes have revealed that azurophil granules are filled with hydrolytic enzymes characteristic of lysosomes, whereas specific granules contain the enzymes alkaline phosphatase and lysozyme.

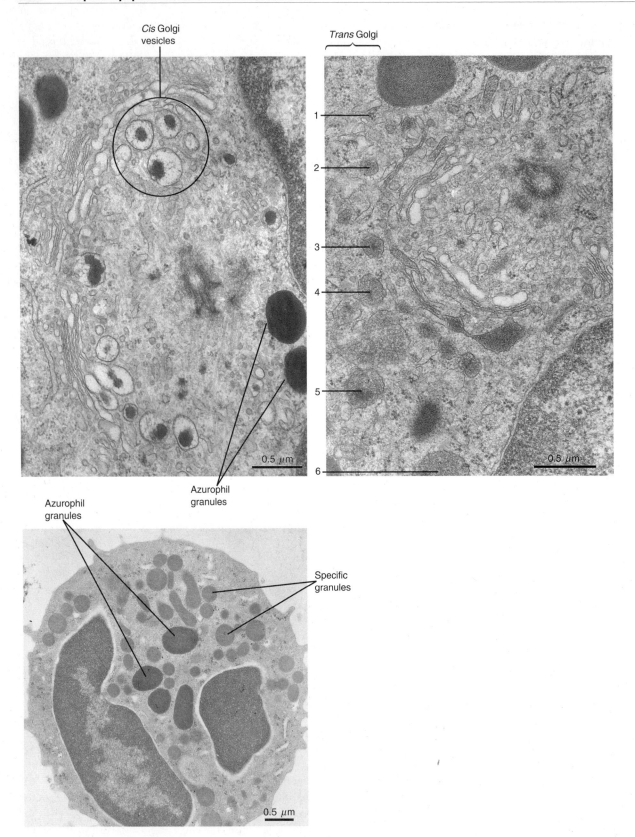

Figure 7-28 The Development of Azurophil and Specific Granules in Neutrophils   (Top left) *Development of azurophil granules in an immature neutrophil. Vesicles destined to form azurophil granules are seen budding from the* cis *face of the Golgi complex. Several mature azurophil granules are also present outside the Golgi zone.* (Top right) *Formation of specific granules from the* trans *face of the Golgi complex. Small granules first accumulate along the* trans *membranes (labeled 1, 2, and 3), and then merge to form larger granules (designated 4, 5, and 6).* (Bottom) *Mature neutrophil showing four azurophil granules and many smaller specific granules in the cytoplasm. Courtesy of D. F. Bainton.*

Electron microscopic examination of developing neutrophils has provided an elegant picture of the role played by the Golgi complex in producing the two types of granules. The first granules to appear in immature cells are the azurophils, which develop from large vacuoles that can be seen budding from the *cis* face of the Golgi cisternae (Figure 7-28, *top right*). These vacuoles have a dense central core of material destined to become the contents of the azurophil granules. After the azurophil granules have matured, smaller vesicles containing a homogeneous granular matrix bud off the *trans* Golgi cisternae and develop into specific granules (Figure 7-28, *top right*). The fact that electron microscopy reveals azurophil granules emerging from the *cis* cisternae and specific granules emerging from the *trans* cisternae provides visual proof that the Golgi complex can discriminate between the various macromolecules it handles, sorting and packaging them in different ways.

CELL SECRETION

Animal and plant cells secrete a multitude of different kinds of enzymes, hormones, neurotransmitters, local mediators, serum proteins, antibodies, and constituents of the extracellular matrix, including cell wall components in plants. In spite of the enormous diversity in the kinds of secretory products involved, the mechanism by which these different substances are secreted from the cell involves a common set of underlying principles.

Secretion Can Be Either Constitutive or Regulated

Two secretory pathways occur in eukaryotic cells. In **constitutive secretion,** secretory and membrane proteins are moved to the cell surface by membrane vesicles in a continuous, ongoing fashion by the process of nonselective bulk flow. The membrane vesicles fuse with the plasma membrane and expel their contents to the cell exterior without the need for any external stimulus. In addition to providing the integral membrane proteins and lipids that are needed by the plasma membrane, the constitutive secretion pathway is used for the unregulated secretion of enzymes, growth factors, and extracellular matrix components.

In contrast to constitutive secretion, which occurs in virtually all cells, the process of **regulated secretion** is restricted to cells that are specialized for the purpose of secreting specific products. Pancreatic cells that secrete digestive enzymes, or anterior pituitary cells that secrete peptide hormones, are typical examples of specialized secretory cells that carry out regulated secretion. Rather than being immediately expelled from the cell, the secretory materials produced by such cells are concentrated and stored in cytoplasmic secretory vesi-

cles; secretion occurs only in response to an appropriate external stimulus.

Vesicles Destined for the Constitutive and Regulated Secretory Pathways Can Be Distinguished from One Another

The existence of two secretory pathways means that the Golgi complex must handle materials destined for the two pathways differently. Proteins destined for the regulated secretory pathway are concentrated and condensed into a compact form as they pass through the Golgi complex. The particular region of the Golgi complex involved in this process varies among cell types and organisms. In guinea pig pancreas, for example, large vesicles at the Golgi periphery, called **condensing vacuoles,** are the usual site for concentrating protein. Other organisms employ the *trans* flattened cisternae and/or the dilated margins of the Golgi stack for the same purpose. In either case, the net result is a highly concentrated secretory product contained in membrane-enclosed vesicles that emerge from the *trans* side of the Golgi complex. As they bud from the *trans* Golgi network, these vesicles are usually coated with a protein called *clathrin* (page 296). The clathrin coat, which is thought to play a role in the budding process, is lost after the vesicles emerge from the Golgi. These small vesicles then fuse with each other to generate larger **secretory vesicles,** which are stored in the cytoplasm until an appropriate signal triggers the expulsion of their contents from the cell. Because they are often characterized by the presence of dense granular contents, secretory vesicles are sometimes referred to as *zymogen* or *secretory granules.*

In contrast to proteins destined for regulated secretion, molecules destined for constitutive secretion are sorted into a separate family of membrane vesicles that are not coated with clathrin as they bud from the Golgi complex. These vesicles, which tend to be smaller than the vesicles involved in regulated secretion, are referred to as **transport vesicles** because they rapidly transport materials to the cell surface. How does the Golgi complex distinguish those proteins that belong in secretory vesicles from those that belong in transport vesicles? Examination of the chemical structures of various secretory proteins has failed to reveal any shared amino acid sequence or carbohydrate chain that might be used as a molecular signal. Yet when DNA sequences coding for different secretory proteins are introduced into the same cell, all proteins destined for regulated secretion are sorted into the same secretory vesicles. Hence some mechanism is clearly capable of recognizing proteins destined for regulated secretion and packaging them into the same secretory vesicles.

Some evidence suggests that this sorting mechanism is based on the tendency of *regulated* secretory proteins to aggregate under the low pH conditions that normally prevail in the *trans* Golgi network. Because molecules destined for *constitutive* secretion do not aggregate under these same low pH conditions, it is possible that a mechanism exists for selectively carrying aggregated proteins to the region of the *trans* Golgi complex where secretory vesicles are formed. The low pH conditions that play a crucial role in this process are produced by an ATP-dependent transporter that pumps protons into the lumen of the *trans* Golgi network.

Cleavage of Precursor Proteins Often Occurs in Immature Secretory Vesicles

Many secreted enzymes and hormones are synthesized as inactive precursors called **proenzymes** or **prohormones** that are converted to their mature, active form by cleavage of the polypeptide chain. In some cases cleavage does not occur until after the protein has been secreted from the cell. For example, the pancreas secretes inactive enzyme precursors, such as trypsinogen and chymotrypsinogen, into the intestines in their inactive forms. Only after secretion are these precursors cleaved to generate the active digestive enzymes, trypsin and chymotrypsin.

In the case of many polypeptide hormones, cleavage occurs before the prohormone has left the cell. To investigate where in the cell this cleavage occurs, experiments have been carried out in which insulin-secreting cells were stained with antibodies that selectively bind either to insulin or to its inactive precursor, *proinsulin* (Figure 7-29). To allow the antibodies to be seen, they were linked to electron-opaque gold particles. In electron micrographs, the antibody against proinsulin was found to stain material present in the Golgi complex and in clathrin-coated vesicles budding from the *trans* Golgi network. In contrast, antibody directed against mature insulin reacts only with mature secretory vesicles. Such observations indicate that proinsulin is cleaved to produce mature insulin as the clathrin-coated immature vesicles are converted to mature secretory vesicles. This implies that the protease responsible for cleaving proinsulin to insulin is packaged into newly forming secretory vesicles along with proinsulin.

Materials Are Expelled from Cells by Exocytosis

The final stage in the secretory pathway occurs when the membrane of a secretory or transport vesicle fuses with the plasma membrane, discharging the contents of the vesicle into the extracellular space (Figure 7-30). This process, called **exocytosis,** occurs spontaneously in constitutive secretion, but must be triggered by an

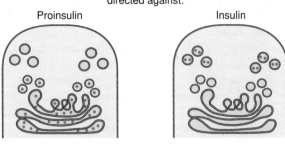

Stained with antibodies directed against:

Proinsulin   Insulin

Figure 7-29 *Identification of the Cellular Site Where Proinsulin Is Cleaved to Insulin* *Insulin-producing cells were stained with antibodies directed against proinsulin or insulin. To permit visualization of the antibody molecules, they were linked to gold particles (colored dots). The antibody against proinsulin stains the Golgi complex as well as clathrin-coated vesicles budding from the* trans *Golgi network. In contrast, the antibody directed against mature insulin reacts only with mature secretory vesicles. This means that proinsulin is cleaved to mature insulin as the clathrin-coated immature vesicles are converted to mature secretory vesicles.*

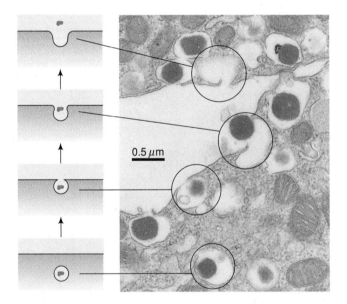

0.5 μm

Figure 7-30 *The Process of Exocytosis* *The electron micrograph of a rat pancreas cell shows secretory vesicles fusing with the plasma membrane and discharging their contents into the extracellular space. Courtesy of L. Orci.*

appropriate external agent, such as a neurotransmitter or hormone, in regulated secretion. Many of the neurotransmitters and hormones that trigger exocytosis share a common mechanism of action: the ability to increase the intracellular concentration of free cytosolic Ca^{2+} by promoting its release from the ER or uptake from outside the cell. The importance of Ca^{2+} has been demonstrated experimentally by treating cells with ionophores that artificially increase the permeability of the plasma membrane to Ca^{2+}. Under such conditions, calcium ions diffuse into the cell from the extracellular fluid and trigger exocytosis. Among the targets of these calcium ions are the **annexins,** a family of Ca^{2+}-binding proteins that promote the fusion of membrane vesicles with the plasma membrane.

In addition to calcium ions, the nucleotides GTP and ATP are also required for exocytosis. Although the exact role of these nucleotides is yet to be unraveled, useful information is beginning to emerge from studies on yeast mutants in which exocytosis fails to occur. Several of these mutants involve abnormalities in genes coding for GTP-binding proteins that are members of a family of proteins called **Rab proteins.** Rab proteins have been implicated not only in the fusion of secretory vesicles with the plasma membrane, but also in a variety of other membrane fusion and budding events, such as fusion of ER membranes with the Golgi complex and the subsequent budding of membrane vesicles from the Golgi region.

During exocytosis, the fusion of secretory vesicles with the plasma membrane typically occurs at a specific location on the cell surface. In glandular cells, for example, discharge occurs only at the region of the cell facing the lumen of the gland. The mechanism by which secretory vesicles are guided to the appropriate region of the cell surface is thought to involve interactions with the cytoskeleton. It has been shown, for example, that drugs that disrupt microtubules also inhibit secretion. Although the exact role played by microtubules during exocytosis is uncertain, it is known in other situations that *motor proteins* such as dynein and kinesin can attach vesicles to microtubules and propel them over long distances by an ATP-dependent mechanism (page 616). A similar type of active vesicle migration may occur in exocytosis as well.

Membrane Components Are Recycled from the Plasma Membrane Back to the Golgi Complex

Since the membranes of secretory vesicle become incorporated into the plasma membrane during exocytosis, the question arises as to how uncontrolled expansion of the plasma membrane is prevented in secretory cells. The answer is to be found in a recycling pathway in which membrane vesicles invaginate and bud off from the plasma membrane, migrate back to the Golgi complex, and fuse with the Golgi cisternae or their associated membranes. Cell secretion can therefore be viewed as a cyclic process in which membrane vesicles continually move back and forth between the Golgi complex and the cell surface (Figure 7-31).

But vesicle recycling in turn raises another potential problem: How do membrane vesicles find the appropriate target membrane prior to initiating fusion? This

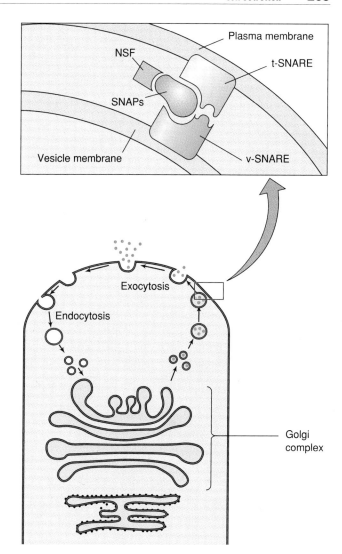

Figure 7-31 *Membrane Recycling between the Golgi Complex and the Plasma Membrane* *In secretory cells, membranes derived from secretory vesicles are incorporated into the plasma membrane during the process of exocytosis. To prevent uncontrolled expansion of the plasma membrane, vesicles invaginate and bud off from the cell surface, migrate back to the Golgi complex, and fuse with Golgi cisternae or their associated membranes. The enlarged area of the diagram illustrates how a vesicle membrane may recognize the proper target membrane prior to membrane fusion. According to this model, a protein in the vesicle membrane (v-SNARE) recognizes a protein in the target membrane (t-SNARE). NSF and SNAPs are proteins that mediate the fusion of the vesicle and target membranes once the two membranes have become attached.*

problem could be quite severe in secretory cells, where membrane vesicles continually pass back and forth between the Golgi complex and the cell surface, between the ER and the Golgi, and between the various Golgi cisternae. What prevents the numerous migrating vesicles from inadvertently fusing with the wrong target membrane? A possible answer has been provided by James Rothman and his colleagues, who discovered that membrane fusion in mammalian cells requires a soluble cytoplasmic protein called *N-ethylmaleimide-sensitive fusion protein* (**NSF**) in combination with several *NSF attachment proteins* (**SNAPs**). The same NSF/SNAPs appear to be involved in the fusion of many different kinds of membranes, suggesting that these proteins are not responsible for directing a particular type of vesicle to its appropriate membrane target. Rothman has proposed that the specificity of membrane fusion is instead provided by membrane proteins called *SNAP receptors* or **SNAREs,** which act as attachment points for SNAPs during membrane fusion (see Figure 7-31). According to his model, every vesicle contains a v-SNARE (vesicle-SNAP receptor) that is designed to interact with a particular type of t-SNARE (target-SNAP receptor), which is only exhibited by the appropriate target membrane. A vesicle may transiently interact with many membranes before its v-SNARE discovers the proper t-SNARE; this recognition event then causes the two membranes to form a stable attachment with each other, setting the stage for membrane fusion.

Bacterial Cells Secrete Materials Directly through the Plasma Membrane

The secretory pathway utilized by eukaryotic cells is not applicable to bacterial cells because bacteria do not contain an ER, Golgi complex, or secretory vesicles. Yet bacteria are still capable of secreting proteins into the external environment. However, instead of bringing secretory proteins to the cell surface via an elaborate system of membrane vesicles, bacterial secretory proteins are directly translocated across the plasma membrane.

Although this appears to be quite different from the secretory process that occurs in eukaryotic cells, bacterial proteins destined for secretion contain signal sequences that cause them to bind to the plasma membrane, just as eukaryotic secretory proteins contain signal sequences that promote their binding to ER membranes. After a bacterial protein becomes bound to the plasma membrane by its signal sequence, the protein is translocated through a protein-translocation channel in the plasma membrane. Again, this process is analogous to the way in which secretory proteins are translocated across the ER membrane in eukaryotic cells. The major difference is that in bacteria, translocation causes the protein to be secreted from the cell because the protein is being moved through the plasma membrane rather than the ER membrane.

LYSOSOMES AND ENDOCYTOSIS

The identification of the lysosome as a distinct cytoplasmic organelle originally emerged from a series of subcellular fractionation experiments carried out by Christian de Duve and his associates in the early 1950s. This accomplishment represents a dramatic example of the exquisite power of subcellular fractionation, for this technique permitted lysosomes to be identified on the basis of their biochemical properties before they had ever been recognized in electron micrographs. It was only after subcellular fractionation experiments provided information as to what to look for, and where to look for it, that lysosomes came to be identified by electron microscopists. We will therefore begin our discussion of lysosomes by examining how they came to be discovered.

Acid Phosphatase Is Localized within a Membrane-Enclosed Organelle

Serendipity, often responsible for important new advances in science, played a significant role in the discovery of lysosomes. De Duve had begun his work with an interest in the effect of insulin on carbohydrate metabolism. He therefore chose to investigate the subcellular localization of glucose 6-phosphatase, the enzyme responsible for releasing glucose into the bloodstream. As a control he monitored the distribution of acid phosphatase, an enzyme that is not directly involved in carbohydrate metabolism. When he homogenized liver tissue in 0.25 M sucrose and isolated subcellular fractions by differential centrifugation, the glucose 6-phosphatase was found to be associated predominantly with the microsomal fraction. This discovery was important in its own right because it was widely believed at the time that microsomes were simply fragments of disrupted mitochondria. The fact that glucose 6-phosphatase occurred in the microsomal fraction, but not the mitochondrial fraction, established that microsomes represent a unique subcellular component.

The data on the distribution of acid phosphatase, on the other hand, were not as clear. Although this enzyme was found to be present in highest concentration in the mitochondrial fraction, the total recovery of acid phosphatase was only about 10 percent of that known to be present in liver based on experiments involving harsher extraction conditions. De Duve's initial reaction was that the assay procedure for acid phosphatase was not working properly. Since it was too late in the day to repeat the measurements, the fractions were put in the refrigerator for storage. When the acid phosphatase activity of this stored material was measured a few days later, the results were tenfold higher than when originally assayed.

Rather than dismissing the initial data as an error, as some might have done, de Duve considered the possibility that the acid phosphatase had been "masked" in the original mitochondrial fraction, and that the enzyme was somehow activated during storage. A clue as to how this might have occurred emerged when he centrifuged the stored mitochondrial fraction. Instead of sedimenting with the mitochondria during centrifugation, the increased acid phosphatase activity appeared in the supernatant. In other words, the tenfold increase in enzyme activity observed upon storage was accompanied by conversion of the acid phosphatase to a freely soluble form that appeared in the supernatant.

Although this discovery was unrelated to de Duve's original interest in insulin (recall that the acid phosphatase assay had only been included as a control), he decided to deviate from his original research because of a feeling that he had stumbled upon something important. He soon discovered that a variety of treatments other than storage can also trigger increases in acid phosphatase activity. Homogenizing in a food blender, freezing and thawing, warming to 37°C, and exposure to detergents or hypotonic solutions all increased the amount of acid phosphatase activity detected in liver homogenates. Since each of these treatments is capable of disrupting membrane vesicles, de Duve postulated that acid phosphatase is localized within a membrane-bound organelle, and that disruption of membrane integrity leads to the release of acid phosphatase from the organelle. Since membranes are impermeable to the substrates employed in assaying for the presence of acid phosphatase, the enzyme can be detected only after it has been released from its membrane-enclosed compartment (Figure 7-32).

Acid Phosphatase and Other Acid Hydrolases Are Packaged Together in Lysosomes

De Duve at first assumed that the mitochondrion was the membrane-enclosed compartment in which acid phosphatase resides, although he was disturbed by the fact that a sizable portion of the acid phosphatase appeared in the microsomal fraction. This problem was soon resolved in an unexpected way when the breakdown of the ultracentrifuge routinely used to prepare the mitochondrial fraction forced one of de Duve's students to use a slower centrifuge. This slower centrifuge was still able to produce a mitochondrial fraction containing large quantities of mitochondria, but they were virtually devoid of acid phosphatase activity. This chance observation suggested that acid phosphatase is localized in some other organelle.

To test this idea, de Duve revised the standard subcellular fractionation scheme to subdivide the mitochondrial fraction into its larger and smaller components based on centrifugation at lower and higher speeds, respectively (Figure 7-33). The fraction containing the larger components was found to be enriched in typical mitochondrial enzymes, such as cytochrome oxidase; it was therefore concluded that this fraction contained purified mitochondria. The fraction containing the smaller components, on the other hand, was enriched in acid phosphatase and four additional enzymes (ribonuclease, deoxyribonuclease, β-glucuronidase, and cathepsin) that can be activated by treatments similar to those that activate acid phosphatase. Because these five enzymes are all hydrolytic enzymes, de Duve concluded that they are packaged together in a unique organelle that is designed to perform an intracellular digestive function. In 1955 he introduced the term **lysosome** (hydrolytic body) to describe this new cytoplasmic organelle.

Although the biochemical evidence for the existence of lysosomes was quite impressive, the nagging problem remained that no one had ever identified such an organelle in electron micrographs. These remaining

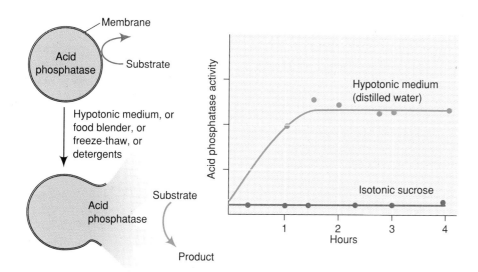

Figure 7-32 Evidence That Acid Phosphatase Resides in a Membrane-Enclosed Vesicle (Left) *Model showing how conditions that disrupt membranes also release acid phosphatase into solution, thereby permitting its activity to be measured.* (Right) *Data from one such experiment illustrate that hypotonic conditions increase the amount of acid phosphatase activity that can be detected in a crude mitochondrial fraction from rat liver.*

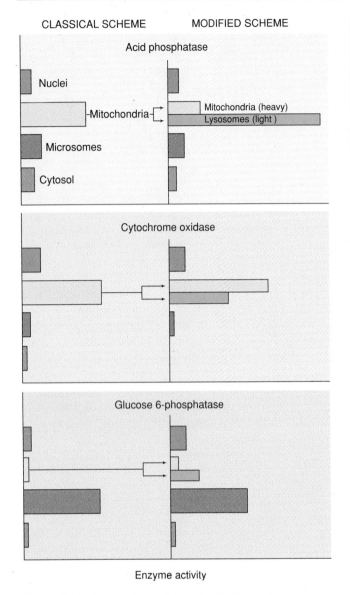

Enzyme activity

Figure 7-33 *Comparison of the Classical Four-Fraction Scheme of Subcellular Fractionation with the Modified Five-Fraction Scheme Introduced by de Duve* *The major difference is the subdivision of the crude mitochondrial fraction into "heavy" and "light" components. Note that the modified scheme enhances the separation of organelles containing cytochrome oxidase (a mitochondrial enzyme) from those containing acid phosphatase (a lysosomal enzyme).*

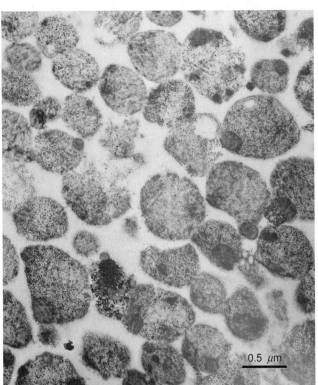

Figure 7-34 *Electron Microscopic Appearance of an Isolated Subcellular Fraction Enriched in Lysosomes* *Courtesy of P. Baudhuin.*

qualms were soon alleviated when Alex Novikoff used the electron microscope to examine subcellular fractions enriched in acid phosphatase. Besides a few mitochondria, he saw large numbers of vesicles measuring several hundred nanometers in diameter that were not present in mitochondrial preparations devoid of acid phosphatase activity (Figure 7-34). Membrane-enclosed vesicles of similar size were subsequently identified in thin sections of liver cells. Although unequivocal identification of these vesicles was hindered by the fact that lysosomes lack a distinctive morphology, this obstacle

was soon overcome with the development of cytochemical staining procedures designed to localize acid phosphatase and other lysosomal enzymes (Figure 7-35).

Subsequent investigations revealed that lysosomes contain several dozen enzymes in addition to the original five described by de Duve, but all of these lysosomal enzymes share an important feature in common; that is, they are all hydrolytic enzymes exhibiting maximal activity at acid pH (Table 7-3). Because of this common feature, such enzymes are called **acid hydrolases.** Included in this category are *proteases* that break down proteins, *nucleases* that break down nucleic acids, *lipases* that break down lipids, and *glycosidases* that break down polysaccharides. Taken together, these enzymes are capable of degrading all the major types of macromolecules found in living cells. In order to avoid their own destruction, such enzymes may possess a unique structure that renders them relatively resistant to the proteases contained within the lysosome.

Not all lysosomes are identical to one another. Examination of lysosomal subfractions separated by isodensity centrifugation has revealed the presence of a heterogeneous array of vesicles of differing enzyme content (Figure 7-36). That all lysosomes are not the same is also reinforced by the observation that lysosomal enzymes occur in vesicles exhibiting a wide variety of

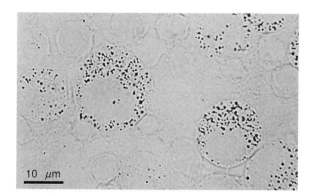

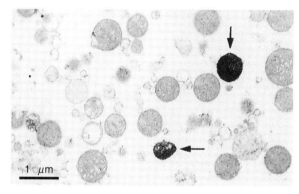

Figure 7-35 *Identification of Lysosomes in Tissue Sections Using Cytochemical Staining for Acid Phosphatase* (Top) *Light micrograph showing localization of acid phosphatase (dark granules) in the lysosomes of neutrophils. (Bottom) Electron micrograph showing that the cytochemical staining reaction for acid phosphatase can be used to distinguish lysosomes (arrows)* from surrounding mitochondria. Cour-*tesy of D. F. Bainton (top) and P. Baudhuin (bottom).*

sizes, shapes, and staining characteristics. We will see shortly that at least some of this diversity reflects differences in functional state and stage of development.

Plant Spherosomes Are Specialized Lysosomes Exhibiting a High Lipid Content

Isolating lysosomes from plant tissues was initially a difficult task because the methods developed for animal tissues do not work with plants. However, modified procedures that were eventually designed led to the purification of plant subcellular fractions enriched in acid hydrolases. When examined under the electron microscope, such fractions are found to contain large numbers of small, membrane-enclosed vesicles. These vesicles resemble **spherosomes**, which are highly refractive spherical particles measuring 0.5–1.0 μm in diameter that are visible when the plant cell cytoplasm is viewed by light microscopy. Cytochemical staining for acid phosphatase in tissue sections has confirmed that spherosomes, as well as larger cytoplasmic vacuoles, contain lysosomal enzymes (Figure 7-37).

Unlike the lysosomes of animal cells, spherosomes stain intensely with fat-soluble dyes and so must have a high

Table 7-3   Selected Lysosomal Enzymes

Enzyme	Natural Substrates
Phosphatases	
Acid phosphatase	Phosphomonoesters
Acid pyrophosphatase	ATP, FAD
Phosphodiesterase	Phosphodiesters
Phosphoprotein phosphatase	Phosphoproteins
Phosphatidic acid phosphatase	Phosphatidic acid
Sulfatases	
Arylsulfatase	Aryl sulfates
Proteases and Peptidases	
Cathepsins (several)	Proteins
Collagenase	Collagen
Arylamidase	Amino acid arylamides
Peptidases	Peptides
Nucleases	
Acid ribonuclease	RNA
Acid deoxyribonuclease	DNA
Lipases	
Triglyceride lipase	Triacylglycerols
Phospholipase	Phospholipids
Esterase	Fatty acid esters
β-Glucocerebrosidase	Glucocerebrosides
Galactocerebrosidase	Galactocerebrosides
Sphingomyelinase	Sphingomyelin
Glycosidases	
α-Glucosidase	Glycogen
β-Glucosidase	Glycoproteins
β-Galactosidase	Glycolipids, glycoproteins
α-Mannosidase	Glycoproteins
α-Fucosidase	Glycoproteins
β-Xylosidase	Glycoproteins
β-Glucocerebrosidase	Glycolipids
α-N-Acetylhexosaminidase	Heparin
β-N-Acetylhexosaminidase	Glycoproteins, glycolipids
Sialidase	Glycoproteins, glycolipids
Lysozyme	Bacterial cell walls
Hyaluronidase	Hyaluronic acid, chondroitin sulfate
β-Glucuronidase	Polysaccharides, glycosaminoglycans, steroid glucuronides

lipid content. The presence of large amounts of lipid in vesicles containing hydrolytic enzymes suggests that the spherosome functions in storing and mobilizing lipid reserves. In addition to their role in lipid metabolism, spherosomes and other hydrolase-containing plant vacuoles are involved in

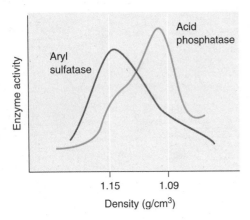

Figure 7-36 *Evidence for Lysosomal Heterogeneity* *A homogenate prepared from fibroblast cells was fractionated by isodensity centrifugation and assayed for the presence of the lysosomal enzymes, acid phosphatase and aryl sulfatase. The fact that these two enzymes are detected in vesicles that differ slightly in density indicates that they are contained in different populations of lysosomes.*

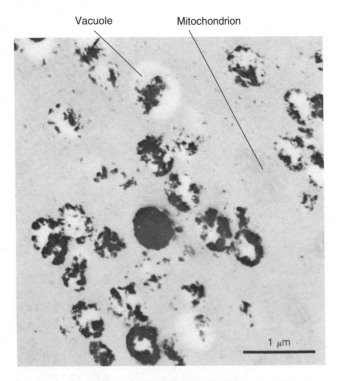

Figure 7-37 *Localization of Acid Phosphatase Activity in Plant Cells Using a Cytochemical Staining Reaction* *The dark deposits, which represent the site of acid phosphatase, are present in smaller vesicles (spherosomes) and larger vacuoles, but not in mitochondria. Courtesy of N. Poux.*

digesting and recycling intracellular constituents in a manner analogous to the animal cell lysosome.

Lysosomes Are Formed by a Pathway That Uses Mannose 6-Phosphate as a Targeting Signal

Since lysosomes are characterized by the presence of a unique set of hydrolytic enzymes, a mechanism must exist for targeting these enzymes to the lysosome when the organelle is being formed. For most lysosomal enzymes, the targeting mechanism involves phosphorylated mannose residues located in N-linked oligosaccharide chains. Lysosomal enzymes in this category are synthesized on ribosomes attached to the rough ER and modified by N-linked glycosylation as they enter the lumen. After being released into the lumen, these proteins pass through transitional vesicles to the *cis* Golgi network. At this point, mannose residues in the N-linked oligosaccharide chains of the lysosomal enzymes are converted to *mannose 6-phosphate* by a two-step enzymatic pathway that occurs in close association with the *cis* face of the Golgi complex. The first step in this sequence is catalyzed by a *phosphotransferase* that selectively recognizes and phosphorylates mannose in lysosomal enzymes. Since lysosomal enzymes do not resemble each other in amino acid sequence, the phosphotransferase must recognize some structural feature that is common to the overall conformation of these proteins.

As newly formed lysosomal enzymes pass through the Golgi complex, the presence of mannose 6-phosphate allows them to be recognized and sorted for transport to lysosomes. The sorting is carried out in the *trans* Golgi network by integral membrane proteins that function as *mannose 6-phosphate receptors*. The receptor sites on these molecules face the lumen of the Golgi complex, where they bind to the mannose 6-phosphate contained within lysosomal enzymes. The presence of mannose 6-phosphate in lysosomal enzymes allows the *trans* Golgi network to distinguish these molecules from other glycoproteins and package them into specialized clathrin-coated vesicles.

After these vesicles have budded off from the *trans* Golgi network, the internal pH of the vesicles is lowered by the action of an ATP-dependent proton pump that transports protons into the vesicle lumen. As the pump acidifies the lumen to about pH 5, the lysosomal enzymes are released from their binding sites on the membrane because the interaction between the phosphorylated mannose groups and their membrane receptors is disrupted at low pH. After the lysosomal enzymes are released into the vesicle lumen as soluble proteins, a lysosomal phosphatase removes the phosphates from the mannose groups to prevent the proteins from rebinding to the mannose 6-phosphate receptors, and the receptors are then returned to the Golgi complex for reutilization (Figure 7-38).

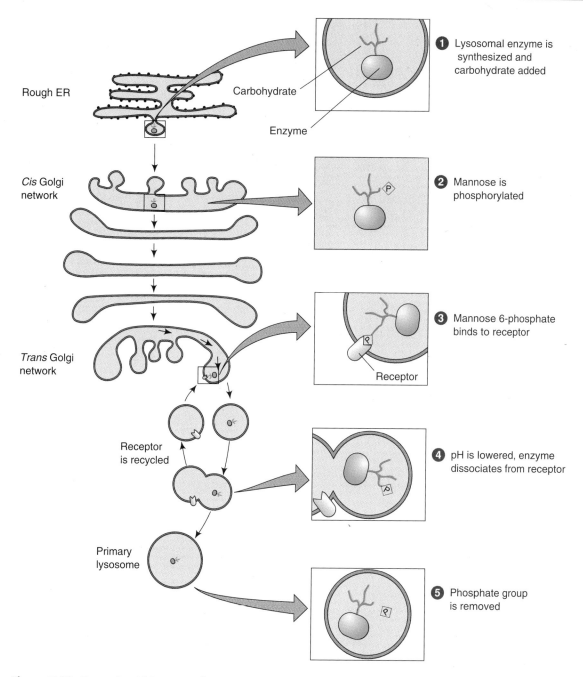

Figure 7-38 *Targeting of Lysosomal Enzymes to Lysosomes by Mannose 6-Phosphate* *Most lysosomal enzymes contain mannose residues in N-linked oligosaccharide chains that are converted to mannose 6-phosphate at the* cis *face of the Golgi complex. As newly formed lysosomal enzymes pass through the Golgi complex, the mannose 6-phosphate binds to membrane receptors that allow the enzymes to be recognized and packaged into specialized vesicles. After the vesicles bud off from the* trans *Golgi network, the internal pH of the vesicles is lowered by an ATP-dependent proton pump, causing the lysosomal enzymes to be released from the receptor sites on the membrane. A lysosomal phosphatase then removes the phosphates from the mannose groups, and the receptors are returned to the Golgi complex.*

Some Proteins Are Targeted to the Lysosome by a Mannose 6-Phosphate Independent Pathway

The importance of mannose 6-phosphate in targeting lysosomal enzymes to their proper destination is dramatically illustrated in patients suffering from the disease *mucolipidosis*, a hereditary disorder character-ized by an inability to phosphorylate mannose. Fibroblasts obtained from patients afflicted with this disease contain large cytoplasmic vesicles containing undigested macromolecules that normally would be degraded by lysosomes. Although such individuals manufacture lysosomal enzymes, the inability to phosphorylate mannose means that most of the enzymes cannot be targeted to lysosomes. Instead, they are secreted into the extracellular space by the normal pathway for constitutive secretion. In spite of this defect, a few lysosomal proteins do manage to become properly packaged into lysosomes in the absence of mannose phosphorylation. Hence the mechanism by which these particular proteins are targeted to lysosomes must be independent of the mannose 6-phosphate recognition system.

Proteins in this category include glycoproteins that reside in the lysosomal membrane as well as lysosomal enzymes that are synthesized as membrane-bound precursors. An example of the latter phenomenon involves acid phosphatase, an enzyme synthesized in the form of a precursor that is retained as an integral membrane protein in the lipid bilayer of the ER membrane. After the membrane-bound precursor has been transported through the Golgi complex to lysosomes, it is cleaved to release the mature form of acid phosphatase into the lysosomal lumen. β-Glucocerebrosidase is another lysosomal enzyme synthesized as an integral membrane protein, although in this case the enzyme remains bound to the membrane after incorporation into lysosomes.

The targeting signal that directs membrane-bound proteins from the ER to the lysosome involves an amino acid sequence that protrudes from the cytosolic surface of the membrane. In the case of acid phosphatase, mutations that change a single tyrosine in this sequence result in an enzyme that accumulates in the plasma membrane rather than lysosomes. Similar results have been obtained in studies involving other integral proteins of the lysosomal membrane, suggesting that this tyrosine is part of the targeting sequence that directs membrane proteins to the lysosome.

Lysosomes Are Involved in Macroautophagy, Microautophagy, Autolysis, and Extracellular Digestion

The discovery that lysosomes contain hydrolytic enzymes suggests that this organelle plays a degradative or digestive role within the cell. One obvious reason for packaging such enzymes inside lysosomes is to protect the cell from digestion by a group of enzymes that in combination could potentially degrade every macromolecule and organelle in the cell. Situations do occur, however, that call for the destruction of cellular constituents. For example, in aging cells or in cells containing damaged organelles, lysosomes can be called upon to selectively degrade small regions of cytoplasm. During this process of **macroautophagy,** organelles targeted for destruction are wrapped in one or more layers of smooth ER membrane. The resulting vesicle, referred to as an *autophagic vacuole,* then fuses with **primary lysosomes,** defined as those lysosomes that are not yet involved in digestive activity. This fusion event generates an *autophagosome* in which digestion

takes place. Autophagosomes can often be recognized by the identifiable remnants of cellular organelles contained within them (Figure 7-39). Macroautophagy is useful to cells because it removes unwanted organelles and permits the chemical building blocks generated during an organelle's destruction to be recycled for other purposes.

Lysosomes also take up and degrade cytosolic proteins by invagination of the lysosomal membrane, creating small vesicles that are internalized within the lysosome and broken down by the organelle's hydrolytic enzymes. This process, referred to as **microautophagy,** is relatively nonselective in the proteins that it degrades. It is thought to provide for the slow, continual recycling of the building blocks used in the construction of protein molecules. However, under stressful conditions nonselective degradation of protein molecules would be detrimental to the cell. For example, during prolonged fasting the continued nonselective degradation of cellular proteins could lead to the depletion of critical enzymes or regulatory proteins. Under these conditions, lysosomes preferentially degrade proteins containing a targeting sequence consisting of glutamine flanked on either side by a tetrapeptide composed of very basic, very acidic, and/or very hydrophobic amino acids. Proteins exhibiting this sequence are presumably targeted for selective degradation because they are dispensable to the cell.

Some situations require the complete dissolution of a cell and its contents, rather than the selective elimination of a few organelles or proteins. Such self-inflicted cell destruction, or **autolysis,** plays an important role in

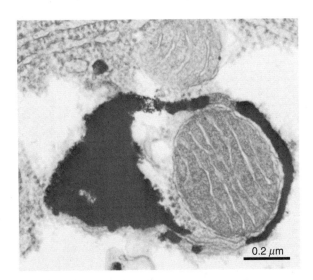

Figure 7-39 *Electron Micrograph of a Mitochondrion Being Digested in an Autophagosome* *The presence of lysosomal enzymes is indicated by the dark material, which was generated by the cytochemical staining reaction for acid phosphatase. Courtesy of M. Locke.*

the development of certain organ systems and body structures. For example, embryonic formation of fingers or toes from undifferentiated blocks of tissue requires selective destruction of the cells located between the newly forming digits. Likewise, metamorphosis of a tadpole into a frog requires resorption of the tadpole's tail. In both cases, tissue destruction is mediated by lysosomal digestion of the constituent cells. Although the mechanism controlling lysosomal enzyme release under such conditions is yet to be unraveled, lysosomes are known to rupture and release their hydrolytic enzymes throughout the cell, triggering a massive breakdown of cellular constituents.

Besides digesting materials that are present inside cells, lysosomal hydrolases can also be secreted into the extracellular space for use in external digestion. During fertilization, for example, penetration of sperm through the outer coatings of the egg cell is aided by the release of lysosomal enzymes from the sperm head (page 686). Situations also occur in which the release of lysosomal enzymes leads to detrimental effects. Cells subjected to physical trauma or microbial infection are a case in point. Lysosomal enzymes are often released in response to such conditions, triggering additional tissue damage and inflammation. The potential danger of lysosomal enzymes is further illustrated by the discovery that an arthritis-like disease can be produced in rabbits by injecting drugs known to disrupt lysosomes. This finding may help to explain why drugs that inhibit lysosomal enzyme release, such as the steroid hormones cortisone and hydrocortisone, are effective in combating severe inflammation. Vitamin A, on the other hand, increases the fragility of the lysosomal membrane and hence promotes the discharge of acid hydrolases. As a result, connective tissue damage and spontaneous bone fractures often occur in individuals who consume excess quantities of vitamin A.

During Phagocytosis, Plasma Membrane Vesicles Containing Particulate Matter Are Brought into the Cell and Fuse with Lysosomes

In addition to their role in degrading intracellular materials, lysosomes also hydrolyze macromolecules that have been brought into the cell by **endocytosis.** Endocytosis refers to any process in which materials are taken into the cell by membrane-bound vesicles that pinch off from the plasma membrane. When the material contained within the vesicles consists mainly of particulate matter, the term **phagocytosis** is employed. Phagocytosis is a property of a relatively limited number of specialized cell types. Amoebae and other unicellular eukaryotes employ phagocytosis to ingest microorganisms and food particles as sources of nutrients. Multicellular animals, on the other hand, contain specialized phagocytic cells called *macrophages* and *neutrophils*

that utilize phagocytosis to ingest and destroy invading bacteria and viruses.

Phagocytosis is triggered by an interaction between receptors on the surface of the phagocytic cell and molecules present on the surface of the material being ingested. At least two groups of receptors have been identified on mammalian phagocytic cells. (1) A group of proteins known as the *complement system* (page 732) circulate in the bloodstream and extracellular fluids, where they bind nonselectively to foreign particles and promote their uptake by binding to *complement receptors* present on the surface of phagocytic cells. (2) Organisms also produce *antibodies* (page 718) that bind selectively to foreign particles, including bacteria and viruses; the coating of antibody molecules promotes phagocytosis of the foreign particles because the antibodies bind to *Fc receptors* carried by phagocytic cells. After a foreign particle has been bound to receptors on the surface of a phagocytic cell, the plasma membrane engulfs and internalizes the particle by wrapping itself around the particle surface (Figure 7-40). The internalized *phagocytic vesicle* or **phagosome** then fuses with lysosomes, forming a larger vesicle called a **secondary lysosome** (Figure 7-41). The lysosomal enzymes hydrolyze the foreign matter originally present in the phagosome, releasing small mole-

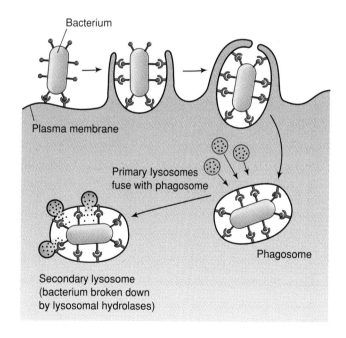

Figure 7-40 *The Mechanism of Phagocytosis During the first stage of phagocytosis, the plasma membrane wraps itself around and engulfs particles that have become bound to cell surface receptors. The internalized phagosome then fuses with lysosomes, forming a secondary lysosome that hydrolyzes the engulfed material. Small molecules produced during the hydrolysis stage can diffuse through the vesicle membrane for use in the cytosol.*

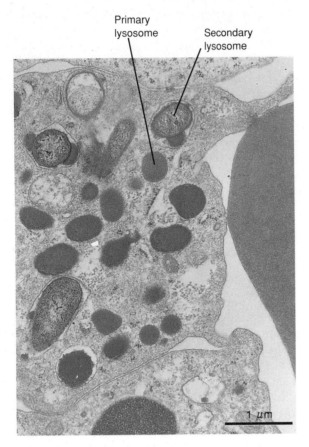

Figure 7-41 *Fusion of a Lysosome with a Phagosome*
Bacteria (color) have been taken up by this cell and reside in phagosomes. Primary lysosomes fuse with these phagosomes to generate secondary lysosomes. Courtesy of D. Zucker-Franklin.

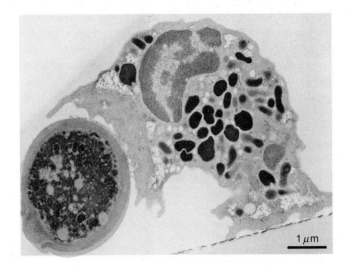

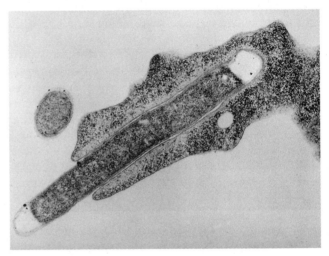

Figure 7-42 *Electron Micrographs of Phagocytosis* (Top) *Transmission electron micrograph showing the phagocytosis of a yeast cell (color) by a neutrophil. (Bottom) Transmission electron micrograph of an amoeba engulfing a bacterium by phagocytosis. Only a portion of the ameba is shown. Courtesy of D. F. Bainton (top) and R. N. Band and H. S. Pankratz (bottom).*

cules that diffuse through the vesicle membrane for use in the cytosol.

Phagocytosis can be an exceptionally impressive event, permitting the uptake and degradation of particles whose sizes approach that of the cell itself (Figure 7-42). A dramatic example is the ingestion and destruction of invading bacteria by neutrophils. As we discussed earlier in the chapter (page 281), the cytoplasm of this cell type contains two kinds of granules: specific granules and azurophil granules (lysosomes). Although specific granules are not lysosomes, they contain hydrolytic enzymes active at neutral or alkaline pH that also degrade macromolecules. When bacteria are ingested by phagocytosis, the specific granules discharge their contents into the phagosomes to initiate the destruction process. A few minutes later the azurophil granules (lysosomes) discharge their contents into the same vesicles, introducing hydrolytic enzymes that quickly complete the degradation of the invading bacteria. The destruction that occurs is often so massive that the neutrophils themselves may die in the process.

During Pinocytosis, Plasma Membrane Vesicles Containing Fluid Are Brought into the Cell and Fuse with Lysosomes

Pinocytosis is a type of endocytosis that involves the nonselective uptake of small droplets of extracellular fluid. In contrast to phagocytosis, the plasma membrane folds inward to form tiny vesicles rather than wrapping itself around an extracellular particle (Figure 7-43, *left*). Any molecules or small particles present in the extracellular fluid are randomly taken into the cell by this process.

The role of lysosomes in mediating the breakdown of materials taken up by pinocytosis was first shown in experiments involving pinocytosis of the enzyme horseradish peroxidase. The advantage of using horseradish

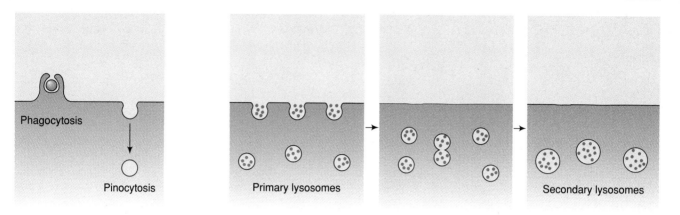

Figure 7-43 *The Process of Pinocytosis* (Left) *During pinocytosis, the plasma membrane folds inward to form tiny pinocytic vesicles, rather than wrapping itself around an extracellular particle as occurs in phagocytosis.* (Right) *Evidence that lysosomes degrade materials present in pinocytic vesicles. In this experiment, cells that had taken up horseradish peroxidase by pinocytosis were stained with dyes that produce a blue color in the presence of horseradish peroxidase, and a red color in the presence of acid phosphatase (a marker for lysosomes). Initially, separate pinocytic vesicles (blue) and lysosomes (red) could be distinguished. But later, purple granules were observed. Since purple is the color that is obtained when blue and red are mixed together, these results indicate that lysosomes have fused with the peroxidase-containing pinocytic vesicles to form secondary lysosomes.*

peroxidase as a marker for pinocytosis is that it can be readily visualized in both electron and light micrographs using cytochemical staining procedures. When cells are incubated in solutions containing horseradish peroxidase, electron micrographs reveal that the enzyme is taken up into newly forming pinocytic vesicles (Figure 7-44). In light micrographs, the location of this enzyme can be determined by a cytochemical procedure that yields a blue reaction product when it reacts with peroxidase. The same tissue sections can also be stained by a procedure that gives a red color in the presence of acid phosphatase, a marker for lysosomes. When cells are examined by these staining procedures half an hour after exposure to horseradish peroxidase, separate red and blue granules can be visualized within the cell (see Figure 7-43, *right*). This indicates that the peroxidase has been taken up by the cell and is present in pinocytic vesicles (blue) that are distinct from lysosomes. When cells are examined at later times, however, purple granules are seen instead of separate red and blue granules, indicating that the lysosomes and peroxidase-containing pinocytic vesicles have fused together to form secondary lysosomes. A day later the color of the granules reverts back to red because the lysosomal enzymes have broken down the peroxidase responsible for the blue color.

In the case of both phagocytosis and pinocytosis, undigested substances can gradually accumulate in a secondary lysosome, converting it to a structure known as a **residual body** (Figure 7-45). Residual bodies usually fuse with the plasma membrane and expel their contents from the cell by the process of exocytosis. When this discharge is slow or absent, the resulting accumulation of residual bodies may contribute to cellular aging.

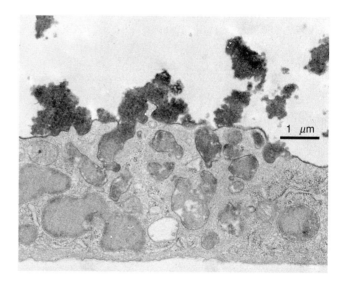

1 μm

Figure 7-44 *Electron Micrograph Showing Pinocytosis of Horseradish Peroxidase by a Mouse Macrophage* The *cytochemical staining procedure employed in preparing this micrograph produces an electron-opaque reaction product wherever horseradish peroxidase is present. The horseradish peroxidase is observed within invaginations of the plasma membrane, which then bud off to generate intracellular pinocytic vesicles. Courtesy of R. M. Steinman.*

Receptor-Mediated Endocytosis Brings Specific Macromolecules into the Cell

Receptor-mediated endocytosis is a specialized type of endocytosis designed to bring specific macromolecules into the cell. More than 50 different proteins, including hormones, growth factors, lymphokines,

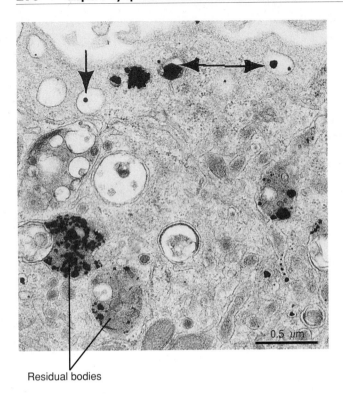

Residual bodies

Figure 7-45 *Electron Micrograph of Residual Bodies In this lung cell, undigestible deposits of colloidal silver initially taken up by pinocytic vesicles (arrows) are seen to accumulate within residual bodies. Courtesy of E. Essner.*

Table 7-4   Some Ligands That Enter Cells by Receptor-Mediated Endocytosis

Hormones
Insulin
Luteinizing hormone (LH)
Follicle-stimulating hormone (FSH)
Growth hormone
Prolactin
Glucagon
Growth factors
Epidermal growth factor
Platelet-derived growth factor
Transforming growth factor β
Nerve growth factor
Lymphokines
Interleukins
Tumor necrosis factor
Interferon
Colony stimulating factor
Nutrients
LDL (cholesterol)
Transferrin (iron)

and nutrients, enter cells in this way (Table 7-4). During receptor-mediated endocytosis, the external ligand first binds to its corresponding plasma membrane receptor; the receptor-ligand complex then becomes concentrated in specific regions of the plasma membrane known as **coated pits** (Figure 7-46). Each coated pit is an infolding of the plasma membrane whose cytoplasmic surface is coated with a polyhedral lattice constructed from the protein **clathrin.** Clathrin molecules consist of three large polypeptide chains and three small polypeptide chains organized into a three-pronged structure called a *triskelion.* Clathrin triskelions polymerize with one another to form the polyhedral lattice seen beneath coated pits (Figure 7-47). Special proteins called *adaptins* help to bind the clathrin molecules to the plasma membrane receptors.

After ligand-receptor complexes have become clustered within a coated pit, the invaginated membrane pinches off and becomes internalized as a **coated vesicle.** The coated vesicle is initially surrounded by a cage of clathrin molecules, but this clathrin coat is quickly shed. Although there is some disagreement as to how the next steps are carried out, the contents of the uncoated vesicles soon accumulate in a membrane network known as the **endosome** compartment. According to one model (Figure 7-48),

the uncoated vesicles first fuse with a network of membrane tubules and vesicles known as *early endosomes.* Carrier vesicles are then employed to transport the receptor-ligand complexes from early endosomes to another membrane compartment known as *late endosomes.* An ATP-dependent proton pump in the late endosomal compartment lowers the pH to about 5, which causes the ligands to dissociate from their receptors. For this reason, late endosomes are also referred to as the *"compartment for uncoupling of receptor and ligand"* or *CURL.*

After they reach the endosomal compartment, receptor-ligand complexes follow one of the four different pathways illustrated in Figure 7-48: (1) The receptor can be returned to the plasma membrane by carrier vesicles while the ligand is delivered to lysosomes for degradation. (2) Both receptor and ligand can be returned to the plasma membrane by carrier vesicles. (3) Both receptor and ligand can be delivered to lysosomes for degradation. (4) The receptor and ligand can be delivered to the opposite side of the cell by carrier vesicles that fuse with the plasma membrane, releasing the ligand into the extracellular space. Because it transports ligands from one side of the cell to the other, this latter process is called **transcytosis.**

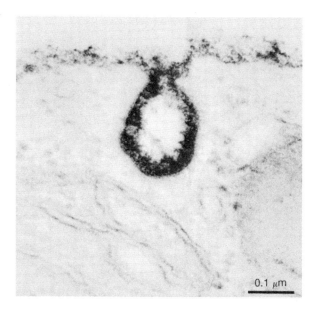
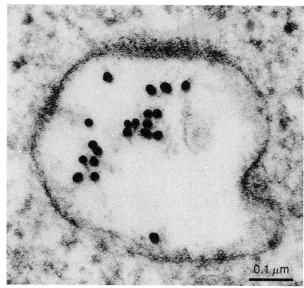

Figure 7-46 *Electron Microscopic Appearance of Coated Pits and Coated Vesicles* (Left) *Electron micrograph of a coated pit invaginating from the plasma membrane. The protein clathrin is present in the dark material that coats the membrane in the invaginated region.* (Right) *Electron micrograph of an internalized coated vesicle containing gold particles taken up from the extracellular space. The fuzzy edge of the coated vesicle consists of clathrin. Courtesy of M. C. Willingham.*

In addition to the clathrin-coated pits that carry out receptor-mediated endocytosis, cells are often equipped with another class of membrane pits called **caveolae.** Caveolae are small, flask-shaped invaginations of the plasma membrane that transport small molecules and ions into cells by a mechanism that has been named **potocytosis** to distinguish it from the closely related process of endocytosis. Unlike the behavior of clathrin-coated pits during endocytosis, caveolae do not pinch off completely to form free cytoplasmic vesicles. Instead, ligands are concentrated within caveolae by caveolar membrane proteins that function either as receptors that bind the ligand, or as enzymes that convert the ligand into an appropriate product. As the ligand is concentrated in the caveolar lumen, the narrow opening that normally connects caveolae to the extracellular space is closed (Figure 7-49). The high concentration of ligand within the caveolae is then thought to drive diffusion of the ligand across the membrane into the cytoplasm.

Receptor-Mediated Endocytosis Is Utilized for Several Purposes, Including Nutrient Delivery and Signal Transduction

Cells utilize receptor-mediated endocytosis for a variety of different purposes. One important role is the delivery of nutrients. If the ligand that binds to a particular plasma membrane receptor is a nutrient, then receptor-mediated endocytosis can be employed to deliver it to the cell interior. Developing egg cells, for example, em-ploy receptor-mediated endocytosis to ingest yolk proteins from the bloodstream, which are then stored in the cell for use as nutrients later in embryonic development (page 677).

A similar function is carried out by proteins that bind nutrients such as *transferrin* (which binds iron), *transcobalamin* (which binds vitamin B_{12}), and *low-density lipoprotein* (which binds cholesterol). The uptake of each of these proteins by receptor-mediated endocytosis allows its bound nutrient to be delivered to the cell interior. The low-density lipoprotein (LDL) is an especially important example because of the role of its bound cholesterol in both lipid metabolism and cardiovascular disease. Since cholesterol is a lipid and therefore insoluble in an aqueous environment, LDL serves as a carrier that transports it through the bloodstream and extracellular fluids. LDL is a protein-containing particle constructed from cholesterol, cholesterol esters, phospholipids, and a protein called *apo-B.* After the LDL particle binds to its plasma membrane receptor and is taken up by receptor-mediated endocytosis, the LDL receptors are recycled to the plasma membrane for reutilization; at the same time the apo-B protein is degraded and the cholesterol and phospholipids are released for use within the cell (see Figure 7-48, *pathway 1*). The released cholesterol is utilized for a variety of metabolic purposes, including the feedback inhibition of excessive cholesterol synthesis by the cell itself. A rare genetic disease, termed *familial hypercholesterolemia,* is caused by an inherited deficit in LDL receptors. Cells of individuals with

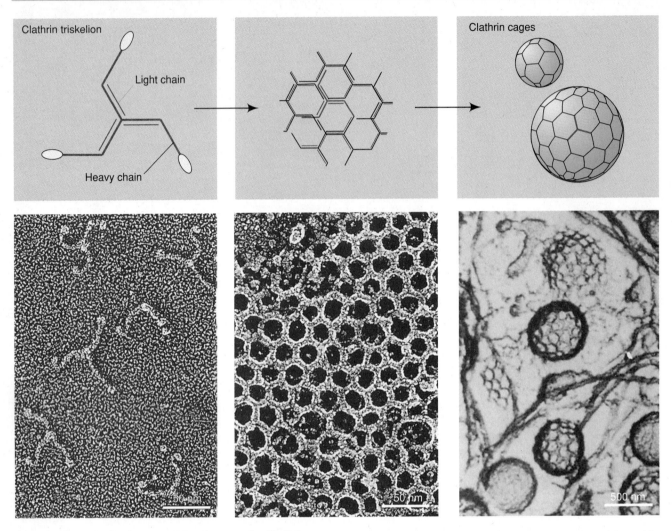

Figure 7-47 *Structures Formed by the Protein Clathrin* (Top) *Diagram illustrating how clathrin triskelions combine with one another to form a polyhedral lattice consisting of hexagonal and pentagonal units. The clathrin cages shown at the right typically surround a membrane vesicle. Clathrin cages contain 12 pentagons and a variable number of hexagons.* (Bottom left) *Electron micrograph of clathrin triskelions visualized by shadowing with platinum.* (Bottom middle) *Freeze-etch electron micrograph of a clathrin lattice in a human carcinoma cell.* (Bottom right) *Electron micrograph of clathrin cages isolated from calf brain. Micrographs courtesy of J. Heuser (left, middle) and N. Hirokawa and J. Heuser (right).*

such a deficiency cannot take up by the cholesterol needed to regulate the synthesis of this lipid. As a result, excessive amounts of cholesterol are formed and secreted into the bloodstream, leading to severe cardiovascular disease.

Receptor-mediated endocytosis is also involved in the process of signal transduction by extracellular ligands such as hormones and growth factors. After an extracellular signaling ligand has bound to its plasma membrane receptor, signals are often terminated by a process called *down-regulation* (page 224); during down-regulation, ligand-receptor complexes are internalized by receptor-mediated endocytosis and degraded, thereby decreasing the number of available receptors (see Figure 7-48, *pathway 3*). In some cases, internalizing the ligand-receptor complex may also

allow the signaling ligand to exert direct effects inside the cell.

In addition to the preceding functions, receptor-mediated endocytosis is involved in removing unneeded enzymes and proteins from the extracellular space and in transporting proteins from one side of a cell layer to the other by transcytosis (see Figure 7-48, *pathway 4*).

Lysosomal Storage Diseases Are Caused by Genetic Defects in Lysosomal Enzymes

We have seen in the preceding sections that lysosomal enzymes degrade a variety of intracellular and extracellular materials. It would therefore be expected that defects in lysosomal function might have severe

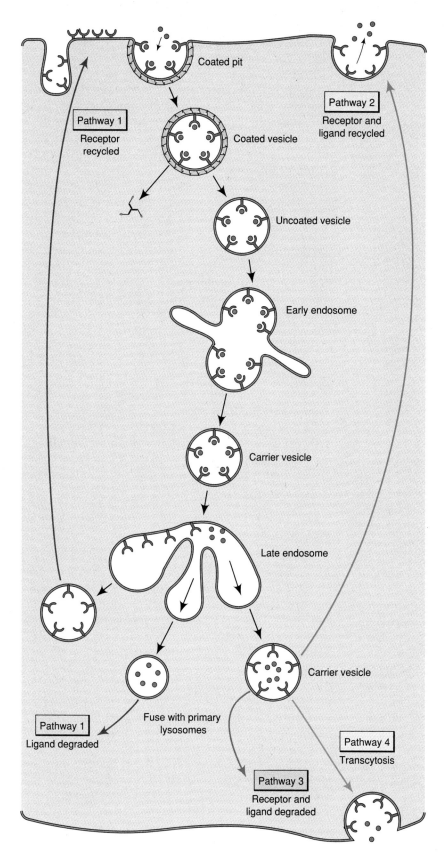

Figure 7-48 Pathways Involved in Receptor-Mediated Endocytosis *(1) The incoming ligand is degraded while its membrane receptor is returned to the plasma membrane; or (2) both receptor and ligand are returned to the plasma membrane, where the ligand is released; or (3) the incoming ligand and its membrane receptor are both degraded; or (4) the incoming ligand and its membrane receptor are transported from one side of the cell to the other, where the ligand is released from the cell (transcytosis).*

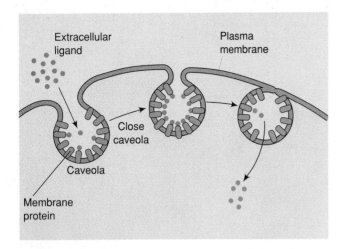

Figure 7-49 A Model Showing How Caveolae May Transport Small Molecules and Ions into Cells by Potocytosis *Extracellular ligands are concentrated in caveolae by binding to caveolar membrane proteins. After the caveolae close, the ligand is released and diffuses across the caveolar membrane and into the cytoplasm. In some cases, enzymes located in the caveolar membrane convert the ligand to a product prior to diffusion of the product into the cytoplasm.*

consequences. The first indication that such defects occur was the discovery that several genetic diseases of young children are characterized by an excessive intracellular accumulation of polysaccharides or lipids. The quantity of polysaccharide or lipid stored is often massive enough to interfere with and even destroy the afflicted cells. Depending on the cell type involved, symptoms such as muscle weakness, skeletal deformities, and mental retardation may result. The first of these diseases to have its underlying mechanism unraveled was *type II glycogenosis,* an illness that causes children to die at an early age with abnormally large amounts of glycogen in the liver, heart, and muscles. In 1963 Henri-Géry Hers discovered that type II glycogenosis is associated with a severe deficiency of the lysosomal enzyme β-glucosidase, which catalyzes the breakdown of glycogen. In the absence of this enzyme, undigested glycogen accumulates within lysosomes.

In addition to providing an explanation for type II glycogenosis, this discovery led Hers to postulate the existence of a wide spectrum of other **lysosomal storage diseases** that correspond to genetic defects in various lysosomal enzymes. He predicted that in each case, the disease symptoms are caused by an abnormal accumulation of the undigested substrates of the defective enzyme. This unifying theory has led to an understanding of the causes of what were once a bewildering array of mysterious diseases (Table 7-5). Some of these diseases involve defects in polysaccharide degradation, but most of them result from an inability to degrade glycolipids. Because glycolipids are

highly concentrated in brain tissue and are important constituents of the myelin sheath surrounding the axon, the symptoms of abnormal glycolipid accumulation usually include severe mental retardation. The glycolipid accumulated in brain cell lysosomes can often be recognized morphologically by its tendency to form unusual layered structures known as *zebra bodies* (Figure 7-50).

Discovering the molecular basis of lysosomal storage diseases has led to the suggestion that such illnesses might be treated by replacing the defective enzymes. Direct injection of missing lysosomal enzymes has been tried with some success in a few instances, although the injected enzymes must be chemically modified to promote their uptake into cells. An alternative approach is to encapsulate the required enzyme in artificial lipid vesicles (liposomes). The encapsulation process protects the enzymes from destruction and has the added advantage that liposomes are actively taken up by cells and fuse with lysosomes. In this way, a missing lysosomal enzyme might be delivered directly to its appropriate site of action. Encouraging results from animal studies suggest that this approach may ultimately be useful for the treatment of inherited enzyme deficiencies in humans. An alternative therapeutic approach is to try to correct or replace the gene coding for the defective lysosomal enzyme using recombinant DNA technology (page 110).

MEMBRANE BIOGENESIS

As we have seen in this and preceding chapters, cells contain a variety of different types of membranes constructed from unique combinations of proteins and lipids. The question therefore arises as to how these various kinds of membranes are assembled. Two basic modes of membrane assembly can theoretically be envisioned. One is that membranes are formed by the spontaneous self-assembly of membrane lipids and proteins. This idea is not completely unrealistic, for it has been shown in laboratory experiments that purified phospholipids and proteins spontaneously form artificial membrane vesicles, or *liposomes,* that share some of the properties of cellular membranes (page 162). The problem with self-assembled liposomes as a model for how membranes are assembled by living cells is that the protein arrangement in artificial membranes almost always lacks the asymmetry inherent to normal cellular membranes.

The alternative to self-assembly is that membranes grow by insertion of lipids and proteins into pre-existing membranes. Since this approach utilizes existing cellular membranes, newly synthesized proteins and lipids can be selectively inserted into one side of the

Table 7-5 Selected Lysosomal Storage Diseases

Disease	Symptom	Substance Accumulated	Enzyme Defect
Type II glycogenosis	Muscle weakness	Glycogen — Gluc–α–Gluc	α-Glucosidase
Niemann-Pick disease	Liver-spleen enlargement Mental retardation	Sphingomyelin — Cer–P–choline	Sphingomyelinase
Gaucher's disease	Liver-spleen enlargement Bone erosion	Glucocerebroside — Cer–β–Gluc	β-Glucosidase
Metachromatic leukodystrophy	Mental retardation	Sulfatide — Cer–β–Gal(OSO$_3^-$)	Sulfatidase
Fabry's disease	Skin rash Kidney failure	Ceramide trihexoside — Cer–β–Gluc–β–Gal–α–Gal	α-Galactosidase
Pseudo-Hurler disease	Liver enlargement Skeletal deformities	Ganglioside (GM$_1$) — Cer–β–Gluc–β–Gal(NeuAcN)–β–GalNAc–β–Gal	β-Galactosidase
Tay-Sachs disease	Mental retardation Blindness Muscular weakness	Ganglioside (GM$_2$) — Cer–β–Gluc–β–Gal(NeuAcN)–β–GalNAc	β-N-Acetylhexos-aminidase

Abbreviations: Gluc = glucose; Cer = ceramide (*N*-acylsphingosine); Gal = galactose; NeuAcN = *N*-acetylneuraminic acid; NAc = *N*-acetyl. Broken lines represent sites of enzymatic cleavage.

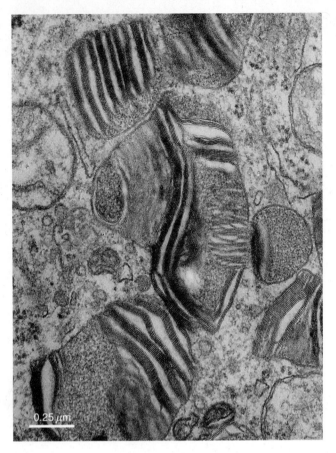

Figure 7-50 *Electron Micrograph of Zebra Bodies* *In this nerve cell the distinctive layered arrangement of the glycolipid that accumulates in defective lysosomes is clearly evident. Courtesy of H. Loeb.*

membrane or the other as membrane growth occurs. The net result is that an asymmetrical arrangement of membrane lipids and proteins is maintained as membrane growth occurs. This ability to maintain membrane asymmetry supports the idea that membranes grow by expansion of intact membranes rather than by self-assembly from isolated lipids and proteins.

How Are Phospholipids Inserted into Cellular Membranes?

If membranes grow by expansion, then mechanisms must exist for inserting newly synthesized lipids and proteins into membranes that are already present in the cell. Most of the phospholipid destined for intracellular membranes is synthesized in the lipid bilayer of the ER, which also contains proteins called **phospholipid translocators (flippases)** that catalyze the movement of phospholipids back and forth across the bilayer (page 261). The formation of these new phospholipid molecules causes the lipid bilayer of the ER to expand. Vesicles then bud off from the ER and fuse with other

membranes, moving phospholipids from the ER to other locations.

Phospholipids can also be moved from the ER to other membrane systems by carrier molecules termed **phospholipid transfer proteins.** Several such proteins have been identified, each specific for a different phospholipid. Phospholipid transfer proteins function by extracting a specific type of phospholipid molecule from one cellular membrane, diffusing through the cytosol, and inserting the phospholipid into another membrane (Figure 7-51).

In prokaryotic cells, which have no endoplasmic reticulum, phospholipids are synthesized directly within the plasma membrane. Insight into the question of how lipid asymmetry is generated in bacterial membranes has been obtained in experiments carried out on the bacterium *Bacillus megaterium* by James Rothman and Eugene Kennedy. In these studies, cells were first incubated with ^{32}P-phosphate for one minute to label newly synthesized molecules of the membrane phospholipid, *phosphatidylethanolamine* (*PE*). The reagent *trinitrobenzenesulfonic acid* (*TNBS*) was then added to the cell suspension. TNBS reacts chemically with PE to form a stable covalent complex. Because TNBS cannot cross the plasma membrane, it combines only with PE molecules that are exposed on the exterior surface of the plasma membrane. After labeling cells with ^{32}P-phosphate and TNBS, the plasma membrane fraction was isolated and the membrane lipids separated from each other by thin-layer chromatography. Because the PE-TNBS complex migrates faster than PE, the PE present on the outer surface of the plasma membrane can be separated from the PE present on the inner side (Figure 7-52). When the separated PE fractions were subsequently assayed for radioactivity, only the PE from the cytoplasmic surface (not combined with TNBS) was found to contain ^{32}P. Hence PE must be synthesized in the cytoplasmic leaflet of the plasma membrane bilayer.

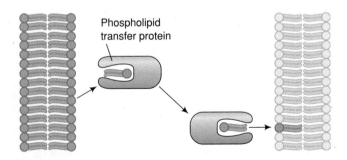

Figure 7-51 *Function of Phospholipid Transfer Proteins* *These carrier proteins can move specific phospholipids from one membrane to another.*

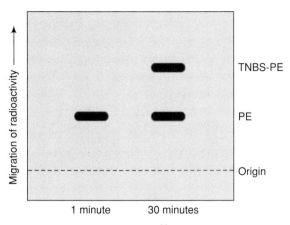

Figure 7-52 *Membrane Phospholipid Synthesis in Bacteria Incubated with* ^{32}*P-Phosphate* *PE represents phosphatidylethanolamine located in the inner (cytoplasmic) half of the plasma membrane bilayer, and TNBS-PE represents phosphatidylethanolamine located in the outer half of the bilayer. In cells incubated with* ^{32}*P-phosphate for 1 minute, only PE from the inner half (not combined with TNBS) contains* ^{32}*P, indicating that PE is synthesized in the inner half of the plasma membrane bilayer. After 30 minutes of labeling, 30 percent of the total radioactivity is detected in the PE-TNBS complex, which is derived from the outer half of the bilayer. The rapid movement of newly synthesized PE from the inner to the outer half of the bilayer indicates that the bacterial plasma membrane must contain catalysts similar to the phospholipid translocators of the eukaryotic ER.*

To determine the ultimate fate of these newly synthesized PE molecules, studies were carried out in which bacterial cells were incubated with ^{32}P-phosphate for half an hour instead of one minute prior to addition of TNBS. Under these conditions, 30 percent of the total radioactivity was detected in the PE-TNBS complex. In other words, 30 percent of the newly synthesized PE now appeared in the outer leaflet of the plasma membrane bilayer. This rapid movement of newly synthesized PE to the outer half of the bilayer is comparable to the rate at which phospholipid translocators catalyze the flip-flop of membrane lipids across the lipid bilayer of the ER. In other words, the bacterial plasma membrane must contain catalysts similar to the phospholipid translocators of the eukaryotic ER. The general absence of such catalysts in other membranes explains the low rate of lipid flip-flop observed in most locations other than the ER and the bacterial plasma membrane.

This model raises a new question, however, and that concerns the mechanism by which lipid asymmetry is established. We noted in an earlier chapter that phospholipids tend to be asymmetrically distrib-

uted in biological membranes, with certain phospholipids present in higher concentration on one side of a membrane and other phospholipids enriched on the opposite side (page 165). If regions where new membrane material is being formed contain phospholipid translocators that catalyze the movement of lipids back and forth across the bilayer, why isn't an equilibrium achieved in which each phospholipid is equally distributed on both sides of the membrane? Although a complete answer to this question is not yet available, at least two factors may contribute to phospholipid asymmetry. (1) Phospholipid translocators are substrate specific, promoting the movement of particular types of phospholipids across the bilayer. Phospholipids that are not efficiently moved by translocators will tend to remain on the side of the bilayer where they were synthesized. (2) Even those phospholipids that are moved across the bilayer by phospholipid translocators often become more concentrated on one side of the bilayer. Such phospholipids may be thermodynamically more stable on one side of the membrane than the other because environmental conditions are different on opposite sides of cellular membranes. Hence thermodynamic equilibrium for each particular phospholipid may cause it to favor one side of the bilayer over the other.

How Are Integral and Peripheral Membrane Proteins Incorporated into Cellular Membranes?

In addressing the question of how proteins become inserted into cellular membranes, we must remember that membranes contain three distinct classes of proteins: integral proteins, peripheral proteins, and lipid-anchored proteins (page 168). The difference between the way in which peripheral and integral membrane proteins are handled is nicely illustrated by experiments carried out by James Rothman and Harvey Lodish using vesicular stomatitis virus (VSV) as a model system. VSV, which causes an upper respiratory disease in farm animals, directs the cells it infects to synthesize several virus-specific proteins. Among them are two proteins that become incorporated into the plasma membrane. One, designated the M protein, is a peripheral membrane protein associated with the cytoplasmic surface of the plasma membrane. The other, designated the VSV glycoprotein, is an integral plasma membrane protein. Studies in which radioactive amino acids were used to label and follow the fate of newly synthesized viral proteins revealed that the M protein is synthesized on free cytoplasmic ribosomes and released in a soluble form into the cytosol. Only later does it become bound to the cytoplasmic surface of the plasma membrane. This arrangement appears to be typical of the way in which peripheral membrane proteins are synthesized and incorporated into membranes.

In contrast to the M protein, the VSV glycoprotein contains a signal sequence that causes it to be synthesized on ribosomes attached to the ER. As it begins to emerge into the ER lumen, the signal sequence is removed and carbohydrate chains are added. After synthesis has been completed, the glycoprotein remains anchored in the membrane bilayer rather than being released into the ER lumen. As is shown in Figure 7-53 (*left*), the glycoprotein is then transported to the plasma membrane by the following pathway: (1) Membrane vesicles containing the VSV glycoprotein bud off from the ER and fuse with the Golgi complex, where the carbohydrate chains of the VSV glycoprotein are further modified. (2) Vesicles containing the VSV glycoprotein emerge from the Golgi complex and migrate to the cell surface. (3) The vesicles then fuse with the plasma membrane, exposing the VSV glycoprotein to the cell exterior. An important point to note about the preceding sequence is that the portion of the VSV glycoprotein containing the carbohydrate chains always faces the lumen of the membrane system in which the protein occurs. When a membrane vesicle containing the VSV glycoprotein fuses with the plasma membrane, this orientation causes the carbohydrate chains to face the cell exterior. Thus the final asymmetric orientation of membrane glycoproteins is established at the time of synthesis by the direction in which the signal sequence inserts the growing polypeptide chain into the ER membrane. Subsequent studies have confirmed that similar pathways are employed for inserting other integral proteins into the plasma membrane.

The preceding process is closely related to the pathway for constitutive secretion, which we discussed earlier in the chapter (page 283). During constitutive secretion, materials are transported via membrane vesicles from the ER to the Golgi complex to the cell surface, where membrane vesicles fuse with the plasma membrane and expel their contents into the extracellular space. At the same time, integral proteins contained within the membranes of such vesicles become incorporated into the plasma membrane. The transport pathway that incorporates integral membrane proteins into the plasma membrane is thus the same pathway that is used for constitutive secretion. In a few cases the pathway for regulated secretion, which also involves the fusion of membrane vesicles with the plasma membrane, may likewise insert integral proteins into the plasma membrane. In prokaryotic cells, where no endoplasmic reticulum is present, fusion of membrane vesicles with the plasma membrane is not necessary. Integral membrane proteins instead contain signal sequences that allow them to be inserted directly into the plasma membrane.

What about integral membrane proteins that are destined for a location other than the plasma membrane? We have already seen that lysosomal membrane proteins contain a special targeting signal that allows them to be sorted into lysosomes (page 292). Targeting signals are presumably present that direct proteins into other membranes as well. Some of the targeting signals involved in directing proteins into the membranes of mitochondria, chloroplasts, peroxisomes, and the nucleus will be described in the chapters devoted to those organelles (Chapters 8, 9, and 10).

How Are Lipid-Anchored Proteins Incorporated into Cellular Membranes?

In the preceding section we described how peripheral and integral membrane proteins become incorporated into cellular membranes. Insertion of lipid-anchored proteins occurs by several different mechanisms, depending upon the type of protein involved.

1. **Glycolipid-anchored membrane proteins.** Many proteins associated with the external surface of the plasma membrane are bound to the membrane by covalent attachment to the glycolipid, **glycosylphosphatidylinositol (GPI).** Proteins in this category are synthesized on ribosomes attached to the rough ER. As we discussed earlier in the chapter, the ER contains enzymes that catalyze the attachment of these newly formed proteins to membrane-bound GPI (see Figure 7-17). As is shown in Figure 7-53 (*right*), the glycolipid-anchored protein is then transported to the plasma membrane by the following pathway: (1) Membrane vesicles containing the glycolipid-anchored protein bud off from the ER and fuse with the Golgi complex. (2) Vesicles containing the glycolipid-anchored protein emerge from the Golgi complex and migrate to the cell surface. (3) The vesicles then fuse with the plasma membrane, exposing the glycolipid-anchored protein to the cell exterior. Since the glycolipid-anchored protein faces the inside of the membrane vesicle, fusion of the vesicle with the plasma membrane will cause the glycolipid-anchored protein to face the cell exterior.

2. **Fatty acid–anchored membrane proteins.** A second class of lipid-anchored proteins includes proteins that are bound to membranes by covalently attached fatty acid chains. This group contains several proteins that are linked to the cytoplasmic surface of the plasma membrane by either *myristic acid,* a 14-carbon saturated fatty acid, or *palmitic acid,* a 16-carbon saturated fatty acid. Proteins in this category are synthesized as soluble cytosol proteins and are then covalently attached to fatty acids embedded in the membrane bilayer.

3. **Prenylated membrane proteins.** A third class of lipid-anchored proteins are bound to membranes by the covalent attachment of isoprenoid lipid groups, a process known as **prenylation.** Prenylated proteins, like fatty acid-anchored proteins, are synthesized as soluble cytosol proteins before being modified by prenylation. Proteins that are to be prenylated contain the C-terminal amino acid sequence *Cys-A-A-X,* which corresponds to cysteine followed by two aliphatic amino acids (valine, leucine, or isoleucine) followed by any amino acid (X) at the C-terminus. During prenylation, a 15-carbon *farnesyl* group or 20-carbon *geranylgeranyl* group is attached to the cysteine residue in the Cys-A-A-X sequence (Figure 7-54). After attachment, the farnesyl or geranylgeranyl group is inserted into the lipid bilayer of the appropriate target membrane. Finally, in some cases a membrane fatty acid such as palmitate is also linked to the prenylated protein.

SUMMARY OF PRINCIPAL POINTS

• The endoplasmic reticulum (ER) exists in two forms: a rough ER containing attached ribosomes and a smooth ER lacking ribosomes. The ER divides the cytoplasm into two compartments: the cytosol and the cisternal space (lumen). The cytosol contains enzymes involved in metabolic pathways, whereas the cisternal space provides a route for the movement of materials through various intracellular compartments and, in some cases, to the cell exterior.

• During subcellular fractionation, the ER is broken into fragments that reseal to form tiny vesicles called microsomes. Smooth microsomal vesicles contain enzymes involved in the metabolism of carbohydrates, lipids, and foreign substances.

• The rough ER plays a central role in the synthesis of secretory proteins, integral membrane proteins, and proteins destined to reside in the lumen of the ER, Golgi complex, and lysosomes. Proteins in these categories contain an N-terminal signal sequence that guides attachment and insertion of the polypeptide chain through the ER membrane as the polypeptide is being synthesized. Secretory proteins and proteins destined to reside in the lumen of the ER, Golgi complex, or

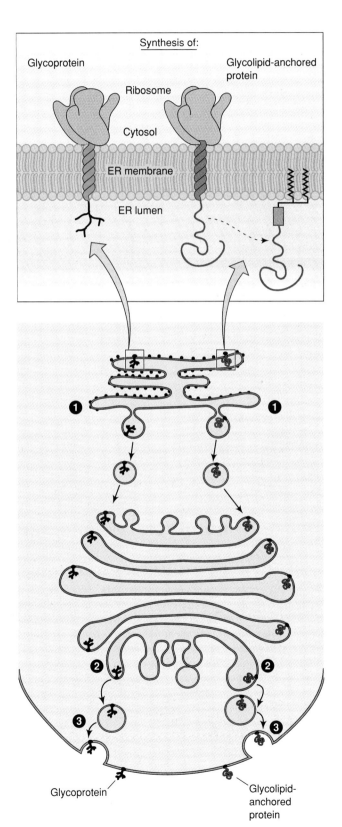

Figure 7-53 *Mechanism by Which Cells Establish the Orientation of Integral Membrane Glycoproteins and Glycolipid-Anchored Proteins in the Plasma Membrane* (Left side) *The pathway for inserting newly formed integral membrane glycoproteins into the plasma membrane involves three stages: (1) membrane vesicles containing newly formed transmembrane glycoproteins bud off from the ER and fuse with the Golgi complex, (2) vesicles containing the membrane glycoprotein emerge from the Golgi complex and migrate to the cell surface, and (3) the vesicles then fuse with the plasma membrane, exposing the glycoprotein to the cell exterior.* (Right side) *The pathway for inserting glycolipid-anchored proteins into the plasma membrane involves a similar three-stage pathway.*

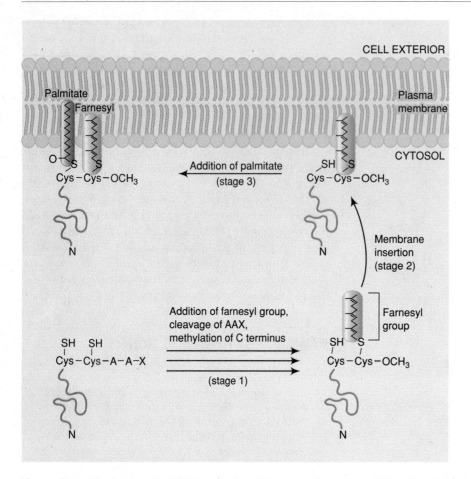

Figure 7-54 *Mechanism by Which Prenylated Proteins Are Inserted into Membranes*
The formation of this type of lipid-anchored membrane protein involves three stages: (1) a 15-carbon farnesyl group (or 20-carbon geranylgeranyl group) is added near the C-terminus of the protein molecule in the cytosol, (2) the farnesyl (or geranylgeranyl) group is inserted into the lipid bilayer of the appropriate target membrane, and (3) in some cases a membrane fatty acid such as palmitate is linked to the prenylated protein.

lysosomes are released into the lumenal space; integral membrane proteins are retained within the bilayer of the ER membrane by stop-transfer sequences that halt translocation of the polypeptide chain through the membrane.

• Proteins synthesized on membrane-attached ribosomes are typically modified by N-linked glycosylation after the growing chain enters the lumen of the ER; in some cases, amino acid hydroxylation or attachment to membrane-bound glycolipid also occurs. In addition, the ER contains enzymes that help newly formed proteins fold into the proper conformation.

• Proteins synthesized in the rough ER are routed to the Golgi complex by membrane vesicles that shuttle back and forth between the two compartments. The Golgi complex consists of a stack of flattened, smooth-membrane-bounded cisternae and associated vesicles. As they pass through the Golgi complex, proteins are subject to both N-linked and O-linked glycosylation. Proteins passing through the Golgi complex contain specific chemical signals that target them to locations such as lysosomes, secretory vesicles, intracellular granules, the ER, or the Golgi itself. In the absence of a specific targeting signal, molecules pass to the cell surface by the process of bulk flow. In addition to its role in protein process-

ing and transport, the Golgi complex synthesizes polysaccharides such as hyaluronate in animal cells and hemicellulose and pectin in plants.

• During constitutive secretion, secretory proteins and integral membrane proteins are moved to the cell surface in a continuous fashion by nonselective bulk flow. In contrast, regulated secretion occurs only in response to external stimuli. In both types of secretion, membrane vesicles fuse with the plasma membrane, discharging their contents into the extracellular space. After this process of exocytosis has taken place, membrane components are recycled back to the Golgi complex by vesicles that bud from the plasma membrane. In bacteria, proteins are secreted by direct translocation through the plasma membrane rather than by fusion of vesicles with the membrane.

• Lysosomes are small vesicles containing acid hydrolases that serve an intracellular digestive function. During lysosome formation, hydrolytic enzymes are targeted to lysosomes by mannose 6-phosphate groups that are recognized by membrane receptors present in the *trans* Golgi network.

• During phagocytosis, the plasma membrane surrounds large particles and brings them into the cell for destruction by lysosomes. In pinocytosis, a small region of the plasma membrane folds inward and buds off as a small vesicle that nonselectively takes in materials dissolved in the extracellular fluid. In receptor-mediated endocytosis, plasma membrane receptors bind to specific extracellular molecules and bring them into the cell by the budding of clathrin-coated vesicles. During potocytosis, caveolae facilitate the uptake of small molecules and ions without pinching off completely to form free cytoplasmic vesicles.

• Membrane growth occurs by the insertion of lipids and proteins into existing membranes. Phospholipids are synthesized by smooth ER membranes (or the plasma membrane in prokaryotic cells) and are incorporated directly into the lipid bilayer. Phospholipid translocators catalyze the movement of phospholipids from one side of a bilayer to the other, whereas phospholipid transfer proteins catalyze the movement of phospholipids from one membrane to another.

• Peripheral membrane proteins are synthesized on free cytoplasmic ribosomes and released into the cytosol prior to associating with membrane surfaces. In contrast, integral membrane proteins are formed on membrane-attached ribosomes and inserted directly into the ER membrane while they are being synthesized. Integral membrane proteins synthesized in the rough ER are passed via membrane vesicles to the membranes of other organelles, including the Golgi complex, lysosomes, and the plasma membrane. Glycolipid-anchored membrane proteins are synthesized in the rough ER and transported by membrane vesicles to the plasma membrane. Fatty acid-anchored and prenylated membrane proteins are synthesized as soluble cytosol proteins, which are then inserted into membranes after addition of lipid groups such as fatty acids, farnesyl groups, and geranylgeranyl groups.

SUGGESTED READINGS

Books

Austen, B. M., and O. M. R. Westwood (1991). *Protein Targeting and Secretion,* IRL Press, Oxford, England.

Holtzman, E. (1989). *Lysosomes,* Plenum, New York.

Pugsley, A. P. (1989). *Protein Targeting,* Academic Press, San Diego.

Articles

Anderson, R. G. W. (1993). Potocytosis of small molecules and ions by caveolae, *Trends Cell Biol.* 3:69-72.

Avran, P., and D. Castle (1992). Protein sorting and secretion granule formation in regulated secretory cells, *Trends Cell Biol.* 2:327-331.

Bishop, W. R., and R. M. Bell (1988). Assembly of phospholipids into cellular membranes: biosynthesis, transmembrane movement and intracellular translocation, *Annu. Rev. Cell Biol.* 4:579-610.

Brodsky, F. M. (1988). Living with clathrin: its role in intracellular membrane traffic, *Science* 242:1396-1402.

Chrispeels, M. J., and B. W. Tague (1991). Protein sorting in the secretory system of plant cells, *Int. Rev. Cytol.* 125:1-45.

Cross, G. A. M. (1990). Glycolipid anchoring of plasma membrane proteins, *Annu. Rev. Cell Biol.* 6:1-39.

Dice, J. F. (1990). Peptide sequences that target cytosolic proteins for lysosomal proteolysis, *Trends Biochem. Sci.* 15:305-308.

Driouich, A., L. Faye, and L. A. Staehelin (1993). The plant Golgi apparatus: a factory for complex polysaccharides and glycoproteins, *Trends Biochem. Sci.* 18:210-214.

Dunn, W. A., Jr. (1994). Autophagy and related mechanisms of lysosome-mediated protein degradation, *Trends Cell Biol.* 4:139-143.

Ferro-Novick, S., and R. Jahn (1994). Vesicle fusion from yeast to man, *Nature* 370:191-193.

Gilmore, R. (1991). The protein translocation apparatus of the rough endoplasmic reticulum, its associated proteins, and the mechanism of translocation, *Curr. Opinion Cell Biol.* 3:580-584.

Goldstein, J. L., M. S. Brown, R. G. W. Anderson, D. W. Russell, and W. J. Schneider (1985). Receptor-mediated endocytosis: concepts emerging from the LDL receptor system, *Annu. Rev. Cell Biol.* 1:1-39.

Griffiths, G., and J. Gruenberg (1991). The arguments for pre-existing early and late endosomes, *Trends Cell Biol.* 1:5-9.

Guengerich, F. P. (1993). Cytochrome P450 enzymes, *Amer. Sci.* 81:440-447.

High, S., and C. J. Stirling (1993). Protein translocation across membranes: common themes in divergent organisms, *Trends Cell Biol.* 3:335-339.

Holtzman, E. (1992). Intracellular targeting and sorting, *BioScience* 42:608-620.

Johnson, A. E. (1993). Protein translocation across the ER membrane: a fluorescent light at the end of the tunnel, *Trends Biochem. Sci.* 18:456-458.

Kornfeld, S., and I. Mellman (1989). The biogenesis of lysosomes, *Annu. Rev. Cell Biol.* 5:483-525.

Low, M. G., and A. R. Saltiel (1988). Structural and functional roles of glycosyl-phosphatidylinositol in membranes, *Science* 239:268-275.

Mellman, I., and K. Simons (1992). The Golgi complex: in vitro veritas? *Cell* 68:829-840.

Meyer, D. I. (1991). Protein translocation into the endoplasmic reticulum: a light at the end of the tunnel, *Trends Cell Biol.* 1:154-159.

Novick, P., and P. Brennwald (1993). Friends and family: The role of the Rab GTPases in vesicular traffic, *Cell* 75:597-601.

Orci, L., M. Ravazzola, M.-J. Storch, R. G. W. Anderson, J.-D. Vassalli, and A. Perreiet (1987). Proteolytic maturation of insulin is a post-Golgi event which occurs in acidifying clathrin-coated secretory vesicles, *Cell* 49:865-868.

Pearce, B. M. F., and M. S. Robinson (1990). Clathrin, adaptors, and sorting, *Annu. Rev. Cell Biol.* 6:151-171.

Pelham, H. R. B. (1989). Control of protein exit from the endoplasmic reticulum, *Annu. Rev. Cell Biol.* 5:1-23.

Pelham, H. R. B. (1991). Recycling of proteins between the endoplasmic reticulum and Golgi complex, *Curr. Opinion Cell Biol.* 3:585-591.

Pfeffer, S. R. (1992). GTP-binding proteins in intracellular transport, *Trends Cell Biol.* 2:41-46.

Rapoport, T. A. (1992). Transport of proteins across the endoplasmic reticulum membrane, *Science* 258:931–936.

Redman, C. M., and Sabatini, D. D. (1966). Vectorial discharge of peptides released by puromycin from attached ribosomes, *Proc. Natl. Acad. Sci. USA* 56:608–615.

Robinson, D. G., and H. Depta (1988). Coated vesicles, *Annu. Rev. Plant Physiol. Plant Mol. Biol.* 39:53–99.

Sandoval, I. V., and O. Bakke (1994) Targeting of membrane proteins to endosomers and lysosomes, *Trends Cell Biol.* 4:292–297.

Shepherd, V. L. (1989). Intracellular pathways and mechanisms of sorting in receptor-mediated endocytosis, *Trends Pharmacol. Sci.* 10:458–462.

Simon, S. M., and G. Blobel (1991). A protein-conducting channel in the endoplasmic reticulum, *Cell* 65:371–380.

Singer, S. J. (1990). The structure and insertion of integral proteins in membranes, *Annu. Rev. Cell Biol.* 6:247–296.

Söllner, T., S. W. Whiteheart, M. Brunner, H. Erdjument-Bromage, S. Geromanos, P. Tempst, and J. E. Rothman (1993). SNAP receptors implicated in vesicle targeting and fusion, *Nature* 362:318–324.

van Deurs, B., P. K. Holm, K. Sandvig, and S. H. Hansen (1993). Are caveolae involved in clathrin-independent endocytosis? *Trends Cell Biol.* 3:249–251.

von Heijne, G. (1990). Protein targeting signals, *Curr. Opinion Cell Biol.* 2:604–608.

Wickner, W. T., and H. F. Lodish (1985). Multiple mechanisms of protein insertion into and across membranes, *Science* 230:400–407.

Wirtz, K. W. A. (1991). Phospholipid transfer proteins, *Annu. Rev. Biochem.* 60:73–99.

Wolin, S. L. (1994). From the elephant to E. coli: SRP-dependent protein targeting, *Cell* 77:787–790.

C h a p t e r 8

Mitochondria and the Capturing of Energy Derived from Food

As we learned in the first two chapters, cells cannot survive without a source of chemical building blocks and energy. These two requirements are related in that chemical building blocks and energy are both present in the food molecules that organisms produce or ingest, and both are simultaneously transformed every time a chemical reaction occurs. In this chapter we will focus primarily on the question of how energy is captured, but it should be kept in mind that the reactions we will be discussing also provide cells with chemical building blocks.

The way in which cells capture energy from ingested food molecules depends upon whether or not oxygen is present. The term **anaerobic** refers to metabolism that doesn't use oxygen, while **aerobic** refers to metabolism that does use oxygen. In anaerobic metabolism, energy-rich food molecules are broken down in the cytosol and a small fraction of the released energy is captured for use by the cell. In aerobic metabolism, the preceding set of reactions is followed by an oxygen-requiring pathway that captures a much larger fraction

of the released energy; this oxygen-requiring process is associated with mitochondria in eukaryotic cells, and with the plasma membrane in prokaryotic cells. We will begin our discussion of how cells capture energy from nutrients by considering what happens during anaerobic metabolism, and will then discuss the events associated with aerobic metabolism.

ANAEROBIC PATHWAYS FOR CAPTURING ENERGY

At the outset of this chapter, it is important to understand that several unifying themes recur repeatedly in all metabolic pathways for capturing energy, aerobic as well as anaerobic. These recurring themes include the use of *oxidation reactions* for releasing chemical energy, the use of *coenzymes* for coupling oxidation-reduction reactions to each other, and the use of *ATP* and related high-energy phosphate compounds as intermediates in energy transfer. Before beginning our discussion of how cells obtain energy from food molecules, we will briefly describe each of these unifying themes.

Energy Is Released by Oxidation Reactions

Substances taken in as food, such as carbohydrates and fats, are energy-rich molecules whose energy can be captured for use by cells. An energy-rich molecule is one that can be *oxidized* in an exergonic reaction that liberates large amounts of free energy. By definition, an **oxidation** reaction involves the removal of electrons. Hence any time a substance gives up electrons, it is being oxidized. Conversely, any substance that gains electrons during a chemical reaction is said to be undergoing **reduction.** Oxidation and reduction always occur together; electrons removed from a substance being oxidized are transferred to another substance, which becomes reduced.

A simple example of an oxidation-reduction reaction is the transfer of a single electron from an iron ion (Fe^{2+}) to a copper ion (Cu^{2+}) as follows:

$$Fe^{2+} + Cu^{2+} \rightarrow Fe^{3+} + Cu^+$$

Reduced   Oxidized   Oxidized   Reduced

As we will see later in the chapter, this simple type of oxidation plays an important role in the final stages of aerobic energy metabolism. However, most biological oxidations involve the transfer not just of electrons, but of *electrons bound to hydrogen atoms.* To illustrate this principle, let us consider the following hypothetical reaction:

$$XH_2 + Y \rightarrow X + YH_2$$

In this reaction, two hydrogen atoms are being removed from substance XH_2 and transferred to substance Y. Each of these hydrogen atoms contains a single electron, so a total of two electrons is being shifted from X to Y. Substance XH_2 is losing these electrons and is therefore undergoing oxidation; in contrast, substance Y is gaining electrons and thus undergoing reduction.

Thus oxidation-reduction reactions involve the transfer of electrons from one substance to another, either alone or as part of a hydrogen atom. Another way of stating this is that the substance being oxidized acts as an *electron donor,* and the substance being reduced functions as an *electron acceptor.* The term "oxidation" was originally coined because molecular oxygen readily acts as an electron acceptor, but it is now clear that many other substances can also serve as electron acceptors in oxidation-reduction reactions. As we will see in this chapter, oxidation-reduction reactions are central to cellular energy metabolism because oxidation is an exergonic process that releases energy from carbohydrates, fats, and other energy-rich molecules taken in as food.

Coenzymes and ATP Play Central Roles in Transferring Energy from One Reaction to Another

Cellular oxidation-reduction reactions are often catalyzed by enzymes that utilize coenzymes as electron donors or acceptors. As we learned in Chapter 2, the coenzymes employed for this purpose are **NAD⁺, NADP⁺, FAD,** and **FMN** (page 57). Enzymes that use FAD or FMN as coenzymes are called *flavoproteins;* in most cases, the FAD or FMN molecule is tightly bound to the protein. In contrast, NAD^+ and $NADP^+$ bind loosely to enzymes; as a result, NAD^+ and $NADP^+$ can diffuse from one enzyme to another after being oxidized or reduced. Despite these differences, NAD^+, $NADP^+$, FAD, and FMN all have an important property in common: Each can accept a pair of electrons from a substrate undergoing oxidation, and then transfer the electrons to another molecule that is being reduced (Figure 8-1). When one of these coenzymes undergoes reduction by accepting electrons from a molecule undergoing oxidation, the free energy of the coenzyme molecule increases. The reason for the increase is that the reduced coenzyme has captured some of the energy released when the substrate was oxidized. The reduced coenzymes NADH, NADPH, $FADH_2$, and $FMNH_2$ are thus high-energy molecules that can transfer energy to other reactions.

ATP is another type of molecule that is encountered repeatedly in the pathways that capture energy from food molecules. In Chapter 2 we learned that the energy released by thermodynamically favorable reactions can be used to drive the formation of ATP from ADP and free phosphate. The energy stored in ATP is then employed for energy-requiring activities such as active transport, the synthesis of macromolecules, and cell motility. As we proceed to discuss the pathways by which cells capture energy under anaerobic and aerobic conditions, we will encounter numerous steps that involve ATP.

Glucose Plays a Central Role as a Source of Both Energy and Chemical Building Blocks

Our discussion of the metabolic reactions that cells utilize to obtain energy will begin with the pathway for oxidizing the six-carbon sugar, **glucose.** Although cells can obtain energy from other organic molecules as well, most of these other substances are converted into molecules that enter at one point or another into the same pathway that is used for oxidizing glucose.

The glucose oxidation pathway performs two basic functions for the cell. First, energy released during the oxidation of glucose is captured by coupling it to the synthesis of ATP. Second, some of the chemical products generated by this pathway are utilized as precursors for synthesizing chemical building blocks needed by cells, such as amino acids, nucleotides, fatty acids, and other sugars. These building blocks are in turn incorporated into proteins, nucleic acids, lipids, and polysaccharides by synthetic pathways that require energy derived from ATP. Thus the pathway for metabolizing glucose plays a central role as a source of both energy and chemical building blocks (Figure 8-2).

Glycolysis Is the First Stage in Extracting Energy from Glucose

The first stage in the pathway for oxidizing glucose involves a set of reactions occurring in the cytosol known as **glycolysis.** This group of reactions is believed to be one of the first energy capturing pathways to have evolved in living cells, appearing in ancient bacteria that lived more than three billion years ago. Since oxygen was not present in the earth's atmosphere at that time, glycolysis originally functioned in an anaerobic environment. Today this pathway occurs in virtually all cells, operating in aerobic as well as anaerobic environments.

The discovery of glycolysis was initially based on observations made by nineteenth-century wine makers, who found that yeast cells can convert sugar to ethanol under anaerobic conditions. However, it was not until the 1930s that the individual steps of glycolysis were elucidated in Germany by Gustav Embden, Otto Meyerhof, and Otto Warburg. These investigators reported that during glycolysis, glucose is converted by a sequential series of 10 reactions into two molecules of the three-carbon compound, *pyruvate.* During this process some of the energy released from the glucose molecule is captured in the form of the energy-rich molecules, ATP and NADH.

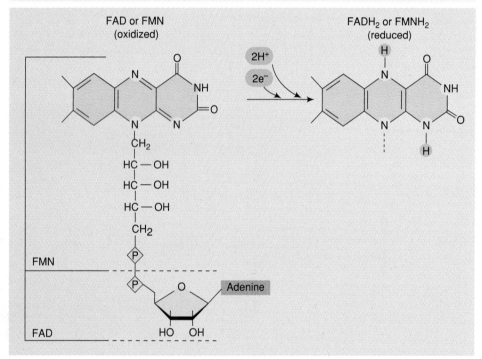

Figure 8-1 *Structures of the Coenzymes NAD⁺, NADP⁺, FAD, and FMN in the Oxidized and Reduced States* One proton (H^+) and two electrons (e^-) are transferred during the reduction of NAD^+ and $NADP^+$, whereas two protons and two electrons are transferred during the reduction of FAD and FMN. In spite of the difference in the number of protons involved, two electrons are always transferred during the oxidation or reduction of these coenzymes.

The ten steps of glycolysis are summarized in Figure 8-3. During the first phase of this pathway (Steps 1–5), two molecules of ATP are consumed in reactions that add phosphate groups to the glucose skeleton, and the sugar's six-carbon skeleton is cleaved in half to produce two molecules of glyceraldehyde 3-phosphate. During the second phase of glycolysis (Steps 6–10) the carbon skeleton is oxidized (in Step 6) and then converted into two molecules of pyruvate; this phase of glycolysis generates two molecules of NADH and four molecules of

ATP. Since two ATP molecules are *consumed* during the first phase of glycolysis and four ATP molecules are *produced* during the second phase, there is a net gain of two molecules of ATP per molecule of glucose metabolized. *Hence the overall result of glycolysis is the conversion of one molecule of glucose to two molecules of pyruvate, accompanied by the net formation of two molecules of ATP and two molecules of NADH.*

The breakdown of glucose does not normally end at Step 10 (the formation of pyruvate) because the conver-

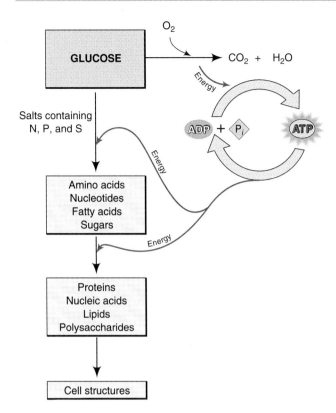

Figure 8-2 *The Central Role Played by Glucose as a Source of Both Energy and Chemical Building Blocks* *Energy released during glucose oxidation is used to drive the synthesis of ATP. The breakdown of glucose also yields compounds that function as precursors for synthesizing amino acids, nitrogenous bases, fatty acids, and other sugars.*

sion of NAD^+ to NADH that accompanies glycolysis would quickly lead to a depletion of NAD^+. The NAD^+ molecule is an essential requirement for glycolysis because it functions as the electron acceptor in Step 6, where the carbon skeleton of glucose is oxidized. The absence of NAD^+ would prevent this reaction from occuring and thus bring glycolysis to a halt. Hence NAD^+ must be regenerated for glycolysis to continue.

Fermentation Replenishes NAD^+ While Reducing Pyruvate to Lactate or Ethanol

Under anaerobic conditions, NAD^+ is regenerated in a reaction in which the pyruvate produced by glycolysis is reduced, accompanied by the oxidation of NADH to NAD^+. When pyruvate is reduced in this way, the overall pathway for converting glucose to a reduced derivative of pyruvate is called **anaerobic glycolysis** or **fermentation.** The actual product generated by the reduction of pyruvate varies in different organisms (Figure 8-4). In yeast, which can survive in the complete absence of oxygen, pyruvate is converted to carbon dioxide and *ethanol* (a type of alcohol). This process, called *alcoholic fermentation,* has important practical applications both

in baking bread and in the production of alcoholic beverages. When bread is made, yeast is added because the carbon dioxide generated by yeast cells during alcoholic fermentation causes the bread to rise, while the ethanol evaporates harmlessly during the baking process. In the beer- and wine-making industries, yeast is employed because the ethanol it produces when fermenting sugar provides these beverages with their alcohol content.

Fermentation is not limited to organisms that live anaerobically. Most animals are aerobic organisms that require oxygen for survival, and yet certain animal tissues still carry out fermentation. A typical example occurs in the skeletal muscle tissue of a person who is exercising vigorously. If the contracting muscle cells consume oxygen faster than it can be supplied from the bloodstream, the cells become temporarily anaerobic. Under such conditions, muscle cells obtain energy from a fermentation pathway in which glucose is broken down to pyruvate by glycolysis; the pyruvate is then reduced to *lactate* (lactic acid), regenerating NAD^+ in the process. The ATP produced by this *lactate fermentation* pathway provides the energy required for muscle contraction. During prolonged exercise the large amounts of lactate that accumulate may cause muscle pain and cramping.

Ethanol and lactate are not the only products that can be formed by fermentation. Bacteria convert pyruvate to a variety of reduced products, including lactate, butyrate (present in rancid butter), and acetone. But no matter what the final product of the pyruvate reduction step happens to be, the important shared feature of all these pyruvate reduction reactions is that they simultaneously oxidize NADH to NAD^+, thereby replenishing the NAD^+ required for Step 6 in glycolysis. Thus fermentation does not change the amounts of NAD^+ and NADH present in cells because the conversion of NAD^+ to NADH, which accompanies the breakdown of glucose to pyruvate, is reversed when pyruvate is reduced during the final stage of fermentation.

The Two ATP Molecules Produced During the Fermentation of Glucose Represent a Relatively Small Energy Yield

Since there is no net production of NADH during the fermentation of glucose, the only by-product is the two molecules of ATP generated by glycolysis. These two ATP molecules represent the total amount of energy captured by fermentation for use by the cell. How much of the total energy theoretically available in the glucose molecule does this represent?

To answer this question, one must first determine how much energy would be released if a molecule of glucose were completely oxidized to CO_2 and H_2O:

$$C_6H_{12}O_6 + 6\,O_2 \rightarrow 6\,CO_2 + 6\,H_2O$$

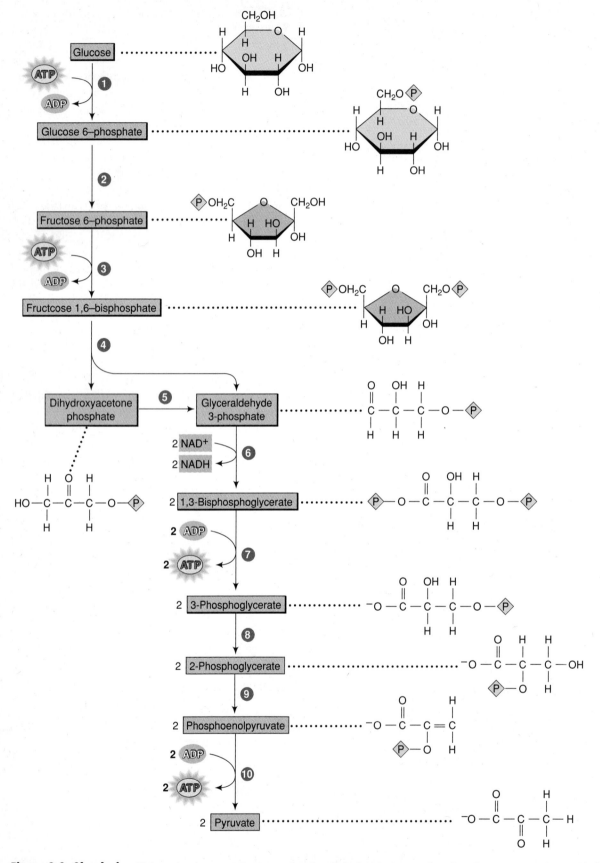

Figure 8-3 Glycolysis   *This pathway converts one molecule of glucose to two molecules of pyruvate accompanied by the net production of two molecules of ATP and two molecules of NADH.*

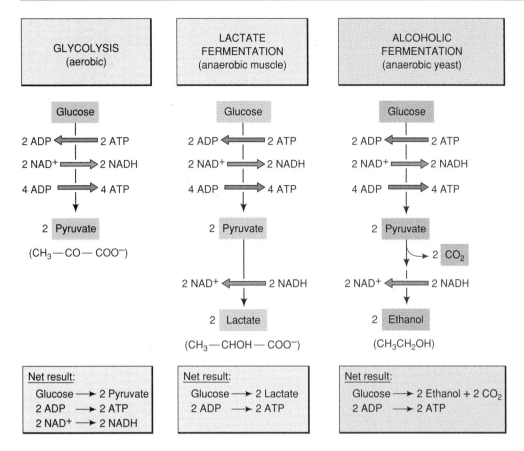

Figure 8-4 *Comparison of Glycolysis and Fermentation* *Glycolysis involves the net oxidation of the carbon skeleton of glucose accompanied by the reduction of NAD^+ to NADH. During fermentation the pyruvate produced by glycolysis is reduced while NADH is oxidized back to NAD^+.*

Using the techniques described in Chapter 2 (page 35), the $\Delta G^{\circ\prime}$ for this reaction can be calculated to be -686 kcal/mol; in other words, 686 kcal of energy is released per mole of glucose oxidized to CO_2 and H_2O under standard conditions. How much of this energy is captured during fermentation? As we showed in Figure 2-25, the $\Delta G^{\circ\prime}$ for the hydrolysis of ATP to ADP under conditions that typically prevail in an intact cell is about -11 kcal/mol. Since two molecules of ATP are made during fermentation, this value is multiplied by 2 to obtain -22 kcal/mol. If 686 kcal/mol of energy is theoretically available in glucose and 22 kcal/mol is captured in the form of two molecules of ATP, then the overall recovery is 22 divided by 686, or about 3 percent. The yield is so small because no *net* oxidation of glucose occurs during fermentation. Oxidation of the carbon skelcton does take place during Step 6 of glycolysis, but this oxidation is later reversed when pyruvate is reduced to lactate or ethanol. Hence there is no change in the overall oxidation state of the carbon skeleton of glucose during fermentation.

The small amount of energy obtained when glucose is metabolized by fermentation is sufficient to meet the energy needs of organisms that can live in the absence of oxygen. However, aerobic organisms derive much larger amounts of energy from glucose by using molecular oxy-

gen to oxidize glucose to carbon dioxide and water. This aerobic process, called **cellular respiration,** utilizes a series of metabolic pathways known as the Krebs cycle, electron transport, and oxidative phosphorylation. The **Krebs cycle** oxidizes the pyruvate generated by glycolysis, and **electron transport** and **oxidative phosphorylation** then use the released energy to drive ATP synthesis. Whereas the enzymes involved in glycolysis and fermentation are located in the cytosol, cellular respiration occurs within the mitochondria of eukaryotic cells, and in association with the plasma membrane of prokaryotic cells. Because we have not yet discussed the structure of mitochondria in any detail, the architecture of this organelle must be described before we can explain how energy is captured during cellular respiration.

ANATOMY OF THE MITOCHONDRION

Mitochondria Were First Discovered and Functionally Described Using Microscopy and Subcellular Fractionation

It is difficult to say who first discovered the mitochondrion. Nineteenth-century light microscopists often noted the presence of small particles in the cytoplasm, but it is not clear how many of these were actually mitochondria. One of the first systematic studies was initiated in 1850 by Rudolph Kölliker, who described the existence of regularly spaced particles in the cytoplasm of muscle cells. He even isolated the particles from shredded muscle tissue and discovered that they swell when placed in water, leading him to conclude that each particle is surrounded by a semipermeable membrane. Similar particles were soon identified in other tissues, and although they were initially given a variety of different names, the term **mitochondrion** ("threadlike granule") eventually came to be universally accepted.

In 1900 Leonor Michaelis made a remarkable discovery while in the process of staining live cells with the dye *Janus green*. Initially this dye stains mitochondria green, but as cells consume oxygen, the color gradually disappears. Such alterations in color are known to indicate a change in the oxidation-reduction state of the dye, suggesting that mitochondria function in oxidation-reduction reactions. Unfortunately, this clue to the physiological role of mitochondria was not generally appreciated because most biologists of that era believed that mitochondria were involved in transmitting hereditary characteristics to differentiating cells. Even Otto Warburg's discovery in 1913 that granules isolated from cell homogenates consume oxygen did not trigger the realization that the granules were mitochondria, and that mitochondria are therefore sites of cellular oxidation.

The uncertainty concerning mitochondrial function was not resolved until methods were developed for isolating and purifying functional mitochondria. In the early 1940s, Albert Claude pioneered the development of the first subcellular fractionation techniques for isolating mitochondria and other organelles (page 137). Although Claude's original procedures involved fractionating cell homogenates in salt solutions that disrupt mitochondrial function, mitochondria isolated in this way were nevertheless found to contain components of the Krebs cycle and the respiratory chain, both of which are part of the cellular respiration pathway. In 1948 functionally active mitochondria were finally isolated by George Hogeboom, Walter Schneider, and George Palade, who used a fractionation medium containing sucrose rather than salt. This breakthrough allowed Eugene Kennedy and Albert Lehninger to demonstrate that isolated mitochondria are capable of carrying out the Krebs cycle, electron transport, and oxidative phosphorylation. Because such observations indicated that mitochondria are the site of the main energy-capturing pathways of eukaryotic cells, the mitochondrion came to be known as the "power plant" of the cell.

Mitochondria Have Outer and Inner Membranes That Define Two Separate Compartments

A few years after these pioneering discoveries, George Palade and Fritiof Sjøstrand independently published the first high-resolution electron micrographs of mitochondria. In place of the relatively formless granule familiar to light microscopists, the mitochondrion suddenly appeared as an organelle exhibiting a complex internal architecture (Figure 8-5). It is encompassed by two membranes, an **outer mitochondrial membrane** that defines the external boundary of the organelle and an **inner mitochondrial membrane** exhibiting numerous folds, known as **cristae,** that project into the mitochondrial interior. The cristae are usually arranged as flat membrane sheets, but in certain organisms they may take the form of tubules, vesicles, prisms, or concentric whorls (Figure 8-6). In negatively stained preparations, the surface of the cristae facing the mitochondrial interior is seen to be covered with small knoblike projections, called F_1 **particles,** which measure about 9 nm in diameter (Figure 8-7). Later in the chapter we will see that these particles are the major site of mitochondrial ATP formation.

The outer and inner mitochondrial membranes divide the mitochondrion into two distinct compartments: the **intermembrane space,** which is located between the outer and inner membranes, and the mitochondrial **matrix,** which is enclosed by the inner mitochondrial membrane.

Mitochondria Can Form Large Interconnected Networks

Mitochondria often occur in close association with other cytoplasmic structures that either utilize the ATP produced by mitochondria or provide mitochondria with energy-rich substrates. Two striking examples of such arrangements are presented in Figure 8-8. On the left, the mitochondria of muscle cells are lined up adjacent to the contractile myofibrils that utilize ATP during muscle contraction. On the right, a mitochondrion is pictured surrounding a lipid droplet that contains fatty acids destined for mitochondrial oxidation.

In thin-section electron micrographs, mitochondria typically appear as oval structures measuring several micrometers in length and 0.5–1.0 μm across. This appearance has fostered the notion that most mitochondria are sausage shaped and that typical cells contain from several hundred to a few thousand of these organelles. However, this view has been challenged by the work of Hans-Peter Hoffman and Charlotte Avers, who showed that the three-dimensional shape of mitochondria within intact cells cannot be determined by looking at a few thin-section micrographs. After examining a complete series of thin sections through an entire yeast cell, they concluded that the oval profiles observed in individual micrographs all represent slices through a single large, extensively branched mitochondrion. This view has been buttressed by the analysis of other eukaryotic cells, where mitochondrial profiles seen in thin-section

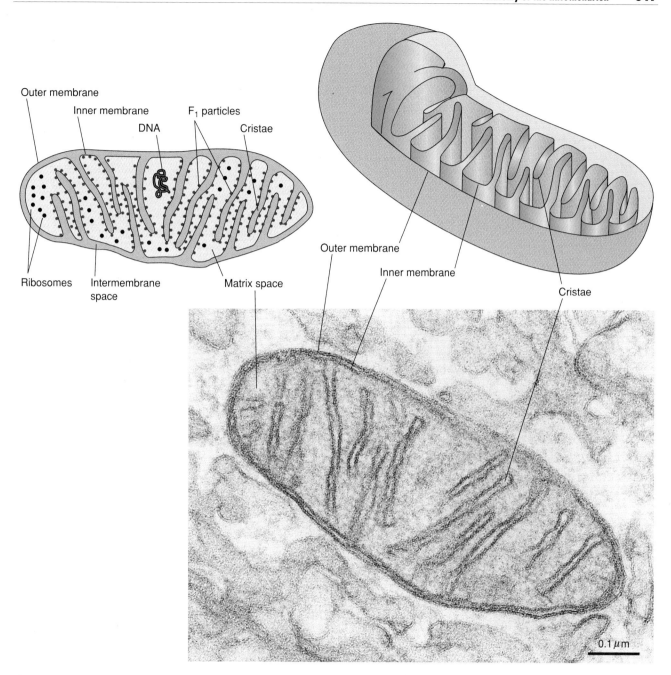

Figure 8-5 *Structural Organization of the Mitochondrion* *Mitochondria are bounded by two membranes that together enclose an inner compartment called the mitochondrial matrix. The inner membrane exhibits numerous folds, called cristae, that project into the interior of the mitochondrion. Micrograph courtesy of B. Tandler.*

micrographs appear to represent portions of larger, interconnected mitochondrial networks (Figure 8-9, page 320). Such results suggest that the number of individual mitochondria present in a cell is considerably smaller than once believed.

The conclusion that mitochondria exist as interconnected networks rather than numerous independent organelles has received further support from studies in which intact living cells were examined by phase-contrast microscopy (Figure 8-10, page 321). Such investigations have revealed that living cells contain large

branched mitochondria in a dynamic state of flux, with segments of one mitochondrion frequently pinching off and fusing with another mitochondrion. These dynamic interactions suggest that the concept of the "number" of mitochondria present in any given cell may in fact be meaningless.

As we learned in Chapter 4, differential centrifugation can be employed to isolate mitochondria from cell homogenates. Microscopic examination of mitochondria isolated in this way usually reveals the presence of numerous small mitochondria of relatively uniform size

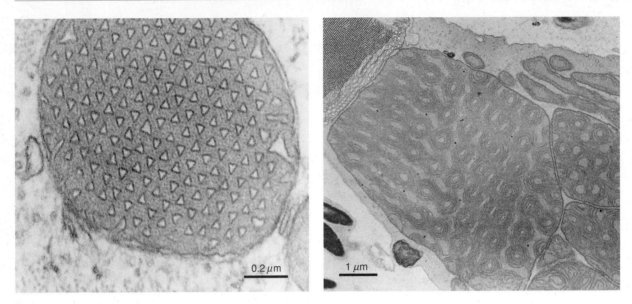

Figure 8-6 Electron Micrographs of Unusual Mitochondrial Cristae (Left) *Prismatic cristae present in mitochondria of nerve cells in the hamster brain. (Right) Concentric cristae in giant mitochondria of photoreceptor cone cells in the eye of the tree shrew. Courtesy of N. B. Rewcastle and A. P. Anzil (left) and T. Samorajski, J. M. Ordy, and J. R. Keefe (right).*

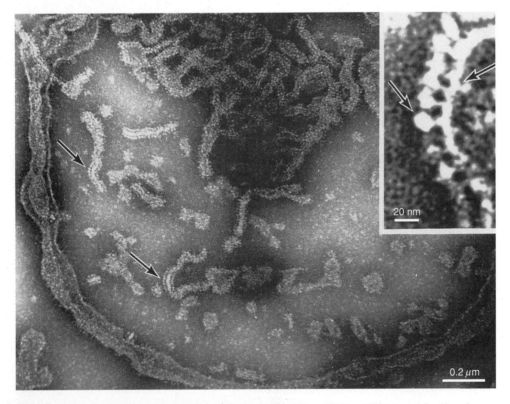

Figure 8-7 Electron Micrograph of Negatively Stained Beef Heart Mitochondria Showing F₁ Particles Protruding from the Surfaces of the Cristae *The inset shows these particles at higher magnification. F_1 particles are the site where ATP is made during oxidative phosphorylation. Courtesy of H. Fernández-Morán.*

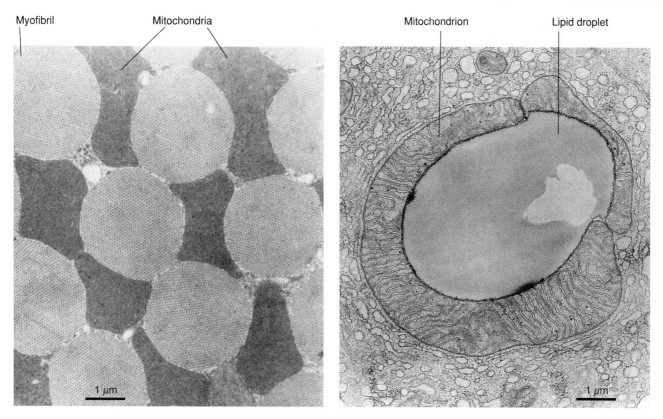

Figure 8-8 *Examples Illustrating the Proximity of Mitochondria to Structures That Either Use ATP or Provide Substrates for Mitochondrial Oxidation Reactions* (Left) *Cross section through insect flight muscle showing the contractile myofibrils surrounded by the darker mitochondria.* (Right) *A pancreatic exocrine cell showing a lipid droplet surrounded by a mitochondrion. Courtesy of B. Sacktor* (left) *and J. D. Jamieson* (right).

(Figure 8-11). However, it is likely that these individual mitochondria are artificially generated when the branched, interconnected mitochondrial network is disrupted during homogenization. This phenomenon resembles what takes place when the endoplasmic reticulum is fragmented into tiny, membrane-bound vesicles (microsomes) during homogenization and subcellular fractionation (page 259).

Mitochondrial Membranes and Compartments Can Be Separated from Each Other for Biochemical Study

The development of methods for isolating mitochondrial membranes and compartments has helped us understand how this organelle functions. The first successful technique for separating inner and outer mitochondrial membranes was developed by Donald Parsons and his colleagues. In this procedure (Figure 8-12), mitochondria are placed in a hypotonic solution or exposed to the detergent *digitonin* until the outer membrane ruptures, releasing the contents of the intermembrane space into solution. The inner membranes, outer membranes, and components of the intermembrane space are then sepa-

rated by centrifugation. Isolated inner and outer membranes can be readily distinguished from each other by electron microscopy; isolated outer membranes look like empty sacs, whereas inner membranes form vesicles called *mitoplasts,* which contain trapped matrix material within them (Figure 8-13). Mitoplasts can be further fractionated into inner membrane and matrix components by exposure to the detergent Lubrol, which disrupts the inner membrane. The disrupted inner membrane usually reseals itself into small *inner membrane vesicles* that contain F_1 particles protruding from the exterior surface.

Biochemical studies utilizing isolated mitochondrial components have provided important information regarding the location of various metabolic activities within the mitochondrion (Table 8-1, page 322). For example, the enzymes of the Krebs cycle and fatty acid oxidation are detected mainly in isolated matrix preparations, whereas electron transport and oxidative phosphorylation are associated with isolated inner membrane vesicles. Mitochondrial membranes and compartments also contain enzymes involved in lipid breakdown, the synthesis of fatty acids and amino acids, and the oxidation of substrates not directly related to cellular respiration, indicating that the mitochondrion performs a variety of functions for the cell.

Figure 8-9 A Model of Interconnected Mitochondrial Networks in a Pig Skin Cell
This model, which was created by examining a complete series of thin sections through the cell, shows that individual mitochondrial profiles observed in thin-section micrographs represent portions of larger, interconnected mitochondrial networks. The model is overlaid on a two-dimensional representation of the cell. Courtesy of M. L. Vorbeck.

The Outer Mitochondrial Membrane Is Permeable to Small Molecules

The outer mitochondrial membrane contains an unusual array of enzymes (see Table 8-1). Several are involved in lipid metabolism, implicating the outer membrane in both the synthesis of membrane phospholipids and in the preliminary breakdown of lipids prior to their oxidation in the mitochondrial matrix. The outer membrane also contains *monoamine oxidase,* an enzyme involved in terminating the action of amine neurotransmitters such as norepinephrine and dopamine (see Table 17-2).

The permeability of the outer mitochondrial membrane has been studied by incubating intact mitochondria in the presence of various substances and then measuring how much of each substance is taken up. Most small molecules, such as salts, sugars, nucleotides, and coenzymes, penetrate rapidly. But their final average concentration within the mitochondrion is usually less than their concentration in the incubation medium, suggesting that only a portion of the mitochondrion is accessible to solutes in the external environment (Figure 8-14). The volume of this readily

permeable compartment, which varies from 20 to 80 percent of the total mitochondrial volume, correlates well with the size of the intermembrane space as visualized by electron microscopy. Such observations suggest that the outer membrane is relatively permeable to small molecules whereas the inner membrane is not; as a result, small molecules readily pass from outside the mitochondrion into the intermembrane space, but not into the matrix. This conclusion has been confirmed by the discovery that the outer membrane contains transmembrane proteins called **porins,** which permit the passage of most molecules smaller than 10,000 daltons. Similar proteins reside in the outer membrane of Gram-negative bacteria and in the outer chloroplast membrane.

Since the inner mitochondrial compartment houses most of the mitochondrion's metabolic activities, the permeability of the outer membrane to many small molecules allows the inner compartment more direct access to substances present in the cytosol. This arrangement is also compatible with theories claiming that mitochondria evolved from symbiotic bacteria hundreds of millions of years ago (page 665). According to

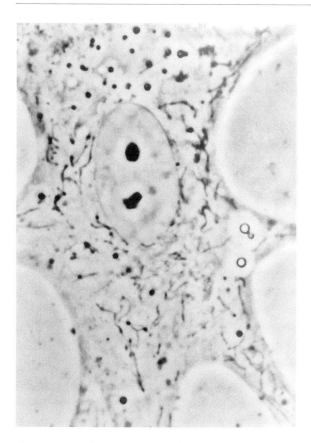

Figure 8-10 *Phase Contrast Micrograph of a Living Fibroblast Showing Interconnected Threadlike Mitochondria in the Cytoplasm* *Courtesy of D. W. Fawcett.*

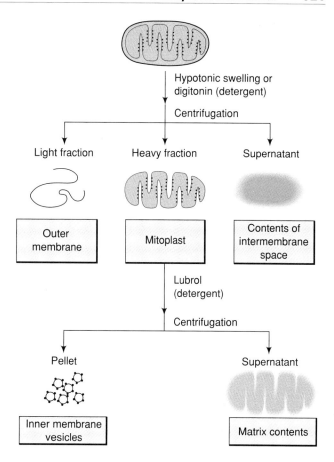

Figure 8-12 *Isolation of Mitochondrial Components*
Mitochondria are first exposed to hypotonic conditions to rupture the outer membrane, allowing the inner membrane plus matrix (mitoplasts) to be isolated by centrifugation. Treating mitoplasts with the detergent Lubrol disrupts the inner membrane and releases the matrix components, yielding inner membrane vesicles that contain F_1 particles protruding from the exterior surface.

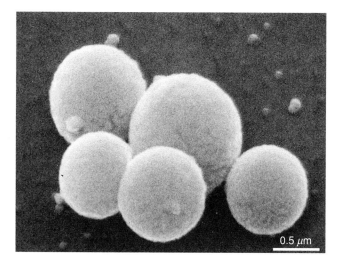

Figure 8-11 *Isolated Mitochondria Viewed by Scanning Electron Microscopy* *Although they appear as separate spherical structures, these individual mitochondria may be artificially produced when the branched, interconnected mitochondrial network is disrupted during cell homogenization and fractionation. Courtesy of C. R. Hackenbrock.*

such theories, the inner mitochondrial membrane and its enclosed matrix are derived from the bacterial plasma membrane and cytoplasm, whereas the outer mitochondrial membrane is either derived from the bacterial outer membrane or was added from the cytoplasmic membranes of the host eukaryotic cell. The presence of porins in both the outer mitochondrial membrane and the outer membrane of Gram-negative bacteria supports the view that the two membranes have a common evolutionary origin.

Because the outer mitochondrial membrane is relatively permeable to most small molecules, the solute composition of the intermembrane space closely mirrors that of the cytosol. However, a few enzymes are selectively localized in the intermembrane space (because of their large size, the enzymes cannot pass through the porin channels). Prominent among these enzymes is *adenylate kinase,* which catalyzes the transfer of the

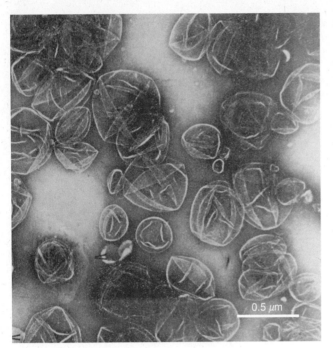
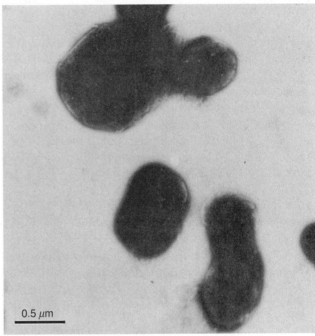

Figure 8-13 *Electron Microscopic Appearance of Isolated Outer and Inner Mitochondrial Membranes* (Left) *Isolated outer membranes exhibit a characteristic "folded-bag" appearance. (Right) Isolated inner membranes form vesicles called mitoplasts, which contain trapped matrix material. Courtesy of B. Chance.*

Table 8-1 Distribution of Selected Components within the Mitochondrion

Outer Membrane	Intermembrane Space	Inner Membrane	Matrix
Cytochrome b_5	Adenylate kinase	NADH dehydrogenase	Pyruvate dehydrogenase
NADH-cytochrome b_5 reductase	Nucleoside	Succinate dehydrogenase	Fatty acid β-oxidation enzymes
Monoamine oxidase	diphosphokinase	Cytochrome $b-c_1$ complex	Krebs cycle enzymes
Fatty acyl-CoA synthetase	Nucleoside	Cytochrome oxidase	DNA polymerase
Glycerolphosphate-acyl	monophosphokinase	Cytochrome c	RNA polymerase
transferase		ATP synthase (F_1–F_0 complex)	Ribosomes
Nucleoside diphosphokinase		Transport carriers (e.g., for	Transfer RNAs
Porin		adenine nucleotides, phosphate,	
		Ca^{2+}, Na^+, etc.)	
Membrane lipid content:		Membrane lipid content:	
Phospholipid/protein = 0.9		Phospholipid/protein = 0.3	
Cardiolipin/phospholipid = 0.03		Cardiolipin/phospholipid = 0.22	
		Ubiquinone	

terminal phosphate group of ATP to AMP, forming two molecules of ADP:

$$AMP + ATP \rightleftharpoons 2\,ADP$$

The equilibrium constant for this reaction is close to 1, which means that adenylate kinase allows ATP, ADP, and AMP to be readily interconverted.

The Mitochondrial Matrix Is the Site Where the Krebs Cycle Occurs

The mitochondrial matrix contains all but one of the enzymes of the Krebs cycle, a metabolic pathway whose role in oxidizing the pyruvate produced by glycolysis will be discussed shortly. The protein concentration of isolated matrix preparations is much higher than is typi-

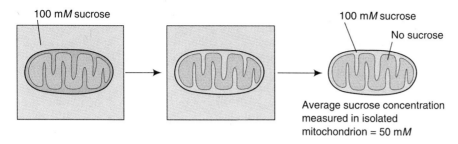

100 mM sucrose

100 mM sucrose

No sucrose

Average sucrose concentration
measured in isolated
mitochondrion = 50 mM

Figure 8-14 *Measurement of Mitochondrial Permeability* (Left) *Mitochondria are placed in a solution containing 100 mM sucrose. (Middle) The sucrose equilibrates across the outer mitochondrial membrane. (Right) Mitochondria are removed from the sucrose solution and their overall sucrose content is measured. If the average sucrose concentration in the isolated mitochondria is found to be 50 mM, it can be concluded that the 100 mM sucrose solution has access to only half of the total mitochondrial fluid volume. In other words, if half the mitochondrial fluid is contained in the intermembrane space and has a sucrose concentration of 100 mM, while the other half of the fluid is in the matrix and has no sucrose, the average sucrose concentration of the entire mitochondrion will be 50 mM.*

cal for enzymes dissolved in an aqueous solution. It has therefore been proposed that many matrix enzymes are anchored within a structural framework and/or are part of large, multiprotein complexes. This idea has received support from experiments in which mitochondria were artificially swollen by suspending them in a hypotonic medium. Under these conditions, the matrix does not randomly disperse; instead, openings form in what appears to be an organized matrix network. Moreover, when the enzymes of the Krebs cycle are isolated under gentle conditions, many remain united together as multiprotein complexes rather than dispersing as separate entities.

In addition to Krebs cycle enzymes, the mitochondrial matrix contains DNA, RNA, ribosomes, and enzymes involved in nucleic acid and protein synthesis. Because these components are involved in mitochondrial growth and division, we will delay their considera-

tion until the topic of mitochondrial biogenesis is covered in Chapter 14.

The Inner Mitochondrial Membrane Is the Main Site of Mitochondrial ATP Formation

The inner mitochondrial membrane is the site of electron transport and oxidative phosphorylation, which are pathways that utilize the energy released during oxidation reactions to drive the formation of ATP. The conclusion that the inner membrane is the major site of these activities rests on biochemical analyses of isolated inner membrane preparations, as well as microscopic examination of tissue sections stained for components involved in these pathways (Figure 8-15). The inner mitochondrial membrane is unique in both appearance and chemical composition. Its unique appearance stems

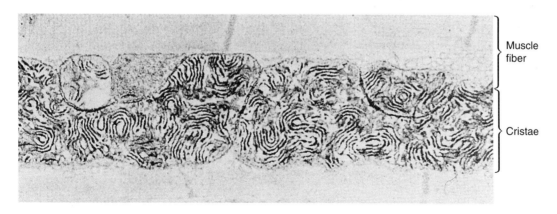

Muscle
fiber

Cristae

Figure 8-15 *Electron Micrograph of a Muscle Cell Stained for Cytochrome Oxidase* The black reaction product, which indicates the presence of cytochrome oxidase, is localized on the mitochondrial cristae, but not in the matrix space or the adjacent muscle fibers. Courtesy of W. A. Anderson.

from the presence of the F_1 particles that protrude from the matrix side of the cristae, thereby imparting a visible asymmetry to the membrane. The inner membrane is also unusual in its high protein-to-lipid ratio, the presence of large amounts of the unusual lipid *cardiolipin,* and its high ratio of unsaturated to saturated phospholipids and the virtual absence of cholesterol, both of which tend to increase membrane fluidity (page 165). The high protein content of the inner membrane reflects the presence of numerous proteins involved in electron transport and oxidative phosphorylation.

Although the role of the inner membrane in electron transport and oxidative phosphorylation will be our primary focus in this chapter, it is important to note that the inner membrane also performs other functions. Since the outer mitochondrial membrane is freely permeable to small molecules, the inner membrane is the only effective permeability barrier between the cytosol and the mitochondrial matrix. In contrast to the freely permeable nature of the outer membrane, the inner membrane is an impermeable barrier to most nucleotides, sugars, and small ions (see Figure 8-14). Small molecules that need to be taken up or expelled by mitochondria are transported across the inner membrane by specific carrier proteins that will be described in more detail later in the chapter.

The inner mitochondrial membrane is also the site of several enzymatic pathways that are not directly related to energy metabolism. A prominent example involves the synthesis of steroid hormones, which are produced as a cooperative effort between mitochondria and the endoplasmic reticulum. The first step in the synthetic pathway is the cleavage of the side chain of cholesterol by an enzyme present in the inner mitochondrial membrane. The product of this reaction is subsequently modified by enzymes in the endoplasmic reticulum, yielding products that are returned to mitochondria and hydroxylated in NADPH-dependent reactions that utilize a membrane-bound cytochrome P-450 (page 261).

HOW MITOCHONDRIA CAPTURE ENERGY

Now that the architecture of the mitochondrion has been described, we are ready to address the question of how this organelle captures energy derived from food. We have already seen that glycolysis, which takes place in the cytosol, converts glucose to two molecules of pyruvate. In the presence of oxygen, pyruvate is then oxidized by mitochondria and the released energy is used to drive the formation of ATP. The first stage in this process occurs in the mitochondrial matrix and begins with the production of a compound called *acetyl CoA.*

Pyruvate and Fatty Acids Are Oxidized to Acetyl CoA in the Mitochondrial Matrix

In cells that function aerobically, pyruvate molecules produced by glycolysis pass through the porin channels of the outer mitochondrial membrane and are transported across the inner mitochondrial membrane into the matrix. Here pyruvate is acted upon by *pyruvate dehydrogenase,* a multiprotein complex consisting of three enzymes, five coenzymes, and two regulatory proteins. These components work together to catalyze the oxidation of pyruvate to acetyl CoA. During this reaction, three events take place: (1) NAD^+ is reduced to NADH, (2) one of the three carbons originally present in pyruvate is released as CO_2, and (3) the remaining two-carbon *acetyl group* is joined to *coenzyme A (CoA),* a derivative of the vitamin pantothenic acid, to generate the final product of the reaction, **acetyl CoA** (Figure 8-16, *Step 1*).

Acetyl CoA is a central molecule in mitochondrial energy metabolism because it is the main input to the Krebs cycle. Before describing what happens when acetyl CoA is metabolized by the Krebs cycle, it should be pointed out that the breakdown of glucose molecules is only one of several possible ways of generating acetyl CoA. When a meal containing carbohydrates is eaten, most of the acetyl CoA produced in mitochondria can be traced to the breakdown of glucose molecules derived from the ingested food. Any excess food that has been consumed is used to create energy-storing molecules that can be broken down later when demands for energy or chemical building blocks require it. Two main groups of energy-storing molecules are produced: Glucose is either stored in the form of polysaccharides such as *glycogen* and *starch* (page 14) or it is converted to fatty acids, which are then stored in fats called *triacylglycerols* (page 15). Fats are the most commonly used mechanism for long-term energy storage because the oxidation of fatty acids yields several times more energy per gram than the oxidation of polysaccharides such as glycogen and starch. If long-term energy reserves were stored in the form of glycogen rather than fat, the body weight of some animals would be increased by as much as 50 percent.

When organisms require energy between meals, glucose-storing polysaccharides are broken down first, followed by the breakdown of fats if necessary. The breakdown of polysaccharides yields glucose molecules that are converted in the cytosol to pyruvate, which is then oxidized to acetyl CoA in the mitochondrial matrix. The hydrolysis of fats yields fatty acids that are also transported into the mitochondrial matrix, where they are broken down to acetyl CoA by a cyclic set of reactions known as the β-**oxidation pathway.** In addition to producing acetyl CoA, this pathway generates one molecule of NADH and one molecule of $FADH_2$ for

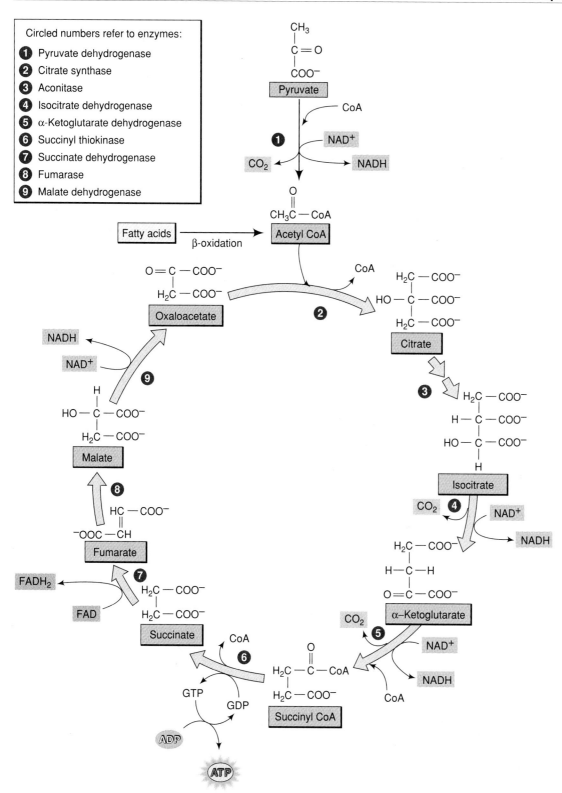

Figure 8-16 *The Krebs Cycle*   *For each molecule of acetyl CoA metabolized, the cycle produces three molecules of NADH, one molecule of FADH₂, one molecule of ATP, and two molecules of CO₂. When pyruvate is used as a source of acetyl CoA for the Krebs cycle, the conversion of pyruvate to acetyl CoA generates an additional molecule of NADH and CO₂. Figure 8-17 shows how fatty acids are broken down to yield acetyl CoA that is metabolized by the Krebs cycle.*

every two fatty acid carbon atoms that are converted to acetyl CoA (Figure 8-17). The fact that the oxidation of both glucose and fatty acids generates acetyl CoA indicates the central importance of this compound in cellular energy metabolism.

The Oxidation of Acetyl CoA in the Krebs Cycle Generates a Small Amount of ATP plus NADH and FADH₂

Acetyl CoA derived from the oxidation of pyruvate and fatty acids serves as the main input to the Krebs cycle, which oxidizes the two-carbon acetyl group of acetyl CoA. Acetyl CoA enters the Krebs cycle by combining with the four-carbon compound *oxaloacetate* to produce the six-carbon compound *citrate*. As is shown in Figure 8-16, citrate is the starting point for a series of reactions that include (1) oxidations accompanied by the reduction of NAD^+ to NADH or FAD to $FADH_2$, (2) cleavages that release CO_2, and (3) one reaction that synthesizes ATP. For each molecule of acetyl CoA that enters the cycle, *a total of three molecules of NADH, one molecule of FADH₂, one molecule of ATP, and two molecules of CO₂ are made.* The release of two carbon atoms in the form of CO_2 results in the conversion of citrate (six carbons) to oxaloacetate (four carbons). This oxaloacetate molecule can then react with another molecule of acetyl CoA and repeat the cycle.

The discovery of this cyclic set of reactions by Sir Hans Krebs in the 1930s is a classic example of elegant biochemical reasoning. By adding various four-, five-, and six-carbon compounds to homogenates of muscle

tissue and analyzing the resulting products, Krebs was able to detect each of the steps involved in the conversion of citrate to oxaloacetate. The crucial discovery that led him to suspect that these reactions are tied together in a cycle involved the use of the enzyme inhibitor *malonate* (see Figure 2-11). Malonate is a competitive inhibitor of the enzyme *succinate dehydrogenase*, which catalyzes the reversible reaction by which succinate is converted to fumarate (Step 7 in Figure 8-16). When Krebs added malonate to muscle tissue, an unexpected result was obtained. The conversion of succinate to fumarate was depressed, but the conversion of fumarate to succinate was not. Since the enzyme succinate dehydrogenase is inhibited when malonate is added, the conversion of fumarate to succinate must occur by some mechanism other than the simple reversal of Step 7. Krebs suggested that this mechanism involves a circular pathway by which fumarate is first converted to oxaloacetate (Steps 8 and 9), which is in turn converted to citrate and then back to succinate (Steps 2 through 6). The idea that oxaloacetate (the end product of the pathway) can be converted to citrate (the starting substrate) suggested to Krebs that the pathway was circular, a conclusion that has subsequently been confirmed by a variety of other experimental approaches. In honor of Krebs this pathway is commonly referred to as the **Krebs cycle,** although the terms *citric acid cycle* and *tricarboxylic acid cycle* are used as well.

Most of the Krebs cycle enzymes are soluble components of the mitochondrial matrix that can be isolated as a single multienzyme complex, suggesting that the substrates and products of the various reactions are passed directly from one enzyme to the next. *Succinate dehydrogenase,* which catalyzes Step 7 of the cycle, is an exception; this particular enzyme is a component of the inner mitochondrial membrane.

NADH and FADH₂ Are Oxidized Back to NAD⁺ and FAD by the Respiratory Chain

Each time a molecule of acetyl CoA is oxidized by the Krebs cycle, its two carbons are completely oxidized to CO_2. What happens to the energy that is released during this process? Although a small portion is directly coupled to the synthesis of ATP in Step 6, most of the energy is trapped in the form of the energy-rich reduced coenzymes, NADH and $FADH_2$. The energy trapped in these reduced coenzymes is subsequently released by oxidizing them back to NAD^+ and FAD, respectively. This oxidation process involves the transfer of electrons from NADH or $FADH_2$ to oxygen, thereby forming water:

$$NADH + \frac{1}{2} O_2 + H^+ \rightarrow NAD^+ + H_2O$$
$$FADH_2 + \frac{1}{2} O_2 \rightarrow FAD + H_2O$$

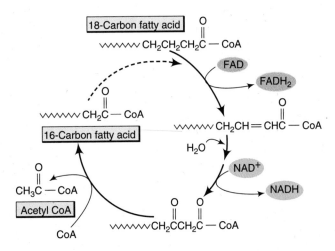

Figure 8-17 Production of Acetyl CoA from Fatty Acids by the β-oxidation Pathway *This example shows how an 18-carbon fatty acid is broken down to a 16-carbon fatty acid, producing one molecule each of acetyl CoA, FADH₂, and NADH. The 16-carbon fatty acid can enter the cycle again and be shortened to a 14-carbon fatty acid, and so forth. Each time a fatty acid enters the cycle, it is shortened by two carbons that are released in the form of acetyl CoA.*

If carried out as written, however, these two reactions would release too much energy at once. Cells therefore utilize a series of electron carriers called the **respiratory chain** to transfer electrons in a stepwise fashion from NADH or $FADH_2$ to oxygen. During this process, free energy is released in a more gradual fashion so that it can be used to drive the formation of ATP. The flow of electrons through the respiratory chain is called **electron transport** (or *electron transfer*), while the accompanying formation of ATP is referred to as **oxidative phosphorylation.** In addition to releasing energy that allows ATP to be formed, the transfer of electrons from NADH and $FADH_2$ to the respiratory chain regenerates the oxidized coenzymes NAD^+ and FAD, which can then be used again in the Krebs cycle.

The Respiratory Chain Is Comprised of Four Different Classes of Molecules

Although electron transport and oxidative phosphorylation are functionally linked to each other in the cell, it is easier to understand how the two processes operate by discussing them separately. We will start with the process of electron transport through the respiratory chain. Most of the electron carriers that make up the respiratory chain absorb light of a particular wavelength, allowing them to be identified by their characteristic absorption spectra (Figure 8-18). Four different kinds of molecules are members of the respiratory chain: *cytochromes, iron-sulfur proteins, flavoproteins,* and *ubiquinones.* Before examining how they are arranged in the respiratory chain, we will briefly describe their basic features.

1. Cytochromes. Cytochromes are a group of proteins that have in common the presence of a **heme** prosthetic group. The heme group is constructed from a *porphyrin* ring and a centrally bound iron atom that is capable of accepting and donating electrons (Figure 8-19). In contrast to coenzymes like NAD^+ and FAD, the oxidation and reduction of the iron atom of the heme group is accomplished by the transfer of a single electron rather than an electron pair. During reduction, the iron atom is converted from its Fe^{3+} state to the Fe^{2+} state by the addition of a single electron; conversely, oxidation involves the conversion of Fe^{2+} to Fe^{3+}.

Cytochromes were first discovered in a series of elegant experiments carried out in the 1920s by David Keilin. In these classic studies, light that had passed through a thin piece of insect muscle or yeast culture was directed through a prism to divide the light into the spectrum of colors of which it is composed. Keilin discovered that the muscle or yeast cells caused two sets of three dark bands to appear in the spectrum, indicating the presence of molecules that absorb light at these

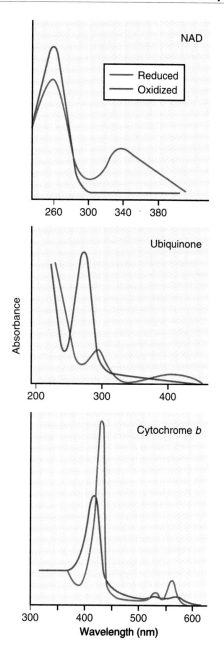

Figure 8-18 *Absorption Spectra of Several Components of the Respiratory Chain* *Each electron carrier involved in the respiratory chain absorbs light maximally at a different wavelength, allowing the carriers to be easily distinguished from each other. The light-absorbing properties of the individual carriers changes during oxidation and reduction, so light absorption can be used to monitor the oxidation-reduction states of the various components of the chain.*

particular wavelengths (Figure 8-20). The molecules responsible for these bands were named *cytochromes a, b,* and *c.* The most significant discovery made by Keilin was that bubbling oxygen through a suspension of yeast causes the absorption bands to disappear, suggesting that the ability of cytochromes to absorb light is abol-

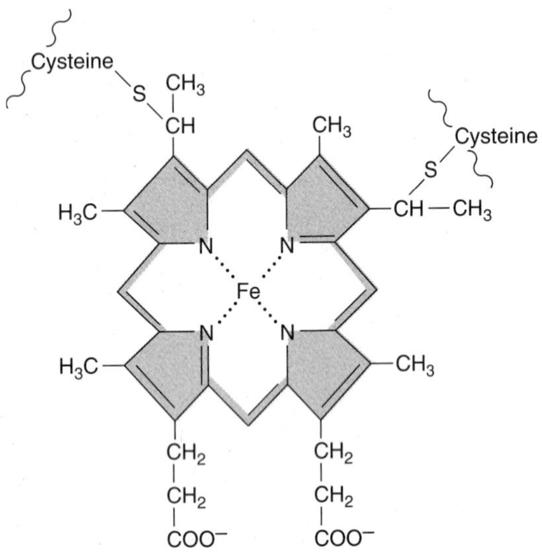

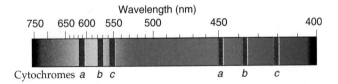

Figure 8-19 *Structure of a Heme Group* *The porphyrin ring (color) contains a series of side chains that vary among different cytochromes. This particular heme group, which occurs in cytochrome c, is covalently attached to the polypeptide chain through two cysteine residues. The heme group is not covalently attached in most other cytochromes.*

ished when cytochromes become oxidized by transferring electrons to oxygen. With remarkable insight, he speculated that cytochromes function in cellular oxidation-reduction pathways in which electrons are transferred from food molecules to oxygen.

Subsequent work by Keilin and others led to the isolation and characterization of the proteins responsible for each of the absorption bands. The individual cytochromes were found to differ from each other in size, absorption spectra, and the type of heme ring they contain. Refined spectroscopic analysis revealed that at least five different cytochromes are present in the mitochondrial respiratory chain: cytochromes b, c, c_1, a, and a_3. All of these proteins, with the exception of cytochrome c, are integral membrane proteins firmly embedded within the inner mitochondrial membrane, making them difficult to extract and purify. Cytochrome c, on the other hand, is a peripheral membrane protein

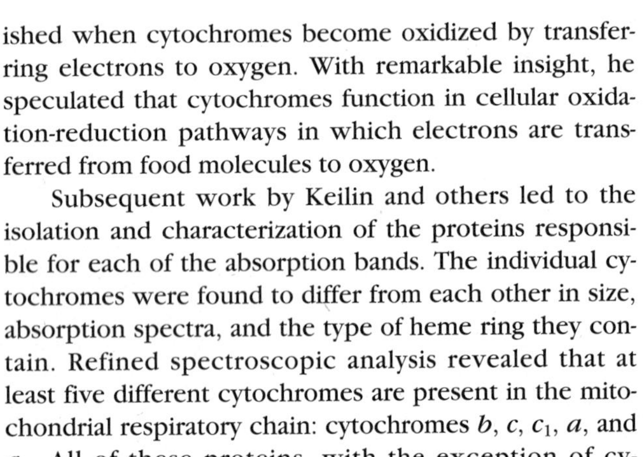

Wavelength (nm)

750 650 600 550   500   450   400

Cytochromes *a* *b* *c*   *a*   *b*   *c*

Figure 8-20 *Absorption Spectrum of Intact Muscle as Seen by Keilin* *Light that had passed through the muscle tissue was directed through a prism to divide the light into a color spectrum. The dark bands represent light absorption by cytochromes a, b, and c. Each cytochrome absorbs light in two different regions of the spectrum.*

that is loosely associated with the surface of the inner membrane that faces the intermembrane space. Because of its weak association with the membrane surface, cytochrome c can dissociate from the membrane and diffuse into the intermembrane space.

Cytochromes a and a_3 differ from the other cytochromes in that they contain bound copper ions in addition to iron. These copper ions transfer single electrons by cycling between the Cu^{2+} and Cu^+ states, just as iron atoms transfer electrons by cycling between the Fe^{3+} and Fe^{2+} states.

2. Iron-sulfur proteins. Cytochrome heme groups do not account for all of the iron atoms in the respiratory chain. The more recently discovered iron-sulfur proteins lack a heme group but instead contain iron bound to the sulfur atom of the amino acid cysteine. Like cytochromes, these **iron-sulfur centers** transfer electrons by cycling between the Fe^{3+} and Fe^{2+} states of the iron atoms.

3. Flavoproteins. Flavoproteins are enzymes that contain FMN or FAD as a tightly bound prosthetic group. Oxidation-reduction reactions involving either FMN or FAD transfer two electrons at a time along with two protons. In some cases a flavoprotein can also be an iron-sulfur protein. For example, the enzyme *NADH dehydrogenase* contains both an FMN group and iron-sulfur centers. The FMN group of this enzyme accepts electrons from NADH and then transfers them to the iron-sulfur centers on the way to the next electron carrier in the respiratory chain.

4. Ubiquinones. In addition to the preceding classes of proteins, the respiratory chain also utilizes a lipid known as **ubiquinone** or **coenzyme Q** as an electron carrier. Ubiquinones exists in several forms that differ in the lengths of their side chains (Figure 8-21). Reduction of ubiquinone involves the addition of two electrons along with two protons, whereas oxidation occurs by the reverse process. Because of its lipid nature, ubiquinone is readily incorporated into the hydrophobic lipid bilayer of the mitochondrial inner membrane.

Redox Potentials Indicate the Energetically Most Favorable Route for Electrons Passing through the Respiratory Chain

Now that we are acquainted with the kinds of electron carriers that make up the respiratory chain, the question arises as to the sequence in which these carriers are arranged. One way of trying to determine this sequence is to compare the electron-donating power of the different carriers using a simple electrical cell. To illustrate how such a cell operates, let us consider an ion like iron that exists in both an oxidized state (Fe^{3+}) and a reduced state (Fe^{2+}). Such a pair, consisting of the oxidized and

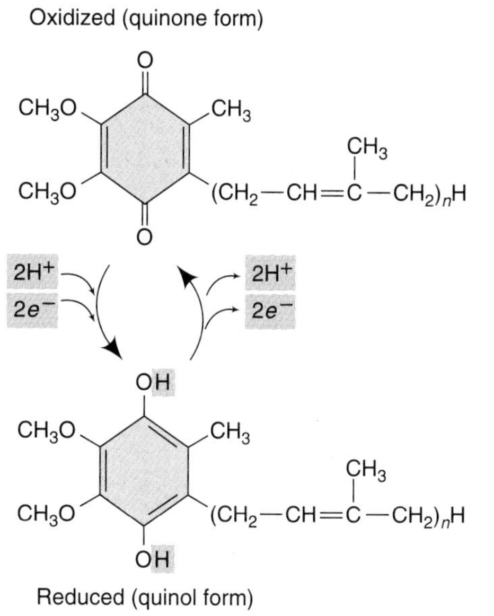

Oxidized (quinone form)

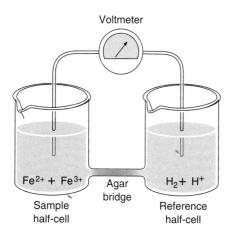

Reduced (quinol form)

Figure 8-21 *Structure of the Oxidized and Reduced Forms of Ubiquinone* *Although the length of the side chain is variable, in most mammalian mitochondria* n = 10.

reduced forms of the same substance, is referred to as a **redox couple.** In the electrical cell illustrated in Figure 8-22, a mixture of Fe^{2+} and Fe^{3+} is placed in the sample half-cell, and a mixture of H^+ and H_2 is placed in the reference half-cell. Electrons are then permitted to flow from one half-cell to the other using a connecting bridge made of agar. If H_2 donates electrons *more readily* than the electron donor being tested, electrons will flow from the reference half-cell to the sample half-cell. Conversely, if H_2 donates electrons *less readily* than the electron donor being tested, electrons will flow in the

Voltmeter

Fe^{2+} + Fe^{3+} Agar   H$_2$ + H$^+$
bridge

Sample   Reference
half-cell   half-cell

Figure 8-22 *An Electrical Cell Used to Measure Standard Redox Potentials* *The agar bridge permits electrons to flow from one half-cell to the other. Electrons flow from the cell containing the stronger electron donor to the cell containing the weaker electron donor. The voltmeter permits the magnitude and direction of the electron flow to be measured.*

opposite direction. In either case, the magnitude and direction of the electron flow can be measured using a voltmeter connected to the two half-cells.

When the experiment is performed under standard conditions (1 *M* concentration of reactants, 25°C, pH 7.0, and 1 atmosphere pressure), the resulting voltage is referred to as the **standard redox potential (E'_0).** A negative value of E'_0 means that electrons flow from the sample half-cell to the reference half-cell, while a positive value of E'_0 indicates that electrons flow in the reverse direction. Such information permits the relative strengths of various redox couples to be easily compared. *The more negative the value of E'_0 for a redox couple, the greater its tendency to donate electrons.* Thus redox couples whose redox potentials are very negative will tend to donate electrons to redox couples whose redox potentials are less negative (or more positive).

The standard redox potentials for most of the electron carriers involved in the respiratory chain are summarized in Table 8-2. This table lists the redox potentials in sequential order beginning with those that have the greatest tendency to donate electrons and ending with those that have the greatest tendency to accept electrons. Such information is useful because it represents the energetically most favorable route that can be taken by electrons passing from one carrier to the next under standard conditions. In fact, the exact amount of free energy that would be released as electrons pass from carrier to carrier can be calculated from the redox potentials using the following equation:

$$\Delta G^{\circ\prime} = -nF\,\Delta E'_0 \tag{8-1}$$

where n is the number of electrons transferred, F is the faraday (23,062 kcal/V/mol), $\Delta E'_0$ is the difference in redox potential between the two redox couples in volts, and $\Delta G^{\circ\prime}$ is the change in free energy that occurs when electrons are passed between the two redox couples. To illustrate the use of this equation, let us consider the oxidation of pyruvate ($E'_0 = -0.19$) by oxygen ($E'_0 = +0.82$). Inserting these values into Equation 8-1 yields

$$\Delta G^{\circ\prime} = -2 \times 23{,}062 \times [0.82 - (-0.19)]$$
$$= -2 \times 23{,}062 \times 1.01$$
$$= -46{,}600 \text{ cal/mol} = -46.6 \text{ kcal/mol}$$

Hence for any two redox couples, Equation 8-1 can be employed to calculate the change in free energy that occurs when electrons are passed from one couple to the other.

Although a consideration of standard redox potentials provides information about the energetically most favorable sequence of electron carriers under standard

Table 8-2   Standard Redox Potentials of Selected Electron Carriers

Oxidized Form	Reduced Form	n	E'_0 (volts)
NAD^+	$NADH + H^+$	2	−0.32
FMN	$FMNH_2$	2	−0.30
FAD	$FADH_2$	2	−0.22
Pyruvate	Lactate	2	−0.19
Ubiquinone	$Ubiquinone-H_2$	2	+0.04
Cytochrome b (Fe^{3+})	Cytochrome b (Fe^{2+})	1	+0.07
Cytochrome c_1 (Fe^{3+})	Cytochrome c_1 (Fe^{2+})	1	+0.23
Cytochrome c (Fe^{3+})	Cytochrome c (Fe^{2+})	1	+0.25
Cytochrome a (Fe^{3+})	Cytochrome a (Fe^{2+})	1	+0.29
Cytochrome a_3 (Fe^{3+})	Cytochrome a_3 (Fe^{2+})	1	+0.55
$\frac{1}{2}O_2 + 2H^+$	H_2O	2	+0.82

Note: n = number of electrons transferred.

conditions, this does not necessarily reflect the sequence of carriers within intact cells, where the concentration of reactants is quite different than the value of 1 *M* specified by standard conditions. Other experimental approaches are therefore needed to determine the actual sequence of electron carriers in the mitochondrial inner membrane.

Difference Spectra Reveal the Sequence of the Respiratory Chain within Intact Mitochondria

The most direct way of studying the behavior of electron carriers while they still reside in mitochondria is to monitor their light-absorbing properties, which change as each carrier becomes oxidized or reduced (see Figure 8-18). Such measurements are usually made by taking separate samples of mitochondria in the reduced and oxidized states and measuring the difference in the light-absorbing properties of one versus the other. The result, called a **difference spectrum**, exhibits a series of peaks and valleys that correspond to the wavelengths at which the absorbance of the respiratory carriers differs between the oxidized and reduced state (Figure 8-23). The amplitude of these peaks and valleys can therefore be used to calculate the relative degree of oxidation of each of the individual carriers. In order to increase the sensitivity and reliability of this method, Britton Chance developed a technique for measuring the absorption of each carrier at two wavelengths: a wavelength at which its absorption change is maximal and a nearby reference wavelength where its absorption does not change upon oxidation-reduction. This dual-wavelength approach permits nonspecific changes in the light-absorbing properties of mitochondria to be canceled out by subtracting the absorbance reading obtained at the refer-

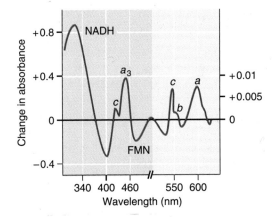

Figure 8-23 *Difference Spectrum of Isolated Rat Liver Mitochondria* *The graph represents the difference in absorbance between mitochondria in a reduced state (excess substrate, no oxygen) and those in an oxidized state (excess oxygen). Most of the respiratory carriers appear as peaks because they absorb more strongly when reduced. One exception is the FMN group of NADH dehydrogenase, which absorbs less in its reduced state and therefore appears as a valley. The prominence of each peak or valley can be used to determine the oxidation-reduction state of the individual carriers. The small letters (a, a_3, b, and c) refer to the various cytochromes.*

ence wavelength from the reading obtained at the point of maximum absorption.

Difference spectra have been used in several ways to help elucidate the sequence of carriers in the respiratory chain. In one approach, oxygen is added to oxygen-depleted mitochondria and the resulting changes in the difference spectra are recorded with time. Under such conditions the first component that becomes oxidized is cytochrome a_3, followed sequentially by cytochromes a, c, and b, FMN, and NADH. These results indicate that cytochrome a_3 is located closest to oxygen in the respiratory chain and that each of the remaining components is located progressively farther away from oxygen in the sequence.

In an alternative approach, difference spectra have been employed in conjunction with inhibitors that block the flow of electrons at specific points in the respiratory chain. All carriers located prior to the site where electron flow is blocked will remain in the reduced state because their electrons cannot be passed on to oxygen; in contrast, carriers located beyond the block remain oxidized after they have donated their electrons because the blockage earlier in the chain prevents them from receiving any more electrons. The sequence in which the carriers are arranged can therefore be deduced by using difference spectra to determine which carriers remain reduced and which are oxidized in the presence of various inhibitors. For example, difference spectra obtained from mitochondria exposed to the inhibitor *antimycin* reveal that one group of carriers (NADH, FMN, ubiquinone, and cytochrome b) are all fully reduced, whereas a second group of carriers (cytochromes c, a, and a_3) become fully oxidized. This pattern suggests that antimycin blocks the flow of electrons between the first group of carriers and the second group; in other words, the first group precedes the second group in the respiratory chain. Similar experiments involving the use of inhibitors that block electron transfer at other points has helped to elucidate the overall sequence of the respiratory chain.

Data obtained from the preceding experimental approaches have led to the conclusion that the respiratory chain is arranged in the sequence illustrated in Figure 8-24. This sequence is compatible with information obtained from redox potentials; that is, the sequence of electron carriers is dictated by the relative ease with which each component accepts and donates electrons. Some of the carriers in this scheme transfer two electrons at a time (NAD$^+$, FAD, FMN, and ubiquinone), whereas others transfer electrons one at a time (the iron-sulfur centers and cytochromes). In the last step of the sequence, where electrons are passed to oxygen to form water, four electrons are required:

$$O_2 + 4e^- + 4H^+ \rightarrow 2H_2O$$

It is crucial that four electrons be transferred during this final step because the reduction of an oxygen molecule by less than four electrons generates substances such as hydrogen peroxide and the superoxide anion (O_2^-), which are toxic to cells.

The Respiratory Chain Is Organized into Four Multiprotein Complexes

Thus far we have been viewing the respiratory chain as a series of individual electron carriers (i.e., NADH, FMN,

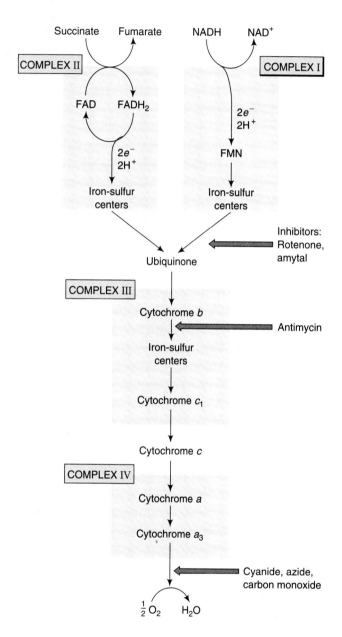

Figure 8-24 Sequence of Electron Carriers in the Mitochondrial Respiratory Chain *Inhibitors that block electron transfer at certain points in the chain have been useful in determining the sequence of electron carriers. The individual electron carriers are organized into four multiprotein complexes. Electrons derived from NADH are transferred first to complex I, then to complex III, and finally to complex IV before being transferred to oxygen to form water. Electrons derived from succinate in the Krebs cycle are first passed to FAD in complex II, and then to complexes III and IV before being transferred to oxygen to form water. Electrons are transferred between the complexes by the diffusible carriers, ubiquinone and cytochrome c.*

ubiquinone, cytochromes b, c, c_1, a, and a_3). However, the behavior of these carriers in mitochondrial extraction experiments suggests that they are organized into multiprotein complexes rather than as individual components. For example, when mitochondrial membranes are extracted under gentle conditions, only ubiquinone and cytochrome c are readily removed. The other electron carriers remain bound to the inner membrane and are not released until the membrane is exposed to detergents or concentrated salt solutions. Such treatments release the remaining electron carriers from the membrane not as individual molecules, but as four multiprotein **respiratory complexes:**

1. Respiratory complex I, also called the *NADH dehydrogenase complex,* catalyzes the transfer of electrons from NADH to ubiquinone.

2. Respiratory complex II, also called the *succinate dehydrogenase complex,* catalyzes the transfer of electrons from succinate in the Krebs cycle to ubiquinone in the respiratory chain.

3. Respiratory complex III, also called the *cytochrome $b-c_1$ complex,* catalyzes the transfer of electrons from ubiquinone to cytochrome c.

4. Respiratory complex IV, also called the *cytochrome oxidase complex,* catalyzes the transfer of electrons from cytochrome c to oxygen.

The respiratory chain can thus be viewed as a group of protein complexes that are functionally linked by the diffusible carriers ubiquinone and cytochrome c. Ubiquinone transfers electrons from both complexes I and II to complex III, whereas cytochrome c transfers

electrons from complex III to complex IV (see Figure 8-24). Because the inner mitochondrial membrane is virtually free of cholesterol and has a high ratio of unsaturated to saturated phospholipids, its fluidity is very high. This raises the possibility that the four respiratory complexes diffuse freely in the inner mitochondrial membrane rather than being physically joined together.

This idea has been tested by Charles Hackenbrock, who fused isolated mitochondrial membranes with artificial phospholipid vesicles to produce membranes with varying phospholipid contents. Freeze-fracture micrographs of these membranes revealed that the average distance between intramembrane particles (respiratory complexes) increases as membrane phospholipid is added, indicating that the particles are free to diffuse within the membrane and are not joined together as a chain of respiratory complexes (Figure 8-25). Metabolic studies of these phospholipid-enriched membranes revealed that the rate of electron transfer is depressed as phospholipid is added, presumably because the average distance separating the complexes has been increased. These results indicate that the respiratory chain is not a physically linked set of electron carriers, but rather a group of independent protein complexes that diffuse freely within the membrane. Such an arrangement suggests that factors that influence the diffusion of the respiratory complexes may play a role in regulating the rate of electron transfer through the chain.

The Respiratory Chain Contains Three Coupling Sites for Oxidative Phosphorylation

Now that we have described the transfer of electrons through the respiratory chain, we can address the ques-

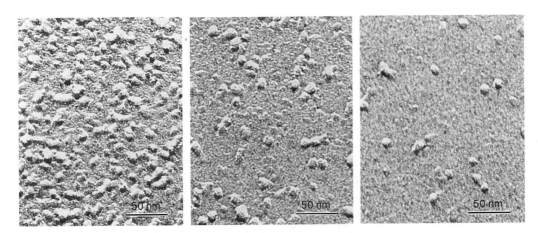

Figure 8-25 Freeze-Fracture Electron Micrographs of Mitochondrial Inner Membranes with Differing Phospholipid Content (Left) *Control membranes with no added phospholipid.* (Middle and right) *As increasing amounts of phospholipid are added, the distance between the intramembrane particles increases. Such observations indicate that the membrane particles are free to diffuse within the membrane rather than being joined together. Courtesy of C. R. Hackenbrock.*

tion of how electron transfer is coupled to ATP synthesis in the inner mitochondrial membrane. This type of ATP formation is called **oxidative phosphorylation** because it accompanies the oxidation of reduced coenzymes by the respiratory chain. The idea that ATP synthesis is coupled to electron transfer was first proposed in the late 1930s by Vladimir Belitzer, who measured the rate of ATP synthesis and oxygen consumption in minced muscle preparations. His results indicated that at least two ATPs are formed per pair of electrons passed through the respiratory chain to oxygen. Subsequent measurements by other investigators led to the conclusion that this *P/O ratio* (number of ATPs formed per oxygen atom reduced) is actually closer to 3. Although some disagreement persists as to the precise value of the P/O ratio, we will use a value of 3 because it simplifies the task of describing the mechanism of oxidative phosphorylation.

The most straightforward interpretation of a P/O ratio of 3 is that three steps in the respiratory chain release enough free energy to allow the formation of ATP. As we saw earlier, knowing the redox potentials of the various members of the respiratory chain allows us to calculate how much energy is released as electrons are passed from one carrier to another. Such calculations reveal that the reactions carried out by complexes I, III, and IV (but not complex II) each yield enough energy for the synthesis of one molecule of ATP per pair of electrons transferred. It has therefore been proposed that each of these three complexes acts as a **coupling site** that couples electron transfer to the synthesis of ATP.

This conclusion has been experimentally confirmed using *site-specific* assays that utilize inhibitors of electron transport in combination with added electron donors and acceptors. For example, when inner membrane vesicles are incubated in the presence of the inhibition *antimycin* to prevent the flow of electrons past cytochrome *b*, NADH can be added to serve as an electron donor and ubiquinone as an electron acceptor. Under such conditions roughly one molecule of ATP is synthesized per pair of electrons transferred from NADH to ubiquinone. This finding confirms the notion that a coupling site for ATP synthesis occurs between NADH and ubiquinone, which corresponds to respiratory complex I. If an artificial electron donor known to reduce cytochrome *c* is employed in place of NADH, roughly one ATP molecule is produced per pair of electrons transferred from cytochrome *c* to oxygen. Hence a coupling site for ATP synthesis must also exist between cytochrome *c* and oxygen, which corresponds to respiratory complex IV. This type of experimental approach has been effective in demonstrating that complexes I, III, and IV each act as a site where the synthesis of a molecule of ATP is coupled to the transfer of a pair of electrons through the respiratory chain.

Complex II, on the other hand, does not couple ATP synthesis to electron transfer, confirming that this complex does not release enough energy to permit ATP formation.

The preceding findings indicate that a pair of electrons donated to the respiratory chain by NADH will encounter *three* coupling sites for ATP synthesis because these electrons pass through complexes I, III, and IV. In contrast, electrons donated by $FADH_2$ encounter only *two* coupling sites for ATP synthesis because the electrons donated by $FADH_2$ pass through complexes II, III, and IV (recall that complex II does not act as a coupling site). Since each coupling site releases enough energy to make about one molecule of ATP per pair of electrons transferred, this means that *the oxidation of NADH by the respiratory chain will generate three molecules of ATP, whereas the oxidation of $FADH_2$ produces only two ATPs.*

The F_1–F_0 Complex Is the Site of ATP Synthesis During Oxidative Phosphorylation

Thus far we have described the sequence of carriers in the respiratory chain and have identified the three coupling sites for ATP formation. But how is the synthesis of ATP actually carried out? Identification of the enzyme responsible for ATP formation was first made possible by membrane reconstitution experiments in which the inner mitochondrial membrane was taken apart and reassembled from isolated components. Such experiments begin with isolated inner membrane vesicles that contain F_1 particles bound to their outer surface. In the presence of NADH, ADP, and inorganic phosphate, these vesicles are capable of transferring electrons from NADH through the respiratory chain to oxygen, and coupling this electron transfer to ATP synthesis.

The first successful disassembly and reconstitution experiments were carried out in the laboratory of Efraim Racker in the 1960s. Inner membrane vesicles containing attached F_1 particles were first disrupted by treatment with urea or shaking with glass beads; the resulting material was then separated into pellet and supernatant fractions by centrifugation (Figure 8-26). Microscopic examination of the pellet revealed the presence of membrane vesicles lacking their normal F_1 particles. These smooth vesicles possessed an intact respiratory chain that could carry out electron transport, but no ATP was synthesized. The supernatant, on the other hand, consisted of isolated F_1 particles floating free in solution. By themselves, these particles carried out neither electron transport nor ATP synthesis. However, they were able to catalyze the hydrolysis of ATP, which is the reverse of ATP synthesis. Since enzymes catalyze chemical reactions in both directions, it was proposed that the F_1 particles normally catalyze ATP synthesis when they are attached to the inner mito-

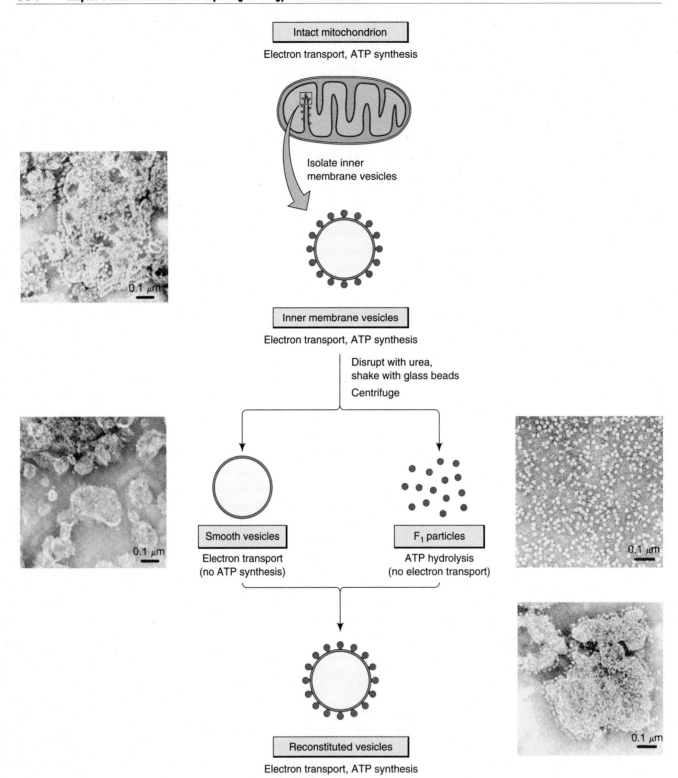

Figure 8-26 *Reconstitution of the Inner Mitochondrial Membrane* *When F_1 particles are removed from inner membrane vesicles, the remaining smooth vesicles carry out electron transport but do not make ATP. Reattaching the F_1 particles restores the ability to synthesize ATP in conjunction with electron transport, suggesting that the F_1 particles are the site of ATP formation. Micrographs courtesy of E. Racker.*

chondrial membrane, but can no longer do so when removed from the membrane because the respiratory chain must be present to provide the energy needed for ATP formation.

This hypothesis was confirmed by adding the supernatant containing the isolated F_1 particles back to the pellet containing the smooth membrane vesicles. Under these conditions, the F_1 particles reattached to the membranes; at the same time, the membranes regained the ability to synthesize ATP. It was therefore concluded that the F_1 particles are the site where ATP synthesis occurs when electrons flow through the respiratory chain.

The reconstitution approach has also been extended to the analysis of individual coupling sites for ATP synthesis. In one such study, Racker's group reconstituted membrane vesicles from a mixture of purified phospholipids, respiratory complex I, F_1 particles, and a group of hydrophobic proteins called F_0 that span the inner mitochondrial membrane and normally function to anchor F_1 to the surface of the membrane. Membrane vesicles generated in this way were found to be capable of coupling ATP synthesis to the flow of electrons from NADH to ubiquinone; a coupling site for ATP formation had thus been reconstituted in an artificially constructed membrane. Comparable results have since been obtained with respiratory complexes III and IV. Because ATP synthesis in such reconstituted membranes requires the presence of both F_1 and F_0, the **F_1-F_0 complex** is commonly referred to as an **ATP synthase**.

An Electrochemical Proton Gradient across the Inner Mitochondrial Membrane Drives ATP Formation

Now that we have identified the components involved in mitochondrial electron transfer as well as oxidative phosphorylation, one crucial question remains: How is the energy which is released during the transfer of electrons through the respiratory chain used to drive the synthesis of ATP during oxidative phosphorylation? Early work on this problem produced so many diverse ideas that an expert in the field commented in the 1960s that *anyone who is not confused about oxidative phosphorylation just does not understand the situation!* At that time, oxidative phosphorylation was widely believed to proceed by the formation of high-energy chemical intermediates. According to this viewpoint, the free energy released during an oxidation-reduction reaction is conserved by linking one of the substrates of the reaction to some unidentified substance to form a high-energy intermediate:

$$A_{red} + B_{ox} + X \rightarrow X \sim A_{ox} + B_{red}$$

where A and B are components of the respiratory chain, and X is an unidentified substance employed to create the high-energy intermediate ($X \sim A_{ox}$). In a subsequent reaction, the high-energy intermediate would be broken down in a reaction that makes ATP:

$$X \sim A_{ox} + ADP + P_i \rightarrow X + A_{ox} + ATP$$

This theory, which was first clearly formulated in 1953 by E. C. Slater, led to a vigorous search for the hypothetical high-energy intermediates. Many claims were made to the effect that the elusive intermediate had been discovered, but all proved to be unfounded.

While the biochemists of the time were feverishly occupied in the hunt for high-energy intermediates, the alarming suggestion was made by the British biochemist Peter Mitchell that they, like the emperor's new clothes, might not exist at all. In 1961 Mitchell proposed an alternative model, which he referred to as **chemiosmotic coupling.** According to this theory, the energy released during the passage of electrons through the respiratory chain is used to pump protons (H^+) from the matrix side of the inner membrane into the intermembrane space (Figure 8-27). The result would be an **electrochemical proton gradient** across the inner mitochondrial membrane in which the proton concentration is higher in the intermembrane space than in the matrix. The electrochemical gradient consists of two components: a difference in electric charge or membrane potential ($\Delta\psi$), and a difference in the concentration of protons (ΔpH). As we learned in Chapter 5, electrochemical gradients store energy that is released when the solute in question moves down its gradient. Mitchell therefore postulated that the synthesis of ATP is driven by the energy that is released when protons move down their electrochemical gradient by passing back across the intermembrane space and into the mitochondrial matrix.

This theory originally met with considerable skepticism, especially since it was proposed in the virtual absence of experimental data. Over the years, however, a large body of evidence has been amassed in its support and Mitchell was eventually awarded a Nobel Prize for his pioneering contributions. Today the chemiosmotic coupling model is a well-verified idea that provides a unifying framework for understanding energy transformations not just in mitochondrial membranes, but in chloroplast and bacterial membranes as well. Because

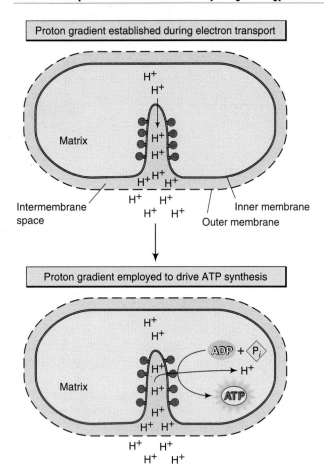

Figure 8-27 *Proton Flow across the Inner Mitochondrial Membrane as Postulated by the Chemiosmotic Theory* (Top) *During the transfer of electrons through the respiratory chain, protons are pumped from the mitochondrial matrix into the intermembrane space, creating a gradient in which the proton concentration is higher in the intermembrane space than in the matrix. The outer mitochondrial membrane is portrayed with a dashed line because it is readily permeable to protons, allowing them to pass out of the mitochondrion.* (Bottom) *The subsequent diffusion of protons down their electrochemical gradient causes them to move through the ATP synthase complex, driving the formation of ATP.*

of the profound impact this theory has had on our thinking, it is important to understand some of the evidence upon which this theory now rests. In the following sections, we will briefly discuss eight of the most important lines of experimental evidence and see what these data tell us about the detailed operation of chemiosmotic coupling.

1) Electron Transfer through the Respiratory Chain Causes Protons to Be Pumped Out of the Mitochondrial Matrix

Shortly after the chemiosmotic model was first proposed, Mitchell and his colleague Jennifer Moyle experi-

mentally verified that the flow of electrons through the respiratory chain is accompanied by the movement of protons across the inner mitochondrial membrane. In these studies, mitochondria were first suspended in a medium in which electron transfer could not occur because oxygen was lacking. The proton concentration (pH) of the medium was then monitored as electron transfer was stimulated by the addition of known quantities of oxygen. Under such conditions the pH of the medium declined rapidly. Because a decline in pH reflects an increase in proton concentration, it was concluded that electron transfer causes the mitochondrial inner membrane to pump protons from the mitochondrial matrix into the external medium (the presence of the outer mitochondrial membrane can be ignored because it is freely permeable to small molecules and ions).

The addition of NADH to the respiratory chain was found to stimulate a greater pH change than the addition of succinate, which generates $FADH_2$ during its oxidation (Figure 8-28). This outcome is exactly what would be predicted if the pH change were involved in ATP synthesis, since NADH yields more ATP when oxidized by the respiratory chain than does $FADH_2$. The magnitude of the pH difference initially suggested to Mitchell that two protons are pumped across the inner membrane at each coupling site for ATP synthesis, but later experiments have suggested that more than two protons per coupling site may be involved.

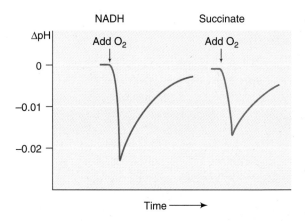

Figure 8-28 *Experiment in Which the pH of the External Medium Was Monitored after Exposing Mitochondria to a Brief Pulse of Oxygen* *Under such conditions, oxidation of NADH or succinate by the respiratory chain is accompanied by a decrease in the pH of the external medium, indicating that protons are being pumped out of the mitochondrion. After the oxygen is consumed, electron transfer through the respiratory chain stops and the pH gradient dissipates. The pH change that occurs during electron transport is greater with NADH than with succinate, which correlates with the fact that the oxidation of NADH yields more ATP than the oxidation of succinate.*

2) Thermodynamic Calculations Indicate That the Proton Gradient Stores Enough Energy to Drive ATP Formation

Thermodynamic calculations indicate that the magnitude of the proton gradient is sufficient to account for the amount of ATP synthesized. In metabolically active mitochondria, the proton gradient across the inner membrane consists of a pH difference of about 1.4 pH units (higher pH on the matrix side) and an electrical potential or $\Delta\psi$ of approximately 0.14 volts (more negative on the matrix side). This electrochemical gradient exerts a force, called the **proton motive force (pmf),** which tends to drive protons back down their concentration gradient. The value of the proton motive force can be calculated by summing the contributions of the membrane potential and the pH gradient using the following equation:

$$\text{pmf} = \Delta\psi + 2.303\, RT\, \Delta\text{pH}\, / F$$
$$\underset{\substack{\text{(membrane} \\ \text{potential)}}}{\qquad} \underset{\text{(pH gradient)}}{\qquad} \tag{8-2}$$

where pmf is the proton motive force in volts, $\Delta\psi$ is the membrane potential in volts, R is the gas constant (1.987 cal/mol-K), T is the absolute temperature, ΔpH is the pH difference across the membrane (pH of the matrix *minus* the pH of the intermembrane space), and F is the faraday (23,062 kcal/V/mol).

For a mitochondrion at 37°C exhibiting an electrical potential of 0.14 volts and a pH gradient of 1.4 units across the inner membrane, Equation 8-2 can be used to calculate the proton motive force as follows:

$$
\begin{aligned}
\text{pmf} &= \Delta\psi + 2.303 RT\, \Delta\text{pH}/F \\
&= 0.14 + 2.303(1.987)(273+37)(1.4)/23{,}062 \\
&= 0.14 + 0.086 \\
&= 0.226 \text{ volts}
\end{aligned}
$$

The proton motive force is an electrical force expressed in volts, just as the redox potential (E'_0) is an electrical force expressed in volts. Therefore the change in free energy that is represented by a proton motive force of 0.226 volts can be calculated by substituting this value as $\Delta E'_0$ into Equation 8-1 (page 329):

$$
\begin{aligned}
\Delta G^{\circ\prime} &= -nF\Delta E'_0 \\
&= -1 \times 23{,}062 \times 0.226 \\
&= -5{,}212 \text{ cal/mol} = -5.2 \text{ kcal/mol}
\end{aligned}
$$

Hence a proton motive force of 0.226 volts across the inner mitochondrial membrane corresponds to a free energy change of about 5.2 kcal/mol of protons. If two protons are pumped across the inner membrane per molecule of ATP synthesized, then $2 \times 5.2 = 10.4$ kcal of energy are available. Current estimates suggest that the

$\Delta G'$ for ATP synthesis under conditions prevailing within the mitochondrion is around 11 kcal/mol, which is quite close to the 10.4 kcal provided by the proton gradient.

Although the preceding calculations indicate that the free-energy content of the proton gradient is within the range required for ATP synthesis, the absolute values of the numbers involved should not be taken too literally. First of all, there is a difference of opinion as to how many protons are actually transported per coupling site. Mitchell's proposal of two protons per site is a minimum value; if the real value is higher, as is suggested by current experimental measurements, then the free-energy content of the proton gradient would be correspondingly higher. Differences of opinion also exist regarding the number of ATP molecules synthesized per coupling site. Before the advent of the chemiosmotic model, it was believed that each coupling site synthesizes exactly one molecule of ATP per pair of electrons transferred. Hence the oxidation of NADH, whose electrons pass through three coupling sites, was thought to be accompanied by the formation of exactly three molecules of ATP. With the chemiosmotic coupling model, however, there is no need to be constrained to integer numbers such as 3.0 because electron transfer establishes a proton gradient rather than making ATP directly. Current measurements suggest that the proton gradient established during NADH oxidation may drive the synthesis of slightly less than three molecules of ATP per molecule of NADH passing a pair of electrons through the respiratory chain to oxygen.

3) Artificially Created pH Gradients Can Drive ATP Synthesis in the Absence of Electron Transport

The fact that a proton gradient is established during electron transport by the respiratory chain and that it contains sufficient free energy to drive ATP formation does not prove that it actually does so. Direct evidence that proton gradients in fact provide the driving force for ATP formation has been obtained by exposing mitochondria or inner membrane vesicles to artificial pH gradients. When mitochondria are suspended in a solution in which the external proton concentration is suddenly increased by the addition of acid, ATP is generated in response to the artificially created proton gradient. Because such artificial pH gradients induce ATP formation even in the absence of oxidizable substrates capable of passing electrons to the respiratory chain, it is evident that ATP synthesis can be induced by proton gradients even in the absence of electron transport.

4) Reconstituted Membrane Vesicles Containing Complexes I, III, or IV Establish Proton Gradients

Experiments involving the reconstitution of membrane vesicles from mixtures of isolated components have

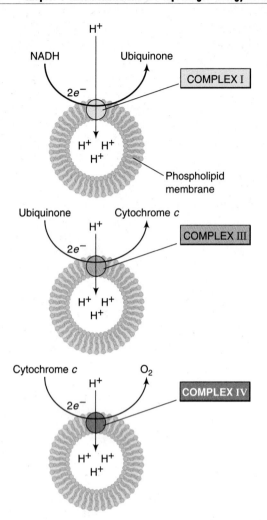

Figure 8-29 *Behavior of Artificial Phospholipid Membranes Reconstituted with Respiratory Chain Complexes* *In membranes vesicles reconstituted with complex I, III, or IV, protons are pumped into the vesicle as electron transport occurs. The inside of such vesicles corresponds to the intermembrane space, while the outside corresponds to the mitochondrial matrix.*

provided considerable support for the chemiosmotic theory. Since respiratory complexes I, III, and IV each has a coupling site for ATP synthesis, the chemiosmotic model predicts that each of these complexes should be capable of pumping protons across the inner mitochondrial membrane. This prediction has been confirmed experimentally by reconstituting artificial phospholipid vesicles containing either complex I, III, or IV. When provided with appropriate oxidizable substrates, each of these three types of vesicles pumps protons across the membrane (Figure 8-29).

Such results support the idea that the ability of each respiratory complex to drive ATP formation stems from its ability to pump protons across the inner mitochondrial membrane. Figure 8-30 presents a model showing how the three complexes are thought to work together to establish a proton gradient, which in turn drives ATP

synthesis by the F_1–F_0 complex. Some disagreement exists as to the exact number of protons pumped across the membrane by each respiratory complex, as well as the mechanism by which proton movement is accomplished. Mitchell originally proposed that the protons being pumped were the protons that are picked up and released by components of the respiratory chain as they undergo oxidation and reduction. For example, ubiquinone picks up two protons along with two electrons when it is reduced, and releases these protons when it is oxidized. Since ubiquinone can move freely in the lipid bilayer, it might pick up protons from the matrix side of the inner membrane and release them on the opposite side of the membrane. This type of mechanism cannot, however, explain proton pumping beyond cytochrome *b*, since none of the known electron carriers after this point releases or receives protons as part of its oxidation-reduction cycle.

5) Uncoupling Agents Abolish both the Proton Gradient and Oxidative Phosphorylation

Additional evidence for the role played by proton gradients in ATP formation has been obtained using agents that abolish such gradients by making membranes leaky to protons. The first such substance to be studied in mitochondrial systems was *dinitrophenol*, a compound known for many years to inhibit oxidative phosphorylation but not electron transfer. Agents of this type, which allow electrons to flow through the respiratory chain but prevent the released energy from driving ATP formation, are known as **uncoupling agents.** In 1963 Mitchell demonstrated that dinitrophenol makes cellular membranes freely permeable to protons by functioning as a proton ionophore (page 182). Hence in the presence of dinitrophenol, membranes are incapable of maintaining a proton gradient. The fact that exposure to dinitrophenol abolishes the ability of mitochondria to make ATP by oxidative phosphorylation suggests that a proton gradient is required for this type of ATP synthesis.

Other ionophores that increase membrane permeability to protons have also been found to uncouple oxidative phosphorylation. Two of the more commonly employed agents are *carbonyl cyanide m-chlorophenylhydrazone (CCCP)* and the peptide antibiotic, *gramicidin* (page 182). These substances completely uncouple oxidative phosphorylation from electron transfer because they abolish both the pH gradient and the membrane potential across the inner mitochondrial membrane. In contrast, ionophores that facilitate the diffusion of ions other than protons influence only one of these two constituents of the electrochemical gradient. *Valinomycin,* for example, dissipates the membrane potential by making membranes freely permeable to potassium ions. Hence the electrical potential created by the presence of a higher proton concentration

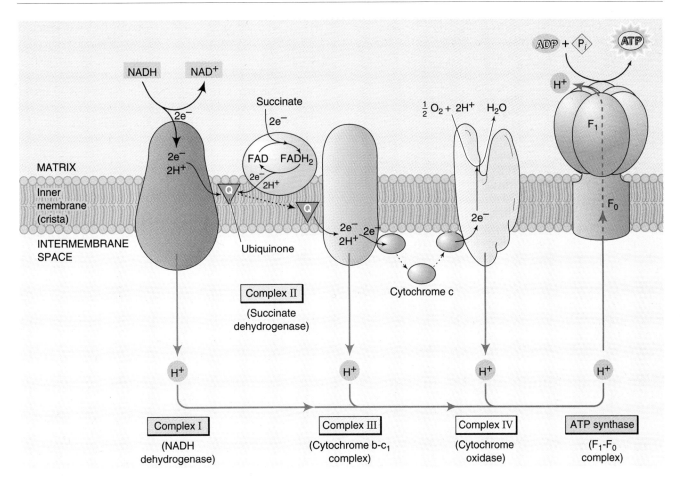

Figure 8-30 *Model of the Respiratory Chain in the Inner Mitochondrial Membrane* *Each respiratory complex pumps protons across the inner membrane, establishing an electrochemical proton gradient. The movement of protons back down this gradient drives the formation of ATP by ATP synthase. The exact number of protons pumped across the membrane by each complex is uncertain.*

outside the mitochondrion is abolished by the valino-mycin-mediated flow of K^+ into the mitochondrion. The ionophore *nigericin,* on the other hand, facilitates a neutral exchange of protons for potassium ions. Because this exchange is electrically neutral, the membrane potential is unaffected but the pH gradient is abolished. In the presence of either valinomycin or nigericin alone, oxidative phosphorylation can continue because either a pH gradient or a membrane potential is still present. When administered together, however, valinomycin and nigericin abolish both components of the electrochemical gradient and ATP synthesis is completely uncoupled from electron flow through the respiratory chain.

In mitochondria that have been uncoupled, the energy released when electrons flow through the respiratory chain is lost as heat instead of being captured for ATP formation. Most mammals contain a special type of fat tissue, called **brown fat,** whose mitochondria are deliberately uncoupled for the purpose of producing heat. The inner membrane of such mitochondria contains an *uncoupling protein* that allows protons that

have been pumped across the membrane to return back through the membrane without driving ATP synthesis. As a result, the energy that would have been used for ATP synthesis is instead released as heat that helps to maintain the organism's body temperature. This process is especially important for maintaining body heat in hibernating animals and in the newborn.

6) Oxidative Phosphorylation Requires an Intact Membrane Defining an Enclosed Compartment

An obvious prediction of the chemiosmotic model is that oxidative phosphorylation requires an intact mitochondrial membrane enclosing a defined compartment. Otherwise, the proton gradient that drives oxidative phosphorylation could not be maintained. This prediction has been verified by the discovery that electron transfer carried out by isolated respiratory carriers cannot be coupled to ATP synthesis unless the carriers are incorporated into membranes that form enclosed vesicles.

7) The Components of the Respiratory Chain Are Asymmetrically Oriented within the Inner Mitochondrial Membrane

The chemiosmotic model requires that protons be selectively pumped in one direction across the inner mitochondrial membrane. It logically follows that the electron carriers that comprise the respiratory chain must be asymmetrically oriented within this membrane; otherwise protons would be randomly pumped in both directions. The topographical organization of the respiratory chain has been studied by exposing inner membrane vesicles to antibodies, enzymes, and labeling reagents designed to interact with various membrane components. Such studies have revealed that some constituents of the respiratory complexes face the matrix side of the inner membrane while others are exposed on the opposite side. This finding confirms the prediction that respiratory chain components are asymmetrically distributed across the inner mitochondrial membrane.

8) ATP Is Synthesized as Protons Flow through the F_1–F_0 Complex

We have already seen that ATP synthesis by the inner mitochondrial membrane is carried out by ATP synthase, an enzyme consisting of F_1 particles attached to the membrane by a group of transmembrane proteins known as F_0. How does a gradient of protons influence the ability of ATP synthase to manufacture ATP? A clue to the role played by F_0 has emerged from the discovery that removing the F_1 particles from the inner mitochondrial membrane causes the membrane to become leaky to protons. These leaks can be sealed by treating F_1-depleted membranes with *oligomycin,* an inhibitor of oxidative phosphorylation that binds to F_0. Since the binding of oligomycin to F_0 stops proton movement through the membrane, it has been concluded that F_0 normally functions as a transmembrane channel through which protons flow down their electrochemical gradient from the intermembrane space to the mitochondrial matrix (see Figure 8-30). According to this view, the ability of oligomycin to inhibit oxidative phosphorylation is based on its capacity to prevent the flow of protons through F_0. This inhibitory mechanism is quite different from that of uncoupling agents, which inhibit oxidative phosphorylation by making the membrane leaky to protons.

Both direct and indirect mechanisms have been proposed to explain how protons passing through the F_0 portion of ATP synthase might drive the synthesis of ATP by F_1. Direct mechanisms invoke the participation of protons in the chemical reaction by which ATP is made, whereas indirect mechanisms envision that pro-

tons passing through F_0 trigger conformational changes in F_1, which in turn promote ATP synthesis. It has been experimentally verified that F_1 undergoes changes in conformation when mitochondrial membranes are exposed to a pH gradient, but it is as yet unknown whether such changes are responsible for triggering ATP synthesis or whether they are simply a by-product of the catalytic mechanism.

A Maximum of 38 Molecules of ATP Are Produced per Molecule of Glucose Oxidized

We have now described the complete metabolic pathway for the aerobic oxidation of glucose. As a net result of glycolysis, the Krebs cycle, and electron transfer through the respiratory chain, the glucose molecule is completely oxidized to six molecules of CO_2; at the same time, six molecules of oxygen are consumed and six molecules of water are produced. The net equation for aerobic glucose oxidation can therefore be written as follows:

$$C_6H_{12}O_6 + 6\ O_2 \rightarrow 6\ CO_2 + 6\ H_2O$$

Earlier in the chapter we saw that this reaction is highly exergonic, with a $\Delta G^{\circ\prime}$ of -686 kcal/mol. The question therefore arises as to how much of the released energy is captured by the cell in the form of ATP.

In order to answer this question, we must first discuss the fate of the NADH molecules that are produced in the cytosol during glycolysis. In eukaryotic cells the main pathway for oxidizing NADH under aerobic conditions is the mitochondrial respiratory chain. But *cytoplasmic* NADH molecules cannot pass their electrons directly to the respiratory chain because mitochondria are usually impermeable to NADH. Instead, the cytoplasmic NADH molecules generated during glycolysis pass their electrons to carrier compounds that, unlike NADH, can pass through mitochondrial membranes; these carrier compounds then transfer their electrons to the respiratory chain. Figure 8-31 summarizes the operation of two such carrier systems, the **malate-aspartate shuttle** and the **glycerol phosphate shuttle**. The malate-aspartate shuttle delivers electrons from cytoplasmic NADH to intramitochondrial NAD^+, whereas the glycerol phosphate shuttle delivers electrons from cytoplasmic NADH to intramitochondrial FAD. Since the oxidation of NADH by the respiratory chain yields three molecules of ATP, while the oxidation of $FADH_2$ generates only two ATPs, the malate-aspartate shuttle is inherently more efficient than the glycerol phosphate shuttle.

When all sources of ATP production are included, the complete aerobic oxidation of one molecule of glucose by eukaryotic cells is found to yield a maximum of 38 molecules of ATP (Table 8-3). Of these 38 ATPs, 2 are

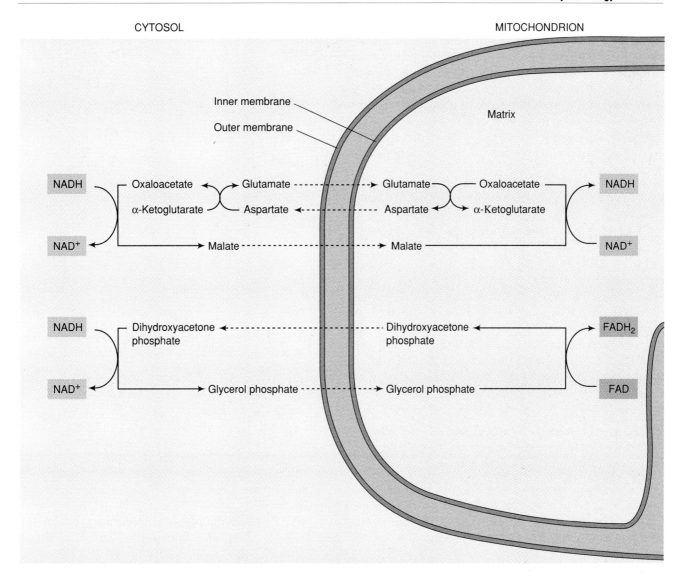

Figure 8-31 *Shuttle Mechanisms for Transferring Reduced Coenzymes from the Cytosol to the Mitochondrial Matrix*
The malate-aspartate shuttle (top) *transfers electrons from NADH in the cytosol to NAD^+ in the matrix, whereas the glycerol phosphate shuttle* (bottom) *transfers electrons from NADH in the cytosol to FAD in the matrix.*

Table 8-3 Calculation of the Maximum ATP Yield During the Aerobic Oxidation of Glucose					
Source	**Yield of CO_2**	**Net ATP Yield From Substrate-Level Phosphorylation**	**Yield of Reduced Coenzyme**	**Number of ATP Molecules Formed per Reduced Coenzyme**	**Maximum ATP Produced**
Glucose → 2 pyruvate			2 NADH ×	3	= 6
		2 ATP			= 2
2 Pyruvate → 2 acetyl CoA	2 CO_2				
			2 NADH ×	3	= 6
Krebs cycle (2 turns)	4 CO_2				
			6 NADH ×	3	= 18
			2 $FADH_2$ ×	2	= 4
		2 ATP			= 2
				Total Yield =	38

made during glycolysis and 2 are made during the Krebs cycle. The term **substrate-level phosphorylation** is used to refer to this type of ATP formation because the synthesis of ATP is part of a chemical reaction in which a substrate is converted to a product. Thus a total of 4 ATPs are made by substrate-level phosphorylation. The remaining 34 ATPs are formed by the process of *oxidative phosphorylation,* which as we have seen is driven by the proton gradient across the inner mitochondrial membrane.

A total of 38 ATP molecules represents a sizable recovery of the energy released during the aerobic oxidation of glucose. Since the $\Delta G'$ for the synthesis of ATP under conditions prevailing a typical cell is about 11 kcal/mol (see Figure 2-25), the total energy captured in the form of ATP is 38 ATPs × 11 kcal/mol = 418 kcal/mol. The complete oxidation of glucose, on the other hand, releases 686 kcal/mol. The overall energy recovery is therefore 418 divided by 686, or about 60 percent. Although the remaining 40 percent of the released energy is lost as heat, the overall efficiency of this process is still very high, especially when compared to the efficiency of most engines and other energy-transducing devices.

Molecules Involved in Mitochondrial Metabolism Are Actively Transported across the Inner Mitochondrial Membrane

The mitochondrion serves as the main site of ATP synthesis for aerobic eukaryotic cells, taking up pyruvate and other oxidizable substrates and delivering ATP to the rest of the cell. These activities require the movement of various small molecules, such as pyruvate, ATP, ADP, and inorganic phosphate, into and out of the mitochondrion. Although the outer mitochondrial membrane is freely permeable to such molecules, the inner membrane acts as a permeability barrier to most small molecules and ions. Without this permeability barrier, mitochondria would not be able to maintain different proton concentrations on opposite sides of the inner membrane, an essential condition for ATP formation. The impermeability of the inner membrane also means, however, that membrane transport systems are required for transporting the various molecules that need to enter and exit the mitochondrion.

Many of these transport systems depend upon the proton gradient that exists across the inner mitochondrial membrane (Figure 8-32). A typical example is the *H⁺-linked pyruvate carrier,* which transports pyruvate into the matrix coupled to the inward flow of protons down their electrochemical gradient. Likewise, the inorganic phosphate required for synthesizing ATP from ADP is transported into the matrix by an *H⁺-linked phosphate carrier* that couples the uptake of phos-

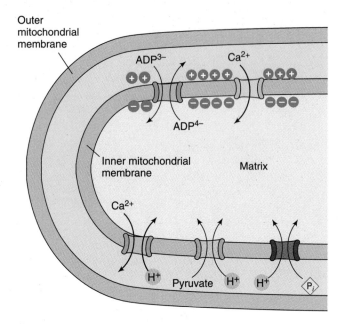

Figure 8-32 *Several Membrane Transport Systems Driven by the Electrochemical Proton Gradient across the Inner Mitochondrial Membrane* *The three transport systems at the bottom of the diagram are driven by the difference in proton concentration, while the two at the top are driven by the membrane potential.*

phate to the inward movement of protons. The high external proton concentration generated by the activity of the respiratory chain drives these two types of active transport.

ATP and ADP are two other crucial substances that need to be transported into and out of the mitochondrion. Mitochondria take up ADP from the cytosol and convert it to ATP, which is then delivered back to the cytosol for use by the rest of the cell. The transport of these two nucleotides across the inner mitochondrial membrane is handled by a transport protein called the *ADP-ATP carrier.* Under normal conditions this carrier transports ADP into the matrix and ATP in the opposite direction, an appropriate arrangement for an organelle that manufactures ATP for the rest of the cell. The ability of this carrier to selectively transport ATP and ADP in opposite directions is based on the existence of the membrane potential across the inner mitochondrial membrane. The ADP molecule has a charge of −3, whereas ATP has a charge of −4. Hence the uptake of a molecule of ADP coupled to the export of a molecule of ATP is equivalent to the net expulsion of one negative charge. Because the membrane potential generated by the proton gradient causes the exterior of the mitochondrion to be positively charged relative to the matrix, the membrane potential serves as a driving force for this type of exchange; in other words, the negative charge being expelled from the matrix is traveling down an electrical charge gradient from an area that is negatively charged to an area that is positively charged. If the membrane potential is abolished by adding an uncoupling agent to dissipate the proton gradient, the selectivity of the ADP-ATP carrier is abolished; ADP and ATP are then transported in both directions at equal rates.

The electrochemical gradient across the inner mitochondrial membrane also plays a role in the transport of calcium ions, which are stored in mitochondria and which regulate the activity of a number of important cytoplasmic proteins (page 219). The uptake of Ca^{2+} across the inner mitochondrial membrane is mediated by a membrane carrier that is driven by the membrane potential. Since the membrane potential generated by the proton gradient causes the interior of the mitochondrion to be negatively charged relative to the exterior, the positively charged calcium ions are pulled across the mitochondrial membrane by electrostatic attraction to the negatively charged matrix side. Conversely, the transport of Ca^{2+} out of the mitochondrion involves a different membrane carrier that mediates the exchange of Ca^{2+} inside the mitochondrion for H^+ outside the mitochondrion. Thus the membrane potential drives the uptake of Ca^{2+} and the pH gradient drives the expulsion of Ca^{2+}. Since the uptake and expulsion of Ca^{2+} from mitochondria are mediated by two different transport systems, regulating the relative activities of these two systems provides cells with a mechanism for influencing the Ca^{2+} concentration in the cytosol.

The preceding examples illustrate the central role played by the electrochemical proton gradient in the transport of small molecules and ions into and out of mitochondria. When energy derived from the electrochemical gradient is employed to drive such transport systems, *it decreases the total amount of energy in the gradient that is available for use in ATP formation*. In other words, the yield of 38 ATPs per glucose molecule oxidized that we calculated earlier (see Table 8-3) simply represents the *maximum amount of ATP* that can be made by the proton gradient. If the proton gradient is partially dissipated by using it to drive active transport, the amount of ATP that can be produced will be correspondingly lower. The net effect is that the energy released during glucose oxidation is used to drive both ATP synthesis and active transport.

Allosteric Regulation Plays an Important Role in Controlling Glucose Oxidation

The oxidation of glucose and the accompanying formation of ATP are carried out by a complex set of metabolic pathways whose activities must be precisely coordinated and regulated. One way in which these pathways are controlled is through the availability of

oxygen. We have seen that in the presence of oxygen, a maximum of 38 ATPs is produced per molecule of glucose metabolized, while only two molecules of ATP are produced in the absence of oxygen. It logically follows that in order to maintain comparable levels of ATP production, glycolysis must proceed 19 times faster in anaerobic cells than in aerobic cells. In other words, a control mechanism is needed that can speed up glycolysis in the absence of oxygen and slow it down when oxygen is present. This type of control, known as the **Pasteur effect,** was first discovered more than a century ago by Louis Pasteur, who reported that exposing anaerobic cells to oxygen causes the rate of glucose consumption to decrease dramatically.

Since molecular oxygen does not participate directly in glycolysis, how does oxygen influence the rate at which glycolysis proceeds? This question has been answered by experiments in which tissues carrying out anaerobic glycolysis were rapidly frozen shortly after the introduction of oxygen. Rapid freezing interrupts metabolism and allows the intermediates formed during glycolysis to be isolated and quantified. The results of such studies have revealed that shortly after oxygen is added to anaerobic tissues, the products formed by the first two steps of glycolysis increase in concentration, whereas the products formed by the steps beyond this point decrease in concentration. Such results indicate that the presence of oxygen inhibits Step 3 in glycolysis, which is catalyzed by the enzyme *phosphofructokinase* (see Figure 8-3).

Subsequent investigations involving purified phosphofructokinase have revealed that this enzyme is subject to allosteric regulation (page 51); ATP acts as an allosteric inhibitor of phosphofructokinase, while AMP serves as an allosteric activator. Using ATP and AMP as allosteric regulators makes good sense because the relative concentrations of these substances is a reflection of a cell's need for chemical energy. If the intracellular concentration of ATP is high relative to its breakdown products ADP and AMP, then there is no need for more ATP to be produced. Under such conditions ATP slows down the metabolism of glucose by acting as an allosteric inhibitor of phosphofructokinase. Conversely, when the intracellular concentration of ATP is low relative to that of ADP and AMP, the rate of glycolysis needs to be increased. Under such conditions AMP acts as an allosteric activator of phosphofructokinase, speeding up glycolysis and thereby fostering the production of more ATP.

Phosphofructokinase is not the only enzyme subject to allosteric control by ATP and its breakdown products (Figure 8-33). When ATP levels are high, the oxidation of pyruvate is slowed down by the ability of ATP to function as an allosteric inhibitor of *pyruvate dehydrogenase*. Conversely, when ATP is depleted and

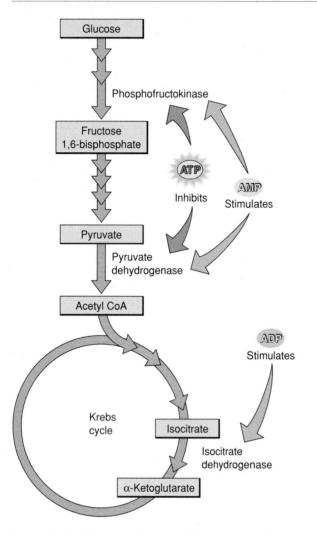

Figure 8-33 Allosteric Control of Several Key Enzymes of Glycolysis and the Krebs Cycle by ATP, ADP, and AMP *High concentrations of ATP act to inhibit this pathway, whereas high concentrations of ADP and AMP are stimulatory.*

chain occurs only when ATP can be made. To illustrate this type of control, let us consider what happens when all of a cell's ADP has been converted to ATP by oxidative phosphorylation. Under such conditions it would be wasteful for electron transfer to continue because the released energy cannot be used to drive the synthesis of any additional ATP. Electron transfer does continue for a brief period of time, but it soon builds up an excessive proton gradient that cannot be dissipated by using it to drive ATP formation. Eventually the proton gradient becomes so steep that no more protons can be pumped, which in turn serves as a negative-feedback mechanism that prevents any further electron transfer through the respiratory chain. This control mechanism thus helps to ensure that the rate of electron transfer is appropriately matched to a cell's need for ATP synthesis.

Changes in the rate of electron transfer are often reflected in the structural appearance of the mitochondrion. For example, the number of mitochondrial cristae can increase dramatically in cells where electron transfer rates are maintained at high levels because of a continued demand for large amounts of ATP. Morphological changes have also been observed in mitochondria subjected to different experimental conditions. Mitochondria incubated in the presence of the substrates required for electron transfer and oxidative phosphorylation normally exhibit a *condensed configuration* in which the matrix space is reduced and the intermembrane space is expanded. But if oxidative phosphorylation is prevented by omitting ADP from the medium, mitochondria assume the *orthodox configuration* in which the intermembrane space is small and the matrix space is distended (Figure 8-34). Such observations indicate that the metabolic state of a mitochondrion is reflected in its morphological appearance.

HOW BACTERIA CAPTURE ENERGY

Although bacteria lack mitochondria, the metabolic pathways they employ for oxidizing food molecules and synthesizing ATP are similar to those found in eukaryotes. Glycolysis occurs in the cytosol of bacterial cells just as it does in eukaryotic cells. However, the absence of mitochondria in bacteria means that cellular respiration must be handled somewhat differently.

The Respiratory Chain and Oxidative Phosphorylation Occur in the Plasma Membrane of Bacterial Cells

Electron transfer through the respiratory chain and oxidative phosphorylation are associated with the inner mitochondrial membrane in eukaryotic cells, so membranes might be expected to carry out these ac-

ADP and AMP levels rise, AMP acts as an allosteric activator of pyruvate dehydrogenase and ADP acts as an allosteric activator of *isocitrate dehydrogenase,* the Krebs cycle enzyme that catalyzes the conversion of isocitrate to α-ketoglutarate. Both of these actions lead to an increased rate of pyruvate oxidation and hence increased ATP production.

Respiratory Control Regulates the Flow of Electrons through the Respiratory Chain

Allosteric control by ATP and its breakdown products is only one of several ways in which the pathways involved in cellular energy metabolism are regulated. Another mechanism that deserves special mention is the phenomenon of **respiratory control,** which helps to ensure that electron transfer through the respiratory

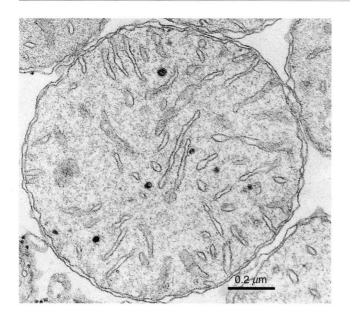
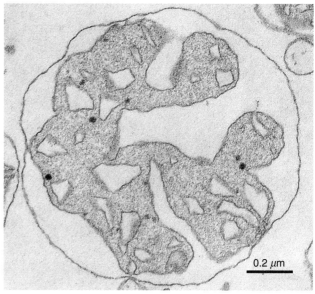

Figure 8-34 *Changes in Mitochondrial Appearance in Different Metabolic States* (Left) *The orthodox configuration, characterized by a small intermembrane space and a distended matrix compartment, is exhibited by mitochondria that are not carrying out oxidative phosphorylation.* (Right) *The condensed configuration, characterized by an expanded intermembrane space and a reduced matrix space, is exhibited by active mitochondria that are carrying out both electron transfer and oxidative phosphorylation. Courtesy of C. R. Hackenbrock.*

tivities in prokaryotic cells as well. This expectation has been verified by experiments in which plasma membrane vesicles isolated from bacteria cells were found to be capable of oxidizing NADH and using the released energy for ATP synthesis. Moreover, electron micrographs reveal that these plasma membrane vesicles are covered with spherical particles which resemble the F_1 particles of the mitochondrial inner membrane (Figure 8-35). These particles, which protrude from the inner surface of the plasma membrane in intact bacterial cells, contain an ATP-synthesizing enzyme complex that is similar to the one found in mitochondrial F_1 particles.

The respiratory chains of bacterial cells, like those of eukaryotes, are comprised of cytochromes, iron-sulfur proteins, flavoproteins, and quinones. However, wide variations are observed among bacteria in the particular electron carriers that are employed and the sequence in which they are organized. In the bacterium *E. coli*, the respiratory chain begins with the transfer of electrons from NADH to the FAD group of a membrane-bound flavoprotein. The electrons next pass through a series of iron-sulfur proteins to a ubiquinone, from which they are transferred through cytochrome b_1 and cytochrome o to oxygen (Figure 8-36). Coupling sites for ATP synthesis have been identified at two points in this chain. The first coupling site for ATP synthesis is associated with the transfer of electrons from NADH to

ubiquinone, and the second site is associated with the transfer of electrons from ubiquinone to oxygen. As we will see in a moment, each of these coupling sites pumps protons across the bacterial plasma membrane, generating a proton gradient that drives the formation of ATP. The respiratory chains of other bacteria exhibit

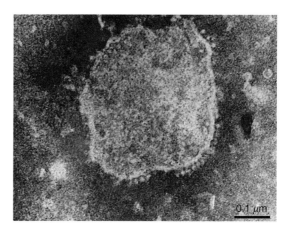

Figure 8-35 *A Negatively Stained Plasma Membrane Fragment from the Bacterium* Mycobacterium *Showing Spherical Particles Protruding from the Membrane Surface* *This plasma membrane fragment has resealed itself into an inverted vesicle whose outer surface corresponds to the inner (cytosolic) surface of the plasma membrane in an intact cell. Courtesy of A. F. Brodie.*

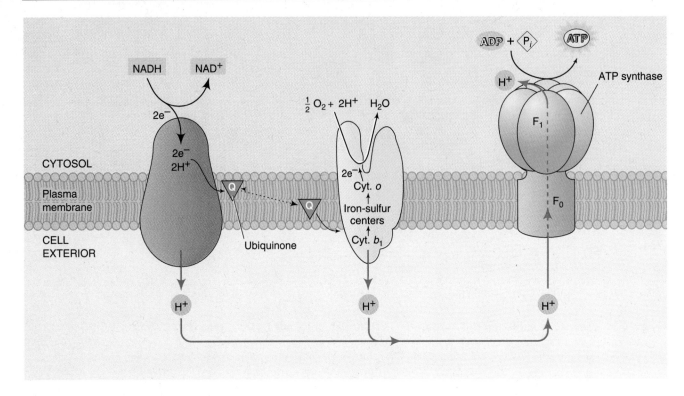

Figure 8-36 *Organization of the Respiratory Chain in the Plasma Membrane of the Bacterium E. coli* *As in the mitochondrial respiratory chain, protons are pumped across the bacterial plasma membrane during electron transfer and the resulting electrochemical proton gradient is employed to drive ATP synthesis by ATP synthase. (Abbreviation: Cyt. = cytochrome.)*

significant variations from that of *E. coli*, both in the arrangement of the carriers and in the location of ATP coupling sites. As a consequence bacterial respiration is a more flexible process than mitochondrial respiration; it can oxidize a wider range of substrates and in some cases can employ other electron acceptors, such as sulfate or nitrate, in place of oxygen.

The study of bacterial mutants in which electron transfer is uncoupled from ATP synthesis has led to the identification of several genes and gene products that are involved in oxidative phosphorylation. For example, the *unc A* gene has been found to code for a polypeptide subunit of the F_1 particle, whereas the *unc B* gene codes for a polypeptide chain that is part of the integral plasma membrane protein, F_0. The discovery that bacteria contain F_1 and F_0 components that are similar to those found in mitochondrial **ATP synthase** suggests that bacteria and mitochondria synthesize ATP by similar mechanisms.

An Electrochemical Proton Gradient across the Bacterial Plasma Membrane Drives Oxidative Phosphorylation

Studies on the mechanism by which ATP synthesis is coupled to electron transfer in bacterial cells have revealed that an electrochemical proton gradient is in-

volved just as in mitochondrial respiration. As electrons flow through the bacterial respiratory chain, the pH inside the cell increases relative to the external pH. This means that protons are being pumped out of the cell, creating a pH gradient as well as a membrane potential. Such an electrochemical gradient has been shown to be capable of driving ATP formation, just as it does in mitochondria.

Another parallel with mitochondria has emerged from the discovery that the proton gradient across the bacterial plasma membrane provides the energy not just for ATP synthesis, but for active transport as well. A typical example involves the H^+-linked carrier that mediates the active transport of lactose across the bacterial plasma membrane (page 191). This lactose carrier couples the uptake of lactose to the movement of protons down their electrochemical gradient, and thus depends upon a proton gradient in which the proton concentration is higher outside the cell than inside. Similar H^+-linked carriers are responsible for the active uptake of a variety of amino acids.

The fact that solute transport and ATP synthesis both depend on the existence of an electrochemical proton gradient is exploited by certain toxins that act on bacterial cells. For example the bacterium *E. coli* produces two toxic proteins, called *colicins E1* and *K,* that kill susceptible bacteria by promoting the diffusion of K^+ through the plasma membrane and hence disrupting the membrane potential. A decrease in the membrane potential component of the electrochemical gradient across the plasma membrane would be expected to inhibit ATP formation as well as solute transport systems driven by the gradient. Such toxins therefore kill bacterial cells by virtue of their ability to inhibit both active transport and the ability to make ATP.

In addition to its role in active transport, we will see in Chapter 13 that the electrochemical proton gradient across the plasma membrane provides the energy that drives the rotation of bacterial flagella. When energy derived from the electrochemical gradient is employed for flagellar rotation or active transport, it of course decreases the total amount of energy in the electrochemical gradient that is available for ATP synthesis.

PEROXISOMES AND RELATED ORGANELLES

We have already seen in this chapter that the main pathways of oxidative metabolism are compartmentalized within mitochondria in eukaryotic cells. However, another type of membrane-enclosed compartment called the *peroxisome* also plays a role in certain oxidative pathways in eukaryotic cells. Unlike mitochondria, which resemble one another in all eukaryotic cell types and are specialized to carry out one major function, peroxisomes are a heterogeneous group of related organelles that appear to be involved in a variety of different functions.

Peroxisomes Were Discovered by Isodensity Centrifugation

In Chapter 7 we saw how Christian de Duve's studies on the distribution of the enzyme acid phosphatase in isolated subcellular fractions led to the discovery of lysosomes. Among the other enzymes monitored by de Duve was an enzyme involved in purine degradation known as *urate oxidase*. In spite of its high concentration in lysosomal fractions isolated by differential centrifugation, some minor differences in its distribution pattern relative to that of other lysosomal enzymes suggested to de Duve that urate oxidase might be localized in some other organelle. When isodensity gradient centrifugation (page 135) was employed to improve the separation between organelles, it became clear that urate oxidase occurs in an organelle whose density differs slightly from that of lysosomes (Figure 8-37, *top*). Shortly thereafter the enzymes *D-amino acid oxidase* and *catalase*, which are involved in the formation and

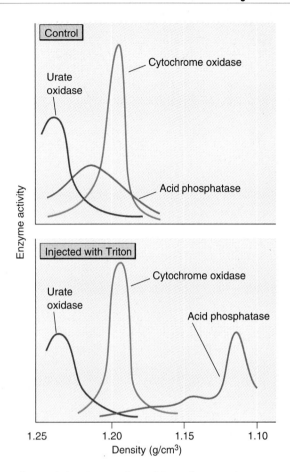

Figure 8-37 *Separation of Peroxisomes, Lysosomes, and Mitochondria by Isodensity Centrifugation* These three organelles are identified by the presence of the enzymes urate oxidase, acid phosphatase, and cytochrome oxidase respectively. (Top) Small differences between the densities of these three organelles permit them to be partially separated from one another by isodensity centrifugation of cell homogenates. (Bottom) Injection of animals with the detergent Triton WR-1339 enhances the separation that can be obtained because this substance is taken up by lysosomes, thereby lowering their density.

breakdown of hydrogen peroxide respectively, were found to behave in the same way as urate oxidase. Because these latter enzymes are involved in the metabolism of *hydrogen peroxide*, the organelle in which they are found was named the **peroxisome.**

The conclusion that peroxisomes and lysosomes are distinct organelles received subsequent support from experiments in which the release of acid phosphatase and catalase from detergent-treated subcellular fractions was compared. If these two enzymes are located in the same organelle, they should be released simultaneously when this organelle is ruptured by adding detergent. However, it was found that ten times more detergent is required to liberate catalase than acid phosphatase, indicating that these two enzymes are located in different organelles (Figure 8-38).

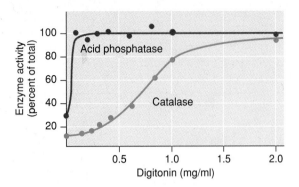

Figure 8-38 *Release of Acid Phosphatase from Lysosomes and Catalase from Peroxisomes by Treatment with the Detergent Digitonin* *The fact that the two enzymes are not released simultaneously verifies that they reside in different organelles.*

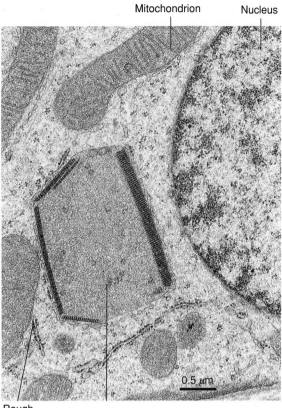

Figure 8-39 *Peroxisome in a Rat Kidney Cell Showing Crystalloid Inclusions* *Both rough endoplasmic reticulum and mitochondria are closely associated with the peroxisome. Courtesy of J. M. Barrett.*

Because of the relatively small difference in density between peroxisomes and lysosomes, isodensity centrifugation does not normally separate these organelles very well. Better separation can be accomplished by isolating peroxisomes from animals that have been first injected with the detergent *Triton WR-1339*. This detergent is selectively accumulated within lysosomes and causes their density to decrease, improving the separation between lysosomes and peroxisomes (see Figure 8-37, *bottom*).

Electron microscopic examination of isolated peroxisomes reveals the presence of small, roughly spherical vesicles that resemble organelles seen in tissue sections by early electron microscopists and given the general name *microbodies*. Such structures vary from 0.2 to 2 μm in diameter and consist of a finely granular matrix surrounded by a single membrane. They often exhibit elaborate crystalloid cores containing the enzyme urate oxidase (Figure 8-39). Microbodies displaying such cores are clearly peroxisomes, but in cases where distinctive cores are not present, peroxisomes are often difficult to distinguish from lysosomes and other small cytoplasmic vesicles. Unequivocal identification requires the use of a cytochemical staining reaction that is specific for the enzyme catalase. One reaction widely used for this purpose is based on the ability of catalase to oxidize *diaminobenzidine* to an electron-opaque reaction product (Figure 8-40).

It should be emphasized that the term *microbody* is a descriptive label that is employed by some electron microscopists any time they see a "small body." Unless such structures have a urate oxidase core or are stained for catalase, they cannot be unequivocally identified as peroxisomes. As we will see shortly, other types of microbodies exist in addition to peroxisomes.

Peroxisomes Contain Enzymes Involved in the Production and Breakdown of Hydrogen Peroxide

Peroxisomes are most abundant in liver and kidney cells in vertebrates, in the leaves and seeds of plants, and in eukaryotic microorganisms such as yeast, protozoa, and fungi. The spectrum of enzymes present in peroxisomes varies considerably among tissue sources (Table 8-4). The only enzyme common to all peroxisomes is catalase, which accounts for up to 15 percent of the peroxisomal protein content. In addition to catalase, hydrogen-peroxide producing oxidases such as urate oxidase, amino acid oxidase, and glycolate oxidase are usually present. These enzymes employ molecular oxygen as an oxidizing agent and generate hydrogen peroxide (H_2O_2) as a product:

$$RH_2 + O_2 \rightarrow R + H_2O_2 \qquad (8\text{-}3)$$

The H_2O_2 generated by reactions of this type is subsequently hydrolyzed by catalase acting in one of two

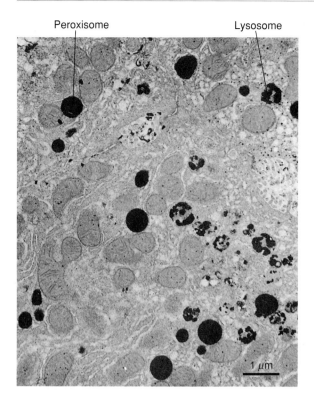

Peroxisome   Lysosome

1 μm

Figure 8-40 *Electron Micrograph of a Rat Liver Cell Stained for Both Acid Phosphatase to Identify Lysosomes and Catalase to Identify Peroxisomes* *Because the staining reaction for acid phosphatase generates a more granular deposit than that produced by the reaction for catalase, lysosomes and peroxisomes are readily distinguished from one another. Courtesy of S. Goldfischer.*

Table 8-4   Main Enzymatic Activities Found in Peroxisomes

Enzymatic Activity	Rat Liver	Plant Seedling	Plant Leaf
1. Catalase	+	+	+
2. Urate oxidase	+	+	+
3. D-Amino acid oxidase	+	+	−
4. L-Amino acid oxidase	±	+	−
5. L-α-Hydroxy acid oxidase	+	+	+
6. Glycolate oxidase	+	+	+
7. Fatty acid β-oxidation	+	+	−
8. Citrate synthase	−	+	−
9. Aconitase		+	−
10. Isocitrate lyase	−	+	−
11. Malate synthase	−	+	−
12. Malate dehydrogenase	−	+	+
13. Glyoxylate reductase	+	−	+
14. Glyoxylate transaminase	−	−	+

Note: (+) = present, (−) = absent, (blank) = not measured.

ways. In the so-called *catalatic mode,* a molecule of H_2O_2 donates its electrons to another molecule of H_2O_2, forming molecules of water and oxygen:

$$H_2O_2 + H_2O_2 \xrightarrow{\text{catalase}} O_2 + 2H_2O \qquad (8\text{-}4)$$

In the *peroxidatic mode,* the electron donor is a substrate other than H_2O_2:

$$RH_2 + H_2O_2 \xrightarrow{\text{catalase}} R + 2H_2O \qquad (8\text{-}5)$$

In both cases, the net result is the breakdown of hydrogen peroxide. The hallmark of peroxisomes is therefore the association of H_2O_2-producing oxidases with the H_2O_2-degrading enzyme, catalase.

Peroxisomes Perform Several Metabolic Functions

Because hydrogen peroxide is toxic, it seems logical to conclude that peroxisomes act to protect cells from exposure to this substance by compartmentalizing the enzymes that produce and destroy it. The flaw in this argument, however, is that many H_2O_2-producing oxidases occur in the cytosol and in mitochondria, suggesting that the peroxisome contributes little to protecting the cell from the effects of H_2O_2. This conclusion is further reinforced by the fact that many cells producing H_2O_2 do not even contain peroxisomes. The role of peroxisomes may therefore be related to metabolic pathways these organelles carry out rather than with the simple need to protect cells from H_2O_2. An examination of the list of enzymes present in peroxisomes suggests several possible functions for this organelle.

1. Inactivation of toxic substances.   One potential role for peroxisomes involves their ability to couple H_2O_2 degradation to the inactivation of toxic compounds. Catalase is the most active enzyme present in the peroxisome, and therefore it degrades H_2O_2 at a much faster rate than H_2O_2 is formed by other peroxisomal reactions. The resulting scarcity of hydrogen peroxide favors the peroxidatic mode of catalase action, in which an alternative electron donor is substituted for the second molecule of H_2O_2 (see Equation 8-5). Electron donors that can be employed for this purpose include a variety of noxious substances such as methanol, ethanol, phenols, nitrites, formaldehyde, and formate. Oxidation of these compounds linked to the breakdown of H_2O_2 results in the detoxification of what are otherwise toxic substances. This detoxification role is compatible with the fact that animal peroxisomes occur mainly in liver and kidney, which are two organs that play prominent roles in protecting organisms from toxic substances.

2. Regulation of oxygen concentration. Because peroxisomal oxidases employ molecular oxygen as an oxidizing agent (see Equation 8-3), the reactions catalyzed by these enzymes may influence oxygen levels within the cell. In liver cells, for example, as much as 20 percent of the total oxygen consumed is accounted for by peroxisomes. Most of the remaining oxygen consumption occurs in mitochondria, where the oxidation of energy-rich molecules by molecular oxygen releases energy that is used for ATP formation. Hence *respiration,* which refers to the oxidation of substances by molecular oxygen, occurs in both mitochondria and peroxisomes. The major difference is that much of the energy released during mitochondrial respiration is captured in the form of ATP, whereas the energy released during peroxisomal respiration is lost entirely as heat.

Peroxisomal respiration also differs from mitochondrial respiration in its sensitivity to oxygen concentration. Mitochondrial respiration occurs at a maximal rate when the oxygen concentration is about 2 percent, with further increases in oxygen producing no further increase in respiratory rate. The rate of peroxisomal respiration, on the other hand, increases in direct proportion to the oxygen concentration (Figure 8-41). This differing behavior gives mitochondria an advantage over peroxisomes in utilizing small amounts of oxygen, but allows peroxisomes to exceed the respiratory ability of mitochondria at high oxygen concentrations and thus protect cells from the potentially toxic effects of large amounts of oxygen.

3. Lipid metabolism. In recent years, a number of enzymes involved in lipid metabolism have been identified in peroxisomes. Especially interesting is the

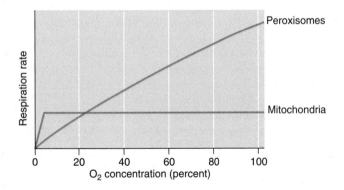

Figure 8-41 *Relationship between Oxygen Concentration and Respiration Rate in Peroxisomes and Mitochondria* *Mitochondria are more efficient than peroxisomes in utilizing small amounts of oxygen, while peroxisomes surpass mitochondria in metabolizing high concentrations of oxygen, which would otherwise be toxic to cells.*

presence of several enzymes involved in the synthesis of *ether phospholipids,* which utilize an ether linkage (C−O−CH$_2$) in place of an acyl linkage (C−O−C=O) to join a fatty acid to one of the oxygen atoms in glycerol. Enzymes involved in the β-oxidation of fatty acids and in the metabolism of dolichol and steroids have also been identified in peroxisomes, suggesting that this organelle may play an important role in the degradation and biosynthesis of various kinds of lipids.

4. Metabolism of nitrogenous bases and carbohydrates. Because uric acid is a degradation product of the purines found in nucleic acids, the localization of urate oxidase and other enzymes of purine metabolism within peroxisomes suggests that this organelle may play a role in the breakdown of nitrogenous bases derived from DNA and RNA. Peroxisomes also have been implicated in the metabolism of various kinds of carbohydrates and organic acids. The combination of enzymes present in peroxisomes suggests that this organelle may be involved in **gluconeogenesis,** which is the process by which carbohydrates are synthesized from lipids and other noncarbohydrate materials. This speculation is reinforced by the fact that the vertebrate cell types containing the largest numbers of peroxisomes are liver and kidney, which are known to be major sites of gluconeogenesis. However, the role of peroxisomes in gluconeogenesis has so far been established only in plant seedlings, where a special type of peroxisome known as a *glyoxysome* is involved.

Glyoxysomes Are Used by Plants to Synthesize Carbohydrates from Lipids

Many plants store large amounts of lipid in their seeds for use as an energy source during germination. At the appropriate time, this lipid is converted to carbohydrate by a pathway that can be divided into two major parts. In the first part, fatty acids are degraded by β-oxidation to yield acetyl CoA. Acetyl CoA is then converted to carbohydrate by a modified version of the Krebs cycle known as the **glyoxylate cycle** (Figure 8-42). In essence, the glyoxylate cycle bypasses the CO$_2$-producing steps of the Krebs cycle with the aid of two enzymes not present in animal cells, *isocitrate lyase* and *malate synthase.* The net result of the glyoxylate cycle is the conversion of two molecules of acetyl CoA to one molecule of succinate, which is a precursor for the synthesis of a variety of carbohydrates.

In the late 1960s Harry Beevers and his associates discovered that in plant seedlings, the enzymes of the glyoxylate cycle and fatty acid β-oxidation are localized in organelles that can be separated by isodensity cen-

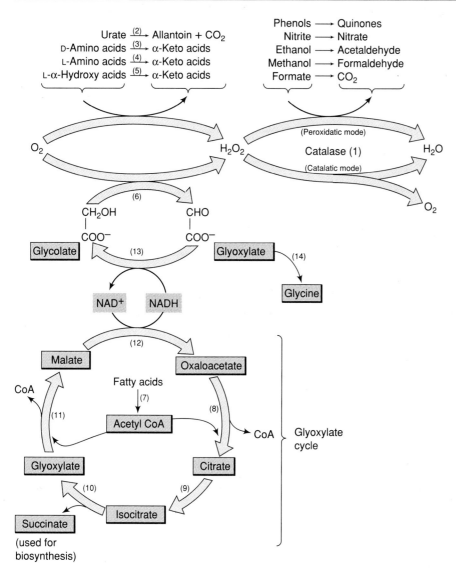

Figure 8-42 *The Main Reactions Occurring in Peroxisomes* *The numbers refer to enzymes listed in Table 8-4. The complete glyoxylate cycle is restricted to the peroxisomes (glyoxysomes) of plant seedlings.*

trifugation from both mitochondria and chloroplasts. These organelles, logically named **glyoxysomes,** also contain catalase and several H₂O₂-producing oxidases. Hence they represent a special type of peroxisome in which the complete set of glyoxylate cycle enzymes is present along with the normal H₂O₂-associated enzymes.

True glyoxysomes bearing the complete set of enzymes required for gluconeogenesis by the glyoxylate cycle have thus far been identified only in plant seedlings. Protozoa and yeast have peroxisomes containing some, but not all, of the glyoxylate cycle enzymes. In such cases the glyoxylate cycle may be carried out by cooperation between mitochondria and peroxisomes, each executing a portion of the cycle. In animals, fatty acid β-oxidation occurs in peroxisomes, but the acetyl CoA produced by this pathway cannot be used for glu-

coneogenesis by the glyoxylate cycle because the glyoxylate cycle does not occur in higher animals.

Glycosomes Are Used by Trypanosomes to Speed Up Glycolysis

Yet another type of peroxisome, called the **glycosome,** has been discovered in a group of parasitic protozoa known as *trypanosomes*. Glycosomes differ from animal peroxisomes and plant glyoxysomes in that enzymes involved in glycolysis account for more than 90 percent of the glycosome's protein content. In spite of this unusual enzyme makeup, the glycosome is thought to be a member of the peroxisome family because it contains catalase as well as enzymes involved in fatty acid β-oxidation.

Although the glycosome does not carry out the complete pathway of glycolysis, it does contain the enzymes needed for the first seven steps of this ten-step pathway (see Figure 8-3). The remainder of the glycolytic pathway occurs in the cytosol. Compartmentalizing glycolysis in this unique way allows glycolysis to be carried out at much faster rates in trypanosomes than in any other type of eukaryotic cell.

It is thus apparent that the peroxisome family is an unusual group of organelles. In contrast to organelles like mitochondria and chloroplasts, which have retained one main function through evolutionary time, the peroxisome family consists of a heterogeneous group of organelles involved in a variety of different functions. Yet these organelles remain related to one another by virtue of the presence of enzymes involved in hydrogen peroxide metabolism and the metabolism of lipids.

SUMMARY OF PRINCIPAL POINTS

- Glycolysis converts glucose to two molecules of pyruvate, accompanied by the net formation of two molecules of ATP and the conversion of two molecules of NAD^+ to NADH. Under anaerobic conditions, pyruvate is converted to a reduced product such as lactate or ethanol accompanied by the oxidation of NADH to NAD^+. This process of fermentation generates two molecules of ATP per molecule of glucose metabolized.

- Under aerobic conditions, glucose is completely oxidized to CO_2 and H_2O by cellular respiration. In eukaryotic cells, respiration occurs in mitochondria. Mitochondria are enclosed by two membranes: a smooth outer membrane and an inner membrane thrown into a series of invaginations known as cristae. The area between the outer and inner membranes is called the intermembrane space, and the region enclosed by the inner membrane is the matrix. The inner mitochondrial membrane is the site of the respiratory chain and oxidative phosphorylation, and contains spherical F_1 particles protruding from its surface.

- In the mitochondrial matrix, pyruvate and fatty acids are oxidized to acetyl CoA, which in turn is metabolized by the Krebs cycle. Each molecule of acetyl CoA that enters the Krebs cycle leads to the production of three molecules of NADH, one molecule of $FADH_2$, one molecule of ATP, and two molecules of CO_2. The NADH and $FADH_2$ produced by this process are then oxidized back to NAD^+ and FAD by the respiratory chain, a group of electron carriers localized within the inner mitochondrial membrane. These carriers are organized into four multiprotein complexes that together catalyze the transfer of electrons from NADH and $FADH_2$ to oxygen.

- The oxidation of a molecule of NADH by the respiratory chain generates a maximum of three molecules of ATP, whereas the oxidation of $FADH_2$ produces only two ATPs. This type of ATP formation, called oxidative phosphorylation, is driven by an electrochemical proton gradient that is established during the flow of electrons through the respiratory chain. During electron transfer, protons are pumped from the mitochondrial matrix into the intermembrane space; these protons then diffuse back into the matrix by passing through the F_1–F_0 complex (ATP synthase), stimulating the complex to synthesize ATP. Uncoupling agents inhibit oxidative phosphorylation by making membranes leaky to protons, thereby dissipating the proton gradient.

- In addition to its role in ATP synthesis, the electrochemical proton gradient drives several mitochondrial transport systems. The pH component of the gradient promotes the uptake of pyruvate and inorganic phosphate, while the membrane potential drives the uptake of ADP coupled to the export of ATP.

- When cellular levels of ATP are high, glycolysis and cellular respiration are inhibited because ATP acts as an allosteric inhibitor of phosphofructokinase and pyruvate dehydrogenase. When ATP levels fall, these pathways are stimulated because AMP, which is generated as the ultimate breakdown product of ATP, serves as an allosteric activator of phosphofructokinase and pyruvate dehydrogenase.

- Although bacteria lack mitochondria, the metabolic pathways by which they synthesize ATP are similar to those occurring in eukaryotes. The bacterial plasma membrane, which exhibits F_1 particles protruding from its inner surface, serves in lieu of the mitochondrial membrane as the site of electron transport and oxidative phosphorylation. The electron carriers that make up bacterial respiratory chains are different from those employed by eukaryotes, but electron transfer through the chain is still accompanied by the formation of an electrochemical proton gradient. This gradient, which exhibits a higher proton concentration outside the cell than inside, is used to drive ATP synthesis as well as certain active transport reactions and flagellar rotation.

- Peroxisomes are eukaryotic organelles characterized by the presence of the H_2O_2-degrading enzyme catalase, a variety of H_2O_2-generating oxidases, and selected enzymes involved in lipid and carbohydrate metabolism. Several possible roles have been considered for peroxisomes, including inactivation of toxic substances, regulation of oxygen concentration, and metabolism of lipids, nitrogenous bases, and carbohydrates. The glyoxylate cycle, which permits carbohydrates to be synthesized from lipids, is present in plant seedling peroxisomes (glyoxysomes). Trypanosomes contain a related organelle called the glycosome, which carries out most of glycolysis.

SUGGESTED READINGS

Books

Fahimi, H. D., and H. Sies, eds. (1987). *Peroxisomes in Biology and Medicine,* Springer-Verlag, New York.

Fiskum, G., ed. (1986). *Mitochondrial Physiology and Pathology,* Van Nostrand Reinhold, New York.

Tzagoloff, A. (1982). *Mitochondria,* Plenum, New York.

Articles

Abrahams, J. P., A. G. W. Leslie, R. Lutter, and J. E. Walker (1994). Structure at 2.8Å resolution of F_1 - ATPase from bovine heart mitochondria, *Nature* 370:621–628.

Calhoun, M. W., J. W. Thomas, and R. B. Gennis (1994). The cytochrome oxidase superfamily of redox-driven proton pumps, *Trends Biochem. Sci.* 19:325-330.

Capaldi, R. A., R. Aggeler, P. Turina, and S. Wilkins (1994). Coupling between catalytic sites and the proton channel in F_1F_0-type ATPases, *Trends Biochem. Sci.* 19:284-289.

Dihanich, M. (1990). The biogenesis and function of eukaryotic porins, *Experientia* 46:146-153.

Ernster, L., and G. Schatz (1981). Mitochondria: a historical review, *J. Cell Biol.* 91:227s-255s.

Hatefi, Y. (1985). The mitochondrial electron transport and oxidative phosphorylation system, *Annu. Rev. Biochem.* 54:1015-1069.

Hinkle, P. C., and R. E. McCarty (1978). How cells make ATP, *Sci. Amer.* 238 (March):104-123.

Mannella, C. A. (1992). The 'ins' and 'outs' of mitochondrial membrane channels, *Trends Biochem. Sci.* 17:315-320.

Mitchell, P. (1979). Keilin's respiratory chain concept and its chemiosmotic consequences, *Science* 206:1148-1159.

Opperdoes, F. R. (1988). Glycosomes may provide clues to the import of peroxisomal proteins, *Trends Biochem. Sci.* 13:255-260.

Racker, E. (1980). From Pasteur to Mitchell: a hundred years of bioenergetics, *Fed. Proc.* 39:210-215.

Senior, A. E. (1988). ATP synthesis by oxidative phosphorylation, *Physiol. Rev.* 68:177-231.

Trumpower, B. L., and R. B. Gennis (1994). Energy transduction by cytochrome complexes in mitochondrial and bacterial respiration: The enzymology of coupling electron transfer reactions to transmembrane proton translocation, *Annu. Rev. Biochem.* 63:675-716.

Vallee, D., and J. Gärtner (1993). Penetrating the peroxisome, *Nature* 361:682-683.

van den Bosch, H., R. B. H. Schutgens, R. J. A. Wanders, and J. M. Tager (1992). Biochemistry of peroxisomes, *Annu. Rev. Biochem.* 61:157-197.

Wainio, W. W. (1985). An assessment of the chemiosmotic hypothesis of mitochondrial energy transduction, *Int. Rev. Cytol.* 96:29-51.

Chapter 9

Chloroplasts and the Capturing of Energy Derived from Sunlight

In Chapter 8 we learned that carbohydrates and fats are excellent food sources because they contain reduced carbon atoms whose oxidation by cellular respiration releases free energy. This also means, however, that maintaining the earth's food supply requires a mechanism for regenerating the reduced carbon atoms that have been oxidized by cellular respiration. Reducing these carbon atoms is ultimately accomplished by a group of reactions known as **photosynthesis**. Photosynthesis captures energy from sunlight and utilizes it to drive the reduction of carbon dioxide, generating oxygen and carbohydrates as final products. The oxygen produced by this process is responsible for the oxygen-rich atmosphere of our planet, while the carbohydrate feeds not only the photosynthesizing organisms themselves, but also all the animals that eat these photosynthesizing organisms, the animals that eat these animals, and so forth through the food chain. Photosynthesis is thus one of the cornerstones upon which life on earth is built.

Organisms that are capable of carrying out photosynthesis are referred to as **autotrophs** ("self-feeding"),

whereas those that cannot photosynthesize and must therefore consume food containing reduced carbon are termed **heterotrophs** ("feeding on others"). Although we most often think of multicellular plants when discussing photosynthesis, single-celled photosynthetic organisms produce more than half of the roughly 40 billion tons of carbon reduced annually by photosynthesis. These microorganisms include eukaryotes such as algae, diatoms, and dinoflagellates, as well as prokaryotes such as photosynthetic bacteria and cyanobacteria (formerly called blue-green algae).

A great deal of effort has been expended in recent years in trying to understand how photosynthetic organisms trap light energy and use it to drive the reduction of carbon. In eukaryotic autotrophs the light-trapping reactions occur in specialized organelles known as chloroplasts, while in prokaryotic autotrophs the comparable reactions take place within membranes dispersed throughout the cytoplasm. But despite this superficial difference, the photosynthetic pathways of prokaryotes and eukaryotes are fundamentally similar. Moreover, many of the events involved in photosynthesis resemble processes that occur during cellular respiration. As we will see in this chapter, the crucial stages of both photosynthesis and respiration occur in membranes, involve electron transfer reactions, and synthesize ATP utilizing an electrochemical proton gradient.

EARLY STUDIES OF PHOTOSYNTHESIS

Before embarking upon a detailed examination of the molecular events that are involved, it will be helpful for us to step back and briefly view the process of photosynthesis as a whole. In this introductory section we will examine some of the early experimental evidence that led to the formulation of the basic equation of photosynthesis and to the idea that photosynthesis consists of separate "light" and "dark" reactions that occur in different regions of the chloroplast.

The Basic Equation of Photosynthesis Was Derived from the Independent Observations of Several Scientists

The original formulation of the chemical equation summarizing photosynthesis required the work of several scientists spanning a period of nearly 200 years. The first major contributor was the Flemish physician Jan Baptista van Helmont, who carried out a simple experiment in which he measured the increase in weight of a willow tree grown in a bucket of soil to which only water was added. After five years the tree was found to have gained 160 pounds, while the soil in which it was planted weighed only 2 ounces less than it had at the beginning of the experiment. Since water was the only ingredient added, van Helmont concluded that the organic matter synthesized by plants is derived from water molecules.

It was not until 100 years later that the English minister Joseph Priestley implicated the gaseous components of air in this process. In 1771 Priestley reported an experiment in which a candle was burned in an enclosed chamber until it spontaneously extinguished. Mice subsequently placed in such a chamber were found to suffocate, leading Priestley to conclude that the burning candle had "injured" the air. But if a small green plant was placed in the chamber, the mice could then breathe and survive normally. It was therefore concluded that plants can "restore" air to a normal state, a monumental realization whose significance was not fully appreciated at the time because it preceded the discovery of the gases oxygen (O_2) and carbon dioxide (CO_2).

News of Priestly's observations stimulated Jan Ingenhousz, a Dutchman serving as court physician to the Austrian empress, to investigate the phenomenon in more detail. A few years later Ingenhousz made two crucial discoveries: Only the green portion of a plant can "restore" air, and light is required during the process. In 1782 a Swiss minister, Jean Senebrier, further observed that the "restoration" of air is accompanied by the uptake of carbon dioxide. Around this same time the discovery of oxygen led to the realization that the restoration of air by green plants is based upon their ability to produce O_2. Taken together, the preceding observations suggested that photosynthesis involves the uptake of CO_2 and the production of organic matter and O_2. But careful measurements of the uptake and production of these substances during photosynthesis led the Swiss chemist Theodore de Saussure to conclude that the amount of organic matter and oxygen a plant produces is greater than the weight of the carbon dioxide it consumes. This discrepancy led de Saussure to rediscover van Helmont's old idea that the synthesis of organic matter by plants also requires water.

Thus by the end of the eighteenth century, it was clear that photosynthesis converts water and carbon dioxide to organic matter and oxygen. In 1845 the German surgeon Julius Robert Mayer, who formulated the law of conservation of energy, suggested that light provides the energy needed to drive the overall process. With this final insight, the overall equation describing photosynthesis could be written as follows:

$$CO_2 + H_2O \xrightarrow[\text{Green Plant}]{\text{Light}} (CH_2O)_n + O_2$$

One of the many types of organic matter that can be produced by this pathway is the carbohydrate *glucose,* in which case n equals 6. This equation is therefore commonly written as:

$$6CO_2 + 6H_2O \xrightarrow[\text{Green Plant}]{\text{Light}} C_6H_{12}O_6 + 6O_2 \qquad (9\text{-}1)$$

The elucidation of the basic equation of photosynthesis illustrates how the independent contributions of many individuals may be required for an important scientific breakthrough. In this particular case the observations of individuals of differing eras (seventeenth through nineteenth centuries), nationalities (Dutch, English, Swiss, and German), and occupations (physicians, ministers, and chemists) were involved. But in spite of the magnitude of the accomplishment, it was only the first of many steps in unraveling the pathway of photosynthesis. As biochemical techniques gradually improved, it became apparent that photosynthesis is not a simple mixing together of carbon dioxide and water to produce oxygen and carbohydrate. Rather it involves a complex series of chemical reactions that are organized into two discrete pathways known as the "light" and "dark" reactions of photosynthesis.

Chloroplasts Carry Out Photosynthesis in Two Stages Called the Light and Dark Reactions

The idea that photosynthesis occurs in two discrete stages first emerged from studies reported in 1905 by F. Blackman. His work revealed that the rate of oxygen production by photosynthetic cells exposed to intense illumination and low CO_2 concentration increases with increasing temperature. The rate of oxygen production by cells maintained under conditions of low light intensity and high CO_2 concentration, on the other hand, does not vary with changing temperature. On the basis of these observations, Blackman proposed that photosynthesis involves two separate pathways: a temperature-insensitive pathway that requires light, and a temperature-sensitive process that requires CO_2. These two pathways were designated the **light** and **dark reactions** of photosynthesis respectively. The light reactions are responsible for absorbing light and producing oxygen, while the dark reactions are involved in **CO_2 fixation** (taking up carbon dioxide from the atmosphere and converting it to organic matter containing reduced carbon).

Additional support for the existence of separate light and dark reactions was provided by the discovery that photosynthesis is stimulated to a greater extent by flashing light than by continuous light. As the length of the dark period between light flashes is increased, the quantity of CO_2 incorporated per light flash rises (Figure 9-1). The most straightforward interpretation of this phenomenon is that the reactions involved in CO_2 fixation continue during the dark periods, driven by energy absorbed by the light reactions during the preceding light period.

A series of classic experiments carried out by Robert Hill in England during the 1930s provided yet

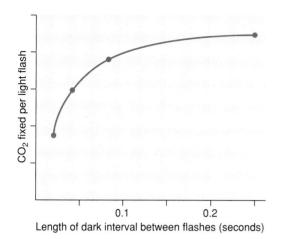

Figure 9-1 *Quantity of CO_2 Fixed by a Green Alga Exposed to Flashing Light* As the length of the dark period between light flashes is increased, the quantity of CO_2 incorporated per light flash rises. Such data suggest that CO_2 fixation continues during the dark periods.

more support for the concept of separate light and dark reactions. Hill was the first to show that isolated chloroplasts evolve oxygen when illuminated. However, because of damage inflicted by the isolation conditions employed, the chloroplasts evolved oxygen only when an artificial electron acceptor was added in place of the normal electron acceptor, CO_2. For instance, addition of the electron acceptor Fe^{3+} permits the following reaction to take place:

$$4Fe^{3+} + 2H_2O \xrightarrow{\text{Light}} 4Fe^{2+} + 4H^+ + O_2$$

This production of oxygen accompanied by the reduction of an artificial electron acceptor such as Fe^{3+} has come to be known as the *Hill reaction*. The existence of the Hill reaction indicates that the light-requiring steps of photosynthesis can occur without the simultaneous fixation of CO_2 by the dark reactions. It also reveals that water must be the source of the oxygen atoms liberated during photosynthesis because oxygen can be produced in the absence of CO_2 fixation.

Although Hill's studies revealed that chloroplasts absorb light and produce oxygen, they provided no information about the CO_2-fixing steps of photosynthesis. In the early 1950s Daniel Arnon and his colleagues showed that chloroplasts isolated under gentler conditions than those employed by Hill are capable of fixing CO_2 as well as absorbing light and producing oxygen. With such gently isolated chloroplasts, an artificial electron acceptor need not be added because carbon dioxide acts as the natural electron acceptor. Hence it was concluded that the complete pathway of photosynthesis occurs within the chloroplast.

In order to determine where in the chloroplast these reactions occur, subsequent studies were carried out in which the soluble components of the chloroplast were isolated by extracting chloroplasts with water. The remaining chloroplast membranes were found to be capable of absorbing light and producing oxygen, but not fixing CO_2; the enzymes required for CO_2 fixation appeared in the water-soluble extract. It was therefore inferred that the chloroplast membranes carry out the light reactions of photosynthesis, while the soluble interior region of the chloroplast is responsible for the dark reactions of CO_2 fixation.

The Light and Dark Reactions of Photosynthesis Are Linked by NADPH and ATP

The question of how the light and dark reactions are linked to each other was at first a puzzle. A clue to the identity of the substances that join these two sets of reactions was obtained in 1951, when Wolf Vishniac and Severo Ochoa discovered that illuminated chloroplasts reduce the coenzyme $NADP^+$ to NADPH. A few years later Arnon reported that illuminated chloroplasts also synthesize ATP. Since ATP and NADPH are both high-energy compounds, it was proposed that they function as intermediates that transfer energy from the light reactions of photosynthesis to the dark reactions.

To test this idea, Arnon added ATP and NADPH to soluble chloroplast extracts known to contain the enzymes that carry out the dark reactions. The result was a marked stimulation of CO_2 fixation. Because maximal stimulation of CO_2 fixation was found to require the presence of NADPH as well as ATP, Arnon concluded that both of these substances are involved in transferring energy from the light reactions to the dark reactions. At about the same time that Arnon was carrying out these experiments, the enzymatic steps involved in the dark reactions were being investigated by Melvin Calvin and his associates. This work led to the discovery that the CO_2-fixing dark reactions are organized into a cyclic pathway that has come to be known as the **Calvin cycle.** Most interesting was the discovery that the Calvin cycle requires both NADPH and ATP. Taken together, the preceding observations suggest that solar energy trapped by the light reactions of photosynthesis is utilized to drive the formation of NADPH and ATP, and that these high-energy compounds then drive the CO_2-fixing Calvin cycle.

The main question left unanswered by this model concerns the mechanism by which energy derived from sunlight promotes the formation of NADPH and ATP. By analogy to mitochondrial respiration, where ATP synthesis is made possible by energy released as electrons flow from reduced coenzymes to oxygen, the idea arose in the late 1950s that ATP synthesis in chloroplasts is mediated by the light-induced flow of electrons from water to the coenzyme $NADP^+$. According to this model, the absorption of light energy triggers the removal of electrons from water, producing oxygen; the transfer of these electrons to $NADP^+$ is then accompanied by the formation of ATP, generating the NADPH and ATP required by the Calvin cycle (Figure 9-2). The idea that the *light reactions produce oxygen as well as ATP and NADPH* has turned out to be basically correct, although crucial questions arise as soon as one begins to probe the details of this model: How is solar energy absorbed? What is the nature of the carriers that transfer electrons from water to $NADP^+$? How is the released energy used to drive ATP formation? How are the relevant pathways organized within the cell? Answering such questions requires us to first examine the structure of the main organelle involved in photosynthesis, the **chloroplast.**

ANATOMY OF THE CHLOROPLAST

Early Light Microscopic Studies Identified the Chloroplast as the Site of Photosynthesis

Because of their relatively large size and conspicuous green color, chloroplasts were one of the first intracellular structures to be discovered by light microscopists. Soon after the invention of the compound microscope in the seventeenth century, van Leeuwenhoek observed the presence of chloroplasts in algae and, by the year

1800, chloroplasts had been identified in the leaves of a wide variety of plant cells. Shortly thereafter it was discovered that chloroplasts rupture when placed in a hypotonic solution, suggesting that they are bounded by a semipermeable membrane.

The first insight into the biological function of chloroplasts was provided in 1894 by the German plant biologist, Thomas Engelmann. At the time it was already known that oxygen is produced during photosynthesis. To investigate the question of where photosynthesis occurs within plant cells, Engelmann utilized a special strain of bacteria known to be attracted to oxygen. Such bacteria were placed in a medium containing *Spirogyra*, a large unicellular alga that exhibits a single spiral chloroplast that winds from one end of the cell to the other (Figure 9-3). When the algal cells were illuminated with a finely focused beam of light and examined by light microscopy, the oxygen-seeking bacteria were seen to congregate near those regions of the cell where the chloroplast had been illuminated. It was therefore concluded that the light-absorbing and oxygen-producing steps of photosynthesis are housed within the chloroplast. This conclusion was an important milestone in cell biology because it was the first time a specific metabolic pathway had been assigned to a particular subcellular organelle.

Chloroplasts Contain Membranes That Define Three Separate Compartments

The realization that chloroplasts are involved in photosynthesis stimulated significant interest in the struc-

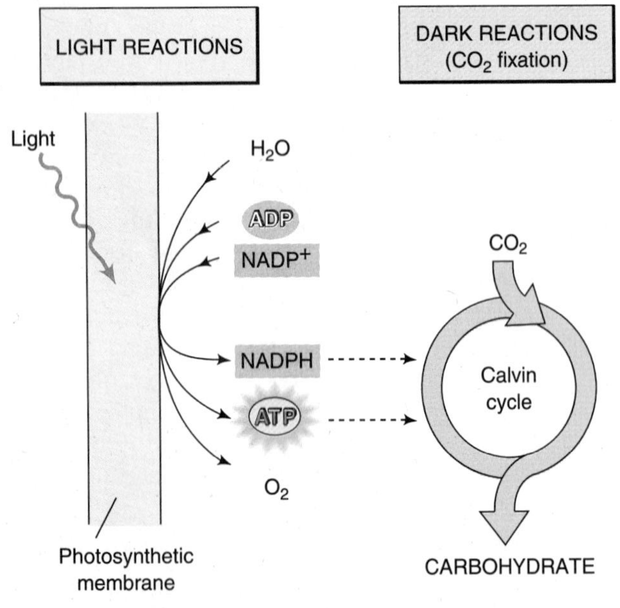

Figure 9-2 Relationship between the Light and Dark Reactions of Photosynthesis   *The ATP and NADPH generated by the light reactions are utilized to drive the energy-requiring dark reactions. The light reactions occur in chloroplast membranes of eukaryotic autotrophs and in comparable cytoplasmic membranes of prokaryotic autotrophs.*

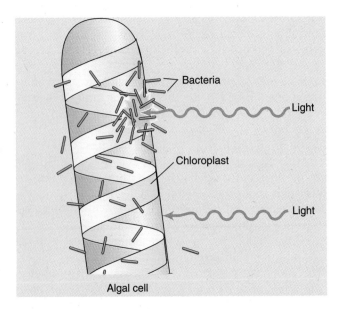

Figure 9-3 Engelmann's Experiment with Spirogyra, a Large Algal Cell Containing a Single Spiral Chloroplast   *Oxygen-seeking bacteria are attracted to the region of the cell where the chloroplast has been illuminated, indicating that the chloroplast produces oxygen.*

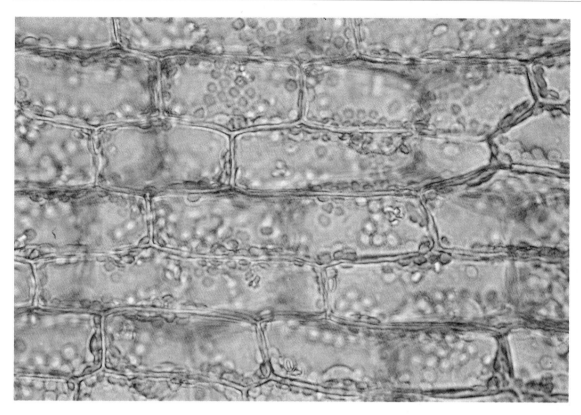

Figure 9-4 *Chloroplasts in Leaf Cells of Elodea Viewed by Light Microscopy* *Courtesy of W. J. Hayden.*

ture of this organelle. When viewed with the light microscope, chloroplasts typically appear as oval or spherical structures exhibiting a prominent green pigmentation (Figure 9-4). They occur in the leaves and other green tissues of higher plants, as well as in algae. The chloroplasts of leaf cells typically measure about 2–4 μm in diameter and up to 10 μm in length, which is roughly twice the size of a typical mitochondrial profile. The chloroplasts of algae may even be larger, and often assume unusual shapes such as spirals, cups, and circular bands that wind around the cell. In algae as few as one or two large chloroplasts are present per cell, but plant cells typically contain several dozen or more. These multiple chloroplasts appear to be largely independent of each other rather than forming an interconnected network, as often occurs in mitochondria (page 316).

Relatively little internal structure can be seen when chloroplasts are viewed with the light microscope, although early light microscopists did detect tiny green granules within the chloroplast, which they named **grana.** The complexity of chloroplast structure did not become evident, however, until thin-sectioning techniques for the electron microscope were developed in the early 1950s. As expected from the ability of hypotonic solutions to rupture chloroplasts, electron micrographs revealed that the organelle is surrounded by a membrane envelope. This **chloroplast envelope** consists of two closely apposed membranes, termed the *outer* and *inner membranes*, which are separated from each other by an *intermembrane space* (Figure 9-5). The chloroplast envelope encloses the interior space, or **stroma,** of the chloroplast, which is analogous to the mitochondrial matrix. Within the stroma is another set of membranes, called **thylakoid membranes**, which are organized as a system of flattened sacs. The internal space bounded by the thylakoid membranes is called the **thylakoid lumen.**

Thylakoid membranes are organized in two different ways, referred to as the stacked and unstacked configurations. In **stacked** or **grana thylakoids,** thylakoid sacs are stacked upon each other like a pile of coins, generating large membrane masses that correspond to the grana observed by light microscopy. Grana stacks are connected to one another by membrane-bound channels referred to as **unstacked thylakoids.** Each unstacked thylakoid is a large flattened sheet that makes connections to many or all of the individual thylakoids of a given granum stack (see Figure 9-5). In this way, the individual thylakoid lumens of each granum stack become interconnected both with one another and with the thylakoid lumens of other granum stacks. Hence the entire thylakoid membrane system of the chloroplast defines a single, enormously complex, membrane-enclosed compartment. Negative staining has revealed that the outer surfaces of thylakoid membranes are studded with spherical **CF$_1$ particles** (Figure 9-6), which resemble the F$_1$ particles observed on mitochon-

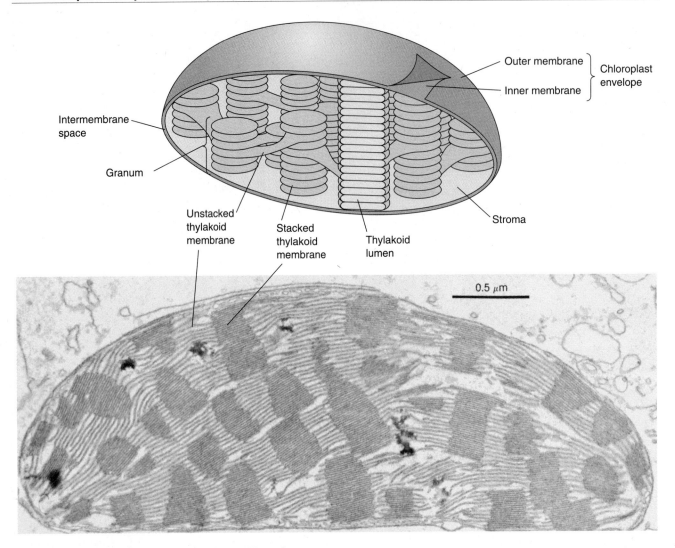

Figure 9-5 *Structural Organization of the Chloroplast* *Chloroplasts are bounded by a double-membrane envelope that encloses an inner compartment called the chloroplast stroma. Stacked and unstacked thylakoid membranes are distributed throughout the chloroplast stroma. Micrograph courtesy of J. Rosado-Alberio.*

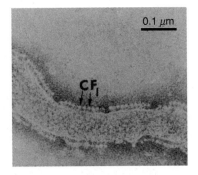

Figure 9-6 *Electron Micrograph of a Negatively Stained Thylakoid Membrane Showing CF$_1$ Particles Protruding from Its Surface* *CF$_1$ particles are the site where ATP is made during the light reactions of photosynthesis. Courtesy of M. P. Garber.*

drial cristae. We will see later in the chapter that CF$_1$ particles play a role in photosynthetic ATP formation that is comparable to the role played by mitochondrial F$_1$ particles in oxidative phosphorylation.

During chloroplast development, points of continuity are occasionally observed between thylakoid membranes and the inner membrane of the chloroplast envelope (Figure 9-7). But such connections are rarely seen in mature chloroplasts, indicating that the thylakoid membranes eventually become separated from the chloroplast envelope. The final organization of the thylakoids varies among different kinds of plant cells. Stacking of thylakoid membranes into grana is observed in the chloroplasts of most higher plant cells and in green algae, but does not occur in other kinds of algae. The chloroplasts of these other algae contain separated

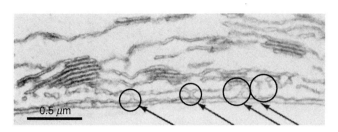

Figure 9-7 Electron Micrograph Showing Connections between the Chloroplast Envelope and Thylakoid Membranes *Points of continuity* (arrows) *between the inner membrane of the envelope and thylakoid membranes are usually observed only during the early stages of chloroplast development. Courtesy of J. Rosado-Alberio.*

thylakoids that run parallel to one another. In red algae the parallel thylakoids are completely separated (Figure 9-8), but in other algae the thylakoids may be grouped into sets of two or three. But even when they are grouped in this way, adjacent thylakoids are not connected to each other as is the case in the grana stacks of higher plants.

The anatomy of a chloroplast resembles that of a mitochondrion in several ways. Both organelles are surrounded by two membranes, and both have a complex system of internal membranes that contain protruding spherical particles involved in ATP synthesis. The major difference is that thylakoid membranes are not usually continuous with the inner membrane of the chloroplast

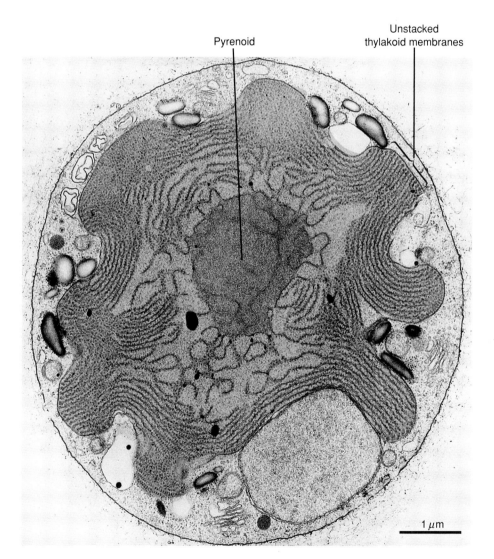

Pyrenoid

Unstacked
thylakoid membranes

Figure 9-8 Electron Micrograph Showing the Single Chloroplast of the Red Alga Porphyridium
The majority of the cell is occupied by a large chloroplast (color) *containing parallel unstacked thylakoids, but no stacked thylakoids. The large granular deposit located in the center of the chloroplast is called a pyrenoid. Courtesy of E. Gantt.*

envelope, while mitochondrial cristae *are* continuous with the inner membrane surrounding the mitochondrion. As a result (Figure 9-9), chloroplasts consist of three separate compartments (intermembrane space, stroma, and thylakoid lumen), whereas mitochondria have only two compartments (intermembrane space and matrix).

The Main Components of the Chloroplast Can Be Isolated for Biochemical Study

The development of techniques for isolating chloroplasts and their various components has played an important role in advancing our understanding of the functional organization of this organelle. Chloroplasts can be isolated from plant cells in one of several different ways. Some procedures utilize harsh homogenization techniques, such as grinding cells with sand using a mortar and pestle, to break the cell wall and release the chloroplasts into solution. Following removal of

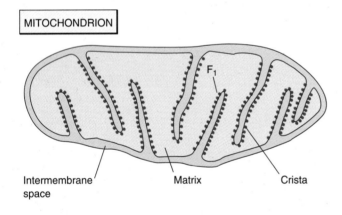

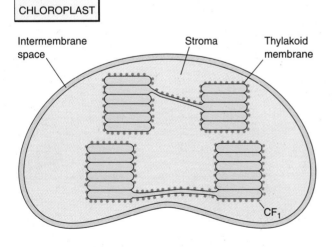

Figure 9-9 *Comparison of the Compartments Found in Mitochondria and Chloroplasts* *Chloroplasts possess three separate compartments (intermembrane space, stroma, and thylakoid lumen), whereas mitochondria have only two compartments (intermembrane space and matrix).*

nuclei by centrifugation at low speed, chloroplasts are collected by centrifuging at higher speeds. A more recent approach to chloroplast isolation employs the enzymes *cellulase* and *pectinase* to break down the cell wall. The advantage of this approach is that the resulting wall-less cells, or **protoplasts,** can be disrupted by gentler techniques prior to isolating chloroplasts by centrifugation.

Chloroplasts isolated under harsh conditions are usually capable of carrying out the light-induced production of oxygen, ATP, and NADPH, but not CO_2 fixation. Electron micrographs of such defective *class II* chloroplasts reveal the presence of little or no stroma, and broken or missing outer envelopes. In contrast, gentler disruption techniques yield completely intact *class I* chloroplasts that are capable of carrying out the complete pathway of photosynthesis, including CO_2 fixation. Class I and II chloroplast preparations have both been useful as starting material for isolating the various membranes and compartments that make up the chloroplast (Figure 9-10). In one commonly used procedure, class I chloroplasts are suspended in a hypotonic solution to rupture the chloroplast envelope, followed by isodensity centrifugation to separate the stroma, outer envelope, and thylakoids from one another. Alternatively, class II chloroplasts lacking the outer envelope and stroma can be used as starting material for separating the thylakoid membranes into stacked and unstacked thylakoids. In this approach chloroplasts are disrupted in a special apparatus, called a *French pressure cell*, that shears the thylakoid membranes by forcing them through a small orifice following rapid decompression. This shearing process breaks the connections between the stacked and unstacked thylakoids, thereby allowing the two membrane fractions to be separated from one another by differential centrifugation. Biochemical studies involving components isolated in these ways have provided a great deal of information concerning the functional organization of the chloroplast.

Chloroplasts Are Enclosed by a Relatively Permeable Outer Membrane and an Impermeable Inner Membrane

The chloroplast envelope separates the cytosol from the interior regions of the chloroplast, where photosynthesis occurs. The permeability of the chloroplast envelope has been studied by incubating chloroplasts in the presence of various solutes and then measuring the amount of solute taken up. Most small molecules enter the chloroplast rapidly, but their final concentration inside the chloroplast is less than in the surrounding medium. As was the case for comparable experiments involving mitochondria (page 320), such results indicate that only a portion of the internal volume of the chloroplast can

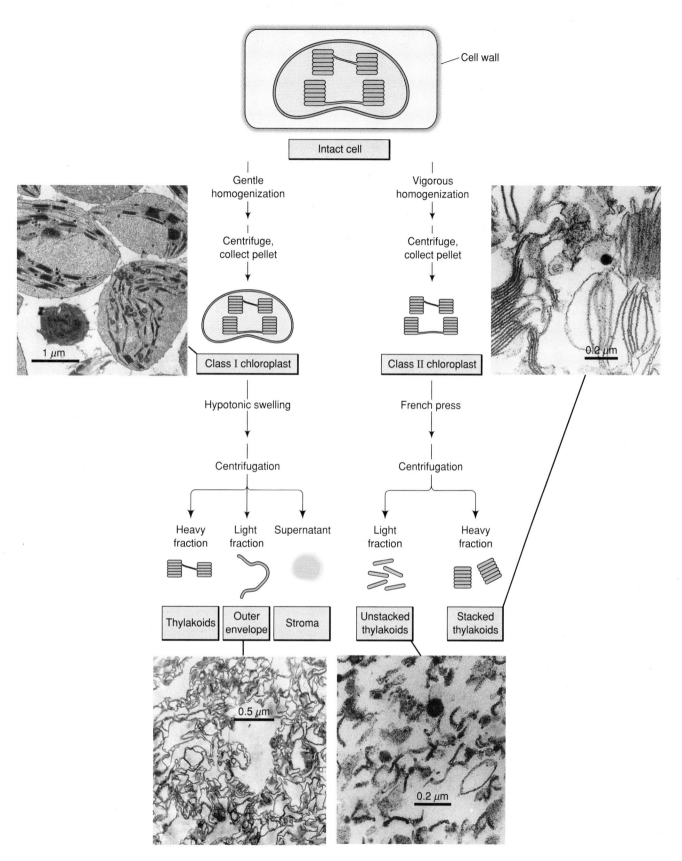

Figure 9-10 *Procedures for Isolating Chloroplasts and Their Components* *Class I chloroplasts are isolated using gentle homogenization techniques and therefore retain the chloroplast envelope; rupturing class I chloroplasts under hypotonic conditions allows thylakoid membranes, the chloroplast envelope, and the components of the stroma to be separated from one another. Class II chloroplasts lacking the outer envelope and stroma are prepared using vigorous homogenization techniques; connections between the thylakoid membranes are then ruptured in a French pressure cell, a procedure that allows stacked and unstacked thylakoids to be separated from each other. Micrographs courtesy of A. A. Benson (left) and R. B. Park (right).*

equilibrate its solutes with the external environment. The volume of this readily permeable compartment correlates with the relative size of the intermembrane space as visualized by electron microscopy, indicating that chloroplasts, like mitochondria, possess an outer membrane that is readily permeable to small molecules and an inner membrane that is not. The permeability of the outer membrane is caused by the presence of transmembrane proteins called **porins,** which permit the passage of molecules and ions smaller than about 10,000 daltons. Similar proteins occur in the outer mitochondrial membrane (page 320), as well as in the outer membrane of Gram-negative bacteria (page 251).

Whereas the outer membrane of the chloroplast envelope is usually smooth, the inner membrane frequently exhibits small folds, distinct from thylakoids, that protrude for short distances into the stroma. In some plants these infoldings create an intricate network of interconnected vesicles and tubules, called the *peripheral reticulum,* which greatly increases the surface area of the inner membrane (Figure 9-11). Studying the properties of the inner and outer membranes has been facilitated by the discovery that the two membranes can be separated from each other by isodensity centrifugation. Isolated inner membranes exhibit a much higher protein-to-lipid ratio than do isolated outer membranes (Table 9-1). This high protein content correlates with the observation that the inner membrane contains a larger number of intramembrane

protein particles when viewed by freeze-fracture electron microscopy.

Many of the proteins present in the inner membrane are enzymes involved in the synthesis of phospholipids and glycolipids. Studies on the intracellular distribution of these lipid-synthesizing enzymes have led to the conclusion that the chloroplast envelope is the principal location of lipid synthesis, not just for chloroplasts but for the entire cell as well. This situation contrasts with that of typical animal cells, where the endoplasmic reticulum is the main site of lipid synthesis. Another group of proteins found in the inner membrane of the chloroplast envelope are membrane transport proteins. Since the outer envelope membrane is freely permeable to most small molecules and ions, the inner membrane is the principal permeability barrier between the cytosol and the chloroplast stroma. Hence small molecules that need to be taken up or expelled by chloroplasts are transported across the inner membrane by specific carrier proteins.

The intermembrane space separating the inner and outer membranes of the envelope typically measures between 2 and 10 nm across. The permeability of the outer membrane makes this intermembrane space freely accessible to the small molecules of the cytosol, while the impermeability of the inner membrane prevents access by molecules present in the stroma. Little is known about the properties of the few proteins that appear to be localized within the intermembrane space.

The Chloroplast Stroma Is the Site of CO₂ Fixation

The chloroplast stroma, which is analogous to the mitochondrial matrix, exhibits a finely granular appearance in electron micrographs. Several kinds of specialized structures can be detected within the stroma. Prominent among these are *starch grains,* which store some of the carbohydrates produced by the dark reactions of photosynthesis (Figure 9-12). Also present are lipid-containing deposits, known as *plastoglobuli,* which accumulate in association with the breakdown of thylakoid membranes (e.g., in aging leaves) and decrease in size and number when new thylakoid membranes are being synthesized (e.g., in dark-adapted cells newly exposed to light). Such correlations suggest that plastoglobuli serve as reservoirs of lipid for thylakoid membrane formation. The stroma of algal chloroplasts often contains large granular deposits referred to as **pyrenoids** (see Figure 9-8). These structures consist almost entirely of crystallized *rubisco (ribulose-bisphos–phate carboxylase),* an enzyme whose role in CO₂ fixation will be discussed later in the chapter. The chloroplasts of higher plants contain related structures, termed **stroma centers,** which are comprised of

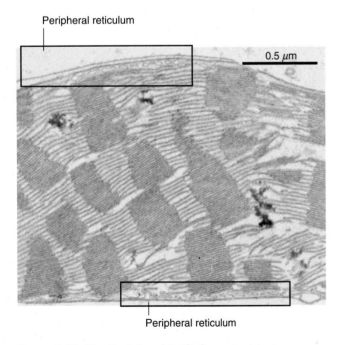

Peripheral reticulum

0.5 μm

Peripheral reticulum

Figure 9-11 *The Peripheral Reticulum* *In this electron micrograph, the peripheral reticulum appears as a network of membrane tubules and vesicles located just beneath the chloroplast envelope (see boxed areas). Courtesy of J. Rosado-Alberio.*

Table 9-1 Lipid Content of Chloroplast Membranes of Spinach Leaves

Lipid	Percent of Total Lipid		
	Outer Membrane	Inner Membrane	Thylakoids
Glycolipids			
Monogalactosyldiacylglycerol	17	55	40
Digalactosyldiacylglycerol	29	29	19
Sulfoquinovosyldiacylglycerol	6	5	5
Phospholipids			
Phosphatidylglycerol	10	9	5
Phosphatidylcholine	32	0	0
Phosphatidylinositol	5	1	1
Light-absorbing pigments			
Chlorophyll	0	0	20
Carotenoids	<1	<1	6
Quinones	0	0	3
Protein/lipid ratio	0.35	0.9	1.5

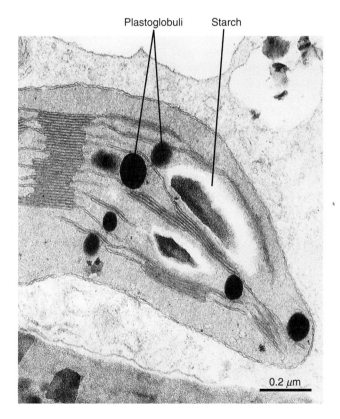

Plastoglobuli   Starch

0.2 μm

Figure 9-12 *Electron Micrograph of a Chloroplast Showing Starch Grains and Plastoglobuli* *Starch grains store carbohydrates produced by the dark reactions of photosynthesis, while plastoglobuli store lipids involved in thylakoid membrane formation. Courtesy of J. V. Possingham.*

tightly packed fibrils 8–9 nm in diameter that are also thought to contain rubisco (Figure 9-13).

Biochemical analyses of isolated stroma preparations have revealed that in addition to rubisco, the stroma contains all of the other enzymes associated with the CO_2-fixing reactions of photosynthesis. The stroma also contains DNA, RNA, ribosomes, and enzymes involved in nucleic acid and protein synthesis. Since these latter components are involved in chloroplast growth and division, we will delay their consideration until the topic of chloroplast biogenesis is covered in Chapter 14.

Thylakoid Membranes Contain Unusual Lipids and Numerous Proteins Involved in the Light Reactions

Thylakoid membranes are unique in both appearance and chemical composition. Their unique appearance stems from the presence of the CF_1 particles that protrude from their outer surfaces, and the fact that thylakoid membranes can be piled upon one another to form grana. The chemical composition of thylakoid membranes is also unusual because phospholipid, which is the main type of lipid in most cellular membranes, accounts for only 5 to 20 percent of the lipid in thylakoid membranes. This situation contrasts markedly with mitochondrial membranes, where more than 90 percent of the membrane lipid is phospholipid, or endoplasmic reticulum membranes, where about 80 percent of the lipid is phospholipid. The bulk of the lipid in thylakoid membranes is accounted for by a variety of

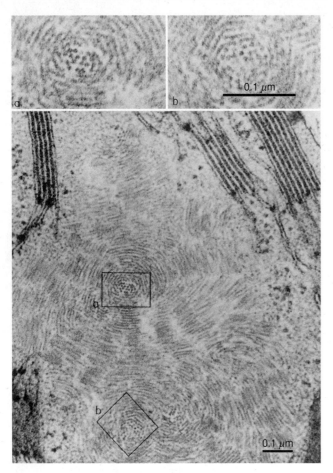

Figure 9-13 *Electron Microscopic Appearance of the Stroma Center* (Bottom) *At lower magnification the stroma center usually appears as a group of fibrils arranged in a circular pattern.* (Top) *At higher magnification the fibrils in the two boxed regions are seen to be composed of granular subunits. Courtesy of B. E. S. Gunning.*

glycolipids and light-absorbing lipid pigments such as *chlorophyll* and *carotenoids*.

In addition to their distinctive lipid makeup, thylakoid membranes are also characterized by an unusually high protein-to-lipid ratio (see Table 9-1). This high protein content reflects the presence of an extensive array of electron carriers, enzymes, and pigment-binding proteins whose role in the light reactions of photosynthesis will be discussed in the following section.

THE LIGHT REACTIONS OF PHOTOSYNTHESIS

Now that we have described the structure of the chloroplast, we are ready to investigate how this organelle captures solar energy and converts it into useful chemical forms. When thylakoid membranes are exposed to light, they produce oxygen, reduce $NADP^+$ to NADPH, and convert ADP to ATP. The ability of isolated thylakoids to carry out these **light reactions** in the ab-

sence of other chloroplast components means that thylakoid membranes are central to the process by which solar energy is captured. In capturing solar energy, thylakoid membranes utilize a group of pigments that contain electrons that become activated upon the absorption of light energy.

Chlorophyll Is the Principal Light-Absorbing Pigment in Thylakoid Membranes

The initial event in photosynthesis is the absorption of solar energy by light-absorbing pigments embedded in thylakoid membranes. These pigments can be isolated by extracting plant tissues with organic solvents to remove lipid molecules. The Russian botanist Mikhail Tswett showed in the early 1900s that the colored pigments present in such extracts can be separated from one another by passing the mixture through columns containing inert adsorbents. Tswett coined the term *chromatography* ("color-writing") for this general type of fractionation procedure, which has subsequently been adapted to the purification of many other kinds of molecules (page 146).

The main photosynthetic pigments identified in plant cell extracts are the **chlorophylls,** a family of lipid pigments that reside in thylakoid membranes and give plants their green color. Evidence that chlorophyll is involved in the light-absorbing step of photosynthesis has been provided by measuring the ability of light of varying wavelengths to stimulate photosynthesis. Such studies have revealed that blue or red light is more effective than green light in promoting photosynthesis in higher plants (Figure 9-14). This pattern matches the absorption spectrum of purified chlorophyll, which absorbs red and blue light but not green light. Since the wavelengths of light that are absorbed by chlorophyll are also the most effective in stimulating photosynthesis, it has been concluded that chlorophyll plays a central role in absorbing the light energy that drives photosynthesis.

The skeleton of the chlorophyll molecule consists of a central *porphyrin ring* to which a long hydrocarbon side chain called *phytol* is attached (Figure 9-15). The alternating double bonds in the porphyrin ring are responsible for absorbing visible light, while the hydrophobic phytol chain permits chlorophyll to be inserted into the lipid bilayer of the thylakoid membrane. Although porphyrin rings also occur in cytochromes, hemoglobin, and myoglobin, the porphyrin ring of chlorophyll is distinguished by several unique features: A *magnesium atom*, rather than an iron atom, is bound to the center of the ring; a cyclopentanone ring is fused to the basic porphyrin structure; and some of the side chains attached to the porphyrin ring differ from those occurring in other types of porphyrins.

Plants contain several types of chlorophyll that differ in their porphyrin side chains. The predominant form, designated *chlorophyll* a, occurs in all photosynthetic eu-

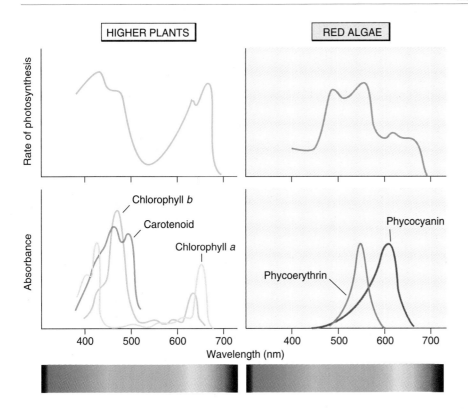

Figure 9-14 *Relationship between the Light-Absorbing Properties of Photosynthetic Pigments and the Ability of Light of Varying Wavelengths to Induce Photosynthesis* (Left) *In higher plants, maximum rates of photosynthesis are observed with those wavelengths of light that are absorbed by chlorophyll and carotenoids, suggesting that these two classes of molecules are the main light-absorbing photosynthetic pigments.* (Right) *In contrast to the situation in higher plants, red algae exhibit maximum rates of photosynthesis when illuminated with light in the 500–600-nm region of the spectrum. Phycoerythrin is the main pigment responsible for absorbing light of this wavelength.*

karyotes and in cyanobacteria. Most photosynthetic cells also contain a second type of chlorophyll (Table 9-2). In higher plants and in green algae the additional molecule is *chlorophyll b,* and in brown algae, diatoms, and dinoflagellates it is *chlorophyll c.* Slightly different forms of chlorophyll, designated *bacteriochlorophylls* and *Chlorobium chlorophyll,* are utilized by purple and green photosynthetic bacteria. The various types of chlorophyll differ in their light-absorbing properties. For example, chlorophyll *a* absorbs light maximally at 420 and 663 nm, while chlorophyll *b* absorbs most strongly at 460 and 645 nm (see Figure 9-14). When chlorophyll molecules reside within chloroplast membranes, their light-absorbing characteristics are altered slightly by interactions between chlorophyll and membrane proteins.

Carotenoids and Other Accessory Pigments Funnel Light Energy to Chlorophyll

In addition to chlorophyll, thylakoid membranes contain several *accessory pigments* that increase the effi-

ciency of photosynthesis by absorbing light in those regions of the spectrum where chlorophyll absorbs light inefficiently. The most widely occurring accessory pigments are the **carotenoids,** a family of lipid molecules constructed from long hydrocarbon chains containing alternating double bonds and ending in substituted cyclohexene rings (see Figure 9-15). Carotenoid molecules absorb light in the violet/blue-green region of the spectrum (400–500 nm), giving them brilliant yellow, orange, and red colors. Carotenoids are responsible for the colors of many vegetables, such as carrots and tomatoes, but in actively photosynthesizing tissues their presence is usually masked by the green color of the more abundant chlorophyll molecules. At the end of the growing season, chlorophyll molecules often break down before carotenoids, giving leaves their brilliant fall colors.

A special class of accessory pigments, known as **phycobilins,** occur in red algae and cyanobacteria. The color of red algae results from the presence of a red phycobilin called *phycoerythrin,* while the color of cyanobacteria is imparted by a blue phycobilin known as *phycocyanin.* Both of these lipid pigments are bound to

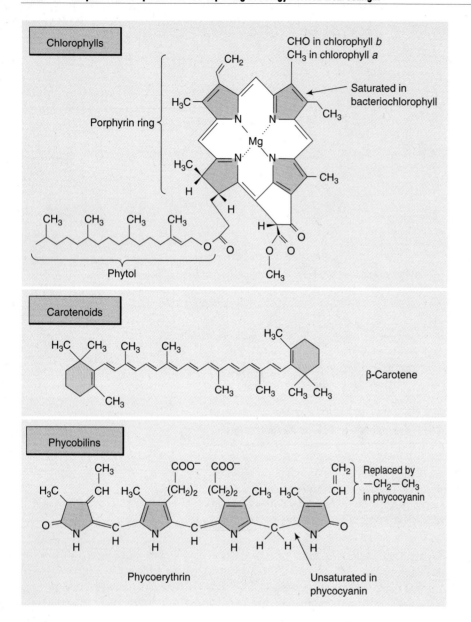

Figure 9-15 *Structures of the Main Photosynthetic Pigments*   *Note the resemblance of the four rings in phycoerythrin (and phycocyanin) to the four rings linked to form the porphyrin ring of chlorophyll.*

protein molecules, forming **phycobilisome granules** that are attached to the surface of thylakoid membranes (Figure 9-16). The phycobilin molecule is constructed from the same four rings that occur in porphyrins, but in phycobilins the rings are linked to form a linear chain rather than a circular structure (see Figure 9-15).

Although accessory pigments increase the amount of light that can be absorbed by photosynthetic cells, they do not substitute for chlorophyll because the light energy absorbed by accessory pigments must be transferred to chlorophyll *a* before it can enter the photosynthetic pathway. This qualification in no way minimizes the importance of the accessory pigments for photosynthesis. In red algae and cyanobacteria, maximum rates of

photosynthesis occur when cells are illuminated with light whose wavelength corresponds to that absorbed by the accessory pigments rather than chlorophyll (see Figure 9-14). In such cases the accessory pigments are more important than chlorophyll in the initial light-absorbing step, even though the absorbed energy is subsequently channeled to chlorophyll.

Photosystems Are Formed from a Mixture of Chlorophylls, Carotenoids, and Proteins

In addition to interacting with accessory pigments, chlorophyll molecules also interact with one another during the light absorbing-step of photosynthesis. This

Table 9-2   Photosynthetic Pigments Found in Different Organisms

	Eukaryotes				Prokaryotes		
	Higher plants	Green algae	Diatoms, brown algae	Red algae	Cyanobacteria	Purple bacteria	Green bacteria
Chlorophylls							
Chlorophyll *a*	+	+	+	+	+		
Chlorophyll *b*	+	+					
Chlorophyll *c*			+				
Bacteriochlorophyll *a*						+	+
Bacteriochlorophyll *b*						+	
Chlorobium chlorophyll							+
Carotenoids	+	+	+	+	+	+	+
Phycobilins				+	+		

Figure 9-16 *Thylakoid Membranes Showing Attached Phycobilisomes* *In this high-magnification electron micrograph of a chloroplast from a red alga, numerous phycobilisomes are seen attached to the unstacked thylakoids. Courtesy of E. Gantt.*

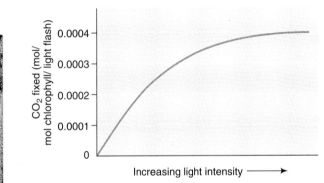

Figure 9-17 *The Effect of Increasing Amounts of Light on the Rate of CO_2 Fixation by Algae* *At maximum rates of CO_2 fixation, about 0.0004 moles of CO_2 are fixed per mole of chlorophyll present. Therefore the fixation of one molecule of CO_2 requires several thousand chlorophyll molecules.*

conclusion first emerged in the early 1930s from experiments in which Robert Emerson and William Arnold exposed algae to brief flashes of light. They expected that light of sufficient intensity to drive photosynthesis at its maximum rate would cause one molecule of carbon dioxide to be fixed per molecule of chlorophyll per light flash. The data revealed, however, that a maximum of about 0.0004 moles of CO_2 are fixed per mole of chlorophyll present (Figure 9-17). This number indicates that the fixation of one molecule of CO_2 requires roughly $1 \div 0.0004 = 2500$ molecules of chlorophyll.

Although these experiments showed that thousands of chlorophyll molecules are required for the fixation of each molecule of CO_2, they did not reveal how much light is absorbed during the process. Light is composed of a stream of particles referred to as **photons**. Cells that are photosynthesizing at maximal efficiency have been found to absorb *about eight photons of light per molecule of CO_2 that is fixed.* Since roughly 2500 chlorophyll molecules are required per molecule of CO_2 fixed, each of the eight photons required for the fixation of a single molecule of CO_2 must have a target of $2500 \div 8 \approx 300$ chlorophyll molecules. In other words, chlorophyll is organized into functional units of several hundred chlorophyll molecules that work together to absorb a photon of light.

Subsequent research has revealed that these functional units, now called **photosystems,** are constructed from a combination of chlorophylls, carotenoids, other lipids, and proteins. Each photosystem consists of two main components: a *light-harvesting complex* and a *reaction-center complex.* The **light-harvesting complex** typically contains several hundred chlorophyll molecules and a variable amount of carotenoid all linked together by protein. When a photon of light is absorbed by a chlorophyll or carotenoid molecule located in the light-harvesting complex, one of its electrons is activated. This excited state is transferred from pigment to pigment until it reaches a special pair of chlorophyll *a* molecules located in the reaction-center complex. The **reaction-center complex** consists of several polypeptide chains associated with chlorophyll *a* and several other lipid molecules that serve as electron donors and acceptors. The bulk of the light-absorbing pigments within a given photosystem thus function as "antenna" pigments that funnel their activation energy to a pair of chlorophyll *a* molecules located in the reaction-center complex (Figure 9-18). This chlorophyll *a* in turn passes its excited electron to a chain of electron carriers called the *photosynthetic electron transfer chain.*

The Photosynthetic Electron Transfer Chain Is Analogous to the Mitochondrial Respiratory Chain

The **photosynthetic electron transfer chain** is a series of electron carriers whose role in photosynthesis is analogous to the role of the mitochondrial respiratory

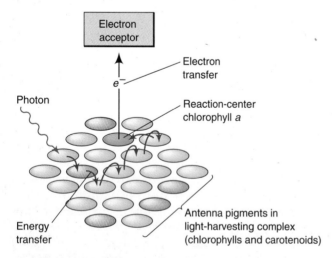

Figure 9-18 *The Relationship between Antenna Pigments and the Reaction-Center Chlorophyll* *Photons of light are initially absorbed by antenna pigments in the light-harvesting complex associated with each photosystem. The antenna pigments funnel the absorbed energy to the reaction-center chlorophyll* a, *activating an electron that is passed to an electron acceptor in the photosynthetic electron transfer chain.*

chain in cellular respiration. It will be recalled that the respiratory chain in the mitochondrial cristae transfers electrons from NADH and $FADH_2$ to oxygen, releasing energy that is used to drive ATP formation. In a similar fashion, the photosynthetic electron transfer chain in the thylakoid membranes carries electrons from water to $NADP^+$. The main difference is that the transfer of electrons from water to $NADP^+$ by thylakoid membranes is a thermodynamically unfavorable process that consumes energy rather than releasing it. Hence the function of the photosystems is to absorb light energy and use it to drive this energetically uphill transfer of electrons from water to $NADP^+$.

Like the mitochondrial respiratory chain, the photosynthetic electron transfer chain utilizes *cytochromes, iron-sulfur proteins, flavoproteins,* and *quinones.* Because the basic properties of these classes of molecules were presented in the preceding chapter (page 327), we will simply list the principal components that participate in photosynthetic electron transfer.

1. **Cytochromes.** Several cytochromes unique to plant cells are involved in photosynthetic electron transfer. Those whose roles have been most clearly defined are *cytochromes b_6* and *f,* which absorb light at 563 and 553 nm, respectively.

2. **Ferredoxin.** In the late 1950s an iron-sulfur protein isolated from plant leaves was found to enhance the rate of $NADP^+$ reduction when added to isolated chloroplasts. This protein, termed *ferredoxin,* contains a single oxidation-reduction site composed of two iron atoms bound to two sulfur atoms. Like other iron-sulfur proteins, ferredoxin transfers electrons by cycling between the Fe^{3+} and Fe^{2+} states. In its oxidized state, ferredoxin exhibits strong absorption bands at 420 and 463 nm.

3. **$NADP^+$ reductase.** The main flavoprotein taking part in photosynthetic electron transfer is $NADP^+$ reductase, an FAD-containing enzyme that catalyzes the transfer of electrons from ferredoxin to $NADP^+$.

4. **Plastocyanin.** The photosynthetic electron transfer chain also includes a copper-containing protein called *plastocyanin.* In its oxidized state this protein absorbs light at a wavelength of 600 nm, giving the molecule a blue color. Upon reduction the absorption band and the blue color disappear. The copper that is bound to plastocyanin transfers electrons by cycling between the Cu^{2+} and Cu^+ states.

5. **Plastoquinone.** The participation of lipid molecules in the photosynthetic electron transfer chain was first suggested by the observation that extraction of chloroplasts with organic solvents inhibits the Hill reaction. The essential lipid was later identified as *plastoquinone,* a small molecule whose structure and absorption spectrum closely resembles that of

ubiquinone. Like ubiquinone, plastoquinone transfers electrons by cycling between its quinone and quinol forms (see Figure 8-21).

6. **NADP⁺.** The final electron acceptor in the photosynthetic electron transfer chain is $NADP^+$, a derivative of NAD^+ that contains an additional phosphate group (see Figure 8-1). $NADP^+$ accepts two electrons during its reduction to NADPH, just as NAD^+ accepts two electrons when being reduced to NADH.

Two Photosystems Are Involved in the Photosynthetic Electron Transfer Chain of Chloroplasts

The preceding types of electron carriers are arranged into a chain that is designed to transfer electrons from water to $NADP^+$. Observations made by Robert Emerson in the early 1940s first suggested that photons of light are absorbed at more than one point in this chain. Emerson's studies were triggered by the observation that oxygen production declines dramatically when chloroplasts are illuminated with light in the far-red region of the spectrum, even though chloroplast pigments still absorb light of these wavelengths. Emerson discovered that the decrease in photosynthetic activity at long wavelengths, termed the *red drop*, can be largely overcome by simultaneous illumination with a second beam of light of a shorter wavelength (Figure 9-19). Moreover, when chloroplasts are simultaneously illuminated with 650- and 700-nm light beams, oxygen is produced much faster than when each beam is used separately and the two rates of oxygen production are added together.

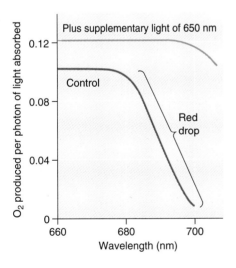

Figure 9-19 *The Red-Drop Effect* *When photosynthetic cells are illuminated with light of varying wavelengths, photosynthetic activity falls sharply for wavelengths above 680 nm, even though photosynthetic pigments absorb light at these longer wavelengths. The drop in photosynthetic activity that occurs between 680 and 700 nm can be overcome by supplementing the light with a second light beam of 650 nm.*

These findings led to the conclusion that photosynthesis involves two light-absorbing systems that absorb light energy of different wavelengths. The light-absorbing system that is more efficient at longer wavelengths is called **photosystem I**, while the system active at shorter wavelengths is referred to as **photosystem II**. The reaction-center chlorophyll of photosystem I is called **P700** because it absorbs light maximally at 700 nm; similarly, the reaction-center chlorophyll of photosystem II is designated **P680** because it absorbs light maximally at 680 nm. The difference in the absorption maxima of the reaction-center chlorophylls of the two photosystems is caused by interactions with surrounding proteins rather than by any inherent differences in the chlorophyll molecules themselves.

Shortly after the existence of two separate photosystems was first suggested by Emerson's experiments, Robert Hill and Fay Bendall proposed that the two photosystems are arranged in a linear sequence rather than as two independent light-absorbing pathways. According to this model, the absorption of light by one photosystem triggers an initial activation of a chlorophyll electron, followed by a further activation of the electron by light energy absorbed by the other photosystem. This proposal was supported by simple calculations that indicated that an electron derived from water requires an energy input equivalent to at least two photons of light in order to achieve an activated state whose free energy is sufficient to reduce $NADP^+$ to NADPH.

The Photosynthetic Electron Transfer Chain Is Organized into Several Protein Complexes That Transfer Electrons from Water to NADP⁺

Now that the photosystems and electron carriers that make up the photosynthetic electron transfer chain have been introduced, we are ready to address the question of how these components are organized in the thylakoid membrane. Many of the experimental approaches employed to elucidate the sequence of the mitochondrial respiratory chain have also been used to investigate the sequence of carriers in the photosynthetic electron transfer chain. These approaches, which were described in Chapter 8, include the following: (1) determining the standard redox potentials of isolated carriers, (2) testing the ability of isolated carriers to oxidize and reduce one another, (3) examining light-induced changes in difference spectra, (4) using specific inhibitors to block electron flow at particular points, (5) utilizing artificial electron donors and acceptors to study electron flow through various regions of the chain, and (6) analyzing genetic mutants defective in particular regions of the photosynthetic chain.

Data obtained from the preceding approaches have led to the model of photosynthetic electron transfer shown in Figure 9-20. Because this diagram resembles a

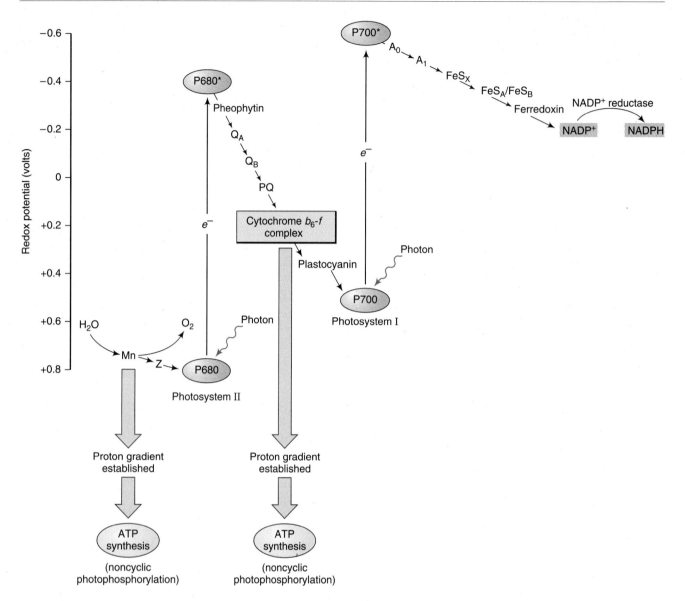

Figure 9-20 *Sequence of Electron Flow through the Photosynthetic Electron Transfer Chain* *Absorption of light energy by the two photosystem reaction centers (P680 and P700) elevates the oxidation-reduction potential, allowing electrons to be transferred from a low free-energy level in H_2O to a higher free-energy level in NADPH. P680* and P700* refer to the excited states that are induced when the absorption of light excites an electron associated with either the P680 or P700 reaction-center chlorophyll a. Q_A and Q_B are plastoquinone molecules bound to photosystem II, while PQ refers to the Q_B plastoquinone molecule after it has diffused out of the photosystem II complex. In two regions of the photosynthetic electron transfer chain, a proton gradient is established that drives ATP formation. Because of the zigzag or Z-shaped pattern of this diagram, it is sometimes referred to as the Z scheme.*

sideways letter Z, it is often called the **Z scheme**. At two points in the Z scheme, a photon of light is absorbed to increase the free-energy level of the electron being transferred. Although the Z scheme is illustrated as a sequence of separate electron carriers, most of its components are arranged in thylakoid membranes as constituents of three protein complexes: *photosystem II*, the *cytochrome b$_6$-f complex*, and *photosystem I*. The existence of these complexes allows us to describe

photosynthetic electron transfer as a process involving four distinct stages.

1) Photosystem II Splits Water and Transfers Its Electrons to Plastoquinone

In spite of its name, **photosystem II** is actually involved in the first stage of photosynthetic electron transfer (the names *photosystem I* and *photosystem II* were assigned chronologically before the functions of the two photosystems had been determined). The reaction center of photosystem II contains at least six integral membrane polypeptides and several tightly bound lipids, including the reaction-center P680 chlorophyll. Also associated with the reaction center are three peripheral membrane proteins and several ions, including four atoms of manganese (Mn). The light-harvesting complex of photosystem II, called **LHCII**, contains several dozen chlorophyll *a* and chlorophyll *b* molecules associated with several carotenoids and polypeptide chains.

Photosystem II transfers electrons from water to the electron carrier *plastoquinone*, liberating oxygen along the way. Energy obtained from photons of light is needed to drive this process because the transfer of electrons from water to plastoquinone cannot occur without an input of energy. During photosynthesis, the light-harvesting complex associated with photosystem II absorbs photons of light and funnels this energy to the reaction-center P680, which consists of a special pair of chlorophyll *a* molecules. This causes an electron associated with a P680 chlorophyll molecule to enter an excited state (P680*). The excited electron is then transferred to a nearby molecule of **pheophytin**, which is a modified molecule of chlorophyll *a* in which two hydrogen atoms replace the central Mg^{2+}. The pheophytin transfers the electron to a plastoquinone molecule called Q_A, which passes it to another plastoquinone called Q_B. After Q_B. has received two electrons (and simultaneously picked up two protons from solution), it diffuses out of the photosystem II complex and is replaced by another, unreduced plastoquinone.

But what happens to the P680 whose electron has been transferred to plastoquinone by this process? The loss of an electron from P680 converts it to a positively charged ion, *P680+*, whose missing electron is then replaced by an electron derived from the splitting of water:

$$2H_2O \rightarrow O_2 + 4H^+ + 4e^-$$

Note, however, that while the water-splitting reaction releases four electrons simultaneously, P680+ can accept only one electron at a time. A clue as to how this discrepancy is overcome has been provided by the work of Pierre Joliot and Bessel Kok, who measured the amount of oxygen produced by photosynthetic membranes after exposure to brief flashes of light. The results indicated that oxygen production reaches a peak after every fourth flash of light. The existence of this pattern led to the proposal that photosystem II splits water using a four-step clocklike mechanism called the **water-oxidizing clock**. The water-oxidizing clock is a cyclic mechanism involving four light-induced electron transfer steps, each transferring a single electron from

the clock to P680+ (Figure 9-21). After four electrons have been transferred from the clock to P680+, the clock replaces its missing electrons by removing four electrons from two molecules of water, releasing oxygen and returning the clock to its initial state. Current evidence suggests that manganese ions, which can assume several different oxidation states from +2 to +7, are a central component of the clock, and that they undergo successive changes in oxidation state that correspond to changes in the clock. During each clock stage a manganese ion transfers an electron to an electron carrier called Z, which then transfers the electron to P680+. Z has recently been identified as a tyrosine residue in one of the polypeptide chains that make up the photosystem II reaction center.

On the basis of the preceding information, the path followed by electrons through photosystem II can be summarized as follows:

$$H_2O \rightarrow Mn \rightarrow Z \rightarrow P680 \rightarrow Pheophytin \rightarrow Q_A \rightarrow Q_B$$

2) The Cytochrome b_6-f Complex Transfers Electrons from Plastoquinone to Plastocyanin

After photosystem II has carried out the light-induced transfer of electrons from water to Q_B, the reduced Q_B molecule leaves photosystem II and diffuses through the lipid bilayer as a free molecule of reduced plastoquinone (PQ), eventually delivering its electrons to the cytochrome b_6-f complex in the membrane. The **cytochrome b_6-f complex** consists of four polypeptide subunits, three of which have iron atoms that transfer electrons by cycling between the Fe^{3+} and Fe^{2+} states. The three iron-containing proteins are cytochrome b_6, cytochrome f, and an iron-sulfur protein. The role of the cytochrome b_6-f complex in photosynthetic electron transfer is to pass electrons from photosystem II to photosystem I. It accomplishes this task by accepting electrons from reduced plastoquinone and passing them to *plastocyanin*, a copper-containing protein located in the thylakoid lumen; plastocyanin in turn passes electrons to photosystem I.

3) Photosystem I Transfers Electrons from Plastocyanin to Ferredoxin

The reaction center of **photosystem I** consists of at least 11 integral membrane polypeptides and several kinds of tightly bound lipid, including the reaction-center P700 chlorophyll. Associated with the reaction center is a light-harvesting complex called **LHCI**, which contains about a hundred molecules of chlorophyll *a* and *b* plus accessory pigments bound to protein. The

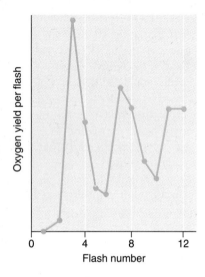

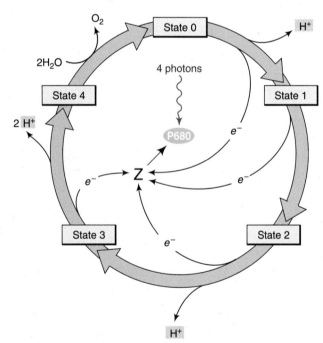

Figure 9-21 *The Water-Oxidizing Clock* (Top) *Oxygen production by photosynthetic membranes exposed to brief flashes of light exhibits a four-step periodicity; that is, oxygen production reaches a maximum value after every four flashes of light. (Bottom) This four-step periodicity is explained by a model called the water-oxidizing clock, which postulates that the oxidation of water by photosystem II involves four light-induced electron transfer steps, each transferring a single electron from the clock to P680⁺. After the four electrons are transferred to P680⁺, the clock replaces its missing electrons by removing four electrons from two molecules of water, releasing oxygen and returning the clock to its initial state.*

function of photosystem I is to transfer electrons from plastocyanin to ferredoxin, a small protein that is loosely associated with the stromal surface of the thylakoid membranes. Because the transfer of electrons from plastocyanin to *ferredoxin* is thermodynamically unfavorable, energy derived from light is needed to drive the process.

The required photons of light are absorbed by the light-harvesting complex of photosystem I, which funnels the absorbed energy to a special pair of chlorophyll *a* molecules that make up the P700 reaction center. As a consequence, an electron associated with one of the P700 chlorophyll molecules enters an excited state and is passed to a series of protein-bound electron carriers called A_0, A_1, FeS_X, FeS_A, and FeS_B. A_0 is a molecule of chlorophyll *a*, A_1 is a quinone called *phylloquinone* (or *vitamin K_1*), and FeS_X, FeS_A, and FeS_B are iron-sulfur centers. From the last iron-sulfur center in this sequence, the electron is transferred to ferredoxin. Meanwhile P700, which was the original source of the excited electron being passed to ferredoxin, receives a replacement electron donated by plastocyanin.

From the preceding information, the path followed by electrons through photosystem I can be summarized as follows:

$$\text{Plastocyanin}(Cu^{2+}) \rightarrow P700 \rightarrow$$
$$A_0 \rightarrow A_1 \rightarrow FeS_X \rightarrow FeS_A/FeS_B \rightarrow \text{Ferredoxin}$$

4) Electrons Are Transferred from Ferredoxin to NADP⁺ by the Enzyme NADP⁺ Reductase

The electrons that ferredoxin receives from photosystem I are usually passed on to $NADP^+$ in a reaction catalyzed by *$NADP^+$ reductase*. Because the reduction of $NADP^+$ to NADPH involves two electrons, two molecules of reduced ferredoxin are required to effect the reduction of $NADP^+$.

ATP Is Synthesized by Both Noncyclic and Cyclic Photophosphorylation

We have now seen how the components of the photosynthetic electron transfer chain work together to transfer electrons from water to $NADP^+$. Although the entire light-driven pathway of electron flow from water to $NADP^+$ is collectively referred to as the "light reactions" of photosynthesis, light is only involved in the light-absorbing steps mediated by photosystems II and I. The absorption of photons of light by these two photosystems provides the energy that drives the transfer of electrons from a relatively low free energy level in water to a much higher free energy level in NADPH.

In addition to capturing energy in the form of NADPH, the light-induced flow of electrons through the photosynthetic electron transfer chain is also used to drive the formation of ATP. This type of ATP synthesis, known as **photophosphorylation**, is analogous to mitochondrial oxidative phosphorylation in that both are driven by an electrochemical proton gradient that is generated as electrons flow through a series of carriers and protons are pumped across a membrane. The main difference is that the energy that drives electron flow through the mitochondrial respiratory chain is derived from the oxidation of energy-rich food molecules, whereas the energy that drives electron flow through the photosynthetic electron transfer chain is derived from photons of light.

Photophosphorylation occurs in two different ways. In the more prevalent type of photophosphorylation, ATP formation is coupled to the flow of electrons from water to $NADP^+$ through the photosynthetic electron transfer chain. As electrons flow through the chain, protons are pumped across the thylakoid membrane; the resulting electrochemical proton gradient then drives ATP formation. Because this process links ATP synthesis to the one-way flow of electrons from water to $NADP^+$, it is called **noncyclic photophosphorylation**.

The alternative pathway for light-induced ATP formation, termed **cyclic photophosphorylation**, involves a circular pathway of electron flow centered around photosystem I. In cyclic photophosphorylation, the light-induced transfer of electrons from photosystem I to ferredoxin occurs as usual. But instead of being employed to reduce $NADP^+$, these electrons are transferred from ferredoxin to the cytochrome b_6-f complex. From there they are returned via plastocyanin to photosystem I, creating a circular pathway of electron flow (Figure 9-22). During electron transfer through the cytochrome b_6-f segment of the pathway, protons are pumped across the thylakoid membrane via a mechanism that involves the reduction and oxidation of plastoquinone; the resulting proton gradient then drives ATP formation. Cyclic photophosphorylation typically accounts for 10 to 20 percent of the total ATP produced by illuminated chloroplasts, although its contribution can be larger when insufficient $NADP^+$ is available to serve as a final electron acceptor.

The main difference between cyclic and noncyclic photophosphorylation is that ATP is the only product of cyclic photophosphorylation, whereas ATP, NADPH, and oxygen are all produced during noncyclic photophosphorylation. The existence of these alternative pathways gives photosynthetic cells flexibility in the way they utilize energy derived from photons of light. When cells require relatively large amounts of ATP, cyclic photophosphorylation is stimulated because it allows ATP to be synthesized as the sole product of light-induced electron transfer. Alternatively, when cells need

to fix CO_2 and convert it to carbohydrates, noncyclic photophosphorylation is favored because it generates NADPH as well as ATP. As we will see shortly, the combination of NADPH and ATP generated by the noncyclic pathway is employed in the dark reactions of photosynthesis to fix CO_2 and reduce it to carbohydrate.

The Photosynthetic Electron Transfer Chain Contains Two Coupling Sites for Photophosphorylation

The conclusion that ATP formation accompanies both the noncyclic and cyclic flow of electrons through the photosynthetic electron transfer chain raises the ques-

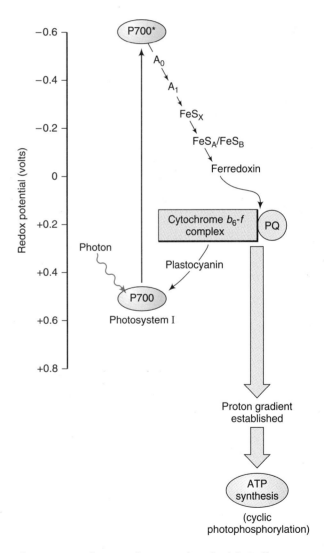

Figure 9-22 *Electron Flow Associated with Cyclic Photophosphorylation* As an alternative to passing its electrons to $NADP^+$ (Figure 9-20), reduced ferredoxin can pass its electrons back to photosystem I. This pathway of cyclic electron flow allows ATP to be made without the net transfer of electrons from water to $NADP^+$.

tion of the number and location of the coupling sites for ATP synthesis. As we learned in Chapter 8, a **coupling site** corresponds to a region in an electron transfer chain where sufficient energy is released to pump protons across a membrane and hence establish a proton gradient that can be used to drive ATP formation.

Evidence for the existence of two coupling sites in photosynthetic electron transfer was first obtained in the laboratories of Achim Trebst and Norman Good, who employed artificial electron donors to probe the ATP-synthesizing capabilities of individual regions of the photosynthetic electron transfer chain. When chloroplasts were incubated under conditions in which water, the normal electron donor, served as a source of electrons for the photosynthetic electron transfer chain, the P/O ratio was found to be about 1.2; that is, about 1.2 molecules of ATP were synthesized per atom of oxygen produced. In contrast, using reduced *dichlorophenolindophenol (DCPIP)* as an electron donor caused the P/O ratio to fall to about 0.6. This halving of the ATP yield suggests that the photosynthetic electron transfer chain contains at least two coupling sites for ATP synthesis, one located before and one located after the site where DCPIP donates electrons to the chain. Since DCPIP donates electrons to plastoquinone, one of the coupling sites must be situated prior to plastoquinone.

Localization of this coupling site was facilitated by the use of artificial electron donors such as potassium iodide (KI) and benzidine, which can substitute for water as a source of electrons feeding the electron transfer chain (Figure 9-23). As with water, electrons derived from these artificial electron donors flow through the entire sequence of carriers and are eventually employed to reduce NADP⁺ to NADPH. In terms of the amount of ATP synthesis that occurs, however, artificial electron donors do not always behave the same as water. When potassium iodide is substituted for water the P/O ratio declines to roughly half its normal value, indicating that water itself must participate in one of the coupling sites for ATP synthesis. An insight into the role played by water has been provided by the observation that, unlike potassium iodide, benzidine produces the same amount of ATP as when water is the electron donor. Benzidine resembles water in that it releases protons upon transferring its electrons to the photosynthetic electron transfer chain, but potassium iodide does not. Taken together, the preceding observations suggest that protons released during the splitting of water play an important role in the first coupling site for photophosphorylation. Given the well-established role of proton gradients in mitochondrial oxidative phosphorylation (see Chapter 8), it is not surprising to find that the release of protons plays a role in the first coupling site of photophosphorylation.

Experiments using other electron donors and acceptors have indicated that plastoquinone, which normally transfers electrons from photosystem II to the cytochrome b_6-f complex, is involved in the second site of photophosphorylation. Since the reduction and oxidation of plastoquinone is accompanied by the uptake and release of protons, plastoquinone is believed to pick up protons from one side of the thylakoid membrane and release them on the opposite side, just as ubiquinone is thought to transport protons across the inner mitochondrial membrane (page 338).

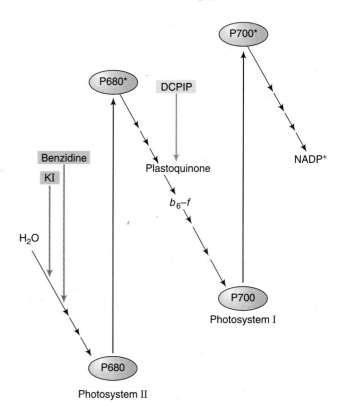

Electron donor	P/O ratio
DCPIP	0.6
KI	0.6
Benzidne	1.2

Figure 9-23 *Experimental Data Supporting the Conclusion That Photophosphorylation Involves Two Coupling Sites* *When DCPIP is employed as an artificial electron donor, the ATP yield is half of that observed when benzidine is used as the electron donor. This halving of the ATP yield suggests that the photosynthetic electron transfer chain contains at least two coupling sites for ATP synthesis, one located before and one located after the site where DCPIP donates electrons to the chain. Potassium iodide (KI) donates electrons to the chain at the same point as benzidine but produces only half as much ATP because KI, unlike benzidine, does not release protons upon oxidation. Hence with KI, ATP synthesis does not occur at the first coupling site because protons cannot be pumped there.*

Membrane Reconstitution Studies Have Revealed That the CF₁-CF₀ Complex Is an ATP Synthase

So far we have described the sequence of the photosynthetic electron transfer chain and have identified two coupling sites for ATP formation. But how is the synthesis of ATP carried out? Identification of the component responsible for ATP synthesis was first made possible by membrane reconstitution experiments whose results paralleled those obtained from the reconstitution of mitochondrial inner membranes (page 333). Such experiments, which were pioneered in the laboratory of Efraim Racker, begin with isolated thylakoid membrane vesicles that contain **CF₁ particles** (which resemble the F_1 particles of mitochondrial inner membranes) bound to their outer surface. In the presence of $NADP^+$, ADP, and inorganic phosphate, isolated thylakoid vesicles carry out the light-induced transfer of electrons from water to $NADP^+$ accompanied by the formation of ATP.

When the CF₁ particles are removed from these thylakoid vesicles, the CF₁-depleted membranes continue to carry out the light-induced transfer of electrons from water to $NADP^+$, but the ability to synthesize ATP is lost (Figure 9-24). In contrast, the isolated CF₁ particles carry out neither electron transfer nor ATP synthesis; they do, however, catalyze the hydrolysis of ATP, which is of course the reverse of ATP synthesis. Since enzymes catalyze chemical reactions in both directions, such observations suggest that CF₁ particles normally catalyze ATP synthesis when they are attached to the thylakoid membrane, but they can no longer do so when removed from the membrane because the photosynthetic electron transfer chain must be present to provide the energy that drives the ATP-forming reaction.

This hypothesis has been confirmed by adding isolated CF₁- particles back to CF₁-depleted thylakoid membranes. Under such conditions, the CF₁ particles reattach to the thylakoid membranes and the membranes regain the ability to couple ATP synthesis to the light-induced flow of electrons from water to $NADP^+$. It has therefore been concluded that the CF₁ particles are the site of ATP formation. The similarity of this organization to that of the mitochondrial inner membrane is quite striking.

The resemblance to mitochondria has been further reinforced by the discovery that CF₁ is normally bound to **CF₀**, a group of hydrophobic proteins that span the thylakoid membrane and anchor CF₁ to the membrane surface. The properties of the **CF₁-CF₀ complex** closely parallel those of the mitochondrial F_1-F_0 complex (page 335). Because the ability of reconstituted thylakoid membranes to carry out ATP synthesis depends on the presence of both CF₁ and CF₀, the CF₁-CF₀ complex is referred to as an **ATP synthase.**

An Electrochemical Proton Gradient across the Thylakoid Membrane Drives ATP Formation

How is the energy that is released during photosynthetic electron transfer used to promote the synthesis of ATP by the CF₁-CF₀ complex? In Chapter 8 we described Peter Mitchell's **chemiosmotic coupling** theory, which states that electrochemical proton gradients link the formation of ATP to electron transfer reactions. In that context, we examined evidence supporting the idea that proton gradients are involved in the synthesis of ATP by mitochondrial inner membranes and bacterial plasma membranes (pages 335 and 346). We will now discuss some of the evidence that supports the idea that ATP synthesis by thylakoid membranes also utilizes an electrochemical proton gradient.

1) Protons Are Pumped into the Thylakoid Lumen During Photosynthetic Electron Transfer

Shortly after Mitchell suggested that electrochemical proton gradients drive ATP synthesis coupled to electron transfer pathways, studies carried out by André Jagendorf supported the proposed role of proton gradients in photophosphorylation. In these experiments, chloroplasts were first placed in a medium in which ATP synthesis could not occur because ADP and inorganic phosphate were lacking. When such chloroplasts were illuminated with bright light, the pH of the external solution was found to increase dramatically. Since an increase in pH reflects a decline in proton concentration, it was concluded that protons are pumped into the thylakoid lumen during the light reactions of photosynthesis. The participation of the resulting proton gradient in ATP synthesis was suggested by a subsequent experiment in which ADP and inorganic phosphate were added back to chloroplasts after the proton gradient had been established and the light had been turned off. Under these conditions, ATP synthesis occurred as the pH of the medium fell back to normal levels (Figure 9-25). It was therefore concluded that ATP synthesis takes place as protons flow down their electrochemical gradient—that is, from the thylakoid lumen into the chloroplast stroma.

2) Artificially Created pH Gradients Can Drive ATP Synthesis in the Absence of Light

Although the preceding observations demonstrated that a proton gradient is established in illuminated chloroplasts and dissipated in chloroplasts synthesizing ATP in the dark, such data do not prove that the proton gradient actually drives ATP formation. To directly test this idea, Jagendorf created an artificial pH gradient by soaking chloroplasts for several hours in the dark in a pH 4 solution and then placing them in a medium of pH 8. Under such conditions the pH of the stroma quickly

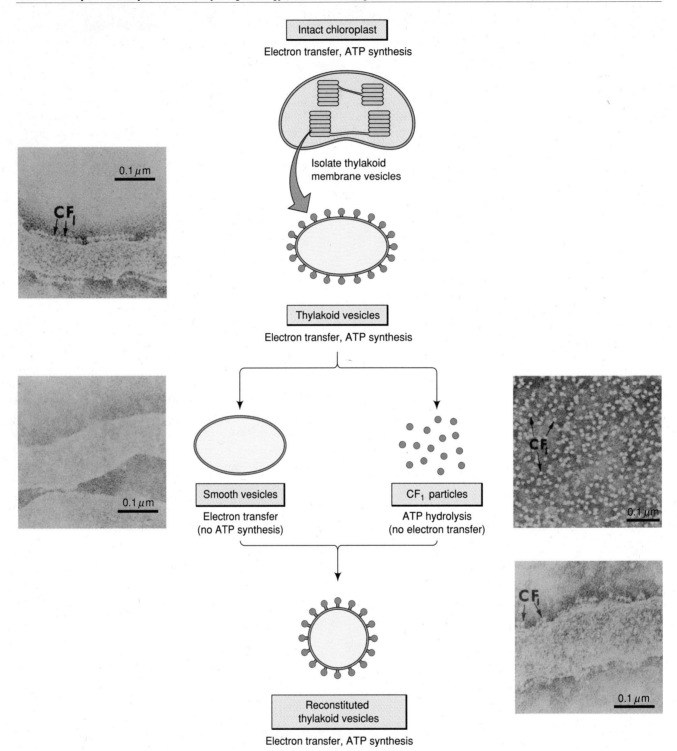

Figure 9-24 *Reconstitution of Thylakoid Membranes* *When CF₁ particles are removed from thylakoid vesicles, the remaining smooth vesicles carry out photosynthetic electron transfer but do not make ATP. Reattaching the CF₁ particles restores the ability to synthesize ATP in conjunction with electron transfer, suggesting that the CF₁ particles are the site of ATP formation. Note the similarity of these reconstitution experiments to those carried out on mitochondrial membranes (see Figure 8-26). Micrographs courtesy of M. P. Garber.*

rose to 8, while the pH within the thylakoid lumen remained at 4. This artificially created pH gradient triggered a rapid burst of ATP synthesis that continued until the pH gradient was dissipated (Figure 9-26). Since the entire experiment was carried out in the dark, it was concluded that an electrochemical proton gradient can drive ATP synthesis in the absence of light-induced electron flow. This experiment, reported in 1966, was the first direct demonstration in any experimental system that a pH gradient can induce ATP formation.

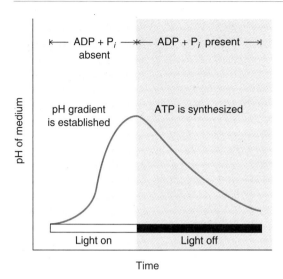

Figure 9-25 *Evidence That Protons Are Pumped into the Thylakoid Lumen During Photosynthetic Electron Transfer* *When isolated chloroplasts are exposed to light, the pH of the external solution rises (i.e., the proton concentration falls), indicating that protons are being pumped into the thylakoid lumen. The resulting proton gradient can drive ATP formation in the dark if ADP and P_i are added.*

3) The Driving Force for Photophosphorylation Is Derived Almost Entirely from ΔpH

The preceding two sets of experiments suggest that the light-induced flow of electrons through the photosynthetic electron transfer chain causes protons to be pumped from the chloroplast stroma into the thylakoid lumen, generating both a pH gradient (ΔpH) in which the pH is lower in the thylakoid lumen than in the stroma, and a membrane potential (Δψ) that is more positive on the side of the thylakoid membrane facing the lumen, where more protons are located. The diffusion of protons back out of the thylakoid lumen and into the stroma then drives ATP formation (Figure 9-27).

Although the artificial gradient of 4 pH units used in Jagendorf's experiments might seem drastic when compared to physiological conditions, actual measurements of chloroplasts stimulated with saturating light intensities have shown that the pH within the thylakoid lumen declines to about 4.5 while the pH of the stroma rises to about 8, yielding a ΔpH of 3.5 pH units. Sequestering the strongly acidic side of the proton gradient within the thylakoid lumen protects the remaining chloroplast components, including the enzymes of CO_2 fixation located in the stroma, from this harsh environment.

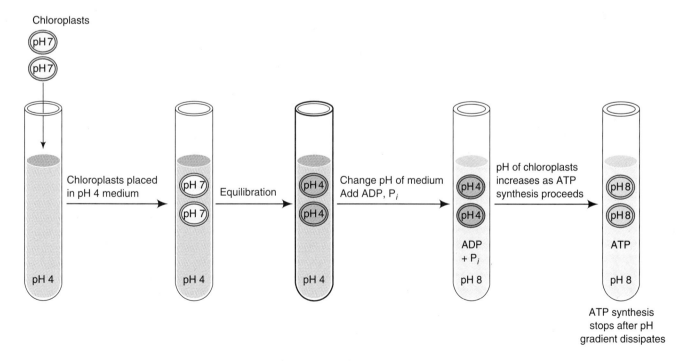

Figure 9-26 *Evidence That a Proton Gradient Can Drive Chloroplast ATP Synthesis in the Absence of Light-Induced Electron Transfer* *An artificial pH gradient was created by soaking chloroplasts for several hours in the dark in a pH 4 solution, and then placing them in a medium of pH 8. The resulting proton gradient triggered a rapid burst of ATP formation, even though the experiment was performed in the dark and hence no light-induced electron transfer could occur.*

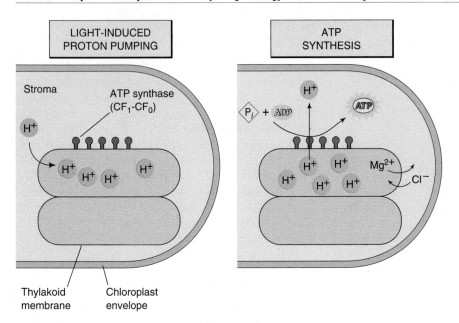

Figure 9-27 *Direction of Proton Flow During Photosynthetic Electron Transfer* (Left) *During light-induced electron transfer, protons are pumped across the thylakoid membrane from the chloroplast stroma into the thylakoid lumen.* (Right) *ATP formation is driven by the subsequent flow of protons down their electrochemical gradient (from the thylakoid lumen through ATP synthase and back into the stroma). The electrochemical gradient consists of both ΔpH (a pH gradient in which the pH is lower in the thylakoid lumen than in the stroma) and Δψ (a membrane potential that is more positive on the side of the thylakoid membrane facing the lumen). Because thylakoid membranes are permeable to Mg^{2+} and Cl^-, the membrane potential causes the positively charged magnesium ions to diffuse out of the thylakoid lumen and the negatively charged chloride ions tend to move in, abolishing most of the membrane potential. Hence the driving force for ATP formation is derived almost entirely from ΔpH.*

Measurements of the electrical potential across the thylakoid membrane (Δψ) have typically yielded values less than 0.01 volts, which is an order of magnitude smaller than the membrane potential across the inner mitochondrial membrane. Hence the driving force for ATP synthesis in chloroplasts derives almost entirely from ΔpH, whereas in mitochondria both ΔpH and Δψ make significant contributions. The reason for the low value of Δψ in chloroplasts is that the thylakoid membrane is more permeable than the inner mitochondrial membrane to ions such as Mg^{2+} and Cl^-. Hence passive diffusion of ions through the thylakoid membrane tends to dissipate the membrane potential component of the electrochemical proton gradient (see Figure 9-27).

4) Thermodynamic Calculations Indicate That the Proton Gradient Stores Enough Energy To Drive Photophosphorylation

Thermodynamic calculations have revealed that the magnitude of the proton gradient across the thylakoid membrane is sufficient to account for the amount of ATP synthesized by photophosphorylation. We have just seen that in illuminated chloroplasts, the proton

gradient across the thylakoid membrane consists of a pH difference (ΔpH) of about 3.5 pH units and a membrane potential (Δψ) of no more than 0.01 volts. Substituting these values into Equation 8-2 (page 337) allows us to calculate the proton motive force across the thylakoid membrane as follows:

$$
\begin{aligned}
\text{pmf} &= \Delta\psi + 2.303RT\,\Delta\text{pH/F} \\
&= 0.01 + 2.303(1.987)(273 + 37)(3.5)/23{,}062 \\
&= 0.01 + 0.215 \\
&= 0.225 \text{ volts}
\end{aligned}
$$

The change in free energy represented by a proton motive force of 0.225 volts can be determined by substituting this value into Equation 8-1 (page 329):

$$
\begin{aligned}
\Delta G^{\circ\prime} &= -n\text{F}\,\Delta E'_0 \\
&= -1 \times 23{,}062 \times 0.225 \\
&= -5{,}189 \text{ cal/mol} \approx -5.2 \text{ kcal/mol}
\end{aligned}
$$

The proton motive force across the thylakoid membrane thus corresponds to a free-energy change of about 5.2 kcal/mol of protons, which is basically the same value that we calculated for the free-energy change across the mitochondrial inner membrane (page 337). Current estimates suggest that the $\Delta G'$ for ATP synthesis under conditions prevailing within the chloroplast is around 14 kcal/mol, which is somewhat higher than the value of 11 kcal/mol that applies to mitochondrial ATP formation. The reason for the difference is that ADP and inorganic phosphate (P_i) are present in lower concentrations in chloroplasts than in mitochondria, shifting the equilibrium point of the reaction and effectively increasing the free-energy requirement. Since we have just calculated that the free-energy change of the proton gradient across the thylakoid membrane is 5.2 kcal/mol of protons, a minimum of three protons must flow back through the thylakoid membrane per molecule of ATP synthesized to provide the 14 kcal/mol required for ATP formation ($3 \times 5.2 = 15.6$ kcal/mol, which is slightly more than the 14 kcal/mol required). Although experimental measurements of the number of protons passing through the thylakoid membrane per molecule of ATP synthesized have yielded values ranging between two and three, these thermodynamic considerations suggest that the correct value is closer to three.

How does this value compare with the number of protons that are transported across the thylakoid membrane during the light-induced flow of electrons from water to NADP? Although experimental estimates vary somewhat, reported values range as high as six protons pumped across the membrane per pair of electrons passing from water to $NADP^+$. If three protons are needed to drive the synthesis of one molecule of ATP, then six protons would provide enough free energy to drive the formation of a maximum of two molecules of ATP per pair of electrons passing from water to $NADP^+$. As we will see later in the chapter, this value is more than adequate to explain the amount of ATP that needs to be synthesized by the light reactions of photosynthesis.

5) Reconstituted Membrane Vesicles Containing a Photosystem and ATP Synthase Establish Proton Gradients and Make ATP

Experiments involving the reconstitution of membrane vesicles from mixtures of isolated thylakoid components have provided considerable support for the chemiosmotic coupling theory. By mixing isolated preparations of photosystem I or II with purified phospholipids, it is possible to construct artificial membrane vesicles that are capable of carrying out the light-induced transport of protons. If ATP synthase (CF_1-CF_0) is included in the reconstitution mixture, the proton gradient can be utilized to drive ATP formation, providing more support for the idea

that photophosphorylation is mediated by formation of an electrochemical proton gradient (Figure 9-28).

Related experiments have provided information concerning the role played by CF_0, the hydrophobic transmembrane protein component of ATP synthase that anchors CF_1 to the membrane surface. Removal of CF_1 from the thylakoid membrane causes the thylakoid membrane to become leaky to protons, suggesting that CF_0 normally serves as a transmembrane channel through which protons flow on their way to CF_1 (Figure 9-29). This overall arrangement, in which the flow of protons through CF_0 leads to the synthesis of ATP by CF_1, is basically the same as what occurs in mitochondria with the analogous proteins, F_1 and F_1 (page 340).

6) Uncoupling Agents Abolish Both the Proton Gradient and Photophosphorylation

Additional support for the role of proton gradients in photophosphorylation has been provided by the use of ionophores that disrupt the electrochemical gradient across the thylakoid membrane. Agents such as grami-

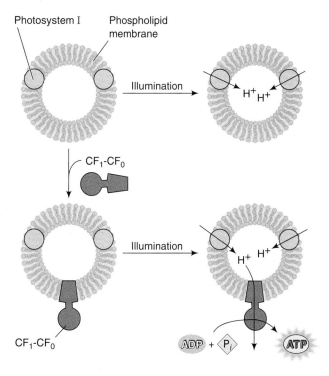

Figure 9-28 Proton Pumping and ATP Synthesis in Artificial Phospholipid Membrane Vesicles Reconstituted with Photosystem I *Membrane vesicles reconstituted with photosystem I pump protons into the vesicle when illuminated, but do not make ATP. Light-induced ATP formation requires the presence of CF_1-CF_0 as well as photosystem I. These results indicate that light-induced electron transfer by the photosystem is associated with the formation of an electrochemical proton gradient, and that the CF_1-CF_0 complex functions as an ATP synthase that couples the proton gradient to the synthesis of ATP.*

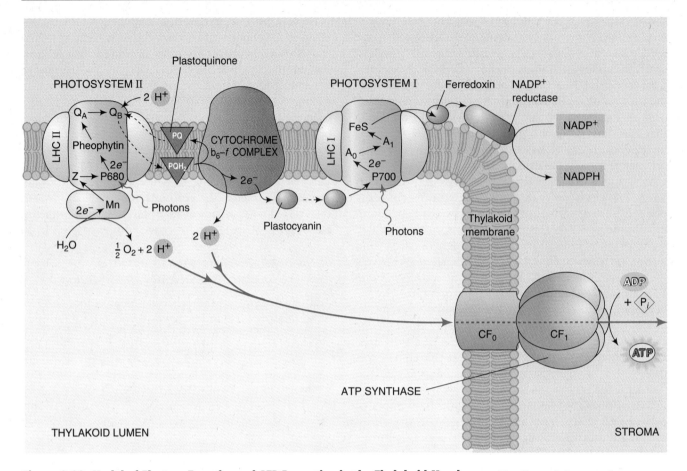

Figure 9-29 *Model of Electron Transfer and ATP Formation in the Thylakoid Membrane* *The flow of electrons from water to NADP+ is accompanied by the pumping of protons into the thylakoid lumen. The subsequent movement of these protons through ATP synthase (CF1-CF0 complex) then drives the formation of ATP. Some uncertainty exists as to the exact number of protons pumped per pair of electrons passed through the electron transfer chain. This model accounts for a total of four protons, but experimental measurements suggest that the number may be as high as six.*

cidin and CCCP, which make membranes freely permeable to protons and thereby collapse both ΔpH and Δψ, uncouple ATP synthesis from electron transfer in chloroplasts as they do in mitochondria (page 338). Ionophores that eradicate only ΔpH (e.g., nigericin) or Δψ (e.g., valinomycin) can also be used in combination to uncouple photophosphorylation. When ionophores are employed to collapse the electrochemical gradient, the light-induced flow of electrons through the photosynthetic chain continues, accompanied by the production of oxygen and the reduction of NADP+ to NADPH. Photophosphorylation, however, is abolished. Such observations indicate that photophosphorylation, like oxidative phosphorylation, depends upon the existence of an electrochemical proton gradient.

7) Photophosphorylation Requires an Intact Membrane Enclosing a Defined Compartment

The chemiosmotic coupling theory predicts that the coupling of ATP synthesis to the flow of electrons through the photosynthetic electron transfer chain

should require an intact thylakoid membrane enclosing a defined compartment, for otherwise a proton gradient could not be maintained. This prediction has been verified by the discovery that isolated thylakoid membrane preparations that are capable of carrying out photophosphorylation always consist of intact membrane vesicles.

8) The Rate of Photophosphorylation Correlates with the Rate of Formation of the Proton Gradient

Further evidence for the involvement of proton gradients in photophosphorylation has been provided by experiments in which the relationship between the magnitude of the proton gradient and the capacity to synthesize ATP has been measured as a function of time after illumination. This approach has revealed that the two processes have identical kinetics; that is, the rate at which the proton gradient forms is closely correlated with the rate at which ATP synthesis increases (Figure 9-30).

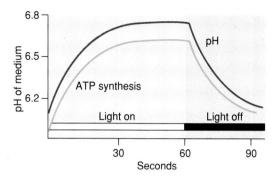

Figure 9-30 *Data Comparing the Rate of ATP Formation and the Rate at Which a Proton Gradient Is Formed in Illuminated Chloroplasts* *The rate at which the proton gradient is formed correlates closely with the rate at which ATP synthesis increases, supporting the idea that the proton gradient drives ATP formation.*

9) The Components of the Photosynthetic Electron Transfer Chain Are Asymmetrically Oriented across the Thylakoid Membrane

Because it requires that protons be selectively pumped in one direction (from the stroma into the thylakoid lumen), the chemiosmotic model predicts that the constituents of the photosynthetic electron transfer chain must be asymmetrically oriented across the thylakoid membrane. One way of testing this prediction involves incubating thylakoid vesicles with antibodies directed against various thylakoid components to determine whether a particular constituent is exposed at the outer or inner membrane surface. It has been shown, for example, that antibodies against ferredoxin and CF_1 readily bind to intact thylakoid vesicles, implying that these two components are exposed on the outer (stromal) membrane surface.

Another approach for investigating the orientation of thylakoid membrane components involves the use of artificial electron donors and acceptors. In general, lipid-soluble electron donors are capable of reducing photosystems I and II when added to intact membrane vesicles, but lipid-insoluble electron donors are not. Since only lipid-soluble electron donors can pass through the membrane to react with components exposed on the inner membrane surface, such observations suggest that the electron-accepting sites of photosystems I and II are preferentially exposed at the inner (lumenal) surface of the thylakoid membrane.

On the basis of these and related experimental approaches, a clearer picture of thylakoid membrane organization has begun to emerge (see Figure 9-29). Although the details of this picture will undoubtedly be subject to future revision, the overall arrangement is clearly consistent with the chemiosmotic coupling model. Electron carriers are asymmetrically distributed

within the membrane in such a way that protons are pumped into the thylakoid lumen each time a pair of electrons flows through the chain. The flow of these protons back down their electrochemical gradient then drives the formation of ATP by the ATP synthase (CF_1-CF_0) complex.

Photosystems I and II Are Spatially Separated in Stacked and Unstacked Thylakoids

The schematic model of the thylakoid membrane illustrated in Figure 9-29 emphasizes that photosynthetic electron transfer and photophosphorylation require the participation of four discrete multiprotein complexes, namely photosystem II, the cytochrome b_6-f complex, photosystem I, and ATP synthase. These protein complexes are functionally linked by three mobile components: *plastoquinone*, which diffuses through the lipid bilayer to transfer electrons from photosystem II to the cytochrome b_6-f complex; *plastocyanin*, which diffuses through the thylakoid lumen to transfer electrons from the cytochrome b_6-f complex to photosystem I; and *protons*, which are pumped across the thylakoid membrane and diffuse through the thylakoid lumen to link the photosynthetic electron carriers to ATP synthase. The existence of these mobile links between the protein complexes means that the complexes themselves need not be contiguous in the thylakoid membrane.

Evidence indicates that these protein complexes are in fact spatially separated between the stacked and unstacked thylakoids. One of the first techniques used for studying this issue was freeze-fracture electron microscopy. In fractured membranes, the two fracture faces are referred to as the *P face* and the *E face* (page 124). The predominant component observed in views of the P face is a family of small particles 8–11 nm in diameter, whereas the E face exhibits a group of larger particles 10–18 nm in diameter (Figure 9-31).

The following lines of evidence suggest that the large particles correspond to photosystem II, and that at least some of the smaller particles correspond to photosystem I. (1) Stacked thylakoid fractions isolated by centrifugation are enriched in photosystem II activity and large E-face particles, while unstacked thylakoid membranes contain more photosystem I activity and smaller P-face particles. (2) Plants grown in the dark for extended periods of time contain modified chloroplasts, called **etioplasts**, which lack chlorophyll, thylakoids, and photosynthetic activity. Upon exposure to light, photosystem I activity develops along with unstacked thylakoids and small P-face particles. Photosystem II activity appears later, accompanied by the formation of grana stacks and large E-face particles. (3) The leaves of some mutant strains of tobacco have large yellow patches that are defi-

P face (stacked)   P face (unstacked)

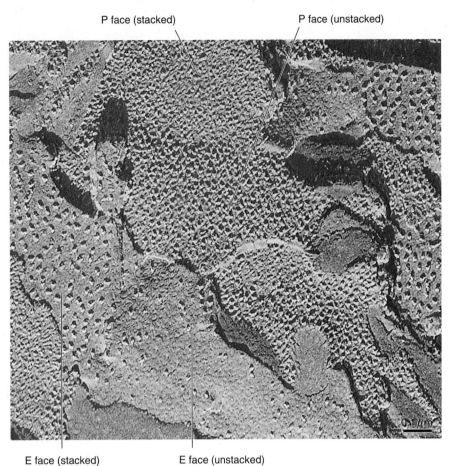

E face (stacked)   E face (unstacked)

Figure 9-31 *Freeze-Fracture Micrographs Showing the Fracture Faces of Stacked and Unstacked Thylakoids* *The protein particles seen in the E face are larger than those observed in the P face. The large particles are thought to correspond to photosystem II, while at least some of the smaller particles correspond to photosystem I. Courtesy of K. D. Allen.*

cient in chlorophyll. The chloroplasts in these yellow areas lack stacked thylakoids, photosystem II activity, and large E-face particles.

The preceding evidence supports the conclusion that photosystem II is preferentially localized in stacked thylakoids and corresponds to the large E-face particles, whereas photosystem I is preferentially localized in unstacked thylakoids and corresponds to some of the smaller P-face particles. This conclusion has received further support from electron microscopic studies using gold-labeled antibodies directed against specific photosystem proteins. Such studies have revealed that photosystem I is completely absent from stacked thylakoids in the regions where membranes are in direct contact with each other. CF₁ particles, which protrude from the surface of the thylakoid membrane, are also absent from these same regions. In contrast, most of the photosystem II complex occurs in the stacked membrane regions, while the cytochrome b_6f complex is found in both stacked and unstacked thylakoids. Hence there is a marked difference in the distribution of the main photosynthetic protein complexes in stacked and unstacked thylakoids (Figure 9-32).

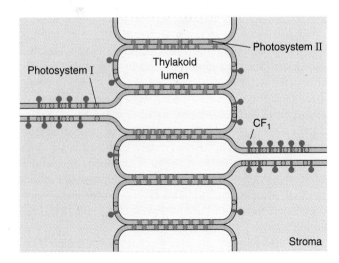

Figure 9-32 *Diagram Comparing the Distribution of CF₁, Photosystem I, and Photosystem II in Stacked and Unstacked Thylakoids* *The distribution of the cytochrome b_6f complex is not shown because it is present in both stacked and unstacked thylakoids.*

LHCII Acts as a Regulator of Photosystem Activity

The association of photosystem II with stacked thylakoids raises questions concerning both the origins and functional significance of such an arrangement. Several observations suggest that the mechanism underlying membrane stacking involves LHCII, which is the light-harvesting complex associated with photosystem II. LHCII is present in reduced amounts in mutant organisms deficient in stacked thylakoids. Moreover, treatment of isolated thylakoid membranes with enzymes known to degrade the protein constituents of LHCII abolishes the ability of thylakoids to form stacks. Finally, artificial phospholipid vesicles reconstituted in the presence of LHCII form membrane stacks, but reconstituted vesicles lacking LHCII do not.

The conclusion that thylakoid stacking is induced by the presence of LHCII does not in itself provide much insight into the functional significance of grana stacking. An important clue has emerged, however, from the discovery that a mutant strain of the green alga *Chlamydomonas* lacking stacked thylakoids also requires a higher than normal light intensity to achieve maximal rates of photosynthesis (Figure 9-33, *top*). Comparable results have been obtained in studies in which dark-adapted pea seedlings were exposed to light to trigger chloroplast development. Under such conditions the formation of grana stacks parallels the appearance of LHCII, and both events correlate with an increased photosynthetic efficiency as measured by a decrease in the amount of light required per pair of electrons transferred (see Figure 9-33, *bottom*). The conclusion that the presence of LHCII and the establishment of grana stacking are involved in regulating light-harvesting efficiency is compatible with the finding that plants grown at low light intensities contain enlarged grana stacks (Figure 9-34).

LHCII has also been shown to function as a regulator of the relative activities of photosystems I and II. Since the flow of electrons from water to NADP⁺ depends on two light-absorbing steps carried out by these photosystems in sequence, the rates at which photosystems I and II operate must be precisely balanced with each other. This control is accomplished in part by a regulatory mechanism involving LHCII. When photosystem II activity is operating faster than photosystem I, the reduced plastoquinone generated by photosystem II accumulates. The increase in the amount of reduced plastoquinone interacting with the cytochrome b_6-f complex causes the latter complex to activate the enzyme *LHCII kinase*, which catalyzes the phosphorylation of proteins associated with LHCII. Phosphorylation of LHCII causes it to dissociate from photosystem II and move from the stacked thylakoids to the unstacked thylakoids, where photosystem I is found. The resulting presence of LHCII in the unstacked thylakoids allows photosystem I to ab-

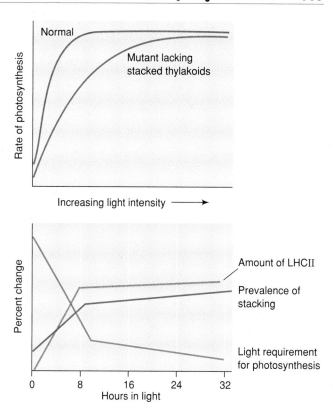

Figure 9-33 *Evidence for a Relationship between Grana Stacking and Photosynthetic Efficiency* (Top) *Effect of varying light intensity on photosynthesis in a mutant strain of* Chlamydomonas *lacking stacked thylakoids. A higher than normal light intensity is required for maximal photosynthetic activity.* (Bottom) *Effect of illumination on pea seedlings that had been kept in the dark for a week. Note that light induces the synthesis of LHCII and the formation of grana stacks. At the same time a decrease occurs in the amount of light required per pair of electrons transferred through the photosynthetic electron transfer chain, indicating that light-trapping has become more efficient.*

sorb more light energy, thereby bringing the activities of photosystems I and II back into balance.

THE CO₂-FIXING DARK REACTIONS

We have now seen how the light reactions of photosynthesis absorb energy derived from photons of light and use this energy to promote the formation of NADPH and ATP. During the next phase of photosynthesis, the energy stored in NADPH and ATP is employed to drive the conversion of CO_2 to carbohydrate. This process of **CO₂ fixation** is commonly referred to as the **dark reactions** of photosynthesis because it can take place in the dark as long as NADPH and ATP are available. However, it should be emphasized that the dark reactions *do not require darkness*; in fact, they proceed perfectly well in the light as long as NADPH and ATP are being provided by the light reactions.

Experiments Using Radioactive CO₂ First Revealed the Role Played by the Calvin Cycle in CO₂ Fixation

When radioactive isotopes first became available to the scientific community after World War II, they provided biologists with a powerful new tool for analyzing the pathway by which CO_2 is converted to carbohydrate during photosynthesis. This opportu-
nity was quickly seized by Melvin Calvin and his collaborators, who exposed green algae to radioactive carbon dioxide ($^{14}CO_2$) and then analyzed the radioactive products by paper chromatography (Figure 9-35). Such experiments revealed that a large number of radioactive products appear in cells that have been exposed to $^{14}CO_2$ for a few minutes or more. When the length of exposure to $^{14}CO_2$ is reduced to

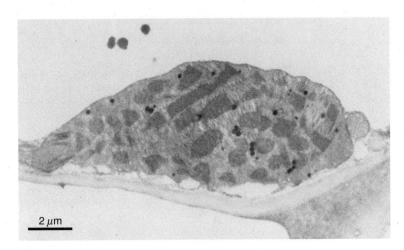

Figure 9-34 *Electron Microscopic Appearance of Chloroplasts in a Plant Grown at Low Light Intensity* *The grana stacks are significantly larger that those observed in plants exposed to normal amounts of light, suggesting that extensive grana stacking improves light-harvesting efficiency. Courtesy of A. Melis.*

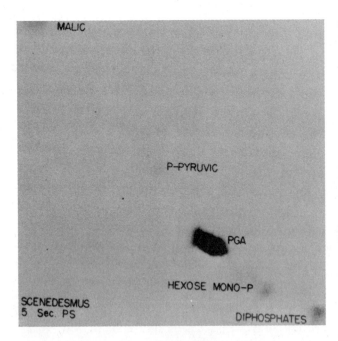

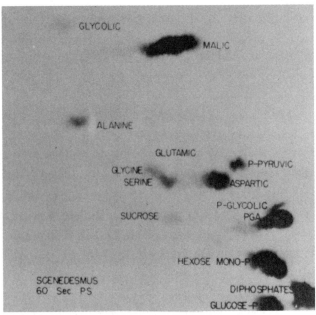

Figure 9-35 *Paper Chromatography of Radioactive Products Extracted from Green Algae Exposed to *$^{14}CO_2$ (Left) *After 5 seconds of incubation in the presence of $^{14}CO_2$, most of the radioactivity is in 3-phosphoglycerate (PGA). (Right) After 60 seconds of incubation with $^{14}CO_2$, radioactivity is detected in a large number of products. Taken together, the two experiments suggested that the initial product of CO_2 fixation is 3-phosphoglycerate. Courtesy of M. Calvin.*

a few seconds, however, the vast majority of the radioactivity appears in a single compound called *3-phosphoglycerate (3-PGA)*.

Because chemical analyses revealed that only one of the three carbon atoms in 3-PGA is radioactive, Calvin first guessed that the dark reactions of photosynthesis link CO_2 to a two-carbon acceptor molecule to form the three-carbon compound, 3-PGA. After several years of searching in vain for such a two-carbon precursor, a new observation revealed the error in this approach; it was discovered that longer exposures to $^{14}CO_2$ cause all the carbon atoms in 3-PGA to become radioactive instead of just one. The most straightforward explanation of this unexpected finding was that $^{14}CO_2$ is initially joined to an acceptor molecule whose carbon skeleton is itself gradually acquiring radioactive carbon atoms. Since all radioactivity initially appears in 3-PGA, the 3-PGA molecule must be giving rise to a radioactive product that in turn combines with $^{14}CO_2$ to produce more 3-PGA. In other words, the pathway for CO_2 fixation is circular. This realization eventually allowed Calvin to work out the details of the circular reaction pathway,

now called the **Calvin cycle**, by which photosynthetic cells fix and reduce CO_2 (Figure 9-36).

One of the crucial insights that permitted Calvin to unravel the organization of this pathway was the realization that the acceptor molecule to which CO_2 is initially joined might contain five carbon atoms rather than two. According to this idea, the reaction of CO_2 with a five-carbon acceptor molecule would generate a transient six-carbon compound that rapidly breaks down to two molecules of 3-PGA. One five-carbon molecule that generates two molecules of 3-PGA upon addition of CO_2 and subsequent cleavage is *ribulose 1,5-bisphosphate (RuBP)*. If RuBP were the true acceptor molecule, one would expect its concentration to increase when CO_2 is removed from the environment surrounding chloroplasts because no more CO_2 would be available to react with the RuBP. At the same time the concentration of 3-PGA in the chloroplast should fall, for 3-PGA is the expected product of the reaction between CO_2 and RuBP. As predicted, when chloroplasts are suddenly switched from an atmosphere containing large amounts of CO_2 to an atmosphere containing little CO_2, the concentration

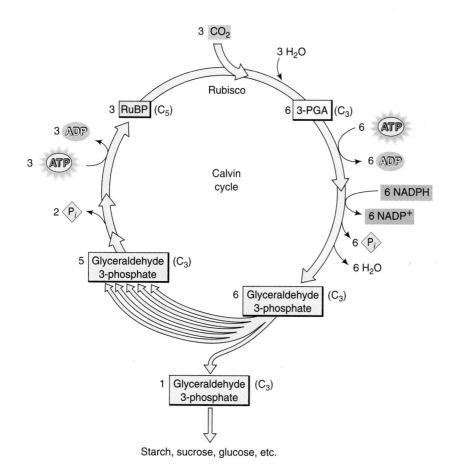

Figure 9-36 Overview of the Calvin Cycle *For every three molecules of CO_2 fixed by this cycle, one molecule of glyceraldehyde 3-phosphate is produced at a net cost of 9 molecules of ATP and 6 molecules of NADPH. The numbers in parentheses (e.g., C_3) indicate the number of carbon atoms present in each molecule.*

of 3-PGA within the chloroplasts falls while the concentration of RuBP increases (Figure 9-37). Such observations have led to the conclusion that RuBP is the five-carbon acceptor molecule that reacts with CO_2, generating two molecules of 3-PGA.

Glyceraldehyde 3-Phosphate Is the Principal Product of CO_2 Fixation by the Calvin Cycle

The discovery that 3-PGA accumulates in plants maintained in darkness implies that the initial CO_2-fixing step of the Calvin cycle does not require energy derived from the light reactions. The main energy requirement occurs during the next two steps of the cycle, where ATP and NADPH generated by the light reactions are utilized to drive the reduction of 3-PGA to *glyceraldehyde 3-phosphate* (see Figure 9-36). Since the presence of reduced carbon gives molecules a high free-energy content, this reduction step permits photosynthesis to produce molecules that serve as energy-rich nutrients.

Once glyceraldehyde 3-phosphate is formed, the succeeding steps in the Calvin cycle are designed to accomplish two main objectives: (1) to regenerate the RuBP molecules that will serve as CO_2 acceptors for subsequent turns of the cycle, and (2) to convert newly fixed carbon atoms into various carbohydrates needed by the cell. These objectives are accomplished by metabolizing glyceraldehyde 3-phosphate in two different ways. First, most of the glyceraldehyde 3-phosphate molecules are metabolized by a complex pathway that regenerates RuBP. In addition, for every five molecules of glyceraldehyde 3-phosphate utilized in the regenera-

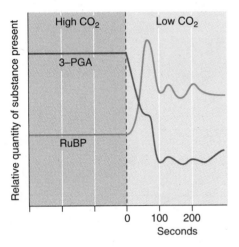

Figure 9-37 *Effects of CO_2 on 3-Phosphoglycerate (3-PGA) and Ribulose 1,5-Bisphosphate (RuBP) Concentration in Intact Chloroplasts* *If RuBP reacts with CO_2 to form 3-PGA, one would expect the concentration of RuBP to increase and the concentration of 3-PGA to fall when the CO_2 concentration in the atmosphere is lowered. The data obtained in these experiments confirm this prediction.*

tion of RuBP, one molecule of glyceraldehyde 3-phosphate is employed to produce carbohydrates for use by the cell. Although glucose is often written as the end product of such metabolism, this is done for the sake of simplicity rather than metabolic accuracy. In fact glucose is only one of many different carbohydrates that are produced from glyceraldehyde 3-phosphate, so glyceraldehyde 3-phosphate itself is properly viewed as the principal product of the Calvin cycle.

Much of the glyceraldehyde 3-phosphate produced by the Calvin cycle is transported out of the chloroplast for use elsewhere in the cell. Since glyceraldehyde 3-phosphate is an intermediate in glycolysis (see Figure 8-3), it can enter the glycolytic pathway and be metabolized much like a molecule of glucose. Glyceraldehyde 3-phosphate is also converted to glucose phosphate, which is a precursor for the synthesis of both *cellulose* and the disaccharide *sucrose*. In the pathway for sucrose formation, glyceraldehyde 3-phosphate is converted to glucose 1-phosphate and fructose 6-phosphate by reactions that resemble the reversal of several steps in glycolysis. The glucose 1-phosphate then reacts with the nucleotide UTP to form the activated sugar derivative *UDP-glucose*, which combines with fructose 6-phosphate to produce sucrose. Sucrose is the main form in which carbohydrate is transferred between cells and tissues in plants, playing a role comparable to the circulating glucose that occurs in the bloodstream of animals.

The glyceraldehyde 3-phosphate that is not exported from chloroplasts is converted to *starch* in the chloroplast stroma, forming starch grains that are readily visible in electron micrographs (see Figure 9-12). Starch is produced by converting glyceraldehyde 3-phosphate to glucose 1-phosphate, which is then reacted with ATP to produce the activated precursor of starch formation, *ADP-glucose*. Starch serves as a glucose storage mechanism in plants just as glycogen stores glucose in animals (page 14). Excess energy captured by chloroplasts during periods of intense illumination is employed to promote the formation of starch, which is then broken down when it is dark to provide the cell with needed glucose.

The Calvin Cycle Requires 18 Molecules of ATP and 12 Molecules of NADPH When Converting CO_2 to Glucose

Although the Calvin cycle generates many different kinds of carbohydrates, for simplicity the six-carbon sugar glucose is usually indicated as the end product in the summary equation of photosynthesis (see Equation 9-1). In order to write a balanced equation for the production of glucose, we need to determine how much ATP and NADPH are required. Figure 9-36 illustrates that a total of 9 molecules of ATP and 6 molecules of

NADPH are consumed during the fixation of 3 molecules of CO_2 by the Calvin cycle (note that ATP is utilized at two different points in the cycle). The net result is the production of one molecule of the *three-carbon compound*, glyceraldehyde 3-phosphate. It follows that the synthesis of glucose, which contains *six* carbons instead of *three*, requires a total of 18 molecules of ATP and 12 molecules of NADPH per 6 molecules of CO_2 fixed by the Calvin cycle.

In addition to ATP and NADPH, Figure 9-36 shows that water is both an input and an output of the Calvin cycle. When these water molecules are included, the net equation for the production of one molecule of glucose by the Calvin cycle becomes:

$$6CO_2 + 18ATP + 12NADPH + 12H^+ + 6H_2O \rightarrow$$
$$C_6H_{12}O_6 + 18ADP + 18P_i + 12NADP^+ + 12H_2O$$

Since water is present on both sides of this expression, it can be simplified to the following net reaction for the production of glucose:

$$6CO_2 + 18ATP + 12NADPH + 12H^+ \rightarrow$$
$$C_6H_{12}O_6 + 18ADP + 18P_i + 12NADP^+ + 6H_2O$$
$$(9\text{-}2)$$

Combining the Light Reactions with the Calvin Cycle Generates the Overall Equation of Photosynthesis

The 18 ATP and 12 NADPH molecules required for the production of a molecule of glucose are provided by the light reactions of photosynthesis. In order to generate 12 molecules of NADPH, the light reactions must transfer 12 pairs of electrons from water to $NADP^+$. The number of ATP molecules that are synthesized as these 12 electron pairs pass through the photosynthetic electron transfer chain has been difficult to determine experimentally. Earlier in the chapter we estimated that the number of protons pumped across the thylakoid membrane by the photosynthetic electron transfer chain is sufficient to drive the synthesis of a maximum of two molecules of ATP per pair of electrons passing from water to $NADP^+$ (page 381). Hence for 12 pairs of electrons, this maximum estimate would yield $12 \times 2 = 24$ molecules of ATP. Since only 18 ATP molecules are required by the Calvin cycle for the production of a molecule of glucose, let us assume for the moment that the light reactions pump at least enough protons to make the required 18 ATP molecules. Given this assumption, the equation for the production of 12 molecules of NADPH by the light reactions can be written as follows:

$$12H_2O + 18ADP + 18P_i + 12NADP^+ \rightarrow$$
$$6O_2 + 18ATP + 12\,NADPH \quad (9\text{-}3)$$

Combining this equation for the light reactions with Equation 9-2, which represents the dark reactions, we obtain:

$$6CO_2 + 6H_2O \rightarrow C_6H_{12}O_6 + 6O_2 \quad (9\text{-}4)$$

This equation of course corresponds to the summary equation of photosynthesis described earlier in the chapter (Equation 9-1).

How efficient is this overall process for using energy captured from sunlight to drive the formation of carbohydrates? The $\Delta G^{\circ\prime}$ for the synthesis of glucose represented by Equation 9-4 is 686 kcal/mol. This means that the light reactions must capture at least 686 kcal of free energy in the form of ATP and NADPH for every mole of glucose to be synthesized. Earlier in the chapter we estimated that about eight photons of light are absorbed per molecule of CO_2 fixed under conditions of maximal photosynthetic efficiency (page 369). Since glucose contains six carbons, a total of about $8 \times 6 = 48$ photons of light are absorbed per molecule of glucose produced. Although the free energy of a photon of light varies with the wavelength of light involved, we can employ 41 kcal/mol of photons, which corresponds to light of 700 nm, as an approximation. The total energy input per mole of glucose formed is therefore

$$48 \text{ mol of photons} \times 41 \text{ kcal/mol} = 1968 \text{ kcal}$$

Since the synthesis of one mole of glucose traps 686 kcal, the overall efficiency of photosynthesis under these conditions is $686 \div 1968 \approx 35$ percent. At shorter wavelengths the energy content of a photon of light is higher, resulting in a corresponding reduction in efficiency. For light of 400 nm, for example, the calculated efficiency drops to about 20 percent. These calculations are of course based on the assumption that the transfer of electrons from water to $NADP^+$ is accompanied by the synthesis of the 18 ATP molecules required by the Calvin cycle. If the actual ATP yield is less than 18 molecules, then extra ATP would need to be produced by cyclic photophosphorylation; since the absorption of more photons of light would be required, the overall efficiency of photosynthesis would be correspondingly lower.

Molecules Involved in Chloroplast Metabolism Are Selectively Transported across the Inner Chloroplast Membrane

In photosynthetic eukaryotic cells, the chloroplast plays a major role in providing reduced carbon and chemical energy for the rest of the cell. Since the inner chloroplast membrane is relatively impermeable to most small molecules, specific membrane carriers are required for transporting many of the substances that need to enter and exit the chloroplast. Prominent among these transport systems is the *triose phosphate translocator*, which transports glyceraldehyde 3-phosphate out of the chloroplast in exchange for either inorganic phosphate or 3-PGA from the cytosol. The main function of this carrier is to enable the chloroplast to export fixed carbon in the form of glyceraldehyde 3-phosphate to the cytosol for use in the synthesis of sucrose, cellulose, and other carbohydrates. The phosphate taken up by chloroplasts in exchange for glyceraldehyde 3-phosphate is employed in the conversion of ADP to ATP.

In addition to this primary role, the triose phosphate translocator is also involved in a shuttle mechanism whose net effect is to transfer electrons and ATP out of the chloroplast for use by metabolic pathways in the cytosol. The shuttle mechanism is based on the ability of the triose phosphate translocator to couple the export of glyceraldehyde 3-phosphate to the uptake of 3-PGA via a process that is driven by the proton gradient across the chloroplast envelope. Because glyceraldehyde 3-phosphate has a net charge of *minus 2* and 3-PGA has a net charge of *minus 3*, the uptake of 3-PGA coupled to the export of glyceraldehyde 3-phosphate is equivalent to the net uptake of one negative charge by the chloroplast. But photosynthetic electron transfer causes the proton concentration to be higher in the cytosol than in the stroma, creating a proton gradient that tends to drive the flow of positively charged protons into the chloroplast. The uptake of each molecule of 3-PGA (charge = -3) by the chloroplast is therefore coupled to the uptake of one proton (charge = $+1$), producing a net uptake of two negative charges; this exactly balances the two negative charges that are being exported from the chloroplast in the form of glyceraldehyde 3-phosphate.

Figure 9-38 illustrates how the proton-driven exchange of 3-PGA for glyceraldehyde 3-phosphate serves as an indirect shuttle mechanism for transferring electrons and ATP out of the chloroplast. The glyceraldehyde 3-phosphate transported out of the chloroplast is oxidized in the cytosol to 3-PGA by two steps in glycolysis that yield NADH and ATP (see Steps 6 and 7 in Figure 8-3). The 3-PGA produced by this process is then returned to the chloroplast by the triose phosphate translocator, where it is reduced back to glyceraldehyde 3-phosphate by steps in the Calvin cycle that utilize ATP and NADPH. Hence the net effect is that NADPH and ATP produced in the chloroplast are transferred to the cytosol in the form of NADH (not NADPH) and ATP.

Light Regulates the Activity of Rubisco and Other Calvin Cycle Enzymes

Since the Calvin cycle requires NADPH and ATP generated by the light reactions of photosynthesis, it is important that the activities of the Calvin cycle and the light reactions be precisely coordinated with each other.

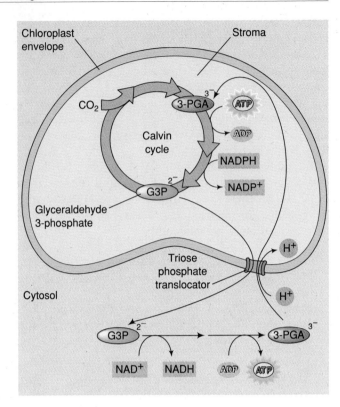

Figure 9-38 *Use of the Triose Phosphate Translocator as a Shuttle Mechanism for Transferring Electrons and ATP Out of the Chloroplast* *The electrons transferred out of the chloroplast are used to drive the reduction of NAD^+ to NADH in the cytosol. The proton gradient established during photosynthetic electron transfer drives the shuttle mechanism by ensuring that the number of charges that enter and leave the chloroplast are equal.*

One site of control involves the first step in the Calvin cycle, where CO_2 and RuBP are joined together to form two molecules of 3-PGA. The enzyme that catalyzes this reaction is called *ribulose 1,5-bisphosphate carboxylase,* or **rubisco**. Rubisco accounts for roughly half the total protein of the chloroplast, and as much as 20 percent of the entire protein content of a typical photosynthesizing plant cell. It is present in all photosynthetic organisms except for a few photosynthetic bacteria, and is easily the most abundant protein in nature; in fact, it has recently been estimated that more than a million grams of rubisco are synthesized on earth every second!

Purified rubisco requires a much higher CO_2 concentration for maximal activity than is required for the operation of the Calvin cycle in intact chloroplasts. This discrepancy initially led to skepticism concerning the physiological relevance of rubisco, but doubts were dispelled when it was demonstrated that mutant organisms lacking rubisco are unable to fix CO_2. It was later shown that purified rubisco does not function as efficiently as expected because it requires conditions that resemble those prevailing within the chloroplast stroma where it normally resides. Two especially important factors are the pH and the Mg^{2+} concentration. In intact chloroplasts we have seen that the light reactions cause protons to be pumped from the stroma into the thylakoid lumen, raising the pH in the stroma to about pH = 8. In addition, the membrane potential produced by the proton gradient causes Mg^{2+} to diffuse from the thylakoid lumen into the stroma (see Figure 9-27). The rubisco molecules residing in the stroma are activated by both the increased pH and the elevated Mg^{2+} concentration.

Two other factors also regulate the activity of rubisco. The first is CO_2, which activates rubisco by binding to one of the enzyme's lysine residues to form a carbamate group (lysine-NH-CO_2). This *carbamylation* of rubisco is catalyzed by the enzyme *rubisco activase*, which requires ATP for maximal activity. Rubisco activase is relatively inactive in dim light or darkness, presumably because it requires ATP made by the light reactions. In darkness, some plants also make an inhibitor of rubisco called *2-carboxyarabinitol-1-phosphate (CA1P)*. Thus at least four different factors influenced by light and darkness control the activity of rubisco, namely pH, Mg^{2+}, carbamylation, and CA1P.

In addition to these effects on rubisco, several other enzymes in the Calvin cycle are also controlled by changing light conditions. These particular Calvin cycle enzymes have disulfide (S−S) bonds that must be reduced to sulfhydryl (−SH) groups in order for the enzymes to exhibit optimal activity. Light indirectly controls the reduction of these disulfide bonds using a mechanism that involves the protein ferredoxin. In the presence of intense illumination, increased amounts of reduced ferredoxin are produced. The enzyme *ferredoxin-thioredoxin reductase* catalyzes the transfer of electrons from reduced ferredoxin to a small protein called *thioredoxin*, thereby reducing a disulfide group in thioredoxin. The reduced thioredoxin in turn activates several Calvin cycle enzymes, as well as chloroplast ATP synthase, by reducing their disulfide bonds to sulfhydryl groups. This mechanism helps to ensure that both photophosphorylation and the Calvin cycle operate at maximum rates under conditions of intense illumination.

PHOTORESPIRATION AND C₄ PLANTS

In 1920, the German biochemist Otto Warburg first noted that CO_2 fixation is inhibited by increasing concentrations of O_2 (Figure 9-39). We now know that this inhibitory effect is caused by **photorespiration**, a light-dependent pathway that consumes O_2 and evolves CO_2 (the opposite of photosynthesis). In the presence of light, photorespiration decreases the efficiency of photosynthesis because it oxidizes reduced carbon atoms produced by the photosynthetic pathway with-

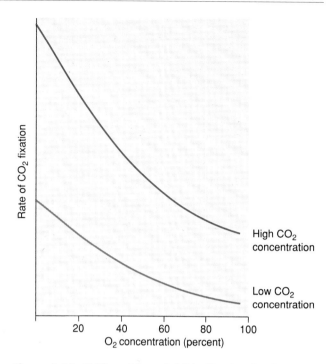

Figure 9-39 Ability of O_2 to Inhibit CO_2 Fixation in Soybean Plants *This inhibitory effect is due to photorespiration, a light-dependent pathway that consumes O_2 and evolves CO_2. The ability of O_2 to inhibit CO_2 fixation is most pronounced at low CO_2 concentrations.*

out capturing the released energy in a useful chemical form such as ATP or reduced coenzymes.

Photorespiration Occurs When O_2 Substitutes for CO_2 in the Reaction Catalyzed by Rubisco

The existence of photorespiration can be explained by an unusual property of the enzyme rubisco. When the O_2 concentration is high and the CO_2 concentration is low, O_2 can substitute for CO_2 as substrate in the reaction catalyzed by rubisco. Under such conditions rubisco catalyzes the interaction of RuBP with O_2, an oxidation reaction that generates phosphoglycolate and 3-PGA as products instead of the two molecules of 3-PGA normally produced by rubisco (Figure 9-40). While the 3-PGA continues through the Calvin cycle as normal, the phosphoglycolate is converted to glycolate, which is then transferred to peroxisomes (glyoxysomes) for further oxidation by molecular oxygen. The final product of peroxisomal metabolism is the amino acid glycine, which is transported to mitochondria; here two molecules of glycine are converted to one molecule of the amino acid serine, accompanied by the release of CO_2. Serine is returned to the peroxisome for conversion to glycerate, which then returns to the chloroplast to enter the Calvin cycle. Photorespiration is thus a cyclic pathway that involves the cooperation of

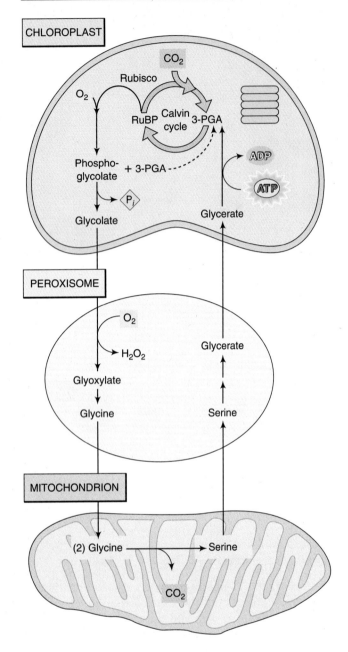

Figure 9-40 *The Process of Photorespiration* *When rubisco acts on O_2 rather than CO_2, RuBP is diverted from the Calvin cycle into a pathway that involves the cooperation of chloroplasts, peroxisomes, and mitochondria. During the process, some of the carbon atoms present in RuBP are oxidized to CO_2 by mitochondria without the released energy being captured in a chemically useful form.*

three organelles: chloroplasts, mitochondria, and peroxisomes. For this reason, these three organelles often occur in close proximity to one another in plant cells (Figure 9-41).

Photorespiration causes some of the carbon atoms present in RuBP to be oxidized to CO_2 by mitochondria without any of the released energy being captured in a chemically useful form. Several factors make this phenomenon a light-

dependent process. First, the RuBP molecules that serve as the starting point for photorespiration are made by the Calvin cycle, which requires ATP and NADPH generated by the light reactions. Second, photorespiration is favored by the presence of O_2, which is produced by the light reactions. And finally, the enzyme rubisco is subject to activation by several light-dependent mechanisms (page 390).

The rate of photorespiration in illuminated plants can be exceptionally high, wasting as much as 50 percent of the reduced carbon generated by photosynthesis. For this reason, attempts are being made to alter the genetic makeup of important agricultural plants so that they will carry out less photorespiration and thus be more efficient at converting atmospheric CO_2 into carbohydrates.

C_4 Plants Minimize Photorespiration by Using a Special Pathway for Concentrating CO_2

Some species of plants have already developed their own mechanisms for overcoming the problem of photorespiration. Because photorespiration is favored when the concentration of CO_2 is low, these plants have evolved a special metabolic pathway that minimizes photorespiration by maintaining a high CO_2 concentration in the cells where photosynthesis occurs. Among the plants exhibiting this property are sugarcane, maize, sorghum, and a variety of tropical grasses and plants. The ability of these plants to handle CO_2 in a special way was first suggested by experiments in which sugarcane and other tropical plants were exposed to an atmosphere containing $^{14}CO_2$. Whereas Calvin's experiments using this approach with photosynthetic algae had shown that 3-PGA is the first radioactive compound to be formed, the comparable experiments with tropical plants revealed that radioactivity initially appears in the four-carbon acids *oxaloacetate* and *malate*. To determine whether these four-carbon acids are intermediates in a unique carbon-fixing pathway, M. Hatch and C. Slack monitored the radioactive products that appear over time in sorghum leaves exposed to $^{14}CO_2$. They discovered that radioactivity turns up first in oxaloacetate and malate, and then sequentially in 3-PGA, six-carbon sugars, and finally sucrose (Figure 9-42).

This pattern suggests the existence of a pathway in which CO_2 is incorporated into four-carbon acids prior to its entry into the Calvin cycle. Plants that use such a pathway are referred to as C_4 **plants** because the *initial* product of CO_2 fixation contains four carbon atoms; in contrast, those plants that depend solely on the Calvin cycle are referred to as C_3 **plants** because the initial product of CO_2 fixation is the three-carbon compound, 3-PGA. Several versions of the C_4 pathway have evolved in different plants, but all serve the same function: to deliver CO_2 in high concentration to the Calvin cycle.

The leaves of C_4 plants exhibit a special type of anatomy that reflects a division of labor between two

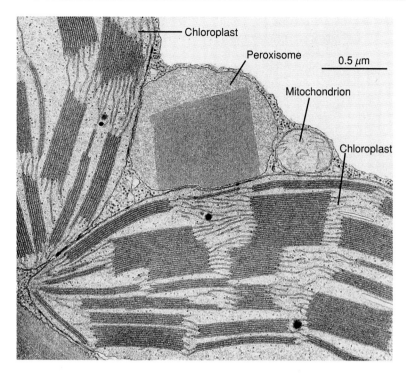

Figure 9-41 *Electron Micrograph of a Tobacco Leaf Cell Showing Two Chloroplasts, a Peroxisome, and a Mitochondrion in Close Proximity* *This close association reflects the functional interactions that occur between these organelles during photorespiration. Courtesy of E. H. Newcomb.*

cell types involved in photosynthesis: *bundle sheath cells* and *mesophyll cells*. **Bundle sheath cells** are large, thick-walled cells that form a prominent layer around the leaf veins of C₄ plants (Figure 9-43); they contain numerous chloroplasts and function as the main site of the light reactions and the Calvin cycle. In contrast, the **mesophyll cells** fill the rest of the leaf interior and function as the initial sites of CO_2 fixation, producing four-carbon (C₄) acids destined for transport to the bundle sheath cells. After arriving in the bundle sheath cells the C₄ acids are degraded, releasing CO_2 for entry into the Calvin cycle.

The initial CO_2-fixing reaction in C₄ plants is carried out in the cytosol of mesophyll cells by the enzyme **phosphoenolpyruvate (PEP) carboxylase**, which joins CO_2 with phosphoenolpyruvate to form the C₄ acid, *oxaloacetate*. Oxaloacetate is then converted to other C₄ acids (malate and aspartate) that are transported to bundle sheath cells and degraded to CO_2 plus pyruvate. The net result is a sequence in which CO_2 is first fixed in mesophyll cells by PEP carboxylase, transferred in the form of C₄ acids to bundle sheath cells, and then released again as free CO_2 (Figure 9-44). Although the value of fixing CO_2 in one cell type only to release it again as CO_2 in another cell type might appear questionable, closer examination reveals that the high efficiency of PEP carboxylase permits this process to function as a CO_2-concentrating mechanism. Unlike the Calvin cycle enzyme rubisco,

which has a low affinity (high K_m) for CO_2 and can confuse O_2 with CO_2, PEP carboxylase has a high affinity (low K_m) for CO_2 and does not confuse O_2 with CO_2. As a result, PEP carboxylase can fix CO_2 when the CO_2 concentration in the plant is too low for rubisco to function effectively. Under these conditions, the PEP carboxylase of mesophyll cells fixes the CO_2 and delivers it in the form of C₄ acids to bundle sheath cells; here the C₄ acids are broken down to release high concentrations of CO_2, which can then be fixed by rubisco and metabolized via the Calvin cycle. Thus the localization of PEP carboxylase in mesophyll cells and rubisco in bundle sheath cells allows the two cell types to cooperate with each other in fixing CO_2 (Figure 9-45).

This arrangement for trapping and delivering CO_2 has a clear adaptive advantage in hot, arid environments, where most C₄ plants flourish. In hot, dry weather, plant leaves prevent excessive water loss by closing their surface pores, or **stomata**, which normally function to permit gas exchange with the atmosphere. Although closing these pores reduces the loss of water through evaporation, it also limits access of atmospheric CO_2 to the leaf cells. As a result, the intracellular concentration of CO_2 falls to levels that are too low to be fixed by the Calvin cycle enzyme rubisco. However, the more efficient C₄ pathway can still fix CO_2 and deliver it in concentrated form to the Calvin cycle, thereby allowing photosynthesis to proceed under conditions that would

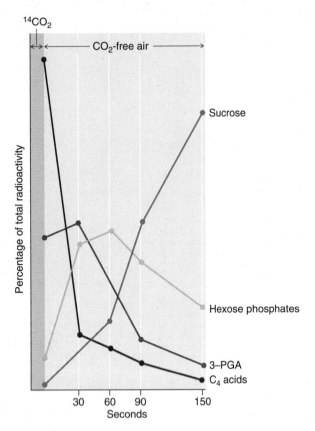

Figure 9-42 *Data Demonstrating the Existence of the C₄ Pathway* *In these experiments, sorghum leaves were briefly exposed to $^{14}CO_2$ and were then transferred to a CO_2-free atmosphere. The data indicate that newly fixed carbon appears first in four-carbon (C₄) acids, and then sequentially in 3-PGA, hexose phosphates, and sucrose.*

otherwise be prohibitory. Moreover, the high concentration of CO_2 delivered to the Calvin cycle by the C₄ pathway prevents O_2 from competing with CO_2 for rubisco; hence photorespiration is absent or barely detectable in C₄ plants. The absence of photorespiration in C₄ plants makes their overall efficiency in harvesting solar energy considerably greater than that of C₃ plants. In fact, it is likely that the C₄ pathway evolved at least in part to overcome the inefficiency of photorespiration.

An appreciation of the photosynthetic efficiency of C₄ plants is sometimes obscured by the fact that one of the steps in the C₄ pathway requires an input of ATP (see Figure 9-44). Although the existence of this step means that the C₄ pathway requires some energy, the ATP that is utilized can be regenerated at little cost in cells that are capable of cyclic photophosphorylation. This easily generated ATP allows C₄ plants to fix CO_2 in a way that avoids the enormous inefficiencies introduced by photorespiration.

CAM Plants Produce Malate at Night and Use It to Provide CO₂ to the Calvin Cycle During the Day

We have just seen that in C₄ plants, CO_2 is fixed by PEP carboxylase in one cell type (mesophyll cells) and then released in another cell type (bundle sheath cells) for refixation by the Calvin cycle. Certain species of plants living in hot arid environments employ a related strategy for increasing the efficiency of CO_2 fixation that does not require the participation of two different cell types. Such plants, called *succulents*, have thick fleshy leaves that are comprised mainly of mesophyll cells. Commonly encountered examples of succulents include the numerous varieties of cactuses and orchids. Because CO_2 fixation in succulents exhibits some unusual features that were first discovered in the plant family *Cras-*

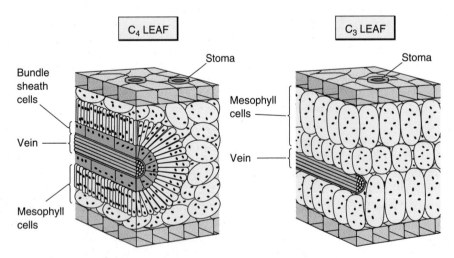

Figure 9-43 *Organization of Leaf Cells in C₄ and C₃ Plants* *A prominent layer of bundle sheath cells surrounds the veins of C₄ leaves, but not C₃ leaves.*

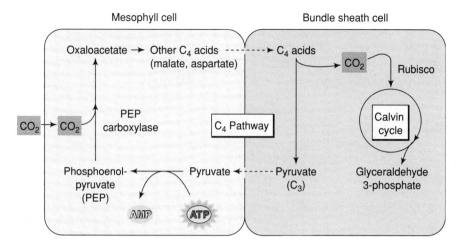

Figure 9-44 *The C₄ Pathway for Fixing CO₂ and Delivering It to the Calvin Cycle*
In C₄ plants, CO₂ is initially fixed in mesophyll cells by PEP carboxylase, generating C₄ acids that are transported to bundle sheath cells; here the CO₂ is released for fixation by the Calvin cycle. In C₃ plants, atmospheric CO₂ is fixed directly by the Calvin cycle in mesophyll cells without the prior production of C₄ acids.

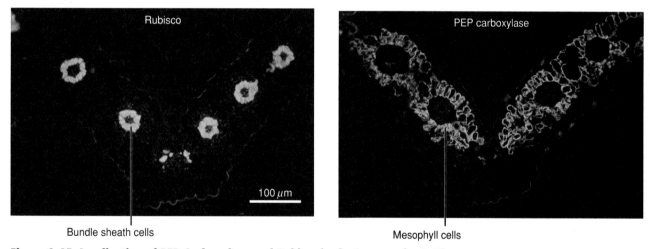

Figure 9-45 *Localization of PEP Carboxylase and Rubisco in the Leaves of a C₄ Plant* (Left) *Staining with fluorescent antibodies directed against rubisco shows that rubisco is located in bundle sheath cells.* (Right) *Staining with fluorescent antibodies directed against PEP carboxylase shows that it is located in mesophyll cells. Courtesy of J. O. Berry.*

sulaceae, this type of CO₂ metabolism is referred to as **crassulacean acid metabolism (CAM)**.

The leaves of CAM plants, like those of other plants, contain stomata that can be opened or closed to control gas and water exchange between the leaf and the surrounding atmosphere. When the weather is hot and dry, CAM plants have a problem obtaining sufficient CO₂ during the daytime because opening the stomata to permit CO₂ entry would allow excessive amounts of water to be lost to the surrounding hot dry air by evaporation. CAM plants therefore open their stomata mainly at night, when the air is cooler and the humidity is higher. The entering CO₂ is fixed by PEP carboxylase located in

the cytosol of the mesophyll cells, generating large amounts of malate, a C₄ acid. The malate that accumulates during the night is temporarily stored in a large vacuole that occupies the central region of the mesophyll cells. When daytime arrives, the stomata are closed and PEP carboxylase is temporarily inactivated. Malate then diffuses from the vacuole into the cytosol, where it is broken down to release CO₂ that is subsequently fixed by the Calvin cycle enzyme rubisco. At night the stomata are again opened, PEP carboxylase is activated, and the cycle is repeated.

Thus CO₂ fixation in CAM plants resembles the comparable process in C₄ plants in that CO₂ is fixed twice, first by PEP carboxylase and then by rubisco. But instead of carrying out the first step in bundle sheath cells and the second in mesophyll cells, CAM plants carry out both steps in the same cell type, the first step at night and the second during the day.

PHOTOSYNTHESIS IN PROKARYOTES

Photosynthetic prokaryotes differ from photosynthetic eukaryotes in that they lack chloroplasts; nonetheless they still utilize membranes to carry out the light reactions of photosynthesis. In *cyanobacteria*, the light reactions occur within intracellular photosynthetic membranes that resemble the thylakoid membranes of higher plants (Figure 9-46). The shape and arrangement of these photosynthetic membranes, which are distributed throughout the cytoplasm, vary significantly among the various kinds of cyanobacteria. Most of the light energy that drives photosynthesis is absorbed by phycobilin pigments present in phycobilisome granules attached to the membranes (see Figure 9-16). The absorbed energy is then passed on to chlorophyll molecules situated within the membranes to which the phycobilisomes are attached. The subsequent light and dark reactions are fundamentally similar to the comparable pathways of eukaryotes.

The Light Reactions in Purple and Green Bacteria Utilize Only One Photosystem

In addition to cyanobacteria, two other kinds of prokaryotes are capable of photosynthesis: *purple*

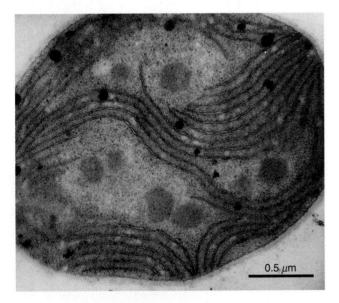

Figure 9-46 Electron Micrograph of a Cyanobacterium
Photosynthetic membranes can be seen in the cell interior. Courtesy of N. J. Lang and B. A. Whitton.

bacteria and *green bacteria*. In these two groups of bacteria, the light reactions also occur in intracellular photosynthetic membranes (Figure 9-47). The photosynthetic membranes of purple bacteria are created by invaginating the plasma membrane into vesicular or tubular channels that occasionally form stacks resembling grana. In contrast, the photosynthetic membranes of green bacteria take the form of spherical cytoplasmic vesicles that appear to be independent of the plasma membrane.

Our understanding of the light reactions in photosynthetic bacteria has been greatly aided by the work of Johann Deisenhofer, Robert Huber, and Hartmut Michel, who utilized X-ray crystallography to determine the three-dimensional structure of the photosynthetic reaction-center complex of the purple bacterium, *Rhodopseudomonas viridis*. As we learned in Chapter 5, this transmembrane protein complex consists of four polypeptide chains: the L and M subunits that span the lipid bilayer, an H subunit bound to the cytosolic membrane surface, and a cytochrome bound to the exterior membrane surface (see Figure 5-26). The L and H subunits contain several prosthetic groups that are involved in the transfer of electrons through the reaction-center complex; these include two molecules of bacteriochlorophyll *b* that comprise the reaction center itself, two additional molecules of bacteriochlorophyll *b*, two bacteriopheophytins, two quinones designated Q_A and Q_B, and a bound atom of iron. Because maximum light absorbance occurs near 870 nm, the reaction center is called *P870*.

When a photon of light is absorbed by the P870 bacteriochlorophyll *b*, an electron is activated and transferred to one of the bacteriopheophytin molecules. From there the electron is passed to Q_A and then Q_B. After Q_B has received two electrons (and picked up two protons from solution), it diffuses out of the reaction-center complex as QH_2 and moves through the lipid bilayer to a *cytochrome b-c_1 complex*. From here the electrons are passed to a cytochrome molecule that transfers them back to P870 (Figure 9-48). Thus the activated electrons that were originally derived from the P870 reaction center are eventually returned to P870, creating a circular pathway of electron flow around a single photosystem. As a result there is no net transfer of electrons from water to $NADP^+$ and hence no accompanying production of oxygen.

During the process of electron flow, ATP is formed by a mechanism that closely resembles the process by which ATP is synthesized in chloroplasts and mitochondria. As electrons are passed from the photosynthetic reaction center to the cytochrome *b-c_1* complex, protons are pumped from the cytosol into the lumen of the photosynthetic membranes. The resulting proton gradient then drives ATP formation utilizing an ATP synthase complex that is similar to the ones utilized by mitochondrial and chloroplast membranes.

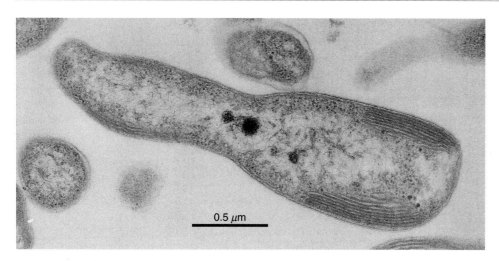

Figure 9-47 Electron Micrograph of a Purple Photosynthetic Bacterium *Photosynthetic membranes are seen lying just beneath the plasma membrane. Courtesy of A. R. Varga.*

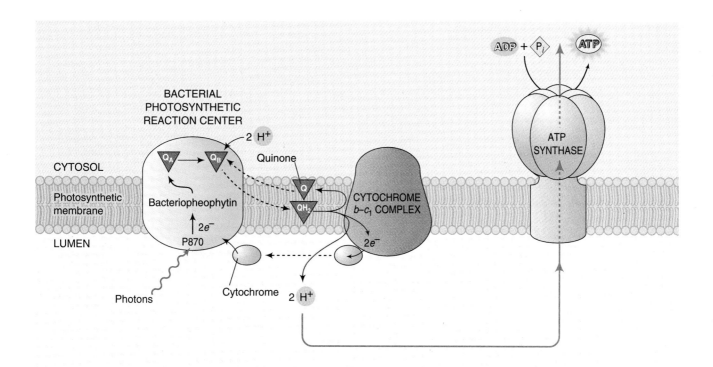

Figure 9-48 Cyclic Pathway of Photosynthetic Electron Transfer in Purple Bacteria *When a photon of light is absorbed by P870, an electron in the reaction-center bacteriochlorophyll b is activated and transferred via bacteriopheophytin to the quinones Q_A and Q_B. After Q_B has received two electrons (and picked up two protons from solution), it diffuses out of the reaction-center complex and through the lipid bilayer to a cytochrome b–c₁ complex, which passes the electrons to a cytochrome molecule that transfers them back to P870. Because the pathway of electron flow is circular, there is no net transfer of electrons from water to NADP⁺ and no accompanying production of oxygen. However, electron flow does lead to the pumping of protons from the cytosol into the lumen of the photosynthetic membranes; the subsequent movement of these protons down their electrochemical gradient (through the ATP synthase and back into the cytosol) drives ATP formation.*

Photosynthetic Bacteria Can Make NADPH Using Electron Donors Other Than Water

Since the preceding pathway of electron flow does not involve the transfer of electrons from water to $NADP^+$, how do purple bacteria generate the NADPH that is required for the carbon-reducing dark reactions of photosynthesis? The answer is that electrons originating in P870 are not always returned to P870 by the cyclic pathway illustrated in Figure 9-48; in some cases, they are instead transferred through a series of carriers to $NADP^+$. When this transfer occurs, the electrons lost from P870 are replaced by an electron donor other than water. One commonly used donor is *hydrogen sulfide* *(H₂S)*, a gas that generates elemental sulfur upon donating electrons to the photosynthetic reaction center. Under such conditions, the overall equation for photosynthesis can be written as follows:

$$6CO_2 + 12H_2S \rightarrow C_6H_{12}O_6 + 6H_2O + 12S$$

The sulfur produced by this process accumulates in large cytoplasmic globules that are eventually excreted from the cell. When substances such as H_2S are substituted for water as electron donors, oxygen is not produced by the light reactions because water molecules are not being split.

Besides employing unusual electron donors in place of water, bacterial photosynthesis also differs from photosynthesis in plants in the kinds of electron acceptors used for the dark reactions. In addition to reducing CO_2, which is the classic electron acceptor for photosynthesis, bacteria also employ electrons derived from the light reactions to reduce hydrogen ions to H_2, and atmospheric nitrogen to ammonia. This latter reaction is especially important because it provides a mechanism for fixing atmospheric nitrogen in a form that can be used by living organisms.

Halobacterium Captures Solar Energy without Fixing CO₂ or Other Electron Acceptors

In addition to the photosynthetic bacteria described so far, the bacterium *Halobacterium halobium* is also capable of capturing solar energy. *Halobacterium* is not photosynthetic in the classic sense because it does not utilize light energy to drive the reduction of carbon or other electron acceptors. However, *Halobacterium* does trap solar energy and utilize it to drive the synthesis of ATP. The crucial component in this process is **bacteriorhodopsin**, a plasma membrane protein related to the rhodopsin molecule that makes vision possible in animals (page 768). Bacteriorhodopsin absorbs light maximally at 560 nm, giving it a vivid purple color. It often congregates into large masses in the plasma membrane, forming "purple patches" that occupy up to half the membrane surface.

In order to investigate how bacteriorhodopsin works, Walther Stoeckenius and Efraim Racker incorpo-

rated this protein along with ATP synthase into artificial phospholipid membranes. When illuminated, these membrane vesicles first generated a proton gradient and then synthesized ATP. Since no electron transfer carriers were present, it was concluded that light causes bacteriorhodopsin to act as a proton pump, producing a proton gradient that is used to drive ATP formation. This phenomenon provides yet another example of the widespread applicability of the chemiosmotic coupling theory, and shows that electrochemical proton gradients can be utilized as energy sources even in the absence of electron transfer.

SUMMARY OF PRINCIPAL POINTS

• Photosynthesis utilizes energy derived from sunlight to drive CO_2 fixation, generating oxygen and carbohydrates as final products. The light reactions of photosynthesis involve the light-induced transfer of electrons from water to $NADP^+$, accompanied by the synthesis of ATP and the production of oxygen. The dark reactions then utilize the NADPH and ATP produced during the light reactions to drive CO_2 fixation by the Calvin cycle.

• The chloroplast is the site of photosynthesis in eukaryotic cells. Chloroplasts are enclosed by a double-membrane envelope consisting of an outer membrane that is relatively permeable to small molecules and an inner membrane that is not. Movement of materials into and out of chloroplasts is controlled by transport systems localized within the inner membrane. The inside of the chloroplast contains a mixture of stacked and unstacked thylakoid membranes that enclose a compartment known as the thylakoid lumen. The compartment between the thylakoids and the chloroplast envelope is the stroma. The light reactions of photosynthesis occur in the thylakoid membranes, while the dark reactions take place in the stroma.

• In the light reactions, electrons are transferred by the photosynthetic electron transfer chain from water to $NADP^+$, accompanied by the release of oxygen and the formation of ATP. The flow of electrons through the chain is driven by photons of light absorbed by two photosystems. During photosynthetic electron transfer, protons are pumped from the stroma into the thylakoid lumen; the protons then diffuse down their electrochemical gradient by passing through the CF_1-CF_0 complex (ATP synthase) and back into the stroma, stimulating the complex to synthesize ATP. Uncoupling agents inhibit ATP formation by making the thylakoid membrane leaky to protons, thereby dissipating the proton gradient.

• ATP synthesis accompanying photosynthetic electron transfer is called photophosphorylation. Noncyclic photophosphorylation accompanies the flow of electrons from water to $NADP^+$ and involves two coupling sites for ATP synthesis: one associated with the oxidation of water by photosystem II, and the other associated with the oxidation of plastoquinone by the cytochrome b_6-f complex. Cyclic photophosphorylation is based on a circular flow of electrons around photosystem I and hence utilizes only the second coupling site for ATP synthesis.

• The dark reactions of photosynthesis utilize NADPH and ATP generated by the light reactions to drive the fixation of CO_2 in the chloroplast stroma. CO_2 fixation is carried out by the Calvin cycle enzyme rubisco, which joins CO_2 to RuBP to produce two molecules of 3-PGA. Energy derived from NADPH and ATP is then used to reduce 3-PGA to glyceraldehyde 3-phosphate, which serves as a precursor for the synthesis of carbohydrates such as sucrose, starch, and glucose.

• The efficiency of photosynthesis can be severely limited by photorespiration, which uses molecular O_2 to oxidize the reduced carbon atoms generated by photosynthesis. Photorespiration occurs when O^2 substitutes for CO_2 in the reaction catalyzed by rubisco; the result is the oxidation of RuBP and the release of some of its carbon atoms as CO_2. Because photorespiration is not coupled to the synthesis of ATP or any other high-energy compound, it wastes energy that has been trapped by photosynthesis. C_4 plants overcome this problem by fixing CO_2 in mesophyll cells using PEP carboxylase, an enzyme that does not confuse O_2 for CO_2. PEP carboxylase generates C_4 acids that are transported to bundle sheath cells, where the C_4 acids are broken down to release CO_2 that is fixed by the Calvin cycle enzyme rubisco. CAM plants utilize a similar two-step pathway for CO_2 fixation, but rather than carrying out the PEP carboxylase step in bundle sheath cells and the rubisco step in mesophyll cells, CAM plants carry out both steps in mesophyll cells, the first step at night and the second during the day.

• Although photosynthetic bacteria lack chloroplasts, they have photosynthetic membranes that carry out the light reactions of photosynthesis. Cyanobacteria utilize a photosynthetic pathway similar to the one employed by higher plants, but purple and green bacteria utilize a different sequence of electron carriers and light-absorbing pigments. Moreover, they do not produce oxygen because electron donors such as hydrogen sulfide are employed in place of water. The dark reactions of bacterial photosynthesis are also unusual in that electrons derived from the light reactions can be used to reduce hydrogen or nitrogen in addition to the typical electron acceptor, CO_2. In *Halobacterium halobium,* solar energy is trapped by the membrane pigment bacteriorhodopsin, generating a proton gradient that is employed to drive ATP formation in the absence of electron transfer or photosynthesis.

SUGGESTED READINGS

Books

Bogorad, L., and I. K. Vasil, eds. (1991). *The Photosynthetic Apparatus: Molecular Biology and Operation*, Academic Press, San Diego.

Briggs, W. R., ed. (1989). *Photosynthesis*, Liss, New York.

Foyer, C. H. (1984). *Photosynthesis*, Wiley, New York.

Hoober, J. K. (1984). *Chloroplasts*, Plenum, New York.

Salisbury, F. B., and C. W. Ross (1992). *Plant Physiology*, 4th Ed., Wadsworth, Belmont, CA, Chs. 10 and 11.

Articles

Barber, J., and B. Andersson (1994). Revealing the blueprint of photosynthesis, *Nature* 370: 31-34.

Bogorad, L. (1981). Chloroplasts, *J. Cell Biol.* 91:256s-270s.

Calvin, M. (1962). The path of carbon in photosynthesis, *Science* 135:879-889.

Deisenhofer, J., O. Epp, K. Miki, R. Huber, and H. Michel (1985). The structure of the protein subunits in the photosynthetic reaction center of *Rhodopseudomonas viridis* at 3 Å resolution, *Nature* 318:618-624.

Douce, R., and J. Joyard (1990). Biochemistry and function of the plastid envelope, *Annu. Rev. Cell Biol.* 6:173-216.

Flügge, U.-I., and H. W. Heldt (1991). Metabolite translocators of the chloroplast envelope, *Annu. Rev. Plant Physiol. Plant Mol. Biol.* 42:129-144.

Golbeck, J. H. (1992). Structure and function of photosystem I, *Annu. Rev. Plant Physiol. Plant Mol. Biol.* 43:293-324.

Govindjee and W. J. Coleman (1990). How plants make oxygen, *Sci. Amer.* 262 (February):50-58.

Hartman, F. C., and M. R. Harpel (1994). Structure, function, regulation, and assembly of D-ribulose-1,5-bisphosphate carboxylase/oxygenase, *Annu. Rev. Biochem.* 63: 197-234.

Hatch, M. D. (1987). C_4 photosynthesis: a unique blend of modified biochemistry, anatomy, and ultrastructure, *Biochim. Biophys. Acta* 895:81-106.

Joliot, P., and B. Kok (1975). Oxygen evolution in photosynthesis, In: *Bioenergetics of Photosynthesis* (Govindjee, ed.), Academic Press, New York, pp. 387-412.

Knaff, D. B. (1988). The photosystem I reaction centre, *Trends Biochem. Sci.* 13:460-461.

Knaff, D. B. (1991). Regulatory phosphorylation of chloroplast antenna proteins, *Trends Biochem. Sci.* 16:82-83.

Krauss, N., et al. (1993). Three-dimensional structure of system I of photosynthesis at 6 Å resolution, *Nature* 361:326-331.

Kühlbrandt, W., D. N. Wang, and Y. Fujiyoshi (1994). Atomic model of plant light-harvesting complex by electron crystallography, *Nature* 367: 614-621.

Miller, K. R., and M. K. Lyon (1985). Do we really know why chloroplast membranes stack? *Trends Biochem. Sci.* 10:219-222.

Murakami, S. (1992). Structural and functional organization of the thylakoid membrane system in photosynthetic apparatus, *J. Electron Microsc.* 41:424-433.

Portis, A. R., Jr. (1992). Regulation of ribulose 1,5-bisphosphate carboxylase/oxygenase activity, *Annu. Rev. Plant Physiol. Plant Mol. Biol.* 43:415-437.

Racker, E., and W. Stoeckenius (1974). Reconstitution of purple membrane vesicles catalyzing light-driven proton uptake and adenosine triphosphate formation, *J. Biol. Chem.* 249:662-663.

Rutherford, A. W. (1989). Photosystem II, the water-splitting enzyme, *Trends Biochem. Sci.* 14:227-232.

Trissl, H.-W., and C. Wilhelm (1993). Why do thylakoid membranes from higher plants form grana stacks?, *Trends Biochem. Sci.* 18: 415-419.

Vermaas, W. (1993). Molecular biological approaches to analyze photosystem II structure and function, *Annu. Rev. Plant Physiol. Plant Mol. Biol.* 44:457-481.

Youvan, D. C., and B. L. Marrs (1987). Molecular mechanisms of photosynthesis, *Sci. Amer.* 256 (June):42-48.

C h a p t e r   1 0

The Nucleus and Transcription of Genetic Information

In Chapter 3, which introduced the topic of information flow within cells, we learned that DNA base sequences code for the amino acid sequences of protein molecules. The transmission of information from DNA to protein involves two distinct stages: the *transcription* of DNA sequences into messenger RNA, and the *translation* of messenger RNA into amino acid sequences during protein synthesis. The mechanisms underlying these two stages in the flow of genetic information are considered in detail in this and the following chapter.

To understand the events that occur during transcription, one first must be acquainted with the way in which DNA is organized within the cell. In eukaryotic organisms the vast bulk of the DNA is localized within the **nucleus,** a large double-membrane enclosed organelle that has long been known to play a key role in directing cellular activities. Although prokaryotic cells lack a true membrane-enclosed nucleus, they too contain a special region in which the genetic material is localized. In this chapter we will begin by examining the organization of the nuclear regions of eukaryotic and prokaryotic cells, and will then describe how the genetic information contained in these regions is transcribed and processed for subsequent use by the protein-synthesizing machinery of the cytoplasm.

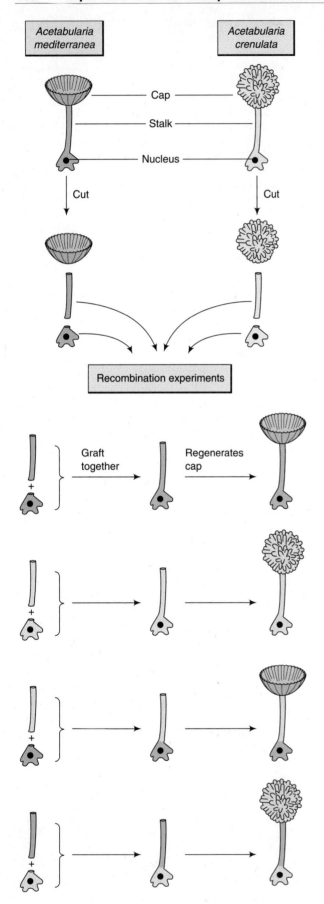

Recombination experiments

Graft together → Regenerates cap →

THE NUCLEUS AND NUCLEOID

The role played by the nucleus in determining the properties of the rest of the cell was appreciated by biologists even before it was known that DNA contains genetic information. One of the first studies to demonstrate the influence exerted by the nucleus was carried out by Joachim Hämmerling and Jean Brachet in the early 1950s on the unicellular alga, *Acetabularia.* Several distinct species of *Acetabularia* can be distinguished, each containing a uniquely shaped cytoplasmic *cap* at one end of the cell. The large size of this single-celled organism and the localization of the nucleus at one end of the cell make it relatively simple to remove the nucleus from one type of cell and transplant it to another. To investigate the influence exerted by the nucleus, Hämmerling and Brachet removed the nucleus from one species of *Acetabularia* and transplanted it to a cell of another species whose nucleus and cap had been removed. The resulting hybrid cell always formed a new cap whose shape was determined by the nucleus it received (Figure 10-1). These findings made it clear that the nucleus contains the genetic information required for controlling activities that occur elsewhere in the cell.

The Nucleus of Eukaryotic Cells Is Bounded by a Double-Membrane Envelope

Although the nucleus is usually pictured as a large spherical membrane-enclosed organelle located near the center of the cell, considerable variations exist in nuclear size, shape, and position. The nucleus typically accounts for about 5 to 10 percent of the total cell volume, but in some cases this value may reach 80 percent or more. Such variations in size are determined both by the amount of DNA present and the physiological state of the cell. Nuclei range from one to several hundred micrometers in diameter, with typical nuclei measuring about 5 to 15 μm across. The nucleus is usually spherical in shape, but it also may be ellipsoid, flattened, lobed, or irregular, depending on the cell type. The position of the nucleus within the cell also varies and is often characteristic for a given type of cell. Although a single nucleus per cell is the general rule, cells that are exceptionally large may contain multiple nuclei. A striking example occurs in the skeletal muscle of vertebrates, where individual muscle cells reach several

Figure 10-1 *Evidence for Nuclear Control of Cytoplasmic Events*   *When a nucleus is transplanted from one species of Acetabularia to a cell of another species whose nucleus and cytoplasmic cap have been removed, the resulting hybrid cell forms a new cap whose shape is determined by the transplanted nucleus.*

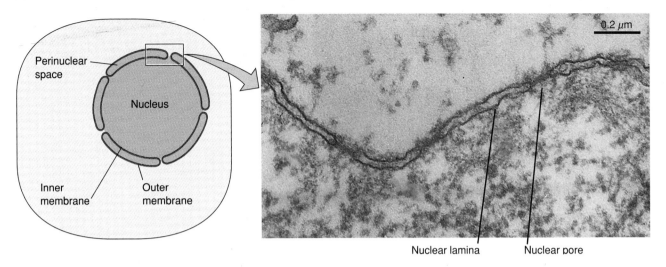

Nuclear lamina   Nuclear pore

Figure 10-2 *The Nuclear Envelope* *The electron micrograph clearly shows the two membranes that make up the nuclear envelope. In several locations, the continuity of the envelope is interrupted by nuclear pore complexes. The fuzzy material adhering to the inner surface of the inner membrane is called the nuclear lamina. Courtesy of E. W. Daniels.*

millimeters or even centimeters in length. These enormous cells usually contain dozens or even hundreds of separate nuclei.

The typical eukaryotic nucleus is comprised of five main structural components: (1) a double-membrane **nuclear envelope** that separates the internal contents of the nucleus from the cytoplasm, (2) a fluidlike **nucleoplasm** that contains the soluble material of the nucleus, (3) one or more small spherical **nucleoli** that are involved in the formation of ribosomes, (4) a **nuclear matrix** that provides a skeletal framework for the nucleus, and (5) DNA-containing fibers that are called **chromatin** fibers when they are dispersed throughout the nucleus or **chromosomes** when they become organized into discrete structures. In the following sections we will examine each of these components in turn.

The Nuclear Envelope Is Supported by the Nuclear Lamina and Contains Numerous Nuclear Pore Complexes

The presence of a nuclear envelope in eukaryotic cells establishes a membrane barrier between the nuclear compartment and the cytoplasm that does not exist in prokaryotes. Since light microscopy reveals only a narrow fuzzy border at the outer surface of the nucleus, little was known about the structure of the nuclear envelope prior to the advent of electron microscopy. Thin-section electron micrographs reveal that the nucleus is bounded by a **nuclear envelope** consisting of two membranes: an *outer nuclear membrane* abutting the cytoplasmic compartment and an *inner nuclear membrane* adjacent to the interior of the nucleus (Figure 10-2). Between the two membranes of

the nuclear envelope is a *perinuclear space* that typically measures from 15 to 30 nm across. The outer nuclear membrane often exhibits points of continuity with the endoplasmic reticulum and may contain ribosomes attached to its outer surface. The connections between the outer nuclear membrane and membranes of the ER make the perinuclear space continuous with the lumen of the ER (see Figure 7-4).

Components of the cytoplasmic cytoskeleton (microtubules, actin filaments, and intermediate filaments) are often associated with the outer surface of the nuclear envelope, helping to anchor the nucleus and to maintain its proper shape. Directly beneath the inner surface of the inner nuclear membrane is a thin layer of fibrous material, the **nuclear lamina,** which acts as a supporting structure for the nuclear envelope. The nuclear lamina is constructed from intermediate filaments made of proteins called *lamins.*

One of the most distinctive features of the nuclear envelope is the presence of specialized openings in the membrane known as **nuclear pore complexes (NPCs).** Each NPC is a wheel-shaped structure measuring about 120 nm in external diameter and exhibiting an eight-sided symmetry (Figure 10-3). The hub of the wheel consists of a cylindrical plug of material called the *central transporter.* Eight *spokes* project outward from the central transporter and attach to *nucleoplasmic* and *cytoplasmic rings* located on opposite sides of the pore complex. The surface of the cytoplasmic ring is often covered with eight *cytoplasmic granules,* while the nucleoplasmic ring is associated with thin filaments that project into the nucleoplasm, forming a cagelike structure called the *NPC basket* (Figure 10-4). In at least some organisms the NPC basket is attached to an inter-

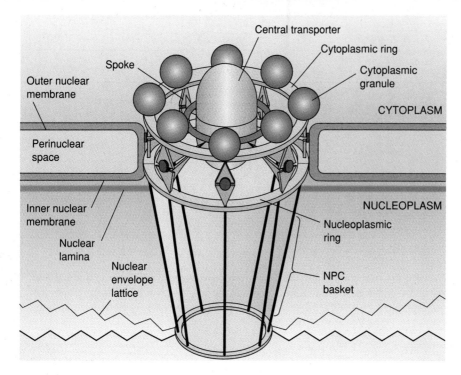

Figure 10-3 *Model of the Nuclear Pore Complex*   *This three-dimensional model illustrates the relationship between the nuclear pore complex and the inner and outer membranes of the nuclear envelope. The eight spokes and eight cytoplasmic granules impart an eight-sided symmetry to the pore complex.*

woven lattice of filaments, called the *nuclear envelope lattice,* which is distinct from the nuclear lamina.

Nuclear Pore Complexes Are Channels through Which Molecules Enter and Exit the Nucleus

Nuclear pore complexes usually occupy from 10 to 25 percent of the total surface area of the nuclear envelope. The first evidence that these structures are involved in transporting substances into and out of the nucleus was obtained in 1965 by Carl Feldherr, who injected gold particles of various sizes into the cytoplasm of amoebae and then examined the cells by thin-section electron microscopy. Shortly after injection, the gold particles were seen passing through the nuclear pore complexes and into the nucleus (Figure 10-5). The rate of particle entry into the nucleus was inversely related to particle diameter; that is, the larger the gold particle, the slower it entered the nucleus. Particles larger than 10 nm in diameter were excluded entirely. Since the overall diameter of a nuclear pore complex is an order of magnitude larger than the 10-nm gold particles, it was concluded that the pore complex contains a small *passive diffusion channel* through which small particles and molecules can freely move. Studying the rates at which radioactive molecules of varying sizes pass into the nucleus has led to the conclusion that the passive diffusion channel has a functional diameter of about 10 nm, which is sufficient to permit the unhin-

dered diffusion of ions and small molecules, including proteins up to about 4000 daltons in size. Proteins in the range of 4000 to 60,000 daltons may also move through the passive diffusion channel, but at rates that are inversely proportional to their sizes. For proteins at the larger end of this range, passive diffusion is too slow to be physiologically important.

Many of the molecules that move either from the nucleus to the cytoplasm or in the opposite direction are too large to pass through the nuclear pore complexes by passive diffusion. Included in this category are RNA molecules that are synthesized in the nucleus and then exported to the cytoplasm, and nuclear proteins that are synthesized in the cytoplasm but function within the nucleus. Although molecules in these two categories are usually too large to move through the passive diffusion channels, electron microscopy has revealed that such molecules still pass through the nuclear pore complexes. Some of the earliest evidence involved the behavior of large RNA-protein complexes that are produced in the nucleus and then transported to the cytoplasm. These *ribonucleoprotein (RNP)* particles often measure 50 nm or more in diameter, which is significantly larger than the 10-nm functional diameter of the passive diffusion channel of the nuclear pore complexes. Yet electron microscopy has shown that such RNP particles leave the nucleus by passing through nuclear pore complexes, becoming compressed and elongated as they do so (Figure 10-6). It therefore has

CYTOPLASMIC FACE

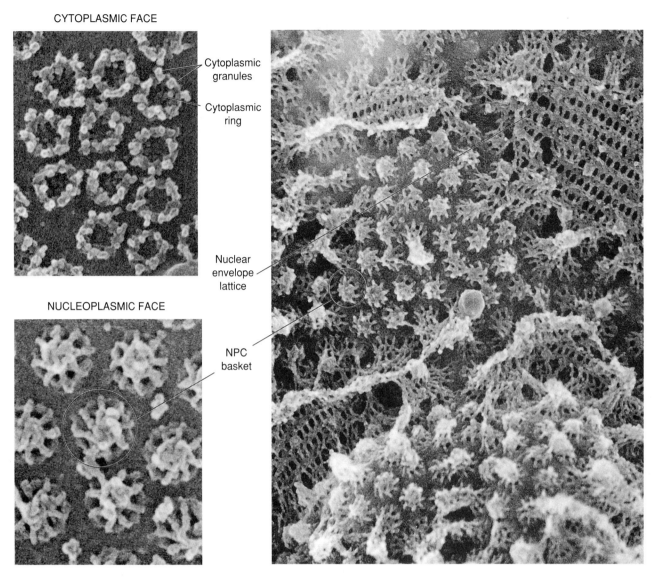

NUCLEOPLASMIC FACE

Cytoplasmic granules

Cytoplasmic ring

Nuclear envelope lattice

NPC basket

Figure 10-4 *Nuclear Pore Complexes Visualized by High-Resolution Scanning Electron Microscopy* (Top left) *Numerous nuclear pore complexes are visible in this view of the cytoplasmic face of a nuclear envelope isolated from amphibian oocytes. The cytoplasmic granules that cover the cytoplasmic ring of each pore complex are clearly evident. (Bottom left) When the nuclear envelope is viewed from the nucleoplasmic side, the NPC basket associated with each pore complex can be seen. (Right) In this lower-magnification view of the nucleoplasmic surface of the nuclear envelope, several patches of nuclear envelope lattice are retained, attached to the innermost aspect of the NPC baskets. Courtesy of M. W. Goldberg.*

been concluded that in addition to the passive diffusion of smaller molecules, the nuclear pore complex also actively transports selected larger molecules through a central *transport channel* (presumably corresponding to the central transporter illustrated in Figure 10-3). The transport channel, which has a functional diameter of about 25 nm, behaves as if it is *gated;* that is, the channel is normally closed but can be opened to accommodate the passage of molecules of certain size, shape, and chemical characteristics.

How does the transport channel recognize molecules that need to be transported into or out of the nucleus and yet are too large to pass through the pore complex by passive diffusion? One molecule that has

served as a model for addressing this question is *nucleoplasmin,* a protein of 165,000 daltons that is synthesized in the cytoplasm and then accumulated within the nucleus. To investigate the mechanism of nucleoplasmin uptake, experiments have been carried out in which gold particles too large to move through the passive diffusion channels were coated with nucleoplasmin and then injected into the cytoplasm of frog oocytes. After such treatment, the gold particles pass rapidly through the nuclear pore complexes and into the nucleus. However, movement of the particles into the nucleus can be prevented by removing a short amino acid sequence called the **nuclear localization sequence (NLS)** from the nucleoplasmin molecule. This is only

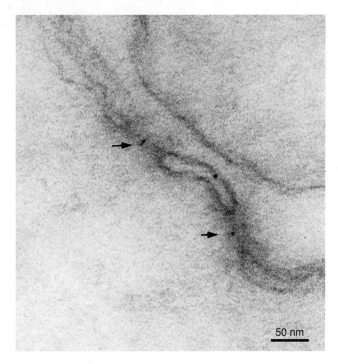

Figure 10-5 *Gold Particles Migrating through the Nuclear Pore Complex* *A cell of the giant amoeba* Chaos *was injected with colloidal gold particles 1 minute before fixation for electron microscopy. In this section two nuclear pore complexes are seen, each containing a centrally located gold particle* (arrows). *Courtesy of C. M. Feldherr.*

one of numerous examples in which a short stretch of amino acids has been found to target a protein to a specific cellular site after it has been synthesized on cytoplasmic ribosomes (page 511). Cytoplasmic proteins containing a nuclear localization sequence are recognized by special NLS-binding proteins, which bind to the sequence and deliver the protein to the nuclear pore complex. After becoming associated with the cytoplasmic surface of the nuclear pore complex, the NLS-containing proteins are actively transported through the transport channel and into the nucleus by a process that utilizes energy derived from the hydrolysis of ATP.

Although nuclear pore complexes are usually observed only in the nuclear envelope, they have also been detected in an unusual class of cytoplasmic membranes known as *annulate lamellae,* which occur in developing egg and sperm cells, and in certain embryonic and tumor cells. Annulate lamellae consist of stacks of several dozen cytoplasmic membranes that contain pore complexes similar to those of the nuclear envelope (Figure 10-7). Since the number of pore complexes present in the nuclear envelope changes with differing states of nuclear activity, it has been suggested that the annulate lamellae might serve as a repository of extra nuclear pore complexes. Such an arrangement would permit new nuclear pore complexes to be

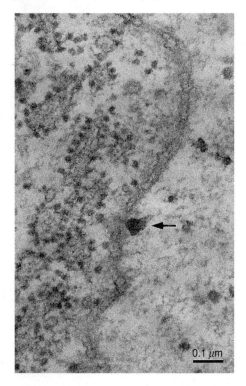 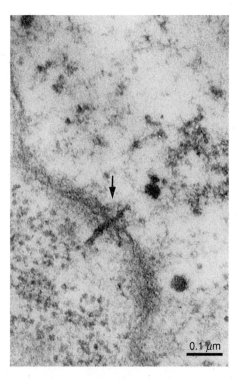

Figure 10-6 *Granules Consisting of RNA and Protein Passing through the Nuclear Pore Complex* (Left) *A spherical granule* (arrow) *residing in the nucleus is stretched out slightly in the direction of the nuclear pore complex.* (Right) *A granule moving through the nuclear pore complex is seen to be compressed into an elongate shape as it passes through. The nucleus is on the right. Courtesy of B. J. Stevens.*

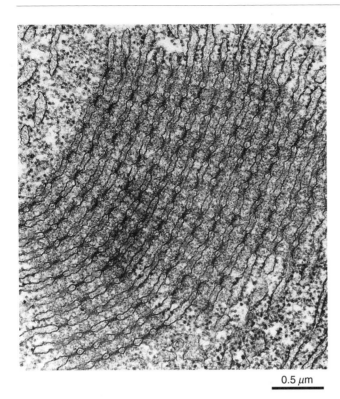

0.5 μm

Figure 10-7 *Annulate Lamellae in a Zebrafish* *This electron micrograph shows an array of annulate lamellae comprised of several dozen cytoplasmic membranes containing pores that resemble the nuclear pore complexes of the nuclear envelope. Courtesy of R. G. Kessel.*

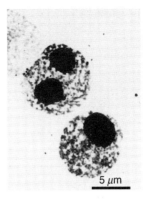

5 μm

Figure 10-8 *The Nucleolus as Seen by Light Microscopy* *The two isolated nuclei visible in this micrograph were stained with silver. Two prominent nucleoli are visible in one nucleus, and one nucleolus can be seen in the other. Courtesy of J. Olert.*

rapidly incorporated into the nuclear envelope as needed. However, it would not explain why annulate lamellae are limited to a few selected cell types, and so it is equally probable that they serve some other, as yet unidentified, function.

The Nucleoplasm Is the Soluble Portion of the Nucleus

The **nucleoplasm** (also called the *nuclear sap*) is an amorphous, fluidlike material that contains the soluble components of the nucleus, making it comparable to the cytosol fraction of the cytoplasm. Molecules residing in the nucleoplasm can be isolated by extracting purified nuclei with dilute aqueous solutions, which yields a mixture of three classes of substances: (1) *small molecules* such as coenzymes, ions, and nucleotides, (2) *proteins,* including enzymes that manufacture the nucleotides required for nucleic acid synthesis, as well as regulatory proteins and enzymes involved in DNA replication and transcription, and (3) *RNA molecules* of various sizes and types, most of which are bound to protein molecules to form *ribonucleoprotein (RNP) complexes.* A special class of small nuclear RNA molecules are found in RNP complexes called *snurps* or *snRNPs (small nuclear RNP parti-*

cles), which carry out splicing reactions involved in the processing of messenger RNA (page 431) and ribosomal RNA (page 481). Other RNP complexes containing larger RNA molecules have been observed passing out through nuclear pore complexes (see Figure 10-6), suggesting that they function in the cytoplasm. Some of these RNP particles are ribosomal subunits, while others contain messenger RNA destined for translation on cytoplasmic ribosomes.

The Nucleolus Contains Granules and Fibrils Involved in Ribosome Formation

One of the most striking structural features of the eukaryotic nucleus is the presence of a large spherical structure known as a **nucleolus** (Figure 10-8). Although most cells contain only one or two nucleoli, the occurrence of several more is not uncommon and in certain situations hundreds or even thousands may be present. The typical nucleolus is a spherical structure measuring several micrometers in diameter, but wide variations in size and shape are observed. Because of their relatively large size, nucleoli are readily visible with the light microscope and hence were first observed more than 200 years ago. Although silver staining techniques gave early light microscopists a striking view of nucleolar structure and even led to the description of a "threadlike" substructure within the nucleolus, it was not until the advent of electron microscopy that the fibrils and granules that are contained within the nucleolus became clearly visible.

In Chapter 11 we will examine the role played by the nucleolus in manufacturing the ribosomal subunits that are required for cytoplasmic protein synthesis. The relationship of the nucleolar fibrils and granules to the process of ribosome formation will be discussed at that time.

The Nuclear Matrix Provides a Supporting Framework for the Nucleus

Roughly 80 to 90 percent of the total nuclear mass is accounted for by chromatin fibers, which consist of DNA and associated protein. Because they occupy the bulk of the nucleus, it might be expected that removing the chromatin fibers would cause the nucleus to collapse into a relatively structureless mass. It was therefore surprising when Ronald Berezney and Donald Coffey first reported in 1974 that an insoluble filament network retaining the overall shape of the nucleus remains behind after more than 95 percent of the chromatin has been removed by a combination of nuclease and detergent treatments (Figure 10-9). This network, which was named the **nuclear matrix,** is analogous to the cytoskeleton residing in the cytoplasm, although the filaments that make up the nuclear matrix differ in both appearance and protein composition from those of the cytoskeleton. Nonetheless several proteins, including actin, are common to the cytoskeleton and nuclear matrix.

Although the function of the nuclear matrix is not completely understood, a close connection between the matrix and chromatin fibers is suggested by the discovery that isolated nuclear matrix preparations always contain small amounts of tightly bound DNA and RNA. Nucleic acid hybridization techniques have revealed that the tightly bound DNA is enriched in sequences that are actively being transcribed into RNA. Moreover, when ³H-thymidine is employed to monitor DNA synthesis in dividing cells, the newly synthesized radioactive DNA is also found to be preferentially associated with the nuclear matrix. The preceding observations suggest that the nuclear matrix is involved in anchoring chromatin fibers at locations where DNA or RNA is being synthesized.

Chromatin Fibers Consist of DNA Associated with Histone and Nonhistone Proteins

The nuclear DNA of eukaryotic cells is bound to a diverse group of proteins, forming a mass of intertwined

Remnant of
nucleolus

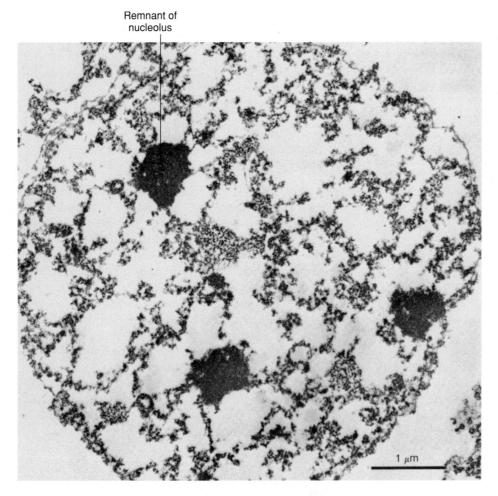

Figure 10-9 *Electron Micrograph of an Isolated Nuclear Matrix* *The nuclear matrix is isolated by subjecting purified nuclei to detergent extraction and nuclease digestion to remove the nuclear envelope, the nucleoplasm, and most of the chromatin fibers. In addition to the nuclear matrix, the remnants of several nucleoli can be seen in this micrograph. Courtesy of R. Berezney.*

chromatin fibers that collectively account for the bulk of the nuclear contents. Although chromatin fibers are normally dispersed throughout the nucleus, during cell division (and in a few other situations) they become condensed into compact structures referred to as **chromosomes.** Depending on the conditions under which they are examined, chromatin fibers usually measure between 10 and 30 nm in diameter. In the intact nucleus they form attachments both to the nuclear matrix and to the inner surface of the nuclear envelope, often adjacent to nuclear pore complexes.

Our current understanding of the molecular organization of chromatin fibers has been made possible by the development of techniques for extracting and purifying chromatin from isolated nuclei. A common approach is to rupture nuclei by exposing them to hypotonic solutions, followed by differential centrifugation to sediment the chromatin fibers. Chromatin can also be isolated by extracting nuclei with solutions of high ionic strength, such as 1.0 M NaCl. In either case, centrifugation is employed to separate the isolated chromatin fibers from other nuclear components. Although some variability is observed in the composition of chromatin fibers isolated from different sources and by different techniques, protein typically accounts for about two-thirds of the total mass, DNA for about one-third, and RNA for no more than a few percent.

Because protein is present in such high concentrations in chromatin preparations, considerable effort has been expended in characterizing these proteins and investigating their roles in chromatin structure and function. Such investigations have led to the identification of two main classes of proteins in chromatin fibers: a small group of basic proteins called *histones* and a more diverse class of proteins referred to as *nonhistones*.

Histones Are Small Basic Proteins That Are Similar in All Eukaryotic Cells

The **histones** are a group of small proteins whose high content of the basic amino acids lysine and arginine gives them a strong positive charge at typical values of intracellular pH. Histones are divided into five main types designated *H1, H2A, H2B, H3,* and *H4* (Table 10-1). Chromatin contains roughly equal numbers of H2A, H2B, H3, and H4 molecules, and about half that number of histone H1 molecules. These proportions are remarkably constant among all eukaryotic cells, regardless of the kind of cell or its physiological state (Figure 10-10). In addition to the five main classes of histones, unusual variants occur in a few cell types. The red blood cells of birds, for example, produce a special histone called *H5* as the cells mature and their chromatin fibers become inactivated.

One of the more remarkable features of histones is the extent to which their amino acid sequences have been conserved over evolutionary time. For example, a

Table 10-1 The Five Major Types of Histones

Type of Histone	Lysine	Arginine	Number of Amino Acids	Size (daltons)
H1	29%	1%	215	23,000
H2A	11%	9%	129	14,000
H2B	16%	6%	125	13,800
H3	10%	13%	135	15,300
H4	11%	14%	102	11,200

Note: Data are for calf thymus histones; molecular weights have been rounded off.

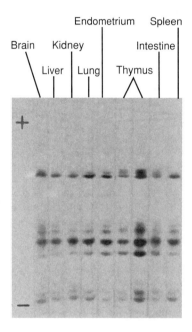

Figure 10-10 *Polyacrylamide Gel Electrophoresis of Histones Isolated From Various Tissues of the Calf* Note the similarity in the types of histones present in each tissue. Courtesy of R. Chalkley.

comparison of the amino acid sequences of histone H4 molecules isolated from pea seedlings and calf thymus reveals that only two out of 102 amino acids are different. This striking similarity in histone H4 sequence in two organisms that diverged from each other millions of years ago indicates that even minor changes in histone structure can have a severe impact on nuclear function.

Although the amino acid sequence of histone molecules is highly conserved, variability in histone structure can be generated in other ways. For example, histone H1 exists in multiple forms, each characterized by a slightly different amino acid sequence. Histones are also subject to reversible chemical modification reactions such as *acetylation, methylation, phosphorylation,* and *ADP ribosylation* (Figure 10-11). These reactions alter the properties of histone molecules in ways that influence chromatin assembly, replication, and transcription.

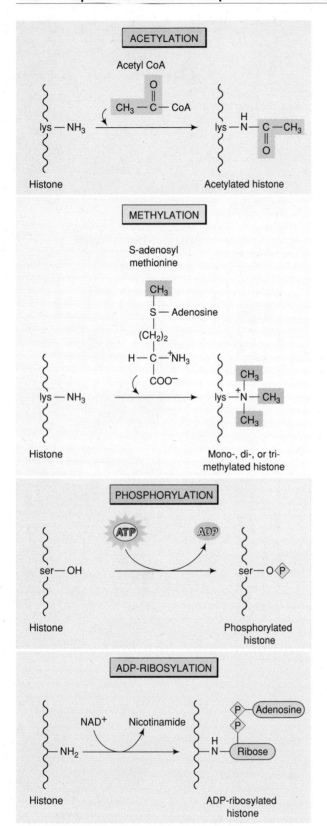

Figure 10-11 *Reactions for the Covalent Modification of Histone Molecules* *The acetyl, methyl, phosphate, and ADP-ribosyl groups attached to histone molecules by these reactions can be removed by appropriate hydrolytic enzymes. The structures of the NAD⁺ and nicotinamide molecules involved in ADP-ribosylation are illustrated in greater detail in Figure 6-18.*

Histones are present in almost all eukaryotic nuclei. However, in the sperm cells of some animals, histones are replaced by small proteins known as **protamines.** Arginine accounts for roughly two-thirds of the amino acids present in protamines, making them extremely basic. X-ray diffraction studies of intact sperm heads have revealed that protamine molecules fit snugly into the minor groove of the DNA double helix. This association is stabilized by ionic bonds between the positively charged arginine residues in protamine and negatively charged phosphate groups in DNA. Such an arrangement creates a stable crystalline structure that efficiently packages and protects the DNA of sperm cells, whose main function is to deliver an intact set of DNA molecules to the egg.

Nonhistone Proteins Are a Heterogeneous Group of DNA-Associated Molecules

In contrast to histones, the **nonhistone proteins** are a diverse group of molecules (Figure 10-12). Whereas histones are present in all cell types in virtually the same concentration relative to the total nuclear DNA content, the nonhistone protein composition of different cells tends to be quite variable. Many nonhistone proteins are enzymes; some of them, such as DNA and RNA polymerases, employ DNA as a substrate, while others are protein-modifying enzymes, such as the protein kinases and acetylases that catalyze the modification of chromatin proteins (Table 10-2).

In addition to enzymes, structural and regulatory proteins have also been identified among the nonhistone proteins. One of the most abundant members of

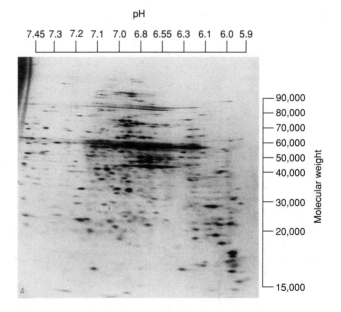

Figure 10-12 *Two-Dimensional Polyacrylamide Gel Electrophoresis of Nonhistone Proteins from a Human Cell* *Note the large number of different proteins. Courtesy of E. H. McConkey.*

Table 10-2   Some Enzymes Identified in the Nonhistone Protein Fraction

Enzyme	Function
Enzymes That Employ Nucleic Acids as Substrates	
DNA polymerases	DNA synthesis
RNA polymerases	RNA synthesis
Nucleases	Cleavage of DNA and/or RNA
DNA ligase	Joining DNA segments during DNA replication and repair
Poly-A polymerase	Addition of poly-A tails to messenger RNA
DNA methylase	Methylation of DNA
Topoisomerases	Relaxation of supercoiled DNA
Helix-destabilizing enzymes	Unwinding of DNA double helix and stabilization of resulting single-stranded DNA
Enzymes That Act on Chromatin Proteins	
Proteases	Protein cleavage
Histone acetylases and deacetylases	Acetylation and deacetylation of histones
Protein kinases	Phosphorylation of histones and nonhistones
Histone methylases	Histone methylation

this class are the **high-mobility group (HMG) proteins,** named because of their rapid electrophoretic mobility. As we will see later in the chapter, HMG proteins are preferentially associated with chromatin fibers whose DNA is being actively transcribed. Among the other nonhistone proteins are proteins that provide the structural backbone of chromosomes (page 414), and proteins called *transcription factors,* which regulate gene transcription (page 422).

The Nucleosome Is the Fundamental Unit of Chromatin Structure

The DNA contained within a typical nucleus would measure a meter or more in length if it were completely extended, whereas the nucleus itself is usually no more than 5–10 μm in diameter. A significant topological problem is therefore encountered when packaging such an enormous length of DNA into a nucleus that is almost a million times smaller in diameter. This task is accomplished largely through the action of nuclear protein molecules, which bind to DNA and fold it into chromatin fibers.

One of the first insights into this folding process emerged in the late 1960s, when X-ray diffraction studies carried out by Maurice Wilkins revealed that purified chromatin fibers exhibit a repeating structure with a periodicity of 10 nm. This repeat structure is not observed with either DNA or histones alone, but it does appear when histones and DNA are mixed together. It was therefore concluded that histones impose a repeating structural organization upon DNA. A clue to the nature of this structure was provided in 1974, when Ada and Donald Olins published electron micrographs of chromatin fibers prepared in a new way. Nuclei swollen by exposure to hypotonic conditions were gently fixed

with formalin and then stained, a sequence that avoids the harsh solvents employed in earlier procedures for preparing chromatin for microscopic examination. Chromatin fibers viewed in this way appear as a series of tiny particles attached to one another by thin filaments (Figure 10-13). This "beads-on-a-string" appearance led to the suggestion that the beads are made of histones and the thin filaments connecting the beads correspond to DNA.

On the basis of electron microscopic evidence alone, it would have been difficult to determine whether the particles are a normal component of chromatin structure or an artifact generated during sample preparation. Fortunately, independent evidence for the existence of a repeating structure in chromatin emerged at about the same time from the studies of Dean Hewish and Leigh Burgoyne, who discovered that rat liver nuclei contain a nuclease that is capable of cleaving the DNA helix in chromatin fibers. After exposing chromatin to this nuclease, the partially degraded DNA was purified to remove chromatin proteins and then analyzed by gel electrophoresis. The result was a distinctive pattern of fragment sizes in which the small-

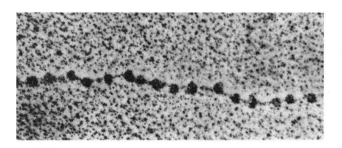

Figure 10-13 *The Beads-on-a-String Appearance of an Isolated 10-nm Chromatin Fiber* Courtesy of J. Griffith.

est piece of DNA measured about 200 base pairs in length, and the remaining fragments were exact multiples of 200 base pairs (Figure 10-14). Since nuclease digestion of protein-free DNA does not generate this pattern of fragments, it was concluded that (1) chromatin proteins are clustered along the DNA molecule in a regular pattern that repeats at intervals of roughly 200 base pairs, and (2) the DNA located between these protein clusters is most susceptible to nuclease digestion, yielding fragments that are multiples of 200 base pairs in length.

The preceding observations raised the question of whether the protein clusters postulated to occur at 200 base-pair intervals correspond to the spherical particles observed in electron micrographs of chromatin fibers. Answering this question required a combination of the nuclease digestion and electron microscopic approaches. Chromatin was therefore briefly exposed to *micrococcal nuclease,* a bacterial enzyme that, like the rat liver nuclease, cleaves chromatin DNA at intervals of 200 base pairs. The fragmented chromatin was then separated into fractions of varying sizes by moving-zone centrifugation and examined by electron microscopy. The smallest fraction was found to contain single spherical particles, the next fraction contained clusters of two particles, the succeeding fraction contained clusters of three particles, and so forth (Figure 10-15). When DNA was isolated from these fractions and analyzed by gel electrophoresis, the DNA from the fraction containing single particles measured 200 base pairs in length, the DNA from the fraction containing clusters of two particles measured 400 base pairs in length, and so

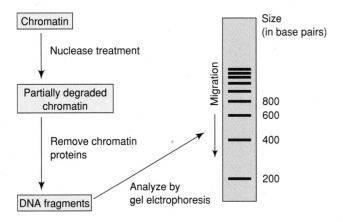

Figure 10-14 *Evidence That Proteins Are Clustered at 200 Base-Pair Intervals along the DNA Molecule in Chromatin Fibers* In these experiments, DNA fragments generated by nuclease digestion of rat liver chromatin were analyzed by gel electrophoresis. The discovery that the DNA fragments are multiples of 200 base pairs suggests that histones are clustered at 200 base-pair intervals along the DNA, thereby conferring a regular pattern of protection against nuclease digestion.

on. It was therefore concluded that the spherical particles observed in electron micrographs are each associated with 200 base pairs of DNA. This basic repeat unit, containing an average of 200 base pairs of DNA associated with a protein particle, is called a **nucleosome.**

An Octamer of Histones Forms the Core of the Nucleosome

The first insights into the molecular organization of the nucleosome emerged from the work of Roger Kornberg and his associates, who developed techniques for assembling nucleosomes from purified mixtures of DNA and protein. They found that chromatin fibers composed of nucleosomes can be generated by mixing together DNA and all five histones. However, when they attempted to use individually purified histones in such experiments, they discovered that nucleosomes could only be assembled if the histones had been purified by gentle techniques that left histone H2A bound to histone H2B, and histone H3 bound to histone H4. When these H3-H4 and H2A-H2B complexes were mixed with DNA, chromatin fibers exhibiting normal nucleosomal structure were reconstituted. Kornberg therefore concluded that histone H3-H4 and H2A-H2B complexes are an integral part of the nucleosome.

To investigate the nature of these histone interactions in more depth, Kornberg and his colleague, Jean Thomas, treated isolated chromatin with a chemical reagent that forms covalent crosslinks between protein molecules that are located next to each other. After treatment with this reagent, the chemically crosslinked proteins were isolated and analyzed by polyacrylamide gel electrophoresis. Protein complexes the size of eight histone molecules were prominent in the gels, suggesting that the nucleosomal particle contains an *octamer* of eight histones. Given the knowledge that histones H3-H4 and histones H2A-H2B each form tight complexes, and that these four histones are present in roughly equivalent amounts in chromatin, it was proposed that the histone octamer contains two H2A-H2B dimers and two H3-H4 dimers.

How is the DNA of a chromatin fiber bound to the histone octamer? An answer to this question first emerged from studies involving pancreatic *DNase I,* an enzyme that, unlike micrococcal nuclease, is capable of cleaving DNA molecules that are bound to a protein surface. Treatment of chromatin with DNase I generates DNA fragments that are multiples of about ten base pairs in length. Since ten base pairs is the approximate length of one complete turn of the DNA double helix, the most straightforward interpretation of these data is that the DNA double helix is wound around the surface of the histone octamer, and that a small region of each turn of the double helix is exposed and potentially subject to cleavage by DNase I (Figure 10-16).

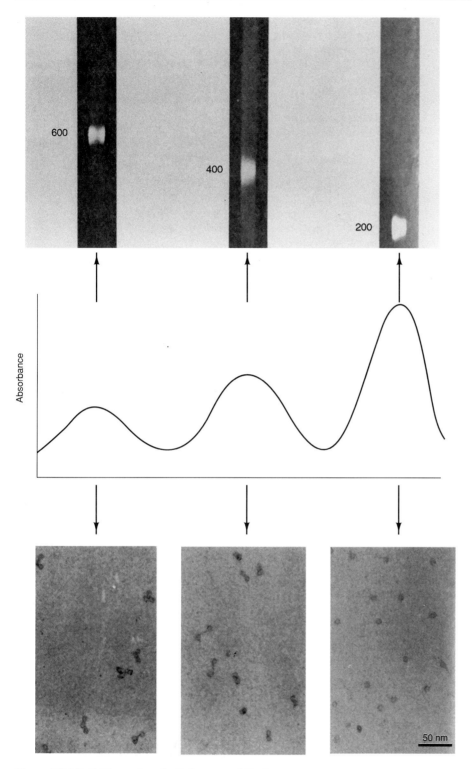

Figure 10-15 *Evidence for the Existence of Nucleosomes*   *Chromatin that had been par-*
tially degraded by treatment with micrococcal nuclease was fractionated by moving-zone
centrifugation (center graph). The individual peaks were then analyzed both by electron
microscopy (bottom) and by gel electrophoresis after removal of chromatin proteins (top).
The peak on the right consists of single protein particles associated with 200 base pairs of
DNA, the middle peak consists of clusters of two particles associated with 400 base pairs of
DNA, and the peak on the left consists of clusters of three particles associated with 600
base pairs of DNA. This indicates that the basic repeat unit in chromatin is a protein parti-
cle associated with 200 base pairs of DNA. Photographs courtesy of R. D. Kornberg.

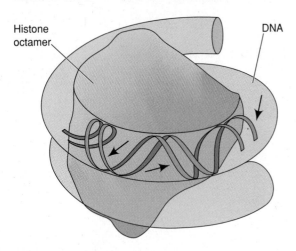

Figure 10-16 *Model of a Nucleosome* *The arrows point to the exposed DNA regions that are susceptible to cleavage by DNase I at approximately ten base-pair intervals.*

One issue that is not addressed by the preceding model concerns the significance of histone H1, which is not part of the octamer. If individual nucleosomes are isolated by briefly digesting chromatin with micrococcal nuclease, histone H1 is still present (along with the four other histones and 200 base pairs of DNA). When digestion is carried out for longer periods, the DNA fragment is further degraded until it reaches a length of about 146 base pairs; during the final stages of the digestion process, histone H1 is released. The remaining particle, consisting of a histone octamer associated with 146 base pairs of DNA, is referred to as a *core particle*. The DNA that is degraded during digestion from 200 to 146 base pairs in length is referred to as *linker DNA* because it joins one nucleosome to the next. Since histone H1 is released upon degradation of the linker DNA, it has been concluded that the histone H1 molecules are closely associated with the linker region. The length of the linker DNA varies somewhat between organisms, but the DNA associated with the core particle always measures close to 146 base pairs. Because of their uniform size, core particles can be readily crystallized for analysis by X-ray diffraction and electron microscopy. Results from such studies have disclosed that the core particle measures about 11 nm in diameter and 5.7 nm in height. The 146 base pairs of DNA associated with the core particle is sufficient to wrap around the particle roughly 1.8 times.

Nucleosomes Are Packed into Higher-Order Structures to Form Chromatin Fibers and Chromosomes

Winding DNA around histone particles allows the DNA molecule to be folded in an orderly fashion so that it can fit into a nucleus whose diameter is orders of magnitude smaller than the extended length of the DNA it contains. To quantify the extent of DNA folding, a **packing ratio** is calculated by determining the total extended length of the DNA and dividing it by the length of the particle or fiber in which the DNA is packaged. A nucleosome, for example, contains 200 base pairs of DNA, which would measure about 70 nm in length if completely extended; since the diameter of a nucleosome is about 10 nm, the packing ratio is roughly $70 \div 10 = 7$.

The formation of nucleosomes is only the first step in the packaging of nuclear DNA. Isolated chromatin fibers exhibiting the "beads-on-a-string" appearance measure about 10 nm in diameter, but the chromatin fibers of intact cells are often closer to 30 nm in diameter (Figure 10-17). In preparations of isolated chromatin, the 10-nm and 30-nm forms of the chromatin fiber can be interconverted by changing the salt concentration of the solution. During the transition from the 10-nm to the 30-nm fiber, the chain of nucleosomes becomes coiled upon itself into a more compact structure known as a **solenoid.** Several models have been proposed to explain how nucleosomes are packed together in the 30-nm solenoid, one of which is illustrated in Figure 10-18. The 30-nm fiber does not occur in chromatin preparations whose histone H1 molecules have been removed, suggesting that histone H1 mediates the packing of nucleosomes into the 30-nm solenoid.

When nucleosomes are packed together to form 30-nm chromatin fibers, the packing ratio increases from 7 to about 40. Under some conditions the 30-nm fiber is folded even further, producing structures with significantly higher packing ratios. The most common example occurs during cell division, when chromatin fibers are packaged into compact chromosomes whose packing ratios approach 10,000. When the structure of chromosomes is described in detail in Chapter 13, we will see that this folding process generates uniquely identifiable chromosome shapes. Although the folding mechanism is not completely understood, chromosomes from which histones have been extracted retain a nonhistone protein backbone whose shape resembles

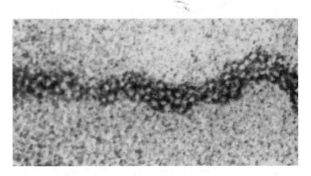

Figure 10-17 *Electron Micrograph of a Negatively Stained 30-nm Chromatin Fiber* *Courtesy of B. Hamkalo.*

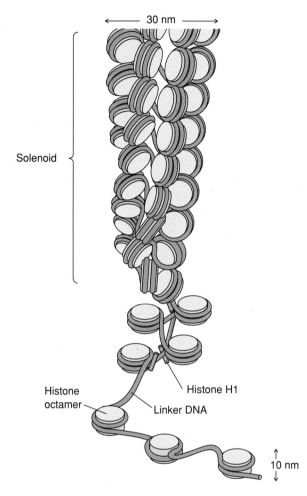

Figure 10-18 *A Model Showing How Nucleosomes May Be Packed in Chromatin Fibers* *Isolated chromatin fibers usually measure about 10 nm in diameter and exhibit a beads-on-a-string appearance. The "beads" correspond to histone octamers wrapped by short stretches of DNA, while the "string" corresponds to the DNA that passes from bead to bead. In intact cells the 10-nm chromatin fiber appears to coil upon itself to form a 30-nm diameter fiber called a solenoid. For clarity, histone H1 is not shown in the 30-nm diameter region of the fiber.*

that of the intact chromosome. This backbone or *scaffold* serves as an anchoring point to which long loops of DNA are attached (Figure 10-19). One of the main constituents of the scaffold is *topoisomerase II,* an enzyme whose role in controlling DNA coiling will be described shortly.

In addition to the chromosomes that form at the time of cell division, chromatin fibers can be packaged into several other types of chromosomes. Classic examples are the giant polytene chromosomes of insects (page 446) and the lampbrush chromosomes formed during meiotic division of certain oocytes (page 675). In spite of striking differences in the appearance of these various types of chromosomes, they all appear to be constructed by folding the same type of 30-nm chro-

matin fiber. The overall picture of eukaryotic chromatin that therefore emerges is of a DNA backbone that undergoes several levels of coiling and folding as a result of its interactions with histones and other proteins.

The Bacterial Nucleoid Contains DNA That Is Extensively Supercoiled and Bound to a Few Basic Proteins

The DNA of bacterial cells is localized in a special region of the cell called the **nucleoid.** Unlike the eukaryotic nucleus, the nucleoid is not membrane-enclosed, nor are nucleoli or a nuclear matrix present. Although no membrane boundary exists, the DNA-containing fibers of the nucleoid are packed together in a way that maintains a distinct boundary between the nucleoid and the rest of the cell (Figure 10-20).

The packaging of bacterial DNA has been most thoroughly studied in *E. coli,* which contains a single circular DNA molecule whose extended length is about a thousand times longer than that of the cell itself. To make the DNA molecule compact enough to fit into the nucleoid, it is twisted so that it coils upon itself. This process of DNA **supercoiling** can be carried out in two different ways. In *positive supercoiling,* the DNA is twisted in a direction that is the *same* as the turns of the DNA double helix; in *negative supercoiling,* the DNA is twisted in the opposite direction. DNA that is supercoiled is much more compact than nonsupercoiled or **relaxed** DNA (Figure 10-21). In bacterial cells, negative supercoiling is employed to make DNA compact enough to fit into the region occupied by the nucleoid.

Supercoiling is controlled by a class of enzymes called **topoisomerases.** A topoisomerase alters the extent of DNA supercoiling by a three-stage mechanism that involves (1) cleaving one or both strands of the DNA double helix, (2) passing one or both DNA strands through the break, and (3) joining the broken ends back together. Two kinds of topoisomerases with opposing effects are utilized to control supercoiling. *Topoisomerase I removes* negative supercoils in a reaction that involves the transient cleavage of one of the two strands of the DNA double helix; *topoisomerase II introduces* negative supercoils (or *removes* positive supercoils) in a reaction that involves the transient cleavage of both DNA strands (Figure 10-22). Introducing negative supercoils is an energy-requiring process that is driven by the hydrolysis of ATP. Since negative supercoils are introduced by topoisomerase II and removed by topoisomerase I, the proper amount of negative supercoiling can be maintained by balancing the activities of these two enzymes.

The DNA of bacterial cells does not have as much bound protein as the DNA of eukaryotic cells, but several types of evidence suggest that some structural protein is present. For example, electron microscopic studies utilizing disrupted bacterial cells have revealed the presence

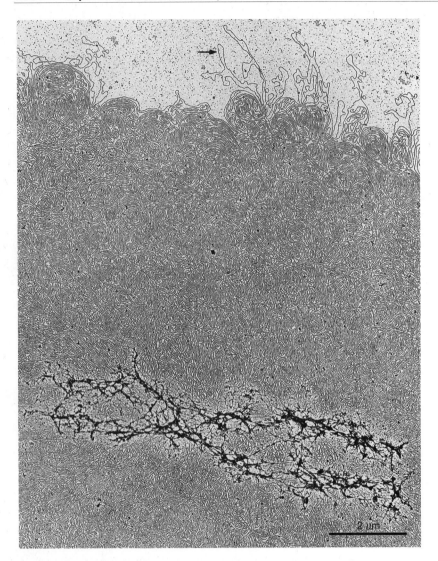

Figure 10-19 *Electron Micrograph Showing the Protein Scaffold That Remains after Removing Histones from Human Chromosomes* *The chromosomal DNA remains attached to the scaffold as a series of long loops. The arrow points to a region where the DNA molecule can be clearly seen. Courtesy of U. K. Laemmli.*

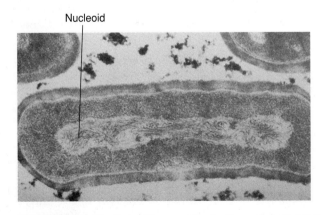

Nucleoid

Figure 10-20 *Electron Micrograph of a Bacterium Exhibiting a Distinct Nucleoid*

of DNA-containing fibrils averaging 12 nm in diameter. Since a naked DNA double helix measures about 2 nm in diameter, these findings suggest that other substances are associated with the DNA. In addition, chemical analyses of DNA-containing fibrils isolated from bacterial cells have revealed the presence of two proteins, called *HU* and *H*, which resemble histones in structure. However, there is no evidence that these proteins create a repeating nucleosomal arrangement such as that occurring in eukaryotes. Nucleosomes may not be required in bacteria because bacterial cells contain several orders of magnitude less DNA than eukaryotic cells, making the topological problem of DNA packaging less severe.

Although nucleosomes appear to be absent, the structural organization of bacterial DNA does share

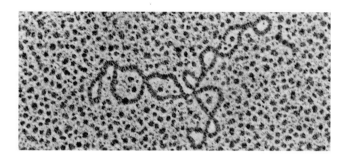

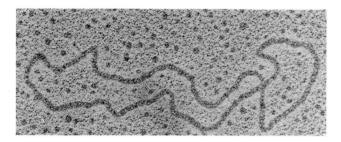

Figure 10-21 *DNA Supercoiling Observed by Electron Microscopy* *These two electron micrographs show a circular DNA molecule of the same virus in the supercoiled* (top) *and relaxed* (bottom) *states. Note that the supercoiled molecule is more compact. Courtesy of J. C. Wang.*

one feature with eukaryotic chromosomes. When bacterial cells are ruptured, the released DNA looks like a series of loops that are anchored to a structural framework located within the nucleoid (Figure 10-23). The circular chromosome of *E. coli* consists of about a hundred of these loops, each containing roughly 40,000 base pairs of negatively supercoiled DNA bound to protein. Because the two ends of each loop are anchored, the supercoiling of individual loops can be altered without influencing the supercoiling of adjacent loops. This arrangement at least superficially resembles the organization of eukaryotic chromosomes, where long loops of DNA are attached to a nonhistone protein scaffold.

GENE TRANSCRIPTION

Now that we have examined the structural organization of the nuclear regions of eukaryotic and prokaryotic cells, we are ready to proceed with our investigation of how genes are transcribed into RNA molecules. While some aspects of this process differ significantly between prokaryotes and eukaryotes, the underlying principles are similar enough to allow us to describe RNA synthesis in these two classes of cells at the same time.

RNA Is Synthesized by Several Different Kinds of RNA Polymerase

The first step in unraveling the molecular mechanism of gene transcription came with the isolation and purification of enzymes called **RNA polymerases,** which catalyze the synthesis of RNA using DNA as a template (page 81). In bacterial cells, a single type of RNA polymerase catalyzes the synthesis of all classes of RNA (messenger RNA, ribosomal RNA, and transfer RNA). Purified preparations of bacterial RNA polymerase can be fractionated by column chromatography into two components: a *core enzyme* consisting of four subunits, and an accessory protein termed the *sigma factor.* In the absence of sigma factor, the core enzyme tends to transcribe both strands of a DNA template randomly. When sigma factor is added, RNA synthesis is restricted to sites known to be transcribed within intact cells. Once RNA synthesis has begun, the sigma factor is released from the core enzyme and can be utilized by a second RNA polymerase molecule to initiate another round of transcription.

In contrast to prokaryotes, eukaryotic cells contain three different kinds of RNA polymerase. The existence of these multiple RNA polymerases first became apparent in the mid-1960s, when it was reported that the base composition of the RNA synthesized by isolated nuclei can be varied by altering the incubation medium. When incubated in a low ionic-strength solution containing Mg^{2+}, nuclei synthesize ribosomal RNA; in contrast, incubation in a Mn^{2+}-containing medium of higher ionic strength leads to the synthesis of messenger RNA. Subsequent studies involving microscopic autoradiography revealed that RNA synthesis occurs in the nucleolus when a Mn^{2+}-containing medium of high ionic strength is employed, and in the remainder of the chromatin when a Mg^{2+}-containing medium of low ionic strength is used (Figure 10-24).

The preceding observations suggest that eukaryotic cells contain multiple forms of RNA polymerase located in different regions of the nucleus. This interpretation was verified in the late 1960s by Robert Roeder and William Rutter, who successfully solubilized the RNA polymerases from eukaryotic nuclei and fractionated them by ion-exchange chromatography into distinct enzymes designated *RNA polymerases I, II, and III.* When nucleoli and chromatin are separated prior to purification of these enzymes, RNA polymerase I activity is detected mainly in the nucleoli, while RNA polymerases II and III are associated with the chromatin (Figure 10-25). In addition to their differing locations within the nucleus, the three RNA polymerases differ in their ionic requirements, sensitivity to inhibitors, and the type of RNA products they synthesize (Table 10-3, page 420). RNA polymerase I preferentially transcribes genes coding for

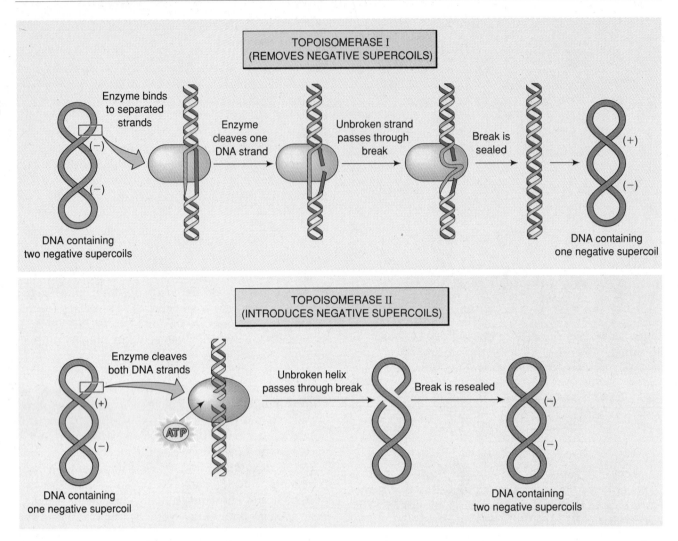

Figure 10-22 *Reactions Catalyzed by Topoisomerases I and II* (Top) *Topoisomerase I removes negative supercoils by transiently cleaving one strand of the DNA double helix and passing the unbroken strand through the break.* (Bottom) *Topoisomerase II introduces negative supercoils by transiently cleaving both strands of the DNA double helix and passing an unbroken region of the DNA double helix through the break.*

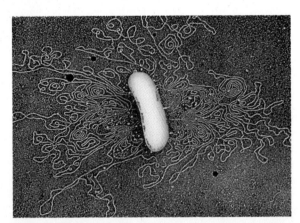

Figure 10-23 *DNA Released from the Nucleoid of a Ruptured Bacterial Cell* *The DNA forms a series of loops that remain attached to a structural framework within the nucleoid. Note the similarity to the organization of eukaryotic DNA, which forms long loops attached to a nonhistone protein scaffold (see Figure 10-19). Courtesy of Dr. Gopal/ SPL/Photo Researchers.*

the large ribosomal RNA molecules, RNA polymerase II transcribes genes coding for messenger RNAs and a few small nuclear RNAs, and RNA polymerase III transcribes genes coding for transfer RNAs, 5S ribosomal RNA, and some small nuclear RNAs. The three enzymes can also be distinguished from one another by their susceptibility to inhibition by *α-amanitin,* a highly toxic peptide isolated from the poisonous mushroom *Amanita phalloides.* RNA polymerase II is very sensitive to inhibition by α-amanitin, RNA polymerase III is moderately sensitive to inhibition, and RNA polymerase I is not inhibited at all (Figure 10-26).

In spite of the differences between the RNA polymerases of prokaryotic and eukaryotic cells, the reactions catalyzed by these enzymes are quite similar. As we will see in the following sections, each RNA polymerase molecule catalyzes a four-step process that involves the (1) binding of RNA polymerase to a DNA promoter site, (2) initiation of RNA synthesis, (3) elon-

Nucleolus

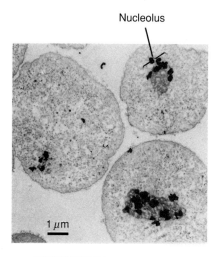

1 μm

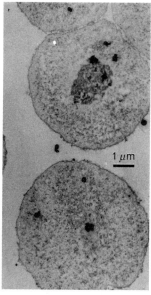

1 μm

Figure 10-24 *Microscopic Autoradiography of Nuclei Incubated with Radioactive RNA Precursors under Different Conditions* (Top) *When nuclei are incubated in a medium of low ionic strength containing Mg²⁺, silver grains are observed mainly over the nucleolus.* (Bottom) *When nuclei are incubated in a medium of higher ionic strength containing Mn²⁺, silver grains are observed mainly over the chromatin. Courtesy of V. G. Allfrey.*

gation of the newly forming RNA chain, and (4) termination of RNA synthesis and release of the completed RNA molecule.

RNA Polymerase Binds to DNA Promoter Sites

The first step in RNA synthesis is the binding of RNA polymerase to a DNA sequence called a **promoter site,** which is located adjacent to the region to be transcribed. The location of promoter sequences is commonly investigated by *DNA footprinting,* a technique that can be used to identify the region to which any DNA-binding protein has become bound. In this procedure, a DNA-binding protein such as RNA polymerase is

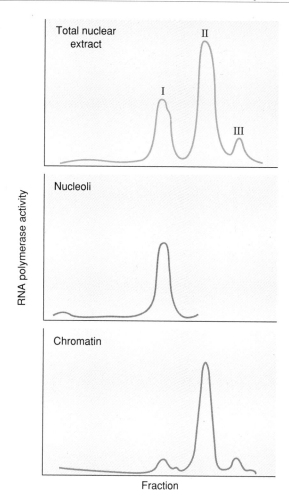

RNA polymerase activity

Total nuclear extract

I

II

III

Nucleoli

Chromatin

Fraction

Figure 10-25 *Chromatography of RNA Polymerases Extracted from Rat Liver Nuclei* *RNA polymerase I is preferentially associated with nucleoli, while RNA polymerases II and III occur mainly in the chromatin.*

allowed to bind to a piece of DNA that has been labeled at one end with ³²P. The DNA, some of which is covered by RNA polymerase, is then briefly digested with DNase I and the resulting DNA fragments are analyzed by gel electrophoresis. Since the region of the DNA molecule covered by RNA polymerase is protected from digestion by DNase I, the pattern of fragments can be used to locate the region to which the protein is bound (Figure 10-27). In addition to DNA footprinting, recombinant DNA techniques have also been useful in studying promoter sequences. Essential sequences within the promoter region have been identified by deleting or adding specific base sequences to cloned genes and then testing the ability of the altered DNA to be transcribed by RNA polymerase.

The preceding techniques have revealed that bacterial promoters typically encompass a region of about 35 base pairs located immediately adjacent to the point where transcription begins. By convention, promoter sequences are described in the 5′ →3′ direction on the DNA strand that lies opposite the template strand. The

Table 10-3   Distinguishing Features of Eukaryotic RNA Polymerases

RNA Polymerase	Location	Optimal Ionic Conditions	Sensitivity to α-Amanitin	Products
I	Nucleolus	Mg^{2+}, low ionic strength	Insensitive	28S, 5.8S, 18S ribosomal RNAs
II	Chromatin	Mn^{2+}, high ionic strength	Very sensitive	Messenger RNA precursors, small nuclear RNAs
III	Chromatin	Mn^{2+}, high ionic strength	Moderately sensitive	Transfer RNA, small nuclear RNAs, 5S ribosomal RNA

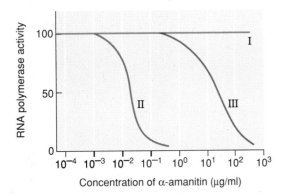

Figure 10-26 Effects of α-Amanitin on Eukaryotic RNA Polymerases I, II, and III *RNA polymerase II is very sensitive to inhibition by α-amanitin, RNA polymerase III is moderately sensitive, and RNA polymerase I is not inhibited at all.*

terms **upstream** and **downstream** are used to refer to DNA sequences that are located toward the 5' (upstream) or 3' (downstream) end of this strand. By comparing the DNA sequences of the promoters for a large number of genes, it has become apparent that two sequences located upstream from the start site recur in almost all bacterial promoters (Figure 10-28). Because these sequences are similar but not always identical in different promoters, they are expressed as an idealized **consensus sequence,** which represents the most common version of the sequence. One consensus sequence, called the *–10 sequence* or *Pribnow box,* is centered 10 bases upstream from the start site and consists of the bases *TATAAT;* the other consensus sequence, called the *–35 sequence,* consists of the bases *T TGACA* and is located 35 bases upstream from the start site. Base substitutions in either of these sequences decrease the efficiency with which RNA polymerase can initiate transcription of the adjoining gene.

In eukaryotic cells, promoter sequences vary with the type of RNA polymerase involved. Promoters for protein-coding genes, which are transcribed by RNA polymerase II, usually contain the consensus sequence *TATAAA* located about 30 bases upstream from the start site for transcription. This sequence, called the **TATA box,** is present in about 80 percent of the genes tran-

scribed by RNA polymerase II; it resembles the –10 sequence of prokaryotic promoters but is located a bit further upstream. Removing the TATA box abolishes the ability of genes containing such a sequence to be transcribed, indicating that the TATA box plays a crucial role in positioning RNA polymerase II for transcription. In addition to (or instead of) the TATA box, 10 to 15 percent of the genes transcribed by RNA polymerase II contain two other kinds of control sequences, typically located between 60 and 120 bases upstream from the transcription start site. One exhibits the consensus sequence CCAAT and is referred to as a **CCAAT box,** while the other has the consensus sequence GGGCG and is called a **GC box** (see Figure 10-28).

Although promoter sequences are usually situated upstream from the start site for transcription, an exception occurs in promoters recognized by RNA polymerase III. This unusual behavior was first reported by Donald Brown and his colleagues, who investigated the ability of RNA polymerase III to synthesize 5S ribosomal RNA, a small RNA molecule that forms part of the ribosome. In these experiments the 5S gene and its surrounding sequences were first cloned using recombinant DNA techniques. Various portions of the gene and/or its surrounding sequences were then removed by nuclease digestion, and the ability of the resulting DNA molecule to be transcribed was tested. It was discovered that removing sequences located upstream from the transcription start site does not inhibit RNA synthesis, nor does removal of sequences from the beginning of the 5S gene itself. However, removal of sequences beyond base 47 abolishes the ability of the 5S gene to be transcribed, suggesting that sequences in the middle of the 5S gene are required for the initiation of transcription. By removing DNA sequences from the other end of the 5S gene and again monitoring the effects on RNA synthesis, the crucial area was pinpointed to a segment about 35 base pairs in length situated in the middle of the gene (Figure 10-29). This special type of promoter is referred to as an **internal control region.**

To confirm its importance for promoting RNA synthesis, the internal control region was removed from the center of the 5S RNA gene and cloned into the middle of other DNA molecules. The recombinant DNA

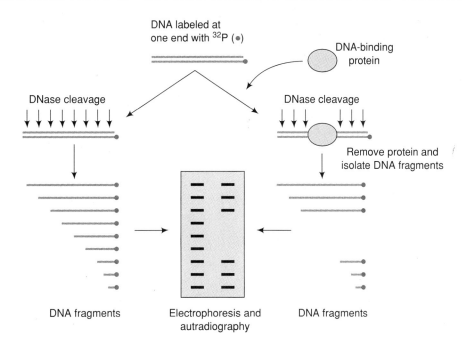

Figure 10-27 *The DNA Footprinting Technique* *This procedure is used to locate the site within a DNA molecule where a particular protein binds. A DNA molecule that has been labeled at one end with ³²P is bound to the protein in question and partially degraded with DNase; the resulting pattern of DNA fragments is then compared to the pattern obtained when DNA is not bound to protein. Since the DNase can cleave DNA only where it is not covered by protein, comparison of the two patterns allows the protein-binding region of the DNA molecule to be located.*

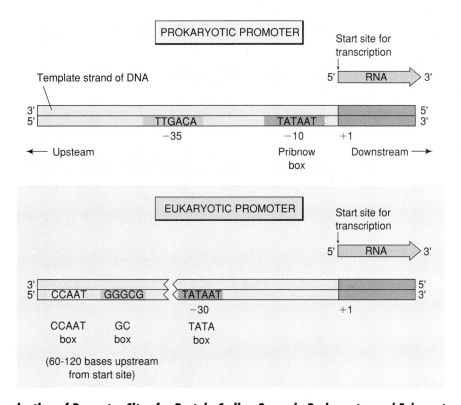

Figure 10-28 *Organization of Promoter Sites for Protein-Coding Genes in Prokaryotes and Eukaryotes* (Top) *Most bacterial promoters contain the consensus sequences TATAAT and TTGACA located about 10 and 35 bases upstream from the transcriptional start site respectively.* (Bottom) *About 80 percent of eukaryotic genes transcribed by RNA polymerase II contain a TATA box about 30 bases upstream from the transcriptional start site. In addition to (or instead of) the TATA box, about 10–15 percent of the genes transcribed by RNA polymerase II contain CCAAT and/or GC boxes.*

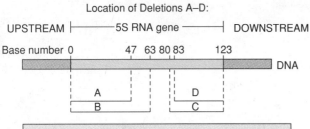

Location of Deletions A–D:

Fragment deleted	Ability of resulting DNA molecule to be transcribed by RNA polymerase III
A	Yes
B	No
C	No
D	Yes

Conclusion: Sequence between 47 and 83 must be present for transcription to occur

Figure 10-29 *Identification of the DNA Region That Promotes Transcription of the 5S RNA Gene by RNA Polymerase III* *Transcription of the 5S RNA gene can be experimentally inhibited by removing sequences from the middle of the gene, but not from the beginning or end of the gene. Such a DNA sequence located in the middle of a gene that promotes the initiation of transcription is called an internal control region.*

molecules formed in this way were found to be transcribed by RNA polymerase III starting at a point located about 50 bases *upstream* from the site at which the fragment was inserted. This behavior is in striking contrast to the promoters for other RNA polymerases, which cause transcription to be initiated *downstream* from the promoter sequence.

Transcription Factors Allow RNA Polymerase to Recognize Promoter Sites and Initiate RNA Synthesis

Both prokaryotic and eukaryotic RNA polymerases require the presence of special accessory proteins, called **transcription factors,** in order to bind to their appropriate promoters and initiate the synthesis of RNA. In bacterial cells, the sigma factor (page 417) plays such a role. Once a sigma factor has enabled an RNA polymerase molecule to bind to an appropriate promoter and initiate RNA synthesis, the sigma factor is released from the enzyme because it plays no role in the actual synthesis of RNA. Bacterial cells produce different kinds of sigma factor, each recognizing a different type of promoter. By altering the type of sigma factor being produced, cells can change the types of genes that are being transcribed into RNA.

In eukaryotic cells, a large number of transcription factors have been identified. Some of these proteins serve as general initiation factors that participate directly with RNA polymerase in initiating the transcription of a broad spectrum of genes. Each type of RNA

polymerase functions with a particular group of such factors. Some of the factors bind to DNA, whereas others bind to RNA polymerase or to other transcription factors. Figure 10-30 illustrates the interactions among the general initiation factors that are involved in initiating transcription by RNA polymerase II. The key transcription factor in this group is *TFIID* (also called *TBP*), a protein that starts the initiation process by binding to the TATA box. This step is followed by the binding of factors TFIIA, TFIIB, RNA polymerase II combined with TFIIF, and TFIIE. The multiprotein complex produced by the interaction of these various components with RNA polymerase is called the *preinitiation complex.*

In addition to requiring this group of factors, the initiation of transcription can also be influenced by proteins that interact with DNA sequences located upstream from the TATA box. For example, transcription factors called *CTF* and *SP1* bind specifically to the CCAAT and GC boxes, respectively, leading to an increase in the rate at which transcription is initiated. Proteins such as *CTF* and *SP1,* which function as upstream transcription factors, typically exhibit a modular structure in which one region of the protein binds to DNA and another region of the protein activates transcription. This arrangement allows one portion of the molecule to bind to an upstream DNA sequence while another region interacts directly or indirectly with proteins involved in the preinitiation complex (see Figure 10-30).

In addition to the CCAAT and GC boxes, another class of DNA sequences known as **response elements** lie within a few hundred bases upstream from pro-

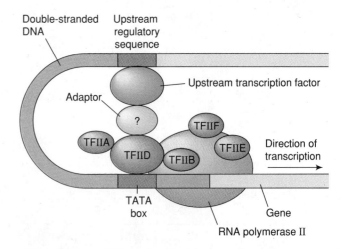

Figure 10-30 *Interactions among the Transcription Factors That Form a Preinitiation Complex with RNA Polymerase II* *This process begins with the binding of TFIID to the TATA box, followed by binding of factors TFIIA, TFIIB, RNA polymerase II combined with TFIIF, and TFIIE. Folding of the DNA molecule allows transcription factors that bind to DNA sequences located upstream from the TATA box to interact with the preinitiation complex.*

moter sites. Unlike the CCAAT and GC sequences, which occur in a broad spectrum of genes, DNA response elements are involved in regulating the activity of specific groups of genes in response to particular signals. As we will see later in the chapter, the binding of *gene-specific transcription factors* to DNA response elements allows the expression of specific genes to be turned on and off.

Enhancers and Silencers Alter the Activity of Promoters That May Be Located Far Away in the DNA Molecule

DNA sequences located near the promoter region are not the only sequences that influence the initiation of transcription. For many eukaryotic genes, the rate at which transcription is initiated can be dramatically increased by the action of an unusual family of sequences called **enhancers.** Although enhancers have no promoter activity of their own, they stimulate transcription when placed either upstream or downstream from a normal promoter site. They are capable of stimulating transcription when situated several thousand bases away from the start site of transcription, and are even active when inserted in reverse orientation in the DNA molecule. Sequences called **silencers** exhibit characteristics that are similar to those of enhancers, but they inhibit rather than stimulate transcription.

How do enhancers and silencers influence transcription when they are located far away from the start site for transcription? Like promoters and their associated upstream regulatory sequences, enhancers contain DNA sequences that bind to specific proteins. Although the enhancer with its bound protein(s) may be located thousands of bases away from the gene it acts upon, DNA molecules are thought to fold in such a way that the enhancer with its bound protein factor is brought into close proximity with the promoter site. This arrangement would allow proteins bound to enhancers or silencers to interact directly with components of the preinitiation complex.

RNA Synthesis Is Usually Initiated with ATP or GTP after the DNA Template Has Been Partially Unwound

Once the preinitiation complex has been formed, the process of RNA synthesis can begin. In genes transcribed by RNA polymerase II, transcription usually begins about 30 bases downstream from the TATA box. If DNA sequences located between the TATA box and the start site for transcription are removed, transcription is still initiated about 30 bases downstream from the TATA box, rather than at the normal start site. Hence the initiation site for transcription is determined by the location of the TATA box. In bacterial cells a similar function is

served by the Pribnow box. In eukaryotic genes transcribed by RNA polymerase III, the location of the internal control region determines the start site for transcription; however, in this case transcription is always initiated about 50 bases *upstream* from the control region.

In order to serve as a template for RNA synthesis, the DNA double helix must be partially unwound so that complementary base-pairing can occur between the template strand and the newly forming RNA chain. The amount of unwinding induced by RNA polymerase has been estimated by carrying out gel electrophoresis of circular DNA molecules that have had varying amounts of RNA polymerase added. Adding RNA polymerase unwinds the DNA and therefore increases the amount of negative supercoiling, which can be detected by a change in electrophoretic mobility of the DNA. Such studies have revealed that the binding of RNA polymerase unwinds about 17 base pairs of DNA. After this short region of DNA double helix has been unwound, the first base in the new RNA chain is incorporated at the start site. This first nucleotide is usually an ATP or GTP, although CTP and UTP are occasionally employed. Unlike the nucleotides incorporated during the subsequent steps of RNA synthesis, the initiating nucleotide retains all three phosphate groups.

Experiments utilizing selective inhibitors have revealed that the binding of RNA polymerase to DNA and the initiation of RNA synthesis are separate processes. The most thoroughly studied inhibitors that interfere with the initiation step are the *rifamycins,* a family of antibiotics that prevent initiation of RNA chains in bacterial cells without affecting the binding of RNA polymerase to DNA. In contrast to the rifamycins, polyanions such as *heparin* inhibit the ability of RNA polymerase to bind to DNA but do not interfere with initiation by RNA polymerase molecules that are already bound to DNA.

Elongation of the RNA Chain Occurs in the 5' → 3' Direction

As soon as the initial ATP or GTP has been incorporated, the RNA chain is *elongated* by the formation of phosphodiester bonds between the first (α) phosphate group of each incoming nucleotide and the 3' hydroxyl of the preceding one (Figure 10-31). During this process, the terminal two phosphates of the incoming nucleotide are released as pyrophosphate (PP_i). Elongation of the nucleotide chain then proceeds in the 5' → 3' direction, just as it does during DNA synthesis. After RNA synthesis begins, transcription factors TFIIB and TFIIE are released from RNA polymerase while another transcription factor, called TFIIS, becomes bound. Once elongation has proceeded far enough into the gene to free up the promoter and initiation sites, a new RNA polymerase molecule can

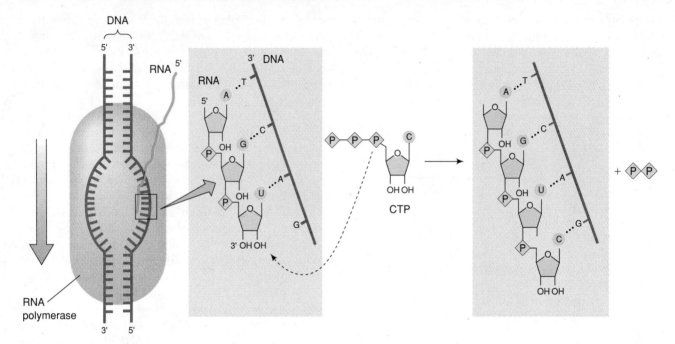

Figure 10-31 *The Elongation Step of RNA Synthesis* *During elongation of an RNA chain, which proceeds in the 5' → 3' direction, a phosphodiester bond is formed between the first (α) phosphate group of an incoming nucleotide and the 3'-terminal hydroxyl group of the growing RNA chain, releasing the two terminal phosphates of the incoming nucleotide as pyrophosphate (PP$_i$).*

bind and begin the synthesis of a second molecule of RNA. In this way the transcription rate of a single gene can be maximized by having several RNA polymerase molecules transcribing it simultaneously.

Because the base sequence of the newly forming RNA chain is determined by complementary base-pairing between each new base and the corresponding base on the template strand of the DNA molecule, RNA polymerase must continue to unwind the DNA double helix as elongation proceeds. The overall length of the unwound region need not be very large, however, because the growing RNA chain is continually peeled off the DNA template as the RNA polymerase moves along. As we discussed earlier in the chapter, the DNA of eukaryotic cells is assembled into chromatin fibers by the presence of nucleosomes. How RNA polymerase unwinds and transcribes DNA that is packaged into nucleosomes is not well understood, but it is known that nucleosomes do exist in regions that are being transcribed. It is possible that nucleosomes are transiently disassembled ahead of the transcribing RNA polymerase and then reform behind the polymerase after it has transcribed that part of the DNA.

The elongation stage of transcription is the target of *actinomycin D*, one of the more widely used inhibitors of RNA synthesis. Actinomycin D selectively binds between adjacent G-C and C-G base pairs, becoming inserted into the DNA double helix in a way that prevents the passage of RNA polymerase (Figure 10-32). For this reason, the transcription of genes enriched in G-C base pairs is most susceptible to inhibition by this antibiotic.

Unlike DNA polymerase (page 72), RNA polymerase has no 3'-exonuclease activity that would allow it to correct mistakes by removing mismatched base pairs as RNA elongation proceeds. For this reason, RNA synthesis is subject to a higher error frequency than DNA synthesis. This does not appear to create a problem, however, because many RNA molecules are usually transcribed from each gene, and so a small number of inaccurate copies can be tolerated. In contrast, only one copy of each DNA molecule is made when DNA is replicated prior to cell division. Since each newly forming cell receives only one set of DNA molecules, it is crucial that the copying process be extremely accurate.

Transcription Can Be Terminated by RNA Sequences That Either Form a Stem-and-Loop Structure or Are Recognized by Rho Factor

The elongation stage of RNA synthesis proceeds until a base sequence is synthesized that signals the end of transcription. In bacterial cells, two classes of termination signals can be distinguished that differ in whether or not they require the participation of a protein called **rho.** RNA molecules that are terminated without the rho factor contain a short self-complementary GC-rich sequence followed by several U residues near the 3' end. Since GC base pairs are held together by three hydrogen bonds and AU base pairs are joined by only two hydrogen bonds (see Figure 3-5), this configuration pro-

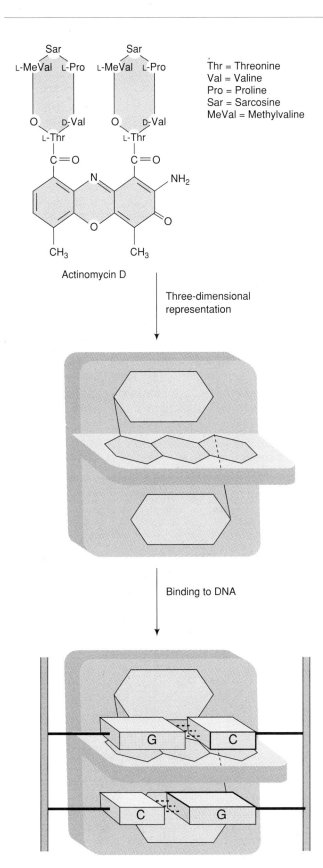

Actinomycin D

Three-dimensional
representation

Binding to DNA

Figure 10-32 *Structure of Actinomycin D* *The actinomycin D molecule selectively binds between adjacent G-C and C-G base pairs in a DNA double helix, preventing the passage of RNA polymerase.*

Thr = Threonine
Val = Valine
Pro = Proline
Sar = Sarcosine
MeVal = Methylvaline

motes termination in the following way: First, the self-complementary GC region forms a tightly base-paired stem-and-loop structure that tends to pull the RNA molecule away from the DNA; then the weaker bonds between the final sequence of U residues and the DNA template are broken, releasing the newly formed RNA molecule (Figure 10-33). Termination of this kind does not appear to require the participation of any special type of termination factor.

In contrast, terminating the synthesis of RNA products that do not form a GC-rich stem-and-loop structure requires the participation of the rho factor. Genes of this type were first discovered in experiments in which purified DNA obtained from bacteriophage lambda was transcribed with purified RNA polymerase. Some genes were found to be transcribed into RNA molecules that are longer than the RNAs produced by intact cells, suggesting that transcription was not terminating properly. This problem could be corrected by adding rho factor, which binds to specific termination sequences located near the 3' end of newly forming RNA molecules. The rho factor then acts as an ATP-dependent unwinding enzyme, moving along the RNA toward its 3' end and unwinding it from the DNA template as it proceeds.

Further insight into the termination process has emerged from the discovery that bacteriophage lambda also codes for the production of an **antitermination protein** that selectively suppresses the termination of certain rho-dependent genes. Suppressing termination permits RNA synthesis to proceed beyond the normal stop site and into adjacent genes whose transcription is essential for phage development. The antitermination protein can only function in bacterial cells that produce a protein called *NusA,* suggesting that the NusA protein may be involved in the regulation of rho-dependent termination.

In contrast to the situation in bacteria, eukaryotic genes do not possess termination sequences that halt transcription at the proper 3' end of each forming mRNA. Transcription proceeds instead for hundreds or even thousands of nucleotides beyond the 3' end of the mature mRNA; as we will see shortly, the proper 3' end of the message is then created by RNA processing reactions (page 428).

PROCESSING OF MESSENGER RNA

Transcription of DNA by RNA polymerase usually produces RNA molecules that are longer than would be expected for messenger RNAs (mRNAs) coding for the products of individual genes. Two factors contribute to the increased length of these initial RNA products. In bacteria, several adjacent genes are often transcribed as an integrated unit, producing a single large transcript called a **polycistronic mRNA.** Polycistronic mRNAs

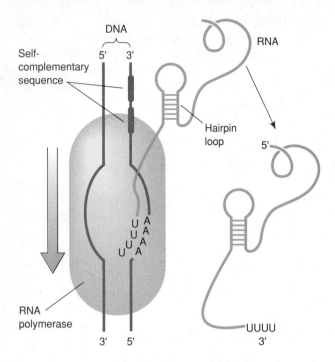

Figure 10-33 *Termination of Transcription in Prokaryotic Genes That Do Not Require the Rho Termination Factor* *A short self-complementary sequence near the end of the gene allows the newly formed RNA molecule to form a stem-and-loop structure that helps to dissociate the RNA from the DNA template.*

ris, who utilized microscopic autoradiography to study the behavior of newly synthesized RNA molecules that had been labeled by exposing cultured cells for brief periods to ^{3}H-uridine. Counting the number of silver grains over the nucleus and cytoplasm at varying times after exposure to radioisotope led Harris to the surprising discovery that the amount of radioactive RNA appearing in the cytoplasm is less than 10 percent of that which had initially become labeled in the nucleus (Figure 10-34). This unexpected result led Harris to conclude that the bulk of the nuclear RNA is degraded in the nucleus without ever entering the cytoplasm.

This remarkable conclusion initially met with considerable skepticism, for it did not seem compatible with the idea that DNA is transcribed into messenger RNA molecules that function in cytoplasmic protein synthesis. But experiments on the behavior of isolated nuclear RNA carried out soon thereafter by James Darnell and his associates provided further support for Harris's contention. In these experiments, cultured human cells were first incubated with ^{3}H-uridine for varying periods of time, and the nuclear RNA was then isolated and analyzed by moving-zone centrifugation (page 135). When labeling periods of only a few minutes were utilized, the radioactive RNA was found to sediment as a heterogeneous array of molecules whose sedimentation coefficients ranged from 20S to 100S. Because of the

typically remain intact during translation, directing the synthesis of several different polypeptides in succession.

The other reason for the existence of RNA molecules that are longer than expected is that many genes are transcribed into precursor RNAs, which must be cleaved into smaller RNA products. This conversion of a precursor RNA into a mature product, called **RNA processing**, may involve the removal, addition, and/or chemical modification of various base sequences. Ribosomal and transfer RNAs are processed in prokaryotic as well as eukaryotic cells, whereas the processing of messenger RNA occurs mainly in eukaryotes. The processing steps associated with the formation of eukaryotic messenger RNA will be described in the following sections, while the processing of ribosomal RNA will be covered in Chapter 11 when we discuss the formation of ribosomes.

Most of the RNA Sequences Synthesized in Eukaryotic Cells Are Degraded in the Nucleus without Entering the Cytoplasm

In bacterial cells, most of the RNA sequences transcribed from DNA act as messages that guide the synthesis of polypeptide chains. The situation in eukaryotic cells is significantly more complex. The first indication of this complexity was reported in 1962 by Henry Har-

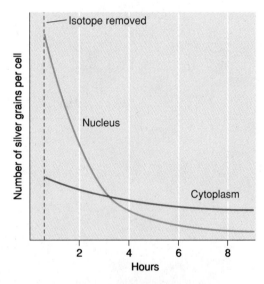

Figure 10-34 *Evidence for the Rapid Degradation of Newly Synthesized Nuclear RNA* *Cultured cells were incubated for 30 minutes with ^{3}H-uridine; the radioisotope was then removed, incubation was continued for varying amounts of time, and the distribution of radioisotope was monitored by microscopic autoradiography. The data reveal that the rapid disappearance of radioactive RNA from the nucleus is not accompanied by a comparable increase in radioactive RNA in the cytoplasm. It was therefore concluded that most of the labeled nuclear RNA is rapidly degraded.*

heterogeneity of the molecules contained in this RNA fraction, they were referred to as **heterogeneous nuclear RNAs (hnRNAs).**

To investigate the fate of these molecules, Darnell performed an experiment in which cells were first incubated briefly with ^{3}H-uridine to label the hnRNA fraction. Actinomycin D was then added to inhibit the formation of any additional radioactive hnRNA, and the cells were reincubated for varying periods of time to determine the fate of the previously labeled hnRNA molecules. Within a few hours the bulk of the radioactivity was found to disappear from the RNA fraction (Figure 10-35), a finding which supported Harris's contention that most newly synthesized RNA is rapidly degraded.

Messenger RNAs Are Derived from Larger RNA Precursors

The realization that most of the radioactivity incorporated into hnRNA is degraded without entering the cytoplasm is compatible with two possible interpretations: hnRNA is either selectively degraded, with certain sequences preferentially preserved for ultimate export from the nucleus, or the degradation is nonselective, and all sequences present in the hnRNA fraction are represented in the small number of RNA molecules that escape degradation and enter the cytoplasm. These alternatives were first investigated in the late 1960s by Ruth Shearer and Brian McCarthy, who employed nucleic acid hybridization techniques to compare the RNA sequences present in the nucleus and cytoplasm. Results from such experiments revealed that many of the base sequences present in hnRNA molecules are not represented in cytoplasmic RNA. It was therefore concluded that a selective degradation of hnRNA occurs,

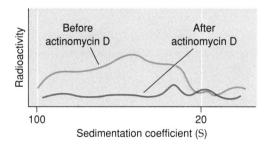

Figure 10-35 *The Behavior of Newly Synthesized hnRNA Molecules* *Cultured human cells were incubated for 5 minutes with radioactive uridine; actinomycin D was then added to prevent the synthesis of any additional radioactive RNA, and incubation was continued for an additional 150 minutes. Total RNA was isolated and analyzed by moving-zone centrifugation. Note that prior to the addition of actinomycin D, the newly synthesized RNA sediments as a heterogeneous family of molecules, most of which are relatively large. Several hours after the addition of actinomycin D, most of this radioactive RNA has disappeared.*

yielding only a specific subset of RNA sequences for transport to the cytoplasm. This conclusion has subsequently been verified by nucleic acid hybridization data from a variety of cell types, which indicate that only 10 to 20 percent of the sequences present in hnRNA are typically represented in cytoplasmic mRNAs.

The preceding observations imply that hnRNA is processed in such a way that certain sequences are preferentially conserved for export to the cytoplasm as mRNAs. One of the first eukaryotic gene products for which this processing was directly demonstrated is the mRNA coding for *β-globin*, a protein produced in large quantities in developing red blood cells. Because of the high concentration of β-globin mRNA in red cells, it can be easily purified and utilized as a template for the synthesis of globin cDNA using reverse transcriptase. In the late 1970s several laboratories utilized globin cDNA molecules as hybridization probes for determining whether globin sequences are present in hnRNA. In these experiments, RNA isolated from cells that had been incubated with ^{3}H-uridine was fractionated by either moving-zone centrifugation or polyacrylamide gel electrophoresis. When the RNA fractions obtained in this manner were hybridized to globin cDNA, it was discovered that β-globin sequences are present in two classes of RNA molecules measuring 15S and 9S in size (Figure 10-36).

To determine the relationship between the 15S and 9S RNAs, experiments were carried out in which cells exposed for a few minutes to ^{3}H-uridine were subsequently reincubated in the absence of radioisotope. Under these conditions, all the radioactive β-globin sequences are eventually converted to the 9S form, which corresponds to the size of mature globin mRNA. Such results suggest that the 9S globin mRNA is derived from a larger, 15S precursor. Subsequent investigations involving a wide variety of other eukaryotic genes have confirmed that *mRNAs are typically derived from larger precursor molecules.* It has therefore been concluded that hnRNAs function as mRNA precursors, or **pre-mRNAs** as they are now generally called.

How are pre-mRNA molecules converted into mRNA? In the following sections we will see that several chemical changes are involved, including packaging the initial pre-mRNA molecule with protein, adding special "cap" and "tail" structures to its two ends, and removing base sequences from its interior. The fact that long stretches of RNA are usually removed and degraded during this process helps to explain why so much of the RNA synthesized in the nucleus never reaches the cytoplasm.

Messenger RNA Precursors Are Packaged with Protein Particles as They Are Being Transcribed

One of the first events to occur during the processing of eukaryotic mRNA is that the newly forming pre-mRNA molecule becomes bound to protein particles. An impor-

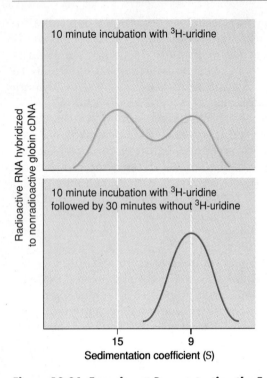

Figure 10-36 *Experiment Demonstrating the Existence of a 15S RNA Precursor to 9S β-Globin Messenger RNA*
(Top) *RNA isolated from cells labeled for 10 minutes with ³H-uridine was fractionated by moving-zone centrifugation and assayed for β-globin mRNA sequences by hybridizing to β-globin cDNA. β-Globin RNA sequences are found in both 15S and 9S RNA molecules. (Bottom) After subsequent incubation of cells in the absence of ³H-uridine, radioactive β-globin sequences are only found in the 9S RNA, indicating that the 15S RNA precursor is processed to form 9S RNA within 30 minutes of its synthesis.*

tant insight into the organization of these protein particles was provided in the mid-1960s by Georgii Georgiev and collaborators, who extracted pre-mRNA molecules along with their tightly bound proteins from purified nuclei. When this material was analyzed by moving-zone centrifugation under conditions designed to prevent ribonuclease activity, Georgiev observed a series of distinct RNA-protein complexes ranging in size from 30S to 200S. Electron micrographs revealed the presence of single particles in the first peak, clusters of two particles in the second peak, and so on (Figure 10-37). However, when the material was briefly digested with ribonuclease prior to centrifugation, only the first peak containing single particles was detected, suggesting that the particles are normally held together by RNA.

These observations led Georgiev to postulate that each newly forming pre-mRNA molecule becomes bound to a series of protein particles, which are called *heterogeneous nuclear ribonucleoprotein particles* or **hnRNPs**. Such an arrangement is reminiscent of the nucleosomal organization of chromatin fibers, although the protein particles employed in the packaging of pre-

mRNA are about twice the diameter of nucleosomes and are associated with a stretch of RNA about 500–600 nucleotides long. At least eight different proteins have been identified in hnRNP particles, none of which are histones.

Messenger RNA Precursors Acquire 5' Caps and 3' Poly-A Tails

Earlier in the chapter we learned that unlike other nucleotides found in RNA chains, the initiating nucleotide retains its three phosphate groups. In the pre-mRNA molecules of eukaryotic cells, this 5' terminal nucleotide is subsequently modified by the addition of a 7-methylguanosine group (Figure 10-38). During the modification reaction the ribose rings of the first, and often the second, bases of the RNA chain become methylated. Together these alterations create a special structure, known as a **5' cap**, which is unique to eukaryotic mRNAs and which plays an important role in the initiation of protein synthesis (page 491). The 5' cap is added early during the process of RNA synthesis, usually before more than a few dozen nucleotides have been incorporated into the newly forming RNA chain.

The 3' end of pre-mRNA molecules is also modified in a special way, in this case by the addition of between 50 and 250 adenine nucleotides to form a structure known as a **poly-A tail.** Since genes lack the long stretches of the base T that would be required to encode such poly-A sequences, it has been concluded that poly A is added to the 3' end of pre-mRNA molecules after transcription has taken place. Direct support for this conclusion has come from the isolation of the enzyme *poly-A polymerase*, which catalyzes the synthesis of poly-A sequences without the requirement for a DNA template. The addition of poly A is part of the process that creates the proper 3' end of eukaryotic mRNA molecules. Unlike bacteria, where specific termination sequences halt transcription at the correct 3' end of newly forming mRNAs, the transcription of eukaryotic pre-mRNAs often proceeds hundreds or even thousands of nucleotides beyond the 3' end of the mature mRNA. A special AAUAAA sequence located slightly upstream from the proper 3' end then signals where the poly-A tail should be added. The RNA chain is cleaved about 20 bases downstream from the AAUAAA sequence and poly-A polymerase catalyzes the formation of the poly-A tail (Figure 10-39). In addition to creating the poly-A tail, these processing events associated with the AAUAAA signal may also help to trigger the termination of transcription.

Although most eukaryotic mRNAs possess poly-A tails, the exact role played by this sequence is not clear. Since histone mRNAs have been found to lack poly-A tails, it is obvious that the presence of poly A is not an absolute requirement. Moreover, the presence of a poly-A

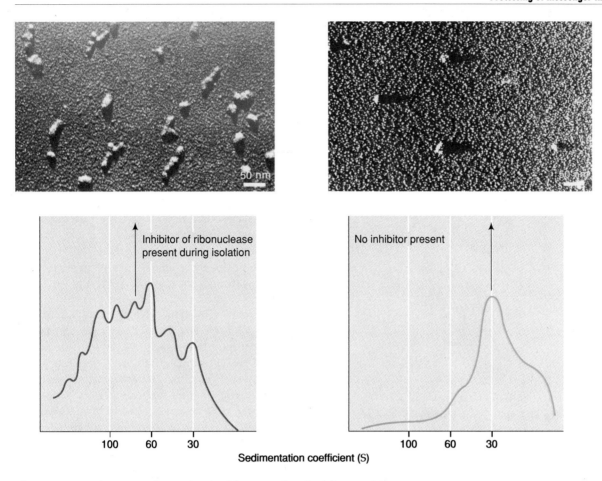

Inhibitor of ribonuclease
present during isolation

No inhibitor present

Sedimentation coefficient (S)

Figure 10-37 *Discovery of Protein Particles Associated with Pre-mRNA*   (Left) *Pre-mRNA was extracted from nuclei along with its tightly bound proteins and analyzed by moving-zone centrifugation in the presence of an inhibitor of ribonuclease. The material sediments as a series of peaks that contain varying numbers of protein particles. (Right) In the absence of the ribonuclease inhibitor, RNA is cleaved and the clusters of protein particles are all converted to single particles. Hence RNA must be joining the particles together. Micrographs courtesy of G. P. Georgiev.*

tail does not in itself determine that a given RNA molecule is destined to become a functional message. For example, in tissues that do not produce hemoglobin, globin sequences have been detected in nuclear RNA molecules that contain poly-A tails. In spite of the presence of poly A, these RNA molecules are not processed into functional mRNAs. Hence the addition of poly A is not in itself the determining factor that governs whether a given mRNA sequence is processed and transported to the cytoplasm.

An alternative possibility is that poly-A tails serve a cytoplasmic function related to the process of protein synthesis. In the next chapter we will see that poly-A tails undergo changes in length that may play a role in regulating the stability of cytoplasmic mRNA molecules.

Eukaryotic Genes Are Interrupted by Introns That Are Removed During RNA Processing

During the processing of pre-mRNA into mRNA, the RNA molecule often undergoes a dramatic decrease in

size. In extreme cases, the length of the final mRNA may be less than 10 percent of the length of the pre-mRNA molecule from which it is derived. It was originally thought that pre-mRNA molecules contain extraneous sequences at both ends of the final mRNA sequence and that this extra material is simply removed during RNA processing. However, this relatively simple view was shown to be incorrect in the late 1970s when a series of surprising discoveries led to a complete reevaluation of our concept of eukaryotic genes. These discoveries relied heavily on the use of **R looping**, a technique in which single-stranded RNA is hybridized to double-stranded DNA under conditions that favor the formation of hybrids between RNA and DNA. When RNA hybridizes to a complementary sequence located in one of the two DNA strands, the other DNA strand is displaced as a single-stranded loop that can be visualized by electron microscopy.

In 1977 Susan Berget, Claire Moore, and Phillip Sharp reported the results of experiments in which this technique was used to study the hybridization proper-

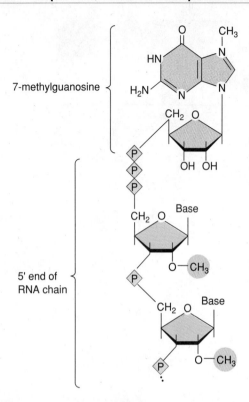

Figure 10-38 *The Cap Structure Located at the 5' End of Eukaryotic Pre-mRNA and mRNA Molecules* *The methyl groups attached to the first two riboses of the RNA chain are not always present.*

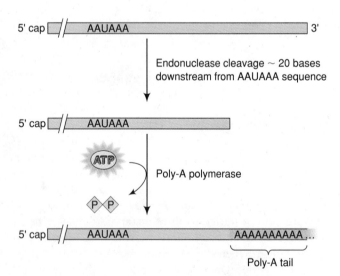

Figure 10-39 *Addition of a Poly-A Tail to Pre-mRNA* *Transcription of eukaryotic pre-mRNAs often proceeds beyond the 3' end of the mature mRNA. An AAUAAA sequence located slightly upstream from the proper 3' end then signals that endonuclease cleavage should occur about 20 bases downstream from the signal site, followed by addition of a poly-A tail catalyzed by poly-A polymerase.*

ties of mRNA molecule produced by adenovirus, a DNA virus that infects a variety of eukaryotic cells. Electron microscopy revealed that hybridizing an adenovirus mRNA molecule to adenovirus DNA results in the formation of several separate R loops (Figure 10-40). If the adenovirus mRNA had been transcribed from a single continuous stretch of DNA, only a single R loop would have been expected. This rather surprising result indicated that the DNA sequences coding for the mRNA are not continuous with each other, but instead are separated by sequences that do not appear in the final message. Such sequences that disrupt the linear continuity of the message-encoding regions of a gene are known as **introns,** while sequences destined to appear in the final mRNA are termed **exons** (see Figure 3-27).

Shortly after the discovery of introns in adenovirus DNA, a similar arrangement was detected in other eukaryotic genes (Table 10-4). Although electron microscopic examination of DNA/RNA hybrids has been a major tool in such studies, restriction mapping has also played an important role (page 99). When the restriction maps of chromosomal genes are compared to the restriction maps of the corresponding cDNAs (made by transcribing a gene's mRNA with reverse transcriptase), significant differences are often observed because of the presence of introns in the gene that are not represented in the final mRNA (Figure 10-41).

The discovery that eukaryotic genes are disrupted by introns that do not appear in the final mRNA raises the question of whether introns are transcribed into pre-mRNA. This question has been addressed by experiments in which pre-mRNA and DNA are mixed together and the resulting hybrids examined by electron microscopy. In contrast to the appearance of hybrids between mRNA and DNA, which exhibit multiple R loops where the DNA molecule contains sequences that are not present in the mRNA, pre-mRNA hybridizes in one continuous stretch to the DNA molecule, forming a single R loop (Figure 10-42). It can therefore be concluded that pre-mRNA molecules represent continuous copies of their corresponding genes, containing introns as well as sequences destined to be part of the final mRNA. Since introns must be removed from pre-mRNA during its conversion to mRNA, this arrangement explains why many of the RNA sequences synthesized in the nucleus never reach the cytoplasm.

RNA-Protein Complexes Called snRNPs Facilitate the Removal of Introns from Pre-mRNA

Introns are removed from pre-mRNA molecules by a process called **RNA splicing.** Proper splicing can be disrupted by altering the base sequence of short stretches located at either end of the intron, indicating that these

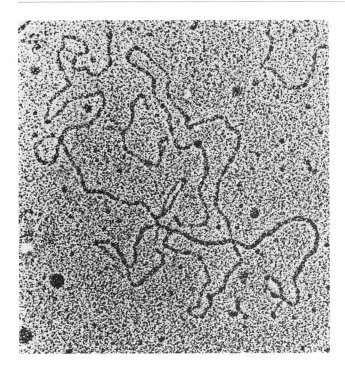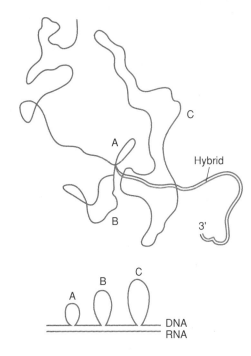

Figure 10-40 R-Looping Experiment Showing That Adenovirus Messenger RNA Is Transcribed from Nonadjacent Regions of Adenovirus DNA   *The electron micrograph pictures a single-stranded adenovirus DNA molecule hybridized to adenovirus mRNA. In the line drawing on the right, loops A, B, and C represent single-stranded DNA that has not hybridized to the RNA. Below the line drawing is a schematic diagram that illustrates why the existence of such loops indicates that the mRNA molecule is encoded by nonadjacent DNA sequences. Micrograph courtesy of P. A. Sharp.*

Table 10-4   A Few Examples of Genes with Introns

Gene	Organism	Number of Introns	Number of Exons
Actin	Drosophila	1	2
β-Globin	Mammals	2	3
Insulin	Chicken, human	2	3
Lysozyme	Chicken	3	4
Actin	Chicken	3	4
Ovalbumin	Chicken	7	8
Collagen	Chicken	50	51

sequences determine the location of the 5' and 3' *splice sites*—that is, the points where the two ends of the intron are cleaved from the pre-mRNA. Base sequence analyses of hundreds of introns have revealed that the 5' end of an intron typically starts with the sequence GU and the 3' end terminates with the sequence AG; in addition, a short stretch of bases adjacent to these GU and AG sequences tends to be similar among different introns.

The base sequence of the remainder of the intron appears to be largely irrelevant to the splicing process. Introns vary from a few dozen to thousands of

nucleotides in length, but most of the intron can be artificially removed without altering the splicing process. One exception is a special sequence located several dozen bases upstream from the 3' end of the intron and referred to as the *branch point*. The branch point plays an important role in the mechanism by which introns are removed. Intron removal typically requires the participation of a group of small nuclear RNAs that are bound to proteins to form complexes known as "**snurps**" or **snRNPs** (small nuclear ribonucleoprotein particles). During RNA splicing, a group of snRNPs bind sequentially to an intron to form a splicing complex known as a **spliceosome** (Figure 10-43). The first step in this process is the binding of an snRNP called *U1*, whose RNA contains a nucleotide sequence that allows it to base-pair with the 5' splice site. A second snRNP, called *U2*, then binds to the branch-point sequence. Finally, another group of snRNPs (*U4/U5/U6*) brings the two ends of the intron together to form a mature spliceosome.

At this stage, the pre-mRNA is cleaved at the 5' splice site and the newly released 5' end of the intron is covalently joined to an adenine residue located at the branch-point sequence, creating a looped structure called a *lariat*. The 3' splice site is then cleaved and the two ends of the exon are joined together, releasing the intron for

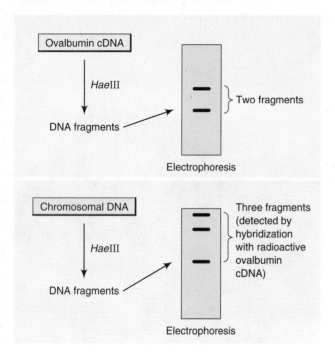

Figure 10-41 *Detection of Introns Using Restriction Enzymes* (Top) *Purified chicken ovalbumin cDNA cleaved with the restriction enzyme HaeIII yields two fragments, indicating the presence of one HaeIII site in the cDNA molecule.* (Bottom) *Treatment of chicken chromosomal DNA with HaeIII generates three fragments containing ovalbumin sequences, thereby indicating the presence of two HaeIII sites in the ovalbumin gene. The extra HaeIII site is situated within an intron and therefore does not appear in the final ovalbumin mRNA molecule (from which ovalbumin cDNA is derived).*

subsequent degradation. Electron micrographs have revealed the presence of snRNPs and mature spliceosomes bound to RNA molecules that are still in the process of being synthesized (Figure 10-44), indicating that introns can be removed before transcription of the pre-mRNA molecule is completed.

The participation of snRNPs is almost always required for intron removal. However, a few kinds of genes, including ribosomal RNA genes in ciliated protozoa and several mitochondrial genes, have been found to be **self-splicing**; that is, intron removal proceeds in the absence of snRNPs or any other protein-containing entity (page 482). Some self-splicing RNAs release the excised intron in a lariat configuration, as is typical of nuclear pre-mRNA splicing, while others utilize a non-lariat mechanism in which both splice sites are cleaved and the intron is released as a linear fragment.

Although most of the protein-coding genes of eukaryotes contain introns that must be removed during the formation of mRNA, not all genes are organized in this fashion. Introns are extremely rare in the genes of prokaryotic cells and are also absent from eukaryotic genes coding for histone molecules. The realization that

many, but not all, eukaryotic genes contain introns raises some interesting questions. At first glance it may seem inefficient and potentially hazardous to interrupt genes with sequences that do not appear to serve any useful function and that are simply destined for removal. One possible explanation for this arrangement has emerged from the discovery that exons often code for different functional regions of polypeptide chains, each of which can independently fold into a separate domain (Figure 10-45). The presence of introns between such exons may facilitate their evolutionary rearrangement into new combinations, creating proteins consisting of new combinations of domains. In an analogous fashion, changes in the way in which exons are spliced together during RNA processing would allow a given gene sequence to code for more than one protein. As we will see later in the chapter, such a mechanism is one of the many ways in which the process of gene expression is regulated.

GENE REGULATION IN PROKARYOTES

A cell's DNA houses the genes coding for every protein the cell is potentially capable of synthesizing. But the need for these proteins varies considerably, depending upon the cell type, its physiological state, the availability of external nutrients, and changing environmental conditions. To respond to these changing needs, cells have evolved a wide range of mechanisms for turning on and off the expression of individual genes. We will begin our examination of these mechanisms by exploring how gene expression is controlled in bacteria.

Genes Involved in Lactose Metabolism Are Organized into an Inducible Operon

Our understanding of how bacteria regulate the expression of individual genes owes a great debt to the pioneering work of Francois Jacob and Jacques Monod on the control of lactose metabolism in *E. coli.* The central enzyme in the pathway for lactose metabolism is *β-galactosidase,* which cleaves the disaccharide lactose into the monosaccharides glucose and galactose. Bacteria that are grown in the absence of lactose do not produce β-galactosidase because they have no need for the enzyme when lactose is absent. However, if lactose is added to the growth medium, β-galactosidase begins to be synthesized (Figure 10-46). This type of regulation, in which a small molecule such as lactose triggers the production of a specific protein, is called *induction,* and the molecule that triggers the process is called an **inducer.**

Through a combination of biochemical and genetic studies on the genes involved in lactose metabolism, Jacob and Monod developed a general model of gene regulation with far-reaching implications. The corner-

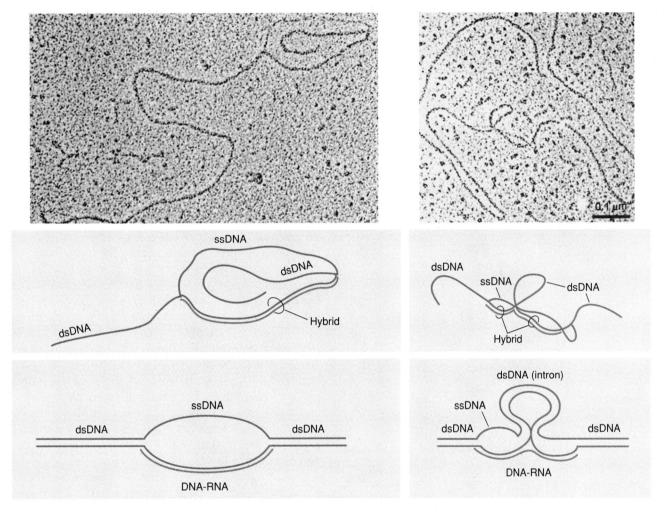

Figure 10-42 *Hybridization of 15S and 9S Globin RNAs to the DNA Gene Coding for β-Globin* *Each electron micrograph is accompanied by a line drawing and a schematic diagram that summarizes what has occurred.* (Left) *The 15S β-globin pre-mRNA hybridizes in one continuous stretch to the β-globin gene.* (Right) *Hybridization of 9S β-globin RNA to the globin gene is interrupted by a large R loop that corresponds to the location of an intron. Abbreviations: dsDNA = double-stranded DNA, ssDNA = single-stranded DNA. Micrographs courtesy of P. Leder.*

stone of this model rested on the discovery that the control of lactose metabolism involves two types of genes: *structural genes* that code for the structure of enzymes involved in lactose uptake and metabolism, and a *regulatory gene* whose product regulates the activity of the structural genes. Three structural genes are involved in lactose metabolism: (1) the *lacZ* gene, which codes for *β-galactosidase,* (2) the *lacY* gene, which codes for a lactose transport protein known as *lactose permease,* and (3) the *lacA* gene, which codes for a lactose modifying enzyme called *galactoside transacetylase.* These three genes, situated adjacent to each other in the bacterial DNA, are only transcribed when an inducer such as lactose is present.

Genetic studies carried out by Jacob and Monod revealed that the ability of inducers to control the transcription of the three structural genes depends upon the presence of a regulatory gene called *lacI.* If the *lacI* gene is deleted, bacteria synthesize β-galactosidase, lac-

tose permease, and galactoside transacetylase regardless of whether an inducer such as lactose is present; hence the *lacI* gene must normally code for a product that *inhibits* expression of the *lacZ, lacY,* and *lacA* genes. The product of a regulatory gene that inhibits the expression of other genes is called a **repressor.**

Subsequent investigations revealed that mutations located immediately upstream from the beginning of the *lacZ* gene can also cause the products of the *lacZ, lacY,* and *lacA* genes to be produced regardless of the presence of inducer. Since mutations in this DNA region, called the **operator,** influence the transcription of all three structural genes, it was concluded that the three structural genes are transcribed as part of an integrated unit. This integrated transcriptional unit is termed the ***lac* operon,** and contains the following five elements in sequence: a promoter (*lacP*), an operator (*lacO*), and the three structural genes (*lacZ, lacY,* and *lacA*).

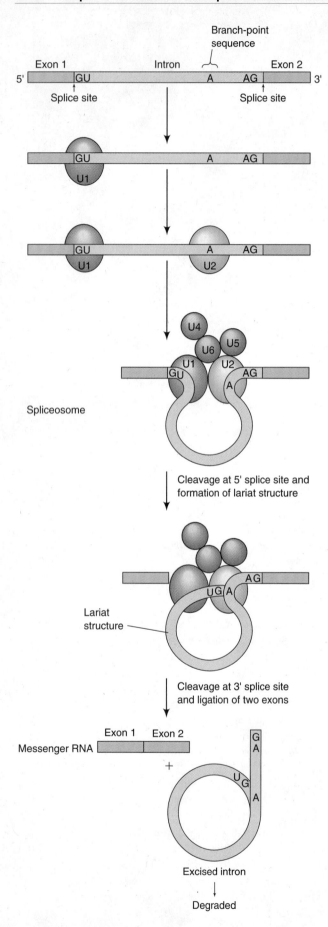

Branch-point sequence

Exon 1   Intron   Exon 2

Spliceosome

Cleavage at 5' splice site and formation of lariat structure

Lariat structure

Cleavage at 3' splice site and ligation of two exons

Messenger RNA   Exon 1   Exon 2

Excised intron

Degraded

The *Lac* Repressor Is a Protein That Contains Binding Sites for Both DNA and Inducers Such as Lactose

Based on the preceding information, Jacob and Monod proposed that the *lac* repressor binds to DNA sequences in the *lac* operator, preventing RNA polymerase from initiating transcription of the adjacent structural genes. It was further postulated that inducers such as lactose act by binding to the *lac* repressor and preventing it from interacting with the *lac* operator, thereby allowing RNA polymerase to initiate transcription of the *lac* genes. When this model was first proposed in 1961, the only known mechanism for recognizing specific base sequences was base-pairing between complementary nucleic acids. Since the preceding model requires that the *lac* repressor be able to specifically recognize and bind to the base sequence of the *lac* operator, it was initially believed that the repressor would turn out to be a nucleic acid.

However, in 1966 Walter Gilbert and Benno Müller-Hill successfully isolated the *lac* repressor and demonstrated that it is a protein molecule. Because a typical bacterial cell contains only 10–20 copies of the repressor, a sensitive assay was required for its identification and isolation. For this purpose Gilbert and Müller-Hill utilized *isopropyl thiogalactoside (IPTG)*, a potent inducer of the *lac* genes. Since it had been postulated by Jacob and Monod that inducers function by binding to the *lac* repressor, various molecules isolated from bacterial cells were tested for their ability to bind to IPTG. This testing was accomplished by placing bacterial macromolecules inside a dialysis membrane bag (page 143) and dialyzing against radioactive IPTG. When IPTG-binding macromolecules are absent from the bag, the radioactive IPTG diffuses back and forth across the membrane until its concentration is equal on both sides. But if macromolecules that bind to IPTG are present, the radioactive IPTG becomes more concentrated within the bag.

Using this approach to identify molecules that bind to IPTG, Gilbert and Müller-Hill eventually purified an IPTG-binding protein that exhibited several properties expected of the *lac* repressor. First, the IPTG-binding protein cannot be detected in mutant bacteria exhibiting a defective *lac* repressor gene. Second, when the

Figure 10-43 *Intron Removal by Spliceosomes*
The pre-mRNA first assembles in a stepwise fashion with the U1 snRNP, U2 snRNP, and U4/U5/U6 snRNPs (along with some non-snRNP splicing factors), forming a mature spliceosome. The pre-mRNA is then cleaved at the 5' splice site and the newly released 5' end is linked to an adenine (A) residue located at the branch-point sequence, creating a looped lariat structure. Next the 3' splice site is cleaved and the two ends of the exon are joined together, releasing the intron for subsequent degradation.

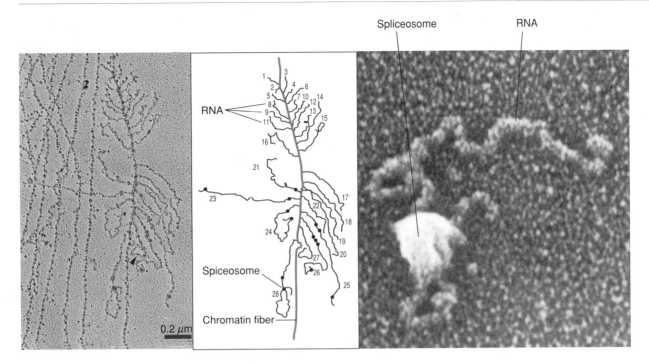

Figure 10-44 *Visualization of Spliceosomes by Electron Microscopy* (Left) *In this electron micrograph of chromatin in the process of being transcribed, newly forming RNA molecules (each numbered separately) can be seen protruding from the central chromatin fiber. The darker granules on the RNA molecules represent snRNPs that are beginning the process of spliceosome assembly. In the case of RNA molecule number 28, a mature spliceosome has formed.* (Right) *Higher magnification electron micrograph of a single RNA molecule showing a spliceosome. Courtesy of A. L. Beyer* (left) *and J. Griffith* (right).

IPTG-binding protein is mixed with DNA samples containing *lac* operon sequences, strong binding between the protein and DNA is observed (Figure 10-47, *left*). Third, the IPTG-binding protein does not bind to other kinds of DNA, or to *lac* operon DNA containing a mutated operator sequence. And finally, the IPTG-binding protein loses its ability to bind to *lac* operator DNA in the presence of IPTG (see Figure 10-47, *right*). Taken together, these properties indicate that the IPTG-binding protein is the *lac* repressor.

The *Lac* Repressor Inhibits Transcription by Binding to *Lac* Operator DNA through a Helix-Turn-Helix Motif

The preceding data demonstrated that the *lac* repressor is a protein molecule whose binding to the *lac* operon is regulated by inducer molecules such as IPTG and lactose. From this information, a relatively complete picture of the *lac* operon could be constructed (Figure 10-48). According to this model, the *lac* repressor normally binds to the operator region of the *lac* operon, inhibiting transcription of the operon by RNA polymerase. When an inducer, such as lactose or IPTG, binds to an allosteric binding site on the *lac* repressor, the conformation of the repressor changes so that it is no longer able to bind to DNA. The *lac* operon can then be transcribed by RNA polymerase.

Soon after this model was first proposed, the development of mutant strains of bacteria that produce excessive amounts of *lac* repressor made it possible to isolate and purify large quantities of the repressor protein. The purified repressor was found to be a tetramer consisting of four identical polypeptide chains, each containing a binding site for an inducer molecule. Although the three-dimensional structure of the repressor has not yet been determined, examination of its amino acid sequence has revealed the presence of a **helix-turn-helix** motif, which occurs in many DNA-binding proteins. The helix-turn-helix motif consists of two regions of α helix separated by a few amino acids that allow the polypeptide chain to bend between the two helices (Figure 10-49). Studies of the behavior of the helix-turn-helix motif in a variety of DNA-binding proteins have led to the conclusion that one α helix, called the *recognition helix,* contains amino acid side chains that recognize specific DNA sequences by forming hydrogen bonds with bases located in the major groove of the DNA double helix. The second helix stabilizes the motif through hydrophobic interactions with the recognition helix.

The availability of purified *lac* repressor molecules also made it possible to isolate the DNA sequence that comprises the *lac* operator. In these studies, purified *lac* repressor molecules were added to DNA preparations containing *lac* operon sequences, and the result-

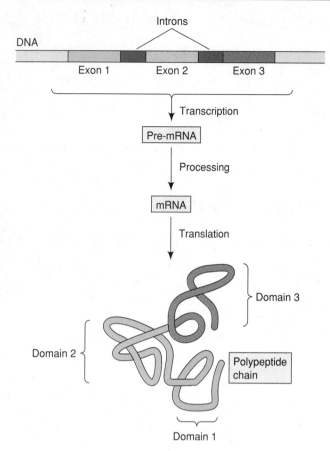

Figure 10-45 *The Proposed Role of Exons in Coding for Protein Domains*   *Each domain represents a separate region of a polypeptide chain that is capable of independently folding into a functional unit.*

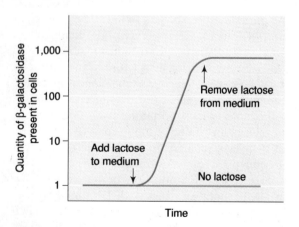

Figure 10-46 *Induction of β-Galactosidase by Lactose in E. Coli*   *Bacteria do not produce β-galactosidase unless lactose is present in the growth medium. This type of regulation, in which a small molecule such as lactose triggers the production of a specific protein, is called induction.*

ing mixture was then digested with deoxyribonuclease. Under such conditions the *lac* repressor binds to the *lac* operator and protects it from digestion; the undigested DNA sequence can then be isolated and analyzed. Experiments of this type have revealed that the

lac operator is roughly two dozen bases in length and is a **palindrome**—that is, a base sequence that is roughly the same when read in the 5' → 3' direction in each DNA strand (Figure 10-50).

The Catabolite Activator Protein (CAP) and Cyclic AMP Exert Positive Control over Operons Susceptible to Catabolite Repression

Because the *lac* repressor inhibits transcription of the *lac* operon, it is said to exert **negative control.** Several years after the *lac* repressor was first isolated, a DNA-binding protein was identified that exerts **positive control** over the *lac* operon; that is, it activates transcription of the *lac* operon when it binds to DNA. This **activator** protein is involved in a control system that allows cells to turn off the synthesis of enzymes that are not needed when glucose is abundant. When glucose is plentiful, enzymes that catalyze the breakdown of other nutrients are unnecessary. Included in this category of degradative or *catabolic* enzymes are such molecules as β-galactosidase, arabinose isomerase, and tryptophanase. The ability of glucose to depress the synthesis of such enzymes is called **catabolite repression.**

An early clue to the mechanism underlying catabolite repression was provided in the late 1960s by Ira Pastan and Robert Perlman, who discovered that exposing bacterial cells to glucose lowers the intracellular concentration of the signaling molecule, *cyclic AMP* (page 211). The exact role played by cyclic AMP did not become clear, however, until Geoffrey Zubay and his associates isolated a DNA-binding protein that binds both to cyclic AMP and to DNA sequences located adjacent to catabolite-repressible genes (genes whose expression is inhibited in the presence of glucose). Mutant bacteria lacking this protein, termed the **catabolite activator protein** or **CAP,** are unable to transcribe catabolite-repressible genes, even in the absence of glucose. Hence the CAP protein must be required for the transcription of such genes.

Subsequent studies revealed the basis for this requirement: The CAP protein, when complexed with cyclic AMP, binds to the promoter sites of catabolite-repressible genes and facilitates the binding of RNA polymerase. Like the *lac* repressor, the CAP protein contains a helix-turn-helix motif that permits it to recognize the DNA sequence to which it binds. Another parallel with the *lac* repressor is that the binding site recognized by the CAP protein is palindromic (see Figure 10-50). Binding of the CAP–cyclic AMP complex to this site brings the CAP protein in close proximity with RNA polymerase, allowing it to stimulate the initiation of transcription (Figure 10-51).

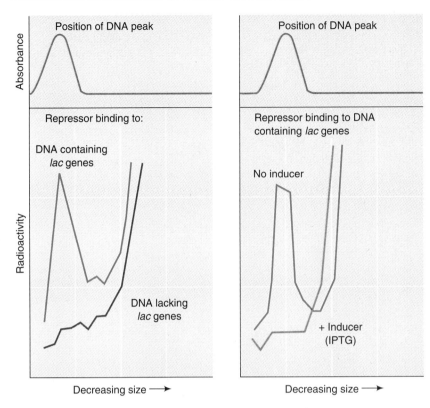

Figure 10-47 *Evidence That the* Lac *Repressor Binds to* Lac *Operon DNA* *Radioactive* lac *repressor was mixed with DNA and then analyzed by moving-zone centrifugation to separate free repressor from repressor bound to DNA. The two top panels show the position of the DNA after centrifugation. (Left) The radioactive repressor only binds to DNA samples that contain the* lac *operon genes. (Right) Binding of the* lac *repressor to* lac *operon DNA is abolished in the presence of the inducer, IPTG.*

Among the many operons subject to control by catabolite repression is the *lac* operon. The *lac* operon is thus subject to both positive and negative control, the former based upon binding of the CAP activator protein to the upstream end of the promoter and the latter based on binding of the *lac* repressor downstream from the promoter (Figure 10-52). This arrangement means that efficient expression of the *lac* operon requires the presence of both cyclic AMP (to activate the CAP protein) and an inducer such as lactose (to inactivate the *lac* repressor).

In the *lac* operon, positive and negative control are exerted by two different regulatory proteins (CAP and the *lac* repressor). It is also possible for positive and negative control to be exerted by the same protein. An example of this phenomenon occurs in the arabinose operon, whose transcription is inhibited by a repressor protein that binds to operator sites located within the operon. When it becomes bound to arabinose, however, the conformation of the repressor protein changes so that it activates rather than inhibits transcription.

Genes Coding for Enzymes Involved in Tryptophan Synthesis Are Organized into a Repressible Operon

Inducible operons such as the *lac* operon are ideally suited to situations in which the presence of a particular substance (in this case, lactose) signals the need for the enzymes encoded by a certain set of genes. In some cases, however, the opposite is true; that is, the presence of a particular substance indicates that the enzymes encoded by a certain set of genes are *not* needed. For example, bacteria placed in a growth medium containing the amino acid tryptophan have no need for enzymes that are involved in the synthesis of tryptophan. The genes coding for these enzymes are therefore organized into an operon whose expression is inhibited when tryptophan is present. Operons that are *inhibited* when a particular compound is added to the cellular environment are called *repressible operons* to distinguish them from inducible operons like *lac*, which are activated when a particular compound is added.

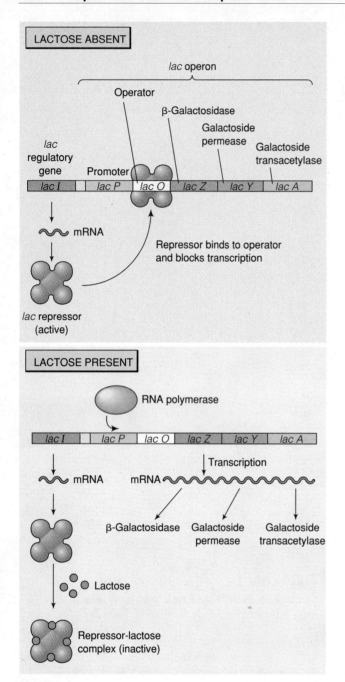

Figure 10-48 *The Effect of Lactose on Transcription of the Lac Operon* (Top) *In the absence of lactose, the lac repressor binds to the* lac *operator DNA, blocking transcription of the* lac *genes by RNA polymerase. (Bottom) In the presence of lactose, the binding of lactose to the* lac *repressor prevents the repressor from binding to* lac *operator DNA. As a result, the* lac *genes can be transcribed by RNA polymerase.*

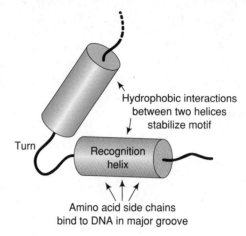

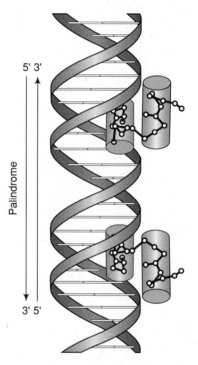

Figure 10-49 *The Helix-Turn-Helix Motif* *The helix-turn-helix motif represents a region of a polypeptide chain that binds to DNA in a sequence-specific manner. The bottom diagram shows how the helix-turn-helix motif binds in the major groove of the DNA double helix. DNA-binding proteins exhibiting this motif bind to DNA sequences that are palindromes, meaning that the sequence of bases in each DNA strand is the same when read in the 5' → 3' direction. Hence two helix-turn-helix motifs can bind to such DNA sites, one at each end of the palindromic sequence.*

The operon that codes for enzymes involved in tryptophan synthesis is called the ***trp* operon.** This operon contains five structural genes whose transcription is regulated by the protein product of a regulatory gene called *trpR*. The *trpR* gene codes for a repressor protein that by itself has no affinity for the *trp* operator and hence no effect on transcription. However, the protein is converted into an active repressor when it binds to tryptophan, which is said to act as a **co-repressor.** The repressor-tryptophan complex binds to an operator DNA sequence that is palindromic, just like the sequences recognized by the *lac* repressor and the CAP

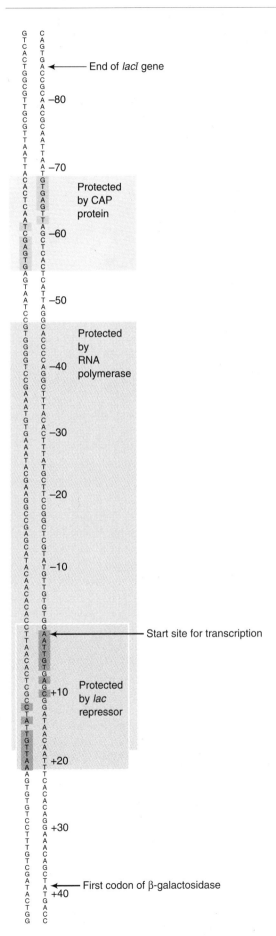

End of *lacI* gene

Protected by CAP protein

Protected by RNA polymerase

Start site for transcription

Protected by *lac* repressor

First codon of β-galactosidase

Figure 10-50 *Nucleotide Sequence of the Promoter and Operator Regions of the Lac Operon* *The binding sites for RNA polymerase, the* lac *repressor, and the CAP protein as determined by DNase I footprinting are marked. The palindromic sequences in the binding sites for the* lac *repressor and the CAP protein are highlighted.*

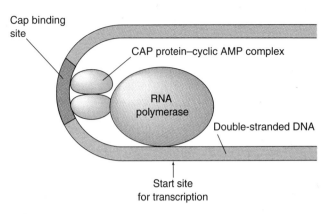

Cap binding site

CAP protein–cyclic AMP complex

RNA polymerase

Double-stranded DNA

Start site for transcription

Figure 10-51 *Activation of Transcription by the CAP Protein-Cyclic AMP Complex* *Binding of the CAP protein–cyclic AMP complex to its DNA binding site causes DNA to bend, bringing the CAP protein close to RNA polymerase.*

protein. Because the *trp* operator site overlaps with the *trp* promoter, binding of the repressor-tryptophan complex to the operator prevents RNA polymerase from binding to DNA and initiating transcription of the *trp* operon.

Like the *lac* repressor and the CAP protein, the *trp* repressor has a DNA-binding site that exhibits the helix-turn-helix motif. This same motif occurs in many of the DNA-binding regulatory proteins produced by bacteria, bacterial viruses, and, as we will see shortly, eukaryotic cells as well. The amino acid sequence of the recognition helix appears to be solely responsible for determining the type of DNA sequence a DNA-binding protein will recognize. In fact, changing a few amino acids in this helix can change a repressor that acts on one set of genes into a repressor that interacts with a different set of genes.

Both DNA Supercoiling and DNA Sequence Rearrangements Can Influence Bacterial Gene Expression

The interaction of repressor and activator proteins with operons is the principal means by which bacteria regulate gene transcription, but several other modes of gene control are also employed. Some of these mechanisms involve changes in the DNA molecule itself. For example, one type of DNA change that has been implicated in the regulation of bacterial gene activity involves DNA supercoiling. During transcription by RNA polymerase, localized unwinding of the DNA double helix must take

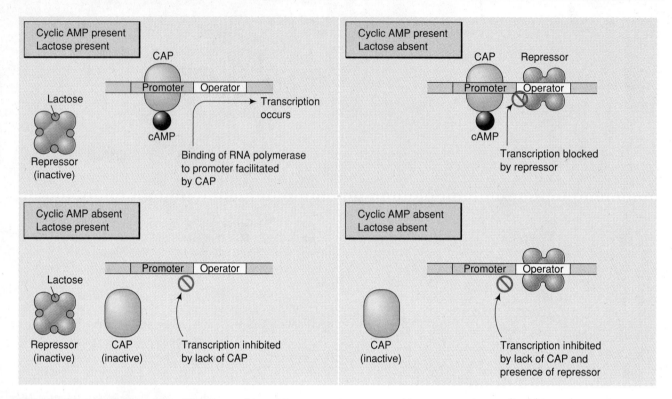

Figure 10-52 *Summary of the Positive and Negative Control Systems Acting on the* Lac *Operon* *Cyclic AMP bound to the CAP protein exerts positive control over transcription of the* lac *operon, whereas the* lac *repressor exerts negative control. Efficient expression of the* lac *operon therefore requires the presence of both cyclic AMP and an inducer such as lactose.*

place in order to allow base-pairing between the DNA template strand and the newly forming RNA chain. Because bacterial DNA is organized into long loops whose ends are anchored and therefore unable to rotate (page 417), localized unwinding creates tension in the double helix and the resulting formation of positive supercoils. The supercoiled state is energetically unfavorable for transcriptional processes and hence tends to slow the rate of transcription.

This obstacle can be overcome by the enzyme topoisomerase II, which catalyzes the formation of negative supercoils (page 415). When transcription takes place in a negatively supercoiled DNA molecule, the tension normally generated by localized unwinding of the double helix is utilized to remove such negative supercoils rather than to generate positive supercoils. The possibility that this phenomenon is involved in regulating the expression of specific genes is suggested by the discovery that inhibitors of topoisomerase II depress the transcription rate of some bacterial genes but not others. It has therefore been proposed that certain regions of the bacterial chromosome have specific sites for interaction with topoisomerase II, and that the negative supercoils generated in these regions facilitate gene transcription.

Rearrangements in DNA base sequence represent another means of altering gene expression. One of the simplest rearrangement mechanisms is observed in the bacterium *Salmonella* during *phase variation*, a phenomenon that involves the alternate expression of two genes coding for differing forms of the bacterial flagellar protein, *flagellin*. Switching between the expression of two alternate genes helps *Salmonella* evade the immune response of the organism it is infecting. The activities of the two flagellin genes, called *H1* and *H2*, are regulated by an invertible control sequence located immediately upstream from the flagellin H2 gene. This 970 base-pair control region contains both a promoter and a sequence that resembles a *transposable element*, which is a DNA sequence that can move from one location to another in DNA (page 95). In cells in which the flagellin H2 gene is active, the control element is oriented so that its promoter sequence is situated at the appropriate position upstream from the H2 gene. RNA polymerase can therefore transcribe both the flagellin H2 gene and an adjacent gene that codes for a repressor protein which inhibits transcription of the flagellin H1 gene (Figure 10-53).

However, in cells in which the H1 gene is active, the control element is inverted so that its promoter sequence is disconnected from the H2 gene. Under these

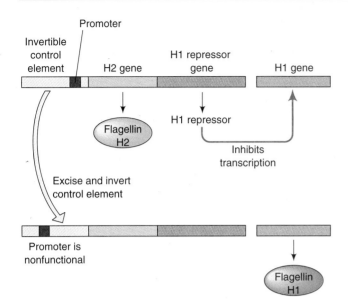

Figure 10-53 *The Mechanism of Phase Variation in Salmonella* *The expression of the H1 and H2 flagellin genes is alternated by inverting the orientation of a control element located adjacent to the H2 gene.*

conditions neither the flagellin H2 gene nor the repressor gene can be transcribed; moreover, the absence of repressor production activates transcription of the flagellin H1 gene. Hence the event that triggers alternate transcription of the flagellin H1 and H2 genes is the excision of the 970 base-pair control segment and its reinsertion in the opposite orientation. The resemblance of this invertible control sequence to a transposable element suggests that the mechanism of excision and reinsertion is analogous to the mechanisms by which transposable elements move from site to site within DNA.

Bacterial Gene Expression Can Be Regulated at Many Other Levels

In addition to the controls described thus far, bacterial gene transcription can be regulated in several other ways. One group of mechanisms involves changes in the enzyme RNA polymerase. For example, when *E. coli* is infected by bacteriophage *T4,* a small number of viral genes are first transcribed by the bacterial RNA polymerase into mRNAs. These messages in turn code for enzymes and proteins that interact with the bacterial RNA polymerase, causing chemical modification of the existing subunits and the addition of several new subunits to the bacterial RNA polymerase. This altered RNA polymerase then selectively transcribes the remaining viral genes. A somewhat different approach is employed by bacteriophage *T7;* the DNA of this virus is transcribed into an mRNA that codes both for protein factors that inactivate the bacterial RNA polymerase and for a new RNA polymerase that selectively transcribes phage genes.

The final prokaryotic control mechanism to be considered in this section is related to the termination of transcription. Earlier in the chapter we learned that the production of antitermination proteins allows transcription to proceed beyond its normal stop point and into adjacent genes that would not otherwise be transcribed (page 425). Control over the process of termination can also be exerted by a phenomenon called *attenuation,* in which the rate at which mRNA is translated into protein determines whether or not the transcription of the gene coding for that message will be prematurely terminated. Because regulation of this type is linked to the process of protein synthesis, it will be described more fully in Chapter 11 when we discuss the control of mRNA translation.

GENE REGULATION IN EUKARYOTES

The regulation of gene expression in eukaryotic cells exhibits some underlying similarities to the comparable events in prokaryotic cells, but major differences are evident as well. One important difference is that genes with related functions are not organized into operons in eukaryotes. In addition, the existence of RNA splicing and the need to transport mRNAs through the nuclear envelope create potential points of control that do not exist in prokaryotes (Figure 10-54). Finally, the DNA of eukaryotic cells is organized into chromatin fibers and contains various types of noncoding sequences that are not present in prokaryotes. We will begin our discussion of eukaryotic gene regulation by describing some of these unusual features of eukaryotic DNA.

Eukaryotic Cells Contain Far More DNA Than Appears to Be Needed for Genetic Coding Purposes

Any discussion of gene regulation in eukaryotic cells requires a basic familiarity with the way in which eukaryotic DNA is organized. For any organism, the total amount of DNA present in a single (haploid) set of chromosomes is known as the **C value.** The C value for eukaryotic cells ranges from about five times greater to more than ten thousand times greater than that of bacterial cells (Table 10-5). This enormous variation in DNA content is not easy to explain based on differences in the complexity of eukaryotic organisms. For example, it is currently estimated that the bacterium *E. coli* contains about 4000 genes and humans contain between 10,000 and 50,000 genes. Therefore at most, humans contain about ten times the number of genes as *E. coli,* and yet human cells possess about a thousand-fold more DNA.

The fact that eukaryotic cells contain far more DNA than appears to be needed for genetic coding is called the **C value paradox.** Comparison of the amount of

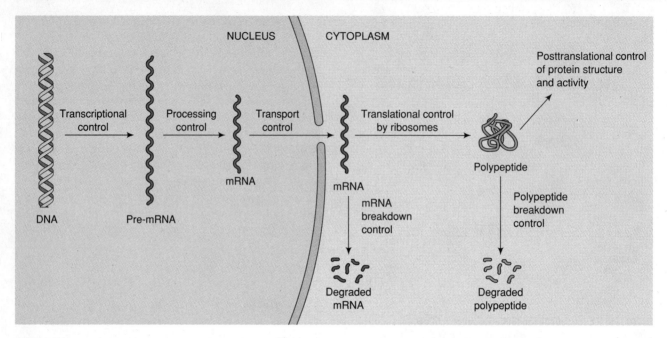

Figure 10-54 *Levels at Which Gene Expression Can Be Regulated in Eukaryotes* *Control can be exerted at the level of gene transcription, RNA processing, RNA transport from nucleus to cytoplasm, translation, mRNA breakdown, protein breakdown, and posttranslational alterations in protein structure and activity.*

Table 10-5   The C Values of Selected Organisms

	DNA Content		
Organism	**Daltons**	**Base Pairs**	**Length**
Prokaryotes			
Escherichia coli	2.8×10^9	4.1×10^6	1.4 mm
Salmonella typhimurium	8.0×10^9	1.1×10^7	3.8 mm
Eukaryotes			
Fungi:			
Yeast (*S. cerevisiae*)	1.2×10^{10}	1.8×10^7	5.95 mm
Neurospora crassa	1.9×10^{10}	2.7×10^7	9.18 mm
Invertebrates:			
Sea urchin	5.0×10^{11}	8.0×10^8	27.2 cm
Fruit fly	1.2×10^{11}	1.8×10^8	5.95 cm
Vertebrates:			
Human	1.9×10^{12}	2.8×10^9	94 cm
Mouse	1.5×10^{12}	2.2×10^9	75 cm
Frog	1.4×10^{13}	2.3×10^{10}	7.65 m
Plants:			
Lily	2.0×10^{14}	3.0×10^{11}	100 m
Maize	4.4×10^{12}	6.6×10^9	2.24 m

DNA found in various eukaryotes has revealed another aspect of the C value paradox: The DNA contents of closely related organisms sometimes differ by as much as two orders of magnitude. In addition, certain plants and amphibians possess ten times as much DNA as human cells, yet these organisms do not have a significantly greater number of genes. A complete explanation for the various aspects of the C value paradox is yet to emerge, but they seem to suggest the existence of large amounts of DNA that perform no obvious genetic function.

DNA Reassociation Studies Reveal That Eukaryotic DNA Contains Both Unique and Repeated Sequences

A significant advance in our understanding of the nature of some of this extra DNA occurred in the late 1960s, when DNA hybridization experiments carried out by Roy Britten and David Kohne led to the discovery of repeated DNA sequences. In these experiments, DNA was broken into small fragments by physical shearing and dissociated into single strands by heating. The temperature was then lowered to permit the single-stranded fragments to hybridize back to one another. The rate of this *DNA reassociation* process depends on the concentration of each individual DNA sequence; the higher the concentration of any given DNA sequence in the initial double-stranded DNA population, the greater the probability that it will collide with a complementary strand to which it can hybridize.

Given these considerations, how would the reassociation properties of different kinds of DNA be expected to compare? As an example, let us consider DNA derived from a bacterial cell and from a typical mammalian cell containing a thousandfold more DNA. If this difference in DNA content reflects a thousandfold difference in the kinds of DNA sequences present, then bacterial DNA should reassociate a thousand times faster than does mammalian DNA. The rationale underlying this prediction is that in mammalian and bacterial DNA samples of equal concentration, any particular DNA sequence should be present in a thousandfold lower concentration in the mammalian DNA sample because there are a thousand times more kinds of sequences present.

When experiments comparing the reassociation kinetics of mammalian and bacterial DNA were actually performed by Britten and Kohne, the results were not exactly as expected. Figure 10-55 summarizes the data obtained when the reassociation kinetics of calf and *E. coli* DNA were compared. In this graph, the percentage of the total DNA that has become reassociated is plotted as a function of the starting concentration of DNA multiplied by the length of time. This parameter of DNA concentration × time, or $C_{o}t$, is employed in

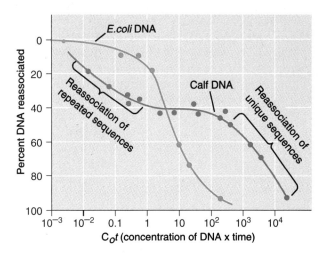

Figure 10-55 *Reassociation Kinetics of Calf and E. Coli DNAs* *The calf DNA that reassociates more rapidly than the bacterial DNA consists of repeated sequences.*

place of time alone because it allows the direct comparison of data obtained from reactions run at different DNA concentrations. When graphed in this way, the data reveal that calf DNA consists of two classes of sequences that reassociate at very different rates. One class of DNA sequence, which accounts for about 40 percent of the calf DNA, reassociates more rapidly (i.e., at a lower $C_{o}t$ value) than bacterial DNA. The most straightforward explanation for this unexpected result is that calf DNA contains **repeated DNA sequences** that are present in multiple copies. The existence of multiple copies increases the relative concentration of the sequences, thereby generating more collisions and a faster rate of reassociation than would be expected if each sequence were present in only a single copy.

The remaining 60 percent of the sequences present in calf DNA reassociate about a thousand times more slowly than *E. coli* DNA, which is the behavior expected of sequences that are present in single copies. This fraction is therefore called **unique-sequence DNA** to distinguish it from the repeated sequences that reassociate more quickly. Unique-sequence DNA is present in one copy per single (haploid) set of chromosomes. Hence in diploid cells containing two sets of chromosomes, two copies of each unique DNA sequence are present. Most protein-coding genes are members of the unique-sequence class, although this does not mean that all unique DNA codes for proteins. In bacterial cells virtually all of the DNA is unique-sequence DNA, whereas eukaryotes exhibit a large variation in the relative proportion of unique and repeated sequences.

Unique and repeated sequences can be easily separated from each other by reassociating DNA at a $C_{o}t$ value that is high enough to permit the reassociation of repeated, but not unique, DNA sequences. This allows the repeated sequences to reassociate into double-stranded DNA while most unique sequences remain single stranded. The single- and double-stranded DNAs are then separated from each other by physical means, such as chromatography on hydroxylapatite columns. Repeated DNA sequences purified in this way exhibit melting temperatures that are lower than that expected from normal DNA molecules of comparable base composition. From studies involving DNA molecules of defined sequence, it is known that a decreased melting temperature occurs when two DNA strands are not properly matched throughout their sequence; for every 1 percent of the bases that are not properly paired, the melting temperature decreases by about 1°C. From this information it has been calculated that typical repeated DNA sequences are about 10 percent mismatched after reassociation, and in some cases the mismatching is even more extensive. The presence of mismatching in this reassociated DNA means that the repeated sequences are not exact copies of each other (Figure 10-56).

Eukaryotic Cells Contain Several Classes of Repeated DNA Sequences

The repeated components of eukaryotic DNA consist of a heterogeneous group of sequences that are repeated to varying extents. A small contribution is made by a few genes, such as those coding for histones and ribosomal RNAs, which are present in hundreds or thousands of copies per cell. The remaining repeated sequences fall into several different categories. The

most highly repeated sequences are called *satellite DNAs* because their unique base sequence and extensive repetition allow them to be detected during iso-density centrifugation as a shoulder or satellite peak that is distinct from the main band of DNA (Figure 10-57). Satellite DNAs typically consist of relatively short sequences that are repeated over and over. In the case of mouse satellite DNA, for example, a simple sequence about a dozen bases long is repeated with minor variations more than ten million times, accounting for roughly 10 percent of the organism's total DNA content.

Satellite DNAs of varying lengths, sequence arrangements, and chromosomal locations have been detected in a variety of animal and plant cells. In spite of this variation, most satellite DNA sequences share three basic features: (1) they consist of multiple copies of the same sequence repeated in tandem; (2) they are not transcribed into RNA; and (3) they are preferentially associated with constitutive heterochromatin (page 450), centromeres (page 528), and telomeres (page 533). Because these chromosome regions tend to be permanently coiled up and are not transcribed into RNA, it has been proposed that satellite DNA sequences play a structural role that helps to maintain chromatin in an inactive state. In addition, we will see in Chapter 12 that the repeated sequences found in telomeres play a special role in DNA replication.

Besides satellite DNAs, the repeated DNA family also contains sequences that are scattered throughout a cell's chromosomes. These repetitive sequences are divided into two main categories called short and long interspersed repeats. *Short interspersed repeats* are

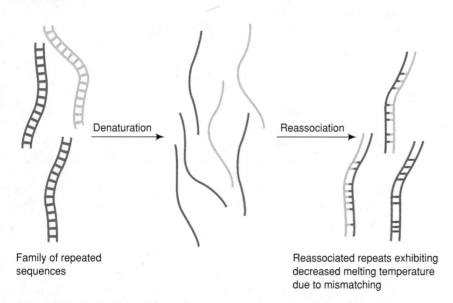

Family of repeated sequences

Denaturation

Reassociation

Reassociated repeats exhibiting decreased melting temperature due to mismatching

Figure 10-56 *Diagram Showing Why Reassociated Repeated DNA Sequences Exhibit a Decreased Melting Temperature* *The magnitude of the decrease in melting temperature indicates the extent of mismatching between the various members of the repeat family.*

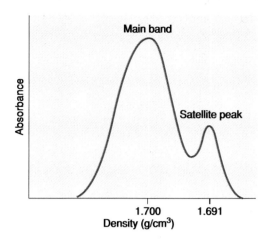

Figure 10-57 *Fractionation of Mouse DNA by Isodensity Centrifugation* *About 10 percent of the total DNA has a base composition that is different enough from the rest of the DNA to cause it to sediment as a separate satellite peak.*

typically 100 to 500 base pairs long, whereas *long interspersed repeats* are several thousand base pairs in length. Both types of interspersed sequences are repeated as many as hundreds of thousands of times. Short interspersed repeats are often found in introns or nontranslated regions of protein-coding genes, where they are transcribed by RNA polymerase II. A selected subset of short interspersed repeats is also transcribed by RNA polymerase III. One prominent class of short interspersed repeat is the *Alu* sequence, which consists of sequences about 300 base pairs long that are repeated almost a million times in human DNA. Although *Alu* sequences account for more than 5 percent of the DNA present in human cells, their function is yet to be clearly established.

The short interspersed repeats that are localized within introns are degraded during mRNA splicing, but the functions of other interspersed repeats have not been clearly established. The possibility has even been raised that some of this repeated DNA represents "selfish" sequences that perform no function at all. According to this view, repeated DNA sequences may have arisen during evolution from viruses or transposable elements that are capable of multiplying and inserting into various chromosomal regions. Such inserted sequences may behave like parasites that are replicated along with the rest of the chromosomal DNA but perform no function for the host cell.

The realization that a major portion of the DNA present in eukaryotic cells consists of repeated sequences whose functional significance is unknown should be kept in mind as we proceed with our discussion of eukaryotic gene regulation. As we will see, a number of differences distinguish the genetic regulatory mechanisms of eukaryotes from those of prokaryotes, and it is possible that at least some repeated DNA

sequences are involved in regulatory processes that are specific to eukaryotes.

The Cytoplasm Influences the Activity of Nuclear Genes

Now that the unique features of eukaryotic DNA have been introduced, we are ready to examine the mechanisms by which the expression of this DNA is controlled. A considerable body of evidence supports the notion that the spectrum of genes expressed in eukaryotic cells is influenced by the cytoplasmic environment in which the nucleus resides. The technique of *nuclear transplantation* has been especially useful in demonstrating this point. In one type of transplantation study, transcriptionally inactive nuclei have been taken out of cells that are synthesizing neither DNA nor RNA and transplanted into egg cells whose nuclei were previously removed. When placed in this new cytoplasmic environment, a nucleus that had previously been inactive will begin to synthesize RNA and, in some cases, DNA. In the most dramatic experiments of this type, carried out on the African clawed toad *Xenopus laevis,* John Gurdon removed the nuclei from cells lining the intestine of swimming tadpoles and transplanted them into egg cells whose nuclei had already been removed. Although the transplanted nuclei were derived from intestinal cells, they were able to direct the development of the egg cell into a complete new toad. Clearly the egg cytoplasm must exert profound effects on nuclear activity if it can induce the nucleus of an intestinal cell to behave in this way.

The technique of cell fusion (page 117) has also provided important insights into the factors involved in the cytoplasmic control of gene activity. Using this approach, Henry Harris and his associates have investigated the biochemical and structural changes that occur after cells containing active nuclei are fused with cells containing inactive nuclei. One cell type that has been extensively studied is the chicken red blood cell, which contains a nucleus that is inactive in both DNA and RNA synthesis. Although the chromatin of the red cell nucleus is normally packed into a tightly condensed mass of fibrils, fusing the red cell with a more active cell type causes the red cell nucleus to swell and its condensed chromatin to uncoil (Figure 10-58). At the same time, nonhistone proteins derived from the active cell type accumulate within the red cell nucleus, which then begins to synthesize RNA.

In such experiments, the new cytoplasmic environment has been shown to activate the transcription of genes that are not normally transcribed in red blood cells. For example, fusing chick red blood cells with mouse cells triggers the production of the enzyme *inosinic acid pyrophosphorylase (IMPase).* Since the mouse cells used in these studies lack the IMPase gene,

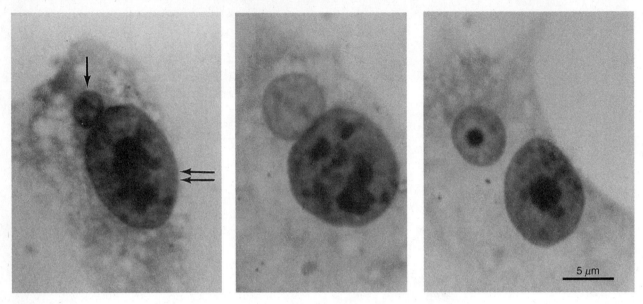

Figure 10-58 *Reactivation of a Red Cell Nucleus after Fusion of Chick Red Blood Cells with Cultured Mouse Cells* (Left) *Twelve hours after fusion the red cell nucleus* (arrow) *still contains condensed, darkly staining chromatin. The double arrow points to the larger mouse cell nucleus.* (Middle) *Two days after fusion the red cell nucleus has swelled and its chromatin has uncoiled.* (Right) *Four days after fusion a nucleolus has appeared within the red cell nucleus. Courtesy of E. Sidebottom.*

and the IMPase gene of chick red cells is normally inactive, one can conclude that some factor present in the mouse cell cytoplasm causes the previously inactive chick IMPase gene to be expressed.

The preceding observations indicate that the pattern of genes expressed in a given cell type is not rigidly or permanently determined within the nucleus itself, but instead is subject to influence by cytoplasmic conditions. In the remainder of the chapter we will investigate some of the mechanisms by which changes in cytoplasmic and environmental conditions can trigger such alterations in gene expression.

The Activity of Individual Genes Can Be Visualized in Polytene Chromosomes

Because the DNA of most eukaryotic cells is dispersed throughout the nucleus as a mass of intertwined chromatin fibers, it is difficult to distinguish the activity of individual genes with a microscope. However, a way around this obstacle has been provided by an unusual situation that occurs in certain insect cells. In flies belonging to the order Diptera, metabolically active tissues such as the salivary glands, intestines, and excretory organs grow by an increase in the size of, rather than the number of, their constituent cells. This process generates giant cells whose volumes may be thousands of times greater than normal. The development of giant cells is accompanied by the formation of enormous chromosomes measuring hundreds of micrometers in

length and several micrometers in width (Figure 10-59). These chromosomes are roughly ten times greater in length and a hundred times greater in width than the chromosomes that appear at the time of cell division in typical eukaryotic cells.

In the early 1960s Hewson Swift utilized DNA staining methods to monitor the increase in DNA content that accompanies the formation of giant chromosomes in the fruit fly *Drosophila*. He discovered that the increase in DNA occurs as a series of about ten doublings, at which point the DNA content has increased roughly a thousandfold ($2^{10} = 1024$). This observation suggests that giant chromosomes are multistranded or **polytene**—that is, composed of a large number of identical DNA molecules generated by successive rounds of replication. Because this replication occurs in a cell that is not dividing, the newly synthesized DNA molecules accumulate in the nucleus and line up in parallel to form a polytene chromosome.

A characteristic feature of polytene chromosomes is the presence of a distinct pattern of darkly staining regions, termed *bands,* which are separated from each other by lighter staining areas known as *interbands.* It has long been known that mutations involving the deletion of specific genes are accompanied by the disappearance of particular bands. This observation led Calvin Bridges to propose in the early 1930s that each band corresponds to a single gene, but more recent evidence indicates that individual bands often contain several genes. Since the polytene chromosomes of insects such as *Drosophila* have a total of about 5000 bands, and since

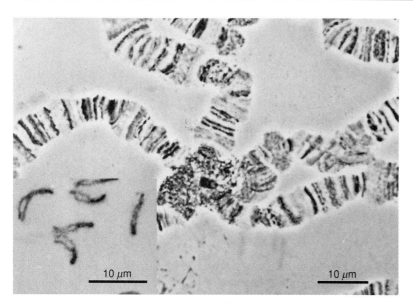

10 μm   10 μm

Figure 10-59 *Polytene Chromosomes of* Drosophila *Observed by Light Microscopy*
The bands of the polytene chromosomes appear as dark regions separated from one another by the interbands. For size comparison, the inset shows typical metaphase chromosomes at the same magnification. Courtesy of J. Gall.

each of these bands contains no more than a few genes, it can be concluded that *Drosophila* probably contains somewhere between 10,000 and 50,000 genes.

Electron microscopy has provided several clues as to how polytene chromosomes are constructed. In thin-section electron micrographs, 30-nm fibers resembling the chromatin fibers of typical eukaryotic cells can be observed in both the band and interband regions. The major difference in the appearance of the two regions is that the chromatin fibers are densely packed in the bands, while in the interbands they are arranged as more loosely packed parallel bundles (Figure 10-60). Based on this information, it has been proposed that polytene chromosomes are constructed from chromatin fibers that are aligned in parallel and pass continuously from one end of the chromosome to the other (Figure 10-61, *top*). According to this model the individual chromatin fibers, each containing a single DNA molecule, are tightly folded and packed together in the band regions and more loosely assembled in the interbands.

Chromosome Puffs and Balbiani Rings Represent Sites of Gene Transcription

The pattern of chromosome banding is relatively constant for each type of polytene chromosome, but certain bands undergo changes in appearance that are specific for a particular tissue or stage of development. Such bands uncoil into a looser, puffed-out configuration known as a chromosome **puff** or, when it becomes exceptionally large, a **Balbiani ring** (Figure 10-62). In 1952 Wolfgang Beerman postulated that puffing is a reflection of the activity of individual genes. This idea was

originally based on the observation that each tissue type exhibits its own characteristic pattern of puffs, and that the puffing pattern changes in a reproducible way as development proceeds.

Beerman supported this hypothesis with a series of experiments in which the expression of a specific trait was correlated with the development of a particular Balbiani ring. These studies focused on the salivary secretions of the midge *Chironomus*. In one species of midge, known as *Ch. pallidivitattus*, the salivary gland produces a secretion that is granular; in the closely related midge, *Ch. tentans*, the salivary gland produces a clear fluid. Upon comparing the salivary gland chromosomes of the two organisms, Beerman noted that a large Balbiani ring present in *Ch. pallidivitattus* is missing in *Ch. tentans*. In order to determine whether this Balbiani ring is responsible for the granular secretion, the two species were allowed to interbreed. The salivary gland secretion of the resulting hybrid midge was found to be about half as granular as the secretion of *Ch. pallidivitattus*, and the Balbiani ring in question was half as large. Beerman therefore concluded that the formation of this Balbiani ring represents transcription of the gene responsible for the granularity of the salivary gland secretion.

The idea that puffs and Balbiani rings represent sites where DNA is transcribed was confirmed in the early 1960s by Claus Pelling, who employed microscopic autoradiography to monitor the incorporation of ^{3}H-uridine into RNA. His studies revealed that newly synthesized RNA is localized almost exclusively in the puffed regions of polytene chromosomes, and that inhibitors of RNA synthesis such as actinomycin D pre-

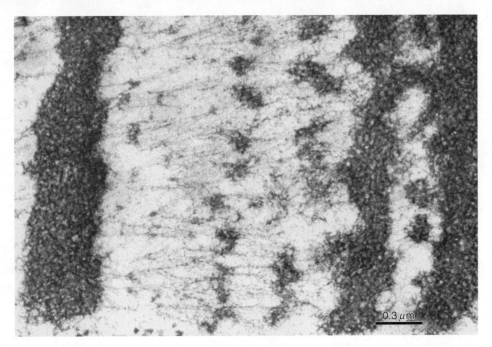

Figure 10-60 *Electron Micrograph Showing the Band and Interband Regions of a Polytene Chromosome*   *The light staining region, which represents an interband, contains a loosely packed array of chromatin fibers. The chromatin fibers are more densely packed in the darker staining regions, which correspond to bands. Courtesy of V. Sorsa.*

vent this radioactive labeling from occurring (Figure 10-63). Efforts to characterize the RNA synthesized in the puffed regions have been aided by the fact that the Balbiani rings of *Chironomus* salivary gland cells are large enough to be isolated by microsurgery. This approach has been elegantly exploited by Bo Lambert, who extracted radioactive RNA from several isolated Balbiani rings and hybridized it back to unlabeled chromosomes. When RNA was isolated from a particular Balbiani ring, it was found to hybridize back only to the DNA of that Balbiani ring. It was therefore concluded that each Balbiani ring synthesizes a specific type of RNA that is uniquely complementary to the DNA stretch found in that Balbiani ring.

Subsequent experiments have revealed that RNAs that hybridize to specific Balbiani rings can also be isolated from the nucleoplasm, where they are present in RNA-protein (RNP) complexes, and from the cytoplasm, where they are associated with ribosomes. Taken together, the preceding observations indicate that each chromosomal puff synthesizes a specific type of RNA that is complexed with protein to form an RNP particle. After release into the nucleoplasm, these RNP particles are transported through the nuclear pore complexes to the cytoplasm (see Figure 10-6). The RNA component of the RNP complex then becomes associated with ribosomes, where it functions as an mRNA that directs the synthesis of a particular polypeptide chain.

Chromosome Puffing Can Be Induced by the Steroid Hormone Ecdysone

Although our understanding of the factors that regulate puff formation is by no means complete, one of the first changes that can be detected at newly forming puffs is the accumulation of nonhistone proteins (Figure 10-64).

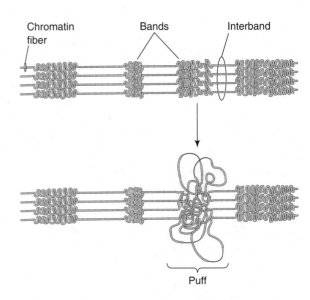

Figure 10-61 *Model of a Polytene Chromosome*   *During gene transcription, the chromatin fibers often uncoil into a visible puff or Balbiani ring.*

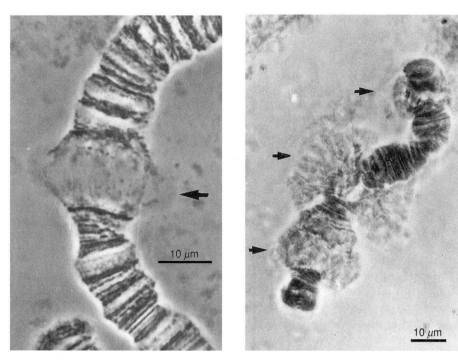

Figure 10-62 *Puffs and Balbiani Rings* (Left) *A polytene chromosome from an insect salivary gland exhibiting one puff.* (Right) *A polytene chromosome from the midge* Chironomus *exhibiting three Balbiani rings. Courtesy of B. Daneholt.*

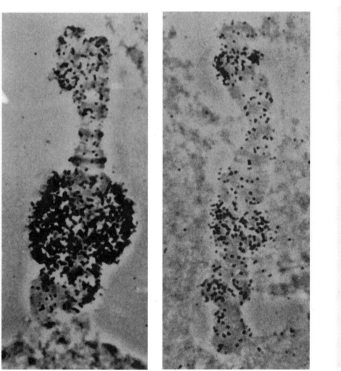

Figure 10-63 *Autoradiograph of a Polytene Chromosome from Cells Incubated with ³H-Uridine to Label Newly Synthesized RNA* (Left) *Numerous silver gains occur over the puffs, indicating that they represent sites of RNA synthesis.* (Right) *In cells treated with actinomycin D to inhibit RNA synthesis, the puffs decrease in size and the incorporation of ³H-uridine is markedly reduced. Courtesy of W. Beerman.*

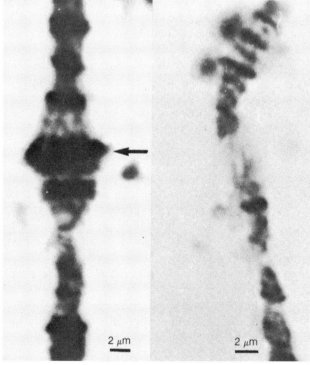

Figure 10-64 *Light Micrograph Showing the Accumulation of Protein in the Puff of a Polytene Chromosome* *The intense staining of the puff observed in the chromosome on the left is abolished when the chromosome is pretreated with an enzyme that breaks down protein molecules (right), indicating that the puff is enriched in protein. Courtesy of Th. K. H. Holt.*

Some nonhistone proteins are common to multiple puffs, while others are specific to certain ones. As nonhistone proteins accumulate, the puff begins to swell and the chromatin fibers uncoil, becoming thinner and less condensed than the typical 30-nm fibers characteristic of a normal band (see Figure 10-61, *bottom*). Shortly after the uncoiling process begins, the synthesis of RNA is initiated.

The normal process of insect molting is one situation in which the molecular events that trigger the puffing process are well established. When an insect molts, a reproducible pattern of chromosomal puffs develops. In the early 1960s Ulrich Clever reported that this same sequence of puffs can be induced experimentally by injecting nonmolting insects with the steroid hormone *ecdysone*. An identical puffing pattern is also observed when isolated salivary glands are incubated with ecdysone (Figure 10-65). The puffing induced by ecdysone occurs in two phases: *early puffs* develop within a few minutes of hormone administration, whereas *late puffs* form several hours later and are accompanied by regression of the early puffs. Inhibitors of protein synthesis block formation of the late but not the early puffs, suggesting that the RNAs synthesized by the early puffs code for the synthesis of proteins required for development of the later puffs. As in the case of steroid hormones that act in other organisms, the effects of ecdysone on gene transcription are mediated by a protein receptor to which the steroid binds. We will see later in the chapter that steroid-hormone receptors bind to specific DNA sequences and activate gene transcription. The ecdysone receptor is thus an example of a nonhistone protein that selectively regulates the transcriptional activity of a specific group of puffs; when ecdysone binds to the ecdysone receptor, the receptor undergoes a change in conformation that allows it to bind to DNA and activate transcription.

Although it is now clear that chromosome puffs represent sites of gene transcription, the absence of visible puff formation does not necessarily indicate the absence of transcription. Autoradiographic studies of cells incubated with ³H-uridine have revealed the presence of newly formed radioactive RNA over bands where obvious puffs are not present, suggesting that these bands also represent sites of transcription. In the case of the genes coding for the histone molecules of *Drosophila*, it has been clearly demonstrated that transcription occurs in banded regions that lack puffs.

The Formation of Heterochromatin Allows Large Segments of DNA to Be Transcriptionally Inactivated

Chromosome puffs are not the only situation in which alterations in eukaryotic gene expression are associated with visible changes in chromatin structure. It was first noted in the early 1920s that certain regions of eukary-

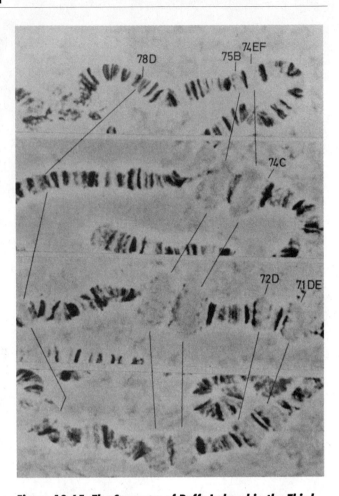

Figure 10-65 *The Sequence of Puffs Induced in the Third Chromosome of Drosophila by the Hormone Ecdysone* Note the changing appearance of the three numbered bands (78D, 75B, and 74EF) after exposure to ecdysone for 1, 2, 4, and 6 hours (top to bottom). Courtesy of M. Ashburner.

otic chromosomes do not uncoil after cell division is completed but instead remain tightly condensed throughout the cell cycle. Subsequent studies have revealed that this condensed form of chromatin is not transcribed into RNA. Such condensed, genetically inactive chromatin is referred to as **heterochromatin,** while the more loosely packed, transcriptionally active form of chromatin is called **euchromatin.** Heterochromatin occurs in two forms known as constitutive heterochromatin and facultative heterochromatin. *Constitutive heterochromatin* contains permanently inactivated DNA sequences that are never destined to be transcribed, such as the satellite DNAs found in chromosomal centromeres. In contrast, *facultative heterochromatin* contains genes that are potentially capable of being transcribed, but have become inactivated in certain cell types as part of normal development.

A dramatic example of facultative heterochromatin occurs in the sex chromosomes of mammals. In the early 1950s Murray Barr and his collaborators observed that the cells of female mammals often exhibit a small, darkly staining mass of chromatin that is not detectable in the comparable cells of males (Figure 10-66). Subsequent studies revealed that this structure, termed a **Barr body,** represents an X chromosome that has been completely converted to heterochromatin. Mary Lyon was the first to suggest that Barr bodies are produced by the random inactivation of one of the two X chromosomes that are normally present in females, leaving only one X chromosome transcriptionally active. This theory therefore predicts that every cell in an adult female has only one active X chromosome, just as in males where only one of the two sex chromosomes is an X chromosome.

Lyon further proposed that once an X chromosome has been inactivated in a particular cell, all descendants of that cell will have the same X chromosome inactivated. Since the initial inactivation process is thought to be random, some of the cells present in a female adult would be expected to derive from a cell lineage in which one X chromosome is inactive, and the rest would be expected to derive from cells in which the other X chromosome is inactive. In support of this hypothesis, Lyon pointed out that female animals containing genes for different coat colors on their two X chromosomes always have mottled coats. A classic example of this phenomenon occurs in the tortoise-shell cat, whose coat consists of patches of yellow and black fur. The tortoise-shell pattern is usually restricted to female cats, and can be readily explained by expression of one X chromosome in the yellow patches of fur and expression of the other X chromosome in the black patches.

Subsequent support for the Lyon hypothesis has been obtained from the direct analysis of proteins known to be encoded by genes localized on the X chro-

mosome. The behavior of one such protein, the enzyme *glucose 6-phosphate dehydrogenase,* has been studied in cultured cells obtained from individuals who synthesize two different forms of this enzyme. Each form of the enzyme is encoded by a single gene located on one of the two X chromosomes. Cell cultures derived from such individuals produce both forms of glucose 6-phosphate dehydrogenase, but clones obtained by growing cultures from isolated single cells produce only one form of the enzyme or the other (Figure 10-67). Hence the original cultures must consist of a mixed population of two cell types, each expressing only one of the two X chromosomes.

Transcriptionally Active Chromatin Is Sensitive to Digestion with DNase I

Although the preceding example shows that the formation of heterochromatin can inactivate large blocks of DNA, such a process represents a relatively coarse con-

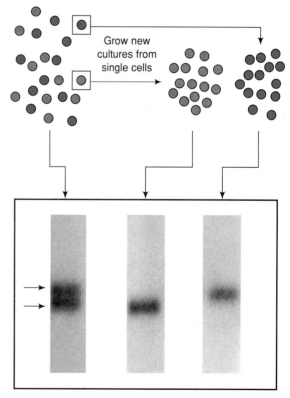

Electrophoresis of
glucose 6-phosphate dehydrogenase

Figure 10-67 *Evidence That the Cells of Human Females Express Only One X Chromosome* *Cells were obtained from an individual who produces two forms of glucose 6-phosphate dehydrogenase (arrows), an enzyme encoded by a gene located on the X chromosome. Although the original cell population produced both forms of the enzyme, clones obtained by growing cultures from single cells produce only one form of the enzyme or the other. Photograph courtesy of R. G. Davidson and H. M. Nitowsky.*

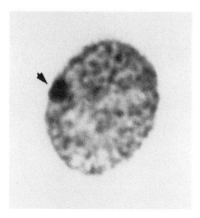

Figure 10-66 *The Barr Body* *The dark staining mass of chromatin (arrow) observed near the edge of the nucleus is a typical Barr body. This structure, which represents an inactivated X chromosome, occurs in the cells of female mammals. Courtesy of G. R. Wilson.*

trol over gene expression. More subtle changes in chromatin structure appear to be associated with the selective activation and inactivation of individual genes. In the mid-1970s Harold Weintraub and his associates demonstrated that actively transcribed genes are more susceptible than inactive genes to digestion with pancreatic DNase I. For example, if chromatin isolated from developing chick red blood cells is briefly exposed to DNase I, DNA sequences coding for globin are destroyed while sequences coding for ovalbumin remain intact (Figure 10-68). These results reflect the fact that the globin gene is expressed in developing red cells but the ovalbumin gene is not. The opposite result is obtained when the same procedure is carried out using chromatin isolated from oviduct, where the ovalbumin gene is expressed and the globin gene is inactive. In this case the ovalbumin gene is more sensitive than the globin gene to DNase I digestion.

Sensitivity to DNase I is exhibited by genes that are being actively transcribed, by genes that have recently been transcribed but are no longer active, and by DNA sequences located adjacent to genes of the preceding types. These observations suggest that DNase I sensitivity is not caused by the process of gene transcription itself, but instead reflects an altered chromatin structure in regions associated with active or potentially active genes. If chromatin structure is altered in the region of active genes, the question arises as to whether this alteration involves the loss of the normal nucleosomal organization of chromatin fibers.

Two lines of evidence indicate that nucleosomes do in fact remain associated with actively transcribed genes. First, electron microscopic studies have revealed that chromatin that is being actively transcribed retains the characteristic beads-on-a-string appearance, with nucleosomes visible both in front of and behind the advancing RNA polymerase molecule (Figure 10-69). Second, we learned earlier in the chapter that micrococcal nuclease degrades chromatin into DNA fragments that are multiples of 200 base pairs in length when nucleosomes are present (page 412). Base sequences contained within both active and inactive genes are found to be cleaved into 200 base-pair fragments, indicating that the DNA of actively transcribed genes remains associated with nucleosomes.

DNase I Hypersensitive Sites Are Short Nucleosome-Free Regions Located Adjacent to Active Genes

Certain sequences located adjacent to active genes are so sensitive to digestion by trace amounts of DNase I that they are referred to as **DNase I hypersensitive sites.** Hypersensitive sites tend to occur up to a few hundred bases upstream from the transcriptional start sites of active genes, and are about tenfold more sensitive to DNase I digestion than the bulk of the DNA associated with these genes. The idea that DNase I hypersensitive sites represent regions that are free of nucleosomes first emerged from studies involving the eukaryotic virus *SV40*. When the SV40 virus infects a host cell, its circular DNA molecule becomes associated with histones and forms typical nucleosomes that can be observed with the electron microscope. However, a small region of the viral DNA remains free of nucleosomes (Figure 10-70). This DNA region consists of an enhancer sequence that contains several DNase I hypersensitive sites.

In many cases, sites that are hypersensitive to digestion with DNase I tend to be extremely sensitive to cleavage with other nucleases as well. This sensitivity is especially notable with nucleases that preferentially de-

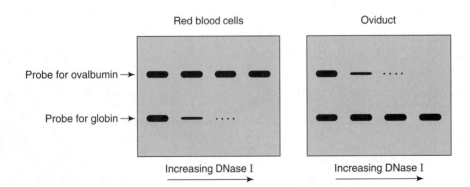

Figure 10-68 *Sensitivity of Active Genes to Digestion With DNase I*   *In these experiments, chromatin was digested with DNase I and the DNA was then isolated, cut with a restriction enzyme, and analyzed by Southern blotting for the presence of sequences that react with probes for either globin or ovalbumin genes. Note that the globin gene is sensitive to DNase I digestion in red blood cells but not in oviduct, whereas the opposite is true for the ovalbumin gene. This shows that the globin gene is active in red blood cells and inactive in oviduct, whereas the ovalbumin gene is inactive in red blood cells and active in oviduct.*

Not transcribed   Transcribed

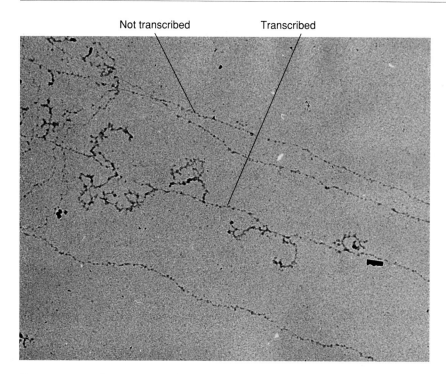

Figure 10-69 *The Presence of Nucleosomes in Chromatin That Is Being Actively Transcribed* This electron micrograph of chromatin isolated from a Drosophila *cell shows a chromatin fiber that is being transcribed as well as a chromatin fiber that is not being transcribed. Note that both types of fibers have the same beads-on-a-string appearance that indicates the presence of nucleosomes. Newly forming RNA molecules can be seen emerging from the transcribed chromatin fiber. Courtesy of O. L. Miller, Jr.*

grade DNA regions exhibiting an altered conformation, suggesting that the DNA configuration in the region of the hypersensitive sites may be different from that of the bulk of the DNA. One possibility is that the DNA located in the hypersensitive sites is highly supercoiled. Negative supercoiling is known to stabilize an unusual structural form of DNA discovered by Alexander Rich and referred to as *Z-DNA*. In the normal or *B-form* of the DNA molecule, the sugar-phosphate backbone is arranged as a smooth right-handed double helix; Z-DNA, in contrast, is a left-handed helix with a zigzag backbone (Figure 10-71). Although Z-DNA was first detected in synthetic DNA molecules consisting of alternating purines and pyrimidines, antibodies directed against Z-DNA also react with some of the DNA found in normal cells. A possible relationship between Z-DNA and transcriptional activity is suggested by the distribution of Z-DNA in the nuclei of ciliated protozoa. These cells contain two kinds of nuclei: a large transcriptionally active macronucleus and a smaller inactive micronucleus. When the cells are stained with antibodies against Z-DNA, only the transcriptionally active macronucleus reacts (Figure 10-72). Z-DNA has also been localized in the puffs of polytene chromosomes, further supporting the idea that this unusual DNA structure may be associated with transcriptional activity.

Changes in Histones, HMG Proteins, and Connections to the Nuclear Matrix Are Observed in Active Chromatin

In addition to an altered sensitivity to DNase I, several other characteristics are commonly associated with the structural organization of actively transcribed chromatin. Given the essential role of histones in chromatin organization, changes in histone properties might be expected in the region of active genes. Evidence that histone acetylation is altered in active chromatin has been obtained from experiments in which chromatin was incubated with DNase I to selectively degrade genes that are transcriptionally active. Such treatment leads to the preferential release of the acetylated form of histones H3 and H4, suggesting that acetylated histones are preferentially associated with the nucleosomes of active genes. Further support for this idea has come from studies involving *sodium butyrate,* a fatty acid that inhibits the removal of histone acetyl groups and hence causes histones to become excessively acetylated. Chromatin isolated from cells treated with sodium butyrate is more susceptible than normal to digestion with DNase I, suggesting that histone acetylation alters nucleosomal structure in a way that enhances the susceptibility of DNA to enzymatic cleavage.

Location of DNase I
hypersensitive sites

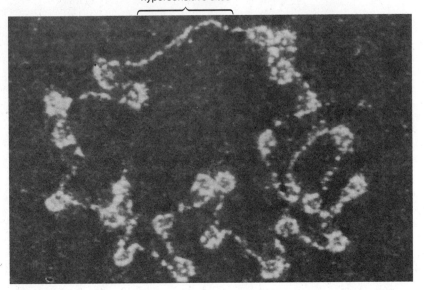

Figure 10-70 *The Circular DNA Molecule of an SV40 Virus in an Infected Host Cell*   *A small region near the top of the molecule lacks nucleosomes. This region corresponds to the location of several DNase I hypersensitive sites.*

In addition to the increased acetylation of histones H3 and H4, another histone alteration associated with transcriptionally active chromatin is the lack of histone H1. Since histone H1 is required for the formation of the 30-nm chromatin fiber (page 415), the absence of histone H1 may help to maintain active chromatin in the form of uncoiled 10-nm fibers. A related feature of transcriptionally active chromatin is its high content of nonhistone HMG proteins (page 411). If HMG proteins are removed from isolated chromatin preparations, the active genes lose their sensitivity to DNase I. Sensitivity to DNase I can be restored by adding back two HMG proteins, *HMG 14* and *HMG 17*. If HMG 14 and 17 are isolated from one tissue and added to HMG-depleted chromatin derived from another tissue, the pattern of DNase I sensitive genes is characteristic of the tissue from which the chromatin, not the HMG proteins, was obtained. For example, if HMG-depleted liver chromatin is mixed with HMG 14 and 17 isolated from brain tissue, the pattern of DNase I sensitive genes in the resulting chromatin resembles that of liver, not brain. This means that chromatin must exhibit tissue-specific differences that are recognized by HMG 14 and 17, allowing the HMG proteins to bind to those genes that are capable of being activated in any given tissue. Binding of HMG 14 and 17 is thought to enhance DNase I sensitivity by helping to uncoil chromatin fibers into a more open configuration, perhaps by displacing histone H1 from the 30-nm fiber.

A final structural characteristic associated with actively transcribed genes is a close relationship with the nuclear matrix. This property of active genes was first

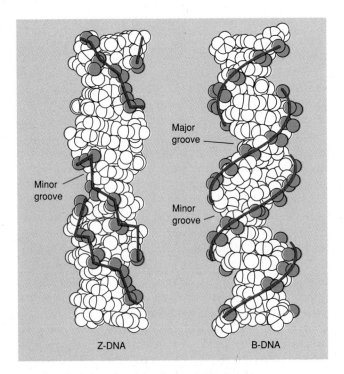

Figure 10-71 *The Structures of B-DNA and Z-DNA*   *In the normal B-form of DNA, the sugar-phosphate backbone forms a smooth right-handed double helix; in Z-DNA, the backbone forms a zigzag left-handed helix. Color is used to highlight the DNA backbone.*

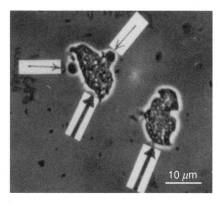

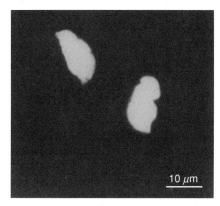

Figure 10-72 *Staining of Z-DNA with Fluorescent Antibodies in the Ciliated Protozoan* Stylonychia (Top) *Phase- contrast micrograph showing the micronuclei* (light arrow) *and macronuclei* (heavy arrow). (Bottom) *Fluorescence microscopy of the same field showing that antibodies directed against Z-DNA bind only to the macronucleus. Courtesy of A. Rich.*

established for the chicken gene coding for ovalbumin, a protein formed only in cells lining the oviduct. In studies carried out by Bert Vogelstein and his collaborators, the nuclear matrix was isolated from chick oviduct cells, which actively transcribe the ovalbumin gene, and from chick liver cells, which do not. When the small amount of DNA that remains associated with the nuclear matrix after isolation was hybridized to cloned ovalbumin gene sequences, it was found that ovalbumin DNA sequences are enriched in nuclear matrix fractions isolated from oviduct, but not from liver. In other words, DNA sequences coding for ovalbumin are closely associated with the nuclear matrix in tissues where the ovalbumin gene is being actively transcribed. Subsequent studies have revealed that transcriptionally active DNA is bound to the nuclear matrix by special base sequences called *matrix attachment regions (MARs)*, which are located at the ends of active chromatin domains.

Although several structural characteristics of transcriptionally active chromatin have now been described, the existence of these structural changes does not address the underlying question of how the DNA se-

quences contained in these regions are initially selected for activation. In the next several sections, we will examine some of the mechanisms that have been implicated in this process.

DNA Methylation Suppresses the Ability of Some Eukaryotic Genes to Be Transcribed

One way of influencing which genes can be transcribed is to introduce alterations directly into the DNA molecule. **DNA methylation** is an example of a DNA alteration that is thought to influence gene activity. Most DNA molecules contain small amounts of a modified base, *5-methylcytosine,* which is generated by enzymatic transfer of a methyl group to cytosine residues that are already present in DNA. Methylated cytosine tends to occur next to guanine, forming the dinucleotide sequence –CG–. The enzyme *DNA methylase,* which catalyzes the methylation reaction, is most efficient at methylating –CG–sequences that are paired with opposing –GC– sequences that are already methylated. Hence, once an opposing pair of –CG– sequences has become methylated, the methylation pattern will tend to be reproduced during DNA replication and passed on from generation to generation (Figure 10-73).

Several lines of evidence suggest that DNA methylation influences gene activity. First, the pattern of DNA methylation in active and inactive genes has been investigated using the restriction enzymes *Msp*I and *Hpa*II. Both of these enzyme cleave the recognition site –CCGG–; however, *Hpa*II works only if the central C is unmethylated, whereas *Msp*I cuts whether the C is methylated or not. By comparing the DNA fragments generated by these two enzymes, it has been concluded that a number of DNA sites exhibit a tissue-specific methylation pattern; that is, they are methylated in some tissues but not in others. In general, such sites are unmethylated in tissues where the gene is active or potentially active, and methylated in tissues where the gene is inactive. For example, C residues located near the 5' end of the globin gene are methylated in tissues that do not produce hemoglobin but are unmethylated in red blood cells.

Independent support for the conclusion that decreased DNA methylation is associated with gene activity has emerged from experiments employing *5-azacytidine,* a cytosine analog that cannot be methylated because it contains a nitrogen atom in place of carbon at the site where methylation normally occurs. Because methylation patterns tend to be inherited, incorporation of 5-azacytidine into DNA triggers an undermethylated state that is maintained for many cell generations after the drug has been removed. Exposing cells to 5-azacytidine has been found to trigger the activation of several kinds of genes, including some that are located within the heterochromatin of inactivated X chromosomes.

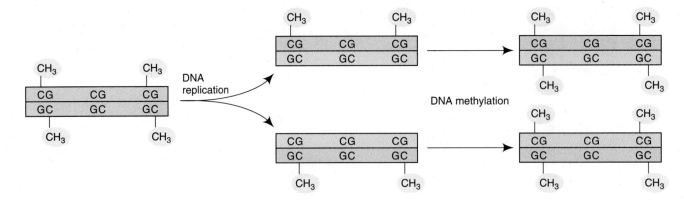

Figure 10-73 *The Inheritance of DNA Methylation Patterns* *After a DNA molecule has been replicated, DNA methylase is most efficient at methylating –CG– sequences that are paired with opposing –GC– sequences that are already methylated. Hence the original methylation pattern is maintained.*

Finally, several studies utilizing cloned genes have revealed that methylated DNA sequences introduced into cells are poorly expressed, whereas the same DNA is actively transcribed if its methyl groups are first removed. Taken together, the preceding observations suggest that DNA methylation suppresses gene transcription. Although the mechanism for this effect is not well understood, it has been proposed that methylation helps to maintain chromatin in the form of condensed 30-nm fibers, which makes DNA less accessible to transcription factors and RNA polymerase than when it is present in uncoiled 10-nm chromatin fibers.

Caution must be taken not to carry such speculations too far, however. Examples are known of methylated DNA sequences that are transcribed and unmethylated sequences that are not. Moreover, the extent of DNA methylation varies significantly among eukaryotes, being prominent in mammals and higher plants but rare in simpler eukaryotes such as *Drosophila* and yeast. Thus instead of representing an essential component of eukaryotic gene regulation, DNA methylation may simply represent one of many factors that contribute to the overall control of gene expression.

Gene Amplification and Deletion Can Alter Gene Expression

The maximum rate at which RNA can be transcribed from any gene is directly related to the number of copies of the gene that is present. Changing the number of gene copies is therefore another type of DNA alteration that can influence gene expression. We have already seen that the production of thousands of copies of the DNA molecules that make up polytene chromosomes facilitates the production of the large quantities of RNA that are needed by giant cells. Although this represents an extreme case in which multiple copies of all gene sequences are produced, a more selective mechanism called **gene amplification** allows cells to produce extra copies of individual genes. For example,

extra genes coding for ribosomal RNAs are produced in egg cells so that the cells can produce the large numbers of ribosomes needed during embryonic development (page 479). In amphibians this process can generate a thousandfold increase in the number of ribosomal genes and the concomitant formation of a thousand or more nucleoli.

Another example of selective amplification involves the genes coding for the chorion proteins that make up the tough outer coat of insect eggs. Like ribosomal RNA genes, chorion genes are preferentially replicated just prior to the time at which their gene product is required in large amounts. In contrast to ribosomal gene amplification, however, the amplified chorion genes remain within the chromosomal DNA.

Despite the evidence that selective amplification of individual genes can occur, it does not appear to be a very widespread phenomenon. DNA hybridization experiments using DNA probes specific for various kinds of protein-coding genes have revealed that most genes are present in only one or a few functional copies. But histone genes represent a notable exception to this generalization. In contrast to the preceding examples, multiple histone genes are a permanent component of the nuclear DNA rather than the product of an amplification process occurring in one cell type at a specific stage of development. It is interesting to note, however, that the number of histone genes varies among different organisms, depending on their maximum need for histone production. In the sea urchin, where embryonic cell division is extremely rapid and histone synthesis must proceed rapidly to package the newly synthesized DNA into nucleosomes, the histone genes are repeated almost a thousandfold. Organisms whose embryos do not divide so rapidly, such as birds and mammals, typically have 10 to 20 copies of the histone genes, and yeast, which does not produce an embryo at all, has only two copies.

In addition to amplifying gene sequences whose products are in great demand, some cells also delete genes whose products are not required. An extreme example of DNA deletion (also called *DNA diminution*) occurs in mammalian red blood cells, which discard their nuclei entirely after adequate amounts of hemoglobin have been made. A less extreme example occurs in a group of tiny crustaceans known as copepods. During the embryonic development of copepods, the heterochromatic regions of their chromosomes are excised and discarded from all cells except those destined to become gametes. In this way, up to half of the organism's total DNA content is removed from its body cells.

DNA Rearrangements Control Mating Type in Yeast and Antibody Production in Animals

We have already learned from examples involving bacterial cells that rearrangements in DNA sequence represent another way of altering gene expression (page 440). In eukaryotes, it has been known for many years that inherited rearrangements in the chromosomal locations of individual genes can exert dramatic effects on gene activity. Only recently has it become apparent, however, that deliberate DNA rearrangements also occur as part of the normal developmental process for regulating selected genes.

A particularly interesting type of DNA rearrangement is involved in controlling the mating process of yeast. Mating normally occurs when cells of two opposite mating types, termed *a* and α, fuse with each other to form a diploid cell. Since mating type is inherited, yeast colonies descended from a single cell might be expected to be of the same type and therefore incapable of mating with one another. This obstacle is overcome, however, by a mechanism that allows yeast cells to convert from one mating type to the other. During the switching process, the gene present at the *mating-type locus* is excised and replaced with a gene of the opposite type. This replacement is made possible by the presence of master copies of the *a* and α genes that are not transcribed. However, duplicate copies of these inactive master genes can be synthesized and inserted in place of the currently existing gene at the mating-type locus (Figure 10-74). This type of regulation is referred to as the *cassette* mechanism because the insertion of different genes into the same location determines which one will be expressed, just as putting different audio cassettes into a tape player determines what type of music will be heard.

A somewhat different type of DNA rearrangement is associated with the production of *antibody* molecules in the white blood cells of higher animals. We will learn in Chapter 16 that the polypeptide chains that comprise an antibody molecule contain two regions: a variable region whose amino acid sequence varies among different kinds of antibodies, and a constant region whose sequence is always the same. The DNA sequences that code for the variable and constant regions

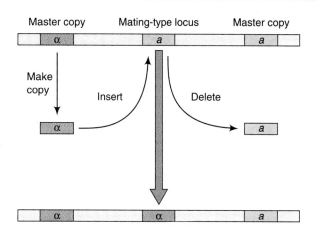

Figure 10-74 *The Cassette Mechanism for Regulating Expression of the Mating-Type Locus in Yeast* *In this type of regulation, copies of different genes (a or α) can be inserted into the mating-type locus, thereby determining which mating-type gene will be transcribed.*

of a given polypeptide chain are not normally adjacent to one another, but are brought together by a specific type of rearrangement that only occurs in cells that produce antibodies. As a result of this rearrangement, multiple kinds of antibodies can be produced by bringing together differing variable regions with the same constant region. The process of rearrangement also activates transcription via a mechanism that involves enhancer sequences (page 423). Enhancer sequences are located immediately upstream from the DNA sequences coding for the constant region, but a promoter sequence is not present in this area and thus transcription does not normally occur. The promoter for gene transcription is located upstream from the variable region, but it is not efficient enough to promote transcription in the absence of an enhancer sequence. Hence prior to DNA rearrangement, the promoter and enhancer sequences of an antibody gene are far apart and transcription does not occur; only after rearrangement are they close enough for transcription to be activated.

DNA Response Elements Allow Nonadjacent Genes to Be Regulated as a Unit

Although gene amplification, deletion, and rearrangement are used to control gene activity in selected situations, they are not very widely employed regulatory mechanisms. The sequence organization and chromosomal location of most eukaryotic genes remains the same whether or not the gene is transcribed. Yet it is well known that certain genes, and often groups of genes, are selectively activated in specific tissues in response to developmental or environmental signals. How are such genes selected for activation?

In bacterial cells, we learned that groups of related genes are organized into operons whose transcription is regulated by repressor and/or activator proteins that bind to DNA sequences located near the start site for transcription. Despite extensive effort, operons resembling those found in bacterial cells have not been detected in eukaryotes. However, DNA sequences that play a role in regulating the transcription of genes in response to specific signals have been identified near the transcriptional start sites of various eukaryotic genes. Because these sequences allow transcription to be regulated in response to a particular type of signal, they are referred to as **response elements.** By placing the same response element next to genes residing at different locations, the transcription of such a group of genes can be regulated by the same signal even though the genes are not adjacent to one another.

The Heat-Shock Response Element Coordinates the Expression of Genes That Are Activated by Elevated Temperatures

One of the first DNA response elements to be identified is a sequence that coordinates the activity of genes whose products protect organisms against excessive heat. If cells growing in culture are briefly warmed by raising the temperature a few degrees, the transcription of several **heat-shock genes** is activated. In cells of *Drosophila*, where this increased transcription can be visualized as chromosomal puffs, the most prominent product of the heat-shock genes is a protein called *hsp70*. The region immediately upstream from the start site of the *hsp70* gene contains several sequences commonly encountered in eukaryotic promoters, such as a TATA box, a CCAAT box, and a GC box. In addition, a seven base consensus sequence has been identified that is unique to those genes whose transcription is activated in response to elevated temperature. This *heat-shock response element* is located 62 bases upstream from the start site for transcription of the *Drosophila hsp70* gene, and in a comparable location in other heat-inducible genes.

In order to investigate the role of this sequence in regulating transcription, Hugh Pelham employed recombinant DNA techniques to transfer the heat-shock consensus sequence from the *hsp70* gene to a position adjacent to a viral gene coding for the enzyme *thymidine kinase*. Transcription of the thymidine kinase gene is not normally induced by heat. But when the hybrid gene was introduced into cells, elevating the temperature triggered the expression of thymidine kinase. In other words, the presence of the heat-shock response element caused the thymidine kinase gene to be activated by heat just like a typical heat-shock gene. Subsequent investigations have revealed that the activation of heat-shock genes is mediated by the binding of a protein called the *heat-shock transcription factor* to the heat-shock response element. The heat-shock transcription factor is present in an inactive form in nonheated

cells, but elevated temperatures cause a change in the structure of the protein which allows it to bind to the heat-shock response element in DNA. The protein is then further modified by phosphorylation, which makes it capable of activating transcription of the adjacent genes.

The principles illustrated by the behavior of the heat-shock genes are commonly encountered in eukaryotic gene regulation. The central step in this type of regulation is the binding of a **gene-specific transcription factor** to an upstream DNA response element that is specific for a particular set of genes. Response elements of the same type are often situated next to genes located on different chromosomes; hence the binding of a gene-specific transcription factor to its corresponding DNA response element wherever the response element may occur allows genes located on different chromosomes to be activated in a coordinated fashion.

Examining the amino acid sequences of a variety of gene-specific transcription factors has led to the discovery of several recurring structural motifs that mediate the binding of such proteins to DNA. In the following sections we will describe some examples of transcription factors that illustrate the most commonly encountered structural motifs.

Homeotic Genes Regulate Embryonic Development by Coding for Transcription Factors Exhibiting a Helix-Turn-Helix Motif

The study of mutations that influence the development of the fruit fly *Drosophila* has provided a wealth of information about the genes that control the formation of animal body plans. One of the most interesting findings to emerge from such studies involves an unusual class of genes known as **homeotic genes.** When mutations occur in one of these genes, a strange thing happens during embryonic development; that is, one part of the body is replaced by a structure that normally occurs somewhere else. For example, mutations in the homeotic gene *Antp* can cause the antennae of the fly to be replaced by a pair of legs (Figure 10-75). Although such a phenomenon might at first glance appear to be no more than an oddity of nature, the fact that legs are formed in the wrong place suggests that the *Antp* gene normally helps to control the formation of the body plan of the developing embryo.

One of the most interesting discoveries to emerge from the study of the *Antp* gene is that it contains a 180 base-pair sequence near its 3' end that resembles a comparable sequence occurring in other homeotic genes. Termed the **homeobox,** this DNA sequence codes for a stretch of 60 amino acids called a **homeodomain.** Homeobox sequences have been detected in organisms as diverse as sea squirts, frogs, mice, and humans—an evolutionary distance of more than 500 million years.

Eye   Leg

Figure 10-75 *The Homeotic Mutant Antp*. *This* Drosophila *mutation causes the antenae of the fly to be replaced by a pair of legs. Courtesy of T. C. Kaufman.*

This widespread occurrence suggests that homeotic genes serve an important developmental function that has been highly conserved during evolution.

Homeotic genes exert their influence over embryonic development by coding for transcription factors that activate or inhibit the transcription of other genes by binding to specific DNA sequences. Since these transcription factors all share the presence of a 60 amino acid homeodomain, the homeodomain is a logical candidate for the site that allows the factors to bind to DNA. This hypothesis has been confirmed by synthesizing the homeodomain region of the *Antp* protein and showing that this 60 amino acid stretch binds to DNA in the identical sequence-specific manner as the intact *Antp* protein.

Like the bacterial *lac* repressor, *trp* repressor, and CAP proteins discussed earlier in the chapter, the homeodomain contains a helix-turn-helix motif. The role of this motif in DNA binding has been directly demonstrated by experiments involving a homeotic gene called *bicoid*, which is required for establishing the proper anterior-posterior polarity during *Drosophila* development. The amino acid lysine is present in position number 9 of the recognition helix of the *bicoid* protein, while glutamine is present in the comparable position in the *Antp* protein. Normally the *bicoid* and *Antp* proteins bind to different DNA sequences. However, if the lysine present at position 9 of the *bicoid* protein is mutated to glutamine, the *bicoid* protein then binds to DNA sequences recognized by the *Antp* protein. Thus, not only does the helix-turn-helix motif mediate DNA binding by homeotic proteins, but single amino acid differences in this motif can alter the DNA sequence to which the protein binds.

Cysteine-Histidine Zinc Fingers Are Present in Transcription Factors in Variable Numbers

Earlier in the chapter, we learned that transcription of the 5S RNA gene by RNA polymerase III requires the presence of an internal control region located within the middle of the gene (page 422). In order for this gene to be efficiently transcribed, a transcription factor called *TFIIIA* must first bind to the internal control region. TFIIIA was one of the first transcription factors to be isolated in pure form. The purified protein contains nine atoms of zinc and nine repeats of a sequence about 30 amino acids long. Although some of the amino acids vary from repeat to repeat, two cysteines always occur near one end of the repeated structure and two histidines near the other. Together the two cysteines and two histidines bind a single atom of zinc, allowing the rest of the sequence to form a loop that projects from the surface of the protein (Figure 10-76, *top middle*). This motif, called a **cysteine-histidine zinc finger,** is repeated nine times in factor TFIIIA.

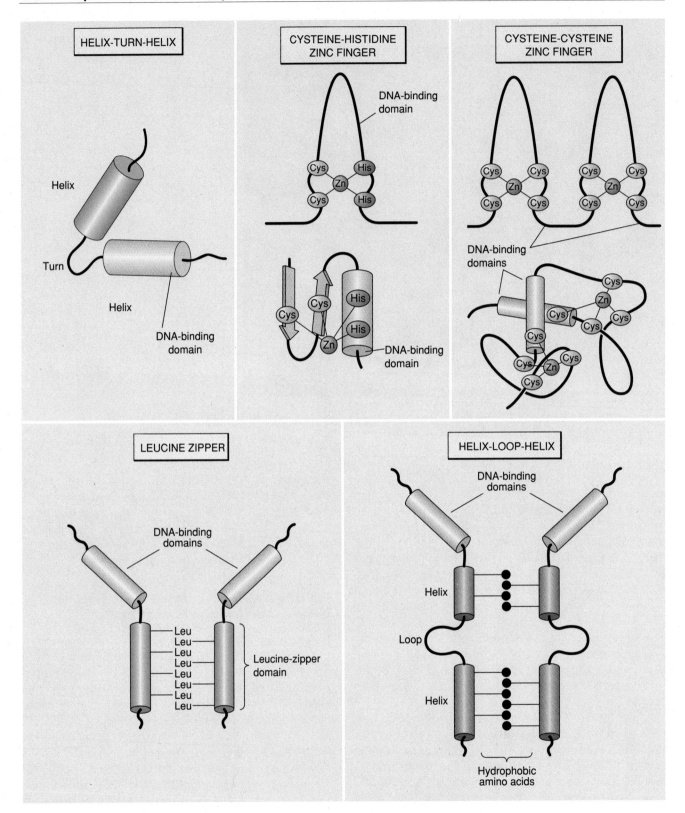

Figure 10-76 DNA-Binding Motifs Encountered in Proteins That Bind to Specific DNA Sequences *The diagrams for the two kinds of zinc fingers include a schematic representation of the polypeptide chain as a well as a model showing how the chain is thought to be folded.*

Soon after the initial identification of zinc fingers in TFIIIA, similar structures were detected in a variety of other transcription factors, including a series of proteins that regulate various aspects of embryonic development. The number of cysteine-histidine zinc fingers varies considerably among transcription factors, ranging from two fingers to several dozen or more. The existence of these repetitive zinc fingers allows a protein molecule to make repeated contacts along the length of a DNA molecule (Figure 10-77). Several lines of evidence support the conclusion that zinc fingers play a crucial role in the mechanism of DNA binding. First, if zinc is removed from TFIIIA, zinc fingers do not form and the protein loses its ability to bind to DNA. Likewise, progressive removal of zinc fingers from TFIIIA has been shown to be accompanied by a parallel loss in DNA-binding ability. A similar dependence on zinc fingers has been demonstrated for the *Kruppel* protein, which is essential for proper thoracic and abdominal development in *Drosophila*. In the *Kruppel* protein, mutation of just one of the cysteine residues involved in zinc binding results in the formation of a mutant fly that looks just like a fly in which the *Kruppel* gene has been completely deleted.

Steroid Hormones Interact with Receptor Proteins Localized in the Nucleus

The next DNA-binding motif that we will be discussing is involved in mediating the action of steroid hormones on gene transcription. Hormones are signaling molecules that alter the metabolism or behavior of target cells containing receptors that bind to the hormone in question. In Chapter 6 we examined how peptide hormones and epinephrine trigger their responses by binding to cell surface receptors. In contrast, steroid hormones pass through the plasma membrane and bind to intracellular receptors. The existence of hormone receptors in steroid target tissues first came to light around 1960 when Elwood Jensen and his colleagues showed that radioactive estrogen injected into rats is concentrated and retained by estrogen target tissues, such as uterus and vagina, but is rapidly lost from nontarget tissues. Such observations suggested the presence in target tissues of receptors that bind to estrogen and prevent its escape back into the circulation.

Shortly thereafter the protein receptors that bind to steroid hormones such as estrogen, glucocorticoids, progesterone, testosterone, and ecdysone were identified. Early studies employing cell fractionation suggested that these steroid hormone receptors are cytoplasmic proteins that only enter the nucleus after they become bound to the appropriate steroid hormone. However, data from more recent investigations suggest that this model is incorrect; most, if not all, steroid hormone receptors appear to be localized in the nucleus. Some of the initial evidence in support of this notion came from studies using the drug *cytochalasin*, which causes cells to expel their nuclei. When assays are carried out for the presence of steroid hormone receptors after cytochalasin treatment, virtually all of the receptor activity is found in the expelled nuclei and none in the remaining cytoplasm. A similar conclusion has been reached in studies where cells are stained with fluorescent monoclonal antibodies directed against steroid hormone receptors. Again, steroid receptors are observed only in the nucleus. Earlier data suggesting a cytoplasmic location for steroid hormone receptors were apparently distorted by the fact that steroid receptors readily leak out of the nucleus during cell homogenization, especially when they do not contain a bound steroid hormone.

In addition to receptors for steroid hormones, the nucleus also contains receptors for the thyroid hormone *thyroxine* and the vitamin A derivative, *retinoic acid*. The nuclear receptors for these two nonsteroid hormones exhibit both structural and functional similarities to the receptors for steroid hormones.

Steroid Hormone Receptors Act as Transcription Factors That Bind to Hormone Response Elements

The fact that steroid hormones bind to nuclear receptors suggests that these hormones influence events occurring inside the nucleus. The first clue to the nature of the nuclear events emerged in the late 1950s, when it was discovered that injecting rats with the steroid hormone hydrocortisone triggers the synthesis of the liver enzymes tyrosine aminotransferase and tryptophan oxygenase. It was soon found that other steroid hormones act in a comparable fashion, stimulating the formation of specific proteins in their respective target tissues (Table 10-6). The ability of steroid hormones to stimulate the formation of new proteins can usually be inhibited by blocking gene transcription with either actinomycin D or the RNA polymerase II inhibitor, α-amanitin. Hence steroid hormones must be inducing the synthesis of new proteins by activating the transcription of their respective genes.

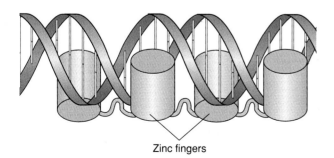

Zinc fingers

Figure 10-77 *Model Showing How Repetitive Cysteine-Histidine Zinc Fingers Bind to DNA* *Note that adjacent fingers make contact with DNA from opposite sides of the double helix.*

Table 10-6 Selected Examples of Proteins Whose Synthesis Is Induced by Steroid Hormones

Hormone/ Target Tissue	Induced Protein
Estrogen:	
Oviduct (hen)	Ovalbumin, conalbumin, ovomucoid, lysozyme
Liver (rooster)	Vitellogenin, very low density lipoprotein
Pituitary (rat)	Prolactin
Glucocorticoids:	
Liver (rat)	Tyrosine aminotransferase, tryptophan oxygenase
Kidney (rat)	Phosphoenolpyruvate carboxylase
Pituitary (rat)	Growth hormone
Progesterone:	
Oviduct (hen)	Avidin, conalbumin, ovomucoid, lysozyme
Uterus (rat)	Uteroglobin
Testosterone:	
Liver (rat)	α2-Microglobulin
Prostate (rat)	Aldolase

Further support for this conclusion has been obtained by using hybridization with cDNA probes to measure the concentration of individual mRNAs in cells stimulated by steroid hormones. Such studies have demonstrated the enhanced production of specific mRNAs in various systems, including estrogen stimulation of ovalbumin production in the hen oviduct and vitellogenin synthesis in amphibian liver; progesterone stimulation of ovalbumin, conalbumin, lysozyme, and avidin formation in the hen oviduct; hydrocortisone stimulation of tyrosine transaminase and tryptophan oxygenase synthesis in mammalian liver; and testosterone stimulation of α2-microglobulin synthesis in mammalian liver.

The preceding data indicate that each steroid hormone selectively activates the transcription of a particular group of genes. The key to this selectivity lies in the ability of steroid hormone receptors to bind to DNA sequences called **hormone response elements,** which are located within a few hundred base pairs upstream from the promoters of genes whose transcription is hormonally regulated. All of the genes whose transcription is activated by a particular steroid hormone contain the same response element, explaining how a signaling hormone can activate the transcription of a selected group of genes. For example, genes activated by estrogen contain a 15 base-pair *estrogen response element* near the upstream end of their promoters, while genes activated by glucocorticoids contain a *glucocorticoid response el-*

ement exhibiting a slightly different base sequence (Figure 10-78). Like many other DNA sequences that are recognized by regulatory proteins, hormone response elements are palindromes. Since the receptors for steroid hormones regulate transcription by binding to DNA response elements, they are in essence acting as gene-specific transcription factors.

The Cysteine-Cysteine Zinc Finger Is the DNA-Binding Motif of Steroid Hormone Receptors

Receptors for steroid hormones typically contain three distinct domains: a domain that binds to the appropriate steroid hormone, a domain that binds to the receptor's DNA response element, and a domain that is responsible for activating transcription. The process by which a hormone causes its receptor to activate transcription involves several distinct steps (Figure 10-79). The first stage is the binding of the hormone to its receptor. Prior to this step, the receptor is unable to bind to DNA or activate transcription because it is complexed with two other proteins called *hsp90* (a heat-shock protein) and *p59*. The binding of a steroid hormone to its receptor displaces these two proteins and unmasks the receptor domains that are required for DNA-binding and transcriptional activation. At this stage two molecules of the steroid hormone-receptor complex join together to form a dimer, which in turn binds to the receptor's DNA response element. Because hormone response elements are palindromes, the two receptor molecules present in the dimer can bind to the two symmetrical sequences that are present at opposite ends of the palindrome. Binding of receptor to the response ele-

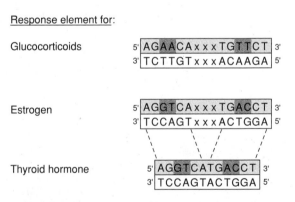

Response element for:

Glucocorticoids   5′ AGAACAxxxTGTTCT 3′
                   3′ TCTTGTxxxACAAGA 5′

Estrogen   5′ AGGTCAxxxTGACCT 3′
                   3′ TCCAGTxxxACTGGA 5′

Thyroid hormone   5′ AGGTCATGACCT 3′
                   3′ TCCAGTACTGGA 5′

Figure 10-78 DNA Sequences of Several Hormone Response Elements *Note that all three sequences are palindromes. The response elements for glucocorticoids and estrogen differ in only two base pairs at each end of the symmetrical sequence. The thyroid hormone element contains the same two symmetrical sequences as the estrogen element, but the three bases that separate the two sequences in the estrogen element are not present. (The dashed lines are included to help the reader line up the comparable regions of the estrogen and thyroid hormone elements.)*

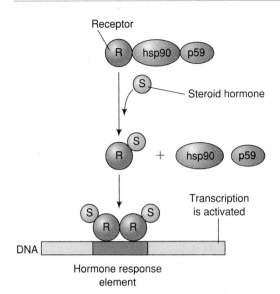

Receptor

Steroid hormone

Transcription
is activated

DNA

Hormone response
element

Figure 10-79 *Pathway by Which Steroid Hormones Activate Gene Transcription* *In the absence of hormone, steroid hormone receptors are complexed with two proteins (hsp90 and p59) that prevent the receptor from interacting with DNA. The binding of a steroid hormone to its corresponding receptor displaces these two proteins, allowing the receptor to form a dimer that binds to the hormone response element and activates transcription.*

ment transforms the DNA at that location into a DNase I hypersensitive site, suggesting that a nucleosome has been displaced (see figure 10-70). Nucleosome removal presumably facilitates transcription by making the promoter region more accessible to RNA polymerase and its associated transcription factors.

Steroid hormone receptors recognize their respective DNA response elements through a DNA-binding motif known as a **cysteine-cysteine zinc finger.** This motif differs from the cysteine-histidine zinc finger discussed earlier in that it utilizes four cysteines rather than two cysteines and two histidines to bind an atom of zinc (see Figure 10-76, *top right*). In contrast to cysteine-histidine fingers, which are present in variable numbers in proteins that contain them, the cysteine-cysteine finger motif of steroid hormone receptors consists of only two adjacent zinc fingers. The importance of this motif for DNA binding has been demonstrated by "finger-swap" experiments in which the finger motifs from two different receptor proteins are exchanged. For example, replacing the finger region of the estrogen receptor with the finger region normally present in the glucocorticoid receptor produces a hybrid receptor that binds to estrogen but recognizes the DNA response element for glucocorticoids.

Although steroid hormone receptors usually activate the transcription of genes located adjacent to hormone response elements, in a few cases transcription is inhibited. A prime example involves the gluco-

corticoid receptor, which can bind to two different kinds of response elements; one type of response element is adjacent to genes whose transcription is activated, and the other is adjacent to genes whose transcription is inhibited. Following binding of the glucocorticoid receptor to the inhibitory type of response element, the receptor depresses the initiation of gene transcription by interfering with the binding of other transcription factors to the promoter region. The existence of separate activating and inhibiting response elements for the same hormone receptor allows a single hormone to stimulate the formation of mRNAs coding for one group of proteins while it depresses the production of mRNAs coding for a different group of proteins.

The Leucine Zipper and Helix-Loop-Helix Motifs Facilitate DNA Binding by Joining Two Polypeptide Chains

In addition to the helix-turn-helix and zinc finger motifs, a third DNA-binding motif that occurs widely in eukaryotic transcription factors is the **leucine zipper.** This structure is built from protein regions in which the amino acid leucine is present as every seventh amino acid. Such regions form an α helix in which the repeating leucine residues occur every two turns on the same side of the helix. Because leucines are hydrophobic amino acids that attract one another, the stretch of leucines exposed on the outer surface of one α helix can interlock with a comparable stretch of leucines on the other α helix, "zippering" the two α helices together (see Figure 10-76, *bottom left*). If the α helices are present in two different polypeptide chains, this process will join the chains into a single protein consisting of two polypeptide subunits.

Adjacent to the leucine zipper of proteins containing this motif is another α-helical region whose high content of basic amino acids makes it ideally suited for binding to DNA, which is acidic. The role of the leucine zipper and this adjacent *basic domain* in DNA binding has been explored in experiments involving two zipper-containing transcription factors called C/EBP and GCN4. If the leucine zipper region from C/EBP is substituted for the leucine zipper of GCN4, the DNA-binding properties of the GCN4 molecule are unaltered. However, if the basic region of C/EBP is substituted for the basic region of GCN4, the resulting hybrid protein exhibits the DNA-binding specificity of C/EBP. In other words, the DNA-binding specificity of the leucine zipper-containing proteins is determined by the amino acid sequence of the basic domain. The leucine zipper itself serves an indirect structural role, holding the two subunits of the protein together and positioning the basic domains in the proper orientation for DNA binding.

Abnormal forms of transcription factors exhibiting the leucine zipper motif have been implicated in the development of certain kinds of cancer. One example involves mutations in genes coding for polypeptides called **Jun** and **Fos,** which are joined together by a leucine zipper to form the **AP1 transcription factor.** The AP1 transcription factor binds to DNA sequences called *AP1 response elements,* which reside next to genes whose expression is associated with cell growth and division. When cells are exposed to extracellular signaling molecules that stimulate cell proliferation, the AP1 transcription factor is activated and this group of genes is transcribed. Mutations in genes coding for either Jun or Fos can lead to the formation of an abnormal AP1 transcription factor that disrupts normal growth control, thereby contributing to the development of cancer (page 799).

Most of the DNA-binding domains that have been identified in eukaryotic transcription factors fit into either the helix-turn-helix, zinc finger, or leucine zipper categories. However, several less frequently encountered DNA-binding domains have also been described. One of these, called the **helix-loop-helix** motif, consists of two α helices separated by a nonhelical loop. The helix-loop-helix motif differs from the helix-turn-helix motif in that it contains hydrophobic amino acids lined up on one side of each α helix. When helix-loop-helix motifs are present in two separate polypeptide chains, these hydrophobic regions attract one another and join the polypeptide chains together in a fashion reminiscent of the leucine zipper (see Figure 10-76, *bottom right*). Like leucine zipper-containing proteins, the helix-loop-helix motif lies adjacent to a region enriched in basic amino acids that serves as the DNA-binding domain.

Transcription Factors Contain Acidic, Glutamine-Rich, or Proline-Rich Regions That Function as Activation Domains

So far our discussion of the structural motifs observed in eukaryotic transcription factors has focused largely on domains involved in DNA binding. After a transcription factor has become bound to its appropriate DNA sequence, it must then act to facilitate (or in a few cases inhibit) the initiation of transcription. In general, a region of the protein molecule that is distinct from the DNA-binding domain is responsible for activating transcription. This region, called the **activation domain,** can be identified by "domain-swap" experiments in which the DNA-binding region of one transcription factor is combined with various regions of a second transcription factor. The resulting hybrid molecule will be able to activate gene transcription only if it contains an activation domain provided by the second transcription factor.

Studies of this type have led to the discovery that activation domains often have a high proportion of acidic amino acids, producing a strong negative charge that is generally clustered on one side of an α helix. Mutations that increase the number of negative charges tend to increase a protein's ability to activate transcription, while mutations that decrease the net negative charge or disrupt the clustering on one side of the α helix diminish the ability to activate transcription. In addition to acidic domains, several other kinds of activating domains have been identified in transcription factors. Some are enriched in the amino acid glutamine, and others contain large amounts of proline. Hence several structural motifs appear to be involved in activating transcription, just as several kinds of motifs have been implicated in DNA binding.

How do activation domains influence transcription? A number of experiments suggest that after a transcription factor has become bound to a DNA response element, the transcription factor's activation domain makes contact with one of the general transcription factors that are involved in formation of the preinitiation complex. A prominent candidate for this target component is TFIID (pages 422), a factor that is required for the transcription of a wide variety of genes both with and without TATA boxes. Although the binding of transcription factors to upstream DNA sequences has been shown to influence the interaction of TFIID with the promoter site, it is not clear whether the activation domain interacts directly with TFIID or with some kind of "adaptor" molecule that in turn influences TFIID (see Figure 10-30).

The Activity of Transcription Factors Is Subject to Control

The central role played by transcription factors in regulating the expression of specific genes raises a final question: How is the activity of a gene-specific transcription factor regulated when cells need to turn a particular set of genes on or off? One approach is to regulate the synthesis of individual transcription factors so that they are only produced in the proper cell type or in response to an appropriate stimulus. The synthesis of a transcription factor, like the synthesis of other proteins, can be regulated at several levels, including the translation of its mRNA, the splicing of its pre-mRNA, and the transcription of its corresponding gene. Current evidence suggests that the latter mechanism is the predominant mode of control. In other words, the synthesis of most transcription factors is controlled by regulating the transcription of the gene that codes for the transcription factor. This, of course, only sets the critical point of gene regulation one stage further back, since the cell must regulate transcription of the genes coding for these transcription factors.

It is not surprising, therefore, that mechanisms have evolved for altering the activity of individual transcription factors without changing the rate at which the genes coding for the transcription factors are transcribed. We have already discussed two examples of such mechanisms: The binding of a steroid hormone to its receptor changes the conformation of the receptor so that it can bind to DNA and activate transcription, and elevated temperature alters the properties of the heat-shock transcription factor so that it can bind to DNA and activate transcription. In addition, protein phosphorylation appears to be an especially widespread mechanism for reversibly altering the properties of transcription factors. We learned in Chapter 6, for example, that cyclic AMP is a widely employed second messenger that mediates the effects of a variety of extracellular signaling molecules. Cyclic AMP stimulates protein kinase A, which in turn catalyzes the phosphorylation of a variety of proteins, including a transcription factor called the **CREB protein.** The CREB protein binds to *cyclic AMP response elements (CRE),* which are located adjacent to genes whose transcription is induced by cyclic AMP. Phosphorylation of the CREB protein stimulates its activation domain, which in turn allows the factor to activate the transcription of cyclic AMP-inducible genes (Figure 10-80) This is only one of several ways in which phosphorylation influences the activity of transcription factors. Phosphorylation can alter the properties of the DNA-binding domain as well as the activation domain, and in some cases inhibits rather than stimulates the ability of a transcription factor to activate transcription.

Control of RNA Splicing and Nuclear Export Represent Additional Ways of Regulating Eukaryotic Gene Expression

Although our discussion of eukaryotic gene regulation has focused predominantly on transcription, this is only the first step in the production of mRNA in eukaryotic cells. The need for RNA splicing and the subsequent export of mRNA through the nuclear pore complexes create additional points of control that do not exist in bacterial cells.

Many situations have been described in which gene expression is controlled by altering the pattern of RNA splicing. This type of regulation was first demonstrated in animal viruses such as adenovirus and SV40, whose DNA is transcribed into RNA molecules that can be spliced in different ways to produce different mRNAs. The ability to splice the same pre-mRNA molecule into differing mRNAs has since been demonstrated for cellular genes as well. The pre-mRNA molecule transcribed from the gene that codes for the lens protein αA_2-*crystalline,* for example, can be spliced in two different ways to produce two distinct mRNAs coding for slightly different forms of crystalline. One of the mRNAs codes for a crystalline chain that is 22 amino acids longer than the crystalline encoded by the alternate mRNA. The extra segment of 22 amino acids is encoded by an exon that is spliced out during processing of the mRNA molecule coding for the shorter form of the crystalline chain (Figure 10-81). In addition to a role in normal gene regulation, alterations in splicing can also be produced by

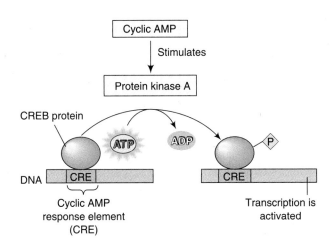

Figure 10-80 *Activation of Gene Transcription by Cyclic AMP* *Genes activated by cyclic AMP possess an upstream cyclic AMP response element (CRE) that binds to a transcription factor called the CREB protein. In the presence of cyclic AMP, protein kinase A catalyzes phosphorylation of the CREB protein, stimulating its activation domain.*

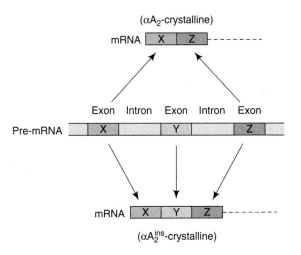

Figure 10-81 *Alternative Splicing of the Pre-mRNA Coding for Lens αA_2-Crystalline* *Two mRNAs coding for different forms of αA_2-crystalline can be spliced from the same pre-mRNA. The mRNA coding for the longer form of αA_2-crystalline contains an extra exon that is removed during splicing of the shorter mRNA.*

mutations. Some forms of *thalassemia,* which is a group of diseases involving hemoglobin abnormalities, are caused by mutations located near the ends of introns. Such mutations disrupt the normal splice sites, leading to abnormal splicing and the production of defective forms of hemoglobin.

Once an RNA molecule has been spliced, the resulting mRNA must become associated with ribosomes before it can be translated into a protein product. In prokaryotic cells, which lack a nuclear envelope, ribosomes have direct access to newly synthesized RNA molecules and thus translation of mRNA can begin before its transcription has been completed. The mRNAs of eukaryotic cells, on the other hand, must pass through nuclear pore complexes and into the cytoplasm before protein synthesis can be initiated. Relatively little is known about the factors that control the export of mRNA from the nucleus, but some evidence indicates that RNA splicing and export may be closely coupled. The most direct support for this statement comes from experiments in which introns have been artificially removed from purified genes and the genes then inserted back into eukaryotic cells. Although such genes are actively transcribed, their transcripts are not exported from the nucleus unless at least one intron is initially present.

In one particular case the regulation of RNA export from the nucleus has been clearly demonstrated. This example involves the export of RNA molecules produced by the virus associated with AIDS (page 743). One type of RNA molecule encoded by the virus is synthesized in the nucleus and remains there until a viral protein called *rev* is produced. After the *rev* protein is formed, the viral RNA molecules that had previously been retained in the nucleus are transported to the cytoplasm.

Thus, we have seen in this chapter that a wide variety of control mechanisms are available in both prokaryotic and eukaryotic cells for determining which genes will be transcribed into functional molecules of mRNA. Once a molecule of mRNA has been produced, the next stage in the process of gene expression is its translation into a polypeptide chain. The role of the ribosome in the translation process, and the ways in which translation is regulated, will be discussed in the following chapter.

SUMMARY OF PRINCIPAL POINTS

• The main components of the eukaryotic nucleus are the nuclear envelope, nucleoplasm, nucleoli, nuclear matrix, and chromatin fibers. The nuclear envelope has numerous nuclear pore complexes that transfer material back and forth between the nucleus and cytoplasm, the nucleoplasm is the soluble phase of the nucleus, nucleoli are involved in ribosome formation, and the nuclear matrix provides a skeletal framework to which chromatin fibers are attached.

• Chromatin fibers consist of long DNA molecules that are bound to histones to form nucleosomes, which are particles consisting of an octamer of histones associated with about 200 base pairs of DNA. Extended chains of nucleosomes form a 10-nm chromatin fiber exhibiting a beads-on-a-string appearance; in intact cells, the 10-nm chromatin fiber is usually coiled upon itself to form a 30-nm fiber. In dividing cells, the 30-nm chromatin fiber is further folded to form a discrete chromosome, which consists of a series of looped domains attached to a nonhistone protein scaffold.

• The nucleoid of bacterial cells contains DNA that is extensively supercoiled and bound to a few basic proteins. Supercoiling is controlled by topoisomerase II, which introduces negative supercoils, and topoisomerase I, which removes them.

• Bacteria utilize one RNA polymerase for synthesizing all types of RNA, whereas eukaryotic cells utilize three different enzymes: RNA polymerase I for synthesizing large ribosomal RNAs, RNA polymerase II for producing mRNAs and a few small nuclear RNAs, and RNA polymerase III for synthesizing tRNAs, 5S rRNA, and the remaining small nuclear RNAs. The first step in gene transcription is the binding of an RNA polymerase to a DNA promoter site, which is usually located immediately upstream from the transcription start site. In eukaryotic cells, the initiation of RNA synthesis requires the participation of numerous transcription factors, including gene-specific factors that bind to upstream DNA sequences called response elements. After RNA synthesis has been initiated, elongation of the RNA chain proceeds in the $5' \rightarrow 3'$ direction until a termination signal is reached. In bacterial cells, transcription is terminated by RNA sequences that either form stem-and-loop structures or are recognized by the rho factor.

• The transcription of most eukaryotic protein-coding genes generates pre-mRNA molecules that are converted into mature mRNAs by a process that involves both the addition of a 5' cap and a 3' poly-A tail, and the removal of introns by a splicing process mediated by snRNPs.

• Regulation of gene expression in prokaryotes involves groups of neighboring genes called operons. The *lac* operon is an inducible operon whose transcription is regulated by the *lac* repressor, which inhibits transcription by binding to the *lac* operator. Inducers of the *lac* operon, such as lactose, bind to the *lac* repressor and prevent it from binding to the *lac* operon. The *lac* operon is also subject to positive regulation by the CAP protein–cyclic AMP complex, which binds upstream from the *lac* promoter. The *trp* operon is a repressible operon whose transcription is inhibited by the tryptophan repressor when the repressor is bound to its corepressor, tryptophan.

• Gene regulation in eukaryotes is complicated by the fact that DNA is packaged into chromatin fibers containing various kinds of noncoding DNA sequences that are not present in prokaryotes. Transcriptionally active chromatin is characterized by an increased sensitivity to DNase I digestion, increased acetylation of histones H3 and H4, a lack of histone H1, the presence of HMG 14 and 17, and a close relationship with the nuclear matrix. Gene expression can be turned off by condensing chromatin fibers into heterochromatin, such as occurs during X chromosome inactivation in female mammals. Changes in gene expression can also be produced by direct changes in the DNA molecule, such as decreased methylation, gene amplification, gene deletion, and gene rearrangements.

• In eukaryotes the expression of nonadjacent genes exhibiting related functions is coordinated by placing the same response element next to genes located at different chromosomal sites. For example, the heat-shock response element coordinates the expression of genes whose expression is activated by elevated temperatures. Heat triggers a change in the structure of the heat-shock transcription factor, permitting it to bind to the heat-shock response element and activate transcription.

• Transcription factors exhibit recurring structural motifs that mediate their binding to specific DNA response elements. One such motif occurs in transcription factors encoded by homeotic genes, which control the formation of an organism's body plan during embryonic development. Homeotic genes code for transcription factors exhibiting a helix-turn-helix motif whose amino acid sequence determines the base sequence to which the protein will bind.

• Another DNA-binding motif is involved in the regulation of gene expression by steroid hormones, which activate (or inhibit) gene transcription by binding to nuclear receptors that exhibit a DNA-binding motif called a zinc finger. After a steroid hormone binds to its corresponding receptor, the receptor functions as a transcription factor that controls transcription by binding via the zinc finger motif to DNA response elements located adjacent to hormone responsive genes.

• In addition to the regulation of gene transcription, eukaryotes also exert control over gene expression by employing alternate pathways of RNA splicing and by controlling the export of RNA to the cytoplasm.

SUGGESTED READINGS

Books

Agutter, P. S. (1991). *Between Nucleus and Cytoplasm,* Chapman and Hall, New York.

Latchman, D. S. (1990). *Gene Regulation: A Eukaryotic Perspective,* Unwin Hyman, London.

Latchman, D. S. (1991). *Eukaryotic Transcription Factors,* Academic Press, San Diego.

McKnight, S. L., and K. R. Yamamoto (1993). *Transcriptional Regulation,* Cold Spring Harbor Laboratory Press, Plainview, NY.

Van Holde, K. E. (1988). *Chromatin,* Springer-Verlag, New York.

Wolffe, A. (1992). *Chromatin Structure and Function,* Academic Press, San Diego.

Articles

Akey, C. W., and M. Radermacher (1993). Architecture of the *Xenopus* nuclear pore complex revealed by three-dimensional cryo-electron microscopy, *J. Cell Biol.* 122:1–19.

Beato, M. (1989). Gene regulation by steroid hormones, *Cell* 56:335–344.

Becker, P. B. (1994). The establishment of active promoters in chromatin, *BioEssays* 16: 541–547.

Borst, P., and D. R. Greaves (1987). Programmed gene rearrangements altering gene expression, *Science* 235:658–667.

Britten, R. J., and D. E. Kohne (1968). Repeated sequences in DNA, *Science* 161:529–540.

Buratowski, S. (1994). The basics of basal transcription by RNA polymerase II, *Cell* 77:1–3.

Darnell, J. E. (1985). RNA, *Sci. Amer.* 253 (October):68–78.

Dingwall, C., and R. Laskey (1992). The nuclear membrane, *Science* 258:942–947.

Felsenfeld, G. (1985). DNA, *Sci. Amer.* 253 (October):58–67.

Forbes, D. J. (1992). Structure and function of the nuclear pore complex, *Annu. Rev. Cell Biol.* 8:495–527.

Gehring, W. J. et al. (1994). Homeodomain-DNA recognition, *Cell* 78:211–223.

Hanes, S. D., and R. Brent (1989). DNA specificity of the bicoid activator protein is determined by homeodomain recognition helix residue 9, *Cell* 57:1275–1283.

Harris, H. (1994). An RNA heresy in the fifties, *Trends Biochem. Sci.* 19:303–305.

Holliday, R. (1989). A different kind of inheritance, *Sci. Amer.* 260 (June):60–73.

Hunter, T., and M. Karin (1992). The regulation of transcription by phosphorylation, *Cell* 70:375–387.

Kohtz, J. D. et al. (1994). Protein-protein interactions and 5'-splice-site recognition in mammalian mRNA precursors, *Nature* 368:119–124.

Kornberg, R. D., and Y. Lorch (1992). Chromatin structure and transcription, *Annu. Rev. Cell Biol.* 8:563–587.

Kornberg, R. D., and A. Klug (1981). The nucleosome, *Sci. Amer.* 244 (February):52–64.

Lamond, A. I. (1993). The spliceosome, *BioEssays* 15:595–603.

Malim, M. H., J. Hauber, S.-Y. Le, J. V. Maizel, and B. R. Cullen (1989). The HIV-1 *rev trans*-activator acts through a structured target sequence to activate nuclear export of unspliced viral mRNA, *Nature* 338:254–257.

McKeown, M. (1992). Alternative mRNA splicing, *Annu. Rev. Cell Biol.* 8:133–155.

McKnight, S. L. (1991). Molecular zippers in gene regulation, *Sci. Amer.* 264 (April):54–64.

Mehlin, H., and B. Daneholt (1993). The Balbiani ring particle: a model for the assembly and export of RNPs from the nucleus?, *Trends Cell Biol.* 3:443–447.

Morse, R. H. (1992). Transcribed chromatin, *Trends Biochem. Sci.* 17:23–26.

Pelham, H. R. B. (1982). A regulatory upstream promoter element in the Drosophila hsp 70 heat-shock gene, *Cell* 30:517–528.

Pruss, G. J., and K. Drlica (1989). DNA supercoiling and prokaryotic transcription, *Cell* 56:521–523.

Ptashne, M. (1989). How gene activators work, *Sci. Amer.* 260 (January):41–47.

Rhodes, D., and A. Klug (1993). Zinc fingers, *Sci. Amer.* 268 (February):56–65.

Roeder, R. G. (1991). The complexities of eukaryotic transcription initiation: regulation of preinitiation complex assembly, *Trends Biochem. Sci.* 16:402–408.

Sakonju, S, D. F. Bogenhagen, and D. D. Brown (1980). A control region in the center of the 5S RNA gene directs specific initiation of transcription: I. The 5' border of the region, *Cell* 19:13–25.

Schwabe, J. W. R., and D. Rhodes (1991). Beyond zinc fingers: hormone receptors have a novel structural motif for DNA recognition, *Trends Biochem. Sci.* 16:291–296.

Selker, E. U. (1990). DNA methylation and chromatin structure: a view from below, *Trends Biochem. Sci.* 15:103–107.

Sharp, P. A. (1994). Split genes and RNA splicing, *Cell* 77:805–815.

Silver, P. A. (1991). How proteins enter the nucleus, *Cell* 64:489–497.

Steitz, J. A. (1988). Snurps, *Sci. Amer.* 258 (June):56–63.

Tsukiyama, T., P. B. Becker, and C. Wu (1994). ATP-dependent nucleosome disruption at a heat-shock promoter mediated by binding of GAGA transcription factor, *Nature* 367:525–532.

Wang, J. C. (1985). DNA topoisomerases, *Annu. Rev. Biochem.* 54:665–697.

Welshons, W. V., M. E. Lieberman, and J. Gorski (1984). Nuclear localization of unoccupied oestogen receptors, *Nature* 307:747–749.

Wickens, M. (1990). How the messenger got its tail: addition of poly(A) in the nucleus, *Trends Biochem. Sci.* 15:277–281.

Workman, J. L., and A. R. Buchman (1993). Multiple functions of nucleosomes and regulatory factors in transcription, *Trends Biochem. Sci.* 18:90–95.

Chapter 11

The Ribosome and Translation of Genetic Information

In the preceding chapter we saw how messenger RNA (mRNA) is generated by the process of gene transcription and, in eukaryotes, RNA processing. During the next stage in the pathway of information flow, the base sequence of an mRNA molecule guides the sequence in which amino acids are incorporated into polypeptide chains. This process of mRNA *translation* occurs on cytoplasmic particles known as ribosomes. Although ribosomes were once viewed as passive structures whose sole purpose was to provide a physical surface for the interaction between mRNA and the other components involved in protein synthesis, it is now evident that the ribosome actively participates in the reaction pathway. In essence, the role of the ribosome in protein synthesis resembles that of a large complicated enzyme.

The ribosome is an unusual "enzyme," however, because it contains more than 50 different proteins and several kinds of RNA. The presence of so many molecules within a single macromolecular complex has made it difficult to unravel the structural organization of the ribosome and to determine the roles played by its components. Yet an understanding of the molecular architecture of the ribosome is essential if we are to understand the mechanism by which proteins are synthesized. We will therefore begin this chapter by exploring what is known about the structure and assembly of ribosomes; we will then discuss the role of this organelle in the process of protein synthesis; and finally, we will conclude by discussing the kinds of control mechanisms that influence the activity and destination of the protein molecules being synthesized by ribosomes.

RIBOSOME STRUCTURE

The Use of Electron Microscopy and Subcellular Fractionation Led to the Discovery of the Ribosome and Its Role in Protein Synthesis

Ribosomes measure about 25 nm in diameter, which is an order of magnitude smaller than the wavelength of visible light and hence beneath the resolving power of the light microscope. For this reason, ribosomes were not discovered by cell biologists until procedures for examining cells by electron microscopy were developed in the early 1950s. Several electron microscopists of this period noted the presence of small granules in the cytoplasm of thin-sectioned cells, but it was the extensive observations of George Palade and his associates that revealed that similar granules are present in the cytoplasm of widely diverse cell types, and are especially numerous in cells that are actively engaged in protein synthesis. Palade and his colleague, Philip Siekevitz, isolated these granules using subcellular fractionation techniques and found that they sediment along with membranes of the endoplasmic reticulum in the *microsomal fraction* (page 138). When chemical analysis revealed that the microsomal fraction is enriched in ribonucleic acid, the term **ribosome** was coined to refer to the granules.

The realization that ribosomes are involved in protein synthesis first emerged from experiments in which cells exposed for brief periods to radioactive amino acids were homogenized and subjected to subcellular fractionation. After such treatment, the highest concentration of newly synthesized radioactive protein was detected in the microsomal fraction. Separating this fraction into its ribosome and membrane components revealed that the radioactive protein is associated predominantly with ribosomes (see Figure 3-21), implicating the ribosome as the site of protein synthesis.

Ribosomes Are Constructed from Large and Small Subunits

Soon after the ribosome was identified as the site of protein synthesis, centrifugation studies showed that isolated ribosomes dissociate into *large* and *small subunits* when the magnesium ion concentration is lowered (Figure 11-1). Prokaryotic ribosomes are constructed from 50S and 30S subunits that join together to form a 70S ribosome, whereas eukaryotic cytoplasmic ribosomes consist of 60S and 40S subunits that combine to form an 80S ribosome. (Recall from our discussion on page 135 that S values correlate roughly with a particle's molecular weight, but cannot be added to one another.) In Chapter 14, we will learn that subunits of slightly different sizes occur in the ribosomes of chloroplasts and mitochondria (see Table 14-4).

Although the idea that ribosomes are composed of two subunits originally emerged from studies involving

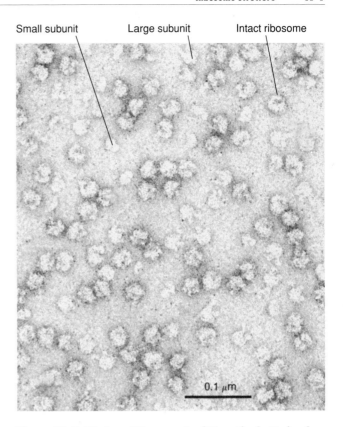

Small subunit   Large subunit   Intact ribosome

0.1 μm

Figure 11-2 *Electron Micrograph of Negatively Stained Ribosomes and Ribosomal Subunits* *Intact ribosomes, isolated large subunits, and isolated small subunits of ribosomes isolated from the bacterium* E. coli *can all be seen in this micrograph. Courtesy of J. A. Lake.*

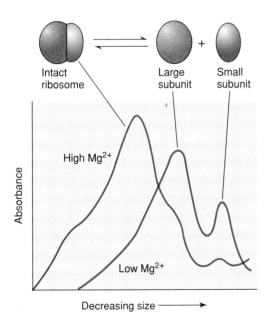

Figure 11-1 *Identification of Ribosomal Subunits by Moving-Zone Centrifugation* *At low Mg²⁺ concentration, intact ribosomes dissociate into large and small subunits.*

moving-zone centrifugation, the existence of ribosomal subunits has subsequently been confirmed by electron microscopy (Figure 11-2). High-magnification micrographs of negatively stained ribosomal subunits reveal that each subunit is an asymmetric particle exhibiting a characteristic pattern of protuberances and ridges. The small subunit consists of four main regions termed the *platform, cleft, head,* and *base,* while the large subunit contains distinctive areas known as the *central protuberance, ridge,* and *stalk* (Figure 11-3). When the two subunits come together to produce an intact ribosome, the central protuberance of the large subunit and the head of the small subunit face each other.

The large and small subunits each consist of a mixture of **ribosomal RNAs (rRNAs)** and **ribosomal proteins** (Figure 11-4). Prokaryotic ribosomes contain a single 16S rRNA molecule in the small subunit and molecules of 23S rRNA and 5S rRNA in the large subunit. Eukaryotic cytoplasmic ribosomes typically have one 18S rRNA molecule in the small subunit and single 28S rRNA, 5S rRNA, and 5.8S rRNA molecules in the large subunit. Ribosomal RNA molecules are characterized by the presence of numerous sequences that can form complementary base pairs with sequences located else-

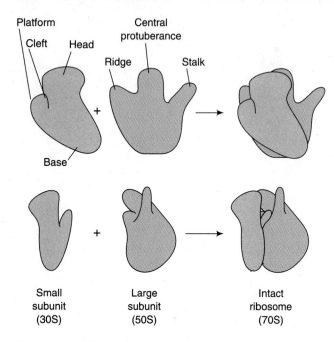

Figure 11-3 *Three-Dimensional Model of an E. coli Ribosome Illustrated from Two Different Angles.*

where in the same molecule. Such interactions cause rRNA molecules to fold into three-dimensional configurations that exhibit many localized double-stranded regions (Figure 11-5).

The number of different proteins present in ribosomes has been more difficult to assess, but electrophoretic and chromatographic fractionation has led to the identification of 21 different small subunit proteins and 32 large subunit proteins in the ribosomes of the bacterium *E. coli*. The small subunit proteins are designated S1 to S21, while the large subunit proteins are called L1 through L34 (the numbering system extends to 34 because of mistakes in early work on ribosomal protein identification). Each protein is present in one copy per ribosome with two exceptions; a protein called *S20* or *L26* is present in two copies, one in the small subunit and one in the large subunit, and four copies of the protein *L7/L12* occur in the large subunit. The latter protein is designated with two numbers because it exists in two forms that differ in their state of acetylation. The number of proteins in eukaryotic ribosomes is significantly higher than in bacterial ribosomes; current estimates suggest that eukaryotic cytoplasmic ribosomes contain about 80–85 ribosomal proteins, compared to 52 different proteins in *E. coli* ribosomes.

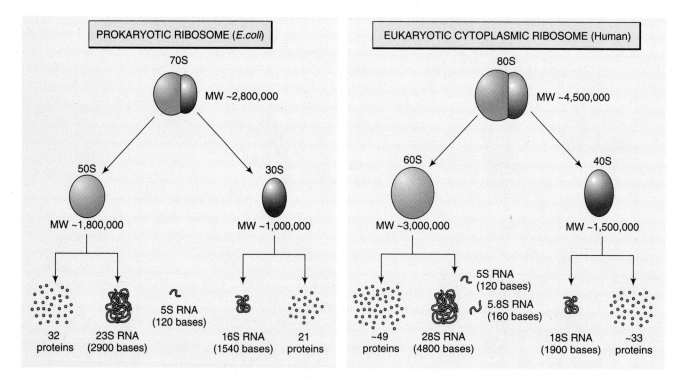

Figure 11-4 *Chemical Composition of Typical Prokaryotic and Eukaryotic Ribosomes* *Ribosomes of the bacterium E. coli contain three types of RNA (23S, 16S, 5S) plus 52 different proteins (one of these 52 proteins occurs in both the large and small subunits, explaining why the values of 32 proteins in the 50S subunit and 21 proteins in the 30S subunit add up to 53). Eukaryotic cytoplasmic ribosomes are significantly larger than prokaryotic ribosomes, possessing four types of RNA (28S, 18S, 5.8S, and 5S) and about 82 different proteins. (Abbreviation: MW = molecular weight.)*

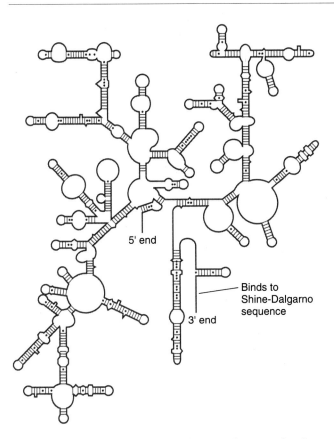

5' end

Binds to
Shine-Dalgarno
sequence

3' end

Figure 11-5 *Folding of the 16S Ribosomal RNA Molecule of E. Coli* *Folding is caused by base-pairing between complementary sequences.*

Ribosomes Can Be Reconstituted from Mixtures of Purified Ribosomal Proteins and RNA

To fully understand the role played by the ribosome in protein synthesis, it is necessary to know how the protein and RNA molecules that comprise this organelle are arranged. One way to explore the molecular organization of the ribosome is by reassembling ribosomes from mixtures of purified ribosomal proteins and RNA. This reconstitution approach was pioneered in the late 1960s by Masayasu Nomura, who showed that the 30S subunit of the *E. coli* ribosome can be reconstituted by mixing together its 21 proteins and 16S RNA. When reconstituted 30S subunits are mixed with normal 50S subunits, the resulting ribosomes function normally in protein synthesis. A few years after these pioneering experiments, the 50S subunit was also reconstituted. Hence all the information required for assembling an intact and functionally competent 70S ribosome is inherent in the RNA and protein components of which it is comprised.

Reconstitution experiments in which ribosomal proteins are added in a defined sequence have revealed that ribosome assembly occurs in a stepwise fashion in which the binding of certain ribosomal proteins to rRNA induces conformational changes that facilitate the binding of other proteins. During the reconstitution process, particles of intermediate size are formed that resemble ribosomal precursor particles found in living cells. For example, when cell-free reconstitution is carried out at low temperatures, a 21S particle is assembled that contains approximately two-thirds of the proteins normally found in the mature 30S subunit. In intact cells, a 21S particle containing the same set of proteins occurs as an intermediate in the ribosome assembly pathway. When 21S particles reconstituted at low temperatures are incubated at higher temperatures in the presence of the remaining ribosomal proteins, they undergo a conformational change that permits them to complete the assembly process.

The reconstitution approach has also been helpful in investigating the functions of individual ribosomal proteins. By omitting or substituting specific proteins during reconstitution, the effects of an individual protein on the properties of the reconstituted ribosome can be determined. For example, ribosomal proteins that function as targets for inhibitors of protein synthesis have been investigated in this way. One such inhibitor is *streptomycin,* an antibiotic that disrupts protein synthesis and causes incorrect amino acids to be incorporated into newly forming polypeptide chains. Mutant bacteria that are resistant to the effects of streptomycin have been employed to identify the ribosomal protein with which this antibiotic interacts. In these experiments, ribosomal proteins isolated from ribosomes of streptomycin-resistant bacteria were substituted one at a time for normal ribosomal proteins during ribosome reconstitution. The inclusion of a single protein, *S12,* from the streptomycin-resistant ribosomes was found to make the reconstituted ribosomes resistant to streptomycin. Hence protein S12 must represent the target of streptomycin.

The Molecular Structure of Ribosomes Has Been Investigated Using a Variety of Chemical, Physical, and Microscopic Techniques

The reconstitution approach has provided important information about the functional roles played by certain ribosomal components, but it tells us relatively little about the location of individual protein and RNA molecules within the ribosome. In order to construct a complete molecular model of the ribosome, the location, structure, and function of each protein and rRNA molecule needs to be known. Although no individual technique is capable of providing all the necessary information, a detailed picture of ribosome architecture is beginning to emerge from data generated by a diverse group of chemical, physical, and microscopic approaches. Some of the more widely used techniques are described in the following sections.

Chemical Crosslinking Detects Neighboring Ribosomal Proteins

One way of investigating the location of individual proteins within the ribosome involves the use of reagents that chemically link neighboring proteins to one another. Reagents of this type are referred to as **crosslinking agents** because they contain two reactive chemical groups, each capable of forming a covalent bond with an amino acid side chain. In a typical crosslinking experiment, ribosomes are first incubated in the presence of a crosslinking agent to link neighboring ribosomal proteins. All of the ribosomal proteins are then extracted and analyzed by polyacrylamide gel electrophoresis. Linked proteins appear as bands that migrate more slowly than normal because their molecular weights are greater than those of typical ribosomal proteins. If the material in these bands is isolated and treated with reagents that disrupt protein crosslinking, the two neighboring proteins can be identified (Figure 11-6).

Neutron Diffraction Yields a Three-Dimensional Map of the Relationships among Ribosomal Proteins

The most complete map of the locations of ribosomal proteins within the ribosome has been produced using the technique of *neutron diffraction*. In this approach, pioneered by Peter Moore and Donald Engelman, ribosomal proteins are isolated from cells that have been grown in the presence of heavy water to introduce atoms of deuterium (^{2}H) into cellular proteins. Ribosomal subunits are then reconstituted from a protein mixture in which two ribosomal proteins contain deuterium and the rest do not. The distance between the two deuterium-labeled proteins is determined by placing the sample in a neutron beam and analyzing the resulting interference pattern. Such interference patterns are altered by the diffraction of neutrons by deuterium, thereby allowing the distance between the two deuterium-labeled proteins to be calculated. By using different combinations of deuterium-labeled proteins during the reconstitution step, it is possible to map the location of each protein within the ribosome (Figure 11-7).

Enzymatic Digestion and Radioactive Labeling Experiments Allow Sites of RNA-Protein Interaction to Be Identified

In addition to information concerning the location of each ribosomal protein within the ribosome, a molecular model of the ribosome requires information con-

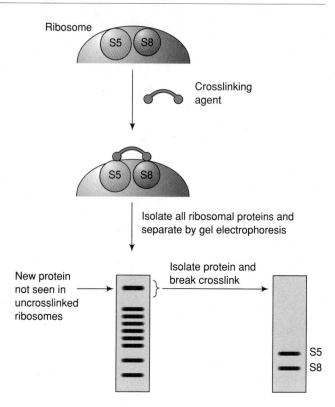

Figure 11-6 The Use of Crosslinking Agents to Detect Neighboring Ribosomal Proteins *Ribosomes are first incubated with a crosslinking agent to link neighboring proteins. When the ribosomal proteins are later analyzed by electrophoresis, any proteins that have become crosslinked will migrate more slowly than normal because their molecular weights are greater than those of typical ribosomal proteins. Such crosslinked proteins are then isolated and treated with reagents that disrupt crosslinking, allowing the individual proteins to be identified.*

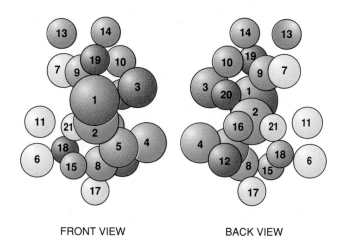

FRONT VIEW   BACK VIEW

Figure 11-7 Spatial Arrangement of the 21 Proteins in the Small Ribosomal Subunit of E. Coli as Mapped by Neutron Diffraction *Front and back views of the same model are shown.*

cerning the sites of interaction between ribosomal proteins and rRNA. One way of obtaining such information is to mix isolated ribosomal proteins with purified rRNAs, followed by digestion with ribonuclease. The regions of the rRNA molecules that contain bound protein are protected from digestion and can be isolated and analyzed. Alternatively, rRNA can be degraded with ribonuclease and the resulting RNA fragments then tested for their ability to bind to ribosomal proteins. Among the information obtained from such approaches is the fact that certain ribosomal proteins bind to the 5' end of the 16S RNA molecule during the early stages of assembly of the small subunit.

Although they are not part of the ribosome itself, other molecules such as messenger and transfer RNAs bind to the ribosome during the process of protein synthesis. The ribosomal components with which they interact have been identified using compounds called **affinity labels,** which are derivatives of substances that normally interact with the ribosome. An affinity label differs from a normal substrate, however, in that it contains a highly reactive side chain that forms covalent linkages with the ribosomal site to which it binds. Affinity labeling compounds have been designed to mimic several of the normal components involved in protein synthesis, including messenger RNA and transfer RNA. Ribosomal proteins that become radioactive after exposure to such affinity labels must be located at or near the site to which the affinity label binds.

Antibodies Directed Against Individual Ribosomal Proteins Permit the Surface of the Ribosome to Be Mapped

In the early 1970s Georg Stöffler and his associates purified the proteins present in the *E. coli* ribosome and injected them into rabbits to trigger the formation of antibodies. These pioneering studies opened the way for the use of specific antibodies for mapping both the topographical and functional organization of the ribosome. Each antibody molecule contains two binding sites for the protein with which it interacts. Hence treating ribosomal subunits with an antibody directed against an individual ribosomal protein will cause two ribosomal subunits containing that protein to be linked together to form pairs. Examining the resulting pairs under the electron microscope provides information as to where the protein is located. For example, treating 30S ribosomal subunits with an antibody directed against protein S14 causes two 30S subunits to be joined at the head region (Figure 11-8). Hence protein S14 must be situated in the head of the 30S subunit.

Models of the Bacterial Ribosome Reveal Where Specific Events Associated with Protein Synthesis Take Place

Information derived from the preceding approaches has allowed a relatively detailed model of the bacterial ribosome to be constructed (Figure 11-9). The locations of most ribosomal proteins are well established because relevant data have been obtained from several independent sources. For example, the proposed existence of a cluster of proteins S4, S5, S8, and S12 on the side of the small subunit facing away from the large subunit is compatible with data obtained from studies using ribosome reconstitution, neutron diffraction, chemical crosslinking, and antibody labeling.

Many of the important functional regions of the ribosome have been identified in such models. For example, it has been established that the small subunit binds messenger and transfer RNAs to the ribosome, while the large subunit catalyzes peptide bond formation. The platform that protrudes from the middle of the small subunit contains the 3' end of the 16S rRNA molecule, which forms part of the binding site for mRNA. Binding sites for transfer RNAs and protein factors involved in the initiation of protein synthesis are located in the cleft adjacent to the platform. The large subunit catalyzes peptide bond formation and the hydrolysis of GTP, which is required for both the binding of transfer RNA and the movement of messenger RNA during protein synthesis. Peptide bond formation occurs adjacent to the central protuberance of the large subunit, whereas GTP hydrolysis is catalyzed by a tetramer of the L7/L12 protein that constitutes the stalk of the large subunit. As the newly forming polypeptide

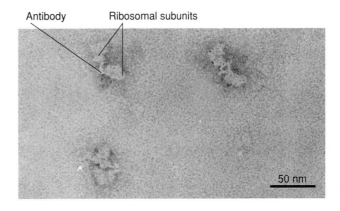

Figure 11-8 *Small Ribosomal Subunits Incubated with an Antibody Directed Against Protein S14* *This negatively stained electron micrograph of* E. coli *ribosomal subunits reveals that the antibody molecule has linked the subunits into pairs. Each antibody molecule is binding to the head region of two subunits. Courtesy of J. A. Lake.*

chain elongates, it moves through a tunnel in the large subunit and emerges from an exit hole located at the end opposite the central protuberance.

RIBOSOME BIOGENESIS

Now that the molecular organization of the ribosome has been described, we are ready to address the question of how ribosomes are manufactured by living cells. Because of the large number of molecules involved, synthesizing ribosomes is a complex task that requires the precise coordination of numerous steps. We will begin by examining the way in which this task is carried out in eukaryotic cells, where a specific region of the nucleus is specialized for the purpose of making ribosomes.

The Nucleolus Is the Site of Ribosome Formation in Eukaryotic Cells

In eukaryotic cells, ribosome formation occurs in the **nucleolus.** Although the nucleolus was originally discovered by light microscopists, it was not until the advent of electron microscopy that the structural components that make up the nucleolus were clearly identified (Figure 11-10). In thin-section electron micrographs, each nucleolus appears as a membrane-free organelle consisting of three components: (1) a *granular component* containing particles 15–20 nm in diameter that account for the bulk of the nucleolus; (2) one or more lightly stained *fibrillar centers* comprised of loosely packed thin fibrils; and (3) a *dense fibrillar component* consisting of a layer of tightly packed,

densely stained fibrils that surround each fibrillar center. The dense fibrillar component contains the proteins *nucleolin* and *fibrillarin,* which are two of the most abundant proteins of the nucleolus.

The first definitive evidence linking the nucleolus with ribosome formation was provided in the early 1960s by Robert Perry, who employed a microbeam of ultraviolet light to destroy the nucleoli of living cells. The treated cells lost their ability to synthesize rRNA, suggesting that the nucleolus is involved in the formation of ribosomes. In later experiments Perry found that low concentrations of actinomycin D inhibit the incorporation of ^{3}H-uridine into rRNA without affecting the formation of other kinds of RNA. Microscopic autoradiography revealed that this treatment selectively prevents the synthesis of RNA by the nucleolus, suggesting that the nucleolus is involved in rRNA formation (Figure 11-11). Additional evidence emerged from studies carried out by Donald Brown and John Gurdon on the African clawed toad, *Xenopus laevis.* Through genetic crosses, it is possible to produce *Xenopus* embryos whose cells lack nucleoli. Brown and Gurdon discovered that such embryos, termed **anucleolar mutants,** are incapable of synthesizing rRNA and therefore die during early development, again implicating the nucleolus in ribosome formation.

If ribosomal RNAs are synthesized in the nucleolus, then the DNA sequences that code for these RNAs would be expected to reside in the nucleolus as well. This prediction has been verified in experiments involving *Xenopus* embryos, where the rRNA genes are extensively amplified. The DNA that codes for rRNA is

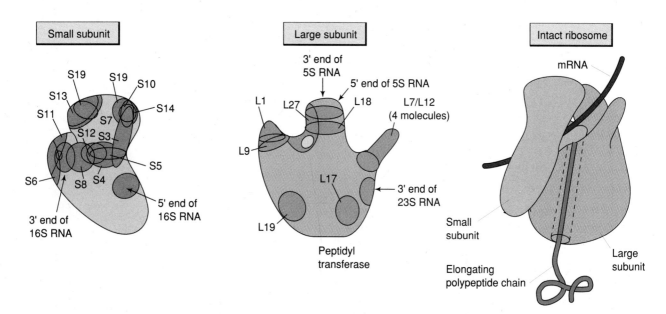

Figure 11-9 *Model of the E. Coli Ribosome Indicating the Location of Several Ribosomal Proteins and RNAs* *The diagram on the right shows the relationship between the ribosome, messenger RNA, and the growing polypeptide chain during the process of protein synthesis.*

Fibrillar center

Dense fibrillar component

Granular component

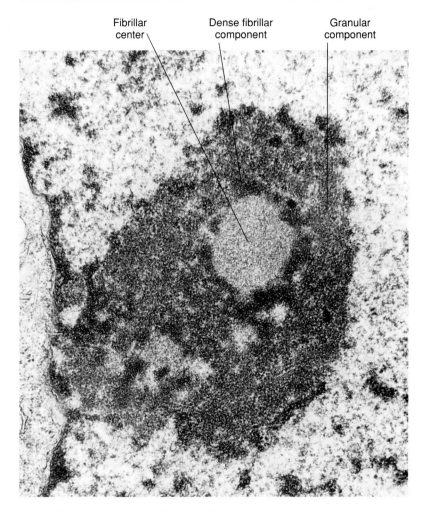

Figure 11-10 *Thin-Section Electron Micrograph of a Nucleolus*   *The three major constituents of the nucleolus (granular component, fibrillar center, and dense fibrillar component) are clearly visible. Courtesy of U. Scheer.*

characterized by a high content of the bases G and C, which makes it denser than the bulk of the nuclear DNA. Hence when nuclear DNA is subjected to isodensity centrifugation, the DNA coding for rRNA forms a *satellite peak* that is separate from the rest of the DNA (Figure 11-12). The satellite peak is missing in DNA samples obtained from anucleolar mutant embryos, indicating that organisms lacking nucleoli also lack the DNA sequences coding for rRNA. The notion that the genes coding for rRNA are located within the nucleolus has received further support from experiments in which radioactive 18S and 28S rRNAs were hybridized directly to tissue sections. When the tissues were examined by autoradiography, the radioactive rRNA was found to hybridize to DNA located within the nucleolus (Figure 11-13). In contrast, radioactive 5S rRNA hybridizes to chromatin fibers situated outside the nucleolus. Hence the genes coding for 18S and 28S rRNAs are localized within the nucleolus, whereas the genes coding for 5S rRNA are not.

The Genes Coding for Ribosomal RNAs Are Extensively Amplified

Cells that are active in protein synthesis may contain hundreds of thousands of ribosomes, which means that large amounts of rRNA must be synthesized. We learned in Chapter 10 that one way of increasing the production of a given type of RNA is to increase the number of copies of the gene encoding that RNA. In the case of ribosomal RNAs, an increase in the number of gene copies is accomplished in two ways. First, the chromosomal DNA of both eukaryotic and prokaryotic cells contains multiple copies of the genes coding for rRNAs. The bacterium *E. coli* contains seven sets of rRNA genes, while eukaryotic cells typically contain between a few hundred and several thousand copies of the 28S, 18S, and 5.8S rRNA genes, and up to 50,000 copies of the 5S rRNA genes.

In cells that need to make large numbers of ribosomes, a second mechanism for increasing the number of

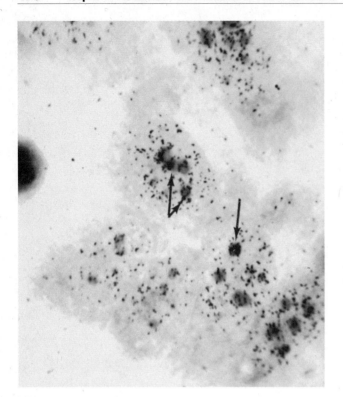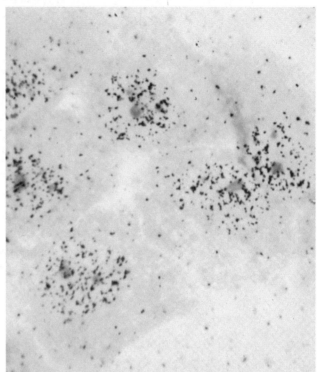

Figure 11-11 *Autoradiographic Evidence Indicating That Ribosomal RNA Synthesis Occurs in Nucleoli* (Left) *Autoradiograph of human cells incubated with 3H-uridine to label newly synthesized RNA. Note the presence of diffuse labeling over the chromatin, as well as the dense labeling of the nucleoli (arrows). (Right) Autoradiograph of cells preincubated with a low concentration of actinomycin D to selectively inhibit the formation of ribosomal RNA. Note the reduction in the amount of radioactivity incorporated into nucleoli. Courtesy of R. P. Perry.*

rRNA gene copies is also utilized; this additional mechanism involves the process of **gene amplification.** The amplification of rRNA genes has been thoroughly studied in developing egg cells (oocytes) of amphibians. During the early growth phase of amphibian oocytes, rRNA genes are amplified more than a thousandfold. Gene amplification is accompanied by the formation of several thousand nucleoli, each containing several hundred copies of the genes coding for 28S, 18S, and 5.8S rRNAs. Hence each oocyte ends up with nearly half a million copies of these rRNA genes, whereas other cells in the same organism contain only a few hundred copies. The enormous number of rRNA genes in the oocyte allows the developing egg cell to produce the quantity of ribosomes that are needed by the embryo after fertilization takes place.

Most of the amplified rRNA genes of the oocyte reside in circular DNA molecules that are separate from the chromosomal DNA. Although it is not clear how the first circular DNA molecules containing rRNA genes are derived, a special type of replication process is responsible for increasing the number of circular DNA molecules once a few have been generated. This mechanism, termed **rolling circle replication,** is comparable to the way in which the DNA of certain single-stranded DNA viruses is replicated. In rolling circle replication, the 3' end of one of the two

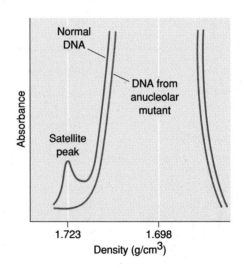

Figure 11-12 *Behavior of DNA Coding for 18S and 28S rRNAs in Xenopus laevis. Isodensity centrifugation of total nuclear DNA reveals the presence of a small satellite peak that contains sequences coding for 18S and 28S rRNAs. Since the satellite peak is not present in DNA obtained from mutant organisms that lack nucleoli, this suggests that the DNA coding for 18S and 28S rRNA is localized within the nucleolus.*

fibrillar regions. Furthermore RNA polymerase I, the enzyme responsible for transcribing the genes for 18S, 5.8S, and 28S rRNA, has been localized in the fibrillar centers of nucleoli using gold-labeled antibodies (page 131) directed against the enzyme. Electron microscopic procedures pioneered by Oscar Miller even allow the genes present in the fibrillar centers to be observed while they are being transcribed. Miller discovered that exposing isolated oocyte nuclei to distilled water causes the nuclei to rupture and the granular outer layers of the nucleoli to disperse, leaving the fibrillar regions readily visible. Staining these isolated fibrillar regions with heavy metals followed by electron microscopic examination reveals that they are comprised of a long *central fiber* from which clusters of shorter *lateral fibrils* emerge (Figure 11-15). In each cluster of lateral fibrils the length of the projecting fibrils increases progressively from one end of the cluster to the other, forming an "arrowhead" pattern. Between these clusters the central fibril is bare.

By treating such preparations with a variety of stains and enzymes, Miller determined that the central

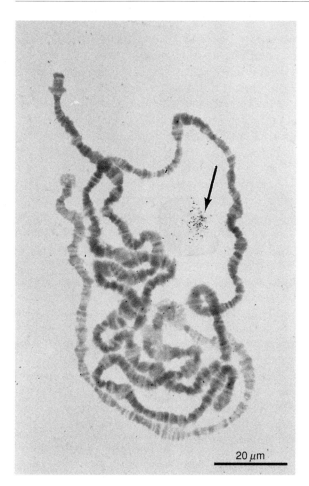

Figure 11-13 *Evidence for the Localization of rRNA Genes in the Nucleolus* Insect salivary gland cells were incubated with radioactive 18S and 28S rRNAs under conditions that promote the hybridization of RNA to DNA. Light microscopic autoradiography reveals that the radioactive RNA hybridizes predominantly with DNA in the nucleolus (arrow), supporting the conclusion that the nucleolus contains the genes coding for 18S and 28S rRNAs. Courtesy of M. L. Pardue.

strands of a circular DNA molecule is extended by a replication process that displaces the 5' end of the same strand from the circle (Figure 11-14). The newly formed single-stranded DNA is then cleaved and serves as a template for the synthesis of the appropriate complementary strand.

The Genes for 18S, 5.8S, and 28S rRNAs Are Transcribed in the Nucleolar Fibrillar Center into a Single Pre-rRNA Molecule

Several lines of evidence suggest that the genes coding for 18S, 5.8S, and 28S rRNAs are transcribed in the fibrillar center of the nucleolus. In cells that have been briefly exposed to ³H-uridine to label newly forming RNA molecules, microscopic autoradiography reveals that nucleolar labeling occurs predominantly over the

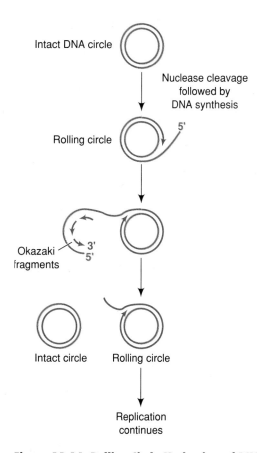

Figure 11-14 *Rolling Circle Mechanism of DNA Replication* As in the case of DNA replication involving linear DNA molecules, replication of one of the two strands involves the formation of discontinuous "Okazaki fragments" that are subsequently joined together (page 71).

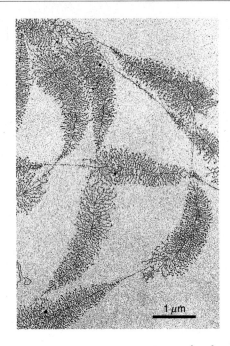

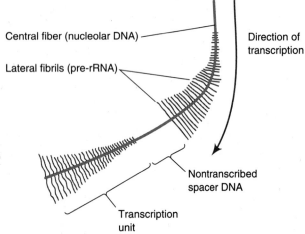

Figure 11-15 *Transcription of Nucleolar DNA Viewed by Electron Microscopy* (Top) *Electron micrograph of the fibrillar region of an oocyte nucleolus visualized after the nucleus has been ruptured by exposure to distilled water.* (Bottom) *Diagram showing the relationship between the central DNA fiber and the lateral fibers, which correspond to pre-rRNA molecules in varying stages of transcription. Micrograph courtesy of O. L. Miller, Jr.*

fiber contains DNA and the lateral fibrils contain RNA. Hence he concluded that the central fiber represents a chromatin fiber in the process of being transcribed, while the lateral fibrils are newly forming RNA molecules. The arrowhead pattern of lateral fibrils corresponds to rRNA molecules in varying stages of transcription. The short fibrils located at one end of each cluster represent RNA molecules whose transcription has just been initiated, while the progressively longer fibrils observed as one proceeds toward the other end of the cluster are RNA molecules whose transcription is further along.

DNA cloning techniques have permitted the DNA sequences present in the central fibers to be isolated and analyzed. The genes coding for 18S, 5.8S and 28S rRNAs are located adjacent to one another in this DNA, with each set repeated multiple times. *Nontranscribed spacer* DNA separates the sets of three rRNA genes from one another. Each group of three genes is transcribed as a unit into a single RNA molecule called **pre-rRNA,** which serves as a precursor of the three ribosomal RNAs. The DNA region that is transcribed into this single pre-rRNA molecule is referred to as a **transcription unit,** which is a general term that refers to a DNA region bounded by a single transcription initiation site and a single transcription termination site. In addition to containing the genes for 18S, 5.8S, and 28S rRNA, the rRNA transcription unit includes *transcribed spacer sequences* that surround each of the three genes. These spacer sequences are removed when the pre-RNA molecule is converted into mature ribosomal rRNAs

The nontranscribed spacer region that lies upstream from each rRNA transcription unit contains promoter sequences to which several transcription factors bind in combination with RNA polymerase I. After transcription has been initiated, RNA polymerase I transcribes the entire transcription unit until it reaches one or more short termination sequences enriched in thymine residues that lie downstream from the 3' end of the 28S rRNA gene.

Pre-rRNA Undergoes Several Cleavage Steps During Its Conversion to Mature Ribosomal RNAs

The pathway involved in the processing of eukaryotic rRNA precursors was first investigated in the early 1960s by Klaus Scherrer and James Darnell, who incubated cells with ³H-uridine for varying periods of time and then analyzed the newly synthesized radioactive RNA by moving-zone centrifugation. When labeling was carried out for relatively short periods of time, a peak of radioactive RNA sedimenting at roughly 45S appeared. With slightly longer labeling periods, radioactive RNA sedimenting at 32S was also detected; finally, after 30 minutes of labeling, radioactivity was detected in 18S and 28S rRNAs as well. Two possible explanations for such data can be envisioned: either the 45S and 32S RNAs are synthesized more rapidly than 28S and 18S RNAs and can therefore be detected sooner, or the 45S and 32S RNAs are detected first because they are precursors of 28S and 18S rRNAs.

To distinguish between these alternatives, experiments were carried out in which cells were incubated for a short period of time with ³H-uridine to allow the 45S RNA, but not the 32S, 28S, or 18S RNAs, to become radioactive. Actinomycin D was then added to block the formation of any additional RNA, and incubation was continued. Under such conditions the radioactive 45S RNA gradually disappeared, while radioactivity in 32S, 28S, and 18S RNAs increased (Figure 11-16, *left*). Since the 32S, 28S, and 18S RNAs are not synthesized in the presence of actinomycin D, it was concluded that the 45S RNA functions as a pre-rRNA that is subsequently converted into these smaller rRNAs.

The steps involved in rRNA processing are more clearly resolved when polyacrylamide gel electrophoresis is employed instead of moving-zone centrifugation (see Figure 11-16, *right*). The information obtained in this way has led to the model of 45S pre-rRNA processing summarized in Figure 11-17. Although the sizes of the various rRNA precursors involved in this pathway differ somewhat among organisms, a similar kind of processing scheme has been observed in a large number of different animal and plant cell types. In such pathways, the initial product generated by transcription of the ribosomal RNA genes is a precursor molecule that ranges in size from 34S to 45S (Table 11-1). During processing, 25 to 50 percent of the RNA present in the initial precursor is removed and destroyed. As in the processing of messenger RNA (page 431), a specific set of small nuclear RNPs (snRNPs) have been implicated in pre-rRNA processing. The first cleavage step appears to be catalyzed by an snRNP that consists of a small nuclear RNA called *U3* bound to several proteins, including *fibrillarin*. Fibrillarin is localized within the dense fibrillar component of the nucleolus, where it forms complexes with several small nuclear RNAs in addition to U3.

Table 11-1 Size of Pre-rRNA in Various Eukaryotes

	Sedimentation Coefficient of Pre-rRNA	Percentage of Precursor Conserved During Processing
Fruit fly	34S	76
Yeast (*S. cerevisiae*)	37S	77
Slime mold	37S	80
Trout	38S	81
Tobacco plant	38S	71
Frog	40S	81
Chicken	45S	57
Mouse	45S	51
Human	45S	51

Besides being cleaved, pre-rRNA molecules are also modified by the addition of methyl groups. The main site of methylation is the 2' hydroxyl group of the sugar ribose, although a few bases are methylated as well. Methylation of rRNA has been studied by incubating cells with radioactive *S-adenosyl methionine*, the substance that serves as the methyl group donor for cellular methylation reactions. When human cells are incubated with radioactive S-adenosyl methionine to

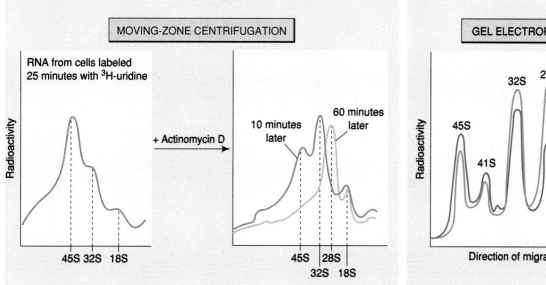
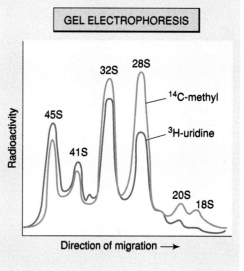

Figure 11-16 *Evidence for the Processing of 45S Pre-rRNA*   (Left) *After human cells have been incubated for 25 minutes with* ³H-uridine, *the bulk of the labeled RNA sediments at 45S when examined by moving-zone centrifugation. The subsequent fate of this 45S RNA was determined taking cells labeled with* ³H-uridine *and adding actinomycin D to block further RNA synthesis. During subsequent incubation, the radioactive 45S RNA is first converted to 32S RNA, and then to 28S and 18S rRNAs. (Right) Polyacrylamide gel electrophoresis provides a clearer picture of 45S pre-rRNA processing than moving-zone centrifugation. Human cells were incubated either with* ³H-uridine *to label newly synthesized RNA, or* ¹⁴C-S-adenosyl methionine, *which serves as a methyl group donor for RNA methylation reactions. Examination of radioactive RNA isolated from the nucleoli of human cells reveals the presence of 41S and 20S intermediates in rRNA processing that were not detected by centrifugation. The distribution of* ¹⁴C-methyl *groups indicates that 45S pre-rRNA is methylated, and that the methyl groups are retained during subsequent rRNA processing.*

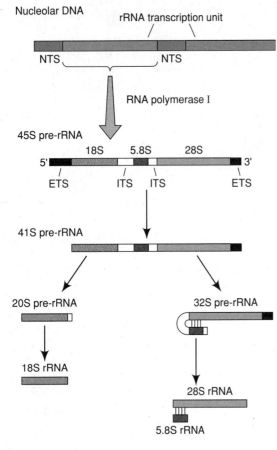

Figure 11-17 *Model of 45S Pre-rRNA Processing* After transcription of the nucleolar rRNA transcription unit by RNA polymerase I, the resulting 45S pre-rRNA molecule undergoes a series of cleavage reactions that lead to the formation of 18S, 28S, and 5.8S rRNAs. The 28S and 5.8S rRNAs are held together by hydrogen bonds between complementary base pairs.

label newly incorporated methyl groups, 45S RNA becomes radioactive first, followed by the 41S, 32S, and 20S RNAs, and finally 28S and 18S RNAs (see Figure 11-16). All of the radioactive methyl groups initially incorporated into the 45S pre-rRNA are retained in the finished rRNA products, indicating that the methylated segments are selectively conserved during rRNA processing. Although the exact role played by RNA methylation is unclear, it may help to guide RNA processing by protecting specific regions of the pre-rRNA molecule from cleavage. When RNA methylation is artificially inhibited by depriving cells of one of the essential components required for transfer of methyl groups, RNA processing is disrupted.

Some Ribosomal RNA Precursors Contain Introns That Are Removed by Self-Splicing

In most eukaryotic organisms, the sequences removed from pre-rRNA molecules during RNA processing are all spacers rather than introns; that is, the sequences being excised are located between genes rather than within a given gene. However, in certain species of the ciliated protozoan *Tetrahymena,* a 413 base-pair intron is located within the gene that codes for the 26S rRNA of the large ribosomal subunit (the 26S rRNA of *Tetrahymena* is comparable to the 28S rRNA molecule of higher eukaryotes). In the early 1980s, Thomas Cech and his colleagues made a surprising discovery about the way in which this intron is removed from the precursor that generates 26S rRNA. The splicing reaction that excises this intron was found to occur in a solution that contains nucleotides and the purified 26S rRNA precursor, but no protein catalyst. To be certain that no *Tetrahymena* enzyme was responsible for catalyzing intron removal, the DNA coding for the 26S rRNA precursor was cloned in bacteria and transcribed under cell-free conditions to generate artificial 26S RNA precursor molecules. The artificially made precursor molecules were again found to catalyze intron removal in the absence of any protein catalyst. This phenomenon, called **self-splicing,** was the first time an RNA molecule had ever been found to catalyze a chemical reaction. RNA molecules that catalyze reactions are now referred to as **ribozymes.**

Shortly after this discovery, self-splicing was also observed in mRNA and rRNA precursors in yeast and fungal mitochondria, in tRNA and rRNA precursors in chloroplasts, and in mRNA precursors manufactured by certain bacterial viruses. Depending on the splicing mechanism involved, these self-splicing RNA sequences are classified into two principal groups (Figure 11-18). *Group I self-splicing introns* are removed by a mechanism that requires free guanosine or one of its phosphrylated derivatives; during splicing, the intron is released as a linear DNA fragment that is then converted to linear and circular pieces. *Group II self-splicing introns* are removed by a mechanism that resembles the spliceosome-mediated removal of introns from premRNA (page 432), releasing the intron in the form of a lariat. However, with self-splicing introns the participation of spliceosomes is not required.

The Formation of Ribosomal Subunits Requires 5S rRNA and Ribosomal Proteins in Addition to 18S, 5.8S, and 28S rRNAs

In addition to the 18S, 5.8S, and 28S rRNAs produced by pre-rRNA processing, ribosome assembly also requires 5S rRNA. Although 18S, 5.8S, and 28S rRNAs are auto-

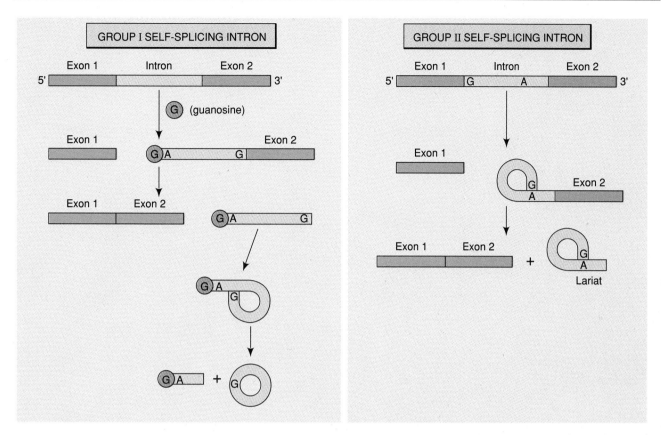

Figure 11-18 *Mechanism of Intron Removal by Group I and Group II Self-Splicing Introns* *Group I introns are released as linear DNA fragments, which are then converted to linear and circular pieces. Group II introns are removed by a process that resembles the spliceosome-mediated removal of introns from pre-mRNA, which releases the intron in the form of a lariat.*

matically produced in equal quantities because they are derived from the same pre-rRNA molecule, 5S RNA is transcribed from a separate set of genes localized outside the nucleolus and then moves into the nucleolus. The 5S rRNA genes are present in larger numbers than the genes coding for 45S pre-rRNA and are transcribed more rapidly; the result is the production of an excess of 5S rRNA molecules, some of which are eventually degraded. The genes coding for 5S rRNA are transcribed by RNA polymerase III, which initiates transcription after transcription factor TFIIIA binds to the 5S rRNA gene's internal control region (page 422). Unlike pre-rRNA, the RNA product generated by transcription of the 5S rRNA gene requires little or no processing. In some organisms a short stretch of nucleotides is removed from its 3' end, while in other organisms 5S rRNA is not processed at all.

So far, our discussion of ribosome formation has focused on the synthesis of rRNAs. To be complete, our picture of ribosome formation must also address the question of how and when ribosomal proteins become associated with the newly forming ribosomal RNAs. One way of exploring this question is to extract isolated nucleoli with solvents that permit RNA to be isolated

with its bound proteins still attached. The analysis of such extracts by moving-zone centrifugation has revealed that nucleoli contain a high concentration of ribonucleoprotein granules sedimenting at 80S and 55S (Figure 11-19). When the RNA molecules present in these particles are purified, 45S and 5S RNAs are found in the 80S particles and 32S and 5S RNAs are detected in the 55S particles. Since the 45S pre-rRNA found in the 80S particle is a precursor for rRNAs present in the large subunit (28S and 5.8S rRNAs) as well as the small subunit (18S rRNA), it can be concluded that the 80S nucleolar particle is a precursor of both the small and large ribosomal subunits. The 55S particle, on the other hand, must function as a precursor of the large ribosomal subunit because it contains the 32S rRNA. The preceding conclusions have been supported by electrophoretic analyses of the protein constituents of the ribonucleoprotein particles; the proteins of the 80S particle resemble those of both the large and small ribosomal subunits, whereas the proteins of the 55S subunit resemble those of the large subunit only.

On the basis of the above information and a variety of other observations, the model of ribosome assembly

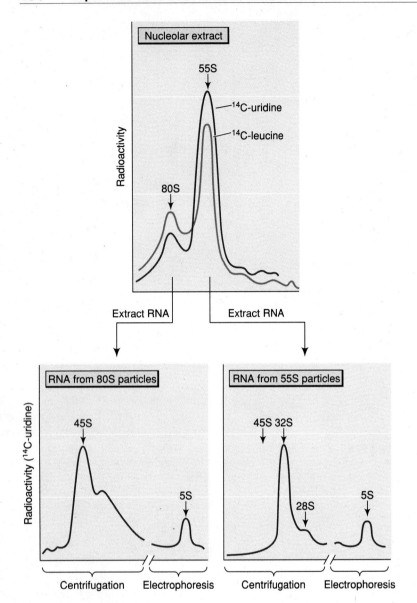

Figure 11-19 *Moving-Zone Centrifugation of a Nucleolar Extract Obtained from Cells Incubated with Either Radioactive Leucine or Uridine* *The two major peaks, sedimenting at 80S and 55S, both contain a mixture of protein (labeled with leucine) and RNA (labeled with uridine). The bottom graphs show the types of RNA that can be extracted from these two peaks.*

summarized in Figure 11-20 has been formulated. According to this model, the newly forming pre-rRNA molecule begins to associate with protein molecules before transcription is complete. Two types of proteins eventually interact with the pre-rRNA; some are involved in rRNA cleavage reactions and are released prior to the formation of mature ribosomal subunits; others are ribosomal proteins that form part of the final ribosome. These ribosomal proteins are synthesized in the cytoplasm and rapidly concentrated within the nucleolus, where they become associated with the pre-rRNA molecule to form the 80S ribonucleoprotein particle. After RNA processing cleaves the 80S particle into precursor

particles for the large and small subunits, the small subunit is quickly formed and exported to the cytoplasm, while the large subunit is produced and exported more slowly. This difference in export rates explains why large ribosomal subunit precursor particles (55S particles) are more readily detected in nucleolar extracts than are precursor particles for the small subunit.

The relationship between the preceding events and the structural organization of the nucleolus has been investigated by microscopic autoradiography. When cells are briefly incubated with ^{3}H-uridine, autoradiography reveals that the newly formed radioactive pre-rRNA is localized in the central regions of the nucleolus, where

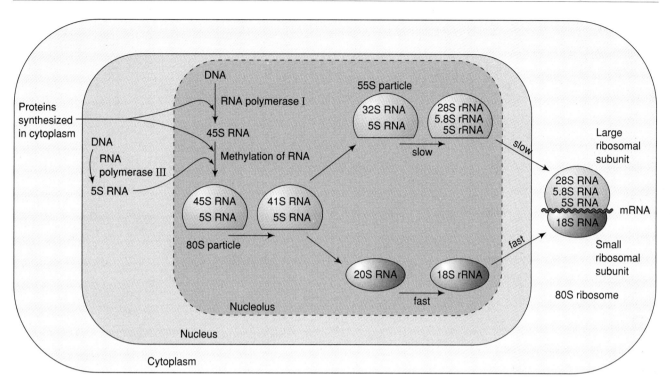

Figure 11-20 *Summary of the Main Steps Involved in Ribosome Formation in Human Cells* *According to this model, 45S pre-rRNA is transcribed from nucleolar DNA, bound to protein, methylated, and combined with 5S RNA to produce an 80S ribonucleoprotein particle. RNA processing then converts the 80S particle into large and small ribosomal subunits, which are exported to the cytoplasm.*

the fibrillar centers and dense fibrillar components are located. If the fate of the newly formed pre-rRNA is investigated by briefly labeling cells with ³H-uridine and then adding actinomycin D to block further RNA synthesis, the radioactivity migrates into the granular component that surrounds the fibrillar centers and dense fibrillar components (Figure 11-21). Hence it appears that pre-rRNA is synthesized in the fibrillar centers, where it becomes associated with protein prior to being processed in the dense fibrillar components. The resulting ribonucleoprotein granules then become part of the granular component of the nucleolus. According to this view, the granules that make up the granular component represent partially and completely processed rRNA molecules bound to ribosomal proteins and perhaps other proteins involved in rRNA processing.

Prokaryotic Cells Manufacture Ribosomes without a Nucleolus

Although bacteria do not contain nucleoli, they manufacture ribosomes by a process that shares many features with ribosome formation in eukaryotes. In both types of cells, the genes coding for rRNAs occur in multiple copies and are transcribed into precursor molecules that are subsequently processed into mature rRNAs. The organization of the rRNA genes does, however, exhibit

some important differences. In contrast to the extensive repetition of the rRNA genes observed in eukaryotes, bacterial rRNA genes are present in a relatively small number of copies. The rRNA genes of *E. coli*, for example, are repeated only seven times. Another difference is that the genes coding for 5S rRNA are transcribed separately from the rest of the rRNA genes in eukaryotes, whereas the bacterial 5S rRNA gene is part of the transcription unit that produces the other two rRNAs. In the rRNA transcription unit of *E. coli*, the three genes are arranged in the order 16S-23S-5S.

Evidence that the 16S, 23S, and 5S rRNA genes are part of the same transcription unit has been obtained from studies of mutant bacteria that are defective in the enzyme *ribonuclease III (RNase III)*. Such cells produce RNA molecules that contain 16S, 23S, and 5S rRNA sequences joined together in a single molecule. This type of precursor is not in cells where RNase III activity is normal, suggesting that the function of RNase III is to cleave the 16S-23S-5S rRNA precursor into smaller molecules. When electron microscopy is employed to examine the transcription of rRNA genes in cells containing normal RNase III, separate clusters of RNA molecules can be seen emerging from the 16S and 23S rRNA genes (Figure 11-22). This arrangement would be expected if the 16S rRNA gene is transcribed first and the newly forming RNA product is released by RNase-III cleavage

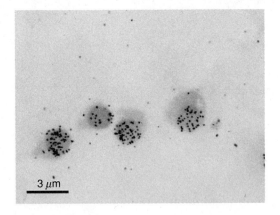

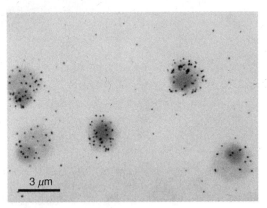

Figure 11-21 *Light Microscopic Autoradiography of Nucleoli Incubated with Radioactive Uridine to Label Newly Synthesized Ribosomal RNA* (Top) *After brief exposure to ³H-uridine, most of the radioactivity is localized near the fibrillar centers. (Bottom) Nucleoli were briefly labeled with ³H-uridine followed by several hours of incubation in the presence of actinomycin D to prevent further synthesis of radioactive RNA. During this subsequent incubation, the radioactivity moves into the granular component surrounding the fibrillar centers. Courtesy of M. Alfert.*

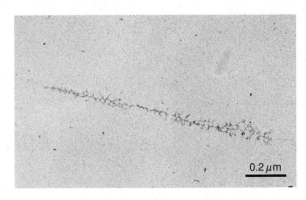

Newly forming RNA molecules

16S rRNA gene   23S rRNA gene

Direction of transcription ⟶

Figure 11-22 *Electron Micrograph of Bacterial Ribosomal RNA Genes in the Process of Being Transcribed* *Separate clusters of RNA molecules are seen emerging from the 16S and 23S rRNA genes because the 16S rRNA gene is transcribed first and the newly forming RNA product is released by RNase-III cleavage before the RNA polymerase proceeds from the 16S rRNA gene into the adjacent 23S rRNA gene. Micrograph courtesy of O. L. Miller, Jr.*

before the RNA polymerase molecule proceeds from the 16S gene into the adjacent 23S gene.

RNase III is not the only enzyme involved in cleaving bacterial rRNA precursors. Bacteria deficient in other types of ribonuclease also fail to produce rRNAs of the proper size. Instead, they accumulate RNA molecules that are slightly larger than mature 16S, 23S, and 5S rRNAs. These observations indicate that bacterial rRNA processing involves more than one kind of ribonuclease (Figure 11-23). In addition to producing rRNAs, processing of the RNA transcribed from the rRNA transcription unit also generates several transfer RNA molecules. One or two of these tRNAs are derived from sequences located between the 16S and 23S genes, and one or two more originate from sequences located downstream from the 5S RNA gene.

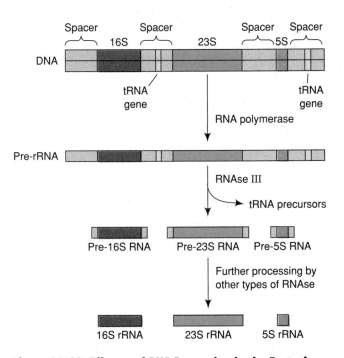

Figure 11-23 *Ribosomal RNA Processing in the Bacterium E. Coli* *Cleavage of the initial pre-rRNA transcript begins before transcription of this set of genes is completed.*

As in the case of eukaryotic ribosome formation, bacterial rRNA becomes methylated and associated with ribosomal proteins during processing. Bacterial pre-rRNA contains relatively few methyl groups compared to the number present in the final 16S, 23S, and 5S rRNA molecules, indicating that methylation of these RNAs occurs after processing and assembly of the ribosomal subunits has begun. During rRNA processing, the binding of ribosomal proteins to rRNA precursors generates two classes of ribonucleoprotein particles. One contains the 23S and 5S precursor RNAs and serves as an intermediate in the formation of the large ribosomal subunit; the other contains the 16S precursor and is an intermediate in the formation of the small ribosomal subunit.

PROTEIN SYNTHESIS

In Chapter 3 we learned that protein synthesis occurs on ribosomes, which translate the base sequences of messenger RNA molecules into the amino acid sequences of newly forming polypeptide chains. During the process, transfer RNA molecules serve as intermediaries that recognize mRNA codons and insert the appropriate amino acids into growing polypeptides. Now that the structure and biogenesis of ribosomes have been described in detail, we can examine how the ribosome facilitates the process of protein synthesis.

Protein Synthesis Can Be Studied in Cell-Free Systems

Our understanding of the molecular events that occur during protein synthesis is derived largely from studies involving *cell-free systems* (page 82). The earliest systems for studying protein synthesis outside of living cells were developed shortly after World War II, when radioactive compounds first became widely available to the scientific community. This development was crucial because access to radioactive amino acids made it possible to detect the tiny amount of protein synthesis that occurs under cell-free conditions.

In the early 1950s Paul Zamecnik and his collaborators established the basic conditions required for the incorporation of radioactive amino acids into polypeptide chains. One of the more extensive studies was carried out by Philip Siekevitz, who showed that homogenates of liver tissue incorporate radioactive amino acids into protein. When such homogenates were fractionated by centrifugation, the newly synthesized radioactive protein was found in the microsomal fraction, which consists of a mixture of ER membranes and ribosomes (page 138). Since this discovery suggested that the microsomal fraction might be involved in protein synthe-

sis, isolated microsomes were incubated with radioactive amino acids to see whether they could synthesize protein in the absence of other cellular components. Efficient rates of protein synthesis were observed only when the mitochondrial and cytosol fractions were added to isolated microsomes. It was later discovered that ATP and GTP can be substituted for the mitochondrial fraction, indicating that the mitochondria simply function as a source of energy. The role played by the cytosol was clarified when the presence of transfer RNAs and aminoacyl-tRNA synthetases in this fraction was discovered a few years later.

The next set of investigations addressed the question of whether the ribosomes or the membranes of the microsomal fraction are the actual site of protein synthesis. In these studies, microsomal membranes were solubilized with detergent and the solution centrifuged to collect membrane-free ribosomes. Ribosomes prepared in this way retain the capacity to carry out protein synthesis, indicating that the ribosome and not the membrane is the essential component. It has therefore been concluded that in spite of the attachment of many ribosomes to membranes (the endoplasmic reticulum in eukaryotic cells and the plasma membrane in prokaryotic cells), membrane attachment is not required for protein synthesis. As we learned in Chapter 7, the main function of membranes is in the intracellular transport and packaging of newly synthesized protein molecules, especially those proteins destined for secretion from the cell or insertion into membranes.

Protein Synthesis Proceeds from the N-Terminus to the C-Terminus

The synthesis of polypeptide chains is an ordered, stepwise process that begins at the N-terminus of the polypeptide chain and involves the sequential addition of amino acids to the growing chain until the C-terminus is reached. The first experimental evidence supporting this view was provided in 1961 by Howard Dintzis, who investigated the synthesis of hemoglobin in developing red blood cells that had been incubated briefly with radioactive amino acids. He reasoned that if the time of incubation with radioisotope is kept relatively brief, then the radioactivity present in completed hemoglobin chains should be concentrated at the end of the molecule that has been most recently synthesized. Amino acids located elsewhere would not be radioactive because they would have been incorporated into the protein before the radioisotope was added. This idea was tested by isolating hemoglobin from cells that had been briefly incubated with radioactive leucine. After digesting the hemoglobin with the protease trypsin, the resulting peptide fragments were analyzed for radioactivity. The highest concentration of radioactivity was detected in the C-terminal end, indicating that amino acids are added to the growing polypeptide chain beginning at the N-terminus and proceeding toward the C-terminus (Figure 11-24).

Translation of Messenger RNA Proceeds in the 5' → 3' Direction

In theory, the base sequence of a messenger RNA molecule can be read in two different directions—from the 5' end to the 3' end or from the 3' end to the 5' end. The first attempts to determine the way in which the base sequence is actually read during protein synthesis involved the use of artificial RNA molecules. A typical example is the synthetic RNA that can be made by adding the base C to the 3' end of poly A:

$$(5') \, A \, A \, A \, A \, A \, A \, A \, A \, A \, A \cdots A \, A \, C \, (3')$$

When added to a cell-free protein-synthesizing system, this RNA stimulates the synthesis of a polypeptide that consists of a stretch of lysine residues with an asparagine at the C-terminal end. Since AAA codes for lysine and AAC codes for asparagine, the RNA must be translated in the 5' → 3' direction. This conclusion has received further support from studies in which the base sequences of naturally occurring mRNAs were compared to the amino acid sequences of the polypeptide chains they encode. In all cases the amino acid sequence of the polypeptide chain corresponds to the order of the codons in mRNA read in the 5' → 3' direction.

Ribosomal Subunits Associate and Dissociate During Protein Synthesis

The fact that ribosomes are constructed from small and large subunits raises the question of whether the subunits come apart and rejoin as part of the normal mechanism for synthesizing proteins. An answer to this question was first provided in 1968 by Raymond Kaempfer, who utilized both radioactive isotopes and the heavy and light isotopes of nitrogen (^{15}N and ^{14}N) to monitor the behavior of bacterial ribosomes during protein synthesis. Ribosomes isolated from cells that had been labeled with both a radioactive isotope and heavy nitrogen were mixed with an excess of nonradioactive ribosomes containing light nitrogen. When the resulting mixture was incubated under conditions that permit protein synthesis to occur, the radioactive heavy ribosomes were quickly replaced by radioactive ribosomes whose density was between that of heavy and light ribosomes (Figure 11-25). It was therefore concluded that the subunits of the heavy ribosomes had dissociated from one another and then randomly reassociated with subunits derived from the nonradioactive light ribosomes. Adding an inhibitor of protein synthesis to the reaction mixture prevents the change in ribosome density from taking place, thereby indicating that ribosomal subunits dissociate and then come back together as part of the normal mechanism by which proteins are made.

At what stage in the process of protein synthesis do ribosomal subunits dissociate and then rejoin? In order to answer this question, we need to take a closer look at how ribosomes translate messenger

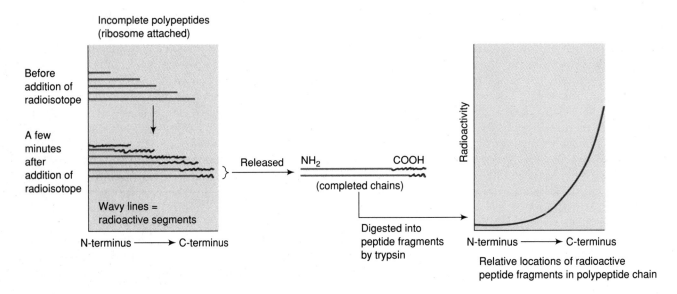

Figure 11-24 Experiment Showing That Protein Synthesis Proceeds from the N-Terminus toward the C-Terminus *After immature red blood cells (reticulocytes) were exposed to radioactive leucine for 4 minutes, completed chains of radioactive hemoglobin were isolated and digested into peptide fragments with trypsin. The highest concentration of radioactivity was found in peptide fragments located near the C-terminal end of hemoglobin.*

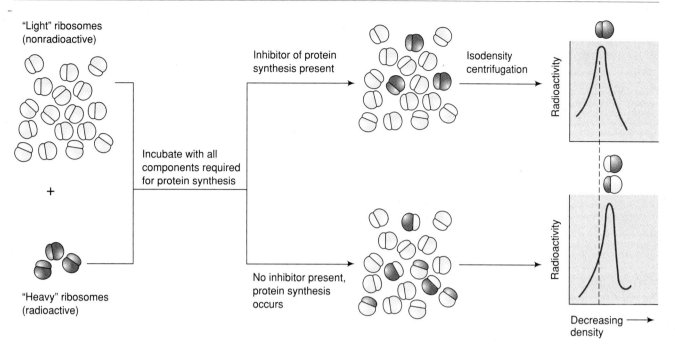

Figure 11-25 *Experiment Showing That Ribosomal Subunits Dissociate and Then Reassociate During Protein Synthesis* Nonradioactive light (^{14}N-containing) ribosomes were mixed with radioactive heavy (^{15}N-containing) ribosomes. (Top) *If protein synthesis is inhibited, the heavy and light ribosomes remain intact.* (Bottom) *If protein synthesis is allowed to occur, radioactive subunits appear in ribosomes whose density is less than that of the original heavy ribosomes. Thus the small and large heavy ribosomal subunits must have come apart and then reassociated with subunits derived from the light ribosomes.*

RNA. In both prokaryotic and eukaryotic cells, the process of mRNA translation can be divided into three distinct phases: (1) an *initiation phase,* in which mRNA is bound to the ribosome and positioned for proper translation; (2) an *elongation phase,* in which amino acids are sequentially joined in an order specified by the arrangement of codons in the mRNA; and (3) a *termination phase,* in which the mRNA and newly formed polypeptide chain are released from the ribosome. In the following sections we will examine each of these phases in turn.

The Initiation of Protein Synthesis Requires an Initiator Transfer RNA

Messenger RNA molecules almost always contain extra bases at their 5' and 3' ends that are not translated into amino acids during protein synthesis (Table 11-2). The 5' untranslated sequence is called the *leader,* while the 3' untranslated region between the end of the coding sequence and the beginning of the poly-A tail (if present) is called the *trailer.* The existence of leader and trailer sequences means that special signals are required for determining where translation of a given mRNA will

Table 11-2   Number of Nucleotides in the Translated and Untranslated Regions of Several Eukaryotic Messenger RNAs

	Complete mRNA	Coding Region	5' Untranslated Leader	3' Untranslated Trailer
α-Globin (rabbit)	551	429	36	86
α-Globin (human)	575	429	37	109
β-Globin (rabbit)	589	444	53	92
β-Globin (human)	628	446	50	132
Ovalbumin (chick)	1859	1164	64	631
Insulin (rat)	443	333	57	53

start and stop. Starting the process of translation requires the participation of a special type of transfer RNA molecule. The general properties of **transfer RNAs (tRNAs)** and their role in protein synthesis have already been described in Chapter 3. Like messenger and ribosomal RNAs, transfer RNAs are transcribed from their corresponding genes in the form of precursor RNAs that are cleaved and trimmed to generate mature tRNAs. Amino acids are then attached to their appropriate tRNAs by a family of enzymes called *aminoacyl-tRNA synthetases* (page 89). The resulting aminoacyl-tRNAs bring amino acids to the ribosome for incorporation during protein synthesis.

The first clue that a special kind of tRNA is associated with the initiation of protein synthesis surfaced in the early 1960s when it was discovered that roughly half the proteins in *E. coli* contain methionine at their N-terminus. This finding was surprising because methionine is a relatively rare amino acid, accounting for no more than a few percent of the amino acids present in bacterial proteins. The importance of this phenomenon became apparent in 1964, when K. Marcker and F. Sanger isolated two different methionine-specific tRNAs from *E. coli*. One, designated *tRNA^Met*, carries a normal methionine destined for insertion into the internal regions of polypeptide chains. The other, called *tRNA^fMet*, carries a methionine that is converted to the derivative *formylmethionine* after it becomes bound to the tRNA (Figure 11-26). In formylmethionine, the amino group of methionine is blocked by the addition of a formyl group and so cannot form a peptide bond with another amino acid; only the carboxyl group is available for bonding to another amino acid. Hence formylmethionine can be situated only at the N-terminal end of a

polypeptide chain, suggesting that tRNA^fMet is involved in initiating protein synthesis. This idea was soon verified by the discovery that bacterial polypeptide chains in the early stages of being synthesized always contain formylmethionine at their N-terminus. Following completion of the polypeptide chain (and in some cases while it is still in the process of being synthesized), the formyl group, and often the methionine itself, is enzymatically removed.

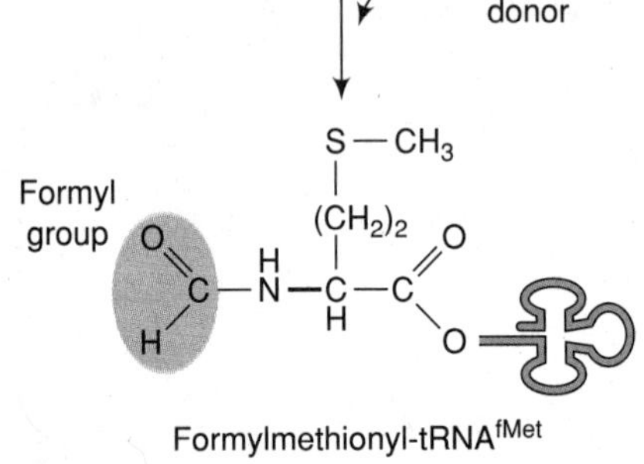

Thus bacteria contain two kinds of tRNA for the amino acid methionine. One, designated *tRNA^fMet*, is an **initiator tRNA** that recognizes the AUG start codon and initiates protein synthesis with the amino acid formylmethionine. The other, called *tRNA^Met*, recognizes internal AUG codons and inserts methionine into polypeptide chains as they are being elongated. Eukaryotic cells also contain two tRNAs that carry methionine, one of which is specialized for initiating protein synthesis. However, formylation is not involved; both forms of eukaryotic methionyl-tRNA carry unmodified methionine.

During Initiation mRNA Binds to the Small Ribosomal Subunit and the AUG Start Codon Becomes Properly Aligned

Initiating and noninitiating tRNAs that carry the amino acid methionine both recognize the same mRNA codon, AUG. Yet only one AUG codon located near the 5' end of each mRNA molecule functions as a *start codon* that initiates protein synthesis (page 87). How is this special codon recognized? Part of the answer involves the mechanism by which mRNA is initially bound to the ribosome. In 1974 J. Shine and L. Dalgarno discovered that many bacterial mRNAs contain a short purine-rich stretch of nucleotides, called the **Shine-Dalgarno sequence,** which is located 5-10 bases upstream from the AUG start codon and whose base sequence is complementary to a base sequence appearing at the 3' end of 16S rRNA (Figure 11-27). This complementary relationship suggested that the initial binding of mRNA to the ribosome is mediated by base-pairing between the Shine-Dalgarno sequence in mRNA and the 3' end of

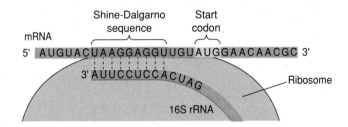

Figure 11-27 Role of the Shine-Dalgarno Sequence in Binding Bacterial mRNAs to the Ribosome *The exact sequence of the purine-rich Shine-Dalgarno region varies somewhat among mRNAs.*

Figure 11-26 Reaction by Which Bacterial Cells Add a Formyl Group to Methionine Bound to tRNA^fMet

16S rRNA. Support for this idea was subsequently provided by experiments in which it was shown that cell-free protein synthesis can be inhibited by the addition of artificial RNAs that are complementary to the Shine-Dalgarno sequence; these artificial RNAs prevent protein synthesis by binding to the Shine-Dalgarno sequence and preventing its interaction with 16S rRNA. Additional evidence for the importance of the interaction between the Shine-Dalgarno sequence and the 3' end of 16S rRNA has emerged from studies of *colicin E3*, a toxic protein produced by certain strains of *E. coli*. Upon entrance into the cytoplasm of susceptible strains of bacteria, colicin E3 catalyzes the removal of a 49 nucleotide fragment from the 3' end of 16S rRNA, thereby creating a ribosome that can no longer initiate protein synthesis.

In contrast to the situation in bacteria, eukaryotic mRNAs do not possess a Shine-Dalgarno sequence. Instead, binding of mRNA to the small ribosomal subunit depends on the methylated cap structure located at the 5' end of eukaryotic mRNAs (page 428). Two kinds of experiments indicate the importance of the cap for efficient translation. One is that mRNAs whose caps have been removed are poorly bound to the small ribosomal subunit and are therefore poorly translated in cell-free protein-synthesizing systems. The other is that adding 7-methyl GMP inhibits the translation of capped messages, presumably because the 7-methyl GMP molecule mimics the cap structure and binds to the ribosomal components that normally interact with the mRNA cap. Such observations suggest that during normal protein synthesis, the capped end of the mRNA binds to the small ribosomal subunit; the mRNA molecule then moves over the surface of the ribosomal subunit until the first AUG codon is encountered. At this point, translation of the coding sequence begins. The notion that translation is initiated at the first AUG codon to be encountered is supported by the discovery that, with few exceptions, extra AUG triplets do not occur between the 5' capped end of eukaryotic mRNAs and the AUG triplet at which translation begins.

Initiation Factors Promote the Interaction between Ribosomal Subunits, mRNA, and Initiator tRNA

Thus far we have seen that the initiation of protein synthesis involves an interaction between mRNA, an initiator tRNA, and the small ribosomal subunit. In prokaryotic cells, this interaction requires the participation of three protein **initiation factors** called IF-1, IF-2, and IF-3 (Figure 11-28). The process by which initiation factors promote the onset of protein synthesis can be divided into three main stages. (1) Initiation factors IF-1, IF-3, and IF-2 bound to GTP become associated with the 30S ribosomal subunit. (2) The 5' end of

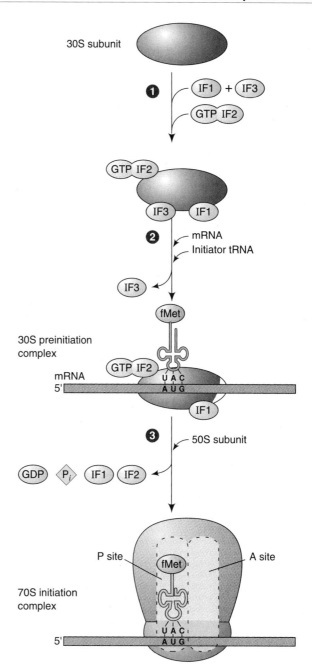

Figure 11-28 *The Initiation of Protein Synthesis in Prokaryotic Cells* *Initiation is a multistep process that requires the participation of the initiation factors IF-1, IF-2, and IF-3 in prokaryotes. In eukaryotes a larger number of initiation factors is involved.*

an mRNA molecule binds to the 30S subunit through its Shine-Dalgarno sequence, aligning the AUG start codon in the proper position for the initiation of protein synthesis. The initiator tRNA with its attached formylmethionine is then bound to the 30S subunit by the action of IF-2, which can distinguish initiator tRNA^fMet from other kinds of tRNA. This attribute of IF-2 helps to explain why AUG start codons pair with the initiator tRNA^fMet, whereas AUG codons located elsewhere in mRNA bind to the noninitiating tRNA^{Met}. After mRNA and the initiator tRNA^{fMet} have become attached to the 30S subunit, IF-3 is released. At this point, the 30S subunit with its associated initiation factors, mRNA, GTP, and initiator tRNA is referred to as a *30S preinitiation complex*. (3) The 50S subunit binds to the 30S preinitiation complex, accompanied by the release of IF-1 and IF-2 as well as the hydrolysis of GTP. The GTP hydrolysis step is carried out by the L7/L12 protein located in the stalk protruding from the 50S subunit. The final complex, consisting of an intact ribosome bound to mRNA and the initiator tRNA^{fMet}, is called a *70S initiation complex*.

Initiation of protein synthesis in eukaryotic cells involves a similar process, although at least six groups of initiation factors (named eIF-1 through eIF-6) are involved rather than three. Eukaryotic factors eIF-1 through eIF-3 are comparable to prokaryotic factors IF-1 through IF-3, while eIF-4 is involved in recognizing the mRNA cap structure and in the hydrolysis of ATP, neither of which occur in prokaryotic cells. During eukaryotic initiation the mRNA cap is bound to the small ribosomal subunit, which then moves downstream along the message until it reaches the first AUG codon. In a few rare instances, the translation of eukaryotic mRNAs can also be initiated at internal sites located far downstream from the 5' cap and initial AUG start site.

More is known about the relationship between specific ribosomal components and the initiation process in prokaryotic cells than in eukaryotes. Experiments in which small ribosomal subunits and initiation factors were incubated in the presence of chemical crosslinking reagents have led to the conclusion that proteins S11, S12, S13, and S19, as well as the 3' end of the 16S rRNA molecule, are located near the region where initiation factors bind to the ribosome. These components are all situated near the head and platform end of the small subunit. In addition, proteins S12 and S21 are required for the binding of the initiator tRNA^{fMet} to the AUG start codon. Although the level of resolution of such studies is still relatively crude, a general picture is beginning to emerge in which the 3' end of 16S rRNA, initiation factors, the initiator tRNA^{fMet}, and the 5' end of the mRNA are closely associated with one another and with a small group of proteins localized in the head and platform region of the small subunit.

In the final initiation complex, the initiator tRNA^{fMet} is positioned with its anticodon paired to the AUG start codon of the mRNA. The formation of this complex ensures that the mRNA is properly aligned for translation to proceed in the proper reading frame. Without such a mechanism an individual mRNA might be translated into three different polypeptides, each encoded by a one-base shift of the reading frame (Figure 11-29). Such improper initiation can be induced experimentally by certain conditions, such as elevated Mg^{2+} concentrations. This effect of Mg^{2+} is sometimes exploited delib-

5' | CAGCUAUGACCAUGAUU | 3'
fMet—Thr—Met— Ile

| CAGCUAUGACCAUGAUU |
Gln—Leu-(stop)-Pro-(stop)

| CAGCUAUGACCAUGAUU |
Ser—Tyr—Asp—His—Asp

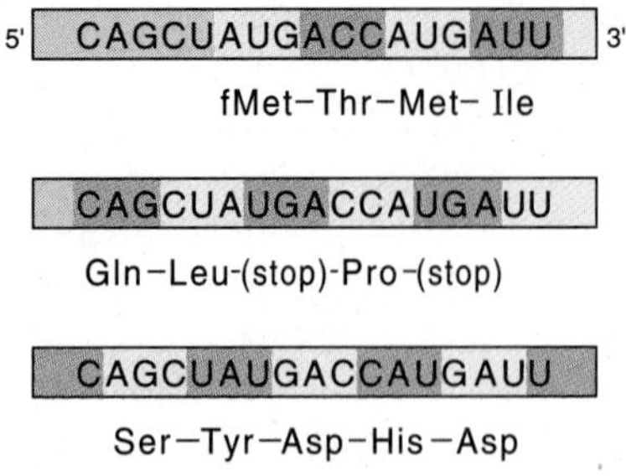

Figure 11-29 *Diagram Showing How the AUG Start Codon Sets the Proper Reading Frame for mRNA Translation* *The top example shows the correct reading frame of the mRNA coding for β-galactosidase. If the initiation codon is not properly recognized, one of the two incorrect reading frames shown in the lower portion of the diagram might be selected instead of the correct one.*

erately, as in forcing ribosomes to initiate the translation of artificial RNAs such as poly U, which would not otherwise be translated because they lack a proper initiation codon.

Ribosomes Contain a P Site That Binds the Growing Polypeptide Chain and an A Site That Binds Incoming Aminoacyl-tRNAs

Once an initiation complex has been formed, protein synthesis proceeds by adding amino acids one at a time in an order specified by codons in mRNA. This process, known as the *elongation phase* of protein synthesis, involves a repeated three-step cycle in which (1) an aminoacyl-tRNA brings an amino acid to the ribosome, (2) the amino acid is joined to the growing polypeptide chain, and (3) the mRNA then advances to expose the next codon. Elongation begins with the binding of an aminoacyl-tRNA whose anticodon is complementary to the codon immediately downstream from the AUG start codon in mRNA. The ribosomal site to which this incoming aminoacyl-tRNA binds is called the **aminoacyl** or **A site,** while the previously bound initiator tRNA is said to occupy the **peptidyl** or **P site.** The latter site is called the peptidyl site because at the comparable stage of ensuing elongation cycles, it will contain the growing polypeptide chain. Binding between mRNA codons and tRNA anticodons occurs in the cleft that lies between the platform and head of the small subunit.

In bacterial cells, the binding of a newly arriving aminoacyl-tRNA to the A site requires GTP bound to a protein **elongation factor** called *EF-Tu* (Figure 11-30, *step 1*). EF-Tu promotes the binding of all aminoacyl-tRNAs *except the initiator tRNA* to the ribosome; hence it ensures that AUG codons located downstream from the start codon do not mistakenly bind to an initiator tRNA^{fMet}. During the binding of each aminoacyl-tRNA to the ribosome, GTP is hydrolyzed to GDP. For many years the hydrolysis of only a single

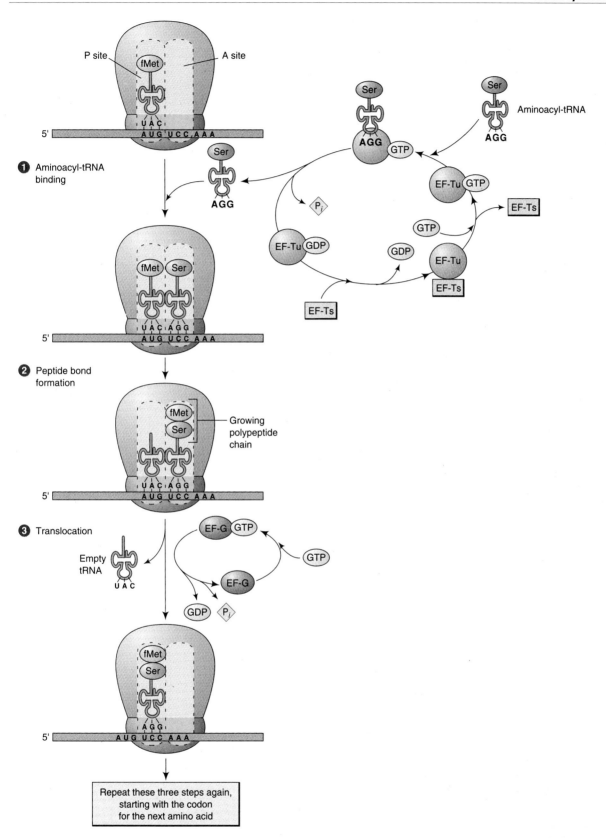

Figure 11-30 *The Elongation Phase of Protein Synthesis* *During elongation, amino acids are added to the growing polypeptide chain by a three-step process that involves (1) binding of the proper aminoacyl-tRNA to the next codon in mRNA, (2) formation of a peptide bond between the amino acid carried by this tRNA and the previously incorporated amino acid, and (3) translocation of the mRNA along the ribosome to bring the next codon into position for translation. This cycle is repeated for each amino acid incorporated into the newly forming polypeptide chain. The elongation factors illustrated in this diagram are those employed by prokaryotic cells; a slightly different set of factors is involved in eukaryotes.*

molecule of GTP was thought to be involved, but recent measurements suggest that two GTPs are hydrolyzed for each amino-acyl-tRNA bound to the ribosome. After GTP hydrolysis has taken place, the resulting EF-Tu/GDP complex is converted back to EF-Tu/GTP through the action of a second elongation factor called *EF-Ts*. A similar cycle occurs in eukaryotic cells, although a slightly different set of elongation factors is involved.

Elongation factors do not recognize individual anti-codons, which means that aminoacyl-tRNAs of all types (other than initiator tRNAs) are indiscriminately brought to the A site of the ribosome. Some mechanism must therefore ensure that only the correct aminoacyl-tRNA is retained by the ribosome for subsequent use during peptide bond formation. If the anti-codon of the incoming aminoacyl-tRNA is not complementary to the mRNA codon exposed at the A site, the aminoacyl-tRNA does not remain bound to the ribosome. If the match is close but not exact, transient binding may take place, but the mismatched aminoacyl-tRNA is eventually rejected. The basis of this error-checking mechanism is not completely understood, but it costs energy because additional GTP hydrolysis often accompanies the transient binding of incorrectly paired aminoacyl-tRNAs.

Although the rules of complementary base-pairing suggest that each tRNA should recognize only one codon in mRNA, many tRNA molecules bind to more than one codon. This phenomenon does not cause misreading of the message, however, because the multiple codons recognized by a single tRNA differ only in the third base. Examination of the genetic code assignments (see Table 3-5) reveals that codons differing in the third base often code for the same amino acid. For example, UUU and UUC both code for phenylalanine; UCU, UCC, UCA, and UCG all code for serine; and so forth. In such cases the same tRNA usually binds to more than one codon. For example, a single tRNA recognizes both the UUU and UUC codons for phenylalanine.

Such considerations led Francis Crick to propose that mRNA and tRNA line up on the ribosome in a way that permits flexibility or "wobble" in the pairing between the third base of the codon and the corresponding base in the anticodon. According to this **wobble hypothesis,** the flexibility in codon-anti-codon binding allows some unexpected base pairs to form (Figure 11-31). The base inosine (I), which is extremely rare in other RNA molecules, often occurs in the wobble position of tRNA anticodons. Although the existence of wobble means that a single tRNA molecule can recognize more than one codon, the codons recognized by a given tRNA always code for the same amino acid.

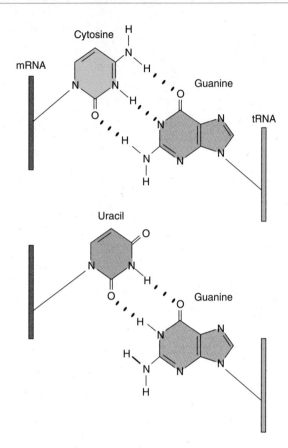

Bases Recognized in Codon (third position only)	Base in Anticodon
U	A
G	C
A or G	U
C or U	G
U, C, or A	I (Inosine)

Figure 11-31 *The Wobble Hypothesis* *The two diagrams illustrate how a slight shift or "wobble" in the position of the base guanine in a tRNA anticodon would permit it to pair with uracil instead of its normal complementary base, cytosine. The table summarizes the base pairs permitted at the third position of a codon by the wobble hypothesis.*

Peptide Bond Formation Is Catalyzed by Ribosomal RNA

Once the appropriate aminoacyl-tRNA has become bound to the A site of the ribosome, the next step in the elongation process is the formation of a peptide bond between the amino group of the amino acid bound at the A site and the carboxyl group by which the initiating amino acid (or growing polypeptide chain) is attached to the tRNA at the P site. The formation of this peptide bond causes the growing polypeptide chain to be transferred from the tRNA located at the P site to the tRNA located at the A site (see Figure 11-30, *step 2*). Peptide bond formation is the only step in protein synthesis that requires neither nonribosomal protein factors nor an outside source of energy such as GTP or ATP. The energy required for driving peptide bond formation is provided by the cleavage of the high-energy bond that joins the amino acid or peptide chain to the tRNA located at the P site (Figure 11-32).

For many years peptide bond formation was thought to be catalyzed by a hypothetical ribosomal protein that was given the name **peptidyl transferase.** However, in 1992 Harry Noller and his colleagues showed that the large subunit of bacterial ribosomes retains peptidyl transferase activity after rigorous treatments designed to destroy and remove all ribosomal proteins. In contrast, peptidyl transferase activity is quickly destroyed when rRNA is degraded by exposing ribosomes to ribonuclease. Such observations suggest that rRNA rather than protein is responsible for catalyzing peptide bond formation. In bacterial ribosomes, peptidyl transferase activity has been localized to the region of the 23S rRNA molecule that is located between the central protuberance and ridge of the large subunit, directly opposite the cleft of the small subunit (see Figure 11-9).

Translocation Advances the Messenger RNA by Three Nucleotides

After peptide bond formation, the P site contains an empty tRNA and the A site contains a *peptidyl-tRNA* (the tRNA to which the growing polypeptide chain is attached). The mRNA now advances relative to the small subunit, bringing the next codon into proper position for translation. During this process of **translocation,** the peptidyl-tRNA moves from the A site to the P site and the empty tRNA is ejected from the ribosome (see Figure 11-30 *step 3*). Translocation requires an elongation factor called *EF-G* (*eEF-2* in eukaryotes), and is accompanied by the hydrolysis of GTP. The binding of EF-G to the ribosome can be inhibited by exposing ribosomes to antibodies directed against proteins L7 and L12, indicating that these related ribosomal proteins are components of the binding site for EF-G.

The net effect of translocation is to bring the next mRNA codon into position for translation during the following elongation cycle. The only difference between succeeding elongation cycles and the first cycle is that an initiator tRNA occupies the P site at the beginning of the first elongation cycle, and peptidyl-tRNA occupies the P site at the beginning of all subsequent cycles. As elongation proceeds, the newly forming polypeptide chain gradually gets longer and eventually protrudes from the surface of the ribosome. Digesting ribosomes with proteases has revealed that a segment of the growing polypeptide chain about 30–35 amino acids long is protected from degradation, indicating that a short stretch of the newly forming polypeptide chain is buried within the ribosome interior.

We have now seen that the elongation phase of protein synthesis consists of three distinct steps: (1) the binding of an aminoacyl-tRNA to its appropriate codon in mRNA, (2) the formation of a peptide bond between the amino acid carried by this tRNA and the previously incorporated amino acid, and (3) translocation of the mRNA along the ribosome so that the next codon is brought into position for transla-

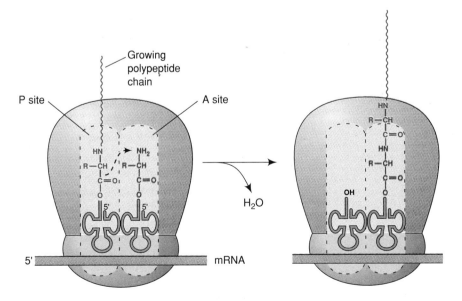

Figure 11-32 *The Peptidyl Transferase Reaction* *In this reaction a peptide bond is formed between the incoming amino acid at the A site and the growing polypeptide chain at the P site. The large subunit rRNA is thought to catalyze this process.*

tion. This cycle is repeated for each amino acid that is incorporated into the newly forming polypeptide. The addition of each amino acid utilizes energy derived from at least five high-energy phosphate bonds: Two high-energy phosphate bonds are consumed when ATP is hydrolyzed to AMP during the formation of aminoacyl-tRNA (see Figure 3-29), and at least three molecules of GTP are hydrolyzed to GDP during each elongation cycle, two in association with EF-Tu during aminoacyl-tRNA binding and one in association with EF-G during translocation.

Protein Synthesis Is Terminated by Release Factors That Recognize Stop Codons

The elongation cycle continues to add amino acids to the growing polypeptide chain until a stop codon (UAA, UAG, or UGA) is reached. Amino acid incorporation ceases at this point because no normally occurring tRNAs contain an anticodon capable of binding to these triplets. In prokaryotes, three protein **release factors** are involved in recognizing stop codons. Factor *RF-1* binds to UAA and UAG, *RF-2* binds to UAA and UGA, and *RF-3* stimulates the activity of both RF-1 and RF-2. In eukaryotes a single release factor appears to recognize all three stop codons. The binding of a release factor to a stop codon located at the A site causes the linkage between the polypeptide chain and the tRNA to be hydrolyzed, releasing the completed polypeptide (Figure 11-33). The ribosomal subunits and tRNA then dissociate from the mRNA for use in subsequent rounds of protein synthesis.

Some insights into the termination mechanism have emerged from the study of mutations in which a codon for a normal amino acid is mutated to a stop codon. For example, mutation of a single DNA base

pair can yield an mRNA in which the codon CAG, which codes for glutamine, is converted to the stop codon UAG. The presence of a stop codon in a position that normally codes for an amino acid will cause the polypeptide chain encoded by the mutated gene to be released prior to completion. This harmful effect can be overcome, however, by the independent occurrence of a **suppressor mutation** that alters the anticodon sequence of a tRNA molecule. An example of a suppressor mutation would be a mutation in the anticodon of a tyrosine tRNA from 3'—AUG—5' to 3'—AUC—5'. Instead of pairing with the normal tyrosine codons UAC and UAU, this mutated tRNA anticodon pairs with the stop codon UAG. In the presence of such a tRNA, the codon UAG specifies the incorporation of tyrosine into the growing polypeptide chain where premature termination would otherwise have occurred. Although the resulting polypeptide will contain a tyrosine residue where a glutamine would normally have resided, such a product has a better chance of being functionally active than a prematurely terminated peptide fragment.

Since suppressor mutations permit a normal terminator codon to be read as if it coded for an amino acid, how is proper termination signaled in cells that produce suppressor tRNAs? The synthesis of most polypeptides is terminated normally in such cells, indicating that a stop codon located in its proper place at the end of a coding sequence still triggers termination while the same codon in an internal location does not. Hence normal termination must involve the recognition of a special sequence or three-dimensional configuration at the end of the mRNA coding sequence in addition to the stop codon itself.

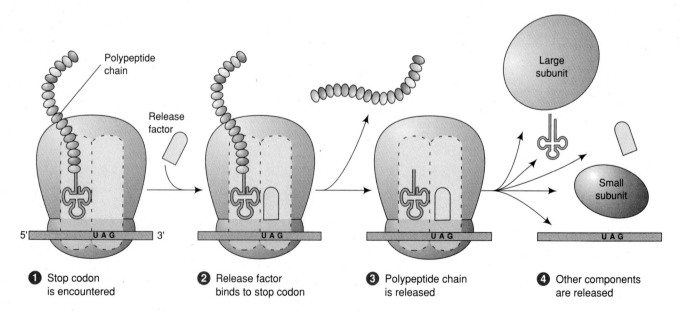

① Stop codon is encountered

② Release factor binds to stop codon

③ Polypeptide chain is released

④ Other components are released

Figure 11-33 *The Termination of Protein Synthesis* *Prokaryotic cells employ several different release factors that vary with the stop codon, whereas eukaryotic cells can utilize the same release factor for all three stop codons.*

Polysomes Consist of Clusters of Ribosomes Simultaneously Translating the Same Messenger RNA

The preceding discussion of the initiation, elongation, and termination phases of protein synthesis has focused upon the interaction of a single ribosome with a single molecule of mRNA. In the early 1960s it was independently discovered in the laboratories of Hans Noll and Alexander Rich that ribosomes function in clusters. The studies carried out in Rich's laboratory utilized immature red blood cells that had been briefly incubated with radioactive amino acids to label newly synthesized proteins. After incubation, the cells were gently broken open and centrifuged at moderate speed to remove nuclei and mitochondria. The resulting supernatant, which contained the cell's ribosomes, was then fractionated by moving-zone centrifugation. Two major peaks of material were observed: a sharp peak sedimenting at 80S and a broader peak sedimenting at 170S. Since eukaryotic ribosomes sediment at 80S, it was expected that newly synthesized radioactive protein would be associated with the 80S peak. Surprisingly, radioactive protein was instead detected mainly in the 170S fraction (Figure 11-34).

Electron microscopic examination revealed that the 80S material consisted of single ribosomes, whereas the 170S material contained clusters of four to six ribosomes (Figure 11-35). In high-magnification micrographs, each ribosome cluster appeared to be held together by a thin strand whose length suggested that it corresponded to a molecule of messenger RNA. If this were the case, then digesting ribosome clusters with ribonuclease under conditions known to degrade mRNA should break the clusters apart, converting them into single 80S ribosomes. As predicted, the 170S peak disappeared and the 80S peak increased in size when such experiments were carried out (see Figure 11-34, bottom).

The preceding experiments suggest that several ribosomes simultaneously translate the same molecule of mRNA, thereby creating a ribosome cluster known as a **polysome.** According to the polysome model, which is illustrated in Figure 11-36, translation begins when a pair of ribosomal subunits forms an initiation complex at the 5' end of an mRNA molecule. As elongation of the new polypeptide proceeds, the mRNA advances along the ribosome and the 5' end of the message soon becomes free again. Another pair of ribosomal subunits then binds to the 5' end of the mRNA and begins a second round of protein synthesis. This process is repeated until ribosomes are spaced along the entire length of the message. Such a mechanism increases the overall efficiency of protein synthesis by permitting multiple ribosomes to translate the same mRNA simultaneously.

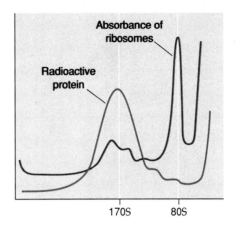

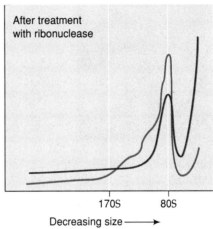

Figure 11-34 *Experiment Demonstrating That Protein Synthesis Occurs on Polysomes* *Immature red blood cells were incubated with radioactive amino acids to label newly forming polypeptide chains. Ribosomes were then isolated and analyzed by moving-zone centrifugation. Absorbance measurements were employed to detect the presence of ribosomes, and measurements of radioactivity were used to detect the presence of newly forming polypeptide chains. (Top) Newly forming radioactive polypeptides are associated with ribosome clusters (polysomes) that sediment at 170S, rather than with individual ribosomes, which sediment at 80S. (Bottom) When isolated ribosomes are treated with ribonuclease prior to moving-zone centrifugation, mRNA is degraded and the ribosome clusters that comprise the 170S peak are converted to individual 80S ribosomes.*

An obvious prediction of the polysome model is that the number of ribosomes present in any given polysome should be directly related to the length of the mRNA being translated. The red blood cells studied in Rich's laboratory are specialized for the production of one major type of protein (hemoglobin), so the single 170S polysome peak containing clusters of four to six ribosomes reflects a relatively simple situation in which one kind of mRNA is being translated. Cells that synthesize a broader spectrum of proteins typically contain an mRNA population that is more heterogeneous in length, so a wider range of

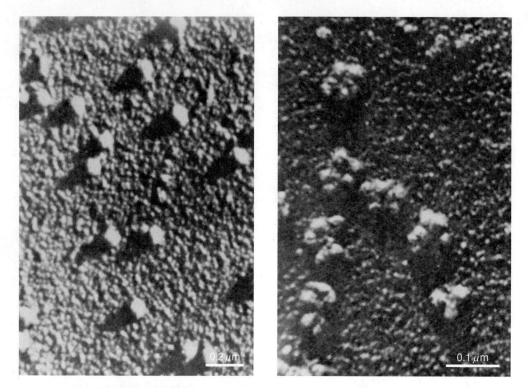

Figure 11-35 *Electron Microscopic Evidence for the Existence of Polysomes* *Ribosomes isolated from immature red blood cells were fractionated by moving-zone centrifugation and examined by electron microscopy. (Left) The material sedimenting at 80S consists of single ribosomes. (Right) The material sedimenting at 170S consists of ribosome clusters (polysomes). Courtesy of A. Rich.*

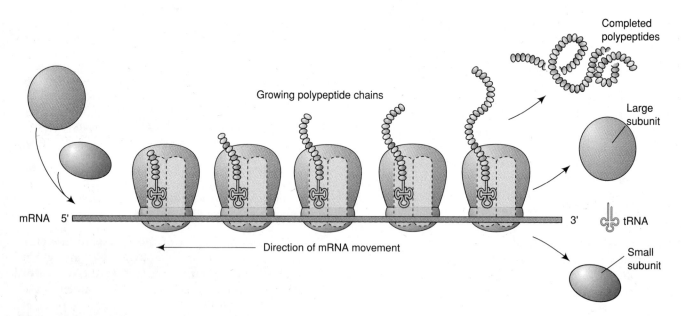

Figure 11-36 *Model Showing How a Polysome Is Created* *As an mRNA molecule advances along a ribosome during translation, the 5' end of the mRNA soon becomes free again, thereby allowing another set of ribosomal subunits to bind to the mRNA and initiate translation. This process allows multiple ribosomes to translate the same mRNA molecule simultaneously.*

polysome sizes is observed (Figure 11-37). In cells that manufacture exceptionally large proteins, polysomes containing 50 or more ribosomes have been detected (Figure 11-38).

Although the polysome concept was initially based on data derived from subcellular fractionation studies, subsequent investigations have revealed that polysomes can also be detected in electron micrographs of tissues that have not been subjected to fractionation procedures. Polysomes visualized in this way often exhibit a considerable degree of order, occurring in highly organized configurations such as spirals or helices (Figure 11-39). In prokaryotic cells, polysomes are formed on mRNA molecules that are still in the process of being transcribed from DNA. As transcription proceeds so does translation, continually freeing the 5' end of the newly forming mRNA for the binding of additional ribosomes (Figure 11-40). This coupling of transcription and translation does not occur in eukaryotic cells because the site of transcription (the nucleus) is separated from the site of translation (the cytoplasm) by the membranes of the nuclear envelope.

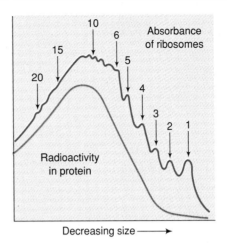

Figure 11-37 *Heterogeneity in the Size of Rat Liver Polysomes* *Rat liver ribosomes were analyzed by moving-zone centrifugation after administration of radioactive amino acids to label newly synthesized protein. The multiple peaks represent polysomes of varying sizes. The number above each peak represents the number of ribosomes present per polysome. Note that most of the newly forming radioactive protein is associated with polysomes containing ten or more ribosomes.*

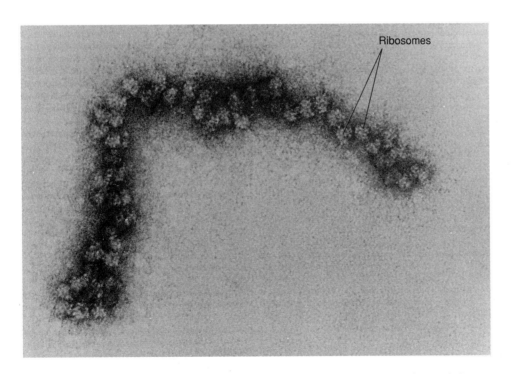

Figure 11-38 *Electron Micrograph of a Giant Polysome Viewed by Negative Staining*
This polysome isolated from chick embryo cells contains more than 50 ribosomes translating the same mRNA molecule. Courtesy of A. Rich.

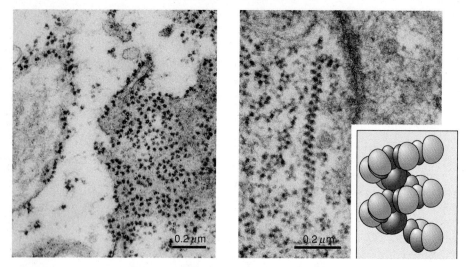

Figure 11-39 *Examples of Some Highly Organized Polysome Configurations* (Left)
*Electron micrograph of radish root cell showing polysomes arranged in spiral patterns.
(Right) A rat intestinal epithelial cell containing polysomes constructed from helically
packed ribosomes. (Inset) A model illustrating how helically packed ribosomes are thought
to be arranged. Courtesy of O. Behnke.*

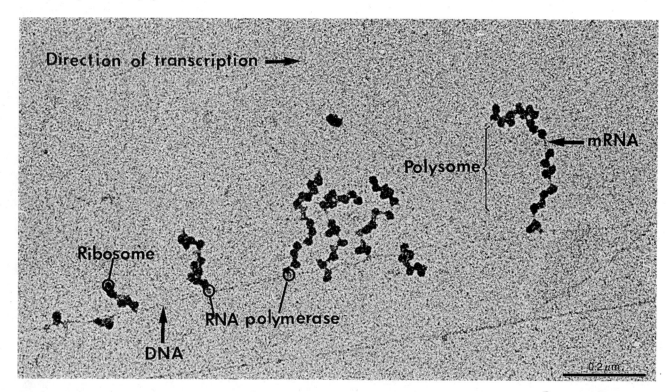

Figure 11-40 *Electron Micrograph Showing the Coupling between Transcription and Translation in Bacterial Cells*
*In this DNA molecule isolated from E. coli, transcription is proceeding from left to right. Before transcription is completed,
ribosomes become associated with the newly forming mRNA molecule and begin translation. Courtesy of O. L. Miller, Jr.*

Inhibitors of Protein Synthesis Are Useful Tools for Studying Translation

Inhibitors of protein synthesis act in a variety of ways
(Table 11-3), and so have provided a wealth of information concerning the mechanisms involved in the translation process. For example, the main evidence for the

existence of separate A and P sites on the ribosome has
come from experiments involving **puromycin**, an inhibitor of protein synthesis whose structure mimics the
end of a tRNA molecule to which an amino acid has
bound. Because of this resemblance, puromycin can
bind to ribosomes in place of an aminoacyl-tRNA and
become incorporated into the growing polypeptide

Table 11-3   Some Commonly Used Inhibitors of Protein Synthesis

Inhibitor	Subunit or Factor Affected	Step(s) Blocked	Reaction Affected	Cell Type Affected
Aurintricarboxylic acid	30S/40S	Initiation	Binding of messenger RNA	Prokaryotes/eukaryotes
Chloramphenicol	50S	Elongation	Peptide-bond formation	Prokaryotes
Colicin E3	30S	Initiation Elongation	Binding of messenger RNA Binding of aminoacyl-tRNA	Prokaryotes
Cycloheximide	60S	Initiation Elongation	Binding of initiator-tRNA Translocation (tRNA release from P site)	Eukaryotes
Diphtheria toxin	eEF-2	Elongation	Translocation	Eukaryotes
Erythromycin	50S	Initiation	Formation of initiation complex	Prokaryotes
Fusidic acid	EF-G/eEF-2	Elongation	Translocation	Prokaryotes/eukaryotes
Kasugamycin	30S	Initiation	Binding of initiator-tRNA	Prokaryotes
Puromycin	50S/60S	Elongation	Peptide-bond formation (triggers chain release)	Prokaryotes/eukaryotes
Spectinomycin	30S	Elongation	Translocation	Prokaryotes
Streptomycin	30S	Initiation Elongation	Binding of initiator tRNA Binding of aminoacyl-tRNA (induces misreading)	Prokaryotes
Tetracycline	30S	Elongation Termination	Binding of aminoacyl-tRNA Binding of RF-1 and RF-2	Prokaryotes

chain. After puromycin has become incorporated, it cannot participate in any subsequent steps; protein synthesis therefore ceases and the polypeptide chain containing puromycin at its C-terminal end is prematurely released from the ribosome.

Figure 11-41 illustrates how this property of puromycin was used to demonstrate the existence of separate A and P sites on the ribosome. In one set of experiments, ribosomes containing an initiator tRNA bound to the P site were incubated with puromycin; under such conditions, puromycin binds to the empty A site and reacts with the initiator tRNA. In a second set of experiments, ribosomes containing an initiator tRNA bound to the P site were first incubated with aminoacyl-tRNAs under conditions that permit a peptide bond to be formed. The result is the formation of a dipeptidyl-tRNA bound to the A site. When puromycin is then added, it cannot bind to the ribosome or react with the dipeptidyl-tRNA. Such results indicate that the dipeptidyl-tRNA is bound to a different ribosome binding site (the A site) than was the original initiator tRNA (the P site). Only after movement of the dipeptidyl-tRNA back to the P site, which requires GTP and elongation factor EF-G, can puromycin again bind to the A site and react with the growing polypeptide chain.

Streptomycin is another antibiotic whose use has shed light on the mechanism of protein synthesis. Streptomycin is an oligosaccharide that interferes with the binding of both initiating and noninitiating tRNAs to the ribosome, thereby inhibiting initiation as well as elongation. In addition, streptomycin causes misreading of the genetic code. For example, when streptomycin is added to ribosomes that are translating the artificial message poly U, the amino acids isoleucine, leucine, serine, tyrosine, and phenylalanine are all incorporated. Normally, poly U codes only for phenylalanine. Earlier in the chapter we learned that mutations in protein S12 render ribosomes resistant to the misreading effects of streptomycin. It has therefore been proposed that protein S12 participates in the error-checking mechanism that normally ensures that the proper aminoacyl-tRNA is bound to the A site of the ribosome.

Some inhibitors of protein synthesis, such as streptomycin, tetracycline, erythromycin, and chloramphenicol, can be used to combat bacterial infections in humans because they selectively inhibit protein synthesis in prokaryotic cells. A few inhibitors, such as cycloheximide and diphtheria toxin, are selective for eukaryotic protein synthesis. **Diphtheria toxin** is of special interest because of its association with the infectious disease diphtheria. Produced by the bacterium *Corynebacterium*

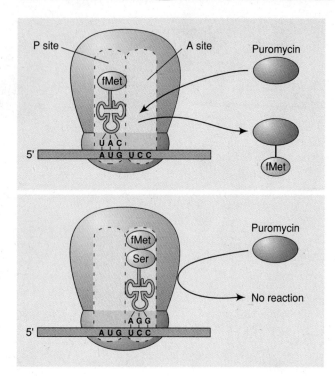

Figure 11-41 *Use of Puromycin to Demonstrate the Existence of Separate A and P Sites* (Top) *Puromycin resembles an aminoacyl-tRNA in chemical structure. As a result, it is capable of binding to the A site and reacting with an amino acid or growing polypeptide chain bound to a tRNA located at the P site. (Bottom) After peptide bond formation has occurred (but before translocation), the growing polypeptide chain occupies the A site. At this stage, puromycin cannot bind to the ribosome because the A site is occupied.*

diphtheriae, diphtheria toxin is a protein that is lethal to nonimmunized humans in doses as small as a few micrograms. It inhibits the translocation step of protein synthesis by catalyzing a reaction in which an *ADP ribose* group derived from an NAD molecule is covalently joined to elongation factor eEF-2. The resulting ADP-ribosylated eEF-2 is unable to participate in translocation. In bacteria the comparable elongation factor (EF-G) is not susceptible to ADP-ribosylation by diphtheria toxin.

Newly Synthesized Polypeptide Chains Are Often Processed by Posttranslational Cleavage Reactions

After an mRNA molecule has been translated into a completed polypeptide chain, the amino acid sequence of the chain may still need to be altered to produce a final protein product. Peptide bond cleavages are often required to remove extra amino acids from the ends of newly formed polypeptides, and to break chains into smaller fragments. Both types of cleavage occur during

the production of insulin, whose mRNA codes for a precursor polypeptide known as *preproinsulin.* The N-terminus of preproinsulin contains a signal sequence that directs the insertion of the growing polypeptide into the lumen of the endoplasmic reticulum (page 265). After removal of the signal sequence, the resulting product, called *proinsulin,* is converted into insulin by a second cleavage reaction that removes a 22 amino acid fragment from the internal region of the molecule, leaving two separate chains held together by disulfide bonds (Figure 11-42, *left*).

Cleavage of a single polypeptide chain may also generate more than one protein product. In such cases, the initial polypeptide generated by mRNA translation is called a **polyprotein.** A particularly striking example of a polyprotein occurs in the brain of vertebrates, where a polypeptide called *pro-opiocortin* serves as a precursor for the hormones corticotropin (ACTH), melanocyte-stimulating hormone (MSH), and β-endorphin (see Figure 11-42, *middle*). In some cases, a more elaborate *protein splicing* mechanism is employed that cleaves peptide fragments from separate regions of a polypeptide chain and joins them together to form a final protein product (see Figure 11-42, *right*).

The Folding of Newly Synthesized Proteins Is Often Facilitated by Enzymes and Chaperones

In this section of the chapter we have focused on the mechanisms involved in manufacturing a polypeptide chain containing the proper sequence of amino acids. We learned in Chapter 1, however, that protein molecules are more than just a linear sequence of amino acids; they must be folded into the proper three-dimensional conformation before they can function properly. Since mRNA translation only determines the linear sequence of amino acids, how do proteins become properly folded? A protein's final three-dimensional shape is ultimately determined by the amino acid sequence of its constituent polypeptide chains, but in order for proper folding to occur, the assistance of two other categories of proteins is often required.

One group of proteins consists of enzymes that catalyze steps in the protein-folding process that might otherwise proceed too slowly. For example, the enzyme *protein disulfide isomerase* catalyzes the formation and breakage of disulfide bonds between cysteine residues located in different parts of a polypeptide chain. This process does not determine the final pattern of disulfide bonds; it simply speeds up the process by which the various combinations of bonds are tested until the most stable arrangement is found. Another enzyme implicated in the folding process, *peptidyl prolyl isomerase,* catalyzes rotation around specific peptide bonds and is

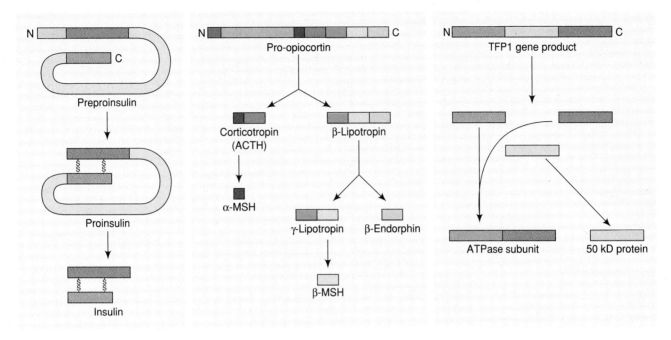

Figure 11-42 *Examples of Posttranslational Cleavage Reactions* *The letters N and C stand for the N-terminus and C-terminus respectively. (Left) The two-step conversion of preproinsulin into insulin. (Middle) The polyprotein pro-opiocortin gives rise to several peptide hormones. (Right) An example of protein splicing.*

thought to promote steps involved in the initial folding and/or rearrangement of protein structures.

The other group of proteins that facilitate polypeptide folding are called **chaperones.** Rather than acting as catalysts, chaperones bind to and stabilize partially folded intermediates during the folding process. Because the synthesis of many chaperones is induced by elevated temperature or other types of stress, they are commonly referred to as *heat shock proteins* or *stress proteins*. However, most of these stress-induced proteins are also produced in the absence of stress and are essential for cell viability under normal conditions.

TRANSLATIONAL AND POSTTRANSLATIONAL CONTROL

In the preceding chapter we described the kinds of control mechanisms that are used by cells to regulate the production of functional messenger RNAs. Once a given messenger RNA has been produced, additional control mechanisms influence the extent to which the message is translated into its corresponding polypeptide product. These *translational control* mechanisms include effects on ribosomes, protein synthesis factors, messenger RNA, and transfer RNA. Once a molecule of messenger RNA has been successfully translated, the resulting protein product is subject to further regulation by various *post-translational control* mechanisms. In the following sections, we will examine some of the more common types of translational and posttranslational control.

Ribosome Production Is Controlled in Several Different Ways

Since translation requires ribosomes, the rate at which ribosomes are produced exerts a profound influence on a cell's ability to translate its mRNAs. We have already learned that oocytes increase their capacity for manufacturing ribosomes by amplifying the number of genes coding for ribosomal RNA (page 477). In cells whose growth is stimulated by hormones, the stimulating hormone often enhances ribosome production by increasing the rate at which the genes coding for rRNA are transcribed. Control can also be exerted at the level of pre-rRNA processing. For example, more than half of the 45S pre-rRNA molecules produced by transcription of the rRNA genes in lymphocytes are normally destroyed rather than being processed into 18S and 28S RNAs; but in lymphocytes that are stimulated to grow and divide by the addition of growth factors, a much larger percentage of the 45S pre-rRNA is processed into mature ribosomal RNA.

In prokaryotic cells, the rate of rRNA synthesis is regulated by the number of active ribosomes. The mechanism that underlies this type of control has been investigated in bacterial cells grown in the absence of an adequate supply of amino acids. Under such conditions the rate of protein synthesis is sharply curtailed, followed by the cessation of rRNA and tRNA formation. During this process, called the **stringent response,** bacteria accumulate two unusual nucleotides called *guanosine tetraphosphate* (**ppGpp**) and *guanosine pentaphosphate* (**pppGpp**). Mutant bacteria unable to regulate rRNA and tRNA synthesis in response to amino acid starvation do not accumulate either ppGpp or pppGpp, suggesting that these two nucleotides are responsible for the inhibition of rRNA and tRNA synthesis.

The formation of ppGpp and pppGpp is mediated by a reaction in which two phosphate groups are transferred from ATP to the 3' hydroxyl group of either GDP or GTP. This reaction is catalyzed by a protein called the *stringency factor,* which is activated by "stalled" ribosomes containing bound mRNA and an A site occupied by a tRNA molecule lacking an attached amino acid. Transfer RNA molecules lacking attached amino acids are prevalent only when amino acids are scarce; hence the conditions that favor the formation of ppGpp and ppGppp only prevail when insufficient amino acids are available for carrying out protein synthesis. Under these conditions ppGpp and ppGppp inhibit the synthesis of rRNA, providing cells with a feedback mechanism for regulating the rate of ribosome formation in response to changes in the ability of the cell to carry out protein synthesis.

Translational Repressors Inhibit Protein Synthesis by Binding to Specific Messenger RNAs

We learned in Chapter 10 that gene transcription is often regulated by proteins that bind to specific DNA sequences. In a comparable fashion, proteins that bind to specific mRNA sites are employed in the control of translation. One striking example involves the expression of the genes that code for ribosomal proteins in *E. coli.* These ribosomal protein genes are distributed among at least ten different operons, each having its own promoter. Since all of the ribosomal proteins are required in roughly equal amounts during the assembly of ribosomes, bacterial cells have evolved a mechanism for coordinating the rate at which the various ribosomal proteins are produced.

Experiments carried out in the laboratory of Masayasu Nomura have shown that this coordination is achieved at the translational level. Most ribosomal protein operons code for at least one protein that functions both as a ribosomal protein and as a **translational repressor** that inhibits translation of the mRNA transcribed from the operon in question (Figure 11-43). The mRNA binding site for these translational repressors typically lies upstream from the AUG start codon and exhibits a stem-and-loop structure. Normally translational repressor molecules do not bind to this mRNA site be-

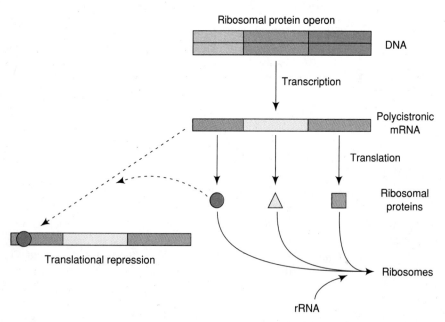

Figure 11-43 *Translational Repression by Ribosomal Proteins in E. Coli*   *Most ribosomal operons code for the synthesis of at least one ribosomal protein that also acts as a translational repressor, inhibiting translation of the mRNA molecule transcribed from that operon. When the concentration of ribosomal RNA is low, such ribosomal proteins bind to mRNA instead, repressing its translation.*

cause they bind more strongly to ribosomal RNA and so are incorporated into newly forming ribosomes. However, when no ribosomal RNA is available for assembling new ribosomes, any excess translational repressor molecules bind instead to the mRNA from which they were translated, inhibiting the further synthesis of ribosomal proteins.

Evidence for translational repression also exists in eukaryotic cells, though not among the mRNAs that code for ribosomal proteins. (The synthesis of ribosomal proteins is not coordinated by translational control in eukaryotes; if a particular ribosomal protein is produced in excessive amounts, the excess is simply degraded.) The best evidence for the existence of a specific translational repressor in eukaryotes involves the synthesis of *ferritin*, an iron-storage protein whose synthesis is stimulated in the presence of iron. The 5' untranslated leader sequence in ferritin mRNA contains a 28 nucleotide segment, called the **iron-responsive element (IRE),** which contains a single stem-and-loop structure that is required for stimulation of ferritin synthesis by iron. If this sequence is experimentally inserted into a gene whose expression is not normally regulated by iron, translation of the resulting mRNA becomes iron sensitive. The mechanism of translational control involves an **IRE-binding protein** that binds to the IRE sequence in ferritin mRNA and blocks the ability of the mRNA to form an initiation complex with ribosomal subunits (Figure 11-44, *top*). The IRE-binding protein remains active as long as iron is absent. In the presence of iron the IRE-binding protein is inactivated, thereby allowing the ferritin mRNA to be translated into ferritin.

In certain kinds of cells, a large fraction of the mRNA population can be translationally repressed. A dramatic example occurs in unfertilized egg cells, where most of the cytoplasmic mRNA is complexed with protein to form **messenger ribonucleoprotein complexes (mRNPs).** Messenger RNA in this form is not associated with polysomes, nor is it readily translated when added to cell-free protein-synthesizing systems. It has therefore been concluded that unfertilized eggs contain inactive mRNA molecules that are "masked" by the presence of bound protein. When egg cells are fertilized, a rapid increase in the rate of protein synthesis occurs. This increase is not prevented by adding actinomycin D, which blocks the synthesis of new mRNAs. Hence the increase in the rate of protein synthesis must result from the "unmasking" of previously inactivated mRNP complexes.

Translation Can Be Regulated by Controlling Messenger RNA Life Span

The average life span of mRNA molecules varies widely. Most bacterial mRNAs are degraded rapidly, exhibiting half-lives in the range of a few minutes. In contrast, the

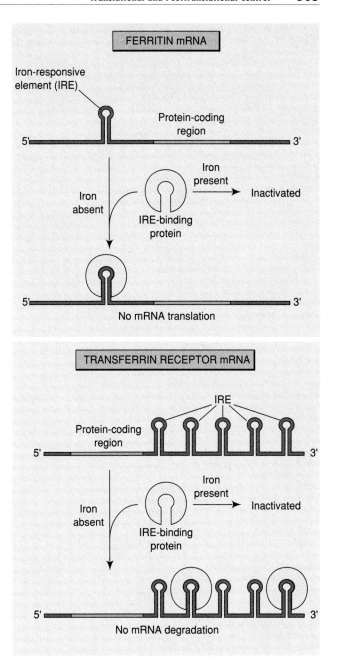

Figure 11-44 *Regulation of Gene Expression by Iron-Responsive Elements in Messenger RNA* (Top) *The binding of IRE-binding protein to the single IRE in ferritin mRNA causes translation of the message to be repressed.* (Bottom) *The binding of IRE-binding proteins to IRE's located in transferrin receptor mRNA causes the message to be stabilized against degradation.*

half-lives of eukaryotic mRNAs typically range from several hours to a few days. Since the rate at which a given mRNA is degraded determines the length of time it is available for translation, alterations in mRNA life span influence the amount of protein a given message can produce.

Messenger RNA degradation is subject to regulation by a variety of factors. For example, the life span of ovalbumin mRNA in the chick oviduct is influenced by the hormone estrogen. A decline in estrogen concentration signals that ovalbumin is no longer needed. In response, the rate of ovalbumin mRNA breakdown increases by an order of magnitude, triggering a dramatic drop in the rate of ovalbumin synthesis. Other hormones known to regulate mRNA life span include growth hormone, which slows the degradation of prolactin mRNA, and cortisone, which slows the degradation of growth hormone mRNA.

Another factor that influences mRNA life span is iron, a metal whose role in regulating the translation of ferritin mRNA was discussed in the preceding section. The uptake of iron into mammalian cells is mediated by a plasma membrane receptor protein known as the **transferrin receptor.** Synthesis of the transferrin receptor is stimulated when iron is scarce and inhibited when iron is abundant. The 3' untranslated region of the transferrin receptor mRNA contains five sets of iron-responsive elements (IREs) that form stem-and-loop structures similar to the single IRE structure observed at the 5' end of ferritin mRNA (see Figure 11-44, *bottom*). In the absence of iron, IRE-binding proteins protect the transferrin receptor mRNA from degradation by binding to its 5' untranslated IRE sequences. When iron is present, the IRE-binding protein is inactivated and so the

mRNA coding for the transferrin receptor is no longer protected from degradation.

The poly-A tail is another factor that influences the life span of eukaryotic mRNA molecules. The most direct evidence implicating poly A has emerged from studies involving mRNA molecules whose poly-A tails have been artificially removed. In one of the more dramatic experiments of this kind, Georges Huez monitored the translation of globin mRNA injected into the cytoplasm of *Xenopus* oocytes. Immediately after injection, globin mRNA whose poly A had been removed was found to be translated at about the same rate as globin mRNA that contained poly A. Within several hours, however, the translation rate of globin mRNA lacking poly A began to decline relative to that of normal globin mRNA (Figure 11-45). Such results indicate that for some messages, the length of the poly-A tail influences mRNA stability. The stabilizing effect of the poly-A tail on mRNA is mediated by a *poly-A binding protein* that binds to the poly-A tail and protects it from degradation.

Antisense RNA Molecules Inhibit the Translation of Complementary Messenger RNAs

Bacterial cells and their viruses contain a class of regulatory genes coding for inhibitory RNA molecules that are complementary to specific mRNAs, allowing the inhibitory RNA to base-pair with its corresponding mRNA. The term **antisense RNA** is used to refer to such regulatory RNA molecules because they are complementary to normal messenger RNAs. An antisense RNA disrupts the function of its target mRNA by preventing it from binding to ribosomes, by blocking its initiation codon, or by making the mRNA more susceptible to degradation.

Artificially produced antisense RNAs are a powerful tool for selectively inhibiting the expression of specific mRNAs. An antisense RNA can be created experimentally by inserting a cloned gene into a specially designed DNA molecule called an *expression vector* (page 104), which allows a cloned gene to be expressed after being introduced into an appropriate host cell. If the cloned gene is inserted *backwards* in the expression vector rather than in its proper orientation, the DNA strand that lies opposite from the normal template strand is transcribed into RNA. The resulting RNA product is therefore an antisense RNA that binds to and inhibits the expression of the mRNA normally produced by the gene. This approach for artificially inhibiting the function of specific mRNAs has been employed for a variety of purposes. For example, a gene coding for the enzyme that breaks down plant cell walls has been inserted backwards into an expression vector and introduced into tomato plants. Inside the cells of the tomato, the vector produces an antisense RNA that in-

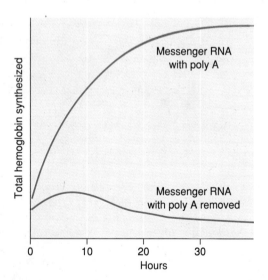

Figure 11-45 *Effect of the Poly-A Tail on mRNA Life Span*
In this experiment, the rate of hemoglobin synthesis was compared in Xenopus *oocytes injected with either normal globin mRNA or globin mRNA whose poly-A tails had been removed. Less hemoglobin is produced by the cells injected with globin mRNA lacking poly-A tails because the absence of the poly A makes the mRNA more susceptible to degradation.*

hibits the enzyme responsible for cell wall breakdown. Such tomatoes ripen more slowly, allowing them to be transported more readily and giving them a longer shelf life without spoilage.

Translation Rates Can Be Increased or Decreased by the Phosphorylation of Protein Synthesis Factors

The use of protein phosphorylation to regulate the rate of mRNA translation was first demonstrated in developing red blood cells, where globin chains are the main product of translation. Protein synthesis in red cells requires the presence of *heme*, an iron-containing prosthetic group that binds to globin chains to form hemoglobin. In cell-free protein-synthesizing systems, the absence of heme causes the rate of protein synthesis to fall; this inhibition can be overcome by adding an excess of initiation factors, indicating that the absence of heme depresses the initiation step of protein synthesis. Heme acts by interfering with the *heme-regulated inhibitor (HRI)*, a protein kinase that catalyzes the phosphorylation and resulting inactivation of eukaryotic initiation factor *eIF-2*. The binding of heme prevents HRI from inactivating eIF-2 by phosphorylation and thus allows protein synthesis to proceed (Figure 11-46).

The effects of protein phosphorylation on translation rates are not restricted to eIF-2. Eukaryotic initiation factor *eIF-4F*, which binds to the 5' mRNA cap, is also regulated by phosphorylation, although in this case phosphorylation activates rather than inhibits the factor. An example of this type of control occurs in cells infected by adenovirus, which inhibits translation in the host cells by triggering the dephosphorylation of eIF-4F. Other initiation and elongation factors involved in eukaryotic protein synthesis are also subject to phosphory-

lation, as are several of the aminoacyl-tRNA synthetases that join amino acids to their respective tRNAs. In most of these cases, the effect of phosphorylation on protein synthesis and the nature of the factors that regulate phosphorylation are yet to be clearly established.

The Availability of Transfer RNAs Can Influence the Translation of Messenger RNA

The population of tRNA molecules found in the cytoplasm varies significantly among cell types, thus raising the question of whether tRNAs play any role in regulating protein synthesis. Some relevant evidence has emerged from studies involving cell-free protein synthesis, where the effects of tRNA availability on mRNA translation can be directly tested. One set of tRNAs whose effects have been examined are those encoded by *bacteriophage T4*. When tRNAs produced by the viral DNA are added to a cell-free protein-synthesizing system derived from bacterial cells, translation of viral mRNA is stimulated to a greater extent than translation of bacterial mRNA. The most straightforward interpretation is that the viral and bacterial mRNAs require a somewhat different spectrum of tRNAs for optimal translation.

Selective effects of tRNA populations on protein synthesis have also been demonstrated during the cell-free translation of globin mRNA. In the presence of high concentrations of tRNA, the α and β chains of hemoglobin are synthesized in roughly equal amounts. But when the tRNA concentration is lowered, the synthesis of α chains is preferentially inhibited (Figure 11-47). This finding suggests that one or more tRNAs is utilized to a greater extent (or even exclusively) by the mRNA coding for α-globin. Hence lowering the concentration

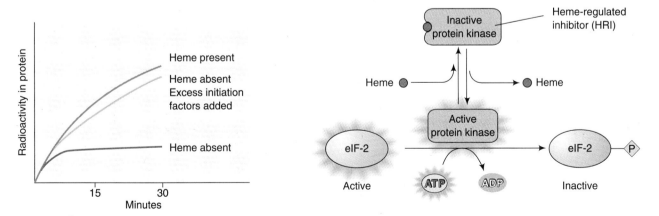

Figure 11-46 *Control of Translation by Phosphorylation of eIF-2 in Eukaryotes* (Left) *Data showing that cell-free protein synthesis is stimulated by heme. The heme requirement can be largely overcome by adding excess initiation factors, which implies that heme acts on the initiation step of protein synthesis.* (Right) *Heme works by blocking the ability of the heme-regulated inhibitor (HRI) to catalyze the phosphorylation and inactivation of eIF-2.*

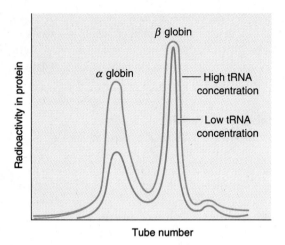

Figure 11-47 *The Effect of tRNA Availability on the Synthesis of α-Globin and β-Globin* *Protein synthesis was carried out in a cell-free system in the presence of varying concentrations of tRNA, and the newly made hemoglobin chains were then separated by ion-exchange chromatography. Note that the synthesis of α-globin chains is preferentially inhibited in the presence of low concentrations of tRNA.*

of such tRNAs preferentially inhibits the translation of α-globin mRNA.

Attenuation Allows Translation Rates to Control the Termination of Transcription

Attenuation is an unusual type of control mechanism because it involves a direct interaction between translation and transcription. The best-understood example of attenuation occurs in the tryptophan (*trp*) operon of *E. coli,* whose regulation has been extensively investigated in the laboratory of Charles Yanofsky. The tryptophan operon, which codes for five enzymes involved in the synthesis of the amino acid tryptophan, contains a typical operator site at which transcription can be blocked by the binding of a tryptophan-repressor complex (page 438). In addition to the operator, another regulatory element called the *attenuator site* occurs in the *trp* operon. Located near the beginning of the first structural gene in the operon, the attenuator site represents a region where transcription is subject to premature termination. Premature termination occurs, however, only if tryptophan is present and the enzymes produced by the tryptophan operon are therefore not needed by the cell.

The mechanism by which the attenuator site senses the presence of tryptophan and triggers premature termination is based on the coupling between transcription and translation that occurs in bacterial cells. As the 5' end of the mRNA encoded by the tryptophan operon is transcribed, it binds to a ribosome and begins to be translated. Two adjacent UGG triplets coding for tryptophan appear early in the mRNA. If the concentration of

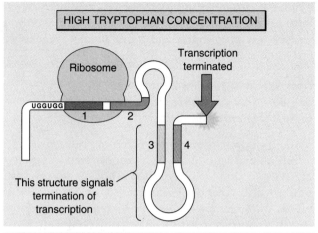

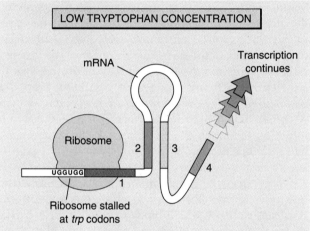

Figure 11-48 *Attenuation in the Tryptophan Operon of E. Coli* (Top) *When the tryptophan concentration is high, translation of the* trp *mRNA proceeds through segment 1. This permits segment 2 to enter the ribosome, which allows segments 3 and 4 to form a stem-and-loop structure that signals RNA polymerase to terminate transcription of the mRNA.* (Bottom) *When the concentration of tryptophan is low, translation of the* trp *mRNA stalls at codons calling for tryptophan. Under these conditions segment 2 forms a stem-and-loop structure with segment 3. Therefore segments 3 and 4 cannot pair, so transcription continues.*

tryptophan in the cell is low, tRNA containing bound tryptophan is not produced in adequate amounts and translation "stalls" when these triplets are encountered. Messenger RNA molecules stalled at this point assume a particular three-dimensional configuration that permits the transcription process to continue (Figure 11-48). But if tryptophan is present in sufficient quantity, stalling does not occur; translation instead proceeds past the two UGG codons, allowing the RNA molecule to assume a configuration that causes transcription to terminate at the attenuator site. The absence of tRNA carrying tryptophan thus serves both to stall translation and, through the resulting effect on mRNA structure, to facilitate transcription of the tryptophan operon by inhibiting premature termination.

A Protein's Biological Activity Is Subject to Control by Posttranslational Alterations in Protein Structure and Conformation

We have now examined many ways in which the flow of genetic information from genes to proteins can be regulated. Once the protein encoded by a particular gene has been produced, the biological activity of the protein is subject to additional types of regulation. In Chapter 2, we learned how allosteric inhibitors and activators bind reversibly to allosteric sites, altering protein conformation in such a way as to inhibit or enhance the molecule's biological activity. We have also seen how the activity of protein molecules can be regulated by the association and dissociation of polypeptide subunits, and by covalent modifications of protein structure such as cleavage of peptide bonds, acetylation, ADP-ribosylation, and phosphorylation (pages 53–55). Since most of these changes are reversible, they provide cells with a wide range of mechanisms for regulating the activity of protein molecules after they have been synthesized.

Differences in Protein Life Span Are an Important Factor in the Control of Gene Expression

The amount of each protein present in a cell is determined by its rate of degradation as well as its rate of synthesis. Hence differences in protein degradation rates can also influence the extent to which the proteins encoded by various genes are expressed in cells. The contributions made by the rates of protein synthesis and degradation in determining the final concentration of any protein are summarized in the equation

$$P = \frac{K_s}{K_d} \tag{11-1}$$

where P is the amount of a given protein present, K_s is the rate order constant for its synthesis, and K_d is the rate order constant for its degradation. Rate constants for degradation are often expressed in the form of a **half-life** ($t_{1/2}$), which corresponds to the amount of time required for half the molecules existing at any moment to be degraded. The value of $t_{1/2}$ can be calculated from K_d by the equation

$$t_{1/2} = \frac{\ln 2}{K_d} \tag{11-2}$$

The half-lives of cellular proteins range from as short as a few minutes to as long as several weeks. Such differences in half-life can exert rather dramatic and surprising effects on the ability of different proteins to respond to changing conditions. A striking example of this phenomenon was uncovered in the laboratory of Robert Schimke in the 1960s. When liver cells are exposed to the steroid hormone cortisone, the concentration of the enzyme *trypto-*

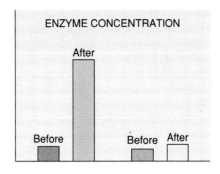

ENZYME CONCENTRATION

After

Before   Before After

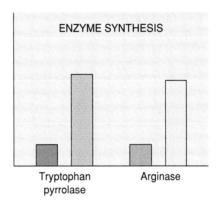

ENZYME SYNTHESIS

Tryptophan   Arginase
pyrrolase

Figure 11-49 *Influence of Protein Half-Life on Enzyme Regulation* *The top graph shows the amount of tryptophan pyrrolase and arginase present in rat liver before and after treatment with cortisone. The bottom graph shows the changes in the rate of enzyme synthesis that occur under the same conditions. Even though the rate of synthesis of these two enzymes is equally stimulated by cortisone, the concentration of tryptophan pyrrolase increases much more dramatically than does the amount of arginase. The explanation for this apparent paradox is that tryptophan pyrrolase has a shorter half-life than arginase.*

phan pyrrolase increases almost tenfold, whereas the enzyme *arginase* increases only slightly (Figure 11-49). At first glance such data would seem to indicate that cortisone selectively stimulates the synthesis of tryptophan pyrrolase. In fact, cortisone stimulates the synthesis of both tryptophan pyrrolase and arginase to about the same extent. The explanation for this apparent paradox is that tryptophan pyrrolase has a much shorter half-life than arginase. Based on the mathematical relationship summarized in Equation 11-1, it can be shown that proteins with short half-lives (large values of K_d) are more sensitive to changes in synthetic rate than proteins with long half-lives. For this reason, enzymes that play an important role in metabolic regulation tend to have short half-lives, allowing their concentration to be rapidly increased or decreased in response to changing conditions.

The factors that determine the rates at which proteins are degraded are not well understood, but it is clear that the process is subject to control. In rats deprived of food, for example, the amount of arginase present in the

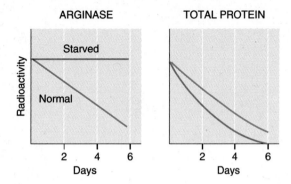

ARGINASE   TOTAL PROTEIN

Figure 11-50 *Regulation of the Degradation Rate of the Liver Enzyme Arginase* *Enzyme degradation rate was measured by injecting rats with a radioactive amino acid to allow proteins to become radioactive, and then measuring the rate at which the radioactivity is lost. The data reveal that the degradation of arginase is inhibited in starved rats, while the degradation rate of total protein is slightly enhanced.*

liver doubles within a few days without any corresponding change in the rate of arginase synthesis. The explanation for the observed increase in arginase is that its degradation rate slows dramatically in starving rats. This effect is highly selective; most proteins are degraded more rapidly, not more slowly, in starving rats (Figure 11-50).

Ubiquitin Targets Selected Proteins for Degradation

The fact that the half-lives of protein molecules are subject to control raises the question of how proteins are selected for degradation. One prominent mechanism for

targeting proteins for destruction is to link them to **ubiquitin,** a small protein that occurs in all eukaryotes. Ubiquitin is conjugated to target proteins by a multistep reaction pathway (Figure 11-51). In an initial ATP-dependent step, ubiquitin is activated by joining it to a *ubiquitin-activating (E1) enzyme.* Activated ubiquitin is then transferred to one of several *ubiquitin-conjugating (E2) enzymes,* which in turn transfer ubiquitin to lysine residues in a target protein. Appropriate target proteins are recognized either by the ubiquitin-conjugating enzyme itself or with the aid of *substrate-recognition (E3) proteins.* Proteins containing bound ubiquitin are subsequently degraded by **proteasomes,** which are large multiprotein complexes that catalyze the ATP-dependent breakdown of a variety of ubiquitin-linked proteins.

How are proteins selected for ubiquitination by the ubiquitin-conjugating system? One signaling factor that influences the rate at which a protein is ubiquitinated is the nature of the amino acid present at its N-terminus. Some N-terminal amino acids cause a protein to be rapidly ubiquitinated and degraded; others make the protein less susceptible to ubiquitination and degradation (Table 11-4). Another signal for ubiquitination is a sequence motif of nine amino acids, called a *destruction box,* which targets proteins containing the motif for destruction. Since eukaryotic cells manufacture a large number of different ubiquitin-conjugating enzymes, it is likely that other signals for ubiquitination are yet to be discovered.

The ubiquitin-dependent pathway is not the only means for degrading proteins. We learned in Chapter 7 that lysosomes take up and degrade cytosolic proteins

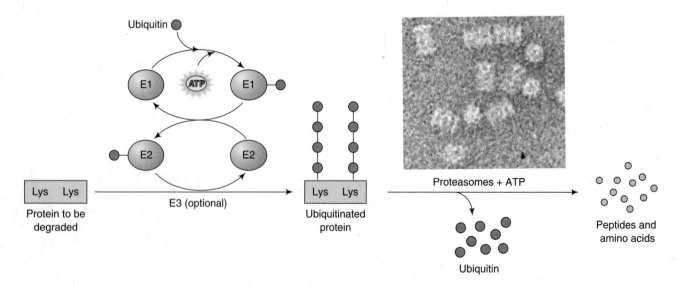

Figure 11-51 *The Ubiquitin-Dependent Pathway for Protein Degradation* *Proteins being targeted for destruction are conjugated to ubiquitin by the sequential action of ubiquitin-activating (E1) and ubiquitin-conjugating (E2) enzymes. Target proteins are recognized either by the E2 enzyme itself, or with the aid of substrate-recognition (E3) proteins. Ubiquitin-linked proteins are subsequently degraded by proteasomes in a reaction that requires ATP. Micrograph courtesy of W. Baumeister.*

Table 11-4 Effect of the N-Terminal Amino Acid on Protein Half-Life

N-Terminal Amino Acid	Protein Half-Life
Arginine	2 min
Asparagine	3 min
Aspartic acid	3 min
Histidine	3 min
Leucine	3 min
Lysine	3 min
Phenylalanine	3 min
Tryptophan	3 min
Glutamine	10 min
Tyrosine	10 min
Glutamic acid	30 min
Isoleucine	30 min
Alanine	> 20 hr
Cysteine	> 20 hr
Glycine	> 20 hr
Methionine	> 20 hr
Proline	> 20 hr
Serine	> 20 hr
Threonine	> 20 hr
Valine	> 20 hr

Note: The same protein (β-galactosidase) was tested with each of the 20 different amino acids artificially added to its N-terminal end. Half-lives were measured in yeast. [Data from A. Varshavsky, 1992. *Cell* 69:725–735.]

by an infolding of the lysosomal membrane, creating small vesicles that are internalized within the lysosome and broken down by the organelle's hydrolytic enzymes. This process of *microautophagy* is relatively nonselective in the proteins it degrades, although under certain conditions specific proteins can be selectively targeted for lysosomal degradation (page 292). In addition, under stressful conditions lysosomes selectively take up and degrade proteins containing a targeting sequence that consists of glutamine flanked on either side by very basic, very acidic, and very hydrophobic amino acids.

PROTEIN TARGETING

Proteins manufactured on cytoplasmic ribosomes are routed to a variety of intracellular and extracellular destinations. In Chapter 7 we learned that a special *signal sequence* occurs at the N-terminus of newly forming polypeptides destined for the endoplasmic reticulum

(ER), Golgi complex, lysosomes, plasma membrane, or secretion from the cell. The signal sequence causes ribosomes making such proteins to become attached to ER membranes and to translocate the newly forming polypeptide into the ER lumen for transport to its proper destination. Polypeptides lacking the signal for ER attachment are synthesized on ribosomes that remain free in the cytosol rather than becoming membrane bound. After release from the ribosome, polypeptides in this class either stay in the cytosol or are routed into the nucleus, mitochondria, chloroplasts, or peroxisomes. In the remainder of the chapter we will discuss the targeting mechanisms that route proteins to these various destinations.

Proteins Are Directed into the Nucleus by a Bipartite Nuclear Localization Sequence Enriched in Basic Amino Acids

In Chapter 10 we learned that the active uptake of proteins into the nucleus is dependent upon the presence of a short stretch of amino acids known as the **nuclear localization sequence (NLS).** The NLS is recognized by cytoplasmic *NLS-binding proteins,* which bind to the sequence and deliver the proteins to the nuclear pore for ATP-dependent transport into the nucleus. Unlike the signal sequences that are utilized to direct polypeptides to the endoplasmic reticulum, the NLS is not removed after the protein enters the nucleus.

Two criteria must be met when attempting to identify nuclear localization sequences. First, is the proposed NLS *necessary* for nuclear uptake of the target protein? This issue is addressed experimentally by deleting or mutating specific amino acids in the suspected NLS and determining whether such changes abolish nuclear uptake of the affected protein. The second question is whether the proposed NLS is *sufficient* to direct a non-nuclear protein into the nucleus. This second issue can be addressed by using recombinant DNA techniques to join the proposed NLS to a protein such as pyruvate kinase, which normally resides in the cytosol. If the sequence in question causes such a cytosolic protein to be transported into the nucleus, then it is clearly functioning as a nuclear localization sequence.

One of the first nuclear localization sequences to be identified using these approaches occurs in the *T antigen,* a protein made by SV40 virus that becomes localized in the nucleus during viral infection of a host cell. The nuclear targeting sequence in the T antigen consists of seven amino acids, five of which are basic:

$$+ \quad + \quad + \quad + \quad +$$
–Pro-Lys-Lys-Lys-Arg-Lys-Val–

Linking the above sequence to pyruvate kinase is sufficient to cause this normally cytosolic protein to become localized in the nucleus. The ability of the sequence to target proteins into the nucleus can be abolished by single amino acid mutations, such as changing the second lysine to threonine or asparagine.

Although the T antigen sequence consists of only a single cluster of basic amino acids, the nuclear localization sequences of other nuclear proteins typically contain two basic clusters separated by a spacer region of about ten amino acids. In proteins containing this *bipartite motif,* neither cluster of basic amino acids is sufficient by itself for nuclear targeting. One of the two basic clusters usually exhibits some resemblance to the SV40 nuclear targeting sequence, but the other does not. This discovery suggests that the T antigen motif simply represents a very efficient version of one of the two basic clusters that normally occur in nuclear proteins. Hence the T antigen basic motif can function alone, whereas other proteins require two basic clusters for nuclear localization.

Because the presence of an NLS is required for the active uptake of nuclear proteins, factors that interfere with the functioning of this sequence can be used by cells to control the movement of proteins into the nucleus. In some nuclear proteins the phosphorylation of serine residues located near the NLS prevents nuclear uptake. When the time is appropriate for nuclear localization of the protein in question, the phosphate groups are removed by a phosphatase and the protein enters the nucleus. In other cases, *nuclear-uptake regulatory proteins (NURPs)* bind to the NLS sequence and prevent it from functioning. When the protein containing the NLS is needed in the nucleus, the NURP is inactivated and nuclear uptake ensues. These examples represent two of a growing list of mechanisms that are used to control the entry of proteins into the nucleus.

Some proteins synthesized in the cytoplasm for transport into the nucleus are bound to RNA. For example, the small nuclear RNA called *U1* is transcribed in the nucleus and then exported to the cytoplasm, where it joins with proteins to form a particle called the *U1 snRNP* (whose role in removing introns from pre-mRNA was described on page 432). Two components are required for the efficient transport of the U1 snRNP into the nucleus. One is the presence of a protein component of the snRNP called *Sm,* and the other is a *trimethylguanosine cap,* which occurs at the 5' end of U1 RNA. Most of the snRNPs involved in removing introns from pre-mRNA are transported into the nucleus using a similar targeting mechanism.

Targeting Sequences Direct Mitochondrial Proteins Made in the Cytosol to Their Proper Locations within the Mitochondrion

Mitochondria contain about 700 different proteins, more than 95 percent of which are synthesized on cytoplasmic ribosomes. The mechanism by which these proteins are transported to their appropriate locations has been studied by incubating purified mitochondria with radioactive cytoplasmic proteins that are destined for mitochondrial uptake. Mitochondria import these proteins in the absence of ongoing protein synthesis, indicating that transport occurs after the newly synthesized polypeptides have been released from the ribosome. Proteins are imported into four different mitochondrial compartments: the matrix, the inner membrane, the intermembrane space, and the outer membrane (Figure 11-52). We will examine transport to each of these locations in turn.

1. Proteins Targeted to the Mitochondrial Matrix. Proteins destined for the mitochondrial matrix are synthesized in the cytosol as precursor molecules that contain a *matrix-targeting signal* at their N-terminus. The matrix-targeting sequence is rich in basic amino acids and in serine and threonine, and is essentially devoid of acidic amino acids. Adding this signal sequence to the N-terminus of proteins that are not normally located in mitochondria is sufficient to cause them to be imported into the mitochondrion.

Polypeptide chains must be in a relatively unfolded state when they are transported into mitochondria. They are kept in this state by binding to one or more chaperones of the *hsp70* family, which use energy derived from ATP hydrolysis to unfold the polypeptide chain and deliver it to a receptor located in the outer mitochondrial membrane. The outer membrane receptor then becomes associated with components in the inner mitochondrial membrane, bringing the inner and outer membranes in close contact and forming a translocation channel that transfers the polypeptide into the mitochondrial matrix. Translocation into the matrix requires the existence of the membrane potential that normally exists across the inner mitochondrial membrane (page 337). As the polypeptide passes into the mitochondrial matrix, its matrix-targeting signal sequence is removed by the action of a matrix peptidase. The polypeptide then binds to another chaperone, which utilizes energy derived from ATP to facilitate proper folding of the protein into its final conformation (see Figure 11-52, *pathway 1*).

2. Proteins Targeted to the Inner Mitochondrial Membrane. Proteins destined for the inner mitochondrial membrane contain matrix-targeting signals that resemble those occurring in matrix proteins. In at least some cases, this signal sequence causes inner membrane proteins to be initially translocated into the matrix. The matrix-targeting signal is then removed, exposing an additional *sorting signal* that consists of about 20 hydrophobic amino acids preceded by several basic residues. This hydrophobic signal directs the insertion of the protein into the inner mitochondrial membrane, where it is retained as an inner membrane protein (see Figure 11-52, *pathway 2A*).

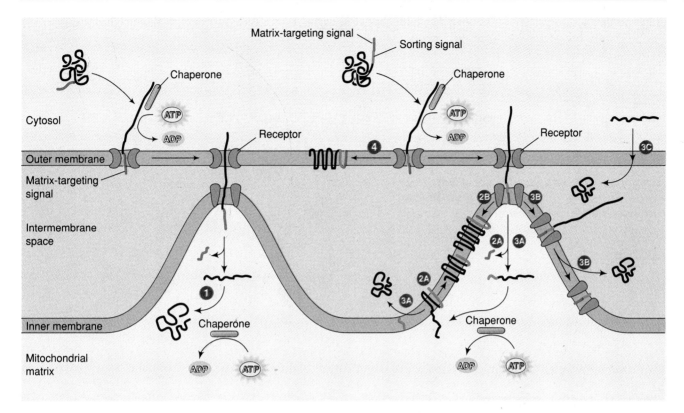

Figure 11-52 *Pathways for Importing Mitochondrial Proteins*   *Matrix proteins are translocated using a single matrix-targeting signal, which is removed after the protein enters the matrix (pathway 1). Two possible transport routes are shown for inner membrane proteins: one that involves passage through the matrix (pathway 2A) and one that does not (pathway 2B). Three possible transport routes are shown for proteins destined for the intermembrane space, one based on passage through the matrix (pathway 3A), one involving movement through the inner membrane (pathway 3B), and a third that involves direct passage across the outer membrane (pathway 3C). Outer membrane proteins are retained in the outer membrane during the first phase of translocation (pathway 4).*

Proteins containing a matrix-targeting signal can also be imported directly into the inner mitochondrial membrane without first being released into the matrix. For example the ADP-ATP carrier, which is involved in transporting ADP and ATP into and out of mitochondria (page 342), is translocated through the outer mitochondrial membrane and then diverted directly from the translocation channel into the inner mitochondrial membrane without entering the matrix or having its targeting sequence removed (see Figure 11-52, *pathway 2B*).

3. Proteins Targeted to the Intermembrane Space. Proteins destined for the intermembrane space are also transported by more than one pathway. Some proteins in this category contain a matrix-targeting signal that permits them to be handled much like the inner membrane proteins described in the preceding section. That is, either they are diverted from the translocation channel directly into the inner membrane or they first pass into the matrix, have their matrix-targeting signal removed, and are then inserted into the inner membrane. In either case, the hydrophobic sequence that directs these proteins into the inner membrane is then re-

moved by a peptidase associated with the outer surface of the inner mitochondrial membrane, releasing the protein into the intermembrane space (see Figure 11-52, *pathways 3A and 3B*).

Yet another mechanism for targeting proteins into the intermembrane space has been demonstrated for cytochrome *c,* which is transferred directly across the outer membrane and into the intermembrane space without requiring a removable signal sequence (see Figure 11-52, *pathway 3C*). After arriving in the intermembrane space, the cytochrome becomes bound to a heme prosthetic group. The folding of the protein chain that accompanies heme addition traps cytochrome *c* within the intermembrane space. During its role in mitochondrial electron transport, cytochrome *c* transiently binds to the outer surface of the inner mitochondrial membrane.

4. Proteins Targeted to the Outer Mitochondrial Membrane.   Proteins that reside in the outer mitochondrial membrane, such as the *porin* molecule that makes the membrane permeable to small molecules and ions (page 320), contain a short matrix-targeting signal followed by a long stretch of hydrophobic amino acids. The hydrophobic amino acids cause the polypeptide chain to be retained in the outer membrane during translocation rather than being transferred all the way through to the matrix. Neither the matrix-targeting signal nor the hydrophobic sequence responsible for retaining porin in the outer membrane is normally removed (see Figure 11-52, *pathway 4*).

Targeting Sequences Direct Chloroplast Proteins Made in the Cytosol to Their Proper Locations within the Chloroplast

As in the case of mitochondria, most chloroplast proteins are synthesized on cytoplasmic ribosomes and then transported into the chloroplast. Energy derived from the hydrolysis of ATP is required for uptake, although a membrane potential is not necessary. Targeting is again accomplished through N-terminal signal sequences that are removed after a protein reaches its proper destination (Figure 11-53). Proteins destined for the chloroplast stroma contain a *stromal-targeting signal* that is relatively rich in serine, threonine, and alanine, and contains few acidic amino acids. Translocation through the chloroplast envelope occurs at contact sites between the outer and inner membranes. After the protein enters the stroma, its stromal-targeting signal is removed by the action of a *stromal peptidase*.

Proteins destined for the thylakoid lumen contain two targeting sequences. The first is a stromal-targeting signal that marks the protein for transport into the stroma, where the sequence is removed by the stromal peptidase. This process exposes a hydrophobic sequence that promotes translocation of the protein across the thylakoid membrane. A peptidase associated with the inner surface of the thylakoid membrane then removes the hydrophobic sequence, thereby releasing the mature form of the protein into the thylakoid lumen. Proteins destined for incorporation into the thylakoid membrane also contain two targeting sequences: a stromal-targeting signal that promotes translocation of the protein into the stroma and a membrane insertion sequence that directs the protein to be incorporated into the thylakoid membrane. Relatively little is known about the targeting signals that direct proteins to the membranes of the chloroplast envelope or the intermembrane space between them.

Targeting Sequences Guide Proteins Made in the Cytosol into Peroxisomes, Glyoxysomes, and Glycosomes

Peroxisomes are cytoplasmic organelles that are bounded by a single membrane and characterized by the presence of enzymes involved in peroxide metabo-

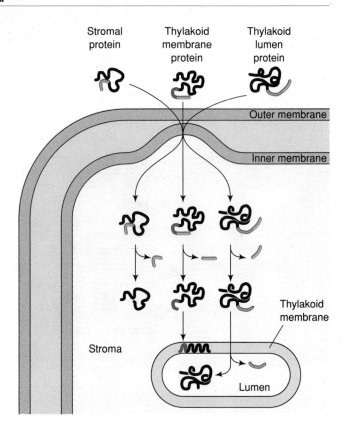

Figure 11-53 *Pathways for Importing Chloroplast Proteins* *Chloroplast proteins are translocated using a stromal-targeting signal that is removed after the protein enters the stroma. Proteins destined for the thylakoid membrane or thylakoid lumen contain an additional sequence that targets the protein to one of these two locations.*

lism and fatty acid oxidation (page 349). Because peroxisomes are often observed in close association with membranes of the ER, it was originally thought that peroxisomal enzymes are synthesized on membrane-attached ribosomes and pass through the lumen of the ER into peroxisomes. This view was later shown to be incorrect by experiments in which tissues were briefly incubated with radioactive amino acids to monitor the route followed by newly synthesized peroxisomal proteins. Subcellular fractionation studies revealed that peroxisomal proteins such as catalase appear first in the cytosol fraction and only later become concentrated in peroxisomes, suggesting that they are synthesized on free ribosomes. Additional support for this conclusion has come from studies showing that messenger RNA isolated from free, but not membrane-bound, ribosomes contains sequences coding for peroxisomal enzymes.

One targeting signal that has been shown to direct the movement of newly synthesized proteins from the cytosol into peroxisomes resides at the C-terminus and consists of the amino acids Ser-Lys-Leu or a closely related sequence. Adding this tripeptide to the C-terminal end of nonperoxisomal proteins is sufficient to cause them to be incorporated into peroxisomes. A similar signal is involved in targeting proteins to glyoxysomes and glycosomes, which are closely related to peroxisomes. Unlike many of the sequences employed for targeting proteins into mitochondria and chloroplasts, the C-terminal tripeptide is not removed from peroxisomal proteins after transport into the peroxisome.

SUMMARY OF PRINCIPAL POINTS

• Protein synthesis occurs on ribosomes, which measure roughly 25 nm in diameter and consist of small and large subunits. Prokaryotic ribosomes are constructed from 30S and 50S subunits that combine to form a 70S ribosome, whereas eukaryotic cytoplasmic ribosomes are constructed from 40S and 60S subunits that combine to form an 80S ribosome. Small ribosomal subunits contain a single molecule of RNA (16S in prokaryotes and 18S in eukaryotes); the larger subunit has two molecules of RNA in prokaryotes (23S and 5S) and three molecules of RNA in eukaryotes (28S, 5.8S, and 5S). Prokaryotic ribosomes contain about 52 different proteins, and eukaryotic cytoplasmic ribosomes typically possess 80–85 proteins.

• Eukaryotic cytoplasmic ribosomes are manufactured in the nucleolus, which contains from several hundred to several thousand copies of the genes coding for 18S, 28S, and 5.8S (but not 5S) rRNAs. These genes are transcribed in the nucleolar fibrillar center into a single pre-rRNA molecule. During the formation of ribosomal subunits, the pre-rRNA molecule is methylated, bound to ribosomal proteins and 5S rRNA, and then cleaved into 18S, 5.8S, and 28S rRNAs. In certain organisms, ribosomal RNA precursors contain introns that are removed by self-splicing.

• The rRNA genes of bacterial cells occur in clusters in which all three ribosomal genes (16S-23S-5S) are closely linked. The rRNA precursor transcribed from these genes is cleaved by RNase III before transcription proceeds from the 16S gene into the adjacent 23S gene.

• Cell-free protein synthesis requires ribosomes, mRNA, tRNAs, aminoacyl-tRNA synthetases, amino acids, protein synthesis factors, and an energy source capable of generating GTP and ATP. Messenger RNA is translated in the 5' → 3' direction, specifying the incorporation of amino acids starting at the N-terminus of the polypeptide chain and proceeding toward the C-terminus.

• Initiation of protein synthesis requires a special initiator tRNA that carries formylmethionine in prokaryotes and methionine in eukaryotes. Initiation factors, GTP, the small ribosomal subunit, mRNA, and the initiator tRNA interact to form a preinitiation complex, which then binds to the large subunit to generate an initiation complex. The Shine-Dalgarno sequence in prokaryotic mRNA and the 5' cap of eukaryotic mRNA help align the mRNA so that protein synthesis is initiated at the AUG start codon.

• During elongation, amino acids are joined together in a linear order specified by codons in mRNA. Elongation involves a three-step reaction cycle in which (1) the aminoacyl-tRNA specified by the next codon in mRNA is bound to the A site of the ribosome, (2) a peptide bond forms between the amino-

acyl-tRNA at the A site and the growing peptide chain (or initiating amino acid) bound to the P site, and (3) translocation advances the mRNA by one codon, accompanied by displacement of the empty tRNA from the ribosome and movement of the peptidyl-tRNA from the A site to the P site. Elongation requires the presence of elongation factors and GTP.

• Termination of protein synthesis occurs when a stop codon (UAA, UAG, or UGA) in mRNA is reached. Stop codons are recognized by release factors that bind to the A site and cause the linkage between the polypeptide chain and the tRNA to be hydrolyzed, releasing the completed polypeptide. The ribosomal subunits and tRNA then dissociate from the mRNA for use in subsequent rounds of protein synthesis.

• Each molecule of mRNA is normally translated by several ribosomes simultaneously, forming a polysome.

• Newly released polypeptide chains are often processed by posttranslational cleavage reactions. The folding of the chain into its proper three-dimensional conformation is facilitated by chaperones as well as enzymes catalyzing steps that might otherwise slow the rate of protein folding.

• Translation is subject to control by (1) translational repressors that bind to mRNA and inhibit translation, (2) factors that influence the life span of individual mRNAs, (3) antisense RNA molecules that inhibit the translation of their corresponding mRNAs, (4) phosphorylation of protein synthesis factors such as eIF-2 and eIF-4F, (5) changes in the availability of tRNAs, and (6) factors that determine the rate of ribosome formation. During attenuation, which depends upon the coupling of transcription and translation that occurs in bacteria, the scarcity of a specific amino acid stalls mRNA translation and prevents the premature termination of transcription.

• Posttranslational control involves the modification of polypeptide chains after they have been synthesized. Among the commonly encountered posttranslational control mechanisms are allosteric regulation, cleavage of peptide bonds, association and dissociation of protein subunits, and chemical modifications such as phosphorylation, acetylation, and ADP-ribosylation.

• Changes in protein degradation rates influence the extent to which various proteins are expressed. Proteins with short half-lives are more sensitive to changes in synthetic rate than proteins with longer half-lives. One mechanism for selecting proteins for degradation involves joining them to ubiquitin followed by digestion by proteasomes.

• Targeting signals route proteins manufactured on free cytoplasmic ribosomes into the nucleus, mitochondria, chloroplasts, and peroxisomes. In proteins destined for mitochondria and chloroplasts, more than one targeting signal may be required to guide the protein to its proper destination within the organelle.

SUGGESTED READINGS

Books

Arnstein, H. R. V., and R. A. Cox (1992). *Protein Biosynthesis,* IRL Press, Oxford, England.

Austen, B. M., and O. M. R. Westwood (1991). *Protein Targeting and Secretion,* IRL Press, Oxford, England.

Graves, D.J., B.L. Martin, and J.H. Wang (1994). *Co- and Post-Translational Modification of Proteins: Chemical Prin-*

ciples and Biological Effects, Oxford University Press, New York.

Hadjiolov, A. A. (1985). *The Nucleolus and Ribosome Biogenesis,* Springer-Verlag, New York.

Hardesty, B., and G. Kramer, eds. (1986). *Structure, Function, and Genetics of Ribosomes,* Springer-Verlag, New York.

Hill, W. E., A. Dahlberg, R. A. Garrett, P. B. Moore, D. Schlessinger, and J. R. Warner, eds. (1990). *The Ribosome: Structure, Function and Evolution,* American Society for Microbiology, Washington, DC.

Ilan, J., ed. (1987). *Translational Regulation of Gene Expression,* Plenum, New York.

Spirin, A. S., ed. (1986). *Ribosome Structure and Protein Biosynthesis,* Benjamin/Cummings, Menlo Park, CA.

Thach, R. E., ed. (1991). *Translationally Regulated Genes in Higher Eukaryotes,* Karger, Basel, Switzerland.

Articles

Agutter, P.S., and D. Prochnow (1994). Nucleocytoplasmic transport, *Biochem. J.* 300: 609-618.

Altmann, M., and H. Trachsel (1993). Regulation of translation initiation and modulation of cellular physiology, *Trends Biochem. Sci.* 18: 429-432.

Cech, T. R. (1986). RNA as an enzyme, *Sci. Amer.* 255 (November):64-75.

Decker, C.J., and R. Parker (1994). Mechanisms of mRNA degradation in eukaryotes, *Trends Biochem. Sci.* 19:336-340.

Dingwall, C., and R. A. Laskey (1991). Nuclear targeting sequences—a consensus, *Trends Biochem. Sci.* 16:478-481.

Draper, D. E. (1989). How do proteins recognize specific RNA sites? New clues from autogenously regulated ribosomal proteins, *Trends Biochem. Sci.* 14:335-338.

Finley, D. (1991). Ubiquitination, *Annu. Rev. Cell Biol.* 7:25-69.

Fournier, M. J., and E. S. Maxwell (1993). The nucleolar snRNAs: catching up with the spliceosomal snRNAs, *Trends Biochem. Sci.* 18:131-135.

Gething, M.-J., and J. Sambrook (1992). Protein folding in the cell, *Nature* 355:33-45.

Glick, B., and G. Schatz (1991). Import of proteins into mitochondria, *Annu. Rev. Genet.* 25:21-44.

Goessens, G. (1984). Nucleolar structure, *Int. Rev. Cytol.* 87:107-158.

Goessling, L. S., S. Daniels-McQueen, M. Bhattacharyya-Pakrasi, J.-J. Lin, and R. E. Thach (1992). Enhanced degradation of the ferritin repressor protein during induction of ferritin messenger RNA translation, *Science* 256:670-673.

Hartl, F.-U., R. Hlodan, and T. Langer (1994). Molecular chaperones in protein folding: the art of avoiding sticky situations, *Trends Biochem. Sci.* 19:20-25.

Hershey, J. W. B. (1988). Protein phosphorylation controls translation rates, *J. Biol. Chem.* 264:20823-20826.

Jentsch, S. (1992). Ubiquitin-dependent protein degradation: a cellular perspective, *Trends Cell Biol.* 2:98-103.

Kane, P. M., et al. (1990). Protein splicing converts the yeast *TFP1* gene product to the 69-kD subunit of the vacuolar H^+-adenosine triphosphatase, *Science* 250:651-657.

Kozak, M. (1992). Regulation of translation in eukaryotic systems, *Annu. Rev. Cell Biol.* 8:197-225.

Lake, J. A. (1981). The ribosome, *Sci. Amer.* 245 (August):84-97.

Lamond, A. I. (1990). The trimethyl-guanosine cap is a nuclear targeting signal for snRNPs, *Trends Biochem. Sci.* 15:451-452.

Liljas, A. (1991). Comparative biochemistry and biophysics of ribosomal proteins, *Int. Rev. Cytol.* 124:103-136.

Moore, P. B. (1988). The ribosome returns, *Nature* 331:223-227.

Nellen, W., and C. Lichtenstein (1993). What makes an mRNA anti-sense-itive? *Trends Biochem. Sci.* 18:419-423.

Nilsen, T. W. (1994). RNA-RNA interactions in the spliceosome: Unraveling the ties that bind, *Cell* 78:1-4.

Noller, H. F. (1991). Ribosomal RNA and translation, *Annu. Rev. Biochem.* 60:191-227.

Noller, H. F., V. Hoffarth, and L. Zimniak (1992). Unusual resistance of peptidyl transferase to protein extraction procedures, *Science* 256:1416-1419.

Nomura, M. (1984). The control of ribosome synthesis, *Sci. Amer.* 250 (January):102-114.

Ross, J. (1989). The turnover of messenger RNA, *Sci. Amer.* 260 (April):48-55.

Scheer, U., M. Thury, and G. Goessens (1993). Structure, function, and assembly of the nucleolus, *Trends Cell Biol.* 3:236-241.

Schmitz, M. L., T. Henkel, and P. A. Baeuerle (1991). Proteins controlling the nuclear uptake of NF-kB, Rel, and dorsal, *Trends Cell Biol.* 1:130-137.

Sollner-Webb, B., and E. B. Mougey (1991). News from the nucleolus: rRNA gene expression, *Trends Biochem. Sci.* 16:58-62.

Stuart, R. A., D. M. Cyr, E. A. Craig, and W. Neupert (1994). Mitochondrial molecular chaperones: their role in protein translocation, *Trends Biochem. Sci.* 19:87-92.

Subramani, S. (1993). Protein import into peroxisomes and biogenesis of the organelle, *Annu. Rev. Cell Biol.* 9: 445-478.

Thach, R. E. (1992). Cap recap: the involvement of eIF-4F in regulating gene expression, *Cell* 68:177-180.

Theg, S. M., and S. V. Scott (1993). Protein import into chloroplasts, *Trends Cell Biol.* 3:186-190

Warner, J. R. (1990). The nucleolus and ribosome formation, *Curr. Opinion Cell Biol.* 2:521-527.

Weijland, A., and A. Parmeggiani (1994). Why do two EF-Tu molecules act in the elongation cycle of protein biosynthesis? *Trends Biochem. Sci.* 19:188-193.

Weintraub, H. M. (1990). Antisense RNA and DNA, *Sci. Amer.* 262 (January):40-46.

Yanofsky, C. (1988). Transcription attenuation, *J. Biol. Chem.* 263:609-612.

Chapter 12

Cell Cycles and Cell Division

The ability to grow and reproduce is a fundamental property of living organisms. Therefore, whether an organism is composed of a single cell or trillions of cells, individual cells must be able to grow and divide. In the preceding two chapters we learned how genetic information guides the synthesis of the protein molecules that carry out most cellular functions. When cells grow and divide, each newly formed cell needs to retain the ability to synthesize all of its essential proteins. This means that the entire set of DNA molecules must be faithfully replicated prior to cell division and then distributed to the two cells created by the division process.

In recent years, much has been learned about the mechanisms that coordinate the events associated with cell division. Prior to each division a cell first passes through a series of preparatory stages, collectively known as the **cell cycle,** which begins when two new cells are formed by the division of a single parental cell and ends when one of these cells divides to form two new cells. In the present chapter we will examine the events associated with the cell cycles of both prokaryotic and eukaryotic cells, emphasizing the mechanisms which ensure that each new cell receives a complete set of genetic instructions.

STUDYING THE CELL CYCLE

Although dividing cells were first observed with the light microscope more than a century ago, it is only in the past few decades that a molecular description of the cell cycle as a whole has begun to emerge. One of the most powerful approaches for studying the cell cycle involves the use of cells that are growing in culture. However, the cells that make up a typical proliferating culture are randomly distributed at various phases of the cell cycle. To identify the events that take place at specific stages of the cycle, specialized procedures are therefore required.

Synchronized Cell Populations Can Be Produced Using Either Induction or Selection Techniques

One approach for investigating the activities that occur at specific points during the cell cycle is to utilize *synchronized* cell populations in which all the cells are at the same stage in the cycle at any given moment. Two kinds of techniques are employed to generate synchronized cell populations. *Induction synchrony* uses conditions that temporarily halt the cell cycle at a particular stage. Temperature changes, pulses of light to photosynthetic cells, and metabolic inhibitors that block specific cell cycle functions are commonly used for this purpose. An example of the latter approach is the *thymidine block* technique, which utilizes high concentrations of thymidine to prevent the synthesis of nucleotides required for DNA synthesis. When cell populations are exposed to high concentrations of thymidine, each cell continues through the cell cycle until it reaches the point at which DNA synthesis would normally occur. When all cells have been blocked at this stage in the cycle, the excess thymidine is removed and the entire cell population begins to synthesize DNA simultaneously.

Although induction techniques have been widely employed for producing synchronized cell populations, the main limitation of this approach is that the conditions employed for temporarily halting the cell cycle may have toxic side effects. This disadvantage is overcome in the other approach to cell synchronization, called *selection synchrony,* in which cells at a particular stage of the cell cycle are selectively removed from a culture and grown as a separate, synchronous population. One such technique exploits the fact that cells in the process of dividing tend to round up and detach from the surface of the culture vessel in which they are growing. This allows dividing cells to be isolated by gently shaking the culture vessel and removing the medium containing the detached cells, leaving behind cells at other phases of the cycle.

Regardless of the method employed to obtain synchronized cell populations, synchronous growth does not persist indefinitely because every cell does not take the same amount of time to proceed through the cell cycle. As a result, synchrony dissipates after a culture has proceeded through several cell cycles and the population must be synchronized again.

Conditional Mutants Are Useful for Identifying Important Events in the Cell Cycle

Another approach for identifying important events in the cell cycle involves the study of mutations that prevent cells from dividing normally. Because cells exhibiting such mutations can be difficult to grow and maintain in culture, biologists often use **temperature-sensitive mutants,** which are cells that exhibit the abnormal trait only when the temperature is elevated slightly above the normal range. The elevated temperature triggers a biochemical defect, such as misfolding of a critical protein, that is not manifested at lower temperatures. Therefore temperature-sensitive mutants can be grown and maintained at normal temperatures and then studied under warmer conditions to investigate the mutant trait.

This approach has been especially useful in studying yeast cells, where several dozen *cell-division cycle (cdc) mutants* blocked at specific stages of the cell cycle have been identified. We will see later in the chapter that proteins encoded by some of these *cdc* genes have been shown to play crucial roles in controlling the progression of cells through the normal cell division cycle.

BACTERIAL DIVISION

Prokaryotes and eukaryotes differ significantly both in the organization of their cell division cycles and in the mechanisms they employ for carrying out many cell cycle activities. Yet in spite of these differences, the cell cycles of these two classes of cells are based upon the same set of objectives. In each case, cell proliferation involves three basic events: *cell growth, chromosome replication,* and *cell division.* We will begin by examin-

ing these events in bacterial cells and will then proceed to a discussion of eukaryotic cells.

The Prokaryotic Cell Cycle Is Constructed from C and D Periods That Can Overlap plus an Optional B Period

In order for cell division to produce two identical cells, the chromosomal DNA must be replicated prior to division and distributed equally to the two new cells. This means that DNA synthesis needs to occur sometime prior to the actual time of cell division. Investigations of the timing of DNA synthesis during the bacterial cell cycle have revealed that the pattern can vary dramatically, depending on the rate at which the cells are growing (Figure 12-1, *left column*). This varying pattern of DNA synthesis has been explained by the *Cooper-Helmstetter model,* which postulates the existence of two time constants within the bacterial cell cycle. One constant, called the *C period,* corresponds to the amount of time required for replication of the chromosomal DNA; the other constant, called the *D period,* refers to the minimum amount of time necessary between the end of DNA replication and the completion of cell division.

Experiments utilizing radioactive precursors to monitor DNA synthesis in bacterial cells have revealed that in cell populations that divide once per hour, DNA replication is restricted to the first 40 minutes. It therefore follows that C = 40 minutes and D = 20 minutes. In bacteria that divide more slowly than once per hour, a gap called the *B period* occurs between cell division and the initiation of DNA synthesis. But what about cells that divide more rapidly than once per hour? Instead of shortening the C or D periods, cell cycles of less than an hour are created by overlapping these two periods. For example, in cells that divide every 40 minutes, a new round of DNA replication is started each time a previous round ends. Although cell division does not occur until 20 minutes after completion of the first round of DNA synthesis, overlapping the C and D periods allows a net cell cycle of 40 minutes to be generated (see Figure 12-1, *middle row*). Cell cycles that are even shorter than 40 minutes can be created by starting a new round of DNA replication before the previous round has been completed (see Figure 12-1, *bottom row*).

Changing Concentrations of ppGpp May Link Growth Rate to the Initiation of DNA Replication

In bacteria the main factor that determines whether cells grow and divide is the presence of an adequate supply of nutrients in the external environment. This

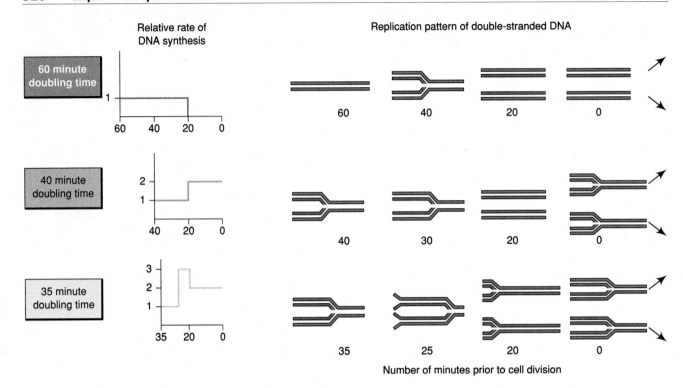

Relative rate of
DNA synthesis

Replication pattern of double-stranded DNA

Number of minutes prior to cell division

Figure 12-1 *DNA Replication During the Cell Cycle of the Bacterium E. Coli* *The three graphs on the left show that the pattern of DNA synthesis during the cell cycle varies dramatically, depending on the rate at which cells are growing and dividing. The line drawings on the right symbolize double-stranded DNA replicating semiconservatively. Although the bacterial chromosome is actually circular and replication is bidirectional, the lines drawn here are representative of what happens in the cell. (Top) In cultures with a doubling time of 60 minutes, DNA replication occupies the first 40 minutes of the cell cycle and then ceases for 20 minutes. (Middle) In cultures with a doubling time of 40 minutes, a new round of DNA replication begins as soon as the preceding one ends. (Bottom) In cultures with doubling times shorter than 40 minutes, a new round of DNA replication begins before the preceding round has been completed.*

means that the frequency of cell division must be closely coordinated with the rate of cell growth. If cells divide too frequently relative to the rate of cell growth, the cells will gradually become smaller and smaller. If cells do not divide fast enough, they will become excessively large.

The mechanism that couples the rate of cell growth to the rate of cell division is thought to involve the nucleotide *guanosine tetraphosphate (ppGpp)*. In Chapter 11 we learned that ppGpp is produced by ribosomes that have become stalled during the process of protein synthesis by a scarcity of amino acids (page 504). Hence the concentration of ppGpp within cells is inversely related to growth rate; that is, ppGpp levels are high when the growth rate is slow and low when the growth rate is fast. How does ppGpp couple the rate of cell growth to the rate of cell division? One suggested mechanism involves *DnaA*, a protein that triggers chromosome replication by opening the DNA double helix at the point where DNA synthesis is to be initiated. In bacteria whose growth rate has been reduced by starving them for essential amino acids, transcription of the gene coding for the DnaA protein is greatly reduced. The ability of changing growth rates to influence expression

of this gene is not observed, however, in mutant cells that have lost the ability to synthesize ppGpp.

It is therefore reasonable to speculate that ppGpp serves as a signal that couples changes in growth rate to the mechanism responsible for initiating DNA replication. According to this view, ppGpp accumulates in slowly growing cells and depresses the transcription of genes whose products are required for the initiation of DNA synthesis, such as the gene coding for DnaA. At faster growth rates, less ppGpp is produced and expression of the genes required for the initiation of DNA synthesis increases.

Bacterial Chromosome Replication Is Bidirectional Starting from a Single Origin

Most bacteria contain a single circular chromosome, measuring about 1-2 mm in circumference, which is comprised of a double-stranded DNA molecule associated with small amounts of protein (page 415). The replication of bacterial chromosomes was first studied in the early 1960s by John Cairns, who incubated *E. coli* with ^{3}H-thymidine to label the chromosomal DNA. When the DNA was isolated and visualized by autoradiography, it

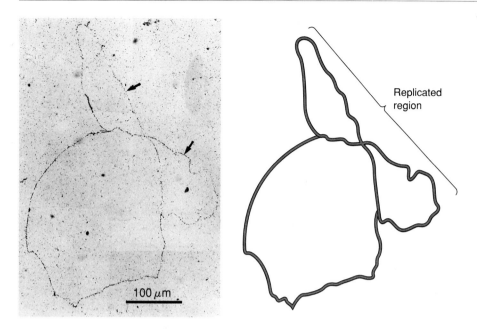

Figure 12-2 *A Replicating Chromosome from the Bacterium* E. Coli *Bacteria were incubated with* ³H-thymidine *and the radioactive chromosomal DNA was then visualized by autoradiography. The two loops in the micrograph* (arrows) *represent the location where the chromosomal DNA has already been replicated. Micrograph courtesy of J. Cairns.*

appeared as an intact circular structure containing a closed loop where the DNA molecule had already unwound and undergone replication (Figure 12-2). It was not clear from such pictures, however, whether the loop was produced by a single replication fork moving in one direction around the circular chromosome or by two replication forks proceeding in opposite directions.

To address this question, bacterial cells were grown for brief periods in the presence of a low concentration of ³H-thymidine, and then transferred to a medium containing a high concentration of ³H-thymidine. If replication proceeds in both directions, autoradiography should reveal an image of a DNA region containing few silver grains (resulting from replication in the presence of a low concentration of ³H-thymidine) surrounded on both sides by regions containing larger numbers of silver grains (resulting from replication in the presence of a high concentration of ³H-thymidine). This pattern is precisely what was observed, indicating that DNA replication is *bidirectional;* that is, it proceeds in both directions around the circular DNA molecule starting from a single point of origin (Figure 12-3).

The initiation of *E. coli* chromosome replication requires a 245 base-pair stretch of DNA called *oriC.* If this DNA segment is artificially removed, the chromosome loses the ability to initiate normal replication. At the appropriate time during the cell cycle, replication is initiated at this site by the DnaA protein mentioned in the preceding section. The DnaA protein binds to *oriC* in an ATP-dependent reaction, separating the two DNA strands and allowing entry of the enzymes required for

DNA replication. The mechanism by which these enzymes then synthesize DNA was discussed in Chapter 3.

Septum Formation Leads to Chromosome Segregation and Cell Division

After DNA replication has been completed, cell division can take place. Cell growth, which occurs throughout the cell cycle, typically causes a cell to double in size by the time it is ready to divide. Most bacterial cells then undergo *binary fission,* which partitions the cell down the middle and generates two identical new cells. A few types of bacteria undergo *asymmetric fission,* which divides cells into two nonidentical cells. A classic example of asymmetric fission occurs in the stalk cells of the bacterium *Caulobacter,* which divide in such a way that each stalk cell gives rise to one stalk cell and one motile cell (Figure 12-4).

The process of cell division is designed to accomplish two objectives: to divide the original cell into two new cells and to distribute a copy of the chromosomal DNA to each of the resulting cells. Bacterial cell division is initiated when a protein called *FtsZ* assembles on the cytoplasmic surface of the plasma membrane, forming a ring around the cell where division will take place. The plasma membrane in this region then grows toward the center of the cell, forming a doubled-layered membrane *septum.* Next the cell wall extends into the space between the two membranes of the septum, generating a *cross wall* that is considerably thicker than a normal cell wall. Finally the thickened wall is cleaved down the middle, and two separate cells are generated. The im-

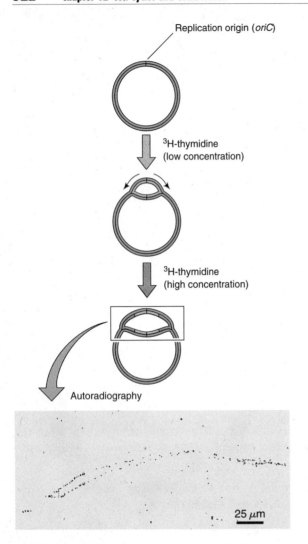

Figure 12-3 *Evidence for Bidirectional Replication of the E. Coli Chromosome* *Bacterial cells were briefly incubated in the presence of a low concentration of ^{3}H-thymidine, followed by incubation with a high concentration of ^{3}H-thymidine. If replication is bidirectional, then the DNA that is replicated first should be less radioactive than the DNA surrounding it on both sides. Autoradiography of the newly replicated chromosomal DNA confirms that labeling is heavier at the two ends than in the middle. Micrograph courtesy of D. M. Prescott.*

portance of the cell wall to the process of cell division has been demonstrated by experiments involving *penicillin*, an antibiotic that inhibits cell wall formation (page 251). Exposing bacteria to penicillin produces cells without walls, called *protoplasts*, that exhibit great difficulty in dividing.

What mechanism ensures that each of the two new cells receives a copy of the bacterial chromosome at the time of cell division? During the process of chromosome replication, the chromosomal DNA is bound to the plasma membrane at multiple sites along the two newly forming DNA molecules, including the replication origin. It has been proposed that as the cell grows,

new plasma membrane is added to the region located between the attachment sites of the two chromosomes, gradually pushing the chromosomes away from each other (Figure 12-5). When the cell is then partitioned down the middle at the time of cell division, the two chromosomes are segregated into separate cells.

Sporulation Is a Special Type of Cell Division That Converts Bacteria into Dormant Endospores

A special type of cell division occurs in several kinds of Gram-positive bacteria when they are exposed to unfavorable environmental conditions that threaten their survival. Such conditions cause these bacteria to form a dormant protected structure called an **endospore** (Figure 12-6). During endospore formation, chromosome replication is followed by migration of one of the two newly formed chromosomes to one end of the cell. A membrane partition called the *spore septum* then forms at that end of the cell, sequestering the chromosome in a small cytoplasmic compartment. Next this region of the cell is surrounded by a thick protective *spore coat*. Finally most of the water is removed, producing a dehydrated endospore that is capable of withstanding extremely harsh conditions and able to remain in a dormant state for many years.

When environmental conditions become more hospitable, endospores are converted back into normal cells. During this *germination* process, the cell water is replenished, the spore coat is discarded, and the endospore uses its stored components to once again assume the characteristics of a normal cell, all within the period of a few hours.

MITOTIC DIVISION

The eukaryotic cell cycle, like its bacterial counterpart, involves cell growth, chromosome replication, and cell division. However, the timing of these events and the mechanisms involved are considerably more complex than in bacteria. The cell cycles of eukaryotes are associated with two kinds of cell division that differ in the way the duplicated chromosomes are distributed to the newly formed cells. The most common type of eukaryotic cell division, called *mitotic division*, generates two cells containing identical sets of chromosomes. The other type of division, called *meiotic division*, generates cells with differing chromosomal makeup for use during sexual reproduction. We will begin our examination of eukaryotic cell cycles by describing the events associated with mitotic division and then will proceed to a discussion of meiotic division.

The Eukaryotic Cell Cycle Is Divided into Four Phases Called G_1, S, G_2, and M

In cells that reproduce by mitotic division, the cell cycle can be subdivided into two general stages: **interphase,**

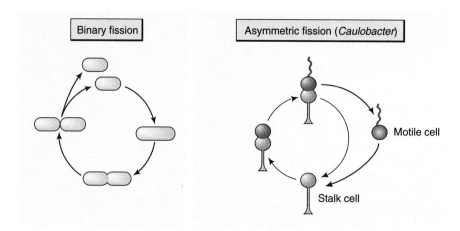

Figure 12-4 *Comparison of Binary Fission and Asymmetric Fission in Bacterial Cells*
Binary fission divides a cell down the middle to form two identical new cells, whereas asymmetric fission divides a cell into two nonidentical cells.

which is the time when the cell grows and replicates its chromosomal DNA, and the mitotic or **M phase**, which is the period when the individual chromosomes become visible and the nucleus and then the rest of the cell divide. Because the microscopic appearance of the chromatin does not change much during interphase, cell biologists originally believed that chromosome replication occurs during M phase. However, in the early 1950s Hewson Swift investigated this issue experimentally by using the Feulgen staining technique (page 128) to measure the quantity of DNA present at various stages of the cell cycle. His data indicated that the amount of DNA doubles during interphase rather than the M phase. Subsequent experiments employing radioactive thymidine to monitor the rate of DNA synthesis revealed that DNA is replicated during a defined period of interphase, which was named the **S phase**. A time gap called the **G₁ phase** separates the S phase from the preceding M phase, and a second gap named the **G₂ phase** separates the end of the S phase from the beginning of the next M phase (Figure 12-7).

The most variable phase of the cell cycle is G₁, which is very brief or absent in rapidly dividing cells but occupies many hours or days in slowly dividing cells. Once a cell leaves G₁ and initiates DNA synthesis, progression through the S, G₂, and M phases proceeds at a relatively fixed rate. In mammalian cells S phase typically occupies about 6–8 hours, G₂ lasts an average of 3–6 hours, and M phase takes no more than an hour or two.

Although the M phase is usually the shortest phase of the cycle, it is comprised of two exceedingly complex events known as *mitosis* and *cytokinesis*. **Mitosis** is the process of nuclear division, which divides the nucleus and chromosomes in a way that creates two new, genetically identical nuclei. At the beginning of mitosis the nuclear envelope breaks down, the chromatin fibers condense into individually recognizable *mitotic chromo-*

somes, and a *spindle* is constructed from microtubules for the purpose of moving the two sets of replicated chromosomes to the two newly forming cells. After the chromosomes have separated, nuclear envelopes form around the two groups of chromosomes to produce two nuclei. During the final stages of mitosis, the companion process of **cytokinesis** (cytoplasmic division) cleaves the original cell into two new cells, each containing one of the nuclei generated by mitosis.

Progression through the Cell Cycle Is Controlled by Cyclin-Dependent Protein Kinases

As a cell progresses through the cell cycle, it encounters a number of control points where decisions are made as to whether or not to proceed to the next stage. The existence of these control points helps to make cell division responsive to external signaling molecules, such as hormones and growth factors, and facilitates the proper coordination of cell cycle events. It would be disastrous, for example, if cell division were to take place before chromosome replication is completed, or if mitotic divisions continued to occur in the absence of cell growth.

The existence of one important control point is suggested by the behavior of cells exposed to conditions that limit cell growth—for example, the absence of nutrients or necessary growth factors. Such conditions cause cells to become arrested at a point during G₁ known as either the **restriction point** or **START**. Cells that have been arrested at this point in G₁ can survive indefinitely without growing and dividing. Because they are no longer progressing through the cell cycle, such G₁-arrested cells are said to reside in the **G₀ state**. When conditions are altered so as to favor growth and division, G₁-arrested cells re-enter the normal cell cycle, although it may take them many hours to regain the capacity to begin DNA synthesis.

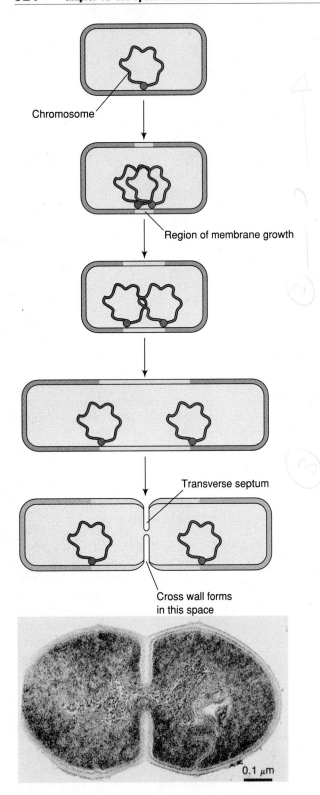

Chromosome

Region of membrane growth

Transverse septum

Cross wall forms
in this space

0.1 μm

Figure 12-5 *Chromosome Segregation During Bacterial Cell Division* *The diagram illustrates how attachment of the replicating bacterial chromosomes to the plasma membrane facilitates the distribution of the chromosomes to the two newly forming cells. The electron micrograph of a dividing* Streptococcus *shows the transverse septum in the process of dividing the cell in two. Micrograph courtesy of M. L. Higgins and L. Daneo-Moore.*

Passing beyond START represents a commitment by the cell to embark upon a pathway that begins with DNA replication and culminates in cell division. Cell fusion experiments provided some early insights into the mechanism that triggers this transition from G_1 into S phase. If S-phase cells are fused with cells in early G_1, the nuclei of the G_1 cells quickly initiate DNA synthesis, even if they would not normally have reached S phase until many hours later. Such observations indicate that S-phase cells contain one or more molecules that trigger the transition from G_1 into S phase. The controlling molecules are not simply the enzymes involved in DNA replication, since these enzymes can be present in high concentration in cells that do not enter into S phase. The nature of the molecule that triggers the G_1-to-S transition has been clarified by genetic studies of the fission yeast *Schizosaccharomyces pombe,* where a gene called *cdc2* is required for this transition. The protein product of the *cdc2* gene is a *protein kinase* that catalyzes the phosphorylation of serine and threonine residues in protein molecules. This particular protein kinase, called *p34^cdc2* or simply **cdc2 kinase,** is present in a wide variety of eukaryotic cells, but is active only when bound to a member of a group of proteins known as **cyclins.**

Besides the G_1-to-S transition, the other major control point in the cell cycle involves the transition from G_2 to M phase. If a cell that has just entered into mitosis is fused with a cell in any stage of interphase, the interphase nucleus begins mitosis prematurely, even if it has not yet replicated its chromosomes. The molecule responsible for triggering entry into mitosis has been isolated from a variety of eukaryotic cell types and is called **maturation promoting factor** or **MPF.** When injected into interphase cells, purified MPF triggers nuclear envelope breakdown, chromosome condensation, and spindle formation, all of which occur during the normal onset of M phase. The MPF molecule consists of two components: *cdc2 kinase* and a *cyclin* that activates the cdc2 kinase. Thus the transition from G_2 to M phase, like the transition from G_1 to S phase, is under the control of a cyclin-dependent protein kinase.

Cyclins that associate with protein kinases to trigger the transition from G_2 to M phase are called *mitotic cyclins,* while cyclins that interact with protein kinases to trigger the transition from G_1 to S phase are called *G_1 cyclins.* In yeast, the same cdc2 kinase is involved in both transitions: the activation of cdc2 kinase by G_1 cyclins initiates the transition from G_1 to S, while the activation of cdc2 kinase by mitotic cyclins controls the transition from G_2 to M phase (Figure 12-8). The cdc2 kinase molecule is presumably directed by the G_1 and mitotic cyclins to catalyze the phosphorylation of different proteins at the G_1-to-S versus G_2-to-M control points.

If progression through critical points in the cell cycle is controlled by cyclin-dependent protein kinases, how is the activity of these kinases in turn regulated? One level of control is exerted by the availability of cyclins, whose concentrations oscillate during the cell cycle. For example, mitotic cyclins gradually increase in

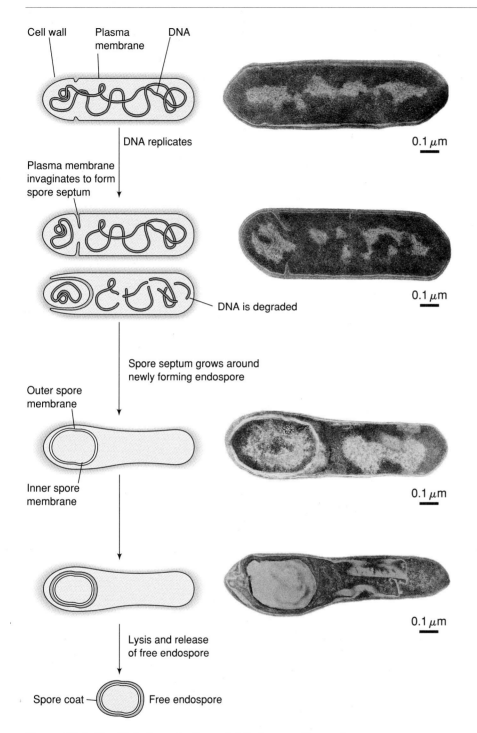

Cell wall   Plasma   DNA
             membrane

DNA replicates

0.1 μm

Plasma membrane
invaginates to form
spore septum

DNA is degraded

0.1 μm

Spore septum grows around
newly forming endospore

Outer spore
membrane

Inner spore
membrane

0.1 μm

0.1 μm

Lysis and release
of free endospore

Spore coat   Free endospore

Figure 12-6 *The Main Steps in Bacterial Endospore Formation* *The plasma membrane first invaginates across one end of the cell to form the double-membrane endospore septum. The septum then grows around the newly forming endospore, enclosing it within two membranes. Finally, a variety of cell wall materials are formed outside the membranes, thereby creating the spore coat. Micrographs courtesy of H. R. Hohl.*

concentration during G_1, S, and G_2, eventually reaching a concentration that is high enough to trigger the onset of M phase (Figure 12-9, *left*). Shortly after entry into M phase, the mitotic cyclins are degraded. During the next cell cycle, mitosis cannot be triggered until the mitotic cyclin concentration is built up again.

Cyclin-dependent protein kinases are also regulated by phosphorylation. When a mitotic cyclin initially binds to cdc2 kinase, thereby forming MPF, this protein complex is inactivated by enzymes that catalyze the phosphorylation of several amino acids in cdc2 kinase. At the end of G_2, a phosphatase produced by the *cdc25*

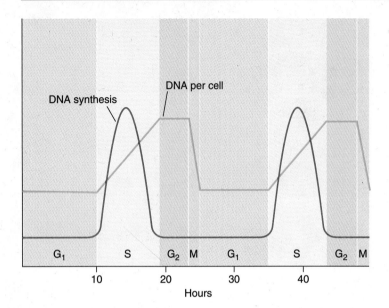

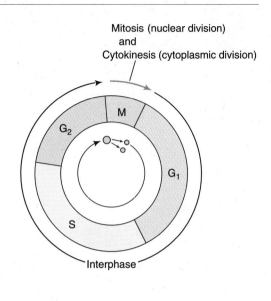

Figure 12-7 *The Eukaryotic Cell Cycle* *The graph on the left shows the rate of DNA synthesis and total DNA content per cell as a function of time in a synchronized cell culture. During S phase, DNA synthesis leads to a doubling of the DNA content per cell. During M phase, the amount of DNA per cell drops in half when the cells divide. The diagram on the right is a schematic view of the cell cycle for such a population.*

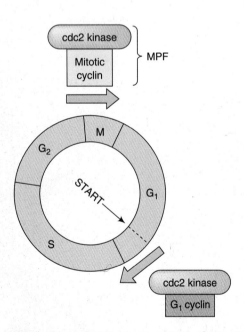

Figure 12-8 *The Role of Cyclin-Dependent Protein Kinases in the Eukaryotic Cell Cycle* *Depending on whether it is bound to a G_1 cyclin or a mitotic cyclin, the cdc2 kinase triggers either the G_1-to-S transition or the G_2-to-M transition.*

gene catalyzes the dephosphorylation of a specific tyrosine residue in cdc2 kinase, activating the kinase and triggering the onset of mitosis (see Figure 12-9, *right*). In cells where DNA replication has been experimentally interrupted, this dephosphorylation and the resulting activation of cdc2 kinase do not occur. Hence the dephosphorylation of cdc2 kinase serves as a control

mechanism that helps to ensure that mitosis does not occur before DNA replication has been completed.

Cell Division in Multicellular Organisms Requires Cell-Specific Growth Factors

Unicellular organisms such as bacteria and yeast live under conditions in which the presence of sufficient nutrients in the external environment is the primary factor that determines whether cells will grow and divide. In multicellular organisms the situation is usually reversed; cells are typically surrounded by nutrient-rich extracellular fluids, but the organism as a whole would be quickly destroyed if every cell were to continually grow and divide just because it had access to adequate nutrients. Cancer is a potentially lethal reminder of what happens when cell growth and division continue unabated without being coordinated with the needs of the organism as a whole (see Chapter 18).

The control of cell growth and division in eukaryotes is mediated by extracellular signaling molecules known as **growth factors.** If animal cells are placed in an artificial culture medium containing only nutrients and vitamins, they normally become arrested in G_1 in spite of the presence of adequate nutrients. Growth and division can be triggered by adding small amounts of blood serum, which contains several protein growth factors. Among them is **platelet-derived growth factor (PDGF),** a protein produced by blood platelets that stimulates the proliferation of connective tissue cells and smooth muscle cells. The PDGF released by platelets at wound sites is instrumental in stimulating

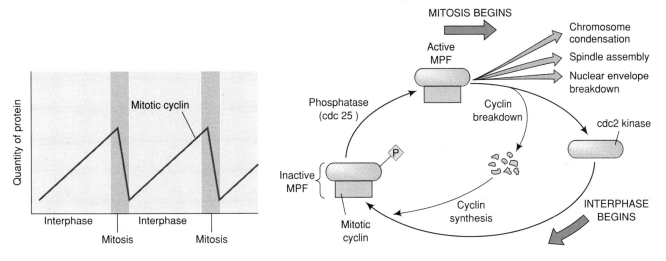

Figure 12-9 *The Role of Mitotic Cyclin and MPF in Triggering the Onset of Mitosis* (Left) *Data obtained from dividing sea urchin eggs show that mitotic cyclins are synthesized during interphase and broken down during mitosis.* (Right) *A schematic model illustrating how mitosis is triggered by MPF. During interphase, mitotic cyclins bind to cdc2 kinase to form MPF, but the MPF molecule is inactivated by phosphorylation. At the end of G_2 a phosphatase produced by the cdc25 gene catalyzes the dephosphorylation of MPF, activating its protein kinase activity and thereby triggering the onset of mitosis. Shortly after mitosis begins, the mitotic cyclins are degraded. During the following interphase, mitotic cyclins must be resynthesized before the cell can divide again.*

the growth of connective tissue that is required for wound healing.

Protein growth factors such as PDGF act by binding to plasma membrane receptors located on the surface of target cells. Different cells have different kinds of receptors and hence differ in the growth factors to which they respond. Growth factors typically interact with receptors that function as **protein-tyrosine kinases,** catalyzing the phosphorylation of tyrosine residues in protein molecules (page 221). Among the growth factors included in this category are *platelet-derived growth factor, epidermal growth factor,* and *nerve growth factor.* The binding of one of these growth factors to its corresponding membrane receptor activates the receptor's protein-tyrosine kinase activity, causing the receptor to phosphorylate itself. This *autophosphorylation* event leads to the activation of a membrane-associated G protein called **Ras,** which in turn triggers the stimulation of a family of protein phosphorylating enzymes called **mitogen-activated protein kinases,** or **MAP kinases** (see Figure 18-18). Activated MAP kinases phosphorylate a variety of proteins involved in stimulating cell growth and division, including the *AP1 transcription factor,* which controls the expression of a set of genes whose products are required for cell proliferation (page 464). As we will learn in Chapter 18, disruption of the preceding pathway for growth-factor induced stimulation of cell growth and division is frequently associated with the development of cancer (page 797).

Although the mechanism by which the activation of Ras and MAP kinases ultimately causes cells to grow and divide is not completely understood, it is clear that an increase in the rate of protein synthesis is necessary if cell growth is to be stimulated. When growth factors are added to cells that have been arrested in G_1, protein synthesis is stimulated and cell growth ensues. The resulting increase in cell mass is followed by entry into S phase, chromosome replication, and cell division. The rate of cell growth and division are usually coordinated so that the appropriate cell size is maintained. It has been shown in amoebae, for example, that division only occurs after the mass of the growing cell has doubled, and that division can be prevented by experimentally removing small portions of cytoplasm. However, this simple relationship does not always hold. Yeast grown in a culture medium depleted of nutrients divide without doubling their mass, yielding cells that are smaller than normal. Similarly, the development of animal embryos usually involves many mitotic divisions in the absence of cell growth, generating smaller and smaller cells. It is clear, therefore, that cell growth can be uncoupled from cell division under certain circumstances.

The existence of separate mechanisms for controlling cell growth and cell division is well illustrated in higher plants, where different signaling molecules control these two processes. Plants synthesize several classes of *plant growth substances* that regulate the growth and differentiation of particular parts of the plant (Table 12-1). Included are *auxins* and *gibberellins,* which stimulate cell enlargement; *cytokinins,* which promote cell division; and *abscisic acid* and the gas *ethylene,* which inhibit growth, especially during winter or the dry season when environmental conditions are unfavorable. If plants cells are placed in a cul-

Table 12-1   Structure of Plant Growth Substances

Class	Structure
Auxins	
Gibberellins	
Cytokinins	
Abscisic acid	
Ethylene	$CH_2 = CH_2$

ture medium containing auxin, the cells grow very large but do not divide. In contrast, when cytokinins are added in the absence of auxin, the cells neither grow nor divide. But when auxin and cytokinins are added together in the proper proportions, the cells grow as well as divide. Moreover, the developmental fate of the dividing cells is determined by the ratio of auxin to cytokinin. When the concentration of cytokinin is higher than that of auxin, the dividing cells tend to develop into shoots; in contrast, an excess of auxin causes the dividing cells to develop into roots.

Eukaryotic Chromosome Replication Is Semiconservative

Now that we have described some of the factors involved in controlling the eukaryotic cell cycle, we are ready to examine the activities that occur as cells prepare to divide. One of the most important events that must take place in preparation for cell division is chromosome replication. Two factors make the process of chromosome replication more complex in eukaryotic cells than in bacteria. First, the DNA content of the eukaryotic nucleus is usually several orders of magnitude greater than that of a bacterial nucleoid (see Table 10-5). Although this increased DNA content is distributed among multiple chromosomes, each individual chromosome still contains from ten to a hundred times more DNA than a typical bacterial chromosome. Second, the

nucleosomal organization of eukaryotic chromatin must be disassembled at the site of DNA replication and reassembled as each new DNA molecule is synthesized.

In spite of these differences, eukaryotes employ a semiconservative mechanism of DNA replication that is similar to the one employed by prokaryotes (page 67). The existence of semiconservative replication in eukaryotic chromosomes was first suggested in the late 1950s by the studies of J. Herbert Taylor, who exposed growing cultures of plant cells to ³H-thymidine and then examined the distribution of radioactivity in each chromosome using light microscopic autoradiography. In eukaryotes, the replication of each chromosome generates two newly synthesized chromosomes that remain attached to each other until the cell divides. As long as they remain attached, the two new chromosomes are referred to as **chromatids,** and the attachment site that holds each pair of chromatids together is called the **centromere.** Taylor discovered that during the first cell division after exposure to ³H-thymidine, both chromatids are radioactive. But during the following cell division, each replicated chromosome consists of one radioactive chromatid and one nonradioactive chromatid. This pattern is exactly what would be expected if each chromosome contains a single DNA double helix that replicates semiconservatively, giving each newly forming chromatid a DNA molecule containing one old DNA strand and one new DNA strand (Figure 12-10).

Eukaryotic Chromosomes Are Replicated Using Multiple Replicons

Earlier in the chapter we learned that DNA replication is initiated at a single site in bacterial chromosomes. In contrast, electron microscopic evidence indicates that eukaryotic chromosomal DNA replication is initiated at multiple sites, creating multiple replication units called **replicons** (Figure 12-11, *top and middle*). Most of these individual replication units fall in the size range of 50,000 to 300,000 base pairs in length, with an average chromosome containing about a thousand replicons. In order to determine the direction in which replication proceeds within each replicon, experiments have been carried out in which cells were first exposed to low concentrations of ³H-thymidine and then to higher concentrations of ³H-thymidine. Under such conditions, the DNA synthesized earlier will be less radioactive than the DNA synthesized later. When DNA isolated from such cells is examined by autoradiography, the DNA in the center of each replicon is found to be less radioactive than the DNA at either end (see Figure 12-11, *bottom*). This pattern is consistent with the theory that replication within each replicon is *bidirectional*, proceeding in both directions away from a single **replication origin.** When the DNA being replicated by one replicon encounters the DNA replicated by an adjacent replicon,

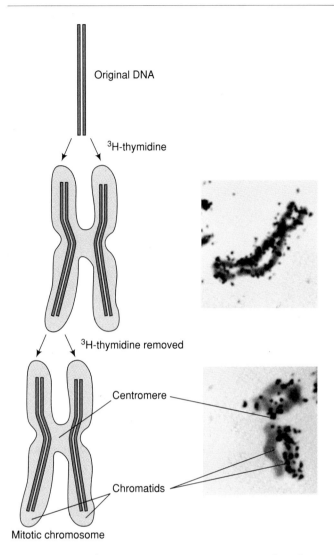

Original DNA

^{3}H-thymidine

^{3}H-thymidine removed

Centromere

Chromatids

Mitotic chromosome

Figure 12-10 *Experimental Evidence Demonstrating the Semiconservative Distribution of Newly Synthesized DNA into the Chromatids of Eukaryotic Chromosomes* *Broad bean cells were incubated for one round of cell division in the presence of ^{3}H-thymidine, followed by one round in the absence of ^{3}H-thymidine. The photographs of mitotic chromosomes present in these cells show that the two chromatids that make up each chromosome are radioactive after the first round of DNA replication, whereas only one of the two chromatids is radioactive after the second round of replication. The diagram shows that such results are compatible with the existence of semiconservative replication. Micrographs courtesy of J. H. Taylor.*

the DNA synthesized by the two replicons is joined together. In this way, DNA synthesized at numerous replication sites is ultimately linked together to form a new chromosomal DNA molecule (Figure 12-12).

The DNA sequences that function as replication origins have been identified in yeast cells by isolating various DNA fragments and inserting them into nonchromosomal circular DNA molecules that lack the ability to replicate. If the inserted DNA fragment gives

the nonchromosomal DNA molecule the ability to replicate within the yeast cell, it is called as an *autonomously replicating sequence,* or *ARS* **element.** The number of *ARS* elements detected in normal yeast chromosomes is similar to the total number of replicons, suggesting that *ARS* sequences function as replication origins. *ARS* elements contain an 11 nucleotide consensus sequence consisting largely of A-T base pairs. Since DNA must be unwound when replication is initiated, the presence of A-T base pairs serves a useful purpose at replication origins because A-T base pairs, held together by two hydrogen bonds, are easier to disrupt than G-C base pairs, which have three hydrogen bonds.

Why does DNA replication involve multiple replicons in eukaryotes but not prokaryotes? Because eukaryotic chromosomes contain more DNA than bacterial chromosomes, it would take eukaryotes much longer to replicate their chromosomes if DNA synthesis were initiated from only a single replication origin. Moreover, the rate at which each replication fork synthesizes DNA is slower in eukaryotes than in bacteria, presumably because the presence of nucleosomes slows down the replication process. Measurements of the length of radioactive DNA synthesized by cells exposed to ^{3}H-thymidine for varying periods of time have revealed that eukaryotic replication forks synthesize DNA at a rate of about 2000 base pairs/minute, compared to 50,000 base pairs/minute in bacteria. Since the average human chromosome contains about 10^8 base pairs of DNA, it would take more than a month to duplicate a chromosome if only a single replication origin were employed.

The ability to increase the speed of chromosome replication by increasing the number of replicons is nicely illustrated by comparing the rates of DNA synthesis in embryonic and adult cells of the fruit fly *Drosophila.* In the developing embryo, S phase is reduced to a few minutes to allow cell division to proceed very rapidly. To replicate chromosomes so quickly, embryonic cells employ a large number of replicons that are simultaneously active and separated by intervals of only a few thousand base pairs. In contrast, adult cells employ fewer replicons spaced at intervals of tens or hundreds of thousands of base pairs, generating an S phase that requires about 10 hours to complete. Since the rate of DNA synthesis at any given replication fork is about the same in embryonic and adult cells, it is clear that the length of the S phase is determined by the number of replicons and the rate at which they are activated rather than the rate at which each replicon synthesizes DNA.

At one time it was thought that eukaryotic DNA replication occurs in association with the nuclear envelope, just as replication of the bacterial chromosome is membrane associated. However, autoradiographic studies of cells briefly exposed to ^{3}H-thymidine have

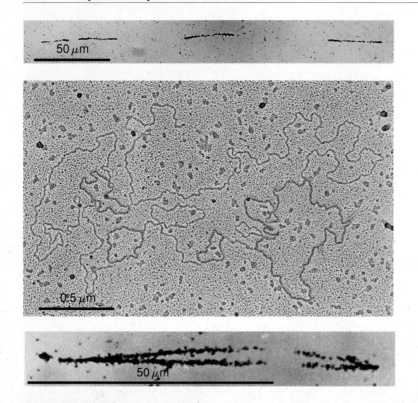

Figure 12-11 *Evidence for Multiple Replication Origins in Eukaryotic DNA*
(Top) *Autoradiograph of a replicating eukaryotic DNA molecule isolated from cells briefly incubated in the presence of ³H-thymidine. Note the presence of several distinct regions of radioactive DNA, each representing a separate replicon.* (Middle) *Electron micrograph of a eukaryotic DNA molecule showing four replicons* (color) *undergoing replication.* (Bottom) *Autoradiographic evidence for the occurrence of bidirectional DNA replication in eukaryotic replicons. Chinese hamster cells were briefly exposed to a low concentration of ³H-thymidine, followed by growth in the presence of a high concentration of ³H-thymidine. Note that the DNA is more radioactive at the two ends of this replicon than in the middle, as would be predicted if replication were bidirectional. Courtesy of J. A. Huberman* (top and bottom) *and D. R. Wolstenholme* (middle).

revealed that newly synthesized radioactive DNA occurs throughout the nucleus, not just adjacent to the nuclear envelope. Subsequent nuclear fractionation studies have shown that this newly formed DNA is associated with the nuclear matrix, leading to the suggestion that replication occurs as the DNA molecule passes through fixed replication sites bound to the nuclear matrix (Figure 12-13). Further support for the idea that eukaryotic DNA replication occurs at fixed matrix locations has been obtained by microscopic examination of cells incubated with DNA precursors that can be visualized by fluorescence microscopy. Such studies have revealed that DNA synthesis is restricted to a few hundred discrete foci within the nucleus, each containing several hundred replicons (Figure 12-14).

Active and Inactive Genes Are Replicated at Different Times During S Phase

If synchronized cell populations are incubated with ³H-thymidine at various times during S phase, the radioac-

tive DNA replicated in early S phase is found to be enriched in the bases G and C, whereas DNA synthesized later in S phase is enriched in the bases A and T. Such observations indicate that different kinds of DNA are replicated at different stages during S phase. The chromosomal locations of the DNA synthesized at these various stages have been identified by incubating cells with *5-bromodeoxyuridine (BrdU),* a substance that is incorporated into DNA in place of thymidine. The presence or absence of BrdU in various regions of mitotic chromosomes is then monitored using appropriate staining procedures or anti-BrdU antibodies. This approach has revealed that different regions of each chromosome are replicated in a defined sequence during S phase (Figure 12-15).

Additional information concerning the order in which DNA sequences are replicated has been obtained by isolating DNA from cells exposed to BrdU at various points during S phase. Because DNA that contains BrdU is denser than normal DNA, it can be separated from the remainder of the DNA by *isodensity centrifugation*

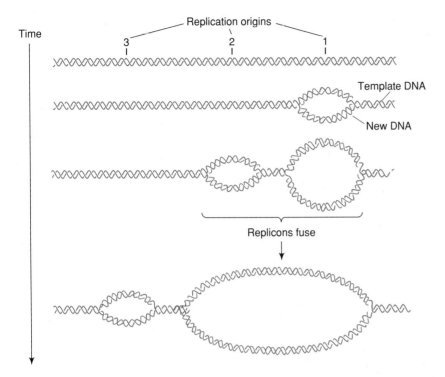

Time

Replication origins

3   2   1

Template DNA

New DNA

Replicons fuse

Figure 12-12 *Model Showing How Replicons Are Joined Together During Chromosomal DNA Replication in Eukaryotes* *In this example, DNA replication is initiated first at replicon 1, followed by replicon 2 and then replicon 3.*

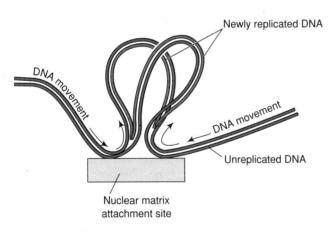

Newly replicated DNA

DNA movement

DNA movement

Unreplicated DNA

Nuclear matrix attachment site

Figure 12-13 *Model Showing How Bidirectional DNA Replication Might Be Achieved by Fixed Replication Sites Associated with the Nuclear Matrix* *The newly replicated DNA emerges as a loop from the center of the replication site as the unreplicated DNA enters from both sides.*

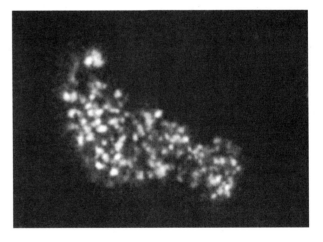

Figure 12-14 *Discrete Sites of DNA Replication within the Nucleus* *Nuclei were briefly incubated in the presence of DNA precursors that make the newly synthesized DNA visible by fluorescence microscopy. Each bright spot represents a region within the nucleus where several hundred replication units (replicons) remain clustered together throughout S phase. Courtesy of R. A. Laskey.*

(page 135). When the purified BrdU-labeled DNA is analyzed by hybridization with a series of DNA probes that are specific for individual genes, it is found that genes which are being actively expressed are replicated relatively early during S phase. If the same gene is analyzed in two different cell types, one in which the gene is active and one in which it is inactive, early replication is observed only in the cell type where the gene is being transcribed.

Certain chromosomal regions are permanently organized into a tightly packed, transcriptionally inactive form known as *heterochromatin* (page 450). In contrast to the early replication of active genes, heterochromatin is replicated late in S phase. Heterochromatin is typically found in chromosomal *centromeres* and in regions at the ends of chromosomes known as *telomeres;* it also comprises the

Early S   Middle S   Late S

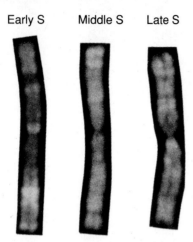

Figure 12-15 *Patterns of DNA Synthesis Visualized in a Mitotic Chromosome Labeled at Different Times During the Previous S Phase* *Replicating DNA was visualized by incubating Chinese hamster cells throughout S phase with BrdU, except for a single hour during either early, middle, or late S, when the BrdU was replaced by thymidine. DNA containing thymidine fluoresces brightly when stained, while DNA made in the presence of BrdU does not. Hence when mitotic chromosomes are visualized by fluorescence microscopy, the bright bands represent DNA synthesized when thymidine was present. Note that DNA located in different regions of the chromosome is synthesized during early, middle, and late S phase. Courtesy of E. Stubblefield.*

bulk of the inactivated X chromosome present in the cells of female mammals. Studies on the replication of X chromosomes have shed light on the question of whether the late replication of heterochromatin is caused by its base sequence or its tightly packed structure. The two X chromosomes found in the cells of a female mammal are similar in base sequence, but one chromosome is active while the other contains heterochromatin and is therefore inactive. The active X chromosome is replicated throughout S phase but the inactive X chromosome, like other forms of heterochromatin, is replicated only during late S phase. Hence the tightly condensed configuration of the inactive X chromosome rather than its base sequence must be responsible for late replication.

Nucleosome Assembly During S Phase Requires the Synthesis of New Histones

Unlike bacterial DNA, the DNA of eukaryotic chromosomes is packaged with histone octamers to form nucleosomes (page 412). Because the total amount of DNA is doubled during chromosome replication, the number of nucleosomes must also double. The histone molecules required for the assembly of new nucleosomes are synthesized mainly during S phase; the production of histones appears to be closely coupled to DNA replication, since exposing cells to inhibitors of DNA synthesis also causes histone synthesis to stop.

The control of histone synthesis is accomplished in at least two ways: The transcription of histone genes is activated during S phase, and the mRNAs transcribed from these histone genes are stable only when DNA synthesis is occurring.

To investigate how newly formed histones are assembled into nucleosomes, experiments have been carried out in which new histones were labeled by growing S-phase cells in the presence of amino acids containing heavy isotopes. Histones synthesized from heavy amino acids are denser than normal, and so can be separated from previously existing histones by isodensity centrifugation. The results of such experiments revealed the presence of two main classes of histone octamers: *new octamers* composed entirely of newly synthesized dense histones, and *old octamers* containing pre-existing histones of normal density. It was therefore concluded that old histone octamers remain largely intact during DNA replication, while the new histones synthesized during S phase are assembled into new octamers.

How are the old and new histone octamers distributed on the two DNA double helices that are formed as replication proceeds? Although studies addressing this issue have not provided an unequivocal answer, the total body of evidence suggests that the old octamers associate with both of the DNA double helices that are produced as the replication fork proceeds, but with a bias toward the leading strand (Figure 12-16). It was originally proposed that old histone octamers are transiently displaced from the DNA double helix by the moving replication fork, and that they quickly rebind to the replicated DNA after the replication fork has passed. This idea has been tested experimentally by incubating a sample of replicating DNA in the presence of a large excess of nonreplicating DNA that lacks nucleosomes. If histone octamers transiently dissociate from the DNA during replication, the dissociated octamers should bind to the excess nonreplicating DNA. However, the pre-

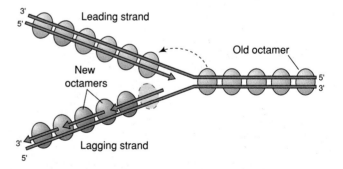

Figure 12-16 *Model Showing the Distribution of Old and New Histone Octamers During Eukaryotic DNA Replication* *Old octamers associate primarily with the leading strand, sliding from the parental double helix ahead of the replication fork to the leading strand behind the fork. New octamers are assembled mainly on the lagging strand.*

dicted transfer of histone octamers to the excess non-replicating DNA was not observed. It therefore has been concluded that histone octamers remain associated with the DNA molecule as the replication fork passes, somehow sliding from the parental DNA helix ahead of the fork to the leading strand behind the fork. At the same time, new histone octamers are assembled, primarily on the lagging strand. During assembly of the new octamers, histones H3 and H4 become bound to the DNA first, followed by histones H2A and H2B.

Telomeres Protect the Ends of Linear Chromosomes from Being Shortened During Replication

Eukaryotic chromosomes, unlike those of bacteria, are usually linear rather than circular. For this reason, a special problem is encountered when the DNA located at the two ends of each linear chromosome is replicated. As we learned in Chapter 3, DNA synthesis on the lagging strand of a replication fork requires the formation of a series of short RNA primers that are later removed,

leaving gaps to be filled by DNA polymerase (page 72). But DNA polymerase can only synthesize DNA in the $5' \rightarrow 3'$ direction, which means that the final gap created at the 3' ends of both strands of a linear chromosome cannot be filled (Figure 12-17, *left*). Hence linear chromosomes are in danger of being shortened at both ends each time replication occurs.

Eukaryotes have solved this problem by placing highly repeated satellite DNA sequences at the terminal ends, or **telomeres**, of each linear chromosome. These special telomeric elements consist of short repeating sequences enriched in the base G in the strand running in the $5' \rightarrow 3'$ direction. The sequence TTAGGG, which occurs at the ends of human chromosomes, is a typical example of such a telomeric or ***TEL*** **sequence** (Figure 12-18). Human telomeres contain between 250 and 1500 copies of the TTAGGG sequence repeated in tandem. At the end of this stretch of repeated sequences, the final TTAGGG folds back and forms unusual G-G base pairs with the preceding TTAGGG sequence (see Figure 12-17, *right*). This folded structure is recognized

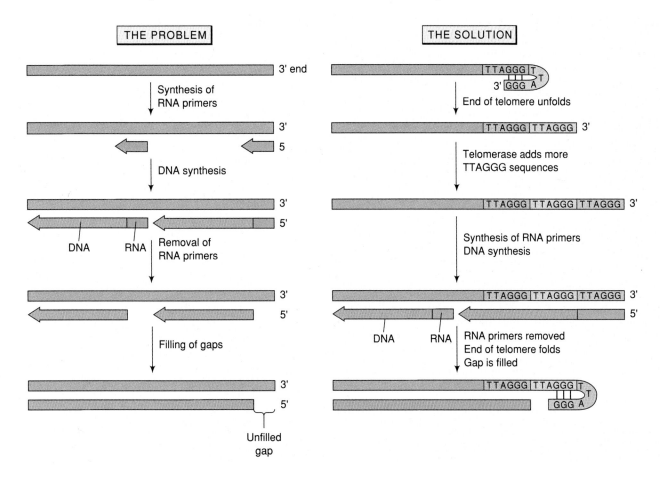

Figure 12-17 *Structure of the Telomere and Its Role in Preventing DNA Shortening* (Left) *When the 3' end of a DNA strand is replicated, an unfilled gap is left at the 5' end of the newly synthesized chain because DNA cannot be synthesized in the $3' \rightarrow 5'$ direction. Linear DNA molecules are therefore in danger of being shortened each time replication occurs.* (Right) *This problem is solved by placing telomeric repeat sequences at the ends of linear chromosomal DNA molecules. Telomerase catalyzes the addition of extra repeat sequences to the ends of the telomeres, thereby counteracting the tendency of the DNA molecule to be shortened during replication.*

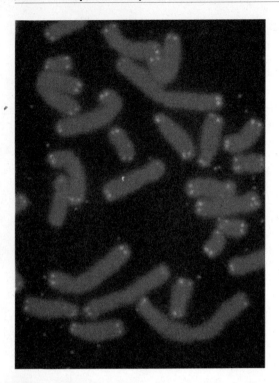

Figure 12-18 *Telomeres of Human Mitotic Chromosomes Stained by a Fluorescent Probe (Yellow) That Recognizes the DNA Sequence TTAGGG* *Each chromosome has been replicated in preparation for cell division, but the resulting chromatids have not yet separated from each other. The bright spot at the end of each chromatid represents the location of the TTAGGG repeat sequence. Courtesy of R. Moyzis.*

by a special enzyme, called **telomerase,** which catalyzes the formation of additional copies of the telomeric repeat sequence. The ability of telomerase to generate additional copies of the repeat sequence compensates for the shortening that occurs at both 3' ends of the chromosome during DNA replication.

Telomerase is an unusual enzyme because it contains RNA as well as protein. In the protozoan *Tetrahymena,* whose telomerase has been well characterized, the RNA part of the enzyme contains the sequence 3'-AACCCC-5', which is complementary to the 5'-TTGGGG-3' repeat sequence that makes up *Tetrahymena* telomeres. The presence of the complementary RNA sequence in the telomerase enzyme molecule suggests that this enzyme uses its own RNA template to create the DNA sequence being added to the ends of telomeres.

The ultimate proof that tandem repeats of telomeric sequences such as TTAGGG or TTGGGG protect the 3' ends of chromosomal DNA from being whittled away during replication has come from studies involving the creation of *yeast artificial chromosomes (YACs).* In order for a linear DNA molecule to function as a chromosome in yeast, it must meet three criteria: (1) the ability to be replicated during the S phase of the cell cycle, (2) the ability to segregate the two newly formed

chromosomal copies to the two cells formed during mitotic division, and (3) the ability to maintain itself from generation to generation. Recombinant DNA techniques have been employed to insert various kinds of sequences into linear DNA molecules to see which sequences perform the preceding functions.

We have already seen that the ability to carry out the first function, namely DNA replication, requires the presence of *ARS* elements that function as replication origins. The ability of chromosomes to segregate properly during mitosis requires a second type of DNA element called a *CEN* **sequence** because it occurs at the *centromere* region where the replicated DNA molecules remain attached prior to mitosis. We will see later in the chapter that *CEN* sequences are involved in attaching chromosomes to the mitotic spindle at the time of cell division. Finally, the third type of sequence element required for a functional chromosome is a *TEL* sequence such as TTAGGG or TTGGGG. When a tandemly repeated *TEL* sequence is added to the end of a linear DNA molecule containing *CEN* and *ARS* sequences, the molecule behaves like a normal chromosome that is properly replicated and distributed for many cell generations. Without *TEL* sequences at its two ends, the chromosome is replicated and distributed properly during mitosis, but eventually disappears from the cell.

M Phase Is Subdivided into Prophase, Metaphase, Anaphase, Telophase, and Cytokinesis

Upon completing the process of chromosome replication, cells exit from S phase and enter into G_2 phase, where final preparations are made for the onset of mitosis. If G_2 cells are fused with S-phase cells, the G_2 nuclei do not initiate a second round of DNA replication, even though the S-phase cells possess all the components required to initiate DNA synthesis. G_2 nuclei must therefore contain some factor that prevents DNA from being replicated again. Once a cell has entered into G_2, it almost always proceeds to mitosis within a few hours. However, certain kinds of cells can be selectively arrested during G_2. Unlike G_1-arrested cells, cells arrested in G_2 are capable of immediately initiating mitosis in response to an appropriate stimulus. Hence they occur in places such as skin, where rapid cell division is required in response to injury. In contrast to G_2-arrested cells, cells arrested during G_1 require many hours to enter mitosis because they must first complete G_1, S, and G_2.

After passing through G_2, cells enter into the M phase of the cell cycle. The purpose of M phase is twofold: to divide the cell into two new cells and to distribute a set of chromosomes to each of the two cells. The main activity of M phase is *mitosis,* which is the process that converts a single nucleus containing two sets of chromosomes into two nuclei, each containing a single set of chromosomes. Mitosis is classically divided into four stages called *prophase, metaphase, anaphase,* and *telophase* (Figures 12-19 and 12-20). During

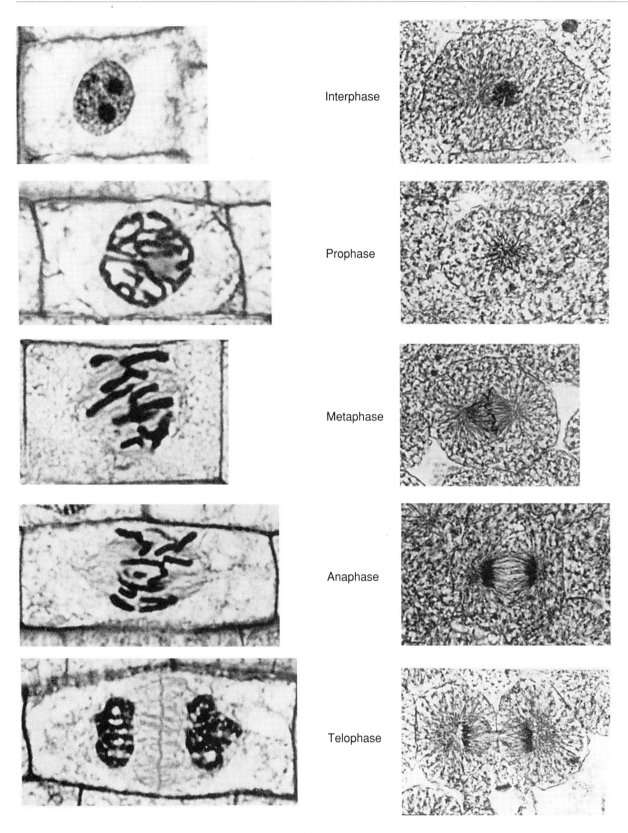

Figure 12-19 *Stages of Mitosis in Animal and Plant Cells* (Left) *Cells from an onion root tip.* (Right) *Cells from newt lung epithelium.*

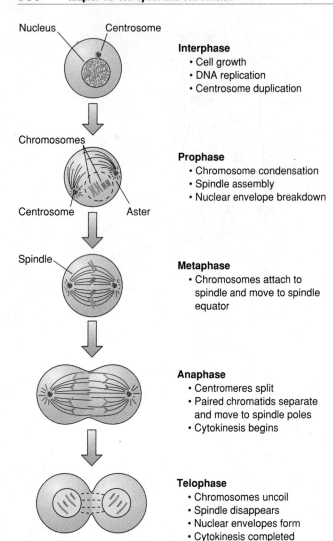

Nucleus   Centrosome

Interphase
• Cell growth
• DNA replication
• Centrosome duplication

Chromosomes

Prophase
• Chromosome condensation
• Spindle assembly
• Nuclear envelope breakdown

Centrosome   Aster

Spindle

Metaphase
• Chromosomes attach to spindle and move to spindle equator

Anaphase
• Centromeres split
• Paired chromatids separate and move to spindle poles
• Cytokinesis begins

Telophase
• Chromosomes uncoil
• Spindle disappears
• Nuclear envelopes form
• Cytokinesis completed

Figure 12-20 *Summary of the Main Events Occurring During Mitosis in an Animal Cell*

prophase the chromosomes condense and become visible as separate entities, the mitotic spindle is assembled from newly forming microtubules, and the nuclear envelope breaks down. At **metaphase** the chromosomes become attached to the spindle and migrate to its equator. During **anaphase** the paired chromatids that make up each chromosome separate and move to opposite poles of the spindle. Finally, during **telophase** the separated chromosomes uncoil into chromatin fibers, the spindle disappears, and new nuclear envelopes form around each set of chromosomes.

While nuclear division is proceeding during mitosis, the companion process of cytokinesis begins dividing the cell into two new cells, each containing one of the nuclei generated by mitosis. Cytokinesis usually begins during anaphase and continues until the end of telophase.

Now that the main activities associated with the M phase of the cell cycle have been introduced, we will

discuss what is known about the mechanisms that underlie these events.

Protein Phosphorylation Initiates Prophase by Triggering Chromosome Condensation and Nuclear Envelope Breakdown

The DNA of interphase nuclei is usually packaged in the form of a diffuse network of intertwined chromatin fibers, making it impossible to identify individual chromosomes. Prophase marks the point in the cell cycle when the replicated chromatin fibers first condense into compact structures that can be recognized as individual chromosomes, each consisting of a pair of chromatids joined together at the centromere. Chromosome condensation is essential because interphase chromatin fibers are so long and intertwined that in an uncompacted form they would become impossibly tangled when attempts were made to distribute the chromosomal DNA during cell division. As the chromosomes condense during early prophase, gene transcription slows down and eventually stops. In animal cells the nucleoli usually disperse at this time; plant cell nucleoli may either remain as discrete entities, undergo partial disruption, or disappear entirely as in animal cells.

Chromosome condensation is triggered by MPF, the regulatory protein that consists of cdc2 kinase bound to a mitotic cyclin (page 524). Histone H1 is one of several proteins known to be phosphorylated by MPF. If MPF is injected into cells that have just emerged from S phase, the phosphorylation of histone H1 and chromosome condensation occur immediately, rather than after the normal G_2 delay of several hours. Such experiments suggest that MPF triggers chromosome condensation by catalyzing the phosphorylation of histone H1, a conclusion that is consistent with other evidence implicating this histone in controlling the higher-order folding of chromatin fibers (page 414).

Nuclear envelope breakdown is another prophase event that has been linked to protein phosphorylation catalyzed by MPF. The inner surface of the nuclear envelope is supported by an underlying *nuclear lamina,* which consists of intermediate filaments constructed from proteins called **lamins** (page 622). At the beginning of prophase the lamin molecules become phosphorylated, a biochemical reaction that triggers lamin depolymerization and hence breakdown of the nuclear lamina. The disappearance of the nuclear lamina in turn destabilizes the nuclear envelope, causing it to disperse into tiny membrane vesicles. The preceding sequence of events is triggered by MPF, which catalyzes the phosphorylation of a specific serine residue in lamins; other sites within lamins are also phosphorylated by protein kinases whose activity is activated by MPF. Hence MPF contributes both directly and indirectly to the phosphorylation of lamins, which in turn triggers the breakdown of the nuclear lamina and the resulting fragmentation of the nuclear envelope.

The Mitotic Spindle Is Generated from Microtubules Whose Growth Is Initiated by Centrosomes

In addition to chromosome condensation and nuclear envelope breakdown, the other main event that occurs during prophase is assembly of the mitotic spindle. The mitotic **spindle** is a structure composed of *microtubules* that is responsible for separating the two sets of chromosomes during mitosis. The properties of microtubules and the mechanisms by which they operate will be described in detail in Chapter 13. To understand the general workings of the mitotic spindle, only a few basic properties of microtubules need to be introduced. Microtubules are long, relatively rigid filaments measuring 25 nm in diameter and up to several millimeters in length. Each microtubule is assembled and disassembled by the addition and removal of subunits comprised of two closely related proteins, *α-tubulin* and *β-tubulin*. One property that is crucial to the functioning of microtubules is that each microtubule has an inherent directionality, both along its surface and at both ends. To identify the direction in which a given microtubule is oriented, one end is called the *plus* end and the other is the *minus* end (page 598).

Pioneering experiments carried out in the laboratory of Sinya Inoué provided one of the first indications that assembly of the mitotic spindle is regulated by changes in the equilibrium between the polymerized and nonpolymerized states of tubulin. In these studies, Inoué showed that exposing cells to low temperature, high pressure, or the drug *colchicine* causes the mitotic spindle to disappear and mitosis to cease. The spindle spontaneously reforms, however, as soon as the temperature is raised, the pressure lowered, or the colchicine removed. This rapid reappearance of the spindle occurs even when inhibitors of protein synthesis are present, indicating that assembly of the spindle is not dependent on the synthesis of new spindle proteins. Such observations led to the conclusion that assembly of the mitotic spindle is based on the polymerization of preexisting tubulin molecules rather than the synthesis of new tubulin.

What causes tubulin to be suddenly polymerized into spindle microtubules at the onset of mitosis? The initiation of spindle assembly is controlled by a specialized cytoplasmic structure known as the **centrosome.** Cells in early interphase contain a single centrosome that consists of a small zone of granular material located to one side of the nucleus. This structure serves as the main site for initiating the polymerization of new microtubules in interphase cells. New microtubules grow by addition of tubulin subunits to their plus ends, while their minus ends remain stabilized by attachment to the centrosome. Embedded within the centrosome of animal cells is a pair of small cylindrical structures called *centrioles,* which are oriented at right angles to each other. Because centrioles are absent in certain cell types, including most higher plant cells, it is clear that they are not essential to the process of mitosis. They do, however, play a role in the formation of cilia and flagella, and so will be discussed in more detail when we cover these cellular appendages in Chapter 13.

The centrosome (including its pair of embedded centrioles if present) is duplicated prior to mitosis, usually during S phase when the chromosomes are being replicated. At the beginning of prophase, the duplicated centrosomes separate from each other and move toward opposite ends of the cell. At this time the ability of each centrosome to initiate the polymerization of microtubules increases dramatically, thereby resulting in the growth of many new microtubules. The enhanced ability to promote microtubule growth is thought to be triggered by the phosphorylation of centrosome proteins by MPF, which is present in high concentration in centrosomes. Some of the microtubules that develop at this stage are relatively short and radiate in all directions from each centrosome, forming structures known as **asters** (see Figure 12-20). Numerous longer microtubules also begin to fill the space separating the two centrosomes. As the centrosomes move toward opposite ends of the cell, these newly forming spindle microtubules lengthen and fill the region between the two migrating centrosomes. By the time the centrosomes reach opposite ends of the cell, the main body of the spindle is almost completely assembled.

At the Beginning of Metaphase, Chromosomes Attach to the Mitotic Spindle through Their Kinetochores

The end of prophase and the beginning of metaphase is marked by the breakdown of the nuclear envelope, which is reduced to a series of small vesicles that disperse throughout the cytoplasm. Nuclear pores soon disappear from these membrane vesicles, which become indistinguishable from membranes of the endoplasmic reticulum. The breakdown of the nuclear envelope allows the spindle microtubules to enter the space previously occupied by the nucleus, thereby bringing the microtubules in contact with the chromosomes. When contact is made between the free plus end of a spindle microtubule and the centromere region of a chromosome, the microtubule binds to the chromosome at the centromere (page 528).

The **centromere** region of a chromosome serves two crucial functions: (1) it holds the two paired chromatids together until the beginning of anaphase, and (2) it serves as the attachment site for spindle microtubules. The centromere is comprised of special DNA se-

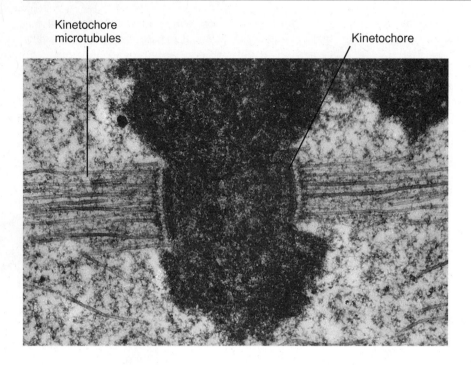

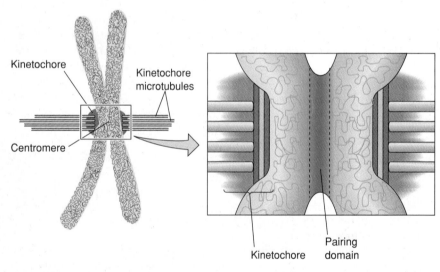

Figure 12-21 *Organization of the Centromere and Kinetochore in a Mitotic Chromosome* (Top) *The electron micrograph shows two kinetochores lying on opposite sides of a mitotic chromosome. Each kinetochore appears as a multilayered structure to which several microtubules are bound.* (Bottom) *The schematic diagrams summarize the relationship between kinetochores, kinetochore microtubules, and the centromere. The region of the centromere where the two chromatids come in close contact is called the pairing domain. Micrograph courtesy of M. Schibler.*

quences, called *CEN* sequences (page 534), that make these two functions possible. Although *CEN* sequences vary both between chromosomes and among different organisms, in higher eukaryotes they typically consist of extensively repeated satellite DNA sequences that can account for several percent of the total chromosomal DNA. Tightly bound to the surface of the centromere is a disk-shaped structure known as a **kinetochore,** which serves as the actual attachment site for microtubules. Each metaphase chromosome contains two kinetochores located on opposite sides of the centromere, one associated with each of the two chromatids. Kinetochores typically measure about 200 nm in diameter and consist of several layers of granular or fibrillar protein to which spindle microtubules can bind (Figure 12-21). Depending on the organism, anywhere from one to several dozen microtubules bind to each kinetochore. The spindle microtubules that become attached to kinetochores are called **kinetochore microtubules,** while the remaining microtubules that span the area between the two ends or *poles* of the spindle are called **polar microtubules** (Figure 12-22).

Later in mitosis, the two chromatids that make up each chromosome are destined to separate from each other and move to opposite poles of the spindle. This means that the two members of each chromatid pair must be attached to opposite spindle poles. When microtubules initially bind to kinetochores, mistakes occasionally cause both members of a chromatid pair to become attached to the same spindle pole, or cause one chromatid to become attached to both poles. However, the most stable configuration occurs when the two kinetochores lying on opposite sides of the centromere become attached to microtubules emerging from opposite spindle poles (Figure 12-23). Apparently the tension generated when microtubules from opposite spindle poles pull a pair of chromatids in opposite directions stabilizes the binding between kinetochores and their respective microtubules.

Metaphase Chromosomes Are Moved to the Spindle Equator by Two Kinds of Forces

When microtubules first become attached to chromosomes at the beginning of metaphase, the chromosomes are randomly distributed throughout the spindle. During the next part of metaphase the chromosomes mi-

grate toward the central region of the spindle, called the *spindle equator.* Movement toward the equator is accomplished through a series of agitated, back-and-forth motions that are generated by at least two different kinds of forces. One force is mediated by the kinetochore microtubules, which tend to move chromosomes toward the pole to which the microtubules are attached. The existence of this "pulling" force can be demonstrated experimentally by using tiny glass microneedles to tear individual chromosomes away from the spindle. A chromosome that has been removed from the spindle remains motionless until new microtubules attach to its kinetochore, at which time the chromosome is drawn back into the spindle.

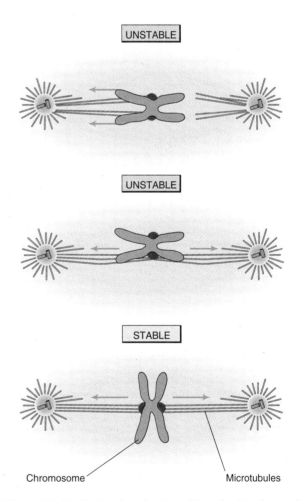

Figure 12-23 *Mechanism for Attaching the Members of a Chromatid Pair to Opposite Spindle Poles* *The two upper configurations represent mistakes in which two paired chromatids are attached to the same spindle pole, or one chromatid is attached to both poles. When such mistakes occur, the linkage between the kinetochore and the spindle microtubules does not persist. But when the two chromatids are pulled in opposite directions* (bottom configuration), *the linkage between the kinetochore and the spindle microtubules is stabilized, thereby ensuring that the chromatids will move to opposite poles during anaphase.*

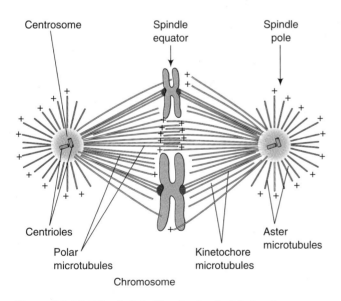

Figure 12-22 *Mitotic Spindle of a Typical Animal Cell* *Spindles contain three kinds of microtubules called aster microtubules, polar microtubules, and kinetochore microtubules. The plus signs indicate the orientation of the plus ends of the microtubules. In most plant cells the spindle does not contain the pairs of centrioles pictured within each centrosome.*

The second force tends to "push" chromosomes away if they approach either spindle pole. The existence of this pushing force has been demonstrated by studies in which one end of a chromosome is broken off using a laser microbeam. Once the broken chromosome fragment has been cut free from its associated centromere and kinetochore, the fragment tends to move away from the nearest spindle pole, even though it is no longer attached to the spindle by microtubules. The nature of the pushing force that propels chromosomes in the absence of microtubule attachments remains to be clearly identified.

The combination of pulling and pushing forces exerted on early metaphase chromosomes drives them to the spindle equator, which is their most stable location. Here the chromosomes line up in random order. Although metaphase chromosomes stop moving at this point, they are in constant tension in both directions. If the kinetochore located on one side of a metaphase chromosome is severed using a laser microbeam, the chromosome promptly migrates toward the opposite spindle pole. Hence metaphase chromosomes remain at the center of the spindle because the forces pulling them toward opposite poles are precisely balanced.

Metaphase Chromosomes Are Intricately Folded Chromatin Fibers

Metaphase is the best time to count and identify a cell's chromosomes because they all line up together across the mitotic spindle. Agents that interfere with the func-

tioning of the spindle, such as the drug colchicine, can be used to generate cell populations that are enriched in metaphase-arrested cells. Microscopic examination of such cells allows individual chromosomes to be identified and classified on the basis of differences in size and shape, generating a chromosomal analysis known as a **karyotype** (Figure 12-24). The most noticeable feature of karyotypes is that the distance between the centromere and the two ends of the chromosome tends to vary for different types of chromosomes.

By the end of metaphase, each chromosome has been compacted to the point where its overall length is about 10,000 times smaller than the extended length of the DNA it contains. Electron microscopic examination has revealed that metaphase chromosomes consist of an elaborately folded mass of fibers that appears to arise from the folding of 30-nm chromatin fibers (Figure 12-25). Current evidence suggests that each of the two chromatids that make up a metaphase chromosome is comprised of a single, intricately folded 30-nm chromatin fiber containing a single linear DNA molecule.

The way in which chromatin fibers are folded to create metaphase chromosomes is highly reproducible. The most striking evidence for this conclusion has emerged from studies of chromosomes stained with various dyes. This approach received a great impetus in the late 1960s when Torbjöern Caspersson discovered that incubating metaphase chromosomes with certain dyes yields a reproducible pattern of stained or fluorescent bands. By altering the conditions under which the chromosomes are prepared and stained, various kinds of

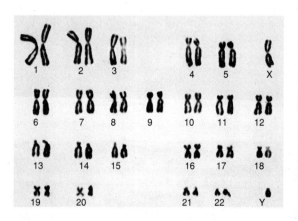
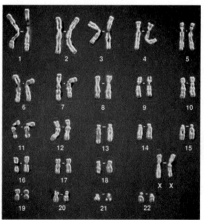

Figure 12-24 A Mitotic Karyotype of Human Chromosomes from Metaphase-Arrested Cells (Left) *This set of chromosomes obtained from the cells of a human male has been stained with a dye that reacts uniformly with the entire body of the chromosome. Human males contain 22 pairs of chromosomes, plus one X and one Y chromosome. The chromosomes in the karyotype have been arranged according to size and centromere position. (Right) This set of human female chromosomes has been stained with dyes that selectively react with certain chromosome regions, creating a unique banding pattern for each type of chromosome. Courtesy of J. F. Gennaro/Photo Researchers* (left) *and CNRI/Science Photo Library/Photo Researchers, Inc.* (right).

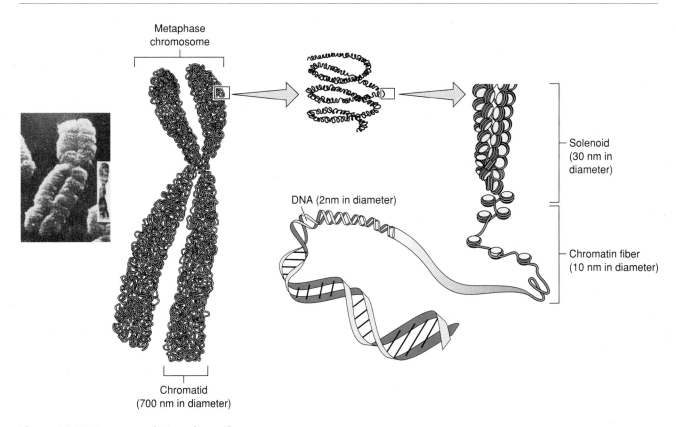

Figure 12-25 *Structure of Metaphase Chromosomes* (Left) *In scanning electron micrographs, metaphase chromosomes appear to be constructed from a tightly packed mass of fibers that are thicker than 10-nm or 30-nm chromatin fibers.* (Right) *A model illustrating how chromatin fibers may be packed in metaphase chromosomes. Micrograph courtesy of C. J. Harrison.*

banding patterns can be produced. Two of the most commonly used techniques stain either *G bands* or *R bands,* which differ mainly in their base composition. G bands contain DNA enriched in A-T base pairs that tends to be replicated late in S phase, whereas R bands contain DNA enriched in G-C base pairs that tends to be replicated earlier in S phase. The banding patterns exhibited by each type of chromosome are so unique that they can be used to identify chromosomes that are difficult to distinguish from one another on the basis of size and centromere position alone (see Figure 12-24, *right*).

The discovery that each type of chromosome exhibits a unique banding pattern suggests the existence of an orderly and reproducible folding of the chromatin fiber within metaphase chromosomes. What guides this folding process? Electron microscopic examination of metaphase chromosomes from which histones and most nonhistone proteins have been removed has revealed the existence of a *scaffold* of nonhistone proteins to which loops of DNA are attached (see Figure 10-19). A chromosome of average size has about 1000 of these *looped domains,* which vary from 20,000 to 100,000 base pairs in length. In interphase cells the ends of these looped domains are anchored to the nuclear matrix, a structure that ex-

hibits biochemical similarities to the scaffold of mitotic chromosomes.

Protein scaffolds isolated from metaphase chromosomes tend to exhibit the same shape as the chromosome from which they were obtained. Such observations suggest that the scaffold defines the pattern of chromosome folding by serving as the attachment site for the looped domains that are distributed throughout the length of the DNA molecule. One of the main constituents of the protein scaffold is topoisomerase II (page 415), an enzyme whose ability to modify DNA supercoiling is presumably required for the proper folding of chromatin fibers in mitotic chromosomes.

Anaphase Is Initiated by the Splitting of Chromosomal Centromeres

At the beginning of anaphase the centromere region of each chromosome splits, allowing the paired chromatids to separate and start moving toward opposite spindle poles. The mechanism responsible for centromere splitting is not completely understood, but several components appear to be involved. One is the enzyme topoisomerase II, which is concentrated near the centromere and which catalyzes changes in DNA supercoiling that are required for chromatid separation. In mutant cells lacking topoisomerase II, the paired chromatids still attempt to separate at the beginning of anaphase, but they are torn apart and damaged instead of separating properly. Another class of proteins implicated in chromatid separation are the *INCENPs* (Inner Centromere Proteins), a group of molecules that are localized in the *pairing domain* of the centromere where the two chromatids are in close physical contact (see Figure 12-21). INCENPs are released into the cytoplasm when anaphase begins, suggesting that they may function as an adhesive that holds the paired chromatids together until the start of anaphase.

Near the onset of anaphase, mitotic cyclin is linked to ubiquitin (page 510) to mark it for destruction. The breakdown of mitotic cyclin leads to the inactivation of cdc2 kinase, which had triggered the onset of mitosis by catalyzing the phosphorylation of specific proteins (page 536). Another change that occurs at this time is a rise in cytosolic Ca^{2+} concentration. An important role for this ion is suggested by the discovery that injecting Ca^{2+} into early metaphase cells causes a premature entry into anaphase. Among the possible targets of Ca^{2+} action is a calmodulin-dependent protein kinase that is present in the mitotic spindle.

Chromosomes Move toward the Spindle Poles During Anaphase A

The splitting of the centromeres that occurs at the beginning of anaphase allows the two chromatids that make up each metaphase chromosome to separate, thereby generating two independent chromosomes. The two chromosomes produced from each chromatid pair then migrate toward opposite spindle poles at a speed of about 1 μm/min. At this rate, it takes no more than a few minutes for the two sets of chromosomes to be segregated to opposite ends of the cell. Chromosome movement is accomplished through two independent mechanisms (Figure 12-26). The first, called **anaphase A,** involves the movement of each set of chromosomes toward the spindle pole to which their kinetochore microtubules are attached. The second, known as **anaphase B,** involves the movement of the two spindle poles away from each other. In some cells anaphase A (chromosome-to-pole movement) and anaphase B (pole-pole separation) occur at the same time, whereas in others anaphase B follows anaphase A.

The mechanisms that underlie these two types of chromosome movement are quite different. The movement of chromosomes toward the spindle poles during anaphase A is accompanied by shortening of the kinetochore microtubules. If cells are exposed to the drug *taxol,* which inhibits microtubule depolymerization, the shortening of the kinetochore microtubules is blocked and the chromosomes do not move toward the spindle poles. Conversely, exposing cells to increased atmospheric pres-

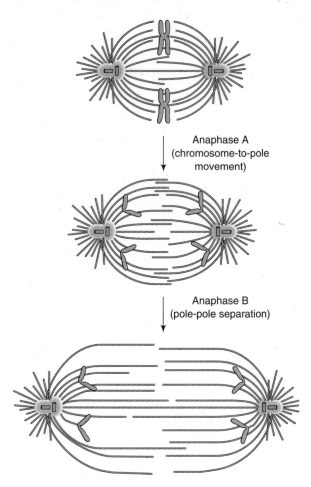

Anaphase A (chromosome-to-pole movement)

Anaphase B (pole-pole separation)

Figure 12-26 *The Two Types of Movement Involved in Chromosome Separation During Anaphase* *Anaphase A involves the movement of chromosomes toward the spindle pole to which they are attached. Anaphase B is the movement of the two spindle poles away from each other. Anaphase A and anaphase B may occur simultaneously.*

sure, which increases the rate of microtubule depolymerization, causes the chromosomes to move more quickly toward the poles. Both observations suggest that microtubule depolymerization is crucial to the anaphase A movement of chromosomes toward the spindle poles.

To investigate which end of the kinetochore microtubule is being depolymerized, metaphase cells have been injected with tubulin molecules linked to a substance that makes them visible using fluorescence microscopy. Initially the labeled tubulin is incorporated into the plus end of the kinetochore microtubules where they attach to chromosomal kinetochores. But these same fluorescent tubulin molecules are quickly released from the kinetochore microtubules when anaphase begins, indicating that the microtubules are being disassembled at their plus ends adjacent to the kinetochore. Such observations suggest that during anaphase A, each kinetochore pulls its attached chromosome toward the spindle pole by advancing along a stationary "track" of kinetochore microtubules, depolymerizing the microtubules as it proceeds (Figure 12-27). This view has received further support from the discovery that kinetochores contain several kinds of **motor proteins,** which are molecules that generate movement by advancing along a surface using energy derived from ATP hydrolysis. A motor protein known as *dynein* (page 609) has been tentatively identified as the protein that pulls chromosomes by advancing along the surface of kinetochore microtubules, but members of another family of motor proteins called *kinesins* (page 617) may also be involved.

Since the preceding model suggests that kinetochore microtubules serve as a stationary track along which the chromosomal kinetochores advance, studies have been carried out to determine whether the kinetochore microtubules are in fact stationary during anaphase A. For example, experiments have been performed in which cells are first injected with fluorescent tubulin molecules, and a laser microbeam is then employed to bleach a small spot on one of the fluorescent kinetochore microtubules (Figure 12-28, *left*). During anaphase A the bleached spots remain stationary, indicating that kinetochore microtubules do not move as the chromosomes migrate toward the spindle poles. In another experiment it has been shown that kinetochore microtubules do not even need to be associated with the spindle pole for chromosome movement to occur. In these studies, a tiny glass needle was used to cut the

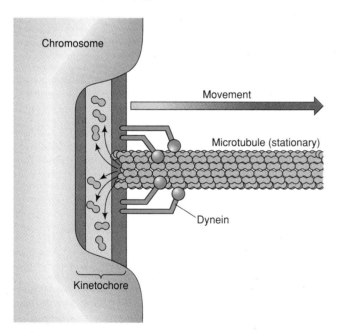

Figure 12-27 *Model for Chromosome Movement toward the Spindle Poles (Anaphase A)* *The kinetochore moves its attached chromosome along a stationary track of kinetochore microtubules, depolymerizing the microtubules as it proceeds. The motor protein responsible for generating chromosome movement has been tentatively identified as dynein, but other motor proteins have not been ruled out. In this diagram, all structures are moving toward the right except for the microtubule, which is stationary.*

mitotic spindle between the chromosomes and one of the spindle poles. Even though the kinetochore microtubules were no longer attached to the spindle pole, the chromosomes nonetheless continued to migrate along the microtubules (Figure 12-28, *right*).

The Spindle Poles Move Away from Each Other During Anaphase B

During anaphase B, which often overlaps anaphase A, the spindle poles move away from each other. The conclusion that anaphase A and anaphase B are based on different mechanisms is supported by the fact that they are inhibited by different treatments. The drug *chloral hydrate,* for example, inhibits pole-pole separation during anaphase B but has no effect on chromosome-to-pole movement during anaphase A.

Pole-pole separation is accompanied by two kinds of changes in the polar microtubules (not to be confused with the changes in the kinetochore microtubules just described). First, polar microtubules are lengthened by the addition of tubulin subunits to the microtubules' plus ends, which reside near the center of the spindle where the microtubules emerging from opposite spindle poles overlap. Second, polar microtubules derived from opposite spindle poles slide apart, forcing the poles away from one another. Microtubule sliding can be induced by exposing isolated spindles to ATP, suggesting that the spindle contains motor proteins that use energy derived from the hydrolysis of ATP to cause the overlapping microtubules to slide away from each other. In the region where the polar microtubules overlap, electron microscopy has revealed the presence of crossbridges that link adjacent microtubules (Figure 12-29). These crossbridges are thought to represent part of a motor mechanism that causes microtubule sliding to occur (Figure 12-30).

During Telophase, Nuclear Envelopes Form around the Two Sets of Chromosomes

The beginning of telophase is marked by the arrival of the two sets of chromosomes at opposite spindle poles. Upon arrival the chromosomes begin to uncoil, dispersing into typical interphase chromatin fibers. These changes are accompanied by the reappearance of nucleoli (if they had disappeared), the resumption of gene transcription, the disappearance of kinetochore microtubules, and the gradual disassembly of the mitotic spindle.

As the chromosomes begin to disperse into chromatin fibers, membrane vesicles that are indistinguishable from membranes of the endoplasmic reticulum begin to condense around the two sets of uncoiling chromosomes. These membrane vesicles gradually coalesce to form two new nuclear envelopes, one surrounding each set of chromosomes. It will be recalled

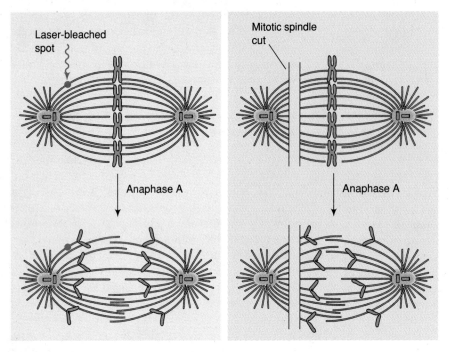

Figure 12-28 *Evidence That Kinetochore Microtubules Remain Stationary as Chromosomes Migrate toward the Spindle Poles* (Left) *Cells were injected with fluorescent tubulin and a laser microbeam was employed to bleach a small spot on one of the kinetochore microtubules. During anaphase A, the bleached spot remains stationary. (Right) A tiny glass needle was used to cut the mitotic spindle between the chromosomes and one of the spindle poles. Even though the kinetochore microtubules are no longer attached to the spindle pole, the chromosomes continue to migrate along the stationary kinetochore microtubules.*

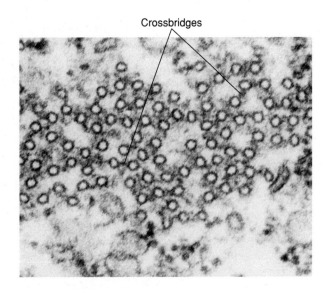

Figure 12-29 *Crossbridges between Spindle Microtubules* *This electron micrograph shows a cross section through the zone where the polar microtubules overlap at the center of the mitotic spindle. Small crossbridges between the microtubules are visible. Courtesy of K. L. McDonald.*

that an increased phosphorylation of lamin molecules triggers the breakdown of the nuclear lamina and the resulting fragmentation of the nuclear envelope at the beginning of prophase. During telophase this process is reversed; lamin phosphorylation decreases dramatically, allowing the nuclear lamina to be reassembled as an underlying support for the nuclear envelope.

Cytokinesis in Animal Cells Is Mediated by a Contractile Ring That Pinches the Cell in Two

At the end of mitosis the process of **cytokinesis** divides the cytoplasm in two, separating the original cell into two new cells. Cytokinesis typically overlaps the final stages of mitosis, beginning during anaphase and proceeding throughout telophase. There are reasons for believing, however, that cytokinesis is not inextricably linked to mitosis. First, a significant time lag may occur between nuclear division and cytokinesis, thereby indicating that the two processes are not tightly coupled to one another. And second, in certain cell types chromosome replication and nuclear division occur in the absence of cytokinesis, generating cells with multiple nuclei.

Cytokinesis usually divides the cell along a plane that passes through the spindle equator, suggesting that

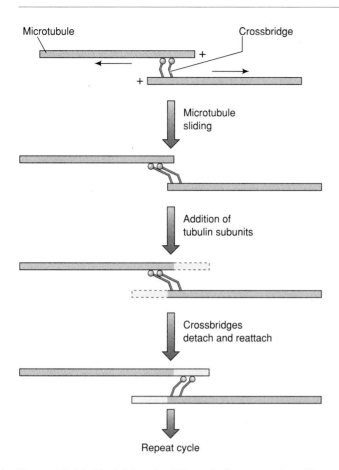

Figure 12-30 *Model for the Microtubule Movements That Cause Pole-Pole Separation (Anaphase B)* *In this model, crossbridges function as "motor molecules" that cause overlapping polar microtubules to slide away from one another. As the microtubules slide apart, tubulin subunits are added to their plus ends to maintain an area of overlap, allowing additional crossbridges to form and trigger further movement.*

the location of the spindle determines where the cytoplasm will be divided. This idea has been investigated experimentally by altering the location of the mitotic spindle using either tiny glass needles or gravitational forces generated by centrifugation. If the mitotic spindle is moved before the end of metaphase, the position of the cleavage plane changes correspondingly. However, if the mitotic spindle is not moved until anaphase, the cleavage plane passes through an area corresponding to the original region occupied by the spindle equator. Hence the site of cytoplasmic division must be programmed by the end of metaphase.

The first signs of cytokinesis in animal cells usually appear toward the end of anaphase, when clusters of dense material accumulate at the equatorial region of the spindle. Soon thereafter a small depression in the plasma membrane forms around the circumference of the cell and a band of actin filaments develops directly beneath this depression or *cleavage furrow* (Figure 12-31). The cleavage furrow then begins to constrict the

cell in half, eventually causing the clusters of dense material located at the equator to coalesce into a single dense structure called the **midbody** (Figure 12-32). Finally, the cleavage furrow completely divides the cytoplasm, thereby generating two new cells.

To determine whether the actin filaments associated with the cleavage furrow generate the force that pinches the cell in two, actin-filament inhibitors such as *cytochalasin* (page 584) have been employed. When cytochalasin is applied to dividing cells with shallow cleavage furrows, the band of actin filaments disappears, cytokinesis stops, and the furrow disappears. Removing the cytochalasin leads to the reappearance of actin filaments and the resumption of cytokinesis. Because this evidence supports the idea that actin filaments generate the force that pinches the cell in two, the band of actin filaments beneath the cleavage furrow has been named the **contractile**

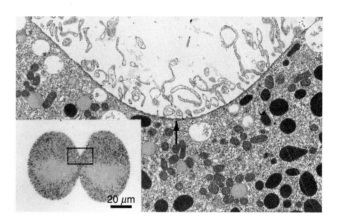

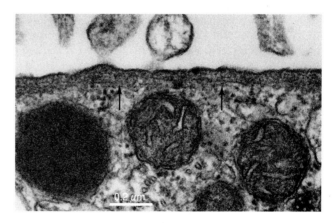

Figure 12-31 *The Cleavage Furrow and the Contractile Ring* (Top) *The inset shows a low-magnification electron micrograph of a dividing sea urchin egg cell after cytokinesis has begun. The enlarged view of the boxed region shows that a dense layer of material* (arrow), *termed the contractile ring, is located beneath the cleavage furrow.* (Bottom) *A higher-magnification electron micrograph of the cleavage furrow, oriented at right angles to the view illustrated in the top micrographs. Numerous actin filaments can be seen within the contractile ring* (arrows). *Courtesy of T. E. Schroeder.*

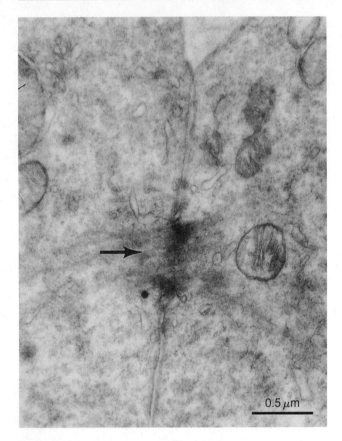

Figure 12-32 *Formation of the Midbody* *This electron micrograph shows an immature rat red blood cell near the end of cytokinesis, when the cleavage furrow has almost completely divided the cell in half. The dense material at the center of the cell (arrow) is the midbody. Courtesy of R. C. Buck.*

ring. Closely associated with the contractile ring is another set of filaments containing the protein *myosin,* which is known to interact with actin to generate the forces responsible for muscle contraction (page 567). Cytokinesis is therefore thought to be driven by an interaction between actin and myosin resembling that which occurs in other contractile systems.

Cytokinesis in Plant Cells Progresses from the Center of the Cell toward the Periphery

Cell division in higher plants differs in several ways from the comparable process in animal cells. For example, the spindle poles in plant cells lack the centrioles and asters that are characteristic of animal cell spindles. But the most prominent difference between division in animal and plant cells involves the mechanism of cytokinesis. Since plant cells are surrounded by a rigid cell wall, they are unable to create a contractile ring at the cell surface that pinches the cell in two. Instead, the cytoplasm is divided by a process that begins in the cell interior and works toward the periphery.

The site where the plant cell is to be divided is determined prior to mitosis by the formation of a band of microtubules that forms a ring around the cell just beneath the plasma membrane. This structure, called the *preprophase band,* marks the site where the new cell wall will be created during cytokinesis. Cytokinesis is initiated in late anaphase by the formation of a layer of membrane vesicles across the equatorial region of the spindle. These vesicles produce a structure, called the **phragmoplast,** which in some ways resembles the midbody of animal cells (Figure 12-33). However, instead of following the formation of the phragmoplast with an infolding of the plasma membrane, the phragmoplast continues to grow outward until it extends across the entire cell. At this point the vesicles of the phragmoplast fuse with one another, generating a double-layered membrane sheet that divides the cell in half. Cell wall material is then deposited between the two membranes to form the **cell plate.** Thus in plant cells cytokinesis progresses from the center of the cell toward the periphery, whereas in animal cells it progresses from the outside of the cell toward the interior (Figure 12-34).

Because the cleavage plane passes through the spindle equator in both animal and plant cells, the two sets of chromosomes situated at the spindle poles are segregated into the two new cells at the time of division. Cytokinesis does not always ensure, however, that the cytoplasm is equally divided. If the spindle forms symmetrically across the cell at the beginning of mitosis, as is typically the case, then cytokinesis will also be symmetrical. But in some cases, cytokinesis is deliberately asymmetric. For example, in the budding yeast *Saccharomyces cerevisiae* the interphase cell produces a small bud that is later separated from the cell by cytokinesis, which divides the cell asymmetrically (Figure 12-35). In contrast, the fission yeast *Schizosaccharomyces pombe* divides by a symmetrical process that involves the formation of a cell plate across the center of the cell.

MEIOTIC DIVISION AND GENETIC RECOMBINATION

During mitotic cell division, the two cells that are produced contain the same number of chromosomes as the original parental cell. Hence mitotic division is employed whenever cells need to proliferate in a way that produces a population of cells that are virtually identical in genetic makeup. In organisms that reproduce sexually, a second type of cell division is also required. Sexual reproduction creates new organisms by the fusion of two cells called **gametes,** one derived from the female parent and the other from the male parent. Female gametes are usually called *eggs,* while male gametes are known as *sperm.* The formation of gametes

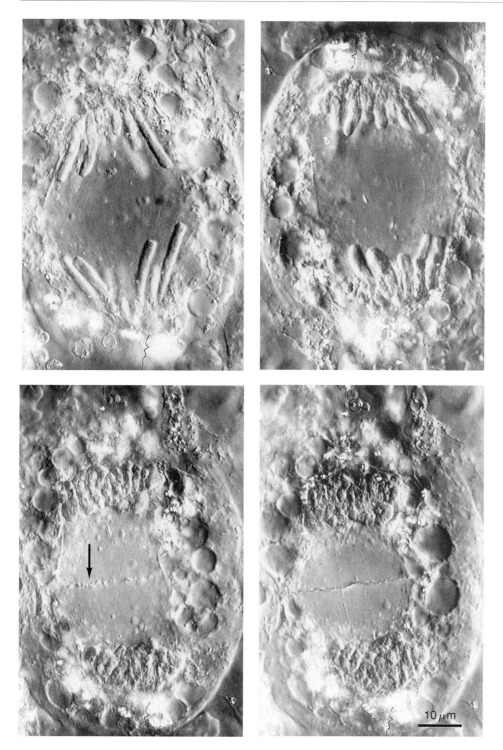

Figure 12-33 *Formation of the Phragmoplast in Dividing Plant Cells* *This series of photographs, taken during late anaphase and telophase, was obtained using the Nomarski interference microscope (page 120). As the chromosomes condense during late anaphase and early telophase, phragmoplast vesicles appear in the center of the cell* (bottom left, see arrow). *At late telophase these vesicles coalesce, forming a double-layered membrane sheet that divides the cell in half* (bottom right). *Courtesy of A. Bajer.*

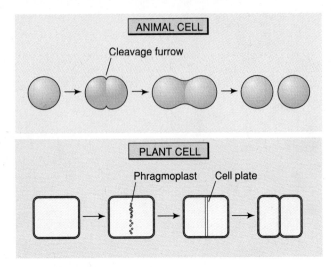

Figure 12-34 *The Difference between Cytokinesis in Animal and Plant Cells* *In animal cells, invagination of the plasma membrane proceeds from the periphery of the cell toward the interior. In plant cells, the phragmoplast initially forms in the center of the cell and then expands toward the periphery.*

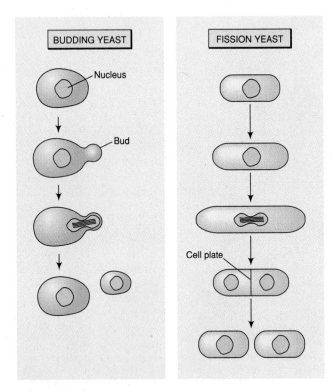

Figure 12-35 *Cytokinesis in Yeast* (Left) *In the budding yeast* Saccharomyces cerevisiae, *a bud forms during interphase which is split off during mitosis by an asymmetric cell division.* (Right) *In the fission yeast* Schizosaccharomyces pombe, *the cell is divided symmetrically during mitosis by a cell plate that forms across the center of the cell. In contrast with mitosis in higher eukaryotes, the nuclear envelope remains intact during mitosis in both kinds of yeast.*

requires a special type of cell division referred to as **meiotic division** or **meiosis.**

Meiosis Converts One Diploid Cell into Four Haploid Cells That Are Genetically Different

When two gametes fuse during sexual reproduction, the new organism that is created must contain the same number of chromosomes as each of the two parents. In eukaryotes the normal chromosome number is usually **diploid,** which means that two copies of each type of chromosome are present. Such chromosome pairs are known as **homologous chromosomes.** Human cells contain two sets of 23 chromosomes, yielding a diploid total of 46. If the fusion of a sperm cell with an egg cell is to produce an offspring with 46 chromosomes, then the sperm and egg must each contain half the normal diploid number of chromosomes—that is, a single set of 23 chromosomes. Cells that contain only a single set of chromosomes are said to be **haploid.**

Sexual reproduction therefore requires a mechanism for generating the haploid gametes that fuse together to form the diploid organism. The mechanism for producing gametes is provided by meiosis, a special type of cell division that creates haploid cells from diploid cells. Meiosis consists of two separate cell divisions occurring in sequence. The first meiotic division, which is preceded by DNA replication, reduces the chromosome number from diploid to haploid; the second meiotic division, which is *not* preceded by DNA replication, separates the paired chromatids that make up each chromosome (Figure 12-36). The net result of the two meiotic divisions is the conversion of one diploid cell into four haploid cells.

Besides reducing the chromosome number from diploid to haploid, meiosis is the primary means of generating the hereditary variability that lies at the heart of evolutionary change. Without genetic variability, organisms would be less able to adapt to changing environments and thus less likely to survive over long periods of time. One way in which meiosis generates genetic variability is by randomly distributing the parental chromosomes among the haploid cells, thereby producing gametes that contain some chromosomes derived from the male parent and others derived from the female parent. A second mechanism employed for increasing genetic variability is the exchange of DNA base sequences between the two members of each pair of homologous chromosomes. This process of genetic **recombination** creates individual chromosomes that have new combinations of paternal and maternal genes. Through the preceding mechanisms, meiosis produces four genetically unique haploid cells containing new combinations of hereditary information.

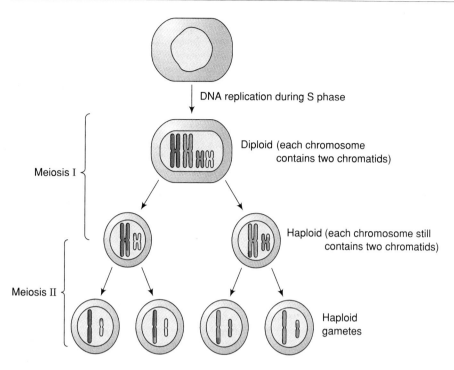

Figure 12-36 *Chromosome Distribution During Meiosis   In this example, two pairs of homologous chromosomes are shown. During meiosis I, the two members of each homologous pair line up next to each other and exchange genetic information; the two members of each pair then separate and move toward opposite spindle poles, leading to a cell division that creates two haploid cells (i.e., cells containing only a single set of chromosomes). During meiosis II, the paired chromatids that make up each chromosome are separated, generating a final total of four haploid cells.*

Homologous Chromosomes Become Paired and Exchange DNA During Prophase I

In order to understand how meiosis converts one diploid cell into four genetically unique haploid cells, it is necessary to examine the steps involved in the first and second meiotic divisions—or *meiosis I* and *meiosis II,* as they are commonly called. Like mitosis, each meiotic division can be subdivided into prophase, metaphase, anaphase, and telophase, but chromosome behavior differs from that during the comparable stages of mitosis (Figure 12-37). At the start of the first meiotic prophase, called *prophase I,* each chromosome is composed of two chromatids that were generated by DNA replication during the previous S phase, just as in mitosis. As we mentioned earlier, each chromosome in diploid cells is a member of a homologous pair in which one chromosome is inherited from the female parent and one from the male parent. During prophase I, homologous chromosomes line up next to each other and exchange genetic information. Because of the complexity of this process, prophase I is subdivided into five periods called *leptotene, zygotene, pachytene, diplotene,* and *diakinesis.*

1. The **leptotene** stage of prophase I begins with the condensation of chromatin fibers into long thin threads.

This condensation superficially resembles what occurs at the beginning of mitosis, except that the chromatin threads are more extended. Separate chromosomes cannot be distinguished from one another, nor are individual chromatids visible. Attachments between the ends of the chromosomes and the nuclear envelope can be seen at this stage.

2. The **zygotene** phase of prophase I is characterized by the development of close pairing, or **synapsis,** between homologous chromosomes. Synapsis is usually initiated by the formation of several points of attachment between the members of each pair of homologous chromosomes, followed by the progressive alignment of the rest of the chromosome until pairing is complete. A small space measuring about 100–200 nm across is typically observed between the two members of each homologous chromosome pair. This space contains a specialized structure, called the *synaptonemal complex,* whose possible role in genetic recombination will be discussed later in the chapter. During zygotene, the paired chromosomes often remain attached to the nuclear envelope at their two ends.

3. The **pachytene** stage begins at the completion of synapsis, and is marked by a dramatic thickening and

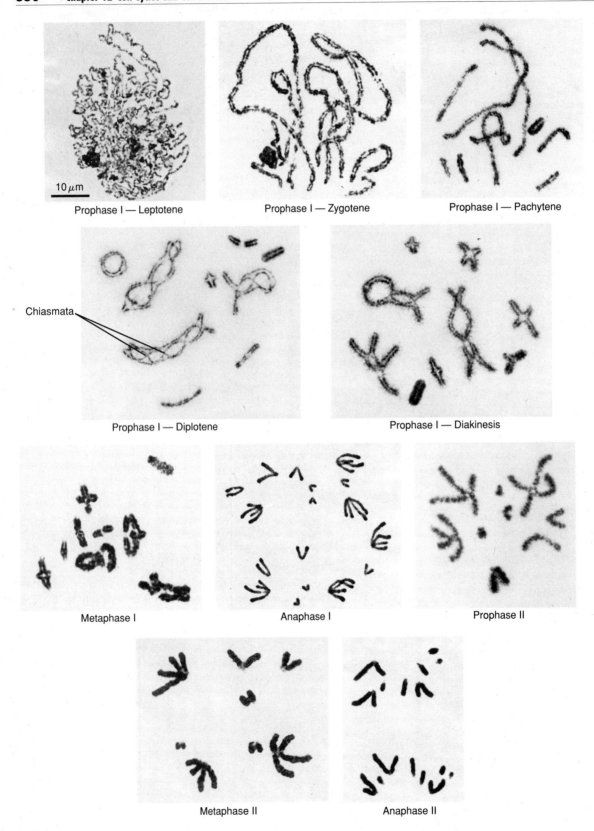

Figure 12-37 *Chromosome Appearance During the Various Stages of Meiosis* *Homologous chromosomes become paired during zygotene, and shorten and thicken at the pachytene stage. During diplotene the four chromatids present in each tetrad can be distinguished; at this stage, the chiasmata are visible. During anaphase I, each newly forming cell receives one member of each pair of homologous chromosomes. At metaphase II, each chromosome consists of a pair of chromatids held together by a centromere. During anaphase II the centromeres split, allowing the chromatids to separate and migrate in opposite directions. Courtesy of B. John.*

shortening of the chromosomes. This process reduces each chromosome to less than a quarter of its previous length. The synaptonemal complex becomes fully developed and the two chromosomes of each homologous pair bind so tightly to one another that they behave as a single unit. This overall unit is referred to as a **tetrad** because it contains a total of *four* chromatids, two derived from each chromosome. At this stage, DNA is exchanged between chromatids by a recombination process whose mechanism will be described shortly. When recombination involves the exchange of chromosome segments between corresponding regions of homologous chromosomes, it is called **crossing over.**

4. During the **diplotene** period of prophase I, the synaptonemal complex disappears and the two homologous chromosomes that make up each tetrad begin to separate. However, the homologous chromosomes remain attached to each other by connections known as **chiasmata** (singular, *chiasma*). Such connections are situated in regions where chromosomes have exchanged DNA segments and hence represent visual evidence that crossing over has occurred (Figure 12-38). As the diplotene stage proceeds, individual chromatids become visible for the first time. It is common for the chromosomes to uncoil at this point and for many nucleoli to form. In certain types of eggs a period of extensive growth ensues in which the cell mass increases by many orders of magnitude. During this growth phase, which can last for several years, the chromosomal and nucleolar DNA is actively transcribed and the chromosomes assume a characteristic morphology consisting of a darkly staining central axis from which numerous loops project. One well-studied example of this kind of meiotic chromosome, the *lampbrush chromosome* of amphibian oocytes, is described in more detail in Chapter 15.

5. During **diakinesis,** which is the final stage of prophase I, the chromosomes return to a highly condensed state. At the same time the chiasmata migrate toward the ends of the chromosomes, a process known as *terminalization* (Figure 12-39). The nucleoli disappear as in the comparable stage of mitosis, and the centrosomes move to opposite sides of the nuclear envelope. Microtubules then begin to form a spindle and the nuclear envelope breaks down and disappears, marking the end of prophase I.

Tetrads Align at the Spindle Equator During Metaphase I

During metaphase I, the tetrads attach via their kinetochores to spindle microtubules and migrate to the spindle equator. At this point the tetrads look quite different from mitotic chromosomes (Figure 12-40). Typical mitotic chromosomes consist of two chromatids joined at the centromere, whereas tetrads are composed of a pair of homologous chromosomes held together by chiasmata (the two chromatids comprising each homolog are usually not distinguishable). In spite of the obvious differences in the organization of meiotic and mitotic metaphase chromosomes, both appear to be comprised of extensively folded 30-nm chromatin fibers.

Because four chromatids occur in each tetrad, four kinetochores are also present. The kinetochores located on the two chromatids derived from each chromosome face in the same direction (rather than in opposite directions, as in mitosis), and therefore they attach to microtubules emanating from the same spindle pole (Figure 12-41). The two kinetochores derived from one chromosome attach to microtubules emanating from one spindle pole, while the two kinetochores derived from the other chromosome attach to microtubules emanating from the opposite spindle pole. This orientation

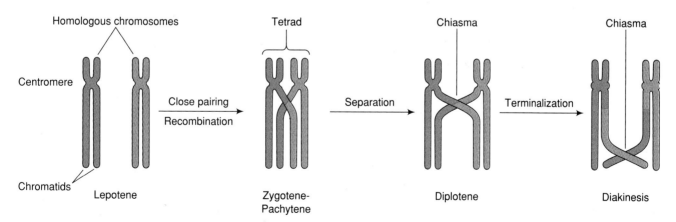

Figure 12-38 *Chromosome Behavior During Prophase I of Meiosis*   *During zygotene and pachytene, homologous chromosomes become tightly paired and the exchange of DNA segments (crossing over) takes place. The homologous chromosomes begin to separate during diplotene, but remain connected by regions (chiasmata) where DNA exchange has occurred. During diakinesis, the chiasmata migrate toward the ends of the chromosomes.*

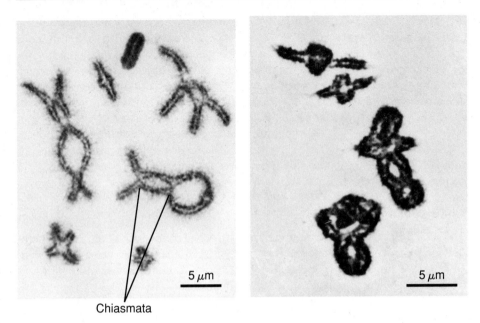

Figure 12-39 *Behavior of Chiasmata During Prophase I of Meiosis* (Left) *The chiasmata, which represent regions of genetic recombination, hold homologous chromosomes together during the diplotene stage of prophase I.* (Right) *At the end of prophase I, the chiasmata move to the ends of the chromosomes by a process called terminalization. Courtesy of B. John.*

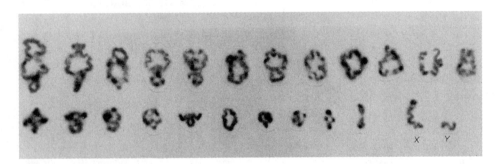

Figure 12-40 *A Meiotic Karyotype of Human Chromosomes from Sperm Cells Arrested During Metaphase I* *Note the difference in appearance between these chromosomes and typical mitotic chromosomes (see Figure 12-24).*

sets the stage for separation of the homologous chromosomes during anaphase.

Homologous Chromosomes Move to Opposite Spindle Poles During Anaphase I

During anaphase I, the members of each pair of homologous chromosomes separate from each other and migrate toward opposite spindle poles. This phenomenon is fundamentally different from the events of mitotic anaphase, when the centromeres split and the two chromatids that comprise each chromosome migrate to opposite poles. In anaphase I the centromeres do *not* split; instead, the pair of chromatids that make up each chromosome remain attached to each other as the homolo-

gous chromosomes separate and move in opposite directions. Because the two members of each pair of homologous chromosomes separate and move toward opposite spindle poles, each pole receives only a haploid set of chromosomes, each of which has two chromatids.

Telophase I and Cytokinesis Produce Two Haploid Cells Whose Chromosomes Consist of Paired Chromatids

The onset of telophase I is marked by the arrival of a haploid set of chromosomes at each spindle pole. Since centromere splitting did not occur during anaphase I, each chromosome still contains two paired chromatids.

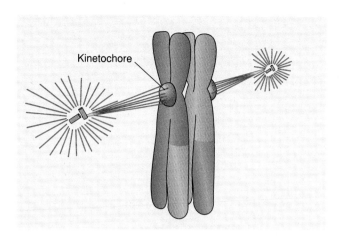

Kinetochore

Figure 12-41 *The Attachment of Homologous Chromosomes to Opposite Spindle Poles During Metaphase I of Meiosis* *Unlike the comparable phase of mitosis (or meiosis II), the kinetochores located on the two chromatids comprising each chromosome face in the same direction, and therefore attach to the same spindle pole. The kinetochores located on the two members of a homologous pair of chromosomes face in opposite directions, and therefore attach to opposite poles. The colored shading within the chromosomes is used to highlight a region where crossing over has occurred.*

After arriving at the spindle poles, the chromosomes uncoil and disperse into chromatin fibers as the spindle disappears. Nuclear envelopes form around the chromosomes and cytokinesis ensues, generating two haploid cells whose chromosomes consist of paired chromatids.

Meiosis II Produces Gametes by a Process That Resembles a Mitotic Division

After meiosis I has been completed, a brief interphase may intervene before meiosis II begins. However, this interphase is *not* accompanied by DNA replication because each chromosome already consists of a pair of replicated chromatids. The purpose of the second meiotic division, like that of a typical mitotic division, is to parcel these paired chromatids into two newly forming cells.

The second meiotic division begins with prophase II, which is marked by chromosome condensation, formation of a new spindle, and breakdown of the nuclear envelope. During the next stage, metaphase II, the chromosomes bind to kinetochore microtubules and migrate to the spindle equator. Like the chromosomes observed during mitosis, each chromosome is composed of two paired chromatids joined at a common centromere, and microtubules attached to opposite sides of the centromere link the chromatids to opposite spindle poles.

The beginning of anaphase II is marked by the splitting of the centromeres, which allows the two

chromatids that make up each metaphase chromosome to separate from each other and migrate to opposite spindle poles. The remaining phases of the second meiotic division consist of telophase II and cytokinesis, which resemble the comparable stages of mitosis. The final result is the formation of gametes containing a haploid set of chromosomes. Because the members of each homologous chromosome pair were randomly distributed to the two cells produced during meiosis I, each of the final gametes contains a random mixture of maternal and paternal chromosomes. Moreover, each of these chromosomes consists of a mixture of maternal and paternal DNA sequences created by the genetic recombination that occurred during prophase I.

Genetic Recombination Occurs in Both Prokaryotes and Eukaryotes

Two unique events are associated with meiosis: reduction of the chromosome number from diploid to haploid, and exchange of genetic information between chromosomes. We have already seen how the behavior of homologous chromosomes during meiosis leads to a reduction of the chromosome number, but the mechanism for exchanging genetic information between chromosomes is considerably more complex and requires further elaboration.

When genetic information is exchanged between two DNA molecules that exhibit extensive sequence similarity, the process is referred to as **homologous recombination.** Although our discussion of homologous recombination is prompted by events occurring in meiosis, similar types of genetic exchange are observed in prokaryotes where meiosis does not take place. For example, certain bacteria undergo a modified type of sexual reproduction termed **conjugation** in which the chromosomal DNA from one bacterium is injected into another bacterium of the same species. Genetic recombination between the two DNA molecules occurs in the recipient cell, generating a chromosome that contains DNA sequences derived from both bacteria. In a related process, termed **genetic transformation,** bacteria take up naked DNA molecules from the external environment. The foreign DNA then undergoes genetic recombination with homologous DNA sequences in the host cell chromosome. Finally, in the process known as **transduction,** viruses incorporate some bacterial DNA sequences while infecting one bacterial cell and then infect a different bacterium, where the released DNA recombines with the chromosome of the newly infected host.

Each of the preceding types of bacterial recombination produces a DNA molecule that contains base sequences derived from two different homologous sources, which is comparable to what occurs during prophase I of meiosis. Our ensuing discussion of homologous recombination mechanisms will therefore draw upon examples involving both prokaryotes and eukaryotes, since the principles involved appear to be similar.

Breakage-and-Exchange Occurs During Homologous Recombination

The notion that genetic material is exchanged between chromosomes during meiosis was first proposed in the early 1900s, when it was initially discovered that homologous chromosomes become tightly paired during prophase I. It was even suggested that chiasmata represent the sites of exchange, although no direct evidence existed at the time to support this idea. It was not until several decades later that genetic studies showed that individual genes switch positions from one homologous chromosome to the other during meiosis. Moreover, the number of such exchanges was found to correlate with the number of chiasmata, thereby supporting the notion that chiasmata represent sites of genetic recombination.

Shortly after the discovery that genes are exchanged between chromosomes during meiosis, two theories were proposed to explain how this is accomplished. The *breakage-and-exchange model* postulated that breaks occur in the DNA molecules of two adjoining chromosomes, followed by exchange and rejoining of the broken segments. In contrast, the *copy-choice model* proposed that genetic recombination occurs while DNA is being replicated. According to the latter view, DNA replication begins by copying a DNA molecule located in one chromosome and then switches at some point to copying the DNA located in the homologous chromosome. The net result would be a new DNA molecule containing information derived from both chromosomes. One of the more obvious predictions made by the copy-choice model is that DNA replication and genetic recombination should occur at the same time. When subsequent studies revealed that DNA replication takes place during S phase but recombination does not occur until prophase I, the copy-choice model had to be rejected.

The first experimental evidence in support of the breakage-and-exchange model was obtained in 1961 by Matthew Meselson and Jean Weigle, who employed bacterial viruses labeled with either the heavy or light isotope of nitrogen. Infection of bacterial cells with two strains of the same virus, one labeled with ^{15}N and the other with ^{14}N, resulted in the production of recombinant virus particles containing genes derived from both viruses. When the DNA from these recombinant viruses was examined, it was found to contain a mixture of ^{15}N and ^{14}N (Figure 12-42). Since these experiments were performed under conditions that prevented any new DNA from being synthesized, the recombinant DNA molecules must have been produced by breaking and rejoining DNA molecules derived from the two original viruses.

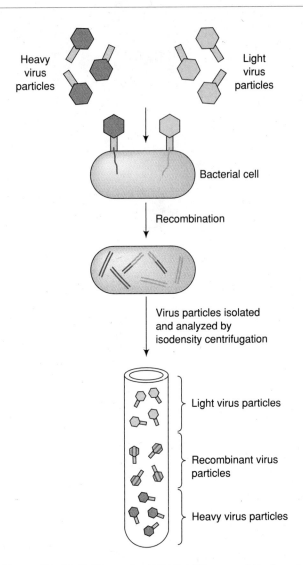

Figure 12-42 Evidence for DNA Breakage-and-Exchange During Genetic Recombination Involving Bacterial Viruses *From* Biochemistry *by C. K. Mathews and K. E. vanHolde. Copyright © 1990 by The Benjamin/Cummings Publishing Company. Reprinted by permission.*

Subsequent experiments involving bacteria whose chromosomes had been labeled with either ^{15}N or ^{14}N revealed that DNA containing a mixture of both isotopes is also produced during genetic recombination between bacterial chromosomes. Moreover, when these recombinant DNA molecules were heated to dissociate them into single strands, a mixture of ^{15}N and ^{14}N was detected within each DNA strand; hence the DNA double helix must have been broken and rejoined during recombination.

A similar conclusion emerged from experiments carried out on eukaryotic cells by J. Herbert Taylor. In these studies, cells were briefly exposed to ^{3}H-thymidine during the S phase preceding the last mitosis prior to meiosis. This labeling produces chromatids containing one radioactive DNA strand per double helix. During the following S phase, DNA replication in the absence of ^{3}H-thymidine generates chromosomes containing one labeled chromatid and one unlabeled chromatid (Figure 12-43). If breakage-and-exchange occurs during the subsequent meiosis, it would be expected to produce DNA molecules containing a mixture of radioactive and nonradioactive segments. When the chromatids of the meiotic cells were examined, some were found to contain a mixture of labeled and unlabeled segments as predicted by the breakage-and-exchange model. Moreover, the frequency of such exchanges was directly proportional to the frequency with which the genes located in these regions underwent genetic recombination. Such observations provided strong support for the notion that genetic recombination in eukaryotic cells, as in prokaryotes, involves a breakage-and-exchange mechanism. These experiments also showed that most DNA exchanges occur between homologous chromosomes rather than between the two paired chromatids of a given chromosome. This selectivity is important because it ensures that genes are exchanged between paternal and maternal chromosomes.

Homologous Recombination Can Lead to Gene Conversion

The conclusion that homologous recombination is based on DNA breakage-and-exchange does not in itself provide much information concerning the underlying molecular mechanisms. One of the simplest breakage-and-exchange models that might be envisioned would involve the cleavage of two homologous DNA molecules at comparable locations, followed by exchange and rejoining of the cut ends. In spite of its attractive simplicity, this model makes predictions that are incompatible with experimental observations. For example, this simple view of breakage-and-exchange implies that genetic recombination should be completely reciprocal; that is, any genes exchanged from one chromosome should appear in the other chromosome, and vice versa. However, in some situations genetic recombination has been found to be *nonreciprocal*.

To illustrate what is meant by nonreciprocal recombination, let us consider a hypothetical situation involving two genes designated P and Q. If one chromosome contains forms of these genes called P1 and Q1, and the other chromosome has alternate forms designated P2 and Q2, genetic exchange would be expected to generate one chromosome with genes P1 and Q2, and a second chromosome with genes P2 and Q1. Although this pattern is the one usually observed, nonreciprocal combinations can occur as well. For example, recombination might generate one chromosome with genes P1 and Q2, and a second chromosome with genes P2 and Q2. In this particular example, the Q1 gene expected on the second chromosome appears to have been "converted" to a Q2 gene. For this reason, nonreciprocal recombination is

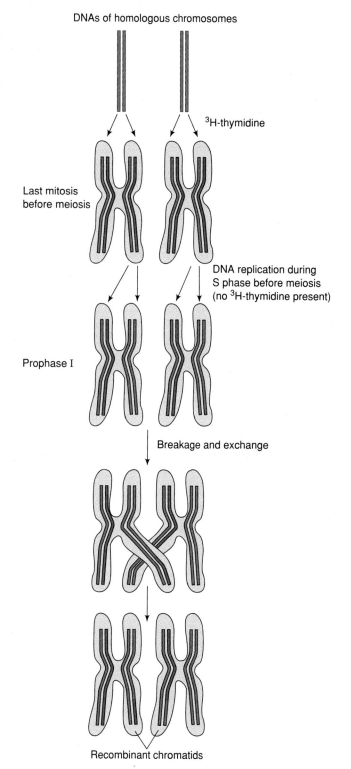

Figure 12-43 Experimental Approach Utilized to Demonstrate the Existence of Breakage-and-Exchange During Genetic Recombination in Eukaryotic Cells *When the chromatids produced during such labeling experiments were examined using autoradiography, some were found to contain a mixture of labeled and unlabeled segments as predicted by the breakage-and-exchange model.*

often referred to as **gene conversion.** Gene conversion is most commonly observed when the recombining genes are located very close to one another on the chromosome. Because the frequency with which two genes recombine is directly related to the distance between them, recombination between closely spaced genes is a rare event. These rare recombination events, and hence gene conversions, are most readily detectable in organisms that reproduce rapidly and generate large numbers of offspring, such as yeast and *Neurospora.*

Neurospora is an especially convenient organism in which to study gene conversion because its meiotic cells are enclosed in a small sac, called an **ascus,** which prevents the cells from moving around and thus allows the lineage of each cell to be easily followed (Figure 12-44). Initially each ascus contains a single diploid cell. Meiotic division of this cell produces four haploid cells that subsequently divide by mitosis, yielding a final total of eight cells. Since the final division is mitotic, it should produce two identical progeny cells for each cell that divides. Yet in a significant number of cases this mitosis produces two cells that are genetically different. Such unexpected results are most often observed with genes that are close to a site of genetic recombination.

How can mitosis produce two cells that are genetically different? The most straightforward explanation is that the chromosome contains one or more genes in which the two strands of the DNA double helix are not entirely complementary. When the two DNA strands in such a noncomplementary region split apart during chromosome replication, the two newly forming DNA molecules will exhibit differing base sequences in this region and thus will represent slightly different genes.

Homologous Recombination Involves the Formation of Single-Strand Exchanges (Holliday Junctions)

The preceding observations suggest that homologous recombination is more complicated than can be explained by a breakage-and-exchange model in which two DNA molecules are simply cleaved and rejoined. To accommodate the complex behavior of genes during homologous recombination, Robin Holliday proposed in 1964 that recombination is based on the exchange of *single* DNA strands between two double-stranded DNA molecules. A subsequent adaptation of Holliday's single-strand exchange model, developed by Matthew Meselson and Charles Radding, is illustrated in Figure 12-45. According to this model, recombination begins with a single-strand break in one DNA molecule (*step 1* in Figure 12-45). The broken strand then "invades" a complementary region of a homologous DNA double helix, displacing one of the two strands (*step 2*). Next the

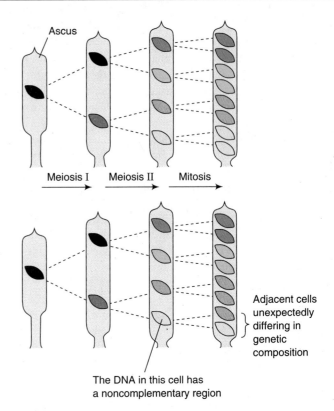

Figure 12-44 *Meiosis in Neurospora* *In the bread mold* Neurospora, *cells undergoing meiotic division are contained within a sac called an ascus. The ascus keeps the individual cells lined up in a row, making it easy to trace each cell's lineage. The separation of homologous chromosomes during meiosis I generates two cells exhibiting different genetic traits. Cells produced during meiosis II may also differ from each other genetically because of the crossing over that occurred during meiosis I. The third division is a simple mitosis, and hence is expected to produce a pair of genetically identical cells for each cell that divides* (top). *Occasionally, however, this terminal mitosis generates a pair of nonidentical cells* (bottom).

process of DNA repair (*step 3*) produces a crossed structure, called a **Holliday junction,** in which a single strand from each DNA double helix crosses over and joins the opposite double helix (*step 4*). Direct support for the existence of Holliday junctions has been obtained by electron microscopy, which has revealed that DNA molecules undergoing homologous recombination appear as DNA double helices joined by single-strand crossovers (Figure 12-46).

Once a Holliday junction has been formed, unwinding and rewinding of the DNA double helices causes the crossover point to move back and forth along the chromosomal DNA (*step 5*). This phenomenon, called *branch migration,* can rapidly increase the length of single-stranded DNA that is exchanged between two DNA molecules. After branch migration has occurred, the Holliday junction is cleaved and the broken DNA

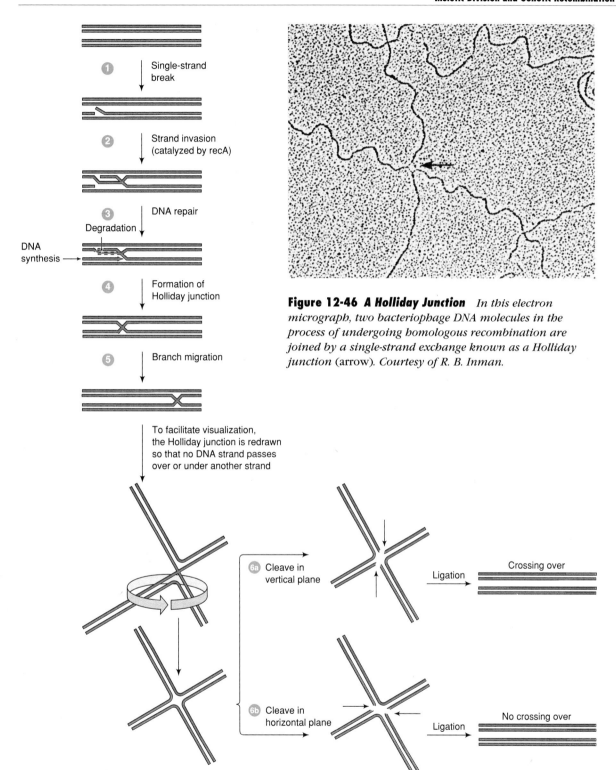

Figure 12-46 A Holliday Junction *In this electron micrograph, two bacteriophage DNA molecules in the process of undergoing homologous recombination are joined by a single-strand exchange known as a Holliday junction (arrow). Courtesy of R. B. Inman.*

Figure 12-45 The Holliday Model for Homologous Recombination as Modified by Meselson and Radding *In this model, recombination begins with a (1) single-strand break in one DNA molecule, followed by (2) invasion of the intact strand by the broken strand (catalyzed by the recA protein). The process of (3) DNA repair then leads to (4) formation of a Holliday junction in which a single strand from each DNA double helix has crossed over and joined the opposite double helix. After (5) DNA unwinding and rewinding causes the crossover point to move (branch migration), cleavage in one plane (6a) creates two DNA molecules that exhibit crossing over, while cleavage in the other plane (6b) generates DNA molecules that do not exhibit crossing over, but contain a noncomplementary region near the site where the Holliday junction had formed. From* Biochemistry *by C. K. Mathews and K. E. vanHolde. Copyright © 1990 by The Benjamin/Cummings Publishing Company. Reprinted by permission.*

strands are joined back together to produce two separate DNA molecules. However, there are two ways in which a Holliday junction can be cut and rejoined. If it is cleaved in one plane, the two DNA molecules that are produced will exhibit crossing over; that is, the chromosomal DNA beyond the point where recombination occurred will have been completely exchanged between the two chromosomes (*step 6a*). If the Holliday junction is cut in the other plane, crossing over does not occur but the DNA molecules exhibit a noncomplementary region near the site where the Holliday junction had formed (*step 6b*).

What is the fate of such noncomplementary regions? If they remain intact, an ensuing mitotic division will separate the mismatched DNA strands and each will serve as a template for the synthesis of a new complementary strand. The net result will be two new DNA molecules with differing base sequences, and hence two cells containing slightly different gene sequences in the affected region. We have just seen that this type of behavior is occasionally observed in *Neurospora,* where two genetically different cell types can arise during the mitosis following meiosis. The alternative fate for a noncomplementary DNA region is for it to be corrected by excision and repair. The net effect of DNA repair would be to convert genes from one form to another; in other words, it would lead to gene conversion.

The identity of the enzymes involved in homologous recombination has been investigated by studying the ability of bacterial extracts to catalyze the formation of Holliday junctions. Such extracts contain a protein, called *recA,* whose presence is crucial to this process. Mutant bacteria with a defective recA gene cannot carry out genetic recombination, nor are extracts prepared from these mutant cells capable of creating Holliday junctions from homologous DNA molecules. Studies using purified preparations of the recA protein have revealed that it catalyzes a reaction in which a single-stranded DNA segment displaces one of the two strands of a DNA double helix; in other words, it catalyzes *step 2* in Figure 12-45. In catalyzing this reaction, the recA protein first coats the single-stranded DNA region. The coated, single-stranded DNA then interacts with a DNA double helix, moving along the target DNA until it reaches a complementary sequence. Here recA denatures the target DNA molecule and promotes hybridization of the single-stranded DNA with its complementary strand.

The Synaptonemal Complex Facilitates Chromosome Pairing and Recombination During Meiosis

During meiosis, homologous recombination is associated with the formation of a zipperlike, protein-containing structure called the **synaptonemal complex.** This complex, which measures about 100–200 nm in thickness, runs lengthwise between homologous chromosomes when they become closely paired during prophase I. It is composed of two dense parallel structures, called *lateral elements,* separated by a less dense *central element* (Figure 12-47). Small spherical structures called *recombination nodules* appear within the synaptonemal complex at the time when meiotic recombination is taking place. Recombination nodules are thought to contain multienzyme complexes that catalyze steps involved in the recombination process.

The formation of synaptonemal complexes begins during the leptotene stage of prophase I, when a lateral element develops along the edge of each chromosome. At zygotene the two chromosomes of each homologous pair come together, leaving a space of about 100 nm between their respective lateral elements. The central element then forms between the lateral elements, generating a fully developed synaptonemal complex. The synaptonemal complex is retained until the beginning of diplotene, when it disassembles as the homologous chromosomes separate. In some cells remnants of the synaptonemal complexes persist through the second meiotic division, forming aggregates known as *polycomplexes* (Figure 12-48).

A number of observations suggest that the synaptonemal complex plays an important role in genetic recombination. First, the synaptonemal complex appears at the time when recombination takes place. Second, its location between the opposed homologous chromosomes corresponds to the region where crossing over occurs. Third, when genetic recombination is prevented by exposing cells to high temperatures or drugs that block DNA synthesis, synaptonemal complexes do not develop. And finally, synaptonemal complexes are absent in organisms that fail to carry out meiotic recombination, such as male fruit flies.

Presumably the role of the synaptonemal complex is to facilitate recombination by maintaining a close pairing between adjacent homologous chromosomes along their entire length. If the synaptonemal complex functions to facilitate recombination, how do cells ensure that such structures form only between homologous chromosomes? Recent observations suggest the existence of a process called *homology searching,* in which a single-strand break in one DNA molecule produces a free strand that "invades" another DNA double helix and checks for the presence of complementary sequences (see *step 2* in Figure 12-45). If extensive homology is not found, the free DNA strand invades another DNA molecule and checks for complementarity, repeating the process until a homologous DNA molecule is detected. Only then does a synaptonemal complex develop, bringing the homologous chromosomes together throughout their length to facilitate the process of recombination.

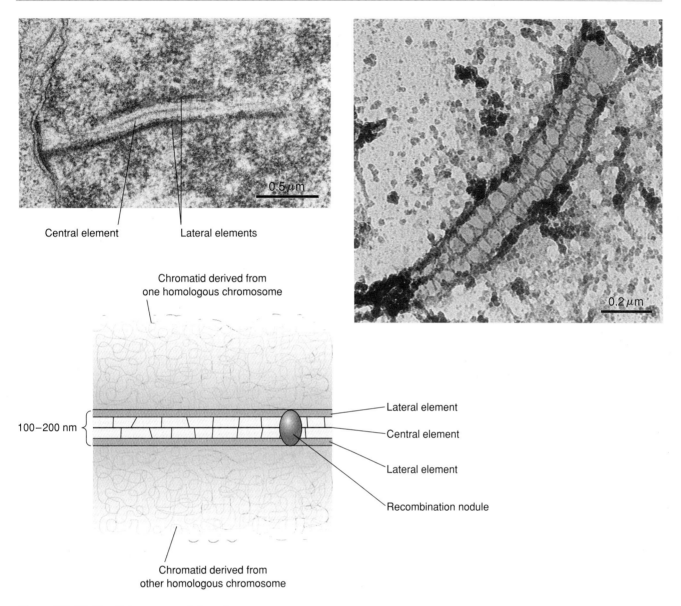

Figure 12-47 *The Synaptonemal Complex* (Upper left) *Thin-section electron micrograph of a synaptonemal complex at relatively low magnification. The central and lateral elements are clearly visible.* (Upper right) *An unsectioned synaptonemal complex* (color) *viewed after removal of adjacent chromosomal fibers by digestion with deoxyribonuclease. The major elements of the synaptonemal complex are clearly visible.* (Bottom left) *A schematic diagram of the synaptonemal complex, which joins a chromatid derived from one homologous chromosome with a chromatid derived from the other member of the homologous pair. Micrographs courtesy of D. E. Comings.*

SUMMARY OF PRINCIPAL POINTS

• The cell cycle involves three main activities: cell growth, chromosome replication, and cell division.

• Bacterial cell cycles consist of a C period when the chromosomal DNA is replicated, a D period that intervenes between the end of DNA replication and cell division, and an optional B period between cell division and the onset of DNA replication. The tetranucleotide ppGpp couples growth rate to DNA replication by controlling the transcription of genes whose products are required for initiating DNA synthesis.

• Bacterial chromosome replication is initiated at a single site on the chromosome and proceeds in both directions around the circular DNA molecule. The replicating chromosome is bound to the plasma membrane, which is thought to facilitate chromosome segregation when septum formation divides the cell in two. Sporulation is a specialized type of cell division that converts bacteria into dormant spores.

• The eukaryotic cell cycle is divided into four phases called G_1, S, G_2, and M. Chromosomal DNA is replicated during S phase, whereas mitosis and cytokinesis occur during M phase. Progression through the cell cycle is regulated by cyclin-dependent protein kinases. The G_1-to-S transition is con-

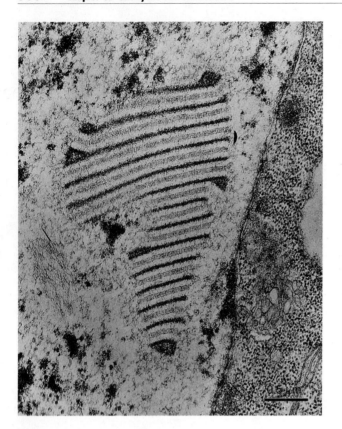

Figure 12-48 Electron Micrograph of a Large Polycomplex in the Nucleus of a Cell from the Mosquito Aedes Aegypti *The polycomplex consists of a series of synaptonemal complexes stacked upon one another. The nuclear envelope lies to the right of the polycomplex. Courtesy of T. F. Roth.*

trolled by a complex between cdc2 kinase and a G_1 cyclin; the G_2-to-M transition is controlled by MPF, which is a complex between cdc2 kinase and a mitotic cyclin.

• In unicellular organisms, the presence of sufficient nutrients is the main factor that determines whether cells grow and divide. In multicellular organisms, cell growth and division are controlled by extracellular growth factors.

• Eukaryotic chromosomal DNA replication is initiated at multiple replicons located within each DNA molecule. Active genes are replicated early during S phase, while inactive genes are replicated later. New nucleosomes are assembled from histones that are synthesized mainly during S phase.

• To function as a chromosome, a linear DNA molecule requires three types of sequences: ARS elements that function as replication origins, centromere (CEN) sequences involved in attaching chromosomes to the mitotic spindle, and telomere (TEL) sequences that protect the ends of chromosomes from being shortened during replication.

• Mitosis is subdivided into prophase, metaphase, anaphase, and telophase. During prophase the chromatin fibers condense into chromosomes, centrosomes trigger the growth of microtubules that form the mitotic spindle, and the nuclear envelope breaks down. These events are triggered by MPF (cdc2 kinase plus mitotic cyclin), which catalyzes the phosphorylation of histone H1, lamins, and perhaps centrosomal proteins.

• During metaphase, chromosomes become attached to spindle microtubules through their kinetochores and are

moved to the spindle equator by a combination of pushing and pulling forces. Each of the two chromatids found in a metaphase chromosome consists of a single, intricately folded chromatin fiber anchored to a protein scaffold.

• At the beginning of anaphase the chromosomal centromeres split, allowing the paired chromatids to separate and move toward opposite spindle poles. Chromosome migration is accomplished by both chromosome-to-pole movements (anaphase A) and pole-pole separation (anaphase B). During anaphase A, each kinetochore advances along a stationary track of kinetochore microtubules, depolymerizing the microtubules as it proceeds. In anaphase B, the polar microtubules elongate and slide away from one another.

• The beginning of telophase is marked by the arrival of the two sets of chromosomes at opposite spindle poles. The chromosomes then disperse into chromatin fibers, nucleoli reappear (if they had disappeared), the spindle breaks down, and nuclear envelopes are formed around the two sets of chromosomes.

• Cytokinesis begins in association with the final stages of anaphase. In animal cells, cytokinesis proceeds from the outside of the cell toward the interior, powered by the contraction of a ring of actin filaments that passes around the circumference of the cell. In plant cells, cytokinesis is mediated by a coalescence of membrane vesicles and deposition of cell wall material that begins in the center of the cell and moves toward the periphery.

• The formation of haploid gametes requires a special type of cell division, called meiosis, which converts one diploid cell into four haploid cells containing new combinations of genetic information. During prophase I, homologous chromosomes become tightly paired with one another and undergo genetic recombination. At metaphase I, the paired homologous chromosomes (tetrads) line up at the spindle equator. Unlike mitosis, the centromeres do not split during anaphase I; instead, the members of each pair of homologous chromosomes separate and move toward opposite spindle poles, giving each pole a haploid set of chromosomes. The final result of the first meiotic division is two haploid cells whose chromosomes consist of paired chromatids. The second meiotic division, which is not preceded by DNA replication, resembles a mitotic division, although the starting chromosome number is haploid rather than diploid.

• Homologous recombination involves breakage-and-exchange between DNA molecules that exhibit extensive sequence homology. Homologous recombination is sometimes accompanied by gene conversion or the formation of DNA molecules whose two strands are not completely complementary to one another. These phenomena can be explained by recombination models that involve the exchange of single DNA strands between double-stranded DNA molecules (Holliday junctions). In eukaryotic cells, the synaptonemal complex facilitates the process of homologous recombination by maintaining a close pairing between homologous chromosomes along their entire length.

SUGGESTED READINGS

Books

Adams, R. L. P. (1991). *DNA Replication,* IRL Press, Oxford, England.

Baserga, R. (1985). *The Biology of Cell Reproduction,* Harvard University Press, Cambridge, MA.

Bray, D. (1992). *Cell Movements,* Garland, New York, Ch. 14.

Cooper, S. (1991). *Bacterial Growth and Division,* Academic Press, San Diego.

Hyams, J. S., and B. R. Brinkley (1989). *Mitosis: Molecules and Mechanisms,* Academic Press, San Diego.

John, B. (1990). *Meiosis,* Cambridge University Press, New York.

Murray, A., and T. Hunt (1993). *The Cell Cycle: An Introduction,* Freeman, New York.

Articles

Baskin, T. I., and W. Z. Cande (1990). The structure and function of the mitotic spindle in flowering plants, *Annu. Rev. Plant Physiol. Plant Mol. Biol.* 41:277-315.

Bollag, R. J., A. S. Waldman, and R. M. Liskay (1989). Homologous recombination in mammalian cells, *Annu. Rev. Genet.* 23:199-225.

Brinkley, B. R., I. Ouspenski, and R. P. Zinkowski (1992). Structure and molecular organization of the centromere-kinetochore complex, *Trends Cell Biol.* 2:15-21.

Burhans, W. C., and J. A. Huberman (1994). DNA replication origins in animal cells: A question of context? *Science* 263:639-640.

Cande, W. Z., and C. J. Hogan (1989). The mechanism of anaphase spindle elongation, *BioEssays* 11:5-9.

Coverley, D., and R. A. Laskey (1994). Regulation of eukaryotic DNA replication, *Annu. Rev. Biochem.* 63:745-776.

Dunphy, W. G. (1994). The decision to enter mitosis, *Trends Cell Biol.* 4:202-207.

Forsburg, S. L., and P. Nurse (1991). Cell cycle regulation in the yeasts *Saccharomyces cerevisiae* and *Schizosaccharomyces pombe, Annu. Rev. Cell Biol.* 7:227-256.

Funnell, B. E. (1993). Participation of the bacterial membrane in DNA replication and chromosome partition, *Trends Cell Biol.* 3:20-25.

Hall, A. (1994). A biochemical function for Ras—at last, *Science* 264:1413-1414.

Holm, C. (1994). Coming undone: How to untangle a chromosome, *Cell* 77:955-957.

Hozák, P., and P. R. Cook (1994). Replication factories, *Trends Cell Biol.* 4:48-52.

Kirschner, M. (1992). The cell cycle then and now, *Trends Biochem. Sci.* 17:281-285.

Koshland, D. (1994). Mitosis: Back to the basics, *Cell* 77:951-954.

McIntosh, J. R., and G. E. Hering (1991). Spindle fiber action and chromosome movement, *Annu. Rev. Cell Biol.* 7:403-426.

McIntosh, J. R., and K. L. McDonald (1989). The mitotic spindle, *Sci. Amer.* 261 (October):48-56.

Meselson, M. S., and C. M. Radding (1975). A general model for genetic recombination, *Proc. Natl. Acad. Sci. USA* 72:358-361.

Moreno, S., and P. Nurse (1994). Regulation of progression through the G1 phase of the cell cycle by the $rum1^+$ gene, *Nature* 367:236-242.

Moyzis, R. K. (1991). The human telomere, *Sci. Amer.* 265 (August):48-55.

Murray, A. W. (1992). Creative blocks: cell-cycle checkpoints and feedback controls, *Nature* 359:599-604.

Murray, A. W., and M. W. Kirschner (1991). What controls the cell cycle, *Sci. Amer.* 264 (March):56-63.

Nigg, E. A. (1993). Cellular substrates of $p34^{cdc2}$ and its companion cyclin-dependent kinases, *Trends Cell Biol.* 3:296-301.

Nishida, E., and Y. Gotoh (1993). The MAP kinase cascade is essential for diverse signal transduction pathways, *Trends Biochem. Sci.* 18:128-131.

O'Farrell, P. H. (1992). Cell cycle control: many ways to skin a cat, *Trends Cell Biol.* 2:159-163.

Pines, J. (1993). Cyclins and cyclin-dependent kinases: take your partners, *Trends Biochem. Sci.* 18:195-197.

Roberge, M. (1992). Checkpoint controls that couple mitosis to completion of DNA replication, *Trends Cell Biol.* 2:277-281.

Roeder, G. A. (1990). Chromosome synapsis and genetic recombination: their roles in meiotic chromosome segregation, *Trends Genet.* 6:385-389.

Saitoh, Y., and U. K. Laemmli (1994). Metaphase chromosome structure: Bonds arise from a differential folding path of the highly AT-rich scaffold, *Cell* 76:609-622.

Sawin, K. E., and J. M. Scholey (1991). Motor proteins in cell division, *Trends Cell Biol.* 1:122-129.

Schulman, I., and K. S. Bloom (1991). Centromeres: an integrated protein/DNA complex required for chromosome movement, *Annu. Rev. Cell Biol.* 7:311-336.

Sherr, C. J. (1993). Mammalian G_1 cyclins, *Cell* 73:1059-1065.

Stahl, F. W. (1987). Genetic recombination, *Sci. Amer.* 256 (February):90-101.

West, S. C. (1994). The processing of recombination intermediates: Mechanistic insights from studies of bacterial proteins, *Cell* 76:9-15.

Zyskind, J. W., and D. W. Smith (1992). DNA replication, the bacterial cell cycle, and cell growth, *Cell* 69:5-8.

Chapter 13

The Cytoskeleton and Cell Motility

During the early part of this century the cell was regarded as a collection of physically independent organelles suspended in an amorphous fluid. However, the advent of electron microscopy and the development of sophisticated staining techniques revealed that the eukaryotic cytoplasm contains a network of interconnected filaments that together impart a dynamic framework to the cell. This framework, called the **cytoskeleton,** performs at least two basic functions. The first, implied by the term "skeleton," is the creation of an internal scaffolding that supports and organizes the cell interior. This structural framework permits cells to assume elaborate shapes that would be unstable in the absence of a supporting lattice, and also organizes and guides interactions among organelles. The second function of the cytoskeleton is its ability to generate movement. Just as the skeleton of the body is associated with muscles that allow the organism to move, the cytoskeleton contains elements that permit both movement of the cell as a whole and movements of intracellular components such as chromosomes, membrane vesicles, granules, and other organelles.

The cytoskeleton is built from three classes of protein filaments known as *actin filaments, microtubules,* and *intermediate filaments.* In some situations a single kind of filament is responsible for a particular function, while in other cases interactions between different filament types are involved. We will begin our examination of the roles played by cytoskeletal filaments with a discussion of actin filaments, whose involvement in muscle contraction was the first aspect of cell motility to be well understood at the molecular level. After describing the roles played by actin filaments in both muscle and nonmuscle cells, we will proceed to a description of microtubules and their role in ciliary and flagellar motion, and then to a discussion of intermediate filaments, which impart mechanical strength to the cytoskeleton. Finally, after describing cytoskeletal interconnections between the three classes of filaments, we conclude the chapter by discussing motility in bacterial cells, which lack a cytoskeleton comparable to that of eukaryotes and yet still have motile flagella.

MUSCLE CONTRACTION

Motility in multicellular animals requires forces of great magnitude, not just for moving the organism as a whole, but for activities such as propelling blood through the circulatory system, air into and out of the respiratory system, and food through the digestive tract. Movements of this type depend on muscle cells, which are specialized for generating motility. The energy that powers muscle contraction is derived from the hydrolysis of ATP. In vertebrates, nearly one-third of the ATP produced by oxidative phosphorylation is utilized for muscle contraction, and this value can reach 90 percent during vigorous physical activity. Hence a major goal underlying the study of muscle motility is to determine how chemical energy stored in ATP is converted to mechanical work during muscle contraction.

Skeletal Muscle Cells Contain Myofibrils Constructed from an Organized Array of Thick and Thin Filaments

Much of what we know about the mechanism of muscle contraction has come from the study of vertebrate *skeletal muscle*, which attaches to the skeleton and is responsible for voluntary movements. Skeletal muscle is ideally suited for studying the molecular basis of contraction because it contains a highly organized, massive array of filaments specialized for generating movement. Skeletal muscle is composed of long cylindrical cells called **muscle fibers,** each measuring 10 to 100 µm in diameter and up to several centimeters in length. Muscle fibers arise during embryonic development from cells known as *myoblasts,* which fuse to form giant cells in which hundreds of nuclei share a common cytoplasm. The muscle fibers arising from the fusion process may extend over a considerable portion of, and in some cases the entire length of, an individual muscle. Contraction of muscle fibers causes the muscle as a whole to shorten, thereby exerting a pulling force on the bones or other structures to which the muscle is attached. If shortening of the muscle is physically prevented during contraction, tension develops instead.

Each muscle fiber is enclosed within a single plasma membrane referred to as the **sarcolemma.** Flattened nuclei are distributed at numerous points along the length of the cell just beneath the sarcolemma. The cytoplasm is occupied largely by cylindrical filament-containing structures called **myofibrils,** which represent the contractile machinery of the muscle cell. The discovery of myofibrils dates back to the early 1800s,

when it was first noted that skeletal muscle exhibits a series of alternating light and dark bands or *striations* when viewed with a light microscope (Figure 13-1). Although some controversy initially arose as to whether the striations are integral to muscle cells or simply reflect a ringlike surface morphology, William Bowman suggested correctly in 1840 that the striations represent bands of intracellular material with differing refractive indices. As light microscopic techniques improved, it became apparent that the light and dark bands reside in cylindrical structures (myofibrils) measuring 1–2 µm in diameter. The myofibrils of each muscle fiber are aligned with their bands in register, giving the cell a striated appearance consisting of alternating light and dark bands called **I bands** and **A bands,** respectively. The letter "I" stands for *isotropic* and the letter "A" stands for *anisotropic,* which are terms that refer to the appearance of these bands when viewed by polarization microscopy (page 120). Running down the middle of each I band is a dense line known as the **Z disk.** The Z disks divide a myofibril into a series of repeating units called **sarcomeres,** each measuring about 2 µm in length (Figure 13-2).

The structures responsible for the repeating pattern of light and dark bands were uncovered in the early 1950s by the pioneering electron microscopy studies of Hugh Huxley and Jean Hanson. Their work revealed that myofibrils are composed of two types of longitudinal filaments: **thin filaments** measuring about 6 nm in diameter and **thick filaments** that are about 15 nm in diameter (Figure 13-3). I bands have a light appearance because they contain only thin filaments, while the A bands are dark because they contain an overlapping

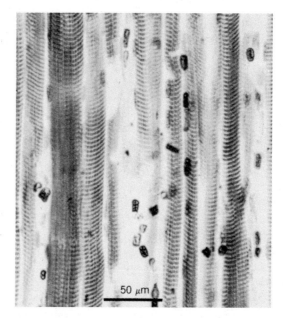

Figure 13-1 *Skeletal Muscle Viewed by Light Microscopy* *The alternating pattern of light and dark bands is clearly apparent. Courtesy of M. H. Ross.*

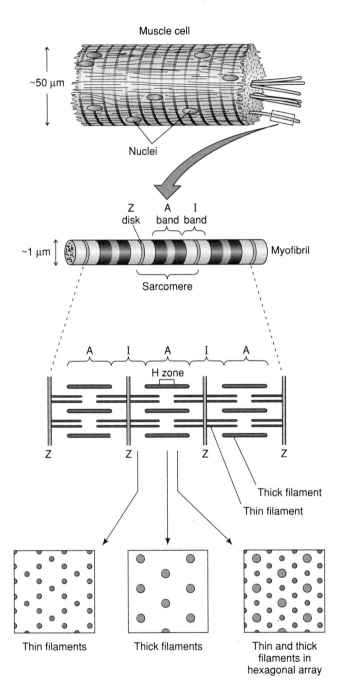

Figure 13-2 Organization of Skeletal Muscle Cells *The cytoplasm of each muscle cell (muscle fiber) contains numerous myofibrils whose cross striations are in register. The light bands (I bands) within the myofibrils contain thin filaments, whereas the dark bands (A bands) contain both thick and thin filaments. The Z disks, which run through the middle of each I band, divide the myofibril into repeating units called sarcomeres. Each A band contains a central region, called the H zone, which lacks thin filaments.*

array of thick and thin filaments. The thick filaments extend throughout the entire length of each A band, whereas the thin filaments are excluded from a central region called the *H zone* (see Figure 13-2). When observed in cross section, the overlapping thick and thin filaments appear as a hexagonal array in which each thick filament is surrounded by six thin filaments. Such an arrangement stipulates that twice as many thin filaments as thick ones are present in each myofibril.

In high-resolution electron micrographs, *crossbridges* between adjacent thick and thin filaments are visible where the two types of filament overlap (Figure 13-4). When thick and thin filaments are isolated from muscle cells and examined separately, the crossbridges protrude from the thick filaments but not the thin filaments, indicating that the crossbridges are constituents of the thick filaments. These crossbridges are the only direct connections linking the thick and thin filaments in intact myofibrils.

Actin and Myosin Are the Main Constituents of Myofibrils

Myofibrils are constructed from protein molecules that account for 50 to 75 percent of the total protein present in a typical skeletal muscle cell. Because myofibril proteins are relatively insoluble in solutions of low ionic strength, the classical approach to their purification involved an initial extraction of muscle with water to remove the soluble proteins of the cytosol, followed by exposure to concentrated salt solutions to extract the components of the myofibril. For many years the only major protein identified in such extracts was **myosin,** a molecule that was postulated to be the major contractile protein in muscle cells more than a century ago. However, in the early 1940s the Hungarian biochemist Albert Szent-Györgyi and his colleagues made the intriguing discovery that the properties of isolated "myosin" depend on how the protein is extracted. If muscle is extracted for a relatively brief period, the extracted myosin has a low viscosity and is unaffected by the addition of ATP. But when a day-long extraction procedure is employed, the myosin is very viscous and its viscosity decreases upon the addition of ATP.

Such observations suggested that extracting muscle for longer periods of time causes a second protein to be isolated along with myosin, and that the behavior of myosin is altered by this additional protein. The second protein, named **actin,** was subsequently purified and shown to bind to myosin, forming an actin-myosin complex whose viscosity is higher than that of pure myosin. Adding ATP causes the actin-myosin complex to dissociate, explaining the viscosity-lowering effect of ATP. The critical importance of these early observations for models of muscle contraction lay in the realization that the binding of actin to myosin is influenced by ATP, the molecule that drives muscle contraction.

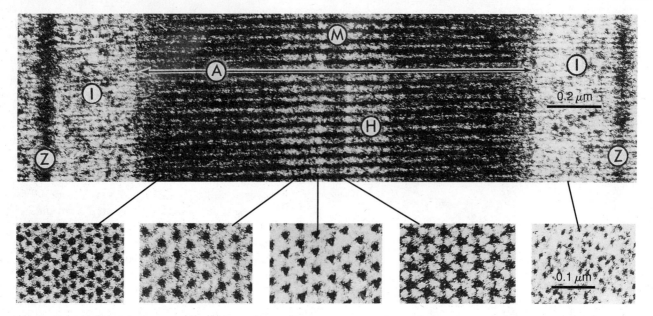

Figure 13-3 *A Myofibril Viewed by Electron Microscopy*   (Top) *Longitudinal view encompassing a single sarcomere. (Abbreviations: Z = Z disk, I = I band, A = A band, H = H zone, and M = M line, which is a small region at the center of the H zone where adjacent thick filaments are linked together.) (Bottom) Cross sections through different regions of the sarcomere show the relative positions of the thin and thick filaments. The first cross section in the series shows thick filaments surrounded by a hexagonal array of thin filaments, which is characteristic of the region in which the thick and thin filaments overlap. Courtesy of F. A. Pepe.*

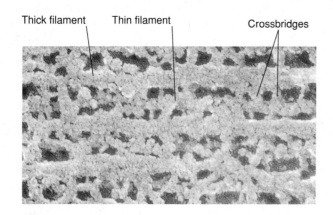

Figure 13-4 *Thick and Thin Filaments in Skeletal Muscle Viewed by Deep-Etch Electron Microscopy*   *Numerous crossbridges linking the two filament types are evident. Courtesy of H. S. Sugi.*

Myosin II Is the Principal Protein of the Thick Filament

Myosin, a large protein with a molecular weight of about 500,000 daltons, accounts for roughly half the protein present in myofibrils. In the early 1930s it was shown that solutions of myosin are *birefringent*, which means that they can rotate polarized light (page 120). Because the A band is also birefringent, such observations led to the proposal that myosin resides within the A band of the myofibril. Further support for this idea

emerged from the discovery that the A band is selectively stained by fluorescent antibodies that bind to myosin. However, these observations did not resolve the issue of which filament type is constructed from myosin because thick and thin filaments are both found in the A band. The issue was resolved when it was shown that extracting myosin from muscle cells results in the disappearance of the thick filaments from the A bands. Hence the thick filaments must be composed of myosin.

The myosin molecule consists of six polypeptide subunits: two identical *heavy chains* with a molecular weight of about 200,000 daltons each, and four *light chains* of about 20,000 daltons each. In electron micrographs, purified myosin looks like a long thin rod containing two *globular heads* protruding at one end. This two-headed type of myosin is called **myosin II** to distinguish it from the smaller, single-headed **myosin I** molecule that is involved in cytoplasmic movements in some nonmuscle cells (Figure 13-5). In the long rod-like portion of the myosin II molecule, two α-helical heavy chains are coiled around each other to form a rigid structure known as a **coiled coil.** The functions of various portions of the myosin molecule have been investigated by using the protease trypsin to cleave the molecule into two fragments called *light meromyosin* and *heavy meromyosin*. The globular heads, which are contained within the heavy meromyosin fragment, can be isolated by further treatment with the protease papain.

Myosin Hydrolyzes ATP, Binds to Actin, and Polymerizes into Filaments

Myosin exhibits three properties that are crucial to an understanding of muscle contraction. The first was discovered in 1939 by the Russian biochemists Vladimir Englehardt and Militsa Ljubimowa, who found that ATP is hydrolyzed to ADP and inorganic phosphate (P_i) in the presence of myosin and Ca^{2+}. Hence myosin is an *ATPase*—that is, an enzyme that catalyzes the hydrolysis of ATP. Subsequent studies utilizing fragments of the myosin molecule revealed the existence of two ATPase sites, one in each of the two globular heads.

The second important property of myosin is its ability to bind to actin. Myosin has two actin-binding sites, one on each of its globular heads. The actin-binding and ATPase activities of myosin are related to each other by the following four-step pathway (Figure 13-6): (1) ATP is hydrolyzed to ADP and P_i by the myosin ATPase site, permitting actin to bind to the myosin. (2) The binding of actin to myosin promotes the release of ADP and P_i from the myosin ATPase site. (3) After releasing ADP and P_i, the myosin ATPase site binds another molecule of ATP. (4) Binding of ATP to the ATPase site triggers the dissociation of actin from myosin. This completes the four-step cycle, which can then be repeated. The net result is the cyclic binding and dissociation of actin and myosin driven by the hydrolysis of ATP.

The third important property of the myosin molecule is its ability to polymerize into filaments. When the ionic strength of a solution of purified myosin is lowered, individual myosin molecules polymerize into filaments that are visible with the electron microscope. The shortest filaments are slightly less than twice the length of a single myosin molecule. Clusters of globular projections protrude from both ends of the filaments, leaving the middle of the shaft bare. The simplest interpretation of this picture is that the short filaments represent bundles of several dozen myosin molecules oriented in opposite directions, with their long tails overlapping in the central bare region and their globular heads forming projections at the two opposite ends (Figure 13-7). The longer filaments produced during the polymerization of myosin approach 1500 nm in length, and resemble the thick filaments isolated directly from

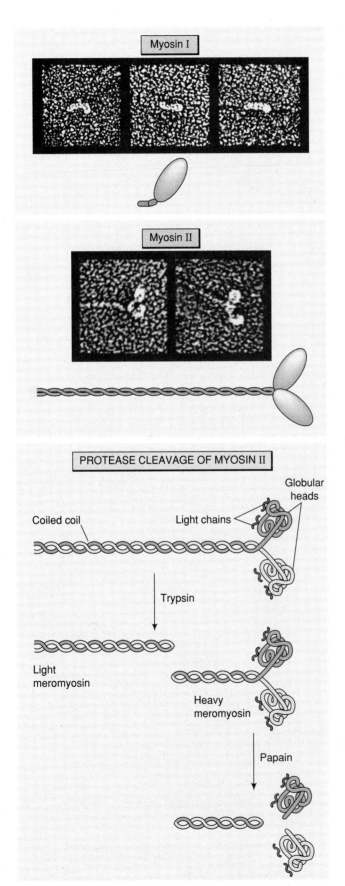

Figure 13-5 *Structure of Two Forms of Myosin* (Top) *Electron micrographs and accompanying diagrams illustrating the structural difference between myosin I, which is involved in movements in nonmuscle cells, and myosin II, which is the main constituent of the thick filaments of muscle cells. Myosin I has a single globular head and a short tail; myosin II has two globular heads and a long tail. (Bottom) Diagram showing how protease cleavage is employed to cleave myosin II into fragments. Micrographs courtesy of R. E. Cheney and J. E. Heuser.*

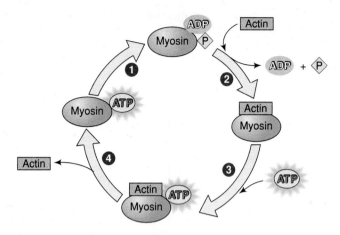

Figure 13-6 *Cyclic Binding and Dissociation of Purified Actin and Myosin Driven by ATP Hydrolysis* *When purified actin and myosin are incubated in the presence of ATP, a four-stage cyclic process is initiated in which (1) ATP is hydrolyzed to ADP and P_i, (2) actin binds to myosin accompanied by the release of ADP and P_i, (3) ATP binds to myosin, and (4) actin dissociates from myosin.*

muscle. These extended filaments exhibit the same central bare region, but the regions covered by the globular projections are longer. Such an arrangement suggests that myosin molecules are added to both ends of the growing filament, always orienting in the same direction as the myosin molecules that have already been incorporated at each end. In other words, the myosin thick filament is a *bipolar structure* assembled from myosin molecules that extend in opposite directions from the central bare region.

Actin Is the Principal Protein of the Thin Filament

About 25 percent of the mass of a myofibril is accounted for by the protein **actin.** The idea that thin filaments are composed of actin first emerged from the discovery that extracting actin from muscle cells causes the thin filaments to disappear. This conclusion was later verified when antibodies against actin were shown to selectively stain the regions of the sarcomere where the thin filaments are located (the I band and part of the A band).

In solutions of low ionic strength, purified actin is a globular polypeptide chain with a molecular weight of about 42,000 daltons. Raising the ionic strength to values closer to those prevailing within living cells causes the globular actin molecule (*G actin*) to polymerize into actin filaments (*F actin*) that resemble the thin filaments observed in skeletal muscle. Thus actin filaments are polymers constructed from monomers of G actin. As long as the actin monomer concentration exceeds a certain level called the *critical concentration,* the polymerization reaction favors the assembly of actin filaments. If

the actin monomer concentration falls below the critical concentration, the rate of actin depolymerization exceeds the rate of polymerization and the actin filaments begin to disassemble. The rates of polymerization and depolymerization are in precise balance only when the actin monomer concentration is exactly equal to the critical concentration.

In high-resolution electron micrographs, actin filaments look as though they are constructed from two helical strands twisted around one another (Figure 13-8, *left*). However, this appearance is misleading because actin cannot be unraveled into two independent strands, nor are actin filaments formed by joining two strands. Instead, actin filaments are constructed from a single chain of actin monomers that are assembled at angles to one another, creating a helical appearance. Filament assembly is carried out by sequentially adding actin monomers to the end of an existing filament, whereas disassembly is accomplished by removing actin monomers from the filament end. During filament assembly, each incoming actin monomer contains a bound ATP molecule that is hydrolyzed shortly after the monomer is added to the growing filament tip.

Actin filaments exhibit a directionality or *polarity* that can be visualized by the *myosin decoration* technique. In this procedure, actin filaments are incubated with myosin fragments such as heavy meromyosin or globular myosin heads. The myosin fragments bind along the entire length of the actin filament, producing a series of lateral projections that look like arrowheads pointing toward one end of the filament (Figure 13-9). The fact that all of the myosin fragments point the same way means that the actin molecules are all oriented in the same direction within the filament, and thus the two ends of the filament can be distinguished from each other. During assembly one end of the actin filament, designated the **plus end,** grows several times faster than the other end, designated the **minus end.** In filaments that have been decorated with myosin fragments, the minus end corresponds to the end toward which the arrowheads point.

When the myosin decoration technique is employed to examine the orientation of thin filaments in intact myofibrils, the arrowheads are consistently found to point away from the Z disk. This means that the thin filaments located on opposite sides of the same sarcomere are oriented in opposite directions, with their minus ends pointing away from the Z disk (Figure 13-10). As we will see shortly, this arrangement is crucial to the mechanism underlying muscle contraction.

Muscle Contraction Is Caused by the Sliding of Thick and Thin Filaments

A crucial breakthrough in our understanding of how actin and myosin produce muscle contraction occurred in the early 1950s when Andrew Huxley and

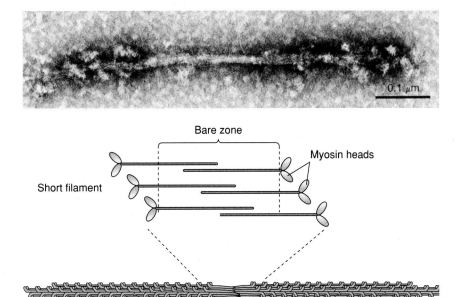

Figure 13-7 *Filaments Produced by the Polymerization of Purified Myosin* (Top) *Electron micrograph of negatively stained short filaments produced during myosin polymerization. The projections at either end are the globular myosin heads, which form crossbridges with the thin filaments in intact sarcomeres.* (Bottom) *Schematic representation of the arrangement of individual myosin molecules in the short and long filaments generated during polymerization. The short filaments consist of two sets of overlapping myosin molecules oriented in opposite directions. Longer filaments are constructed by the addition of myosin molecules to both ends, with each newly added molecule always oriented in the same direction as the molecules already present at each end. Micrograph courtesy of H. E. Huxley.*

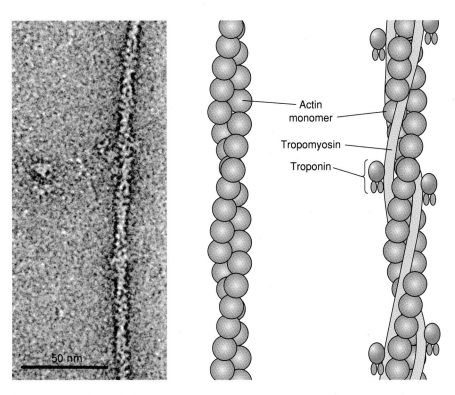

Figure 13-8 *The Actin Thin Filament* (Left) *Electron micrograph of a negatively stained actin filament.* (Middle) *Model of an actin filament showing the helical arrangement of actin monomers.* (Right) *Location of tropomyosin and troponin on the actin filaments of muscle cells. Micrograph courtesy of H. E. Huxley.*

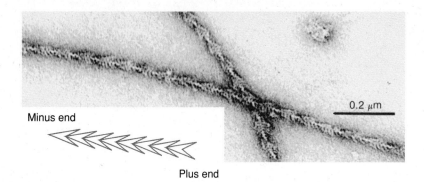

Minus end

Plus end

0.2 μm

Figure 13-9 *The Myosin Decoration Technique*   *Isolated actin filaments were incubated with heavy meromyosin. The arrowheads point toward the minus end of the actin filament. Micrograph courtesy of H. E. Huxley.*

Rolf Niedergerke, and Hugh Huxley and Jean Hanson, independently analyzed and interpreted the banding patterns exhibited by skeletal muscle during various states of contraction. Their measurements revealed that even though individual sarcomeres shorten by as much as 50 percent during muscle contraction, the length of the A bands does not change. The entire reduction in sarcomere length is accounted for by shortening of the I bands, which disappear entirely in fully contracted muscle. The simplest interpretation of this phenomenon is that the thin filaments of the I band slide into the region occupied by the A band, decreasing the overall length of the I band (Figure 13-11). The net result is a shortening of the sarcomere, and hence a decrease in the length of the muscle as a whole. Direct support for this **sliding filament model** soon emerged; electron microscopic studies revealed that the length of the thin and thick filaments does not change during muscle contraction, but the amount of

overlap between the two types of filaments increases. When muscles relax, the overlap between the two types of filaments decreases.

The conclusion that muscle contraction is produced by thin and thick filaments sliding over each other raises the question of what causes filament sliding to occur. The most obvious possibility is that the crossbridges connecting the thin and thick filaments somehow drive the sliding process. If each crossbridge contributes to the sliding force, then the total force generated by a given myofibril should be proportional to the total number of crossbridges. This prediction was rigorously tested in the 1960s in the laboratory of Andrew Huxley, where muscles stretched to differing sarcomere lengths were stimulated to contract and the resulting tension measured. The rationale underlying this approach was that the number of crossbridges that can be formed decreases as the sarcomere is stretched because the degree of overlap between the thick and thin filaments de-

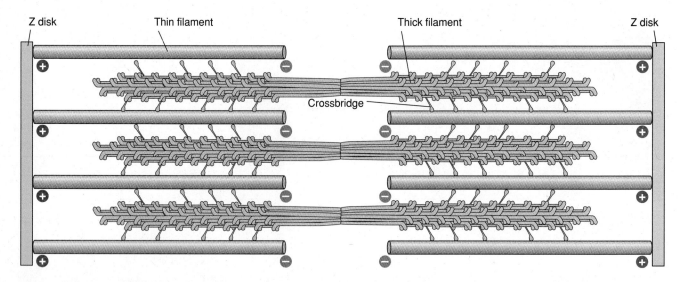

Figure 13-10 *Polarity of the Thin and Thick Filaments in an Intact Sarcomere*   *The polarity of both types of filaments reverses at the center of the sarcomere. This arrangement allows the thin filaments located on opposite sides of the sarcomere to move in opposite directions during muscle contraction.*

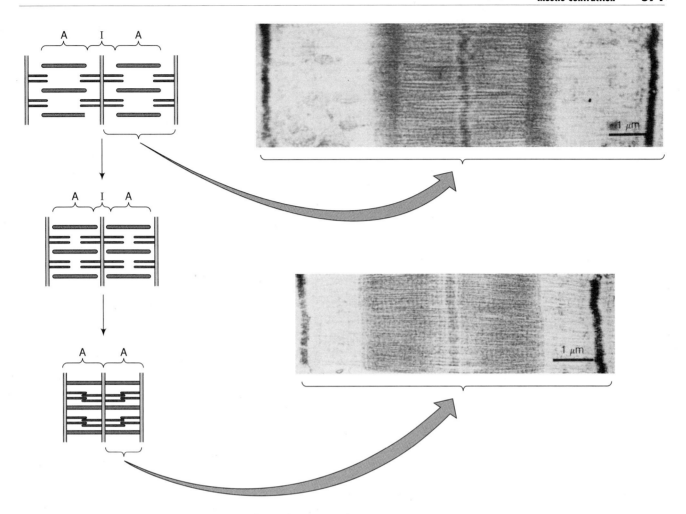

Figure 13-11 *The Sliding Filament Model of Muscle Contraction* *The diagrams on the left illustrate the changes in fila-ment overlap that occur as the sarcomere shortens. The electron micrographs on the right illustrate the shortening of the I band that is produced by filament sliding. Micrographs courtesy of H. E. Huxley.*

creases. As expected, when muscle fibers were stretched to the point where the thin and thick filaments no longer overlap and thus cannot form crossbridges (a sarcomere length of about 3.7 μm), the muscle fiber does not contract when stimulated. But if the sarcomere length is shortened slightly to permit a small amount of overlap between the thick and thin filaments, the muscle fiber regains the capacity to produce contractile tension upon stimulation. As the sarcomere length is progressively set at shorter values, allowing more overlap and hence more crossbridges between thick and thin filaments, the tension produced upon stimulation gradually increases (Figure 13-12). After reaching an optimum configuration in which all crossbridges are in use (a sarcomere length of about 2.2 μm), further shortening of the sarcomere decreases tension because the overlapping thin filaments begin to disrupt the existing crossbridges, and the thick filaments collide with the Z disk.

The sliding filament mechanism requires that the crossbridges located on one side of the sarcomere pull their attached thin filaments in one direction to move them toward the center of the sarcomere, while the crossbridges in the other half of the sarcomere pull their attached thin filaments in the opposite direction to move them toward the center. The physical basis for this differing behavior resides in the construction of the thin and thick filaments. As we have seen, thick filaments are bipolar structures composed of myosin molecules whose orientation is opposite in the two halves of the filament, and the thin filaments located in the two halves of the sarcomere exhibit opposite polarity (see Figure 13-10). Hence the polarity of both thick and thin filaments is opposite in the two sides of the sarcomere, thereby allowing the formation of crossbridges to move the two sets of thin filaments in opposite directions.

Filament Sliding Is Powered by ATP-Driven Changes in the Myosin Head

How do the crossbridges cause the thick and thin filaments to slide over one another? The key to the mechanism lies in the fact that *the crossbridges are the*

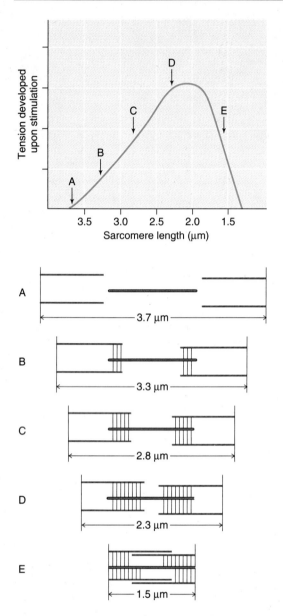

Figure 13-12 *Tension Developed by Isolated Frog Muscle Maintained at Different Sarcomere Lengths* (A) *When the thick and thin filaments do not overlap, no tension can be produced upon stimulation.* (B, C, D) *As the degree of overlap and hence the number of crossbridges increases, the tension produced upon stimulation increases.* (E) *When the sarcomere shortens beyond its optimum configuration, tension decreases because of disruption of crossbridges by overlapping thin filaments, and collision of the thick filaments with the Z disk.*

globular heads of the myosin molecules. Earlier in the chapter we learned that each myosin head has a site that hydrolyzes ATP and a site that binds to actin, an arrangement that allows myosin to participate in an ATP driven attachment-detachment cycle with actin (see Figure 13-6). Since myosin heads correspond to the crossbridges seen in electron micrographs, this is the same as saying that ATP induces the successive formation and breakage of crossbridges between the thick (myosin) and thin

(actin) filaments. A clue as to how the repeated formation and breakage of crossbridges might cause the thin filament to slide relative to the thick filament has come from X-ray diffraction and electron microscopic studies of contracting muscle. These studies have revealed that the configuration of the myosin crossbridges is altered during the process of muscle contraction; prior to contraction the crossbridges protrude at right angles from the thick filament, while during contraction they lie at a 45° angle. This change in spatial orientation allows one to postulate that breakage and reformation of crossbridges occurs in a progressive fashion, with each new crossbridge binding to a site located further along the thin filament than the site from which it was just detached. A successive series of such events, multiplied over thousands of crossbridges, would cause the thin filaments to slide and the myofibril to shorten.

Figure 13-13 summarizes this model and shows how energy released during ATP hydrolysis might trigger the conformational changes that are required for the progressive formation of crossbridges. The model begins with the myosin heads, which carry bound ATP, protruding at right angles from the thick filament. During step 1 of the contraction cycle, ATP is hydrolyzed to ADP and P_i, both of which initially remain bound to the myosin head. In step 2 the P_i dissociates from myosin, triggering the binding of the myosin head to the adjacent actin filament. During step 3, designated the *power stroke,* the myosin head undergoes a conformational change that drives the sliding of the attached actin filament. This conformational change causes the myosin head to protrude from the thick filament at a 45° angle rather than the original right angle. Finally in step 4, ADP is released from the myosin head and a new molecule of ATP is bound, triggering the detachment of the myosin head from actin so that it can initiate another cycle of attachment and sliding. Rapid repetition of this cycle results in a progressive sliding of the thin filament along the thick filament.

T Tubules Transmit Incoming Action Potentials to the Interior of the Skeletal Muscle Cell

The preceding model illustrates how the hydrolysis of ATP drives filament sliding, but it does not indicate how the process is controlled. In fact, the model makes it appear as if muscles will continue to contract as long as ATP is present. Yet this clearly cannot be the case because exquisite control of muscle contraction is required during complex coordinated movements, which occur in the face of relatively constant intracellular ATP concentrations.

It is not ATP, but rather the presence or absence of incoming nerve impulses that determines whether a muscle will contract. In Chapter 6 we learned that electrical signals are carried along nerve cells in the form of

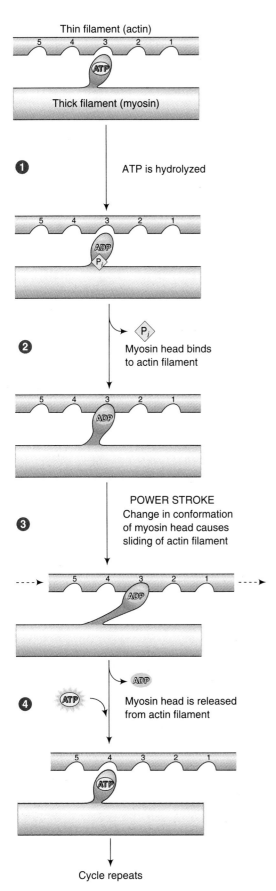

Thin filament (actin)

Thick filament (myosin)

1 ATP is hydrolyzed

2 Myosin head binds to actin filament

3 POWER STROKE
Change in conformation of myosin head causes sliding of actin filament

4 Myosin head is released from actin filament

Cycle repeats

transient changes in membrane potential called *action potentials* (page 199). When an action potential reaches the end of a nerve cell axon, it triggers the release of a neurotransmitter that diffuses across the synapse and binds to a plasma membrane receptor in the adjacent target cell, in this case a muscle cell. The axons that control skeletal muscle contraction have many tiny branches that lie in shallow depressions on the muscle cell surface, maintaining a gap of about 50 nm between the nerve and muscle cell. At these **neuromuscular junctions** the sarcolemma exhibits a series of deep infoldings called *junctional folds* (Figure 13-14). The neurotransmitter *acetylcholine* (page 757) is released from the nerve endings and diffuses across the 50-nm gap or *synapse,* binding to acetylcholine receptors present in the membranes of the junctional folds. The acetylcholine receptors function as neurotransmitter-gated ion channels, opening in response to acetylcholine and permitting an increased flow of Na^+ into the muscle cell (page 210). If the resulting depolarization exceeds a threshold value, an action potential is triggered and then propagated along the entire sarcolemma.

For many years biologists puzzled over the question of how an action potential occurring at the sarcolemma triggers the nearly simultaneous contraction of every myofibril in the cell interior. Simple diffusion of a chemical signal from the cell surface through the cytoplasm would be too slow to account for such rapid contraction. An important clue was provided in 1958 by Andrew Huxley and Robert Taylor, who used a micropipette to apply a small electric current to selected regions on the surface of a frog muscle cell. The tiny current was sufficient to cause a local depolarization of the plasma membrane wherever it was applied, but the magnitude of the depolarization was below the threshold required for triggering an action potential across the entire muscle cell. When applied to most regions of the cell surface, the small current failed to trigger contraction. However, Huxley and Taylor discov-

Figure 13-13 *The Sliding Filament Model of Muscle Contraction* *Prior to contraction, myosin heads containing bound ATP protrude at right angles from the thick filament. (1) A contraction cycle begins with the hydrolysis of ATP, generating ADP and P_i molecules that remain bound to the myosin head. (2) The dissociation of P_i from myosin is accompanied by binding of the myosin head to the adjacent actin filament. (3) During the ensuing power stroke, the myosin head undergoes a conformational change that drives the sliding of the attached actin filament. At this stage, the myosin heads form crossbridges that lie at a 45° angle. (4) In the final step, ATP binds to the myosin head as ADP is released, triggering detachment of the myosin head from actin. The net result is a ratcheting mechanism in which myosin heads induce filament sliding by progressively engaging successive sites on the adjacent actin filament. Note that the sequence of interactions between myosin and actin closely parallels the cyclic binding and dissociation that is observed when purified actin and myosin are incubated with ATP (see Figure 13-6).*

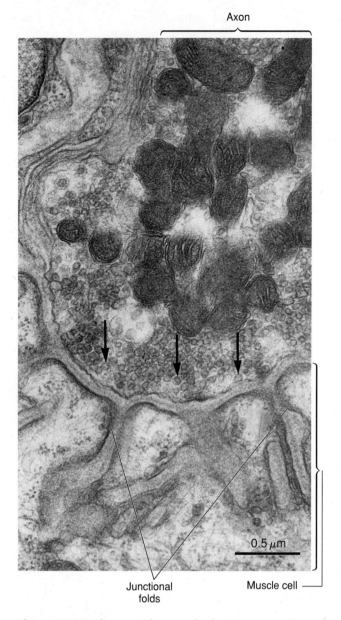

Axon

0.5 μm

Junctional folds   Muscle cell

Figure 13-14 *Electron Micrograph of a Neuromuscular Junction* *The axon occupies the upper part of the micrograph; it contains many small vesicles (arrows) filled with the neurotransmitter acetylcholine. The muscle cell occupies the lower part of the micrograph. Its sarcolemma is thrown into several deep junctional folds. Courtesy of E. G. Gray.*

colemma and the T tubules are occasionally observed, suggesting that T tubules represent invaginations of the plasma membrane and that the lumens of the T tubules are continuous with the outside of the cell. Hugh Huxley tested this idea by incubating frog muscle in a solution of ferritin molecules, whose large size and electron density make them easy to visualize with the electron microscope. It was found that the ferritin molecules accumulated quickly within the lumens of the T tubules but did not appear elsewhere in the muscle (Figure 13-17). Since ferritin is too large to pass through the plasma membrane, the simplest explanation is that T tubules are continuous with the plasma membrane and their lumens open directly to the outside of the cell. This physical continuity between the plasma membrane and T tubules provides a direct pathway for conducting membrane depolarization from the cell surface to the cell interior.

Further support for the idea that T tubules are responsible for spreading the depolarization signal to the cell interior came when Huxley carried out similar experiments in lizard muscle, where T tubules are located near the junctions between the A and I bands, rather than in the center of the I bands. In this case contraction occurred only when the tiny current was applied near the junction between the A and I bands, and the I band shortened only on its stimulated side. Thus once again the site where contraction could be induced corresponded to the location of T tubules.

T Tubules Trigger Muscle Contraction by Promoting the Release of Ca^{2+} from the Sarcoplasmic Reticulum

How is filament sliding triggered once T tubules have transmitted the signal for muscle contraction to the cell interior? The idea that calcium ions are involved first received serious consideration in the 1940s when L. V. Heilbrunn and F. J. Wiercinski injected various substances into muscle cells and found Ca^{2+} to be the only biological agent that triggered contraction. Later evidence emerged from studies in which muscle cells were exposed to substances whose color or fluorescence varies in direct relationship to changes in free Ca^{2+} concentration. Monitoring the behavior of such indicators revealed that the Ca^{2+} concentration in the cytosol rises quickly after a muscle cell is stimulated to contract (Figure 13-18).

Where does this Ca^{2+} come from? The Ca^{2+} concentration in the cytoplasm of muscle cells is regulated by the **sarcoplasmic reticulum,** a system of flattened membrane channels covering the surface of each myofibril that is comparable to the endoplasmic reticulum of other cell types (see Figure 13-16). Subcellular fractionation experiments have revealed that isolated sarcoplasmic reticulum vesicles take up Ca^{2+} from the surrounding medium by an

ered the existence of sensitive spots, always located over an I band, whose stimulation caused the adjacent I band to contract (Figure 13-15).

This phenomenon suggested that the signal for muscle contraction is conducted from the sarcolemma to the interior of the cell along structures located near the I bands. In electron micrographs, a highly developed system of membrane channels called **transverse tubules (T tubules)** can be seen passing from the cell surface into the cell interior at the center of each I band (Figure 13-16). Points of continuity between the sar-

Before current ⟶ After current

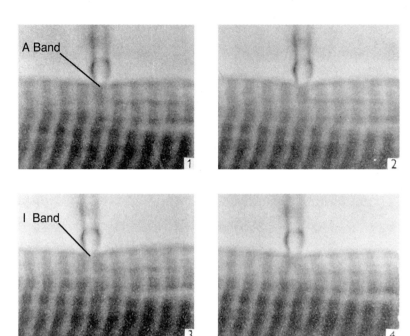

Figure 13-15 *Evidence That the Surface of a Muscle Cell Exhibits Regions That Are Particularly Sensitive to Electrical Stimulation* In these experiments, a tiny electrical current was applied to a muscle cell through a pipet placed above an A band (1 and 2) or an I band (3 and 4). Micrographs 1 and 3 show the muscle before current is applied; micrographs 2 and 4 were taken while current is being applied. Contraction occurs only when current is applied to an I band (micrograph 4), and the I band shortens only in the region directly adjacent to the site of stimulation.

active transport process that is driven by the hydrolysis of ATP. The **calcium pump** responsible for this uptake is so efficient that it can lower the Ca^{2+} concentration of the surrounding medium to less than 10^{-7} *M*. Because the permeability of the sarcoplasmic reticulum to Ca^{2+} is normally quite low, the calcium ions taken up by active transport cannot pass back through the membrane by simple diffusion. Hence Ca^{2+} is effectively trapped inside the sarcoplasmic reticulum lumen, where many of the calcium ions bind to the Ca^{2+}-binding protein *calsequestrin*.

This situation can be altered, however, by signals that are transmitted by T tubules. As we have discussed, T tubules usually pass into the interior of muscle cells along the middle of the I bands, where the Z disks are located. Near each Z disk, the flattened channels of sarcoplasmic reticulum that cover each myofibril are enlarged, forming *terminal cisternae* (see Figure 13-16). The T tubules of vertebrate muscle cells pass between the two terminal cisternae of adjacent sarcomeres, creating a configuration called a **triad** because it consists of three elements: two terminal cisternae surrounding a T tubule (Figure 13-19). In the region of a triad, the membranes of the T tubule and the adjacent terminal cisternae are separated by a space of no more than

10–20 nm. This close proximity forms the basis for signal transmission during muscle excitation. When an action potential is transmitted by T tubules from the cell surface to the center of a triad, the membrane depolarization triggers a transient increase in the permeability of the terminal cisternae to Ca^{2+}. As a result, calcium ions diffuse from the terminal cisternae of the sarcoplasmic reticulum into the cytosol.

Troponin and Tropomyosin Make Filament Sliding Sensitive to Ca^{2+}

The final missing link in our model of muscle contraction concerns the question of how the release of Ca^{2+} from the sarcoplasmic reticulum causes myofibrils to contract. An answer to this question first emerged from studies carried out by Setsuro Ebashi and his colleagues in the early 1960s. The impetus for this work was the discovery that mixtures of actin and myosin prepared in different ways vary in how they respond to calcium ions. Adding Ca^{2+} to crude preparations of actin and myosin enhances the formation of actin-myosin complexes, whereas the interaction between purified actin and purified myosin is unaffected by Ca^{2+}.

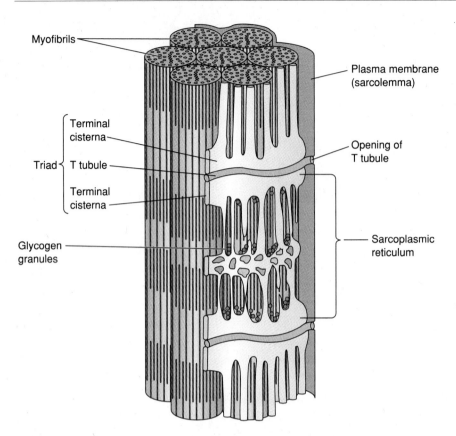

Figure 13-16 *Three-Dimensional Model Showing the Organization of T Tubules and Sarcoplasmic Reticulum in Vertebrate Skeletal Muscle* Note that the T tubules pass from the cell surface into the cell interior by running down the middle of the I bands, where they pass between flattened channels of sarcoplasmic reticulum known as terminal cisternae.

Ebashi correctly interpreted this unresponsiveness as an indication that purified actin and myosin lack one or more of the components required for conferring Ca^{2+} sensitivity upon the actin-myosin interaction. By extracting proteins from muscle tissue whose actin and myosin had already been removed, he was able to isolate two proteins that restore Ca^{2+} sensitivity to purified actin and myosin. The two proteins, *tropomyosin* and *troponin,* are both associated with the thin filament and together they account for about 10 percent of the total myofibrillar protein. **Tropomyosin** is a dimer of 70,000 daltons comprised of α and β subunits of approximately equal molecular weight. **Troponin** is a large globular protein consisting of three polypeptide subunits: *troponin-T,* which binds to tropomyosin, *troponin-I,* which attaches to actin and inhibits its binding to myosin, and *troponin-C,* which binds Ca^{2+}. Ferritin-labeled antibodies directed against troponin have been shown to bind to sites situated every 40 nm along the thin filament, which also happens to be the length of a single tropomyosin molecule. This observation supports the model of thin filament organization illustrated in Figure 13-8 (*right*).

How do tropomyosin and troponin mediate the responsiveness of muscle contraction to Ca^{2+}?

Tropomyosin normally binds to actin in a way that blocks the sites where myosin crossbridges attach to the actin filament. By blocking the formation of crossbridges between myosin and actin, tropomyosin keeps the myofibril in a relaxed state. When Ca^{2+} is released from the sarcoplasmic reticulum, the Ca^{2+} concentration in the area surrounding the myofibrils rises above 10^{-6} *M,* causing calcium ions to bind to troponin. As a result, troponin undergoes a change in conformation that allows it to displace tropomyosin from its position blocking the myosin-binding sites on the actin filament. The myosin heads can then bind to actin, forming crossbridges and setting in motion the contraction cycle depicted in Figure 13-13.

When nerve stimulation of the muscle ceases, depolarization signals are no longer transmitted by T tubules and the sarcoplasmic reticulum regains its impermeability to Ca^{2+}. The activity of the calcium pump quickly lowers the Ca^{2+} concentration in the cytosol to about 10^{-7} *M,* which causes Ca^{2+} to dissociate from troponin. Troponin, in turn, changes conformation, permitting tropomyosin to move back into a position where the myosin-binding sites on actin are again blocked. Not only does this prevent new crossbridges from forming, but old crossbridges dissociate in the presence of ATP (see Figure 13-13), allowing the muscle to relax.

Ferritin granules

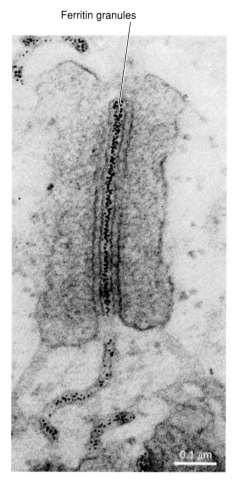

0.1 μm

Figure 13-17 *Evidence That the T Tubule Lumen Is Continuous with the Cell Exterior* *Muscle cells were incubated with ferritin molecules and then examined with the electron microscope. The ferritin molecules quickly accumulate within the lumens of the T tubules but not elsewhere in the muscle. Courtesy of H. E. Huxley.*

Phosphocreatine Replenishes ATP During Muscle Contraction

We have now seen that ATP plays two crucial roles in skeletal muscle contraction. First, ATP hydrolysis is required for the formation and breakage of crossbridges underlying filament sliding. And second, the uptake of Ca^{2+} by the sarcoplasmic reticulum depends on hydrolysis of ATP by the calcium pump. A mechanism for maintaining adequate ATP supplies is therefore crucial for muscle contraction.

It has been known since the early 1930s that the ATP concentration in actively contracting muscle remains virtually constant until physical exhaustion sets in. Moreover, inhibitors of glycolysis or cellular respiration block neither muscle contraction nor maintenance of ATP levels over the short term, pointing to the existence of an energy source in muscle other than glycolysis, respiration, and ATP. In the early 1930s it was

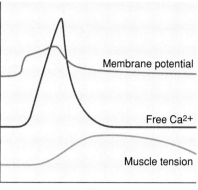

Membrane potential

Free Ca^{2+}

Muscle tension

Time

Figure 13-18 *Change in Free Ca^{2+} Concentration in the Cytosol of a Skeletal Muscle Cell after Depolarization of the Plasma Membrane* *Membrane depolarization triggers a rapid increase in free cytosolic Ca^{2+}; the subsequent fall in free cytosolic Ca^{2+} is due to the binding of Ca^{2+} to troponin, which stimulates muscle contraction (tension) by activating crossbridge formation and filament sliding.*

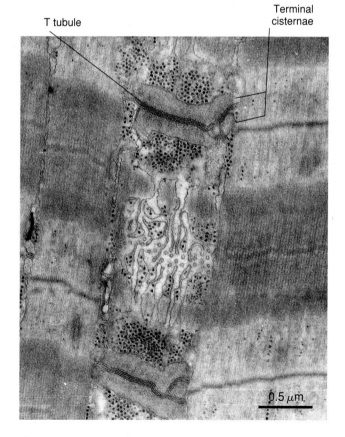

T tubule

Terminal cisternae

0.5 μm

Figure 13-19 *The Triad Configuration in Skeletal Muscle* *In this electron micrograph, the triads are highlighted with color for emphasis. Each triad consists of a T tubule surrounded on both sides by terminal cisternae of the sarcoplasmic reticulum. Courtesy of L. D. Peachey.*

discovered that instead of ATP, the high-energy molecule **phosphocreatine** decreases in concentration in contracting muscle (Figure 13-20). This finding led some investigators to erroneously conclude that phosphocreatine rather than ATP is the energy source that powers the contractile machinery. However, phosphocreatine was later found to donate its phosphate group to ADP in a reaction catalyzed by the enzyme *creatine kinase;* as a result, it was proposed that phosphocreatine functions not as the direct energy source for muscle contraction but as an energy-storing molecule that replenishes ATP during muscle contraction by donating its phosphate group to ADP.

Although this idea first surfaced in the early 1930s, supporting evidence did not emerge until D. F. Cain and R. E. Davies studied the effects of creatine kinase inhibitors 30 years later. Their studies revealed that muscle contraction continues even after creatine kinase has been inhibited. But under such conditions, ATP levels decline while the phosphocreatine concentration re-mains constant, which is the opposite of what occurs during normal muscle contraction. Such results are exactly what would be expected if ATP serves as the direct energy source for contraction, and phosphocreatine normally acts to replenish ATP.

Muscles That Frequently Contract Contain Large Numbers of Mitochondria

Since muscle contraction requires a replenishable supply of ATP, muscle cells typically contain numerous mitochondria, which are the main site of ATP production in aerobic eukaryotic cells (see Chapter 8). Muscle mitochondria tend to be arranged in regular patterns adjacent to the myofibrils, where they may occupy as much as 50 percent of the total cytoplasmic volume (Figure 13-21). The flight muscles of insects, for example, exhibit repeating patterns in which one, two, or three mitochondria are lined up adjacent to each sarcomere. Muscles designed for long periods of continuous use also contain large amounts of **myoglobin,** a cytoplasmic oxygen-binding protein designed to concentrate oxygen from the bloodstream and store it in the muscle cell for use during periods of intense mitochondrial activity. The high content of myoglobin and mitochondrial cytochromes imparts a reddish color to muscle cells of this type, so they are referred to as *red muscle.* The major source of fuel in red muscle is stored fat, whose fatty acids are oxidized by mitochondria to produce ATP (page 324). If ATP levels exceed what is needed to drive muscle contraction, the excess energy is stored by using ATP to drive the conversion of creatine to phosphocreatine; the phosphocreatine is then available to convert ADP to ATP if a future episode of muscle contraction depletes the existing ATP supply.

Muscles that are not called upon for prolonged use typically contain fewer, irregularly spaced mitochondria. Because of their reduced content of myoglobin and cytochromes, muscles of this type exhibit little color and so are termed *white muscle.* The major stored fuel in white muscle is glycogen, which is broken down to glucose molecules that are subsequently metabolized by glycolysis. Since glycolysis produces much less ATP than mitochondrial respiration, white muscles are not capable of prolonged use. Common examples of red and white muscle are the "dark" and "white" meat of chickens and turkeys. The red muscle of the legs is designed for continual use in walking and standing, whereas the white muscle of the breast is used only intermittently.

During prolonged muscle contraction, the rate of mitochondrial oxygen consumption may exceed the rate at which oxygen can be delivered to muscle cells by the bloodstream. After depleting intracellular supplies of oxygen bound to myoglobin, such muscles become anaerobic and are forced to shift to anaerobic glycolysis for ATP

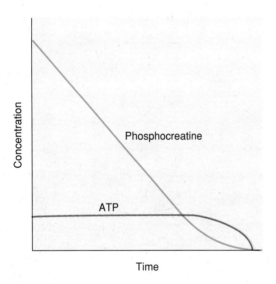

Figure 13-20 *The Role of Phosphocreatine in Muscle Contraction* (Top) *In actively contracting muscle, the phosphocreatine concentration declines while ATP remains relatively constant.* (Bottom) *The explanation for this phenomenon is that phosphocreatine replenishes ATP by donating its phosphate group to ADP.*

Mitochondrion

Figure 13-21 *Arrangement of Mitochondria in Skeletal Muscle* *In this electron micrograph of a frog skeletal muscle cell, the mitochondria are lined up in a regular pattern adjacent to the sarcomeres. Courtesy of C. Franzini-Armstrong.*

production, using stored glycogen and blood glucose as sources of fuel. Because of the low ATP yield of anaerobic glycolysis, glycogen is rapidly depleted and the cells produce large quantities of lactate, the normal end product of anaerobic glycolysis (page 313). Lactate buildup can cause muscle pain and cramping, while glycogen depletion limits the amount of ATP that can be synthesized. Both factors place severe limits on the length of time contraction can be sustained. In white muscle, the scarcity of mitochondria and the low myoglobin content means that glucose must be metabolized almost entirely by anaerobic glycolysis, explaining why, unlike red muscle, white muscle is not capable of prolonged activity.

Several Proteins Help to Maintain the Structural Integrity and Metabolic Efficiency of the Myofibril

The proteins that are directly involved in muscle contraction (myosin, actin, tropomyosin, and troponin) together account for roughly 80 percent of the protein content of myofibrils. The remaining 20 percent includes about a dozen different proteins that function to maintain the structural integrity and metabolic efficiency of the myofibril. The most abundant of these minor proteins is *titin,* so named because its titanic molecular weight of about 3 million daltons makes it one of the largest polypeptides known to exist in nature. Titin is a long flexible molecule that measures almost 1 μm in length; it

extends from the Z disk through the I band to the middle of the A band. By forming an elastic connection between the Z disk and thick filaments, titin plays an important role in maintaining the structural integrity of the myofibril when crossbridges between actin and myosin are not present, as can occur in relaxed or stretched muscle.

In addition to titin, several other proteins are present in the myofibril. Among them are (1) *α-actinin,* which crosslinks the thin filaments to each other in the region of the Z disk; (2) *nebulin,* which is anchored to the Z disk and attaches to the thin filaments; (3) the *M protein,* which links thick filaments together in the region of the *M line;* (4) the *C protein,* which forms bands at regular intervals across the thick filaments and may help to hold them together, and (5) *desmin,* an intermediate filament protein that joins the Z disks of adjacent myofibrils to one another. In addition to these structural proteins, the central region of the A band contains high concentrations of creatine kinase, an enzyme whose role in replenishing ATP during filament sliding has already been discussed.

Defects in Skeletal Muscle Cells Lead to Myasthenia Gravis and Muscular Dystrophy

Several muscle disorders have captured the interest of cell biologists because of the insights they provide concerning the workings of normal muscle. One such malady is **myasthenia gravis,** a disease characterized by progressive weakness and muscle fatigue. Current evidence suggests that the loss of muscle function is caused by a decline in the number of acetylcholine receptors in the sarcolemma. For example, quantitative measurements using radioactive acetylcholine have shown that patients with the disease have a decreased number of receptors that can bind acetylcholine. Such individuals have also been found to produce antibodies that bind to their own acetylcholine receptors. Hence for some reason, people with myasthenia gravis appear to manufacture antibodies that bind to and inactivate the acetylcholine receptors needed by their own muscle cells for initiating contraction.

More difficulty has been encountered studying the cellular defects underlying **muscular dystrophy,** a family of diseases characterized by progressive degeneration of skeletal muscle cells. The most common and debilitating form of the disease, *Duchenne muscular dystrophy,* is inherited by a gene carried on the X chromosome. Early research focused on the question of whether the primary defect is in the muscle cells (*myogenic hypothesis*) or in the nerve cells innervating the muscles (*neurogenic hypothesis*). Because one of the two X chromosomes is randomly inactivated in each cell of human females (page 451), these two possibilities can be distinguished by examining muscle function in women inheriting one normal X chromosome and one X chromosome with the muscular dystrophy gene. Since either a normal or an abnormal X chromosome will be active in any given nerve cell, the neurogenic hypothesis predicts that muscle cells innervated by abnormal nerve cells will be dystrophic and muscle cells innervated by normal nerve cells will be normal. In contrast, the myogenic hypothesis predicts that all muscle cells will be equally affected by the disease because each muscle cell contains multiple nuclei, some presumably expressing an X chromosome that is normal and others expressing an X chromosome carrying the muscular dystrophy gene. The experimental evidence supports the myogenic hypothesis. Women inheriting one normal and one abnormal X chromosome have muscle cells that are all equally affected by the disease, rather than separate populations of normal and dystrophic muscle cells.

The gene that is mutated in individuals with Duchenne muscular dystrophy codes for an abnormal form of an actin-binding protein called **dystrophin.** The amino acid sequence of dystrophin is similar to that of *spectrin,* a protein that helps stabilize and restrict the mobility of some integral proteins in the red blood cell plasma membrane (page 174). Dystrophin is associated with the inner surface of the muscle cell plasma membrane, where it binds to transmembrane glycoproteins that anchor the exterior of the cell to the extracellular matrix. Recent studies have shown that muscular dystrophy can be cured in mice by using recombinant DNA techniques to insert a normal copy of the dystrophin gene into mouse embryos carrying the abnormal form of the dystrophin gene.

Cardiac Muscle Is Composed of Cells That Are Joined Together by Intercalated Disks

Cardiac muscle, which occurs only in the heart, resembles skeletal muscle in that its constituent cells contain sarcomeres, T tubules, and a modified sarcoplasmic reticulum. Because heart muscle must continually contract, each cardiac muscle cell contains a large number of mitochondria with densely packed cristae. Glycogen granules and fat droplets lie nearby to supply the mitochondria with the energy-rich materials needed for ATP synthesis. Instead of the long, multinucleated cells present in skeletal muscle, cardiac muscle is divided by membrane partitions called **intercalated disks** into separate cells containing single nuclei (Figure 13-22). Intercalated discs are specialized regions of the plasma membrane enriched in the proteins *α-actinin* and *vinculin,* which link the membrane to actin filaments located in the cytoplasm. Intercalated disks contain numerous gap junctions (page 240) whose ability to connect cells electrically is important for conducting impulses from cell to cell and thus coordinating the contraction of the heart as a whole.

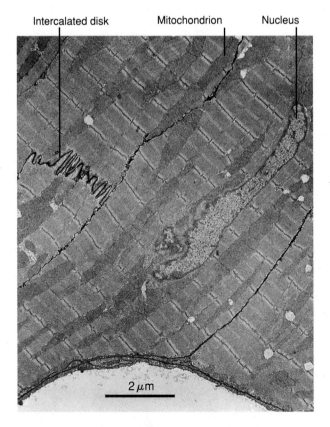

Intercalated disk   Mitochondrion   Nucleus

2 μm

Figure 13-22 *Electron Micrograph of Cardiac Muscle* *An intercalated disk is clearly visible between adjacent cells. Courtesy of J. R. Sommer.*

Though the rate of cardiac muscle contraction is regulated by special nerves, cardiac muscle cells are also able to contract repetitively in the absence of external stimulation or control.

The mechanism of cardiac muscle contraction is similar to that of skeletal muscle. Like skeletal muscle, cardiac muscle contains actin, myosin, tropomyosin, and troponin. Spontaneous action potentials occur in special nodes of tissue and are transmitted by a system of T tubules, causing an influx of Ca^{2+} into the cytoplasm. However, signal transmission by T tubules is slower in cardiac muscle than in skeletal muscle; as a result, heart muscle remains depolarized and contracted 20 to 50 times longer than skeletal muscle, a necessary prerequisite to coordinating the contraction of all the cells in a given region of the heart. Without this extended period of contraction, the first cells to receive an impulse would complete their contraction before the last cells had begun, a situation that would lead to an uncoordinated heartbeat and ineffective pumping of blood.

Cardiac muscle is thus specialized for a particular type of contraction. Its gap junctions permit electrical impulses to be rapidly transmitted from cell to cell, numerous mitochondria provide the ATP that is required for continuous contraction, and a prolonged depolarization phase permits the heart to contract as a coordinated unit.

Smooth Muscle Cells Lack the Highly Organized Array of Filaments Seen in Skeletal and Cardiac Muscle

In addition to skeletal and cardiac muscle, vertebrates contain a third type of muscle tissue called *smooth muscle* because it lacks the striations seen in skeletal and cardiac muscle. Smooth muscle occurs in the walls of internal organs such as the intestinal and genital tracts, and in glands and arteries. Smooth muscle contraction is much slower and more prolonged than contraction in skeletal muscle; examples include contraction of the uterus during delivery of the fetus, of the intestines during digestion of food, and of the iris and ciliary body during entry of light into the eye. The contraction of smooth muscle can be triggered by hormones, as when uterine contractions are induced by the hormone *oxytocin,* or by input from the nervous system.

The appearance and chemical composition of smooth muscle differs from that of skeletal muscle in several ways. The first and most obvious difference is that smooth muscle cells lack the striated banding pattern exhibited by skeletal muscle cells (Figure 13-23). In fact, it is this "smooth" appearance that gives smooth muscle its name. Smooth muscle cells are spindle shaped, about 100-fold smaller than typical skeletal muscle cells, and possess only a single nucleus; smooth muscle also has less sarcoplasmic reticulum than skele-

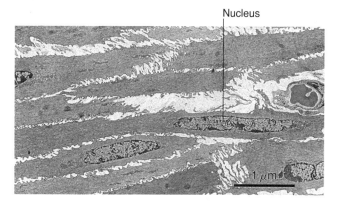

Nucleus

1 μm

Figure 13-23 *Electron Micrograph of Smooth Muscle Cells* *Unlike skeletal and cardiac muscle, smooth muscle cells lack striations. Courtesy of G. Gabella.*

tal muscle, and T tubules are completely missing. The ratio of actin to myosin in smooth muscle is about 15 to 1, which far exceeds the 2 to 1 ratio typically observed in skeletal muscle.

As in skeletal muscle, the actin and myosin of smooth muscle cells reside in thin and thick filaments, respectively. However, these filaments are not arranged in regular arrays to form sarcomeres; instead, the filaments lie in parallel groups that appear to be randomly dispersed throughout the cytoplasm. The thin filaments are linked by the protein α-actinin to **dense bodies** (Figure 13-24), which are distributed throughout the cytoplasm and serve as attachment sites for the thin filaments, just as Z disks anchor the thin filaments of skeletal muscle. The thin filaments of smooth muscle cells are also anchored to the plasma membrane by structures resembling *focal adhesions,* a type of cell junction that binds to actin filaments in other cell types (page 240).

Myosin Light-Chain Kinase Controls Smooth Muscle Contraction

Although the thick and thin filaments of smooth muscle are not arranged in highly organized arrays as in skeletal muscle, the contractile mechanism utilized by smooth muscle cells is similar in principle to that of skeletal muscle. In both cases, contraction is regulated by calcium ions. During smooth muscle contraction, the entry of calcium ions into the cytosol is usually triggered by membrane depolarization. The Ca^{2+} may come from outside the cell, or from the sarcoplasmic reticulum and/or mitochondria. Because the absence of T tubules decreases the rate at which the signal can be delivered to the thick and thin filaments, a slower, more prolonged contraction occurs.

Smooth muscle lacks both troponin and tropomyosin, so Ca^{2+} must act in a way that differs from its role in skeletal muscle. A key event in the control of

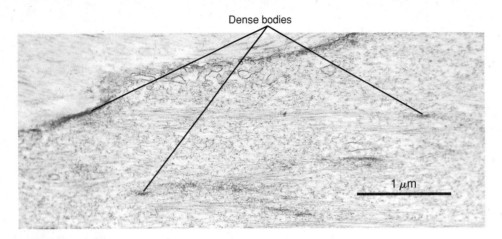

Dense bodies

1 μm

Figure 13-24 *Smooth Muscle Cell Showing Dense Bodies* *To facilitate visualization of the dense bodies, most of the actin and myosin filaments have been removed prior to electron microscopy. Courtesy of P. Cooke.*

smooth muscle contraction by Ca^{2+} is the phosphorylation of myosin light chains by the enzyme **myosin light-chain kinase** (Figure 13-25). The effect of Ca^{2+} on this enzyme is mediated by the Ca^{2+}-binding protein, *calmodulin* (page 219). When the concentration of calcium ions in the cytosol rises, Ca^{2+} binds to calmodulin; the resulting Ca^{2+}-calmodulin complex activates myosin light-chain kinase, which in turn phosphorylates a specific serine residue in one of the myosin light chains. Phosphorylation of the light chain induces a conformational change that permits the myosin molecule to form crossbridges with actin, and hence contraction ensues. When the intracellular concentration of Ca^{2+} falls, myosin light-chain kinase is inactivated and a second enzyme, **myosin light-chain phosphatase,** removes the phosphate group from the myosin light chain. Since the dephosphorylated myosin molecule can no longer bind to actin, the muscle cell relaxes.

In most cases, smooth muscle contraction is initiated by depolarization of the sarcolemma caused by nerve axons lying on the cell surface. However, the contraction of certain types of smooth muscle is induced by hormones. For example, smooth muscle in the uterus contracts in response to the hormone oxytocin, which triggers the entry of Ca^{2+} into the cytosol. The calcium ions then trigger myosin phosphorylation just as they do when an action potential is the initiating stimulus.

ACTIN FILAMENTS IN NONMUSCLE CELLS

Thin filaments measuring about 6 nm in diameter are a prominent component of the cytoplasm of most eukaryotic cells (Figure 13-26). Because of their small size, these structures were originally called *microfilaments.* But subsequent studies revealed that microfilaments in-

cubated with heavy meromyosin exhibit an arrowhead staining pattern that is similar to the pattern observed with muscle thin filaments (see Figure 13-9). In addition microfilaments, like the thin filaments of muscle, can be stained by fluorescent antibodies directed against the protein actin. Thus the microfilaments of nonmuscle cells and the thin filaments of muscle are closely related structures, both of which can be referred to as **actin filaments.**

We will see in this part of the chapter that the actin filaments of nonmuscle cells are involved in a variety of activities, including cytoplasmic streaming, cell crawling, cytokinesis, cell shape changes, and cell surface events such as endocytosis and secretion. The unifying feature shared by these apparently diverse events is that they all exploit the ability of actin filaments to provide structural support and generate movement.

Actin-Binding Proteins Influence the Structural Organization of Actin Filaments

Actin is one of the most abundant proteins of eukaryotic cells, accounting for as much as 10 to 15 percent of the total protein mass. While some unicellular eukaryotes have only a single type of actin, higher eukaryotes produce several forms of actin encoded by a family of related genes. Nonmuscle cells utilize two main types of actin, called *β-actin* and *γ-actin,* whose amino acid sequences differ slightly from that of muscle actin *(α-actin).* The actin of nonmuscle cells exists in a variety of structural forms, including free monomers, individual actin filaments, regularly crosslinked filament bundles, and less regularly crosslinked filament networks that behave like gels (Figure 13-27).

The ability of actin to assume these various states is controlled by several dozen **actin-binding proteins** that interact with actin and influence its properties

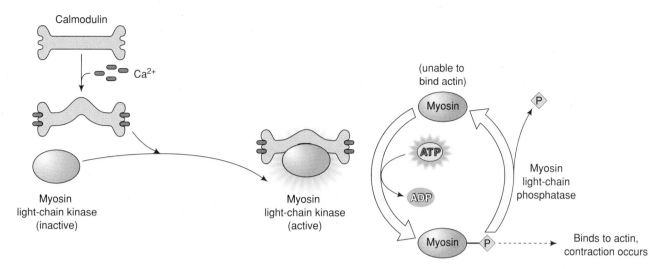

Figure 13-25 *Control of Smooth Muscle Contraction by Myosin Light-Chain Kinase* *When the Ca²⁺ concentration in the cytosol rises, Ca²⁺ binds to calmodulin and the Ca²⁺-calmodulin complex activates myosin light-chain kinase. The activated kinase catalyzes the phosphorylation of a myosin light chain, thereby permitting the myosin molecule to interact with actin and initiate contraction. When the Ca²⁺ concentration falls, the phosphate group on the myosin light chain is removed by myosin light-chain phosphatase. The dephosphorylated myosin molecule can no longer bind to actin, so the muscle cell relaxes.*

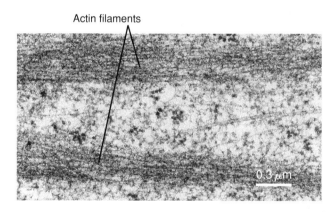

Figure 13-26 *Electron Micrograph Showing Actin Filaments in a Mouse Fibroblast* *The micrograph shows two bundles of actin filaments; each filament measures about 6 nm in diameter. Courtesy of R. D. Goldman.*

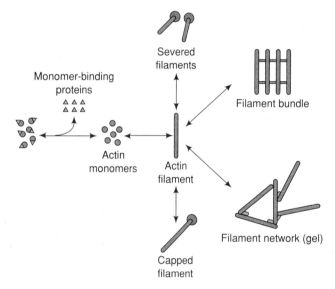

Figure 13-27 *Interrelationships between the Main Structural Forms of Actin* *Actin-binding proteins (purple) are responsible for converting actin filaments from one form to another.*

(Table 13-1). Many actin-binding proteins contain regulatory sites that allow their interaction with actin to be controlled by either Ca²⁺, the Ca²⁺-calmodulin complex, or phosphoinositides such as PIP₂ (page 218). Based on their differing functions, actin-binding proteins can be divided into seven broad categories. (1) *Monomer-binding proteins* bind to monomeric actin, preventing it from polymerizing into actin filaments. (2) *Bundling proteins* create parallel bundles of actin filaments by forming crosslinks between adjacent filaments. (3) *Gelating proteins* also form crosslinks between adjacent actin filaments, but in a less regular fashion than bundling proteins. The result is a loosely interconnected network of filaments that exhibits a gelatinous consistency. (4) *Capping and severing proteins* bind

preferentially to one end of actin filaments (usually the plus end), and thus prevent further addition of monomers. Proteins in this class also tend to bind to actin monomers located internally within an actin filament, causing the filament to break. (5) *Motor proteins* cause movement by making and breaking connections along an actin filament. Myosins I and II are classic examples of proteins in this category. (6) *Regulatory proteins* influence the interaction between actin and other actin-binding proteins. For example, tropomyosin regu-

Table 13-1 Examples of Actin-Binding Proteins

Monomer-binding proteins

Profilin

ADF (actin depolymerizing factor)

Thymosin-β_4

Bundling proteins

α-Actinin

Fimbrin

Villin

Fascin

Synapsin

Gelating proteins

Filamin

α-Actinin

Spectrin

Capping and severing proteins

Gelsolin

Fragmin

Severin

Villin

Motor proteins

Myosin I

Myosin II

Proteins controlling actin interactions

Tropomyosin

Anchoring proteins

Spectrin

Ponticulin

Note: Significant overlap exists between several of these categories. For this reason, proteins that exert multiple effects on actin may be listed in more than one category.

lates the interaction between actin and myosin. (7) *Anchoring proteins* link actin to membranes and other cytoskeletal filaments.

In addition to actin-binding proteins, several drugs employed for experimental and therapeutic purposes also bind to actin and influence its structural organization. One commonly employed group of compounds are the **cytochalasins,** a family of related substances produced by fungi that bind to the plus ends of actin filaments and block polymerization; at high concentrations, they also break actin filaments into fragments. **Phalloidin,** which is produced by certain poisonous mushrooms, promotes rather than blocks actin polymerization. Drugs like cytochalasins and phalloidin are useful tools when trying to determine whether a particular process involves actin filaments. For example, the discovery that certain types of movements, such as cell crawling, are inhibited by both

classes of drugs indicates that actin polymerization and depolymerization both play a role in such activities.

Actin Filaments of Nonmuscle Cells Are Often Anchored to the Plasma Membrane

In muscle cells, actin filaments are arranged in a highly organized array within the myofibrils. The distribution of actin filaments in nonmuscle cells is more variable, depending both on the cell type and its functional state. Nonmuscle actin filaments often occur in loose networks surrounding the nucleus, in bundles that pass through the cytoplasm, and in dense networks located beneath the plasma membrane. The actin-containing structure that most closely resembles the myofibril occurs in cultured fibroblasts, which develop parallel bundles of actin filaments called **stress fibers** that are large enough to be seen with the light microscope. Stress fibers either span the length of the cell or converge on several focal points (Figure 13-28). At one end, stress fibers are attached to a plasma membrane anchoring site called a *focal adhesion,* which is a special type of cell junction that binds the cell to an external surface (page 240). The opposite end of the stress fiber is linked either to other cytoskeletal filaments or to another focal adhesion. Stress fibers resemble small myofibrils in thin-section electron micrographs and contain proteins that are normally associated with myofibrils, such as actin, myosin, tropomyosin, and α-actinin. Such observations suggest that stress fibers function as tiny intracellular "muscles" that utilize actin-myosin interactions to generate forces that pull against the plasma membrane.

Stress fibers represent only one of several ways in which actin filaments can become associated with the plasma membrane. For example, the cells of certain kinds of tissues are held together by *adherens junctions,* which connect to actin filaments lying beneath the plasma membrane (page 239). In epithelial cells these junctions take the form of extensive zones called *adhesion belts* that completely encompass entire cells. In addition to their association with cell junctions, actin filaments are often highly concentrated in the cytoplasm just beneath the plasma membrane. The actin filaments located in this outermost region of the cell, called the **cell cortex,** add mechanical strength to the cell surface and are involved in various types of movement.

Cytoplasmic Streaming Is Driven by Interactions between Actin Filaments and Myosin

When observed by light microscopy, the cytoplasm of most living cells appears to be in constant motion. Some of the activity reflects random *Brownian movement* generated by the thermal energy of the water molecules in

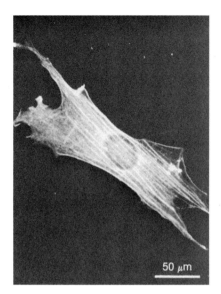 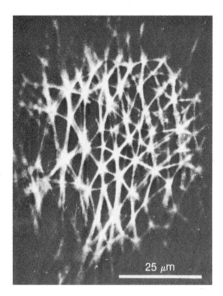 

Figure 13-28 *Fluorescence Micrographs of Nonmuscle Cells Stained with Antibodies Directed against Actin and Myosin*   (Left) *A fibroblast stained with fluorescent antibodies directed against actin exhibits bundles of actin filaments called stress fibers spanning the length of the cell.* (Middle) *Example of a cell stained for actin in which the stress fibers converge on several focal points.* (Right) *A cell stained with fluorescent antibodies directed against myosin reveals the presence of myosin in stress fibers that resemble those observed when cells are stained with antibodies directed against actin. The presence of both actin and myosin in stress fibers suggests that these structures function like tiny intracellular "muscles." Courtesy of E. Lazarides* (left, middle) *and T. D. Pollard* (right).

which the cytoplasmic constituents are dissolved or suspended. But many cells manifest a regular pattern of cytoplasmic flow, or **cytoplasmic streaming,** that is too organized to be random. Cytoplasmic streaming in plant cells follows a circular path around a central vacuole and is therefore referred to as *cyclosis* ("circling"). During cyclosis the outermost or *cortical* region of the cytoplasm is relatively immobile and gelatinous, while the inner moving area of cytoplasm is more fluid (Figure 13-29). Parallel rows of actin filaments lie at the interface between the moving and nonmoving cytoplasmic zones, anchored to stationary chloroplasts located in the cortical cytoplasm (Figure 13-30). The myosin decoration technique (page 568) has revealed that the actin filaments are all oriented in the same direction, with their plus ends pointing in the direction of streaming. Since myosin molecules move toward the plus end of actin filaments during muscle contraction, this arrangement suggests that the movement of myosin molecules along the immobile actin filaments may be responsible for cytoplasmic streaming.

This idea has been tested in two ways. First, it has been shown that disrupting actin filaments by treating cells with the drug *cytochalasin* leads to the cessation of cytoplasmic streaming. Second, the role played by actin and myosin has been examined directly using an experimental assay for measuring cytoplasmic streaming. In this technique, giant cells of the green alga *Nitella* are cut open and latex beads coated with various molecules are placed on the network of actin filaments. If the latex beads are coated with myosin, they

move along the actin filaments by an ATP-driven process that propels the beads in the direction in which cytoplasmic streaming would have normally occurred. This means that myosin bound to the surface of membrane vesicles or other organelles would be able to move these structures around the cell by advancing along the actin filaments.

Cytoplasmic streaming in animal cells, which lack the rigid cell wall found in plants, differs from cyclosis in that the pressure produced by streaming can induce changes in cell shape that trigger movements of the cell as a whole. Because of its tendency to change directions, streaming of this type is referred to as *shuttle streaming.* When focused in a particular direction,

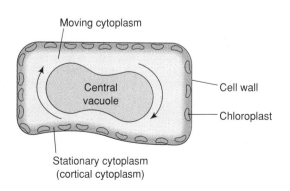

Figure 13-29 *Cytoplasmic Streaming in Plant Cells*   *This type of streaming, where cytoplasm moves in a circular path around a central vacuole, is called cyclosis.*

Figure 13-30 *Scanning Electron Micrograph Showing Chloroplasts and Associated Actin Filaments in a Green Alga* *The chloroplasts are viewed from inside the cell. Parallel rows of actin filaments run across the surface of each chloroplast. These actin filaments lie at the interface between the inner zone of moving cytoplasm and the outer cortical cytoplasm where the stationary chloroplasts are located. Courtesy of N. K. Wessels.*

shuttle streaming leads to a kind of cell locomotion that is described in the following section.

Cell Crawling in Amoebae Depends on Cytoplasmic Streaming Driven by Forces Generated at the Front of the Cell

The ability of cells to move from one location to another depends on two principal mechanisms. One kind of movement, to be described later in the chapter, involves motile appendages called cilia and flagella that allow cells in a fluid environment to *swim* through the liquid in which they are suspended. The other type of cell locomotion is based on cytoplasmic movements that permit cells to move over a **substratum** (a solid surface over which a cell moves or upon which a cell grows). This kind of locomotion is often called *amoeboid movement* because it was first studied in amoebae, but we will refer to it simply as **cell crawling.** In addition to occurring in amoebae and related unicellular eukaryotes, cell crawling is exhibited by certain blood cells, embryonic cells, fibroblasts, cancer cells, and cells growing in tissue culture. Although the mechanisms underlying cell crawling differ somewhat among various cell types, actin filaments always play a crucial role.

The first type of cell crawling to be thoroughly investigated occurs in freshwater amoebae. It was initially

proposed more than a hundred years ago that cell crawling in amoebae is driven by cytoplasmic contractions. This idea was strengthened when it was shown that in the presence of glycerol (which makes the plasma membrane permeable to small molecules added to the external medium), amoebae contract when ATP is introduced. A clue to how cytoplasmic contractions produce cell movement emerged from the discovery that locomotion in amoebae is associated with the formation and retraction of large blunt-ended cytoplasmic projections termed **pseudopodia** (Figure 13-31). Cells move in the direction of the advancing pseudopodium, with the rear portion or "tail" of the cell being pulled forward as the cell advances. The fluid cytoplasm located in the cell interior, termed the **endoplasm,** flows from the tail toward the advancing pseudopodium. As it reaches the pseudopodium, the stream of flowing endoplasm is diverted toward the sides of the cell, where it becomes transformed into a more rigid form of cytoplasm called **ectoplasm.** Meanwhile, the ectoplasm located at the rear of the cell is being broken down to provide new endoplasm for the forward flow. Thus a cyclic interconversion of ectoplasm and endoplasm is established whose net effect is to propel the cell forward (Figure 13-32). The process of converting cytoplasm from a rigid gelatinous consistency (ectoplasm) to a more fluid state (endoplasm) is referred to as a *gel-sol transition,* whereas the opposite process is a *sol-gel transition.*

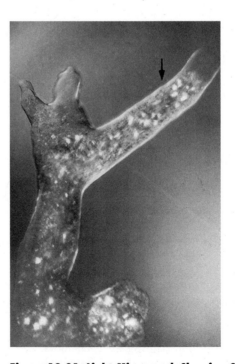

Figure 13-31 *Light Micrograph Showing One Large and Several Smaller Pseudopodia in a Giant Amoeba* *The large pseudopodium* (arrow) *is advancing while the others are retracting. Courtesy of R. D. Allen.*

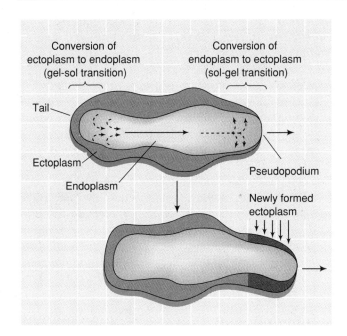

Figure 13-32 *Cytoplasmic Streaming and Cell Locomotion in Amoebae* *Endoplasm flows from the tail to the advancing pseudopodium, where it undergoes a sol-gel transition that converts it to ectoplasm. Meanwhile, ectoplasm located at the rear of the cell undergoes a gel-sol transition, thereby providing new endoplasm for the forward flow. The net effect of this cyclic interconversion of ectoplasm and endoplasm is to propel the cell forward.*

Where in the cell does the crucial event driving the interconversion of ectoplasm and endoplasm take place? The first idea to gain prominence was that the driving force occurs at the rear of the cell, thereby generating a positive pressure gradient that propels the endoplasm toward areas of lower pressure at the front of the cell. This hypothesis was widely accepted until the early 1960s, when Robert Allen was led to question it by an accident that occurred while he was examining an amoeba contained within a glass capillary tube. Using an unfamiliar microscope, he mistakenly moved the lens the wrong way, breaking the tube. To his surprise the amoeba cytoplasm continued to stream, even though the plasma membrane had been ruptured. According to the positive-pressure theory, the streaming should have stopped because any pressure within the cell would have been released upon disruption of the plasma membrane.

Allen cast further doubt on the positive-pressure theory when he later demonstrated that applying suction to the rear end of an amoeba with a small micropipette does not prevent the forward flow of cytoplasm toward the advancing pseudopodia. If the flow of endoplasm toward the pseudopodia is caused by pressure generated at the rear of the cell, then the application of suction to reduce the pressure should have slowed, stopped, or even reversed the normal di-

rection of streaming. On the basis of these observations, Allen concluded that amoeboid movement does not involve propulsion of the endoplasm by pressure generated in the tail.

The obvious alternative is that the endoplasmic stream is pulled forward by forces generated at the front of the cell. Insight into how this mechanism might operate has emerged from electron microscopic studies carried out on isolated amoeba cytoplasm, which is an attractive model for study because its movements can be regulated by the direct addition of substances to the incubation medium. In the absence of an added energy source, the movement of isolated cytoplasm quickly comes to a halt. If ATP and Ca^{2+} are then added, cytoplasmic streaming resumes and structures resembling pseudopodia begin to appear, even though the plasma membrane is missing. When preparations treated in this way are examined with the electron microscope, the most striking morphological change observed after ATP and Ca^{2+} have been added is the formation of large arrays of actin filaments. Such observations have fostered the hypothesis that the formation of actin filament arrays at the front of the cell is involved in pulling the endoplasm forward and converting it to ectoplasm. At the same time, fragmentation of the actin network located at the rear of the cell presumably triggers a gel-sol transition that converts the ectoplasm back to endoplasm.

Actin-Binding Proteins Mediate Sol-Gel Transitions, Gel-Sol Transitions, and the Contraction of Actin Gels

The preceding observations suggest that changes in the behavior of actin filaments play an important role in cell crawling. To investigate the factors that might control this changing behavior, biologists have turned to the study of actin outside of living cells. Cytoplasmic extracts prepared from cells that are capable of crawling typically contain large amounts of actin and numerous actin-binding proteins. Studies of actin-containing cell extracts and purified actin solutions have demonstrated the existence of three kinds of physical changes that are potentially relevant to cell crawling:

1. *Sol-gel transitions.* Actin solutions can be converted into gels by actin-binding proteins that form crosslinks between actin filaments. For example, tiny amounts of the gelating proteins **filamin** or **α-actinin** (page 579) will convert a freely flowing solution of actin filaments into a solid gel.

2. *Gel-sol transitions.* Actin gels can be converted back into fluid solutions by actin-binding proteins that break actin filaments. For example, small amounts of the actin-severing protein **gelsolin** will convert an actin gel into a free-flowing solution.

3. *Contraction of actin gels.* Actin gels are also capable of contracting. If a small glass tube containing an actin gel is exposed to a Ca^{2+}-containing solution at one end, the gel contracts at that end. The contractile response only occurs if myosin is present in the gel. Either double-headed myosin II or single-headed myosin I is capable of inducing contraction. Both forms of myosin are present in nonmuscle cells.

Cell Crawling in Multicellular Animals Is Based on Repeated Cycles of Extension, Attachment, and Contraction

Cell crawling in multicellular animals differs in several ways from the comparable process in amoebae, but similar changes in the properties of actin molecules are involved. In cultured animal cells, the events associated with cell crawling can be subdivided into three stages: (1) *extension* of the leading edge of the cell, (2) *attachment* of the leading edge to the substratum, and (3) *contraction* of the cytoplasm, pulling the cell forward (Figure 13-33). We will briefly examine each of the three stages in turn.

Stage 1: Actin Polymerization Drives the Formation of Cell Surface Extensions

The first stage in the three-step cell crawling cycle involves the formation of transient cell surface extensions that protrude from one side of the cell. In migrating fibroblasts, most of these protrusions take the form of broad flattened pseudopodia called **lamellipodia.** In other cell types, the cytoplasmic extensions may be short and thin (**microspikes**), long and thin (**filopodia**), or thick and cylindrical (**lobopodia**). Regardless of their shape, all of these protrusions are enriched in actin filaments (Figure 13-34). If cells are treated with cytochalasin to disrupt the actin filaments, cell surface extensions do not form.

Insight into the mechanism underlying the formation of cell surface extensions has come from the study of blood platelets, which form numerous microspikes in response to tissue damage or other types of stimulation. Prior to activation, the concentration of nonpolymerized actin in the cytoplasm is about 2 mg/ml, which is twentyfold higher than the concentration at which purified actin spontaneously polymerizes into filaments in a test tube. Hence some factor(s) present in the cytoplasm of nonactivated platelets must prevent actin polymerization from occurring. At least three proteins have been implicated in this process: *profilin, actin depolymerizing factor (ADF),* and *thymosin-β_4.* All three proteins bind to actin monomers and inhibit their polymerization into filaments. During platelet activation, actin is released from its complex with these proteins so that it can polymerize.

When actin polymerizes, the growing filaments are oriented with their fast-growing plus ends pointing toward the plasma membrane. Hence the addition of actin monomers occurs directly beneath the plasma membrane, where it exerts a pressure that propels the plasma membrane outward. A dramatic example of this phenomenon occurs in invertebrate sperm cells, where the rapid polymerization of actin filaments drives the formation of a long thin cytoplasmic extension that passes through the egg coat and makes contact with the egg cell plasma membrane (page 686).

During cell crawling, a state is quickly established in which the addition of actin monomers to the plus end of each filament is balanced by a loss of actin monomers from its minus end. This phenomenon, called **treadmilling,** has been demonstrated in crawling cells injected with fluorescent actin monomers. After waiting for the fluorescent monomers to become incorporated throughout the cell's actin filaments, a laser beam is employed to bleach a small spot near the plus end of an actin filament located at the leading edge of the cell. As time passes, the bleached spot gradually moves farther and farther away from the leading edge. Thus the actin filament appears to be behaving like a treadmill, adding monomers at the plus end beneath the plasma membrane and losing them from the minus end in the cell interior (Figure 13-35). The treadmilling process, which accompanies the forward movement of the cell's leading edge, is associated with the hydrolysis of one ATP molecule for each actin monomer added to the plus end of an actin filament.

Treadmilling of actin filaments is not restricted to cell crawling; it is a general property of actin filaments that can be demonstrated in a test tube using purified actin. Earlier in the chapter we learned that actin polymerizes into filaments only when actin monomers are present at a certain minimum *critical concentration* (page 568). If the prevailing actin concentration is less than the critical concentration, actin filaments depolymerize rather than polymerize. Treadmilling takes place because the critical concentration for the plus end of an actin filament is lower than for the minus end. Therefore when the prevailing actin concentration is higher than the critical concentration for the plus end but lower than the critical concentration for the minus end, treadmilling will occur; that is, the actin filament will add monomers to its plus end while losing them from its minus end.

Stage 2: The Leading Edge of the Cell Attaches to the Substratum by Focal Adhesions

During the initial stages of cell crawling, multiple cell surface extensions form at one end of the cell. When

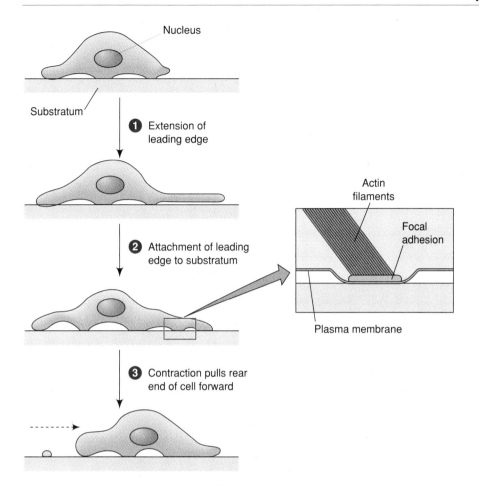

Figure 13-33 *The Three Steps Involved in Cell Crawling in Cultured Animal Cells* *During the first step, extension of the leading edge is driven by actin polymerization. The leading edge of the cell then attaches to the substratum via focal adhesions. Finally, cytoplasmic contraction pulls the rear of the cell forward. The forces generated at this stage are so great that the tail end of the cell actually ruptures, leaving a small tip of cytoplasm bound to the substratum by focal adhesions.*

one of these extensions adheres tightly to the substratum, it becomes dominant and the cell moves in its direction. In cultured fibroblasts the dominant protrusion is a broad, flattened lamellipodium whose leading edge is characterized by fluttering or undulating movements referred to as **ruffling.** Adhesion of the lamellipodium to the underlying substratum occurs in discrete regions of the plasma membrane, termed **focal adhesions,** that measure 1–2 μm long and about 0.5 μm wide. Focal adhesions are enriched in fibronectin receptors that anchor the plasma membrane to fibronectin molecules located in the extracellular matrix. At the inner surface of the plasma membrane, the fibronectin receptors attach to a plaque containing the proteins vinculin and talin (see Figure 6-50, *right*). Emerging from the plaques are bundles of actin filaments (stress fibers) that pass toward the nuclear region, where they mesh with a diffuse network of cytoskeletal filaments.

Stage 3: A Wave of Cytoplasmic Contraction Pulls the Rear of the Cell Forward

During the third step in cell crawling, the trailing end of the cell is rapidly pulled forward as its attachment to the substratum is broken. The forces generated at this stage are so great that the tail end of the cell actually ruptures, leaving behind a small tip of cytoplasm bound to the substratum by focal adhesions. During the process of pulling the tail forward, localized domains of cytoplasm contract in a way that at least superficially resembles contraction in smooth muscle cells. Since actin gels are known to contract when Ca^{2+} is added, it has been proposed that localized changes in Ca^{2+} concentration trigger the wave of cytoplasmic contraction. Support for this idea has come from studies involving amoebae stained with dyes that fluoresce when bound to Ca^{2+}; as the amoebae move, localized areas of increased fluorescence indicate that the Ca^{2+} concentra-

Actin bundles
in filopodia

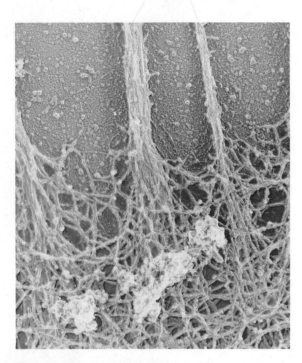

Figure 13-34 *Deep-Etch Electron Micrograph Showing Actin Bundles in Filopodia* *This view of the periphery of a macrophage shows two prominent actin bundles contained within filopodia that extend from the cell surface. The actin filaments in the filopodia merge with a network of actin filaments lying just beneath the plasma membrane. Courtesy of J. Hartwig.*

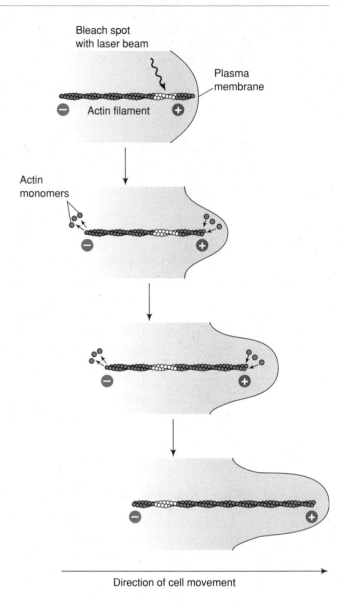

Direction of cell movement

Figure 13-35 *Evidence for Actin Treadmilling* *A laser beam is employed to bleach a small spot on an actin filament located near the leading edge of a moving cell. As the cell crawls, the bleached spot gradually moves further and further back into the cell, indicating that actin monomers are being added to the plus end of the filament and removed from the minus end.*

tion is elevated at the tip of the pseudopodium and at the tail end of the cell.

Because the concentration of myosin is much lower in nonmuscle cells than in smooth muscle, it seems likely that factors in addition to actin-myosin interactions are involved in the cytoplasmic contractions that occur in crawling cells. One possibility is suggested by the discovery that actin-containing gels can be induced to contract by adding agents that break actin filaments. Such observations indicate that actin-severing proteins may play a role in generating localized cytoplasmic contractions during cell crawling.

Crawling Cells Are Guided by Chemical Attractants and Repellents

The direction in which cells move is determined by substances in their surroundings that act as either *attractants* or *repellents*. The ability of a cell to move toward attractants or away from repellents is referred to as **chemotaxis.** A striking example of chemotaxis occurs in the cellular slime mold *Dictyostelium discoideum,* an organism that exists as both free-living individual cells and as multicellular aggregates. In the

presence of adequate nutrients, *Dictyostelium* cells live and function as separate entities. But if the food supply becomes depleted, some of the cells begin releasing a chemical attractant that has been identified as *cyclic AMP,* a molecule which also functions as a chemical messenger in higher organisms (page 211). Cyclic AMP binds to cell surface receptors in the plasma membranes of neighboring slime mold cells, causing the cells to move toward the source of the cyclic AMP. As a result, the cells gradually come together and coalesce into a multicellular aggregate that resembles a tiny worm.

How does cyclic AMP influence the direction in which *Dictyostelium* cells migrate? Within a minute of being exposed to cyclic AMP, cells begin to extend pseudopodia in the direction of the cyclic AMP source. The pseudopodia contain newly formed actin filaments whose assembly is thought to be initiated by membrane proteins like *ponticulin,* which has an actin-binding site that is exposed on the inner surface of the plasma membrane. Binding of cyclic AMP to its cell surface receptor may stimulate ponticulin to initiate the assembly of actin filaments, thereby triggering the formation of pseudopodia.

In higher organisms, a prominent example of chemotaxis is the attraction of phagocytic white blood cells to invading bacteria. The main attractants for this type of chemotaxis are small *formylated peptides* that contain the amino acid formylmethionine at their N-terminus. Because formylmethionine is employed to initiate protein synthesis in bacteria (page 490) but not in animal cells, it is a useful indicator of the presence of bacterial cells. Like the action of cyclic AMP on *Dictyostelium,* formylated peptides bind to plasma membrane receptors of phagocytic white blood cells, triggering actin polymerization and the resulting formation of cell surface extensions (in this case, microspikes). The microspikes quickly become localized at the end of the cell that points toward the area where the concentration of formylated peptides is the highest, and the cell begins to migrate in that direction.

In addition to chemical signals, physical factors also influence cell migration. In 1954 Michael Abercrombie first noted that movement of cells in tissue culture is inhibited when physical contact is made with adjacent cells. This phenomenon, called *contact inhibition of movement,* prevents cells from overlapping and piling up on one another, although in cancer cells such control is often absent (page 784). Cell migration is further influenced by physical contact between cells and the underlying substratum. In general, cells tend to move toward areas of the substratum to which they adhere most strongly. Movement is guided by long slender filopodia that explore the substratum in the areas immediately surrounding the cell (Figure 13-36). If the filopodia make contact with a favorable surface, lamellipodia are extended in the same direction and locomotion proceeds.

Microvilli and Stereocilia Are Supported by Bundles of Crosslinked Actin Filaments

The actin-rich extensions of the cell surface that arise during cell crawling are transient structures that appear for brief periods of time and then retract. Actin filaments are also involved in the formation of **microvilli,** which are relatively permanent, fingerlike projections of the cell surface that occur in cells requiring a large surface area. The structure of microvilli has been extensively studied in the epithelial cells that line the intestines; such cells are covered with thousands of microvilli that dramatically increase the plasma membrane surface area available for the absorption of nutrients. Each microvillus measures 1–2 µm in length and about 0.1 µm in diameter, and contains a supporting bundle of several dozen actin filaments oriented parallel to its long axis (Figure 13-37). Decoration with heavy meromyosin has revealed that the actin filaments are all oriented in the same direction, with their plus ends pointing toward the tip of the microvillus. Hence microvilli, like the cell surface extensions involved in cell crawling, are formed by the addition of actin monomers directly beneath the plasma membrane, exerting a pressure that propels the plasma membrane outward.

The actin filament bundles that form the core of each microvillus are held together by the crosslinking proteins **fimbrin** and **villin.** The central role played by villin in the formation of microvilli has been demonstrated by experiments in which DNA sequences coding for villin were introduced into fibroblasts. Cells treated in this way begin to synthesize villin and then develop microvilli—a quite remarkable phenomenon because fibroblasts do not normally have microvilli. Thus the presence of the protein villin appears to be sufficient to cause microvilli to be produced.

In addition to being crosslinked by fimbrin and villin, the actin filaments that make up the core of the microvillus are also stabilized by two kinds of attachments to the plasma membrane. At the tip of each microvillus, the plus ends of the actin filaments are embedded in a dense cap of material that attaches to the overlying plasma membrane; and along the length of the actin filament bundles, lateral arms link the surface of the actin filaments to the surrounding plasma membrane. These lateral arms consist of a complex between the Ca^{2+}-binding protein calmodulin (page 219) and myosin I. At the base of each microvillus the bundle of actin filaments extends into the **terminal web** (see Figure 13-37), a network of filaments consisting mainly of myosin II, spectrin, and tropomyosin. The presence of myosin within both the terminal web and the lateral arms suggests that some type of contractile movement may occur in microvilli. In support of this idea, isolated terminal web preparations have been found to contract upon the addition of ATP, thereby causing the associated microvilli to fan out. Thus contraction of the terminal web may function within the cell to elicit movements of the microvilli that increase their overall effectiveness as an absorptive surface.

Epithelial cells located in the inner ear contain surface projections called **stereocilia,** which are closely related to microvilli but are several times larger in both length and diameter (Figure 13-38). The term "stereocilia" is an unfortunate name for these structures because

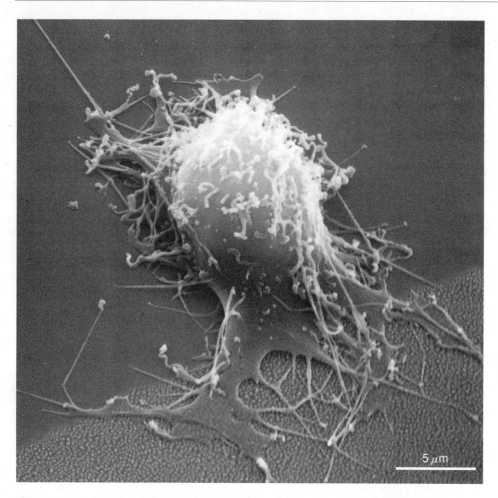

Figure 13-36 *Scanning Electron Micrograph of a Mouse Fibroblast Showing Numerous Filopodia Extending from the Cell Surface* *Courtesy of G. Albrecht-Buehler.*

they are not related to true cilia either in structure or function. Stereocilia are supported by bundles of actin filaments and are sensitive to tiny movements caused either by sound vibrations or by the movement of fluid in the semicircular canals. Such movements alter the permeability of ion channels in the plasma membrane, triggering electrical signals that are sent to the brain.

Actin Filaments Influence the Mobility of Plasma Membrane Proteins

Another role played by the actin filaments located beneath the plasma membrane relates to their ability to influence the mobility of membrane proteins. Some of the earliest evidence pointing to such a role emerged from studies involving *concanavalin A,* a carbohydrate-binding protein that binds to plasma membrane glycoproteins called *lectin receptors.* Lectin receptors are normally free to diffuse laterally in the membrane and so are distributed randomly over the cell surface. When cells are exposed to concanavalin A, lectin receptors containing bound concanavalin A first collect into small patches, and the patches then coalesce into a larger immobile aggre-

gate called a *cap.* During this **capping** process, actin filaments and actin-binding proteins accumulate underneath the cap, suggesting that the movements of the lectin receptors are being guided by interactions with actin filaments located just beneath the plasma membrane. Further support for this idea has come from the discovery that capping is inhibited in cells exposed to cytochalasin, which disrupts actin filaments (page 584).

The influence of cytoskeletal filaments on the mobility of plasma membrane proteins has also been investigated in red blood cells, which have a meshwork of filaments associated with the cytoplasmic surface of the plasma membrane (page 174). The bulk of the meshwork consists of the protein **spectrin** crosslinked by short actin filaments. The spectrin-actin network is linked to the plasma membrane by the proteins ankyrin and the band 4.1 protein. Several lines of evidence suggest that the actin-spectrin network functions in regulating the movement of membrane proteins. First, treatments that cause spectrin to aggregate into clumps, such as exposing membranes to antibodies against spectrin, cause membrane proteins such as glycophorin to clump together. Additional evidence comes from exper-

Microvilli

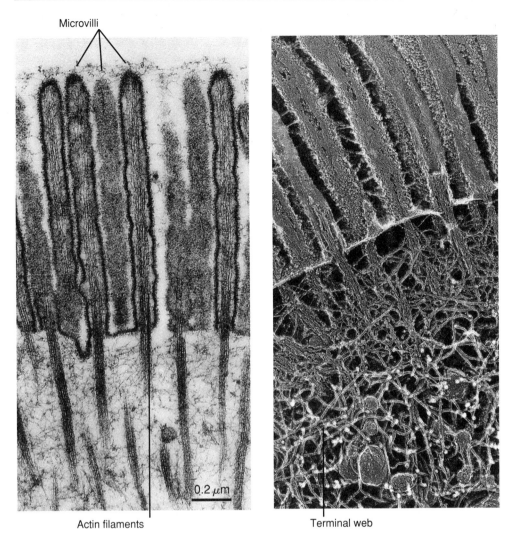

Actin filaments   Terminal web

0.2 μm

Figure 13-37 *Two Views of Microvilli Found in Intestinal Lining Cells* (Left) *In thin-section electron micrographs, each microvillus is seen to contain a core of actin filaments that extends into the underlying cytoplasm.* (Right) *In deep-etch micrographs, a filament network called the terminal web is seen in the cytoplasm directly beneath the microvilli. Courtesy of M. S. Mooseker (left) and J. Heuser (right).*

iments involving artificial membrane vesicles reconstituted from purified phospholipids and the band 3 protein, a molecule that is normally visible as intramembrane particles in freeze-fracture micrographs. In normal red cell membranes, the band 3 particles clump together when exposed to low pH; in reconstituted membranes, clumping has been found to require the presence of spectrin and actin. Such observations reinforce the conclusion that the actin-spectrin network influences the mobility of membrane proteins.

Cell Cleavage by the Contractile Ring Is Based on an Interaction between Actin and Myosin

During the process of cell division in animal cells, the cell is pinched in two by a *contractile ring* of filaments that lies just beneath the plasma membrane (see Figure

12-31). The contractile ring is composed of actin filaments associated with actin-binding proteins such as myosin II, α-actinin, and filamin. Experiments utilizing the drug cytochalasin have revealed that the actin filaments are essential to the functioning of the contractile ring. In cells treated with cytochalasin, the actin filaments disappear from the contractile ring and cell cleavage halts. Myosin is also essential for cytokinesis. If antibodies directed against myosin are injected into dividing cells, the contractile ring fails to operate because the antibodies interfere with myosin function. Additional evidence for the importance of myosin has come from the discovery that *Dictyostelium* mutants lacking myosin are incapable of carrying out cytokinesis. Taken together, the preceding observations suggest that constriction of the contractile ring is produced by an interaction between actin and myosin.

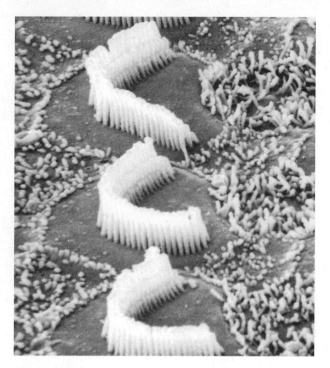

Figure 13-38 *Scanning Electron Micrograph of Epithelial Cells of the Inner Ear Showing Several Rows of Stereocilia*   *Stereocilia are related to microvilli but are much larger. They sense tiny movements caused by sound vibrations or fluid movement in the semicircular canals. Courtesy of Professors Richard Kessel and Randy Kardon, University of Iowa.*

Actin Filaments Influence the Shape of Animal Cells

The process of cell cleavage illustrates that actin filaments lying directly beneath the plasma membrane can exert dramatic effects on cell shape. Such effects are not restricted to cell division. During embryonic development, many instances arise in which groups of cells undergo shape changes. For example, a flat layer of cells may be transformed into a rounded mass, or a group of cells may need to fold in or protrude out from the surface of a tissue mass. Figure 13-39 illustrates how the contraction of actin filaments located in the appropriate regions could produce such changes in cell and tissue shape.

Most animal cells are roughly spherical, which is the shape that would be expected if they were comprised of a fluid cytoplasm bounded by a flexible membrane. But some cells adopt highly asymmetrical shapes that require the presence of actin filaments. For example, in cells that are highly elongated, bundles of actin filaments tend to be oriented parallel to the long axis of the cell; if these filament bundles are disrupted by exposing cells to cytochalasin, the cells lose their asymmetrical shape and become rounded (Figure 13-40). Changes in cell shape also appear to be

mediated by actin-rich cell surface extensions. A vivid illustration of this phenomenon occurs when freshly isolated animal cells exhibiting a spherical shape are placed in culture. Upon making contact with the surface of the culture vessel, a dramatic outgrowth of actin filament-containing microvilli occurs. These microvilli then elongate into filopodia that gradually spread out, causing the cell to become thinly flattened over the vessel surface (Figure 13-41).

Phagocytosis and Exocytosis Require the Participation of Actin Filaments

The related processes of phagocytosis (page 293) and exocytosis (page 284) are additional examples of cell surface activities that involve the participation of actin filaments. When a phagocytic cell binds to a particle that is to be engulfed, actin filaments accumulate directly beneath the plasma membrane near the bound particle. Cytoplasmic extensions then surround the particle and engulf it. This process can be inhibited by exposing cells to cytochalasin, suggesting that actin filaments are involved in particle uptake. However, caution is needed in interpreting experiments involving the effects of cytochalasin on cell surface activities because a commonly used form of this drug (*cytochalasin B*) exerts more than one effect on cells. In addition to disrupting actin filaments, cytochalasin B binds to the plasma membrane and inhibits certain types of solute transport. The possibility therefore arises that cytochalasin is inhibiting phagocytosis through a direct action on the plasma membrane rather than by disrupting actin filaments. To distinguish between these alternatives, investigators can employ an alternative form of the drug, called *cytochalasin D,* which disrupts actin filaments without binding to the plasma membrane. Like cytochalasin B, cytochalasin D has been found to inhibit phagocytosis, supporting the conclusion that actin filaments are involved in this process.

In many cell types the process of cell secretion (exocytosis) is also inhibited by cytochalasins, although in a few cases cytochalasin has been reported to stimulate rather than inhibit secretion. A possible explanation for this discrepancy has emerged from studies involving the discharge of secretory granules from exocrine cells of the pancreas, where low concentrations of cytochalasin stimulate secretion and high concentrations inhibit secretion. The lower concentrations of cytochalasin used in these studies do not disrupt actin filaments, but they do permit cytochalasin to bind to the plasma membrane and inhibit active transport. Only at higher concentrations does cytochalasin disrupt actin filaments. Hence cytochalasin appears to stimulate exocytosis at low concentrations when the plasma membrane is its primary target, and to inhibit exocytosis at higher concentrations when actin fila-

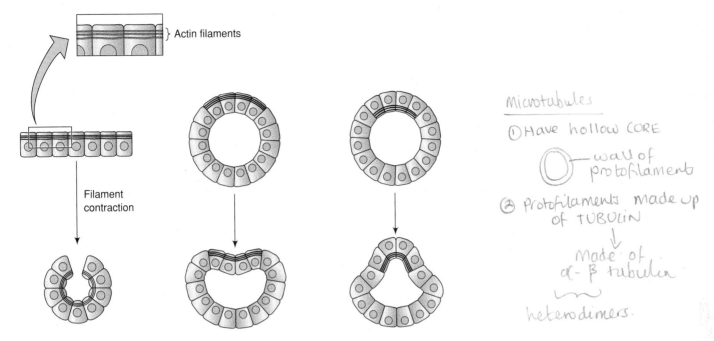

Figure 13-39 *Diagram Illustrating How the Contraction of Actin Filaments Can Produce Changes in Tissue Shape* (Left) *A flat layer of cells can be is transformed into a rounded spherical mass by the contraction of actin filaments located at the same end of each cell. (Middle and Right) Filament contraction in a small group of cells can cause the cells to either fold in (invaginate) or protrude out (evaginate) from the surface of a spherical mass.*

ments are disrupted. The latter finding suggests that as in the case of phagocytosis, actin filaments participate in the process of exocytosis.

MICROTUBULES

In addition to actin filaments, the cytoplasm of most eukaryotic cells contains a second group of cytoskeletal filaments known as **microtubules.** Microtubules play important roles in the motility of cilia and flagella, in the movements of chromosomes and organelles, in controlling the distribution of cell surface components, and in maintaining cell shape and internal organization. These diverse activities are made possible by two underlying properties of the microtubule: (1) its long, thin shape, which facilitates structural roles such as anchoring, guiding, orienting, and supporting other cellular constituents, and (2) its ability to generate movement, which is utilized in moving subcellular components as well as the cell as a whole.

Microtubules Are Long, Hollow Cylinders Constructed from the Protein Tubulin

Microtubules are long, tubelike structures measuring about 25 nm in diameter and up to several millimeters in length. Because individual microtubules are too thin

to be observed with the light microscope, their widespread occurrence in eukaryotic cells was unknown prior to the advent of electron microscopy. Early electron microscopic studies, which employed osmium fixation, revealed the presence of **microtubules** within several subcellular structures, including cilia, flagella, centrioles, basal bodies, and the mitotic spindle. But the extensive network of microtubules that pervades the cytoplasm of most eukaryotic cells was not detected until the early 1960s, when the introduction of gentler fixation techniques using glutaraldehyde permitted visualization of cytoplasmic microtubules that had been disrupted by the harsher osmium fixation procedures employed earlier (Figure 13-42).

Microtubules possess a hollow core that is surrounded by a wall of linear strands called *protofilaments* (Figure 13-43). The walls of most microtubules have 13 protofilaments, but the number is occasionally somewhat larger or smaller. Protofilaments are made of the protein **tubulin,** which in turn consists of two polypeptide chains called *α-tubulin* and *β-tubulin*. Molecules like tubulin, which contain two nonidentical subunits, are referred to as *heterodimers.* Microtubules can be dispersed into protofilaments under conditions in which the α and β chains of the tubulin heterodimer remain joined to each other. This means that each protofilament must contain both α-tubulin and β-tubulin. Additional information concerning microtubule structure has been obtained from X-

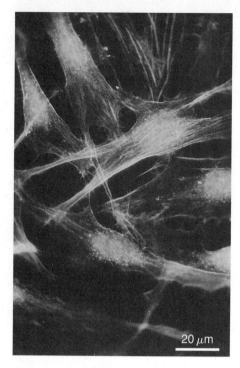
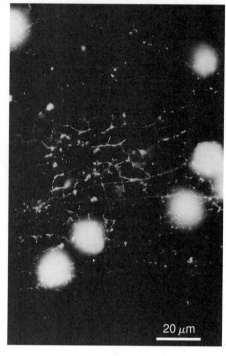

For heterodimers to → microtubules. they need the 'ringform' large tubulin for polymerisation.

Figure 13-40 *Effects of Cytochalasin on Cell Morphology* (Left) *Normal fibroblasts growing in cell culture stained with fluorescent antibodies directed against actin.* (Right) *Appearance of cells after treatment with cytochalasin. The bundles of actin filaments have disappeared and the cells have rounded up. Courtesy of R. Norberg.*

Microtubules have MAPs (protein) as well as tubulin

ray and optical diffraction analyses, which indicate that the subunits of the microtubule wall are packed together with helical geometry. A model of the microtubule wall satisfying the requirements for both a helical arrangement of tubulin subunits and the presence of α-tubulin and β-tubulin within the same protofilament is presented in Figure 13-44.

If you ↑ Pa & ↓ T°C, Microtubules sort of disappear

Microtubule Assembly Is Promoted by Microtubule-Associated Proteins (MAPs)

One of the first insights into the question of how cells make microtubules emerged in the early 1960s from the laboratory of Shinya Inoué, where polarization light microscopy was employed to monitor the behavior of the microtubules that make up the mitotic spindle (page 537). Inoué discovered that exposing cells to low temperature, high pressure, or the drug *colchicine* causes the spindle microtubules to disappear. As soon as the temperature is raised, the pressure lowered, or the colchicine removed, the microtubules spontaneously reappear. These findings suggested that microtubules exist in equilibrium with free tubulin, and that microtubule assembly-disassembly is regulated by factors that influence the equilibrium between polymerized and nonpolymerized tubulin (Figure 13-45, *top*).

Subsequent insights into the assembly process emerged from studies involving the formation of microtubules in cell extracts. The first successful experiments of this type were carried out in the early 1970s by Richard Weisenberg, who discovered that microtubules assemble spontaneously in cell extracts that have been warmed to 37°C in the absence of Ca^{2+} and the presence of GTP. Microtubules formed under such conditions contain several types of proteins in addition to tubulin. These **microtubule-associated proteins,** or **MAPs,** are a diverse group of molecules that account for as much as 10 to 15 percent of the microtubule mass (Table 13-2).

An early clue to the role played by some of these microtubule-associated proteins was provided by an experiment in which microtubules were disassembled by lowering the temperature. Under such conditions, two forms of tubulin are released into solution: the tubulin heterodimer and larger tubulin "ring forms" that look like small, curved protofilaments when viewed by electron microscopy. The tubulin rings readily reassemble into microtubules upon warming, but tubulin heterodimers do not form microtubules unless the ring forms or microtubule fragments are added to act as initiation sites for tubulin polymerization. Subsequent analyses revealed that the isolated ring forms, but not the heterodimers, contain MAPs associated

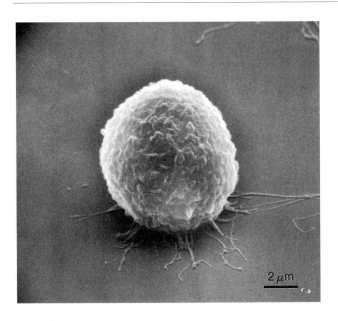
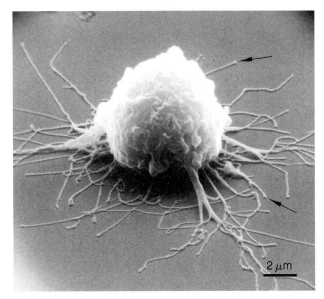
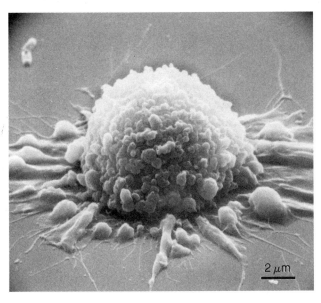
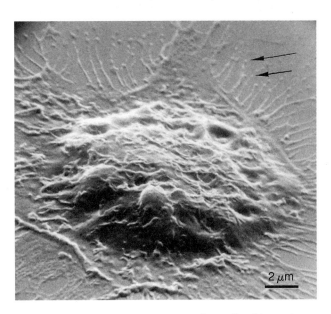

Figure 13-41 *Scanning Electron Micrographs Showing the Flattening and Spreading That Occurs When Fibroblasts Are Placed in Tissue Culture* (Top left) *Spherical cell showing early stages of filopodial outgrowth.* (Top right) *A slightly later stage in which the filopodia* (arrows) *are beginning to advance.* (Bottom left) *As flattening begins, numerous blebs (small rounded protrusions) appear on the cell surface.* (Bottom right) *In the final stages of flattening, cytoplasm is seen flowing along the filopodia in droplets* (arrows). *Courtesy of R. Rajaraman.*

with them, suggesting that MAPs promote the polymerization of tubulin into microtubules. Direct support for this conclusion came when it was shown that microtubule formation is stimulated by adding MAPs to purified tubulin heterodimers. The main proteins involved in promoting microtubule assembly have been identified as members of the *MAP1, MAP2,* and *tau* protein families. These proteins can be phosphorylated at multiple sites by protein kinases, which allows the assembly and properties of microtubules to be controlled by protein phosphorylation.

A Microtubule Is a Polar Structure That Grows More Rapidly at Its Plus End

One question to arise from the preceding observations concerns the issue of whether tubulin is added to one or both ends of a microtubule during the assembly process. This question was first investigated by Joel Rosenbaum and his associates, who incubated fragments of radioactively labeled microtubules with unlabeled tubulin under conditions designed to promote tubulin polymerization. When the resulting mixture was

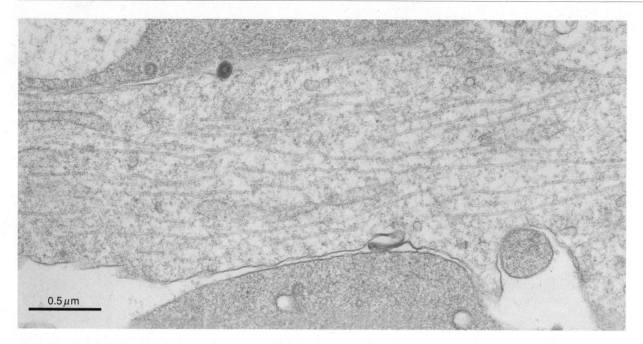

0.5 μm

Figure 13-42 *Thin-Section Electron Micrograph Showing a Group of Cytoplasmic Microtubules in a Sertoli Cell from Rat Testis* *Courtesy of L. D. Russell.*

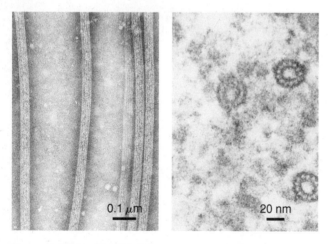

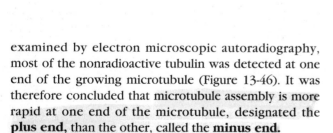

0.1 μm   20 nm

Figure 13-43 *Microtubules Visualized by High-Power Electron Microscopy* (Left) *Longitudinal view of five microtubules. Close examination of each microtubule wall reveals the presence of tiny linear strands called protofilaments.* (Right) *Cross section showing the 13 protofilaments of the microtubule wall surrounding the hollow core. Courtesy of A. Klug (left) and P. R. Burton (right).*

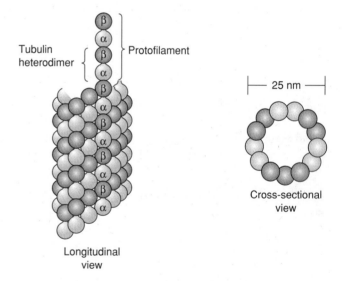

Figure 13-44 *Schematic Model of the Microtubule* *The microtubule is a hollow cylinder whose wall is composed of 13 protofilaments. Each protofilament is constructed from alternating α-tubulin and β-tubulin molecules. The tubulin subunits of the microtubule wall are packed together with helical geometry.*

examined by electron microscopic autoradiography, most of the nonradioactive tubulin was detected at one end of the growing microtubule (Figure 13-46). It was therefore concluded that microtubule assembly is more rapid at one end of the microtubule, designated the **plus end,** than the other, called the **minus end.**

As was the case for actin filaments, the *critical concentration* (page 568) for the plus end of a microtubule is lower than for the minus end. This means that when

the prevailing tubulin concentration is higher than the critical concentration for the plus end, but lower than the critical concentration for the minus end, **treadmilling** can occur; that is, the microtubule will add tubulin heterodimers to its plus end while losing them from its minus end (see Figure 13-45, *bottom*). Although treadmilling is readily observed with isolated microtubules, studies designed to detect microtubule

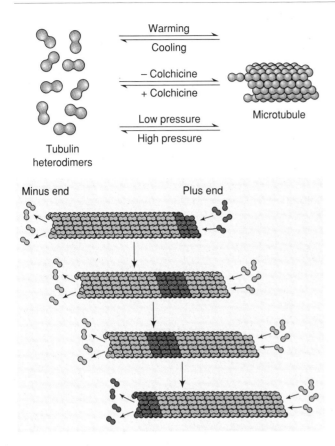

Figure 13-45 *Microtubule Assembly and Disassembly*
(Top) *Microtubule assembly is governed by an equilibrium between free tubulin heterodimers and polymerized tubulin. Low temperature, high pressure, and the drug colchicine all promote microtubule breakdown by shifting the equilibrium in the direction of free tubulin.* (Bottom) *Microtubule assembly occurs more readily at the plus end of a microtubule than at the minus end. When the tubulin concentration is higher than the critical concentration for the plus end but lower than the critical concentration for the minus end, the microtubule can add tubulin heterodimers to its plus end while losing them from its minus end. This phenomenon, called treadmilling, is readily demonstrated with isolated microtubules, but the extent to which it occurs in living cells is uncertain.*

treadmilling in living cells have yielded conflicting results. The experimental approach used to investigate this issue is similar to the one employed to study treadmilling of actin filaments (page 588); the microtubules of intact cells are first stained with light-sensitive dyes, and a laser beam is then employed to bleach a small spot on the microtubule. In most experiments the bleached spot appears to remain stationary, suggesting that treadmilling is not taking place. However, when spindle microtubules are stained with the proper dye and examined in this way, the bleached spot appears to move slowly along the microtubule, suggesting that some treadmilling may actually be taking place.

Table 13-2   Examples of Microtubule-Associated Proteins

Assembly-promoting proteins
  MAP1A
  MAP1B
  MAP2A
  MAP2B
  MAP2C
  Tau proteins
Motor proteins
  Axonemal dynein
  Cytoplasmic dynein (MAP1C)
  Kinesin
Stabilization proteins
  STOP proteins

The discovery that microtubules possess plus and minus ends that behave differently means that microtubules, like actin filaments, **exhibit a directionality or** *polarity.* Determining the direction in which individual microtubules are oriented within cells requires a procedure comparable to the myosin decoration technique used for studying actin filament polarity (page 568). In the procedure devised for investigating microtubule polarity, called the *hook decoration technique,* tubulin is added to cells that have been made permeable to macromolecules by exposing the cells to detergents. Under such conditions, tubulin attaches to the walls of existing microtubules in an aberrant fashion, creating hook-shaped protrusions that are visible with the electron microscope. The hooks always curve in a clockwise direction when viewed from the plus end of the microtubule and in a counterclockwise direction when viewed from the minus end, thereby allowing the orientation of the microtubule to be determined.

GTP Hydrolysis Contributes to Dynamic Instability of Microtubules

In addition to the requirement for MAPs, the assembly of microtubules also involves GTP. The tubulin heterodimer has two binding sites for GTP, one located on α-tubulin and the other on β-tubulin. Shortly after each tubulin heterodimer is added to the growing tip of a microtubule, the GTP bound to the β-tubulin is hydrolyzed to GDP. Experiments carried out by Tim Mitchison and Marc Kirschner have led to the notion that the GTP hydrolysis step controls the balance between microtubule assembly and disassembly. These investigators monitored the effects of different concentrations of free tubulin on the behavior of individual microtubules under cell-free conditions. Surprisingly, they found that lower-

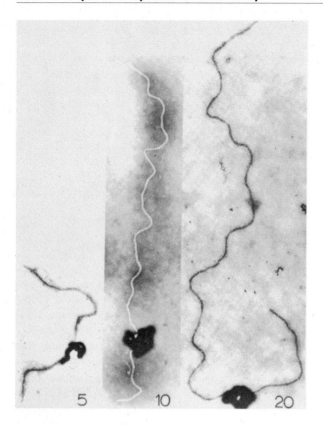

Figure 13-46 *Evidence That Microtubules Grow Mainly at One End* *Radioactive microtubule fragments and unlabeled tubulin heterodimers were incubated together for varying periods of time (as indicated by the number of minutes shown in the lower right corner of each micrograph). The samples were then examined by microscopic autoradiography, which causes large dark spots (silver grains) to appear at sites where radioactivity is located. The results show that as the microtubules grow in length, the radioactive tubulin consistently appears near one end of the microtubule, in the same relative position as in the original preparation of radioactive microtubule fragments. This pattern suggests that unlabeled tubulin is being added to only one end of the growing microtubule. The wavy appearance of the microtubules is caused by the fixation procedure. Courtesy of J. L. Rosenbaum.*

↓ tubulin = ↓ microtubules, BUT
              remainder grow longer!

ing the tubulin concentration causes some microtubules to disappear while the remaining ones grow longer.

To explain this unexpected discovery, Mitchison and Kirschner proposed that microtubules exhibit **dynamic instability** when the tubulin concentration is low. Dynamic instability means that microtubules can readily switch between two phases, a growing phase and a shrinking phase. The growing phase is promoted by the presence of GTP and a high concentration of unpolymerized tubulin. Under such conditions, tubulin-GTP is added to the growing microtubule tip faster than the subsequent GTP hydrolysis step can be carried out. The result is the formation of a short region at the tip of the microtubule where β-tubulin retains bound GTP.

↑ GTP = ↑ growth.

The presence of such a region, called a **GTP cap,** creates a *stable* microtubule tip that promotes the further addition of tubulin heterodimers (Figure 13-47, *top*). If the rate of microtubule elongation decreases because the concentration of free tubulin falls, the rate of GTP hydrolysis catches up and eventually produces a microtubule tip in which the GTP bound to β-tubulin has been hydrolyzed to GDP. The result is an *unstable* microtubule tip that favors the loss of tubulin heterodimers and hence microtubule depolymerization.

Some of the most dramatic evidence for dynamic instability has emerged from studies in which the behavior of individual microtubules is observed directly by light microscopy. When their length is continuously monitored, most microtubules are found to undergo alternating cycles of growth and shrinkage, mainly at the plus end but to some extent at the minus end as well. The switch from microtubule growth to shrinkage, called *microtubule catastrophe,* can cause the microtubule to disappear completely, or it may be reversed by an abrupt transition back to the growing phase, a phenomenon termed *microtubule rescue.* The frequency of microtubule catastrophe is inversely related to the tubulin concentration. High tubulin concentrations make catastrophe less likely to occur, and any catastrophes that do occur are more likely to be reversed by rescue events (see Figure 13-47, *bottom*). At any tubulin concentration, catastrophe is more likely and rescue is less likely at the plus end of a microtubule than at the minus end. In other words, dynamic instability is most pronounced at the plus end.

Microtubule assembly is iniated at the centrosomes.

Centrosomes Serve as Initiation Sites for Microtubule Assembly in Animal Cells

MTOCs are like support cells - allow assembly?

Within intact cells, microtubules are assembled at specific locations and at specific times. The control of where and when microtubules are formed is exerted in large part by **microtubule-organizing centers (MTOCs),** which are intracellular structures that function as initiation sites for microtubule assembly. MTOCs allow microtubule assembly to occur at a lower concentration of free tubulin than would occur in the absence of such initiation sites. Microtubules therefore assemble preferentially at MTOCs rather than randomly throughout the cytoplasm. In animal cells the main MTOC is the **centrosome,** a small zone of granular material located adjacent to the nucleus. The importance of the centrosome as an organizing center for cytoplasmic microtubules can be clearly demonstrated by staining cells with fluorescent antibodies directed against tubulin. When the cells are examined by light microscopy, the main network of cytoplasmic microtubules is seen to be emerging from the centrosome (Figure 13-48).

Embedded within the centrosome is a pair of small cylindrical structures called *centrioles.* While centrioles

[handwritten at top: Centrioles- function to form cilia & Flagella. Without 𝛾-tubulin = NO assembly!]

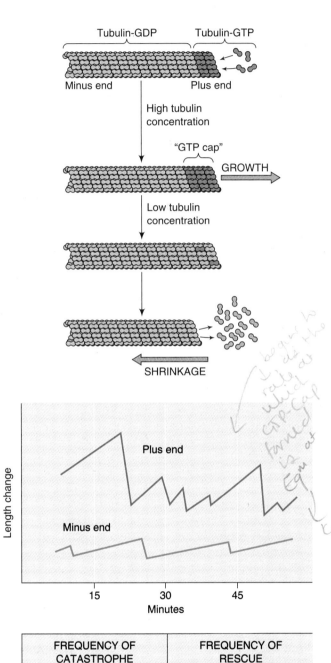

[handwritten notes to right of large graph: begin to do the rate at which GTP-cap formed is at Eqm]

[handwritten below: kinda like saturation!]

Figure 13-47 *The GTP Cap and Its Role in the Dynamic Instability of Microtubules* (Top) *A model illustrating the postulated role of the GTP cap. When the tubulin concentration is high, tubulin-GTP is added to the microtubule tip faster than the incorporated GTP can be hydrolyzed. The resulting GTP cap stabilizes the microtubule tip and promotes further growth. At lower tubulin concentrations the rate of growth decreases, thereby allowing GTP hydrolysis to catch up. This creates an unstable tip (no GTP cap) that favors microtubule depolymerization. (Bottom) Data revealing the existence of dynamic instability. The large graph shows that in an individual microtubule observed by light microscopy, growing and shrinking phases alternate with each other. The plus and minus ends grow and shrink independently. The two smaller graphs show that the frequency of microtubule catastrophe and rescue depends on the tubulin concentration. Catastrophe, defined as the switch from growth to shrinkage, becomes less frequent at high tubulin concentrations. Rescue, defined as the switch from shrinkage to growth, is more frequent at high tubulin concentrations.*

are known to function in the formation of cilia and flagella (page 612), it is the diffuse granular material of the centrosome rather than the centrioles that initiates the assembly of microtubules (Figure 13-49). **Each growing microtubule is anchored to the centrosome at its minus end, which prevents the minus end from being easily disassembled. The plus end protrudes into the cytoplasm and serves as the main site of microtubule growth and shrinkage.** An unusual form of tubulin called γ-*tubulin* has recently been identified in centrosomes. Microtubule assembly is strongly inhibited in cells whose γ-tubulin gene has been disrupted, suggesting that γ-tubulin may initiate the process of microtubule assembly within the centrosome.

The importance of the centrosome in determining when and where microtubules are assembled has been demonstrated in experiments carried out by Richard Weisenberg on clam eggs. Weisenberg compared the ability of components isolated from dividing and nondividing clam eggs to induce the polymerization of tubulin into the microtubules that make up the mitotic spindle. Homogenates from dividing eggs were found to contain material that stimulates microtubule formation when incubated with tubulin. In electron micrographs, this material was seen to consist of centrosomes. When comparable experiments were performed using nondividing clam eggs, the homogenates did not trigger the polymerization of tubulin into microtubules. But when tubulin derived from nondividing eggs was incubated with centrosomes isolated from dividing eggs, the tubulin from nondividing eggs polymerized into microtubules just as effectively as tubulin from dividing eggs. Hence the properties of the centrosome rather than of the tubulin molecule appear to be responsible for triggering the initiation of microtubule formation.

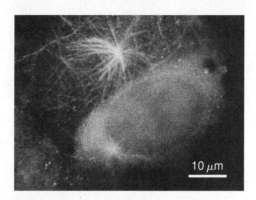

Figure 13-48 *Microtubules Emerging from the Centrosome* *Microtubules were visualized by staining a mouse fibroblast with fluorescent antibodies directed against tubulin. The star-shaped configuration represents microtubules radiating in various directions from the centrosome. Courtesy of B. R. Brinkley.*

Ways to make microtubules more stable (when dynamic instability not occurring)

Microtubule Stability Is Influenced by Various Enzymes, MAPs, and Drugs

We have just seen that newly forming cytoplasmic microtubules have a stable minus end embedded in the centrosome and a free plus end that exhibits dynamic instability. When changes in cell shape, motility, or internal organization are taking place, dynamic instability tends to persist in the microtubule population. But in cells where such changes are not occurring, microtubules often undergo molecular alterations that convert them into relatively stable structures. One mechanism implicated in fostering microtubule stability involves enzymes that catalyze the *acetylation* or the *detyrosination* of α-tubulin after it has been incorporated into microtubules. Tubulin acetylation is catalyzed by the enzyme *tubulin acetyltransferase,* which transfers an acetyl group to a specific lysine in α-tubulin. Detyrosination is carried out by *tubulin detyrosinase,* which removes the tyrosine residue located at the C-terminal end of α-tubulin.

The relationship between these protein modification reactions and microtubule stability has been investigated by exposing cells to drugs that promote microtubule breakdown. Since the few microtubules that survive such treatments are found to be enriched in α-tubulin that has been acetylated and detyrosinated, it appears that acetylation and detyrosination may help to stabilize microtubules against depolymerization. However, microtubules assembled under cell-free conditions from acetylated or detyrosinated tubulin are no more stable than microtubules assembled from unmodified tubulin. Hence neither acetylation nor detyrosination is sufficient by itself to ensure microtubule stability. It is more likely that these modifications serve to target microtubules for the binding of microtubule-associated proteins (MAPs) such as the *STOP proteins,*

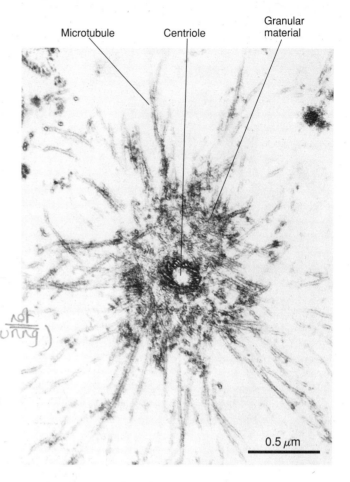

Figure 13-49 *Electron Micrograph of a Centrosome* *The microtubules are emerging from the granular material surrounding the centriole. Courtesy of R. W. Weisenberg.*

which stabilize microtubules against disassembly. In electron micrographs of intact cells the plus ends of cytoplasmic microtubules sometimes appear to be covered by an extra coating of protein, suggesting the possible presence of MAPs that prevent microtubules from depolymerizing.

Microtubule stability is also influenced by a variety of drugs that are used experimentally to interfere with microtubule function. The first such drug to be discovered was **colchicine,** a substance isolated from the autumn crocus and used for medicinal purposes for hundreds of years. Colchicine binds strongly to the tubulin heterodimer, inhibiting tubulin polymerization and promoting the breakdown of microtubules that have already been formed. Other drugs that promote microtubule breakdown include *colcemid, vinblastine,* and *vincristine.* **Taxol,** a toxic substance isolated from the yew tree, differs from the preceding drugs in that it stabilizes microtubules instead of promoting their breakdown. Cells exposed to taxol accumulate large masses of microtubules whose presence interferes with

① Colchicine — involved in breakdown inhibits polymerisat[ion] ② Taxol- opposite it ↑es stability, which = too much microtubules - interfere with a variety of cell functions. Because the microtubules of the mitotic spindle are essential to the process of cell division, drugs that interfere with microtubule assembly and function are sometimes used in cancer patients to inhibit the uncontrolled proliferation of cancer cells (see Table 18-7).

The Axoneme of Eukaryotic Cilia and Flagella Consists of a 9 + 2 Array of Linked Microtubules

Among the various structures that contain microtubules, eukaryotic **cilia** and **flagella** are probably the best understood in terms of microtubule organization and function. Eukaryotic cilia and flagella are motile, membrane-enclosed appendages that project from the surface of certain kinds of cells (Figure 13-50). Depending on the cell type involved, these motile appendages perform one of two alternative functions: (1) cells that are firmly anchored in place employ ciliary motion to move fluids across their surfaces; and (2) cells that are not anchored, such as sperm cells or unicellular organisms, use the movement of cilia or flagella to propel themselves through the fluid medium in which they are suspended.

Although cilia and flagella are closely related structures, the two can be distinguished from each other on the basis of differences in size, number, and pattern of movement. Flagella are the larger of the two organelles, measuring 150 μm or more in overall length; they are present in small numbers per cell and move in a regular pla-

nar wave that propels liquid parallel to the flagellar axis (Figure 13-51). Cilia are shorter (5–10 μm average length), more numerous, and move in a more complex pattern whose net result is to propel fluid across the cell surface. In spite of these differences, the underlying substructure and mechanism of movement is quite similar in cilia and flagella, and most generalizations concerning the behavior of one apply to the other. However, this similarity does not extend to bacterial flagella, whose unique properties will be described later in the chapter.

Cilia and flagella both contain a regularly arranged backbone of linked microtubules, termed an **axoneme,** surrounded by the plasma membrane. Axonemes typically consist of nine outer *doublet tubules* surrounding two *central tubules,* a pattern referred to as the "*9 + 2 arrangement*" (Figures 13-52 and 13-53). The walls of the two central microtubules contain the usual 13 protofilaments, but the doublet tubules are more complicated because each doublet consists of two microtubules physically joined to each other. One of these, termed the *A-tubule,* has a complete wall of 13 protofilaments, whereas the adjoining *B-tubule* contains only 10 or 11 protofilaments because it shares a portion of the A-tubule wall (Figure 13-54). Projecting from each A-tubule are two sets of *dynein arms* that recur every 16–22 nm along the axoneme's length.

The orderly 9 + 2 arrangement of the axoneme is maintained by an intricate system of connections that link the microtubules together. The central microtubules are surrounded by a *central sheath,* while the outer doublets are connected to each other by *nexin links.* In addition, the A-tubules of the outer doublets are connected to the central region by prominent *radial spokes* that terminate in thickened knobs lying adjacent to the central sheath.

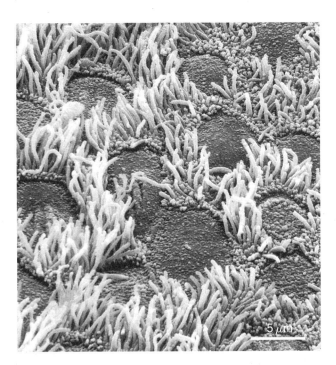

Figure 13-50 *Scanning Electron Micrograph Showing Numerous Cilia Emerging from the Surface of Epithelial Cells Lining the Bronchioles of the Respiratory System* Courtesy of A. T. Mariassy.

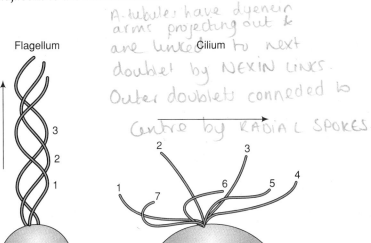

Flagellum   Cilium

Figure 13-51 *Comparison of the Movement Patterns of Cilia and Flagella* *Numbers indicate successive steps in the movement of a single flagellum or cilium. The direction of liquid flow induced by these two types of movement is shown by the arrows.*

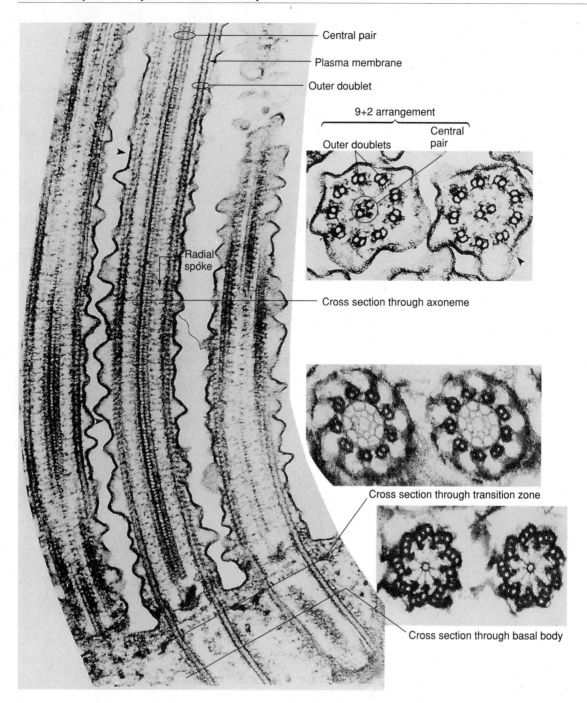

Figure 13-52 *Thin-Section Electron Micrographs of Eukaryotic Cilia* (Left) *In this longitudinal section through three cilia, several of the microtubules that make up the axoneme can be seen.* (Right) *Cross sections at three different levels show the difference between the organization of the ciliary axoneme, the transition zone, and the basal body. Courtesy of W. L. Dentler.*

Identification of the molecules that make up the various components of the axoneme has been facilitated by the development of techniques for the step-wise degradation of cilia. This general approach, pioneered by Ian Gibbons, begins with the removal of cilia or flagella from the cell surface by gentle shearing or an appropriate chemical treatment. After low-speed centrifugation to remove intact cells, Gibbons extracted the isolated cilia with detergent to remove the plasma membrane. The exposed axonemes were then treated with *EDTA*, a compound that binds to divalent cations such as Ca^{2+} and Mg^{2+}. This treatment caused the protein **dynein** to be released into solution. Microscopic examination of the axonemes at this stage revealed that the two arms on each doublet and the pair of central tubules had all been lost, suggesting that one or more of

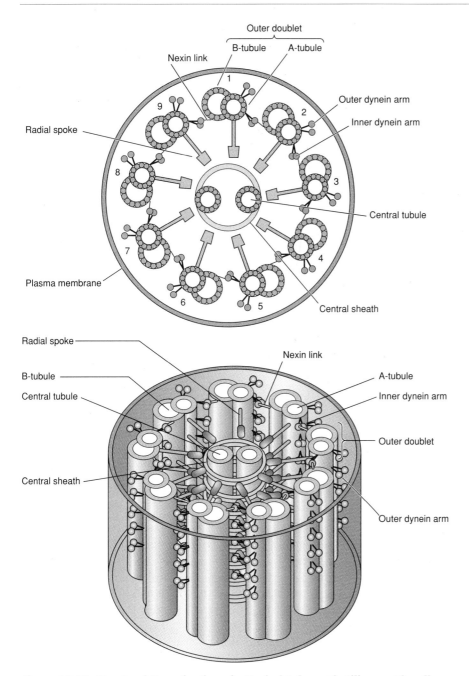

Figure 13-53 *Structural Organization of a Typical Eukaryotic Cilium or Flagellum*
(Top) *In cross-sectional view, the axoneme is seen to consist of nine outer doublet tubules surrounding two central tubules (the "9 + 2 arrangement"). Two sets of dynein arms project from the A-tubule of each outer doublet. The 9 + 2 microtubule array is held together by the central sheath, radial spokes, and nexin links.* (Bottom) *Three-dimensional view of the axoneme. To enhance the visibility of the internal regions, several tubules, arms, and bridges have been omitted.*

these structures is constructed from dynein. To find out which of these components is made of dynein, Gibbons added purified dynein back to EDTA-extracted axonemes and then examined the resulting material by electron microscopy. Under these conditions the arms, but not the central microtubules, were found to reappear, indicating that the arms extending from the outer doublets are made of the protein dynein (Figure 13-55).

As we will see shortly, these dynein arms are responsible for ciliary and flagellar motility.

At least half a dozen proteins in addition to dynein and tubulin have been identified in extracts derived from isolated axonemes. Most of these proteins are involved in holding together the 9 + 2 microtubule array. Included in this category are the proteins that make up the nexin links, the spokes, and the central sheath. In

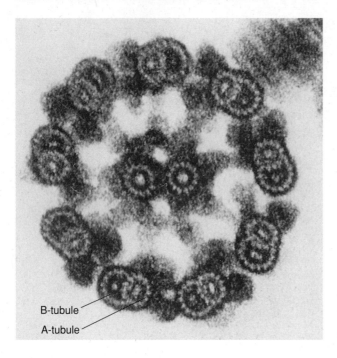

B-tubule
A-tubule

50 nm

Figure 13-54 *High-Resolution Electron Micrograph Showing the 9 + 2 Arrangement of Microtubules in a Flagellar Axoneme* *The protofilaments that make up each microtubule are clearly visible. In the outer doublets the A-tubule contains 13 protofilaments, whereas the wall of the B-tubule contains only 11 protofilaments because it shares part of the A-tubule wall. Courtesy of R. W. Linck.*

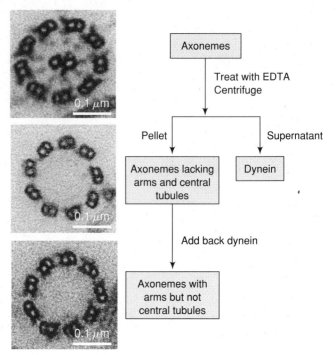

Figure 13-55 *Evidence Showing That the Arms on the Outer Doublets of Ciliary Microtubules Are Composed of the Protein Dynein* *Cilia were first removed from the cell surface by gentle shearing. After low-speed centrifugation to remove intact cells, the isolated cilia were extracted with detergent to remove the plasma membranes. The exposed axonemes were then treated as outlined in the diagram. Micrographs courtesy of I. R. Gibbons.*

addition, a filamentous protein called *tektin* runs lengthwise along individual protofilaments and may help to strengthen the microtubules.

The Motility of Eukaryotic Cilia and Flagella Is Based on Microtubule Sliding

Several lines of evidence indicate that eukaryotic cilia and flagella are intrinsically motile. Perhaps the most striking is the finding that flagella removed from living cells by laser microbeam irradiation continue to move until their supply of stored ATP is exhausted (Figure 13-56). Even purified axonemes isolated from detached cilia are motile if provided with ATP as an energy source.

Although the patterns of ciliary and flagellar movement are relatively complex (see Figure 13-51), the basic motion underlying such patterns is a simple *bending* of the axoneme at appropriate points along its length. One of the earliest theories proposed to explain axonemal bending was that microtubules contract in the presence of ATP, causing the axoneme to bend toward the side of the cilium where the microtubules have contracted. Such a simple mechanism was effectively ruled out, however, by a classic series of studies

carried out by Peter Satir in the 1960s. Satir reasoned that if microtubules contract, the doublets on one side of a bending cilium should be shorter and thicker than those on the opposite side. But when Satir examined sections through the tips of bending cilia, this expectation was not confirmed. Instead, the microtubules located on the inside of the ciliary bend were found to project further into the ciliary tip than the microtubules on the opposite side of the bend. Figure 13-57 shows that such an observation is compatible with the idea that microtubules slide past one another during ciliary bending. Additional support for this conclusion was obtained by measuring the distance between adjacent radial spokes viewed in longitudinal section. Because the radial spokes are fixed to the doublet walls, the distance between spokes should decrease if individual microtubules contract (get shorter). However, the distance between radial spokes was found to be similar in straight and bent regions of the axoneme, indicating that individual microtubules do not contract.

A radically different approach taken by Keith Summers and Ian Gibbons provided independent support for the idea that ciliary microtubules slide past another rather than contract. These investigators discovered that digesting sperm tail axonemes with the

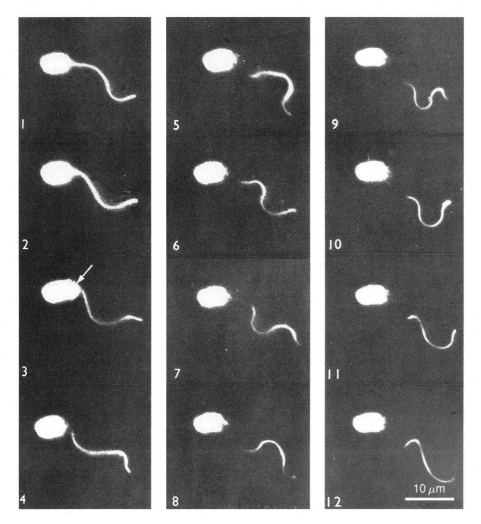

Figure 13-56 Evidence That Eukaryotic Flagella Are Inherently Motile *This series of 12 light micrographs shows the continued movement of a flagellum after it has been detached from the cell with a laser microbeam. (1,2) Intact flagellum prior to detachment. (3,4) A laser beam is directed to the region indicated by the arrow, severing the flagellum from the cell. (5–12) Following detachment, the flagellum continues to move. Courtesy of M. Holwill.*

protease trypsin selectively degrades the radial spokes and nexin links. Instead of bending in response to the addition of ATP, such axonemes extend to several times their original length and then fall apart (Figure 13-58). This ATP-dependent increase in length is most compatible with the theory that microtubules slide past each other during ciliary movement, and that the destruction of the radial spokes and nexin links by trypsin simply exaggerates the amount of sliding that can occur by destroying the connections that normally restrain the sliding process.

If the energy released by ATP hydrolysis causes microtubule sliding, how does microtubule sliding in turn cause the axoneme to bend? The preceding experiments suggest that the radial spokes and nexin links provide physical constraints that limit the amount of sliding that can take place in an intact ax-

oneme. This means that when ATP hydrolysis occurs in a normal cilium or flagellum, the microtubules cannot slide freely past each other; instead, the sliding forces generated by ATP hydrolysis are thought to create localized tension that causes the microtubules to bend. Support for this idea has come from studies involving mutant cells whose cilia lack both radial spokes and the central-tubule/central-sheath complex. Trypsin-treated axonemes obtained from such cilia elongate and fall apart just like normal axonemes, confirming that microtubule sliding occurs. However, the mutant cilia are incapable of normal bending movements, indicating that the radial spokes and the central-tubule/central-sheath complex are required for converting microtubule sliding into axonemal bending. A role for the nexin links has also been inferred from experiments in which mutant axonemes lacking

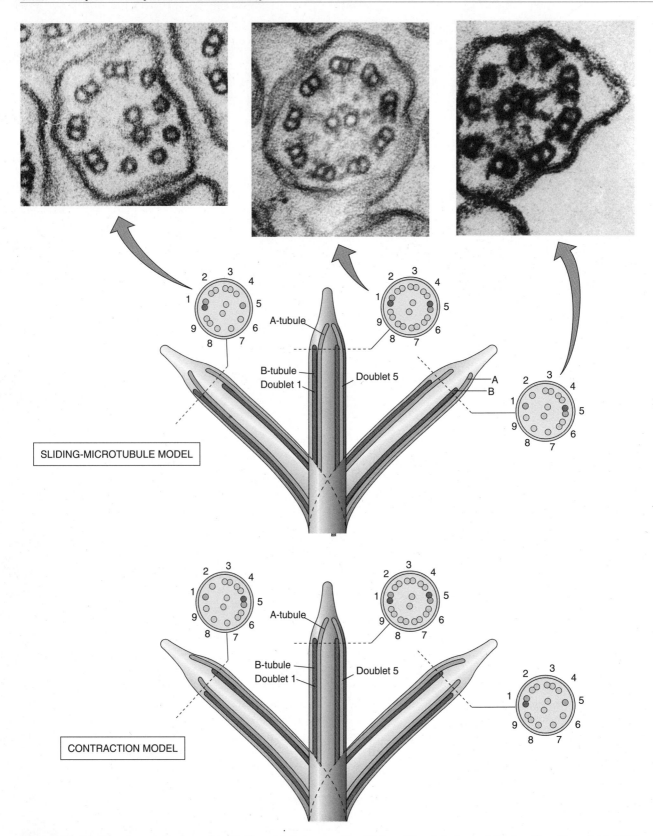

Figure 13-57 Experiment Designed to Distinguish between the Sliding-Microtubule and Contraction Theories of Ciliary Motion *The discovery that the A-tubule of each doublet extends further into the tip of the cilium than the B-tubule provided a convenient way to identify the tip end of each doublet. The sliding-microtubule theory predicts that the B-tubules will be missing on the outer side of the bend, whereas the contraction theory predicts the opposite result. Electron micrographs* (top) *showing sections through the tips of bending cilia are consistent with the predictions of the sliding-microtubule theory. Micrographs courtesy of P. Satir.*

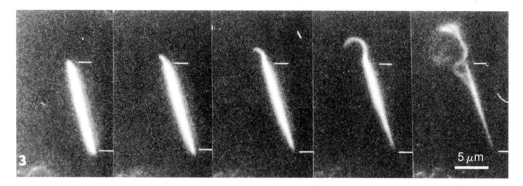

Figure 13-58 *Dark-Field Light Micrographs Showing the Behavior of Trypsin-Treated Axonemes Exposed to ATP* *The successive photographs in the series were taken at intervals of 10–30 seconds. The white lines indicate the initial position of each end of the axoneme. The dramatic lengthening of the axoneme under these conditions provides strong evidence for the occurrence of microtubule sliding. Courtesy of I. R. Gibbons.*

radial spokes but containing nexin links were exposed to ATP without trypsin treatment. Under such conditions the axonemes do not fall apart, suggesting that the nexin links may act to restrain microtubule sliding.

Dynein Is a Motor Protein That Causes Microtubule Sliding

Since microtubule sliding underlies the motility of cilia and flagella, it is important to understand what causes microtubules to slide. Many lines of evidence suggest that sliding is produced by **dynein,** which makes up the inner and outer arms that protrude from the A-tubule of each doublet. Among the evidence implicating dynein in microtubule sliding is the following: (1) Axonemes that have had their dynein arms removed lose the ability to move when ATP is added. Axonemal motility can be restored by adding back purified dynein. (2) Purified dynein is capable of hydrolyzing ATP, suggesting that dynein utilizes the energy derived from ATP hydrolysis to drive microtubule sliding. (3) Mutants of the green alga *Chlamydomonas* that lack both the inner and outer dynein arms have nonmotile flagella. Mutants in which only the inner or the outer arms are missing exhibit reduced flagellar motility, indicating that both the inner and outer arms contribute to microtubule sliding. (4) Nonmotile cilia occur in certain sensory organs, such as the inner ear. These nonmotile cilia lack dynein arms. (5) Finally, some humans suffer from an inherited disease that is associated with both respiratory abnormalities and sterility. Such individuals have nonmotile cilia and flagella that lack dynein arms; the absence of functional cilia on cells lining the passageways into the lung increases the susceptibility to respiratory infections, while the lack of motile flagella on sperm cells leads to sterility.

Dynein is a protein that consists of one, two, or three large polypeptide chains (*heavy chains*) associated with several smaller polypeptides. In electron mi-

crographs these alternate forms of dynein exhibit either one, two, or three *globular heads* attached to a common base. In at least some kinds of axonemes, the inner and outer dynein arms are constructed from different forms of dynein. The flagella of *Chlamydomonas,* for example, appear to contain a triple-headed outer dynein arm and a double-headed inner dynein arm. Each globular head contains a site where ATP is hydrolyzed. ATP hydrolysis powers a cycle in which the dynein arms on one A-tubule bind to and then detach from sites located on the adjacent B-tubule (Figure 13-59). This attachment-detachment cycle allows the dynein head groups to "walk" along an adjacent microtubule, just as myosin head groups move along actin filaments. Because of its ability to produce movement, dynein (like myosin and other movement-inducing proteins) is referred to as a **motor protein.**

The directionality of dynein-induced movement is such that dynein arms attached to the A-tubule of one doublet always push the adjacent doublet toward the tip of the axoneme. In order for this pushing movement to cause a cilium or flagellum to bend, the activity of the dynein arms on opposite sides of the axoneme must be coordinated so that arms on one side of the axoneme are active while those on the opposite side are detached. The discovery that the two central tubules rotate during complex bending movements raises the possibility that coordination is achieved by regulatory signals sent from the central tubules via the radial spokes to the dynein arms. This view is also compatible with the finding that mutant flagella containing defective radial spokes are nonmotile. It should be noted, however, that the sperm cells of certain kinds of invertebrates and fish have flagella that lack the central pair of tubules and yet are still motile. Hence the central tubules may play a role in coordinating complex three-dimensional bending sequences, but they are not an absolute requirement for axonemal motility.

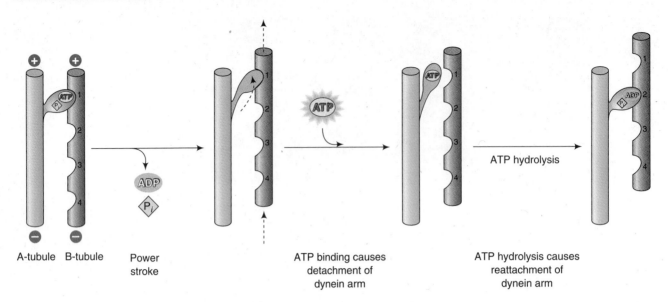

A-tubule B-tubule Power stroke ATP binding causes detachment of dynein arm ATP hydrolysis causes reattachment of dynein arm

ATP hydrolysis

Figure 13-59 *A Model Showing How Dynein Causes Microtubule Sliding in an Axoneme* *The dynein arms projecting from the A-tubule on one doublet interact with the B-tubule of an adjacent doublet. Note that dynein produces microtubule sliding by a mechanism that resembles the process by which myosin head groups cause actin filaments to slide (see Figure 13-13).*

The mechanisms responsible for initiating microtubule sliding are not as well understood as the mechanism that induces the sliding of actin filaments during muscle contraction, but Ca^{2+} has again been implicated. Changes in ciliary motility that occur in response to changing environmental conditions are typically accompanied by alterations in cytosolic Ca^{2+} concentration. The importance of Ca^{2+} has been well documented in the protozoan *Paramecium,* whose ciliary movements change rapidly when the swimming cell encounters a physical obstacle. In response to encountering such a barrier, Ca^{2+} enters the cell from the surrounding environment and the cilia begin to move in the opposite direction. To determine whether the change in cytosolic Ca^{2+} is responsible for the change in ciliary behavior, motility has been studied in *permeabilized* cells whose plasma membranes have been removed to allow direct entry of Ca^{2+} from the medium into the cytosol. When ATP is added to such cells to stimulate ciliary motility, the cells swim forward when the Ca^{2+} concentration is less than 10^{-6} *M.* But raising the Ca^{2+} concentration to 10^{-6} *M* or higher reverses the direction of ciliary motion and the cells swim backward, thus demonstrating that changes in the intracellular Ca^{2+} concentration control the direction in which the cells swim.

Cilia and Flagella Grow by Assembling Microtubules at the Axonemal Tip

Cilia and flagella only occur in certain cell types, where their number and location are precisely controlled. It is therefore important to ask how the assembly of these complex organelles is initiated and regulated. Some early insights into the process of flagellar growth emerged from studies carried out by Joel Rosenbaum and his associates on *Chlamydomonas,* a green alga that normally contains two flagella. The dynamics of flagellar growth were investigated by briefly shearing live cells in a homogenizer to break off the existing flagella and then incubating the cells in the presence of radioactive amino acids under conditions that permit the flagella to regenerate. Electron microscopic autoradiographs of the newly forming flagella revealed that radioactivity is most concentrated over the flagellar tips, indicating that flagella are assembled by the addition of tubulin and other axonemal proteins to the tip, rather than to the base or middle, of the growing axoneme.

One of the more interesting findings to emerge from these studies involved cells in which only one of the two flagella had been amputated during homogenization. In such cases the initial growth of the missing flagellum is accompanied by a shortening of the existing flagellum; when the two flagella reach the same length, both elongate until the proper size is reached (Figure 13-60). Presumably the initial shortening of the existing flagellum provides tubulin subunits to the newly forming flagellum, but the mechanism that coordinates the lengths of the two flagella is not well understood.

The microtubules that make up a ciliary or flagellar axoneme are always oriented with their minus ends at the base of the axoneme and their plus ends at the tip. The discovery that cilia and flagella grow by microtubule assembly at the axonemal tip is therefore consistent with the fact that isolated microtubules tend to grow more rapidly at their plus ends (page 598). However, this conclusion may not apply to the two central microtubules. If isolated flagella are incubated in the

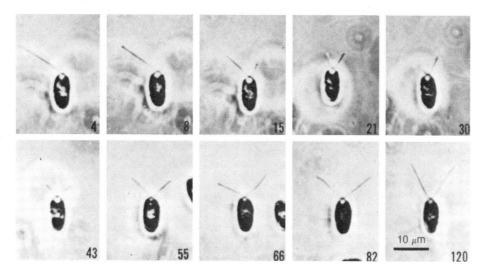

Figure 13-60 *Effects of Flagellar Amputation in* **Chlamydomonas** *After amputation of one of the two flagella, regeneration of the amputated flagellum is accompanied by a dramatic shortening of the intact flagellum. This sequence is shown beginning 4 minutes after amputation of the right flagellum. Courtesy of J. L. Rosenbaum.*

presence of tubulin and then examined microscopically to determine where microtubule growth occurs, the outer doublets, but not the two central microtubules, acquire newly added tubulin subunits at the tip of the flagellum. This difference in behavior may be explained by the finding that central microtubules are normally covered by a cap structure that attaches their tips to the overlying plasma membrane. Addition of tubulin to the tips of the central microtubules only occurs if the cap is artificially removed. Hence in intact cells, the presence of the cap may mean that the central microtubules are assembled at the base rather than at the tip.

Basal Bodies Initiate the Assembly of Cilia and Flagella

The discovery that cilia and flagella grow by assembling microtubules at the axonemal tip raises a related question: How is the assembly of an axoneme initiated in a location where no cilium or flagellum currently exists? The answer to this question involves a specialized structure called the **basal body,** which is located at the base of every eukaryotic cilium or flagellum. The basal body is a cylindrical structure containing nine sets of *triplet* microtubules in place of the doublet microtubules found in the axoneme (Figure 13-61). Of the three tubules that comprise each triplet only the *A-tubule* has a complete circular wall of 13 protofilaments. The *B-tubule* is a horseshoe-shaped structure that shares part of the wall of the A-tubule, and the *C-tubule* is a horseshoe-shaped structure sharing part of the wall of the B-tubule. The A- and B-tubules are continuous with the analogous tubules of the cilium or flagellum, but C-tubules are restricted to the basal body.

Another difference between the basal body and the cilium or flagellum to which it attaches is that the basal body lacks the two central microtubules. A *transition zone* of variable length occurs between the basal body C-tubules and the beginning of the two central microtubules of the axoneme. Just beneath the central microtubules, a dense partition called the *basal plate* crosses the transition zone. Small *transitional fibers* help to anchor the basal body to the plasma membrane, and larger, striated *rootlet fibers* pass from the basal body to the cell interior. The rootlet fibers are composed of the protein *centrin,* a molecule that contracts when it binds Ca^{2+}. Contraction of the rootlet fibers may help to control the orientation and behavior of the attached cilium or flagellum.

The role played by the basal body in initiating the assembly of cilia and flagella has been established in several ways. One approach involves exposing cells to gentle shearing forces, which causes flagella to break off at the transition zone between the basal body and the flagellar axoneme, followed by spontaneous resealing of the plasma membrane. Shortly thereafter the cells begin to form new flagella, but only at sites where basal bodies remain beneath the plasma membrane. It is also possible to purify basal bodies from cells whose cilia or flagella have been removed. When incubated with tubulin, isolated basal bodies initiate the assembly of microtubule bundles.

For many years, a controversy has existed as to whether or not basal bodies contain DNA. Because of their small size and close association with other potential sources of intracellular DNA, staining basal bodies for the presence of DNA tended to yield results that were ambiguous. In 1989, an experiment carried out in the labo-

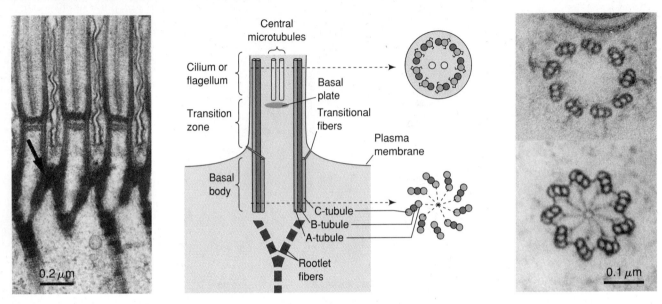

Figure 13-61 *The Basal Body and Its Relationship to a Cilium or Flagellum* *The electron micrograph on the left is a longitudinal section through a row of cilia showing the basal bodies (arrow). The electron micrographs on the right are cross sections through a basal body at two levels, the top one closer to the cilium and the lower one closer to the point where the basal body interfaces with the cell interior. In addition to the characteristic triplet arrangement of the microtubules, note the cartwheel spokes that occur in the lower section. Micrographs courtesy of F. D. Warner.*

ratory of David Luck appeared to resolve the issue. Using recombinant DNA techniques, several genes required for flagellar development in *Chlamydomonas* were cloned. When cells were hybridized with the cloned DNA, the DNA localized to two tiny spots that corresponded to the location of the cell's two basal bodies. Although this finding seemed to indicate that basal bodies contain a specific set of flagellar genes, other investigators have failed to detect DNA in *Chlamydomonas* basal bodies using techniques that should be sensitive enough to detect these genes. Thus the controversy concerning the existence of basal body DNA still continues.

Basal Bodies Arise from Centrioles

If eukaryotic cilia and flagella arise from basal bodies, then where do basal bodies come from? Basal bodies closely resemble **centrioles,** which are microtubule-containing structures embedded within the centrosomes of animal cells. Each centrosome contains two centrioles oriented at right angles to each other (Figure 13-62). A centriole is a short, cylindrical structure whose walls, like those of basal bodies, are composed of nine sets of triplet microtubules. During the first stage in the formation of a cilium or flagellum, a centriole migrates to the cell surface and makes contact with the plasma membrane. The A- and B-tubules of the centriole then act as initiation sites for microtubule assembly, initiating polymerization of tubules that form the nine outer doublets of the axoneme. *After the process of tubule assembly has begun, the centriole is referred to as a basal body.*

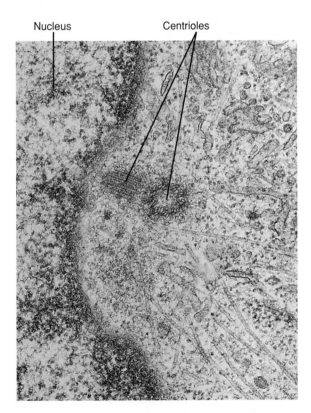

Figure 13-62 *Electron Micrograph of Two Centrioles Oriented at Right Angles* *One centriole appears in cross section and the other in longitudinal section. Courtesy of M. McGill.*

The role of centrioles in initiating the assembly of new cilia and flagella raises the question of how the cell's supply of centrioles is maintained. At least three different mechanisms exist for producing centrioles. The first, typical of cells requiring only a few centrioles, involves the development of new centrioles at right angles to existing centrioles. The process begins with the appearance of a centriole precursor, or *procentriole,* next to an existing centriole. In the newly forming procentriole, the A-tubules are the first element to form. The B-tubules then grow from the A-tubule walls, followed by growth of the C-tubules from the B-tubule walls (Figure 13-63). Finally the triplet tubules elongate, generating a mature centriole lying at right angles to the original centriole. This type of centriole replication occurs prior to mitosis in dividing animal cells, converting the two centrioles normally present in the centrosome to four centrioles. At the beginning of prophase the centrosome splits, taking two centrioles to each pole of the newly forming mitotic spindle. This process ensures that during cell cleavage, the two newly formed cells will each receive a pair of centrioles.

Alternative modes of centriole formation do not require an intimate association between newly forming and previously existing centrioles. One such mechanism occurs when cells that contain a small number of centrioles (usually a single pair) are called upon to produce large numbers of cilia. Under these conditions, dozens of centrioles begin to form next to elongated masses of granular material called *deuterosomes,* which function as organizing centers for the microtubules of the newly developing centrioles (Figure 13-64). As the centrioles are assembled, the deuterosomes diminish in size and eventually disappear. The centrioles then migrate to the cell surface and induce the assembly of cilia.

The remaining mechanism of centriole formation occurs in cells where centrioles are initially absent. This situation is most commonly encountered in plant cells, which usually lack centrioles. Some plant sperm are an exception to the rule because they require centrioles for initiating the formation of flagella. During the differentiation of such sperm, centrioles arise adjacent to dense accumulations of granular material that function as microtubule-organizing centers. In some plants this mechanism leads to the formation of thousands of centrioles, generating a large mass called a *blepharoplast* that is readily visible with the light microscope.

Ordered Microtubule Arrays Occur in Sensory Cilia and in Several Organelles Found in Unicellular Eukaryotes

A variety of sensory organs, including eyes and olfactory organs, contain nonmotile cilia. The axonemes of sensory cilia typically lack dynein arms, and may exhibit other unusual features as well. For example the outer segment of the retinal rod cell, discussed in detail in Chapter 17, is derived from a single cilium whose outer membrane has undergone an elaborate series of invaginations (see Figure 17-25). In this and other sensory cells, cilia serve to increase the plasma membrane surface area available for the reception of incoming stimuli. What role, if any, the axonemal microtubules play in the reception and transmission of sensory signals remains to be determined.

Cilia and flagella are not the only organelles comprised of ordered arrays of microtubules (Figure 13-65). Certain protozoa have long thin projections known as *axopods* protruding from the cell surface. These structures, which are involved in cellular locomotion and in attaching cells to various surfaces, contain an axonemal backbone constructed from spiral coils of microtubules. In some other protozoa, movements are produced by intracellular bundles of crosslinked microtubules called *axostyles.* The crosslinks that hold together adjacent microtubules resemble dynein arms both in appearance

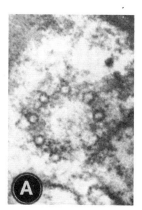 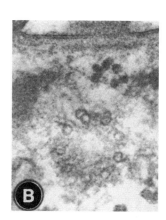 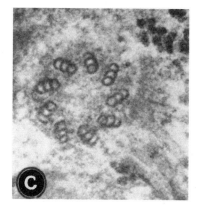 

Figure 13-63 *Formation of New Centrioles Adjacent to Existing Centrioles* (A) *Cross section of an early stage in procentriole development, showing the formation of the A-tubules. (B) In the next stage the B-tubules are seen growing from the A-tubule walls. (C) Formation of the C-tubules is then completed. (D) A newly developing basal body is seen in longitudinal section adjacent to an existing basal body. The newly forming basal body is about one-third grown.*

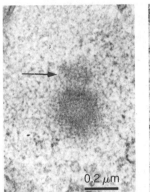
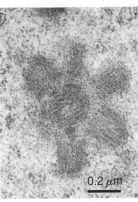

0.2 μm   0.2 μm

Figure 13-64 Development of Centrioles Adjacent to Deuterosomes (Left) *A proecentriole (arrow) is seen forming adjacent to a deuterosome.* (Right) *Six newly forming procentrioles are arranged symmetrically around a deuterosome that is partly hidden by an overlying procentriole. Courtesy of S. P. Sorokin.*

and in their ability to undergo ATP-dependent changes in conformation. Additional examples of organized microtubular arrays occurring in unicellular eukaryotes are *cortical fibers,* which induce rapid changes in cell shape, and the *cytopharyngeal basket,* which is a complex microtubular organelle through which some single-cell organisms feed. In contrast to cilia and flagella, where microtubule doublets are a prominent structural feature, the preceding organelles are comprised of collections of single microtubules. The term "axoneme" is nevertheless employed in referring to such microtubule arrays because, like their counterparts in cilia and flagella, they consist of an ordered arrangement of microtubules held together by intertubule linkages.

Microtubules Are Involved in Moving Cytoplasmic Organelles and Other Intracellular Components

Microtubules can be divided into two broad categories. One includes the relatively stable microtubules found in cilia, flagella, and other axonemal structures. The other is a more loosely organized, dynamically changing network of cytoplasmic microtubules. In contrast to axonemal structures, which usually function to move either the cell as a whole or fluids surrounding the cell, cytoplasmic microtubules are involved in moving intracellular organelles and in determining and maintaining cell shape. The most widely used approach for implicating microtubules in such activities has been to determine whether the activity in question can be inhibited by treating cells with drugs that promote microtubule breakdown, such as colchicine, vinblastine, and vincristine.

The movement of chromosomes during mitosis and meiosis was the first type of intracellular movement to be clearly linked to cytoplasmic microtubules. At the

time of cell division, the microtubules that make up the mitotic spindle move the chromosomes to opposite ends of the cell. At least two mechanisms underlie these movements: (1) Each chromosomal kinetochore moves its attached chromosome along a stationary track of microtubules in a process driven by motor proteins, perhaps including dynein (page 542); and (2) microtubules derived from opposite spindle poles elongate and slide apart, pushing the two poles away from each other (page 543).

In addition to their well-documented role in moving chromosomes, microtubules have been implicated in movements exhibited by a variety of cytoplasmic organelles, including mitochondria, lysosomes, pigment granules, lipid droplets, and other small vesicles. When cultured cells are examined by light microscopy, such organelles often exhibit sudden, rapid movements in a particular direction interspersed with periods of immobility. These **saltatory movements** tend to occur in regions enriched in microtubules and can be inhibited by colchicine, suggesting that microtubules are involved. Microtubules have also been implicated in the movement of secretory vesicles toward the cell surface at the time of cell secretion. In cells specialized for secretion, microtubules are concentrated near the Golgi complex and in the area through which secretory vesicles pass prior to discharge at the cell surface. Disrupting these microtubules with colchicine has been shown to inhibit the secretion of hormones by endocrine cells, plasma proteins by liver cells, and enzymes by salivary gland and pancreatic cells.

The Involvement of Microtubules in Axonal Transport Can Be Observed Using Video-Enhanced Light Microscopy

Some of the best evidence for the role played by microtubules in moving cytoplasmic organelles has come from studies involving nerve cells. A typical nerve cell consists of a central *cell body* and a series of long cytoplasmic extensions that fall into two categories: *dendrites* and *axons* (page 749). **Dendrites,** which usually protrude from the cell body in large numbers, tend to be highly branched and relatively short (less than a millimeter in length). **Axons** are typically unbranched, longer than dendrites (up to a meter or more in length), and only one or two axons emerge from a given cell body. Dendrites are specialized for receiving signals from other cells, whereas axons are involved in sending signals. Both dendrites and axons contain numerous microtubules, but they are arranged in different ways. The microtubules found in dendrites are oriented in both directions, some with their plus ends pointing away from the cell body and some oriented in the reverse direction (Figure 13-66). Axonal microtubules are all arranged with their plus ends pointing away from the cell body.

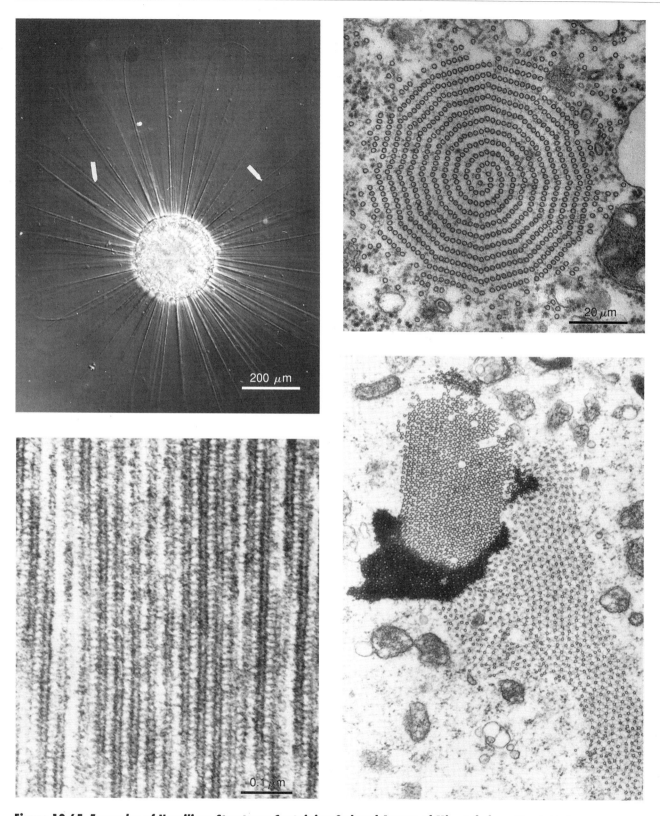

Figure 13-65 *Examples of Nonciliary Structures Containing Ordered Arrays of Microtubules*   (Top left) *A protozoan exhibiting numerous axopods protruding from the cell surface.* (Top right) *Cross section through an axopod showing the spiral arrangement of microtubules.* (Bottom left) *Longitudinal section through an axostyle showing an array of parallel microtubules held together by crosslinks between the adjacent microtubule walls.* (Bottom right) *Cross section of a cytopharyngeal basket in a ciliated protozoan showing the organized array of microtubules held together by crosslinks between adjacent microtubules. Courtesy of L. E. Roth* (top left, top right), *L. G. Tilney* (bottom left), *and J. B. Tucker* (bottom right).

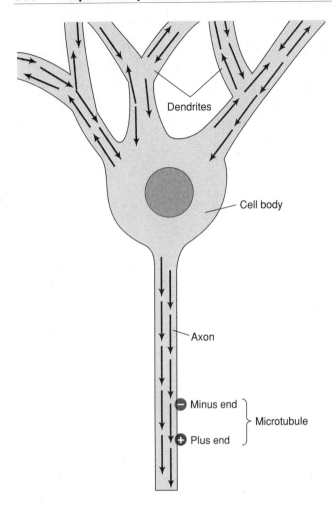

Figure 13-66 *Organization of Microtubules in Dendrites and Axons* *Each arrow represents a microtubule. The microtubules that occur in axons are all oriented with their plus ends pointing away from the cell body, whereas the microtubules found in dendrites point both away from and toward the cell body.*

Because the tip of an axon is far removed from the cell body, the question arises as to how an axon's supply of essential molecules and organelles is maintained. Simple diffusion is inadequate; even small molecules like glucose would take months or years to diffuse from the cell body to the end of the axon. Instead, materials are actively propelled through the axon cytoplasm by active processes known as *slow* and *fast axonal transport*. **Slow axonal transport** moves proteins and cytoskeletal filaments down the axon (away from the cell body) at a rate of about 1–5 mm per day. **Fast axonal transport** is about a hundred times faster and moves membrane vesicles and organelles down the axon (*anterograde transport*) as well as back toward the cell body (*retrograde transport*).

Since fast axonal transport involves movements of organelles and vesicles that can be seen with a microscope, it has been easier to investigate than slow transport. In high-resolution electron micrographs of the

axon cytoplasm, membrane vesicles appear to be connected to microtubules (Figure 13-67). Although this arrangement suggests that microtubules are responsible for moving organelles through the axon, electron microscopy does not allow such movements to be observed directly. In order to view the role played by microtubules, a special type of light microscopy is required. Conventional light microscopy is incapable of resolving objects smaller than 200 nm in diameter and so cannot be used to study the behavior of individual microtubules, which measure about 25 nm in diameter. To overcome this obstacle, Robert Allen developed a modified type of differential interference microscopy known as *Allen video-enhanced contrast microscopy,* or simply **AVEC microscopy.** AVEC microscopy takes advantage of the fact that a television camera can detect subtle differences in contrast better than the human eye. Moreover, video cameras generate images in an electronic form that can be analyzed and enhanced by computer processing. As a result, AVEC microscopy allows the visualization of structures that are an order of magnitude smaller than can be seen with a conventional light microscope. In addition, the specimen does not need to be fixed and stained as is required with electron microscopy, so dynamic events can be observed while they are happening.

The use of AVEC microscopy has provided some striking insights into the mechanism of fast axonal transport. The most illuminating studies have involved squid nerve cells, whose axons are large enough to be dissected away from the cell body for analysis. If a squid axon is broken open under gentle conditions, the axonal cytoplasm (or *axoplasm*) is released as a gelatinous cylinder of material that retains the ability to carry out axonal transport in the presence of added ATP. When the axoplasm is examined by AVEC microscopy, membrane vesicles can be seen moving along microtubule surfaces at rates approaching 2 μm/sec, which is comparable to the speed of fast axonal transport in intact axons (Figure 13-68). Moreover, vesicles are observed moving in both directions along individual microtubules. This means that even though the microtubules of an intact axon are all oriented with their plus ends pointing in the same direction, attached organelles can be moved in both the anterograde and retrograde directions.

(heptad repeat!)

Kinesin and Dynein Propel Vesicles in Opposite Directions along Axonal Microtubules

What kinds of motor proteins are responsible for the observed movement of vesicles and organelles back and forth along axonal microtubules? A partial answer to this question has emerged from studies using *AMP-PNP,* a nonhydrolyzable analog of ATP. When AMP-PNP is

Kinesin - acts as another motor protein.
Kinesin added to beads = (−) → (+) end, ∴ away from centrosome.

kinesin
⊥   ←
axon ⟶
dynein

Centrosome

Microtubules **617**

Crossbridges

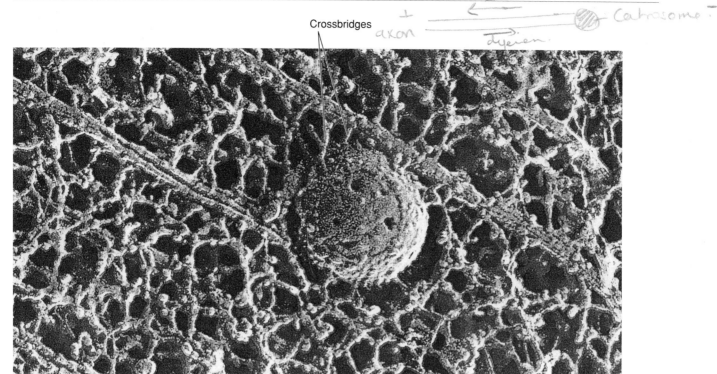

Figure 13-67 Deep-Etch Electron Micrograph Showing a Vesicle Attached to a Microtubule in a Crayfish Axon
Short crossbridges connect the membrane vesicle to the microtubule. Courtesy of N. Hirokawa.

added to isolated axoplasm, it interferes with the ATP hydrolysis step involved in fast axonal transport. As a result, the transport of membrane vesicles stops, leaving the vesicles attached to the microtubule surface.

Examination of microtubules treated in this way has led to the discovery of a microtubule-associated protein called **kinesin.** If kinesin is added to a mixture containing microtubules and either membrane vesicles or plastic beads, the vesicles or beads attach to the microtubules and move along their surfaces. Hence kinesin is functioning as a motor protein that moves organelles and particles along a microtubule surface. To determine whether movement occurs in the minus-to-plus or plus-to-minus direction (or both), experiments have been performed using microtubules growing from isolated centrosomes. In such preparations, the plus ends of the microtubules radiate outward from the centrosome (page 600). When a mixture of kinesin and plastic beads is added, the beads attach to the microtubules and migrate in the minus-to-plus direction (away from the centrosome). Since the plus ends of axonal microtubules point toward the axon tip, the preceding observations indicate that kinesin must be responsible for transporting vesicles from the cell body to the axon tip (the anterograde direction).

Kinesin consists of two light chains and two heavy chains combined together to form a long thin molecule exhibiting two globular heads at one end. Each of the globular heads contains an ATP-binding site, creating a molecule whose organization is reminiscent of myosin II.

After kinesin attaches to an organelle that is to be transported, the head groups of the kinesin molecule "walk" along the surface of a microtubule using energy derived from the hydrolysis of ATP (Figure 13-69). The discovery of kinesin in nerve axons has been followed by the identification of related proteins in other cell types, indicating that kinesin belongs to a family of motor proteins that function in various cell activities. (Surprisingly, at least one recently discovered kinesin-like protein moves in the plus-to-minus direction rather than in the minus-to-plus direction that appears to be more typical for members of the kinesin family.)

Since the kinesin molecules found in nerve axons generate movement in the minus-to-plus direction along the microtubule surface, they cannot be responsible for retrograde axonal transport, which involves movement in the plus-to-minus direction. Retrograde transport is thought to be carried out by *cytoplasmic dynein,* an axonal protein that is closely related to the *axonemal dynein* found in cilia and flagella (page 603). The movement of cytoplasmic dynein along the microtubule surface is directed toward the minus end, so it can transport vesicles and organelles from the tip of the axon back toward the cell body.

Cytoplasmic Microtubules Influence the Distribution of Cell Surface Components

Earlier in the chapter we learned that actin filaments lying directly beneath the cell surface influence the mo-

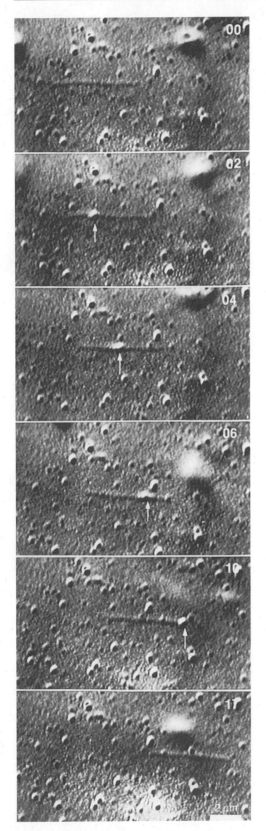

Figure 13-68 Particle Movement along a Microtubule Viewed by AVEC Microscopy *A large particle* (arrow) *can be seen attaching to the microtubule, moving along toward the right, and then being released. Seventeen seconds elapsed during this sequence. Courtesy of D. G. Weiss.*

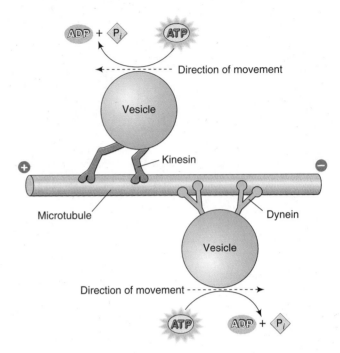

Figure 13-69 Movement of Kinesin and Cytoplasmic Dynein along a Microtubule *These two motor proteins transport vesicles and organelles in opposite directions along microtubules. Kinesin moves toward the plus end of the microtubule while cytoplasmic dynein moves toward the minus end.*

bility of plasma membrane proteins (page 592). Cytoplasmic microtubules located near the plasma membrane may also influence the distribution of cell surface components. One striking set of observations has emerged from the work of Richard Berlin and his colleagues on the behavior of phagocytic white blood cells. During phagocytosis, up to 50 percent of the plasma membrane is involved in forming phagocytic vesicles that bud off from the cell surface and migrate into the cell interior. But in spite of the large amount of membrane removed from the cell surface, the rate of active transport remains relatively constant because transport proteins are selectively excluded from those regions of the plasma membrane that invaginate to form phagocytic vesicles.

Not all membrane proteins behave in the same way, however. When Berlin investigated the distribution of membrane receptors for the lectin concanavalin A (page 592), he found that *more than 50 percent* of the lectin receptors are removed from the cell surface during phagocytosis. Thus in contrast to transport proteins, which are selectively excluded from plasma membrane regions that invaginate to form phagocytic vesicles, lectin receptors are preferentially included in such regions. This selectivity can be abolished, however, by disrupting cytoplasmic microtubules with the drug colchicine; transport proteins and lectin receptors then disappear from the cell surface at the same rate during phagocytosis. These observations suggest that the selective inclusion or exclusion of membrane components from phagocytic vesicles requires the participation of microtubules, which are known to lie beneath phagocytic regions of the plasma membrane.

Microtubules also influence the arrangement of cellulose microfibrils used in the construction of plant cell walls (page 245). In plant cells, microtubules lying directly beneath the plasma membrane tend to be oriented in the same direction as cellulose microfibrils located in the cell wall directly outside the plasma membrane. If microtubules are disrupted by exposing dividing cells to colchicine, the cell walls that are formed contain disordered arrays of microfibrils rather than the parallel bundles normally observed. It is not well understood how microtubules, which are separated from the cell wall by the plasma membrane, influence the orientation of newly forming cellulose microfibrils. However, a good hypothesis is that contact between microtubules and the overlying plasma membrane causes cellulose-synthesizing enzymes localized in the membrane to become oriented parallel to the underlying microtubules.

Microtubules Influence Cell Shape and Internal Organization

One of the first functions ascribed to cytoplasmic microtubules when they were initially discovered in the early 1960s was a skeletal role in determining and maintaining the various shapes assumed by animal cells. Support for this idea came from reports indicating that elongated or flattened cells assume a more spherical shape when their microtubules are disrupted with drugs, low temperature, or high pressure (Figure 13-70). When the treatments are stopped, microtubules reassemble and the original asymmetric shapes are again established. Microtubules responsible for maintaining cell shape have been clearly identified in *blood platelets,* whose flattened shape depends upon a *marginal band* of coiled microtubules that pushes against the plasma membrane. If the marginal band is disrupted, platelets assume a spherical shape. The marginal band of microtubules contributes to the resiliency of blood platelets, which tend to be deformed during their passage through small blood vessels.

The influence of microtubules on cell shape operates by a different mechanism in plant cells, where shape is maintained by a rigid cell wall. As we learned in the preceding section, microtubules lying beneath the plasma membrane influence the direction in which cellulose microfibrils are oriented in the cell wall. Because cellulose microfibrils are inelastic, their orientation influences the direction in which cells can expand

during cell growth; cells therefore tend to elongate at right angles to the direction in which the innermost cellulose microfibrils are oriented.

In addition to influencing the shape of the cell as a whole, microtubules also affect the shape and location of cytoplasmic organelles. For example, the position and organization of both the endoplasmic reticulum and the Golgi complex appear to be influenced by a surrounding network of microtubules. The effects of microtubules have been easiest to demonstrate for the Golgi complex, which is usually located in a discrete region adjacent to the nucleus. When cells are treated with drugs that disrupt cytoplasmic microtubules, membranes derived from the Golgi complex become dispersed throughout the cytoplasm.

INTERMEDIATE FILAMENTS

Besides actin filaments and microtubules, most eukaryotic cells contain a third group of cytoskeletal elements called **intermediate filaments** because their average diameter of roughly 10 nm makes them thicker than actin filaments and thinner than microtubules. Unlike actin filaments and microtubules, which are constructed from globular proteins, intermediate filaments are built from long, rod-shaped proteins. Intermediate filaments are typically the most stable component of the cytoskeleton and the most difficult to solubilize. They are stronger than actin filaments or microtubules, allowing them to function in situations where mechanical strength and support are of prime importance.

Intermediate Filament Proteins Are Grouped into Six Main Classes

Actin filaments and microtubules are built from protein subunits (actin and tubulin) that are similar in all cell types. In contrast, intermediate filaments found in differing cell types and intracellular locations can be constructed from different proteins, although the proteins are related to one another and share some common structural features. Intermediate filament proteins are currently divided into six classes based on similarities in amino acid sequence (Table 13-3). Proteins in all six classes are built on the same plan; they are all *dimers* consisting of two intertwined polypeptide chains, and each chain contains a long central stretch whose amino acid sequence is similar among members of the group and which is flanked at its two ends by regions of variable length and sequence. In the central region, the two intertwined polypeptide chains are in the α-helical configuration and coil around each other, forming a rodlike coiled-coil structure similar to that exhibited by myosin II (page 566).

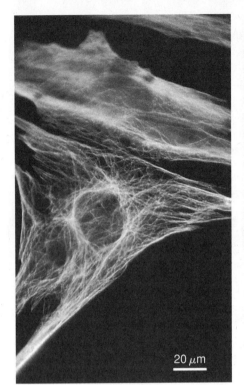
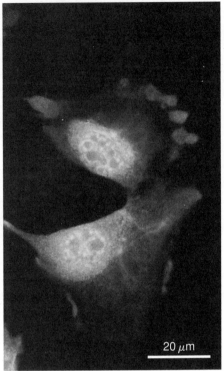

Figure 13-70 *Changes in Cell Shape Following Microtubule Disassembly* (Left) *Mouse fibroblasts in culture stained with a fluorescent antibody directed against tubulin. (Right) The same cell type after treatment with the drug colcemid, which breaks down microtubules. Note that the disappearance of cytoplasmic microtubules is accompanied by a rounding up of the cell. Courtesy of B. R. Brinkley.*

Table 13-3	Intermediate Filament Proteins	
Class	**Examples**	**Source**
Type I	Acidic keratins	Epithelial cells
Type II	Neutral/basic keratins	Epithelial cells
Type III	Vimentin	Fibroblasts, vascular smooth muscle cells
	Desmin	Muscle cells
	Peripherin	Nerve cells
	Glial fibrillary acidic protein	Glial cells
Type IV	Neurofilament proteins	Nerve cells
Type V	Lamins A, B, C	Nuclear lamina of all cell types
Type VI	Nestin	Central nervous system stem cells

During the assembly of intermediate filaments, two dimers oriented in opposite directions come together to form a tetramer (Figure 13-71). The tetramer then polymerizes to form protofilaments, which assemble into intermediate filaments. As a general rule, an intermediate filament protein that is a member of any one of the six classes can assemble by itself, or with proteins that are members of the same class, to form filaments; in contrast, proteins that are members of different classes do not usually assemble with each other. Keratin is an exception to the rule. Keratin filaments are constructed from tetramers that contain a dimer of a type I keratin (acidic keratin) mixed with a dimer of type II keratin (neutral/basic keratin).

The assembly and disassembly of intermediate filaments is controlled by the phosphorylation of sites located in the N-terminal region of the filament proteins. A good example involves the *nuclear lamina,* a dense network of intermediate filaments that underlies the inner surface of the nuclear envelope. The intermediate filament proteins used in the construction of the nuclear lamina are called **lamins.** In Chapter 12 we described how lamins are phosphorylated at the beginning of mitosis, causing the nuclear lamina to depolymerize and the nuclear envelope to break down

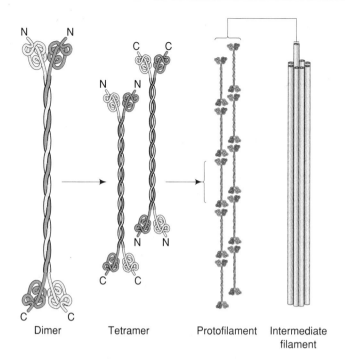

Dimer   Tetramer   Protofilament   Intermediate filament

Figure 13-71 *Assembly of Intermediate Filaments*
Intermediate filament proteins are dimers comprised of two intertwined polypeptide chains that form a coiled coil. Two dimers oriented in opposite directions come together to form a tetramer, which then polymerizes into protofilaments. Intermediate filaments are thought to contain about eight protofilaments.

(page 536). At the end of mitosis, the phosphorylation of lamins decreases, allowing the nuclear lamina to be reassembled as a support for the nuclear envelope. The assembly of other kinds of intermediate filaments is influenced by phosphorylation in a similar fashion; phosphorylation hinders the ability of intermediate filament proteins to assemble into filaments and promotes the breakdown of filaments that are already formed.

Intermediate Filaments Provide Mechanical Strength and Support

The function of intermediate filaments has been most clearly established for keratin filaments, which occur in the epithelial cell layers that cover organ and body surfaces. In Chapter 5 we saw how epithelial cells are held together by mechanical junctions called *desmosomes* (page 238). Desmosomes are especially frequent in such tissues as skin and intestines, where considerable mechanical stress is encountered. Keratin filaments radiate from the inner surfaces of desmosomes into the underlying cytoplasm and around the nucleus, anchoring to the nuclear envelope in the regions of the nuclear pore complexes. The result is the formation of an interconnected filament network that reinforces the mechanical

strength of the epithelial cell layer (Figure 13-72). The network of keratin filaments is especially prominent in the outer cell layers of the skin, where after the cells die it persists as a protective covering at the outer skin surface. Similar keratin networks are involved in the formation of hair and nails.

The importance of keratin filaments in maintaining the integrity of the skin is demonstrated by an unusual group of human diseases called *epidermolysis bullosa,* or more simply, *blistering skin disease.* For people suffering from this inherited disorder, the simple act of turning a doorknob or lifting a book can be a dangerous experience. At the slightest touch, the outer layer of skin cells bursts apart, generating a cluster of painful blisters. The inherited defect has been traced to a mutation in a keratin gene that results in the production of a keratin molecule that cannot assemble properly into intermediate filaments.

Like keratin filaments, other types of intermediate filaments also form networks that provide mechanical support for cells: (1) The *vimentin* filaments found in connective tissue cells form elaborate networks that run from the nucleus to the plasma membrane, where they anchor either to plasma membrane proteins or to micro-

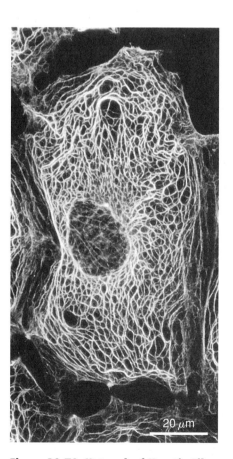

20 μm

Figure 13-72 *Network of Keratin Filaments in a Cultured Epithelial Cell* *The cell was stained with a fluorescent antibody directed against keratin. Courtesy of W. W. Franke.*

tubules closely associated with the cell surface. (2) The *desmin* filaments found in skeletal muscle join the Z disks of adjacent myofibrils together and link them to the plasma membrane. The desmin filaments of smooth muscle serve a comparable function, linking together the *dense bodies* that anchor the actin filaments in these cells (page 581). (3) The *neurofilaments* of nerve cells occur predominantly in axons, where they provide mechanical support to the axon and help to protect this long slender extension of nerve cell cytoplasm from breaking. (4) The *lamin* filaments that make up the nuclear lamina provide support to the nuclear envelope. In addition, recent evidence suggests that the nuclear lamina may make contact with cytoplasmic intermediate filaments in the regions of the nuclear pores. If this is the case, then the nuclear lamina and cytoplasmic intermediate filaments form an interconnected supporting network for the entire cell.

CYTOSKELETAL INTERCONNECTIONS

For convenience, the organization of this chapter has focused on actin filaments, microtubules, and intermediate filaments as separate components of the cytoskeleton. In reality, the cytoskeleton is a dynamically interconnected network of filaments, and many of the activities we have discussed require the participation of more than one type of cytoskeletal component. In this section we will briefly examine some evidence for the existence of cytoskeletal functions that involve an interplay between multiple types of filaments.

Axon Growth Requires Both Actin Filaments and Microtubules

A classic example of cooperative interactions between actin filaments and microtubules occurs in embryonic nerve cells at the time when they begin to develop axons. At the growing tip of a newly forming axon, membrane ruffling and cytoplasmic movements are prominent. If actin filaments are disrupted at this stage of development by exposing cells to cytochalasin, membrane ruffling and cell movements at the axon tip cease. This prevents further elongation of the axon, but the axon remains intact. In contrast, disrupting microtubules by treating cells with colchicine causes the developing axon to retract and disappear, even though ruffling of the plasma membrane continues. Hence the process of axon elongation requires the participation of both actin filaments and microtubules. Actin filaments are needed for membrane ruffling and cell movements at the axon tip, which are essential for the elongation process. As elongation proceeds, microtubules assemble into a skeletal framework that supports the growing axon and whose integrity is required for maintaining the elongated state.

Eukaryotic Cytoplasm Is Supported and Organized by an Underlying Network of Interconnected Cytoskeletal Filaments

Although conventional thin-section electron microscopy has provided a vast amount of information concerning subcellular architecture, slicing specimens into thin sections precludes an in-depth view of the three-dimensional organization of the cell's contents. One way around this obstacle is to use a high-voltage electron microscope, whose increased penetrating power makes it possible to examine intact cells without the need for prior sectioning. Using this approach, Keith Porter and his associates reported in the early 1970s that cells contain a meshwork of interconnected filaments dispersed throughout the cytoplasm and attached to the inner surface of the plasma membrane. Subsequent studies have revealed that this meshwork consists of an interconnected network of microtubules, intermediate filaments, actin filaments, and some thinner filaments containing accessory proteins such as spectrin (Figure 13-73).

The presence of an interconnected network of filaments within the cytoplasm gives cells a potential mechanism for organizing and facilitating interactions

0.2 μm

Figure 13-73 *High-Voltage Electron Micrograph Showing the Dense Interconnected Meshwork of Filaments in the Cytoplasm of a Cultured Toad Cell* *Courtesy of P. C. Bridgeman.*

among various cytoplasmic components. For example, it has long been contended that metabolic pathways occurring in the cytoplasm are too efficient to be based upon random interactions between enzymes and substrates floating freely in solution. Recent evidence suggests that some of these enzymes may be loosely bound to the cytoskeletal network, perhaps oriented in ways that facilitate the operation of metabolic pathways. Metabolic enzymes are not the only "free" components of the cell cytoplasm that may in reality be associated with the cytoskeleton. We learned in Chapter 7 that eukaryotic ribosomes occur in two forms: attached to the endoplasmic reticulum and free in the cytoplasm. Careful examination of high-voltage electron micrographs suggests, however, that many of the cell's "free" ribosomes may in fact be bound to the cytoskeleton.

The functional significance of the bound ribosomes has been investigated in the laboratory of Sheldon Penman, where gentle detergent extraction techniques were developed for removing membranes and soluble proteins without destroying the cytoskeleton. Analysis of the resulting fractions revealed that the ribosomes present in the soluble fraction represent single ribosomes that are not involved in protein synthesis. The ribosomes associated with the cytoskeleton, on the other hand, occur in the form of polysomes that are actively synthesizing proteins. Such observations suggest that "free" cytoplasmic ribosomes may become attached to the cytoskeleton when they engage in protein synthesis.

Another striking discovery to emerge from studies on detergent-extracted cells concerns the relationship between the cytoskeleton and the plasma membrane. Since detergent treatment solubilizes membrane lipids, it might be expected to cause the release of plasma membrane proteins as well. However, electron microscopic observations reveal the presence of a *surface lamina* at the outer boundary of detergent-extracted cells that retains many plasma membrane proteins (Figure 13-74). In addition, the overall shape of a cell remains largely unaltered after detergent extraction, indicating that the surface lamina and the underlying cytoskeleton form an interconnected network that holds the cell together.

BACTERIAL FLAGELLA

In contrast to eukaryotes, bacterial cells lack a cytoskeletal network of microtubules, actin filaments, and intermediate filaments. But bacteria do share with eukaryotes the ability to form motile appendages that allow cells to swim through fluid environments. Although they are referred to as *flagella,* we will see that the motile appendages of bacterial cells do not resemble eukaryotic flagella in either chemical composition, structure, or mechanism of action.

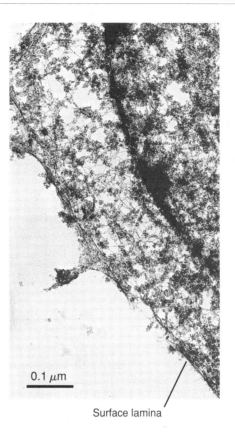

0.1 μm

Surface lamina

Figure 13-74 *Thin-Section Electron Micrograph of a Cultured Human Cell Extracted with Detergent* Note that in spite of the removal of the plasma membrane by the detergent, a distinct outer boundary termed the surface lamina remains at the cell surface. Courtesy of S. Penman.

Bacterial Flagella Are Smaller and Less Complex Than Eukaryotic Flagella

A bacterial flagellum is composed of three distinct elements: a long helical *filament* making up the body of the flagellum, a short *hook* located between the filament and the cell surface, and a *basal body* embedded within the cell envelope (Figure 13-75). The plasma membrane stops at the basal body, rather than enveloping the flagellum as it does in eukaryotic cells. The basal body, which bears no resemblance to the similarly named structure of eukaryotes, consists of a central *rod* surrounded by a series of *rings* (Figure 13-76). Four rings occur in the flagella of Gram-negative bacteria, two associated with the plasma membrane and two associated with the outer membrane (page 250). The flagella of Gram-positive bacteria, which lack an outer membrane, only have the two rings associated with the plasma membrane.

The body of a bacterial flagellum is much smaller and simpler in organization that the comparable structure of eukaryotic flagella. Earlier in the chapter we learned that the axoneme of eukaryotic flagella contains a 9 + 2 array of interconnected microtubules, with each microtubule measuring about 25 nm in diameter. In

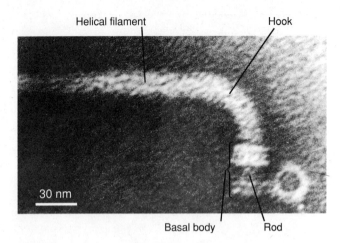

Helical filament   Hook

30 nm

Basal body   Rod

Figure 13-75 High-Magnification Electron Micrograph of a Negatively Stained Bacterial Flagellum *Part of the filament, the curved hook, and the basal body are clearly visible. Courtesy of J. Adler.*

contrast, the diameter of an entire bacterial flagellum is only about 15 nm. Hence a bacterial flagellum is comprised of a single filament whose diameter is smaller than that of an individual microtubule.

Bacterial Flagella Are Assembled by Adding Flagellin Monomers to the Flagellar Tip

Bacterial flagella are constructed from **flagellin,** a protein whose monomers are packed together with helical geometry to form a filament containing a hollow central channel (Figure 13-77). Isolated flagellar filaments can be induced to undergo reversible cycles of breakdown and repolymerization that are analogous in a superficial sense to the depolymerization/polymerization cycles observed with eukaryotic microtubules and actin filaments. In the case of bacterial flagella, warming promotes the depolymerization of isolated filaments into flagellin monomers and cooling promotes filament assembly, provided that small fragments of intact flagella are present to serve as initiation sites for polymerization. Unlike the situation with actin and tubulin, neither ATP nor GTP hydrolysis accompanies the polymerization of flagellin into filaments.

How are bacterial flagella assembled in intact cells? Are flagellin monomers added to the growing flagellum at the base, at its tip, or throughout its length? To distinguish among these alternatives, Tetsuo Iino performed a series of experiments in which bacterial cells were grown in the presence of *fluorophenylalanine,* an amino acid analog that causes the formation of abnormally curly flagella. When cells with partially formed flagella were switched to a growth medium containing fluorophenylalanine, the flagella developed curly tips, an indication that growth must be occurring at the tip (Figure 13-78). Subsequent autoradiographic experiments employing radioactive amino acids confirmed that flagellar growth occurs by addition of newly synthesized flagellin monomers to the tip of the flagellum.

A perplexing issue raised by the preceding conclusion concerns the question of how flagellin monomers reach the flagellar tip. Since bacterial flagella are not surrounded

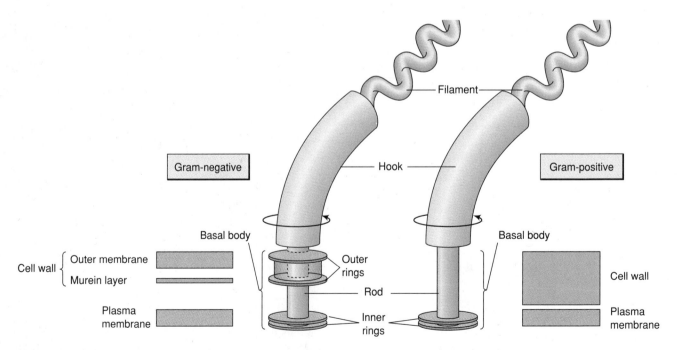

Gram-negative

Gram-positive

Filament

Hook

Basal body

Basal body

Cell wall { Outer membrane
Murein layer

Plasma membrane

Outer rings

Rod

Inner rings

Cell wall

Plasma membrane

Figure 13-76 Organization of the Flagellum in Gram-Negative and Gram-Positive Bacteria *The arrows indicate the direction in which flagella rotate to produce straight swimming. Rotation in the opposite direction produces tumbling. Regions of the flagellum that rotate are highlighted in green.*

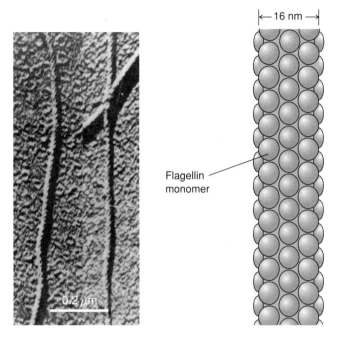

← 16 nm →

Flagellin
monomer

Figure 13-77 *The Filament of the Bacterial Flagellum*
(Left) *Electron micrograph of a shadowed filament from a
bacterial flagellum, showing the helical packing of flagellin
subunits.* (Right) *Model depicting the arrangement of flagellin
subunits in the filament. The filament has a hollow core.*

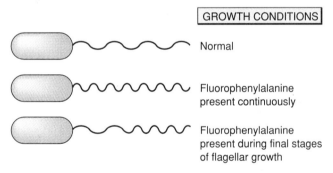

GROWTH CONDITIONS

Normal

Fluorophenylalanine
present continuously

Fluorophenylalanine
present during final stages
of flagellar growth

**Figure 13-78 *Evidence That the Growth of Bacterial
Flagella Occurs at the Tip*** *When bacteria are grown in the
presence of fluorophenylalanine, they form abnormally
"curly" flagella. If fluorophenylalanine is added to cells
whose flagella are partially formed, only the tips of the
flagella become curly. Hence growth must be occurring at
the flagellar tip.*

by a plasma membrane, the growing tip of the flagellum is
directly exposed to the cell exterior. To determine
whether flagellin is simply secreted into the external environment prior to incorporation into the flagellar tip, two
strains of bacteria producing slightly different forms of flagellin were grown in the same culture flask. If flagellin is
secreted into the external medium, one might expect the
flagella of each strain to incorporate some flagellin from
the other strain. Since such an exchange is not observed,
it can be concluded that each cell type transports flagellin
monomers directly to the tips of its own flagella. It has

been hypothesized that flagellin is transported through
the hollow central channel of the flagellum, which is just
large enough to accommodate flagellin monomers.

Bacterial Flagella Are Rotated by a Motor That Is Powered by a Proton Gradient

The movement of bacterial flagella is quite different from
that of eukaryotic flagella. When swimming bacteria are
photographed under the microscope using a high-speed
movie camera, their flagella appear to be rotating in a
screwlike fashion. Two possible explanations for such an
observation can be envisioned. One possibility is that the
flagellum is a rigid helix rotated by a circular movement of
its base. Alternatively, the propagation of helical waves
along the length of the filament might cause it to bend in
such a fashion that it appears to be rotating, when in fact it
is only bending. In order to distinguish between these possibilities, Michael Silverman and Melvin Simon designed an
experiment in which live bacteria were placed on a glass
slide to which antibodies against flagellin had been bound.
The presence of the antibodies caused the bacteria to become attached to the slide via their flagella. When observed under the light microscope, the bacteria were
found to be spinning in circles (Figure 13-79). Hence flagellar motility must involve rotation of the flagella relative
to the cell body rather than waves of helical bending.

The realization that a bacterial flagellum behaves
like a rotating rod suggests that the forces which drive
its movement are exerted at its base where the flagellum
lies buried within the plasma membrane. Such an
arrangement is quite different from that occurring in eukaryotic cells, where movement is driven by the interaction of ATP with dynein arms located along the entire
length of the flagellum. This difference might help to explain why eukaryotic flagella are enclosed by the plasma
membrane and bacterial flagella are not. The membrane
surrounding eukaryotic flagella prevents ATP from es-
caping into the environment, but such an enclosure is
unnecessary in bacterial cells because movement is
driven at the flagellar base rather than along its length.

The prime candidate for the "motor" at the base of
bacterial flagella is the set of inner rings that are associated with the basal body (see Figure 13-76). The outer
rings, which occur only in Gram-negative bacteria, appear to function as a nonrotating seal through which
the central rod of the rotating flagellum passes. In contrast to the situation in eukaryotes, ATP hydrolysis is not
the immediate driving force for the motor. Evidence for
such a conclusion first emerged from the discovery that
bacteria in which oxidative phosphorylation has been
blocked, either by chemical inhibitors or the presence
of a specific mutation, lose the ability to rotate their flagella even though they can still make ATP by glycolysis.

Subsequent studies have shown that the proton
gradient generated during oxidative phosphorylation

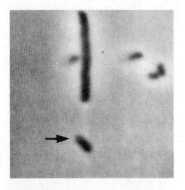

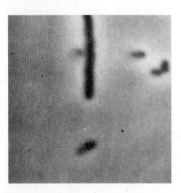

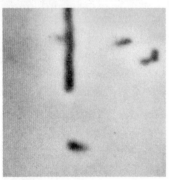

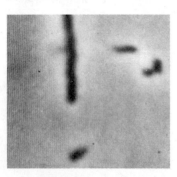

rather than ATP is the energy source for flagellar rotation. In Chapter 8 we learned that protons are pumped out of the bacterial cell during oxidative phosphorylation, creating an electrochemical proton gradient across the plasma membrane (page 346). If a proton gradient is created artificially by adding protons to the external medium in which bacteria are suspended, flagellar rotation is stimulated. Moreover, the velocity of rotation is directly related to the magnitude of the proton gradient; the steeper the proton gradient, the faster the flagella rotate. It therefore has been concluded that the proton gradient across the bacterial plasma membrane drives flagellar rotation. The mechanism by which this process operates has not been clearly established, but it has been proposed that protons flowing across the membrane in the region of the inner rings induce conformational changes in the rings that cause the flagellar rod to rotate.

The Direction of Bacterial Swimming Is Determined by the Balance between Clockwise and Counterclockwise Flagellar Rotation

How does the rotation of bacterial flagella allow bacteria to move in a particular direction? In the absence of attracting or repelling substances in the environment, bacteria swim in a random pattern in which short periods of swimming in smooth straight lines are interrupted by brief episodes of "tumbling" that lead to random changes in direction. Studies on the behavior of mutant bacteria that either never tumble or always tumble have led to the discovery that *swimming in a straight line is associated with counterclockwise rotation of flagella, whereas tumbling is produced by clockwise rotation.* The explanation for this difference is based on the fact that bacteria have multiple flagella located adjacent to one another (Figure 13-80). Counterclockwise rotation promotes packing of the flagella into organized bundles whose individual members move in unison to propel the cell. In contrast, clockwise rotation causes the bundles to fly apart into individual flagella whose uncoordinated movements give rise to random tumbling.

The direction in which bacteria move is determined by substances in their surroundings that function as either *attractants* or *repellents*. Recent observations suggest that attractants act by suppressing clockwise rotation and its associated tumbling movements. When bacteria move toward an area where the concentration of an attractant is higher, their straight swimming is interrupted less and less by random tumbling and they

Figure 13-79 *Evidence That Bacterial Flagella Rotate* *In this experiment, a bacterial cell* (arrow) *was attached to a microscope slide by its flagellum. The sequence of pictures shows the bacterial cell spinning in circles, indicating that flagellar motion is based upon rotation of the flagellum relative to the cell body. Courtesy of M. Simon.*

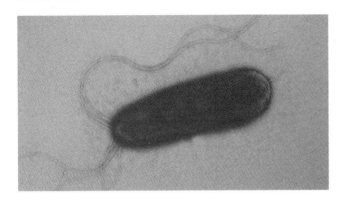

Figure 13-80 *Electron Micrograph of a Bacterial Cell with Multiple Flagella* *Courtesy of K. S. Kim, NYU Medical Center.*

continue moving toward the attractant. The opposite occurs with repellents, which promote clockwise rotation of flagella and enhance tumbling.

Chemotactic Signals Are Relayed by a Pathway That Involves Protein Phosphorylation and Methylation

Bacteria detect the presence of chemical attractants or repellents using *chemotaxis receptor proteins* located in the plasma membrane. The bacterium *E. coli* has four types of chemotaxis receptors known as *Tar, Tsr, Tap,* and *Trg* (Table 13-4). Each receptor recognizes a small group of chemical substances, either as free molecules or when bound to specific binding-proteins present in the periplasmic space (page 251). For example, aspartic acid binds directly to the Tar receptor, but maltose binds to Tar only after the maltose molecule has become associated with a periplasmic maltose-binding protein.

As a group, the four types of chemotaxis receptors recognize several dozen attractants and repellents. Although it might seem as if a cell's survival should require the ability to detect more than a few dozen compounds, it would be superfluous to have receptors for every compound that might be useful or toxic to a cell. For example, environments that contain decaying foodstuffs will contain all the amino acids normally generated by the degradation of protein molecules, so the presence of receptors for only a few different amino acids is adequate for sensing the presence of such an area.

Table 13-4   Chemotaxis Receptors in *E. coli*

Receptor	Attractants	Repellents
Tar	Aspartic acid, maltose	Ni^{2+}
Tsr	Serine, leucine	Protons
Tap	Dipeptides	
Trg	Galactose, glucose, ribose	

After a chemical attractant or repellent binds to its membrane receptor, the signal must be relayed to the flagellar motor. A central component in the signaling pathway is *CheA kinase,* a cytoplasmic protein kinase that catalyzes phosphorylation of the protein *CheY* (Figure 13-81). Phosphorylated CheY interacts with the flagellar motor to promote clockwise rotation and hence tumbling movements. When a chemical attractant binds to its receptor, the CheA kinase is inhibited. As a result, the phosphorylation of CheY is inhibited and tumbling movements are suppressed. In the absence of tumbling movements, bacteria tend to swim straight toward the source of the attractant.

Protein methylation also plays an important role in the chemotactic signaling pathway. It has been known for many years that bacteria exposed to inhibitors of protein methylation can swim only in straight lines. In contrast, bacteria lacking the enzyme that removes protein methyl groups (and hence containing proteins that are excessively methylated) are only able to tumble. The reason for these effects became apparent when it was discovered that after an attractant has bound to its receptor and triggered an alteration in flagellar rotation via CheA kinase, the receptor slowly becomes methylated. Methylation depresses the ability of the receptor to transmit signals to CheA kinase. Hence flagellar rotation is initially altered in response to an attractant, but then slowly returns to normal, even though the attractant is still present. This phenomenon, called *adaptation,* increases the sensitivity of chemotaxis. It means that once a bacterium has moved to an area where an attractant is present in higher concentration than in its original environment, the cell soon "adapts" to the new environment. It is then capable of detecting an environment in which the attractant concentration is even higher.

SUMMARY OF PRINCIPAL POINTS

• The cytoplasm of eukaryotic cells contains an interconnected cytoskeletal network consisting of 6-nm actin filaments, 10-nm intermediate filaments, and 25-nm microtubules. The cytoskeleton serves as a scaffolding that supports and organizes the cell interior, and as a source of motile elements that permit movement of both intracellular components and the cell as a whole.

• Skeletal muscle cells contain striated myofibrils that exhibit an alternating pattern of I bands, which contain thin filaments, and A bands, which contain a mixture of thick and thin filaments. Thick filaments are composed of myosin II and thin filaments are constructed from actin, a globular polypeptide that polymerizes into filaments exhibiting a fast-growing plus end and a slow-growing minus end. During contraction, shortening of the sarcomeres is accomplished by sliding of the thin filaments over the thick filaments. The driving force for filament sliding is the formation and breakage of myosin crossbridges powered by the hydrolysis of ATP.

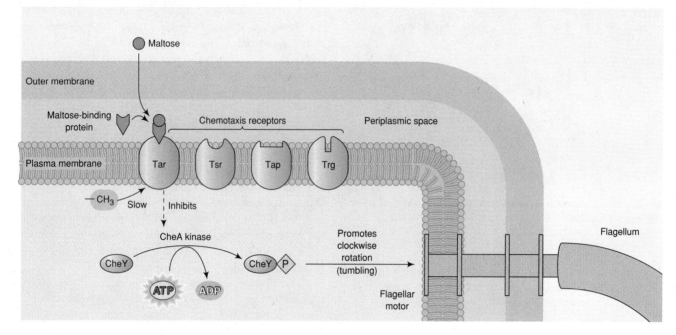

Figure 13-81 *The Main Components Involved in Bacterial Chemotaxis* *This example shows how the chemical attractant maltose influences flagellar rotation in a Gram-negative bacterium. Maltose binds to a periplasmic maltose-binding protein, generating a complex that interacts with the Tar chemotaxis receptor. The Tar receptor then inhibits CheA kinase, blocking the phosphorylation of CheY and thereby inhibiting clockwise rotation of the flagellar motor. Because inhibiting clockwise rotation suppresses tumbling movements, the bacteria swim straight toward the source of the maltose. Binding of maltose to the Tar receptor also activates receptor methylation, which gradually depresses the ability of the receptor to transmit signals to CheA kinase.*

• In stimulated muscle cells, an action potential spreads from the sarcolemma to the T tubules, triggering the release of Ca^{2+} from the sarcoplasmic reticulum. The binding of Ca^{2+} to troponin causes tropomyosin to be displaced from its normal position blocking the myosin-binding sites on the thin filaments. Phosphocreatine serves as an energy reservoir for muscle contraction, donating its high-energy phosphate group to ADP to form ATP.

• Smooth muscle contraction is triggered by the Ca^{2+}-induced activation of myosin light-chain kinase, which in turn catalyzes the phosphorylation of myosin. Phosphorylated myosin then binds to actin and initiates contraction.

• In nonmuscle cells, actin filaments are present in high concentration beneath the plasma membrane and make direct connections to plasma membrane structures such as adherens junctions and focal adhesions. Cytoplasmic streaming in plant cells is driven by the movement of myosin molecules along rows of immobilized actin filaments.

• Cell crawling is a three-stage process involving extension of the leading edge of the cell driven by actin polymerization, attachment of the leading edge to the substratum, and contractions of the cytoplasm that pull the cell forward. The interaction of actin-binding proteins with actin can produce the gel-sol transitions, sol-gel transitions, and actin gel contractions that accompany cell crawling. Chemical attractants in the surrounding environment guide cell crawling by triggering the formation of cell surface extensions that point toward the source of the attractant.

• Bundles of crosslinked actin filaments support cell surface extensions such as microvilli, which increase the surface area of the plasma membrane, and stereocilia, which are used in certain sensory organs to detect tiny movements. In addition, actin fila-

ments pinch the cell in two during cell division, influence the mobility of plasma membrane proteins, mediate changes in cell and tissue shape, and participate in phagocytosis and exocytosis.

• Microtubules are long, hollow filaments that occur both dispersed through the cytoplasm and as organized arrays localized within cilia, flagella, centrioles, basal bodies, and the mitotic spindle. Microtubules are made of tubulin, a protein heterodimer that reversibly assembles into microtubules exhibiting a fast-growing plus end and a slower growing minus end. The plus ends of cytoplasmic microtubules tends to exhibit dynamic instability, while the minus ends are embedded in microtubule-organizing centers known as centrosomes.

• Eukaryotic cilia and flagella function either to propel fluids across the cell surface or to move the cell as a whole. The axoneme of these organelles consists of an ordered array of nine outer doublet tubules surrounding two central tubules (the 9 + 2 arrangement). Ciliary and flagellar movements are generated by microtubule sliding produced by ATP-driven interactions between the dynein arms on one microtubule doublet and the B-tubule of the adjacent doublet.

• The assembly of eukaryotic cilia and flagella is initiated by basal bodies, which are derived from centrioles. Centrioles are produced either by replication of individual centrioles or from masses of granular material that function as microtubule-organizing centers.

• A variety of sense organs contain nonmotile cilia whose axonemes lack dynein arms. Axonemes that differ from those typically encountered in cilia and flagella also occur in the axopods, axostyles, cortical fibers, and cytopharyngeal basket of unicellular eukaryotes.

• Cytoplasmic microtubules are involved in moving both chromosomes and cytoplasmic organelles. Kinesin moves organelles along microtubules in the minus-to-plus direction, whereas dynein moves organelles in the plus-to-minus direction. Cytoplasmic microtubules also influence cell shape and the distribution of cytoplasmic organelles and cell surface components.

• Intermediate filaments are assembled from six classes of long, rod-shaped proteins. The assembly of intermediate filaments is controlled by phosphorylation, which hinders filament formation and promotes the disassembly of filaments that have already formed. The function of intermediate filaments has been clearly established for keratin filaments, which reinforce the mechanical strength of epithelial cell layers. Other intermediate filaments that play mechanical support roles include vimentin filaments of connective tissue cells, desmin filaments of muscle cells, neurofilaments of nerve cell axons, and lamin filaments of the nuclear lamina.

• The bacterial flagellum is a rigid helix that is rotated at its base using energy derived from a proton gradient across the plasma membrane. Counterclockwise rotation of bacterial flagella produces straight swimming, whereas clockwise rotation causes tumbling. The binding of chemical attractants to cell surface receptors promotes straight swimming by inhibiting a cytoplasmic protein kinase that phosphorylates proteins which influence the flagellar motor.

SUGGESTED READINGS

Books

Amos, L. A., and W. B. Amos (1991). *Molecules of the Cytoskeleton,* Guilford Press, New York.

Bagshaw, C. R. (1992). *Muscle Contraction,* 2nd Ed., Chapman and Hall, New York.

Bershadsky, A. D., and J. M. Vasiliev (1988). *Cytoskeleton,* Plenum, New York.

Bray, D. (1992). *Cell Movements,* Garland, New York.

Hyams, J. S., and C. W. Lloyd, eds. (1993). *Microtubules,* Wiley-Liss, New York.

Kreis, T., and R. Vale, eds. (1993). *Guidebook to the Cytoskeletal and Motor Proteins,* Oxford University Press, New York.

Preston, T. M., C. A. King, and J. S. Hyams (1990). *The Cytoskeleton and Cell Motility,* Chapman and Hall, New York.

Warner, F. D., Satir, P., and I. R. Gibbons, eds. (1988). *Cell Movement, Volume 1: The Dynein ATPases,* Liss, New York.

Warner, F. D., and J. R. McIntosh, eds. (1989). *Cell Movement, Volume 2: Kinesin, Dynein, and Microtubule Dynamics,* Liss, New York.

Articles

Allen, R. D. (1987). The microtubule as an intracellular engine, *Sci. Amer.* 256 (February):42–47.

Archer, J., and F. Solomon (1994). Deconstructing the microtubule-organizing center. *Cell* 76:589–591.

Bretscher, A. (1991). Microfilament structure and function in the cortical cytoskeleton, *Annu. Rev. Cell Biol.* 7:337–374.

Bourret, R. B., K. A. Borkovich, and M. I. Simon (1991). Signal transduction pathways involving protein phosphorylation in prokaryotes, *Annu. Rev. Biochem.* 60:401–441.

Bray, D. (1993). Towards the molecular physiology of cell movements, *Trends Cell Biol.* 3:363–365 (the entire issue of this journal is devoted to the topic of cell motility).

Bridgman, P. C., and T. S. Reese (1984). The structure of cytoplasm in directly frozen cultured cells. I. Filamentous meshworks and the cytoplasmic ground substance, *J. Cell Biol.* 99:1655–1668.

Cox, G. A., et al. (1993). Overexpression of dystrophin in transgenic *mdx* mice eliminates dystrophic symptoms without toxicity, *Nature* 364:725–729.

Fuchs, E., and K. Weber (1994). Intermediate filaments: structure, dynamics, function, and disease, *Annu. Rev. Biochem.* 63:345–382.

Fuchs, E., Y. Chan, A. S. Paller, and Q - C. Yu (1994). Cracks in the foundation: keratin filaments and genetic disease, *Trends Cell Biol.* 4:321–326.

Gibbons, I. R. (1981). Cilia and flagella of eukaryotes, *J. Cell Biol.* 91:107s–124s.

Glover, D. M., C. Gonzalez, and J. W. Raff (1993). The centrosome, *Sci. Amer.* 268 (June):62–68.

Hall, J. L. Z. Ramanis, and D. J. L. Luck (1989). Basal body/centriolar DNA: molecular genetic studies in Chlamydomonas, *Cell* 59:12–132.

Hammer, J. A., III (1991). Novel myosins, *Trends Cell Biol.* 1:50–56.

Jiang, M. Y., and M. P. Sheetz (1994). Mechanics of myosin motor: force and step size, *BioEssays* 16:531–532.

Johnson, K. A., and J. L. Rosenbaum (1990). The basal bodies of *Chlamydomonas reinhardtii* do not contain immunologically detectable DNA, *Cell* 62:615–619.

Kalt, A., and M. Schliwa (1993). Molecular components of the centrosome, *Trends Cell Biol.* 3:118–128.

Luna, E. Z., and A. L. Hitt (1992). Cytoskeleton-plasma membrane interactions, *Science* 258:955–964

Mitchison, T., and M. W. Kirschner (1984). Dynamic instability of microtubule growth, *Nature* 312:237–242.

Oakley, B. R. (1992). γTubulin: the microtubule organizer? *Trends Cell Biol.* 2:1–5.

Rayment, I., et al. (1993). Structure of the actin-myosin complex and its implications for muscle contraction, *Science* 261:58–65.

Shimmen, T., and E. Yokota (1994). Physiological and biochemical aspects of cytoplasmic streaming, *Int. Rev. Cytol.* 155:97–140.

Stossel, T. P. (1994). The machinery of cell crawling, *Sci. Amer.* 271 (September): 54–63.

Svoboda, K., C. F. Schmidt, B. J. Schnapp, and S. M. Block (1993). Direct observation of kinesin stepping by optical trapping interferometry, *Nature* 365:721–727.

Tamm, S. (1994). Ca^{2+} channels and signalling in cilia and flagella, *Trends Cell Biol.* 4:305–310.

Taylor, E. W. (1993). Molecular muscle, *Science* 261:35–36.

Vandekerckhove, J. (1990). Actin-binding proteins, *Curr. Opinion Cell Biol.* 2:41–50.

Vale, R. D., T. S. Reese, and M. P. Sheetz (1985). Identification of a novel force-generating protein, kinesin, involved in microtubule-based motility, *Cell* 42:39–50.

Vallee, R. B., and G. S. Bloom (1991). Mechanisms of fast and slow axonal transport, *Annu. Rev. Neurosci.* 14:59–92.

Walker, R. A., et al. (1988). Dynamic instability of individual microtubules analyzed by video light microscopy: rate constants and transition frequencies, *J. Cell Biol.* 107:1437–1448.

Walker, R. A., and M. P. Sheetz (1993). Cytoplasmic microtubule-associated motors, *Annu. Rev. Biochem.* 62:429–451.

Chapter 14

Evolution of Cells and Genetics of Cell Organelles

Cells are comprised of an intricate collection of molecules and organelles, each designed to perform one or more functions. One of the most profound questions that can be posed about this extraordinary arrangement is how did it arise? When did cells first come into existence, and how did they acquire the elaborate set of organelles found in contemporary eukaryotes? Such questions are difficult to investigate experimentally. Most of what we know about cells is derived from the investigative approach; cells are broken open, molecules and organelles are isolated for study, and conclusions are drawn about the functions of these components in the intact cell. But living examples of ancient organisms are not available for laboratory investigation, so a comparable approach cannot be used to study the evolutionary origins of present-day cells.

But it is still possible to design experiments that shed light on the issue of how cells *might* have arisen and evolved in the distant past. Although such studies are incapable of providing a definitive answer to the question of how cells first arose on earth, they do indicate what kinds of events *would have been possible,* and at the same, may shed light on features of contemporary cells that would be hard to explain without an evolutionary perspective. In this chapter we will see what the experimental approach has revealed about two of the most profound questions to be contemplated by biologists: How did the first cells come into existence on earth, and how did the diverse array of organelles contained in contemporary eukaryotic cells arise?

THE FIRST CELLS

In order to create a living cell, one first needs the proper collection of organic molecules; these molecules must then be assembled in precisely the right way to form an intact cell. We will begin the chapter by examining how such events might have occurred on the primitive earth before cells as we know them today came into existence.

Biologists Once Believed That Cells Routinely Arise by Spontaneous Generation

The idea that living organisms arise spontaneously from nonliving matter is thousands of years old, dating back to writings of Aristotle around 350 B.C. As recently as the early 1800s, this theory still appeared to be supported by experiments in which nutrient solutions left exposed to the air were found to become overgrown with microorganisms. However, in 1861 the French microbiologist Louis Pasteur dealt the idea of spontaneous generation a mortal blow by showing that when the air and nutrient broth in a culture flask are sterilized to destroy all microorganisms, no living cells appear in the broth as long as the neck of the flask is kept constricted. But if the neck of the flask is opened to permit entry of unsterile air, microorganisms begin to grow. Pasteur therefore concluded that the cells proliferating in the flask arise from the growth and division of cells that enter the broth from contaminated air. The idea that cells arise only from the growth and division of preexisting cells received further support from biologists who examined growing tissues under the microscope and observed that new cells always arise from the division of cells that already exist. In 1855 the German pathologist Rudolph Virchow summarized these observations in the famous statement *omnis cellula e cellula* ("all cells come from cells").

The basic validity of the idea that cells normally arise only from previously existing cells is now unquestioned, but ironically this concept impeded the development of ideas concerning the origin of the first living cells. The biggest obstacle to explaining how cells initially arose on earth is the need to invoke the occurrence of a step in which cells assemble spontaneously from nonliving matter, and Pasteur's experiments made this concept fairly disreputable. However, in recent years it has become apparent that macromolecules often assemble spontaneously into multimolecular functional entities. Among the many examples of **self-assembly** that have been observed are the assembly of RNA and proteins into ribosomes, tubulin into microtubules, actin into actin filaments, lipids and proteins into membranes, polypeptide chains into multienzyme complexes, and nucleic acids and proteins into viruses.

Once the self-assembly of macromolecules into more complex structures was shown to be a widely occurring phenomenon, the idea that cells might also have assembled spontaneously at one point in evolutionary history gained some theoretical credibility. For Pasteur's failure to observe spontaneous assembly of nonliving matter into living cells does not mean that such an event could never have occurred on earth; it only means that it did not occur under the laboratory conditions employed by Pasteur. Hence we have come almost full circle, from the initial view that cells arise spontaneously by self-organization of nonliving matter, through Pasteur's refutation of this concept, and back to spontaneous self-assembly as a theoretically viable notion.

Fossil Evidence Suggests That Cells First Appeared on Earth About 3.5 Billion Years Ago

Before we can construct plausible models of how cells might have arisen through spontaneous self-assembly, it is necessary to know something about the conditions that prevailed on earth when the first cells arose. The disciplines of astronomy, astrophysics, nuclear physics, geology, geochemistry, and geophysics have together accumulated an enormous mass of data bearing on the age of the earth and the conditions prevailing at the time of its origin. These independent approaches all support the conclusion that the earth is about 4.5 billion years old. The oldest fossil imprints of living cells have been discovered in sedimentary rocks estimated to be about 3.5 billion years old (Figure 14-1). Hence living cells appeared on earth sometime during the first billion years of the planet's existence.

How did these first cells arise? Scientific answers to this question fall into two major categories: Cells either arose from molecules that were present on the primi-

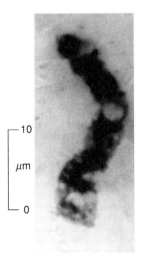

Figure 14-1 *The Fossilized Remains of Primitive Cells*
The outlines of multiple cells are apparent in this section through a rock from Western Australia estimated to be about 3.5 billion years old. Courtesy of J. W. Schopf.

tive earth, or the first cells came to earth from somewhere else in the universe. The hypothesis that living cells were brought to earth from outer space was first seriously proposed in the nineteenth century, and still continues to be discussed. For example, in 1981 Francis Crick argued that the billion years that intervened between the origin of the planet earth and the development of the first cells was not enough time for the evolution of something as complex as a living cell; instead, he proposed that cells may have been sent to earth from a distant planet aboard a spaceship.

The main shortcoming of the idea that cells came to earth from outer space is that it assumes the existence of life elsewhere in the universe, an assumption for which there is no scientific evidence. In contrast, we do have evidence for the presence of cells on earth billions of years ago. So until proof to the contrary is obtained, the simplest assumption is that these cells originated on earth itself, and that the cells found in contemporary organisms are descendants of the cells that initially arose on earth billions of years ago. At first glance, the conditions that prevailed on earth when these first cells appeared seem incompatible with life as we know it. During the first billion years of earth's existence, the planet was largely submerged under water that boiled and steamed beneath an atmosphere of toxic gases. Insignificant amounts of oxygen were present, and the absence of an ozone layer allowed massive amounts of ultraviolet radiation from the sun to reach the earth's surface. The planet was repeatedly bombarded by huge meteorites, and volcanic eruptions, torrential rains, and enormous lightning discharges were frequent. If we are to understand how living cells first arose on earth, our hypotheses must take into account these apparently hostile conditions.

Organic Molecules Form Spontaneously under Simulated Primitive Earth Conditions

In order for cells as we know them to arise on the primitive earth, the organic molecules found in cells first need to be produced. A pioneering insight into how organic molecules might have been formed on the primitive earth was provided in the late 1920s by Aleksandr Oparin, a Russian biochemist to whom we owe many of our current ideas on the evolutionary origin of cells. Oparin suggested that rather than being inhospitable for the evolution of the first cells, the chemical and physical conditions on the primitive earth were in fact favorable for the spontaneous formation of the organic molecules required by cells. According to his hypothesis, the presence of energy sources such as lightning, heat, and radiant energy from the sun could have caused the gases in the primitive atmosphere to react with one another to form small organic molecules.

The first experimental support for this theory was provided in 1953 by Stanley Miller, who attempted to

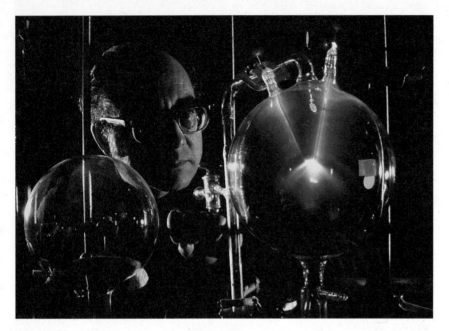

Figure 14-2 *Apparatus Designed by Stanley Miller to Simulate Conditions on the Primitive Earth* *A mixture of gases thought to exist in the early earth atmosphere was heated and subjected to electrical sparking in a sealed glass flask. After a few days, a reddish gooey material appaered within the flask. Chemical analysis of the material revealed the presence of organic molecules such as amino acids and fatty acids. Courtesy of R. Ressmeyer.*

simulate the conditions on the primitive earth in a sealed glass flask (Figure 14-2) To mimic the primordial earth atmosphere, he filled the flask with methane (CH_4), ammonia (NH_3), and hydrogen (H_2). To represent the ocean, he added water. And to simulate the energy sources present on the primitive earth, Miller subjected the mixture to electric sparking as a substitute for lightning and used a heating coil to keep the water boiling. Within a few days, a reddish goo covered the inside of the flask. To his delight, Miller found that the gooey material contained a variety of biologically important organic molecules, including amino acids and fatty acids.

These results seemed to support Oparin's theory that organic molecules could have arisen spontaneously under the conditions that prevailed on the primitive earth. However, caution is required in interpreting such experiments because the exact composition of the primitive earth atmosphere is unknown. Many scientists do not believe that reduced gases such as methane and ammonia were present in high concentrations. They instead suggest that the atmosphere was more oxidized, and thus consisted of a mixture of gases such as carbon monoxide, carbon dioxide, nitrogen, and hydrogen. Subsequent experiments of the Miller type have therefore been carried out on this and many other gas mixtures, using a variety of energy sources (Table 14-1). In each case biologically important organic molecules have been formed, including virtually all the major types of small molecules required by living cells.

Hence experimental evidence indicates that the molecules needed by living cells could have formed

Table 14-1 Results Obtained from Some Primitive Earth Simulation Experiments

Gas Mixture	Energy Source	Products
CH_4, NH_3, H_2, H_2O	Electric discharge	Amino acids, urea, organic acids, fatty acids
CO_2, CO, N_2, NH_3, H_2, H_2O	Electric discharge	Amino acids
CH_4, CO_2, CO, NH_3, N_2, H_2, H_2O	X-rays	Amino acids
HCHO, CH_3CHO, CH_2OH-$CHOH$-CHO, $Ca(OH)_2$	Heat	Sugars
CH_4, NH_3, H_2O	β-Radiation	Adenine
CH_4, NH_3, H_2O	UV-radiation	Amino acids, fatty acids

Note: This list is only a partial representation of the kinds of organic molecules that can be produced by exposing various mixtures of gases to energy sources such as those that may have been prevalent on the primitive earth.

spontaneously on the primitive earth. But simple chemical building blocks are not enough. Cells are constructed from polymers such as proteins and nucleic acids, which are synthesized by joining together monomer building blocks. The question therefore arises as to how polymers first formed on the primitive earth. **Condensation reactions,** in which monomers are joined together to form polymers accompanied by the splitting out of water, are not energetically favorable. Consider, for example, the condensation reaction for joining amino acids:

$$\text{amino acid}_1 + \text{amino acid}_2 \rightleftharpoons$$
$$\text{amino acid}_1 - \text{amino acid}_2 + H_2O$$

The equilibrium for the above reaction lies far to the left, and thus it produces little product. This result is typical for condensation reactions because the reverse reaction involving the hydrolytic breakdown of a polymer tends to be favored energetically over the forward reaction by which the polymer is formed.

There are two possible ways of overcoming this difficulty. One is to carry out the reaction under anhydrous conditions, thus removing the water required by the reverse hydrolysis reaction. Sidney Fox has shown that under such conditions, amino acids are readily polymerized into proteinlike molecules by moderate heating. The second alternative involves the use of a *condensing agent,* which is a compound whose free energy can shift the equilibrium in the direction of polymer formation. An example of a condensing agent formed under simulated primitive earth conditions is *polyphosphate,* a family of molecules of differing length consisting of linear chains of phosphate groups covalently joined together. Under laboratory conditions, the energy released during the breakdown of polyphosphates has been shown to facilitate the formation of simple polypeptide and nucleic acid polymers. It is interesting to note that contemporary cells utilize nucleoside triphosphates such as ATP for this purpose. Such triphosphates contain within their structures a "polyphosphate" sequence that is three phosphate groups long.

As an alternative to polyphosphates, Christian de Duve has proposed that sulfur-based compounds called *thioesters* may have played a central role as condensing agents on the primitive earth. Thioesters are high-energy compounds formed by the reaction between organic acids and *thiols* (substances containing an –SH group):

$$\underset{\text{acid}}{R_1C\overset{\displaystyle O}{\overset{\|}{-}}OH} + \underset{\text{thiol}}{HS-R_2} \longrightarrow \underset{\text{thioester}}{R_1C\overset{\displaystyle O}{\overset{\|}{-}}SR_2} + H_2O$$

The formation of thioesters requires a hot acidic environment, which exists in hot springs, or *hydrothermal vents,* that occur at the ocean bottom. Thioesters are high-energy compounds that are capable of driving condensation reactions. For example, amino acids that have been reacted with thiols to form thioesters can then join together in a reaction that, in contrast to the joining together of free amino acids, is energetically favorable. In contemporary cells, thioesters are employed as high-energy intermediates in a variety of metabolic pathways. A prominent example is *acetyl CoA,* a thioester that is formed by joining a two-carbon acetyl group to the thiol compound, *coenzyme A* (page 324).

Catalysis and Information Transfer Were Essential for Development of the First Cells

We have now seen that simple organic molecules and polymers can be formed spontaneously under various simulated primitive earth conditions. But do such molecules carry out any of the functions required by living cells? Two molecular functions are especially crucial. One is catalysis—that is, the ability to make chemical reactions proceed more quickly (page 37). The other is the ability to store, reproduce, and transmit information. Let us briefly examine how each of these molecular properties might have arisen on the primitive earth.

How Did the First Catalysts Arise?

It is known from the study of organic chemistry that simple organic acids and bases can catalyze reactions involving the uptake or loss of protons. It follows that simple amino acid polymers containing acidic or basic side chains could have served as proton catalysts on the primitive earth. The catalytic properties of such amino acid polymers have been experimentally investigated by Sidney Fox and his associates. As a model system they studied the properties of polymers called **proteinoids,** which can be synthesized in the laboratory by heating mixtures of amino acids in the absence of water. Proteinoids have molecular weights of thousands of daltons and contain all the commonly occurring amino acids. Instead of a random mixture of amino acid polymers, a relatively small number of proteinoids tend to predominate. The amino acid compositions of these proteinoids resemble those of natural proteins and the amino acids are arranged in specific nonrandom sequences, suggesting that the amino acid polymerization reaction spontaneously generates a certain degree of order.

The most interesting property of proteinoids, however, is their ability to catalyze chemical reactions when provided with appropriate substrates. Among the reactions catalyzed by proteinoids are hydrolysis, decarboxylation, oxidation-reduction, and amination. Given the diversity of the kinds of reactions catalyzed, it is conceivable that a primitive kind of metabolism could have been established using such catalysts. Thus both catalysis and metabolism could have arisen spontaneously on the primitive earth before life as we known it appeared.

How Did the Property of Information Transfer Arise?

In contemporary cells, the property of information transfer is associated with nucleic acids. The instructions that guide most cellular activities are encoded in the nucleotide sequences of DNA molecules, which are reproduced every time a cell divides. But the replication of DNA base sequence information, as well as its transcription into RNA, requires protein catalysts (i.e., enzymes). The question therefore arises as to whether the ability to transfer base sequence information between nucleic acids could have arisen on the primitive earth prior to the advent of such protein catalysts.

Leslie Orgel and his colleagues have shown that enzymatic catalysts are not an absolute requirement for the transfer of base sequence information. If high-energy nucleotides containing the bases A, G, C, or U are mixed together in the absence of any template or enzyme, the nucleotides join together spontaneously to form RNA chains consisting of a random mixture of the four bases. But if an RNA molecule containing only the base G is added as a template, RNA chains containing the base C are preferentially synthesized (Figure 14-3). Conversely, runs of C residues in template RNA molecules are copied into RNA products containing mainly G residues. In a similar fashion, the base A in a template RNA molecule can direct the incorporation of the base U into an RNA product, and the base U can direct the incorporation of the base A. Because the preceding reactions occur with no enzymatic catalyst present, they demonstrate that the ability of nucleic acids to transmit base sequence information could have arisen spontaneously on the primitive earth without the need for protein catalysts.

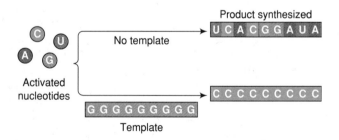

Figure 14-3 *Transfer of Base Sequence Information in the Absence of Enzymatic Catalysts* *In this experiment, nucleotides that had been chemically activated to drive their polymerization into nucleic acids were incubated in the presence or absence of a template RNA molecule containing only the base G. In the presence of the template, the base C is selectively incorporated into the product.*

Organic Molecules Spontaneously Generate Boundaries That Separate Them from the Surrounding Environment

Having established that organic molecules capable of carrying out catalysis and information transfer could have arisen spontaneously on the primitive earth, we now arrive at the crucial question of how these molecules came together to form the first primitive cells. Several models have been developed to show how groups of molecules might have established boundaries that separate them from the surrounding environment, allowing the molecules to work together as a functional unit. One of the first experimental systems to be investigated involves coacervate droplets, which have been extensively studied by Oparin and his colleagues. The phenomenon of **coacervation** occurs in concentrated solutions of macromolecules, which tend to separate spontaneously into two distinct phases. The phase that separates from the rest of the solution consists of small *coacervate droplets* ranging from 1 to 100 μm in diameter. The concentration of macromolecules in the droplets is several orders of magnitude higher than in the surrounding solution. Oparin has shown that coacervates made from concentrated solutions of phosphorylase, an enzyme involved in carbohydrate metabolism, take up glucose phosphate and convert it to the polysaccharide starch. Accumulation of the newly synthesized starch causes the coacervate droplets to grow and divide, providing a model for how the first primitive cells might have evolved.

The coacervate model is not without its limitations, however. Coacervation requires a high concentration of macromolecules, which seems unlikely in the vast oceans that covered the primitive earth. It is conceivable, however, that organic molecules became concentrated by evaporation of water in small lagoons that formed at the peripheries of the primitive seas. Another problem is that coacervates are generally unstable, and tend to break up upon standing. Furthermore, the boundary of coacervate droplets is not a selective barrier, so substances concentrated within the droplet eventually leak back into the external environment. Finally, the use of proteins such as phosphorylase to make coacervate droplets may not be applicable to conditions on the primitive earth because phosphorylase is made by living cells, which did not exist at the time. If one wants to create a model for how molecules came together on the primitive earth, it would be more realistic to utilize molecules that could have occurred on earth before cells appeared.

Such a model system has been worked out in Sidney Fox's laboratory using proteinoids synthesized by heating amino acids in the absence of water. When proteinoids are dissolved in water and allowed to cool slowly, small vesicles a few micrometers in diameter are formed (Figure 14-4). In contrast to coacervates, these

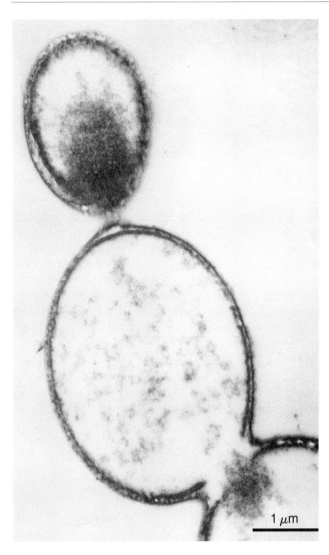

Figure 14-4 *Electron Micrograph of Several Proteinoid Microspheres Showing a Double-Layered Outer Boundary*
This barrier influences the passage of materials into and out of the vesicle. Courtesy of S. Fox.

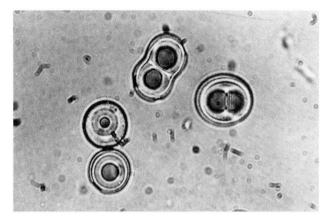

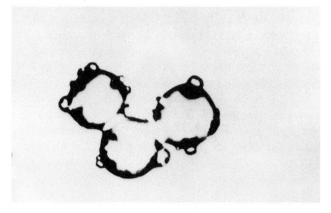

Figure 14-5 *Proteinoid Microspheres in the Process of Dividing* *The microspheres in the top photomicrograph appear to be dividing in half, while those in the bottom photomicrograph are forming tiny buds. Courtesy of S. Fox.*

microspheres are relatively uniform in size and shape and are stable for long periods of time. Microspheres are bounded by a semipermeable barrier that influences the passage of materials into and out of the vesicle. Microscopically this boundary bears a superficial resemblance to the plasma membrane of contemporary cells, although it lacks the lipid that forms the backbone of cellular membranes.

When microspheres are formed from solutions of proteinoids that exhibit catalytic activity, the resulting vesicles can carry out a primitive type of metabolism. If these vesicles are exposed to certain ions or changes in pH, they divide to form more microspheres; if a solution of microspheres is allowed to sit for several weeks, new microspheres are created by a budding process resembling that which occurs in yeast (Figure 14-5). Hence several properties we associate with living cells are exhibited by these unique vesicles which can be

formed under simulated primitive earth conditions. It is also intriguing to note that ancient fossils resembling microspheres have been detected in geological formations estimated to be billions of years old.

Although the study of microspheres has shed light on how cells might have acquired certain properties, it is a gigantic leap from a microsphere to a living cell. Microspheres are no more than small vesicles containing mediocre catalysts that carry out a primitive type of metabolism. The crucial shortcoming of microspheres is that they have no replicating genetic system that would allow them to gradually evolve into more efficient metabolic systems. If a particular microsphere happened by chance to synthesize a new proteinoid whose catalytic properties offered a significant advance over previously produced proteinoids, the lack of a genetic system would prevent the new proteinoid from being replicated and passed on to future microspheres.

In contemporary cells this genetic function is carried out by nucleic acid molecules whose base sequences code for the amino acid sequences of proteins. If a change in base sequence occurs that happens to code for a protein that functions more effectively, that mutation will be replicated and passed on to the next generation of cells. But how did the ability of nucleic acids to code for the amino acid sequence of protein molecules arise in the first place? Since contemporary cells utilize genes made of nucleic acid to guide the synthesis of proteins, but require proteins (enzymes) for the synthesis of nucleic acids, this question is like asking which came first, the chicken or the egg? As one might imagine, there are two schools of thought on the matter.

Did Proteins Arise before Nucleic Acids?

The behavior of proteinoid microspheres has led Sidney Fox to advocate the idea that proteins evolved before nucleic acids. Fox has proposed that conditions on the primitive earth could have led to the formation of microspheres containing proteinoids that catalyzed a primitive type of metabolism. Like microspheres created in the laboratory, microspheres arising on the primitive earth would have been capable of growth and division. Once they arose, proteinoid-based microspheres might have dominated evolution because of their metabolic and physical advantages over noncompartmentalized organic molecules. According to this model, the original unit of biological replication was intact microspheres rather than individual nucleic acids. Only later did microspheres evolve a genetic system in which the information coding for the amino acid sequences of protein molecules came to be encoded in the base sequence of nucleic acids.

A clue as to how the latter step might have occurred is provided by Fox's discovery that certain proteinoids bind preferentially to specific nucleic acids. For example, proteinoids enriched in the amino acid lysine bind better to RNA molecules containing the base C than they do to RNA molecules containing the base G, whereas the opposite is true for proteinoids enriched in the amino acid arginine. The ability of proteinoids to bind specific bases has led Fox to speculate that primitive proteins might have facilitated the synthesis of nucleic acids with particular base sequences. Later in evolution the nucleic acids generated by this process might have acquired the ability to replicate themselves and to direct the synthesis of proteins exhibiting catalytic activity. Once nucleic acids acquired the ability to replicate, gradual evolutionary improvement became possible because any change in nucleic acid base sequence that leads to an improved function can be replicated and passed on.

Did Nucleic Acids Arise before Proteins?

The preceding view of early cellular evolution illustrates how molecules that function as catalysts might have arisen on the primitive earth before molecules that replicate genetic information. An alternative solution to the problem of whether catalysis preceded information transfer would be a molecule that is both catalytic and informational. In the early 1980s, Thomas Cech and his colleagues made the surprising discovery that RNA molecules can function as catalysts (page 482). This discovery has led to the speculation that RNA might have been able to replicate itself on the primitive earth in the absence of protein catalysts. In other words, the answer to the chicken and egg paradox might be that RNA served as both gene and catalyst, egg and chicken.

According to this theory, early cellular evolution was dominated by an "RNA world" in which RNA molecules catalyzed their own replication. Any change in base sequence that happened to improve the ability of an RNA molecule to replicate itself would cause more of that particular RNA molecule to be produced. As a result, RNA molecules with an enhanced ability to replicate gradually predominated. Let us now suppose that one such self-replicating nucleic acid exhibited the ability to catalyze the formation of peptide bonds between amino acids, generating protein molecules. This assumption is not far-fetched because current evidence suggests that peptide bond formation in contemporary ribosomes is catalyzed by ribosomal RNA (page 494). We can also postulate the spontaneous formation of a group of self-replicating RNAs that were capable of binding to amino acids. Finally, all we need for the creation of a primitive genetic system for guiding protein synthesis is to utilize other spontaneously arising RNA molecules as crude "messages" that specify the order in which amino acids are polymerized.

The operation of such a hypothetical primitive system is illustrated in Figure 14-6, where it is seen that the base sequences of the primitive RNA "messages" are recognized by base sequences in the RNA molecules that carry amino acids, thereby causing the attached amino acids to line up in a specific order prior to being joined together. The net result is that the sequence of amino acids in a polypeptide chain is determined by the base sequence of the RNA message, which is in turn capable of self-replication. The information coding for any particular protein could therefore be passed on from generation to generation. If a randomly arising RNA molecule happened by chance to code for a protein that facilitated the operation of such a system, it would enhance the system's chances for survival and would tend to persist as evolution proceeded. Eventually the role of storing genetic information was taken over by DNA, which could have been copied from RNA by a primitive enzyme resembling reverse transcriptase (page 75).

Although the ability of RNA molecules to act as both informational templates and catalysts makes the preceding model quite attractive, upon closer examination some serious problems emerge. RNA is a relatively unstable molecule that is difficult to synthesize in the

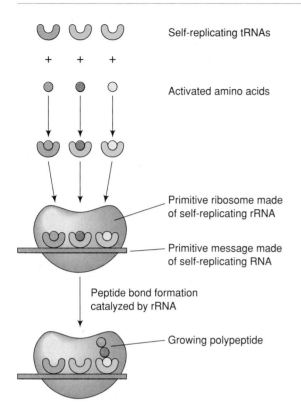

Self-replicating tRNAs

+ + +

Activated amino acids

Primitive ribosome made
of self-replicating rRNA

Primitive message made
of self-replicating RNA

Peptide bond formation
catalyzed by rRNA

Growing polypeptide

Figure 14-6 *Hypothetical Model of an RNA-Based Protein-Synthesizing System* *This model postulates that ancient self-replicating RNA molecules could have acquired some of the properties of present-day transfer RNAs, ribosomal RNAs, and messenger RNAs. As a result, the information contained in an RNA molecule might have been able to specify the amino acid sequence of a polypeptide chain without the need for protein catalysts.*

laboratory under the best of conditions, let alone under conditions that mimic the primitive earth. Moreover, the catalytic properties of RNA molecules are quite primitive when compared to those of proteins. The idea that significant quantities of self-replicating RNA occurred on the primitive earth without aid from some other type of catalyst seems implausible. Thus in spite of the likely importance of RNA in early cellular evolution, it is still possible that primitive protein catalysts were needed to set the stage for the "RNA world."

In the final analysis, our ideas about how organic molecules on the primitive earth evolved into the first self-replicating cells are confined largely to speculation. Laboratory experimentation has told us that the organic molecules utilized by living cells could have formed spontaneously under primitive earth conditions, could have acquired the ability to catalyze chemical reactions and transfer information, and could have created boundaries that separated them from the surrounding environment. But even if such experiments are an accurate reflection of what occurred on the primitive earth, the ensuing steps that led to the formation of living cells as

we know them today have never been simulated in the laboratory.

Ancient Cells Branched into Eubacteria, Archaebacteria, and Eukaryotes Early in Evolution

In spite of the gap in our understanding of how molecules on the primitive earth came together to form the first cells, there is a general consensus from the fossil evidence that the first cells appeared on earth about three and a half billion years ago. These ancient fossilized cells exhibit a relatively simple structure, resembling some present-day prokaryotes in size and shape. It was once believed that primitive prokaryotes inhabited the earth for several billion years before giving rise to eukaryotic cells, which are structurally and genetically more complex. But this view, which was based largely on fossil evidence, has had to be revised as a result of data emerging from nucleic acid sequencing studies. Sequencing studies are useful because a cell's evolutionary history is recorded in the amino acid sequences of its proteins and the nucleotide sequences of its DNA and RNA. Cells that are closely related to one another in terms of evolutionary history have nucleic acid molecules that contain similar base sequences, while distantly related cells exhibit fewer similarities in base sequence.

Sequence analysis of ribosomal RNAs (rRNAs) has been especially useful in studying evolutionary relationships because all cells contain rRNA. The work of Carl Woese and C. Fred Fox on rRNA sequences has been instrumental in modifying our views of cellular evolution. Their studies have revealed that all living cells can be placed into one of three groups, each containing rRNAs that are distinctly different from the rRNAs of the other two groups. These three groups of cells are referred to as **eubacteria, archaebacteria,** and **eukaryotes.** In addition to differences in their rRNAs, the cells in each of the three categories can be grouped together on the basis of other molecular properties, including distinctive transfer RNA modification patterns, RNA polymerases, and sensitivity to inhibitors of protein and nucleic acid synthesis.

Of the three classes of cells, eubacteria and eukaryotes are most familiar. The eubacteria include most of the commonly encountered bacteria, while the eukaryotes encompass the cells of all animals, plants, fungi, and protists (protozoa and algae). In contrast, the third class of cells—the archaebacteria—consists of an unusual group of prokaryotic organisms that thrive in conditions that would be fatal to most other cells. The archaebacteria can be subdivided into three main groups: *Methanobacteria,* which obtain energy by converting carbon dioxide and hydrogen to methane; *Halobacteria,* which thrive in salty environments; and

Sulfobacteria, which obtain energy from sulfur-containing compounds. The various archaebacteria can live at temperatures above the boiling point of water, at the enormous pressures of the ocean depths, in lakes that are saltier than the Dead Sea, and in solutions that are more alkaline than household ammonia or more acidic than gastric juices.

The ability to survive in such hostile environments may be a remnant of traits that were required on the primitive earth, where the environment was more extreme than it is today. However, it should not be inferred that archaebacteria are the most ancient of the three classes of cells. Judging by the sequences of their rRNAs, eubacteria appear to be the oldest of the three groups of cells, emerging perhaps half a billion years before the archaebacteria and eukaryotes split from one another (Figure 14-7).

Although differences in rRNA sequence were instrumental in dividing cells into eubacteria, archaebacteria, and eukaryotes, the rRNAs of these three groups also exhibit some striking similarities. For example, the locations of complementary sequences that generate stem-and-loop structures responsible for rRNA folding are similar in the three groups, thereby giving their rRNAs comparable three-dimensional structures. This similarity suggests that eubacteria, archaebacteria, and eukaryotes evolved from a common ancestral cell, although the identity of the shared ancestor remains a mystery.

How Did the Internal Membranes and Organelles of Eukaryotic Cells Arise?

Contemporary eubacteria and archaebacteria are smaller and structurally simpler than eukaryotes. Because neither eubacteria nor archaebacteria have a nuclear envelope, both are designated as prokaryotes; they also lack membrane-bound organelles such as mitochondria, chloroplasts, endoplasmic reticulum, and the Golgi complex (see Table 1-1). When the eukaryotic cell lineage first diverged from archaebacteria billions of years ago, the initial ancestral eukaryote, which has been called an **urkaryote,** did not immediately acquire all these internal membranes. It required perhaps a billion or more years for the urkaryote to first evolve into a **protoeukaryote,** which is a cell that contains an enveloped nucleus and primitive internal membranes, but is devoid of mitochondria and chloroplasts. Some contemporary protists still exhibit these primitive traits; for example, *Giardia* is an single-celled protist that contains a nucleus and cytoplasmic membranes, but no mitochondria or chloroplasts. During the next billion or so years the protoeukaryote acquired mitochondria and chloroplasts, eventually giving rise to animals, plants, fungi, and modern-day protists.

Although the way in which eukaryotic cells acquired their various organelles is not completely understood, the most extensive information available concerns the origins of mitochondria and chloroplasts. Studies on the properties of these two organelles in contemporary cells have revealed that both organelles have primitive genetic systems that provide clues to their evolutionary origins. We will therefore turn to a discussion of the genes of mitochondria and chloroplasts to learn both how these genes function in present-day cells and to see what they tell us about the evolutionary history of eukaryotes.

GENETIC SYSTEMS OF MITOCHONDRIA AND CHLOROPLASTS

Mutations Affecting Chloroplast Pigmentation Led to the Discovery of Chloroplast Genes

Although most of the genes in eukaryotic cells occur in the nuclear DNA, the idea that a few genes reside in the cytoplasm was first raised by two reports appearing in 1909, one by Carl Correns and the other by Erwin Baur. Both men were studying the inheritance of mutations that prevent chloroplasts from developing their normal green pigmentation. Such mutations produce plants exhibiting a mixture of green, colorless, and variegated (green and colorless) regions. Correns carried out an experiment using mutants of the four-o'clock flower *Mirabilis jalapa* in which flowers derived from normal regions of the plant were fertilized with pollen obtained from colorless regions. He found that the source of the pollen had no effect on the experimental results; flowers from normal green areas produced only green plants, even if the fertilizing pollen was taken from the mutant colorless segments. Conversely, flowers from the colorless areas always gave rise to colorless plants, even when normal pollen was employed. Because the properties of the offspring were always determined by the female parent, this phenomenon was referred to as **maternal inheritance** (Figure 14-8, *top*). Such a pattern is different from the behavior of nuclear genes, which are contributed equally by the male and female parents. It was quickly realized that maternal inheritance might be accounted for by genes located in the cytoplasm, since the cytoplasm of the fertilized egg is contributed almost solely by the female parent.

Slightly different results were obtained by Baur, who chose the geranium *Pelargonium* as his experimental material. Starting with one green and one colorless parent, he found some offspring to be variegated, others pure green, and the rest colorless. The green and colorless offspring both bred true, meaning that each had received genes *from only one of the two parents.* This type of inheritance, in which the genes for a given trait can be contributed solely by the male parent or solely by the female parent, is referred to as **biparental**

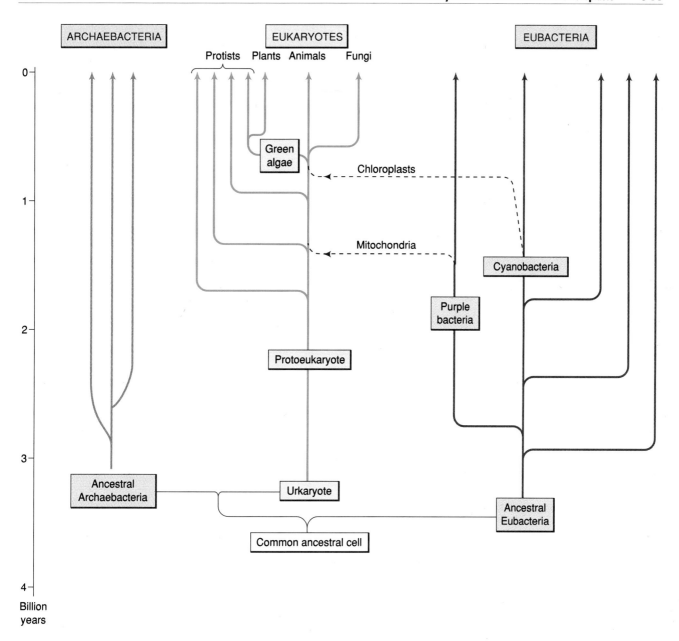

Figure 14-7 *Proposed Evolutionary Lineage of Eukaryotes, Archaebacteria, and Eubacteria*   *This scheme is based largely on information derived from comparison of rRNA sequences. The proposed role of purple bacteria and cyanobacteria in the origins of mitochondria and chloroplasts is discussed later in the chapter (page 665).*

inheritance (see Figure 14-8, *bottom*). Genes inherited in either the maternal or biparental fashion do not behave as if they are located in the nucleus because in nuclear inheritance, the male and female parents *both contribute genes to the offspring*. It therefore was concluded that the maternal and biparental patterns of inheritance reflect the presence of genes in the cytoplasm. Since both Correns and Baur were studying genes that influence chloroplast pigmentation, the most logical location for the genes would be the chloroplast itself.

In the years since these early discoveries, numerous examples of cytoplasmic inheritance involving chloroplast pigmentation have been described. Most, but not all, exhibit maternal inheritance like that observed by Correns. One unusual and illuminating case is the loss of pigmentation produced by the *iojap* mutation in maize. Some *iojap* plants lack green pigment entirely, while others are green with white striping. The *iojap* mutation maps to a nuclear chromosome, but after chloroplasts lose their pigmentation under the influence of the mutation, the nuclear *iojap* gene is no longer required to maintain the unpigmented state. The most straightforward interpretation of this phenomenon is that the nuclear *iojap* gene induces a cytoplasmic mutation that, once established, no longer requires the *iojap* gene for its inheritance. Electron microscopic examination of *iojap* maize has revealed that in areas

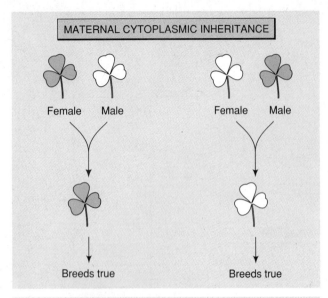

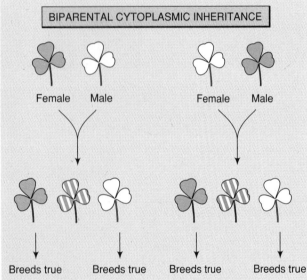

Figure 14-8 *Maternal and Biparental Patterns of Cytoplasmic Inheritance* *The green shading represents green pigmentation in plants, while the striped green pattern symbolizes variegated plants. In maternal cytoplasmic inheritance, the traits inherited by the offspring are determined entirely by the female parent. In biparental inheritance, traits are inherited from either the male or female parent, or both.*

where normal green tissue lies adjacent to mutant colorless tissue, individual cells contain a mixture of two types of chloroplasts: normal pigmented ones and abnormal chloroplasts lacking chlorophyll-containing thylakoids (Figure 14-9). The coexistence of these two kinds of chloroplasts within *the same cell* (and therefore under the influence of the same nuclear genes) provides dramatic proof of the fact that some properties of the chloroplast are not under nuclear control.

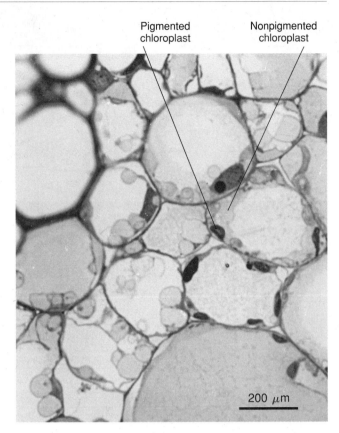

Figure 14-9 *Light Micrograph of* lojap *Maize Showing Cells Containing Both Pigmented and Nonpigmented Chloroplasts* *Note that both kinds of chloroplasts are present in the same cell. Courtesy of E. H. Coe.*

The *Petite* and *Poky* Mutations Led to the Discovery of Mitochondrial Genes

The existence of mitochondrial genes was first suggested by the discovery of the *petite* mutation of yeast by Boris Ephrussi and his colleagues in 1949. *Petite* yeast are incapable of carrying out electron transport and oxidative phosphorylation because, among other defects, they lack the necessary cytochromes. Associated with these biochemical deficiencies is a drastic reduction in the number of mitochondria and mitochondrial cristae (Figure 14-10), but the mutation is not lethal because the cells are still able to produce ATP by glycolysis. Since the ATP yield of glycolysis is quite small compared to that of mitochondrial respiration, *petite* yeast grow slowly and produce colonies that are smaller than normal. It is for this reason that the mutant yeast are called *petites*.

A variety of different mutations can produce *petite* yeast. Some involve nuclear chromosomes, but in most cases the *petite* state is cytoplasmically inherited. Cytoplasmic *petite* mutations fall into two classes: *neutral* and *suppressive*. When yeast bearing a *neutral petite* mutation are crossed with a normal strain, the resulting progeny are always normal; moreover, these normal

PETITE YEAST   NORMAL YEAST

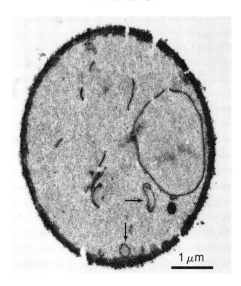
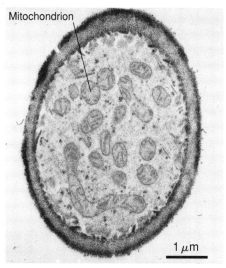

Figure 14-10 *Electron Micrographs of Normal and* Petite Yeast   (Left) *Petite yeast have few mitochondria with poorly developed internal structure* (arrows). (Right) *Normal yeast contain numerous mitochondria with well-developed cristae. Courtesy of Y. Yotsuyanagi.*

progeny continue to give rise to future generations of normal offspring. In other words, the *petite* character disappears altogether. In contrast, organisms exhibiting *suppressive petite* mutations produce offspring that are all *petite*. Neither inheritance pattern resembles the behavior of nuclear genes, where the traits of both parents are ultimately expressed.

Several other features also distinguish cytoplasmic *petite* mutations from typical nuclear mutations. First, the cytoplasmically inherited *petite* condition occurs spontaneously in roughly 1 percent of the cells of a typical yeast colony, while the spontaneous mutation rate of most nuclear genes is much lower (about 0.01 percent). Second, some agents that are not particularly effective as mutagens for nuclear genes, such as *ethidium bromide* and *5-fluorouracil*, can convert virtually an entire population of yeast into *petites* within one generation of growth. Mutagens with close to 100 percent efficiency are unheard of when dealing with nuclear genes. Finally, the cytoplasmic *petite* mutation is also unusual in that it never reverts to the normal state, whereas nuclear mutations usually revert at low but finite frequencies. This apparent irreversibility of the *petite* state led Ephrussi to suggest that it involves the loss of a cytoplasmic hereditary factor.

In 1952 Mary and Herschel Mitchell discovered a mutant strain of the bread mold *Neurospora* that is analogous in many ways to *petite* yeast. Cells of this *poky Neurospora* mutant are slow growing, lack cytochromes *b*, *a*, and a_3, have an excess of cytochrome *c*, and exhibit abnormal mitochondrial cristae. Crosses between *poky* "males" and normal "females" yield all normal offspring, whereas the offspring of the reciprocal

cross express only the *poky* trait. Because the "female" parent contributes most of the cytoplasm during mating, this pattern of maternal inheritance suggests that the gene governing the *poky* trait is located in the cytoplasm. However, some nuclear genes produce a condition that is similar to *poky*, indicating that mitochondrial behavior is under the influence of both nuclear and cytoplasmic genes.

In addition to *poky*, several other slow-growing strains of *Neurospora* have been found to exhibit maternally inherited mitochondrial defects. One such strain, identified in Edward Tatum's laboratory, was used to provide the first direct evidence that cytoplasmic genes affecting mitochondrial traits are located within the mitochondrion itself. In these studies, Elaine Diacumakos injected purified nuclei or mitochondria from slow-growing mutants into cells of a normal strain. Injection of nuclei produced no detectable change in the recipient cells, but introduction of mitochondria derived from the mutant strain caused defects in growth rate and cytochrome content similar to those exhibited by the mutant. It was therefore concluded that mitochondria contain the gene responsible for the mutant condition.

Mitochondria and Chloroplasts Contain Their Own DNA

The genetic evidence indicating that mitochondria and chloroplasts contain genes distinct from those located in the nucleus was overshadowed for many years by the classical notion that DNA resides only in the nucleus. At best, biologists underrated the significance of cytoplasmic inheritance; at worst, they disregarded it entirely. But in the early 1960s the situation was abruptly changed when the presence of DNA in both mitochondria and chloroplasts was discovered. The study of cytoplasmic genes suddenly became respectable because of the demonstrated existence of DNA molecules with which these genes could be associated.

The idea that DNA might exist outside the nucleus dates back to the early 1920s when Ernst Bresslau and Luigi Scremin used the newly developed Feulgen stain (page 128) to localize DNA in a parasitic protozoan known as a trypanosome. This organism, which is responsible for African sleeping sickness and several other diseases, has a large cytoplasmic organelle called a **kinetoplast** that stains intensely with Feulgen dye. Because the connection between DNA and genes was not recognized at the time, the significance of kinetoplast DNA remained unappreciated for many years. When modern electron microscopic techniques were eventually developed in the 1950s, it became apparent that the kinetoplast has the characteristics of a mitochondrion, including separate outer and inner membranes, cristae invaginating from the inner membrane, and a finely granular matrix (Figure 14-11). The Feulgen-positive reaction observed 30 years earlier by Bresslau and Scremin was found to be associated with a mass of fine fibrils present in the kinetoplast (mitochondrial) matrix.

The presence of DNA in trypanosome kinetoplasts, thought by some to be a biological oddity, was soon reaffirmed by the discovery of DNA-containing fibrils in the chloroplasts and mitochondria of a wide variety of organisms. The first such report was that of Hans Ris and Walter Plaut, who showed in 1962 that chloroplasts from the green alga *Chlamydomonas* react positively in the Feulgen staining reaction for DNA. Electron microscopic examination subsequently revealed the presence of fibrils 3 nm in diameter scattered throughout the chloroplast stroma. These fibrils resemble the DNA-containing fibrils found in the nucleoid of prokaryotic cells and can be destroyed by treatment with DNase, indicating that they are composed of DNA. Shortly after the discovery of DNA-containing fibrils in chloroplasts, Margit and Sylvan Nass reported that fibrils susceptible to digestion with DNase are present in the mitochondrial matrix of a variety of cell types. With osmium fixation the mitochondrial DNA-containing fibrils appear as large aggregates 20–40 nm in diameter (Figure 14-12), but following a rinse in uranyl acetate the clumps disperse to reveal individual 3-nm fibrils.

In spite of the microscopic evidence indicating the existence of mitochondrial and chloroplast DNA, the idea was slow to be accepted. Part of the problem can be traced to the inability of the Feulgen reaction to detect tiny amounts of DNA, which led to conflicting data as to whether or not mitochondria and chloroplasts stain with the Feulgen reagent. Fortunately, a solution

Feulgen-positive
material

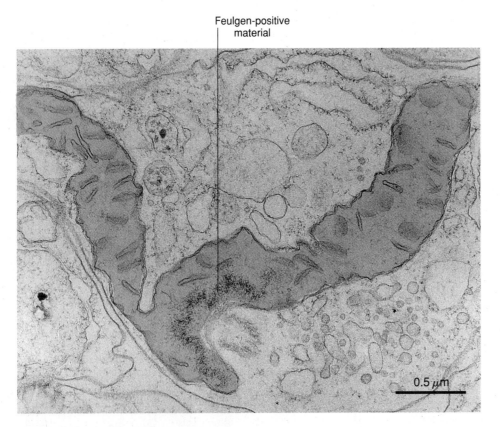

0.5 μm

Figure 14-11 *Electron Micrograph of a Kinetoplast* *The V-shaped kinetoplast (color) contains granular material that can be stained by the Feulgen reaction, which means that it contains DNA. Courtesy of P. R. Burton.*

DNA fibrils

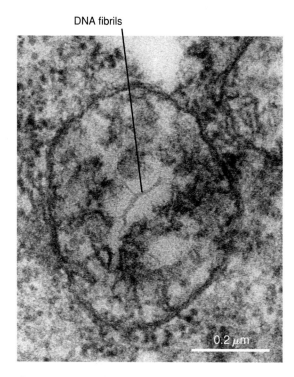

Figure 14-12 *Electron Micrograph Showing DNA-Containing Fibrils in a Sea Urchin Mitochondrion* *Courtesy of M. M. K. Nass.*

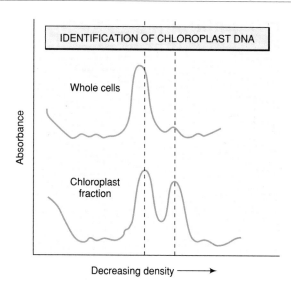

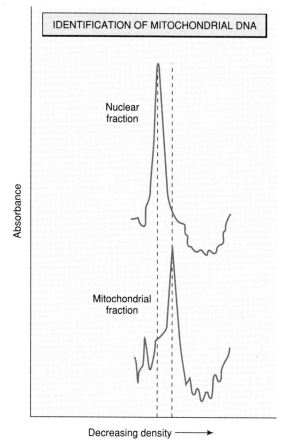

Figure 14-13 *Detection of Chloroplast and Mitochondrial DNAs by Isodensity Centrifugation* (Top) *DNA samples isolated from whole cells and from the purified chloroplast fraction of* Chlamydomonas *were fractionated by cesium chloride isodensity centrifugation. The fact that the lower-density DNA peak is selectively enriched in the chloroplast fraction identifies it as chloroplast DNA.* (Bottom) *DNA samples isolated from nuclear and mitochondrial fractions of* Neurospora *were analyzed by cesium chloride isodensity centrifugation. The lower-density DNA that predominates in the mitochondrial fraction is mitochondrial DNA.*

to the problem was eventually provided by the technique of cesium chloride isodensity centrifugation (page 136), which permitted mitochondrial and chloroplast DNAs to be identified on the basis of their unique densities. In one of the first studies of this type, centrifugation in cesium chloride was employed to compare the DNA obtained from whole cell extracts and purified chloroplasts of *Chlamydomonas* cells. The total cellular DNA was found to separate into two peaks: a large peak with an average density of 1.726 g/cm^3 and a smaller peak with a density of 1.702 g/cm^3. When DNA isolated from purified chloroplasts was analyzed, it was found to be enriched in the DNA of lower density, suggesting that the small amount of low-density DNA detected in total cellular DNA preparations represents chloroplast DNA (Figure 14-13, *top*). Shortly thereafter David Luck and Edward Reich used the same approach to identify *Neurospora* mitochondrial DNA, which is also less dense than the corresponding nuclear DNA (Figure 14-13, *bottom*).

Isodensity centrifugation in cesium chloride has turned out to be such a powerful tool that in some cases nuclear, mitochondrial, and chloroplast DNA from the same cell type can be separated from each other simultaneously. Because isodensity centrifugation provides a way to separate mitochondrial and chloroplast DNAs from nuclear DNA, it has opened the way to the chemical, physical, and genetic characterization of these organellar DNAs.

Why Is Cytoplasmic Inheritance Usually Maternal?

The development of biochemical techniques for studying the DNA of mitochondria and chloroplasts has provided new insights into the question of why cytoplasmic inheritance is usually maternal. The classical explanation has been that the male gamete (e.g., sperm cell) contains little cytoplasm, and so the cytoplasm present in the cell produced by fusion of male and female gametes is derived mainly from the egg. Although this rationale may be correct for organisms in which the male gamete contains little cytoplasm, such a relationship does not always hold. In the green alga *Chlamydomonas,* the male and female gametes are of equal size and contribute equal amounts of cytoplasm when they fuse. Yet inheritance of some chloroplast traits still follows the maternal pattern.

An explanation for this paradox has been proposed by Ruth Sager and her associates, who have monitored the fate of the chloroplast DNAs derived from the male and female parents in *Chlamydomonas.* By labeling the chloroplast DNA of one parent with ^{14}N and that of the other parent with ^{15}N, they were able to show that only one type of chloroplast DNA survives after fusion of the male and female gametes, and that its density always corresponds to that of the female parent. In other words, the paternal chloroplast DNA is selectively destroyed after fusion of the two gametes. The mechanism that marks the paternal DNA for destruction appears to be related to DNA methylation. The chloroplast DNA of maternal origin is methylated in the female gametes prior to gamete fusion, but the paternal chloroplast DNA is not. After gamete fusion, the nonmethylated paternal DNA is selectively degraded just before the two parental chloroplasts fuse. Thus maternal inheritance in *Chlamydomonas* results not from the absence of paternally derived cytoplasm, but from the selective destruction of nonmethylated paternal chloroplast DNA.

Mitochondrial DNA Is Usually Circular and Present in Multiple Copies per Mitochondrion

The amount of mitochondrial DNA found in eukaryotic cells varies considerably. In animals, mitochondrial DNA typically accounts for 0.1 to 1.0 percent of a cell's total DNA content, but in eukaryotic microorganisms such as yeast, the value can be as high as 15 percent. In amphibian oocytes, which have a massive cytoplasmic volume and numerous mitochondria, 90 percent or more of the cell's DNA may be mitochondrial in origin. The DNA located in mitochondria exhibits several features that distinguish it from nuclear DNA. One difference is that mitochondrial DNA lacks histones and hence is not packaged into nucleosomes. Mitochondrial DNA is also characterized by a unique base composition that usually

COMPARISON OF MITOCHONDRIAL (M) AND NUCLEAR (N) DNAs

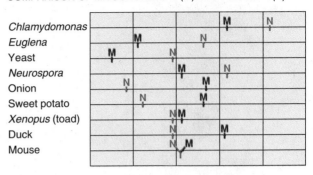

COMPARISON OF CHLOROPLAST (C) AND NUCLEAR (N) DNAs

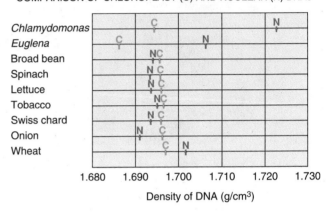

Figure 14-14 *Densities of Nuclear, Mitochondrial, and Chloroplast DNAs in Selected Organisms* *In most organisms, the difference in density between nuclear and mitochondrial or chloroplast DNA is sufficient to allow the DNAs to be separated from each other by cesium chloride isodensity centrifugation.*

permits it to be separated from nuclear DNA by isodensity centrifugation (Figure 14-14).

Electron microscopic studies of purified DNA preparations have revealed another difference between mitochondrial and nuclear DNAs; that is, mitochondrial DNA is typically circular (Figure 14-15). The mitochondrial DNA circles of animal cells are the smallest, averaging about 15,000–20,000 base pairs in size. Mitochondrial DNA circles are somewhat larger in fungi and protists, usually falling in the range of 20,000–100,000 base pairs, and plant mitochondrial DNAs are by far the largest, ranging from about 200,000 to more than a million base pairs in length (Table 14-2). Due to the tendency of large circular DNAs to break during isolation, many reports in the early literature suggest that plant mitochondrial DNA is linear rather than circular. Today we know that such reports can be attributed largely to the breakage of large circular DNA molecules during isolation. However a few unicellular eukaryotes, such as the protozoans *Paramecium* and *Tetrahymena,* appear to have genuine linear molecules of mitochondrial DNA.

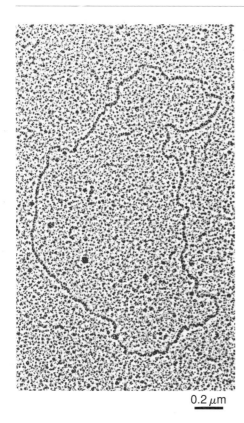

0.2 μm

Figure 14-15 *Electron Micrograph of a Circular Mouse Mitochondrial DNA Molecule* *Most mitochondrial DNAs are circular. Courtesy of M. M. K. Nass.*

Table 14-2 Properties of Mitochondrial and Chloroplast DNAs

Organism	Approximate Size in Base Pairs	Shape
Mitochondrial DNA		
Animals		
Human	16,600	Circular
Chicken	17,000	Circular
Fruitfly (*Drosophila*)	19,000	Circular
Fungi		
Bread mold (*Neurospora*)	60,000	Circular
Yeast (*Saccharomyces*)	78,000	Circular
Protists		
Acanthamoeba	41,000	Circular
Paramecium	45,000	Linear
Tetrahymena	45,000	Linear
Higher Plants		
Turnip	218,000*	Circular
Wheat	430,000*	Circular
Maize	570,000*	Circular
Chloroplast DNA		
Higher Plants		
Tobacco	156,000	Circular
Liverwort	121,000	Circular

*Sizes listed for higher plant mitochondrial DNAs refer to the size of the "master chromosome."

In animals, protists, and fungi, a typical mitochondrion contains between a few dozen and several hundred molecules of DNA, all of which are identical. But in plants the situation is considerably more complex. Instead of a single type of DNA molecule, most plant mitochondria contain a bewildering array of DNA circles of differing sizes. In the mid-1980s, Jeffrey Palmer uncovered the explanation for this diversity while working with the DNA of *Brassica campestris,* a member of the cabbage family that contains three kinds of circular mitochondrial DNAs. The largest DNA circle contains 218,000 base pairs, while the smaller molecules have 135,000 and 83,000 base pairs, respectively. When Palmer examined the base sequence of the largest DNA circle, he discovered the presence of two short repeated sequences lying between DNA stretches measuring 135,000 and 83,000 in length. He therefore proposed that the large DNA circle is a *master chromosome* that, through genetic recombination between the two repeated sequences, produces the smaller-sized circles (Figure 14-16). Other mitochondrial DNAs have subsequently been found to exhibit a similar organization. For example, the largest DNA circle in maize mitochondria contains six repeated sequences scattered around the molecule, allowing recombina-

tion to generate a diverse array of smaller circular DNAs.

Although the existence of mitochondrial DNA has now been well documented, this does not constitute proof that mitochondrial DNA contains genes governing mitochondrial traits. The first direct evidence for an association between a cytoplasmically inherited trait and mitochondrial DNA was provided by Piotr Slonimski and co-workers, who demonstrated that mitochondrial DNA obtained from yeast bearing the *petite* mutation has a different density than the mitochondrial DNA of normal yeast (Figure 14-17). The molecular basis for this density shift was studied in several other laboratories, employing *petites* induced by treatment with ethidium bromide as a model system. These studies revealed that ethidium bromide stimulates the breakdown of mitochondrial DNA and inhibits mitochondrial DNA replication (Figure 14-18). Long-term exposure to ethidium bromide eventually causes the complete destruction of mitochondrial DNA, but if treatment is stopped before all the DNA has been degraded, the remaining DNA re-

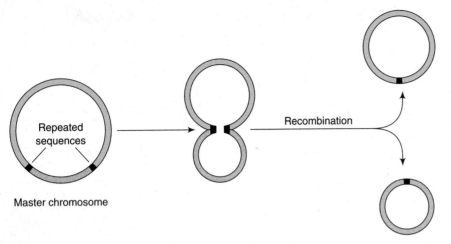

Figure 14-16 *Model for Creating Multiple Kinds of Mitochondrial DNA Circles in Plants* *According to this model, the mitochondria of each type of plant contain a large circular DNA molecule, called a master chromosome, which contains two or more repeated sequences. Genetic recombination between the repeated sequences generates smaller circles containing various segments of the master chromosome.*

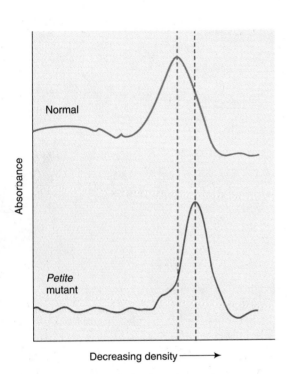

Figure 14-17 *Cesium Chloride Isodensity Centrifugation of Mitochondrial DNAs from Normal and Petite Strains of Yeast* *The average density of mitochondrial DNA is decreased in the* petite *mutant, indicating a major change in base sequence.*

sumes replication until the total content of mitochondrial DNA reaches normal levels. But the new mitochondrial DNA is grossly abnormal, containing few of the sequences present in the original mitochondrial DNA. The presence of abnormal mitochondrial DNA in *petite* yeast argues strongly that this DNA is responsible for the abnormalities in mitochondrial structure and respiratory activity observed in *petite* mutants.

Chloroplast DNA Is Also Circular and Present in Multiple Copies

Early studies of chloroplast DNA utilized isodensity centrifugation as the main tool for its isolation. Although this approach works well with unicellular plants (see Figure 14-13), in higher plants most of the isolated DNA components initially believed to be chloroplast DNA turned out not to be chloroplast DNA at all. These early experiments, in which DNA prepared from isolated chloroplasts was analyzed by cesium chloride isodensity centrifugation, revealed the presence of three DNA components: a major component with a density of about 1.696 g/cm that was thought to represent contaminating nuclear DNA, and two minor DNA components denser than nuclear DNA that were thought to represent chloroplast DNA. For several years, similar reports confirming the existence of these denser chloroplast DNAs appeared. Then around 1968 such reports ceased; higher plant chloroplasts were found to contain a single type of chloroplast DNA exhibiting a density of about 1.696 g/cm. Thus in the original reports the main component actually represented the real chloroplast DNA, while the minor components probably reflected contamination by mitochondrial and/or bacterial DNAs. This example illustrates an interesting point about how scientific ideas sometimes develop. Biologists were initially reluctant to accept electron microscopic evidence for the existence of mitochondrial and chloroplast DNA because of the possibility of artifacts, and were convinced only after the isolation of these DNAs by isodensity centrifugation; yet the centrifugation data that initially convinced them of the existence of chloroplast DNA was itself an artifact!

In early electron micrographs, chloroplast DNA often appeared as a mass of linear fragments. But in 1971 the chloroplast DNA of the alga *Euglena* was isolated as a single large circle, suggesting that the linear fragments seen in chloroplast DNA preparations are produced by the breakage of circular molecules during isolation. Virtually all chloroplast DNA molecules are now known to be circular. Current estimates suggest that a typical plant leaf cell contains about 10,000 chloroplast DNA circles distributed among 50 to 100 chloroplasts, giving each chloroplast between 100 and 200 DNA molecules. Depending on the organism, chloroplast DNA contains anywhere from 70,000 to more than 500,000 base pairs, with an average of 150,000 base pairs being typical for the chloroplasts of higher plants (see Table 14-2).

As in the case of mitochondria, the presence of DNA in chloroplasts is not indisputable evidence that this DNA contains genes governing chloroplast traits. But experimental evidence for such an association does exist. An early set of observations involved cells of the alga *Euglena* that were exposed to ultraviolet light to induce cytoplasmically inherited defects in chloroplast development. Following treatment, nuclear and mitochondrial DNAs were unaffected but the chloroplast DNA disappeared.

Independent support for the existence of circular DNA in chloroplasts has come from the genetic studies of Ruth Sager, who employed the antibiotic *streptomycin* to induce mutations in chloroplast genes of the green alga *Chlamydomonas*. For some unknown reason, streptomycin is not an effective mutagen for nuclear genes, or for genes of bacteria or viruses. But it causes a variety of chloroplast mutations, including loss of photosynthetic activity and resistance to the inhibitory effects of various antibiotics, including streptomycin itself. The availability of this vast array of mutants permitted Sager to map the location of many cytoplasmic genes affecting the chloroplast. The creation of a genetic map generally requires that genes undergo recombination (page 553), for it is the frequency with which maternal and paternal genes recombine that allows their distances from each other on the chromosome to be calculated. Since cytoplasmic inheritance in most organisms, including *Chlamydomonas*, is entirely maternal, it is not normally possible to measure the rates at which maternal and paternal cytoplasmic genes recombine with one another. But this obstacle was overcome in *Chlamydomonas* by irradiating the female parent prior to mating, which causes the progeny to receive cytoplasmic genes from both parents rather than just the mother. Analysis of the relative rates of recombination of cytoplasmic genes in such matings allowed Sager to construct a genetic map, which showed that the chloroplast genes are arranged on a circular rather than a linear DNA molecule (Figure 14-19).

Mitochondrial and Chloroplast DNA Replicate by a Semiconservative Mechanism Involving the Formation of D Loops

The realization that mitochondria and chloroplasts contain their own DNA raises the question of how the DNA in these organelles is replicated. If cells are incubated in the presence of ^{3}H-thymidine to label newly synthesized DNA, microscopic autoradiography reveals that DNA

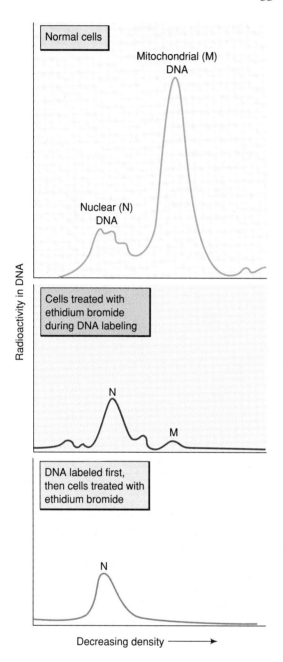

Figure 14-18 *Effects of Ethidium Bromide on Replication and Stability of Yeast DNA* *All graphs represent cesium chloride isodensity centrifugation of total yeast DNA. (Top) In normal yeast cultures incubated with a radioactive DNA precursor, both nuclear and mitochondrial DNA are synthesized. (Middle) When ethidium bromide is included during the labeling period, the synthesis of mitochondrial DNA is selectively inhibited. (Bottom) When yeast are first incubated with a radioactive DNA precursor to label nuclear and mitochondrial DNAs and ethidium bromide is then added, the labeled mitochondrial DNA is degraded while the nuclear DNA remains unaffected.*

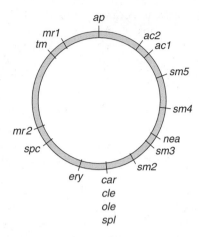

Figure 14-19 *Genetic Map Showing the Circular Arrangement of Genes in the Chloroplast DNA of* Chlamydomonas *This map was created by measuring the rates at which various cytoplasmic genes were transmitted together from parent to offspring. Many of the genes indicated on the map code for resistance to antibiotics that inhibit chloroplast protein synthesis.*

synthesis takes place inside both mitochondria and chloroplasts (Figure 14-20). The replication of mitochondrial and chloroplast DNA is carried out by DNA polymerases that operate independently of nuclear DNA polymerases, thereby allowing mitochondrial and chloroplast DNA to be synthesized anytime during the cell cycle rather than just during S phase when the nuclear DNA replicates.

To determine whether mitochondrial DNA replication involves a semiconservative mechanism like that employed by bacterial and nuclear DNAs, Edward Reich and David Luck used the $^{14}N/^{15}N$ labeling approach (page 67) to study mitochondrial DNA replication. Cells of the bread mold *Neurospora* were initially grown in the presence of ^{15}N, and were then switched to media containing ^{14}N. The data obtained in these experiments were complicated by the fact that it takes a considerable amount of time for the ^{14}N in the growth medium to replace all the old ^{15}N-containing nucleotides in the cell. But after this delay, mitochondrial DNA molecules containing one ^{14}N-containing strand and one ^{15}N-containing strand were detected, and after a subsequent round of replication, mitochondrial DNA composed almost entirely of ^{14}N was found. This pattern is what would be expected from DNA replication that is semiconservative (see Figure 3-7).

Shortly thereafter Kwen-Sheng Chiang and Noboru Sueoka used the same approach to study chloroplast DNA replication. They showed that after transferring *Chlamydomonas* cultures from ^{15}N- to ^{14}N-containing media, chloroplast DNA of intermediate density appears after one round of replication, and equal amounts of intermediate and light density chloroplast DNA appear after two rounds (Figure 14-21). Again, this pattern is exactly what would be expected of semiconservative DNA replication.

Electron microscopy has also been a valuable tool for investigating the mechanism of mitochondrial and chloroplast DNA replication. In 1971 Jerome Vinograd and Piet Borst independently discovered that mitochondrial DNA molecules often exhibit a small loop displaced from the main circle (Figure 14-22, *left*). Denaturation of such DNA molecules causes a small fragment of single-stranded DNA to be released, suggesting that the loop is caused by displacement of one of the two strands of the double helix by a small DNA fragment. The loop is therefore referred to as a *displacement loop* or **D loop**. In addition to relatively short D loops, longer loops have also been observed in replicating mitochondrial DNA (see Figure 14-22, *right*). In some cases the loops are single stranded, while in others they are partially or completely double stranded. A current model of mitochondrial DNA replication consistent with the existence of these various forms is presented in Figure 14-23. According to this model, DNA replication is initiated at different locations on the two DNA strands. Because the two DNA strands differ in density in most mitochondrial DNAs, they are referred to as the *heavy* (H) and *light* (L) strands. The synthesis of a new H strand is initiated first, using an RNA primer as in the replication of nuclear DNAs (page 72). After synthesis of the new H strand has proceeded partway around the circle, forming a visible D loop, synthesis of the new L strand is initiated at a second origin. Each of the two new strands is then completed by continuous synthesis in the 5' → 3' direction.

Chloroplast DNA replication resembles mitochondrial DNA replication in that DNA synthesis is initiated at different sites on the two DNA strands. However, replication of the two chloroplast DNA strands is initiated at about the same time, leading to the formation of two visible D loops (see Figure 14-23). Both strands are then completed by continuous 5' → 3' synthesis.

Mitochondrial DNA Codes for Ribosomal RNAs, Transfer RNAs, and a Small Group of Mitochondrial Polypeptides

Several lines of evidence support the conclusion that mitochondrial DNA serves as a template for the transcription of mitochondrial RNAs. First, isolated mitochondria incorporate radioactive precursors into RNA in a reaction whose dependence on a DNA template is revealed by its sensitivity to inhibition by actinomycin D (page 424) or DNase. Second, mitochondria contain their own RNA polymerase, an enzyme that can be distinguished from nuclear RNA polymerases by its differing susceptibility to inhibitors. Third, the discovery that ethidium bromide inhibits the synthesis of mitochondrial, but not nuclear, RNA supports the conclusion that independent RNA synthetic pathways exist in nuclei and mitochondria.

How are the genes that code for mitochondrial RNAs identified in a mitochondrial DNA molecule? Early

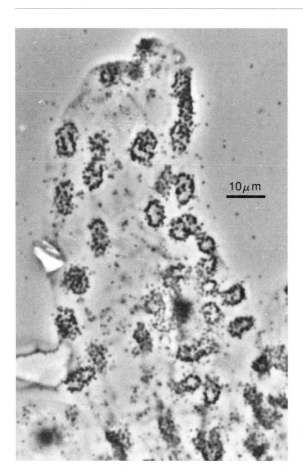 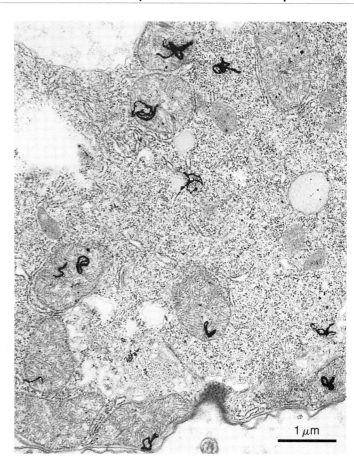

Figure 14-20 *Microscopic Autoradiographs Showing That DNA Synthesis Occurs in Both Chloroplasts and Mitochondria*
(Left) *Light microscopic autoradiograph of a spinach leaf cell following incubation with radioactive thymidine. The silver grains indicating the location of radioactive material are localized primarily over chloroplasts.* (Right) *Electron microscopic autoradiography of the ciliated protozoan* Tetrahymena *following incubation with radioactive thymidine. The silver grains are located primarily over mitochondria. Courtesy of J. V. Possingham* (left) *and R. Charret and J. André* (right).

studies utilized nucleic acid hybridization techniques to study the interaction between mitochondrial RNAs and DNA. For example, tRNA molecules isolated from mitochondria have been linked to *ferritin,* an electron-opaque protein that is visible under the electron microscope. When ferritin-labeled tRNAs are hybridized to mitochondrial DNA, the tRNAs bind to complementary regions along the DNA molecule (Figure 14-24). In this way the locations of mitochondrial genes coding for various tRNAs have been determined.

Additional information about the location of mitochondrial genes has emerged from the use of DNA sequencing techniques. Because mitochondrial DNAs are relatively small compared to nuclear DNAs, it has been possible to determine the complete base sequence of the mitochondrial DNAs obtained from a variety of organisms. Figure 14-25 (page 655) illustrates the map of human mitochondrial DNA that has emerged from such studies. It shows that the human mitochondrial DNA molecule contains a total of 37 genes falling into three categories. (1) Two genes code for 16S and 22S rRNAs, which are components of mitochondrial ribosomes. (2) Twenty-two genes code for mitochondrial tRNAs,

which differ in base sequence from the corresponding tRNAs in the cytosol. (3) Thirteen genes code for polypeptides that become incorporated into the inner mitochondrial membrane, where they function as components of the respiratory chain and ATP synthase complexes (see Figure 8-30). Seven of the polypeptides are subunits of NADH dehydrogenase (complex I of the respiratory chain), one is cytochrome *b* (a member of complex III), three are subunits of cytochrome oxidase (complex IV), and two are components of ATP synthase (the F_0-F_1 complex).

The 37 genes found in human mitochondrial DNA are not all located on the same DNA strand. The genes coding for the two rRNAs and most of the tRNAs and polypeptide chains occur on the H strand, but genes for eight of the tRNAs and one of the polypeptide chains are situated on the L strand. The only region of the DNA molecule that does not code for a product occurs near the origin of replication on the H strand, where the initial D loop is formed.

The preceding gene arrangement is typical of animal mitochondrial DNAs, which are significantly smaller than the mitochondrial DNAs of most protists, fungi,

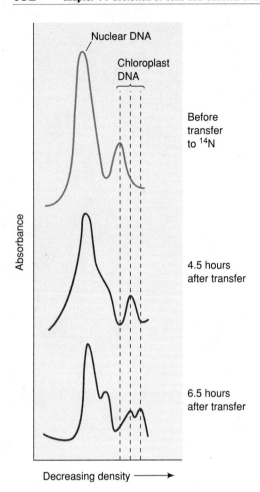

Figure 14-21 *Semiconservative DNA Replication In Chlamydomonas Chloroplasts* *The graphs represent cesium chloride isodensity gradient centrifugation of total cell DNA before and after transfer to ^{14}N-containing media. (Top) The top graph shows the position of the heavy ^{15}N-containing chloroplast DNA in the isodensity gradient. (Middle) After 4.5 hours in ^{14}N-containing media, the density of the chloroplast DNA has shifted to an intermediate value that is between that of ^{14}N-containing and ^{14}N-containing chloroplast DNA. (Bottom) After 6.5 hours in ^{14}N-containing media, equal amounts of intermediate and light density chloroplast DNA appear. This stepwise shift in DNA density is consistent with semiconservative replication.*

and plants. But larger mitochondrial DNAs do not necessarily contain a larger number of genes. In the mitochondrial DNA of yeast, for example, long stretches of noncoding sequences separate the mitochondrial genes. Some yeast mitochondrial genes are also interrupted by the presence of introns. Because of the presence of these various kinds of noncoding sequences, less than a third of the yeast mitochondrial DNA molecule appears to code for functional gene products. But in spite of these differences in gene organization between yeast and human mitochondrial DNAs, all mitochondrial DNAs examined thus far have a basic set of genes in common; included in this group are at least 2 genes

coding for mitochondrial rRNAs, at least 22 genes coding for tRNAs, and genes coding for cytochrome *b*, cytochrome oxidase subunits, and subunits of mitochondrial ATP synthase (Table 14-3).

Chloroplast DNA Codes for More Than a Hundred RNAs and Polypeptides Involved in Chloroplast Gene Expression and Photosynthesis

Chloroplast DNA, like mitochondrial DNA, codes for a variety of different RNAs and polypeptides. The location of individual genes in chloroplast DNA molecules has been investigated both by hybridizing chloroplast RNA to chloroplast DNA, and by determining the base sequence of chloroplast DNA. The resulting data have revealed that chloroplast DNA contains 120 or more genes falling into three distinct groups. (1) The first group consists of genes coding for products involved in the transcription and translation of chloroplast DNA. Included in this category are genes coding for 4 rRNAs, 30 or more tRNAs, 20 ribosomal proteins, and 4 subunits of chloroplast RNA polymerase. (2) The second group consists of genes coding for polypeptides involved in photosynthesis. Included are at least 28 thylakoid membrane polypeptides and the large subunit of the soluble protein *rubisco,* which is involved in the fixation of carbon dioxide by the Calvin cycle (page 390). The thylakoid membrane polypeptides encoded by chloroplast DNA include subunits of ATP synthase (the CF_0-CF_1 complex) and components involved in photosynthetic electron transfer by the cytochrome b_6f complex and photosystems I and II. (3) The remaining genes identified in chloroplast DNA are referred to as *unassigned reading frames* because they consist of base sequences whose codons specify an uninterrupted stretch of amino acids, but the identity and function of the corresponding proteins are as yet unknown.

Both Strands of Human Mitochondrial DNA Are Completely Transcribed into Single RNA Molecules

The transcription of mitochondrial and chloroplast genes differs in a number of ways from the transcription of nuclear genes described in Chapter 10. The earliest studies of mitochondrial DNA transcription were carried out by Guiseppe Attardi and his collaborators, who incubated human mitochondria with radioactive RNA precursors and then tested the ability of the newly formed radioactive RNA to hybridize to the H and L strands of mitochondrial DNA. Unexpectedly, the RNA was found to hybridize to almost all the sequences in *both* DNA strands (in contrast to nuclear gene transcription, where RNA is usually copied from only one DNA strand for any given gene). When longer labeling periods were used, most of the radioactive RNA complementary to the L strand disappeared. These findings suggested that both

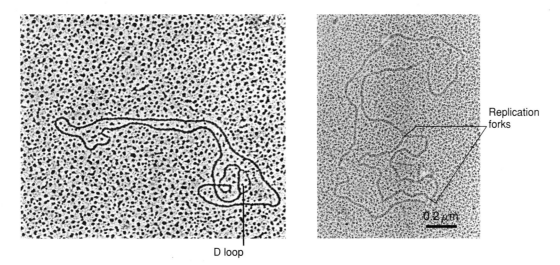

Figure 14-22 *Electron Micrographs of Replicating Mitochondrial DNA* (Left) *In this micrograph of a mitochondrial DNA molecule, a small D loop is present where the DNA is beginning to replicate.* (Right) *This mitochondrial DNA molecule shows a longer DNA segment that has just been replicated, forming a large loop. Courtesy of D. L. Robberson (left) and K. Koike and D. R. Wolstenholme (right).*

mitochondrial DNA strands are completely transcribed, but most of the RNA copied from the L strand is quickly destroyed in the mitochondrion.

Subsequent studies have revealed that transcription is initiated at a promoter located in the D loop area of the DNA molecule, leading to the formation of two complete RNA transcripts, one copied from the H strand and the other copied from the L strand. Transcription is catalyzed by a mitochondrial RNA polymerase that is encoded by nuclear DNA, although it differs from the RNA polymerases that catalyze nuclear gene transcription. After transcription, the RNA molecules generated by copying the H and L strands are enzymatically cut into their respective rRNAs, tRNAs, and mRNAs. A clue to the mechanism underlying the cleavage step may be provided by the organization of genes within the mitochondrial DNA molecule. In almost all cases, genes coding for rRNAs and mRNAs are separated from each other by genes coding for tRNAs (see Figure 14-25). Thus it appears as if tRNAs may serve as biological punctuation marks that determine where the RNA should be cut. Some support for this idea has come from the discovery of *mitochondrial RNase P,* an enzyme that cuts mitochondrial RNA transcripts at the 5' end of tRNA sequences. The mitochondrial mRNAs generated by cleavage of the initial mitochondrial RNA transcript are subsequently modified by addition of poly-A tails to their 3' ends, but they do not acquire the 5' cap structure found in eukaryotic cytoplasmic mRNAs.

Mitochondrial gene transcription in yeast exhibits several features that distinguish it from the comparable process in mammals. In yeast, all mitochondrial genes other than a single tRNA gene are sequestered on one of the two DNA strands, and transcription is initiated at multiple sites throughout this strand rather than at a single promoter as in mammals. Another difference is that, unlike mammalian mitochondrial DNA, some of the protein-coding genes of yeast mitochondrial DNA possess introns. For at least a few of these genes, the intron sequence has been found to play a direct role in its own removal; in these cases, part of the intron is initially translated along with the exon sequences to form a hybrid protein called a **maturase.** The maturase then catalyzes the removal of the intron sequences from the mRNA precursor, yielding a mature mRNA.

Like transcription in mitochondria, gene transcription in chloroplasts differs in several ways from transcription in the nucleus. Transcription of chloroplast genes is catalyzed by a chloroplast RNA polymerase that resembles bacterial RNA polymerase more closely than eukaryotic nuclear RNA polymerases; several subunits of this chloroplast RNA polymerase are encoded by genes residing in chloroplast DNA. Transcription by chloroplast RNA polymerase is initiated at multiple sites throughout the chloroplast DNA, using promoter sites that closely resemble the promoters employed by bacterial cells. Some the genes transcribed by chloroplast RNA polymerase contain introns that are excised by a process comparable to the one employed by mitochondrial maturases.

In addition to intron removal, the base sequence of mitochondrial and chloroplast mRNAs can also be altered by a mechanism called **RNA editing.** During RNA editing, anywhere from a single nucleotide to hundreds of nucleotides in an RNA transcript may be

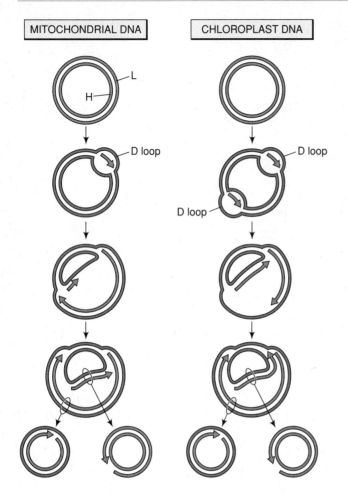

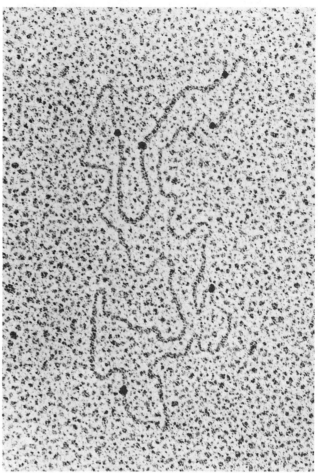

Figure 14-23 *Models of Mitochondrial and Chloroplast DNA Replication* (Left) *During the initiation of mitochondrial DNA replication, the L strand serves as a template for the synthesis of the beginning of a new H strand, forming a small D loop. After synthesis of the new H strand has proceeded partway around the circle, synthesis of the new L strand (using the original H stand as template) is initiated at a second origin. Both strands are then completed by continuous synthesis in the 5' → 3' direction. (Right) In chloroplast DNA replication, synthesis of the two strands is initiated at different locations at about the same time, leading to the formation of two D loops. Both strands are then*

Figure 14-24 *Electron Micrograph of Mitochondrial DNA Hybridized to Ferritin-Labeled Mitochondrial tRNAs* *The large black granules correspond to the locations where the tRNAs have hybridized to the mitochondrial DNA. Courtesy of N. Davidson.*

inserted, deleted, or chemically altered. Such changes often create new initiation and/or stop codons, and can change the reading frame of the message. In some cases, the information required for altering bases in the proper location is provided by short *guide RNAs* that are complementary to the region of the mRNA being altered. Although the most dramatic examples of RNA editing have been observed in mitochondrial and chloroplast mRNAs, small editing changes in the base sequence of mRNAs transcribed from nuclear genes also occur.

Mitochondria and Chloroplasts Synthesize Proteins Using a Unique Set of Ribosomes, Transfer RNAs, and Protein Synthesis Factors

The mRNAs transcribed from mitochondrial and chloroplast DNAs are translated into polypeptides within the two organelles using protein-synthesizing machinery that differs in several ways from that present in the cytosol. Protein synthesis in these organelles involves unique kinds of ribosomes, tRNAs, and protein-synthesis factors, and also differs from cytoplasmic protein synthesis in its susceptibility to inhibitors. We will briefly consider each of these differences in turn.

1. Ribosomes. Mitochondrial ribosomes differ from cytoplasmic ribosomes in size and chemical composition (Table 14-4). The ribosomes of mammalian mitochondria sediment at 55–60S and contain

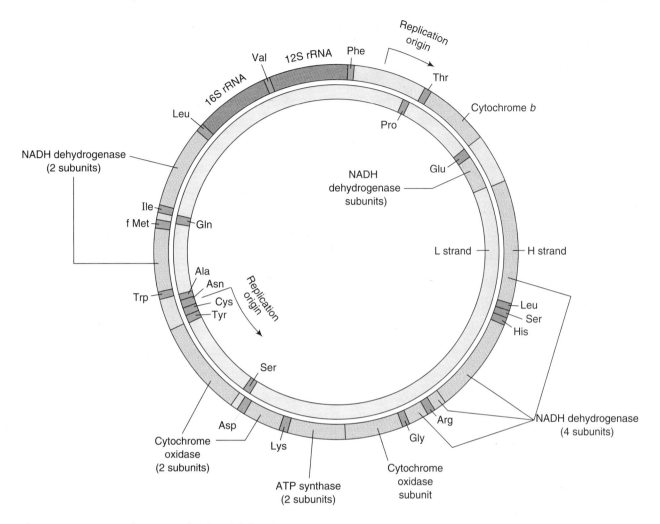

Figure 14-25 *Map of Human Mitochondrial DNA* *The tRNA genes are designed by three-letter abbreviations corresponding to the amino acid to which each tRNA binds. Human mitochondrial DNA codes for 22 tRNAs, 2 rRNAs, and 13 polypeptides.*

Table 14-3 Genes Identified in Mitochondrial DNA

Gene	Human	Yeast	Plants
Ribosomal RNAs			
Large subunit	16S	21S	26S
Small subunit	12S	15S	18S
5S RNA	–	–	+
Transfer RNAs	22	24	~30
Ribosomal proteins	0	1	+
RNase P (RNA component)	–	+	?
NADH dehydrogenase subunits	7	0*	6?
Cytochrome *b*	+	+	+
Cytochrome oxidase subunits	3	3	3
ATP synthase subunits	2	3	4

Note: (+) = present, (–) = absent.

*The mitochondrial DNAs of other fungi, such as *Neurospora*, encode six subunits of NADH dehydrogenase.

12S (small subunit) and 16S (large subunit) rRNAs. The mitochondrial ribosomes of unicellular eukaryotes and plants are somewhat larger, sedimenting in the range of 70–80S. Mitochondrial ribosomes in higher plants, for example, sediment at about 78S and contain 18S rRNA in the small subunit and 26S rRNA plus 5S rRNA in the large subunit. All mitochondrial rRNAs, including the 5S rRNA that occurs only in plant mitochondria, differ from the corresponding rRNAs of cytoplasmic ribosomes. Unlike eukaryotic cytoplasmic ribosomes, none of the mitochondrial ribosomes examined thus far contain 5.8S rRNA hydrogen-bonded to the larger rRNA.

Chloroplast ribosomes exhibit less variability in size than those of mitochondria, sedimenting near 70S and containing subunits of 50S and 30S. Like prokaryotic ribosomes, the large subunit of chloroplast ribosomes contains rRNAs sedimenting at about 5S and 23S, and the small subunit exhibits a single rRNA molecule sedimenting at 16S.

Table 14-4 Sizes of Various Types of Ribosomes

Source	Intact Ribosome	Ribosomal Subunits	Ribosomal RNAs
Cytoplasm (eukaryotes)	80S	(large) 60S (small) 40S	28S 18S, 5.8S, 5S
Cytoplasm (prokaryotes)	70S	(large) 50S (small) 30S	23S 16S, 5S
Mitochondria (mammalian)	55–60S	(large) 45S (small) 35S	16S 12S
Mitochondria (yeast)	75S	(large) 53S (small) 35S	21S 14S
Mitochondria (higher plants)	78S	(large) 60S (small) 45S	26S 18S, 5S
Chloroplasts	70S	(large) 50S (small) 30S	23S 16S, 5S

Note: The data represent "average" values for various classes of organisms.

2. Transfer RNAs. Mitochondria and chloroplasts contain tRNAs and aminoacyl-tRNA synthetases that are unique to the two organelles. The aminoacyl-tRNA synthetases are encoded by nuclear DNA, but the tRNAs are transcribed from either mitochondrial or chloroplast DNA and differ in sequence from the corresponding tRNAs found in the cytosol. Taking into account the existence of wobble, which allows a single tRNA to recognize more than one codon (page 494), at least 32 different tRNAs are required for translation of all possible mRNA codons. Although genes for about 30 different tRNAs have been identified in chloroplasts and plant mitochondria, fewer are present in the mitochondria of animal cells. Mammalian mitochondrial DNA codes for only 22 tRNAs, and no evidence exists for the uptake of any additional tRNAs from the cytosol. Thus it appears as if the wobble rules have been relaxed in animal mitochondria so that a single tRNA can recognize as many as four different codons specifying the

same amino acid, rather than the normal maximum of three (see Figure 11-31).

Another unusual feature of mitochondrial and chloroplast tRNAs is the presence of the tRNA for formylmethionine. This type of tRNA, and the enzyme *transformylase* that formylates the methionine bound to the tRNA, were once thought to be unique to bacterial cells, where formylmethionine is employed in the initiation of protein synthesis (page 490). The discovery of these components in mitochondria and chloroplasts suggests that protein synthesis in the two organelles exhibits some similarities to protein synthesis in bacteria. The possible significance of these similarities will become evident later in the chapter when we discuss the evolutionary origins of mitochondria and chloroplasts (page 664).

3. Protein Synthesis Factors. In addition to the components already described, protein synthesis requires an appropriate set of initiation, elongation, and termination factors. The protein synthesis factors of mitochondria and chloroplasts are structurally distinct from their counterparts in the cytosol. Again, there is a close resemblance to the analogous constituents of prokaryotic cells; several protein synthesis factors present in mitochondria are functionally interchangeable with those obtained from bacterial cells but not interchangeable with their counterparts in the eukaryotic cytosol.

4. Inhibitors of Protein Synthesis. It has been known for many years that protein synthesis in isolated chloroplasts can be blocked by **chloramphenicol,** a drug that inhibits protein synthesis by bacteria but not by eukaryotic cytoplasmic ribosomes. Other inhibitors that are selective for bacterial protein synthesis also have been found to inhibit protein synthesis by mitochondria and chloroplasts, whereas inhibitors selective for eukaryotic cytoplasmic ribosomes, such as **cycloheximide,** are inactive against mitochondrial and chloroplast ribosomes (Table 14-5). Thus the pattern of susceptibility to inhibitors, the similarity in protein synthesis factors, and the utilization of formylmethionine all indicate a similar-

Table 14-5 Effects of Protein Synthesis Inhibitors on Various Types of Ribosomes

Source of Ribosomes	Puromycin	Chloramphenicol	Cycloheximide
Prokaryotic	+	+	−
Eukaryotic			
Cytoplasmic	+	−	+
Mitochondrial	+	+	−
Chloroplast	+	+	−

Note: (+) = inhibited, (−) = no effect.

ity between the protein synthetic systems of mitochondria, chloroplasts, and bacteria.

Mitochondria Employ a Slightly Altered Genetic Code

Examining the base sequence of mitochondrial and chloroplast DNAs has revealed that during mitochondrial (but not chloroplast) protein synthesis, mRNA is translated into polypeptide chains using a genetic code that is slightly different from the universal code used elsewhere by prokaryotic and eukaryotic cells. For example, human mitochondria read the codon AUA as methionine instead of isoleucine, and use the normal stop codon, UGA, to code for the amino acid tryptophan (Table 14-6). Differences in codon usage can occur even among the mitochondria of different organisms. The mitochondria of yeast, but not mammals, read the four codons beginning with CU as threonine rather than leucine, whereas the mitochondria of mammals, but not yeast, employ AGA and AGG as stop signals rather than as codons for the amino acid arginine. Why mitochondria have evolved these unusual patterns while chloroplasts continue to employ the normal genetic code is not completely understood, but the reason is thought to be related to the differing evolutionary histories of the two organelles.

Mitochondria and Chloroplasts Synthesize Polypeptides Whose Genes Reside in Mitochondrial and Chloroplast DNA

We have now seen that mitochondria and chloroplasts contain DNA sequences coding for a limited number of polypeptides, and that the two organelles also possess the enzymes and protein-synthesizing machinery required for transcribing and translating these gene se-

quences into polypeptide chains. Direct evidence that such polypeptides are actually synthesized within mitochondria and chloroplasts has been obtained by incubating isolated mitochondria or chloroplasts with radioactive amino acids, followed by identification of the radioactive products. The main limitation to this approach is that because of the vast number of cytoplasmic ribosomes in most cells, even a small carryover of cytoplasmic ribosomes into preparations of isolated mitochondria or chloroplasts can confuse the results. An alternative approach is to exploit the selectivity of protein synthesis inhibitors, searching for polypeptides whose synthesis is inhibited by chloramphenicol but unaffected by cycloheximide.

One mitochondrial component whose synthesis has been studied in this way is *cytochrome oxidase,* a protein complex consisting of seven polypeptides residing in the inner mitochondrial membrane. When cells are incubated with cycloheximide, which blocks cytoplasmic but not mitochondrial protein synthesis, only three of the seven subunits of cytochrome oxidase are synthesized. Conversely, if cells are incubated with chloramphenicol to block mitochondrial but not cytoplasmic protein synthesis, the other four subunits are synthesized. Hence of the seven subunits that make up cytochrome oxidase, three are manufactured in mitochondria and four in the cytoplasm. The three subunits synthesized in mitochondria correspond to the three polypeptides whose genes reside in mitochondrial DNA (see Table 14-3). Similar experiments have revealed that mitochondria also synthesize the other polypeptides that are encoded in mitochondrial DNA; included in this category are cytochrome *b* and subunits of the ATP synthase complex.

Chloroplast protein synthesis has been studied using techniques resembling those employed with mitochondria. The first product of chloroplast protein synthesis to be identified was the large subunit of *rubisco,* the CO_2-fixing enzyme of the Calvin cycle that consists of a large catalytic subunit and a smaller regulatory subunit. Synthesis of the small rubisco subunit is selectively inhibited in the presence of cycloheximide, whereas formation of the large subunit is blocked by chloramphenicol. Thus the small subunit appears to be made in the cytoplasm and the large subunit in the chloroplast. Verification of this conclusion has come from experiments showing that isolated chloroplasts synthesize the large but not the small rubisco subunit (Figure 14-26). Moreover the gene coding for the large rubisco subunit has been identified in chloroplast DNA, thereby providing further support for the idea that this polypeptide is synthesized in chloroplasts.

Although rubisco is a soluble protein located in the chloroplast stroma, many polypeptides synthesized by chloroplasts have turned out to be thylakoid membrane proteins. Among them are subunits of protein complexes involved in photosynthetic electron transfer

Table 14-6 Examples of Altered Codon Usage in Mitochondria

Codon	Normal Usage*	Mitochondrial Usage	
		Mammals	Yeast
AUA	Ile	Met	Met
UGA	Stop	Trp	Trp
AGA, AGG	Arg	Stop	Arg
CUU,CUC,CUA, CUG	Leu	Leu	Thr

*"Normal usage" refers to codon assignments in mRNA transcribed from eukaryotic nuclear DNA, prokaryotic DNA, and chloroplast DNA.

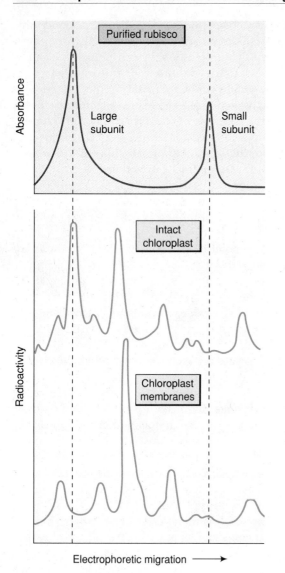

Figure 14-26 *Evidence That Isolated Chloroplasts Synthesize the Large but Not the Small Subunit of Rubisco* (Top) *SDS-polyacrylamide gel electrophoresis of purified rubisco. Absorbance measurements were used to detect the presence of the large and small subunits.* (Bottom) *After incubating isolated chloroplasts with radioactive amino acids, newly synthesized radioactive proteins present in the chloroplast were detected by gel electrophoresis. The results show that: (1) chloroplasts synthesize the large but not the small subunit of rubisco, and (2) many of the polypeptides synthesized by the chloroplast (other than the large rubisco subunit) are components of chloroplast membranes.*

(photosystem II, the cytochrome b_6-f complex, and photosystem I) and subunits of the ATP-synthesizing CF_0-CF_1 complex. Chloroplasts also synthesize some of the polypeptides found in chloroplast ribosomes and chloroplast RNA polymerase. As in the case of mitochondria, all the polypeptides synthesized by chloroplasts are encoded by chloroplast DNA.

Nuclear Genes Cooperate with Mitochondrial and Chloroplast Genes in Making Mitochondrial and Chloroplast Proteins

Mitochondria and chloroplasts carry less than 10 percent of the genetic information required to assemble either a mitochondrion or a chloroplast. Most of the polypeptides needed by mitochondria and chloroplasts are encoded by nuclear genes, synthesized on free cytoplasmic ribosomes, released into the cytosol, and imported into mitochondria and chloroplasts using the targeting mechanisms described in Chapter 11 (pages 512–514). In essence, polypeptides destined for uptake by mitochondria and chloroplasts contain signal sequences that bind to receptor proteins located in the outer membrane of either the mitochondrion or chloroplast. After binding, the polypeptide chain is inserted into or transferred across the mitochondrial or chloroplast membranes, depending on its ultimate location in the organelle (see Figures 11-52 and 11-53). The signal sequences are then cleaved by proteases to generate mature protein molecules.

Since some genes coding for mitochondrial and chloroplast polypeptides occur in the nucleus while others reside within the two organelles, the question arises as to how the expression of the genes located in the organelles is coordinated with those residing in the nucleus. This issue is especially important for proteins that require polypeptides encoded by genes located in two different compartments. For example, cytochrome oxidase requires three polypeptides encoded by mitochondrial DNA and four polypeptides encoded by nuclear DNA, and rubisco requires one polypeptide encoded by chloroplast DNA and one polypeptide encoded by nuclear DNA. The mechanism that coordinates the synthesis of polypeptides made in separate locations but destined for assembly into the same protein is not well understood. In chloroplasts, some insights have emerged from studies involving the synthesis of the large and small subunits of rubisco. If an inhibitor of cytoplasmic protein synthesis is added to intact cells to block formation of the small rubisco subunit by cytoplasmic ribosomes, synthesis of the large rubisco subunit by chloroplasts eventually slows down as well, even though the inhibitor has no direct effect on protein synthesis by isolated chloroplasts. One possible interpretation of these results is that the small rubisco subunit is manufactured on cytoplasmic ribosomes and then enters the chloroplasts, where it stimulates synthesis of the large rubisco subunit.

Why are some mitochondrial and chloroplast polypeptides synthesized within the organelles themselves, while others are manufactured in the cytoplasm and then imported? The answer to this question may lie in the evolutionary history of mitochondria and chloroplasts. We will see shortly that these two organelles are thought to have evolved from bacteria that were ingested by eukaryotic cells a billion or more years ago. As time progressed and the ingested bacteria evolved into mitochondria or chloroplasts, most of the bacterial genes were either lost or transferred to the nucleus. But why weren't all the chloroplast and mitochondrial genes relocated to the nucleus? The answer may be related to the fact that many polypeptides encoded by chloroplast and mitochondrial DNA are hydrophobic molecules that reside in either the inner mitochondrial membrane or the thylakoid membrane. Because it is thermodynamically unfavorable to place a hydrophobic molecule in an aqueous environment, it would be disadvantageous to synthesize hydrophobic polypeptides on free cytoplasmic ribosomes and then release them into the cytosol prior to transport into the mitochondrion or chloroplast. Such polypeptides are therefore synthesized on ribosomes attached to the inner mitochondrial or thylakoid membrane (Figure 14-27); this arrangement allows the polypeptides to be inserted directly into the membrane as they are being synthesized, just as ribosomes bound to the ER insert their newly synthesized polypeptides directly into the ER membrane (page 265).

In most cases, the gene that codes for a given mitochondrial or chloroplast polypeptide occurs either in the organelle or in the nucleus, but not both. However, exceptions do occur. Using DNA hybridization techniques, sequences resembling those that occur in mitochondrial DNA have been identified in the nuclei of a variety of organisms, including insects, mammals, sea urchins, and yeast. However, most of these nuclear sequences contain fragments of different mitochondrial genes that are fused together, making them genetically inactive. The gene coding for a subunit of mitochondrial F_0-F_1 ATP synthase in *Neurospora* has also been detected in both nuclear and mitochondrial DNAs, but in this case the nuclear rather than the mitochondrial copy is expressed. Thus when similar sequences are present in both nuclear and mitochondrial DNA, in some instances the nuclear sequences are functional and in other cases the mitochondrial sequences are functional.

New Mitochondria Arise by Growth and Division of Existing Mitochondria or Promitochondria

The occurrence of DNA, RNA, and protein synthesis within mitochondria and chloroplasts raises the question of the degree to which these two organelles control their own growth and development. In early studies, three mechanisms were proposed to explain the way in which new mitochondria might originate: (1) growth and division of existing mitochondria; (2) complete *de novo* ("from the beginning") formation of new mitochondria from a mixture of the required proteins, nucleic acids, and lipids; and (3) assembly of new mitochondria from other membrane systems, such as the plasma membrane, nuclear envelope, or endoplasmic reticulum.

Microscopic observations have tended to support the idea that new mitochondria arise by growth and division. Mitochondria in the process of dividing have been observed in living cells viewed under the phase microscope, and membrane partitions dividing mitochondria in half can be detected with the electron microscope (Figure 14-28). In the green alga *Micromonas*,

Figure 14-28 *Electron Micrograph Showing Dividing Mitochondria within Fat Cells* Each mitochondrion is constricted into two compartments by a transecting membrane. Courtesy of W. J. Larsen.

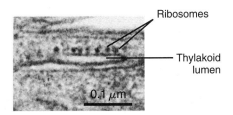

Ribosomes

Thylakoid lumen

0.1 μm

Figure 14-27 *Electron Micrograph Showing Chloroplast Ribosomes Attached to the Outer Surface of a Thylakoid Membrane* These ribosomes synthesize polypeptides that are inserted directly into the thylakoid membrane. Courtesy of H. Falk.

which contains only a single mitochondrion and a single chloroplast, each of the two organelles can be seen dividing in half as the cell divides.

But the preceding microscopic observations do not rule out the possibility that mitochondria also arise by other mechanisms. In the early 1960s, David Luck designed a series of experiments involving *Neurospora* that were aimed at determining whether such alternative mechanisms exist. First, *Neurospora* cultures were incubated with ^{3}H-choline, a phospholipid precursor that becomes incorporated into mitochondrial membranes. The radioactive cells were then transferred to a nonradioactive medium and allowed to grow for varying periods of time. Finally, the cells were examined by microscopic autoradiography and the silver grains were counted to determine the distribution of radioactivity. It was reasoned that one of the following three patterns would emerge: (1) If new mitochondria are formed by growth and division of existing mitochondria, each doubling in the number of mitochondria after the ^{3}H-choline is removed should be accompanied by a 50 percent reduction in the number of autoradiographic grains per mitochondrion. (2) If *de novo* synthesis occurs, then new mitochondria formed after removal of ^{3}H-choline should be nonradioactive, while the radioactivity in the preexisting mitochondria would remain the same. (3) If mitochondria develop from other membranes, then the preexisting radioactive mitochondria should retain their radioactivity, and new mitochondria should at first also be radioactive (because the membranes from which they develop would be radioactive). But as time passes, the absence of ^{3}H-choline from the medium would result in a progressive decrease in the amount of radioactivity in the cellular membranes from which the new mitochondria develop. Therefore the average amount of radioactivity per labeled mitochondrion would decrease gradually. Table 14-7 shows that the data obtained in the studies followed the first pattern, suggesting that new mitochondria arise by the growth and division of existing mitochondria.

In a related set of experiments, Luck exploited the fact that the buoyant density of mitochondria can be altered by growing cells in media containing different concentrations of choline, a lipid precursor that is in-

corporated into mitochondrial membranes. Growth in the presence of high choline concentrations results in the formation of mitochondria of low buoyant density, whereas growth in the presence of low choline concentrations yields mitochondria of high buoyant density. Luck therefore grew cells in a medium containing a low concentration of choline, and then switched them to a medium containing a high concentration of choline. He reasoned that if mitochondria are formed either *de novo* or from other membranes, the high-density mitochondria originally manufactured in the presence of low choline concentrations should remain undisturbed, while a new population of low-density mitochondria would appear as a result of subsequent growth in the presence of a high choline concentration. But if mitochondria arise by growth and division, low-density membrane material should be added to preexisting high-density mitochondria as they grow, and so the mitochondria would be expected to gradually become less dense. This result is exactly what occurred (Figure 14-29), providing added support for the idea that new mitochondria arise by growth and division.

Although new mitochondria are usually formed by growth and division, in some cases mitochondria develop in cells that appear to have no preexisting mitochondria. One such example occurs in yeast, which can survive under either aerobic or anaerobic conditions. When switched to anaerobic conditions, the cells meet all their energy requirements through glycolysis and so mitochondria gradually disappear. If the cells are then exposed to oxygen, the mitochondria rapidly reappear. Although this phenomenon was once considered to be evidence for the *de novo* formation of mitochondria, careful electron microscopic examination of anaerobically grown yeast has revealed the presence of small double-membrane-enclosed vesicles about 1 μm in diameter. Though lacking cristae and cytochromes, these vesicles do contain mitochondrial DNA. When anaerobic yeast are exposed to oxygen, the inner membrane of the vesicles invaginates and forms cristae that acquire newly synthesized cytochromes and other components of the respiratory chain. Because of their ability to differentiate into mitochondria, these small double-membrane vesicles are called **promitochondria.**

Table 14-7 Average Number of Silver Grains per Labeled Mitochondrion after Exposure of *Neurospora* to ^{3}H-Choline

Number of Generations after Labeling	Actual Data	Pattern Predicted by		
		Growth and Division	**De Novo Synthesis**	**Assembly from Nonmitochondrial Membranes**
0	2.0	2.0	2	Intermediate between
1	1.1	1.0	2	growth and division
2	0.5	0.5	2	and *de novo*
3	0.25	0.25	2	synthesis

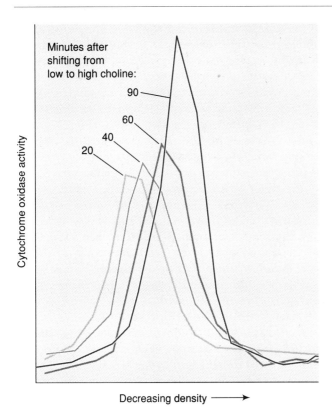

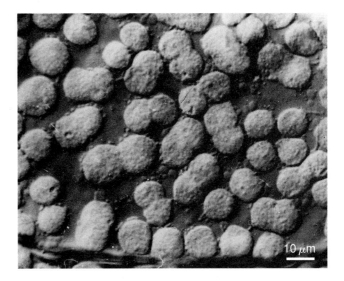

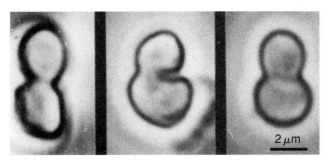

Figure 14-29 *Isodensity Centrifugation of Neurospora Mitochondria Isolated from Cells Shifted from a Medium of Low Choline Concentration to a Medium of High Choline Concentration* Mitochondria were detected by monitoring for the presence of the mitochondrial enzyme cytochrome oxidase. Growth in the presence of low choline concentration yields mitochondria of high buoyant density, whereas growth in the presence of high choline concentration yields mitochondria of low density. The gradual decrease in mitochondrial density observed in these studies is consistent with the conclusion that new mitochondria arise by growth and division of existing mitochondria.

Figure 14-30 *Evidence for Chloroplast Division in Higher Plants* (Top) In a living spinach leaf viewed by Nomarski interference microscopy, chloroplasts are seen in various stages of division, from dumbbell shaped to fully separated. (Bottom) Phase contrast micrographs of isolated spinach chloroplasts reveal several chloroplasts in the process of dividing. Courtesy of J. V. Possingham (top) and R. M. Leech (bottom).

New Chloroplasts Arise by Growth and Division of Existing Chloroplasts or Proplastids

As in the case of mitochondria, microscopic evidence suggests that chloroplasts usually arise by growth and division of existing chloroplasts. In algae, which contain only one or two large chloroplasts per cell, the fate of individual chloroplasts is easy to watch. More than a hundred years ago, light microscopists discovered that each chloroplast in cells of the green alga *Spirogyra* divides in half prior to cell division. Similar observations have been reported for other types of algae, and films of the process of chloroplast division have even been made. In higher plants the large number of chloroplasts per cell makes it harder to follow the fate of individual organelles, but microscopic observations occasionally reveal chloroplasts in the process of dividing (Figure 14-

30). Chloroplasts isolated from higher plant tissues have also been reported to divide when placed in an artificial growth medium, confirming both that higher plant chloroplasts divide and that they can do so in the absence of other cellular components.

Not all plant cells contain chloroplasts. The various cell types found in higher plants all arise from a rapidly dividing, undifferentiated tissue called **meristem.** Meristem cells lack chloroplasts, but they have small organelles called **proplastids** that contain chloroplast DNA. Proplastids measure 1 μm or less in diameter and consist of an undifferentiated stroma surrounded by a double-membrane envelope (Figure 14-31). Depending on where they occur in the plant and how much light they receive, proplastids develop into different kinds of **plastids** designed to serve different functions (Figure 14-32). Some proplastids differentiate into *amyloplasts,* which are starch-filled particles that predominate in starchy vegetables such as potatoes. Other proplastids acquire red, orange, or yellow pigments, forming *chromoplasts* that give flowers and fruits their distinctive colors. Proplastids also develop into organelles that store protein (*proteinoplasts*) or lipids (*elaioplasts*). Since proplastids, amyloplasts, proteinoplasts, and elaioplasts are all colorless, they are collectively referred to as *leukoplasts* (leuko, *white*).

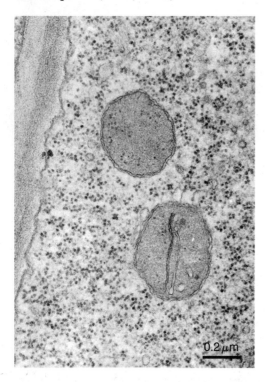

Figure 14-31 *Electron Micrograph of Proplastids in a Root Tip Cell of the Bean* **Phaseolus** *A few internal membranes can be seen in one of the proplastids. Micrograph by W. P. Wergin, courtesy of E. H. Newcomb.*

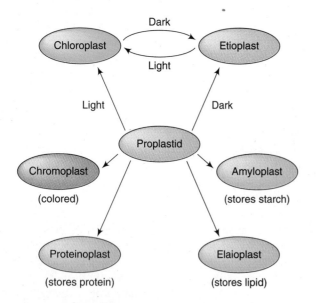

Figure 14-32 Pathways of Proplastid Differentiation *The developmental fate of a proplastid depends on where it occurs in the plant and how much light it receives.*

Although proplastids can develop into a variety of organelles, their main fate in photosynthetic tissues is to evolve into chloroplasts. The conversion of proplastids into chloroplasts requires light, which triggers an enlargement of the proplastid and the formation of membrane tubules that project from the inner membrane into the stroma. The tubules then spread into flat sheets that line up in parallel to form thylakoids. At the same time, chlorophyll and other components of the photosynthetic electron transfer chain are produced and incorporated into the thylakoids.

If developing plant seedlings are allowed to grow in the dark, a different sequence of events takes place. The proplastids still enlarge and membrane sheets invaginate from the inner membrane, but these sheets condense into a set of interconnected membrane tubules known as a **prolamellar body.** The membranes of a typical prolamellar body are arranged as highly ordered, hexagonal arrays of membrane tubules (Figure 14-33). At this stage the plastid is called an **etioplast.** Besides developing from proplastids in dark-grown seedlings, etioplasts also arise when green plants containing mature chloroplasts are transferred from light to darkness. Under such conditions, the thylakoid membranes break up into tubules that fuse together to produce a typical prolamellar body, converting the chloroplast into an etioplast.

Regardless of whether they arise from proplastids in dark-grown seedlings or from mature chloroplasts in plants switched from light to darkness, etioplasts serve the same function: They act as chloroplast precursors that can develop into functional chloroplasts when ex-

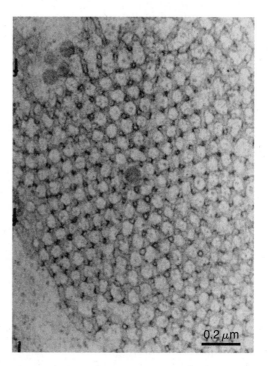

Figure 14-33 *Electron Micrograph of a Prolamellar Body* *In this etioplast of a dark-grown plant seedling, the hexagonal arrangement of membrane tubules that make up the prolamellar body is clearly evident. Courtesy of T. E. Weier.*

posed to light. Within a few hours after an etioplast is illuminated, the prolamellar body becomes transformed into an irregular mass of tubules; membrane sheets then grow out from the tubules and develop into thylakoid membranes (Figure 14-34). This conversion of an etioplast into a chloroplast is accompanied by the formation of components of the photosynthetic electron transfer chain, such as chlorophyll. Etioplasts contain high concentrations of *protochlorophyllide,* which is a metabolic precursor of chlorophyll. When etioplasts are illuminated, the absorption of light by protochlorophyllide triggers its rapid conversion to chlorophyll. After a lag period of several hours, more protochlorophyllide is produced for conversion into additional chlorophyll. This lag period can be abolished, however, by providing cells with *5-aminolevulinate,* a precursor of protochlorophyllide (Figure 14-35). Such findings suggest that the component being produced during the lag period is *ALA synthetase,* the enzyme that catalyzes the formation of 5-aminolevulinate.

Induction of ALA synthetase does not require the continuous presence of light. If etiolated leaves are exposed to red light for a few minutes and then returned to darkness for several hours, chlorophyll synthesis occurs without lag upon subsequent illumination. This means that once the induction of ALA synthetase and the resulting formation of 5-aminolevulinate are triggered by a brief exposure to red light, these processes continue to occur in the ensuing darkness. The induction process triggered by red light can be reversed by a subsequent exposure to far-red light (light whose wavelength is slightly longer than that of red light). This pattern of activation by red light and inhibition by far-red light is not unique to the induction of chlorophyll synthesis, but is shared by a variety of light-induced events in plants, including the formation of Calvin cycle enzymes. The light-absorbing pigment that mediates these effects is a protein called **phytochrome.** Phytochrome exists in two interconvertible forms: an inactive form that absorbs light maximally at 665 nm (red) and an active form that absorbs light maximally at 725 nm (far-red). Illuminating inactive phytochrome with red light converts it to the active form, while illuminating the active form of phytochrome with far-red light converts it back to inactive phytochrome:

$$\text{Phytochrome} \underset{\text{Far-red light (725 nm)}}{\overset{\text{Red light (665 nm)}}{\rightleftharpoons}} \text{Phytochrome}$$
$$\text{(inactive)} \qquad\qquad\qquad \text{(active)}$$

This interconversion between the active and inactive forms of phytochrome explains why treatment with red light triggers the induction of chlorophyll synthesis while exposure to far-red light halts the process. Because treating cells with actinomycin D to inhibit RNA synthesis has been found to block most of the light-induced activation of chlorophyll synthesis, activated phytochrome is thought to exert at least some of its effects by stimulating the transcription of specific genes.

ENDOSYMBIOSIS AND THE ORIGIN OF EUKARYOTIC ORGANELLES

Earlier in the chapter we learned that cellular life first appeared on earth over three billion years ago and then

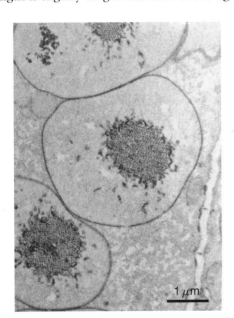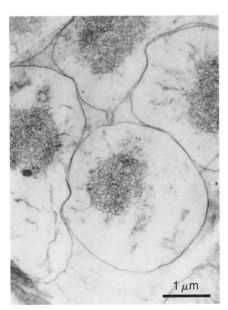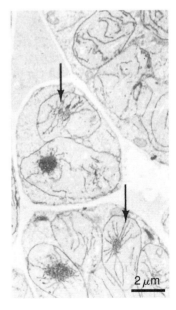

Figure 14-34 *The Conversion of an Etioplast into a Chloroplast* (Left) *Thin-section electron micrograph showing etioplasts in the leaf of a dark-grown bean plant. The regular arrangement of the prolamellar body is clearly evident.* (Middle) *In etioplasts exposed to red light for 10 seconds, the prolamellar body is beginning to disperse.* (Right) *In etioplasts exposed to red light for 5 hours, thylakoid membranes* (arrows) *have begun to grow out from the prolamellar body. Courtesy of L. Bogorad.*

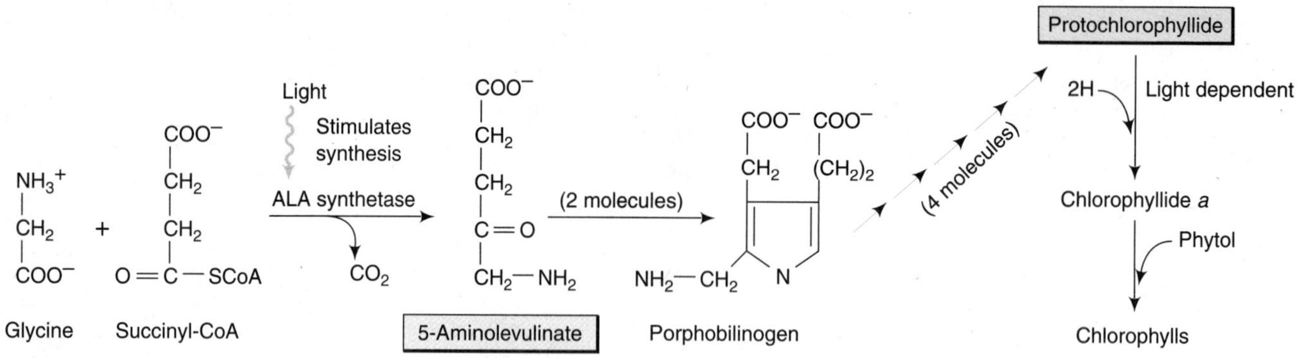

Figure 14-35 *Outline of the Pathway for Chlorophyll Formation* *Protochlorophyllide is the main pigment found in etioplasts. Light induces the conversion of protochlorophyllide to chlorophyll and also stimulates formation of the enzyme ALA synthetase, leading to the production of more protochlorophyllide.*

diverged into three lineages: eubacteria, archaebacteria, and eukaryotes. Eubacteria and archaebacteria are both prokaryotic cell types, retaining a relatively simple internal structure. Eukaryotes, on the other hand, evolved a complex system of internal membranes, including the nuclear envelope, endoplasmic reticulum, Golgi complex, peroxisomes, mitochondria and chloroplasts. Now that we have discussed the genetic systems of mitochondria and chloroplasts in some detail, we are in a better position to discuss current theories regarding the evolutionary origins of these organelles.

Mitochondria and Chloroplasts Exhibit Similarities to Bacterial Cells

The debate on the evolutionary origins of mitochondria and chloroplasts has a long history. As early as 1890, R. Altmann referred to mitochondria as primitive "organisms" analogous to bacteria, suggesting that they can exist as free-living forms outside the host cytoplasm. In 1910 C. Mereschovsky extended the idea, proposing that chloroplasts originally arose from a symbiotic association between primitive photosynthetic bacteria and nonphotosynthetic nucleated cells. A few years later J. F. Wallin claimed to support these theories with the demonstration that mitochondria can grow outside of living cells. However, his data were questionable and the whole notion of a bacterial origin of mitochondria and chloroplasts soon fell into disrepute.

After several decades of ridicule and neglect, these ideas regained prominence when it was discovered in the early 1960s that mitochondria and chloroplasts contain their own DNA. Especially influential was the gradual realization that nucleic acid and protein synthesis in mitochondria and chloroplasts exhibit many similarities to the comparable processes in bacterial cells. Among the similarities are the following:

1. The DNA of chloroplasts and mitochondria, like that of bacterial cells, is usually circular and is not complexed with histones, as is eukaryotic nuclear DNA.

2. Like bacterial ribosomes, mitochondrial and chloroplast ribosomes are usually smaller than the 80S cytoplasmic ribosomes found in eukaryotic cells. For example, chloroplast 70S ribosomes closely resemble bacterial ribosomes in size, and active ribosomes have been artificially assembled containing the small subunit of *Euglena* chloroplast ribosomes and the large subunit of *E. coli* ribosomes.

3. The base sequences of mitochondrial and chloroplast rRNAs resemble bacterial rRNAs more closely than eukaryotic cytoplasmic rRNAs.

4. Protein synthesis is initiated by formylmethionine in mitochondria, chloroplasts, and bacteria, but not in the cytoplasm of eukaryotes.

5. Antibiotics that block protein synthesis by bacterial ribosomes but not by eukaryotic cytoplasmic ribosomes (e.g., chloramphenicol) also inhibit chloroplast and mitochondrial protein synthesis. Conversely, inhibitors that are selective for eukaryotic cytoplasmic protein synthesis (e.g., cycloheximide) exert no effect on chloroplast or mitochondrial protein synthesis.

6. Protein synthesis initiation and elongation factors are interchangeable between mitochondria and bacteria, but not between mitochondria and the eukaryotic cytosol.

7. Certain viral messenger RNAs are efficiently translated by protein-synthesizing systems prepared from bacteria and mitochondria, but not by the cytoplasmic protein-synthesizing apparatus of eukaryotic cells.

8. Bacterial and mitochondrial RNA polymerases are inhibited by the antibiotic *rifampicin,* but eukaryotic nuclear RNA polymerases are not.

9. Eukaryotic cells contain distinct mitochondrial and cytoplasmic forms of *superoxide dismutase,* an enzyme that catalyzes the breakdown of the superoxide anion. The enzyme occurring in mitochondria resembles bacterial superoxide dismutase more closely than it does the eukaryotic cytoplasmic enzyme.

The preceding list of similarities between mitochondria, chloroplasts, and bacterial cells is not intended to imply the absence of differences. The ribosomes of animal mitochondria are significantly smaller than 70S bacterial ribosomes, and the subunits of animal mitochondrial ribosomes and bacterial ribosomes are not functionally interchangeable. Yet in spite of such qualifications, nucleic acid and protein synthesis in mitochondria and chloroplasts exhibits more similarities to the comparable processes occurring in bacterial cells than to the events taking place in the nucleus and cytoplasm of eukaryotic cells.

In addition to biochemical similarities, mitochondria and chloroplasts also resemble bacterial cells in size and shape. Moreover, the plasma membrane of bacteria and the inner membrane of mitochondria both possess protruding spheres that contain the ATP synthase involved in oxidative phosphorylation, and cyanobacteria have intracellular photosynthetic membranes that resemble the thylakoid membranes of chloroplasts.

The Endosymbiont Theory States That Mitochondria and Chloroplasts Evolved from Ancient Bacteria

The similarities between mitochondria, chloroplasts, and bacterial cells have led biologists to formulate the **endosymbiont theory,** which views mitochondria and chloroplasts as having evolved from bacteria that were ingested by primitive cells a billion or more years ago. An overview of the events that might have occurred is provided in Figure 14-36. This scenario begins by assuming that the absence of oxygen in the primitive atmosphere caused the first cells arising on earth to be anaerobic. Some of these primitive anaerobic cells subsequently developed pigments capable of absorbing light energy, thereby allowing them to trap energy from sunlight and use it to drive other energy-requiring reactions. In this way, the earliest photosynthetic cells arose. The first photosynthetic cells probably employed hydrogen sulfide or molecular hydrogen as electron donors, but later acquired the capacity to utilize electrons from water, releasing oxygen into the atmosphere.

As oxygen accumulated in the primitive earth atmosphere, some anaerobic cells may have adapted by moving to oxygen-free environments such as muddy flats, where they exist today as anaerobic bacteria. Others evolved into aerobic bacteria by developing the oxygen-utilizing pathways of electron transport and oxidative phosphorylation. At this point, the stage was set for the evolution of eukaryotic cells as we know them today. It has been hypothesized that the primitive ancestor of

today's eukaryotic cells, called a *protoeukaryote,* developed two features that distinguished it from other cells existing at the time: large size and the ability to ingest nutrients from the environment by phagocytosis. According to the endosymbiont theory, the large size and phagocytic capacity of the ancient protoeukaryote allowed it to ingest bacteria that later evolved into mitochondria and chloroplasts. Let us briefly consider how this evolution might have occurred.

Mitochondria Appear to Have Evolved from an Ancient Purple Eubacterium

The first step in the evolution of mitochondria is thought to have occurred when a primitive protoeukaryote ingested smaller aerobic bacteria by the process of phagocytosis. Some scientists believe that prior to this event, protoeukaryotes were anaerobic and depended entirely on glycolysis for energy. The ingested aerobic bacterium, with its ability to carry out electron transport and oxidative phosphorylation, would therefore provide larger amounts of useful energy than the host protoeukaryote could produce by glycolysis alone. In turn, the host cell provided protection and nutrients to the bacteria residing in its cytoplasm. As a result, the protoeukaryote and the ingested bacteria developed a stable *symbiotic relationship*— that is, a relationship that was mutually beneficial. As the host cell and its cytoplasmic bacteria adapted to living with one another over billions of years, the bacteria gradually lost functions that were no longer needed in their new cytoplasmic environment and eventually evolved into mitochondria. To determine what kind of ingested bacterium might have been involved in this scenario, the base sequence of contemporary mitochondrial rRNAs has been compared to the base sequences of various present-day bacterial rRNAs. The closest match occurs with *purple bacteria,* thereby implying that the ingested bacterium that gave rise to mitochondria was an ancient purple eubacterium.

Critics of the preceding scenario have claimed that it is inaccurate to view the eukaryotic cell as a primitive anaerobic cytoplasm containing aerobic mitochondria. They point out that the eukaryotic cytoplasm must be considered fundamentally aerobic because it contains oxygen-requiring enzymes such as superoxide dismutase and cytochrome P-450. The presence of oxygen-requiring features in eukaryotic cytoplasm suggests that the ancestral protoeukaryote was aerobic rather than anaerobic. But if the protoeukaryote were already capable of aerobic metabolism, what would be the advantage of acquiring an aerobic symbiont? A possible answer is that the aerobic metabolism of the primitive protoeukaryote was less efficient at trapping energy than the electron transport/oxidative phosphorylation pathway utilized by the ingested purple eubacteria. Hence the acquisition of aerobic bacteria capable of

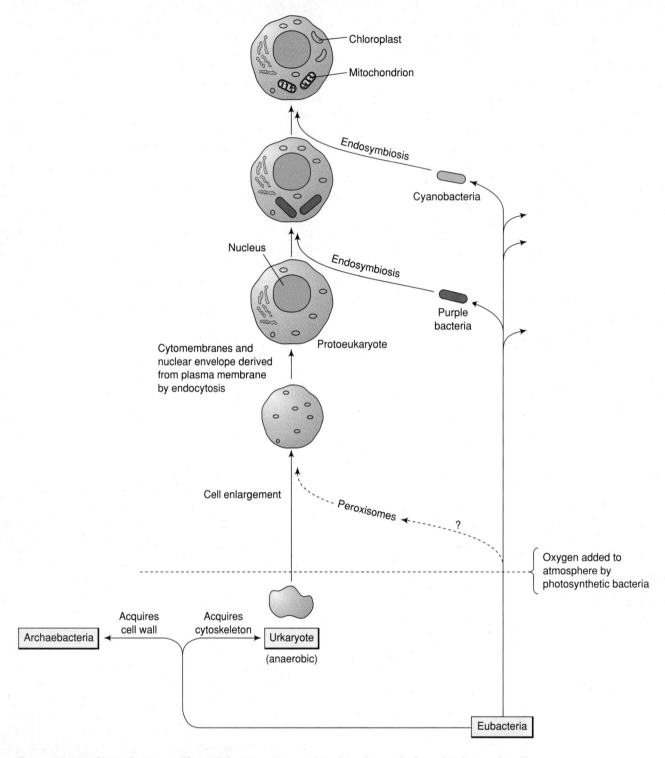

Figure 14-36 *The Main Events That Might Have Occurred During the Evolution of Eukaryotic Cells* *Considerable evidence exists for an endosymbiotic origin for mitochondria and chloroplasts. Endosymbiosis has also been considered as a possible mechanism for the origins of peroxisomes and cilia/flagella (not shown). The cytomembrane system, including the nuclear envelope, is thought to have arisen from infoldings of the plasma membrane produced by endocytosis.*

coupling electron transport to oxidative phosphorylation would have been advantageous even if the host protoeukaryote were already aerobic.

Chloroplasts Appear to Have Evolved from an Ancient Cyanobacterial Type of Eubacterium

Subsequent to ingesting the purple bacteria that were destined to evolve into mitochondria, ancient eukaryotes appear to have been involved in a second symbiotic event—this time leading to the development of chloroplasts. Because chloroplasts do not occur in all eukaryotic cells, this second symbiosis must have been restricted to a subgroup of ancient eukaryotic cells that eventually gave rise to the plant kingdom, and the ingested organisms must have been ancient photosynthetic bacteria that gave the host eukaryote the ability to photosynthesize. With time, the ingested photosynthetic bacteria gradually lost functions that were unnecessary in their new cytoplasmic environment, and thus the bacteria eventually evolved into chloroplasts as we know them today. To determine what kind of photosynthetic bacterium might have been involved in such a scenario, the base sequences of chloroplast rRNAs have been compared to the base sequences of various present-day bacterial rRNAs. Since the closest match occurs with members of the *cyanobacterial* group of eubacteria, it has been suggested that an ancient type of cyanobacterium gave rise to chloroplasts.

Because the chloroplasts found in different kinds of plants exhibit a variety of biochemical differences, the question has arisen as to whether chloroplasts evolved from more than one type of symbiotic bacterium. The main distinguishing feature between differing types of chloroplasts involves the light-absorbing pigments used during photosynthesis. The major chloroplast pigments are chlorophylls *a* and *b* in higher plants and green algae, chlorophylls *a* and *c* in brown algae, dinoflagellates, and diatoms, and chlorophyll *a* and phycobilins in red algae. Does this mean that the chloroplasts of the three groups of plants arose from different symbiotic bacteria? Examination of the photosynthetic pigments employed by present-day photosynthetic bacteria has revealed that *cyanobacteria* utilize chlorophyll *a* and phycobilins for photosynthesis, and a closely related group of bacteria called *prochlorophytes* use chlorophylls *a* and *b*. The rRNAs of both of types of bacteria are closely related to the rRNAs of contemporary chloroplasts, making either (or both) potential candidates for the ancient symbiont that evolved into chloroplasts.

Although the endosymbiont theory is based mainly on the biochemical similarities that are observed among mitochondria, chloroplasts, and bacteria, support of another kind is provided by present-day symbiotic events that resemble what may have occurred in the distant past. The search for relevant symbiotic events has led to the discovery that algae, dinoflagellates, diatoms, and photosynthetic bacteria reside as symbionts in the cytoplasm of cells occurring in more than 150 different kinds of invertebrates and protists (Figure 14-37). In many cases the cell wall of the ingested symbiont is no longer present, and in a few instances the reduction in structure goes even further, with only the chloroplasts of the symbiont remaining.

A striking example of the latter type of symbiosis occurs in certain marine slugs and related mollusks, where the cells lining the animal's digestive tract contain clearly identifiable chloroplasts. These chloroplasts originate from the green algae the mollusks feed upon, and continue to carry out photosynthesis long after being incorporated into the animal's cells. The carbohydrates produced by the photosynthetic process are even distributed as a source of nutrients to the rest of the organism. Although the chloroplasts do not grow and divide, they continue to function in the animal cell cytoplasm for many months. This discovery that chloroplasts derived from ingested algae are taken up by animal cells and assume a stable symbiotic relationship provides support for the idea that chloroplasts could have had their evolutionary origins in endosymbiotic associations between ancient photosynthetic bacteria and ancestral eukaryotes.

Did Peroxisomes, Cilia, and Flagella Also Arise by Endosymbiosis?

Although the evidence for an endosymbiotic origin is strongest in the case of mitochondria and chloroplasts, endosymbiosis has also been considered as a potential route for the origin of other organelles. One possibility involves peroxisomes, a family of membrane-bound organelles containing enzymes involved in the metabolism of hydrogen peroxide, lipids, and carbohydrates (page 349). Peroxisomes lack DNA and other components of a genetic system, so there is little direct evidence to support the idea that they arose by symbiosis. However, the chemical makeup of peroxisomal membranes is distinct from that of other cytoplasmic membranes, and new peroxisomes always arise by growth and division of existing peroxisomes, which is reminiscent of the behavior of mitochondria and chloroplasts. Thus it is conceivable that peroxisomes are remnants of ancient organisms that lost their original genetic material.

This scenario might help to explain the unusual metabolic functions carried out by peroxisomes. Most vertebrate cells get by perfectly well without peroxisomes, and even when they are present, the necessity of many of their enzymes is questionable. Humans that inherit genetic deficiencies in the peroxisomal enzyme catalase, for example, exhibit no obvious disease symptoms. The peroxisomal enzyme urate oxidase is absent in many organisms, including humans, and the function of the peroxisomal enzyme that oxidizes D-amino acids

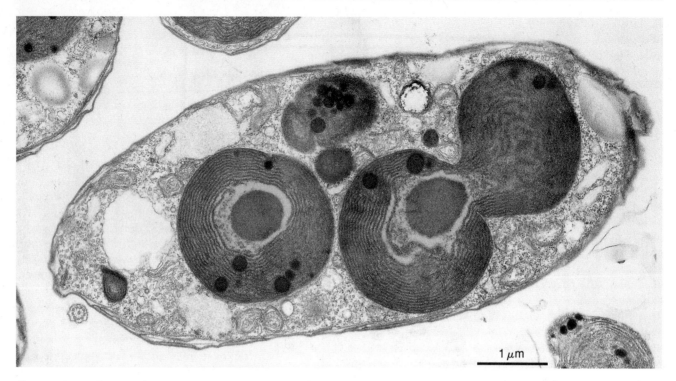

Figure 14-37 *Electron Micrograph Showing Photosynthetic Cyanobacteria Residing as Symbionts in the Cytoplasm of the Flagellated Protozoan* **Cyanophora**   *The symbiotic cyanobacteria are highlighted in color. Courtesy of W. T. Hall.*

is somewhat puzzling because it is the L-amino acids that are normally found in protein molecules. Finally, most of the metabolic events occurring in peroxisomes can take place elsewhere in eukaryotic cells.

The preceding considerations raise the possibility that present-day peroxisomes are remnants of ancient bacteria that invaded the eukaryotic cell lineage before the symbiotic events associated with the evolution of mitochondria and chloroplasts. Experimental support for this theory has come from studies in which the amino acid sequence of the peroxisomal enzyme *thiolase* was compared with the sequence of the same enzyme in bacteria; the sequence similarities observed in these studies were found to be compatible with the idea that peroxisomes originally evolved from endosymbiotic bacteria. The endosymbiotic event is thought to date back to the time when oxygen first appeared on earth. Oxygen can be toxic to cells because it reacts with organic molecules to form hydrogen peroxide. To protect against oxygen toxicity, an ancestral eukaryotic cell might have incorporated an ancient bacterium whose enzymes could couple the formation and breakdown of hydrogen peroxide to the oxidation of intermediates involved in carbohydrate metabolism. This set of metabolic pathways, now carried out by contemporary peroxisomes, is referred to as *peroxisomal respiration* (page 350).

Unfortunately, using peroxisomal respiration to oxidize carbohydrates also has a disadvantage: the energy released during peroxisomal respiration is dissipated as heat rather than being conserved in a useful chemical form. Thus when mitochondria appeared later in evolution, their ability to link oxidation to the formation of ATP permitted them to prevail over peroxisomes. As a result, peroxisomes may have gradually lost many of their enzymes and become less important to cellular metabolism, disappearing from most vertebrate cells. But the fact that peroxisomes now seem to be unnecessary for the survival of individual vertebrate cells does not mean that they are unimportant to organisms as a whole. Several lethal diseases observed in human infants are associated with missing or defective peroxisomes, suggesting that at least some peroxisomal functions are essential for normal development. For example, individuals with *Zellweger syndrome* are unable to produce peroxisomes and therefore exhibit fatal defects in pathways for lipid β-oxidation and ether lipid biosynthesis that normally occur in peroxisomes (page 350). These abnormalities bring about damage to the nervous system, liver, and kidneys that is usually fatal within a year after birth.

Besides peroxisomes, an endosymbiotic origin has also been considered for eukaryotic cilia and flagella. The main proponent of this view, Lynn Margulis, has hypothesized that cilia and flagella arose in the distant past from thin, undulating bacteria of the *spirochete* family. According to her theory, spirochetes attached to the surface of ancient eukaryotic cells in order to take advantage of food leaking through the host cell plasma membrane. Eventually the undulating spirochetes began to propel the host cell through the aqueous environment, thereby creating a mutually beneficial symbiotic relationship. Such relationships involving contemporary cells are well documented. For example, cells of the protist *Mixotricha paradoxa,* which live in the hindgut of Australian termites, are propelled by thousands of spirochetes attached to the cell surface.

The Cytomembrane System of Eukaryotic Cells May Have Arisen from Infoldings of the Plasma Membrane

In addition to the organelles whose origins have already been discussed, a variety of other membrane-bounded compartments also occur in eukaryotic cells. Included are the membrane sacs and tubules that make up the endoplasmic reticulum, Golgi complex, lysosomes, secretion vesicles, endosomes, storage and transport vesicles, and nuclear envelope. The preceding structures are united into a single **cytomembrane system** by their ability to establish connections with each other and with the plasma membrane through the budding and fusion of membrane vesicles. The existence of these interconnections between cytomembranes and the plasma membrane raises the possibility that cytomembranes first arose in ancient eukaryotes from infoldings of the plasma membrane.

What might have triggered the formation of cytomembranes from the plasma membrane? Thomas Cavalier-Smith has proposed that the key event occurred when an ancient eubacterial cell lost the ability to synthesize muramic acid, which is an essential component of the eubacterial cell wall (page 249). Without a cell wall for support, the eubacterium would have been more vulnerable to disruption by external forces. The wall-less cells eventually evolved two strategies for dealing with this problem and so diverged into two lines of descent. One line of descent was comprised of cells that acquired the capacity to produce a cell wall that does not require muramic acid; this group of cells became the archaebacteria. The other line of descent involved cells that maintained their structural integrity by developing an internal cytoskeleton instead of a cell wall. Such cells initiated the eukaryotic cell lineage.

The absence of a cell wall in these ancestral eukaryotes would have exposed the plasma membrane directly to the surrounding environment, allowing endocytosis and exocytosis to develop as mechanisms for moving materials into and out of cells. During endocytosis, infoldings of the plasma membrane would have created membrane channels and vesicles that budded off and entered the cell interior. Gradually these internalized membranes could have given rise to various kinds of cytomembranes, including the nuclear envelope. Thus a membrane-enclosed nucleus could have developed in the eukaryotic lineage before the inges-

tion of symbionts destined to evolve into mitochondria and chloroplasts. In support of this idea, Cavalier-Smith has pointed to the existence of a contemporary eukaryote called *Giardia,* which is an anaerobic single-celled eukaryote that contains a nucleus but no mitochondria or chloroplasts. Thus *Giardia* may be a descendant of an ancient eukaryote that contained a nucleus but had not yet ingested the bacteria that would evolve into mitochondria and chloroplasts.

In the final analysis, many of our ideas about how eukaryotic cells acquired their complex array of organelles over billions of years of evolution must remain speculative because the events under consideration are inaccessible to direct laboratory experimentation. But in spite of this reservation, it is impressive that a relatively coherent picture of the transition from prokaryotes to eukaryotes can be created using two relatively simple concepts: (1) the formation of cytomembranes from infoldings of the plasma membrane produced by the process of endocytosis; and (2) the endosymbiosis of mitochondria, chloroplasts, and perhaps peroxisomes and/or cilia and flagella. Although other ideas have also been put forth to explain how eukaryotic organelles might have arisen, one of the strengths of the proposed roles of endocytosis and endosymbiosis in the evolutionary origins of eukaryotic organelles is that they both involve activities that can be observed in contemporary cells.

SUMMARY OF PRINCIPAL POINTS

- Experiments involving simulated primitive earth conditions have shown that the organic molecules required by living cells could have arisen spontaneously on the primitive earth and acquired both catalytic and information-transferring properties. Proteinoids synthesized under such conditions spontaneously form vesicles called microspheres that carry out primitive metabolic functions.
- The ability of RNA to function as both a catalyst and an information-transferring molecule has led to the idea that self-replicating RNA molecules dominated early cellular evolution. However, it is still possible that primitive protein catalysts were needed to set the stage for this "RNA world."
- Fossil evidence suggests that the first cells appeared on earth about 3.5 billion years ago and diverged into three major lineages: eubacteria, archaebacteria, and eukaryotes.
- Clues to the evolutionary origins of eukaryotic organelles have come from the discovery that mitochondria and chloroplasts contain their own DNA molecules, which are transcribed into ribosomal, transfer, and messenger RNAs. Mitochondrial DNAs typically code for at least 2 mitochondrial rRNAs, 22 mitochondrial tRNAs, cytochrome *b,* and subunits of cytochrome oxidase, mitochondrial ATP synthase, and in many cases, NADH dehydrogenase. Chloroplast DNAs typically code for 120 or more products, including 4 rRNAs, 30 or more tRNAs, 20 ribosomal proteins, 4 subunits of chloroplast RNA polymerase, the large subunit of rubisco, and several

dozen polypeptide components of the cytochrome b_6f complex, photosystems I and II, and chloroplast ATP synthase.

• Genetic information encoded in mitochondrial and chloroplast DNAs is transcribed and translated using RNA polymerases, ribosomes, and transfer RNAs that are different from those employed in the transcription and translation of information encoded in nuclear genes. Mitochondria also employ a genetic code that is slightly different from the universal code employed elsewhere by eukaryotic and prokaryotic cells. Identification of the proteins synthesized by mitochondria and chloroplasts has been aided by the fact that chloramphenicol selectively inhibits protein synthesis by mitochondrial and chloroplast ribosomes, and cycloheximide selectively inhibits protein synthesis by eukaryotic cytoplasmic ribosomes.

• Although the existence of mitochondrial and chloroplast genes allow these organelles to manufacture some of their own polypeptides, most mitochondrial and chloroplast polypeptides are encoded by nuclear genes, synthesized on free cytoplasmic ribosomes, released into the cytosol, and imported into mitochondria and chloroplasts.

• New mitochondria typically arise by the growth and division of existing mitochondria or promitochondria. New chloroplasts arise by growth and division of either existing chloroplasts or precursor organelles known as proplastids and etioplasts.

• According to the endosymbiont theory, mitochondria and chloroplasts arose from eubacteria that were ingested by ancestral eukaryotes a billion or more years ago. This theory is supported by the discovery that the genetic systems of mitochondria and chloroplasts exhibit many similarities to those of bacterial cells. Further support for the concept of endosymbiosis is provided by present-day examples of symbiotic relationships between unicellular photosynthetic organisms and animal cells. Through comparisons of rRNA base sequences in mitochondria, chloroplasts, and contemporary bacteria, it has been concluded that mitochondria most likely originated from an ancient purple eubacterium, and chloroplasts apparently evolved from an ancient cyanobacterial type of eubacterium.

• Although the evidence is less compelling, the possibility of an endosymbiotic origin has also been considered for eukaryotic cilia, flagella, and peroxisomes.

• The cytomembranes of eukaryotic cells (including the nuclear envelope) are thought to have evolved from infoldings of the plasma membrane produced by endocytosis, an event that may have occurred in an ancient eukaryote prior to the acquisition of mitochondria and chloroplasts.

SUGGESTED READINGS

Books

Boffey, S. A., and D. Lloyd (1988). *The Division and Segregation of Organelles,* Cambridge University Press, Cambridge, England.

Crick, F. (1982). *Life Itself: Its Origin and Nature,* Simon and Schuster, New York.

Day, W. (1984). *Genesis on Planet Earth,* 2nd Ed., Yale University Press, New Haven, CT.

de Duve, C. (1991). *Blueprint for a Cell: The Nature and Origin of Life,* Carolina Biological Supply, Burlington, NC.

Fox, S. (1988). *The Emergence of Life: Darwinian Evolution from the Inside,* Basic Books, New York, NY.

Gesteland, R. F., and J. F. Atkins, eds. (1993). *The RNA World,* Cold Spring Harbor Laboratory Press, Cold Spring Harbor, NY.

Gillham, N. (1994). *Organelle Genes and Genomes,* Oxford University Press, NY.

Margulis, L. (1993). *Symbiosis in Cell Evolution,* 2nd Ed., Freeman, New York.

Articles

Attardi, G., and G. Schatz (1988). Biogenesis of mitochondria, *Annu. Rev. Cell Biol.* 4:289-333.

Cavalier-Smith, T. (1987). The origin of eukaryotic and archaebacterial cells, *Ann. NY Acad. Sci.* 503:17-54.

Clayton, D. A. (1991). Replication and transcription of vertebrate mitochondrial DNA, *Annu. Rev. Cell Biol.* 7:453-478.

Gray, M. W. (1989). The evolutionary origins of organelles, *Trends Genet.* 5:294-299.

Grivell, L. A. (1983). Mitochondrial DNA, *Sci. Amer.* 248 (March):78-89.

Horgan, J. (1991) In the beginning. . . , *Sci. Amer.* 264 (February):116-125.

Igual, J. C., C. Gonzalez-Bosch, J. Dopazo, and J. E. Perez-Ortin (1992). Phylogenetic analysis of the thiolase family. Implications for the evolutionary origin of peroxisomes, *J. Mol. Evol.* 35:147-155.

Joyce, G. F. (1989). RNA evolution and the origins of life, *Nature* 338:217-224.

Kabnick, K. S., and D. A. Peattie (1991). *Giardia:* a missing link between prokaryotes and eukaryotes, *Amer. Sci.* 79:34-43.

Miller, S. L. (1987). Which organic compounds could have occurred on the prebiotic earth? *Cold Spring Harbor Symp. Quant. Biol.* 52:17-27.

O'Neill, L., M. Murphy, and R. B. Gallagher (1994). What are we? Where did we come from? Where are we going?, *Science* 263:181-183.

Orgel, L. E. (1994). The origin of life on earth, *Sci. Amer.* 271 (October): 76-83.

Palmer, J. D. (1990). Contrasting modes and tempos of genome evolution in land plant organelles, *Trends Genet.* 6:115-120.

Rochaix, J.-D. (1992). Post-transcriptional steps in the expression of chloroplast genes, *Annu. Rev. Cell Biol.* 8:1-28.

Schuster, W., and A. Brennicke (1994). The plant mitochondrial genome: Physical structure, information content, RNA editing, and gene migration to the nucleus, *Annu. Rev. Plant Physiol. Plant Mol. Biol.* 45:61-78.

Taylor, W. C. (1989). Regulatory interactions between nuclear and plastid genomes, *Annu. Rev. Plant Physiol. Plant Mol. Biol.* 40:211-233.

Vidal, G. (1984). The oldest eukaryotic cells, *Sci. Amer.* 244 (February):48-57.

Woese, C. R. (1981). Archaebacteria, *Sci. Amer.* 244 (June):96-122.

Woese, C. R., O. Kandler, M. L. Wheelis (1990). Towards a natural system of organisms: proposal for the domains Archaea, Bacteria, and Eucarya, *Proc. Natl. Acad. Sci. USA* 87:4576-4579.

Part 3

*Formation of
Specialized Cells*

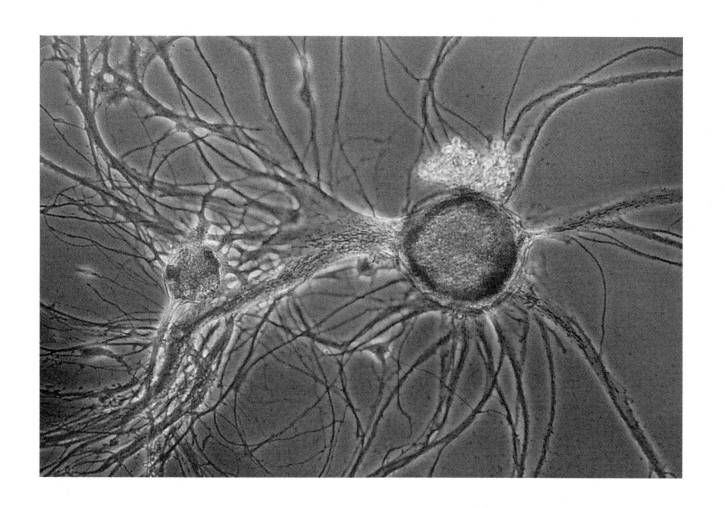

Chapter 15

Gametes, Fertilization, and Early Development

One of the most awe-inspiring of all biological events is triggered every time an egg cell is fertilized. Through a complex series of divisions, differentiations, and interactions, a single fertilized egg gives rise to trillions of new cells of differing types, organized in precise patterns to form the tissues and organs of the newly forming organism. Given the enormity of the task, it seems almost miraculous that each type of fertilized egg cell—be it a frog egg, a sea urchin egg, a human egg, an egg derived from a flowering plant, or any other type of egg—manages to generate a complete new organism of exactly the right type.

Long ago it was believed that every fertilized egg contains a miniature version of a complete adult organism that simply grows in size as development proceeds. But we now know that embryonic development is considerably more complex than this naive picture implies. No visible sign of a mature organism exists within a fertilized egg; instead, the innumerable steps involved in producing a new organism are programmed within the egg cell in ways that are just beginning to be deciphered. In the present chapter we will examine some of the mechanisms that are involved in this developmental process, looking first at how a fertilized egg is created from the fusion of male and female gametes and then examining the principles that guide the development of the fertilized egg into a new organism.

GAMETOGENESIS

In organisms that reproduce sexually, offspring are created through the fusion of haploid cells known as male and female **gametes.** The female gametes of animals and plants are referred to as *ova* or *eggs,* whereas male gametes are called *sperm.* Sperm cells are usually quite small and may be inherently motile, whereas egg cells tend to be quite large and nonmotile. For example, in sea urchins the volume of an egg cell is more than 10,000 times greater than that of a sperm cell; in birds and amphibians, which have massive yolky eggs, the size difference is even greater. But in spite of their differing sizes, sperm and egg cells bring equal amounts of nuclear DNA to the offspring; the increased cytoplasmic volume of the egg does mean, however, that the egg contributes most of the cytoplasmic organelles and cytoplasmic DNA to the offspring.

During the process of gamete formation, or **gametogenesis,** diploid cells known as **primordial germ cells** are transformed by meiosis and accompanying cell differentiation into haploid egg and sperm cells (Figure 15-1). Since the reduction in chromosome number from haploid to diploid that occurs during meiosis has already been described in Chapter 12, we will restrict our discussion of gametogenesis to the morphological and biochemical changes that accompany the formation of egg and sperm cells.

Oogenesis Is Delayed at Prophase I to Allow Time for Oocyte Growth

In most animals, the process of female gamete formation or **oogenesis** takes place in the ovary and begins with a *proliferation phase,* in which the female primordial germ cells (*oogonia*) divide by mitosis to increase their numbers. In humans this proliferation phase is completed by the end of embryonic development, and therefore all the germ cells capable of differentiating into eggs for a woman's entire lifetime are present by the time of birth. Other organisms, such as amphibians and birds, continue to produce new oogonia by mitotic division throughout most of their adult lives.

After mitotic proliferation the oogonia embark upon the process of meiosis, at which point they are renamed *oocytes.* Meiosis is accompanied by genetic recombination and reduction of the chromosome number to the haploid level. In nonmammalian oocytes, meiosis also involves a dramatic increase in cell size. To provide time for this *growth phase* of oogenesis, meiosis is delayed during late pachytene or early diplotene of prophase I. The length of the delay varies widely among species, from a few days to several years. This variability reflects differences in how much growth is required, which is in turn dictated by differences in the way eggs develop after fertilization. In vertebrates whose eggs develop outside the organism (e.g., amphibians and birds), the developing oocyte must grow to a size that is large enough to store all the nutrients required for embryonic development. Hence the growth

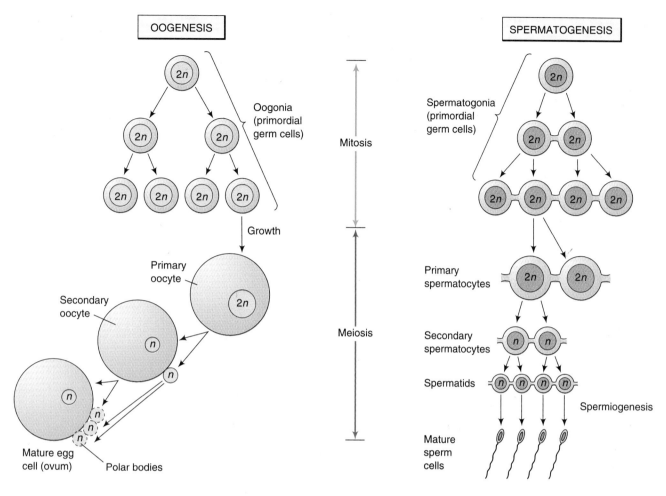

Figure 15-1 *Comparison of Female and Male Gametogenesis in Animals* *During oogenesis, only one of the four haploid products of meiosis yields a functional gamete. Also note that during spermatogenesis, the developing gametes remain connected to one another by cytoplasmic bridges.*

phase of amphibian oocytes produces up to a thousandfold increase in cell mass accompanied by more than a tenfold increase in cell diameter. Oocyte growth is even more dramatic in birds, where it generates gigantic cells like chicken and ostrich eggs. In contrast, mammalian oocytes grow relatively little during oogenesis because the fertilized egg develops within the female reproductive tract and hence does not need to store many nutrients.

Lampbrush Chromosomes and Multiple Nucleoli Produce RNA That Is Stored in the Egg for Use after Fertilization

In oocytes that undergo an extensive growth phase, cell growth is accompanied by a dramatic enlargement of the nucleus and a marked increase in the rate of gene transcription. The intense transcriptional activity causes diplotene chromosomes to acquire an unusual appearance that is unique to this particular stage of meiosis. The term **lampbrush chromosome** was coined during the late nineteenth century to refer to such chromosomes because their shape is reminiscent of the brushes

used at that time to clean the chimneys of oil-burning lamps. Each lampbrush chromosome consists of a long central backbone from which numerous loops project (Figure 15-2). The loops, which occur in pairs projecting from opposite sides of each chromosome, originate from beadlike swellings in the central backbone known as *chromomeres*.

The relationship between the loops and the central backbone of lampbrush chromosomes has been extensively investigated in the laboratories of H. Callan and J. Gall. In one set of experiments, isolated lampbrush chromosomes were stretched until a gap was created in the central backbone at one of the chromomeres. After such treatment, the pair of loops that normally emerge from the chromomere could be seen spanning the break. As is illustrated in Figure 15-3, this finding suggests that the central backbone consists of two closely joined fibers, each giving rise to one of the loops. Since lampbrush chromosomes occur at a stage when each chromosome is still comprised of two chromatids, it has been concluded that the two fibers that make up the central backbone represent chromatids.

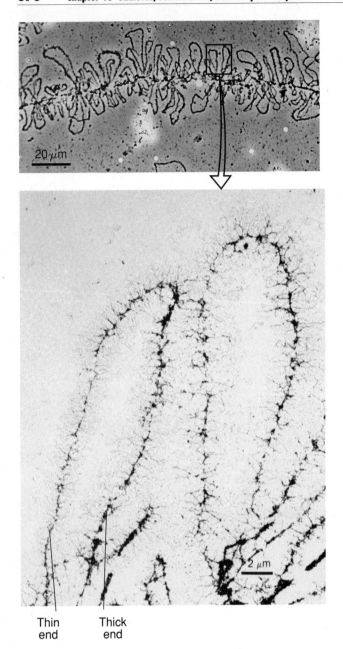

20 μm

2 μm

Thin
end

Thick
end

Figure 15-2 *Microscopic Appearance of Lampbrush Chromosomes* (Top) *Phase-contrast microscopy of a lampbrush chromosome isolated from a newt. Note the numerous loops extending from the central axis.* (Bottom) *Electron micrograph of individual lampbrush chromosome loops, showing tiny RNA-containing fibrils emerging from each loop. The fibrils occur in gradually increasing concentration as one proceeds around the loop, giving the loop a "thin" end and a "thick" end. Courtesy of J. G. Gall (top) and O. L. Miller, Jr. (bottom).*

Most of the loops emerging from lampbrush chromosomes are covered with thin RNA-containing fibrils that occur in gradually increasing concentration as one proceeds around the loop, giving the loop a "thin" end and a "thick" end (see Figure 15-2). The idea that these filaments represent newly forming RNA molecules is supported by autoradiographic studies of cells incu-

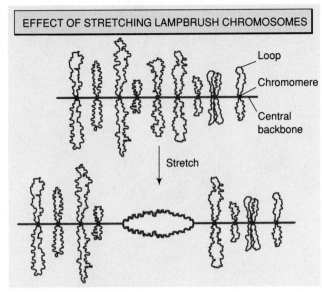

EFFECT OF STRETCHING LAMPBRUSH CHROMOSOMES

Loop

Chromomere

Central
backbone

Stretch

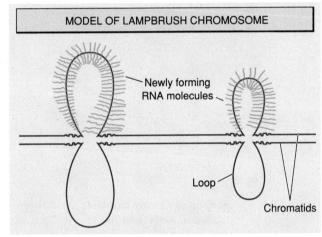

MODEL OF LAMPBRUSH CHROMOSOME

Newly forming
RNA molecules

Loop

Chromatids

Figure 15-3 *Organization of Lampbrush Chromosomes* (Top) *When lampbrush chromosomes are stretched, a gap occurs in the backbone at one of the chromomeres. The pair of loops that normally emerge from the chromomere can be seen spanning the break.* (Bottom) *The preceding observations suggest that the central backbone of each lampbrush chromosome is comprised of two chromatids and that each pair of loops contains one loop derived from each chromatid. Normally the two chromatids are too closely joined in the region of the backbone to be distinguished from each other.*

bated with³H-uridine, which have revealed that the loops represent active sites of RNA synthesis. The most straightforward interpretation of the asymmetrical arrangement of RNA around the loop is that the thin end represents the site at which gene transcription is being initiated; the newly forming RNA molecules are therefore small in this region, but as one proceeds around the loop, transcription has proceeded further and the RNA molecules are therefore larger. Thus each loop behaves like a DNA **transcription unit**—that is, a region bounded by an initiation site for transcription at one end and a termination site at the other end.

The RNA manufactured by lampbrush chromosomes consists mainly of mRNA molecules that are bound to proteins to form inactive ribonucleoprotein (RNP) particles, which are then stored in the oocyte for use after fertilization. The proteins involved in the storage mechanism include transcription factors such as *FRGY2,* which stimulates the expression of genes that are selectively transcribed during the growth phase of oogenesis. After these genes are transcribed, FRGY2 binds to the resulting mRNA molecules and prevents them from being translated in the oocyte.

In addition to its role in synthesizing and storing mRNA molecules for use after fertilization, the growth phase of oogenesis is also a time when large numbers of ribosomes are produced. Because the nucleus must supply ribosomal RNA to an oocyte whose cytoplasmic volume is much greater than that of most typical cells, a severe strain is placed on ribosomal gene transcription in the nucleolus. To accommodate this increased demand, the genes coding for rRNA are selectively amplified (page 477). The result can be a thousandfold or more increase in the number of rRNA genes, accompanied by a comparable increase in the number of nucleoli.

Yolk Platelets Store Nutrients for Use by the Embryo after Fertilization

During the growth phase of oogenesis, oocytes often store vast quantities of protein, lipid, and carbohydrate destined for later use by the embryo. Though some of this stored material is manufactured by the oocyte itself, much of it can be derived from external sources. For example in amphibians and birds, where the egg provides virtually all the nutrients needed during early embryonic development, the liver plays a central role in supplying the required nutrients to the developing oocyte. In response to stimulation by female sex hormones (i.e., estrogen), the liver manufactures large amounts of **vitellogenin,** a macromolecular complex consisting of the phosphoprotein *phosvitin* joined to the lipoprotein *lipovitellin.* Vitellogenin is secreted into the bloodstream and passes to the ovary, where growing oocytes incorporate it by receptor-mediated endocytosis (page 295). Endocytic vesicles containing vitellogenin accumulate within the oocyte and eventually fuse together, producing nutrient-containing vesicles known as **yolk platelets** (Figure 15-4). The term "yolk" does not refer to a single type of nutrient molecule, but rather to the varying mixtures of proteins, lipids, and polysaccharides that are stored in platelets. Besides stored nutrients, yolk platelets also contain hydrolytic enzymes that are activated upon fertilization, thereby allowing the yolk to be digested into chemical building blocks for use by the developing embryo.

The liver is not the only source of nutrients stored in developing oocytes. Additional materials can also be provided by accessory cells called **follicle cells,** which surround most animal oocytes (Figure 15-5). The surfaces of both oocytes and follicle cells are covered with numerous microvilli, increasing the area of contact between the two cell types and thereby facilitating the transport of nutrient materials from the follicle cells to the oocyte. The nutritional role of follicle cells is not limited to oocytes that store nutrients in yolk platelets. Mammalian oocytes, for example, do not produce yolk platelets because the developing embryo will receive its nourishment directly from the mother's bloodstream and hence has relatively little need for food reserves; yet most of the nutrients utilized by the growing mammalian oocyte are still provided by follicle cells.

Another type of accessory cell, termed the **nurse cell,** supports oocyte growth in invertebrates such as insects, annelids, and mollusks. Unlike follicle cells, nurse cells are derived from the same cell lineage as the oocyte itself. For example, in fruit flies four mitotic divisions of an oogonium yield one oocyte and 15 nurse cells. Nurse cells are connected to the developing oocyte by cytoplasmic bridges that allow a variety of nutrients to be transferred. The nutritional contribution of nurse cells to oocyte growth is generally greater than that of follicle cells; in some cases, an entire nurse cell may even be engulfed by an oocyte.

Not all oocytes depend on other tissues to manufacture their stored nutrients. In some amphibians, birds, and invertebrates, yolk proteins are synthesized in the rough endoplasmic reticulum of the oocyte itself and are then transported to the Golgi complex for packaging into platelets (Figure 15-6). Besides yolk platelets, the endoplasmic reticulum is also involved in producing membrane-enclosed **pigment granules** and **cortical granules** that become localized near the periphery of the cell, creating a region called the oocyte *cortex* (Figure 15-7). The cortical granules contain proteases that are destined to play an important role in the early stages of fertilization (page 688).

Oocytes Develop into Asymmetric Cells That Acquire Various Kinds of Protective Coats

The biochemical and morphological changes that occur in oocytes exhibiting an extensive growth phase tend to impart an obvious asymmetry to the cell (Figure 15-8). One end of the oocyte, designated the **vegetal pole,** acquires most of the yolk platelets and stored nutrients. The opposite or **animal pole** of the cell contains little in the way of stored nutrients, but it contains the nucleus and is enriched in ribosomes, endoplasmic reticulum, mitochondria, and pigment granules. Even particular kinds of mRNAs become localized to specific regions of the cytoplasm. Studies involving the use of inhibitors of cytoskeletal function have revealed that microtubules are involved in moving specific mRNAs to particular regions of the cell and that actin filaments

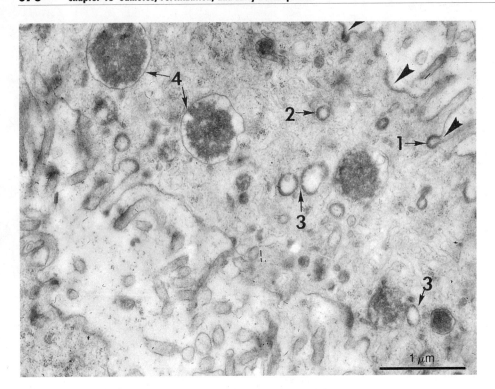

Figure 15-4 *Formation of Yolk Platelets via Pinocytosis* *In this electron micrograph of an oocyte from the milkweed bug, the arrowheads in the upper-right-hand corner point to regions where pinocytosis is occurring at the cell surface. The numbered arrows point to sequential stages in which the pinocytic vesicles (1) invaginate, (2) bud off from the cell surface, (3) aggregate, and (4) fuse together to form yolk platelets. Courtesy of R. G. Kessel.*

help to maintain this asymmetric localization once it has been established.

As oocytes enlarge and develop an asymmetric distribution of cytoplasmic constituents, changes in the organization of the cell surface also occur. Most striking is the formation of an external coat of polysaccharide-rich material known as the **vitelline envelope** in nonmammalian oocytes or the **zona pellucida** ("transparent zone") in mammalian oocytes. This envelope, which protects the egg from chemical and physical injury, often remains intact well past the time of fertilization. When the external envelope is produced by the oocyte itself, it is referred to as a *primary coat;* when formed by the follicle cells, it is called a *secondary coat;* and when produced by the oviduct or other maternal tissue, it is termed a *tertiary coat.* The appearance of these various layers differs significantly among species. Especially variable is the tertiary coat, which ranges from the jelly layer of invertebrate and amphibian eggs to the egg white and external shell of chicken eggs. Impervious shells such as those found in chicken eggs are added after fertilization so that they will not interfere with the entry of sperm.

The Activation of MPF Triggers the Maturation Phase of Oogenesis

After the growth phase of oogenesis has been completed, the oocyte is ready to resume meiosis. Rather than entering into this *maturation phase* of oogenesis spontaneously, the oocyte usually remains arrested in prophase I until it is triggered by an appropriate hormone. In amphibians, the resumption of meiosis is triggered by the steroid hormone *progesterone,* which influences the activity of a protein kinase called **maturation promoting factor** or **MPF.** In Chapter 12, we learned that MPF controls *mitotic* cell division by triggering the transition from G_2 to M phase (page 524); MPF also controls *meiotic* cell division by triggering the transition from prophase I to metaphase I. Progesterone exerts its control over MPF by stimulating the formation of $p39^{mos}$, a protein kinase that inhibits the breakdown of *cyclin,* which is a subunit of the MPF molecule. By inhibiting the breakdown of cyclin, $p39^{mos}$ increases the amount of MPF in the cell, thereby allowing oocytes that had been arrested in prophase I to resume meiosis.

In response to the progesterone-induced activation of MPF, the nuclear envelope breaks down into tiny vesicles, spindle microtubules are assembled, and the first meiotic division, which had been delayed at prophase I to permit the growth phase to occur, is completed. The cytoplasmic cleavage associated with the first meiotic division is highly asymmetrical; instead of dividing the oocyte into two equal cells, the cleavage furrow pinches off one set of chromosomes into a small cytoplasmic protrusion called the **polar body,** which forms near the animal pole (Figure 15-9). The resulting

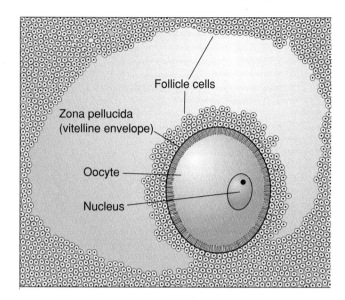

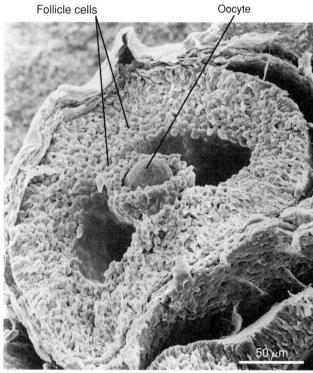

Figure 15-5 ***The Relationship between Follicle Cells and Oocytes*** (Top) *The diagram illustrates the relationship between an oocyte and its surrounding follicle cells in a typical mammalian ovary.* (Bottom) *In this scanning electron micrograph of a rat ovary, the oocyte and follicle cells are clearly visible. Micrograph courtesy of P. Bagavandoss.*

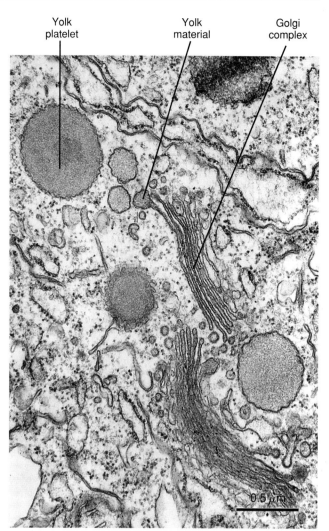

Figure 15-6 ***Formation of Yolk Platelets by the Golgi Complex*** *In this electron micrograph of a jellyfish oocyte, developing yolk platelets are seen closely associated with two Golgi complexes. The dilated ends of the Golgi cisternae also contain yolk material. Courtesy of R. G. Kessel.*

cell is referred to as a *secondary oocyte* to distinguish it from the *primary oocytes* that have not yet completed the first meiotic division (see Figure 15-1, *left*). The secondary oocyte then goes through a second meiotic division, producing another polar body and a mature **egg cell;** at this stage the first polar body may either divide, remain quiescent, or disintegrate. The polar bodies appear to play no subsequent role in development and

eventually disappear. The advantage of having only one of the four haploid products of meiosis develop into an egg is that the cytoplasm that would otherwise have been distributed among four cells is instead concentrated into one cell, thereby maximizing the content of stored nutrients in each egg.

In some organisms the maturation sequence proceeds rapidly to completion; in others, however, it halts at an intermediate stage and is not completed until after fertilization. In vertebrate eggs, for example, the second meiotic division is often arrested at metaphase II until fertilization takes place. Recent evidence suggests that the molecule responsible for this metaphase arrest is p39[mos], the same protein kinase that regulates the transition from prophase to metaphase; p39[mos] inhibits the transition from metaphase to anaphase by preventing

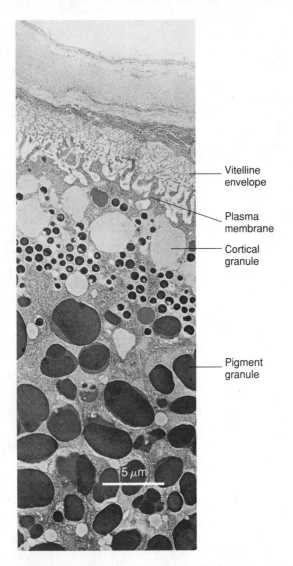

Figure 15-7 *Electron Micrograph of the Surface Region (Cortex) of an Amphibian Oocyte* *Numerous cortical granules and pigment granules are seen lying directly beneath the plasma membrane. Courtesy of J. N. Dumont.*

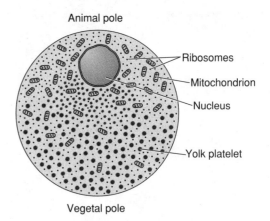

Figure 15-8 *Asymmetric Organization of an Amphibian Oocyte* *The vegetal pole contains most of the yolk platelets and stored nutrients; the animal pole contains the nucleus and is enriched in ribosomes, mitochondria, and pigment granules.*

Spermatogenesis Begins with the Formation of Haploid Spermatids That Remain Connected to One Another by Cytoplasmic Bridges

Male gametogenesis differs in several ways from the process by which eggs are formed during oogenesis. The male gametes of animals, called **sperm cells,** are produced in the testes by the process of **spermatogenesis.** In mammals, spermatogenesis occurs in close association with **Sertoli cells** that provide protection and nutritional support for the developing sperm; in fact, the various stages of sperm development usually occur while the newly forming sperm lie embedded within recesses of the Sertoli cell surface (Figure 15-10). Spermatogenesis begins with mitotic division of the male primordial germ cells or *spermatogonia,* which continue to proliferate throughout most of the organism's adult lifetime, thereby allowing sexually mature males to continually produce large numbers of new sperm cells.

Some of the cells generated during the mitotic proliferation of spermatogonia eventually differentiate into *primary spermatocytes,* which then embark upon meiosis. The first meiotic division transforms primary spermatocytes into *secondary spermatocytes,* which are subsequently converted by the second meiotic division into **spermatids.** In contrast to the comparable phase of oogenesis, no dramatic growth or morphological changes accompany meiosis in male gametogenesis. However, the mitotic and meiotic divisions that lead to the formation of spermatids are accompanied by an unusual type of cytokinesis that fails to completely divide the cytoplasm; as a result, the spermatids remain connected to each other by cytoplasmic bridges (see Fig-

the degradation of cyclin that normally occurs at the onset of anaphase (page 542).

The mature egg cell produced by the process of meiosis is a highly differentiated cell that is specialized for the task of producing a new organism in much the same sense that a muscle cell is specialized to contract, or a red blood cell is specialized to transport oxygen. This inherent specialization of the egg is vividly demonstrated by the observation that even in the absence of fertilization by a sperm cell, many kinds of animal eggs can be stimulated to develop into a complete embryo by artificial treatments as simple as a pinprick. Hence everything needed for programming the early stages of development must already be present in the egg.

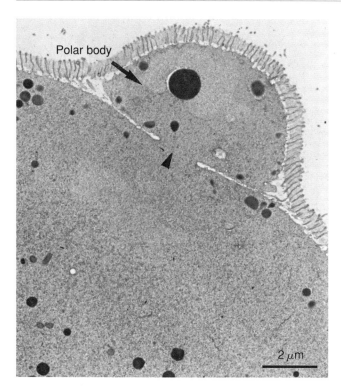

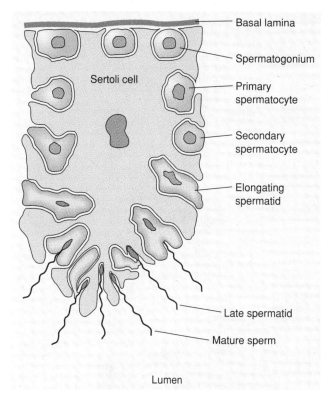

Figure 15-9 *Electron Micrograph of a Polar Body Forming at the Surface of an Egg Cell* A small region of cytoplasmic continuity (arrowhead) *still connects the developing polar body to the egg cell. Courtesy of E. Anderson.*

ure 15-1). Unlike female gametogenesis, where meiosis produces egg cells that are functionally mature, the spermatids produced by meiosis in males must undergo further differentiation to create functional sperm (Figure 15-11). The process of transforming spermatids into mature sperm cells is called **spermiogenesis.** During spermiogenesis the spermatid loses most of its cytoplasm and the remainder of the cell differentiates into two structures, the *head* and the *tail.*

The Sperm Head Contains a Haploid Nucleus and an Acrosome

During the early stages of spermiogenesis, the chromatin fibers of the spermatid become tightly packed into a condensed mass, leading to a marked reduction in nuclear volume (Figure 15-12). Chromatin condensation is accompanied by the removal of most nonhistone proteins and replacement of the normal histones with either special arginine-rich histones, phosphorylated histones, or protamines (page 410). X-ray diffraction analyses have revealed that the chromatin fibers of the developing sperm cell are packed in a highly ordered, almost crystalline configuration. As a result the DNA becomes metabolically inert, unable to act as a template for either RNA or DNA synthesis. Packaging the sperm cell DNA in

Figure 15-10 *Relationship between Sertoli Cells and Developing Sperm Cells in the Mammalian Testis* Sertoli *cells line the seminiferous tubules that make up the testis. Embedded within recesses of the Sertoli cell surface are the developing sperm cells. Mature sperm are released from the Sertoli cell into the lumen of the tubule.*

such a compact, inert form serves at least two functions: It protects the genes from physical or chemical damage and, by reducing nuclear volume, it creates a smaller sperm cell that requires less energy to propel.

The compacted nucleus eventually becomes localized to one end of the sperm cell, forming a region known as the sperm *head* (see Figure 15-11). At the tip of the sperm head, a specialized vesicle called the **acrosome** forms between the nucleus and the overlying plasma membrane. The acrosome develops from small Golgi-derived vesicles that fuse together (Figure 15-13), yielding a large membrane-enclosed organelle of spherical, conical, or elongated shape. After the acrosome has been formed, the Golgi complex moves from its location adjacent to the nucleus toward the tail region of the spermatid, where it is expelled from the cell. The acrosome contains a variety of hydrolytic enzymes, including acid phosphatase, proteases, and hyaluronidase, and thus it represents a specialized type of lysosome. As we will see shortly, the hydrolytic acrosomal enzymes are involved in digesting a path through the egg coats during fertilization.

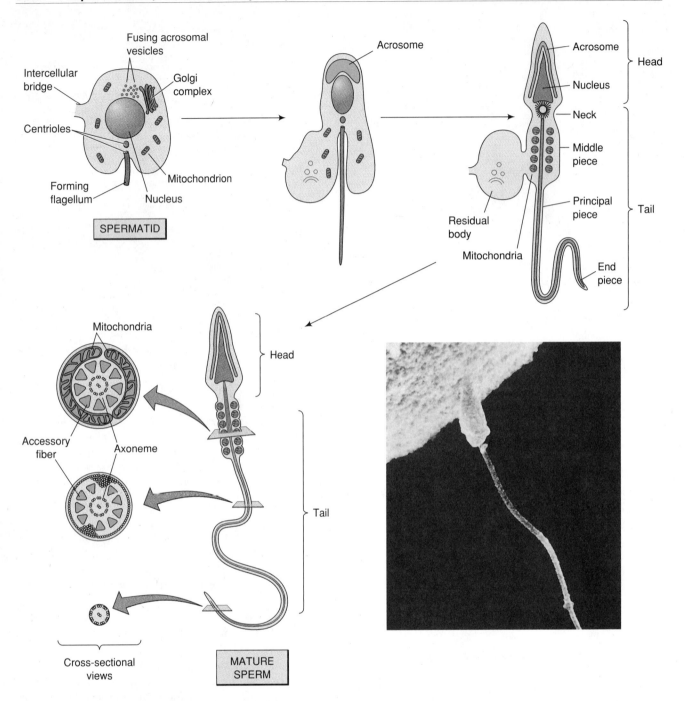

Figure 15-11 *Main Stages in the Process of Spermiogenesis* *During the initial phase of spermiogenesis, the chromatin fibers of the spermatid are tightly packed together and the compacted nucleus becomes localized to one end of the cell, forming the sperm head. As the sperm tail flagellum is assembled at the opposite end of the cell, mitochondria migrate to a position located at the base of the flagellar axoneme. Most of the remaining cytoplasm forms a protrusion called the residual body, which is pinched off and discarded. The scanning electron micrograph at the lower right shows a mature sperm cell making contact with a secondary oocyte. Micrograph courtesy of D. W. Fawcett and E. Phillips.*

The Sperm Tail Consists of a Flagellum and Associated Mitochondria

In most species, the sperm cell is propelled by a long flagellum that forms a structure called the sperm *tail*. Assembly of the flagellum is under the control of one of the cell's centrioles. Spermatids usually contain two centrioles: a *proximal centriole* adjacent to the nucleus and a *distal centriole* situated at right angles to the proximal centriole. During spermiogenesis, the distal centriole initiates the assembly of a flagellar axoneme that forms the core of the developing sperm tail. In addition to the typical "9 + 2" array of microtubules (page 603), sperm tail axonemes sometimes

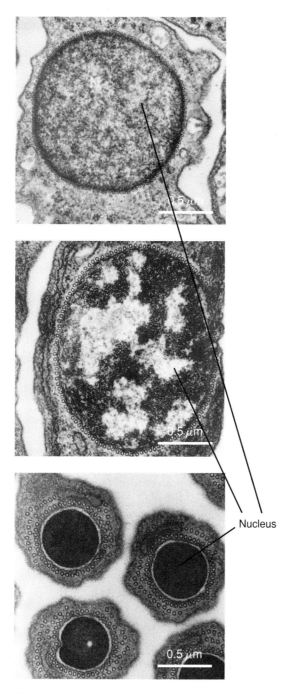

N

0.5 μm

0.5 μm

0.5 μm

Nucleus

Figure 15-12 *Electron Micrographs Illustrating the Chromatin Condensation That Occurs During Formation of the Sperm Head* (Top) *During early spermatid development, the chromatin fibers are evenly dispersed throughout the nucleus.* (Middle) *Shortly thereafter, the chromatin fibers begin to condense into clumps.* (Bottom) *In the final stages of sperm maturation the chromatin has become tightly packed into an amorphous mass, leading to a marked reduction in nuclear volume. Note that the magnification in all three micrographs is roughly the same. Courtesy of W. A. Anderson.*

contain nine thick *accessory fibers,* one adjacent to each of the outer doublets (Figure 15-14). Accessory fibers tend to occur in species where fertilization takes place within the organism's reproductive tract rather than externally, suggesting that they may provide added strength or protection needed when the flagellum propels the sperm through the viscous environment of the reproductive tract. Isolated accessory fibers consist of several polypeptides that do not resemble tubulin, actin, or any of the motor proteins commonly associated with microtubules and actin filaments, reinforcing the idea that the accessory fibers are reinforcing structures rather than motile elements.

While the sperm tail flagellum is being assembled, mitochondria migrate to a position directly beneath the nucleus. Here they form a *middle piece* that surrounds the base of the flagellar axoneme (see Figure 15-11). In many organisms the mitochondria line up end to end, forming a helical sheath around the flagellar axoneme. The mitochondria of some insect and invertebrate sperm actually fuse together to form a composite mass called the *nebenkern,* which then separates into two bodies of equal size located on either side of the developing flagellum (Figure 15-15). The close relationship that tends to exist between mitochondria and the sperm tail flagellum gives the flagellum direct access to mitochondrial ATP, which in turn drives the flagellar movements that propel the sperm.

As the mitochondria take up their position in the middle piece adjacent to the developing flagellum, the remaining cytoplasm of the spermatid becomes localized in a large cellular protrusion called the *residual body.* The entire residual body is eventually pinched off and discarded, yielding a mature sperm cell that lacks many of the organelles initially present in the spermatid cytoplasm. The sperm cell produced by this process is uniquely suited for the task of transporting the paternal genes to the egg. The sperm head is designed for protecting these genes and for penetrating the egg, while the sperm tail, with its flagellum and associated mitochondria, provides motility. Most other cellular components, which would be extraneous to the function of a sperm cell, have been eliminated.

Gametogenesis in Flowering Plants Involves the Production of Eggs and Pollen Grains by the Ovary and Anther

Gametogenesis in plants exhibits a number of unique features that distinguish it from the comparable process in animals. The male and female reproductive organs of flowering plants are termed the **anther** and **ovary,** respectively; both occur within the same plant

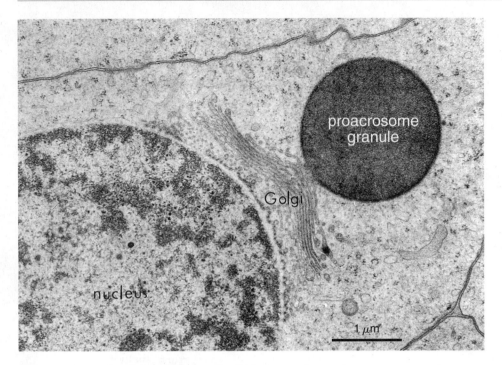

Figure 15-13 *Role of the Golgi Complex in Acrosome Formation*   *In this electron micrograph of a developing grasshopper sperm cell, vesicles derived from the Golgi complex are fusing with one another to form a structure called a proacrosome granule. The proacrosome granule subsequently differentiates into the mature acrosome. Courtesy of D. M. Phillips.*

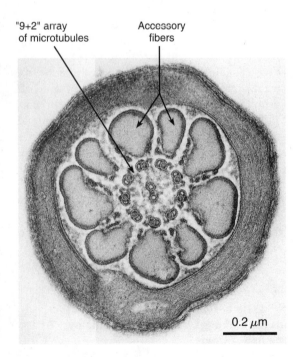

Figure 15-14 *Electron Micrograph of a Guinea Pig Sperm Tail Observed in Cross Section*   *Nine large accessory fibers are seen surrounding a central "9 + 2" array of microtubules. Courtesy of D. S. Friend.*

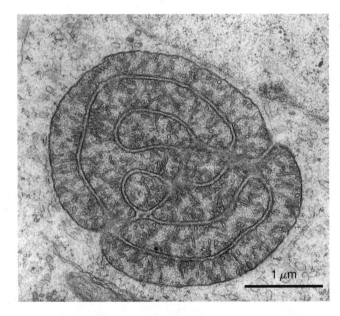

Figure 15-15 *Electron Micrograph of a Four-Layered Nebenkern in an Insect Spermatid*   *This type of structure, which occurs in certain insect and invertebrate sperm, consists of a group of mitochondria that have aggregated together to form a composite mass. Courtesy of S. A. Pratt.*

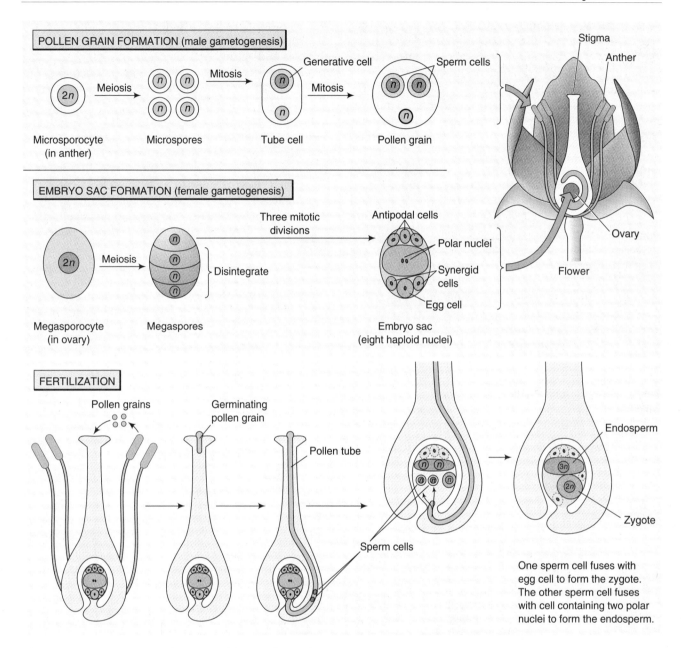

Figure 15-16 *Gametogenesis and Fertilization in Flowering Plants* (Top) *During male gametogenesis, meiotic division of a microsporocyte yields four haploid microspores, each of which develops into a pollen grain. The haploid nucleus of each pollen grain divides mitotically, yielding a tube cell plus a generative cell that is retained within the cytoplasm of the tube cell; the generative cell eventually undergoes another mitotic division to produce two haploid sperm cells. (Middle) During female gametogenesis, meiotic division of a megasporocyte yields four haploid megaspores, three of which disintegrate. The remaining megaspore undergoes three mitotic divisions, yielding an embryo sac containing one egg cell, two polar nuclei, two synergid cells, and three antipodal cells. (Bottom) During fertilization, pollen grains land on the stigma and develop a pollen tube that penetrates into the ovary, where it releases the two sperm cells. One fuses with the egg to form a diploid zygote and the nucleus of the other fuses with the two polar nuclei to produce the triploid endosperm.*

and often within the same flower. Male gametogenesis begins with meiotic division of diploid cells called **microsporocytes** located in the anthers (Figure 15-16, *top*). Each of the haploid **microspores** generated by meiosis becomes encased in a thick spore coat, form-

ing a **pollen grain.** The haploid nucleus of the pollen grain then divides mitotically, yielding a **tube cell** that controls the development of the pollen grain, and a **generative cell** that resides within the cytoplasm of the tube cell. Eventually the generative cell undergoes another mitosis, yielding two haploid **sperm cells** that remain within the tube cell cytoplasm. The mature pollen grain is a metabolically inert structure, protected from the environment by its hard impervious coat. Unlike the male gametes of animals, the sperm cells found in pollen grains are not inherently motile; instead, they depend on movements of the pollen grain produced by external forces such as wind or insects. However, the sperm cells produced by lower plants, such as algae, mosses, and ferns, do resemble the motile sperm of animals, although they often have multiple flagella instead of just one.

The formation of female gametes in flowering plants begins with the meiotic division of diploid **megasporocytes** located in the ovary. Each megasporocyte generates four haploid **megaspores**, three of which disintegrate. The remaining haploid megaspore undergoes three mitotic divisions, yielding an **embryo sac** that contains eight haploid nuclei that mature to form one **egg** cell, two **polar nuclei** suspended in a common cytoplasm, two **synergid** cells, and three **antipodal** cells (see Figure 15-16, *middle*). Compared to the eggs of animal cells, the plant egg is a relatively small cell that lacks elaborate outer coats and stores of nutrients. The polar nuclei play a special role in fertilization to be described later in the chapter, the synergids function in transporting nutrients to the egg, and the antipodal cells usually die without contributing to further development.

FERTILIZATION

The net result of gametogenesis in animals is the production of two haploid cell types, the sperm and the egg. During **fertilization** the sperm and egg fuse, forming a cell in which the diploid chromosome number is reestablished. In mammals, birds, and reptiles the fertilization process occurs within the female reproductive tract, while in amphibians, fish, and invertebrates it takes place after the eggs have been released into the external environment. In either case, fertilization involves several distinct events: an initial contact between sperm and egg, penetration of the egg coats by the sperm, fusion of the sperm and egg plasma membranes, and finally, activation of the egg.

Contact between Sperm and Egg Triggers the Acrosomal Reaction and Penetration of the Egg Coats

Most sperm cells are propelled to the vicinity of the egg by their motile flagella; sperm that do not have flagella either travel by cell crawling (page 588) or are passively carried by the motion of the fluid in which they are suspended. In marine animals that discharge

their sperm and eggs directly into the water, special mechanisms are required both to help the sperm find the egg and to ensure that eggs are not fertilized by sperm produced by the wrong type of organism. The existence of such mechanisms has been well established in sea urchins. The jelly coat of sea urchin eggs contains a small peptide called **resact** that acts as a chemical attractant that diffuses into the surrounding seawater, selectively binding to receptors on the surface of sperm cells originating from the same species of sea urchin. Resact binds to a sperm cell receptor that appears to function as a *guanylyl cyclase* that catalyzes the formation of the second messenger, cyclic GMP (page 222). Binding of resact to its receptor guides the swimming movements of the sperm cell and causes it to migrate selectively toward the egg emitting the chemical attractant.

When a sea urchin sperm makes contact with an egg cell, a sulfated polysaccharide present in the egg jelly binds to a receptor in the sperm plasma membrane. This binding causes Ca^{2+} channels in the sperm membrane to open and Ca^{2+} diffuses into the cell; it also activates a plasma membrane Na^+-H^+ *exchanger* that pumps H^+ out of the sperm cell in exchange for Na^+, causing the pH inside the sperm to rise. The increase in cytosolic Ca^{2+} and rise in pH trigger a dramatic response in the sperm cell known as the **acrosomal reaction.** First, the calcium ions cause the acrosome to fuse with the sperm plasma membrane and expel its hydrolytic enzymes to the cell exterior, where the released enzymes digest a pathway through the outer coatings of the egg (Figures 15-17 and 15-18). In many marine invertebrates, release of the acrosomal enzymes is quickly followed by the formation of a long cytoplasmic protrusion, called the **acrosomal process,** from the region directly in front of the sperm head. Formation of the acrosomal process is driven by the polymerization of actin molecules that had been stored in the region between the acrosome and the nucleus. Actin polymerization is triggered by the rise in intracellular pH, which may act by promoting the dissociation of unpolymerized actin from actin-binding proteins that otherwise prevent the actin from polymerizing.

The acrosomal process quickly traverses the outer egg coats and makes contact with the outer layer of the egg's vitelline envelope. At this point a protein called **bindin,** which is located on the surface of the acrosomal process, binds to specific receptors located on the surface of the vitelline envelope. The interaction between sperm bindin and its vitelline envelope receptor is species-specific; that is, the receptors on the vitelline envelope only recognize bindin molecules produced by sperm of the same sea urchin species, ensuring that fertilization does not occur with sperm from the wrong type of organism. The receptor with which bindin in-

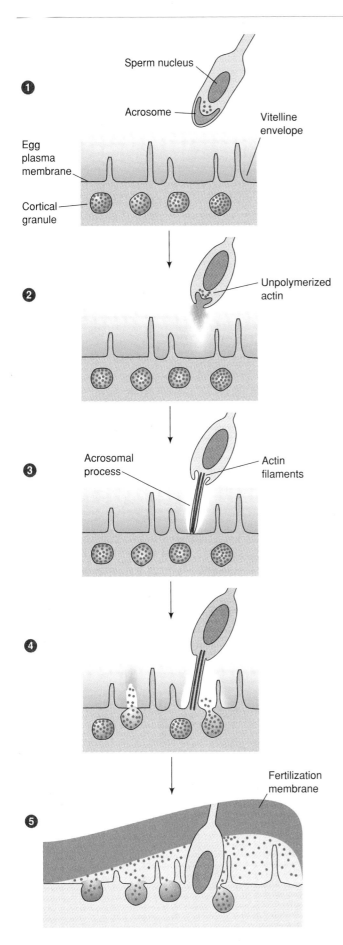

1
Sperm nucleus

Acrosome

Vitelline envelope

Egg plasma membrane

Cortical granule

2
Unpolymerized actin

3
Acrosomal process

Actin filaments

4

Fertilization membrane

5

teracts is thought to be a glycoprotein because treatment of egg cells with lectins (page 234) that bind to glycoproteins prevents fertilization from occurring. Such observations may one day yield practical benefits in the development of new methods of human birth control. For example, if women could be immunized against either the sperm protein or the egg receptor involved in sperm-egg binding, the resulting antibodies might prevent fertilization without interfering with normal hormone function as existing contraceptive pills do.

In vertebrates, where fertilization occurs within the female reproductive tract, the sequence of events leading to sperm-egg adhesion differs slightly from that employed by sea urchins. Sperm cells are deposited directly within the reproductive tract and are propelled to the vicinity of the egg by their flagella, although sperm migration may be facilitated by external forces such as contraction of the oviduct and movements of the cilia lining its walls. The actual meeting of sperm and egg appears to occur largely by chance in such organisms; hence sperm cells must be produced in vast excess to ensure that sufficient numbers reach the egg. In mammals, for example, millions of sperm are deposited in the female reproductive tract during copulation, but no more than a few hundred manage to reach even the general vicinity of the egg.

The events that occur in mammals after sperm cells reach the egg have been most extensively studied in mice. When a mouse sperm encounters an egg, it binds to a glycoprotein called *ZP3* located in the egg's thick outer coat, the zona pellucida (page 678). At least three proteins located in the sperm plasma membrane are involved in binding the sperm cell to ZP3. Binding of the sperm cell to ZP3 in turn triggers the acrosomal reaction. The hydrolytic enzymes released during the acrosomal reaction allow the sperm to bore through the zona pellucida and make direct contact with the egg plasma membrane. This sequence of events differs slightly from that employed by sea urchins; in sea urchins the acrosomal reaction occurs *before* the sperm binds to the vitelline envelope

Figure 15-17 *Steps Involved in Sperm-Egg Fusion in Sea Urchins* *(1) Physical contact between the sperm cell and the vitelline envelope of the egg triggers the acrosomal reaction, which consists of (2) discharge of the sperm cell's acrosomal enzymes and (3) formation of the acrosomal process, which is driven by the polymerization of actin filaments. (4) The subsequent fusion of the sperm and egg cell plasma membranes leads to cortical granule discharge and (5) formation of the fertilization membrane. In organisms whose sperm do not form an acrosomal process, digestion of the vitelline envelope (or zona pellucida) by enzymes released from the acrosome permits the entire sperm head to penetrate the vitelline envelope and fuse with the egg plasma membrane.*

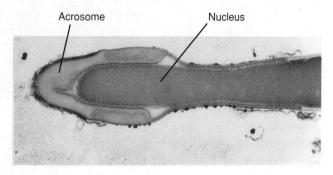

Acrosome   Nucleus

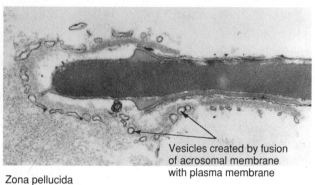

Zona pellucida

Vesicles created by fusion of acrosomal membrane with plasma membrane

Figure 15-18 *The Acrosome Reaction in a Hamster Sperm Cell*   (Top) *Electron micrograph of a hamster sperm cell showing the intact acrosome.* (Bottom) *On contact with the zona pellucida of the egg, the membrane surrounding the acrosome fuses with the sperm cell plasma membrane, creating a series of membrane vesicles. The loss in the continuity of the acrosomal membrane allows the acrosomal contents to diffuse out of the sperm cell and make contact with the zona pellucida. An acrosomal process does not form in vertebrates such as the hamster, but the enzymes released from the acrosome digest a path through the zona pellucida, thereby allowing the sperm to make direct contact with the egg plasma membrane. Courtesy of R. Yanagemachi.*

Following the Fusion of Sperm and Egg, a Transient Depolarization of the Plasma Membrane Prevents Polyspermy

As we have just seen, the binding of a sperm cell to appropriate receptors in the vitelline envelope or zona pellucida is followed by digestion of the envelope or zona in the region of the sperm head, thereby allowing the sperm to make direct contact with the egg plasma membrane. Upon contact, the plasma membranes of the sperm and egg fuse together. Membrane fusion appears to be mediated by specialized "fusogenic" proteins that actively promote the fusion of lipid bilayers. Bindin may be an example of such

a fusogenic protein, since it has been shown that purified bindin causes artificial phospholipid vesicles to fuse together.

The fusion of sperm and egg marks the onset of fertilization and triggers a dramatic sequence of events within the fertilized egg or **zygote.** The earliest events to occur are those designed to prevent **polyspermy**—that is, the fertilization of an egg by more than one sperm. Establishing a quick block to polyspermy is crucial because hundreds or even thousands of sperm may be bound to the outer coatings of the egg when the first sperm fuses with the egg plasma membrane. Experiments carried out by Larinda Jaffe suggest that the initial block to polyspermy is caused by changes in the membrane potential of the egg. Jaffe discovered that the electrical potential across the plasma membrane of the sea urchin egg increases from its resting value of -60 mV to a value approaching $+20$ mV within a few seconds after fertilization. This membrane *depolarization* is caused by a transient influx of Na^+ into the egg (page 199). When Jaffe used an electric current to artificially raise the membrane potential of unfertilized eggs to $+5$ mV, the eggs remained unfertilized, even in the presence of a vast excess of sperm. Fertilization only took place after the electric current was turned off, indicating that changing the membrane potential from a negative value to a positive value (which is what happens immediately after fertilization) is capable of preventing the egg from being fertilized by additional sperm.

An Increased Ca²⁺ Concentration in the Egg Cytosol Triggers the Cortical Reaction, Which Leads to a Permanent Block to Polyspermy

The rapid block to polyspermy created by depolarization of the plasma membrane is transitory in nature, lasting no more than a few minutes. But by the time the membrane potential returns to normal, a more permanent block to polyspermy has been established by the **cortical reaction** that occurs in response to sperm-egg fusion. In the cortical reaction, the cortical granules located near the periphery of the egg cell cytoplasm fuse with the overlying plasma membrane and discharge their contents into the space between the plasma membrane and the vitelline envelope (Figure 15-19). Cortical granule discharge originates at the site of sperm-egg fusion and quickly passes over the entire egg surface, causing a fluid-filled space to form between the plasma membrane and the vitelline envelope.

Among the components discharged from the cortical granules is an enzyme that destroys the sperm receptors of the vitelline envelope, thereby triggering the release of bound sperm and preventing the binding of new sperm (Figure 15-20). Other components discharged from the cortical granules change the vitelline

(which is comparable to the zona pellucida of mammalian eggs), whereas the acrosomal reaction occurs *after* the sperm binds to the zona pellucida in mammals.

envelope from a soft elastic structure into a tough rigid envelope called the **fertilization membrane,** which serves as a barrier to further sperm penetration (Figure 15-21). In spite of its name, the fertilization membrane is not a cellular membrane in the classical sense of a lipid bilayer containing embedded proteins. Instead the fertilization membrane is a hardened form of the vitelline envelope that is produced by the formation of covalent bonds between tyrosine residues located in the glycoproteins that make up the envelope. The formation of these bonds is catalyzed by *ovoperoxidase,* an enzyme that is discharged from cortical granules at the time of fertilization.

The cortical reaction can be artificially induced in unfertilized eggs by exposing them to agents that increase the intracellular concentration of free Ca^{2+}, suggesting that calcium ions trigger the early events associated with fertilization. Further support for this idea has emerged from studies involving **aequorin,** a dye that luminesces in the presence of Ca^{2+}. Aequorin injected into unfertilized eggs is barely luminescent, indicating a low internal concentration of free Ca^{2+}. But the luminescence increases more than four orders of magnitude within a few seconds after fertilization, indicating a dramatic increase in free Ca^{2+} concentration within the egg cell (Figure 15-22). In sea urchin eggs, the calcium ions are derived from the endoplasmic reticulum by a process that utilizes the phosphoinositide signaling pathway (page 218). As we learned in Chapter 6, the phosphoinositide signaling pathway is based on the activation of plasma membrane G proteins that stimulate the synthesis of the second messengers inositol trisphosphate (IP_3) and diacylglycerol (see Figure 6-23). The IP_3 in turn increases the permeability of ER membranes to Ca^{2+}, allowing Ca^{2+} to diffuse from the ER lumen into the cytosol. The increase in free Ca^{2+} that follows the fertilization of sea urchin eggs appears to involve a similar mechanism; that is,

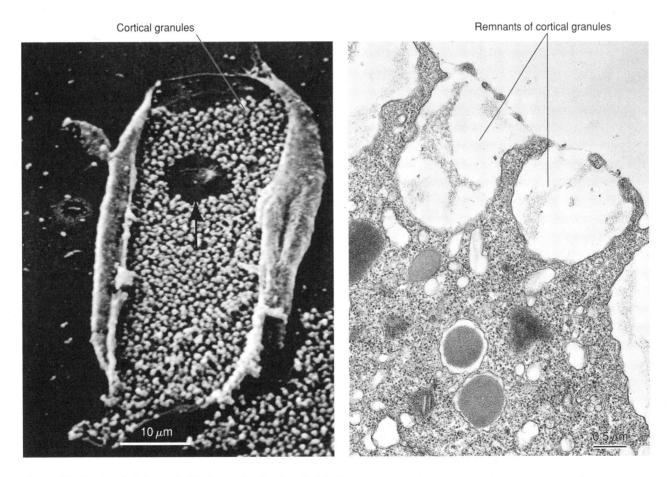

Cortical granules

Remnants of cortical granules

10 μm

0.5 μm

Figure 15-19 *Cortical Granule Discharge in the Sea Urchin Egg* (Left) *In this scanning electron micrograph of an egg surface, the dark area (arrow) represents the region where a sperm cell has just entered the egg. The cortical granules have already discharged their contents and disappeared from the area immediately surrounding the point of sperm entry.* (Right) *A thin-section electron micrograph of a comparable region of another sea urchin egg reveals the remnants of two cortical granules that have just fused with the plasma membrane and discharged their contents. Courtesy of G. Schatten (left) and E. Anderson (right).*

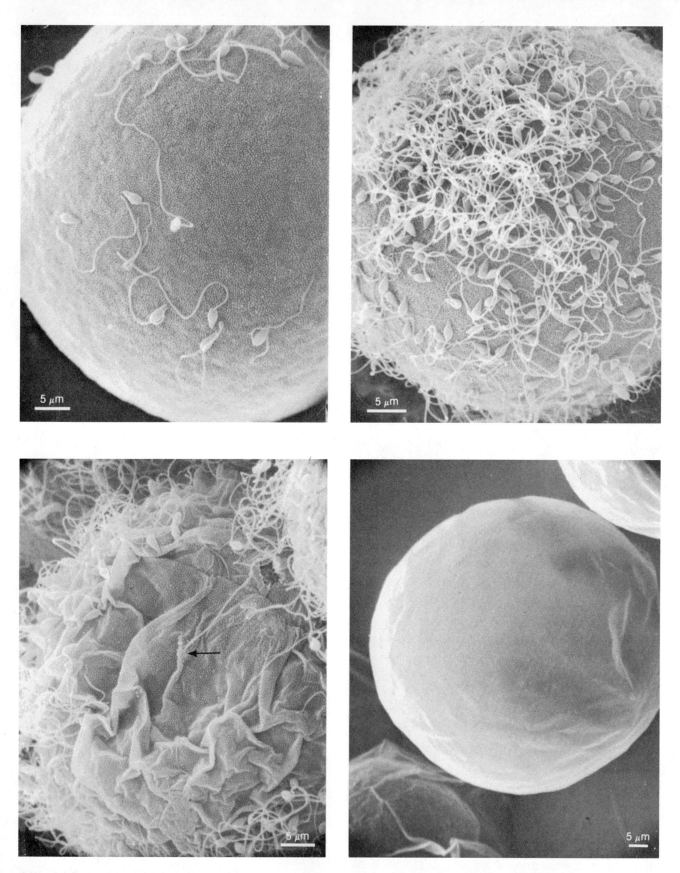

Figure 15-20 *Scanning Electron Micrographs of a Sea Urchin Egg During the First Few Minutes after Fertilization* (Top left) *At the time of fertilization, several sperm cells are seen scattered across the egg surface. (Top right) At 15 seconds after fertilization the number of sperm on the egg surface is still increasing. (Bottom left) At 30 seconds a circular zone lacking bound sperm has developed around the point of sperm entry (the arrow points to the tail of the entering sperm). This sperm-free zone corresponds to the area of cortical-granule breakdown. (Bottom right) Within 3 minutes the hardened fertilization membrane has formed, and no sperm remain bound to the egg surface. Courtesy of M. Tegner.*

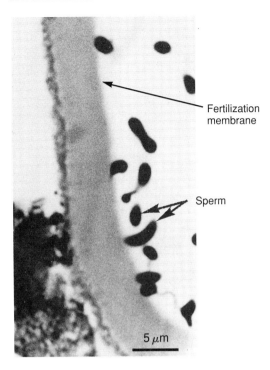

Figure 15-21 *The Fertilization Membrane of a Recently Fertilized Egg Cell Viewed by Light Microscopy* *Note that the sperm cells surrounding the egg are unable to penetrate the barrier created by the fertilization membrane. Courtesy of R. D. Grey.*

the binding of a sperm cell to the egg plasma membrane leads to activation of a G protein that stimulates the formation of IP_3.

Movements of the Sperm Nucleus and the Egg Cytoplasm Occur Shortly after Fertilization

Fusion of the sperm and egg plasma membranes creates a small region of cytoplasmic continuity between sperm and egg. This continuity allows the egg cytoplasm to creep up between the nucleus and the plasma membrane of the sperm, creating a structure called the *fertilization cone* (Figure 15-23). When the egg cytoplasm makes contact with the sperm head, the sperm nucleus is drawn into the egg cell and its compact chromatin begins to uncoil and disperse. As these changes in chromatin structure are taking place, the highly basic chromatin proteins of the sperm nucleus are replaced with less basic proteins derived from the egg cytoplasm.

After the sperm nucleus enters the egg interior, its nuclear envelope disintegrates into small vesicles, momentarily leaving the sperm chromatin without an enclosing membrane. But a new nuclear envelope soon forms around the sperm chromatin, creating a **male pronucleus.** The male pronucleus then migrates toward the nucleus of the egg, which is usually termed the **female pronucleus** at this stage. Movement of the male

pronucleus appears to be guided by microtubules that emanate from the sperm centrioles, which enter the egg along with the sperm nucleus. Upon making contact the two pronuclei usually fuse, forming a diploid nucleus (see Figure 15-23, *bottom left*). The pair of sperm centrioles then divides to produce two centrosomes (page 537), each containing a pair of embedded centrioles; the two centrosomes are destined to become the poles of the mitotic spindle that forms when the egg begins to divide.

Fertilization also triggers major rearrangements in the organization of the egg cell cytoplasm. These cytoplasmic changes are easiest to demonstrate in eggs that contain pigment granules because the shifting pigment distribution can be readily observed. For example, in some amphibian eggs the outer (cortical) layer of cytoplasm at the animal pole is darkly pigmented while the underlying cytoplasm contains fewer pigment granules and therefore appears gray. Shortly after fertilization, the outer layer of cytoplasm rotates about 30° relative to the inner cytoplasm, exposing a gray region of inner cytoplasm called the **gray crescent** (Figure 15-24). Rotation of the outer cortex appears to be mediated by a parallel array of microtubules located between the cortical and inner cytoplasm in the vegetal region of the egg. If eggs are treated with microtubule inhibitors such as colchicine, cytoplasmic rotation does not occur. We will see shortly that cytoplasmic rearrangements such as the one that creates the gray crescent play important roles in controlling cell differentiation as the embryo develops.

An Increase in the pH of the Fertilized Egg Activates Protein Synthesis and DNA Replication

Most of the events described thus far represent early responses to fertilization that are initiated within the first few minutes after fusion of sperm and egg. These initial events are soon followed by the activation of protein synthesis and the initiation of DNA replication in the fertilized egg (Figure 15-25). Changes in protein synthesis are especially prominent in sea urchins, where less than 1 percent of the egg's ribosomes are active prior to fertilization. The rate of protein synthesis begins to increase about 5 minutes after fertilization, and within half an hour protein synthesis is occurring at a rate that is almost an order of magnitude faster than in unfertilized eggs (Figure 15-26). This dramatic increase in protein synthesis is not blocked by inhibitors of RNA synthesis, indicating that the mRNAs being translated must have been present in the unfertilized egg in an inactive state. Several explanations have been proposed to explain the activation of mRNA translation that occurs after fertilization. Among the more prominent possibilities are (1) activation of mRNA through the removal of inhibitory "masking"

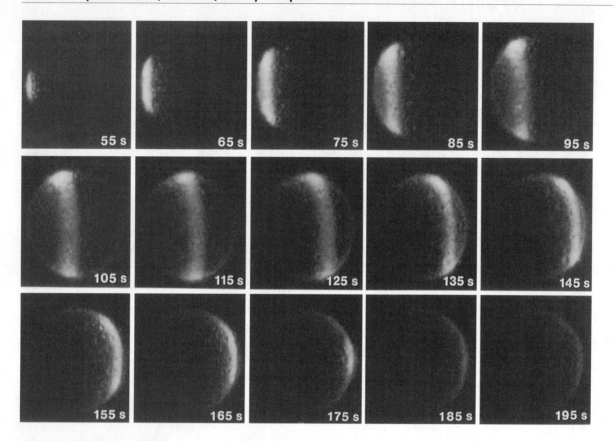

Figure 15-22 *Transient Increase in Free Ca²⁺ Concentration That Occurs in an Egg Cell Immediately after Fertilization*
In this experiment an egg of the medaka fish has been injected with aequorin, a dye that emits light when bound to free Ca²⁺ in the cytosol. The amount of time that has elapsed after fertilization is indicated in seconds in each photograph. These pictures show that in the first few minutes after fertilization, a transient wave of increased Ca²⁺ concentration passes across the egg, starting from the point of sperm entry on the left. Courtesy of Y. Hiramoto.

proteins that prevent translation, (2) addition of a 5' methylated cap structure to uncapped mRNAs, and (3) activation of initiation factors involved in protein synthesis. Most oocytes appear to use more than one of these mechanisms to activate mRNA translation after fertilization.

The activation of protein synthesis and DNA replication that occurs in fertilized eggs is triggered by a rise in intracellular pH. The pH of an unfertilized sea urchin egg is about 6.8, an unusually low value that is thought to keep cellular metabolism relatively inactive. If the pH of an unfertilized egg is artificially raised by exposing it to ammonia, protein synthesis and DNA replication are both activated. These responses occur without increasing the internal Ca²⁺ concentration or triggering the cortical reaction, suggesting that an increased pH stimulates protein and DNA synthesis by a mechanism that differs from the way in which the early responses to fertilization are triggered. Further support for the proposed role of pH in activating egg metabolism has come from the discovery that the intracellular pH of the sea urchin egg increases from 6.8 to 7.3 during the first minute or two after fusion of sperm and egg. This increase is caused by the activation of a plasma mem-

brane *Na⁺-H⁺ exchange carrier* that takes up sodium ions from the external environment in exchange for protons being pumped out of the cell, causing a decrease in the intracellular proton concentration (and hence an increase in intracellular pH). The Na⁺-H⁺ carrier is thought to be activated by protein kinase C, a component of the phosphoinositide signaling pathway that is stimulated by binding of the sperm cell to the egg plasma membrane (Figure 15-27).

Two Fertilization Events Occur in Flowering Plants, Producing a Diploid Zygote and Triploid Endosperm

Fertilization in flowering plants exhibits several unique features that distinguish it from the comparable process in animals. One prominent difference is that the male gametes of flowering plants are not inherently motile; plants must instead rely on environmental forces such as wind, rain, and insects to carry pollen grains from the anther and bring them into contact with the **stigma,** a sticky projection of the ovary specialized for trapping pollen. When pollen grains

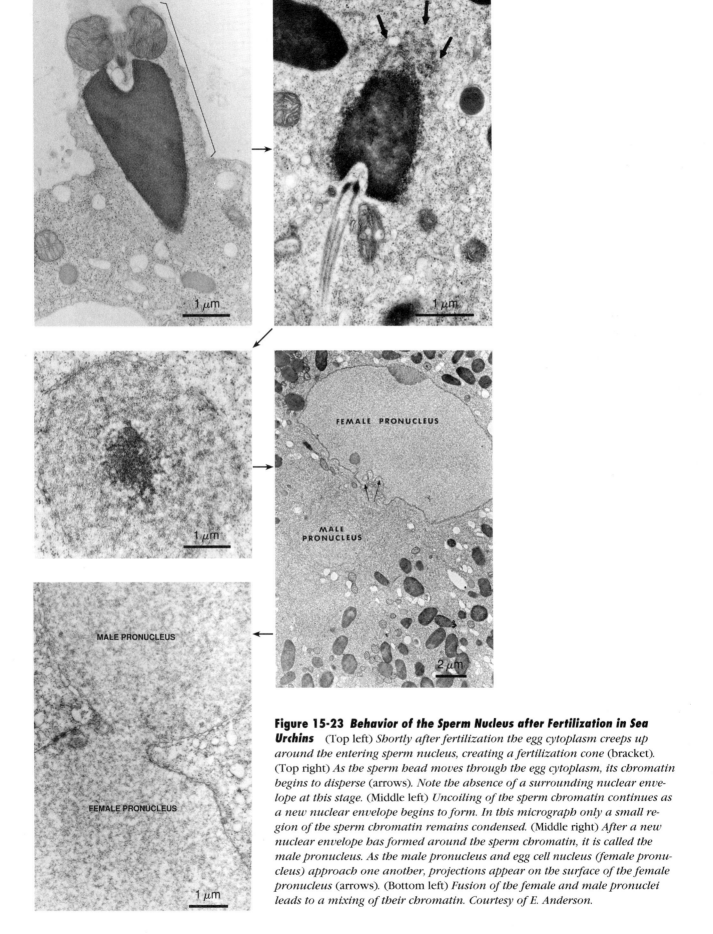

Figure 15-23 *Behavior of the Sperm Nucleus after Fertilization in Sea Urchins* (Top left) *Shortly after fertilization the egg cytoplasm creeps up around the entering sperm nucleus, creating a fertilization cone* (bracket). (Top right) *As the sperm head moves through the egg cytoplasm, its chromatin begins to disperse* (arrows). *Note the absence of a surrounding nuclear envelope at this stage.* (Middle left) *Uncoiling of the sperm chromatin continues as a new nuclear envelope begins to form. In this micrograph only a small region of the sperm chromatin remains condensed.* (Middle right) *After a new nuclear envelope has formed around the sperm chromatin, it is called the male pronucleus. As the male pronucleus and egg cell nucleus (female pronucleus) approach one another, projections appear on the surface of the female pronucleus* (arrows). (Bottom left) *Fusion of the female and male pronuclei leads to a mixing of their chromatin. Courtesy of E. Anderson.*

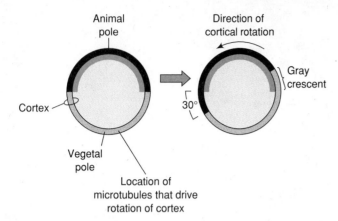

Figure 15-24 *Movement of Cortical Cytoplasm in a Fertilized Frog Egg* *The view on the left represents a schematic cross section through the egg shortly after fertilization. The cortex in the vegetal hemisphere is unpigmented and relatively transparent; in contrast, the cortex in the animal hemisphere is heavily pigmented and the underlying cytoplasm is somewhat less pigmented, appearing gray. Prior to the first cleavage, microtubules located near the vegetal pole drive the rotation of the cortical cytoplasm about 30° relative to the underlying cytoplasm. The rotation exposes a region of the underlying gray cytoplasm called the gray crescent.*

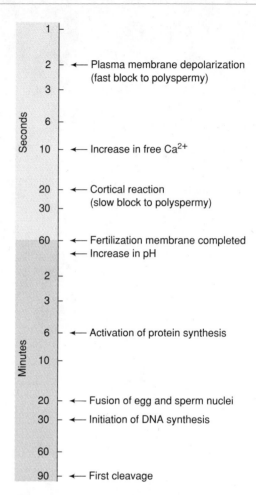

Figure 15-25 *Responses Triggered by Fertilization of a Sea Urchin Egg* *The increase in free Ca²⁺, first detected about 10 seconds after fertilization, triggers the cortical reaction, which in turn creates the slow block to polyspermy. The increase in pH, which occurs within 1–2 minutes after fertilization, leads to the activation of protein and DNA synthesis.*

make contact with the stigma, proteins derived from the pollen wall interact with proteins located on the surface of the stigma. If this interaction signals that the appropriate type of pollen has landed on the stigma, the tube cell of the pollen grain is activated, triggering a burst in respiration, synthesis of RNA, rupture of the thick spore coat, and the gradual outgrowth of a cytoplasmic projection termed the **pollen tube** (see Figure 15-16, *bottom*). The pollen tube, still surrounded by the plasma membrane and a primary cell wall, penetrates the tissues of the stigma and enters the ovary. Since pollen grains that have been experimentally germinated in the absence of external nutrients still produce pollen tubes that grow to normal length, it has been concluded that all the materials required for growth of the pollen tube are present in the initial pollen grain.

When the embryo sac is finally reached, the pollen tube penetrates one of the two synergids and releases the two sperm cells. One fuses with the egg, forming a diploid zygote as in animal fertilization. A second fertilization event is carried out by the other sperm nucleus, which fuses with the two polar nuclei. The net result of the second event is a *triploid* nucleus (three sets of chromosomes) destined to form a specialized nutritive tissue known as **endosperm.**

Once the diploid zygote has been formed, it divides mitotically to generate the plant embryo. The en-

dosperm also undergoes extensive growth and division, accumulating vast stores of starch and protein from other regions of the parent plant. The result is a rich nutritive tissue designed to nourish the embryo. Development is usually arrested at an early stage, yielding a *seed* composed of the diploid embryo surrounded by the triploid endosperm, enclosed in tissues that form the seed coat. The seed may remain in an arrested state for long periods of time, waiting for appropriate conditions to trigger its germination and subsequent development into a new plant.

FORMATION OF THE EMBRYO

After fertilization has taken place, the fertilized egg embarks upon a series of divisions, differentiations, and in-

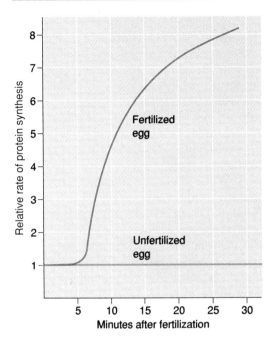

Figure 15-26 *Stimulation of Protein Synthesis Observed in Sea Urchin Eggs in Response to Fertilization* *Since this dramatic increase cannot be blocked by inhibitors of RNA synthesis, it must represent translation of mRNAs that had been stored in the egg in an inactive state.*

teractions that lead to the formation of a new organism. Although it is beyond the scope of this chapter to describe the numerous variations in developmental patterns that occur during the formation of animal and plant embryos, a relatively small number of recurring themes are routinely encountered during early development. In discussing these themes we will focus on examples involving animal development, but similar principles apply to plants as well. The main exception is cell migration, which plays an important role in animal but not plant development, presumably because plant cell walls prevent such movements from occurring.

Cleavage Converts the Fertilized Egg into a Multicellular Blastula

Depending on the organism involved, an animal egg may be at any stage of maturation when fertilization occurs (e.g., primary oocyte, metaphase I, secondary oocyte, metaphase II, or at the end of meiosis). In eggs where maturation is not yet complete, fertilization triggers the resumption of meiosis. The mechanism responsible for initiating the completion of meiosis has been extensively studied in amphibian eggs, which are

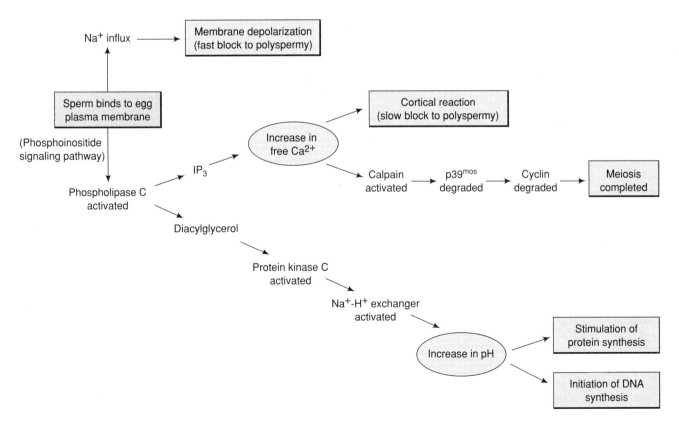

Figure 15-27 *Mechanisms Involved in Eliciting the Main Responses to Fertilization in Sea Urchin Eggs* *The fusion of sperm and egg elicits two main responses in the egg plasma membrane: an increase in Na+ permeability that triggers membrane depolarization (the fast block to polyspermy), and activation of the phosphoinositide signaling pathway, which generates the second messengers IP3 and diacylglycerol (DAG). IP3 triggers the increase in free Ca2+ that elicits the cortical reaction (the slow block to polyspermy), while DAG is responsible for the increase in pH that has been implicated in the activation of protein and DNA synthesis.*

arrested at metaphase II until fertilization occurs. As we learned earlier in the chapter, p39^{mos} maintains metaphase arrest by preventing the degradation of mitotic cyclin, a molecule whose breakdown is normally associated with the transition from metaphase to anaphase (page 679). For the cell cycle to begin again, p39^{mos} needs to be inactivated. The inactivation process is triggered by the increase in cytosolic Ca^{2+} that occurs immediately after fertilization. The increased Ca^{2+} concentration stimulates a Ca^{2+} dependent protease called **calpain,** which in turn degrades p39^{mos}. Once p39^{mos} has been degraded, cyclin can be broken down and the cell is able to complete meiosis (see Figure 15-27).

After completing meiosis, the fertilized egg enters a phase of rapid mitotic division known as **cleavage.** Little or no cell growth occurs during the cleavage stage; the successive mitotic divisions simply partition the egg into smaller and smaller cells, converting the egg into a hollow spherical mass of small cells termed a **blastula.** Because cell division does not need to wait for cell growth, the cells of a cleavage-stage embryo divide faster than any cells in an adult organism. A single fertilized frog egg, for example, divides into 37,000 cells in less than two days. The rapid pace of cell division is accomplished mainly by eliminating the G$_1$ phase of the cell cycle, which is unnecessary in the absence of cell growth. As the cleavage stage of development continues, a G$_1$ phase is gradually introduced, so that by the end of cleavage a typical cell cycle is present and cell growth accompanies cell division.

The cleavage of most eggs (mammalian eggs being the main exception) proceeds normally in the presence of inhibitors of RNA synthesis, indicating that the metabolic events that occur during cleavage depend only on mRNAs stored in the egg. Because the egg cell is asymmetrically organized, the cytoplasmic divisions that accompany cleavage produce cells that begin to inherit differing types of cytoplasm and mRNAs. Additional differences among the cells generated during cleavage also occur because eggs with large amounts of yolk tend to divide unequally, generating larger and more slowly dividing cells near the vegetal pole. Hence the asymmetric distribution of cytoplasmic components that initially characterized the fertilized egg causes the cells formed during the cleavage stage to differ in their properties.

Gastrulation Converts a Blastula into a Three-Layered Gastrula Composed of Ectoderm, Mesoderm, and Endoderm

At the end of the cleavage stage, the cells of the blastula embark upon a series of elaborate movements that alter their spatial relationships with one another (Figures 15-28 and 15-29). Although the details of this process vary among organisms, a universal feature in vertebrate eggs is the formation of an opening in the blastula surface called the **blastopore,** which allows cells to migrate into the interior of the blastula. Migration of cells through the blastopore produces a **gastrula** composed of three cell layers called the **ectoderm** (outer layer), the **mesoderm** (middle layer), and the **endoderm** (inner layer). Ectoderm and endoderm are destined to form sheets of cells called **epithelia** that cover the external and internal surfaces of the organism; ectoderm produces epithelia covering external surfaces (skin and related glands) as well as neural tissue, whereas endoderm differentiates into epithelia covering internal surfaces (gastrointestinal tract and related glands). Finally, mesoderm develops into a diffuse spongework of *mesenchyme* cells that give rise to supporting tissues such as muscle, cartilage, bone, blood, and connective tissue.

The new cellular relationships that have been created in the gastrula trigger a combination of cell-cell interactions, metabolic changes, and cell movements that lead to the formation of tissues and organs. The process by which embryonic cells give rise to the complex assortment of specialized cell types that make up these tissues and organs is called **differentiation.** Differentiated cells can be distinguished from one another by differences in cellular architecture as well as differences in the products they make. For example, red blood cells, nerve cells, and lymphocytes can be easily recognized both by their microscopic appearances and by the products they manufacture; red cells make hemoglobin, nerve cells produce neurotransmitters, and lymphocytes make antibodies.

Many of the cells present during early cleavage are potentially capable of differentiating into almost any specialized cell type, but at some point the developmental fate of each cell becomes committed by a process called **determination.** Determination of a cell's fate can be accomplished in two different ways; one involves the inheritance of cytoplasmic constituents, and the other is based on interactions between neighboring cells. In the following sections, we will explore each of these mechanisms in turn.

Cytoplasmic Segregation of Morphogenic Determinants Can Influence How Cells Differentiate

The influence of the cytoplasm on cell differentiation has been demonstrated in a variety of egg types. A striking example occurs in snail eggs, where cell differentiation is profoundly affected by the **polar lobe,** a large cytoplasmic protrusion temporarily formed at the vegetal pole during the first few cleavages (Figure 15-30). If the polar lobe cytoplasm is surgically removed, an em-

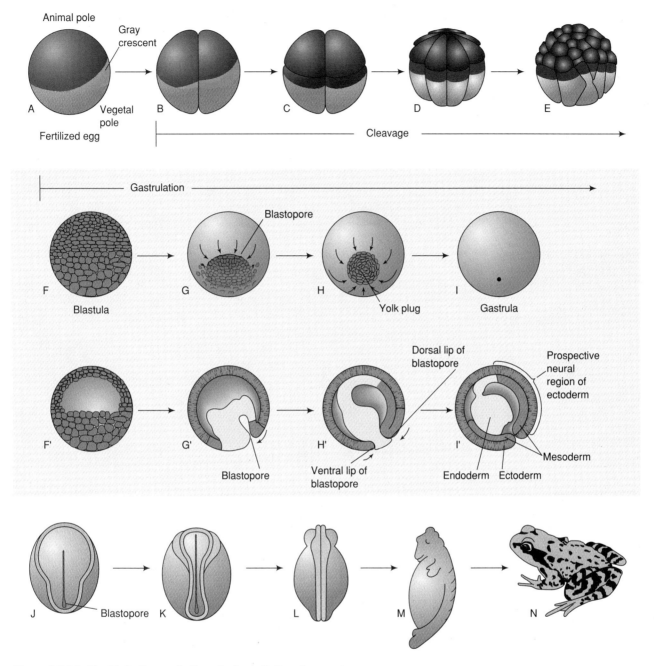

Figure 15-28 The Main Stages in Frog Embryonic Development *Gastrulation is shown in both surface view (F through I) and cross-sectional view (F' through I') to illustrate how cell movements produce a three-layered embryo. Cells that are destined to migrate through the blastopore are shaded red in views G' through I' to facilitate the tracing of their movements.*

bryo with a defective foot, eye, and shell develops. In amphibian eggs, cell differentiation is strongly influenced by the cortical cytoplasm located in the region of the gray crescent (page 691). If the gray crescent is injured by pricking, the resulting embryo develops a grossly abnormal nervous system. The preceding observations suggest that the egg cytoplasm contains substances that control the differentiation of particular cell types. Such substances, called **morphogenic determinants,** are asymmetrically distributed in the egg. Hence

as the dividing cells of the embryo incorporate different regions of egg cytoplasm, the morphogenic determinants become segregated into different cell types.

Some of the earliest evidence for the influence of cytoplasmic determinants on the behavior of individual cells was provided by the studies of Theodor Boveri on embryonic development in the roundworm *Ascaris*. In a fertilized *Ascaris* egg, the first cleavage divides the egg into one cell that contains cytoplasm derived from the animal pole and a second cell containing cytoplasm

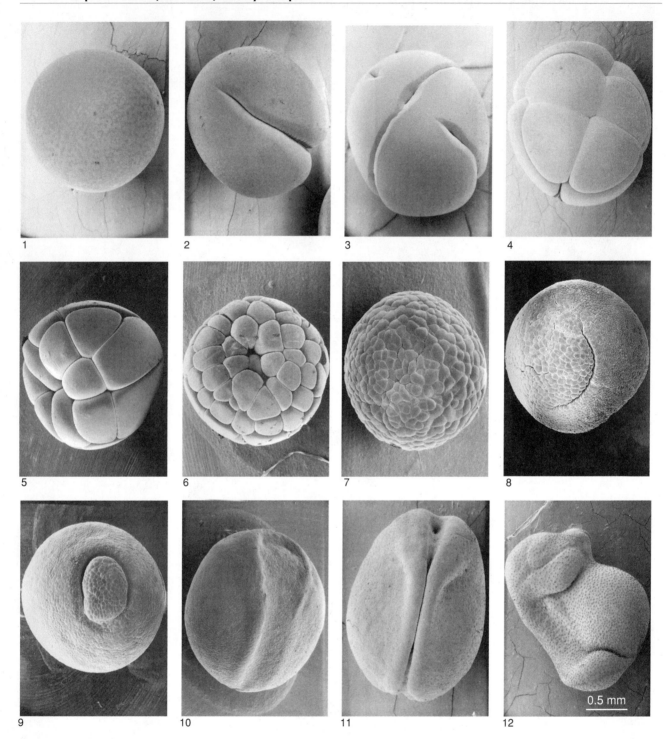

Figure 15-29 Development of a Frog Egg Viewed by Scanning Electron Microscopy *(1) An unfertilized egg. (2) Two-cell stage. (3) Four-cell stage. (4) Eight-cell stage. (5) Sixteen-cell stage. (6) The 32- to 64-cell stage. (7) Late blastula. (8) Gastrula with blastopore. (9) The yolk-plug stage showing the yolk plug surrounded by the blastopore. (10) An early neurula stage. (11) Late neurula stage showing the folding that gives rise to the neural tube. (12). Tailbud stage. Compare these photographs with the comparable stages illustrated in Figure 15-28. Courtesy of R. G. Kessel.*

derived from the vegetal pole. As the cell containing cytoplasm derived from the animal pole prepares for its next division, the heterochromatic regions of its chromosomes are shed into the cytoplasm and degenerate, while the remaining chromosomal regions break up into numerous small chromosomes. This process, called *chromosome diminution,* yields cells that have lost a portion of the normal chromosomal material that they no longer need. To investigate whether the cytoplasm is responsible for triggering this process, Boveri carried

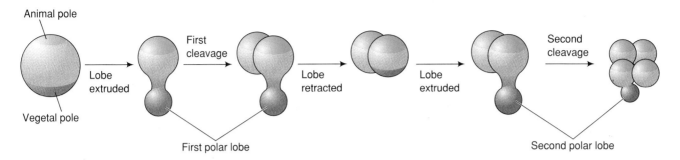

Figure 15-30 *Polar Lobe Formation in Mollusk Eggs* *The first polar lobe is formed prior to the first cleavage and is subsequently resorbed into the cytoplasm of one of the two cells produced by this division. The second polar lobe is formed by this same cell prior to the second cleavage and is again resorbed into one of the two cells produced by cell division. As a result, the vegetal cytoplasm segregated into the polar lobe is selectively passed to one of the four cells generated by the second cleavage. Color shading is employed to help illustrate how the cytoplasm derived from the animal and vegetal poles is distributed during the first two cleavages.*

out an experiment in which *Ascaris* eggs were centrifuged prior to cleavage in order to shift the orientation of the mitotic spindle relative to the cytoplasm. In this way, eggs were created in which the first cleavage plane ran from the animal pole to the vegetal pole rather than across the equator of the egg (Figure 15-31). When cleavage occurs in this direction, the two resulting cells each receive a mixture of both animal and vegetal cytoplasm, and chromosome diminution does not occur in either cell. Hence the kind of cytoplasm a cell receives determines whether or not the process of chromosome diminution will occur.

Additional evidence for the effect of cytoplasmic determinants on cell differentiation has come from studies involving germ cell development in the fruit fly, *Drosophila*. In insect embryos, the cells destined to become primordial germ cells initially arise from the posterior end of the egg. To determine whether the cytoplasm in this region of the egg is responsible for triggering germ cell differentiation, experiments have been carried out in which a tiny bit of cytoplasm is removed from the posterior end of one *Drosophila* egg and injected into the opposite (anterior) end of another egg. Under such conditions, germ cells develop in the anterior region of the injected egg where they normally would not have done so. The injected cytoplasm derived from the posterior end of the egg contains distinctive *polar granules* composed of RNA and protein whose presence has been correlated with the ability to form germ cells.

Cytoplasmic granules have also been identified as morphogenic determinants in the roundworm, *Caenorhabditis elegans*. This particular organism has become an important model for studying embryonic development because the mature adult contains only 959 cells; this makes it possible to trace out every cell division, cell migration, and cell death leading from the fertilized egg to the adult roundworm. Shortly after

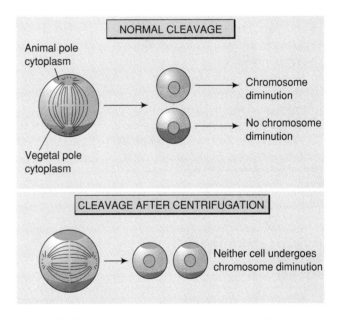

Figure 15-31 *Role of the Cytoplasm in Controlling Chromosome Diminution in* Ascaris *Eggs* (Top) *During normal cleavage, the cell receiving cytoplasm derived from the animal pole undergoes chromosome diminution.* (Bottom) *If the egg is centrifuged prior to the first cleavage to reorient the mitotic spindle, the cleavage creates two cells that each receive a mixture of animal and vegetal cytoplasm. Under such conditions neither cell undergoes chromosome diminution, indicating that the type of cytoplasm a cell receives determines whether or not the process of chromosome diminution will occur.*

fertilization of a *Caenorhabditis* egg, cytoplasmic granules called *P granules* migrate to one end of the egg cell. As a result, only one of the two cells produced by the first cleavage receives P granules. Prior to the next cleavage, the P granules again migrate to one end of the cell in which they reside; hence the granules are again distributed to only one of the two cells generated during cell division. A similar phenomenon is repeated during the next two cell divisions, so only one cell resulting from the first four cell divisions ends up receiving P granules. The cell that receives the P granules is destined to give rise to the sperm and egg cells of the adult. The cytoskeleton plays an important role in the P-granule movements that underlie the preceding events; if actin filaments are disrupted by treating cells with the inhibitor *cytochalasin* (page 584), granule movements are prevented and the normal pattern of sperm and egg cell formation is disrupted.

Although the examples described thus far involve embryonic cells, cytoplasmic control of cell differentiation has also been observed in adult organisms. One extensively studied example involves the grasshopper *neuroblast,* a cell type whose every division leads to the formation of one differentiated ganglion cell and a new neuroblast destined to divide again and repeat the process. Since dividing neuroblasts are always oriented in the same direction, it is possible to predict which regions of the neuroblast cytoplasm will give rise to the new neuroblast and the ganglion cell, respectively. In an elegant study, J. Gordon Carlson inserted a fine needle into a grasshopper neuroblast to rotate the mitotic spindle so that the chromosomes normally destined for the ganglion cell occupied the cytoplasmic region destined to become incorporated into the neuroblast, and vice versa. Under such conditions, the developmental fate of each cell was determined by the type of cytoplasm it received, not the source of its chromosomes.

Inductive Interactions between Neighboring Cells Can Also Influence How Cells Differentiate

In addition to cytoplasmic control of cell differentiation, the other way in which a cell's developmental fate can be determined is through interactions with neighboring cells. In many regions of a developing embryo, the presence of a particular cell type causes adjacent neighboring cells to differentiate in a specific way. This process, called **induction,** was first demonstrated by experiments in which cells from one embryo were grafted to a new location in another embryo. For example, when cells normally destined to become nervous tissue are removed from an amphibian embryo during early gastrula stage and transplanted to another embryo in a region destined to become epidermis, the grafted cells evolve into epidermis instead of nervous tissue. Likewise, prospective epidermal cells grafted to an area fated to become neural tissue will differentiate into neural tissue. Such observations indicate that the surrounding cells cause the grafted cells to differentiate in a particular way.

But this ability of cells to differentiate in alternative ways under differing conditions is a transient property.

If the same experiment is performed on a late gastrula (about two days later), the results are quite different. Prospective neural tissue placed in the region of prospective epidermal tissue no longer differentiates into epidermis like its neighbors, but instead forms neural tissue as it would normally do. Conversely, prospective epidermis placed in the region of prospective neural tissue develops into epidermis.

What has happened between early and late gastrula that has caused the developmental fate of prospective neural and epidermal cells to become determined? During this period, cells lying directly above the blastopore invaginate into the interior of the embryo to produce a multilayered structure in which some of the invaginated cells lie directly under the prospective neural cells (see Figure 15-28). To find out whether the presence of these newly invaginated cells determines the fate of the overlying prospective neural tissue, Hans Spemann and Hilde Mangold performed a classic series of transplantation experiments on newt embryos in the early 1920s in which the dorsal (upper) lip of the blastopore from one gastrula was grafted to a different position in another gastrula. In order to distinguish cells of the donor from those of the host, two strains of newt with differing pigmentation were employed. The surprising result was that the grafted blastopore material triggered formation of an entire new embryo joined to the host embryo like a Siamese twin (Figure 15-32). The pigmentation of the

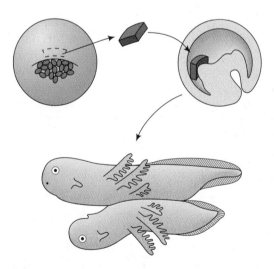

Figure 15-32 Experiment Demonstrating the Existence of Primary Embryonic Induction *When cells obtained from the dorsal lip of the blastopore of a darkly pigmented newt gastrula were transplanted to a nonpigmented newt gastrula, a second embryo developed that was joined to the host embryo like a Siamese twin. Since the Siamese twins are both nonpigmented, the pigmented dorsal lip cells of the graft must have induced the nonpigmented cells of the host to differentiate into the second embryo.*

Figure 15-34 *A Frog That Developed from a Fertilized Egg Whose Nucleus Had Been Replaced with a Nucleus Obtained from an Intestinal Epithelial Cell* *The fact that the frog is normal in all respects indicates that the intestinal nucleus contains all the genes required for directing normal development and producing all the specialized cell types of the adult organism. Courtesy of J. B. Gurdon.*

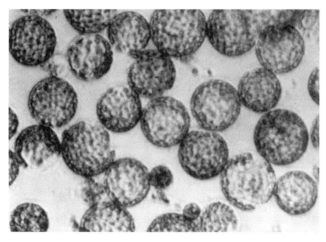

Figure 15-35 *Living Protoplasts Prepared from the Mesophyll Tissue of a Tobacco Plant and Viewed by Light Microscopy* *The removal of the cell wall allows the plant cells to assume a spherical shape. Under appropriate conditions an individual protoplast can develop into a complete new plant. Courtesy of D. A. Evans.*

In 1971 Itaru Takebe and his colleagues improved the technique for growing plants from single, differentiated plant cells. First they employed the enzymes *pectinase* and *cellulase* to digest away the plant cell wall, generating a suspension of wall-less cells called **protoplasts** (Figure 15-35). The protoplasts were then placed in an appropriate culture medium, where they synthesize new cell walls and begin to divide. Under these conditions each protoplast gives rise to a small clump of undifferentiated cells, termed a *callus,* that eventually develops shoots and roots. At this stage the embryonic

plant can be placed in soil, where it will develop into a mature plant identical to the original parent from which the protoplast was derived.

The realization that differentiated plant and animal cells retain all the genes required for programming the development of a complete new organism implies that some mechanism other than a permanent change in the genes themselves must be responsible for cell differentiation. Since differentiated cells contain different kinds of proteins but the same set of genes, it can be concluded that the pattern of gene expression is changing; that is, a different set of genes is being expressed in the form of functional proteins in each differentiated cell type.

Gene Expression in Differentiated Cells Is Controlled by a Variety of Molecular Mechanisms

The observations described in the preceding section imply that cell differentiation is accomplished by turning on and turning off the expression of specific genes. How is this accomplished at the molecular level? In Chapters 3, 10, and 11, we learned that numerous steps are involved in the pathway by which genetic information stored in DNA base sequences is first transcribed into mRNA molecules and then translated into proteins. The first step in this pathway, which involves the transcription of genes by RNA polymerase, is one of the main targets for regulating the pattern of gene expression. Genes that are transcribed into RNA in any given cell type fall into two general categories: (1) **housekeeping genes,** which are expressed in virtually all cell types because they code for proteins involved in structural and metabolic functions essential to most cells, and (2) **tissue-specific genes,** which are expressed in only certain cell types because they code for proteins involved in the specialized functions of particular kinds of differentiated cells.

During cell differentiation, the expression of the appropriate tissue-specific genes must be activated. In many cases, this activation occurs at the level of gene transcription. We learned in Chapter 10, for example, that the changes in the pattern of gene transcription that accompany the development of certain insects can be monitored by observing the puffing patterns of polytene chromosomes (page 447). Such observations have revealed that the transcription of particular genes is turned on (and turned off) at appropriate stages of development. Other examples of tissue-specific genes whose transcription is known to be activated during development include the ovalbumin gene in the chick oviduct, and the globin gene in developing red blood cells. As we discussed in Chapter 10, the molecular mechanisms involved in turning on and off the transcription of specific genes usually involves the participation of *gene-specific transcription factors* (pages 458–465).

Once a particular gene has been transcribed, control of its expression can still be exerted at one of the many succeeding steps that intervene between gene transcription and formation of the final protein product encoded by the gene. First, the initial RNA transcript must undergo posttranscriptional processing steps such as RNA splicing, addition of poly-A, 5' capping, and transport to the cytoplasm. All of these steps are potential sites at which control of gene expression can be exerted. The splicing step is especially important because RNA precursors can often be spliced in more than one way, allowing the same precursor to generate different mRNAs (page 465). Alternative splicing pathways usually generate mRNAs coding for different forms of the same protein, but in some cases mRNAs coding for distinctly different proteins can be spliced from the same RNA precursor. For example, in thyroid cells the splicing of a particular RNA precursor yields an mRNA coding for the hormone *calcitonin,* whereas splicing of the same RNA precursor in nerve cells produces an mRNA coding for a neuropeptide called *CGRP*

After mature mRNA molecules have been produced and transported to the cytoplasm, their expression is subject to further regulation by translational control mechanisms such as changes in translational repressors, protein synthesis factors, and mRNA life span (pages 504–507). And even after an mRNA molecule has been translated, the functional activity of the resulting polypeptide can be controlled by posttranslational events such as cleavage of peptide bonds, modification of amino acid side chains, association with other polypeptide chains, and alterations in the rate of protein degradation (pages 509–511). Thus a large number of transcriptional, posttranscriptional, translational, and posttranslational control mechanisms can be invoked to regulate gene expression during development.

Changes in Gene Expression Can Determine the Developmental Fate of Differentiating Cells

Many of the genes whose transcription is activated during cell differentiation generate products that are the result of differentiation rather than its cause. For instance, the activation of globin gene transcription in red blood cells is one of the last steps to occur after the cell has already begun differentiating into a red blood cell. But in some cases, transcription of a particular gene actually determines a cell's developmental fate. An example of such a gene is the *lin-12* gene of the roundworm *Caenorhabditis.* In developing *Caenorhabditis* embryos, a pair of neighboring cells has been identified that differentiate in different ways; one cell gives rise to a uterine anchor cell, while the other develops into a ventral uterine precursor cell. In mutant organisms in which the *lin-12* gene is not transcribed, both cells develop into anchor cells; conversely, in mutants where the *lin-12* gene is transcribed at excessively high rates, the two cells both develop into ventral uterine precursors. These two cells are just one of several pairs of cells in different parts of the body whose developmental fates are influenced by transcription of the *lin-12* gene.

The *Notch* gene in *Drosophila* is another example of a gene whose transcription influences the pathway of cell differentiation. Soon after gastrulation in *Drosophila,* a group of about 1800 ectodermal cells differentiates into two cell types, neuroblasts and precursors of the hypodermis. But in mutant organisms that are deficient in *Notch* gene transcription, these cells develop solely into neuroblasts instead of a mixture of neuroblasts and hypodermal precursors. Genes that govern a cell's developmental fate have also been identified in vertebrates. In vertebrate muscle cells, transcription of the *MyoD* gene is activated as a result of inductive interactions that occur in the developing embryo. *MyoD* codes for a transcription factor that binds to DNA sequences adjacent to muscle-specific genes, thereby activating transcription. Although *MyoD* is normally expressed only in muscle cells, it has been artificially inserted into other cell types using a viral cloning vector that promotes its transcription. When this approach is employed to promote transcription of the *MyoD* gene in pigment cells, nerve cells, fat cells, fibroblasts, and liver cells, each of these differentiated cell types begins making muscle-specific proteins.

Cell-Cell Recognition and Adhesion Play a Central Role in Morphogenesis

Cell differentiation does not occur in isolation, but as a coordinated part of the development of an organism as a whole. Hence embryonic development requires not just the differentiation of cells into specialized cell types, but also the spatial ordering of differentiated cells into tissues and organs. This process of arranging cells into multicellular patterns is called **morphogenesis** ("development of form"). During morphogenesis, cells often change positions relative to one another and become associated with other types of cells. In order for this process to create the proper multicellular patterns, cells must have the ability to selectively recognize and adhere to cells of the appropriate type. As we learned in Chapter 6, this recognition and adhesion is made possible by cell surface glycoproteins such as *cadherins* and *N-CAMs*, which mediate cell-cell interactions in which adhesive molecules present on the cell surface of one cell bind to adhesive molecules of the same type present on another cell (page 233).

The importance of selective cell-cell adhesion to the process of tissue and organ formation was first suggested by the studies of Aaron Moscona on embryonic cells that had been artificially dissociated from one another and then recombined. He discovered that when cells of two different types are dissociated and then intermixed, they first bind randomly to each other; eventually, however, they become arranged into two groups, a central clump composed of cells of one type and an outer layer composed of cells of the second type (Figure 15-36). Such behavior mimics what often occurs during normal organ development, when a layer of one kind of cells forms around a mass composed of another type of cell.

In analyzing this sorting-out behavior, Malcolm Steinberg subsequently uncovered an interesting pattern. If a mixture of cell types A and B is found to produce aggregates in which the A cells surround the B cells, and a mixture of cell types B and C forms aggregates in which B cells surround C cells, then it can be predicted that a mixture of cell types A and C will yield aggregates in which the A cells surround the C cells. As an example, let us consider the behavior of heart, cartilage, and liver cells. In a mixture of heart and cartilage cells, the heart cells take up the external position; in a mixture of liver and heart cells, the liver cells take up the external position. It can therefore be predicted that liver cells will aggregate external to cartilage cells. Using this approach, Steinberg has arranged cells into a hierarchical table in which each cell type aggregates external to those cell types listed before it in the table and internal to those cell types listed after it (Table 15-1).

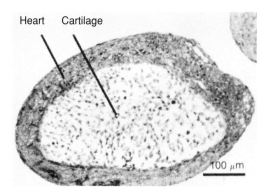

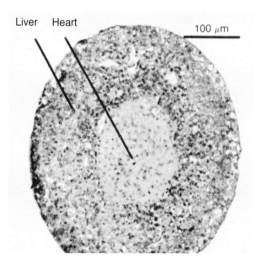

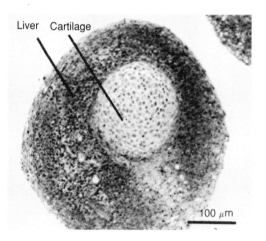

Figure 15-36 *Aggregation Patterns of Liver, Heart, and Cartilage Cells* (Top) *When cartilage and heart cells are mixed together and allowed to reaggregate, the heart cells end up on the exterior.* (Middle) *When liver and heart cells are mixed together, the heart cells now end up in the interior.* (Bottom) *In a mixture of cartilage and liver cells, the liver cells end up on the exterior. Data from such studies have yielded the information summarized in Table 15-1. Courtesy of M. S. Steinberg.*

Table 15-1 Several Cell Types Organized According to Pattern of Aggregation

Epidermis

Cartilage

Pigmented epithelium (eye)

Myocardium (heart)

Nervous tissue (neural tube)

Liver

Note: The cells of each type tend to aggregate internal to the cell types situated lower in the list. Thus epidermal cells would aggregate *inside* all the other cell types listed, whereas liver cells would aggregate *outside* all the other cell types.

The preceding observations suggest that selective adhesion between cells of the same type may play an important role in tissue and organ formation. Direct support for this idea has emerged from experiments involving the cell adhesion glycoprotein N-CAM, which is detectable on all neurons in the peripheral and central nervous systems. N-CAM has been implicated in neural development by the discovery that exposing embryonic cells to antibodies against N-CAM disrupts the orderly formation of neural tissues. Moreover, N-CAM and other cell adhesion molecules appear and disappear in various regions of the embryo during normal development, further supporting the idea that these molecules are involved in guiding tissue and organ formation.

In addition to guiding the formation of tissues and organs, adhesion between cells is also required for the expression of certain metabolic functions. An example

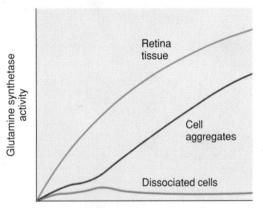

Figure 15-37 *Relationship between Cell-Cell Contacts and Enzyme Induction in Retinal Cells* *In intact retina tissue, exposure to the hormone hydrocortisone stimulates synthesis of the enzyme glutamine synthetase (which catalyzes formation of the amino acid, glutamine). The response to hormone treatment does not occur in dissociated retina cells unless they are allowed to aggregate into clusters.*

of this phenomenon has been observed in embryonic retina cells, where the enzyme *glutamine synthetase* is normally synthesized in response to stimulation by the hormone hydrocortisone. Hormone stimulation of glutamine synthetase does not occur in isolated retina cells that have been separated from one another, but if the isolated cells are first allowed to aggregate into multicellular clusters, formation of the enzyme then becomes sensitive to the addition of hormone (Figure 15-37).

Several potential mechanisms may underlie the ability of cell-cell adhesion to influence tissue function. One is that contact between adjacent cell surfaces might cause changes in plasma membrane receptors that trigger metabolic alterations within the cells themselves. Another possibility is that cell-cell adhesion stimulates the formation of gap junctions, permitting the direct passage of molecules and/or electrical signals between adjacent cells (page 240). Such interactions would presumably facilitate the development of tissues and organs by enhancing communication between neighboring cells.

Cell Migration and Changes in Cell Shape Also Contribute to Morphogenesis

In animal morphogenesis, cell movements and changes in cell shape play important roles in early embryonic development. The first contribution of cell migration occurs during gastrula formation, when a group of cells migrates from the surface of the blastula to its interior cavity, converting a hollow ball of cells into a multilayered sphere. Cells migrate over much longer distances later in development. For example, cells located above the developing spinal cord in a region called the *neural crest* are destined to migrate to diverse regions of the embryo, ultimately differentiating into cartilage, nerve cells, glial cells, pigment cells, and adrenal cells. Long-distance migration also occurs with primordial germ cells, which originate near the developing gastrointestinal tract and later migrate to their final position in the ovary or testis.

How are the preceding kinds of movements guided so that cells end up in the proper location? One component that helps to guide and orient migrating cells is the extracellular matrix. Some of the evidence supporting this idea has come from studies involving amphibian embryos, where cells located near the animal pole of the blastula are surrounded by an extracellular matrix enriched in fibronectin. If antibodies against fibronectin are injected into the blastula, the orderly cell migrations that normally transform the blastula into a gastrula are obliterated. Another component of the matrix that influences cell migration is the basal lamina (page 231). Migration of neural crest cells, for example, is triggered by the breakdown of the surrounding basal lamina. Interactions between migrating cells and extracellular matrix components such as fibronectin and laminin are mediated by cell surface receptors called *integrins* (page 232). In *Drosophila,* mutations involving the genes coding for integrins have been shown to cause disruptions in morphogenesis that are often lethal.

Cellular movements can also be caused by changes in cell shape, even when cells remain in the same positions relative to their neighbors. This phenomenon was discussed in Chapter 13, when we described the role of actin filaments in determining cell shape. If each cell in a flat sheet of cells were to have its actin filaments contract on the same side of the cell, the sheet would be converted into a rounded mass (see Figure 13-39, *left*). Likewise, constriction of a selected group of cells within a spherical mass of cells would cause the involved portion of the tissue to either fold in (*invaginate*) or protrude out (*evaginate*), depending on where the constriction were to occur (see Figure 13-39, *middle and right*). An example of the latter phenomenon occurs in the developing chick oviduct, where glands form by evagination of cells from the oviduct wall. Joan Warren and Norman Wessels have shown that development of these evaginations is preceded by the formation of bands of actin filaments across the inner ends of the cells destined to evaginate. Contraction of the actin filaments is thought to constrict the inner ends of the cells, causing the group of cells to evaginate. Support for such an interpretation has come from studies showing that the inhibitor *cytochalasin* causes both the actin filaments to disappear and the evaginating cell mass to recede back into the oviduct.

Morphogen Gradients Allow Cells to Locate Their Positions within the Developing Embryo

Although selective cell-cell adhesion and guided cell migration are crucial to normal morphogenesis, these two phenomena do not explain how embryos produce exactly the right type of organ in each location in the embryo. For example, it is one thing to produce a bone; it is another to produce the right type of bone (e.g., leg, arm, or skull bone) in the proper location in the organism. In a growing arm bone, for example, how does the embryo know where the fingers belong? If placed at the shoulder end of the arm bone, finger bones would not be very useful!

The solution to this problem requires that cells be able to detect where they are located within the embryo, and that they be able to use this positional information to guide their own differentiation and behavior. One of the most commonly invoked explanations for how cells locate their positions within the embryo hypothesizes the existence of gradients of diffusible molecules called **morphogens.** According to this view, morphogens are synthesized and secreted at particular locations in the embryo and then diffuse into the surrounding tissues, creating gradients of gradually decreasing concentration. By using appropriate receptor molecules to "sense" the concentration of a particular morphogen, cells can assess how far they are located from the source of the morphogen.

The hypothesis that morphogen gradients impart positional information during embryonic development has been extensively investigated in developing chick limbs. The legs and wings of the chick embryo begin forming as small masses of mesodermal tissue, called *limb buds,* that protrude from the body of the embryo. As they grow longer, the limb buds develop appropriate bones, including several *digits* (analogous to fingers or toes) at the terminal end of each limb. The positional information that guides the formation of the limb is generated by a small region of mesodermal tissue, called the **zone of polarizing activity (ZPA),** which is located near the posterior junction of the limb bud with the body wall (Figure 15-38). If some of this polarizing tissue is grafted to the anterior side of another limb bud, the limb produces twice as many digits as normal. It has therefore been proposed that the polarizing mesoderm guides limb development by secreting a morphogen that diffuses from the posterior to the anterior end of the limb, thereby creating a gradient of decreasing morphogen concentration. As the concentration of morphogen decreases to various threshold values at different points in the limb, the cells respond by making specific digits.

Recent evidence suggests that the morphogen involved in guiding the process of limb development is **retinoic acid** (a derivative of vitamin A). One reason for suspecting retinoic acid is that abnormalities in fetal limb development can be induced experimentally by injecting retinoic acids into pregnant animals. Additional evidence has emerged from studies in which beads soaked in retinoic acid were grafted into the anterior margin of chick limb buds. The result was a doubling of the normal number of digits, which is exactly what happens when polarizing mesoderm is grafted to the anterior side of a limb bud whose own posterior mesoderm is intact. Moreover, in limb buds whose polarizing mesoderm has been surgically removed, normal limb development can be reestablished by implanting retinoic acid–soaked beads in place of the missing posterior mesoderm.

How do morphogens such as retinoic acid exert their effects on target cells? Retinoic acid binds to a nuclear receptor that resembles the receptors for steroid and thyroid hormones. As we learned in Chapter 10, such receptors function as transcription factors that activate the transcription of specific genes by binding to specific DNA sequences located upstream from the promoter region (page 461). Hence gradients of retinoic acid are thought to exert their action on target cells by activating the transcription of particular groups of genes.

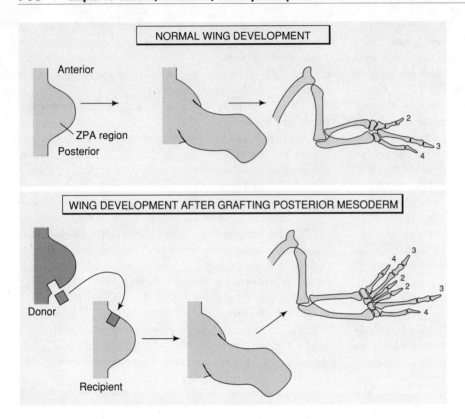

NORMAL WING DEVELOPMENT

Anterior

ZPA region
Posterior

WING DEVELOPMENT AFTER GRAFTING POSTERIOR MESODERM

Donor

Recipient

Figure 15-38 *Role of the Posterior Mesoderm in Imparting Position Information During Wing Formation in a Chick Embryo*   (Top) *During normal development, the chick wing forms three digits that are numbered 2, 3, and 4 because they correspond to the three middle fingers of a five-finger hand.* (Bottom) *If a piece of posterior mesoderm from the ZPA region (zone of polarizing activity) is grafted to the anterior side of another limb bud, the limb produces twice the normal number of digits arranged in the form of a mirror-image duplication.*

Homeotic Genes Control Development in Different Regions of the Embryo

In addition to the role played by diffusible morphogens, another way of providing positional information involves the activity of **homeotic genes,** which govern the specification of body parts in different regions of the embryo. As we learned in Chapter 10, homeotic genes code for a family of transcription factors that bind to DNA in a sequence-specific manner (page 458). Small differences in the amino acid sequence of the helix-turn-helix motif allows the transcription factors encoded by different homeotic genes to activate the transcription of different sets of genes. In this way, each homeotic gene controls the activity of a group of genes involved in specifying the formation of the body parts needed in a particular region of the embryo.

The behavior of homeotic genes has been thoroughly investigated in *Drosophila,* where a small group of homeotic genes controls differentiation in the major regions of the body. Homeotic genes called *lab* and *Dfd* control activities that give specific regions of the head

their characteristic identities, the homeotic genes *Scr, Antp,* and *Ubx* control the development of different segments of the thorax, and the homeotic genes *abdA* and *AbdB* specify different regions of the abdomen (Figure 15-39). Because homeotic genes are responsible for specifying the formation of body parts in different regions of the embryo, mutations involving these genes often have bizarre effects. A mutation in the homeotic gene *Antp,* for instance, can cause the antennae emerging from the head of a fly to be replaced by a pair of legs (see Figure 10-75). The reason for this strange effect is that the *Antp* gene is normally active in the thorax, where it specifies the production of legs; but in its mutant configuration the *Antp* gene is inverted on the chromosome, bringing it under the control of a promoter that is active in the head. The resulting expression of the *Antp* gene in the head region, where it does not normally act, leads to an unexpected pair of legs emerging from the eye sockets.

An equally bizarre mutation involves the homeotic gene *Ubx,* which governs the formation of the thoracic region adjacent to the area where the wings normally

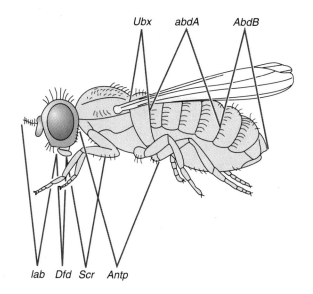

Figure 15-39 *Homeotic Genes in Drosophila Each homeotic gene controls the formation of body parts located in different regions of the embryo. The genes called* lab *and* Dfd *specify regions of the head;* Scr, Antp, *and* Ubx *control segments of the thorax; and* abdA *and* AbdB *specify regions of the abdomen.*

form. When the *Ubx* gene is deleted, the thoracic segment normally controlled by *Ubx* comes under the influence of the adjoining thoracic region, where the wings form. As a result, both regions of thorax produce pairs of wings, generating a total of four wings instead of the normal two.

Programmed Cell Death Is a Normal Part of Morphogenesis

Although cell death might seem to be an undesirable phenomenon, the destruction of specific cells at appropriate times plays an important role in normal embryonic development. During the formation of human limbs, for example, cell death causes the tissue located between the developing fingers and toes to break down and disappear. If these cells were not destroyed, humans would have webbed hands and feet resembling the feet of a duck. Another example occurs in the developing nervous system, where nerve cells are initially overproduced and then die in particular locations, thereby shaping the final pattern of the brain. And finally in the developing immune system, lymphocytes that would mistakenly attack an individual's own tissues are selectively destroyed (page 739).

The preceding examples represent a few of the numerous situations in which a phenomenon known as *programmed cell death* leads to the destruction of specifically targeted cells. Cells that undergo programmed cell death are usually destroyed by a rapid, active process that has been named **apoptosis.** During

apoptosis, cells quickly shrink and shed tiny membrane vesicles that are ingested and destroyed by neighboring cells. This scenario, which takes only a few minutes, contrasts markedly with cell death arising from injury, where cells gradually swell over a period of several hours and then burst (a phenomenon called **lysis**).

In recent years, the genes that control apoptosis have begun to be identified. One of the better understood examples occurs in the roundworm *Caenorhabditis*, which contains only 959 cells. During embryonic development a total of 1090 cells are actually produced, but 131 of them die. The activities of two genes called *ced-3* and *ced-4* are involved in programming the death of these cells; mutations that inactivate either *ced-3* or *ced-4* permit most of the 131 cells that would normally die during development to survive instead. Another gene, called *ced-9,* codes for a product that inhibits the activities of *ced-3* and *ced-4,* thereby helping to ensure that only the proper cells are programmed for cell death. Mutations that inactivate *ced-9* are lethal to the organism because they allow *ced-3* and *ced-4* to destroy cells that normally should survive. It can therefore be concluded that the normal function of *ced-9* is to prevent *ced-3* and *ced-4* from killing those cells that are supposed to live.

Genetic Factors and Oxidative Damage Contribute to Cell Death Associated with Aging

Cell death is not restricted to embryonic development. As organisms grow older, cells begin to deteriorate and die. The possibility that the process of cell aging and death is under genetic control was first suggested in 1961 when Leonard Hayflick reported that normal human fibroblasts have a built-in limit to the number of times they can proliferate. His experiments revealed that fibroblasts taken from an embryo and grown in culture divide about 50 times before they deteriorate and die (Figure 15-40). In contrast, fibroblasts taken from adults multiply only 15–30 times before dying. And fibroblasts isolated from young children suffering from *Werner's syndrome,* a rare disease that causes youngsters to age prematurely, divide only 2–10 times in culture. Further evidence for a relationship between aging and a cell's proliferative capacity came with the discovery that the number of times a cell can divide in culture is related to the life span of the organism. Thus cells of the Galápagos tortoise, whose maximum life span is about 175 years, divide more than a 100 times in culture before dying, whereas cells obtained from mice, whose life expectancy is only a few years, divide fewer than 30 times (Figure 15-41).

Since different kinds of organisms have cells that exhibit differing life spans in culture, the cellular events associated with aging must to some extent be genetically

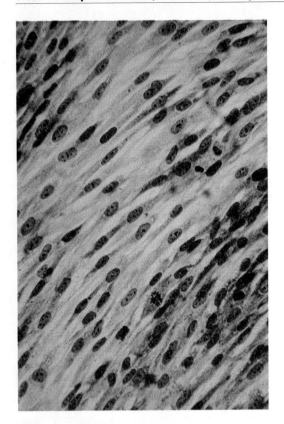

Figure 15-40 *Microscopic Appearance of Young and Old Human Fibroblasts Maintained in Tissue Culture* (Left) *Young fibroblasts that have divided a relatively small number of times in culture exhibit a thin, elongated shape.* (Right) *After dividing about 50 times in culture, the cells undergo a variety of degenerative changes and eventually die. Note the striking difference in morphology between old and young cells. Courtesy of L. Hayflick.*

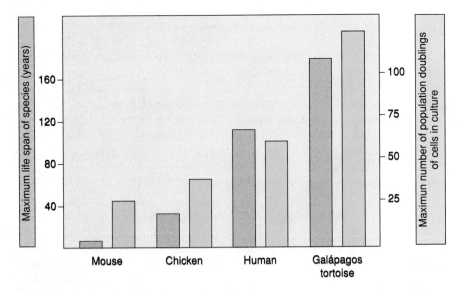

Figure 15-41 *Correlation between an Organism's Maximum Life Span and the Number of Times Its Cells Divide in Culture before Dying* *The cells of organisms with long life spans tend to proliferate longer in culture than the cells of organisms with short life spans, suggesting that aging is to some extent genetically programmed in an organism's cells.*

determined. However, random cell damage has also been implicated in cellular aging. Many biological reactions generate highly reactive compounds, called **free radicals,** which contain an unpaired electron that allows them to oxidize, and thereby damage, nucleic acids, proteins, and lipids. Numerous observations suggest that such oxidative damage contributes to cellular aging. One set of experiments involves long-lived strains of *Drosophila,* which are generated by selecting and breeding flies that live longer than normal. These long-lived strains of *Drosophila* have been found to produce an unusually active version of *superoxide dismutase,* an enzyme that protects cells from oxidative damage by breaking down a potentially dangerous free radical called *superoxide.* Comparable findings have emerged from studies involving the roundworm *Caenorhabditis,* where mutational inactivation of a single gene called *age-1* almost doubles the life span of the organism. Inactivation of the *age-1* gene leads to enhanced production of the enzymes superoxide dismutase and catalase, both of which protect cells from oxidative damage.

If oxidative damage plays an important role in cellular aging, where in the cell does the critical damage occur? An interesting insight into this issue has emerged from studies in which nuclei obtained from young cultured cells were fused with old cells whose nuclei had been removed. Such cells, with their "young" nuclei and "old" cytoplasm, continue to divide like young cells; on the other hand, cells constructed of "young" cytoplasm and "old" nuclei exhibit the limited proliferative capacity that characterizes old cells. These results suggest that the aging of cultured cells is caused largely by events occurring in the nucleus, although they do not define the nature of the nuclear events that are involved. An obvious possibility is accumulated damage to the nuclear DNA and/or proteins. But other kinds of alterations have also been observed in the nuclei of aging cells. For example, changes occur in the telomere sequences that cap both ends of every chromosome and protect them from losing genetic information (page 533). As cells proliferate and age, a tiny bit of each telomere is lost each time the chromosomal DNA is replicated. The loss of telomere sequences may eventually interfere with DNA replication in aging cells, thereby preventing the cells from proliferating. Interestingly, telomere length does not decrease in dividing spermatogonia, which produce the sperm cells that are involved in creating new embryos.

The evidence discussed in this section is consistent with the conclusion that genetic factors and random oxidative damage are both involved in cellular aging and the loss of cellular proliferative capacity. Further insights into the aging process may eventually be gained by studying the properties of cells that do not appear to age and die at all. One cell type in this category is the cancer cell. As we will learn in Chapter 18, cancer cells

are immortal when placed in culture, continuing to divide for as long as they are provided with adequate nutrients. Although cancer is certainly detrimental to the organism as a whole, something learned from studying these immortal cells may one day help us to prevent the losses in cell function that accompany the normal process of aging.

SUMMARY OF PRINCIPAL POINTS

• During oogenesis in animals, meiotic division of an oogonium produces one haploid egg cell and three polar bodies. Meiosis is delayed at prophase I for a period of cell growth that allows the oocyte to produce and store mRNAs, ribosomes, and nutrients for use after fertilization. The growth phase is also accompanied by the formation of cortical granules, exterior egg coats, and the establishment of a cytoplasmic asymmetry that creates distinct animal and vegetal poles.

• During spermatogenesis in animals, meiotic division of spermatogonia yields haploid spermatids that subsequently differentiate into sperm cells. Mature sperm possess a head, which contains the nucleus and acrosome, and a motile tail, which consists of a long flagellum partially enveloped by mitochondria.

• In flowering plants, meiotic division of a megasporocyte is followed by three mitotic divisions of one of the resulting megaspores, yielding one egg cell, two polar nuclei, two synergid cells, and three antipodal cells. Male gametes are formed by meiotic division of microsporocytes, yielding haploid microspores that develop thick spore coats and mature into pollen grains. The haploid nucleus of each pollen grain divides mitotically, yielding a tube cell and a generative cell that eventually undergoes another mitosis, producing two haploid sperm cells.

• During fertilization the sperm and egg fuse, reestablishing the diploid chromosome number. Contact between sperm and egg causes the acrosomal vesicle to discharge its enzymes into the extracellular space, where the enzymes digest a pathway through the egg coats. In many marine invertebrates, discharge of the acrosomal contents is followed by the formation of an acrosomal process that helps the sperm penetrate through to the egg plasma membrane.

• Immediately after fertilization, a transient depolarization of the egg plasma membrane establishes a quick block to polyspermy. A subsequent increase in free Ca^{2+} triggers the cortical reaction, which triggers an formation of a fertilization membrane that creates a long-term block to polyspermy. Finally, an increase in intracellular pH triggers an activation of protein synthesis and the initiation of DNA replication.

• During fertilization in higher plants, pollen grains bind to the stigma and develop a pollen tube that penetrates into the ovary. Upon reaching the embryo sac, one sperm cell from the pollen grain fuses with the egg to form a diploid zygote, and another sperm cell fuses with the two polar nuclei, yielding the triploid endosperm.

• The cleavage stage of animal embryonic development divides the fertilized egg into smaller and smaller cells, yielding a spherical mass of cells termed a blastula. When cleavage ends, cells located above the blastopore migrate into the blastula interior, producing a three-layered gastrula composed of ectoderm, mesoderm, and endoderm.

• The developmental fate of individual cells begins to be determined during cleavage and early gastrulation. One mechanism for determining a cell's developmental fate involves the inheritance of morphogenic determinants located in the cytoplasm. Because of the asymmetric organization of the egg cytoplasm, cells derived from successive divisions of the egg contain dissimilar cytoplasms. A cell's developmental fate can also be determined by inductive interactions between neighboring cells. During primary embryonic induction, for example, ectoderm is induced to differentiate into neural tissue by cells that have invaginated into the gastrula from the dorsal lip of the blastopore. Inductive interactions can be mediated by cell-cell contact, cell-matrix contact, or diffusion of signaling molecules between cells.

• After a cell's developmental fate has been determined, the actual process of cell differentiation is accomplished by changing the pattern of genes that are being expressed. At the molecular level, a large number of transcriptional, posttranscriptional, translational, and posttranslational control mechanisms are invoked to accomplish this regulation of gene expression.

• In addition to cell differentiation, embryonic development also requires the spatial ordering of cells into tissues and organs. During this process, selective cell-cell recognition and adhesion permit cells of differing types to arrange themselves into organized patterns. Cell movements, which are often guided by interactions involving the extracellular matrix, also play an important role during morphogenesis.

• Morphogen gradients allow cells to determine their relative positions within the developing embryo. During limb development, for example, cells located near the posterior margin of the limb bud produce the morphogen retinoic acid, thereby forming a gradient that determines where the various limb bones will be formed. Positional information is also provided by homeotic genes, which govern the specification of body parts in different regions of the embryo. Each homeotic gene produces a transcription factor that controls the activity of a group of genes involved in specifying a certain set of body parts. Mutations in homeotic genes often have bizarre effects, such as the formation of an extra set of wings or legs emerging from the head of a fly.

• Normal morphogenesis is accompanied by the programmed death of selected cells. Cell deterioration and death are also associated with aging, which is caused by a combination of genetic factors and random oxidative damage.

SUGGESTED READINGS

Books

Edelman, G. M. (1988). *Topobiology: An Introduction to Molecular Embryology,* Basic Books, New York.

Gilbert, S. F. (1991). Developmental Biology, 3rd Ed., Sinauer Associates, Sunderland, MA.

Longo, F. J. (1987). *Fertilization,* Chapman and Hall, New York.

Russell, S. D., and C. Dumas, eds. (1992). *Sexual Reproduction in Flowering Plants,* International Review of Cytology, Vol. 140, Academic Press, San Diego.

Articles

Bedinger, P. A., K. J. Hardeman, and C. A. Loukides (1994). Travelling in style: the cell biology of pollen, *Trends Cell Biol.* 4:132–138.

Brickell, P. M., and C. Tickle (1989). Morphogens in chick limb development, *BioEssays* 11:145–149.

Carlson, J. G. (1952). Microdissection studies of the dividing neuroblast of the grasshopper, *Chortophaga viridifasciata* (De Geer), *Chromosoma* 5:199–220.

DeRobertis, E. M., G. Oliver, and C. V. E. Wright (1990). Homeobox genes and the vertebrate body plan, *Sci. Amer.* 263 (July):46–52.

Edelman, G. M. (1989). Topobiology, *Sci. Amer.* 260 (May):76–88.

Garbers, D. L. (1989). Molecular basis of fertilization, *Annu. Rev. Biochem.* 58:719–742.

Greenwald, I. S., P. W. Sternberg, and H. R. Horvitz (1983). The *lin-12* locus specifies cell fates in *Caenorhabditis elegans, Cell* 34:435–444.

Gurdon, J. B. (1962). Adult frogs derived from the nuclei of single somatic cells, *Develop. Biol.* 4:256–273.

Gurdon, J. B. (1992). The generation of diversity and pattern in animal development, *Cell* 68:185–199.

Hayflick, L. (1980). The cell biology of human aging, *Sci. Amer.* 242 (January):58–65.

Hengartner, M. O., R. E. Ellis, and H. R. Horvitz (1992). *Caenorhabditis elegans* gene *ced-9* protects cells from programmed cell death, *Nature* 356:494–499.

Hill, D. P., and S. Strome (1990). Brief cytochalasin-induced disruption of microfilaments during a critical interval in 1-cell *C. elegans* embryos alters the partitioning of developmental instructions to the 2-cell embryo, *Development* 108:159–172.

Jaffe, L. A. (1976). Fast block to polyspermy in sea urchins is electrically mediated, *Nature* 261:68–71.

Lawrence, P. A., and G. Morata (1994). Homeobox genes: Their function in Drosophila segmentation and pattern formation, *Cell* 78:181–189.

Martin, S. J., D. R. Green, and T. G. Cotter (1994). Dicing with death: dissecting the components of the apoptosis machinery, *Trends Biochem. Sci.* 19: 26–30.

Montell, D. J. (1994). Moving right along: regulation of cell migration during *Drosophila* development, *Trends Genet.* 10:59–62.

Orr, W. C., and R. S. Sohal (1994). Extension of life-span by overexpression of superoxide dismutase and catalase in *Drosophila melanogaster, Science* 263:1128–1130.

Rose, M. R., T. J. Nusbaum, and J. E. Fleming (1992). *Drosophila* with postponed aging as a model for aging research, *Lab. Anim. Care* 42:114–118.

Rusting, R. L. (1992). Why do we age? *Sci. Amer.* 267 (December):130–141.

Sardet, C., A. McDougall, and E. Houliston (1994). Cytoplasmic domains in eggs, *Trends Cell Biol.* 4:166–172.

Shepard, J. F. (1982). The regeneration of potato plants from leaf-cell protoplasts, *Sci. Amer.* 246 (May): 154–166.

Sidhu, K. S., and S. S. Guraya (1991). Current concepts in gamete receptors for fertilization in mammals, *Int. Rev. Cytol.* 127:253–288.

Steinberg, M. S. (1970). Does differential adhesion govern self-assembly processes in histogenesis? Equilibrium configurations and the emergence of a hierarchy among populations of embryonic cells, *J. Exp. Zool.* 173:395–434.

Sulston, J. E., E. Schierenberg, J. G. White, and J. N. Thomson (1983). The embryonic cell lineage of the nematode *Caenorhabditis elegans, Develop. Biol.* 100:64–119.

Tafuri, S. R., and A. P. Wolffe (1993). Dual roles for transcription and translation factors in the RNA storage particles of *Xenopus* oocytes, *Trends Cell Biol.* 3:94–98.

Thomsen, G., et al. (1990). Activins are expressed early in *Xenopus* embryogenesis and can induce axial mesoderm and anterior structures, *Cell* 63:485–493.

Tickle, C., J. Lee, and G. Eichele (1985). A quantitative analysis of the effect of all-*trans*-retinoic acid on the pattern of chick wing development, *Dev. Biol.* 109:82–95.

Ward, G. E., C. J. Brokaw, D. L. Garbers, and V. D. Vacquier (1985). Chemotaxis of *Arbacia punctulata* spermatozoa to resact, a peptide from the egg jelly layer, *J. Cell Biol.* 101:2324–2329.

Wasserman, P. (1988). Fertilization in mammals, *Sci. Amer.* 259 (December):78–84.

Watanabe, N., G. F. Vande Woude, Y. Ikawa, and N. Sagata (1989). Specific proteolysis of the *c-mos* proto-oncogene product by calpain on fertilization of *Xenopus* eggs, *Nature* 342:505–511.

Weigel, D., and E. M. Meyerowitz (1994). The ABCs of floral homeotic genes, *Cell* 78:203–209.

Weintraub, H. et al (1989). Activation of muscle-specific genes in pigment, nerve, fat, liver, and fibroblast cell lines by forced expression of MyoD, *Proc. Natl. Acad. Sci. USA* 86:5434–5438.

Chapter 16

Lymphocytes and the Immune Response

Animals fight a constant battle to protect themselves from infection by potentially harmful agents such as bacteria, viruses, fungi, and parasites. The defensive strategies employed in this battle are most sophisticated in higher vertebrates, which have developed an immune system that can selectively destroy or inactivate foreign molecules and cells without harming the host's own molecules or normal cells. Moreover, the immune system retains a memory of each attack, allowing it to respond more efficiently the next time it encounters the same invader. That is why individuals rarely suffer more than once from infectious diseases such as chicken pox, mumps, or whooping cough.

The ability of the immune system to defend an organism against foreign invaders is based on the properties of specialized white blood cells called *lymphocytes.* Lymphocytes attack foreign cells and molecules in two different ways: by secreting antibody molecules that bind to specific molecules and by killing cells directly. In this chapter we will see how these two defensive strategies allow animals to protect themselves against millions of potentially harmful agents.

OVERVIEW OF THE IMMUNE RESPONSE

Antigens Are Substances That Elicit an Immune Response

Any molecule that is capable of provoking an immune response is referred to as an **antigen.** Several factors influence the ability of substances to act as antigens. First, a molecule must usually be recognized as being foreign. The more a foreign substance differs in structure from substances found in an animal's own tissues, the greater the intensity of the immune response mounted against that foreign substance. The second factor that influences a molecule's ability to function as an antigen is its size and shape; large, complex molecules tend to be better antigens than small, simple molecules. Finally, antigens must be susceptible to degradation and processing by cells. As a consequence, stainless steel pins and plastic valves can be implanted into humans without eliciting an immune response because these materials cannot be degraded by cells.

When an animal is invaded by a foreign microorganism, the immune system is simultaneously exposed to a diverse array of new antigens, including proteins, lipopolysaccharides, polysaccharides, and lipids. The larger antigens typically exhibit several surface regions called **epitopes,** each of which elicits its own separate immune response. The number of epitopes present in any given molecule is related to the molecule's size; proteins, for example, typically have about one epitope for every 50 amino acids. Thus the immune response provoked by an invading microorganism is a mixture of many separate immune responses directed against a multitude of foreign antigens and against numerous epitopes exhibited by each antigen.

Antigens Are Processed by Antigen-Presenting Cells

In order to trigger an efficient immune response, antigens must first be processed by an **antigen-presenting cell (APC)** that degrades the antigen and "presents" the resulting antigen fragments to other cells involved in the immune response. **Macrophages** are among the most commonly encountered type of APC; other examples include *dendritic cells* in the spleen and *Kupffer cells* in the liver. Macrophages are large round cells measuring about 15–20 μm in diameter that reside mainly in the spleen, bone marrow, and lymph nodes. Immature macrophages called *monocytes* also circulate in the bloodstream, where they account for about 5 percent of the white blood cells. Macrophages readily engulf foreign particles and cells by the process of phagocytosis (page 293). Although all macrophages are avid phagocytes, only some can process foreign antigens in a way that stimulates an immune response. The macrophages that perform this function carry specialized plasma membrane glycoproteins called **major histocompatibility complex (MHC) class II molecules.** MHC class II molecules reside only on the surfaces of cells involved in immune responses.

The processing of foreign antigens by macrophages and other APCs involves four main stages (Figure 16-1, *left*). First, antigens are taken into the cell by a relatively nonspecific form of endocytosis that brings all mole-

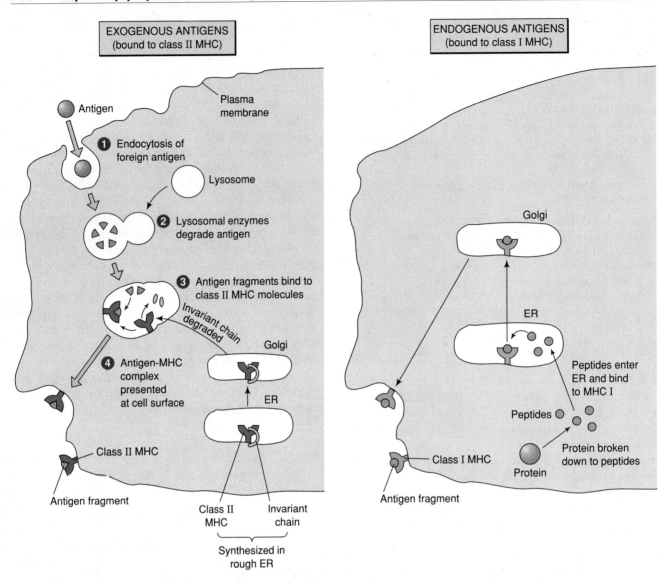

EXOGENOUS ANTIGENS
(bound to class II MHC)

Antigen

Plasma membrane

1 Endocytosis of foreign antigen

Lysosome

2 Lysosomal enzymes degrade antigen

3 Antigen fragments bind to class II MHC molecules

Invariant chain degraded

Golgi

4 Antigen-MHC complex presented at cell surface

ER

Class II MHC

Antigen fragment

Class II MHC

Invariant chain

Synthesized in rough ER

ENDOGENOUS ANTIGENS
(bound to class I MHC)

Golgi

ER

Peptides enter ER and bind to MHC I

Peptides

Protein broken down to peptides

Class I MHC

Protein

Antigen fragment

Figure 16-1 *Pathways for Processing Exogenous and Endogenous Antigens* (Left) *Exogenous antigens are taken up by macrophages and other antigen-processing cells via the process of endocytosis. The internalized antigen is then broken into fragments, which are bound to MHC class II molecules and transported to the cell surface for presentation to other cells of the immune system. The invariant chain is a polypeptide that transiently associates with class II MHC molecules, preventing class II molecules from binding to peptide fragments in the ER or Golgi complex. (Right) Endogenous antigens are molecules that originate within the host cell. Endogenous antigens are cleaved into fragments that are transported into the ER lumen, where they bind to class I MHC molecules that carry the antigen fragments to the cell surface.*

cules dissolved in the extracellular fluid into the cell. Next, the internalized endocytic vesicles fuse with lysosomes, whose hydrolytic enzymes degrade the foreign antigens into small fragments; in the case of ingested protein antigens, the process yields peptide fragments about 10–20 amino acids long. Third, the resulting fragments become bound to MHC class II molecules, which possess an *antigen-binding cleft* that is specifically designed to bind antigen fragments. And finally, the foreign antigen–MHC II complex is transported to the cell surface for "presentation" to other cells of the immune system. The importance of the preceding pathway has been demonstrated by treating cells with inhibitors that

block endocytosis or the lysosomal degradation of antigens; under such conditions, an immune response is inhibited.

The pathway that was just described is used in the processing of *exogenous antigens*—that is, antigens that have been taken up from outside the cell. Immune responses can also be triggered by *endogenous antigens* that originate within the cells of the host organism (see Figure 16-1, *right*). For example, cells that have become infected by a virus usually manufacture foreign proteins encoded by viral genes. Fragments derived from the viral proteins become bound to a different type of MHC molecule called an **MHC class I**

molecule; unlike class II MHC molecules, which are restricted to the immune system, class I molecules are expressed by almost all nucleated cells. Peptide antigens destined to bind to class I MHC molecules are produced by protein cleavage in the cytosol and are then transported into the ER lumen, where the fragments bind to class I molecules associated with the ER membrane. The resulting antigen–MHC I complex is then transported to the cell surface, where its presence can trigger an immune response against the infected cell. (Peptide fragments derived from a cell's own normal proteins can also bind to MHC class I molecules and be presented at the cell surface, but such "self" antigen–MHC complexes do not usually elicit an immune response.)

Class I and class II MHC molecules are both synthesized in the rough ER. Yet peptide fragments that originate inside the cell and move into the lumen of the ER bind to class I, but not class II, MHC molecules. This selectivity is made possible by the *invariant chain,* a polypeptide that associates with class II MHC molecules in the ER and prevents them from binding to peptide fragments (see Figure 16-1, *left*). After class II MHC molecules have been transported to endocytic vesicles, the invariant chain is degraded by lysosomal enzymes and the class II molecule is then free to bind antigen fragments.

B Cells Produce Antibodies and T Cells Are Responsible for Cell-Mediated Immunity

After an antigen has been processed and presented on the surface of a macrophage or other APC, an immune response is triggered in a specialized class of white blood cells called **lymphocytes.** Lymphocytes are small round cells ranging from 7 to 15 μm in diameter that occur in large numbers in lymphoid organs such as the spleen, lymph nodes, and thymus. They also account for about 25 percent of the white blood cells circulating in the bloodstream. Lymphocytes in the blood can pass through capillary walls and enter into most tissues, where they ultimately make their way to a system of lymphatic vessels that transfer the cells back to the venous side of the circulatory system. Humans contain a trillion or more lymphocytes, giving the immune system a total mass equal to that of a large organ such as the liver.

Lymphocytes are subdivided into two principal classes, **B cells** and **T cells,** both of which arise from the same type of undifferentiated *stem cell* located in the bone marrow (Figure 16-2). Some of these stem cells remain in the bone marrow and differentiate into B cells, whereas others migrate to the thymus and develop into T cells. Because they represent the sites where lymphocytes first arise from stem cells, the thymus and bone marrow are called *primary lymphoid organs.* Lymphocytes do not become completely

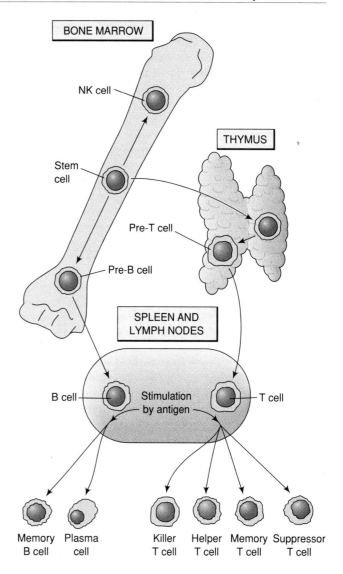

Figure 16-2 Developmental Origins of B Cells, T Cells, and NK Cells *All three types of lymphocytes develop from the same undifferentiated stem cell, which initially resides in the bone marrow. Some of these stem cells remain in the bone marrow and develop into B cells and NK cells, whereas others migrate to the thymus and develop into T cells. Maturation of T cells and B cells into functionally competent cells occurs in secondary lymphoid organs such as lymph nodes and spleen, and requires the presence of antigen.*

functional until they migrate to the *secondary lymphoid organs,* mainly the lymph nodes and spleen.

B cells and T cells are involved in two different types of immunity. B cells develop into cells that produce **antibodies,** which are proteins that circulate in the bloodstream and penetrate into extracellular fluids, where they bind to the foreign antigen that induced the immune response. In contrast, T cells are involved in **cell-mediated immunity,** a process in which T cells bind to and kill cells that exhibit foreign antigens on their surface. This type of immunity is responsible for killing virus-infected cells, and for rejecting skin grafts and organ transplants. Several different kinds of B cells and T cells are involved in antibody formation and cell-mediated immunity. The functions of these various cell types—including *memory B cells, plasma cells, helper T cells, killer T cells, memory T cells,* and *suppressor T cells*—will be described later in the chapter when we discuss the mechanism of antibody formation and cell-mediated immunity in detail. In addition to the various types of B and T cells, a small fraction of the total lymphocyte population consists of **natural killer (NK) cells** that develop in the bone marrow but are neither B cells nor T cells (see Figure 16-2). Natural killer cells are capable of destroying tumor cells and virus-infected cells in the absence of either antibody formation or a typical cell-mediated immune response.

ANTIBODY STRUCTURE AND DIVERSITY

One of the most remarkable features of the immune system is that it can be stimulated by a virtually endless array of foreign antigens, and yet in each case, the immune response is precisely targeted against the specific antigen(s) that elicited the response. This selectivity is so exquisite that the immune system is capable of distinguishing between two foreign proteins that differ in only a single amino acid. As a first step in describing how the immune system responds to so many antigens with such extraordinary specificity, we will examine the structure of antibody molecules.

Antibodies Are Constructed from Light and Heavy Chains

As mentioned previously, antibodies (also called **immunoglobulins**) are proteins that are synthesized and secreted into body fluids in response to foreign antigens. Each antibody molecule is designed to bind specifically to the particular antigen that stimulated its synthesis. Antibodies are present in highest concentration in blood plasma, but also occur in saliva, tears, milk, and secretions of the respiratory and intestinal tracts. Antibody molecules are subdivided into five major classes, called *IgG, IgA, IgM, IgD,* and *IgE,* which differ in structural features, localization, and function (Table 16-1). Of the five classes, IgG is the main type of antibody found in the bloodstream.

Early studies of antibody structure were carried out in the late 1950s by Rodney Porter, who discovered that treating a crude mixture of antibodies with the protease *papain* yields three fragments. Two of these, called *Fab fragments,* are the same size and can bind to antigen. The remaining fragment, termed the *Fc frag-*

Table 16-1 Some Properties of the Five Major Classes of Antibodies

Class	Type of Heavy Chain	Type of Light Chain	Forms*	Percent of Total Antibodies in Blood	Location and Function
IgG	γ	κ or λ	Monomer	80	Main type of antibody found in blood; binds to phagocytic cells and activates complement when bound to antigen
IgA	α	κ or λ	Monomer or dimer	13	Main type of antibody found in secretions of of the digestive, respiratory, and urinary tracts, including saliva, tears, and breast milk
IgM	μ	κ or λ	Monomer or pentamer	6	Occurs in blood and on the surface of B cells, where it acts as an antigen receptor; binds and activates complement
IgD	δ	κ or λ	Monomer	0.2	Found on the surface of many B cells, where it plays a role in B-cell activation
IgE	ϵ	κ or λ	Monomer	0.002	Produced in response to parasitic infections, where it promotes the release of inflammatory agents from mast cells and basophils; involved in allergic reactions such as asthma and hay fever

*The immunoglobulin monomer consists of two heavy chains and two light chains. Some immunoglobulins also exist in the form of a dimer (two four-chain units) or a pentamer (five four-chain units).

ment, does not react with antigen but is capable of binding to *complement,* which is a group of proteins that help antibodies destroy invading microorganisms (page 732). The Fc fragment readily forms crystals ("Fc" stands for *crystallizable* fragment), which means that the Fc fragments derived from a heterogeneous mixture of antibodies are virtually identical in structure; Fab fragments, on the other hand, do not form crystals, indicating that the Fab portion of the molecule differs from antibody to antibody.

Another approach to investigating antibody structure was taken by Gerald Edelman, who used mercaptoethanol (which cleaves disulfide bonds) to dissociate antibodies into their constituent polypeptide chains. Such treatment releases two *heavy chains* of about 50,000 daltons and two *light chains* of about 25,000 daltons. To determine the relationship of the heavy and light chains to the Fab and Fc fragments, Porter ran a series of tests utilizing antiserum specifically directed against either the Fab or Fc fragments. Isolated light chains were found to react only with antiserum directed against Fab fragments, whereas heavy chains reacted with antiserum directed against both Fab and Fc fragments. It was therefore concluded that the Fab fragment contains the entire light chain and part of the heavy chain, and the Fc fragment contains the remainder of the heavy chain.

The preceding data led to the four-chain model of antibody architecture illustrated in Figure 16-3. According to this model, antibodies are Y-shaped structures whose two arms correspond to the Fab fragments and whose stem corresponds to the Fc fragment. Each Y-shaped antibody has two identical *antigen-binding sites,* one located in each of the two arms. Electron microscopic observations have confirmed the underlying Y-shaped architecture of antibodies and have revealed that the stem is joined to the arms by a flexible hinge region that allows the two arms of the "Y" to rotate (Figure 16-4).

Each of the five classes of antibody is characterized by a distinctive type of heavy chain; the heavy chains that occur in IgG, IgA, IgM, IgD, and IgE molecules are called *gamma (γ), alpha (α), mu (μ), delta (δ),* and *epsilon (ε)* chains, respectively. In addition to the five types of heavy chains, antibodies contain two kinds of light chains known as *kappa (κ)* and *lambda (λ)* chains. Although κ and λ chains occur in all five classes of antibodies, any given antibody molecule always has either two identical κ chains or two identical λ chains. Likewise, the two heavy chains that make up a basic four-chain antibody unit are always identical. Thus a four-chain unit consisting of two identical light chains and two identical heavy chains is a shared characteristic of all antibodies, regardless of whether they are members of the IgG, IgA, IgM, IgD, or IgE class. IgA mole-

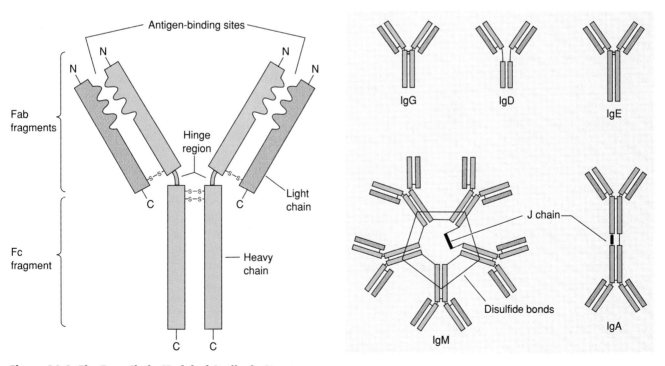

Figure 16-3 *The Four-Chain Model of Antibody Structure* (Left) *An IgG molecule is constructed from two identical heavy chains and two identical light chains. The enzyme papain cleaves the molecule at the two hinge regions, yielding two Fab fragments and a single Fc fragment.* (Right) *Schematic illustration of the five classes of antibody molecules. IgM and IgA are shown in their pentamer and dimer forms respectively. A special polypeptide called a J chain facilitates the formation of IgM pentamers and IgA dimers.*

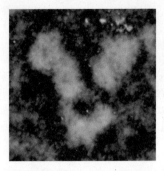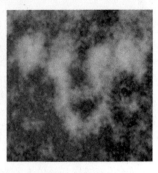

Figure 16-4 *Electron Micrographs of IgG Molecules Demonstrating the Flexibility of the Hinge Region* *The molecule on the left exhibits the typical Y-shaped configuration. In the molecule on the right the two arms have rotated toward the stem, giving the molecule a T-shaped appearance. Courtesy of D. W. Metzger.*

cules also exist as dimers containing two of these four-chain units, and IgM molecules occur as pentamers constructed from five of the four-chain units. A special polypeptide called a *J chain* facilitates the formation of IgA dimers and IgM pentamers (see Figure 16-3).

Light Chains and Heavy Chains Are Composed of Constant and Variable Regions

Blood plasma contains a complex mixture of antibodies directed against thousands of different antigens, making it difficult to investigate the structure of individual antibody molecules in normal blood samples. The first successful attempts to overcome this problem involved patients with *multiple myeloma,* a type of cancer characterized by the uncontrolled proliferation of antibody-producing cells. The urine of myeloma patients contains large amounts of an abnormal substance known as *Bence-Jones protein,* which consists of IgG light chains that have been synthesized in excess by the patient's tumor and excreted into the urine. Each patient produces a single type of Bence-Jones protein corresponding to the light chains of a single antibody molecule. The first detailed information about the amino acid sequence of Bence-Jones proteins was reported in 1965 by Norbert Hilschmann and Lyman Craig, who showed that Bence-Jones light chains obtained from two different myeloma patients exhibited differing amino acid sequences. Comparison of the two sequences revealed an unexpected pattern destined to have a dramatic impact on the field of immunology: *The differences between the amino acid sequences of the two light chains were all located at one end of the molecule.* A similar pattern was subsequently observed for other kinds of light chains as well; in all cases, light chains exhibit both a **constant region,** whose amino acid sequence is the same in all antibodies, and a **variable region,** whose sequence varies among antibodies that bind to different antigens (Figure 16-5).

In addition to the Bence-Jones light chains that appear in the urine, the blood of myeloma patients contains large quantities of the complete IgG molecule produced by the myeloma tumor cells. This *myeloma protein* is the equivalent of a homogeneous population of normal IgG antibodies directed against a single antigen. Amino acid sequencing of myeloma proteins has revealed that antibody heavy chains, like light chains, possess both variable and constant regions. The variable regions, which always occur at the N-terminal end, are about 115 amino acids long in heavy chains and 107 amino acids long in light chains, whereas the constant regions contain about 330 amino acids in heavy chains and 107 amino acids in light chains.

Sequencing studies have revealed that the constant region of the IgG heavy chain is comprised of three domains (C_H1, C_H2, and C_H3), each about 110 amino acids long, whose amino acid sequences resemble each other (and also resemble the sequence of the constant region of the light chain). The existence of these related domains suggests that the gene coding for the heavy-chain constant region evolved from the repeated duplication of a primordial gene that initially coded for a polypeptide about 110 amino acids long. The constant domains of the heavy and light chains carry out functions that are shared widely among antibodies, regardless of the particular antigen to which an antibody binds. As we will see later in the chapter, such functions include interacting with components of the complement system (page 732) and anchoring antibody molecules to the plasma membrane.

Antigen-Binding Sites Are Formed by the Hypervariable Regions of Heavy and Light Chains

The differences in antibody structure that allow each antibody to bind selectively to a particular antigen reside in the variable regions of the light and heavy chains. An early clue to the relationship between these variable regions and an antibody's antigen-binding site was provided by amino acid sequencing studies of myeloma proteins, which revealed that the sequence variability observed among myeloma proteins is not uniformly distributed throughout the variable regions of the heavy and light chains. Instead, the variation in amino acid sequence is confined largely to three small **hypervariable regions** located in the heavy chain, and three comparable regions located in the light chain. The three hypervariable regions are designated *CDR1, CDR2,* and *CDR3* starting from the N-terminus of each chain (the term "CDR" stands for "complementarity determining region"). Ranging from about 5 to 15 amino acids in length, the three hypervariable regions are located approximately 30, 50, and 95 amino acids from the N-terminus.

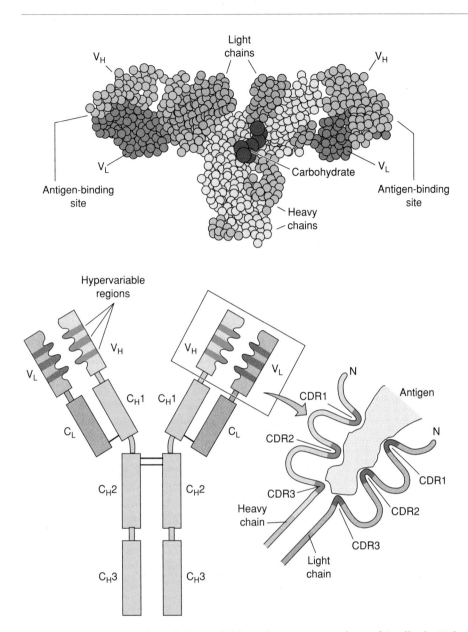

Figure 16-5 *Organization of the Variable and Constant Regions of Antibody Molecules*
(Top) *A space-filling model of an IgG molecule deduced from X-ray crystallography. Each of the small spheres represents an amino acid. Rotation of the molecule at the hinge region is responsible for its T-shaped appearance.* (Bottom) *Schematic diagram illustrating the organization of the heavy and light chains of an IgG molecule. Each light chain is composed of a constant domain (C_L) whose sequence is the same from molecule to molecule, and a variable domain (V_L) whose sequence varies between antibodies. Each heavy chain is composed of one variable domain (V_H) and three constant domains (C_H1, C_H2, and C_H3). Most of the sequence variability exhibited by antibodies is restricted to three hypervariable regions located within the V_L domain and three hypervariable regions located within the V_H domain. The heavy and light chains are folded so that these hypervariable regions (CDR1, CDR2, CDR3) come together to form the antigen-binding site.*

Since the sequence differences between antibody molecules are confined mainly to the hypervariable regions of the light and heavy chains, Elvin Kabat proposed in 1970 that the hypervariable regions come together during the folding of the heavy and light chains to form the antigen-binding site. This hypothesis was first tested in experiments that utilized radioactive antigens containing highly reactive chemical groups. When such antigens bind to their corresponding antibodies, the radioactive chemical groups of the antigen form covalent bonds with neighboring amino acids in the antibody molecule. Chemical analyses revealed that the amino acids that become radioactive are located in the hypervariable regions of the antibody molecule.

Further support for the idea that the antigen-binding site is created by the hypervariable regions came from studies employing X-ray crystallography to investigate the three-dimensional structure of antibodies bound to their corresponding antigens. Such analyses have confirmed that the hypervariable regions of the heavy and light chains are folded together as shown in Figure 16-5 to form the antigen-binding site.

Clonal Selection Allows Specific Antibodies to Be Formed in Response to Each Antigen

Every time an organism encounters a new foreign antigen, it produces antibody molecules whose antigen-binding sites are specifically designed to bind to the triggering antigen. The exquisite specificity of this response was first demonstrated in the early 1900s by Karl Landsteiner, who investigated the ability of small molecules to act as antigens. Small molecules are usually incapable of triggering antibody formation when administered by themselves, but antibody production can be elicited by attaching the small molecule, called a *hapten,* to a larger *carrier* molecule. Landsteiner discovered that every time he synthesized a new hapten by introducing a minor change in its structure, animals injected with the resulting hapten-carrier complex would produce a new antibody that was uniquely specific for the newly synthesized hapten. Thus it gradually became apparent from Landsteiner's work that animals can produce a virtually endless array of unique antibodies, including antibodies directed against substances that the organism or its ancestors could never have encountered before in nature.

One of the first theories that attempted to explain this remarkable specificity of the immune system was proposed by Linus Pauling in 1940. According to his *instructive theory* of antibody formation, antigens interact with antibodies as they are being synthesized, directing the folding of the newly forming antibody chain into a structure that specifically binds to the stimulating antigen. An alternative to the instructive theory, called the *clonal selection theory,* was developed by Niels Jerne and Macfarlane Burnet. According to the clonal selection theory, animals possess millions of different lymphocytes, each producing a different antibody; a triggering antigen simply binds to those lymphocytes that already make an antibody that is specific for the antigen in question, causing these particular lymphocytes to proliferate and make more of the same type of antibody. Thus the clonal selection theory postulates that an antigen *selects* cells that are already making a particular antibody rather than *instructing* cells how to make such an antibody.

The instructive theory is easy to test experimentally because it postulates that an antigen must be present during antibody folding to guide the antibody into its proper three-dimensional shape. Experiments have

therefore been performed in which antibodies are unfolded by treating them with a denaturing agent, which causes the antibody molecule to lose its ability to bind to its corresponding antigen. When the denaturing agent is subsequently removed, antibody molecules spontaneously regain the ability to bind to the appropriate antigen, *even if the antigen is absent during the renaturation process.* It has therefore been concluded that the ability of an antibody to bind to its appropriate antigen is encoded in the antibody's amino acid sequence rather than being instructed by the antigen during antibody folding. Such a conclusion is incompatible with the instructive theory of antibody formation, but it does not contradict the clonal selection theory.

The central premise of the clonal selection theory has also been evaluated experimentally. The clonal selection theory postulates that each antibody-producing cell and its offspring make a single type of antibody directed against a particular antigen. This prediction has been tested by isolating single lymphocytes and analyzing the types of antibodies made by individual cells and their descendants. Such experiments have revealed that *each antibody-producing cell makes only a single type of antibody, and that cells derived from the proliferation of any single lymphocyte all make the same type of antibody.* This observation and numerous others have provided strong confirmation for the idea that the specificity of the immune response is based on **clonal selection.** In other words, animals are able to produce antibodies directed against a vast array of different antigens because every organism possesses millions of different lymphocytes, each capable of making a unique antibody before it has ever encountered the corresponding antigen. The role played by a stimulating antigen is simply to trigger the proliferation of those lymphocytes that are already manufacturing the appropriate antibody.

DNA Segments Coding for the Variable and Constant Regions of Antibody Molecules Are Rearranged in Antibody-Producing Cells

The notion that organisms spontaneously make millions of different antibodies raises a fundamental genetic paradox: If every antibody molecule were to be encoded by a separate gene, then virtually all of an individual's DNA would be occupied by the millions of required antibody genes. One simple way of reducing the number of required genes is to assemble antibodies from different combinations of heavy and light chains. For example, if a hypothetical organism were to possess 1000 different light-chain genes and 1000 different heavy-chain genes, assembling the various heavy and light chains in all possible combinations would generate $1000 \times 1000 = 1,000,000$ different kinds of antibody molecules. Hence in theory, a million different antibodies could be specified by only 2000 genes, which would require less than 1 percent of an organism's DNA. This idea that the genes coding for all the antibodies an organism can manufacture are separately inherited in an individual's DNA is called the *germ line theory*, for it assumes that the genes coding for every antibody molecule are inherited through the gametes (sperm and egg) that develop from the primordial germ cells (page 674).

Another potential way of producing millions of different antibodies is based on the premise that individuals inherit a relatively small number of antibody genes, and that additional diversity in antibody structure is generated during lymphocyte development by recombining and/or mutating these genes. The terms *somatic recombination* and *somatic mutation* are used to refer to such theories. The main difference between the germ line and somatic recombination/mutation theories comes down to the question of whether the information coding for a million different antibodies is transmitted from parent to offspring through the sperm and egg, or is created by DNA alterations that take place during the subsequent development of an organism's antibody-producing cells.

The first major breakthrough in evaluating these theories occurred in 1976 when Susumu Tonegawa reported that the variable and constant regions of antibody molecules are encoded by separate regions of DNA, termed **V** and **C gene segments,** whose location within the DNA *differs in embryonic and antibody-producing cells.* This conclusion emerged from studies in which mRNA coding for an antibody light chain was hybridized to DNA fragments generated by treating DNA obtained from either embryonic cells or mature antibody-producing cells with restriction enzymes. Surprisingly, the light-chain mRNA was found to hybridize to two different fragments of embryonic DNA, but to only one fragment of the DNA obtained from mature antibody-producing cells (Figure 16-6). Of the two embryonic DNA fragments that hybridized to light-chain mRNA, one was found to possess DNA sequences coding for the light-chain variable region and the other contained sequences coding for the constant region. It was therefore concluded that the gene segments coding for the variable and constant regions of antibody light chains are separated from each other in embryonic DNA, but are brought together during the development of antibody-producing cells. Subsequent studies have revealed that the DNA sequences coding for the variable and constant regions of heavy chains undergo similar rearrangements.

Recombination Involving Multiple V, D, and J Segments Is a Major Contributor to Antibody Diversity

Although the preceding studies showed that DNA sequences coding for the variable (V) and constant (C) regions of antibody molecules are rearranged during the

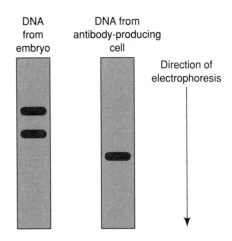

Figure 16-6 *Evidence for DNA Rearrangement in Antibody-Producing Cells* (Left) *Embryonic DNA was cleaved with a restriction enzyme and the resulting fragments were then separated by electrophoresis. When radioactive light-chain mRNA was hybridized to the separated DNA fragments, two different fragments became radioactive. (Right) When a similar experiment was carried out using DNA derived from mature antibody-producing cells, the radioactive light-chain mRNA hybridized to a single DNA fragment whose size differed from that of either of the original fragments. The reason for this altered behavior is that two gene segments coding for portions of the light chain are physically separated from each other in embryonic DNA (and hence occur in two different restriction fragments), but are brought together during the development of antibody-producing cells (and hence appear in the same restriction fragment).*

development of antibody-producing cells, they did not address the question of how many V gene segments are involved. This information is crucial for determining how animals are able to generate millions of different antibodies exhibiting different antigen-binding sites. The development of recombinant DNA technology has made it possible to address this question by comparing the DNA base sequences of cloned antibody genes derived from embryonic cells and from mature antibody-producing cells. Such studies have revealed that human DNA contains about 80 different V segments coding for light-chain variable regions, and 100–200 V segments coding for heavy-chain variable regions. This number of segments could account for a maximum of only $80 \times 200 = 16,000$ different combinations of light chains and heavy chains. Thus the number of different genes inherited through the germ cells cannot by itself explain how humans manufacture a million or more different antibodies.

The first clue as to how the additional diversity is generated came from a detailed examination of the base sequences of V gene segments. These analyses revealed that V segments are not long enough to code for the entire variable region of a heavy or light chain. The missing bases, which are omitted from the 3' end of each V gene segment, code for about 10–15 amino acids located at the end of the variable region that lies adjacent to the constant region in the polypeptide chain. Instead of being part of the V segment, the DNA coding for this short stretch of amino acids is organized as a series of "joining" or **J gene segments** located adjacent to the constant (C) gene segment in embryonic DNA (Figure 16-7). Human λ light-chain genes have four J segments, κ light-chain genes have five J segments, and heavy-chain genes contain at least six J segments. (J segments should not be confused with *J chains*, which are unrelated polypeptides that help hold together IgA dimers and IgM pentamers as shown in Figure 16-3.)

In addition to the V and J segments, a third type of sequence called a "diversity" or **D gene segment** also codes for part of the variable region of heavy chains. *Thus the variable region of a heavy chain is encoded by three separate DNA segments: a V segment, a D segment, and a J segment.* Human heavy chain genes have about 100–200 V segments, more than 20 D segments, and at least 6 J segments (see Figure 16-7). During development of an antibody-producing cell, genetic recombination allows any of the heavy-chain V segments to be spliced to any heavy-chain D segment and any heavy-chain J segment (Figure 16-8). Hence at least $200 \times 20 \times 6 = 24,000$ different heavy-chain variable regions can be formed.

Unlike heavy-chain genes, the variable regions of light-chain genes do not have D segments. *Therefore the variable region of a light chain is encoded by only two separate DNA segments: a V segment and a J segment.* Human κ light-chain genes have about 80 V segments and five J segments (see Figure 16-7), generating a total of $80 \times 5 = 400$ different κ light-chain variable re-

gions. The random combination of heavy chains containing 24,000 different variable regions with light chains containing 400 different variable regions could generate as many as $24,000 \times 400 = 9,600,000$ different antibodies. Additional diversity can also be produced by assembling antibodies with λ light chains instead of κ light chains, although human λ light-chain genes have only two different V segments and thus contribute relatively little variability to antibody structure.

Antibody Diversity Is Further Increased by Inexact Joining of Gene Segments and Somatic Mutation

We have now seen that random recombination between multiple V and J (or V, D, and J) gene segments is capable of introducing an enormous amount of diversity into the structure of antibody molecules. A clue to the mechanism involved in joining the recombining segments together first emerged from the discovery that short complementary sequences, called "recombination" or **R sequences,** are located at the 3' end of each V sequence, on both sides of each D sequence, and at the 5' end of each J sequence. The presence of these complementary sequences guides the formation of DNA loops that bring the joining segments together, which in turn allows the DNA between the recombining segments to be excised (Figure 16-9, *top*).

During the process of segment joining, a small amount of imprecision is introduced that further increases antibody diversity. One reason for the imprecision is that the joining position within any given V or J segment is not precisely defined. For example, the joining of the V and J segments of light chain genes may occur at several different nucleotide positions located

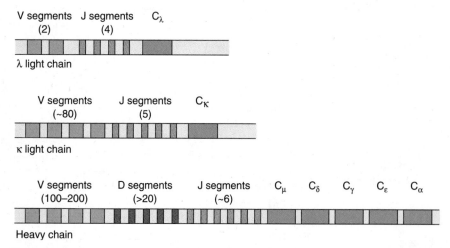

Figure 16-7 *Organization of the Genes Coding for Antibody Light and Heavy Chains in Humans*
The various segments are not drawn to scale. The designations C_μ, C_δ, C_γ, C_ϵ, and C_α represent the constant gene segments coding for the constant chains of IgM, IgD, IgG, IgE, and IgA antibodies, respectively. (To simplify the diagram, only one type of C segment is shown for each of the five kinds of heavy chains.) C_λ and C_κ correspond to the constant gene segments for λ and κ chains.

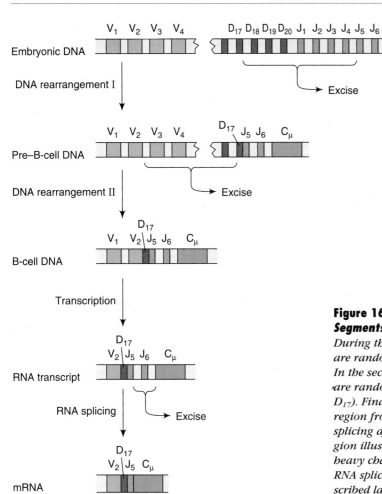

Figure 16-8 *Random Recombination between V, D, and J Segments During the Formation of an Antibody Heavy Chain* *During the first DNA rearrangement, some D and J segments are randomly deleted (in this case bringing D_{17} adjacent to J_5). In the second DNA rearrangement, some V and D segments are randomly deleted (in this case bringing V_2 adjacent to D_{17}). Finally, the sequences separating the newly formed V-D-J region from the C (constant) region are removed by RNA splicing after transcription has taken place. The constant region illustrated in this particular example (C_μ) codes for the heavy chain of an IgM molecule. DNA rearrangements and RNA splicing events involving the constant region are described later in the chapter.*

near the codon for amino acid number 95. A shift of one or two bases in the position of the joining site can change the amino acid being specified at this point in the base sequence (see Figure 16-9, *bottom*). In addition, a variable number of nucleotides is often lost or added to the splice sites during segment joining. Taken together, the variability in splice site location and the addition and removal of nucleotides during splicing leads to a significant increase in the diversity of antibody structure.

Recombination between V and J (or V, D, and J) segments has a major impact on an antibody's antigen-binding properties because the D and J segments code for portions of hypervariable region CDR3, which is part of an antibody's antigen-binding site (see Figure 16-5). In contrast, hypervariable regions CDR1 and CDR2 are encoded by base sequences located far from the V-J and V-D-J splice sites; hence the random joining of gene segments does not increase the sequence diversity of CDR1 and CDR2, which also comprise part of the antigen-binding site. Yet changes in the amino acid sequence of the CDR1 and CDR2 regions do occur under certain circumstances. This realization first emerged from studies involving **hybridomas,** which are permanent cell lines derived from single antibody-producing

lymphocytes (page 143). Because they are derived from single cells, hybridomas produce a single type of antibody directed against one particular antigen. But if a hybridoma is continually exposed to its corresponding antigen, a large number of mutations begin to occur at the CDR1 and CDR2 sites as the cells proliferate. The discovery that the mutation rate at these locations is about a million times higher than for other protein sites suggests that the process is driven by a specific mutagenic mechanism. Since antigens selectively stimulate the proliferation of lymphocytes whose antibodies are targeted against the stimulating antigen, cells exhibiting mutations that enhance the binding affinity between antibody and antigen will proliferate more rapidly and will gradually tend to predominate.

Thus we see that the ability of the immune system to produce an extraordinarily diverse spectrum of antibodies is based on several mechanisms, including (1) the inheritance of multiple light and heavy chain gene segments through the germ cells, (2) the random joining of V-J and V-D-J sequences through genetic recombination, (3) the imprecision of the joining process, (4) the random association of the resulting heavy and light chains, and (5) the mutation of CDR1 and CDR2 sequences during an immune response. The final result is

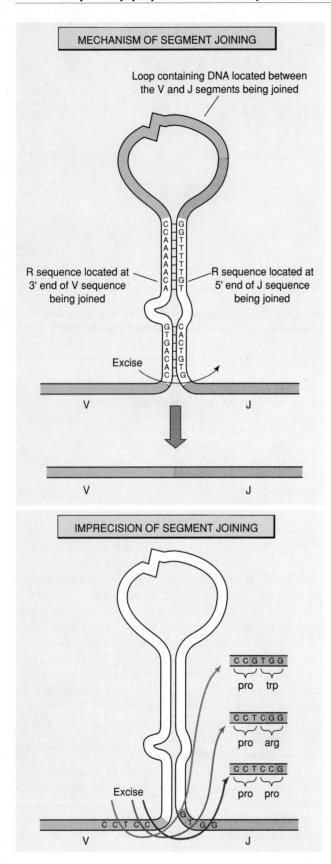

MECHANISM OF SEGMENT JOINING

Loop containing DNA located between the V and J segments being joined

R sequence located at 3' end of V sequence being joined

R sequence located at 5' end of J sequence being joined

Excise

V   J

V   J

IMPRECISION OF SEGMENT JOINING

C C G T G G
pro   trp

C C T C G G
pro   arg

C C T C C G
pro   pro

Excise

V   J

the production of an extremely diverse group of antibodies capable of binding to virtually any conceivable foreign antigen.

All Antibodies Produced by a Given B Cell Have the Same Antigen-Binding Site

During somatic recombination, each developing B cell generates only one functional light chain V-J segment and one functional heavy chain V-D-J segment. In producing the single V-J combination for the light chain, a cell must choose between the light-chain genes on the maternal and paternal chromosomes, and select either the λ or κ light-chain gene. Likewise, in producing the single V-D-J combination for the heavy chain, a cell must again choose between the heavy-chain genes located on the maternal and paternal chromosome. The net result is that each cell generates one active V-J segment coding for a light chain and one active V-D-J segment coding for a heavy chain; hence the antibodies manufactured by the cell will contain a single type of variable region in all of its heavy chains, and a single type of variable region in all of its light chains. This arrangement ensures that every antibody produced by the cell has the same antigen-binding site.

After they have been created by somatic recombination, the light-chain V-J segment and heavy-chain V-D-J segment remain separated from their corresponding constant (C) segments in the chromosomal DNA of mature antibody-producing cells. As a result, transcription of both the light- and heavy-chain genes yields RNA transcripts in which the variable and constant coding regions are separated by extra bases that must be removed by RNA splicing (see Figure 16-8). Other intervening sequences (introns) are also removed from the transcripts during RNA processing.

Secreted and Membrane-Bound Forms of the Same Antibody Are Made by Altering the RNA Transcribed from the Heavy-Chain Gene

Each class of antibodies can be made in two different forms: one bound to the plasma membrane and the

Figure 16-9 *The Mechanism of Segment Joining* (Top) *Complementary R sequences located adjacent to V and J segments guide the formation of a stem-and-loop structure that brings the joining segments together. A similar mechanism is employed for V-D and D-J joining.* (Bottom) *A small shift in the exact position of the joining site can change the amino acid being specified, thereby increasing the diversity of antibody structure.*

other secreted from the cell. The membrane-bound form of an antibody molecule functions as an *antigen receptor,* while the secreted form is produced after cells have been stimulated by exposure to antigen. Membrane-bound and secreted forms of an antibody molecule differ only in the sequence of the C-terminal region of the heavy chain. The C-terminus of secreted antibodies is relatively hydrophilic, whereas the same region of membrane-bound antibodies contains a hydrophobic segment that anchors the molecule to the lipid bilayer of the plasma membrane.

Cells can switch between the production of the two forms of antibody by altering the way in which the RNA transcript of the heavy-chain gene is processed. As shown in Figure 16-10, the mRNA coding for membrane-bound antibody is processed from a long RNA transcript that includes the sequences specifying the C-terminal hydrophobic domain that anchors the protein to the plasma membrane, whereas the mRNA coding for the secreted antibody is processed from a shorter RNA transcript that lacks these sequences. Although this difference determines whether antibodies will be secreted or bound to the cell surface, both types of antibody still exhibit the same antigen-binding site.

CELL INTERACTIONS DURING ANTIBODY FORMATION

The clonal selection theory is based on the idea that an organism has millions of different lymphocytes, each making a different kind of antibody, and that antigens simply stimulate the proliferation of those cells that happen to manufacture an antibody directed against the antigen in question. Now that we have seen how organisms generate millions of different lymphocytes producing millions of different antibodies, we are ready to explore the issue of how the appropriate cells are stimulated to divide and secrete antibody in response to a stimulating antigen.

Antibodies Are Made by B Cells with Assistance from Helper T Cells

Two pioneering discoveries reported in the 1950s laid the foundation for our current understanding of the cellular interactions involved in antibody production. First, Robert Good, Jacques Miller, and Byron Waksman independently reported that surgical removal of the thymus from newborn mice and rabbits yields animals that are unable to produce antibodies or reject skin grafts. Such

observations suggested that thymus-derived lymphocytes—that is, T cells—are required for antibody formation as well as for cell-mediated immunity.

The second discovery, reported in 1954 by Bruce Glick, involved a lymphoid organ called the *bursa of Fabricius,* which is the place where B cells develop in birds. (In mammals, which have no bursa of Fabricius, the comparable stages of B-cell development occur in the bone marrow.) Glick discovered that surgical removal of the bursa of Fabricius from newborn chickens blocked their ability to produce antibodies but had no effect on cell-mediated immunity. Such observations suggested that bursa-derived lymphocytes—that is, B cells—are required for antibody production.

Taken together, the preceding two sets of experiments suggested that antibody production requires the participation of both B cells and T cells. Further support for this conclusion came from the studies of Henry Claman and his co-workers, who studied mice whose immune systems had been destroyed by exposing animals

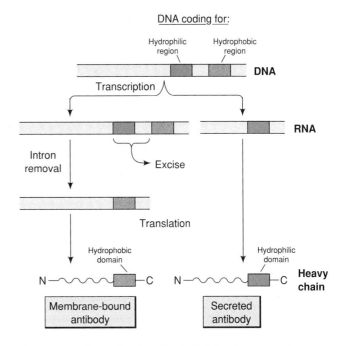

Figure 16-10 *Mechanism for Switching between the Production of Membrane-Bound and Secreted Antibodies* *By changing the length and splicing pattern of the RNA transcript produced from the heavy-chain gene, cells produce either the membrane-bound or secreted form of the same antibody molecule. The mRNA coding for the membrane-bound form is processed from a long RNA transcript that includes the hydrophobic domain that anchors the protein to the plasma membrane, whereas the mRNA coding for the secreted antibody is processed from a shorter RNA transcript lacking these sequences.*

to X-rays. They found that X-irradiated animals lose the ability to produce antibodies, and that antibody formation cannot be restored by injecting the animals with either thymus lymphocytes (T cells) or bone marrow lymphocytes (B cells) derived from healthy mice. But the irradiated animals do regain the ability to produce antibodies if they are injected with a mixture of *both* T cells and B cells derived from healthy animals.

The conclusion that T cells and B cells are both required for antibody production raises the question of which cell type is actually making the antibodies. This issue was first investigated by studies in which Miller and his colleague Graham Mitchell removed the thymus from mice of one strain (the "host"), thereby eliminating their ability to produce T cells. Since T cells (as well as B cells) are required for antibody formation, the animals lacking a thymus gland could not produce antibodies. The host animals were then injected with thymus cells derived from another strain of mice (the "donor"). As expected, the ability of the host animals to produce antibodies in response to a triggering antigen was restored.

To find out whether the antibodies were being synthesized by cells derived from the host or the donor, Mitchell and Miller took advantage of the fact that the cells of each mouse strain carry their own unique cell surface *MHC molecules* (page 715). An experiment was therefore performed in which a host animal that had been injected with donor lymphocytes was stimulated to produce antibodies by exposure to a specific antigen (Figure 16-11). Once antibody formation had been activated, antibody-producing lymphocytes were removed from the spleen. To determine whether these antibody-producing cells were of host or donor origin, they were treated with antibodies directed against the MHC molecules of either the host or donor strain to selectively destroy either the host or the donor lymphocytes. Destroying the host lymphocytes was found to abolish antibody formation, but destroying the donor lymphocytes had no effect; hence the antibody-producing cells must have been host lymphocytes. Since the host lacked a thymus and so had no T cells of its own, the antibody-producing cells could not be T cells. And yet T cells must play a role in antibody formation because the host could not be activated to make antibodies in the first place unless it first received T cells from the donor. It was therefore concluded that T cells are not antibody-producing cells, but somehow "help" to activate the process of antibody formation; T cells that carry out this helping function have come to be known as **helper T cells.**

If T cells are not responsible for making antibodies, then the logical candidate is the B cell. Direct support for the idea that B cells are the actual site of antibody formation came from reconstitution experiments in which normal T cells were mixed with B cells derived from animals bearing an abnormal chromosome. The

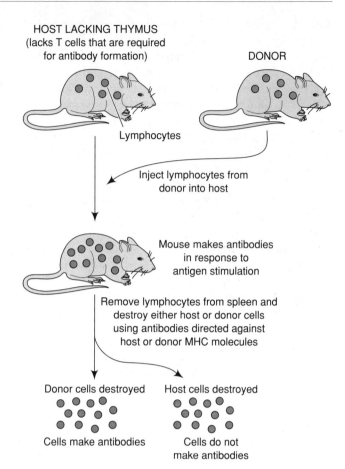

Figure 16-11 Experiment Exploring the Role of Thymus Lymphocytes (T Cells) in Antibody Formation *Since T cells are required for antibody formation, the host mouse lacking a thymus (and hence lacking T cells) cannot make antibodies. The experiment shows that when the host receives an injection of lymphocytes from a normal donor animal, the host's own lymphocytes start producing antibodies. Since the host has no T cells of its own, it is evident that T cells cannot be the site of antibody formation.*

mixture was then injected into irradiated animals incapable of producing antibodies. When antibody-producing cells were later removed from the host, they were found to contain the marker chromosome. However, the reciprocal experiment using T cells possessing the marker chromosome and B cells lacking the marker yielded antibody-forming cells that exhibit normal chromosomes. Taken together, these observations indicate that the B cell must be the site of antibody synthesis.

Helper T Cells Bind to Antigen–MHC II Complexes on the Surface of Antigen-Presenting Cells

When lymphocytes are exposed to a foreign antigen, the induction of antibody formation does not take place unless a third cell type is present in addition to B cells and helper T cells. The third cell type, called an *anti-*

gen-presenting cell (APC), was defined earlier in the chapter as a cell that exhibits the capacity to degrade foreign antigens and "present" the resulting antigen fragments on its surface in combination with MHC class II molecules (page 715). One of the most commonly encountered antigen-presenting cells is the macrophage. The involvement of macrophages in antibody formation was first demonstrated by experiments in which isolated macrophages and lymphocytes were placed on opposite sides of an artificial membrane. When a foreign antigen was added to both cell populations, no antibody synthesis occurred. But when the cells were first mixed together and then exposed to antigen, antibody formation did take place. It was therefore concluded that direct contact between macrophages and lymphocytes is required during the induction of antibody formation.

Why do lymphocytes need to make contact with macrophages before antibodies can be produced? We learned earlier in the chapter that macrophages ingest and degrade foreign antigens, generating antigen fragments that are presented at the cell surface in combination with MHC class II molecules. In order to initiate an immune response, the antigen–MHC II complex must first be recognized by helper T cells that bind to the macrophage surface. Two cell surface components expressed by helper T cells are involved in this step. One is called a **T-cell antigen receptor (TCR)** because its function is to bind to the antigen fragment being presented by the antigen-presenting cell. A TCR molecule is a plasma membrane protein comprised of two nonidentical polypeptide chains. Two major types of TCRs have been identified; the *alpha-beta receptor,* which consists of an alpha chain and a beta chain, and the *gamma-delta receptor,* which is constructed from a gamma chain and a delta chain. T cells possessing alpha-beta receptors are involved in recognizing foreign peptides, whereas T cells carrying gamma-delta receptors appear to recognize molecules produced by host tissues under abnormal conditions (e.g., tissues that have been

overheated or undergone abrasion). Like antibody light and heavy chains, each TCR chain has an N-terminal variable region and a C-terminal constant region. TCRs also resemble antibodies in another way: The genes coding for TCR chains have multiple V, J, and D segments that randomly recombine during T-cell development, creating millions of different sequences. Organisms can therefore produce millions of different TCRs, each capable of binding to a particular antigen, just as they make millions of different antibodies. A further similarity between TCRs and antibodies is that each T cell makes only a single type of TCR, just as each B cell makes only a single type of antibody. This means that each T cell is specifically targeted against a single antigen.

In addition to TCR molecules, a second cell surface component is also required by T cells before they can bind to antigen–MHC II complexes presented by macrophages and other antigen-presenting cells. The additional component is **CD4,** a cell surface glycoprotein that occurs in close association with the TCR molecules of helper T cells (Figure 16-12, *left*). Each CD4 molecule has a binding site that specifically recognizes an MHC class II molecule. This arrangement allows the TCR-CD4 complex to mediate the binding of helper T cells to antigen-MHC II complexes; the TCR component is responsible for recognizing the antigen fragment, while the CD4 glycoprotein binds to the class II MHC molecule.

Interleukins Stimulate the Proliferation of Helper T Cells

For any given antigen that has been administered to an animal to induce an immune response, only a few helper T cells are likely to carry a TCR that can bind the stimulating antigen. When these few T cells bind to the antigen presented on a macrophage surface, the resulting cell-cell interaction stimulates the proliferation of

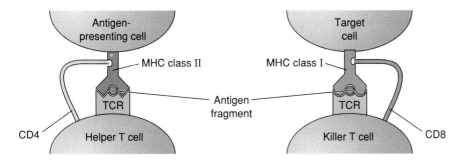

Figure 16-12 *Mechanism of Antigen Binding by T-Cell Antigen Receptors (TCRs)*
(Left) *The TCRs of helper T cells act in conjunction with the cell surface glycoprotein CD4, creating a TCR-CD4 complex that mediates the binding of helper T cells to antigen–MHC II complexes.* (Right) *The TCRs of killer T cells act in conjunction with the cell-surface glycoprotein CD8, creating a TCR-CD8 complex that mediates the binding of killer T cells to antigen–MHC I complexes.*

the bound helper T cells. The net result is an increase in the number of T cells that recognize the triggering antigen. Hence once again we encounter the phenomenon of clonal selection—that is, the ability of an antigen to selectively stimulate the proliferation of lymphocytes targeted against that antigen.

How does the binding of a helper T cell to an antigen fragment presented by a macrophage trigger the proliferation of the bound T cell? Control over the process is exerted largely by **lymphokines,** a group of proteins that are secreted by lymphocytes and that function as local mediators that control cell growth and differentiation. (Because many lymphokines are also produced by cell types other than lymphocytes, they are sometimes referred to as *cytokines.*) Prominent among the lymphokines that regulate antibody induction are a group of related proteins known as **interleukins.** At least two interleukins are involved in stimulating the proliferation of helper T cells that have become bound to antigen–MHC II complexes presented by macrophages (Figure 16-13, *top*). First, the binding of a helper T cell to a macrophage causes the macrophage to secrete *interleukin-1 (IL-1),* a lymphokine that stimulates the bound helper T cells to release other lymphokines. In response, the helper T cells synthesize both *interleukin-2 (IL-2)* and cell surface receptors for IL-2. Once the newly formed IL-2 receptors become incorporated into the plasma membrane of the helper T cell, the IL-2 produced by the T cell binds to the receptors and stimulates the cell to divide. The T cell can then leave the surface of the macrophage and still continue to proliferate because it now produces IL-2 molecules that bind to and stimulate its own cell surface IL-2 receptors.

B Cells Are Activated by Helper T Cells through a Mechanism That Also Involves Interleukins

We have now seen how foreign antigens stimulate the production of helper T cells whose TCRs are targeted against the stimulating antigen. But antibody molecules are produced by B cells, not T cells, so how does the proliferation of a clone of antigen-specific helper T cells lead to the production of the proper antibody by B cells? According to the clonal selection theory, the ultimate effect of a foreign antigen is to stimulate the proliferation of one or a few B cells programmed to make antibodies against the triggering antigen. This implies that B cells, like T cells, contain antigen receptors that allow them to recognize the presence of a specific antigen. The idea that B cells contain specific antigen receptors on their surfaces was first suggested by affinity chromatography experiments in which a heterogeneous population of lymphocytes was passed through an inert matrix to which a given antigen had been coupled (Figure 16-14).

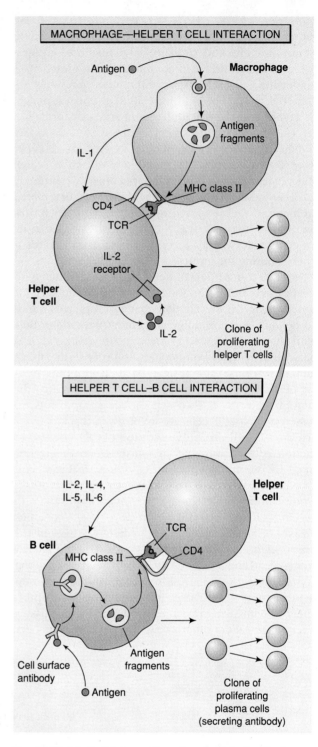

Figure 16-13 *Cell Interactions Involved in Antibody Formation* (Top) *During the first stage, helper T cells bind to antigen–MHC II complexes presented by macrophages or other antigen-presenting cells. This interaction causes the macrophage to produce IL-1, which stimulates the helper T cells to produce IL-2 and IL-2 receptors, which in turn stimulates the helper T cell to divide.* (Bottom) *The activated helper T cells then bind to B cells containing the same antigen bound to MHC II. Interleukins secreted by the bound helper T cells stimulate the B cells to divide and differentiate into a clone of antibody-secreting plasma cells.*

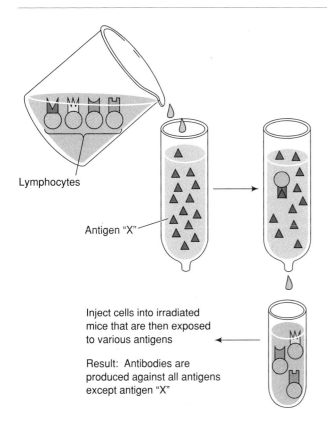

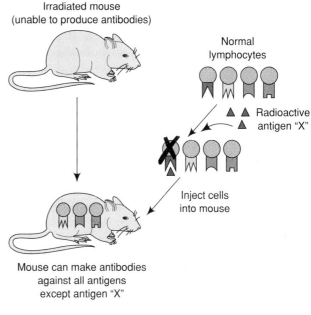

Figure 16-15 *"Antigen Suicide" Experiment   When a highly radioactive antigen is added to a population of lymphocytes, those cells that bind to the antigen are killed by the radioactivity ("antigen suicide"). As a result, the lymphocyte population can no longer produce antibodies against the antigen in question. If either B cells or T cells are exposed to a radioactive antigen prior to injecting a mixture of B and T cells into mice whose own immune system has been crippled, antibody formation against the antigen in question is abolished. Therefore, B cells and T cells must both contain antigen receptors.*

Figure 16-14 *Evidence for the Existence of Antigen Receptors on the Surface of Lymphocytes   A heterogeneous population of lymphocytes is passed through an inert matrix to which antigen "X" has been coupled. Cells that have receptors for antigen "X" bind to the matrix. The cell population that does not bind to the matrix is capable of making antibodies against all antigens except antigen "X."*

The cells that did not bind to the matrix-bound antigen were then injected into irradiated mice. Such animals were found to be capable of making antibodies targeted against any antigen except the original antigen that had been bound to the matrix. It was therefore concluded that those lymphocytes that make antibodies directed against the original antigen had been removed during affinity chromatography because they contain antigen receptors that bound them to the matrix-associated antigen.

These experiments did not indicate, however, whether the cells containing the antigen receptor were B cells or helper T cells, or both. To investigate this question an "antigen suicide" technique was devised in which highly radioactive antigens were bound to lymphocytes, killing them by radiation damage. Such experiments revealed that treating either B cells or T cells with a radioactive antigen prior to injecting the cells into an irradiated animal inhibits the ability of the animal to produce antibodies directed against that particular antigen (Figure 16-15). Hence B cells as well as T cells must possess antigen receptors.

The antigen receptors carried by T cells are cell surface *TCR molecules* (page 729), whose role in binding T cells to antigens presented by macrophages has already been described. The antigen receptors of B cells, on the other hand, are cell surface *antibody molecules*, which differ from secreted antibodies in that they possess a hydrophobic C-terminal segment that anchors them to the plasma membrane (page 727). Each B cell contains between 20,000 and 200,000 cell surface antibody molecules, all exhibiting the same antigen-binding site as the secreted antibodies made by the cell in question; the only difference is the presence of the hydrophobic stretch of amino acids that anchors the cell surface form of the antibody molecule to the plasma membrane.

The binding of an antigen to its corresponding antibody on the surface of a B cell causes the antigen to be taken into the cell by the process of receptor-mediated endocytosis (page 295). The internalized antigen is then broken down into fragments that are transported back to the cell surface bound to MHC class II molecules. The resulting antigen–MHC II complex on the B-cell surface is recognized by those helper T cells whose TCRs are targeted against the stimulating antigen (see Figure 16-13, *bottom*). The binding of these helper T cells to the surface of B cells causes the helper T cells to secrete several interleukins—mainly IL-2, IL-4, IL-5, and IL-6—directly onto the surface of the B cell. The interaction of these interleukins with the B cell causes the B cell to begin dividing.

Proliferating B Cells Develop into Plasma Cells and Memory Cells

Once B cells have been stimulated to divide by helper T cells, most of the proliferating B cells enlarge, develop an extensive endoplasmic reticulum, and begin secreting large amounts of antibody. The resulting antibody-secreting cells are called **plasma cells** (Figure 16-16). As we learned earlier, the onset of antibody secretion involves a change in RNA processing that generates the secreted form of the antibody molecule instead of the membrane-bound form (page 727). Initially the differentiating cells manufacture IgM antibodies, but within a few days they switch over to the production of another antibody class (usually IgG, but in some cases IgA or IgE). Figure 16-17 illustrates how this process of **class switching** is carried out. In B cells that have never been stimulated by an appropriate antigen, the gene segments coding for the constant regions of the five classes of heavy chains are situated adjacent to one another; the segment coding for the μ chain, which occurs in IgM molecules, is located first in the sequence. Class switching is accomplished by looping out and excising some of these heavy-chain constant gene segments, bringing new constant gene segments into position to be expressed. Looping and excision is facilitated by short complementary sequences, called *switch (S) regions,* which guide the formation of stem-and-loop structures that bring the joining segments together.

Plasma cells are not the only cell type produced when B cells are stimulated to divide by a triggering antigen in conjunction with helper T cells. A small fraction of the proliferating B-cell population is set aside as a reserve population of cells directed against the stimulating antigen. Such cells, called **memory B cells,** are indistinguishable in appearance from unstimulated lymphocytes and do not secrete antibody. But if the organism is exposed to the same antigen a second time, the reserve population of antigen-specific memory cells proliferates and differentiates into antibody-secreting plasma cells, thereby allowing the *secondary response* to a given antigen to occur more rapidly and produce more antibody than the initial or *primary response* (Figure 16-18). The effectiveness of the secondary response explains why humans rarely get such diseases as chicken pox or mumps more than once.

The Complement System Assists Antibodies in Defending Organisms against Bacterial Infections

We have now seen how antigens trigger a cooperative interaction between macrophages, helper T cells, and B cells that leads to the secretion of antibodies directed against the stimulating antigen. But what function do the newly secreted antibody molecules actually perform? In the case of soluble antigens, antibodies circulating in the bloodstream and extracellular fluid can bind to the antigen and block its action. For example, the bacterium *Clostridium tetani* produces a highly toxic protein called *tetanus toxin* that circulates throughout the body and interferes with nerve cell function. Because tetanus toxin is a foreign protein, it elicits the formation of antibodies that in turn bind to the toxin and block its effects. Hence people can be immunized against tetanus by injecting them with an inactivated form of the tetanus toxin, which stimulates the formation of antibodies that protect the individual against the native toxin.

In addition to their ability to inactivate soluble antigens, antibodies also play a role in the destruction of invading microorganisms such as bacteria. The ability of antibodies to trigger the destruction of a bacterial cell requires the participation of the **complement system,** a group of about 20 soluble proteins that circulate in the bloodstream and extracellular fluids. Most complement proteins are named with the letter C followed by a number that represents the order in which they were discovered (e.g., C1, C2, C3 . . .), but a few are designated with other letters (e.g., H, B, and D). The complement system operates by two routes known as the classical and alternative pathways. The *classical pathway* requires that the immune system first produce antibodies directed against foreign antigens exposed on the surface of invading bacteria. After the antibodies attach to the bacterial cell surface, complement component C1 binds to the heavy-chain constant region of the antibody molecules. The bound C1 then triggers a cascade of events involving the binding and cleavage of complement components C4, C2, and C3 (Figure 16-19). The main products of these reactions are an enzyme complex called *C4b-C2b* and the protein fragment *C3b,* which together initiate the assembly of the "late complement components" (C5 through C9). This assembly sequence triggers the aggregation of C9 molecules into doughnut-shaped structures called **membrane-attack complexes,** which form pores in the bacterial plasma membrane. The pores allow small molecules to pass freely into and out of the cell, but proteins and other macromolecules remain trapped inside. Osmosis therefore drives the flow of water into the bacterium, causing it to swell and burst.

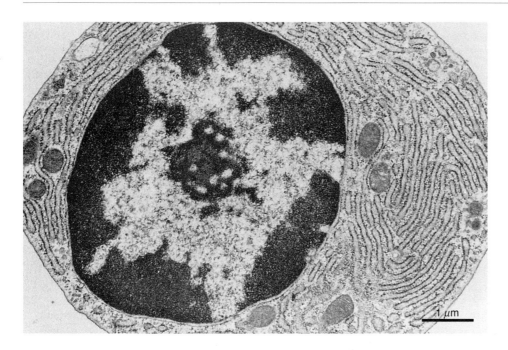

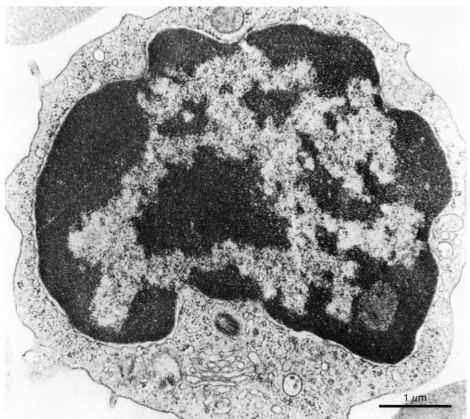

Figure 16-16 *Electron Microscopic Appearance of Plasma Cells and T Cells* (Top) *Plasma cells are characterized by the presence of a well-developed rough ER, which synthesizes antibody molecules destined for secretion from the cell.* (Bottom) *In contrast, T cells have little rough endoplasmic reticulum, although they do possess a large number of free cytoplasmic ribosomes. Courtesy of D. Fawcett.*

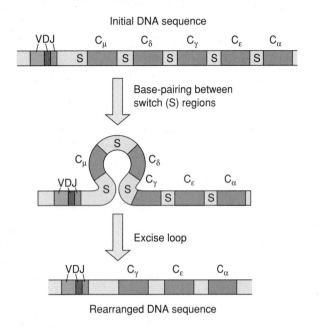

Initial DNA sequence

Base-pairing between switch (S) regions

Excise loop

Rearranged DNA sequence

Figure 16-17 *Mechanism of Class Switching This example shows how differentiating plasma cells switch from the production of IgM antibodies (which have a C_μ heavy chain) to the production of IgG antibodies (which have a C_γ heavy chain). Complementary base-pairing between switch (S) regions causes some of the C segments to be looped out and excised, bringing the C_γ segment into position to be expressed. To simplify the diagram, only one type of C segment is shown for each of the five kinds of heavy chains.*

The complement system can also attack bacteria in the absence of antibodies using a mechanism called the *alternative pathway* (see Figure 16-19). The alternative pathway is initiated by complement protein C3. The C3 molecules that normally circulate in the bloodstream are always breaking down at a slow rate, generating C3b fragments that readily bind to the surface of most cells. Upon binding to a cell, the C3b fragment is usually inactivated by a complement component called *factor H.* However, certain kinds of cell surfaces inhibit the activity of factor H; for example, the lipopolysaccharides found in Gram-negative bacterial cell walls effectively prevent factor H from inactivating C3b. Under such conditions, C3b binds to complement *factor B,* which is in turn cleaved by complement *factor D.* The net result is the formation of an enzyme complex that promotes the binding of more C3 and its cleavage to C3b. Eventually enough C3b accumulates to trigger the assembly of the "late complement components" into a membrane-attack complex, just as in the classical pathway.

The protein fragments that are generated during both the classical and alternative pathways serve several functions in addition to facilitating the formation of membrane-attack complexes. The C3b fragment, for example, binds to receptors carried by phagocytic cells such as *macrophages* and *neutrophils.* Therefore when C3b molecules are present on the surface of a bacterial cell, phagocytic cells bind to the C3b and engulf the

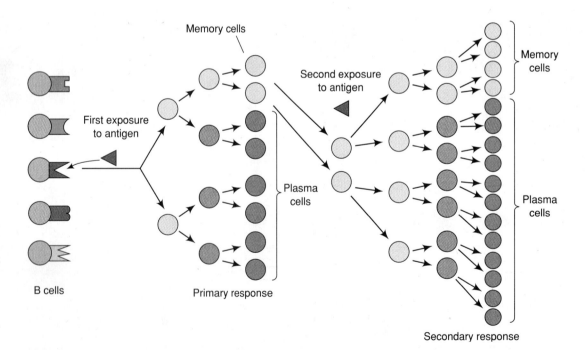

Figure 16-18 *Response of B Cells to Antigen Stimulation During the first exposure to an antigen, B cells specific for that antigen are stimulated to divide, generating antibody-producing plasma cells. In addition, a small number of proliferating cells are set aside as memory cells. During a second exposure to the same antigen, these memory cells proliferate and differentiate into plasma cells, thereby allowing the secondary response to occur more rapidly and produce more antibody than the primary response.*

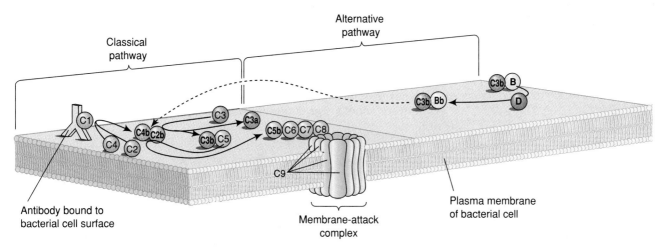

Figure 16-19 *The Classical and Alternative Pathways of Complement Activation* *The classical pathway is initiated by the binding of complement factor C1 to antibody molecules attached to the surface of a bacterial cell. C1 triggers the binding and cleavage of C4, C2, and C3, yielding the enzyme complex C4b-C2b and the protein fragment C3b. Together these components initiate the assembly of the "late complement components" (C5 through C9), culminating in the formation of membrane-attack complexes. In the alternative pathway of complement activation, the binding of C3b to a cell surface leads to the formation of an enzyme complex called C3b-Bb, which functions like C4b-C2b in triggering the formation of membrane-attack complexes.*

bacterium. Another protein fragment, called C3a, is also generated during the cleavage of complement protein C3. The C3a fragment binds to *mast cells,* stimulating them to release two substances. One substance is a chemotactic factor that attracts phagocytes to the vicinity of the infecting bacteria, and the other is *histamine,* a small molecule that increases the permeability of nearby capillaries. The net effect is to allow white blood cells and more antibodies to enter the tissue, creating a local *inflammatory response* at the site of the bacterial infection.

CELL-MEDIATED IMMUNITY

We have now seen how foreign antigens stimulate the immune system to manufacture antibodies that neutralize the toxic effects of soluble antigens and promote the destruction of bacteria. But some microorganisms are able to evade this type of immune response by entering into the cells of the host. The bacteria that cause tuberculosis and leprosy, fungi that reside in the cytoplasm of host cells, and viruses are all examples of infectious agents that attempt to evade the immune system by residing inside cells. In order to defend against such infectious agents, animals have developed a second type of immune response known as **cell-mediated immunity.** In cell-mediated immunity, lymphocytes attack infected cells directly rather than secreting antibodies. In addition to destroying cells that harbor infectious agents such as viruses, bacteria, and fungi, cell-mediated immu-

nity is employed in the destruction of cancer cells and foreign tissue grafts.

Killer T Cells Are Responsible for Cell-Mediated Immunity

The question of which type of lymphocyte is responsible for cell-mediated immunity was first explored in studies that exploited the fact that T cells, but not B cells, contain a cell surface component called *Thy-1.* If a mixture of B and T cells is incubated with antibodies directed against Thy-1 in the presence of complement, the T cells are selectively destroyed. The discovery that such treatments also inhibit the ability of lymphocyte populations to carry out cell-mediated immunity has led to the conclusion that cell-mediated immunity requires the presence of T cells. In contrast, the removal of B cells from lymphocyte populations does not interfere with cell-mediated immune responses. Further support for the proposed role of T cells in cell-mediated immunity has come from studies involving the "nude" mutation in mice. **Nude mice,** which are born without a thymus and thus lack T cells, are unable to carry out any cell-mediated immune reactions (or produce antibodies).

Although the preceding observations implicate T cells in cell-mediated immunity, they do not identify the type of T cell involved. T cells are subdivided into several different categories whose members carry different cell surface proteins. For example, we have already mentioned that helper T cells are characterized by the presence of the cell surface glycoprotein CD4 (page 729). The identity of the T cell population involved in cell-mediated immunity has been investigated using an assay known as *cell-mediated lympholysis*. In this procedure, which allows cell-mediated immunity to be monitored in a test tube, lymphocytes from one strain of mouse are mixed with *target cells* that have been isolated from another strain of mouse and labeled with radioactive chromium (^{51}Cr). Normally ^{51}Cr diffuses out of the target cells very slowly. But if any of the added lymphocytes recognize the target cells as foreign, a cell-mediated immune response is triggered and the radioactive target cells are attacked and killed. The extent of cell killing can be monitored by measuring the release of radioactivity from the dying cells into the medium.

To determine which kind of lymphocyte is responsible for killing the target cells, the preceding type of experiment has been carried out using lymphocytes that were first exposed to antibodies targeted against various T-cell subpopulations. It was discovered that the most effective way of preventing lymphocytes from killing target cells is to expose them to antibodies directed against a cell surface glycoprotein called **CD8**; it was therefore concluded that T cells exhibiting the CD8 glycoprotein are responsible for killing targeted cells during cell-mediated immunity. This type of T cell is known as a *cytotoxic* or **killer T cell.** The surface of a killer T cell is characterized by the presence of CD8 but not CD4, whereas helper T cells carry CD4 but not CD8.

Killer T Cells Bind to Antigen–MHC I Complexes on Target Cells

The mechanism by which killer T cells recognize target cells as being foreign or abnormal was first investigated in the 1970s by Rolf Zinkernagel and Peter Doherty, who studied the interaction between killer T cells and virus-infected cells in mice. Cells that have become infected by viruses often manufacture viral proteins that can be recognized as foreign by killer T cells of the host organism. However, Doherty and Zinkernagel observed that T cells targeted against the viral antigens could only recognize and kill virus-infected cells that came from the same inbred strain of mouse as the T cells (Figure 16-20). This finding suggested that in addition to recognizing viral antigens, killer T cells simultaneously recognize an antigen that is specific to the host.

Subsequent studies revealed that the host antigen being recognized by killer T cells is a class I MHC molecule, which appears on the surface of almost all cell types. As we learned earlier in the chapter, virus-infected cells often present viral protein fragments at the cell surface in combination with MHC class I molecules (page 716). The viral antigen–MHC I complex is recognized by TCRs located on the surface of killer T cells; any given TCR simultaneously recognizes both a foreign antigen fragment and several amino acids located in the

antigen-binding cleft of the MHC molecule. Hence it recognizes both the foreign antigen and the organism's own MHC molecule. The TCRs of killer T cells closely resemble the comparable antigen receptors of helper T cells, although the TCRs of killer T cells are associated with CD8 glycoprotein molecules instead of the CD4 glycoproteins carried by helper T cells. The CD8 glycoprotein has a binding site for class I MHC molecules; thus the TCR-CD8 complex of killer T cells mediates the binding of killer cells to antigen–MHC I complexes, just as the TCR-CD4 complex of helper T cells mediates the binding of helper cells to antigen–MHC II complexes (see Figure 16-12, *right*).

A T cell that has just bound to an antigen–MHC I complex on a target cell is properly referred to as a *pre–killer T cell* because it is not yet capable of destroying the cell being attacked. The conversion of a pre–killer cell into a killer cell requires the participation of lymphokines produced by helper T cells; among the factors involved in this process are the interleukins *IL-2, IL-4,* and *IL-6,* and other lymphokines called *interferons*

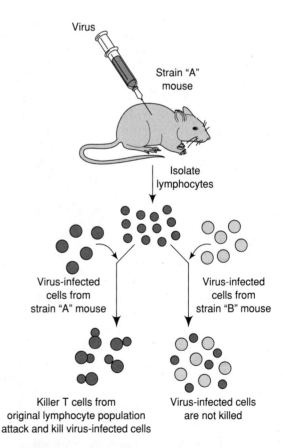

Figure 16-20 Experiment Showing That Killer T Cells Targeted against Virus-Infected Cells Selectively Kill Cells Derived from the Same Strain of Mouse as the T Cells *The reason for this specificity is that the killer T cells recognize part of the host's MHC class I molecules along with the viral antigens.*

(Figure 16-21). In response to stimulation by these lymphokines, the pre–killer T cell proliferates and differentiates into a clone of mature killer T cells, increasing the number of T cells targeted against the triggering antigen. Some of the pre–killer T cells may also be set aside as **memory T cells,** which function as a reserve population of cells directed against the stimulating antigen (similar to the function of the memory B cells already discussed).

Killer T Cells Destroy Their Target Cells by Secreting the Pore-Forming Protein, Perforin

The binding of a killer T cell to a cell exhibiting a foreign antigen–MHC I complex sets in motion a series of events that leads to the destruction of the target cell. Cell death is usually caused by **perforin,** a protein stored in membrane-bound vesicles located in the cytoplasm of killer T cells. Shortly after a killer T cell binds to a target cell, the storage vesicles accumulate near the site where the two cells have made contact. The concentration of free Ca^{2+} within the killer T cell then increases, which causes the vesicles to fuse with the plasma membrane and discharge perforin to the cell exterior and onto the surface of the target cell. The secreted perforin molecules quickly insert themselves into the plasma membrane of the target cell, where they polymerize into doughnut-shaped structures that form pores through the plasma membrane (Figure 16-22). These pores make the target cell leaky to small ions and molecules, thereby causing the cells to swell and burst from the excess water that moves into the cell by osmosis. This scenario closely resembles the mechanism by which cells are destroyed by membrane-attack complexes generated by the complement system (page 732).

Since perforin is a potentially lethal molecule, its effects must be carefully targeted so that surrounding cells are not inadvertently killed. One safeguard built into the system is that only nonpolymerized perforin molecules can be inserted into membranes. If any free perforin molecules diffuse away from the target cell surface, they quickly polymerize into aggregates that can no longer become incorporated into membranes. Hence the chance of accidental injury to neighboring cells is minimized.

Grafted Tissues Are Rejected by Cell-Mediated Immune Responses Directed against Foreign MHC Molecules

In addition to attacking cells that have been invaded by viruses or other infectious agents, cell-mediated immunity is employed to destroy cell types that an organism views as being foreign or abnormal. A prominent example occurs during the transplantation of tissues or organs from one individual to another. **Allografts** are

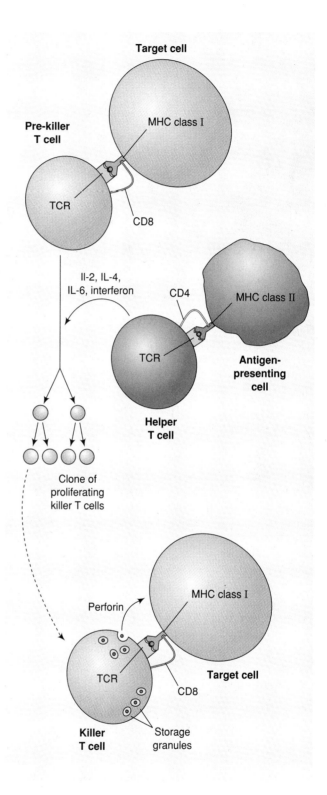

Figure 16-21 *Activation of Killer T Cells* *During the first step in the activation of killer T-cells, an antigen–MHC I complex on a target cell is recognized by the TCR-CD8 complex of a pre-killer T cell. Lymphokines produced by helper-T cells then cause the pre-killer T cell to proliferate and differentiate into a clone of mature killer T cells.*

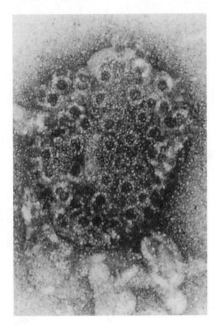
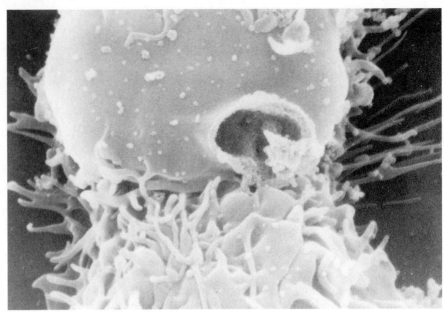

Figure 16-22 *Destruction of Target Cells Displaying Foreign Antigens by Killer T Cells*   (Left) *The surface of this target cell is covered with circular pores composed of perforin, a protein secreted by killer T cells. (Right) A killer T cell (the bottom cell) is shown tightly bound to a target cell in this scanning electron micrograph. Secretion of perforin by the killer T cell causes the target cell to become leaky, and eventually to rupture. Courtesy of E. R. Podack (left) and G. Kaplan (right).*

tissues or organs transplanted between genetically different members of the same species—for example, the transplantation of a kidney or heart from one person to another (other than between identical twins). Unless *immunosuppressive drugs* are employed to inhibit the immune system, allografts are usually destroyed by cell-mediated immunity within a few weeks. The drugs most commonly employed to prevent the rejection of transplanted organs are *corticosteroids* and *cyclosporin*, both of which depress the production of IL-2 and other lymphokines involved in the activation of helper and killer T cells. **Xenografts** are tissues or organs that have been transplanted between different species—for example, the transplantation of a baboon liver into a human infant. Because the antigenic foreignness of a xenograft is much greater than that of an allograft, xenografts trigger an intense immune reaction that is difficult to suppress.

Tissue and organ grafts are destroyed largely through the action of killer T cells that recognize as foreign the class I and II MHC molecules expressed by the graft. Because the genes coding for MHC molecules are extremely variable among individuals, the MHC molecules of any two people are likely to exhibit many differences in amino acid sequence and hence trigger a strong cell-mediated immune response. In addition, the MHC molecules carried on the surface of grafted cells contain peptide fragments derived from the processing of proteins made by the foreign cell. Thus killer T cells may also be responding to a variety of foreign peptides being presented by the foreign MHC molecules. This

means that numerous T-cell clones carrying different TCRs can recognize grafted cells as being foreign, triggering an intense immune response. As a consequence, the number of T cells responding to a foreign tissue graft is usually hundreds of times greater than the number of T cells that respond to a typical viral or bacterial antigen.

NK Cells Destroy Cancer Cells by a Mechanism That Does Not Involve the Recognition of Antigen–MHC Complexes

Organ transplant patients who have been treated with immunosuppressive drugs often exhibit an increased risk of developing cancer. As a result, it has been proposed that the immune system may play a role in the destruction of cancer cells. Although it was initially thought that killer T cells perform this function, studies involving nude mice failed to support the proposed involvement of T cells in rejecting cancer cells. As discussed previously, nude mice lack a thymus and so cannot produce T cells, and yet such animals are no more susceptible to developing cancer than are normal mice. However, the discovery that mice deficient in *natural killer (NK) cells* (page 718) do exhibit an increased incidence of cancer suggests that NK cells are involved in the immune rejection of cancer cells.

Natural killer cells have been shown to destroy numerous kinds of cancer cells, as well as certain virus-infected cells, using a cell-killing mechanism that resembles the way in which killer T cells destroy their target cells. NK cells and killer T cells both adhere to target cells and kill them by secreting the pore-forming protein, perforin, but the mechanism for recognizing target cells is different in the two cases. Whereas killer T cells utilize TCRs and CD8 glycoproteins to recognize foreign antigen fragments bound to MHC class I molecules on a targeted cell, NK cells have neither TCRs nor CD8 molecules (nor do they have CD4 glycoproteins or produce antibodies). Thus NK cells must recognize the abnormal antigens found on cancer cells by some mechanism that differs from the way in which T cells and B cells recognize antigens.

Interferons Provide an Early Nonspecific Immunity to Viral Infections

Another type of immunity that differs from that provided by T cells and B cells involves the production of secreted proteins called **interferons.** As mentioned previously, interferons are among the stimulatory factors that mediate the activation of killer T cells (see Figure 16-21). In addition to their role in lymphocyte activation, interferons are also synthesized and secreted by a variety of vertebrate cell types in response to viral infection. Interferon production usually reaches its maximum level during the first few days of a viral infection, before sufficient time has passed for an effective primary immune response to be mounted. During this early period, interferons provide a nonspecific type of immunity that inhibits the reproduction of a broad spectrum of viruses.

Interferons secreted by virus-infected cells act by binding to plasma membrane receptors of neighboring cells, where they trigger the formation of two enzymes that help to make cells resistant to viral infection. One enzyme, called *oligoadenylate synthetase,* catalyzes the synthesis of *2,5-A,* a short oligonucleotide consisting of adenosine residues joined together by 2',5'-phosphodiester bonds; the 2,5-A in turn activates an RNA-degrading enzyme called *RNase L,* which inhibits viral protein synthesis by cleaving viral mRNAs. The second enzyme induced by interferon is a *protein kinase* that catalyzes the phosphorylation of *eIF-2,* an initiation factor required for eukaryotic protein synthesis (page 507). Phosphorylation inhibits the activity of eIF-2, thereby slowing down the rate at which the cell can translate viral mRNAs.

TOLERANCE, AUTOIMMUNITY, AND AIDS

The ability of the immune system to attack a diverse array of antigens without harming the host's own normal tissues allows animals to mount a sophisticated defense against infectious agents. But it also means that

the failure of the immune system to function properly can trigger a variety of devastating illnesses. In the final section of the chapter, we will explore how some of the more commonly encountered abnormalities of immune function arise.

Tolerance Helps the Immune System Distinguish between Self and Nonself

The realization that an immune response can be directed against millions of different antigens raises a profound question: How are organisms prevented from harming themselves by making antibodies or directing cell-mediated immune responses against their own antigens and normal cells? In other words, how does the immune system distinguish what is "self" from what is "nonself"? An early clue was provided in the 1940s by Ray Owen, who discovered that nonidentical twin cattle occasionally exchange blood cells during embryonic development and retain these "foreign" cells into adulthood. If they had been injected directly into adult cattle, such foreign cells would have been attacked and destroyed by the immune system.

The preceding observations led Macfarlane Burnet to propose that antigens encountered by the immune system during embryonic development are recognized as being self-antigens, inducing a permanent state of immunological **tolerance.** Subsequent support for this idea was provided by Peter Medawar, who injected embryos of one mouse strain with tissues derived from another strain. When the animals reached adulthood, they were unable to reject skin grafts from mice whose tissues they had encountered during embryonic life, although their immune responses were otherwise normal.

What causes animals to become "tolerant" of antigens encountered during embryonic development? Burnet and his colleague Frank Fenner suggested a mechanisms called the **clonal deletion theory,** which postulated that lymphocytes carrying antigen receptors targeted against an organism's own molecules are eliminated during embryonic development when the developing lymphocytes encounter these components. Direct evidence in support of this theory was eventually obtained by Harald von Boehmer and Pawel Kisielow, who devised a technique for monitoring the development of individual T cells carrying different antigen receptors (TCRs). Normally it would be impossible to monitor the developmental fate of the few T cells that happened to carry any one particular receptor because such cells would represent only a tiny fraction of the total T-cell population. Von Boehmer and Kisielow circumvented this obstacle by using mice suffering from *severe combined immune deficiency (SCID),* a genetically inherited disease that makes it impossible for an animal to manufacture any TCRs of its own. Genetic engineering techniques were employed to introduce a single functional TCR gene into a SCID mouse embryo, creating a transgenic mouse that produced a uniform population of T cells all carrying the same antigen receptor (Figure 16-23).

To test the clonal deletion theory, a transgenic mouse was created using a TCR gene coding for an antigen receptor targeted against one of the animal's own antigens. T cells carrying this antigen receptor would be potentially harmful to the organism because they could carry out an immune response targeted against one of the host's own components. When such an experiment was performed, immature T cells bearing a TCR directed against the host's own antigen were found to appear in the animal's thymus during early embryonic development; however, as predicted by the clonal deletion theory, such cells were eliminated before they matured into functional T cells.

Experiments involving transgenic mice carrying a variety of different TCRs have led to the following conclusions regarding T-cell formation (Figure 16-24). During early embryonic development a heterogeneous population of immature T cells expressing billions of different TCRs arises in the thymus. These immature cells, which serve as precursors for both helper T cells and killer T cells, carry both CD4 and CD8 on their surface. Before maturation is completed, cells that would be potentially harmful to the organism are destroyed;

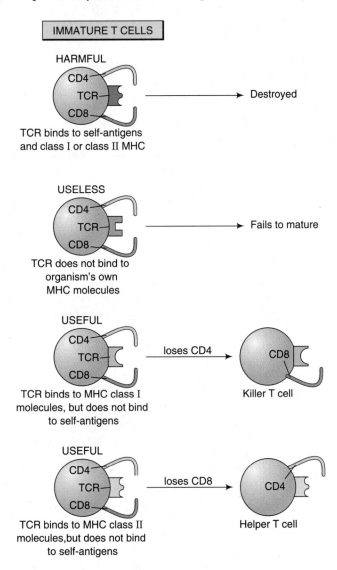

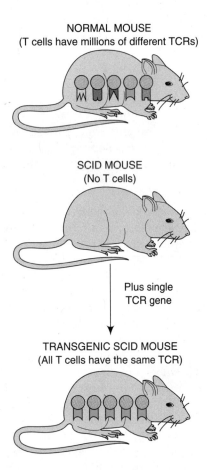

Figure 16-23 _The Use of Transgenic Mice for Studying T-Cell Development_ (Top) _Normal mice possess millions of T cells carrying different TCRs._ (Middle) _Mice suffering from SCID do not make any TCRs, and thus do not produce T cells._ (Bottom) _By using genetic engineering techniques to introduce a single functional TCR gene into a SCID mouse embryo, a transgenic mouse can be created in which all T cells have the same identical antigen receptor. It is then easy to follow the developmental fate of T cells containing that particular TCR._

Figure 16-24 _Fate of T Cells During Embryonic Development_ _Immature T cells carry both CD4 and CD8; hence they serve as precursors of helper T cells as well as killer T cells. Potentially harmful T cells that recognize the organism's own antigen fragments are destroyed, while useless cells that cannot recognize the organism's own intact MHC molecules fail to mature. In contrast, useful cells that recognize the organism's own intact MHC molecules expressed on cell surfaces (but not self-antigen fragments bound to MHC molecules) are selected for maturation into either helper T cells or killer T cells._

such cells contain TCRs that bind to self-antigen fragments associated with either class I or class II MHC molecules. By deleting such cells before they can mature into helper or killer T cells, the organism protects itself from future attack by its own immune system. In contrast, immature lymphocytes whose TCRs bind to the organism's own intact MHC molecules expressed on cell surfaces (but not to self-antigen fragments bound to MHC molecules) are selected for further maturation. Depending on whether the TCR binds to a class I or class II MHC molecule, the selected cell either loses CD4 and develops into a killer T cell containing CD8, or it loses CD8 and develops into a helper T cell containing CD4.

Although the existence of clonal deletion is now well established, it represents only one of several mechanisms that appear to be involved in immune tolerance. An additional mechanism has been implicated in the type of tolerance that can be induced by exposing animals to either very high or very low doses of a particular antigen (Figure 16-25). Clonal deletion cannot be responsible for this type of tolerance because moderate doses of the antigens employed in such experiments have been found to trigger a conventional immune response. In some cases tolerance induced by low doses of antigen appears to be caused by **suppressor T cells,** which are special T cells that inhibit the ability of other lymphocytes to respond to antigen stimulation. Evidence for the proposed role of suppressor T cells has come from studies in which lymphocytes were isolated from animals that had been made tolerant to a particular antigen by exposing them to low doses of that antigen. When lymphocytes isolated from the tolerant animal are injected into a second animal, the recipient becomes tolerant to the antigen in question; thus the recipient animal must have received cells that suppress the activity of its own lymphocytes targeted against the triggering antigen.

Another type of tolerance, called **anergy** ("without working"), occurs when organisms are exposed to antigens under conditions that do not trigger an immune response. For example, it has been known for many years that antigens injected into animals elicit the strongest immune reaction when mixed with an *adjuvant* ("helping substance") consisting of mineral oil, saline solution, detergent, and heat-killed bacteria. If a foreign antigen is injected into an animal without adjuvant, the organism often becomes tolerant; in other words, not only does the animal fail to make antibodies to the antigen, but subsequent injections of the same antigen mixed with adjuvant (which would normally have elicited an immune response) also fail to trigger antibody formation. The apparent explanation for this phenomenon is that lymphocytes shut themselves down instead of becoming activated if they encounter a foreign antigen that has not been properly presented by an antigen-presenting cell. Thus in the absence of an adjuvant, certain

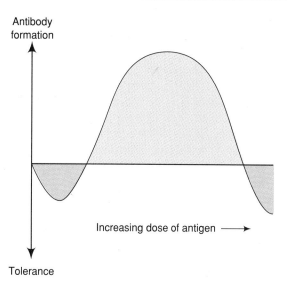

Figure 16-25 *Effect of Antigen Concentration on the Induction of Antibody Synthesis* *Moderate doses of antigen trigger a normal antibody response, but exposing animals to very high or very low doses of the same antigen can induce a permanent state of immunological tolerance.*

antigens may not be taken up or properly prepared by antigen-processing cells; when this occurs and lymphocytes encounter the unprocessed antigen directly, the cells that bind to the antigen become inactivated.

The Failure of Immunological Tolerance Can Trigger Autoimmune Diseases

Immune tolerance is designed to delete or suppress lymphocytes that are potentially harmful because their antigen receptors are targeted against an organism's own molecules and cells. When the mechanisms responsible for tolerance fail, immune responses directed against normal healthy tissues can trigger the onset of severe **autoimmune diseases.** Several dozen autoimmune diseases have been identified in humans, including *multiple sclerosis,* in which the myelin surrounding nerve axons is attacked; *rheumatoid arthritis,* in which the joints are targeted; *type I juvenile diabetes,* in which insulin-producing cells are destroyed; *myasthenia gravis,* in which the acetylcholine receptor is targeted; and *systemic lupus erythematosis,* in which DNA, blood vessels, skin, and kidneys are attacked.

Some important insights into the nature of autoimmunity have emerged from studies involving *experimental autoimmune encephalomyelitis (EAE),* an autoimmune disorder of animals that resembles multiple sclerosis in humans. EAE can be induced by injecting animals with a basic protein derived from the myelin sheath that surrounds nerve axons (page 751). The injected protein triggers an immune response directed against myelin, leading to a widespread paralysis that is often fatal. To determine what type of immune response is responsible for this disease, Irun Cohen and Avraham Ben-Nun isolated lymphocytes from rats afflicted with EAE and cultured them in the presence of myelin basic protein to stimulate the proliferation of lymphocytes containing antigen receptors that bind to this antigen. The proliferating lymphocytes were then isolated and identified as helper T cells whose antigen receptors are specific for myelin basic protein. When such T cells were injected back into normal rats, the animals developed EAE.

The discovery that EAE can be induced by T cells whose receptors are targeted against a single antigen raises the possibility that the disease might be treated by finding a way to inactivate the disease-causing lymphocytes. This prospect has been investigated by taking a T-cell clone directed against myelin basic protein, inactivating the cells by X-irradiation, and then injecting the irradiated cells back into a normal animal (Figure 16-26). Not only do the animals fail to develop EAE, but future injections of myelin basic protein also fail to produce the disease. The explanation for the perma-

nent resistance is that the injected T cells acted as an antigen, triggering an immune response directed against the antigen-binding site of the TCRs carried by these T cells; hence the animals acquired lasting protection against any lymphocyte whose antigen receptor binds to myelin basic protein. In such cases, where the antigen-binding site of a TCR (or antibody molecule) functions as an antigen, the regions of the antigen-binding site that elicit the immune response are collectively referred to as an **idiotype,** and the immune response is called an *anti-idiotypic response.* Since organisms have millions of different TCRs exhibiting millions of potential idiotypes, the concentration of any given idiotype is usually too small to induce tolerance during embryonic development. Thus in theory, it should be possible to treat people suffering from any autoimmune problem by vaccinating them with T cells exhibiting the idiotype responsible for their disease; if successful, such treatment would result in the production of anti-idiotypic T cells that prevent disease by binding to the TCRs on any disease-causing lymphocytes.

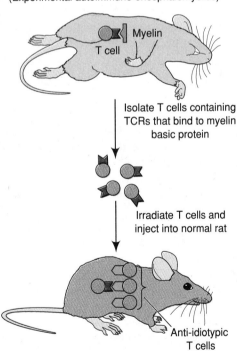

EAE RAT
(Experimental autoimmune encephalomyelitis)

Myelin
T cell

Isolate T cells containing
TCRs that bind to myelin
basic protein

Irradiate T cells and
inject into normal rat

Anti-idiotypic
T cells

Rat develops anti-idiotypic T cells
that prevent disease by binding to the TCRs
on disease-causing lymphocytes

Figure 16-26 *Vaccination against Autoimmunity* *In this experiment, T cells responsible for the autoimmune disease EAE were isolated from rats, inactivated by irradiation, and injected into a healthy rat. The recipient animal developed anti-idiotypic T cells that bound to the TCRs on the disease-causing lymphocytes, thereby making the animal permanently immune to EAE.*

Acquired Immunodeficiency Syndrome (AIDS) Is Caused by a Depletion of Helper T Cells Associated with HIV Infection

In the early 1980s, physicians noticed a growing epidemic of a new human illness characterized by both an unusual susceptibility to rare infectious diseases like *Pneumocystis pneumonia* and the frequent occurrence of an usual type of cancer called *Kaposi's sarcoma.* Since the immune system would normally be expected to prevent the kinds of infections seen with this disease, a defect in immunity was suspected. This suspicion was confirmed by the discovery that afflicted individuals have dramatically reduced numbers of helper T cells. As we have seen, helper T cells are involved in both cell-mediated immunity and antibody formation, so depletion of this cell type has a devastating effect on the immune system. Because the immune deficiency seemed to be acquired through the exchange of body fluids such as blood and semen, the disease was named **acquired immunodeficiency syndrome,** or **AIDS.**

A few years after the disease was first described, it was discovered that lymphocytes obtained from AIDS patients are infected with an RNA virus which was named the **human immunodeficiency virus (HIV).** Because it uses reverse transcriptase to copy its RNA into DNA, HIV is classified as a retrovirus (page 74). HIV consists of an RNA-protein *core* surrounded by a lipid bilayer *envelope* derived from the plasma membranes of human cells (Figure 16-27). Projecting from the surface of the viral envelope are numerous protein "spikes" that play a crucial role when HIV binds to and enters target cells. The outer region of each spike is constructed from the viral glycoprotein *gp120,* which attaches to the glycoprotein *gp41* embedded in the viral envelope. The discovery that gp120 binds tightly to CD4 molecules has helped to explain how HIV attacks the immune system. Infection begins with the binding of viral gp120 to a target cell bearing CD4, which is carried mainly on the surface of helper T cells. The lipid bilayer of the virus then fuses with the plasma membrane of the target cell, the viral core enters into the cytoplasm, and the viral RNA is uncoated and copied by reverse transcriptase into DNA sequences that become integrated into the host-cell DNA. The integrated viral DNA sequences are subsequently transcribed into RNA, viral proteins are synthesized, and the infected cell produces viral particles that bud from the cell surface and infect other helper T cells (Figure 16-28). The main site of infection is the lymph nodes and spleen, where a large fraction of the body's helper T-cell population is infected by HIV. Though a variety of mechanisms have been proposed to explain how HIV infection eventually leads to the depletion of helper T cells, the mechanisms involved in the process are still unknown; it has even been proposed that other factors in addition to HIV infection may be required for T-cell depletion and the onset of AIDS.

Individuals who have become infected with HIV attempt to fight the virus by producing antibodies and mounting cell-mediated immune responses against a variety of HIV components. In fact, it is the presence of antibodies directed against HIV that allows people who have been infected with HIV to be diagnosed. But neither the antibodies nor the cell-mediated immune responses directed against HIV appear to confer protection against AIDS. One reason for this failure may be that the T cells needed for mounting an adequate immune response are killed or inactivated by the virus. Another complicating factor is the genetic variability of HIV. The reverse transcriptase that catalyzes HIV replication lacks the proofreading mechanism used by DNA polymerases to correct the base-pairing mistakes that normally occur during cellular DNA replication (page 72). Moreover, HIV reverse transcriptase seems to be even more error prone than the comparable enzyme of

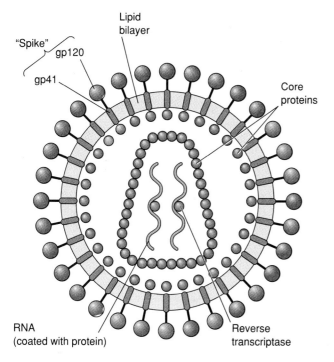

Figure 16-27 *Model Showing the Major Features of HIV*
The virus consists of an RNA-protein core surrounded by a lipid bilayer envelope. The protein gp120, located in the "spikes" that protrude from the envelope, bind to CD4 molecules carried on the surface of helper T cells.

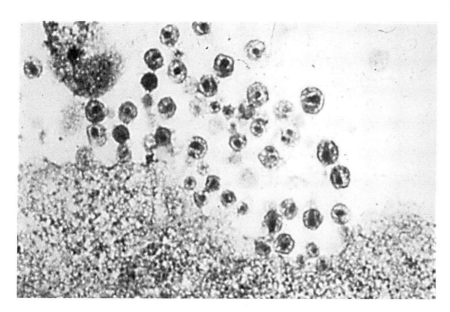

Figure 16-28 *The Surface of an HIV-Infected Helper T Cell* *Numerous HIV particles are seen budding from the plasma membrane. Courtesy of A. Fauci.*

other retroviruses. As a consequence, HIV mutates very quickly, introducing about one error for every 2000 nucleotides incorporated. So any immune response directed by the host against a particular HIV antigen may quickly become obsolete as the viral gene coding for that antigen undergoes mutation. The rapid mutation of HIV creates a similar problem for investigators who have been trying to develop a vaccine designed to prevent HIV infection, since the virus may mutate to evade the protective effect conferred by any vaccine.

Attempts to treat AIDS with antiviral drugs have likewise met with limited success. The most widely used treatment for AIDS is *azidothymidine (AZT)*, a nucleoside that can be incorporated into DNA in place of thymidine. Because AZT lacks the 3' hydroxyl group found in the normal nucleotides that are incorporated into DNA, elongation of the DNA chain is halted once AZT has been incorporated. AZT has turned out to be a useful inhibitor of HIV replication because, unlike mammalian DNA polymerases, the reverse transcriptase of HIV actually prefers AZT to deoxythymidine. Nonetheless, AZT is still very toxic to human DNA polymerase and the rapid mutation rate of HIV leads to the quick appearance of mutated forms of reverse transcriptase that are resistant to AZT. In an attempt to circumvent this problem, test-tube experiments have been carried out in which HIV-infected cells are simultaneously exposed to several different drugs that all inhibit reverse transcriptase. Under such conditions, HIV sometimes mutates to a point where it is incapable of replicating. Experimental trials are currently under way to determine whether similar drug combinations will be therapeutically beneficial when administered directly to AIDS patients.

SUMMARY OF PRINCIPAL POINTS

• Antigens are molecules that stimulate antibody formation or a cell-mediated immune response. To elicit either type of immune response, antigens must be processed into fragments that are presented at the cell surface in combination with either a class I or II MHC molecule.

• Lymphocytes are subdivided into two principal classes, called B cells and T cells. B cells develop in the bone marrow (or bursa of Fabricius in birds), while T cells develop in the thymus. B cells produce antibodies, which are proteins that circulate in body fluids and bind to foreign antigens. T cells assist in the formation of antibodies and carry out cell-mediated immune reactions in which T cells bind to and kill cells directly.

• Antibody molecules are built from light and heavy chains, each of which has an N-terminal variable region and a C-terminal constant region. The variable region of both light and heavy chains exhibits three hypervariable segments that come together to form the antigen-binding site.

• Animals have millions of different B cells, each making a single type of antibody specific for a particular antigen. The role of a triggering antigen is to bind to those B cells that make an antibody specific for the antigen in question, causing these particular cells to proliferate and secrete antibody.

• The ability of an organism to produce millions of different antibodies is based on several mechanisms, including (1) the inheritance of multiple light- and heavy-chain gene segments through the germ cells, (2) the random joining of V-J and V-D-J sequences through genetic recombination, (3) the imprecision of the joining process, (4) the random association of the resulting heavy and light chains, and (5) the mutation of CDR1 and CDR2 sequences during an immune response.

• The formation of antibodies involves cooperation between three cell types: B cells, helper T cells, and antigen-presenting cells such as macrophages. During the first stage of antibody induction, the T-cell antigen receptors (TCRs) of helper T cells bind to antigen fragments presented in combination with class II MHC molecules on the macrophage surface. This interaction triggers the production of interleukins that stimulate proliferation of the helper T cells. In the second phase of antibody induction, antigens bind to antibody molecules located on the B-cell surface, causing the antigen to be taken up, processed, and transported back to the cell surface bound to class II MHC molecules. Helper T cells then bind to the antigen–MHC II complex on the B-cell surface and secrete interleukins that stimulate the B cell to proliferate into antibody-producing plasma cells. Some of the responding B cells are also set aside as memory cells in preparation for any subsequent exposure to the same antigen.

• Antibodies function by binding to their target antigens, blocking any toxic effects that the antigen may have exhibited. Antibodies also bind to antigens located on the surface of invading microorganisms, triggering activation of the complement system and the resulting formation of membrane-attack complexes that cause the microorganisms to swell and burst.

• Cell-mediated immunity is carried out by killer T cells, which attack target cells that harbor foreign infectious agents. Animals have millions of different T cells, each carrying a TCR that is specific for a different antigen. Killer T cells bind to antigen–MHC I complexes on the surface of target cells and secrete the protein perforin, which creates pores in the plasma membrane of the targeted cell. Besides attacking cells that harbor viruses and other infectious agents, killer T cells destroy tissue and organ grafts by recognizing the class I and II MHC molecules expressed by the grafted cells as being foreign. In a related type of cell-mediated immunity, natural killer (NK) cells destroy cancer cells by a mechanism that does not involve recognition of antigen–MHC complexes.

• Interferons produced by virus-infected cells bind to plasma membrane receptors of neighboring cells, triggering the formation of enzymes that provide a relatively nonspecific type of immunity to viral infection.

• To prevent organisms from carrying out immune responses against their own molecules and cells, lymphocytes targeted against self-antigens are destroyed during embryonic development, thereby creating a state of immunological tolerance for those antigens. Tolerance to individual antigens can also be induced by suppressor T cells, which inhibit the ability of other lymphocytes to respond to antigen stimulation. The failure of immune tolerance can trigger autoimmune diseases in which immune responses are targeted against an individual's own tissues.

• AIDS is a severe immunodeficiency disease characterized by the depletion of helper T cells. The lymphocytes of AIDS patients are infected with HIV, a retrovirus containing an envelope protein (gp120) that binds to the CD4 glycoprotein carried by helper T cells. Devising a strategy for preventing or treating AIDS has been difficult because the reverse transcriptase that catalyzes HIV replication is especially prone to errors and hence the virus mutates rapidly.

SUGGESTED READINGS

Books

Abbas, A. K., A. H. Lichtman, and J. S. Pober (1991). *Cellular and Molecular Immunology,* Saunders, Philadelphia.

Austyn, J. M. (1989). *Antigen-Presenting Cells,* IRL Press, Oxford, England.

Golub, E. S., and D. R. Green (1991). *Immunology: A Synthesis,* 2nd Ed., Sinauer Associates, Sunderland, MA.

Hamblin, A. S. (1988). *Lymphokines,* IRL Press, Oxford, England.

Mak, T. W., and H. Wigzell (1991). *AIDS: 10 Years Later, FASEB Journal,* Vol. 5, No. 10.

Owen, M. J., and J. R. Lamb (1988). *Immune Recognition,* IRL Press, Oxford, England.

Ritter, M. A., and I. N. Crispe (1992). *The Thymus,* IRL Press, Oxford, England.

Tizard, I. R. (1992). *Immunology: An Introduction,* 3rd Ed., Saunders, Philadelphia.

Articles

Ada, G. L., and G. Nossal (1987). The clonal-selection theory, *Sci. Amer.* 257 (August):62-69.

Berzofsky, J. A., S. J. Brett, H. Z. Streicher, and H. Takahashi (1988). Antigen processing for presentation to T lymphocytes: function, mechanisms, and implications for the T-cell repertoire, *Immunol. Rev.* 106: 5-31.

Brodsky, F. M. (1992). Antigen processing and presentation: close encounters in the endocytic pathway, *Trends Cell Biol.* 2:109-115.

Brown, J. H., et al. (1993). Three-dimensional structure of the human class II histocompatibility antigen HLA-DR1, *Nature* 364:33-39.

Chow, Y.-K., et al. (1993). Use of evolutionary limitations of HIV-1 multidrug resistance to optimize therapy, *Nature* 361:650-654.

Cohen, I. R. (1988). The self, the world and autoimmunity, *Sci. Amer.* 258 (April):52-60.

DeMars, R., and T. Spies (1992). New genes in the MHC that encode proteins for antigen processing, *Trends Cell Biol.* 2:81-86.

Engelhard, V. H. (1994). How cells process antigens, *Sci. Amer.* 271 (August): 54-61.

Gallo, R. C., and L. Montagnier (1988). AIDS in 1988, *Sci. Amer.* 259 (October):40-48. (This entire issue of *Scientific American* is devoted to AIDS.)

Gellert, M. (1992). V(D)J recombination gets a break, *Trends Genet.* 8:408-412.

Germaine, R.N. (1994). MHC-dependent antigen processing and peptide presentation: providing ligands for T lymphocyte activation, *Cell* 76:287-299.

Greene, W. C. (1993). AIDS and the immune system, *Sci. Amer.* 269 (September):98-105.

Grey, H. M., A. Settee, and S. Buus (1989). How T cells see antigen, *Sci. Amer.* 261 (November):56-64.

Janeway, C. A., Jr. (1993). How the immune system recognizes invaders, *Sci. Amer.* 269 (September):72-79.

Johnson, H. M., F. W. Bazer, B. E. Szente, and M. A. Jarpe (1994). How interferons fight disease, *Sci. Amer.* 270 (May): 68-75.

Kägl, D., et al. (1994). Cytotoxicity mediated by T cells and natural killer cells is greatly impaired in perforin-deficient mice, *Nature* 369: 31-37.

Marrack, P., and J. W. Kappler (1993). How the immune system recognizes the body, *Sci. Amer.* 269 (September):80-89.

Miller, J. F. A. P. (1994). The thymus: Maestro of the immune system, *BioEssays* 16: 509-513.

Pascual, V., and J. D. Capra (1991). Human immunoglobulin heavy-chain variable region genes: organization, polymorphism, and expression, *Adv. Immunol.* 49:1-74.

Podack, E. R., and A. Kupfer (1991). T-cell effector functions: mechanisms for delivery of cytotoxicity and help, *Annu. Rev. Cell Biol.* 7:749-504.

Rennie, J. (1990). The body against itself, *Sci. Amer.* 263 (December):106-115.

Schwartz, R. H. (1993). T cell anergy, *Sci. Amer.* 269 (August):62-71.

Sheriff, S., et al. (1987). Three-dimensional structure of an antibody-antigen complex, *Proc. Natl. Acad. Sci. USA* 84:8075-8079.

Silverton, E. W., M. A. Navia, and D. R. Davies (1977). Three-dimensional structure of an intact human immunoglobulin, *Proc. Natl. Acad. Sci. USA* 74:5140-5144.

Smith, K. A. (1990). Interleukin-2, *Sci. Amer.* 262 (March):50-57.

Steinman, L. (1993). Autoimmune disease, *Sci. Amer.* 269 (September):106-114.

Temin, H. M., and D. P. Bolognesi (1993). Where has HIV been hiding? *Nature* 362:292-293.

Teyton, L., and P. A. Peterson (1992). Invariant chain—a regulator of antigen presentation, *Trends Cell Biol.* 2:52-56.

Tonegawa, S. (1985). The molecules of the immune system, *Sci. Amer.* 253 (October):122-131.

van Bleek, G. M., and S. G. Nathenson (1992). Presentation of antigenic peptides by MHC class I molecules, *Trends Cell Biol.* 2:202-207.

von Boehmer, H., and P. Kisielow (1991). How the immune system learns about self, *Sci. Amer.* 265 (October):74-81.

Weiss, A., and D.R. Littman (1994). Signal transduction by lymphocyte antigen receptors, *Cell* 76:263-274.

Weiss, R. A. (1993). How does HIV cause AIDS?, *Science* 260: 1273-1279.

Weissman, I. L., and M. D. Cooper (1993). How the immune system develops, *Sci. Amer.* 269 (September):64-71.

Young, J. D.-E., and Z. A. Cohn (1988). How killer cells kill, *Sci. Amer.* 258 (January):38-44.

Chapter 17

Neurons and Synaptic Signaling

Animals are comprised of a vast number of cells whose activities must be precisely coordinated with each other. Because the cells that need to interact are often separated by long distances, animals have developed a nervous system that allows virtually instantaneous communication among the various regions of the body. Evolution of the nervous system has reached its zenith in higher vertebrates, where billions of nerve cells form trillions of connections with one another. These elaborate cellular networks receive sensory information from both inside and outside the body, integrate and analyze the information, and then respond by triggering appropriate responses. As the chapter unfolds, we will see that in spite of its inherent complexity, the nervous system communicates using two cellular phenomena that are well understood: changes in the electrical properties of the plasma membrane, and the release of chemical neurotransmitters that transmit signals from one cell to the next. As background for this chapter, the descriptions of membrane potentials, action potentials, and plasma membrane receptors included in Chapter 6 (pages 195–221) should first be reviewed.

NEURON STRUCTURE

To facilitate communication between distant regions of the body that is both rapid and precisely targeted, the cells of the nervous system have evolved a highly specialized morphology that makes them look quite different from any other cell type. We will therefore begin our discussion of neural communication by describing the architectural organization of nervous tissue, emphasizing the relationship between nerve cell structure and the mechanism by which nerve cells communicate.

The Discovery of Neurons Was Hindered by Their Unusual Morphology

Shortly after Schleiden and Schwann first formulated the cell theory in 1839 (page 4), microscopic examination of a variety of different tissues provided growing support for their contention that all living tissues are composed of cells. But a notable exception was encountered in the nervous system, where little evidence for the presence of individual cells was initially detected. Instead, early microscopic studies of nervous tissue revealed only hazy outlines that looked like a network of connecting channels running between spherical cell bodies. Biologists therefore tended to view the nervous system as a physically continuous network of cell bodies and cytoplasmic extensions without boundaries that would define individual cells.

A major impediment to understanding the true organization of nervous tissue no doubt lay in the unusual size and shape of typical nerve cells, whose long, branching cytoplasmic extensions make individual cells difficult to visualize in their entirety. This obstacle was finally overcome in 1873 by an impoverished Italian physician, Camillo Golgi, who was experimenting with different fixing and staining methods for nervous tissue. Working by candlelight in his kitchen, Golgi discovered that exposing chromate-hardened brain tissue to silver nitrate reveals the complete outline of individual nerve cells, including their branched cytoplasmic extensions (Figure 17-1). But the significance of this remarkable advance was not fully appreciated until several years later, when the Spanish neurobiologist Santiago Ramón y Cajal used Golgi's technique to trace the cytoplasmic extensions of individual nerve cells. These studies led Ramón y Cajal to conclude that each nerve cell body has its own set of cytoplasmic extensions that are not continuous with those of other cell bodies. In other words, it appeared as if each cell body and its associated cytoplasmic extensions represented a single nerve cell, or **neuron.**

Ramón y Cajal's conclusion that nervous tissue is comprised of individual neurons did not go unchallenged, however. Golgi and many others still believed that the cytoplasmic extensions of nerve cell bodies are all continuous with one another, and a heated controversy developed that was exacerbated when the two major protagonists, Golgi and Ramón y Cajal, shared the Nobel Prize in 1906. Although the theory that nervous tissue is composed of individual neurons gradually gained prominence, it took almost 50 years before the advent of electron microscopy finally allowed cell biologists to detect the membrane boundaries that separate the cytoplasmic extensions of one neuron from those of its neighbors. Thus more than a hundred years intervened between Schleiden and Schwann's formulation of the cell theory and definitive proof for the idea that nervous tissue is composed of individual cells, each bounded by its own plasma membrane.

Axons and Dendrites Are Cytoplasmic Extensions of the Cell Body That Send and Receive Electrical Signals

A typical neuron is composed of two principal regions: a *cell body* and a series of thin cytoplasmic extensions

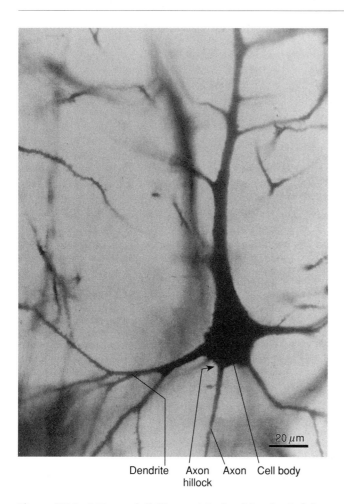

Dendrite   Axon   Axon   Cell body
            hillock

Figure 17-1 *A Nerve Cell (Neuron) Stained by the Golgi Method* *The cell body, numerous dendrites, an axon, and the axon hillock are clearly visible. Courtesy of A. Peters.*

emerging from the cell body called *dendrites* and *axons* (Figure 17-2). The **cell body** contains a centrally placed nucleus that often exhibits a prominent nucleolus. Under the light microscope, the cytoplasm of the cell body appears to be filled with large clumps of material that stain intensely with basic dyes. For many years this material, named *Nissl substance* after its discoverer Franz Nissl, was believed to be an artifact generated during sample preparation. However, electron microscopic studies eventually revealed that the Nissl substance corresponds to dense masses of ribosomes and endoplasmic reticulum that occur in high concentration in the cell body. In addition, the cell body usually contains an extensive Golgi complex and large numbers of mitochondria and lysosomes.

The **dendrites** and **axons** that project from the cell body can be distinguished from each other in several ways (Table 17-1). The most important difference is functional; dendrites are specialized for receiving signals from other cells, whereas axons are involved primarily in sending signals. Dendrites are highly branched, relatively short (less than a millimeter in length), emerge in large numbers from a single cell body, and exhibit small protuberances called dendritic

spines that are specialized for receiving signals from other cells. Axons often have relatively few branches, are longer than dendrites (up to a meter or more in length), and only one or two axons is present per cell. The enormous length of some axons allows neurons in the central nervous system to make direct contact with cells located at the end of a person's arms or legs. Axons emerge from a spherical region of the cell body called the *axon hillock,* which is characterized by a reduced content of free and membrane-bound ribosomes.

The cytoplasm of dendrites and axons often contains numerous vesicles and mitochondria, but the most prevalent components are cytoskeletal filaments that provide mechanical support and facilitate the movement of proteins and organelles. The cytoskeletal elements involved in these activities include microtubules, neurofilaments, and actin filaments (Figure 17-3). **Microtubules** (also called *neurotubules*) are present in high concentration in both axons and dendrites; in axons they are oriented with their plus ends (the end where microtubule assembly occurs) pointing away from the cell body, whereas in dendrites they are oriented in both directions—that is, with their plus ends pointing both away from and toward the cell body (see Figure 13-66). **Neurofilaments,** which are a type of intermediate filament (page 619), occur in highest concentration in axons, where they are organized into longitudinal crosslinked arrays that provide mechanical support and help keep the axon from breaking. Finally, **actin filaments** form a network beneath the plasma membrane of the entire neuron.

Glial Cells Play a Supporting Role and Produce Myelin Sheaths

Supporting cells called **glial cells** occur in large numbers in the nervous system, filling the spaces between neuron cell bodies and surrounding the axons and dendrites. In the mammalian brain, glia outnumber neurons by a ratio of at least ten to one. Two types of glial cells predominate in the *central nervous system* (brain and spinal cord). The first, called an **astrocyte**, is a star-shaped cell possessing long cytoplasmic extensions that contain large numbers of intermediate filaments composed of *glial acidic fibrillary protein.* The great mechanical strength of intermediate filaments suggests that astrocytes provide structural support for the nervous system.

The other type of glial cell found in the central nervous system is the **oligodendrocyte,** which produces insulating sheaths that allow axons to transmit electrical signals quickly and efficiently. The presence of such sheaths was first noted in the 1860s by Rudolph Virchow while he was investigating the microscopic organization of different regions of the brain. When viewed with the naked eye, the vertebrate brain and spinal cord

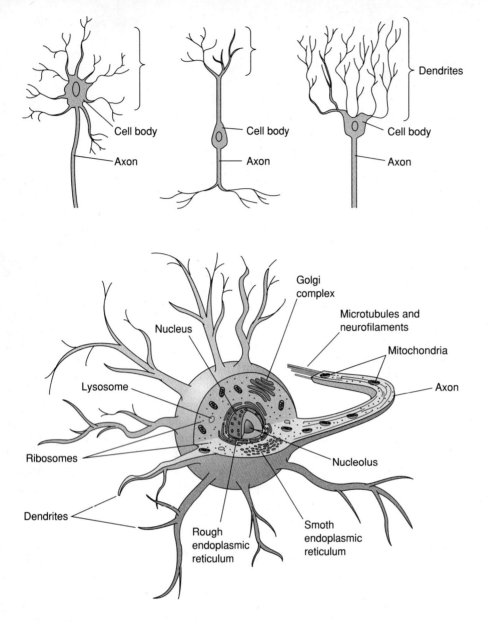

Figure 17-2 Organization of the Neuron (Top) *Schematic diagram showing several arrangements of the cell body, axon, and dendrites.* (Bottom) *Diagram of a single neuron showing the various organelles located in the cell body.*

Table 17-1	Distinguishing Features of Typical Dendrites and Axons	
Dendrites	**Axons**
Many per cell	Never more than one or two
Relatively short (< 1 mm)	May be very long (1 meter or more)
Diameter lessens with distance from the cell body	Diameter relatively constant
Many branches	Few branches
Contain spines	Smooth surface
Never myelinated	Often myelinated
Microtubules usually outnumber neurofilaments	Neurofilaments usually outnumber microtubules
Microtubules oriented both toward and away from cell body	Microtubules oriented with minus end toward cell body
Ribosomes present	Ribosomes absent
Receive signals	Send signals

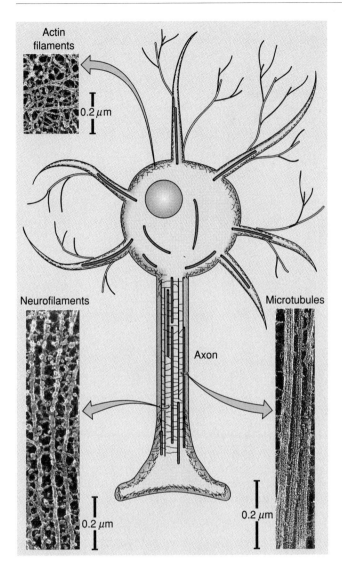

Figure 17-3 *The Neuronal Cytoskeleton* *Microtubules and neurofilaments are concentrated in axons and dendrites, where they are oriented in parallel. Actin filaments are arranged in a meshlike network beneath the plasma membrane of the entire neuron. Micrographs courtesy of N. Hirokawa.*

exhibit differently colored regions referred to as *gray* and *white matter,* respectively. Upon microscopic examination, Virchow discovered that white matter contains a closely packed array of axons covered by a white material which he named **myelin** (from the Greek word for "marrow") because it is concentrated in the core, or marrow, of the brain. Subsequent studies revealed that myelin consists of sheaths of stacked membranes that surround and insulate the axons of many (but not all) neurons, increasing the speed at which electrical signals are transmitted. In the central nervous system, myelin is produced by oligodendrocytes, which wrap their plasma membranes around axons to produce a multilayered membrane sheath; in peripheral nerves, the comparable function is carried out by another kind

of glial cell, the **Schwann cell** (see Figure 5-3). The final thickness of a myelin sheath is directly related to axon diameter; the largest axons have myelin sheaths with the greatest number of membrane layers, whereas small axons have fewer layers, or even none at all.

In the late 1870s the French pathologist Louis-Antoine Ranvier first noted that the myelin sheath is periodically interrupted, creating **nodes of Ranvier** where one glial cell terminates and the next one begins. As each node is approached, the layers of myelin terminate one by one until the plasma membrane of the axon is exposed (Figure 17-4). We will see later in the chapter that these bare areas of plasma membrane are directly involved in the mechanism by which electrical signals are propagated along an axon.

Axonal Transport Moves Materials Back and Forth along the Axon

Because the nucleus and biosynthetic machinery of the neuron are localized in the cell body, far removed from distant regions of the axon, the question arises as to how the axon's supply of essential molecules and organelles is maintained. Simple diffusion is not an adequate explanation, because even small molecules such as glucose would take months or years to diffuse the length of a typical axon. To solve this problem, neurons have developed mechanisms for transporting materials down the axon at rates that vary with the substance being transported. The transport process was first detected in 1948 by Paul Weiss and Helen Hiscoe, who found that constricting an axon causes a swelling to appear just before the point of obstruction (Figure 17-5, *top*). When the constriction is removed, the accumulated material can be seen migrating down the axon at a rate of about 1–5 mm per day. Weiss initially proposed that the flow of material is caused by bulk cytoplasmic movements propelled by contractile waves moving down the axon surface. Although the postulated contractile waves have been detected in neurons growing in culture (where bulk cytoplasmic flow may contribute to axon lengthening), comparable evidence for bulk cytoplasmic flow in the nongrowing axons of mature neurons has not been forthcoming. Indeed, it would be hard to explain what happens to the flowing cytoplasm that would continually be arriving at the axon tip in mature cells.

It has subsequently been shown that movement of materials down nongrowing axons involves the transport of substances through the cytoplasm rather than bulk movement of the cytoplasm itself. Two broad classes of transport have been detected. The first, called **slow axonal transport,** moves proteins and cytoskeletal filaments down the axon at rates of about 1–5 mm per day. Although microtubule sliding has been implicated in this type of transport, the force-generating mechanism has not been clearly identified.

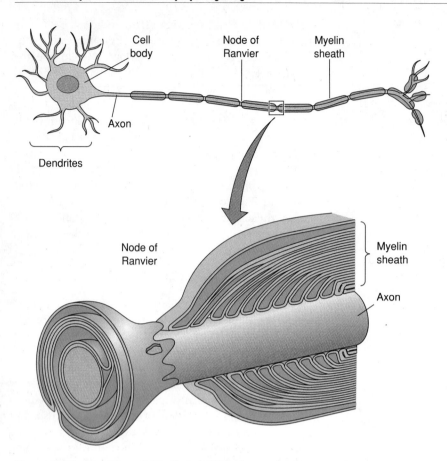

Figure 17-4 *Structure of the Node of Ranvier* *The diagram of the entire neuron* (top) *shows how the nodes of Ranvier interrupt the myelin sheath. In the enlarged diagram at the bottom, the cutaway portion* (right side) *shows how the layers of myelin terminate one by one as the node is approached, exposing the plasma membrane of the axon.*

The other type of transport is called **fast axonal transport** because it moves materials up to a hundred times faster than slow axonal transport. Fast transport was first detected by injecting radioactive substances into the cell body and then monitoring the appearance of radioactivity at various points along the axon (see Figure 17-5, *bottom*). Such experiments revealed that whereas slow axonal transport moves certain molecules down the axon at rates of about 1–5 mm per day, other materials are simultaneously propelled down the same axon by fast transport at rates of 10–500 mm per day (Figure 17-6). At these rates, it would take about a day for fast transport and a year for slow transport to move a cellular component from the cell body to the tip of an axon 35 cm long.

In addition to differing in speed, fast and slow transport can be distinguished by differences in the materials being transported. Slow transport mediates the movement of proteins and cytoskeletal filaments, whereas fast transport propels membrane-bound organelles such as mitochondria and membrane vesicles. Another difference is that slow transport moves substances only toward the axon tip (the *anterograde* direction), but fast transport can move materials toward the axon tip as

well as back toward the cell body (the *retrograde* direction). Fast transport in the anterograde direction moves mitochondria and neurotransmitter-filled vesicles toward the axon tip; fast transport in the retrograde transport moves other membrane-bound vesicles back to the cell body. As we learned in Chapter 13, fast transport is mediated by the motor proteins *kinesin* and *dynein* (pages 616–617). Vesicles linked to axonal microtubules by kinesin move in the anterograde direction, whereas vesicles attached to microtubules by dynein migrate in the retrograde direction.

TRANSMISSION OF NERVE IMPULSES

Neurons transmit signals by changing the electrical potential across their plasma membranes. We explained in Chapter 6 that the main type of electrical signal used by nerve and muscle cells is the **action potential** (pages 199–202). An action potential is caused by stimuli that trigger a transient increase in membrane permeability to Na^+, thereby allowing a tiny amount of Na^+ to diffuse

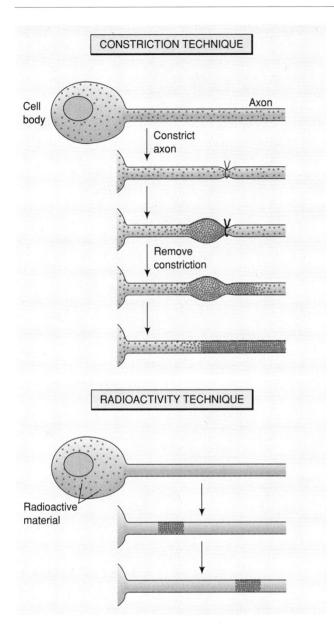

CONSTRICTION TECHNIQUE

Cell body

Axon

Constrict axon

Remove constriction

RADIOACTIVITY TECHNIQUE

Radioactive material

Figure 17-5 Two Methods for Detecting Axonal Transport (Top) *Placing a constriction in an axon causes material to accumulate just before the constriction point. When the constriction is removed, the mass of accumulated material can be observed migrating down the axon. (Bottom) Radioactive substances injected into a cell body are incorporated into cellular molecules that gradually migrate down the axon. The moving radioactivity can be detected either by autoradiography or by cutting the axon into segments and directly measuring the radioactivity in each segment (see Figure 17-6).*

into the cell. The resulting small depolarization causes a few voltage-gated Na^+ channels to open, which in turn permits an additional influx of Na^+ that depolarizes the membrane even further and thus causes more Na^+ channels to open. As this cycle repeats, the rapidly increasing depolarization of the plasma membrane eventually triggers the opening of voltage-gated K^+ channels; K^+

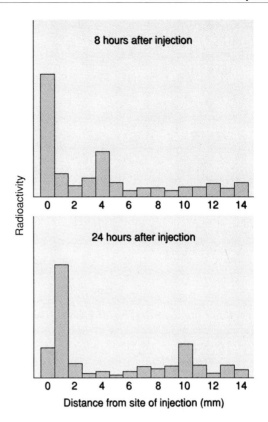

8 hours after injection

24 hours after injection

Radioactivity

Distance from site of injection (mm)

Figure 17-6 Evidence for the Existence of Both Slow and Fast Axonal Transport in the Same Axon *After injecting radioactive amino acids into the cell bodies of crayfish neurons, the axons were cut into 1-mm sections and the radioactivity in each segment was measured. After 24 hours, one peak of radioactivity has migrated about 10 mm down the axon (fast axonal transport) while a second, slower-moving peak of radioactivity has migrated only 1 mm from the cell body (slow axonal transport).*

therefore diffuses out of the cell, reestablishing the normal membrane potential.

Although action potentials play a prominent role in neural communication, they are only one aspect of the mechanism by which electrical signals are transmitted through the nervous system. In this segment of the chapter, we will see how action potentials and other types of electrical signals are propagated along the neuron surface and then transmitted from one neuron to another.

Axon Diameter and Myelination Influence the Rate at Which Action Potentials Are Propagated

Once an action potential has been triggered at a given site on a nerve cell plasma membrane, the signal must be propagated to the end of the axon so that it can be transmitted to another cell. The way in which the electrical signal is propagated along the plasma membrane varies among different types of neurons. Let us begin by considering the behavior of neurons with long axons (>1 mm). To initiate an action potential, a stimulus must

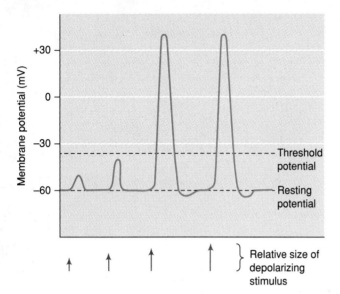

Figure 17-7 Evidence for the Existence of a Threshold Potential *If a small electric current is applied to a neuron to trigger a tiny depolarization, an action potential is not elicited. An action potential is triggered only after the depolarizing stimulus has exceeded a critical value called the threshold potential. Once the threshold potential has been reached, further increases in the magnitude of the depolarizing stimulus do not affect the size of the action potential.*

first depolarize a site on the neuron plasma membrane to a critical value termed the **threshold potential** (Figure 17-7). An increase in stimulus strength beyond this point has no further effect on the magnitude of the action potential. The reason subthreshold stimuli do not bring about action potentials is that the Na^+ influx caused by small depolarizations is less than the normal rate of K^+ efflux, allowing the latter to return the membrane potential to its resting value. Hence an action potential can only be triggered by an initial depolarization that is large enough to produce a rate of Na^+ influx that is greater than the initial rate of K^+ efflux.

Once an initial action potential has been triggered, a local flow of electric current is induced that alters the membrane potential in surrounding areas of the plasma membrane; during this process, positive ions associated with the inner surface of the plasma membrane at the site of the action potential diffuse to adjacent, negatively charged regions of the inner membrane surface, creating a flow of current that depolarizes the surrounding membrane and triggers new action potentials (Figure 17-8). This cycle is repeated again and again along the membrane surface, creating a series of action potentials that travel away from the original site of excitation. Since most action potentials are initiated at or near the cell body, impulses tend to travel away from this region and toward the end of the axon. Each newly initiated action potential is equivalent in magnitude to the original one, so the signal that eventually reaches the tip of the axon is virtually identical to the originating signal.

The velocity of action potential propagation is influenced by two principal factors: axon diameter and myelination. Large-diameter axons provide less electrical resistance to the flow of current, so action potentials move along the plasma membrane more rapidly. In quantitative terms, this phenomenon yields a propagation velocity that is roughly proportional to the square root of the axon diameter (Figure 17-9, *inset*).

Myelination, on the other hand, influences propagation velocity by virtue of its insulating effect. In unmyelinated axons, the ions carrying the local currents that trigger new action potentials can leak back across the plasma membrane, thereby reducing the flow of current to adjacent membrane regions and hence decreasing the effectiveness of action potential propagation. By reducing ion leakage, the presence of a myelin sheath increases the speed at which areas adjacent to the initial action potential are brought to threshold. Because action potentials can only occur where the plasma membrane is exposed to extracellular fluid (which contains the Na^+ needed for an action potential), the presence of a myelin sheath also influences propagation velocity by preventing action potentials from occurring anywhere except at the nodes of Ranvier, where myelin is absent. Action potentials in myelinated axons therefore jump from node to node, a phenomenon known as *saltatory conduction.* The time required for a depolarizing current to flow from node to node is small compared to the time needed to generate a new action potential at each node, so propagation velocity increases with increasing distance between nodes. The distance between nodes, which are spread out along the axon, increases with axon diameter, leading to faster propagation in larger axons (see Figure 17-9). Myelination permits large axons to propagate action potentials at velocities in excess of 100 meters per second, which is an order of magnitude faster than the velocity of impulse conduction in unmyelinated nerve axons of similar diameter. The importance of myelination for normal signal transmission is vividly demonstrated by the disease *multiple sclerosis,* an autoimmune disorder in which destruction of myelin in various regions of the brain causes speech and vision defects, tremors, and paralysis (Figure 17-10).

Although the propagation of action potentials is an efficient means of transmitting electrical signals in neurons with long axons, most neurons in the central nervous system have axons measuring less than a millimeter in length and do not utilize (nor can they generate) action potentials for signal transmission. If a neuron of this type is partially depolarized, either by an external stimulus or by a chemical signal released from another neuron, the localized change in membrane potential is passively conducted across the rest of the plasma membrane by local current flow without ever triggering an action potential. However, the leakage of ions through the plasma membrane causes local currents to dissipate rapidly and hence signal transmission is limited to distances of about a millimeter. But in

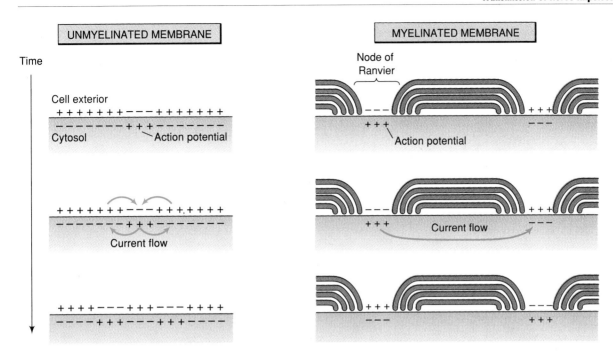

Figure 17-8 Propagation of Action Potentials   (Left) *After an initial action potential has been triggered in an unmyelinated axon, a flow of current is induced that depolarizes the surrounding membrane, causing action potentials to spread along the axon.* (Right) *In myelinated axons, action potentials can only occur where the plasma membrane is exposed to the extracellular fluid. Therefore action potentials jump from one node of Ranvier to the next.*

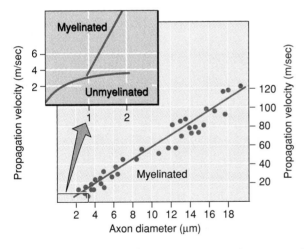

Figure 17-9 Relationship between Axon Diameter and the Velocity at Which Nerve Impulses Are Propagated   *In unmyelinated axons, velocity increases with the square root of axon diameter (inset). Myelinated axons propagate impulses much faster because the action potential jumps from one node of Ranvier to the next. Since the spacing between nodes increases linearly with axon diameter, velocity in myelinated nerves increases linearly with diameter.*

spite of this limitation, the transmission of local currents has the advantage of being equally suitable for propagating membrane hyperpolarizations and depolarizations. The existence of this mechanism for transmitting signals means that a **nerve impulse** can be any change in membrane potential that is transmitted along nerve cells, including, but not restricted to, action potentials.

Neurons Communicate with Each Other at Specialized Junctions Called Synapses

Electrical impulses that travel along nerve cells have an inherent directionality that is imposed by the organization of the cell's dendrites and axons. Dendrites are designed to receive signals from sensory cells or the axons of other neurons, whereas axons transmit signals to other neurons or to appropriate target cells. The directionality of signal transmission is imposed by cell-cell junctions called **synapses,** which are specialized for transmitting nerve impulses from one cell to another. Most synapses transmit signals in one direction only, from the *presynaptic* cell to the *postsynaptic cell.* An individual neuron may be involved in hundreds or even thousands of different synapses, functioning as the presynaptic cell at some synapses and the postsynaptic cell at others. While many synapses transmit signals from the axon of one neuron to the dendrite of a second neuron, others may transmit signals from axon to cell body, axon to axon, or even dendrite to dendrite. Synapses also transmit signals from nerve axons to other cell types, such as secretory cells and muscle cells. The synapses that link nerve axons to muscle cells, known as *neuromuscular junctions,* have already been described in Chapter 13 (page 573).

Synapses are of two fundamentally different types: *chemical* and *electrical* (Figure 17-11). **Chemical synapses,** which are the most common, transmit signals by releasing chemical *neurotransmitters* that diffuse from the presynaptic cell to the postsynaptic cell. A space of 20 to 30 nm, known as the **synaptic cleft,** separates the plasma membranes of the presynaptic

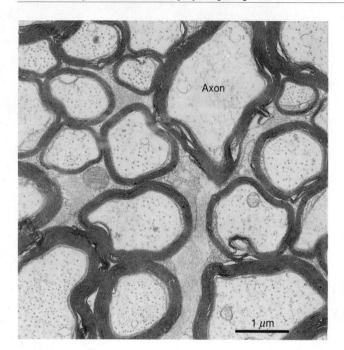 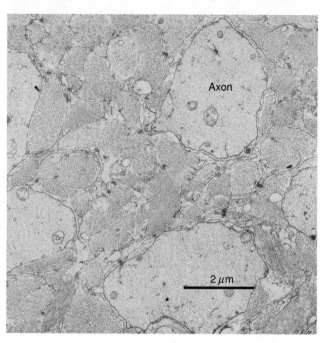

Figure 17-10 *Disruption of the Myelin Sheath in Multiple Sclerosis* (Left) *Thin-section electron micrograph showing myelinated axons in a normal individual.* (Right) *Electron micrograph of nervous tissue from an individual with multiple sclerosis, showing axons lacking their normal myelin sheath. Courtesy of C. S. Raines.*

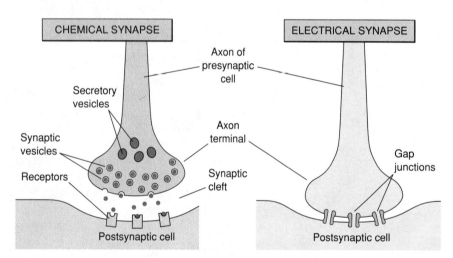

Figure 17-11 *Structure of Chemical and Electrical Synapses* *In chemical synapses, signals are transmitted by chemical neurotransmitters that are discharged by exocytosis from the presynaptic cell, diffuse across the synaptic cleft, and bind to receptors on the surface of the postsynaptic cell. In electrical synapses, signals are transmitted by ions that flow directly through gap junctions that link the presynaptic and postsynaptic cells.*

and postsynaptic cells in the region of a chemical synapse. The tip of the axon is expanded near the synapse to form a bulblike structure called an **axon terminal.** Axon terminals are characterized by the absence of a myelin sheath and the presence of numerous vesicles that contain neurotransmitters destined to be released from the axon upon arrival of an electrical impulse (Figure 17-12). These neurotransmitter-filled vesicles are of two general types: small **synaptic vesicles** about 50 nm in diameter that are

localized directly beneath the plasma membrane at the axon tip, and larger **secretory vesicles** measuring 90–250 nm that are more diffusely distributed in the axon terminal.

In **electrical synapses,** which are relatively rare, the presynaptic and postsynaptic cells are physically joined by *gap junctions* (page 240) that allow direct electrical communication between the cytoplasms of the two cells. At such locations, the plasma membranes of the presynaptic and postsynaptic cells are separated

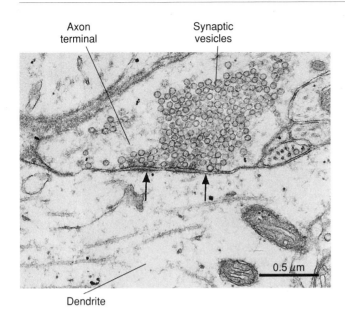

Figure 17-12 *Electron Micrograph of a Chemical Synapse* *The axon terminal contains numerous small synaptic vesicles. The arrows point to the synaptic cleft. Courtesy of S. G. Waxman.*

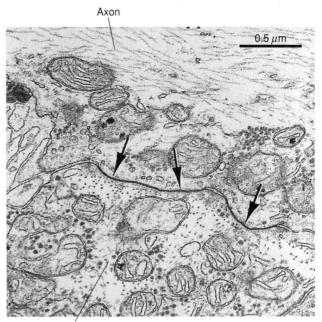

Figure 17-13 *Electron Micrograph of an Electrical Synapse* *This micrograph shows an axon and dendrite whose plasma membranes are very closely apposed where the two cells are joined by an electrical synapse (arrows). Numerous gap junctions, too small to be seen at this magnification, link the two cells in the region of the synapse. Courtesy of S. G. Waxman.*

by a space of about 2 nm, which is roughly one-tenth the distance of the synaptic cleft in chemical synapses (Figure 17-13). The flow of ions through the gap junctions of electrical synapses allows membrane depolarization to pass directly from the presynaptic to the postsynaptic cell. This direct electrical coupling permits signals to be transmitted across the synapse more rapidly than at chemical synapses, where time is required for the diffusion of chemical neurotransmitters across the synaptic cleft.

Neurons Employ a Variety of Different Chemicals as Neurotransmitters

When a nerve impulse reaches a synapse, the signal must be transmitted to the adjacent neuron. Signals move across synapses so rapidly that for many years synaptic transmission was thought to be a purely electrical phenomenon. Although this explanation has turned out to be correct for electrical synapses, most synapses are of the chemical type, where signals are carried across the synaptic cleft by the diffusion of molecules called **neurotransmitters.** Two criteria must be met before a substance can qualify as a neurotransmitter. First, the molecule in question must be released from nerve endings upon stimulation, and second, it must be shown that administering the molecule to the postsynaptic cell evokes the same response as a normal nerve impulse. Based on these criteria, several dozen molecules have been identified as neurotransmitters, including the amines *acetylcholine, norepinephrine, dopamine, serotonin,* and *histamine;*

the amino acids *glutamate, γ-aminobutyric acid (GABA),* and *glycine;* the peptides *enkephalin, β-endorphin, neurotensin, somatostatin,* and *substance P;* and the purine nucleotide, *ATP* (Table 17-2).

To determine which type of neurotransmitter is released at any given synapse, cells can be stained with fluorescent antibodies designed to detect the presence of particular compounds. A surprising conclusion to emerge from such studies is that many neurons utilize more than one kind of neurotransmitter. In numerous cases, a single axon terminal has been found to store both a small molecule neurotransmitter—such as acetylcholine, dopamine, serotonin, norepinephrine, GABA, or glycine—along with a peptide neurotransmitter like enkephalin, neurotensin, somatostatin, or substance P. Another common arrangement is the presence of two or more peptide neurotransmitters in the same axon terminal, and in a few cases, two small molecule neurotransmitters (e.g., serotonin and GABA) have been detected together.

Neurotransmitters Are Stored in Vesicles That Discharge Their Contents into the Synaptic Cleft

During the transmission of signals across a chemical synapse, neurotransmitters are released from the axon terminal. An early clue to the mechanism underlying

Table 17-2 Examples of Neurotransmitters

Neurotransmitter	Structure	Major Site of Action
Amines		
Acetylcholine	$CH_3C\!-\!O(CH_2)_2\overset{+}{N}(CH_3)_3$ (with $=O$ on carbonyl)	Neuromuscular junction and brain
Norepinephrine		Central nervous system and sympathetic nervous system
Dopamine		Brain and sympathetic nervous system
Serotonin		Brain
Histamine		Hypothalamus
Amino Acids		
γ-Aminobutyric acid (GABA)	$HOOC(CH_2)_3NH_2$	Central nervous system
Glycine	$HOOCCH_2NH_2$	Spinal cord
Glutamate	$HOOC(CH_2)_2CHCOOH$ (with NH_2)	Central nervous system
Peptides		
Thyrotropin releasing factor (TRF)	3 amino acids	Hypothalamus
Leu-enkephalin	5 amino acids	Brain
Met-enkephalin	5 amino acids	Brain
Gonadotropin releasing factor (GnRF)	10 amino acids	Hypothalamus
Somatostatin	14 amino acids	Hypothalamus
β-Endorphin	31 amino acids	Pituitary
Substance P	13 amino acids	Brain
Neurotensin	10 amino acids	Brain
Calcitonin gene-related peptide (CGRP)	37 amino acids	Brain
Vasoactive intestinal peptide (VIP)	28 amino acids	Brain, intestine
Purines		
ATP		Brain

Note: This table is only a partial listing of neurotransmitters. The number of peptides released from nerve cells and implicated in chemical signaling in the nervous system is significantly longer than the list presented here.

neurotransmitter release was provided in the 1950s by Bernhard Katz, who discovered that the postsynaptic membrane of resting frog muscle undergoes continual spontaneous depolarizations measuring about 0.5 mV. These **miniature postsynaptic potentials** are orders of magnitude smaller than the larger depolarizations, called **excitatory postsynaptic potentials,** that occur in the postsynaptic membrane when the presynaptic neuron is actually stimulated. But careful examination of the size of the depolarizations produced by normal nerve stimulation revealed them to be exact multiples of the spontaneous miniature potentials, leading Katz to conclude that neurotransmitter is released from the presynaptic neuron in discrete units or *quanta,* rather than in continuously graded amounts.

At about the same time as the quantal release theory was being formulated, electron microscopists discovered the presence of membrane vesicles in axon terminals, and subcellular fractionation studies revealed the presence of neurotransmitters in these vesicles. The preceding observations led to the speculation that each vesicle contains a "quantum" of neurotransmitter that is released from the cell upon nerve stimulation. According to this theory, miniature postsynaptic potentials are caused by the spontaneous expulsion of the contents of a few vesicles from the axon terminal, while the excitatory postsynaptic potentials that occur during normal nerve stimulation are produced by the release of hundreds of vesicles triggered by the arrival of an action potential. As would be predicted by this hypothesis, excessive stimulation of nerve cells has been shown to deplete the vesicles that normally reside in the axon terminal. Presynaptic vesicles have even been observed in the process of fusing with the plasma membrane, providing direct support for the conclusion that the neurotransmitters stored in these vesicles are discharged into the synaptic cleft (Figure 17-14).

Neurotransmitter Release Is Regulated by the Entry of Calcium Ions into the Axon Terminal

How does an action potential arriving at an axon terminal trigger the discharge of neurotransmitter-filled vesicles? Like exocytosis in other cell types (page 284), neurotransmitter discharge is regulated by calcium ions. The plasma membrane of an unstimulated neuron is relatively impermeable to Ca^{2+}, whose concentration is higher outside the cell (about $10^{-3} M$) than inside (about $10^{-7} M$) When an action potential reaches the axon terminal, the membrane depolarization causes voltage-gated Ca^{2+} channels in the presynaptic plasma membrane to open. Calcium ions then diffuse into the cytosol and trigger the exocytosis of neurotransmitter-filled vesicles. Experimental support for the preceding scenario has come from studies using *aequorin,* a protein that emits light when exposed to Ca^{2+}. If aequorin is introduced into nerve

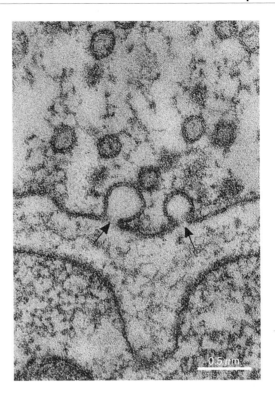

Figure 17-14 *Electron Micrograph Showing Fusion of Synaptic Vesicles with the Presynaptic Plasma Membrane* *The arrows point to two synaptic vesicles in the axon terminal* (top) *discharging their contents into the synaptic cleft. Courtesy of J. E. Heuser.*

cells, light is emitted at the axon terminal each time the cell transmits a nerve impulse, indicating an increase in the cytosolic Ca^{2+} concentration. The proposed role of these calcium ions in triggering the release of neurotransmitter has been verified by showing that neurotransmitter discharge can be induced by injecting Ca^{2+} directly into axon terminals.

Compared to exocytosis in other cell types, Ca^{2+}-induced neurotransmitter release into the synaptic cleft is extremely rapid. The speed of the process is based on the existence of two populations of synaptic vesicles: (1) a *releasable pool* of vesicles bound to the inner surface of the plasma membrane and poised for neurotransmitter release, and (2) a *reserve pool* of vesicles attached to the cytoskeleton by the protein **synapsin.** To explain the speed of neurotransmitter release, it has been proposed that the releasable vesicles are bound to the plasma membrane in a way that aligns a channel-forming protein in the vesicle membrane with a comparable protein in the plasma membrane (Figure 17-15). According to this model, the channel formed by the two proteins would normally be closed, but an increase in Ca^{2+} concentration triggers a conformational change that allows the channel to open and release neurotransmitter directly into the synaptic cleft.

The vesicles that make up the releasable pool represent a small fraction of the total synaptic vesicle population. It has therefore been suggested that controlling

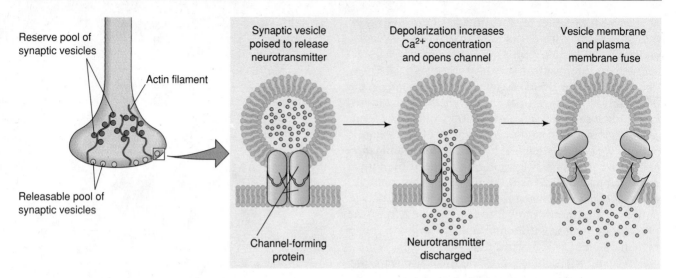

Figure 17-15 *A Model That Has Been Proposed to Explain the Speed of Neurotransmitter Release   Synaptic vesicles are divided into two populations: a reserve pool of vesicles bound to the actin cytoskeleton and a releasable pool of vesicles bound to the presynaptic plasma membrane. It has been suggested that the releasable vesicles are bound to the plasma membrane in a way that aligns a channel-forming protein in the vesicle membrane with a comparable protein in the plasma membrane; calcium ions that enter the axon upon membrane depolarization trigger a conformational change in the channel-forming protein that opens the channel and allows neurotransmitter to diffuse into the synaptic cleft.*

the relative number of vesicles in the releasable pool might be employed by nerve cells as a mechanism for regulating the efficiency of synaptic communication. It is known that the entry of calcium ions into the axon terminal during synaptic transmission stimulates a Ca^{2+}-calmodulin–dependent protein kinase which in turn catalyzes the phosphorylation of *synapsin*. The phosphorylation of synapsin disrupts its ability to link synaptic vesicles to the cytoskeleton, allowing more vesicles to join the releasable pool. Hence the number of vesicles that can discharge neurotransmitter into the synaptic cleft is elevated in neurons that have been repeatedly stimulated, making synaptic transmission more efficient.

Synaptic Vesicles Are Recycled after Fusing with the Presynaptic Membrane

Since neurotransmitter release is accompanied by the fusion of synaptic vesicles with the presynaptic membrane, a mechanism must exist for preventing uncontrolled expansion of the plasma membrane during synaptic transmission. The discovery that the number of vesicles present in the axon terminal does not decrease significantly during normal stimulation of nerve cells suggests that the membrane material being added to the plasma membrane is continually being recycled to form more vesicles. The recycling process can be observed microscopically by incubating neurons in a medium containing *horseradish peroxidase,* a tracer molecule that is easily visualized. Nerves stimulated in the presence of horseradish peroxidase collect the tracer in new vesicles that form by invagination of the plasma

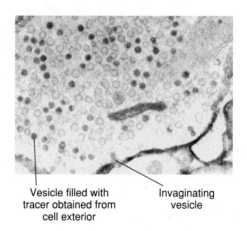

Figure 17-16 *Evidence for Recycling of Synaptic Vesicles   Neurons were incubated in a medium containing horseradish peroxidase, a tracer molecule that forms an electron-opaque reaction product. The dark synaptic vesicles are filled with the tracer, indicating that these vesicles must have incorporated the horseradish peroxidase when they invaginated from the plasma membrane. Courtesy of A. B. Harris.*

membrane (Figure 17-16). These vesicles are of a special type called *coated vesicles,* which contain a protein coat made of *clathrin* (page 296). The coated vesicles coalesce into larger structures, which are then converted back into new synaptic vesicles (Figure 17-17). This recycling process does not reutilize vesicle membranes indefinitely, however. In axon terminals where neurotransmitter discharge is actively taking place, some of the horseradish peroxidase appears in lysosomes that are returned to the cell body by retrograde

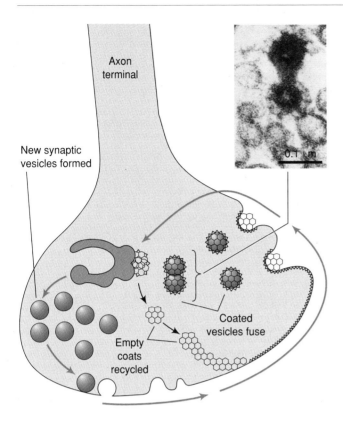

Figure 17-17 *Synaptic Vesicle Recycling* *Membrane material added to the plasma membrane during exocytosis of synaptic vesicles is retrieved by endocytosis of coated vesicles that then fuse together, lose their coats, and ultimately subdivide into new synaptic vesicles. The electron micrograph shows the fusion of two coated vesicles with each other. Micrograph courtesy of J. E. Heuser.*

transport. Such vesicles are replaced by new synaptic vesicles that are delivered to the axon terminal by anterograde transport.

Neurotransmitters Bind to Postsynaptic Receptors That Mediate Either Fast or Slow Chemical Transmission

Neurotransmitters that have been released from an axon terminal diffuse across the synaptic cleft in a few tenths of a millisecond and bind to specific receptors located in the postsynaptic plasma membrane. Binding of a neurotransmitter to its corresponding receptor usually brings about a change in the permeability and hence electrical potential of the postsynaptic membrane. *Excitatory* neurotransmitters trigger depolarization of the postsynaptic membrane, whereas *inhibitory* neurotransmitters induce membrane hyperpolarization.

Chemical neurotransmission can be divided into two principal categories based on differences in the type of receptor involved (Figure 17-18). The more familiar category is **fast chemical transmission,** which utilizes receptors that function as *neurotransmitter-*

gated ion channels. The binding of a neurotransmitter molecule to such an ion channel receptor alters the permeability of the channel, triggering an immediate change in ion flow across the postsynaptic membrane and a resulting alteration in membrane potential. Because the neurotransmitter receptor alters membrane permeability directly, fast chemical transmission requires only a few milliseconds to transmit a signal from one cell to another. The neurotransmitters most commonly employed for fast transmission are *acetylcholine* and *glutamate* for excitatory signals, and *glycine* and *GABA* for inhibitory signals.

Many synapses also utilize a second signaling mechanism called **slow chemical transmission** because it is slower in both the onset of the response, which requires hundreds of milliseconds, and in the duration of the response, which may last for seconds, minutes, or hours. The receptors employed for slow chemical transmission utilize **G proteins** (page 211) to exert their effects on the postsynaptic cell. G protein-linked receptors themselves are not ion channels, but they can influence ion channels or alter the levels of intracellular second messengers through the action of intermediary G proteins. Most of the *amine* and *peptide* neurotransmitters exert their effects on postsynaptic cells through G protein-linked receptors.

In addition to differing in the type of receptor employed and the speed of the response, slow and fast transmission also differ in the way the neurotransmitters are stored and released in the axon terminal. Fast transmitters are synthesized in the axon terminal and packaged into small *synaptic vesicles* that are secreted from sites specialized for rapid release. In contrast, slow neurotransmitters are usually synthesized and packaged into larger *secretory vesicles* in the cell body; the vesicles are then transported to the axon terminal, where their contents are secreted from many sites. Some neurotransmitters are stored in both kinds of vesicles. Many amines, for example, are stored in both secretory and synaptic vesicles, but they function mainly in slow transmission because their main targets are G protein-linked receptors.

Acetylcholine Is Involved in Both Fast Excitatory Transmission and Slow Inhibitory Transmission

Fast excitatory transmission is most frequently mediated by either acetylcholine or glutamate. Acetylcholine is employed at the neuromuscular junction to trigger the contraction of skeletal muscle (page 573), while glutamate is the main excitatory neurotransmitter in the central nervous system. Acetylcholine and glutamate both mediate fast excitatory transmission by binding to and opening neurotransmitter-gated cation channels in the postsynaptic membrane. Since the electrochemical gradient is steeper for Na^+ than for K^+, the influx of Na^+ exceeds the efflux of K^+, leading to an inward flow of sodium ions. This depolarizes the postsynaptic membrane to its threshold, triggering an action potential. The best characterized neurotransmitter-gated ion channel is the **nicotinic acetylcholine receptor,** which occurs in the plasma membrane of vertebrate skeletal muscle cells and in the electric organs of certain fish. The three-dimensional structure of this ion channel receptor and the mechanism by which its permeability properties are influenced by acetylcholine were described in Chapter 6 (pages 208–211).

While the interaction of acetylcholine with nicotinic receptors is a classic example of fast excitatory transmission, this is not the only way that acetylcholine acts. Acetylcholine released by certain neurons that innervate heart muscle triggers a slow inhibitory transmission event that leads to a decrease in the rate at which the heart beats. The effects of acetylcholine on skeletal and heart muscle differ because the two tissues have different kinds of acetylcholine receptors. Instead of nicotinic receptors, heart muscle utilizes **muscarinic acetylcholine receptors.** The binding of acetylcholine to muscarinic receptors activates a G protein that causes K^+ channels to open. Since the concentration of K^+ is higher inside the cell than outside, K^+ diffuses out of the muscle cell and the membrane hyperpolarizes (the membrane potential becomes more negative). This hyperpolarization of the plasma membrane inhibits muscle contraction, which is normally triggered by membrane *depolarization.*

Once a neurotransmitter has bound to its receptor and triggered an alteration in the postsynaptic cell, its action must be terminated to prevent the target cell from remaining in a continued state of excitation (or inhibition). In the case of acetylcholine, the enzyme *acetylcholinesterase* residing in the synaptic cleft quickly degrades the neurotransmitter to choline and acetate. Because acetylcholine was the first neurotransmitter to be studied in detail, inactivation by enzymatic degradation was once thought to terminate the action of other neurotransmitters as well. Subsequent research has revealed, however, that enzymatic inactivation is the exception rather than the rule. The most common mechanism for terminating the action of a neurotransmitter is to transport it back into the axon terminal from which it was released, a phenomenon known as *neurotransmitter reuptake.* The reuptake process is highly selective; for example, norepinephrine-releasing neurons take up only norepinephrine, serotonin-releasing neurons take up only serotonin, and so forth.

GABA and Glycine Mediate Fast Inhibitory Transmission

The binding of acetylcholine to muscarinic receptors illustrates the phenomenon of *slow* inhibitory transmission because membrane hyperpolarization is induced by a G protein rather than by the direct action of the neurotransmitter itself. A different type of inhibitory mechanism is employed by the amino acid neurotransmitters *γ-aminobutyric acid (GABA)* and *glycine,* which are the main inhibitory neurotransmitters of the nervous system. Instead of binding to receptors that ac-

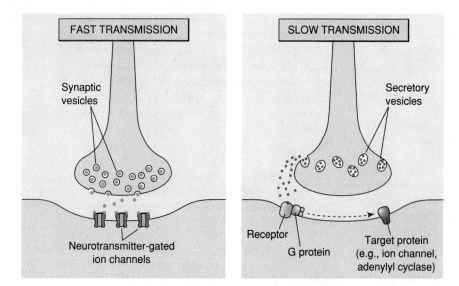

Figure 17-18 Fast and Slow Chemical Transmission (Left) *Fast transmission utilizes small-molecule neurotransmitters that are released from synaptic vesicles, diffuse across the synaptic cleft, and act directly on ion channels in the postsynaptic membrane.* (Right) *During slow transmission other neurotransmitters, including peptides and most amines, are released from large secretory vesicles and then act on G protein-linked receptors in the postsynaptic membrane. Both types of vesicles, and hence both types of transmission, can occur in the same axon terminal.*

tivate G proteins, GABA and glycine function by directly opening Cl⁻ channels. Since Cl⁻ is present in higher concentration outside the cell than inside, negatively charged chloride ions flow into the neuron and hyperpolarize the plasma membrane. The net result is a *fast* inhibitory response because the neurotransmitters are acting directly on a neurotransmitter-gated ion channel. Inhibitory transmitters are commonly employed by one neuron to control the release of neurotransmitters by another neuron. In this phenomenon, called *presynaptic inhibition,* the axons of inhibitory neurons terminate on the axons of excitatory neurons (Figure 17-19). When an inhibitory neuron is stimulated, it hyperpolarizes the axon of the excitatory neuron by releasing an inhibitory neurotransmitter such as GABA or glycine. After the excitatory axon has been hyperpolarized, action potentials that reach the axon terminal are less effective at depolarizing the membrane and triggering neurotransmitter release.

The neurotransmitter-gated Cl⁻ channels that function as receptors for glycine and GABA exhibit a structural resemblance both to each other and to the nicotinic acetylcholine receptor. All three receptors are constructed from five subunits that surround a central hydrophilic channel. Since the glycine and GABA receptors function as channels for negative ions and the acetylcholine receptor acts as a channel for positive ions, one might expect differences in the charged amino acids lining the three kinds of ion channels. The important differences appear to be located at the two ends of each channel (Figure 17-20). Negatively charged amino acids surround both ends of the channel in the

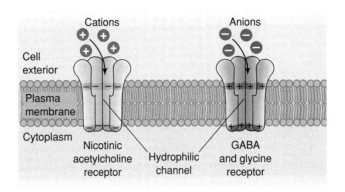

Figure 17-20 *Schematic Representation of Neurotransmitter-Gated Ion Channels* In order to visualize the central hydrophilic channel, only four of the five subunits comprising each of these ion channels are shown. Negatively charged amino acids at both ends of the channel facilitate the movement of positive ions through the nicotinic acetylcholine receptor, whereas positively charged amino acids at both ends of the channel facilitate the movement of negative ions through the GABA and glycine receptors.

acetylcholine receptor, whereas positively charged amino acids surround both ends of the glycine and GABA receptors. Presumably the negatively charged amino acids in the acetylcholine receptor facilitate the movement of positively charged sodium and potassium ions through the channel, and the positively charged amino acids in the GABA and glycine receptors facilitate the movement of negatively charged chloride ions through the channel.

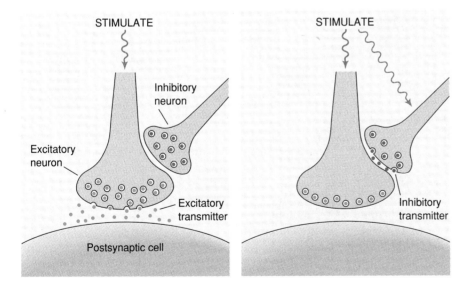

Figure 17-19 *Presynaptic Inhibition* This phenomenon occurs in situations where the axon of an inhibitory neuron terminates on the axon of an excitatory neuron. (Left) When only the excitatory neuron is stimulated, excitatory neurotransmitter is released from its axon terminal and the postsynaptic cell is activated. (Right) When the excitatory and inhibitory neurons are both stimulated, the inhibitory transmitter opens Cl⁻ (or K⁺) channels in the presynaptic membrane of the excitatory neuron, hyperpolarizing the membrane and thereby reducing the ability of the excitatory neuron to release neurotransmitter.

Several Neurotransmitters Bind to Receptors That Influence Adenylyl Cyclase

Slow synaptic transmission exhibits a number of similarities to hormonal communication by non-neural cells. As we learned in Chapter 6, the binding of hormones to plasma membrane receptors often leads to the formation of intracellular second messengers such as cyclic AMP (page 211). Several lines of evidence point to the conclusion that cyclic AMP also mediates some of the actions of certain neurotransmitters, including dopamine, serotonin, and norepinephrine. First, subcellular fractionation experiments involving homogenates of brain tissue have shown that the *synaptosome fraction,* which is enriched in synaptic membranes, contains large amounts of *adenylyl cyclase,* the enzyme that catalyzes cyclic AMP formation. Second, electrical stimulation of nervous tissue has been found to elevate cyclic AMP levels in postsynaptic cells, as does the direct application of neurotransmitters like dopamine, norepinephrine, and serotonin. And finally, direct administration of cyclic AMP to neurons elicits changes in membrane potential similar to those produced by these neurotransmitters alone.

One of the first neurotransmitter-controlled adenylyl cyclase systems to be detected in neurons is associated with the action of **dopamine.** Dopamine is a particularly interesting neurotransmitter because it has been implicated in a variety of medical problems. For example, the tremors and uncontrollable body movements that occur in Parkinson's disease are associated with the degeneration of neurons that release dopamine. This association has fostered the idea that individuals afflicted with Parkinson's disease might improve if the amount of dopamine in the brain could be restored to normal. Since dopamine cannot enter the brain from the circulation, patients have been treated instead with a dopamine precursor called *levo-dihydrophenylalanine* (L-DOPA). Although administration of L-DOPA does relieve some of the symptoms of Parkinson's disease, progression of the illness unfortunately continues. In contrast to Parkinson's disease, which involves the underproduction of dopamine, patients with certain types of schizophrenia produce too much dopamine. Therefore schizophrenia is often treated with such drugs as chlorpromazine (Thorazine), which blocks dopamine receptors. Another type of dopamine abnormality is associated with the use of cocaine, which blocks the uptake system responsible for transporting dopamine back into the axon terminal from which it was released. When the uptake system is blocked, dopamine persists in the synaptic cleft for an abnormally long period of time and continues to act upon the postsynaptic cell.

The medical importance of dopamine has led to considerable interest in the mechanism of action of dopamine receptors. Using brain regions rich in dopamine-secreting neurons as starting material, Paul Greengard was the first investigator to isolate adenylyl cyclase preparations whose activity is stimulated by dopamine. The ability of dopamine to activate adenylyl cyclase is inhibited by antischizophrenic drugs known to block dopamine receptors, whereas drugs that mimic the actions of dopamine are found to stimulate adenylyl cyclase activity. Such observations suggest that at least some of the effects of dopamine on nerve cells are mediated by the binding of dopamine to receptors that trigger the activation of adenylyl cyclase, leading to an enhanced production of cyclic AMP.

Cyclic AMP-Dependent Phosphorylation of a Potassium Channel Is Associated with a Simple Type of Learning

The role of cyclic AMP in synaptic transmission has been extensively studied in *Aplysia californica,* a marine mollusk possessing a relatively simple nervous system that is capable of simple types of learning. In response to a gentle stimulus, such as a mild electric shock to the tail, *Aplysia* reflexively withdraws its gill. Repeating the stimulus several times in succession causes the gill-withdrawal reflex to become less intense as the organism becomes accustomed (*habituated*) to the stimulus. However, if the habituated organism is then subjected to a strong noxious stimulus, such as a sharp blow to the head or tail, it responds to the next gentle electric shock by vigorously withdrawing its gill. This suddenly increased sensitivity, called *sensitization,* represents a simple type of learning—that is, a change in an organism's behavior based on previous experience.

The sensitization of *Aplysia* to electric shock is mediated by a special group of regulatory neurons, called **facilitator neurons,** whose axons make synaptic connections with the axons of the sensory cells that trigger the gill-withdrawal reflex (Figure 17-21). Electric shock causes the facilitator neurons to release serotonin, which activates adenylyl cyclase and stimulates cyclic AMP formation in the axon of the sensory neuron. Cyclic AMP in turn activates protein kinase A, which phosphorylates a K^+ channel and thereby causes it to close. Since the repolarization phase at the end of a normal action potential depends on an outward flow of K^+, the reduced efflux of K^+ prolongs the action potential by keeping the membrane depolarized for a longer period of time. Therefore Ca^{2+} channels activated by the action potential remain open longer and more neurotransmitter is released. The net effect is that release of serotonin from the facilitator neurons causes the signal transmitted by the sensory neurons to become stronger with time.

Although the gill-withdrawal reflex is an extremely simple type of learning, it illustrates an important general principle: Learning is associated with changes in the efficiency of synaptic communication between

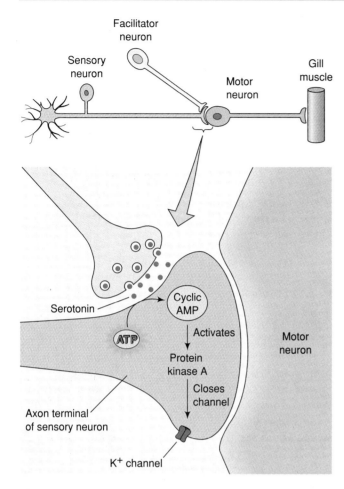

Figure 17-21 *Mechanism by Which Serotonin Released from Facilitator Neurons Enhances the Gill-Withdrawal Reflex in Aplysia* *Serotonin released by the facilitator neuron stimulates the production of cyclic AMP in the sensory neuron, triggering the closing of K⁺ channels. When the sensory neuron is subsequently stimulated and an action potential arrives at its axon terminal, the closed K⁺ channels prolong the action potential, thereby causing the sensory cell to discharge more neurotransmitter and hence activate the motor neuron more effectively.*

nerve cells. The phosphorylation of K⁺ channels is but one of many cellular changes that appear to be involved in changing the efficiency of synaptic communication during learning.

Enkephalins and Endorphins Inhibit Pain-Signaling Neurons by Binding to Opiate Receptors

The milky fluid derived from seedpods of the opium poppy has been known since ancient Greek times to relieve pain and induce a sense of well-being. *Morphine,* the major active ingredient in this fluid, was first isolated in 1803 and gained widespread medical use before its toxic and addictive properties came to be appreciated. Many attempts have since been made to produce

painkillers with the potency, but not the addictive properties, of morphine. For example, two methyl groups were added to morphine in the laboratories of the Bayer company in the 1890s, producing a derivative called *heroin* that was initially touted as a nonaddictive painkiller. In spite of this failure and others like it, the fact that *opiates* (morphine and its derivatives) are the most effective drugs for the treatment of severe chronic pain has motivated continued interest in such substances.

It has long been suspected that opiates produce their effects by binding to specific receptors in nerve cell membranes. Historically this belief was based on the structural specificity of opiates. Not only do small changes in the chemical structure of opiates lead to dramatic changes in potency, but only one of the two possible stereoisomers (mirror-image forms) of any given opiate is pharmacologically active. The existence of **opiate receptors** was finally demonstrated in the early 1970s, when Avram Goldstein, Solomon Snyder, Eric Simon, and Lars Terenius independently demonstrated that radioactive opiates bind to opiate receptors present in membrane fragments derived from brain tissue. The biological relevance of this binding was demonstrated by showing that (1) antagonists that block the pharmacological effects of opiates inhibit the binding of opiates to their receptors, and (2) the ability of various opiates to bind to opiate receptors correlates with the physiological potency of each drug (Figure 17-22).

Studies on the distribution of opiate binding sites in the nervous system has revealed that large numbers of opiate receptors reside in the areas of the spinal cord

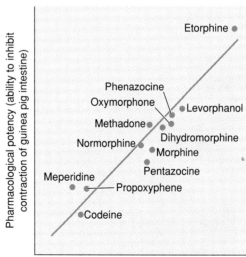

Figure 17-22 *Relationship between the Pharmacological Potencies of Various Opiates and Their Ability to Bind to Opiate Receptors* *Opiates that bind most strongly to opiate receptors also exhibit the greatest pharmacological potency, thereby suggesting that opiates act by binding to receptors.*

and brain that are associated with the perception of dull chronic pain. The brain's limbic system, which mediates emotional behavior, also contains many opiate receptors. Hence it is not surprising that drugs that selectively interact with such pathways relieve pain and induce euphoria. But why are opiate receptors present in the first place? Since they are clearly not designed to interact with morphine, the existence of opiate receptors suggests that the body must manufacture its own morphinelike substances that normally bind to the receptors. This possibility was confirmed in 1975 when John Hughes and Hans Kosterlitz isolated two small peptides from brain tissue that mimic the ability of morphine to inhibit the contraction of intestinal muscle, and whose activities are blocked by antagonists that interfere with the action of morphine. The two peptides, each consisting of five amino acids, were named *Met-enkephalin* (Tyr-Gly-Gly-Phe-Met) and *Leu-enkephalin* (Tyr-Gly-Gly-Phe-Leu). Both **enkephalins** were found to bind to opiate receptors. In addition, more than a dozen larger peptide hormones called **endorphins** have been isolated from nervous tissue and shown to bind to opiate receptors.

Analysis of the enkephalin content of various regions of the nervous system has revealed that the distribution of enkephalins closely parallels that of opiate receptors. Moreover, fluorescent antibodies directed against enkephalins selectively stain nerve endings (Figure 17-23), suggesting that enkephalins function as neurotransmitters. Support for this conclusion has come from the discovery that applying enkephalins to certain neurons slows down synaptic transmission, as would be expected for an inhibitory neurotransmitter. Although the mechanism by which enkephalins inhibit synaptic transmission is not completely understood, opiate receptors are known to be G protein–linked receptors.

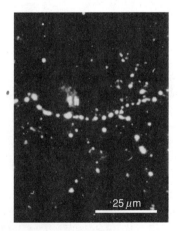

Figure 17-23 Nervous Tissue Stained with Fluorescent Antibodies That Bind to Enkephalin *The antibodies stain nerve endings located in discrete areas of the brain, suggesting that enkephalins function as neurotransmitters. Courtesy of S. H. Snyder.*

One of the first effects of enkephalin binding to an opiate receptor is the activation of a G protein that opens a plasma membrane K^+ channel; since the K^+ concentration is higher inside the cell than outside, K^+ diffuses out of the cell and the membrane hyperpolarizes. Opiate receptors can also close plasma membrane Ca^{2+} channels and inhibit adenylyl cyclase. Closing of Ca^{2+} channels is thought to inhibit neurotransmitter release, but the significance of adenylyl cyclase inhibition is not well understood.

Nitric Oxide Is a Novel Type of Neurotransmitter That Acts by Stimulating Cyclic GMP Formation

Nitric oxide (NO) is a toxic, short-lived gas molecule that at first glance appears to be an unlikely candidate for a neurotransmitter. (Nitric oxide should not be confused with nitrous oxide, or N_2O, which is the "laughing gas" used for anesthesia.) Nitric oxide is produced by the enzyme *NO synthase,* which converts the amino acid arginine to nitric oxide and citrulline. The first clearly defined role of nitric oxide in neural signaling involves the control of blood vessel dilation. It has been known for many years that acetylcholine dilates blood vessels by causing the smooth muscle layers of the vessels to relax. But the acetylcholine receptors that mediate this effect do not reside in the muscle cells themselves; instead, acetylcholine binds to receptors located in the thin layer of *endothelial cells* that lines the inner surface of blood vessels. If the endothelial cells are removed, the vessels no longer relax when acetylcholine is added.

Figure 17-24 (*top*) illustrates how the binding of acetylcholine to endothelial cells triggers the relaxation of the adjacent muscle cells. This complex multistep pathway can be divided into six stages. (1) Acetylcholine binds to G protein–linked receptors that activate the phosphoinositide signaling pathway, causing inositol trisphosphate (IP_3) to be produced by the endothelial cells. (2) IP_3 triggers the release of stored calcium ions from the endoplasmic reticulum, raising the Ca^{2+} concentration in the cytosol. (3) The calcium ions bind to calmodulin, forming a Ca^{2+}-calmodulin complex that stimulates NO synthase to produce nitric oxide. (4) Nitric oxide is a gas that readily diffuses through biological membranes, allowing it to pass from the endothelial cells into the adjacent smooth muscle cells. (5) Upon entering the smooth muscle cells, nitric oxide activates the enzyme *guanylyl cyclase,* which catalyzes the formation of cyclic GMP. (6) The increased cyclic GMP concentration activates protein kinase G, which induces muscle relaxation by catalyzing the phosphorylation of appropriate muscle proteins.

In addition to its role in dilating blood vessels, nitric oxide has been implicated in the mechanism of neural signaling by glutamate, the most commonly employed

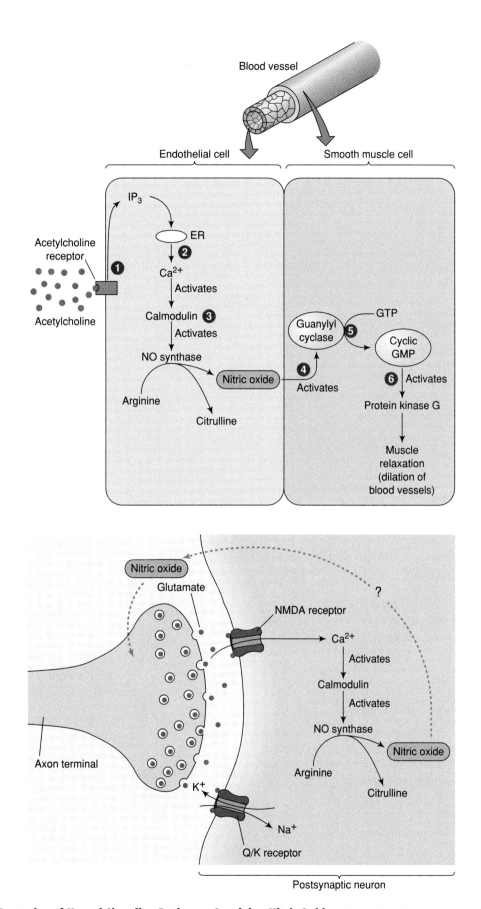

Figure 17-24 _Examples of Neural Signaling Pathways Involving Nitric Oxide_   (Top) _The binding of acetylcholine to endothelial cells triggers the production of nitric oxide, which diffuses into the adjacent smooth muscle cells and stimulates guanylyl cyclase, thereby leading to muscle relaxation._ (Bottom) _The binding of glutamate to NMDA receptors during synaptic transmission triggers the production of nitric oxide in the postsynaptic neuron. It has been postulated that the nitric oxide diffuses back into the presynaptic axon terminal and signals the axon to increase its output of neurotransmitters._

neurotransmitter in the brain. Glutamate interacts with several different classes of receptors, some functioning as ion channels and others linked to G proteins. One type of ion channel, called the *NMDA receptor,* is highly permeable to Ca^{2+} as well as to Na^+ and K^+. Hence the binding of glutamate to NMDA receptors allows calcium ions to enter the postsynaptic cell and bind to calmodulin, forming a Ca^{2+}-calmodulin complex that stimulates NO synthase to produce nitric oxide (see Figure 17-24, *bottom*).

Recent evidence suggests that the nitric oxide formed by this pathway may be involved in the process of **long-term potentiation,** or **LTP,** which is thought to be one of the primary means for storing memories. LTP refers to the increase in the efficiency of synaptic transmission that occurs when repeated stimulation of a presynaptic neuron leads to the repetitive transmission of signals across a synapse to a postsynaptic neuron. Such repetition causes the synaptic connection between the two cells to become strengthened or "potentiated"; that is, a bigger response occurs in the postsynaptic cell the next time signals are sent. Recent experiments have shown that inhibitors of NO synthase prevent LTP from occurring in slices of rat brain, and that rats injected with inhibitors of NO synthase lose the ability to learn certain types of behavior. Such observations suggest that nitric oxide may be involved in establishing LTP. One possible role is that the nitric oxide formed in the postsynaptic neuron during synaptic transmission may diffuse back into the presynaptic axon terminal, signaling the axon to increase its output of neurotransmitters.

The realization that nitric oxide functions in neural signaling raises the question of whether any other gases are similarly involved. One potential candidate is *carbon monoxide,* a gas that shares with nitric oxide the ability to dilate blood vessels by stimulating guanylyl cyclase. Several regions of the brain contain the enzyme *heme oxygenase,* which catalyzes the formation of carbon monoxide during the breakdown of heme. The production of carbon monoxide by brain cells and its ability to stimulate cyclic GMP formation indicates that carbon monoxide might play a signaling role comparable to that of nitric oxide.

DETECTING STIMULI AND TRIGGERING RESPONSES

We have now seen that the transmission of nerve impulses involves changes in membrane potential that are passed from neuron to neuron using electrical and chemical synapses. But if nerve impulses are to carry in-

formation that is useful to the organism as a whole, neurons must be able to do more than just communicate with one another. The nervous system must also receive sensory information from both inside and outside the body and use the information that has been received to trigger appropriate responses. In this segment of the chapter we will briefly discuss a few examples that illustrate how the nervous system detects stimuli and triggers responses.

Vertebrate Photoreceptors Detect Light Using a Mechanism That Involves Cyclic GMP

The ability of the nervous system to perceive its internal and external environments depends on specialized **sensory cells** that can be grouped into four principal types: *photoreceptors* that are sensitive to light, *mechanoreceptors* that respond to physical pressure or movement (e.g., sound and touch), *chemoreceptors* that detect chemicals (e.g., odor and taste), and *thermoreceptors* that monitor temperature. In each case the sensory cell transforms the stimulus being perceived into a common form—namely, a change in membrane potential.

To illustrate how sensory cells convert external stimuli into changes in membrane potential, let us consider photoreceptors as an example. The vertebrate retina is comprised of two kinds of photoreceptor cells called **rods** and **cones** because of the distinctive shapes of their light-sensitive tips (Figure 17-25). Rod cells are more sensitive to dim light but cannot distinguish colors; cone cells are responsible for color vision, but function only in bright light. In both rods and cones, light absorption occurs in a specialized region of the cell termed the *outer segment.* Because it is attached to the rest of the cell by a narrow stalk containing a typical ciliary axoneme, the outer segment can be considered to be an elaborately modified cilium. The outer segment of each photoreceptor cell contains hundreds of flattened *membrane disks* that store a light-absorbing pigment called **rhodopsin.** Each rhodopsin molecule consists of a protein called *opsin* covalently joined to a light-absorbing derivative of vitamin A known as *11-cis-retinal.* In addition to rhodopsin, cone cells contain blue, green, and red pigments that mediate color perception.

When a photon of light strikes 11-*cis*-retinal, it alters the conformation of the rhodopsin molecule (Figure 17-26). The altered rhodopsin in turn activates a G protein called G_t (or *transducin*), which stimulates the breakdown of cyclic GMP by *cyclic GMP phosphodiesterase.* In photoreceptor cells, cyclic GMP controls the permeability of plasma membrane *cyclic GMP-gated Na^+ channels;* when illumination triggers a decrease in cyclic GMP concentration, the lack of cyclic GMP causes these Na^+ channels to close. The resulting decrease in the flow of Na^+ into the cell leads to an increased negative charge

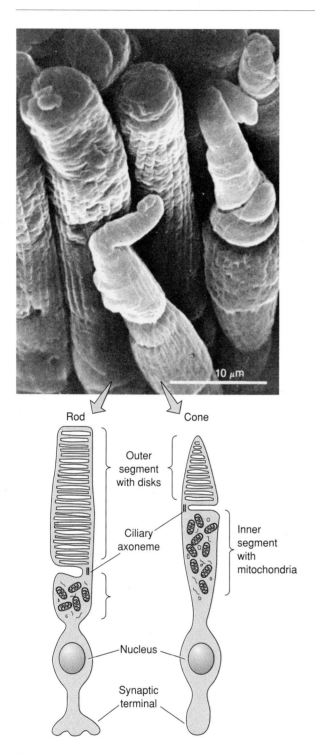

Figure 17-25 Rod and Cone Cells (Top) *Scanning electron micrograph of rod and cone cells in the retina of the mud puppy. The outer segments of the rod cells are cylindrical in shape, whereas the outer segments of the cone cells are conical. (Bottom) Schematic diagram of typical rod and cone cells. The outer segment contains tightly packed membrane disks that store rhodopsin and other photoreceptor pigments. Micrograph courtesy of E. R. Lewis, Y. Y. Zeevi, and F. S. Werblin.*

inside the cell—that is, a hyperpolarization of the plasma membrane. Hence in vertebrate photoreception, the presence of light is translated into membrane hyperpolarization rather than depolarization. Absorption of a single photon of light by rhodopsin can block the transit of millions of sodium ions across the outer segment plasma membrane. This signal amplification is made possible by the fact that one rhodopsin molecule activates thousands of G_t molecules, and each phosphodiesterase molecule activated by G_t can in turn degrade thousands of cyclic GMP molecules per second.

Sensory Stimuli Alter the Membrane Potential of Sensory Cells by a Variety of Different Mechanisms

Some of the principles illustrated by photoreceptor cells apply to the reception of sensory stimuli by other cells as well. Olfactory cells, for example, perceive odors by a mechanism that also involves G proteins. The binding of an odor molecule to an olfactory cell receptor activates a G protein called G_{olf}, which in turn stimulates adenylyl cyclase to produce cyclic AMP. The cyclic AMP triggers membrane depolarization by opening a plasma membrane cation channel.

G proteins are not always involved in sensory reception, however. Taste bud cells, for example, receive salty sensations through a direct mechanism in which sodium ions in the cell's external environment diffuse through the Na^+ channels in the plasma membrane, depolarizing the taste cell directly. The epithelial *hair cells* found in the inner ear, which are the sensory cells for balance and hearing, act as mechanoreceptors that perceive movements using a mechanism that involves neither G proteins nor plasma membrane receptors. Hair cells exhibit huge microvilli called *stereocilia* (page 591) that are sensitive to tiny movements caused either by sound vibrations or by the movement of fluid in the semicircular canals. These movements trigger membrane depolarization by opening cation channels in the plasma membrane of the stereocilia. The channels open so rapidly that they are thought to be acting as *mechanically gated ion channels*—that is, ion channels that are directly opened by the mechanical movements of the stereocilia.

Responses Triggered by the Nervous System Include Muscular Contraction, Glandular Secretion, and Neurosecretion

Stimuli detected by sensory cells trigger nerve impulses that are transmitted to other neurons by synaptic communication. But in order for these impulses to influence the rest of the organism, some neurons must be capable of triggering responses other than exciting or inhibiting another neuron. Such responses can be grouped into three main categories: muscle contraction, glandular secretion, and neurosecretion.

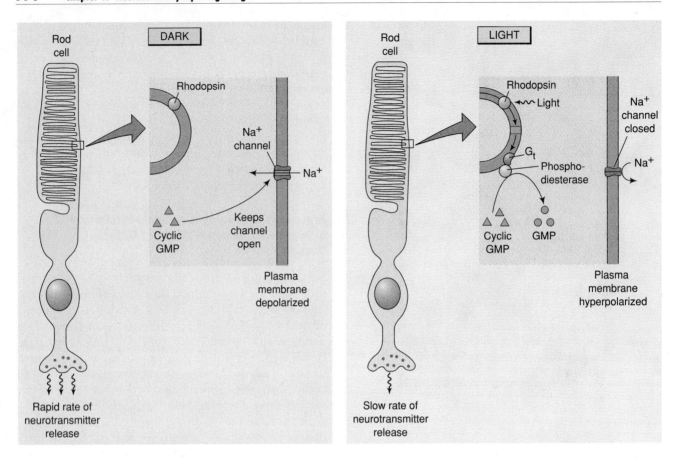

Figure 17-26 *The Effect of Light on Rod Cells* (Left) *In the dark, cyclic GMP keeps plasma membrane Na⁺ channels open and the membrane is partially depolarized, thereby leading to neurotransmitter release from the rod cell.* (Right) *Light induces a change in the conformation of rhodopsin, triggering the activation of a G protein (G$_t$) that stimulates the breakdown of cyclic GMP by phosphodiesterase. The decreased cyclic GMP concentration causes the Na⁺ channels to close, leading to hyperpolarization of the plasma membrane and a reduced rate of neurotransmitter release from the rod cell.*

We have already discussed the mechanism by which nerve impulses stimulate *muscle contraction* (pages 564–582). In addition to initiating the contraction process, nerve impulses are also involved in regulating muscle contraction. For example, cardiac muscle and certain smooth muscles exhibit spontaneous rhythmic contractions that do not require nerve impulses for initiation. But the rate of such contractions is under neural control. In the case of the heart, neurons releasing norepinephrine increase membrane excitability and therefore speed the rate of contraction, whereas neurons releasing acetylcholine decrease membrane excitability and therefore slow the rate.

Nerve cells also activate (or inhibit) many types of *glandular secretion*, such as the secretion of saliva from salivary glands or epinephrine from the adrenal medulla. Cells of the adrenal medulla are developmentally related to neurons but fail to sprout axons; instead of utilizing their released epinephrine as a neurotransmitter, they secrete it as a hormone into the bloodstream. Epinephrine circulating throughout the organism exerts effects similar to those produced by neurons that utilize norepinephrine as a neurotransmit-

ter. For example, circulating epinephrine and neurons utilizing norepinephrine as a neurotransmitter can both trigger increases in heart rate, blood pressure, breathing rate, and blood glucose concentration.

The close relationship between neurotransmitters and hormones is underscored by the observation that some neurons secrete hormones directly into the bloodstream by a process called **neurosecretion.** A prominent example occurs in the hypothalamus, where certain neurons secrete peptide *releasing factors* into the circulation that travel to the anterior pituitary gland and stimulate the release of other hormones (Figure 17-27). Among the peptides secreted by hypothalamic neurons are (1) *thyrotropin releasing factor (TRF),* which promotes the release of thyrotropin from the anterior pituitary, (2) *gonadotropin releasing factor (GnRF),* which promotes the release of the hormones FSH and LH from the anterior pituitary, (3) *corticotropin releasing factor (CRF),* which promotes the release of the hormone ACTH from the anterior pituitary, (4) *growth hormone releasing factor,* which promotes the release of growth hormone from the anterior pituitary, and (5) *somatostatin,* which inhibits the release of growth hor-

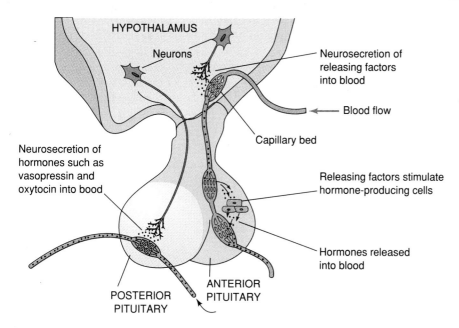

HYPOTHALAMUS
Neurons
Neurosecretion of releasing factors into blood
Blood flow
Capillary bed
Neurosecretion of hormones such as vasopressin and oxytocin into bood
Releasing factors stimulate hormone-producing cells
Hormones released into blood
POSTERIOR PITUITARY
ANTERIOR PITUITARY

Figure 17-27 Neurosecretion by Neurons Located in the Hypothalamus *The neuron on the left has an axon that passes into the posterior pituitary, where it secretes hormones like vasopressin and oxytocin directly into the bloodstream. The neuron on the right secretes peptide releasing factors into capillaries that carry them to the anterior pituitary, where they induce the release of other hormones into the bloodstream.*

mone from the anterior pituitary. Other neurons located in the hypothalamus have axons that pass directly into the posterior pituitary, where they secrete the hormones *vasopressin* and *oxytocin* into the bloodstream. Neurosecretion is accomplished by the same basic mechanism as neurotransmitter release. In both cases the arrival of a wave of membrane depolarization at the nerve ending triggers exocytosis of small vesicles. The only difference is that in typical neurons the vesicles release a neurotransmitter that interacts with the adjacent neuron, whereas in neurosecretory neurons the vesicles release peptide hormones and releasing factors that enter the circulation.

NEURON GROWTH AND DEVELOPMENT

The nervous system of higher vertebrates is an extraordinarily complex network consisting of billions or even trillions of interconnected cells situated in the brain, spinal cord, and peripheral nerves. In order for this system to function properly in analyzing sensory information and triggering appropriate responses, these billions of cells must be properly connected to one another. In the concluding section of the chapter we will briefly consider how neurons establish these connections during embryonic development.

The Growth Cone Directs the Outgrowth of Neurites

In vertebrates, neurons arise during early embryonic development from cells that separate from the primitive

ectoderm shortly after gastrulation. These cells, called *neuroblasts,* stop dividing and begin to develop cytoplasmic extensions known as **neurites,** which are destined to become axons and dendrites. Much has been learned about the development of neurites by studying cultures of embryonic neuroblasts induced to form neurites by appropriate incubation conditions. Experiments of this sort were first carried out in the early 1900s by Ross Harrison, one of the pioneers of cell culture techniques. By carefully observing the development of cultured neuroblasts, he was the first to show that the axon is a direct outgrowth of the neuron rather than an independently formed entity.

Shortly before Harrison's pioneering work, Ramón y Cajal had applied the Golgi silver-stain technique to elongating axons and discovered that the tip of each growing axon contains an expanded region, which he named the **growth cone.** Although the cells being examined were fixed and stained, and hence nonliving, Ramón y Cajal suggested that the growth cone represents an area of dynamic change at the growing tip. This hypothesis was soon confirmed by Harrison, who showed that the neurites of cultured nerve cells exhibit growth cones that are in constant motion. In recent years, higher-resolution microscopic techniques have revealed that the growth cone is covered with slender motile filopodia that protrude and wave about (Figure 17-28). These filopodia behave as if they are "sensing" the surrounding environment and determining the direction in which the neurite should grow. In some cases it has been demonstrated that growth cones turn in the direction in which their filopo-

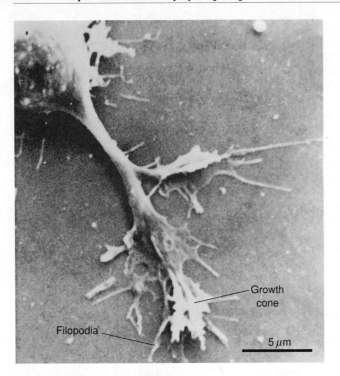

Growth
cone

Filopodia

5 μm

Figure 17-28 *Scanning Electron Micrograph Showing the Growth Cone of a Neuron in Culture* Note the presence of numerous filopodia. Courtesy of S. M. Rothman.

dia adhere best; as they adhere to new surfaces, the filopodia exert tension on the rest of the axon, pulling it along. If the connection between such a growth cone and its attached axon is experimentally severed, the growth cone continues to move without the axon.

Axon elongation driven by movements of the growth cone requires the cooperation of actin filaments and microtubules. Norman Wessells has shown that treating elongating axons with cytochalasin to disrupt actin filaments causes the filopodia to become immobile and retract, halting axon elongation within a few minutes. In contrast, disrupting microtubules with colchicine does not alter filopodial appearance or motility, nor does it exert an immediate effect on axon elongation. Within a half hour of colchicine treatment, however, the axon begins to shorten and, although filopodial activity remains unaltered, the axon eventually collapses back into the nerve cell body. The contrasting effects of these two drugs on axon elongation suggest that actin filaments are required for growth cone movement and axon elongation, whereas microtubules serve as a skeletal framework whose integrity is essential for maintaining the elongated state (Figure 17-29).

Neurite Growth Is Stimulated by Nerve Growth Factor as Well as a Variety of Other Proteins

The growth and development of nerve cells is controlled by a variety of growth factors. The most thoroughly studied example, called **nerve growth factor**

(NGF), was discovered in the early 1950s by Rita Levi-Montalcini and Viktor Hamburger. NGF differs from most growth factors in that it does not actually promote cell proliferation; NGF instead stimulates neurite outgrowth and maintains cell viability. The main targets of NGF are *sensory neurons,* which carry nerve impulses from the periphery to the central nervous system, and *sympathetic neurons,* which are responsible for the regulation of involuntary actions such as breathing and the heartbeat. Sensory neurons are most sensitive to NGF during embryonic development, whereas sympathetic neurons are under the influence of NGF during both embryonic and postnatal development. The selectivity of NGF has been dramatically demonstrated by injecting newborn animals with antibodies directed against NGF; the result is the selective degeneration of the sympathetic nervous system. Both sympathetic and sensory neurons survive poorly when cultured in the absence of NGF, but other cells grow well without it. When NGF is added to a culture of sensory or sympathetic neurons, the cells respond with a striking and massive outgrowth of neurites (Figure 17-30).

Like many other growth factors, NGF binds to a plasma membrane receptor that functions as a protein-tyrosine kinase (page 221). Activation of the tyrosine kinase by NGF is thought to stimulate cell growth through a cascade of events initiated by the phosphorylation of one or more intracellular proteins. Although NGF was the first growth factor shown to promote the growth of nerve cells and foster the development of neurites, a variety of other molecules have subsequently been implicated as well. Included are growth factors such as *insulin-like growth factor II (IGF-II),* extracellular matrix proteins such as *laminin, fibronectin,* and *collagen* (pages 226–232), and cell adhesion molecules such as *N-CAMs* and *cadherins* (page 233).

Axons Are Guided to Their Proper Destination by Cell-Cell Contacts, Matrix Molecules, and Diffusible Substances

Once neurites begin to elongate under the influence of NGF or other appropriate growth factors, they need to follow the proper pathway to link up with the appropriate target cells. The problem is especially pronounced for axons, which must often grow for many centimeters toward a specific region of termination and then make contact with the proper target cell, where they form a synapse at the correct location on the cell body, dendrite, or axon. One of the first attempts to investigate this process was carried out by Roger Sperry using the amphibian visual system as a model. In the visual system, axons emerge from cell bodies located in the retina and grow toward a particular region of the brain called the *optic tectum.* Axons leaving the ventral (lower) portion

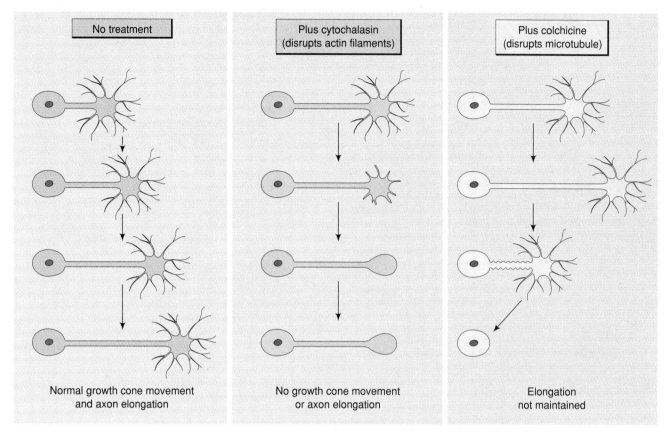

Figure 17-29 ***Effects of Cytochalasin and Colchicine on Axon Elongation in Cultured Neurons*** *The effects of cytochalasin suggest that actin filaments are required for growth cone movement and axon elongation, while the effects of colchicine suggest that microtubules provide a skeletal framework that maintains the elongated state.*

WITHOUT NERVE GROWTH FACTOR   WITH NERVE GROWTH FACTOR

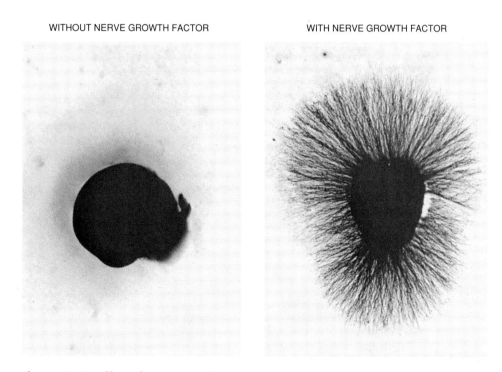

Figure 17-30 ***Effect of Nerve Growth Factor (NGF) on Cultured Nerve Cells*** (Left) *An untreated mass of nerve cells.* (Right) *Nerve cells cultured in the presence of NGF, which induces a massive outgrowth of neurites. Courtesy of R. Levi-Montalcini.*

of the retina migrate into the dorsal (upper) area of the optic tectum, whereas axons originating in the dorsal retina grow toward the ventral tectum. Sperry's approach was to cut the optic nerve, rotate the eye 180°, and then allow the nerves to regenerate. In spite of the disruption, the damaged axons sorted out correctly and migrated to their proper locations.

To explain this remarkable result, Sperry proposed that neurons carry chemical identification tags on their surfaces that permit them to recognize the cells with which they are destined to make synaptic contact. As a test of the hypothesis, Stephen Roth isolated cells from the dorsal and ventral portions of the retina and monitored their ability to adhere to tissue removed from the dorsal or ventral half of the optic tectum. As predicted by Sperry, cells derived from the dorsal retina adhere preferentially to the ventral tectum, and ventral retinal cells bind selectively to the dorsal tectum. The specificity of this cell-cell interaction suggests that neurons carry surface molecules that help guide them to their appropriate targets.

Sperry's pioneering work on the retinotectal system has prompted a vigorous investigation of the mechanisms that guide axons along their proper pathways. In grasshopper embryos, where the developmental trajectories of individual axons have been precisely mapped, several kinds of cell-cell interactions that guide axon extension have been identified. For example, when the axons of developing sensory neurons begin to elongate, their growth cones initially move along epithelial surfaces; but at specific locations, the direction of growth cone extension suddenly changes. These abrupt changes occur at sites where the filopodia of growth cones make contact with specific immature neurons called *guidepost cells.* If the guidepost cells are destroyed with a laser microbeam, the axons wander off in inappropriate directions.

Growth cones also employ the axons of neighboring neurons as guiding scaffolds upon which to extend. In an elegant series of studies investing this phenomenon, Corey Goodman and his associates have found that the growth cone of a particular grasshopper neuron always migrates along the same bundle of axons. If this axon bundle is destroyed before the growth cone of the identified neuron reaches it, the growth cone stops growing or wanders randomly in the vicinity of the missing bundle, even though other axon bundles are within reach of its filopodia. Subsequent investigations have revealed that a family of cell surface glycoproteins called **fasciclins** are involved in the process by which growth cones recognize appropriate axon bundles. Axon bundles differ in the quantities and types of fasciclin they contain; if the ability of an organism to produce a particular type of fasciclin is disrupted by mutation, axon bundles that would have normally contained the missing fasciclin fail to develop.

Fasciclins are but one of several dozen molecules that appear to influence the direction in which axons grow. These guidance factors fall into three main categories: cell surface proteins, extracellular matrix proteins, and diffusible proteins. The long list of molecules implicated in guiding axon outgrowth suggests that the final wiring pattern of the nervous system is achieved not by a single recognition mechanism but by a sequential pathway of recognition steps that gradually guides a growing axon to its proper destination. Some of these guidance interactions involve factors that repel rather than attract growth cones. For example, growth cones of axons derived from the central nervous system have been found to withdraw and move in another direction when they make contact with axons derived from peripheral nerves.

Once axons have entered the appropriate target area, growth cones are able to distinguish among potential postsynaptic partners. The discrimination process involves molecules carried by the target cells as well as chemical gradients that define the positions of given cells within the target tissue. When an axon finally makes contact with the appropriate cell, it induces the formation of a synapse. In the case of skeletal muscle cells, contact with an axon causes acetylcholine receptors that had been diffusely distributed across the muscle cell plasma membrane to form high-density clusters near the site of contact. A protein in the axon terminal called *agrin* has been implicated in triggering this event. If purified agrin is added to cultures of embryonic muscle cells, it causes the formation of multiple clusters of acetylcholine receptors in the absence of nerve cells.

Most Mature Neurons Lose the Capacity to Divide

In higher vertebrates, most neurons lose the capacity to divide after they begin to form axons and dendrites. If neurons are destroyed after this stage, they cannot be replaced. However, injured neurons can sometimes regenerate new axons if their existing axons are cut or damaged. In such an event, the old axon degenerates beyond the point of injury because it has become physically separated from the cell body upon which it depends for nutrition. The cell body then synthesizes new axoplasm and the axon stub begins growing again. The newly forming axon, elongating at a rate of a few millimeters per day, eventually reaches its original destination, where under favorable conditions it may even reestablish proper synaptic connections.

Although most neurons do not divide after a certain stage of development has been reached, tumors of proliferating neuroblasts occasionally arise. These tumors, called **neuroblastomas,** grow and divide readily in culture, and can be induced to differentiate into cells possessing dendrites, axons, and the ability to synthesize neurotransmitters. Since normal neurons cannot be propagated in culture, the availability of neuroblastoma cells exhibiting many attributes of mature neurons, but capable of growth in culture, has provided biologists with an alternative way of studying nerve cells.

SUMMARY OF PRINCIPAL POINTS

• A neuron consists of a cell body exhibiting long cytoplasmic extensions called axons and dendrites. Dendrites are shorter than axons, exhibit an extensively branched pattern, lack a myelin sheath, and are specialized for receiving signals. Axons are less numerous than dendrites, extend from the cell body for longer distances, often exhibit a myelin sheath, and are specialized for sending signals.

• Slow axonal transport moves proteins and cytoskeletal filaments down the axon at rates of about 1–5 mm per day, whereas fast axonal transport moves vesicles and mitochondria at rates of 10–500 mm per day. Fast transport is mediated by the motor proteins kinesin and dynein, which link organelles to microtubules. Kinesin propels materials down the axon, while dynein moves in the opposite direction.

• Neurons transmit signals by changing the electrical potential across their plasma membranes. Action potentials are caused by stimuli that trigger a transient increase in membrane permeability to Na^+. The resulting diffusion of Na^+ into the cell depolarizes the plasma membrane. Depolarization triggers the opening of K^+ channels, allowing an efflux of K^+ that reestablishes the resting membrane potential.

• Myelination speeds the propagation of action potentials by preventing the leakage of local currents through the plasma membrane and by forcing action potentials to jump from one node of Ranvier to the next.

• Nerve impulses are transmitted from cell to cell at specialized junctions called synapses. At electrical synapses, signals are transmitted by ions that flow directly through gap junctions that link the presynaptic and postsynaptic membranes. At chemical synapses, signals are transmitted by chemical neurotransmitters that are released from the presynaptic cell, diffuse across the synaptic cleft, and bind to receptors on the postsynaptic cell.

• Neurotransmitters are stored in vesicles whose contents are discharged from the axon terminal by exocytosis when the plasma membrane is depolarized. Exocytosis is triggered by calcium ions, which enter the cell when depolarization causes plasma membrane Ca^{2+} channels to open.

• Neurotransmitters involved in fast chemical transmission bind to receptors that function as neurotransmitter-gated ion channels, whereas neurotransmitters involved in slow chemical transmission bind to G protein-linked receptors, activating G proteins that influence ion channels or the production of second messengers such as cyclic AMP or IP_3.

• When acetylcholine binds to nicotinic acetylcholine receptors, fast excitatory transmission ensues because the nicotinic receptor functions as a cation channel that is opened directly by acetylcholine. When acetylcholine binds to muscarinic acetylcholine receptors, slow inhibitory transmission ensues because the muscarinic receptor activates a G protein that causes K^+ channels to close, thereby triggering membrane hyperpolarization.

• GABA and glycine are fast inhibitory transmitters that bind to receptors which function as Cl^- channels. Binding of neurotransmitter opens the channels, causing an influx of Cl^- that hyperpolarizes the plasma membrane.

• Dopamine, serotonin, and norepinephrine bind to receptors that regulate the activity of adenylyl cyclase. In sensory neurons the cyclic AMP produced by adenylyl cyclase acts by closing K^+ channels, which prolongs the action potential by keeping the membrane depolarized for longer periods of time.

• Enkephalins and endorphins are inhibitory neurotransmitters that inhibit pain-signaling pathways by binding to opiate receptors, which are G protein-linked receptors that open K^+ channels, close Ca^{2+} channels, and inhibit adenylyl cyclase.

• The binding of acetylcholine or glutamate to the plasma membrane of certain cells triggers an increase in cytosolic Ca^{2+} concentration, which can stimulate the formation of nitric oxide. The nitric oxide diffuses out of the cell and influences neighboring cells by activating the enzyme guanylyl cyclase.

• Environmental stimuli are detected by sensory cells whose membrane potential is altered by changes in temperature, pressure, chemicals, or light. In vertebrate photoreceptors light is absorbed by rhodopsin, which activates a G protein that stimulates cyclic GMP degradation by phosphodiesterase. The resulting decrease in cyclic GMP concentration causes Na^+ channels to close, thereby hyperpolarizing the plasma membrane.

• During development of the nervous system, axon growth is guided by cell-cell contacts, matrix molecules, and diffusible substances. Once neurons acquire axons and dendrites, they usually lose the capacity to divide.

SUGGESTED READINGS

Books

Burgoyne, R. D., ed. (1991). *The Neuronal Cytoskeleton,* Wiley-Liss, New York.

Hall, Z. W., ed. (1992). *An Introduction to Molecular Neurobiology,* Sinauer, Sunderland, MA.

Kandel, E. R., J. H. Schwartz, and T. M. Jessell (1993). *Principles of Neural Science,* 3rd Ed., Elsevier, New York.

Keynes, R. D., and D. J. Aldley (1992). *Nerve and Muscle*, 2nd Ed., Cambridge University Press, New York.

Articles

Bixby, J. L., and W. A. Harris (1991). Molecular mechanisms of axon growth and guidance, *Annu. Rev. Cell Biol.* 7:117–159.

Bradshaw, R. A., T. L. Blundell, R. Lapatto, N. Q. McDonald, and J. Murray-Rust (1993). Nerve growth factor revisited, *Trends Biochem. Sci.* 18:48–52.

Bredt, D. S., and S. H. Snyder (1994). Nitric oxide: A physiologic messenger molecule, *Annu. Rev. Biochem.* 63:175–195.

De Camilli, P., F. Benfenati, F. Valtorta, and P. Greengard (1990). The synapsins, *Annu. Rev. Cell Biol.* 6:433–460.

Di Chiara, G., and R. A. North (1992). Neurobiology of opiate abuse, *Trends Pharmacol. Sci.* 13:185–193.

Dunant, Y., and M. Israel (1985). The release of acetylcholine, *Sci. Amer.* 252 (April):58–66.

Ferro-Novick, S., and R. Jahn (1994). Vesicle fusion from yeast to man, *Nature,* 370:191–193.

Fesce, R., F. Grohovaz, F. Valtorta, and J. Meldolesi (1994). Neurotransmitter release: fusion or 'kiss-and-run'?, *Trends Cell Biol.* 4:1–4.

Golding, D. W. (1994). A pattern confirmed and refined—synaptic, nonsynaptic and parasynaptic exocytosis, *BioEssays* 16:503–508.

Goodman, C. S. (1994). The likeness of being: Phylogenetically conserved molecular mechanisms of growth cone guidance, *Cell* 78:353–356.

Gottlieb, D. I. (1988). GABAergic neurons, *Sci. Amer.* 258 (February):82–89.

Greengard, P., F. Valtorta, A. J. Czernik, and F. Benfenati (1993). Synaptic vesicle phosphoproteins and regulation of synaptic function, *Science* 259:780–785.

Grenninglok, G., E. J. Rehm, and C. S. Goodman (1991). Genetic analysis of growth cone guidance in Drosophila: Fasciclin II functions as a neuronal recognition molecule, *Cell* 67:45–57.

Hargrave, P. A., H. E. Hamm, and K. P. Hofmann (1993). Interaction of rhodopsin with the G-protein, transducin, *BioEssays* 15:43–50.

Kahil, R. E. (1989). Synapse formation in the developing brain, *Sci. Amer.* 261 (December):76–85.

Levi-Montalcini, R., and P. Calissano (1986). Nerve growth factor as a paradigm for other polypeptide growth factors, *Trends Neurosci.* 9:473–477.

O'Conner, V., G. J. Augustine, and H. Betz (1994). Synaptic vesicle exocytosis: molecules and models, *Cell* 76:785–787

Raper, J. A., M. J. Bastiani, and C. S. Goodman (1984). Pathfinding by neuronal growth cones in grasshopper embryos. IV. The effects of ablating the A and P axons upon the behavior of the G growth cone, *J. Neurosci.* 4:2329–2345.

Schnapf, J. L., and D. A. Baylor (1987). How photoreceptor cells respond to light, *Sci. Amer.* 256 (April):40–47.

Shepherd, G. M. (1991). Sensory transduction: entering the mainstream of membrane signaling, *Cell* 67:845–851.

Snyder, S. H. (1984). Drug and neurotransmitter receptors in the brain, *Science* 224:22–31.

Snyder, S. H., and D. S. Bredt (1992). Biological roles of nitric oxide, *Sci. Amer.* 266 (May):68–77.

Südhof, T. C., P. De Camilli, H. Niemann, and R. Jahn (1993). Membrane fusion machinery: Insights from synaptic proteins, *Cell* 75:1–4.

Chapter 18

Cancer Cells and Growth Control

It is hard to imagine anyone familiar with the events occurring inside living cells who does not feel a sense of awe at the complexities involved. Given the vast number of activities that must be coordinated during the lifetime of each cell, it is perhaps not surprising to find that malfunctions occasionally arise. Of the many diseases that result from aberrations in cell function, cancer is among the most prominent. More than one out of every four people in the United States is now expected to develop cancer, making it the most common cause of death other than cardiovascular disease.

When biologists consider the problem of cancer, three underlying questions quickly come to mind: *What is cancer? What causes cancer? Can cancer be prevented or cured?* Although our answers for these questions are still incomplete, enormous progress has been made in recent years and there is reason to believe that this dreaded disease will eventually be brought under control. In the present chapter we will explore some of the experimental advances that provide the basis for such optimism. As we do so, it will become apparent that an understanding of the behavior of cancer cells requires an intimate knowledge of how normal cells behave and, conversely, investigating the biology of cancer cells has deepened our understanding of normal cells.

WHAT IS CANCER?

The term *cancer,* which means "crab" in Latin, was coined by Hippocrates in the fifth century B.C. to describe diseases in which tissues grow and spread unrestrained throughout the body, eventually choking off life. Cancers can originate in almost any tissue of the body. Depending on the cell type involved, they are grouped into three main categories. (1) **Carcinomas,** which are the most common types of cancer, arise from the epithelial cells that cover external and internal body surfaces. Lung, breast, and colon cancer are the most frequent cancers of this type. (2) **Sarcomas** originate in supporting tissues of mesodermal origin, such as bone, cartilage, fat, connective tissue, and muscle. (3) **Lymphomas** and **leukemias** arise from cells

of blood and lymphatic origin. The term *leukemia* is employed when the cancer cells circulate in large numbers in the bloodstream rather than growing mainly as solid masses of tissue.

Tumors Arise When the Rate of Cell Division Exceeds the Rate of Cell Differentiation and Loss

Cancer is based on a loss of normal growth control that produces a growing tissue mass known as a **tumor** or **neoplasm.** Uncontrolled growth does not mean that tumor cells always divide more rapidly than normal cells. The crucial issue is not the rate of cell division, but rather the relationship between the rate of cell division and the rate of cell differentiation and loss. In normal tissues the rate of cell division and the rate of cell differentiation and loss are kept in precise balance. For example, cells located in the basal layer of the skin divide at exactly the rate that is needed to replace the cells that are continually differentiating and being shed from the surface of the skin. A similar phenomenon occurs in bone marrow, where new blood cells are produced to replace aging cells that must be destroyed, and in the lining of the gastrointestinal tract, where new epithelial cells are produced to replace the cells that are continually shed. In each of these situations, the rate of cell division is carefully balanced with the rate of cell differentiation and loss so that no net accumulation of new cells occurs.

In contrast, the rate of cell division in tumors *exceeds* the rate of cell loss, and thus the number of dividing cells increases. If cell division is relatively rapid, the tumor will grow quickly in size; if the cells divide more slowly, tumor growth will be slower. But regardless of the growth rate, tumors continually increase in size because new cells are being produced in greater numbers than needed. As more and more of these dividing cells accumulate, the normal organization of the tissue gradually becomes disrupted (Figure 18-1).

Malignant Tumors Are Capable of Spreading by Invasion and Metastasis

Depending upon their pattern of growth, tumors are classified as either benign or malignant. **Benign** tumors grow in a confined local area and so are rarely life threatening. **Malignant** tumors are a more serious problem because they can invade surrounding tissues and enter into the circulatory system, allowing them to spread to distant parts of the body by the process of **metastasis.** The term **cancer** refers to any malignant tumor—that is, a tumor that can spread by invasion and metastasis. Hundreds of different kinds of tumors, malignant as well as benign, arise in humans. Table 18-1 lists some of the more common tumors and shows how they

Normal   Malignant

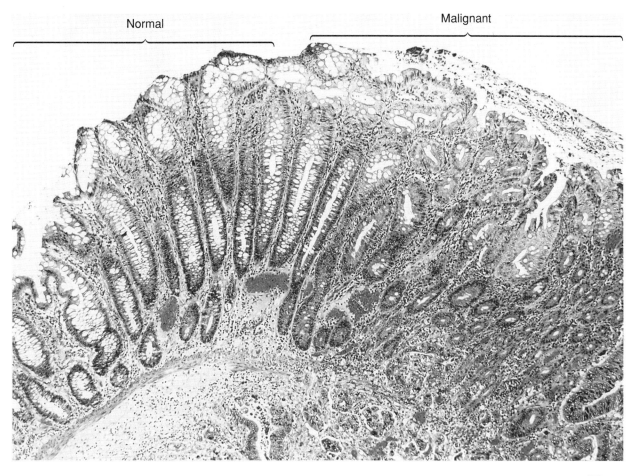

150 μm

Figure 18-1 *Light Micrograph of a Colon Carcinoma*   *The left side shows normal colon tissue covered by an epithelium containing numerous tubular mucous glands. On the right side, proliferating cancer cells derived from the epithelium have disrupted the organized pattern of mucous glands and are invading into the underlying tissue. Courtesy of G. D. Abrams.*

are named by adding the suffix *-oma* to the name of the cell type involved.

By the time a person is diagnosed as having cancer, the malignant cells have often spread to other parts of the body. Although surgeons can usually remove the original tumor, metastases are more difficult to detect and treat. For this reason, malignant tumors are almost always more hazardous than benign ones, although exceptions to the rule do occur. Basal cell carcinoma, for example, is a type of skin cancer that tends to be relatively innocuous because it rarely metastasizes and the tumors are easy to detect and remove. Some benign tumors, on the other hand, occur in surgically inaccessible locations, such as the brain, and are therefore potentially life threatening.

Since cancer would be easy to treat if metastasis could be prevented, considerable attention has been paid to studying the properties of cancer cells that make metastasis possible. Such studies have revealed

that metastasis is a multistep process that can be divided into three distinct stages (Figure 18-2).

Stage 1: Cancer Cells Invade Surrounding Tissues and Vessels

Unlike the cells of benign tumors, cancer cells tend to wander off from the main tumor mass and invade surrounding tissues. During the process, the cells often pass through barriers to normal cell migration such as the basal lamina and the walls of blood and lymphatic vessels. Several properties of the cancer cell facilitate this invasive behavior. First, cell-cell adhesiveness is diminished among cancer cells, and hence the forces that hold together a growing tumor mass are decreased. Second, cancer cells tend to be more mobile than their normal counterparts, increasing the likelihood that they will migrate away from the tumor mass. And finally, cancer cells can secrete proteases that digest a path through the extracellular matrix.

Table 18-1 Some of the Main Types of Cancer	
Tissue of Origin	**Name of Tumor**
Carcinomas (about 90% of all cancers)	
Skin	Basal cell carcinoma
	Squamous cell carcinoma
Lung	Pulmonary adenocarcinoma
Breast	Mammary adenocarcinoma
Stomach	Gastric adenocarcinoma
Colon	Colon adenocarcinoma
Uterus	Uterine endometrial carcinoma
Prostate	Prostatic adenocarcinoma
Ovary	Ovarian adenocarcinoma
Pancreas	Pancreatic adenocarcinoma
Urinary bladder	Urinary bladder adenocarcinoma
Liver	Hepatocarcinoma
Sarcomas (about 5% of all cancers)	
Bone	Osteosarcoma
Cartilage	Chondrosarcoma
Fat	Liposarcoma
Smooth muscle	Leiomyosarcoma
Skeletal muscle	Rhabdomyosarcoma
Connective tissue	Fibrosarcoma
Blood vessels	Hemangiosarcoma
Nerve sheath	Neurogenic sarcoma
Meninges	Meningiosarcoma
Lymphomas/Leukemias (about 5% of all cancers)	
Red blood cells	Erythrocytic leukemia
Bone marrow cells	Myeloma or
	myelocytic leukemia
White blood cells	Lymphoma or
	lymphocytic leukemia

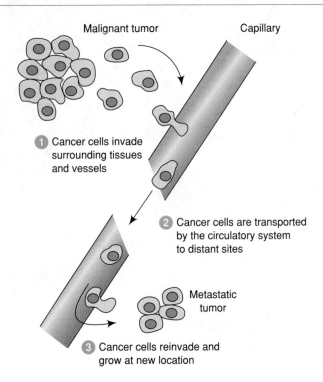

Figure 18-2 *Stages in the Process of Metastasis* *Only a small fraction of the cells in a primary malignant tumor can successfully carry out all three steps involved in metastasis: (1) invasion of surrounding tissues and vessels, (2) transport via the circulatory system, and (3) reinvasion and growth at a distant site.*

Stage 3: Cancer Cells Selectively Reinvade and Grow at Particular Sites

Some of the cancer cells that survive in the circulatory system eventually form clumps that become arrested in the capillaries of another tissue or organ. The arrested cells then penetrate through the capillary walls and invade the surrounding tissue, where they grow and divide. Only a tiny fraction of the cancer cells that initially enter the bloodstream successfully reach this final stage of metastasis. The question therefore arises as to whether the few cells that do succeed represent a random subgroup of the original population, or a set of special cells that are uniquely suited for metastasis.

To address this question, Isaiah Fidler studied the behavior of mouse melanoma cells injected into the circulatory system of healthy mice (Figure 18-3). A few weeks after the injections, metastases were detected in a variety of locations, including the lungs. Cells from the lung metastases were removed and injected into another mouse, leading to the production of more lung metastases. By repeating the procedure many times, Fidler eventually obtained a population of cancer cells that metastasized to the lung much more frequently than did the original tumor cell population. He therefore concluded that the initial mouse melanoma consisted of a heterogeneous population of cells with

Stage 2: Cancer Cells Are Transported to Distant Sites by the Circulatory System

Once cancer cells have broken through the walls of blood or lymphatic vessels, they are carried by the circulatory system to distant parts of the body. A malignant tumor weighing only a few grams may release several million cancer cells per day into the bloodstream, although most cancer cells do not thrive in this environment. Experimental studies in which radioactive cancer cells are injected into the bloodstream of laboratory animals have shown that less than one in a thousand radioactive cells survives for more than a few weeks. Such findings suggest that the bloodstream is a relatively inhospitable place for most cancer cells and that only a tiny number typically survive the trip to potential sites of metastasis.

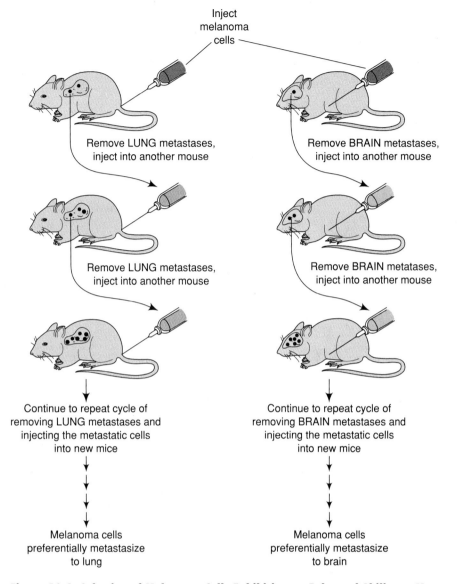

Inject melanoma cells

Remove LUNG metastases, inject into another mouse

Remove LUNG metastases, inject into another mouse

Continue to repeat cycle of removing LUNG metastases and injecting the metastatic cells into new mice

Melanoma cells preferentially metastasize to lung

Remove BRAIN metastases, inject into another mouse

Remove BRAIN metastases, inject into another mouse

Continue to repeat cycle of removing BRAIN metastases and injecting the metastatic cells into new mice

Melanoma cells preferentially metastasize to brain

Figure 18-3 *Selection of Melanoma Cells Exhibiting an Enhanced Ability to Metastasize* (Left) *Mouse melanoma cells were injected into the tail vein of a mouse and the resulting lung metastases were then removed and injected into another mouse. After repeating this cycle ten times, the final population of melanoma cells was found to produce many more lung metastases than the original cell population.* (Right) *A similar experiment was performed in which brain metastases were removed and injected into another mouse. After this cycle was repeated ten times, the final population of melanoma cells was found to metastasize preferentially to the brain.*

differing metastatic capabilities, and that the experimental protocol selected for cells that are particularly well suited for metastasizing.

Besides differing in their ability to metastasize, do cancer cells also vary in the sites to which they tend to spread? To investigate this possibility, mouse melanoma cells were injected into normal mice and metastases were isolated from the brain rather than the lungs. Repeating the same process many times in succession led to the isolation of a population of melanoma cells that preferentially metastasized to the brain. Hence malig-

nant tumors must consist of cells that differ not only in their capacity to metastasize, but also in the sites to which they tend to metastasize.

The Immune System Can Inhibit the Process of Metastasis

Does the body have any defense mechanisms that can slow or hinder the process of metastasis? Experiments carried out by Michael Feldman and Lea Eisenbach suggest that in some cases, the immune system may be able to suppress metastasis by attacking cancer cells that exhibit certain cell surface molecules. These studies involved two strains of mouse lung cancer cells, one called *D122* that metastasizes with high frequency, and another called *A9* that rarely metastasizes. As we learned in Chapter 16, the ability of the immune system to recognize cells as being foreign or abnormal depends on the presence of cell surface glycoproteins called *MHC molecules* (page 735). When Feldman and Eisenbach examined the MHC molecules carried by the two lines of tumor cells, they discovered a striking difference: A9 cells carry two types of MHC (called H-2K and H-2D), whereas the D122 cells express only one form (H-2D).

The discovery that D122 and A9 cells carry different cell surface MHC molecules raises the question of whether the differing metastatic potential of the two kinds of cells is related to the ability of the immune system to recognize and attack the two cell types. This possibility was investigated by injecting A9 and D122 cells into separate groups of animals and monitoring the production of *killer T cells* (page 735), which are components of the immune system that normally function to destroy foreign and abnormal tissues. The animals were found to produce numerous killer T cells targeted against the A9 cancer cells, but few killer T cells appeared in response to D122 cells.

Why do killer T cells attack A9 cells more readily than D122 cells? The most obvious possibility is that the immune system recognizes the H-2K MHC molecules, which are carried by A9 but not D122 cells. To investigate this possibility, cloned DNA containing H-2K gene sequences was introduced into D122 cells (Figure 18-4). As predicted, the altered D122 cells expressing the H-2K gene exhibited a reduced capacity to metastasize when injected into mice. However, the primary

tumor at the site of injection grew normally, implying that individual tumor cells attempting to metastasize through the circulatory system are more susceptible to immune attack than are tumor masses residing outside the circulation.

A Group of Characteristic Traits Are Shared by Cancer Cells

The ability to spread by invasion and metastasis is the one unique characteristic that distinguishes cancer from other kinds of cellular growth, but it is not the only unusual feature of cancer cells. Over the past several decades, studies involving hundreds of different malignancies have revealed that cancer cells tend to exhibit a group of shared traits, although none of these properties is necessarily exhibited by all cancer cells. In the following sections, we will briefly describe these frequently observed traits, which together constitute what is commonly referred to as the *profile of the cancer cell.*

Cancer Cells Have a Distinctive Appearance

The cells of malignant tumors usually have a distinctive appearance that can be recognized by microscopic examination. Although no single morphological trait is sufficient for distinguishing a cancer cell from a normal cell, some frequently encountered structural features are quite useful in diagnosing the presence of a malignancy. Prominent among these features is the tendency of malignant cells to undergo **anaplasia,** a process involving the loss of cell differentiation and the disruption of the proper orientation of cells to one another. Cancer cells also tend to exhibit large, irregularly shaped nuclei, prominent nucleoli, and a cell surface covered

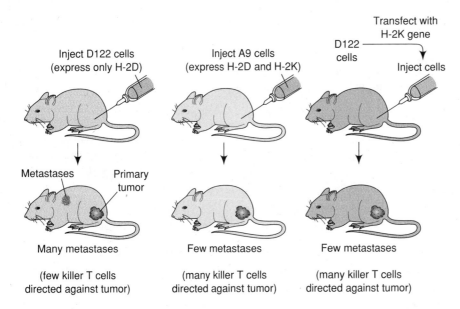

Figure 18-4 *Effects of Immune Recognition on the Metastatic Competence of Cancer Cells* (Left) *D122 lung cancer cells, which express cell surface H-2D molecules, generate numerous metastases. (Middle) A9 lung cancer cells, which express both H-2D and H-2K molecules, metastasize poorly. (Right) If the H-2K gene is introduced into D122 cells, the cells lose their ability to metastasize. Since H-2K is a cell surface MHC molecule recognized by the immune system, such studies suggest that immune recognition influences the ability of cancer cells to metastasize.*

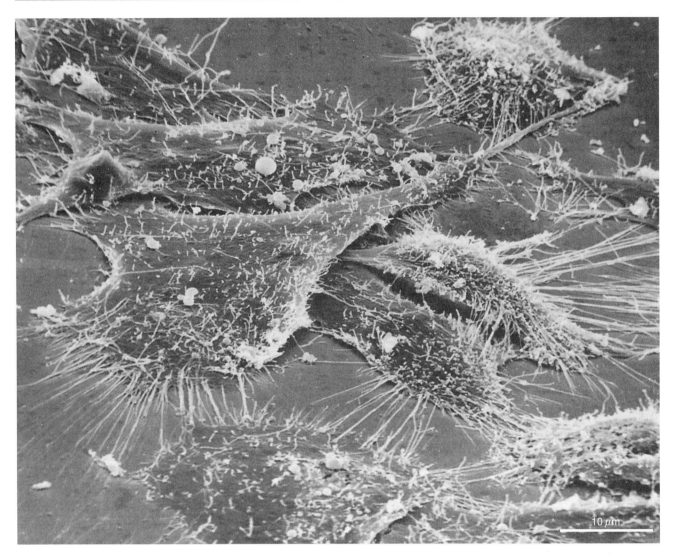

Figure 18-5 *Scanning Electron Micrograph of Several Poorly Differentiated Esophageal Cancer Cells* *Note the numerous radiating lamellipodia and fine microvilli covering the surface of these cells. Courtesy of K. M. Robinson.*

with microvilli and lamellipodia (Figure 18-5). The number of cells undergoing mitosis is usually elevated, and abnormal mitoses (Figure 18-6) and multinucleated giant cells are also encountered.

Changes in the appearance of cancer cells can be so striking that it is possible to diagnose the presence of a malignancy by examining a few isolated cells. This approach has been employed to great advantage in the **Pap smear,** a screening procedure for uterine cancer developed in the 1930s by George Papanicolaou. When doing a Pap smear, a tiny sample of a woman's vaginal secretions is obtained and examined under the microscope. If the cells in the fluid are found to exhibit unusual features, such as large irregular nuclei or prominent variations in cell size and shape (Figure 18-7), it is a sign that a uterine malignancy may be present and that further tests need to be done. Because Pap smears permit uterine cancer to be detected in its early stages

before metastasis has occurred, they have saved the lives of hundreds of thousands of women.

Cancer Cells Produce Tumors When Injected into Laboratory Animals

Perhaps the most distinctive feature of cancer cells is their ability to produce tumors when injected into an appropriate organism. For animal tumors this criterion is applied by injecting cells into normal animals of the same inbred strain. But with human cells the situation is more complicated. Injecting cancer cells into humans for testing purposes is clearly unethical, and the injection of human tumor cells into laboratory animals is not reliable because the animal's immune system may reject the human cells simply because they are foreign. One way around the problem is to inject human cells into **nude mice,** which are immunologically deficient mice lacking a normal thymus gland. Because the thymus

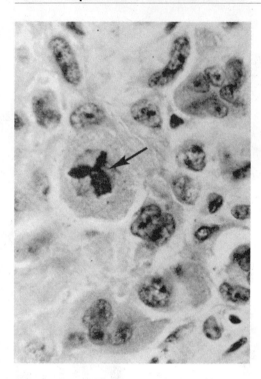

Figure 18-6 *Light Micrograph Showing a Population of Cancer Cells Exhibiting a Marked Variation in Size and Shape* *The prominent cell in the center has an abnormal tripolar mitotic spindle* (arrow).

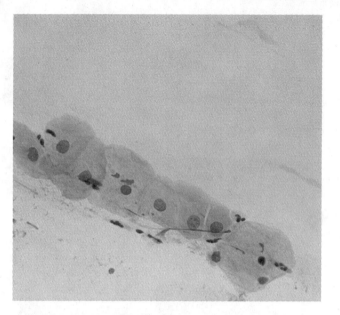

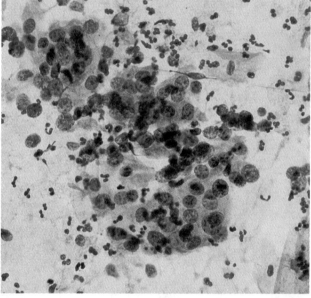

30 μm

Figure 18-7 *Examples of Normal and Abnormal Pap Smears* (Top) *In a normal Pap smear, the cells are of uniform size and contain small spherical nuclei.* (Bottom) *In this abnormal Pap smear, marked variations in cell size and shape are evident, and the nuclei are larger relative to the size of the cells. The abnormalities exhibited by these isolated cells suggest that they may be derived from a uterine cancer, and hence further examination of the uterus should be carried out. Courtesy of G. D. Abrams.*

gland produces T cells that are required for cell-mediated immunity (page 717), nude mice are incapable of rejecting foreign cells. Hence human cancer cells injected into nude mice usually grow into tumors without immunological rejection.

Cancer Cells Are Immortal in Culture

Normal cells typically exhibit a limited life span when grown in culture. Human fibroblasts, for example, multiply for about 50–60 generations and then deteriorate and die (see Figure 15-40). Malignant cells, on the other hand, behave as if they are immortal. Human *HeLa* cells, which were obtained from a uterine carcinoma in 1953, have divided more than ten thousand times in culture without signs of deterioration.

Cancer Cells Grow to High Densities in Culture

When normal cells are placed in culture, they divide until the surface of the culture vessel is covered by a single layer of cells. When this *monolayer stage* is reached, cell movement and cell division usually cease. In the early 1950s, Michael Abercrombie and Joan Heaysman introduced the term *contact inhibition* to refer to the decrease in cell motility that occurs when cells make contact with one another in culture. The same term has also been used to refer to the inhibition

of cell division that occurs when culture conditions become crowded. Because of the confusion that can result from the double meaning of *contact inhibition*, we will use the phrase *density-dependent inhibition of growth* to refer to the inhibition of cell division that occurs in crowded cultures.

Malignant cells are usually less susceptible to density-dependent inhibition of growth than their normal counterparts. Instead of stopping at the monolayer stage, malignant cells continue to divide and pile up on top of each other, forming multilayered aggregates. The relationship between this tendency to grow to high densities in culture and the ability to form tumors has been investigated by Stuart Aaronson and George Todaro using fibroblast cell lines derived from mouse embryos. Through manipulations of the culture conditions under which the cells were grown, cell lines differing in their susceptibility to density-dependent inhibition of growth were obtained. One type of cell line was produced by growing cells under uncrowded conditions; every time the cell population increased and crowding was imminent, the cells were diluted and transferred to a new culture flask. The cells obtained in this way were found to be very sensitive to density-dependent inhibition of growth. Another set of cell lines was established by continually growing cells in overcrowded conditions. Such cell populations became less susceptible to density-dependent inhibition of growth, reaching much higher population densities before growth ceased. When the various cell populations were tested for their ability to produce tumors in mice, tumor-forming ability was found to be directly related to the loss of density-dependent growth control; that is, cells capable of growing to the highest population densities in culture were most effective at forming tumors in animals (Figure 18-8).

Why are cancer cells less susceptible to density-dependent inhibition of growth than normal cells? Although the term *contact inhibition* was used for many years to refer to the tendency of cells to stop dividing when culture conditions become crowded, there is little evidence to support the idea that cell-cell contact is actually responsible. The degree of crowding tolerated before cell division stops in normal cells is subject to considerable variation, depending on the culture conditions employed. For example, if the serum concentration in the growth medium is too low, cells will stop dividing before cell-cell contacts have become established. Conversely, cell cultures that have stopped dividing under crowded conditions will start proliferating again if additional serum is added. Such observations suggest that density-dependent growth inhibition is caused by depletion of nutrients or growth factors rather than by direct cell-cell contact.

Some cancer cells overcome this limitation by manufacturing their own growth factors, which allows them to continue proliferating after the growth factors initially present in the medium have been depleted. One class of growth factors secreted by cancer cells are called *transforming growth factors (TGFs)* because they cause normal cells to acquire some of the growth characteristics of cancer cells. Two distinct types of TGF have been characterized: *TGF-α*, a small polypeptide that acts by binding to the cell surface receptor for epidermal growth factor, and *TGF-β*, a larger protein consisting of two polypeptide subunits. Stimulation of cell division requires the combination of both TGF-β and TGF-α (or epidermal growth factor). Although the production of TGFs may help to explain certain aspects of cancer cell growth, it does not provide a complete explanation of the malignant state because certain normal cells produce TGFs as well.

Cancer Cell Growth is Anchorage-Independent

Most normal cells do not grow well when they are suspended in a liquid medium or a semisolid material such as soft agar. But when provided with an appropriate surface to which they can adhere, the cells will attach to the surface, spread out, and commence growth. This type of growth is said to be *anchorage-dependent*. In contrast to the behavior of normal cells, most cancer cells grow well when they are suspended in a liquid or semisolid medium. Cancer cell growth is therefore characterized as being *anchorage-independent*.

Experiments carried out in the laboratory of Robert Pollack suggest that the ability to grow without anchorage in culture is closely correlated with a cell's ability to form malignant tumors. In these studies a series of fibroblast cell lines were tested for their ability to induce tumors in nude mice. The original cell line exhibited many of the typical characteristics of cancer cells: decreased density-dependent inhibition of growth, low serum requirement for growth, expression of new cell surface antigens (page 787), and anchorage-independent growth. By isolating single cells from the original cell population and growing a series of clones, Pollack generated separate cell lines that had lost one or more of these properties. Of the four properties studied, only anchorage-independent growth was consistently retained by cell lines capable of forming tumors in animals.

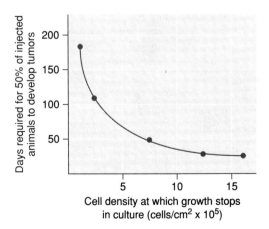

Figure 18-8 Tumor-Forming Ability of Several Mouse Fibroblast Cell Lines That Differ in Their Susceptibility to Density-Dependent Inhibition of Growth *Each point represents a different cell line. The data show that cell lines that grow to higher densities in culture produce tumors more rapidly when injected into mice.*

But the connection between anchorage-independent growth and tumor formation is not without its exceptions. Some human cell lines exposed to cancer-causing chemicals exhibit anchorage-independent growth, yet are unable to induce tumors when injected into nude mice. Conversely, mouse 3T3 fibroblasts, which are anchorage-dependent and normally unable to induce tumor formation, gain the capacity to produce tumors when they are attached to glass beads prior to implantation. Such observations suggest that in spite of its general correlation with tumorigenicity, anchorage-independent growth in culture is not an absolute prerequisite for tumor formation.

Cancer Cells Lack Normal Cell Cycle Controls

When normal cells are exposed to suboptimal growth conditions such as inadequate nutrients or growth factors, they stop growing at a specific point in the cell cycle called the *restriction point* or *START,* which occurs near the end of G_1 (page 523). Cells held up at this stage of the cell cycle are said to be in the G_0 *state.* In contrast to normal cells, cancer cells continue to grow and divide under conditions of high cell density, low serum concentration, or suboptimal nutrient concentration that would cause normal cells to stop at the restriction point. If the nutritional deprivation is severe, cancer cells eventually die at random points in the cell cycle rather than entering into G_0.

This difference in the behavior of normal and cancer cells has been used as a basis for devising ways to kill cancer cells selectively. Arthur Pardee and his associates have constructed a simple scheme employing a combination of caffeine, which blocks normal cells at the restriction point, and cytosine arabinoside, which interferes with DNA synthesis and therefore kills cells during S phase. When cultures of normal cells are exposed to caffeine followed by cytosine arabinoside, the caffeine halts cells in G_1, thereby preventing them from entering S phase and being killed by cytosine arabinoside. If the two drugs are later removed, the cells are released from restriction point control and begin dividing again.

In contrast, when cancer cells are exposed to the same drug combination, caffeine does not stop cells in G_1 because of the absence of restriction point control. The malignant cells therefore proceed into S phase and are killed by cytosine arabinoside. The discovery that cultured cancer cells can be killed by treatments that do not harm normal cells raises the possibility that a similar strategy might eventually be applied to cancer treatment in human patients.

Cancer Cells Exhibit Cell Surface Alterations

Changes in the composition of the plasma membrane are almost universally observed in cancer cells, although the significance of such changes is often difficult to assess because similar alterations tend to occur in normal cells that have been stimulated to divide. For example, active transport systems for the uptake of sugars, amino acids, and nucleosides are frequently activated in tumor cells, but a similar increase in transport activity is observed in normal cells stimulated to divide by the addition of appropriate nutrients or growth factors. It is therefore important to determine which membrane changes play unique roles in cancer cells and which changes are characteristic of dividing cells in general. Among the cell surface changes that appear to be particularly distinctive and important for the behavior of cancer cells are those that influence the properties of adhesiveness, agglutinability, cell-cell communication, and antigenic composition.

The first of these properties, namely the decreased adhesiveness of cancer cells, was mentioned earlier in the chapter when we discussed the ability of cancer cells to wander off and invade surrounding tissues. Cancer cells do not stick to one another like normal cells because they are deficient in the cell surface glycoprotein, *fibronectin* (page 230). In addition to its role in promoting cell-cell adhesion, fibronectin also influences cell shape and motility. Addition of fibronectin to malignant cells restores cell adhesion, flattened cell morphology, and contact inhibition of movement. It does not, however, restore normal growth control (i.e., density-dependent inhibition of growth), indicating that depletion of fibronectin is not responsible for the uncontrolled growth of cancer cells.

A second cell surface property altered in cancer cells is their enhanced tendency to clump together or *agglutinate* when exposed to lectins (Figure 18-9). **Lectins** are proteins that contain multiple carbohydrate-binding sites, and therefore a single lectin molecule can bind simultaneously to carbohydrate groups exposed on the surface of more than one cell (page 234). As a result, lectin molecules can create cell-cell linkages that cause cell clumping. The most straightforward explanation for the enhanced agglutinability of tumor cells would be an increased number of lectin receptors, but careful measurements have led to the surprising conclusion that the total number of lectin receptors is similar in normal and cancer cells. What does differ, however, is receptor mobility. In cancer cells, an altered association between lectin receptors and the cytoskeleton allows lateral diffusion of the receptors to occur more readily, thereby leading to enhanced cell clumping in the presence of lectins.

Another cell surface trait commonly exhibited by cancer cells is a decline in the number of gap junctions, suggesting a deficiency in cell-cell communication. Support for this conclusion has come from experiments showing that fluorescent dyes injected into normal cells move rapidly into surrounding cells

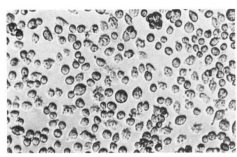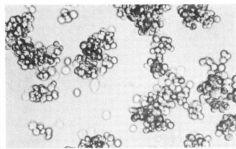

Figure 18-9 *Behavior of Normal and Cancer Cells Exposed to the Lectin, Concanavalin A*
(Left) *Normal cells exposed to concanavalin A tend to remain separated from one another.*
(Right) *Cancer cells exposed to concanavalin A exhibit an enhanced tendency to adhere to*
each other and form clumps. Courtesy of L. Sachs.

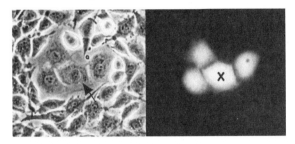

Figure 18-10 *Evidence for Decreased Cell-Cell*
Communication in Cancer Cells (Left) *Phase-contrast*
micrograph showing four normal liver cells surrounded by
numerous malignant cells in culture. The arrow points to
where fluorescent dye was injected into one of the four
normal cells. (Right) *Fluorescence microscopy of the same*
cells shortly thereafter reveals that the dye has diffused into
the three adjacent normal cells but not into the surrounding
malignant cells. Courtesy of W. R. Loewenstein.

that are normal but not into cells that are malignant
(Figure 18-10). Several observations suggest that the inability of cancer cells to communicate through gap
junctions plays a role in the loss of normal growth control. For example, Werner Loewenstein has carried out
a series of investigations in which normal cells were
fused with malignant cells that had lost the ability to
form gap junctions, thereby creating hybrid cells exhibiting both gap junctions and normal growth control.
But the gap junctions eventually disappeared from
some of the hybrid cells. When this happened the cells
reverted to uncontrolled growth, suggesting that normal growth control may depend on the ability of cells
to communicate through gap junctions.

The final cell surface property altered in malignant
cells is their antigenic composition. Although this area
of research has had a controversial history, it is now
well established that some cancers exhibit *tumor-specific cell-surface antigens.* Part of the difficulty in reaching a consensus on this issue has come from the fact
that cancer cells often exhibit a variety of antigenic
changes. In human melanomas, for example, at least
three classes of antigens have been identified. Antigens
of the first type are specific both for melanomas and for

the individual from whom a particular melanoma is obtained. Antigens of the second type are specific for
melanomas, but not for the individual from whom the
tumor is obtained. Antigens of the two preceding types
are not detectable on normal cells and can therefore be
classified as tumor specific. Antigens of the third type
are present on both normal and melanoma cells, although their concentration in melanoma cells is greater.

The existence of tumor-specific antigens (the first
two classes described above) raises the question of why
individuals afflicted with cancer do not reject their own
tumors. According to the **immune surveillance
theory**, immune rejection of newly forming malignant
cells is a normally occurring event in healthy individuals,
and cancer simply reflects the occasional failure of a sufficient immune response to be mounted against aberrant
cells. If correct, this theory suggests that one approach to
treating cancer patients might involve attempts to stimulate an individual's own immune system to destroy the
person's cancer cells (page 808).

Cancer Cells Secrete Proteases, Embryonic Proteins, and Proteins That Stimulate Angiogenesis

Because they would be potentially useful for the diagnosis of cancer, a great deal of effort has been expended
in searching for molecular "markers" that are produced
only by cancer cells. Unfortunately none of the molecules identified thus far have turned out to be ab-
solutely unique to cancer cells, although a number of
them have provided some important insights into the
behavior of malignant tumors. Prominent among these
are proteases, embryonic proteins, and molecules that
stimulate blood vessel formation.

It has been known for many years that malignant tumors often secrete proteases. Although protein-digesting enzymes are also secreted by certain kinds of
normal cells, the enhanced production of such enzymes
by cancer cells may contribute to some of the properties of malignant tumors. For example, the decrease in
adhesiveness and loss of anchorage dependence that

are frequently exhibited by cancer cells may be caused by protease digestion of cell surface components. As discussed earlier in the chapter, secretion of proteases might also provide a partial explanation for the invasive properties of malignant tumors. The main protease secreted by cancer cells is *plasminogen activator,* an enzyme that catalyzes the cleavage of *plasminogen* to *plasmin,* which itself is a protease:

$$\text{Plasminogen} \xrightarrow[\text{activator}]{\substack{\text{Cleaved by} \\ \text{plasminogen}}} \text{Plasmin}$$
$$\text{(inactive precursor)} \qquad\qquad \text{(active protease)}$$

Because plasminogen is present in high concentration in extracellular fluids, its conversion to plasmin leads to a large increase in the amount of protease activity in the region surrounding a malignant tumor.

In addition to proteases, some cancer cells manufacture and secrete proteins that are usually produced only by embryos. For example, *α-fetoprotein,* a protein made by embryonic liver cells, is found in only trace amounts in normal adults, but its concentration in the bloodstream increases dramatically in patients with liver cancer. *Carcinoembryonic antigen (CEA),* a glycoprotein produced in the embryonic digestive tract, and fetal hormones such as *chorionic gonadotropin* and *placental lactogen,* are also secreted by certain human tumors. Testing for embryonic markers such as α-fetoprotein and carcinoembryonic antigen in the bloodstream has been used diagnostically to monitor for the presence of particular kinds of cancer, but the fact that these substances are made by only a few tumor types has limited the usefulness of this approach.

Finally, some tumors also secrete growth factors that stimulate the formation of blood vessels. The presence of an adequate blood supply is a critical factor in determining a tumor's ability to grow and invade, so production of substances that stimulate the process of blood vessel formation, or **angiogenesis**, can have a profound influence on tumor behavior. It has been shown, for example, that tumors will not grow beyond a few millimeters in diameter in the absence of a newly forming blood supply. Therefore inhibitors of angiogenesis might be potentially useful tools for inhibiting tumor growth.

Cancer Cells Exhibit Chromosomal Abnormalities

It has been known for many years that cancer cells exhibit genetic alterations that are often reflected in abnormalities in chromosome number and appearance. A classical example is the *Philadelphia chromosome,* an abnormally shaped chromosome that is produced by the translocation of a piece of chromosome 22 to chromosome 9 (Figure 18-11). The Philadelphia chromosome occurs in nearly 90 percent of individuals suffering from chronic granulocytic leukemia. In recent years the genetic changes exhibited by cancer cells have been pinpointed to the level of individual genes, and the role played by the altered genes in causing cancer is beginning to emerge. Two classes of genes are now known to play crucial roles in the development of cancer: a group of altered genes whose presence is associated with malignancy, and a group of genes whose absence (or loss of function) is associated with malignancy. The roles of these two classes of genes will be described in the next section of the chapter, where we focus on the causes of cancer.

WHAT CAUSES CANCER?

Although cancer is commonly perceived as a disease that strikes randomly and without warning, this misconception ignores the results of thousands of investigations on the causes of cancer, some of which date back more than 200 years. The inescapable conclusion emerging from these investigations is that most human cancers are caused by identifiable factors that can be divided into four major groups: *chemicals, radiation, viruses,* and *heredity.*

Chemicals Can Cause Cancer

The first suggestion that chemicals can cause cancer dates back to 1761, when a London doctor named John Hill noted that people who use snuff suffer an abnormally high incidence of nasal cancer. A few years later another British physician, Percival Pott, observed an elevated incidence of cancer of the scrotum among men who had served as chimney sweepers in their youth. It was common practice at the time to employ young boys to clean chimney flues because they fit into narrow spaces more readily than adults. Pott speculated that the chimney soot became dissolved in the natural oils of the scrotum, irritating the skin and eventually triggering the development of cancer. This theory led to the discovery that cancer of the scrotum could be prevented among chimney sweepers by the judicious use of protective clothing and regular bathing practices.

In the years since these pioneering discoveries, it has become increasingly apparent that exposure to certain chemicals can cause cancer. Unfortunately, the ability of a particular chemical to cause cancer often has become obvious only after large numbers of cancer cases are observed in people exposed to the chemical on a regular basis. At the turn of the century, for example, an elevated incidence of skin cancer was noted among workers in the coal tar industry, and an increased incidence of bladder cancer was observed in

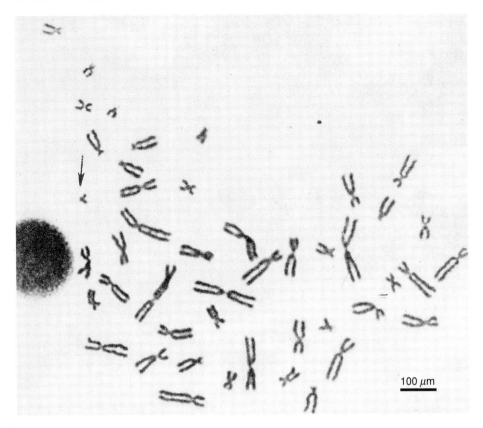

Figure 18-11 *Metaphase Chromosomes from a Cancer Cell Obtained from a Patient with Chronic Granulocytic Leukemia* *The arrow points to the Philadelphia chromosome, which is the remnant of chromosome 22 that remains after the rest of chromosome 22 has been translocated to chromosome 9. Courtesy of P. C. Nowell.*

workers employed by the newly emerging aniline dye industry. As the industrial revolution moved into the twentieth century, the list of known **carcinogens** (cancer-causing agents) became more and more extensive (Table 18-2). Millions of workers are now known to be exposed to chemical carcinogens on a daily basis, although exposure to chemical carcinogens is not always work-related or involuntary. More than one-third of the adults in the United States voluntarily smoke cigarettes, even though tobacco smoke contains over 30 carcinogens and has been implicated in more than one out of every four cancer deaths. This voluntary habit is responsible for 5–10 times as many cancer cases as occupational exposure to all other carcinogens combined.

Chemical Carcinogenesis Involves Initiation and Promotion Stages

The idea that malignant tumors arise by a multistep process was first proposed in the early 1940s by Peyton Rous to explain a phenomenon he had encountered when studying the ability of coal tar to cause cancer in rabbits. Rous had observed that repeated application of coal tar to rabbit skin causes tumors to develop, but the tumors disappear when application of the carcinogen is

stopped. Subsequent treatment of the skin with an irritant such as turpentine, which does not normally cause cancer itself, induces the tumors to reappear again. This pattern suggested to Rous that the coal tar and turpentine are playing two different roles, which he termed **initiation** and **promotion.** According to his theory, initiation causes normal cells to become irreversibly altered to a preneoplastic state, and promotion then stimulates the preneoplastic cells to divide and form tumors.

Independent support for the existence of distinct initiation and promotion phases was obtained by Isaac Berenblum, who investigated the induction of skin cancer in mice (Figure 18-12) Berenblum found that painting the skin of a mouse a single time with methylcholanthrene rarely causes tumors to develop. But subsequent applications of *croton oil*, which is not carcinogenic itself, triggers the formation of multiple tumors once the skin has been exposed to methylcholanthrene. Hence methylcholanthrene is acting as an initiator, whereas the croton oil functions as a promoter.

The mechanisms underlying initiation and promotion are quite different. Initiation is a quick, irreversible process that appears to involve permanent changes in a cell's DNA. Support for this view has come from data showing that the carcinogenic potency of various

Table 18-2 Some of the Chemical Carcinogens Present in the Environment (Air, Food, Water, Workplace, etc.)

Carcinogen	Type of Cancer Induced
Acrylonitrile	Colon, lung
4-Aminodiphenyl	Bladder
Aniline derivatives	Bladder
Arsenic compounds	Lung, skin
Asbestos	Lung, mesothelium
Benzene	Leukemia
Cadmium salts	Prostate, lung
Carbon tetrachloride	Liver
Chromium and chromates	Lung, nasal sinuses
Diethylstilbestrol (DES)	Uterus, vagina
Lead	Kidney
Mustard gas	Lung, larynx
α-Naphthylamine	Bladder
Nickel	Lung, nose
Organochloride pesticides	Liver
Polychlorinated biphenyls	Liver
Radon	Lung
Soot and tars	Skin, lung, bladder
Vinyl chloride	Liver, lung, brain
Wood and leather dust	Nasal sinuses
Tobacco smoke, which contains the following:	Lung, oral cavity, larynx, esophagus, stomach, pancreas, others

Aminostilbene, arsenic, benz[*a*]anthracene, benz[*a*]pyrene, benzene, benzo[*b*]fluoranthene, benzo[*c*]phenanthrene, benzo[*j*]fluoranthene, cadmium, chrysene, dibenz[*a,c*]anthracene, dibenzo[*a,e*]fluoranthene, dibenz[*a,h*]acridine, dibenz[*a,j*]acridine, dibenzo[*c,g*]carbazone, N-dibutylnitrosamine, 2,3-dimethylchrysene, indeno[*1,2,3-c,d*]pyrene, S-methylchrysene, S-methylfluoranthene, α-naphthylamine, nickel compounds, N-nitrosodimethylamine, N-nitrosomethylethylamine, polonium-210, N-nitrosodiethylamine, N-nitrosonornicotine, N-nitrosoanabasine, N-nitrosopiperidine

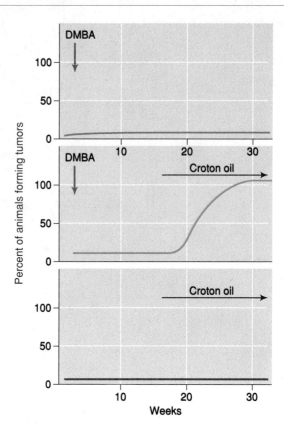

Figure 18-12 *Evidence for the Existence of Initiation and Promotion Phases in Chemical Carcinogenesis* (Top) *Mice treated with a single dose of dimethylbenzanthracene (DMBA) do not form tumors.* (Middle) *Painting the skin of such animals twice a week with croton oil after the DMBA treatment leads to the appearance of skin tumors.* (Bottom) *Croton oil alone does not produce skin tumors. These data are consistent with the conclusion that DMBA is an initiator and croton oil is a promoter.*

chemicals correlates with their ability to bind to DNA (Figure 18-13), and the discovery that most carcinogens are capable of inducing mutations. The ability of carcinogens to act as mutagens provides a simple explanation for the observed ability of carcinogenic chemicals to trigger a stable, inheritable change in a cell's properties.

In contrast to initiation, promotion is a gradual, partially reversible process that requires prolonged exposure to promoting agents. If a cell that has already undergone initiation is exposed to a promoting agent, the cell is stimulated to divide and the number of genetically damaged cells increases. As the damaged cells continue to divide, a gradual selection for cells exhibiting enhanced growth rate and invasive properties occurs, leading to the formation of a malignant tumor. The long period that can be occupied by the promotion phase may explain why cancer often does not develop until many years after exposure to a carcinogenic agent.

Some insights into the mechanism of action of promoting agents have come from studies of **phorbol esters,** a family of tumor promoters present in croton oil (an oil derived from seeds of the tropical plant, *Croton tiglium*). Phorbol esters bind to the plasma membrane and activate protein kinase C, a component of the phosphoinositide signaling pathway whose activity is normally controlled by the second messenger, diacylglycerol (page 221). The activation of protein kinase C (by phorbol esters or diacylglycerol) leads to the

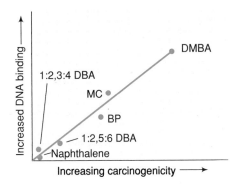

Figure 18-13 *Relationship between Carcinogenic Potency and DNA-Binding Ability* *Six carcinogens were tested for their ability to bind to DNA and to cause cancer when applied to mouse skin. (Abbreviations: DBA = dibenzanthracene, BP = benzpyrene, MC = 20-methylcholanthrene, DMBA = 9,10-dimethyl-1:2-benzanthracene.)*

phosphorylation of a variety of target proteins and eventually to the activation of the *AP1 transcription factor* (page 464), which turns on the transcription of genes involved in stimulating cell proliferation. Thus phorbol esters stimulate cell proliferation via a mechanism that resembles the way in which cell proliferation is normally regulated by signaling molecules that activate the phosphoinositide pathway.

Radiation Can Cause Cancer

Energy that travels through space is referred to as **radiation.** Natural sources of radiation to which humans are normally exposed include ultraviolet rays from the sun, cosmic rays from outer space, and emissions from naturally occurring radioactive elements. Medical, industrial, and military activities have created additional sources of high-energy radiation, mainly in the form of X-rays and radioactivity. Although the various types of radiation to which we are exposed differ in energy and wavelength, several share the ability to cause cancer.

The ability of sunlight to cause cancer was first deduced from the observation that skin cancer is most prevalent in people who spend long hours in the sun, especially in tropical regions where the sunlight is very intense. Only exposed areas of the body tend to be affected, indicating that cancer risk can be minimized by covering the skin. Risk is likewise reduced in individuals who use lotions containing chemical *sunscreens* that absorb ultraviolet radiation. Ultraviolet radiation is also absorbed by normal skin pigmentation, and for this reason dark-skinned individuals usually have lower rates of skin cancer than fair-skinned individuals. Exposure to sunlight rarely causes any type of malignancy other than skin cancer because ultraviolet radiation is too weak to pass through the skin and into the interior of the body. Fortu-

nately the most common kind of skin cancer rarely metastasizes, and its superficial location makes it relatively easy to remove surgically.

X-rays pose a more serious cancer hazard because they are strong enough to penetrate through the skin and reach internal organs. Shortly after the discovery of X-rays by Wilhelm Roentgen in 1895, people working with this type of radiation were found to exhibit an increased incidence of cancer. The suspicion that X-rays cause cancer was soon confirmed by studies revealing that animals exposed to X-rays have an increased risk of developing cancer—a risk directly related to the dose of radiation received. Human exposure to this potentially hazardous form of radiation could be minimized if X-rays served no useful function, but in many situations the health benefits to be gained from the use of medical X-rays far outweigh the risk of inducing cancer. Such is not always the case, however. In countries where medical X-rays have been employed for the treatment of superficial skin conditions of the head and neck, such as ringworm and acne, the rate of thyroid cancer is much higher than in countries where the practice is not widespread. Thus the practical benefits to be gained from every X-ray should be prudently weighed against the increased cancer risk.

Many radioactive elements emit radiation whose carcinogenic effects resemble those of X-rays. One of the first scientists to work with radioactivity was Marie Curie, the co-discoverer of the radioactive elements polonium and radium. She later died of a form of leukemia that appeared to be caused by her extensive exposure to radioactivity. A striking illustration of the danger posed by exposure to radioactive elements occurred during the 1920s in a group of women employed by a New Jersey factory that produced watch dials that glow in the dark. The luminescent paint used in painting the dials contained radium, and was applied with a fine-tipped brush that the employees frequently wetted with their tongues. During the process, minute quantities of radium were inadvertently ingested. Several years later these women developed an alarmingly high rate of bone cancer caused by the radioactive radium that had gradually become concentrated in their bones.

A more horrifying example of radiation-induced cancer occurred in Japan after the atomic bombs were exploded over Hiroshima and Nagasaki in 1945. The massive radioactive fallout produced by these explosions was followed years later by dramatic increases in the incidence of leukemia, lymphomas, and cancers of the thyroid, breast, uterus, and gastrointestinal tract. In the United States individuals exposed to the radioactive fallout produced by the nuclear bombs tested in Nevada in the 1950s also suffered an increased incidence of leukemia several decades later.

Radiation-induced carcinogenesis resembles chemical carcinogenesis in its basic mode of action. Like most chemical carcinogens, radiation is mutagenic and is therefore thought to initiate malignant transformation by causing DNA damage. Many years usually intervene between exposure to an initiating dose of radiation and the appearance of a malignancy, suggesting that subsequent exposure to promoting agents plays a role in stimulating radiation-damaged cells to divide and form tumors.

Viruses Cause a Variety of Different Animal Cancers

The idea that viruses can trigger the development of cancer was first proposed in the early 1900s. Although the ability of chemicals and radiation to cause cancer was already appreciated, the possibility that infectious agents might also play a role was not seriously considered because cancer does not tend to behave like a contagious disease. However, in 1908 the Danish scientists V. Ellerman and O. Bang reported that leukemia can be transmitted to healthy chickens by injecting them with blood extracts obtained from leukemic chickens. Several years later Peyton Rous discovered that sarcomas can also be transmitted between chickens by injection of filtered tumor extracts. Rous tentatively concluded that the chicken sarcoma is transmitted by a virus, and he eventually isolated several tumor viruses from chickens brought to him by local farmers.

Despite the clarity of Rous's data, his work was initially greeted with skepticism and many years passed before the existence of **oncogenic** (cancer-causing) **viruses** came to be widely accepted. Convincing proof required the demonstration that viruses can trigger the development of a variety of different kinds of cancer. In 1933 Richard Shope demonstrated that a skin cancer known as cutaneous papilloma can be transmitted between rabbits using tumor extracts containing no intact cells, suggesting the possible presence of an infectious cancer-causing virus. In the following year, Baldwin Lucké observed that kidney tumors in New England frogs can also be transmitted by a filterable agent. And a few years later, John Bittner reported that breast cancer in mice is transmitted from mother to offspring by a virus present in the mother's milk. In the following decades, dozens of other DNA and RNA viruses were implicated in animal and plant carcinogenesis (Table 18-3). Most of these oncogenic viruses are selective in the hosts they infect and the types of tumors they cause, although there are exceptions to this rule. Polyoma virus, for example, infects a variety of mammals and can induce more than 20 different kinds of tumors, including cancer of the liver, kidney, lung, skin, bone, blood vessels, nervous tissues, and connective tissues.

A common feature of oncogenic viruses is the ability to conceal themselves in infected cells in a hidden or *latent* form in which no new virus particles are produced or released. Latent viruses do not become active until the cell is exposed to appropriate triggering conditions, such as radiation, chemical carcinogens, hormones, or even other viruses. This type of behavior is illustrated by the *feline leukemia virus,* which can be harbored by otherwise healthy cats for many years. Only upon exposure to a physically stressful situation, such as a relatively mild respiratory infection, does the latent virus become activated and cause cancer. At the same time, large numbers of new virus particles are produced and released, thereby triggering a potential cancer epidemic among neighboring cats.

Viruses Have Been Implicated in Only a Few Types of Human Cancer

In contrast to the relatively large number of viruses that cause cancer in animals, only a few viruses have been implicated in producing cancer in humans. The first evidence for the existence of a human tumor virus was provided in the late 1950s by Denis Burkitt, a British surgeon working in Africa. Burkitt noted that a large number of children coming to him for medical care suffered from massive tumors of the jaw. This kind of cancer, now known as *Burkitt's lymphoma,* is most prevalent in areas where mosquito-transmitted infections are common. The epidemic nature of the disease led Burkitt to propose that it is transmitted by a mosquito-borne infectious agent.

A few years later, electron microscopists examining tumor tissue obtained from patients with Burkitt's lymphoma discovered a virus residing in the tumor cells (Figure 18-14). The virus, identified as a member of the herpes group, has come to be called the *Epstein-Barr virus (EBV)* in recognition of the scientists who discovered it. Unlike viruses of animal or plant origin, it is difficult to prove that a virus like EBV causes cancer in humans because ethical considerations prevent testing the hypothesis directly by injecting the virus back into normal individuals. But in spite of the lack of direct proof, several lines of evidence are consistent with the conclusion that the Epstein-Barr virus causes Burkitt's lymphoma: (1) DNA sequences and proteins encoded by the Epstein-Barr virus can be detected in tumor cells obtained from patients with Burkitt's lymphoma, but not in normal cells from the same individuals; (2) addition of purified Epstein-Barr virus to cultures of normal human lymphocytes causes the cells to undergo malignant transformation; and (3) injecting the Epstein-Barr virus into monkeys induces the formation of lymphomas.

Besides its role in Burkitt's lymphoma, the Epstein-Barr virus has also been implicated in the causation of nasopharyngeal cancer in South China. Even more

Table 18-3 Examples of Tumor Viruses

Class	Examples	Tumors Induced	Organism
DNA Viruses			
Herpesviruses	Lucké virus	Kidney adenocarcinoma	Frogs
	Epstein-Barr virus (EBV)	Burkitt's lymphoma, nasopharyngeal carcinoma	Humans
	Marek's disease virus	Lymphoma	Chickens
Papovaviruses	Shope papilloma virus	Papillomas	Rabbits
	SV-40	Subcutaneous, kidney and lung sarcomas	Hamsters
	Polyoma	Liver, kidney, lung, bone, blood vessels, nervous tissue, connective tissues	Mice
	Human papillomaviruses	Cervical cancer	Humans
Adenoviruses	Human adenoviruses	Subcutaneous, intraperitoneal, intracranial	Hamsters
RNA Viruses			
B-type viruses	Bittner mammary tumor virus	Mammary carcinoma	Mice
C-type viruses	Rous sarcoma virus	Sarcomas	Birds, mammals
	Murine leukemia viruses (Gross, Moloney, Friend, Rauscher, and others)	Leukemia	Mice
	Feline leukemia virus	Leukemia	Cats
	Murine sarcoma virus	Sarcoma	Mice
	Feline sarcoma virus	Sarcoma	Cats
	Avian leukemia viruses (avian myeloblastosis and others)	Leukemia	Chickens
	Human T-cell leukemia virus	Leukemias/lymphomas	Humans
Plant viruses	Wound tumor virus	Roots and stems	Plants

interesting is the discovery that in the United States, where more than 90 percent of the population has been exposed to EBV, the major disease induced by the virus is not cancer but *infectious mononucleosis,* a self-limiting proliferation of white blood cells that causes relatively harmless flulike symptoms. The reason why EBV should trigger the development of deadly forms of cancer in one country and only a flulike illness in another is not well understood.

In addition to the Epstein-Barr virus, several other viruses have been linked to human cancers. Among these associations are the *hepatitis-B virus* with liver cancer, *human papilloma viruses* with uterine cervical cancer, and the *human T-cell leukemia virus* with acute T-cell leukemia. The T-cell leukemia virus is particularly interesting because it is structurally related to HIV, the virus

associated with AIDS (page 742). Both viruses act by infecting white blood cells; the major difference is that infection by the T-cell leukemia virus stimulates abnormal proliferation of the infected lymphocytes (i.e., cancer), whereas infection by HIV leads to lymphocyte depletion.

Oncogenic Viruses Insert Their Genetic Information into the Chromosomal DNA of Infected Cells

The mechanism by which oncogenic viruses cause cells to become malignant depends to a certain extent on whether they are DNA or RNA viruses. In the case of DNA cancer viruses, entrance of the virus into the cell is followed by transcription of the viral DNA into viral messenger RNA molecules that are then translated into viral proteins. Replication of both cellular and viral DNA

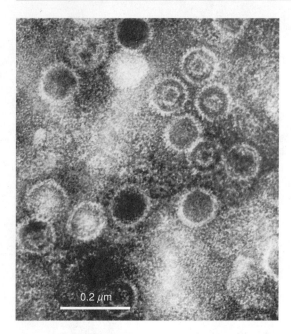

0.2 μm

Figure 18-14 *An Electron Micrograph of Negatively Stained Epstein-Barr Virus Particles* *Such particles have been isolated from cancer cells obtained from patients with Burkitt's lymphoma. Courtesy of M. A. Epstein.*

ensues, followed by cell division. After several weeks of virus-stimulated cell division, one or more copies of the viral DNA molecule usually become integrated into the host chromosomal DNA. At this point the viral genetic information becomes a permanent part of the cell's genetic material and the cell begins to replicate the viral DNA sequences as part of its own DNA. During the entire scenario, no new virus particles need to be produced or released.

In contrast to DNA viruses, RNA tumor viruses cannot insert their genes directly into the host cell's chromosomes because they store their genetic information in the form of RNA rather than DNA. The mechanism by which this limitation is overcome was first revealed in 1970, when Howard Temin and David Baltimore independently discovered that RNA viruses contain a virus-encoded enzyme, termed **reverse transcriptase,** which catalyzes the synthesis of DNA using viral RNA as a template (page 75). The reverse transcriptase reaction generates a DNA molecule called a **provirus,** which is integrated into the host's chromosomal DNA and replicated along with it (Figure 18-15). Single-stranded RNA viruses that employ the reverse transcriptase–mediated pathway for viral replication and integration are known as **retroviruses.**

Thus for both DNA and RNA tumor viruses, the final result is usually the insertion of viral genetic information into the host cell chromosomal DNA. The integrated viral genes can remain inactive for varying periods of time, but they are eventually transcribed into products that cause the cell to become malignant.

Viral Oncogenes Are Altered Versions of Normal Cellular Genes

The fact that oncogenic viruses contain a relatively small number of genes has facilitated the identification of the viral genes that cause cells to become malignant. The first cancer-causing gene to be identified occurs in the *Rous sarcoma virus,* a small RNA virus that produces sarcomas in chickens. In the early 1970s Peter Vogt isolated mutant forms of the Rous virus that are incapable of causing cancer, even though they can still infect cells and undergo replication. Since the only missing function in these *transformation-defective mutants* is the ability to cause malignant transformation, it was concluded that the gene responsible for inducing malignancy is missing or defective. When the genetic composition of transformation-defective mutants was examined, a single gene was found to be absent. Because the missing gene normally allows the virus to induce sarcoma formation, it was named *src.* Genes like *src,* which are capable of inducing malignant transformation, are referred to as **oncogenes.**

Subsequent investigations have led to the surprising discovery that the *src* gene is not restricted to cancer cells. Using nucleic acid hybridization techniques, it has been shown that DNA sequences that are homologous to, but not identical with, the Rous *src* gene can be detected in normal cells of a wide variety of organisms, including salmon, mice, cows, birds, and humans. Since the evolutionary divergence of this group of organisms occurred hundreds of millions of years ago, it can be concluded that *src* gene sequences have been conserved for a large part of evolutionary history and hence must play an important role in normal cell function.

The unexpected discovery that cells contain DNA sequences that are closely related to viral oncogenes has been substantiated by studies on a variety of other tumor viruses. Oncogenes from more than two dozen cancer viruses have now been identified, and in each case they resemble genes present in normal cells. The term **proto-oncogene** has been introduced to refer to these normal cellular genes that closely resemble oncogenes. The resemblance of viral oncogenes to proto-oncogenes suggests that viral oncogenes may have originally been derived from normal cellular genes. According to this theory (Figure 18-16), the first step in the creation of retroviral oncogenes occurred millions of years ago when ancient viruses infected cells and became integrated in the host chromosomal DNA adjacent to normal cellular proto-oncogenes. When the integrated proviral DNA was later transcribed to regenerate new viral RNA molecules, the adjacent proto-oncogene sequences might have accidentally been transcribed as well. In this way, a viral RNA molecule containing normal proto-oncogene sequences could

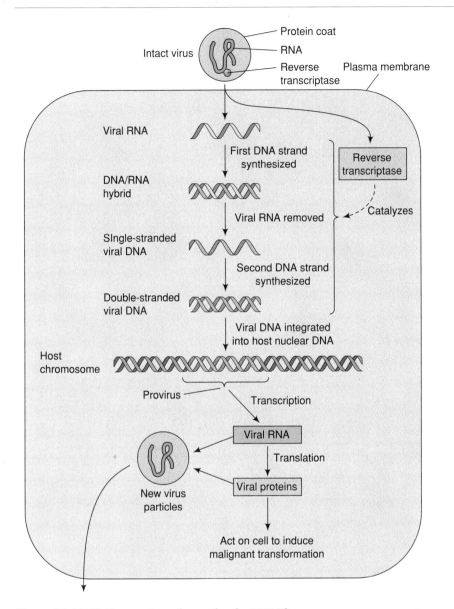

Figure 18-15 *Malignant Transformation by RNA Viruses* *The viral RNA is first copied by reverse transcriptase into a DNA molecule that is then integrated into the host chromosome. Only after integration is the viral genetic information transcribed into proteins that induce malignant transformation.*

have been created. Since a proto-oncogene would initially serve no useful purpose for a virus, it would be free to mutate during subsequent cycles of viral infection. Such mutation would eventually convert the proto-oncogene into an oncogene. This hypothetical scenario could explain why present-day viral oncogenes resemble altered versions of normal cellular genes.

Cellular Oncogenes Arise from the Mutation of Proto-Oncogenes

The realization that oncogenic viruses contain genes that cause cells to become malignant raises the ques-

tion of whether genetic alterations are also involved in nonvirus-induced cancers. The ability of many carcinogens to act as mutagens provides one reason for believing that genetic changes play a role in nonviral carcinogenesis. But the most compelling evidence has come from *gene transfer experiments* in which DNA sequences derived from tumor cells are artificially introduced into normal cells. The first experiments of this type, carried out in the laboratories of Robert Weinberg and Geoffrey Cooper in the early 1980s, revealed that DNA isolated from human bladder cancer cells can cause normal cells to become malignant (Figure 18-17). Gene-cloning procedures were then employed to isolate the cancer-causing gene, which

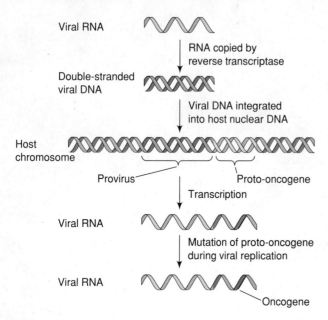

Viral RNA

RNA copied by
reverse transcriptase

Double-stranded
viral DNA

Viral DNA integrated
into host nuclear DNA

Host
chromosome

Provirus   Proto-oncogene

Transcription

Viral RNA

Mutation of proto-oncogene
during viral replication

Viral RNA

Oncogene

Figure 18-16 *A Hypothetical Model for the Origin of Oncogenes in RNA Tumor Viruses* *According to this model, ancient RNA viruses that infected normal cells millions of years ago became integrated in the host chromosome adjacent to proto-oncogenes. When the integrated proviral RNA was later transcribed, the cellular proto-oncogene might have been copied along with the viral genes. Subsequent mutation of the incorporated proto-oncogene could then convert it into an oncogene.*

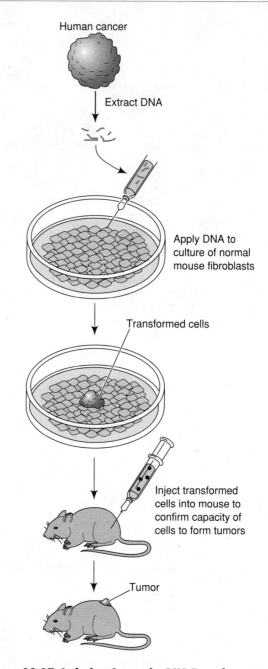

Human cancer

Extract DNA

Apply DNA to
culture of normal
mouse fibroblasts

Transformed cells

Inject transformed
cells into mouse to
confirm capacity of
cells to form tumors

Tumor

Figure 18-17 *Inducing Cancer by DNA Transfer* *When DNA isolated from human cancer cells is introduced into normal mouse fibroblasts, some of the normal fibroblasts undergo malignant transformation. Gene-cloning techniques can be used to isolate the gene sequences that are responsible for causing this transformation.*

turned out to be a member of a family of oncogenes called *ras.* Similar studies involving DNA isolated from a variety of different tumor types have uncovered the existence of other human oncogenes as well (Table 18-4). The term *cellular oncogene* is employed to distinguish this group of cancer-causing genes from viral oncogenes.

Since they are not brought into a cell by viral infection, the question arises as to how cellular oncogenes are created. Recent research suggests that cellular oncogenes are derived from normal proto-oncogenes by at least five different mechanisms:

1. **Point Mutation.** The simplest mechanism for converting a proto-oncogene into an oncogene involves a single base-pair substitution, or *point mutation*, in DNA that creates an oncogene coding for a protein product that differs in only a single amino acid from the protein encoded by the normal proto-oncogene. This phenomenon is frequently encountered in *ras* oncogenes, where point mutations at several locations in the corresponding proto-oncogene generate *ras* oncogenes coding for Ras proteins that contain single amino acid substitutions. [It is common practice to name oncogenes using a three-letter italicized abbreviation (e.g., *ras* oncogene) and to designate the protein

encoded by the oncogene using a capital letter without italics (e.g., Ras protein).]

2. **Local DNA Rearrangement.** The second mechanism for creating oncogenes is based on DNA rearrangements that cause either deletions or base-sequence exchanges between proto-oncogenes and surrounding genes. The *trk* oncogene, for example, is generated by a local DNA rearrangement that disrupts the 5' end of the

Table 18-4   Examples of Oncogenes Detected in Human Cancers

Oncogene	Tumor Type	Produced by
abl	Chronic myelogenous leukemia	Translocation
bcl-2	B-cell lymphoma	Translocation
erbB-1	Squamous carcinoma	Amplification
erbB-2	Carcinoma of breast, ovary, stomach	Amplification
gsp	Thyroid carcinoma	Point mutation
c-myc	Burkitt's lymphoma Breast, lung carcinoma	Translocation Amplification
N-myc	Neuroblastoma Lung carcinoma	Amplification
L-myc	Lung carcinoma	Amplification
H-ras	Colon, pancreas, lung carcinomas; melanoma	Point mutation
K-ras	Leukemias; thyroid carcinoma; melanoma	Point mutation
N-ras	Genitourinary, thyroid carcinoma; melanoma	Point mutation
ret	Thyroid carcinoma	Rearrangement
K-sam	Stomach carcinoma	Amplification
trk	Thyroid carcinoma	Rearrangement

trk proto-oncogene, thereby leading to the production of a *fusion protein*—that is, a hybrid protein in which the N-terminal end of the Trk protein is replaced by an amino acid sequence encoded by a neighboring gene.

3. **Insertional Mutagenesis.** The discovery of the third mechanism for creating oncogenes emerged from the surprising finding that some cancer-causing retroviruses lack oncogenes. These particular viruses cause cancer by integrating a DNA copy of their genetic information into a host chromosome in a region where a proto-oncogene is located, disrupting the structure of the host proto-oncogene and thereby converting it into an oncogene.

4. **Gene Amplification.** The fourth mechanism for creating oncogenes utilizes *gene amplification* (page 456) to increase the number of copies of a particular proto-oncogene. The increased number of gene copies leads to overproduction of the normal protein encoded by the proto-oncogene rather than formation of an abnormal protein product.

5. **Chromosomal Translocation.** The final mechanism for creating an oncogene involves *chromosomal translocation,* a process in which a portion of one

chromosome is physically removed and joined to another chromosome. A classic example occurs in Burkitt's lymphoma, where a piece of chromosome 8 is often translocated to chromosome 14. As a result, the *myc* proto-oncogene is transferred from its normal location on chromosome 8 to chromosome 14, where it becomes situated near the genes coding for antibody molecules. Because antibody genes are extremely active in lymphocytes, the transcriptional regulation of the adjacent *myc* proto-oncogene is disturbed, resulting in an abnormal pattern of synthesis of the Myc protein product.

Oncogenes Code for Proteins Involved in Normal Growth Control

We have now seen that alterations in normal genes called proto-oncogenes can convert them into oncogenes that code for proteins that are either abnormal in structure or are produced in inappropriate amounts. But how do these oncogene-encoded proteins cause cancer? The first clue came from studies involving the *src* oncogene, whose protein product was identified in 1977 by Ray Erikson. In these studies, rabbits were first immunized with chicken sarcoma tissue to produce antibodies directed against the cancer cells; the resulting antibodies were then shown to bind to a protein in sarcoma cells called $p60^{v\text{-}src}$. Because $p60^{v\text{-}src}$ could not be detected in normal cells or cells infected with transformation-defective strains of the Rous virus, it was concluded that $p60^{v\text{-}src}$ is the product of the *src* oncogene. Subsequent studies revealed that $p60^{v\text{-}src}$ catalyzes the phosphorylation of the amino acid tyrosine in protein molecules, and hence is a **protein-tyrosine kinase.**

Shortly after the discovery of $p60^{v\text{-}src}$, a similar enzyme was identified in normal cells. To distinguish it from the viral enzyme, the cellular protein-tyrosine kinase is called $p60^{c\text{-}src}$ (the c stands for "cellular"). The existence of $p60^{c\text{-}src}$ in normal cells indicates that the Rous virus triggers the production of a viral protein-tyrosine kinase that closely resembles a normal cellular enzyme of the same type. However, the amount of viral protein-tyrosine kinase activity detected in Rous-infected cells is generally 30- to 50-fold higher than the normally prevailing levels of a cell's own protein-tyrosine kinase activity. This finding implies that malignant transformation by the Rous virus is associated with the excessive phosphorylation of cellular proteins by the viral enzyme.

In the years following the discovery of $p60^{v\text{-}src}$, the proteins encoded by a variety of other oncogenes have been identified as well (Table 18-5). Although oncogene-encoded proteins exhibit a variety of different functions, most fall into one of five basic categories: *growth factors, receptor and nonreceptor protein-tyrosine kinases, membrane-associated G proteins, protein-serine/threonine kinases,* and *transcription*

Table 18-5   Main Classes of Oncogenes Categorized by the Nature of Their Protein Products

Nature of Protein Product	Examples of Oncogenes	Comments
Growth factors	*sis*	Platelet-derived growth factor (PDGF)
Protein-tyrosine kinases	*erbB*	Membrane receptor for epidermal growth factor (EGF)
	fms	Membrane receptor for colony-stimulating factor-1 (CSF-1)
	src, yes, fgr	Membrane nonreceptor protein-tyrosine kinases
Membrane-associated G proteins	*ras*	Membrane-associated GTP-binding protein
	gsp	G_s (α subunit)
	gip	G_i (α subunit)
Protein-serine/threonine kinases	*raf, mos*	Cytoplasmic protein-serine/threonine kinases
Transcription factors	*jun, fos*	Components of AP1 transcription factor
	erbA	Thyroid hormone receptor

factors. The characteristics of these five groups of oncogene-encoded proteins are briefly described below.

1. Growth factors.   The growth and division of normal cells is regulated by polypeptide growth factors that induce the proliferation of specific cell types (page 526). It is therefore easy to envision how an oncogene that codes for an abnormal form of a growth factor (or too much growth factor) might trigger uncontrolled cell proliferation. Among the oncogenes that function in this way are *sis,* which codes for one of the polypeptide chains of platelet-derived growth factor (PDGF), and *int-2, hst,* and *fgf-5,* which code for fibroblast growth factors.

2. Receptor and nonreceptor protein-tyrosine kinases.   Following the discovery that the *src* oncogene codes for a protein-tyrosine kinase, more than 20 other oncogenes have also been found to code for protein-tyrosine kinases. These oncogene-encoded tyrosine kinases can be subdivided into two broad families: *receptor protein-tyrosine kinases* and *nonreceptor protein-tyrosine kinases*. **Receptor protein-tyrosine kinases** are transmembrane proteins that contain a growth factor receptor domain exposed on the outer surface of the plasma membrane and a tyrosine kinase catalytic domain at the inner surface of the plasma membrane (see Figure 6-26, *top*). In a normal receptor of this type, binding of the appropriate growth factor to the receptor site activates the protein-tyrosine kinase domain, triggering a series of events that leads to the stimulation of cell growth and division. Oncogenes can code for abnormal receptor protein-tyrosine kinases in which the growth factor binding site is disrupted, leading to unregulated activity of the protein-tyrosine kinase site.

In contrast to receptor protein-tyrosine kinases, which span the plasma membrane, **nonreceptor protein-tyrosine kinases** are usually bound to the membrane's cytoplasmic surface or free in the cytosol. The protein encoded by the *src* gene is a prominent example of a nonreceptor protein-tyrosine kinase. Oncogene-encoded nonreceptor kinases often exhibit excessive, unregulated protein-tyrosine kinase activity.

3. Membrane-associated G proteins.   Another group of oncogenes code for plasma membrane **G proteins** that are involved in transmitting cell-surface regulatory signals (page 211). In human cancers, the most commonly encountered oncogenes of this type are members of the *ras* family. *Ras* oncogenes typically arise from point mutations in the *ras* proto-oncogene; the result is the production of mutant **Ras** G proteins that retain bound GTP instead of hydrolyzing it to GDP, thereby keeping the Ras protein in its activated form. As we will see shortly, the Ras protein is a plasma membrane G protein that plays a central role in the transmission of signals from external growth factors to the cell interior.

4. Protein-serine/threonine kinases.   Protein-tyrosine kinases are not the only protein phosphorylating enzymes to be implicated in the mechanism of cancer induction. Most of the protein kinase activity exhibited by mammalian cells catalyzes the phosphorylation of the amino acids serine and threonine, not tyrosine. These **protein-serine/threonine kinases** are soluble cytoplasmic proteins that, like protein-tyrosine kinases, can also be encoded by oncogenes. A prominent oncogene in this group is the *raf* oncogene, which codes for a protein-serine/threonine kinase that transmits signals from plasma membrane Ras proteins to the cell interior.

5. Transcription factors. During the control of normal cell growth and division, extracellular signaling molecules activate intracellular pathways that in turn trigger changes in the expression of genes located in the nucleus. It is not surprising, therefore, that some oncogenes code for proteins that function within the nucleus, predominantly in the regulation of gene transcription. The evidence for such a role has been most clearly demonstrated for the oncogenes *jun* and *fos,* which code for proteins that make up the *AP1 transcription factor* (page 464). The AP1 factor regulates the expression of a group of genes that are involved in stimulating cell proliferation. The *myc* oncogene, which is associated with several kinds of human cancer, also appears to code for a transcription factor.

Although most oncogenes code for proteins that fall into one of the five preceding categories, a small number do not fit into this scheme. Among them are oncogenes coding for plasma membrane receptors that lack protein kinase activity, proteins associated with the cytoskeleton, and cytoplasmic regulatory proteins. Like other oncogene-encoded proteins, these molecules are components of normal pathways for regulating cell growth and division.

Oncogenes Act at Several Different Points in the Same Growth Control Pathway

The realization that oncogenes code for abnormal versions (or abnormal quantities) of proteins involved in normal growth control pathways has greatly enhanced our understanding of how cancer arises. Although the precise sequence of steps leading from the expression of any particular oncogene to the onset of malignancy is yet to be completely elucidated, the general outline of at least one such pathway is beginning to emerge. This pathway, illustrated in Figure 18-18, involves the regulation of cell proliferation by growth factors such as platelet-derived growth factor (PDGF) and epidermal growth factor (EGF), which bind to plasma membrane receptors exhibiting protein-tyrosine kinase activity. The binding of one of these growth factors to its corresponding membrane receptor activates the receptor's protein-tyrosine kinase activity, causing the receptor to phosphorylate itself. The phosphate groups attached to the receptor by this *autophosphorylation* event then serve as binding sites for a variety of cytoplasmic proteins. Among the molecules that bind to these sites are several *adaptor proteins,* which in turn activate the membrane-associated G protein, Ras. Upon activation the Ras protein stimulates Raf, a cytoplasmic protein-serine/threonine kinase that phosphorylates and activates another protein-serine/threonine kinase (Mek), which phosphorylates and activates yet other protein-serine/threonine kinases called *MAP kinases* (page 527).

Activated MAP kinases then phosphorylate a variety of proteins involved in the stimulation of cell growth and division. Among the proteins that become phosphorylated is the Jun protein, a component of the nuclear AP1 transcription factor; activation of AP1 turns on the transcription of genes involved in stimulating cell proliferation.

The preceding model shows how oncogenes coding for growth factors, protein-tyrosine kinases, membrane-associated G proteins, protein-serine/threonine kinases, and transcription factors all act at different points in the same growth control pathway. *By producing abnormal versions (or abnormal quantities) of proteins of these five types, oncogenes can disrupt normal growth control and trigger excessive cell proliferation.* For the sake of simplicity, the growth control mechanism summarized in Figure 18-18 focuses on only one of several target proteins activated by receptor protein-tyrosine kinases, namely the Ras protein. In addition to activating Ras, protein-tyrosine kinases also trigger the phosphorylation and activation of several other proteins, including *phospholipase C.* As we learned in Chapter 6, activation of phospholipase C stimulates the formation of the second messengers diacylglycerol and inositol trisphosphate (see Figure 6-23), which in turn signal a variety of cha[...] behavior. Besides p[...] oncogene-encode[...] phosphorylate oth[...] the behavior of canc[...] kinases phosphorylate[...] cellular matrix (e.g., *vin[...]* nomenon that may co[...] adhesiveness of cancer cells[...] tyrosine kinases is the gap ju[...] whose phosphorylation leads to[...] munication. Such examples illustra[...] havior of an oncogene-encoded pr[...] might contribute to numerous properti[...] malignant growth.

Cancer Can Be Induced by the Loss of Tumor Suppressor Genes That Normally Inhibit Cell Proliferation

We have now seen how the presence of an oncogene can stimulate cell growth and division, thereby fostering the development of malignancy. A second class of genes, called **tumor suppressor genes,** have also been implicated in the causation of cancer. But in contrast to oncogenes, whose *presence* contributes to malignancy, it is the *absence* or loss of function of a tumor suppressor gene that leads to malignancy. The term "tumor suppressor gene" implies that the normal function of genes of this type is to restrain cell growth and division. In other words, tumor suppressor genes act as brakes on the process of cell proliferation, whereas oncogenes function more like an accelerator.

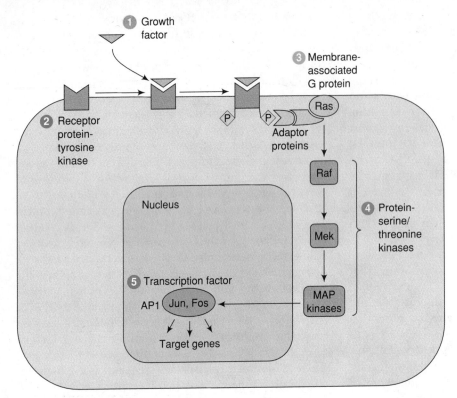

Figure 18-18 *Some of the Steps in Normal Growth Control That Can Be Disrupted by Oncogenes*
This model summarizes the pathway by which growth factors such as PDGF and EGF stimulate cell proliferation through activation of the membrane-associated G protein, Ras. Activation of Ras triggers the phosphorylation of a series of cytoplasmic protein-serine/threonine kinases, thereby leading to phosphorylation of the nuclear AP1 transcription factor, which in turn activates genes involved in stimulating cell proliferation. Oncogenes code for members of all five classes of proteins involved in this pathway: (1) growth factors, (2) receptor protein-tyrosine kinases, (3) membrane-associated G proteins, (4) protein-serine/threonine kinases, and (5) transcription factors. By coding for abnormal versions of these proteins (or by producing too much or too little of these proteins), oncogenes disrupt normal growth control and trigger excessive cell proliferation.

The first tumor suppressor gene to be identified is involved in *hereditary retinoblastoma,* a rare type of eye cancer that develops in young children who have a family history of the disease. Such children inherit a chromosomal deletion in a specific region of one copy of chromosome 13. Although the deletion occurs in all cells, only a few cells in the retina actually become malignant because the initial deletion in chromosome 13 does not cause cancer by itself; for cancer to develop, a subsequent mutation must also occur in the same region of the second copy of chromosome 13. It has therefore been concluded that chromosome 13 contains a gene that normally functions to inhibit cell division, and that deletion or disruption of both copies of the gene must occur before cancer will develop.

The gene lost in hereditary retinoblastoma, called *RB-1,* is a tumor suppressor gene that codes for the nuclear protein *pRB,* a transcriptional repressor that inhibits expression of a group of genes whose products are required for cell proliferation. Hence a lack of pRB resulting from the loss or disruption of both copies of the *RB-1*

gene removes a normal restraining mechanism that functions to inhibit cell growth and division. Independent support for the idea that a loss of pRB can lead to uncontrolled proliferation has come from studies involving DNA tumor viruses such as adenovirus, SV40, and human papillomavirus. Each of these viruses has an oncogene that codes for a protein product that binds to and inactivates pRB, thereby interfering with the ability of pRB to repress transcription and inhibit cell proliferation. Hence cancers triggered by a loss of pRB can be induced in at least two different ways: through mutations that delete or disrupt both copies of the *RB-1* gene, and through viral-encoded proteins that bind to and inactivate pRB.

Following the discovery of the *RB-1* gene, several other tumor suppressor genes have also been implicated in the causation of human cancer. One of the most intriguing is *p53,* a tumor suppressor gene whose protein product is inactivated or deleted in about half of all human malignancies. The p53 protein is a nuclear transcription factor that turns on the activity of genes that arrest cells in the G_1 phase of the cell cycle and, under some circumstances, trigger cell death. Normally the production of the p53 protein is stimulated by treatments that cause DNA damage, such as exposure to ultraviolet light or DNA-damaging drugs. Hence p53 appears to act like a molecular policeman that checks the cell for DNA damage and prevents the cell from proliferating if damage is detected. The loss of p53 function presumably contributes to malignancy by allowing the survival and reproduction of cells in which DNA damage has led to the production of oncogenes and/or the loss of other tumor suppressor genes.

Human Cancers Develop by the Stepwise Accumulation of Mutations Involving Both Oncogenes and Tumor Suppressor Genes

Although abnormalities in the *p53* gene have been detected in a majority of all human cancers, this does not mean that the loss of *p53* by itself is responsible for these malignancies. The most frequent types of human cancer, including colon, lung, and breast cancer, are produced by multiple mutations that involve the inactivation of tumor suppressor genes as well as the conversion of proto-oncogenes into oncogenes. In other words, creating a cancer cell requires that the brakes on cell growth (tumor suppressor genes) be released at the same time as the accelerators for cell growth (oncogenes) are being activated.

In human colon cancer, the mutational steps that convert a normal epithelial cell into a cancer cell have been investigated by examining the genetic alterations exhibited by tumor samples obtained from a variety of patients. The results suggest that colon cancer is often characterized by the presence of a *ras* oncogene as well as deletions of specific portions of chromosomes 5, 17, and 18. The chromosome deletions appear to be responsible for the loss of tumor suppressor genes; in the case of chromosome 17, for example, the lost gene has been identified as *p53*. The most rapidly growing, metastatic colon cancers tend to exhibit all four genetic alterations (i.e., the *ras* oncogene plus the three chromosomal deletions). In contrast, benign tumors have acquired only one or two of the changes. This pattern suggests that cancers develop by a stepwise accumulation of mutations affecting both proto-oncogenes and tumor suppressor genes, and that cells acquiring the most mutations produce the most dangerous malignancies.

The number of mutations associated with colon cancer is too high to be easily accounted for by normal rates of spontaneous mutation. A possible explanation for the high mutation rate has emerged from the discovery that some colon cancers exhibit abnormalities in genes whose disruption leads to the accumulation of mutations in other genes. One such gene, called *hMSH2*, codes for a protein than normally functions in *mismatch repair,* a type of DNA repair mechanism that detects and corrects

base pairs that are improperly hydrogen-bonded (page 94). Mutations in *hMSH2* lead to an increase in the spontaneous mutation rate for other genes, a phenomenon that may help to explain how colon cancers acquire multiple mutations involving both proto-oncogenes and tumor suppressor genes.

An Increased Risk of Developing Cancer Can Be Inherited

The discovery of oncogenes and tumor suppressor genes has provided strong evidence for the role played by genetic changes in converting a normal cell into a cancer cell. As we discussed earlier in the chapter, one way of inducing these genetic changes is through exposure to carcinogenic agents such as radiation, chemicals, and viruses. Alternatively, some of the genetic alterations that are involved in producing cancer can be inherited. Since multiple mutations are usually required to produce a malignancy, a genetic defect that is inherited tends to increase an individual's risk of developing cancer rather than causing cancer by itself. Table 18-6 summarizes some of the inherited defects that are known to increase a person's risk of developing cancer. Although most of these inherited conditions are

Table 18-6	Genetically Inherited Disorders Associated with an Increased Risk of Developing Cancer
Genetic Disorder	**Type of Cancer**
Familial breast cancer	Breast
Xeroderma pigmentosum	Skin
Familial adenomatous polyposis	Colon, rectum
Hereditary retinoblastoma	Retinoblastoma
Familial hydronephrosis	Kidney
Fibrocystic pulmonary dysplasia	Lung
Bloom's syndrome	Leukemia, intestines
von Recklinghausen's neurofibromatosis	Fibrosarcoma, neuroma
Ataxia-telangiectasia	Leukemia, stomach, brain
Paget's disease of bone	Bone
Fanconi's anaplastic anemia	Leukemia, liver, skin
Bruton's agammaglobulinemia	Leukemia
Down syndrome	Leukemia
Phaeochromocytoma	Adrenal medulla
Tylosis with esophageal cancer	Esophagus
Chediak-Higashi syndrome	Retina
Multiple endocrine adenomatosis	Pancreas, pituitary, adrenal cortex, parathyroid

relatively rare, they have provided some important insights into the mechanisms of carcinogenesis. For example, individuals with *xeroderma pigmentosum* inherit a defect in one of the enzymes involved in DNA repair. Because the skin cells of such persons are less able to repair DNA damage resulting from exposure to sunlight, the risk of developing skin cancer is increased.

Insights into the genetic mechanisms that influence cancer risk have also emerged from studies of *familial adenomatous polyposis coli (APC)*, an inherited condition associated with an elevated risk of developing colon and rectal cancer. Cells isolated from individuals inheriting this condition do not form tumors when injected into nude mice, indicating that the cells are not malignant. But if the same cells are first grown in culture in the presence of a phorbol ester, which functions as a promoting agent, tumors do form when the cells are subsequently injected into nude mice. Since phorbol esters by themselves do not normally cause cells to become malignant, it has been concluded that the cells of individuals who inherit APC have already undergone initiation; hence treatment with a promoting agent is sufficient to cause malignancy. The defective gene responsible for APC has recently been identified and tentatively classified as a tumor suppressor gene.

In terms of the total number of individuals affected, the most important hereditary cancer risk is associated with breast cancer. The probability that the average woman in the United States will eventually develop breast cancer is about 1 in 10, but among individuals who have a blood relative with the disease, the risk is closer to 1 in 5. One of the altered genes that increases a woman's chances of developing breast cancer has recently been identified, which means that genetic tests will soon be available for detecting women who are at greatest risk.

Malignancy Is Not Always an Irreversible Change

The discovery of the prominent role played by genetic damage in the development of cancer has helped foster the view that malignancy is an irreversible change. Yet some evidence suggests that malignant cells can, under appropriate conditions, revert to nonmalignant behavior. An example of such reversion occurs in **crown gall disease,** a malignant tumor of plants caused by infection with bacteria that transmit the cancer-inducing *Ti plasmid* (page 107). The plasmid DNA is integrated into the plant cell chromosomal DNA, where it codes for the synthesis of enzymes involved in the synthesis of the plant growth substances, auxin and cytokinin (page 527). In studies carried out by Armin Braun and his colleagues, crown gall tumor cells were grown in tissue culture in the presence of growth factors that promote the development of structures resembling plant shoots. The tumor shoots were then grafted to the cut stems of healthy plants, where they responded to the new environment by developing into stems that produced leaves, flowers,

and even seeds capable of generating new plants. No sign of cancer was apparent. Yet when cells taken from the newly growing leaves were placed in culture, they once again began to grow like tumors. These cells had therefore retained the genetic alteration specifying the malignant state, but the expression of this information apparently depends on environmental conditions.

The reversibility of the malignant state is not restricted to plant tumors. A dramatic example of a potentially reversible animal malignancy is the mouse **teratocarcinoma,** a cancer that develops from primordial germ cells in the ovary or testes. This type of cancer, which also occurs in humans, contains differentiated cell types such as muscle, bone, cartilage, nerve, and skin, intermixed with undifferentiated malignant cells called *embryonal carcinoma cells* (Figure 18-19). In an elegant study carried out in the laboratory of Beatrice Mintz, embryonal carcinoma cells were isolated from a teratocarcinoma occurring in one strain of mouse and injected into a normal blastula obtained from another type of mouse. When the blastulae were reimplanted into pregnant animals and allowed to grow to maturity, they developed into normal adult mice exhibiting no signs of cancer (Figure 18-20). Moreover, genetic analysis revealed that many of the normal cells in these mice had descended from the original cancer cells. Thus the original embryonal carcinoma cells must have differentiated into nonmalignant cells when placed in the environment of a normal developing embryo.

Further support for the idea that malignancy is potentially reversible has come from studies carried out by Robert McKinnell on the Lucké adenocarcinoma of frogs. In these experiments, nuclei isolated from adenocarcinoma cells were transplanted into normal frog eggs whose own nuclei had previously been removed. The eggs were found to develop into normal tadpoles and then adult frogs, even though the nucleus of every cell in these adult frogs had descended from a cancer cell nucleus. Since the Lucké tumor is caused by a DNA tumor virus and thus involves genetic changes, these genetic alterations can apparently be overridden when nuclei are placed in an appropriate cytoplasmic environment.

The preceding experiments indicate that even when genetic damage lies at the heart of carcinogenesis, changing environmental conditions can override the expression of information specifying malignant traits. This potential for controlling the expression of the malignant state raises the possibility of new approaches for treating cancer. Instead of restricting cancer therapy to methods for killing or surgically removing cancer cells, it may eventually be possible to devise methods for converting malignant cells into nonmalignant ones.

CAN CANCER BE PREVENTED OR CURED?

Most cancer research is motivated by the belief that a better understanding of the disease will ultimately lead

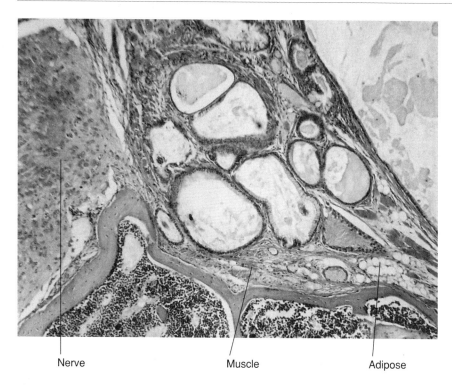

Nerve   Muscle   Adipose

Figure 18-19 *Light Micrograph of a Mouse Testicular Teratocarcinoma Among the different tissues present within the tumor are nerve tissue, adipose tissue, and skeletal muscle. Each of these differentiated tissues is derived from the embryonal carcinoma cells of the tumor itself. Courtesy of L. C. Stevens.*

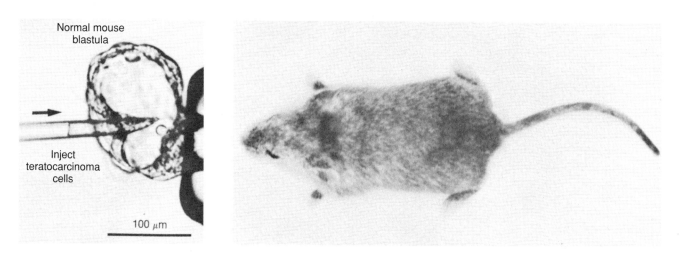

Figure 18-20 *Fate of Teratocarcinoma Cells Injected into a Mouse Blastula* (Left) *Single teratocarcinoma cells being injected into a mouse blastula.* (Right) *Adult mouse that developed from a blastula injected with teratocarcinoma cells. The dark patches of coat color are normal skin that arose from the teratocarcinoma cells, indicating that the injected cancer cells differentiate into nonmalignant cells when inserted into a developing embryo. Courtesy of B. Mintz and K. Illmensee.*

to improved methods for its prevention and treatment. Until recently, research on cancer treatments received more attention than efforts to prevent cancer, but strategies for cancer prevention are beginning to attract more interest now that we know that most cancers are induced by potentially identifiable environmental carcinogens.

Epidemiological Data Allow Potential Carcinogens to Be Identified in Exposed Human Populations

The first step in preventing cancer is to identify the agents that cause cancer. To take a simple example, the discovery that tobacco smoke causes lung cancer provides a straightforward strategy for preventing most cancers of this type; that is, simply avoid tobacco smoke. Likewise, the discovery of the carcinogenic properties of X-rays and sunlight suggests some relatively simple approaches for reducing radiation-induced cancers. Individuals should just avoid unnecessary medical X-rays and use protective skin lotions during prolonged exposure to sunlight.

The greatest impact in avoiding other kinds of cancer-causing agents could be made by eliminating the strongest chemical carcinogens currently found in our food, air, water, clothing, and drugs. But in order to remove these carcinogens from the environment, one must first have a reliable way of identifying them. One way of detecting potential human carcinogens, called the *epidemiological* approach, is based on a comparison of cancer rates among various groups of people exposed to differing environmental conditions. A striking finding to emerge from epidemiological studies is that cancers occur with differing frequencies in different parts of the world. For example, stomach cancer is especially frequent in Japan, breast cancer is prominent in the United States, rectal cancer rates are high in Denmark, and esophageal cancer is prevalent in Iran. In theory, differences in either hereditary or environmental factors might be responsible for the differences in cancer incidence observed in different countries. Cancer rates in people who have moved from one country to another suggest that of the two factors, environment is more important. The most striking data have been collected on Japanese immigrants to the United States. In Japan, the incidence of stomach cancer is greater and the incidence of colon cancer is lower than in the United States. But the difference gradually disappears when Japanese families move to the United States (Figure 18-21), indicating that the risk of developing the two kinds of cancer is determined largely by environmental factors.

Epidemiological data have played an important role in identifying some of the environmental factors that may cause cancer. The most striking statistics have been obtained for lung cancer, a disease that has increased over tenfold in frequency since 1930 (Figure 18-22). When the environmental factors responsible for the epidemic of lung cancer were investigated, it was discovered that virtually all lung cancer patients share one thing in common: They smoke cigarettes. Furthermore, heavy smokers develop lung cancer more frequently than light smokers, and long-term smokers develop lung cancer more frequently than do short-term smokers. Such epidemiological correlations suggest that lung cancer is caused by cigarette smoking, a conclusion that has been verified by the discovery that cigarette smoke contains several dozen carcinogenic chemicals (see Table 18-2).

The main drawback to the epidemiological approach is that it only allows potential carcinogens to be identified after large numbers of cancer cases have appeared. The usefulness of epidemiology is also limited by the long periods of time that usually intervene between exposure to a carcinogen and the development of cancer. In the case of cigarette smoking and lung cancer, for example, more than 20 years elapsed before the increase in cigarette smoking was reflected in an increased rate of lung cancer.

The Ames Test Is a Rapid Screening Method for Identifying Potential Carcinogens

Since the epidemiological approach only allows suspected carcinogens to be identified after large numbers of people have developed cancer, methods are needed for identifying potential carcinogenic hazards before they do any damage. For this reason, the cancer-causing ability of many chemicals has been assessed in laboratory tests. Until recently, such testing was done almost entirely by administering potential carcinogens to laboratory animals. But animal experiments usually take several years to complete and the

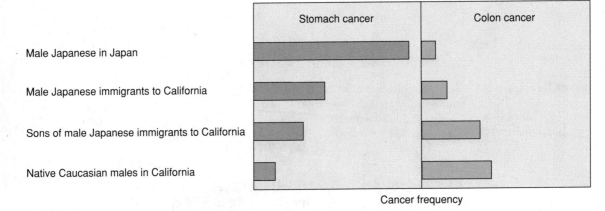

Figure 18-21 *Comparison of the Frequency of Stomach and Colon Cancer in Japan, in the United States, and in Japanese Immigrants to the United States* *When Japanese individuals move to the United States, their susceptibility to developing stomach and lung cancer changes to reflect the rates for such cancers in the United States. These data suggest that environmental factors play a prominent role in causing cancer.*

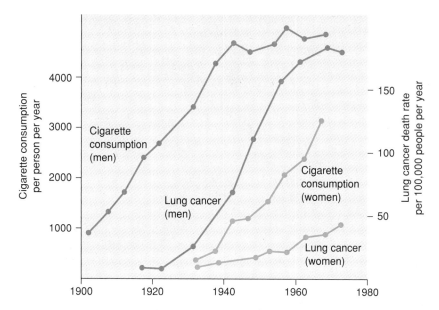

Figure 18-22 *Correlation between Smoking Rates and Lung Cancer in Great Britain*
The data show that a period of about 25 years intervened between an increase in cigarette smoking and an increased incidence of lung cancer. Cigarette smoking and lung cancer are more recent phenomena in women than in men.

data they yield can be difficult to extrapolate to humans. Therefore animal testing by itself is impractical for assessing the dangers posed by the thousands of chemicals present in the environment.

To circumvent this problem, Bruce Ames has devised a rapid screening procedure for the preliminary identification of potential cancer-causing agents. This procedure, which is based on the rationale that most carcinogens act as mutagens, measures the ability of substances to induce mutations in a special strain of bacteria that lack the ability to synthesize the amino acid histidine (Figure 18-23). A sample of bacteria is placed in a culture dish containing the substance to be tested, and the bacteria are then provided with a growth medium that lacks histidine. Normally such bacteria will not grow in the absence of histidine. But if the substance being tested is mutagenic, some of the random mutations that occur will restore the ability of the cells to synthesize histidine. Each bacterium acquiring such a mutation will grow into a visible colony. The total number of colonies is therefore a direct measure of the mutagenic potency of the substance being tested.

The usefulness of this approach for detecting potential carcinogens is hampered by the fact that many of the chemicals to which humans are exposed only become carcinogenic after they have been modified in the liver. As a consequence, chemicals being tested in the Ames procedure are first incubated with a liver homogenate so that such modification reactions can take place before the assay for mutagenicity. When tested in this way, the mutagenic potency of substances assayed by the Ames procedure has been found to correlate well with carcinogenic potency in animals (Figure

18-24). Although it is far from foolproof, the Ames test therefore provides a rapid and inexpensive screening procedure for the preliminary identification of potential carcinogens.

Minimizing Exposure to Carcinogens Requires Changes in Life-Style

Identifying potential carcinogens by epidemiological analysis and/or laboratory testing is only the first step in assessing the danger posed by substances present in the environment. Once a particular carcinogen has been pinpointed, the question arises as to how much of the substance can be tolerated before a clear hazard exists. This question is an extremely difficult one to answer experimentally because small doses of carcinogens produce very few tumors, and therefore it is difficult to detect them statistically. Yet a thorough understanding of the effects of low doses of carcinogenic agents is very important because of its relevance for public policy decisions. Some investigators believe that a threshold level exists for each carcinogen, and that doses below the threshold level do not cause cancer. If this were true, it would significantly influence our views about the regulation of environmental carcinogens. Unfortunately, the lack of reliable data on the effects of low doses of carcinogens leaves the point unresolved, and it would therefore be prudent to consider any dose of carcinogen as a cancer risk until proved otherwise.

In 1958 the United States Congress passed the Delaney Amendment, which instructed the Food and Drug Administration (FDA) to prohibit food additives known to cause cancer in humans or laboratory animals. Under

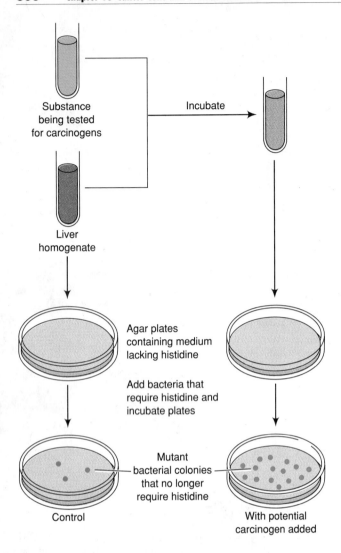

Figure 18-23 *The Ames Test for Identifying Potential Carcinogens* *This procedure, which is based on the rationale that most carcinogens act as mutagens, measures the ability of potential carcinogens to induce mutations in a strain of bacteria that lack the ability to synthesize the amino acid histidine. Each bacterial cell that has mutated to a form in which it no longer requires histidine will grow into a colony that can be counted. The number of colonies is therefore related to the mutagenic potency of the substance being tested. Chemicals being investigated in the Ames test are first incubated with a liver homogenate because many of the chemicals to which humans are exposed only become carcinogenic after they have undergone biochemical modification in the liver.*

this act the FDA has banned substances such as cyclamate and Red Dyes #2 and #4. However, each case has been accompanied by a long and controversial debate about the adequacy of the data. Such debates will no doubt continue because enormous pressure is put on Congress by groups whose economic interests would be adversely affected by publicizing the risks associated with particular substances. One need only consider the example of the tobacco industry, whose political and economic power has hampered regulation of a product implicated in the deaths of several hundred thousand cancer victims a year. Treatment of these cancer victims costs society billions of dollars annually, not to mention the billions of dollars in lost earnings and productivity. A society that condones such a state of affairs does not appear to be committed to the notion of preventing cancer.

The tobacco problem also illustrates the point that millions of people are not willing to modify their lifestyle in order to reduce the risk of developing cancer. If people were to voluntarily stop smoking, government regulation of tobacco would become an irrelevant issue. And tobacco is not the only life-style factor that influences cancer risk. Extensive meat consumption, for example, has been correlated with an increased risk of developing colon cancer (Figure 18-25), and animal studies have pointed to the high fat content of meat as the probable causative factor. Hence careful attention to personal habits such as smoking and diet, along with avoidance of occupations that involve exposure to strong carcinogens, would allow the average person to significantly reduce his or her risk of developing cancer.

Dietary Antioxidants May Help to Counteract the Effects of Some Carcinogens

No matter how seriously society pursues the goal of preventing exposure to cancer-causing agents, it is impossible to eliminate every environmental carcinogen. Even if all human-made carcinogens were to be banned from production, natural carcinogens still exist. Environmental sources of radiation, such as cosmic rays and sunlight, pose a cancer risk, as do naturally occurring carcinogenic chemicals in the environment. Even natural foods, such as vegetables and fruits, contain substances that may be carcinogenic. For example, the *aflatoxins,* a group of poisonous substances secreted by the mold *Aspergillus,* are potent inducers of liver cancer. *Aspergillus* grows readily on nuts and grains in humid tropical climates, causing an exceptionally high incidence of liver cancer in such areas.

Since it is not feasible to prevent exposure to every known carcinogen, the question arises as to whether anything can be done to counteract the toxic effects of carcinogens that we can't avoid. One possible approach is suggested by the discovery that many naturally occurring carcinogens share one property in common: They interact with other molecules to produce highly reactive substances called **free radicals,** which are missing an electron. In an attempt to replace the missing electron, free radicals attack other molecules and strip away an electron, thereby creating additional free radicals, which in turn repeat the process. The net result is a chain reaction that can damage a variety of cellular molecules, including DNA. One way of preventing this

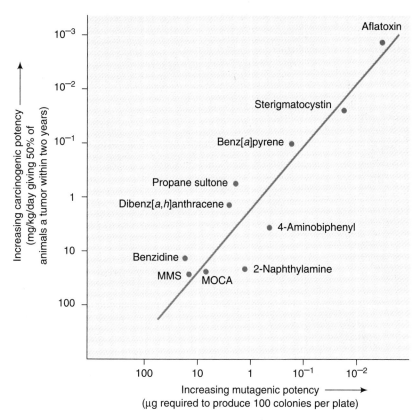

Figure 18-24 *Correlation between Carcinogenic Potency in Animals and Mutagenic Potency in the Ames Test for Ten Different Substances*   *The data reveal that substances which are strong mutagens also tend to be strong carcinogens. (Abbreviations: MOCA = 4-4' methylene-bis2-chloroanaline, MMS = methyl methanesulfonate.)*

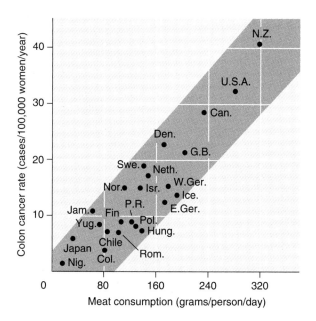

Figure 18-25 *Correlation between Meat Consumption and Colon Cancer*   *Each point represents a different country. The high fat content of meat is thought to be responsible for the elevated risk of developing colon cancer.*

chain reaction involves the use of **antioxidants,** which are molecules that give up electrons to free radicals without becoming harmful themselves. Several vitamin components of the diet can act as antioxidants, most notably vitamins A, C, and E. Epidemiological studies suggest that dietary consumption of such vitamins is correlated with a reduced risk of developing certain kinds of cancer, indicating that proper doses of dietary antioxidants may help to counteract the effects of some carcinogens.

Surgery, Radiation, and Chemotherapy Are the Conventional Treatments for Cancer

Because we do not yet know how to prevent many kinds of cancer, effective methods for treating cancer are badly needed. The three treatment approaches that have been most successful to date are surgery, chemotherapy, and radiation. *Surgery* is utilized to remove tumors that are localized to a particular region of the body. Surgical excision is most effective when the cancer has been diagnosed early enough to minimize the chances that metastasis has already occurred. Unfortunately, cancers involving internal organs often do not produce symptoms until they are disseminated throughout the body, and consequently the effectiveness of the surgical approach is reduced. It is for this reason that early detection of cancer is so important.

Treating cancer with *radiation* is based on the fact that cells engaged in DNA synthesis or mitosis are particularly sensitive to the killing effects of X-irradiation. Hence radiation is an effective way of killing cells that are actively proliferating. The main problem with this approach is that normal dividing cells, such as blood-forming cells in the bone marrow or cells lining the gastrointestinal tract, are destroyed along with cancer cells, thereby limiting the dose of radiation that can be safely employed. Moreover, an additional risk is posed by the fact that radiation itself is carcinogenic. But in spite of these limitations, radiation is quite effective in treating certain kinds of cancer, including Hodgkin's disease (a kind of lymphoma), skin cancer, and specific forms of testicular and bone cancer.

The third approach for treating cancer, called *chemotherapy,* involves the use of drugs that (like radiation) are designed to kill proliferating cells. Because drugs are carried throughout the body by the circulatory system, chemotherapy is particularly well suited for the treatment of tumors that have already metastasized. Antitumor drugs can be divided into at least six different categories on the basis of their varying mechanisms of action (Table 18-7). The major problem with such drugs is similar to the problem encountered with radiation treatment; chemotherapeutic drugs inhibit the division of normal cells as well as cancer cells, producing toxic side effects like diarrhea (caused by destruction of

cells of the intestinal lining), loss of hair (caused by destruction of hair follicle cells), and susceptibility to infections (caused by destruction of blood cells).

For several types of cancer, chemotherapy has been successful in restoring a normal life expectancy to a significant proportion of treated patients. Included in this category are Burkitt's lymphoma, choriocarcinoma, acute lymphocytic leukemia, Hodgkin's disease, lymphomas, mycosis fungoides, Wilms' tumor, Ewing's sarcoma, rhabdomyosarcoma, retinoblastoma, and embryonal testicular tumors. The cure rates for these drug-sensitive tumors range from a low of around 20 percent (metastatic embryonal testicular cancer) to a high of 75 percent (metastatic choriocarcinoma). In some instances drugs alone are responsible for the improved survival rate, whereas in others chemotherapy is used in conjunction with surgery or radiation. The effectiveness of chemotherapy can occasionally be improved by utilizing several drugs in combination instead of a single agent alone. Drug combinations are more effective than individual drugs because the increase in tumor-killing ability observed when several drugs are used together tends to exceed the increase in toxic side effects.

Immunotherapy Is an Experimental Treatment Designed to Destroy Cancer Cells Selectively

Although the use of surgery, radiation, and chemotherapy have led to increased survival rates for certain types of cancer, many of the major cancer killers, including lung and intestinal cancers, often do not respond well

Table 18-7 Some Drugs Used in Cancer Chemotherapy

Class	Examples	Mechanism of Action
1. Antimetabolites	Methotrexate 5-Fluorouracil 6-Mercaptopurine	Inhibit enzymatic pathways for biosynthesis of nucleic acids by substituting for normal substrates
2. Antibiotics (substances produced by microorganisms)	Actinomycin D Adriamycin Daunorubicin	Bind to DNA
3. Alkylating agents	Nitrogen mustard Chlorambucil Cyclophosphamide Imidazole carboximides	Crosslink DNA
4. Mitotic inhibitors	Vincristine Vinblastine Taxol	Interfere with mitotic spindle
5. Hormones	Estrogen (for prostate cancer) Cortisone Progesterone Androgens	Inhibit growth of hormone-sensitive cells by interacting with hormone receptors
6. Miscellaneous agents	L-Asparaginase	Hydrolyzes asparagine

to such treatments. In trying to overcome this problem and find the ultimate cure for cancer, scientists have been looking for a "magic bullet" that will seek out and destroy cancer cells without damaging normal cells in the process. Although it is difficult to know when (or if) this goal will be attained, recent experiments suggest that one possible way of achieving such selectivity is to exploit the ability of the immune system to recognize and destroy cancer cells.

The rationale underlying this approach, called *immunotherapy,* is that tumor cells tend to exhibit cell surface antigens that should make them recognizable by the immune system. Several pioneering experiments carried out on human cancer patients by Steven Rosenberg have involved attempts to utilize a person's own lymphocytes to destroy cancer cells. Initially, lymphocytes were isolated from the blood of cancer patients and treated with *interleukin-2,* which stimulates natural killer lymphocytes (page 738) to recognize and destroy cancer cells. When these cells, called *lymphokine-activated killer (LAK)* cells, were injected back into the patients from whom they were obtained, tumor regression occurred in about 20 percent of the patients.

This modest success led to the search for lymphocytes that were even more effective at killing cancer cells. Rosenberg reasoned that if the body were mounting its own immune response against cancer, the tumor itself would most likely have the highest concentration of lymphocytes that were specifically targeted against the cancer cells. Lymphocytes were therefore isolated from the tumors of cancer patients and grown in culture in the presence of interleukin-2 to stimulate the growth and cancer-destroying properties of the cells. The result was the isolation of a population of killer T cells (page 735) that were specifically targeted against the patient's tumor. When these cells, called *tumor-infiltrating lymphocytes (TILs),* were injected back into the patients from whom they were obtained, tumor regression occurred in about 50 percent of the cases. Moreover, comparable experiments carried out in animals revealed that TILs are 50 to 100 times more effective than LAK cells in inducing tumor regression. In other words, one million TILs are as effective at inducing tumor regression as 100 million LAK cells. The long term goal of this experimental approach is to make the TILs even more effective at killing cancer cells by using recombinant DNA techniques to insert genes designed to enhance the therapeutic potency of the TILs. For example, a protein produced by macrophages called *tumor necrosis factor (TNF)* is especially effective at promoting the destruction of cancer cells. If the TNF gene were first inserted into the TILs before injecting them back into the patient, the cells might be even more effective at destroying the person's tumor.

This strategy is just one of many currently being tested in the hopes of finding ways to promote the immune destruction of cancer cells. Although it is too soon to know which strategy will be most effective, it is exciting to see cancer therapies being developed that are based on a logical rationale for selectively destroying cancer cells. Since the immune system is known to be capable of recognizing and killing specific kinds of cells, this type of selectivity is exactly what is needed if we are ever to create a magic bullet for curing cancer.

SUMMARY OF PRINCIPAL POINTS

• Tumors exhibit uncontrolled growth in which the rate of cell division exceeds the rate of cell differentiation and loss, thereby leading to a progressive increase in cell numbers. Benign tumors grow locally and are rarely life threatening, whereas malignant tumors are more dangerous because they are capable of invasion and metastasis. Only a small proportion of the cells in a malignant tumor can successfully metastasize. In some cases the immune system may help to defend against metastasis.

• Although no single property is universally encountered, cancer cells typically (1) have a distinctive appearance; (2) produce tumors in laboratory animals; (3) are immortal in culture; (4) grow to high densities in culture; (5) are anchorage-independent; (6) lack normal cell cycle controls; (7) exhibit cell surface alterations involving decreased adhesiveness, increased agglutinability, decreased cell-cell communication, and tumor-specific antigens; (8) secrete proteases, embryonic proteins, and proteins that stimulate angiogenesis; and (9) exhibit genetic alterations.

• Chemical carcinogenesis usually involves separate initiation and promotion stages. Initiators act as mutagens, creating preneoplastic cells exhibiting irreversible changes in their DNA, whereas promoting agents stimulate preneoplastic cells to divide and form tumors.

• Radiation-induced cancers are caused by sunlight, which produces skin cancer, and by X-rays and radioactive elements, which induce cancers of internal organs. Like most chemical carcinogens, radiation is mutagenic and hence initiates malignant transformation by inducing DNA alterations.

• Viruses cause a variety of different animal and plant tumors, including a few types of human cancer. Oncogenic DNA viruses integrate their DNA directly into the host cell chromosome, whereas oncogenic RNA viruses employ reverse transcriptase to copy their genetic information into a DNA molecule that is then integrated into the host's chromosomal DNA.

• Genes that cause cancer, called oncogenes, have been identified in both oncogenic viruses and in nonviral human cancers. Oncogenes are closely related to normal cellular genes known as proto-oncogenes. Proto-oncogenes can be converted into oncogenes by point mutation, local DNA rearrangements, insertional mutagenesis, gene amplification, and chromosomal translocation. The resulting oncogenes code for protein products that are either abnormal in structure or are produced in inappropriate amounts. Oncogenes code for proteins involved in normal growth control, including growth factors, receptor and nonreceptor protein-tyrosine kinases, membrane-associated G proteins, protein-serine/threonine kinases, and transcription factors.

• Cancer can also be induced by the loss of tumor suppressor genes, which normally function to inhibit cell proliferation. The common human cancers arise by the stepwise accumulation of mutations involving both the formation of oncogenes and the loss or disruption of tumor suppressor genes.

• Some of the genetic alterations involved in producing cancer can be inherited. Since multiple mutations are usually required to create a malignancy, a genetic defect that is inherited tends to increase an individual's risk of developing cancer rather than causing cancer by itself.

• Although the role played by genetic damage suggests that the development of malignancy is an irreversible change, certain kinds of cancer cells (crown gall, teratocarcinoma, Lucké adenocarcinoma) can revert to nonmalignant behavior under appropriate conditions.

• Potential environmental carcinogens are identified by analysis of epidemiological data and by laboratory testing of chemicals either in animals or in the Ames test. Minimizing exposure to carcinogens often requires changes in life-style, such as avoiding tobacco smoke. Because many carcinogens act by generating free radicals, dietary antioxidants that destroy free radicals, such as vitamins A, C, and E, may help to counteract some of the toxic effects of environmental carcinogens.

• Treating cancer patients with surgery, radiation, and chemotherapy has led to increased survival rates for certain kinds of cancer, but many malignancies do not respond well to such treatments. Recent experimentation is attempting to exploit the ability of the immune system to recognize and kill tumor cells.

SUGGESTED READINGS

Books

Brugge, J., T. Curran, E. Harlow, and F. McCormick, eds. (1991). *Origins of Human Cancer: A Comprehensive Review,* Cold Spring Harbor Laboratory Press, Cold Spring Harbor, NY.

Cooper, G. M. (1990). *Oncogenes,* Jones and Bartlett, Boston.

Evans, C. W. (1991). *The Metastatic Cell: Behaviour and Biochemistry,* Chapman and Hall, New York.

Franks, L. M., and N. M. Teich, eds. (1991). *An Introduction to the Cellular and Molecular Biology of Cancer,* 2nd Ed., Oxford University Press, New York.

Kupchella, C. E. (1987). *Dimensions of Cancer,* Wadsworth, Belmont, CA.

Oppenheimer, S. (1985). *Cancer: A Biological and Clinical Introduction,* 2nd Ed., Jones and Bartlett, Boston.

Ruddon, R. W. (1987). *Cancer Biology,* 2nd Ed., Oxford University Press, New York.

Articles

Ames, B. N., W. E. Durston, E. Yamasaki, and F. D. Lee (1973). Carcinogens as mutagens: a simple test system combining liver homogenates for activation and bacteria for detection, *Proc. Natl. Acad. Sci. USA* 70:2281-2285.

Bishop, J. M. (1994). Misguided cells: The genesis of human cancer, *Biol. Bull.* 186:1-8.

Boon, T. (1993). Teaching the immune system to fight cancer, *Sci. Amer.* 266 (March):82-89.

Bryant, P. J. (1993). Towards the cellular functions of tumour suppressors, *Trends Cell Biol.* 3:31-35.

Burkitt, D. (1962). A children's cancer dependent on climatic factors, *Nature* 194:232-234.

Collett, M. S., and R. L. Erikson (1978). Protein kinase activity associated with the avian sarcoma virus src gene product, *Proc. Natl. Acad. Sci. USA* 75:2021-2024.

Crews, C. M., and R. L. Erikson (1993). Extracellular signals and reversible protein phosphorylation: What to Mek of it all, *Cell* 74:215-217.

Croce, C. M., and G. Klein (1985). Chromosome translocations and human cancer, *Sci. Amer.* 252 (March):54-60.

Culotta, E., and D. E. Koshland, Jr. (1993). p53 sweeps through cancer research, *Science* 262:1958-1961.

Fearon, E. R., and B. Vogelstein (1990). A genetic model for colorectal tumorigenesis, *Cell* 61:759-767.

Feldman, M., and L. Eisenbach (1988). What makes a tumor cell metastatic? *Sci. Amer.* 259 (November):60-68.

Fidler, I. J., and M. L. Kripke (1977). Metastasis results from preexisting variant cells within a malignant tumor, *Science* 197:893-895.

Fishel, R., et al. (1993). The human mutator gene homolog *MSH2* and its association with hereditary nonpolyposis colon cancer, *Cell* 76:1027-1038.

Hall, A. (1994). A biochemical function for Ras—at last, *Science* 264:1413-1414.

Helin, K., and E. Harlow (1993). The retinoblastoma protein as a transcriptional repressor, *Trends Cell Biol.* 3:43-46.

Henderson, B. E., R. K. Ross, and M. C. Pike (1991). Toward the primary prevention of cancer, *Science* 254:1131-1138.

Hunter, T. (1991). Cooperation between oncogenes, *Cell* 64:249-270.

Jiricny, J. (1994). Colon cancer and DNA repair: have mismatches met their match? *Trends Genet.* 10:164-168.

Levene, A. J. (1993). The tumor suppressor genes, *Annu. Rev. Biochem.* 62:623-651.

Liotta, L. A. (1992). Cancer cell invasion and metastasis, *Sci. Amer.* 266 (February):54-63.

Miki, Y., et al. (1994). A strong candidate for the breast and ovarian cancer susceptibility gene *BRCA1*, *Science* 266:66-71.

Mintz, B., and R. A. Fleischman (1981). Teratocarcinomas and other neoplasms as developmental defects in gene expression, *Adv. Cancer Res.* 34:211-278.

Old, L. J. (1988). Tumor necrosis factor, *Sci. Amer.* 258 (May):59-75.

Papadopoulos, N., et al. (1994). Mutation of a *mutL* homolog in hereditary colon cancer, *Science* 263:1625-1629.

Prendergast, G. C., and J. B. Gibbs (1994). Ras regulatory interactions: Novel targets for anti-cancer intervention? *BioEssays* 16:187-191.

Rosenberg, S. A. (1990). Adoptive immunotherapy for cancer, *Sci. Amer.* 262 (May):62-69.

Rotter, V., O. Foord, and N. Navot (1993). In search of the functions of normal p53 protein, *Trends Cell Biol.* 3:46-49.

Sachs, L. (1986). Growth, differentiation and the reversal of malignancy, *Sci. Amer.* 254 (January):40-47.

Schlessinger, J. (1993). How receptor tyrosine kinases activate Ras, *Trends Biochem. Sci.* 18:273-275.

Sherr, C. J. (1994). The ins and outs of RB: coupling gene expression to the cell cycle clock, *Trends Cell Biol.* 4:15-18.

Sugden, B. (1993). How some retroviruses got their oncogenes, *Trends Biochem. Sci.* 18:233-235.

Tanaka, K., and R. D. Wood (1994). Xeroderma pigmentosum and nucleotide excision repair of DNA, *Trends Biochem. Sci.* 19:83-86.

Weinberg, R. A. (1983). A molecular basis of cancer, *Sci. Amer.* 249 (November):126-142.

Weinberg, R. A. (1988). Finding the anti-oncogene, *Sci. Amer.* 259 (September):44-51.

Glossary

(Numbers in parentheses refer to the pages on which these terms are discussed.)

A band Region of a striated muscle myofibril that appears as a dark band when viewed by microscopy; contains thick myosin filaments and those regions of the thin actin filaments that overlap the thick filaments. (564)

absorbance Amount of light absorbed by a solution. (140)

absorption spectrum Graph showing the amount of light of different wavelengths absorbed by a solution. (140)

acetyl CoA High-energy two-carbon compound generated by the breakdown of glucose and fatty acids; main input to the Krebs cycle. (324)

acetylcholine Neurotransmitter that triggers skeletal muscle contraction by binding to nicotinic acetylcholine receptors and slows the heartbeat by binding to muscarinic acetylcholine receptors. (208, 218)

acid hydrolases Degradative enzymes exhibiting maximal activity at acid pH; reside mainly in lysosomes in eukaryotes. (288)

acquired immunodeficiency syndrome *See* AIDS.

acrosomal process Long cytoplasmic protrusion formed by invertebrate sperm cells for the purpose of penetrating through egg coats. (686)

acrosomal reaction Discharge of acrosomal enzymes following contact between sperm and egg. (686)

acrosome Vesicle in the sperm head containing enzymes that digest a path through the egg coats during fertilization. (681)

actin Protein that polymerizes into actin filaments that comprise part of the cytoskeleton of most eukaryotic cells. (565)

actin-binding proteins Family of proteins that trigger changes in actin polymerization, crosslink actin filaments into bundles and networks, initiate and control movements along actin filaments, and link actin filaments to other structures. (582)

actin filaments Cytoskeletal filaments of eukaryotic cells measuring about 6 nm in diameter and composed of the protein actin; also refers to the thin filaments of muscle cells. (9, 582)

α-actinin Actin-crosslinking protein present in high concentration near the Z disk of skeletal muscle myofibrils. (587)

action potential Change in membrane potential involving an initial depolarization followed by a rapid return to the normal resting potential; utilized for cell-cell signaling by neurons and muscle cells. (199, 752)

activation domain Region of a transcription factor, distinct from the DNA-binding domain, that is responsible for activating transcription. (464)

activation energy Amount of energy required to bring chemical reactants to the transition state, thereby allowing the reaction to proceed. (39)

activator protein Protein that activates the transcription of specific genes by binding to DNA; the opposite of a repressor protein. (436)

active site Region of an enzyme that binds to a specific substrate and catalyzes its chemical conversion; also known as the catalytic site. (41)

active transport Movement of molecules or ions across cell membranes against an electrochemical gradient accompanied by the expenditure of energy; carried out by membrane transport proteins. (180)

activin Peptide growth factor that induces the formation of mesoderm in developing embryos. (702)

adenine (A) Purine base found in DNA and RNA; forms a complementary base pair with thymine or uracil by hydrogen bonding. (19)

adenosine diphosphate (ADP) Molecule constructed from the purine base adenine, the five-carbon sugar ribose, and two phosphate groups. (56)

adenosine triphosphate (ATP) Molecule constructed from the purine base adenine, the five-carbon sugar ribose, and three phosphate groups; employed by cells for capturing and storing energy. (56)

adenylyl cyclase Plasma membrane enzyme that catalyzes the conversion of ATP to cyclic AMP; involved in transmitting signals from external ligands to the cell interior. (211)

adherens junctions Plaque-bearing cell junctions that connect to actin filaments and stabilize cells against mechanical stress; include adhesion belts and focal adhesions. (239)

adhesion belt Type of adherens junction that completely encompasses entire cells. (240)

ADP *See* adenosine diphosphate.

ADP-ribosylation Transfer of an ADP-ribose group to a protein molecule. (214)

adrenergic receptor Receptor that binds epinephrine. (213)

adsorption chromatography Type of chromatography in which the molecules being purified are adsorbed to the column matrix by binding forces other than electric charge. (147)

aequorin Molecule that emits light when bound to calcium ions. (689)

aerobic Metabolism that uses oxygen. (310)

affinity chromatography Type of chromatography in which the matrix contains chemical groups that bind selectively to the molecule being purified. (147)

affinity label Molecule containing a highly reactive side chain that causes the compound to become covalently linked to neighboring molecules. (475)

AIDS (acquired immunodeficiency syndrome) Loss of immune function caused by a depletion of helper T cells that occurs in individuals infected with the RNA virus, HIV. (742)

allograft Tissue or organ transplanted between genetically different members of the same species. (737)

allosteric regulation Altering the activity of a protein (usually an enzyme) by the reversible binding of a small molecule to a regulatory region called the allosteric site. (51)

allosteric site Region of a protein molecule designed to bind selectively to a small molecule that regulates the protein's activity. (51)

alpha helix (α helix) Type of protein secondary structure in which hydrogen bonding between the —C═O group of one peptide bond and the —NH group of another peptide bond causes the polypeptide chain to twist into a spiral. (23)

amino acid Organic acid containing a basic amino group ($-NH_3^+$), an acidic carboxyl group ($-COO^-$), and a side chain of varying chemical structure. (14, 20)

aminoacyl site (A site) Ribosomal site that binds incoming aminoacyl-tRNAs. *Also see* peptidyl site. (492)

aminoacyl-tRNA Transfer RNA molecule containing the appropriate amino acid attached to its 3' end. (89)

aminoacyl-tRNA synthetase Enzyme that catalyzes the attachment of the proper amino acid to the 3' end of its corresponding transfer RNA. (89)

amoeboid movement *See* cell crawling.

amphipathic Exhibiting both hydrophilic and hydrophobic properties. (16, 162)

anaerobic Metabolism that doesn't use oxygen. (310)

anaerobic glycolysis *See* fermentation.

anaphase Stage during mitosis (or meiosis) when the chromatids (or homologous chromosomes) separate and move to opposite spindle poles. (536, 541)

anaphase A Movement of chromosomes toward the spindle poles during anaphase. (542)

anaphase B Movement of the spindle poles away from each other during anaphase. (542)

anaplasia Loss of cell differentiation. (782)

anergy Immune tolerance following exposure to an antigen under conditions that do not trigger an immune response. (741)

angiogenesis Process of forming blood vessels. (788)

animal pole Region of an egg cell containing the nucleus and pigment granules, but few yolk platelets. *Also see* vegetal pole. (677)

anion Negatively charged molecule or atom. (144)

ankyrin Peripheral protein of the red cell membrane that links spectrin to the integral membrane protein, band 3. (174)

annexins Family of Ca^{2+}-binding proteins that promote the binding of membranes to each other. (285)

ANP *See* atrial natriuretic peptide.

anther Organ that produces pollen grains in flowering plants. (683)

antibody Protein produced by lymphocytes that binds selectively to a specific antigen; also called immunoglobulin. (130, 717)

anticodon Three-base sequence in a tRNA molecule that forms complementary base pairs with a codon in mRNA. (89)

antigen Substance that induces an immune response. (130, 715)

antigen-presenting cell (APC) Cells that degrade antigens and present the resulting fragments to other cells involved in the immune response; examples include macrophages, dendritic cells, and Kupffer cells. (715)

antioxidant Substance that donates electrons to free radicals without becoming harmful itself. (807)

antipodal cells One of several cell types produced by mitotic division of megaspores during female gametogenesis in flowering plants. (686)

antiport Membrane protein that transports two different ions or molecules in opposite directions across a membrane. (180)

antisense RNA RNA molecule whose base sequence is complementary to that of a specific mRNA. (506)

antitermination protein Protein that suppresses the termination of transcription for selected genes, thereby allowing transcription to proceed into adjacent genes. (425)

anucleolar mutant Lethal mutation characterized by both an inability to synthesize ribosomal RNA and the absence of nucleoli. (476)

AP endonuclease Repair enzyme that cleaves double-stranded DNA adjacent to regions of the backbone that lack attached bases. (94)

AP1 transcription factor Protein that regulates the activity of genes involved in cell proliferation; composed of polypeptides called Jun and Fos. (464)

APC *See* antigen-presenting cell.

apoprotein A protein lacking its normal prosthetic group. (47)

apoptosis Type of cell death in which cells quickly shrink and shed tiny membrane vesicles. (709)

archaebacteria Unusual group of prokaryotic organisms that thrive under harsh environmental conditions that would be fatal to most other organisms; one of the two main subdivisions of the prokaryotes, the other being eubacteria. (639)

***ARS* element** DNA sequence that functions as an origin for DNA replication. (529)

ascus Small sac enclosing the cells produced by meiosis in fungi such as *Neurospora*. (556)

asters Clusters of short microtubules that radiate in all directions from each centrosome at the beginning of mitosis or meiosis. (537)

astrocyte Type of glial cell found in the brain and spinal cord. (749)

ATP *See* adenosine triphosphate.

ATP synthase Enzyme complex that synthesizes ATP using energy derived from the flow of protons across mitochondrial inner membranes, chloroplast thylakoid membranes, or bacterial plasma membranes. (335, 346, 377)

atrial natriuretic peptide (ANP) Peptide hormone that lowers blood pressure by relaxing blood vessels and promoting Na^+ and water excretion by the kidney; binds to membrane receptors exhibiting guanylyl cyclase activity. (222)

attenuation Mechanism for regulating gene expression based on the premature termination of transcription. (508)

autocrine Type of signaling in which a secreted molecule acts on the same cell that produced the molecule. (205)

autoimmune disease Disease caused by an immune response directed against an organism's own cells or molecules. (741)

autolysis Rupturing of a cell caused by its own lysosomal enzymes. (292)

autophagy *See* macroautophagy and microautophagy.

autoradiography Procedure for detecting radioactive molecules by overlaying a sample with photographic film. *Also see* microscopic autoradiography. (98)

autotroph Organism that can synthesize its own organic molecules from nonreduced carbon-containing compounds such as CO_2; all photosynthetic organisms are autotrophs. (355)

AVEC microscopy Type of light microscopy that uses a television camera and computer processing of the image to visualize structures that are an order of magnitude smaller than can be seen with a conventional light microscope. (616)

axon Cytoplasmic extension emerging from a nerve cell body that transmits electrical signals to other nerve cells, often over very long distances. (199, 614, 749)

axon terminal Expanded area at the tip of a nerve axon, usually containing numerous neurotransmitter-filled vesicles. (756)

axoneme Group of interconnected microtubules that form the backbone of eukaryotic cilia and flagella; typically arranged as nine outer doublet tubules surrounding a pair of central tubules, a pattern called the "9 + 2 arrangement." (603)

B cell See B lymphocyte.

B lymphocyte (B cell) Type of lymphocyte that arises in the bone marrow and produces antibodies; nonstimulated B cells produce cell-surface antibodies that function as antigen receptors; activated B cells differentiate into plasma cells, which secrete antibodies into the bloodstream. (717)

bacilli (singular, *bacillus*) Rod-shaped bacteria. (248)

bacteriophage Any virus that infects bacterial cells; also called phage. (28)

bacteriorhodopsin Light-absorbing membrane protein that pumps protons across the bacterial plasma membrane, establishing a proton gradient that can be used to drive ATP synthesis. (189, 398).

Balbiani ring An exceptionally large chromosome puff. (447)

band 3 protein Red cell membrane protein that functions as an anion exchanger. (184)

band 4.1 protein Red cell membrane protein that links glycophorin to spectrin. (174)

Barr body Small, darkly staining mass of nuclear chromatin observed in the cells of female mammals; represents a transcriptionally inactivated X chromosome. (451)

basal body Structure comprised of nine sets of triplet microtubules located at the base of eukaryotic cilia and flagella. (611)

basal lamina Thin sheet of specialized extracellular matrix material that separates epithelial cells from underlying supporting tissues. (231)

base analog Substance whose chemical structure closely resembles one of the normal bases found in DNA. (92)

base-modifying agent Substance that chemically alters the structure of bases found in DNA or RNA, thereby changing their base-pairing properties. (92)

benign (tumor) Tumor that is incapable of invasion and metastasis. (778)

beta sheet (β sheet) Type of protein secondary structure in which two different regions of the same or different polypeptide chains are held together by hydrogen bonds between the —C=O group of one peptide bond and the —NH group of another peptide bond, creating an extended sheetlike structure. (23)

bindin Protein located on the surface of the acrosomal process that binds sea urchin sperm to the vitelline envelope of the egg. (686)

biparental inheritance Pattern of inheritance in which a gene governing a particular trait is contributed either solely by the male parent or solely by the female parent; such genes usually reside within mitochondria or chloroplasts. (640)

blastopore Opening in the surface of a blastula that allows cells to migrate into the interior of the developing embryo. (696)

blastula Hollow ball of small cells generated by cleavage of a fertilized animal egg. (696)

brefelden A Drug that disrupts Golgi membranes. (278)

bright-field light microscope Light microscope in which visible light is transmitted through the specimen. (119)

brown fat Fat tissue containing mitochondria that are uncoupled for the purpose of producing heat. (339)

bundle sheath cells Cells that form a prominent layer around the leaf veins of C_4 plants; site of the light reactions and Calvin cycle in such plants. (393)

C gene segment DNA segment coding for the constant region of an antibody light or heavy chain. (723)

C-terminus End of a polypeptide chain that contains the last amino acid to be incorporated during mRNA translation; usually retains a free carboxyl group. (21, 487)

C value Total amount of DNA present in a single (haploid) set of chromosomes. (441)

C value paradox Presence of more DNA in eukaryotic cells than appears to be needed for genetic coding. (441)

C_3 plants Plants that depend solely on the Calvin cycle for CO_2 fixation. (392)

C_4 plants Plants that utilize PEP carboxylase in mesophyll cells to carry out the initial fixation of CO_2; the CO_2 is subsequently released in bundle sheath cells and fixed again by the Calvin cycle. (392)

cadherins Family of plasma membrane glycoproteins that mediate Ca^{2+}-dependent adhesion between cells. (233)

calcium pump Membrane protein that actively transports Ca^{2+} out of the cell (or into the ER lumen) using energy derived from ATP hydrolysis. (223, 575)

calmodulin Ca^{2+}-binding protein that mediates many of the regulatory effects of calcium ions in eukaryotic cells. (219)

calmodulin-dependent multiprotein kinase Protein kinase that is activated by Ca^{2+}-calmodulin and that in turn catalyzes the phosphorylation of various target proteins. (220)

calpain Ca^{2+}-dependent protease involved in releasing fertilized eggs from metaphase arrest. (696)

Calvin cycle Cyclic pathway used by photosynthetic organisms to fix CO_2 and convert it into carbohydrates; represents the only CO_2-fixing pathway in C_3 plants, and the secondary pathway in C_4 plants. (357, 387)

CAM plants Plants that utilize PEP carboxylase to fix CO_2 at night, generating the C_4 acid, malate; the malate is then broken down during the day to release CO_2, which is fixed by the Calvin cycle. (395)

cancer Any malignant tumor. (778)

5' cap Methylated structure located at the 5' end of eukaryotic mRNAs; created by adding 7-methylguanosine and methylating the ribose rings of the first, and often the second, bases of the RNA chain. (428)

CAP See catabolite activator protein.

capping Movement of cell surface molecules to one region of the plasma membrane (170, 592); also, addition of a methylated cap to the 5' end of eukaryotic mRNAs. (428).

capsule Layer of gelatinous material bound to the outer surface of a bacterial cell. (251)

carbohydrates Class of organic molecules characterized by a 1:2:1 ratio of carbon, hydrogen, and oxygen; includes sugars such as glucose, and polysaccharides such as glycogen, starch, and cellulose. (14)

carcinogen Any agent that causes cancer. (789)

carcinoma Any cancer of epithelial cell origin. (778)

carotenoids Lipids constructed from hydrocarbon chains containing alternating double bonds and ending in substituted cyclohexene rings; function as accessory pigments in photosynthesis. (367)

carrier proteins Membrane transport proteins that function through alternating conformational changes. (184)

catabolite activator protein (CAP) Bacterial protein that binds cyclic AMP and then activates the transcription of catabolite-repressible genes. (436)

catabolite repression Ability of glucose to inhibit the expression of genes coding for enzymes involved in the metabolism of other nutrients. (436)

catalyst Agent that speeds up a chemical reaction without being consumed in the process. (37)

catalytic receptor Membrane protein exhibiting a catalytic domain whose activity is controlled by the binding of a signaling ligand to the protein's receptor site. (221)

catalytic subunit Enzyme subunit containing the enzyme's active site. (52)

cation Positively charged molecule or atom. (144)

caveolae Small, flask-shaped invaginations of the plasma membrane implicated in the transport of small molecules and ions into cells. *Also see* potocytosis. (297)

CCAAT box Eukaryotic promoter sequence located about 60–120 bases upstream from the transcriptional start site of certain genes transcribed by RNA polymerase II. (420)

CD4 Plasma membrane glycoprotein that functions in conjunction with a T-cell antigen receptor (TCR) to bind helper T cells to antigen–MHC II complexes. (729)

CD8 Plasma membrane glycoprotein that functions in conjunction with a T-cell antigen receptor (TCR) to bind killer T cells to antigen–MHC I complexes. (736)

cdc2 kinase Protein kinase activated by cyclins that controls passage through various stages of the eukaryotic cell cycle. (524)

cDNA (complementary DNA) Single-stranded DNA copied from mRNA by reverse transcriptase. (75, 100)

cDNA library Group of recombinant DNA clones produced by copying the entire mRNA population of a particular cell type with reverse transcriptase and then cloning the resulting cDNAs. (100)

cell body Central region of a neuron containing the nucleus and cytoplasmic organelles, but excluding axons and dendrites. (749)

cell cortex Cytoplasmic region just beneath the plasma membrane, usually enriched in actin filaments. (584)

cell crawling Cell locomotion generated by cytoplasmic movements that propel cells across solid surfaces; also called amoeboid movement. (586)

cell cycle Stages involved in preparing for and carrying out cell division; consists of phases called G_1, S, G_2, and M in eukaryotes, or C, D, and B in prokaryotes. (518, 522)

cell envelope Multilayered structure surrounding bacterial cells; includes the plasma membrane, murein wall, outer membrane (in Gram-negative cells), and capsule. (248)

cell fusion Fusion of two cells to form a single, hybrid cell. (117)

cell junctions Specialized connections between the plasma membranes of adjoining cells that function in sealing, adhesion, and communication. *Also see* tight junction, plaque-bearing junction, and gap junction. (236)

cell plate Newly deposited cell wall material that divides the cell in half during plant cell division. (546)

cell theory Theory stating that all organisms are composed of cells, that all cells exhibit similarities in structure and function, and that all cells arise from the division of pre-existing cells. (4)

cell wall Rigid structure of varying chemical composition located external to the plasma membrane in plants, algae, fungi, and bacteria. (5, 243, 248)

cell-mediated immunity Process by which killer T cells destroy target cells exhibiting foreign antigens on their surface. (735)

cellular respiration Process by which cells oxidize organic molecules to carbon dioxide and water in the presence of oxygen; involves the Krebs cycle, electron transport, oxidative phosphorylation, and fatty acid β-oxidation. (315)

cellulose Polysaccharide found in plant cell walls that is composed of glucose molecules joined by $\beta(1 \rightarrow 4)$ glycosidic bonds. (15, 243)

CEN sequence Extensively repeated DNA sequence located at the centromere of eukaryotic chromosomes. (534)

centriole Structure comprised of nine sets of microtubules embedded within the centrosome of animal cells; resembles the basal body of eukaryotic cilia and flagella. (612)

centromere Region of a replicated eukaryotic chromosome where the two chromatids remain joined. (528, 537)

centrosome Small zone of granular material, usually located adjacent to the nucleus, that serves as a microtubule-organizing center; a pair of centrioles is embedded within the centrosome of animal cells. (537, 600)

CF$_0$ Group of hydrophobic membrane proteins that anchor CF$_1$ to the thylakoid membrane and that serve as the proton channel of the chloroplast ATP synthase. (377)

CF$_1$ particle Spherical protein subunit protruding from thylakoid membranes that contains the ATP-synthesizing site of chloroplast ATP synthase. (359, 377)

CF$_1$–CF$_0$ complex The ATP synthase of thylakoid membranes. (377)

chaperones Proteins that bind to and stabilize unfolded polypeptide chains. (269, 503)

chemical synapse Synapse at which signals are transmitted by neurotransmitters that diffuse from the presynaptic cell to the postsynaptic cell. (755)

chemiosmotic coupling Theory postulating that electron transfer pathways establish proton gradients across membranes and that the energy stored in such gradients is then used to drive ATP synthesis; applies to both oxidative phosphorylation and photophosphorylation. (335, 377)

chemotaxis Movement of cells either toward a chemical attractant or away from a chemical repellent. (590)

chiasmata (singular, *chiasma*) Connections between homologous chromosomes produced by crossing over during prophase I of meiosis. (551)

chloramphenicol Drug that inhibits protein synthesis by bacterial, chloroplast, and mitochondrial ribosomes, but not by eukaryotic cytoplasmic ribosomes. (656)

chlorophylls Lipids containing a central porphyrin ring attached to a long hydrocarbon side chain called phytol; function as the principal light-absorbing pigments of thylakoid membranes, where they donate electrons to the photosynthetic electron transfer chain. (366)

chloroplast Double-membrane-enclosed cytoplasmic organelle of plants and algae that carries out both the light and dark reactions of photosynthesis. (9, 358)

chloroplast envelope Double-membrane envelope that forms the outer boundary of the chloroplast. (359)

chromatids Two newly replicated eukaryotic chromosomes joined by a centromere. (528)

chromatin DNA-protein fibers found in interphase nuclei of eukaryotic cells; constructed from nucleosomes, each of which consists of an octamer of histones associated with about 200 base pairs of DNA. (409)

chromatography Group of techniques that separate molecules based on differences in the way they become distributed between two phases. (146)

chromosome A single DNA molecule, complexed with protein, that carries part of the genetic information of a cell; becomes condensed into a compact structure at the time of mitosis or meiosis. (409)

cilia (singular, *cilium*) Motile, membrane-enclosed appendages that project from the surface of some eukaryotic cells and that possess a central axoneme exhibiting a 9 + 2 array of microtubules; shorter and more numerous than closely related organelles called flagella. (9, 603)

cis **face** Region of the Golgi complex that receives newly synthesized proteins from the endoplasmic reticulum. (274)

cis **Golgi network** Network of Golgi channels and vesicles located adjacent to the *cis* face of the Golgi complex. (274)

cisternae Flattened membrane compartments. (273)

cisternal space Internal space bounded by the membranes of the endoplasmic reticulum or Golgi complex. *Also see* ER lumen. (258)

class switching Process by which developing plasma cells switch from the production of IgM antibodies to the production of another antibody class, usually IgG. (732)

clathrin Protein that forms a polyhedral lattice around coated pits and related vesicles. (296)

cleavage Stage of rapid mitotic division that follows fertilization of an egg. (696)

clonal deletion theory Theory postulating that lymphocytes carrying antigen receptors targeted against an organism's own antigens are eliminated during embryonic development, thereby preventing autoimmunity. (739)

clonal selection Process by which an antigen selectively stimulates the proliferation of those lymphocytes that possess antigen receptors targeted against the stimulating antigen. (722)

clone Genetically identical population of cells or molecules arising from the proliferation of a single cell or molecule. (102, 116)

cloning vector DNA molecule capable of rapid replication that can be used for cloning DNA fragments in an appropriate host cell. (101)

CO$_2$ fixation Conversion of atmospheric carbon dioxide into molecules containing reduced carbon, such as carbohydrates. (385)

coacervation Spontaneous separation of a solution of macromolecules into a phase containing small droplets whose macromolecule concentration is higher than that of the surrounding solution. (636)

coated pit Infolding of the plasma membrane whose cytoplasmic surface is coated with clathrin. (296)

coated vesicle Cytoplasmic vesicle coated with clathrin that is formed when coated pits pinch off from the plasma membrane; involved in receptor-mediated endocytosis. (296)

cocci (singular, *coccus*) Spherical bacteria. (248)

codon Three-base sequence in mRNA that either codes for an amino acid or signals the termination of protein synthesis. (86)

coenzyme Small organic molecule that binds reversibly to an enzyme and participates directly in the catalytic process. (47)

coenzyme Q *See* ubiquinone.

coiled coil Protein motif in which two α-helical polypeptide chains coil around each other to form a rigid rodlike structure; examples occur in intermediate filaments and in myosin. (566)

colchicine Substance that inhibits cell division by binding to tubulin heterodimers, thereby inhibiting the assembly of spindle microtubules and promoting the breakdown of existing microtubules. (602)

collagen fibers Extremely strong fibers measuring several micrometers in diameter found in the extracellular matrix; constructed from collagen fibrils that are in turn composed of collagen molecules lined up in a staggered array. (226)

collagens Family of related proteins present in high concentration in the extracellular matrix; constructed from three intertwined polypeptide chains that contain unusually large amounts of the amino acids glycine, hydroxyproline, and hydroxylysine. (226)

colony hybridization Technique in which bacterial colonies are hybridized with a radioactive DNA probe to find out which colonies contain a particular recombinant DNA sequence. (98)

competitive inhibitor Inhibitor of enzyme activity that binds reversibly to an enzyme's active site in direct competition with the normal substrate. (49)

complement system Group of proteins circulating in the bloodstream that assist the immune system in the destruction of invading bacteria. (732)

complementary (base-pairing) Ability of guanine to form a hydrogen-bonded base pair with cytosine, and adenine to form a hydrogen-bonded base pair with thymine or uracil. (66)

complementary DNA *See* cDNA.

condensation reaction Chemical reaction in which monomers are joined to form polymers accompanied by the release of water. (635)

condensing vacuoles Large vesicles formed by the fusion of smaller vesicles at the *trans* face of the Golgi complex; involved in concentrating proteins destined for secretion from the cell. (283)

cones Photoreceptor cells responsible for color vision in the vertebrate retina. (768)

confocal scanning microscope Type of fluorescence light microscope that eliminates blurring by focusing the illuminating beam on a single plane within the specimen. (121)

conformation Three-dimensional shape of a protein or other macromolecule. (26)

conjugation Union between two cells, usually bacteria or unicellular eukaryotes, in which chromosomal DNA is injected from one cell into the other. (553)

connexin Plasma membrane protein involved in the construction of gap junctions. (242)

connexon Basic organizational unit of gap junctions; consists of six connexin molecules surrounding a central aqueous channel. (241)

consensus sequence The most common version of a DNA base sequence that occurs in slightly different forms at different locations. (420)

constant region Region of an antibody or TCR molecule where the amino acid sequence is the same in all antibody or TCR molecules. (720)

constitutive secretion Exocytosis of materials from a cell in a continuous, ongoing fashion. *Also see* regulated secretion. (283)

contractile ring Band of actin filaments that forms beneath the plasma membrane of dividing animal cells and mediates the process of cytokinesis. (545)

co-repressor Small molecule that binds to an inactive repressor protein, converting it into an active repressor. (438)

cortical granules Vesicles located beneath the plasma membrane of an egg cell; contain enzymes destined to be discharged from the cell after fertilization. (677)

cortical reaction Discharge of the contents of the cortical granule from an egg cell triggered by sperm-egg fusion; involved in establishing a permanent block to polyspermy. (688)

cosmid Cloning vector that can be packaged into bacteriophage particles and utilized for cloning large DNA fragments. (101)

coupling site Segment of an electron transfer pathway that couples electron transfer to ATP synthesis. (333, 376)

covalent bond Strong chemical bond created by the sharing of a pair of electrons between two atoms. (25)

crassulacean acid metabolism. *See* CAM plants.

CREB protein Eukaryotic transcription factor that activates the transcription of cyclic AMP-inducible genes by binding to cyclic AMP response elements. (465)

cristae Infoldings of the inner mitochondrial membrane that project into the mitochondrial interior; the site of electron transport and oxidative phosphorylation. (316)

crossing over Exchange of chromosome segments between corresponding regions of homologous chromosomes. (551)

crosslinking agent Substance that contains two reactive chemical groups, thereby allowing it to join two neighboring macromolecules. (474)

crown gall disease Malignant tumor of plants caused by bacteria carrying the Ti plasmid. (802)

cyclic AMP (3',5'-cyclic adenosine monophosphate) Nucleotide employed as a second messenger in a wide variety of signaling pathways; often exerts its effects by activating protein kinase A. (211)

cyclic AMP-dependent protein kinase *See* protein kinase A.

cyclic GMP (3',5'-cyclic guanosine monophosphate) Nucleotide employed as a second messenger in selected signaling pathways; capable of both activating protein kinase G and opening cyclic GMP-gated Na^+ channels. (222)

cyclic photophosphorylation ATP synthesis coupled to the light-induced circular flow of electrons around photosystem I. (375)

cyclins Family of proteins that activate protein kinases involved in regulating progression through the eukaryotic cell cycle. (524)

cycloheximide Drug that inhibits protein synthesis by eukaryotic cytoplasmic ribosomes, but not by chloroplast, mitochondrial, or bacterial ribosomes. (656)

cysteine-cysteine zinc finger *See* zinc finger.

cysteine-histidine zinc finger *See* zinc finger.

cytochalasins Substances that interfere with actin polymerization and disrupt actin filaments. (584)

cytochemical procedures Family of techniques for utilizing microscopy to localize specific molecules and metabolic activities within cells. (128)

cytochrome b_6-f complex Component of the photosynthetic electron transfer chain that transfers electrons from plastoquinone to plastocyanin. (373)

cytochrome P-448 Mixed-function oxidase of smooth ER membranes that oxidizes polycyclic hydrocarbons; absorbs light maximally at 448 nm. (262)

cytochrome P-450 Mixed-function oxidase of smooth ER membranes that hydroxylates toxic substances, steroids, and fatty acids; absorbs light maximally at 450 nm. (262)

cytochromes Proteins containing a heme prosthetic group that are involved in electron transfer pathways. (327, 370)

cytokinesis Cytoplasmic division accompanying mitosis or meiosis. (544)

cytomembrane system System of eukaryotic cytoplasmic

membranes that establish transient connections with one another; includes the endoplasmic reticulum, Golgi complex, lysosomes, secretion vesicles, endosomes, storage and transport vesicles, and the nuclear envelope. (669)

cytoplasm Region of a cell outside the nucleus and enclosed by the plasma membrane; contains a variety of membrane-enclosed organelles and cytoskeletal filaments in eukaryotic cells. (8)

cytoplasmic streaming Bulk flow of the cytoplasm driven by interactions between actin filaments and actin-binding proteins such as myosin. (585)

cytosine (C) Pyrimidine base found in DNA and RNA; forms a complementary base pair with guanine by hydrogen bonding. (19)

cytoskeleton Network of actin filaments, intermediate filaments, and microtubules residing within the cytoplasm of eukaryotic cells. (9, 563)

cytosol Soluble portion of the cytoplasm remaining after organelles, membranes, and cytoskeletal filaments have been removed. (9, 138, 258)

D gene segment (diversity segment) DNA segment coding for a short portion of the variable region of an antibody heavy chain. (724)

D loop (displacement loop) Loop generated when DNA replication is initiated on one of the two strands of a mitochondrial or chloroplast DNA molecule. (650)

DAG *See* diacylglycerol.

dark-field microscope Light microscope in which the light reaching the observer has been scattered by interacting with the specimen. (121)

dark reactions Portion of the photosynthetic pathway in which energy stored in NADPH and ATP is used to drive CO_2 fixation; can occur in the dark, but does not require darkness. (385)

deamination Removal of an amino group, thereby altering the base-pairing properties of bases in DNA and RNA. (91)

deep-etching Prolonged use of a vacuum to remove water from the surface of a freeze-fractured specimen, thereby yielding a striking three-dimensional view of intracellular structures. (125)

dehydration Removal of water. (13)

denaturation Disruption of the normal three-dimensional structure of a protein or nucleic acid; in the case of nucleic acids, usually refers to the disruption of complementary base-pairing. (26, 96)

dendrite Cytoplasmic extension emerging from a nerve cell body that receives electrical signals from other nerve cells. (614, 749)

dense bodies Attachment sites for actin thin filaments found in the cytoplasm of smooth muscle cells. (581)

deoxyribonucleic acid *See* DNA.

deoxyribose Five-carbon sugar employed in the construction of DNA; lacks oxygen on its 2' carbon atom. (19)

dephosphorylation Removal of a phosphate group. (54)

depolarization Change in membrane potential to a less-negative value. (199)

depurination Removal of adenine or guanine from a nucleic acid by breaking the bond that joins either of these bases to deoxyribose. (91)

desmosome Type of plaque-bearing cell junction that is linked to intermediate filaments; joins epithelial cells together to stabilize cell layers against mechanical stress. (238)

determination Process by which a cell becomes committed to differentiate into a particular cell type. (696)

diacylglycerol (DAG) Second messenger produced by the phosphoinositide signaling pathway that functions by activating protein kinase C. (218, 221)

diakinesis Final stage of prophase I of meiosis; associated with chromosome condensation, disappearance of nucleoli, breakdown of the nuclear envelope, and initiation of spindle formation. (551)

dialysis Technique for separating large and small molecules; uses artificial membranes containing pores that permit the passage of water and small molecules, but not large molecules. (143)

difference spectrum Graph showing the difference in the light-absorbing properties of an electron carrier in the oxidized and reduced states. (330)

differential centrifugation Type of velocity centrifugation in which a series of centrifugations are carried out at successively increasing speeds. (135)

differential interference microscope Instrument that increases the contrast of interference microscopy by recombining the split beam in polarized light; also called Nomarski interference microscope. (120)

differentiation Process by which cells acquire specialized structural and functional traits. (696)

diphtheria toxin Lethal protein produced by the bacterium that causes diphtheria; inhibits protein synthesis in infected cells by catalyzing the ADP ribosylation of elongation factor eEF-2. (501)

diploid Containing two sets of chromosomes. (548)

diplotene Stage during prophase I of meiosis when the chiasmata connecting each pair of homologous chromosomes become visible. (551)

dissociation constant (K_d) Ligand concentration at which the binding of a ligand to its receptor is half-maximal. (207)

disulfide bond A covalent —S-S— bond formed between the —SH groups of two cysteines; can occur either within or between polypeptide chains. (25)

DNA (deoxyribonucleic acid) Polymer constructed from nucleotides consisting of deoxyribose phosphate linked to either adenine, thymine, cytosine, or guanine; forms a double helix held together by complementary base-pairing between adenine and thymine, and between cytosine and guanine. (19, 65)

DNA cloning Use of recombinant DNA technology to create identical copies of a particular DNA sequence. (100)

DNA fingerprinting Technique for identifying individuals based on small differences in DNA fragment patterns detected by electrophoresis. *Also see* VNTRs. (108)

DNA glycosylases Enzymes involved in DNA repair that remove deaminated bases, alkylated bases, and bases with abnormal rings. (94)

DNA helicases Enzymes that unwind the DNA double helix, driven by energy derived from ATP hydrolysis. (73)

DNA ligase Enzyme that joins two DNA fragments by catalyzing the ATP-dependent formation of a phosphodiester bond between the 3' end of one nucleic acid chain and the 5' end of another. (71)

DNA methylation Enzyme-catalyzed transfer of a methyl group to a base in DNA; employed in both mismatch repair in bacteria and in the control of gene expression in eukaryotes. (94, 455)

DNA polymerases Enzymes involved in DNA replication that catalyze DNA synthesis in the 5' → 3' direction, using an existing DNA strand as template. (69)

DNA primase Enzyme that initiates DNA replication by synthesizing short RNA primers that are complementary to the DNA template. (72)

DNA probe Single-stranded radioactive DNA molecule that is used in hybridization experiments to identify nucleic acids containing sequences that are complementary to the probe. (97)

DNase I hypersensitive site Short, nucleosome-free region of DNA located adjacent to an active gene that is about tenfold more sensitive to DNase I digestion than the DNA of the active gene itself. (452)

dolichol phosphate Long-chain alcohol embedded in the ER membrane that serves as an attachment site for newly forming oligosaccharide chains prior to their transfer to newly forming polypeptide chains. (270)

domains Localized regions of compact folding in polypeptide chains; often associated with specific functions. (24)

dopamine Neurotransmitter that binds to receptors that, in some cases, activate adenylyl cyclase. (764)

double helix Two intertwined helical chains of a DNA molecule, held together by complementary base-pairing between adenine and thymine and between cytosine and guanine. (66)

downstream Located toward the 3' end of the DNA strand lying opposite the strand that serves as the template for transcription. (420)

dynamic instability Property exhibited by microtubules when the tubulin concentration is low, characterized by rapid fluctuation between a growing phase and a shrinking phase. (600)

dynein Motor protein that moves along the surface of microtubules in the plus-to-minus direction driven by energy derived from ATP hydrolysis. (604, 609)

dystrophin Protein associated with the inner surface of muscle cell plasma membranes that causes muscular dystrophy when mutated. (580)

E face (exoplasmic face) During freeze-fracturing, the fracture face derived from the half of a membrane that lies away from the cytoplasm. *Also see* P face. (124)

ectoderm Outer cell layer of a gastrula; gives rise to epithelia that cover external body surfaces (skin and related glands) and to neural tissue. (696)

ectoplasm Rigid gelatinous region of cytoplasm located beneath the plasma membrane at the sides and rear of crawling amoebae. *Also see* endoplasm. (586)

egg cell The female haploid gamete. (679, 686)

elastin Protein subunit of the elastic fibers that impart elasticity and flexibility to the extracellular matrix. (229)

electrical synapse Synapse at which the presynaptic and postsynaptic cells are joined by gap junctions that allow direct electrical communication between the two cells. (756)

electrochemical equilibrium Condition in which no net diffusion of an ion across a membrane takes place because the electrochemical gradient driving diffusion in one direction is exactly balanced by the electrochemical gradient driving diffusion in the opposite direction. (196)

electrochemical gradient Sum of the electric charge gradient and the concentration gradient for an ion distributed on two sides of a membrane. (181)

electrochemical proton gradient Unequal concentration of protons on two sides of a membrane, creating a gradient consisting of a difference in both electric charge (membrane potential) and proton concentration; established by electron transfer pathways for the purpose of driving ATP formation. (335)

electron microscope Microscope that uses an electron beam to visualize specimens and that therefore has much higher resolving power than a light microscope. (122)

electron transport Stepwise flow of electrons from reduced coenzymes to oxygen mediated by a series of electron carriers known as the respiratory chain. (315, 327)

electron transport chain *See* respiratory chain.

electrophoresis Utilization of an electric field to separate charged molecules; often carried out in gels made of polyacrylamide or agarose. (144)

electrostatic bond (ionic bond) Attractive force between a positively charged chemical group and a negatively charged chemical group. (25)

ELISA (enzyme-linked immunosorbent assay) Technique for detecting tiny amounts of specific proteins utilizing antibodies linked to enzymes that catalyze the formation of colored products. (142)

elongation factors Proteins that catalyze steps involved in the elongation phase of protein synthesis; examples include EF-Tu and EF-Ts. (492)

embryo sac Structure produced during female gametogenesis in flowering plants; contains eight haploid nuclei that mature to form one egg, two polar nuclei, two synergid cells, and three antipodal cells. (686)

endergonic Reaction exhibiting a positive value of ΔG, which means that it cannot proceed in the direction specified without an input of energy. (34)

endocrine Type of signaling in which a hormone molecule is secreted into the circulatory system so that it can act on distant cells. (203)

endocytosis Uptake of material into cells via membrane vesicles that pinch off from the plasma membrane. *Also see* phagocytosis, pinocytosis, and receptor-mediated endocytosis. (293)

endoderm Inner cell layer of a gastrula; gives rise to epithelia that line the gastrointestinal tract and related glands. (696)

endonucleases Enzymes that cut nucleic acids internally within the polynucleotide chain. *Also see* exonucleases. (98)

endoplasm Fluid cytoplasm located in the interior of crawling amoebae. *Also see* ectoplasm. (586)

endoplasmic reticulum (ER) System of interconnected membrane channels and vesicles located in the cytoplasm of most eukaryotic cells; involved in the metabolism of carbohydrates, lipids, and drugs, and in the processing and transport of newly synthesized proteins. (5, 256)

endorphins Peptide hormones that inhibit pain-signaling neurons in the brain by binding to opiate receptors. (766)

endosomes Network of cytoplasmic membrane vesicles and tubules that fuse with endocytic vesicles that have been brought into a cell by receptor-mediated endocytosis. (296)

endosperm Nutritive tissue containing three sets of chromosomes that is produced when a sperm nucleus fuses with two polar nuclei during fertilization in flowering plants. (694)

endospore Dormant, dehydrated cell protected by a thick spore coat; adapted to withstand harsh environmental conditions. (522)

endosymbiont theory Theory postulating that mitochondria and chloroplasts arose from ancient bacteria that were ingested by ancestral eukaryotic cells a billion or more years ago. (665)

energy Capacity to do work. (33)

enhancers DNA sequences that stimulate transcription when placed either upstream or downstream from a promoter site. (423)

enkephalins Small peptide neurotransmitters that inhibit pain-signaling neurons in the brain by binding to opiate receptors. (766)

entropy Randomness or disorder. (34)

enzyme kinetics Study of the factors that influence the rate of enzyme-catalyzed reactions. (43)

enzyme-linked immunosorbent assay *See* ELISA.

enzyme-substrate complex Transient complex formed when an enzyme binds to its substrate(s) during an enzyme-catalyzed reaction. (41)

enzymes Catalysts manufactured by cells to speed up chemical reactions; almost always made of protein, but certain RNA molecules also function as catalysts. *Also see* ribozyme. (37)

epithelium (plural, *epithelia*) Sheet of cells covering an internal or external body surface. (696)

epitope Specific region within an antigen molecule that is capable of eliciting its own unique immune response. (715)

equilibrium Point in a chemical reaction at which no further net change in the concentrations of reactants or products is observed; similarly, the point in membrane transport at which no further net change in solute concentration is observed on either side of the membrane. (35, 175)

equilibrium constant (K_{eq}) Mathematical value reflecting the ratio of product concentration to reactant concentration after a chemical reaction has reached equilibrium. (35)

equilibrium potential (Nernst potential) Magnitude and polarity of the membrane potential that exactly counterbalances the tendency of an ion to diffuse down its concentration gradient. (197)

ER *See* endoplasmic reticulum.

ER lumen Internal space bounded by membranes of the endoplasmic reticulum. *Also see* cisternal space. (258)

etioplast Organelle related to chloroplasts that is found in plants grown in the dark; lacks thylakoid membranes and photosynthetic activity, but develops into functional chloroplasts in the light. (383, 662)

eubacteria One of the two main subdivisions of the prokaryotes (the other being archaebacteria); includes most of the commonly encountered bacteria. (639)

euchromatin Loosely packed configuration of chromatin fibers; contains genes that are transcriptionally active. (450)

eukaryotic cells Organisms whose cells possess nuclei bounded by a double-membrane envelope and organelles that compartmentalize the cytoplasm; found in all animals, plants, algae, fungi, and protozoa. (8, 639)

excinuclease DNA repair enzyme that removes thymine dimers. (94)

excision repair DNA repair mechanism that removes and replaces abnormal bases. (93)

excitatory postsynaptic potential Depolarization of a postsynaptic membrane in response to stimulation of the presynaptic cell. (759)

exergonic Reaction exhibiting a negative value of ΔG, which means that the reaction releases energy and is therefore thermodynamically favorable. (34)

exocytosis Discharge of material from cells via membrane vesicles that fuse with the plasma membrane, discharging the vesicle's contents into the extracellular space. (284)

exons Nucleotide sequences in a gene or its initial RNA transcript that code for the gene's polypeptide product; separated from each other by noncoding sequences called introns. (84, 430)

exonucleases Enzymes that degrade nucleic acids by removing nucleotides from one end of the chain or the other. *Also see* endonucleases. (98)

expression vector A DNA cloning vector that is designed to promote the expression of cloned genes in an appropriate host cell. (104)

extensins Glycoproteins found in plant cell walls that resemble collagen in their high content of hydroxyproline. (244)

extracellular matrix Material secreted by animal cells that fills the spaces between neighboring cells; consists of a mixture of structural proteins (e.g. collagen, elastin) and adhesive proteins (e.g. fibronectin, laminin) embedded in a gelatinous matrix composed of glycosaminoglycans and proteoglycans. (224)

F_0 Group of hydrophobic membrane proteins that anchor F_1 to the inner mitochondrial membrane and that serve as the proton channel of the mitochondrial ATP synthase. (335)

F_1 particle Spherical protein subunit protruding from mitochondrial inner membranes that contains the ATP-synthesizing site of mitochondrial ATP synthase. (316)

F_1–F_0 complex The ATP synthase of mitochondrial inner membranes. (335)

facilitated diffusion Mechanism for increasing the rate of solute transport across cell membranes that does not require energy because the ion or molecule being transported is moving down an electrochemical gradient; carried out by membrane transport proteins. (180)

facilitator neuron Neuron whose axon forms a synapse with another axon that enhances the effect of action potentials arriving at the terminus of the second axon. (764)

FAD and **FADH$_2$** Oxidized and reduced forms of flavin adenine dinucleotide, a coenzyme involved in various cellular oxidation-reduction reactions, including a key step in the Krebs cycle. (58, 311)

fasciclins Cell-surface glycoproteins of nerve axons that are recognized by the growth cones of other axons, thereby influencing the direction in which axons grow. (774)

fast axonal transport Microtubule-mediated movement of vesicles and organelles back and forth along a nerve cell axon at rates of 10–500 mm per day. *Also see* slow axonal transport. (616, 752)

fast chemical transmission Synaptic transmission mediated by neurotransmitters that bind to ion channels in the postsynaptic membrane, triggering an immediate change in membrane potential. (761)

fats *See* triacylglycerols.

fatty acids Linear carbon chains containing an acidic carboxyl group ($-COO^-$) at one end; usually possess an even number of carbon atoms, up to a maximum of about 24. (14, 15)

feedback inhibition Control mechanism in which the product of a metabolic pathway inhibits an enzyme catalyzing an early step in the formation of that product. (51)

female pronucleus Designation for the nucleus of an egg cell after a sperm nucleus has entered the egg. (691)

fermentation Anaerobic breakdown of glucose to a reduced derivative of pyruvate, such as lactate or ethanol; also called anaerobic glycolysis. (313)

ferredoxin Iron-sulfur protein involved in the transfer of electrons from photosystem I to $NADP^+$ during the light reactions of photosynthesis. (370)

fertilization Fusion of a sperm cell with an egg cell, forming a diploid zygote. (686)

fertilization membrane Tough rigid structure created from the vitelline envelope after fertilization; serves as a barrier to further sperm penetration. (689)

Feulgen reaction Staining procedure for detecting DNA by light microscopy. (128)

fibroblast Main cell type involved in producing the extracellular matrix of animal tissues. (224)

fibronectin Glycoprotein component of the extracellular matrix that is also loosely associated with the cell surface; binds cells to the extracellular matrix and guides cell migration. (230)

filamin Protein that crosslinks actin filaments, thereby converting actin solutions into gels. (587)

filopodia (singular, *filopodium*) Long, thin protrusions supported by actin filaments that transiently emerge from the surface of eukaryotic cells. (588)

fimbrin Protein that crosslinks the actin filament bundles of microvilli. *Also see* villin. (591)

first law of thermodynamics Law stating that energy can be neither created nor destroyed. (33)

fixatives Chemicals that kill cells while preserving their structural appearance for microscopic examination. (119)

flagella (singular, *flagellum*) Motile, membrane-enclosed appendages that project from the surface of some eukaryotic cells and that possess a central axoneme exhibiting a 9 + 2 array of microtubules; longer and less numerous than closely related organelles called cilia. (9, 603)

flagellin Protein subunit of bacterial flagella. (624)

flavin adenine dinucleotide *See* FAD.

flavin mononucleotide *See* FMN.

flavoprotein Any enzyme that utilizes either FAD or FMN as a prosthetic group. (328)

flip-flop Movement of a phospholipid molecule from one side of a lipid bilayer to the other; also called transverse diffusion. (162)

flippase *See* phospholipid translocator.

fluid mosaic model Currently accepted model of membrane organization that views membranes as consisting of a fluid lipid bilayer and three classes of membrane proteins (integral, peripheral, and lipid-anchored) that can diffuse laterally. (160)

fluorescence microscopy Light microscopic technique for visualizing fluorescent substances within cells, usually flu-

orescent antibodies or other fluorescent stains. (121)

FMN and FMNH$_2$ Oxidized and reduced forms of flavin mononucleotide, a coenzyme involved in oxidation-reduction reactions. (58, 311)

focal adhesion A small, localized type of adherens junction that attaches cells to the extracellular matrix and other solid surfaces; linked to actin filaments. (240, 589)

follicle cells Accessory cells that surround and provide nutrients to animal oocytes. (677)

Fos One of two polypeptide subunits of the AP1 transcription factor, which regulates the activity of genes involved in cell proliferation. *Also see* Jun. (464)

frameshift mutation Insertion or deletion of one or more bases into a gene, thereby altering the reading of all codons beyond the point of insertion or deletion. (85)

free energy Energy that can be harnessed to do useful work; symbolized by the letter *G*. (34)

free radical Highly reactive molecule containing an unpaired electron. (711, 806)

freeze-etching Modification of the freeze-fracturing technique in which frozen, fractured tissue is exposed to a vacuum to remove water and thereby accentuate surface detail. *Also see* deep-etching. (125)

freeze-fracturing Technique for preparing specimens for electron microscopy in which tissue is rapidly frozen and struck with a sharp knife edge to fracture the tissue; fracture planes often pass through the interior of lipid bilayers, allowing the internal organization of cell membranes to be examined. (124)

G proteins Family of membrane-associated GTP-binding proteins that are involved in signal transduction. (211, 761, 798)

G$_0$ state Designation applied to eukaryotic cells that have become arrested in the G$_1$ phase of the cell cycle and thus are no longer proliferating. (523)

G$_1$ phase Stage of the eukaryotic cell cycle situated between the end of cell division (M phase) and the initiation of DNA synthesis (S phase). (523)

G$_2$ phase Stage of the eukaryotic cell cycle situated between the end of DNA synthesis (S phase) and the start of cell division (M phase). (523)

gamete Haploid cell produced by meiosis that fuses with a gamete produced by an organism of the opposite sex to form a new organism. *Also see* egg cell and sperm cell. (546, 674)

gametogenesis Process of forming gametes. (674)

gap junction Type of cell junction that permits small molecules to pass back and forth between adjoining cells. (240)

gas-liquid chromatography (GLC) Type of chromatography in which molecules in a gas phase are passed over a matrix coated with a heavy liquid; useful for separating molecules that are readily volatilized, such as fatty acids, steroids, and other lipids. (148)

gastrula Stage of embryonic development when migration of cells through the blastopore creates three cell layers: ectoderm, mesoderm, and endoderm. (696)

GC box Eukaryotic promoter sequence consisting of the consensus sequence GGGCG; located about 60–120 bases upstream from the transcriptional start site of certain genes transcribed by RNA polymerase II. (420)

gel filtration chromatography Type of chromatography in which molecules are passed through a column of beads containing microscopic pores; separates molecules based on differences in size. (147)

gelsolin Protein that breaks actin filaments, converting actin gels into free-flowing solutions. (587)

gene Nucleotide sequence in DNA that codes for a functional product, usually a polypeptide chain, but in some cases ribosomal RNA, transfer RNA, or small nuclear RNA. (62, 76)

gene amplification Mechanism for creating extra copies of individual genes by selectively replicating specific DNA sequences. (456, 478)

gene conversion Phenomenon in which genes undergo nonreciprocal recombination during meiosis, thereby leading to a situation in which one of the recombining genes ends up on both chromosomes of a homologous pair rather than being exchanged from one chromosome to the other. (556)

generative cell Haploid cell within a pollen grain that subsequently divides to generate two sperm cells. (685)

gene-specific transcription factor Transcription factor that activates (or inhibits) the transcription of a particular set of genes by binding to a specific DNA response element. (458)

genetic transformation Change in the hereditary properties of a cell brought about by the uptake of foreign DNA. (64, 553)

genomic library Group of recombinant DNA clones produced by breaking nuclear DNA into fragments using restriction enzymes or physical shearing, and then cloning the entire set of fragments in an appropriate cloning vector. (100)

germ cells *See* primordial germ cells.

GLC *See* gas-liquid chromatography.

glial cells Cells that surround, support, and insulate neurons. (749)

gluconeogenesis Synthesis of carbohydrates from lipids or other noncarbohydrate molecules. (350)

glucose Six-carbon sugar that plays a central role in cellular metabolism as a source of both energy and chemical building blocks; stored in the form of the polysaccharides glycogen in animals and starch in plants. (14, 311)

glycerol phosphate shuttle Mechanism for carrying electrons derived from cytoplasmic NADH into the mitochondrion, where the electrons are delivered to FAD in the respiratory chain. (340)

glycocalyx Carbohydrate-rich zone located at the outer boundary of many animal cells. (232)

glycogen Polysaccharide composed of glucose molecules joined by $\alpha(1 \rightarrow 4)$, and to a lesser extent $\alpha(1 \rightarrow 6)$, glycosidic bonds; main storage form of glucose in animals. (14)

glycogen phosphorylase Enzyme that catalyzes the breakdown of glycogen. (216)

glycogen synthase Enzyme that catalyzes the synthesis of glycogen. (216)

glycolipids Lipids that contain a carbohydrate group in place of the phosphate group found in phospholipids; one of three main classes of lipids present in cell membranes, the others being phospholipids and steroids. (16, 161)

glycolysis Metabolic pathway that converts glucose to two molecules of pyruvate, accompanied by the net production of two molecules of ATP and the reduction of two molecules of NAD^+ to NADH. (311)

glycoproteins Proteins possessing covalently attached carbohydrate groups. (172, 269)

glycosaminoglycans Polysaccharides constructed from a repeating disaccharide unit containing one sugar with an amino group and one sugar with a negatively charged sulfate or carboxyl group. (224)

glycosome Specialized type of peroxisome containing most of the enzymes involved in glycolysis; found in parasitic protozoa. (351)

glycosylphosphatidylinositol (GPI) Membrane glycolipid that acts as an attachment site for many lipid-anchored membrane proteins. (304)

glyoxylate cycle Modified version of the Krebs cycle occurring in plant glyoxysomes; converts two molecules of acetyl CoA to one molecule of succinate, thereby permitting the synthesis of carbohydrates from lipids. (350)

glyoxysome Special type of peroxisome found in plant seedlings; carries out the glyoxylate cycle and fatty acid β-oxidation. (351)

Goldman equation Modification of the Nernst equation that determines the value of the resting membrane potential by adding together the contributions of multiple ions of differing permeabilities. (198)

Golgi complex Stack of flattened membrane vesicles involved in glycosylating, sorting, and packaging proteins destined for lysosomes, the plasma membrane, and secretion from the cell; also involved in synthesizing polysaccharides such as hyaluronate in animal cells and hemicelluloses and pectins in plants cells. (8, 272)

GPI *See* glycosylphosphatidylinositol.

Gram-negative Bacteria that do not retain the purple dye used in the Gram-staining procedure; such cells have a thin murein wall surrounded by an outer membrane. (248)

Gram-positive Bacteria that retain the purple dye used in the Gram-staining procedure; such cells possess a thick wall composed of murein and teichoic acids. (248)

grana (singular, *granum*) Stacks of thylakoid membranes located within chloroplasts. (359)

gray crescent Lightly pigmented region near the equator of amphibian eggs that is produced by rotation of the outer layer of cytoplasm after fertilization. (691)

growth cone Expanded, motile region at the tip of a growing axon. (771)

growth factors Proteins that stimulate the growth and division of target cells by binding to plasma membrane receptors. (526)

GTP *See* guanosine triphosphate.

GTP cap Short region at the end of a microtubule where β-tubulin retains bound GTP, thereby stabilizing the tip and promoting the further addition of tubulin heterodimers. (600)

guanine (G) Purine base found in DNA and RNA; forms a complementary base pair with cytosine by hydrogen bonding. (19)

guanosine tetraphosphate and **pentaphosphate** *See* ppGpp and pppGpp.

guanosine triphosphate (GTP) Molecule constructed from the purine base guanosine, the five-carbon sugar ribose, and three phosphate groups; involved in protein synthesis, microtubule assembly, and the activation of G proteins. (56)

guanylyl cyclase Enzyme that catalyzes the conversion of GTP to cyclic GMP; involved in transmitting signals from external ligands to the cell interior. (222)

half-life (t$_{1/2}$) Amount of time required for half the molecules existing at any moment to be degraded. (509)

haploid Containing one set of chromosomes. (548)

heat-shock genes Family of genes whose transcription is activated when cells are exposed to elevated temperatures and whose products help to protect organisms against excessive heat. (458)

helix-destabilizing proteins *See* single-stranded DNA-binding proteins.

helix-loop-helix DNA-binding motif that is closely related to the helix-turn-helix motif but contains hydrophobic amino acids lined up on one side of each α helix. (464)

helix-turn-helix DNA-binding motif found in many proteins that regulate gene transcription; consists of two regions of α helix separated by a few amino acids that allow the polypeptide chain to bend between the two helices. (435)

helper T cell Type of T lymphocyte that is activated by binding to antigen–MHC II complexes and in turn activates both B cells and killer T cells. (728)

heme Prosthetic group containing a porphyrin ring bound to a central iron atom that can accept and donate electrons; occurs in cytochromes and most oxygen carrier proteins. (327)

hemicelluloses Plant cell wall polysaccharides that are constructed from various five- and six-carbon sugars; bind to the surface of cellulose microfibrils. (244)

hemidesmosome Type of plaque-bearing cell junction that is linked to intermediate filaments; attaches epithelial cells to the basal lamina. (239)

heterochromatin Densely packed configuration of chromatin fibers; contains genes that are transcriptionally inactive. (450)

heterogeneous nuclear RNAs (hnRNAs) Originally identified as a class of nuclear RNAs of varying sizes that are rapidly broken down; now known to contain pre-mRNA. (427)

heterotroph Organism that requires nutrients containing reduced carbon because it cannot carry out photosynthesis. (355)

high-mobility group proteins *See* HMG proteins.

high-performance liquid chromatography (HPLC) Type of chromatography in which the molecules being separated are passed through a column matrix at rapid speed using high pressure pumps. (148)

histones Small basic proteins of five major types (H1, H2A, H2B, H3, and H4) that are bound to the nuclear DNA of all eukaryotes; a histone octamer containing two molecules each of histones H2A, H2B, H3, and H4 forms the core of nucleosomes. (409)

HIV (human immunodeficiency virus) RNA retrovirus associated with the human disease, AIDS; disables the immune system by infecting helper T cells. (742)

HMG proteins (high-mobility group proteins) Group of nonhistone proteins that are preferentially associated with chromatin fibers whose genes are being actively transcribed. (411)

hnRNAs *See* heterogeneous nuclear RNAs.

hnRNPs Protein particles involved in packaging nuclear pre-mRNAs. (428)

Holliday junction Crossed structure produced when two DNA molecules are joined together by single-strand crossovers during genetic recombination. (556)

homeobox DNA sequence about 180 base pairs long found in homeotic genes; codes for a DNA-binding protein domain called a homeodomain. (458)

homeodomain DNA-binding region about 60 amino acids long found in transcription factors encoded by homeotic genes; contains a helix-turn-helix motif. (458)

homeotic genes Genes that control the formation of the body plan during embryonic development; code for transcription factors possessing homeodomains. (458, 708)

homogenate Suspension of cell organelles produced by disrupting cells or tissues using physical means such as blending or grinding. (138)

homologous chromosomes Chromosomes that pair with each other and exchange genetic information during meiosis. (548)

homologous recombination Exchange of genetic information between two DNA molecules exhibiting extensive sequence homology. (553)

hormone Signaling molecule that is secreted into the bloodstream by endocrine cells and acts on target cells possessing receptors for the hormone. (203)

hormone response elements DNA regulatory sequences that bind hormone receptors; located upstream from the promoter sites of genes whose transcription is hormonally regulated. (462)

housekeeping genes Genes expressed in most cell types because they code for proteins involved in essential functions. *Also see* tissue-specific genes. (703)

HPLC *See* high-performance liquid chromatography.

hyaluronate Glycosaminoglycan found in high concentration in the extracellular matrix where cells are actively proliferating or migrating, and in the joints between moving bones. (225)

hybrid cell Cell whose nucleus contains chromosomes derived from two different cell types that were joined by cell fusion. (117)

hybridoma Clone of hybrid cells produced by fusing an antibody-producing B lymphocyte with a tumor cell that grows well in culture; each hybridoma produces a single type of monoclonal antibody. (143, 725)

hydrogen bond Noncovalent bond formed between a hydrogen atom in a donor group such as —NH or —OH, and an oxygen or nitrogen atom in an acceptor group. (11, 25)

hydrolysis Reaction in which a chemical bond is broken by the addition of water. (13)

hydropathy plot Graph that identifies the location of hydrophobic amino acid clusters within the primary sequence of a protein molecule; used to locate the membrane-spanning regions of integral membrane proteins. (168)

hydrophilic Readily soluble in water. *Also see* polar. (16, 156)

hydrophobic Poorly soluble in water. *Also see* nonpolar. (15, 16, 156)

hydrophobic effect Tendency of hydrophobic regions of a molecule in aqueous solution to interact with each other, thereby minimizing their interactions with water. (25)

hyperpolarized Exhibiting a membrane potential that is more negative than the normal membrane potential. (199)

hypertonic Exhibiting a solute concentration that is higher than that inside cells. (176)

hypervariable regions Short stretches within the variable region of antibody light and heavy chains that account for most of the sequence variability between antibody molecules; involved in formation of the antigen-binding site. (720)

hypotonic Exhibiting a solute concentration that is lower than that inside cells. (176)

I band Region of a striated muscle myofibril that appears as a light band when viewed by microscopy; contains those regions of the thin actin filaments that do not overlap the thick myosin filaments. (564)

idiotype Region of the antigen-binding site (of a TCR or antibody molecule) that acts as an antigen, thereby eliciting an immune response. (742)

immune surveillance theory Theory postulating that immune rejection of newly forming cancer cells occurs routinely in normal individuals. (787)

immunoaffinity chromatography Type of chromatography in which the matrix is coupled to an antibody that binds specifically to the molecule being purified. (147)

immunoglobulin *See* antibody.

in situ hybridization Nucleic acid hybridization carried out on nucleic acids that still reside within cells. (98)

in vitro Involving isolated cells or subcellular components. (114)

in vivo Involving cells that still reside within intact organisms. (114)

induced-fit model Model postulating that the binding between an enzyme and its substrate triggers alterations in the shape of the enzyme as well as the substrate. (41)

inducer Molecule that induces the synthesis of specific proteins. (432)

induction Process by which a group of cells cause neighboring cells to differentiate in a particular way during embryonic development. (700)

initiation factors Group of proteins that promote the binding of ribosomal subunits to mRNA and initiator tRNA, thereby initiating the process of protein synthesis. (491)

initiation stage of carcinogenesis Irreversible conversion of a cell to a preneoplastic state by agents that induce DNA alterations. (789)

initiator tRNA Special transfer RNA that recognizes the AUG start codon; carries formylmethionine in prokaryotes and methionine in eukaryotes. (490)

inner mitochondrial membrane Innermost of the two membranes that enclose mitochondria; exhibits numerous folds, called cristae, that represent the site of electron transport and oxidative phosphorylation. (316)

inositol trisphosphate (IP$_3$) Second messenger utilized by the phosphoinositide signaling pathway; triggers the release of Ca^{2+} from the ER lumen into the cytosol. (218)

insertion sequence Transposable genetic element less than a few thousand bases long that contains only genes involved in moving the insertion sequence itself. (95)

integral protein Amphipathic membrane protein whose hydrophobic regions are embedded within the lipid bilayer and hydrophilic regions are exposed at the membrane's surfaces. (168)

integrins Family of plasma membrane receptors that bind to extracellular matrix components at the outer membrane surface and interact with cytoskeletal components at the inner membrane surface; includes receptors for fibronectin, laminin, and collagen. (232)

intercalated disks Membrane partitions, enriched in gap junctions, that divide cardiac muscle into separate cells containing single nuclei. (580)

intercalating agent Molecule that inserts itself between adjacent bases in DNA, distorting the shape of the double helix. (92)

interference Process by which two or more light waves combine to reinforce or cancel each other, producing a wave equal to the sum of the two combining waves. (119)

interference microscopy Type of light microscopy in which contrast is enhanced by splitting the light beam, passing one of the resulting beams through the specimen, and then recombining the beams to form an image. *Also see* differential interference microscope. (120)

interferons Group of lymphokines that are involved in activating killer T cells and inhibiting the reproduction of viruses. (739)

interleukins Group of lymphokines that are involved in the activation of both B cells and killer T cells. (730)

intermediate filaments Eukaryotic cytoskeletal filaments measuring about 10 nm in diameter that are characterized by enormous strength; constructed from a family of proteins exhibiting a coiled-coil motif. (9, 619)

intermembrane space Space between the two membranes that enclose a mitochondrion or chloroplast. (316, 359)

internal control region Promoter site located within the middle of a gene rather than upstream from the transcriptional start site; observed in the gene coding for 5S rRNA. (420)

interphase Stage of the eukaryotic cell cycle situated between successive M phases; composed of G$_1$, S, and G$_2$ phases. (522)

introns Noncoding nucleotide sequences that are interspersed between the coding regions (exons) of a gene or its initial RNA transcript; removed by RNA splicing. (84, 430)

ion channel Transmembrane protein that facilitates the diffusion of small ions across the membrane; usually can be opened or closed in response to appropriate stimuli. (184)

ion channel receptor Transmembrane protein that acts as an ion channel whose permeability can be controlled by the binding of a signaling ligand to a receptor site on the protein. (208)

ion-exchange chromatography Type of chromatography in which the matrix contains charged side chains that form transient ionic bonds with the molecules being purified. (147)

ionic bond *See* electrostatic bond.

ionizing radiation Radiation that removes electrons from the molecules with which it interacts. (93)

ionophores Small molecules that increase membrane permeability to specific ions; function either as mobile carriers (e.g., valinomycin) or channels (e.g., gramicidin). (182)

IP$_3$ *See* inositol trisphosphate.

IRE *See* iron-responsive element.

IRE-binding protein Protein that binds to IRE sequences residing in mRNA, thereby repressing translation or stabilizing the mRNA against degradation. (505)

iron-responsive element (IRE) Short nucleotide sequence found in mRNAs whose translation or stability is controlled by iron; binding site for an IRE-binding protein. (505)

iron-sulfur center Protein site containing iron bound to the sulfur atom of the amino acid cysteine. (328)

iron-sulfur proteins Proteins containing iron-sulfur centers that function as electron carriers in electron transfer pathways such as the respiratory chain and the photosynthetic electron transfer chain. (328)

irreversible inhibitor Molecule that binds covalently to an enzyme, permanently disrupting its catalytic activity. (48)

isodensity centrifugation Centrifugation employing a gradient whose density overlaps the density of the components being studied, allowing molecules and organelles to be separated based on differences in density. (135)

isoelectric focusing Electrophoresis performed in a gradient of varying pH; separates molecules based on differences in their isoelectric point (the pH at which each molecule carries no net charge). (146)

isotonic Exhibiting a solute concentration that is the same as that inside cells. (176)

isotopes Different forms of the same chemical element that have the same number of electrons but differing atomic weights. (141)

isozymes Variant forms of an enzyme that catalyze the same reaction. (55)

J gene segment (joining segment) DNA segment coding for a short stretch of amino acids located at the end of the variable region of an antibody heavy or light chain, directly adjacent to the constant region. (724)

Jun One of two polypeptide subunits of the AP1 transcription factor, which regulates the activity of genes involved in cell proliferation. *Also see* Fos. (464)

K_d *See* dissociation constant.
K_{eq} *See* equilibrium constant.
K_m *See* Michaelis constant.

karyotype Photograph of the complete set of chromosomes for a particular cell type, organized as homologous pairs arranged on the basis of differences in size and shape. (540)

kb *See* kilobases.

killer T cell Type of T lymphocyte that binds to antigen-MHC I complexes on target cells and then destroys such cells by secreting the pore-forming protein, perforin. (736)

kilobases (kb) One thousand base pairs. (99)

kinesin Motor protein that moves organelles and particles toward the plus end of microtubules using energy derived from ATP hydrolysis; belongs to a family of related motor proteins, some of which move in the opposite direction along microtubules. (617)

kinetochore Disk-shaped structure located at the centromere of a metaphase chromosome; attaches the chromosome to spindle microtubules. (538)

kinetochore microtubules Spindle microtubules that attach to chromosomal kinetochores. (539)

kinetoplast Large mitochondrion found in certain parasitic protozoa; stained intensely by the Feulgen reaction for DNA. (644)

Krebs cycle Cyclic metabolic pathway that oxidizes acetyl CoA to CO_2, generating ATP, NADH, and $FADH_2$. (326)

lac **operon** Group of adjoining bacterial genes that code for enzymes involved in lactose metabolism and whose transcription is inhibited by the *lac* repressor. (433)

lac **repressor** Protein that inhibits transcription of the *lac* operon by binding to the *lac* operator. (434)

lagging strand DNA strand that is elongated in the $3' \rightarrow 5'$ direction during DNA replication through the joining of short, discontinuous fragments each synthesized in the $5' \rightarrow 3'$ direction. *Also see* leading strand. (71)

lamellipodia (singular, *lamellipodium*) Flattened protrusions that transiently emerge from the surface of eukaryotic cells during cell crawling; supported by actin filaments. (588)

laminin Adhesive glycoprotein of the extracellular matrix, localized predominantly in the basal lamina. (231)

lamins Protein subunits of the intermediate filaments that make up the nuclear lamina. (536, 620)

lampbrush chromosome Chromosome comprised of two chromatids containing numerous DNA loops that are active in RNA synthesis; appears during prophase I of meiosis in oocytes that undergo an extensive growth phase. (675)

lateral diffusion Diffusion of a membrane lipid or protein in the plane of the membrane. (162)

leading strand DNA strand that is synthesized as a continuous chain in the $5' \rightarrow 3'$ direction during DNA replication. *Also see* lagging strand. (71)

leaflet One of the two sides of a lipid bilayer. (162)

lectins Carbohydrate-binding proteins that promote cell-cell adhesion by binding to sugar groups exposed at the outer cell surface. (234, 786)

leptotene First stage of prophase I of meiosis; characterized by condensation of chromatin fibers into visible chromosomes. (549)

leucine zipper DNA-binding motif found in many transcription factors; involves two α helices "zippered" together by hydrophobic interactions between leucine residues. (463)

leukemia Any malignancy of blood or lymphatic cells in which the cancer cells circulate in large numbers in the bloodstream. (778)

LHCI Light-harvesting complex of photosystem I. (373)

LHCII Light-harvesting complex of photosystem II. (373)

ligand Any small molecule that binds to a larger molecule. (206)

light reactions Portion of the photosynthetic pathway in which the absorption of light drives the flow of electrons from water to $NADP^+$, generating oxygen, NADPH, and ATP. (366)

light-harvesting complex (LHC) Collection of light-absorbing pigments, usually chlorophylls and carotenoids, linked together by proteins; photosystem component that absorbs photons of light and funnels the energy to the reaction-center complex. (370)

lignins Polymerized aromatic alcohols present in high concentration in the cell walls of woody plant tissues, where they contribute to the hardening of the wall. (244)

Lineweaver-Burk plot Graph in which the reciprocal of the initial reaction velocity ($1/v_i$) is plotted against the reciprocal of the substrate concentration ($1/[S]$) for an enzyme-catalyzed reaction; useful in determining V_{max} and K_m. (46)

linkers Short synthetic DNAs of known sequence that are covalently joined to the ends of DNA molecules being cloned. (101)

lipid bilayer Universal feature of cell membranes, consisting of a phospholipid sheet two molecules thick in which the hydrophobic tails of the two layers face each other and the hydrophilic head groups face the membrane's surfaces; glycolipids and steroids can also be incorporated into such a structure. (157)

lipid-anchored protein Membrane protein situated outside the lipid bilayer, but covalently bound to lipid molecules residing within the bilayer. (169)

lipids Chemically diverse group of biological molecules that are readily soluble in nonpolar solvents but poorly soluble in water. (15)

lipopolysaccharides Molecules consisting of complex sugar chains with fatty acids attached at one end; found in the outer membrane of Gram-negative bacteria. (250)

liposomes Lipid bilayer vesicles that are produced by exposing a phospholipid-water suspension to ultrasonic vibrations. (162)

lobopodia (singular, *lobopodium*) Thick, cylindrical protrusions supported by actin filaments that transiently emerge from the surface of eukaryotic cells. (588)

local mediator Signaling molecule secreted into the extracellular fluid, where it acts on neighboring cells located within a range of a few millimeters. (205)

lock-and-key theory Theory postulating that the three-dimensional shape of an enzyme's active site is complementary to the shape of the substrate with which it interacts. *Also see* induced-fit model. (41)

long-term potentiation (LTP) Increase in synaptic efficiency that occurs following repeated transmission of signals across a synapse. (768)

LTP *See* long-term potentiation.

lymphocytes Class of white blood cells responsible for both antibody production and cell-mediated immunity. *Also see* B lymphocyte and T lymphocyte. (717)

lymphokines Family of lymphocyte-secreted proteins that control the growth and differentiation of other lymphocytes. (730)

lymphoma Malignant tumor of white blood cells that grows as a solid mass of tissue. (778)

lysis Rupture of the plasma membrane, leading to cell death. (30, 709)

lysogenic virus Virus whose genetic information becomes incorporated into the host cell chromosomal DNA, thereby allowing the virus to persist for an indefinite period of time without destroying the host cell. (29)

lysosomal storage diseases Group of inherited diseases caused by lysosomal enzyme defects that cause the undigested substrate of the defective enzyme to accumulate within cells. (300)

lysosome Membrane-enclosed cytoplasmic organelle containing hydrolytic enzymes that exhibit maximal activity at acid pH. (8, 287)

M phase Stage of the eukaryotic cell cycle when the individual chromosomes become visible and the nucleus and then the rest of the cell divide. (523)

macroautophagy Process by which lysosomes degrade damaged or aging organelles. (292)

macromolecules Biological polymers with molecular weights ranging from several thousand to a million or more, constructed by linking together smaller building blocks; include proteins, nucleic acids, and polysaccharides. (14)

macrophage Cell that degrades antigens and presents the resulting fragments to helper T cells. (715)

major histocompatibility complex *See* MHC class I molecule and MHC class II molecule.

malate-aspartate shuttle Mechanism for carrying electrons derived from cytoplasmic NADH into the mitochondrion, where the electrons are delivered to NAD^+ for subsequent oxidation by the respiratory chain. (340)

male pronucleus Nucleus formed when a sperm cell fertilizes an egg and the sperm chromatin becomes enclosed in a new nuclear envelope within the egg cytoplasm. (691)

malignant (tumor) Any tumor capable of spreading by invasion and metastasis. (778)

MAP kinases (mitogen-activated protein kinases) Family of protein-serine/threonine kinases activated in cells that have been stimulated by protein growth factors. (527)

MAPs *See* microtubule-associated proteins.

maternal inheritance Pattern of inheritance in which the gene governing a particular trait is contributed solely by the female parent; such genes usually reside within mitochondria or chloroplasts. (640)

matrix *See* mitochondrial matrix or extracellular matrix.

maturase Mitochondrial or chloroplast protein translated from an mRNA in which part of an intron is translated along with the exon sequences; the maturase then catalyzes removal of the intron sequences from the mRNA. (653)

maturation promoting factor *See* MPF.

maximum velocity (V_{max}) Maximum rate that an enzyme-catalyzed reaction can attain after the enzyme has become saturated with substrate. (43)

megaspore Haploid cell produced by meiotic division of megasporocytes in flowering plants; undergoes mitotic divisions to produce an egg cell as well as several other cell types. (686)

megasporocyte Primordial germ cell in the ovary of a flowering plant; undergoes meiosis to produce haploid megaspores. (686)

meiosis Series of two cell divisions preceded by a single round of DNA replication; converts a single diploid cell into four haploid cells. (548)

meiotic division *See* meiosis.

melting temperature (T_m) Temperature at which the transition from double-stranded to single-stranded DNA is halfway complete when DNA is denatured by increasing the temperature. (96)

membrane potential Voltage across a membrane; the membrane potential of the plasma membrane is usually negative, which means that the inner surface of the membrane is negative relative to the outer surface. (181, 195)

membrane-attack complex Components of the complement system that form pores in the plasma membrane of bacterial cells being attacked by the immune system. (732)

memory B cells Reserve population of B cells directed against a specific antigen that rapidly proliferate into antibody-secreting plasma cells when the same antigen is encountered again. (732)

memory T cells Reserve population of T cells directed against a specific antigen that rapidly proliferate into killer T cells when the same antigen is encountered again. (737)

meristem Rapidly dividing, undifferentiated cells that give rise to most of the differentiated cell types of multicellular plants. (661)

mesoderm Middle cell layer of a gastrula; gives rise to supporting tissues such as muscle, cartilage, bone, blood, and connective tissue. (696)

mesophyll cells Cells found in the interior of plant leaves; cells in which PEP carboxylase fixes CO_2 in C_4 plants, generating C_4 acids that are transported to bundle sheath cells. (393)

messenger ribonucleoprotein complex (mRNP) Messenger RNA complexed with proteins that inhibit its translation. (505)

messenger RNA (mRNA) RNA molecule that codes for the synthesis of one or more polypeptide chains. (81)

metaphase Stage during mitosis or meiosis when the chromosomes become attached to the spindle and migrate to its equator. (536, 537, 551)

metastasis Spreading of cancer cells to distant parts of the body via the circulatory system. (778)

MHC class I molecule Plasma membrane glycoprotein expressed on the surface of most nucleated cells; binds to endogenous antigen fragments derived from molecules originating within the cell, forming antigen–MHC I complexes that are recognized by killer T cells. (716)

MHC class II molecule Plasma membrane glycoprotein expressed on the surface of cells involved in the immune response, mainly B lymphocytes and antigen-presenting cells such as macrophages; binds to antigen fragments derived from exogenous molecules, forming antigen–MHC II complexes that are recognized by helper T cells. (715)

Michaelis constant (K_m) Substrate concentration at which the rate of an enzyme-catalyzed reaction is equal to one-half the maximum velocity. (43)

Michaelis-Menten equation Equation describing the relationship between K_m and V_{max} for an enzyme-catalyzed reaction. (45)

microautophagy Process in which lysosomes take up and degrade cytosolic proteins by invagination of the lysosomal membrane. (292)

microfilaments Alternative name for the actin filaments of nonmuscle cells. (583)

microscopic autoradiography Procedure in which specimens being examined by light or electron microscopy are overlaid with a photographic emulsion to permit detection of radioactive molecules. (132)

microsomal fraction One of the fractions collected by differential centrifugation during subcellular fractionation; contains microsomes plus free ribosomes. (138)

microsomes Membrane vesicles formed from fragments of the endoplasmic reticulum during cell homogenization. (259)

microspheres Small vesicles that form when proteinoids are dissolved in hot water and allowed to cool slowly; experimental model for studying how primitive cells might have evolved. (637)

microspikes Thin protrusions shorter than filopodia that transiently emerge from the surface of eukaryotic cells; supported by actin filaments. (588)

microspore Haploid cell produced by meiotic division of microsporocytes in flowering plants; becomes encased in a thick spore coat to form a pollen grain. (685)

microsporocyte Primordial germ cell in the anther of a flowering plant; undergoes meiosis to produce haploid microspores. (685)

microtubule-associated proteins (MAPs) Proteins that bind to and influence the properties of microtubules. (596)

microtubule-organizing centers (MTOCs) Structures that initiate microtubule assembly, the primary example being the centrosome. (600)

microtubules Hollow cytoskeletal tubules 25 nm in diameter composed of the protein tubulin; occur in the cilia, flagella, mitotic spindle, and cytoplasm of eukaryotic cells, where they contribute to cell shape and motility. (9, 595)

microvilli Relatively permanent, fingerlike projections of the plasma membrane that increase the area of the cell surface; supported by bundles of crosslinked actin filaments. (10, 591)

midbody Layer of dense material that forms near the spindle equator during cytokinesis in animal cells. (545)

middle lamella First layer of the plant cell wall to be synthesized; ends up farthest away from the plasma membrane, where it functions to hold adjacent cells together. (245)

miniature postsynaptic potentials Tiny depolarizations of a postsynaptic membrane caused by the spontaneous release of neurotransmitter from the presynaptic axon. (759)

minus end Slow-growing end of a microtubule or actin filament. (568, 598)

mismatch repair DNA repair mechanism that detects and corrects base pairs that are improperly hydrogen bonded. (94)

mitochondrial fraction One of the fractions collected by differential centrifugation during subcellular fractionation; enriched in mitochondria, but also contains lysosomes and peroxisomes. (138)

mitochondrial matrix Space enclosed by the inner mitochondrial membrane; site of the Krebs cycle. (316)

mitochondrion Double-membrane-enclosed cytoplasmic organelle of eukaryotic cells that carries out the Krebs cycle, electron transport, and oxidative phosphorylation. (9, 316)

mitogen-activated protein kinases *See* MAP kinases.

mitosis Nuclear division that yields two nuclei containing identical sets of chromosomes; consists of prophase, metaphase, anaphase, and telophase. (523)

mitotic division Type of eukaryotic cell division that generates two cells containing identical sets of chromosomes. (522)

mitotic spindle *See* spindle.

mixed-function oxidase Enzyme catalyzing an oxidation reaction in which one atom of oxygen appears in the product and the other appears in water. (262)

monoclonal antibody Homogeneous population of antibody molecules produced by a hybridoma. (143)

monomers Small molecules that are linked together to form larger molecules called polymers. (14)

monosaccharides Simple sugars; can be linked together to form oligosaccharides and polysaccharides. (14)

morphogenesis Process by which cells become organized into tissues and organs. (705)

morphogenic determinants Cytoplasmic substances that are asymmetrically distributed during cell division, thereby influencing how the resulting cells will differentiate. (697)

morphogens Diffusible molecules that are produced at specific locations in developing embryos and create gradients that allow cells to locate their position within the embryo. (707)

motif Region of protein secondary structure consisting of small segments of α helix and/or β sheet connected by looped regions of varying length. (23)

motor protein Molecule that generates movement by advancing along a surface using energy derived from ATP hydrolysis. *Also see* myosin, kinesin, and dynein. (543, 609)

moving-zone centrifugation Type of velocity centrifugation in which the sample is applied as a narrow layer at the top of the centrifuge tube, and centrifugation is stopped before the particles reach the bottom of the tube; separates organelles and molecules based mainly on differences in size. *Also see* differential centrifugation. (135)

MPF (maturation promoting factor) Protein kinase consisting of cdc2 kinase bound to mitotic cyclin; controls the transition from G_2 phase to M phase during mitosis, and the transition from prophase I to metaphase I during meiosis. (524, 678)

mRNA *See* messenger RNA.

mRNP *See* messenger ribonucleoprotein complex.

MTOC *See* microtubule-organizing center.

murein Polysaccharide-peptide complex that forms the backbone of bacterial cell walls; consists of alternating N-acetylglucosamine and N-acetylmuramic acid residues crosslinked by peptide bridges; also called peptidoglycan. (249)

muscarinic acetylcholine receptor Plasma membrane receptor for acetylcholine that causes K+ channels to open; the resulting diffusion of K+ out of the cell hyperpolarizes the plasma membrane. *Also see* nicotinic acetylcholine receptor. (218, 762)

muscle fibers Long, cylindrical cells found in skeletal muscle, measuring up to several centimeters in length and possessing hundreds of nuclei. (564)

muscular dystrophy Family of diseases characterized by progressive degeneration of skeletal muscle cells; the most common form is caused by a defect in the gene coding for dystrophin. (580)

mutagen Any agent that causes mutations. (92)

mutation Change in the base sequence of a DNA molecule. (75, 90)

myasthenia gravis Disease involving muscle weakness caused by the production of antibodies that bind to and inactivate acetylcholine receptors. (579)

myelin Sheath of stacked membranes that surround and insulate the axons of many neurons; produced by oligodendrocytes and Schwann cells. (158, 751)

myofibrils Cylindrical structures composed of highly organized arrays of thin actin filaments and thick myosin filaments; found in the cytoplasm of skeletal muscle cells. (564)

myoglobin Oxygen-binding protein found in the cytoplasm of skeletal muscle cells. (578)

myosin Motor protein that advances along actin filaments using energy derived from ATP hydrolysis; the two-headed myosin II molecule occurs in the thick filaments of skeletal muscle, whereas single-headed myosin I is involved in cytoplasmic movements in nonmuscle cells. (565)

myosin light-chain kinase Enzyme that phosphorylates myosin light chains, thereby triggering smooth muscle contraction. (582)

myosin light-chain phosphatase Enzyme that removes phosphate groups from myosin light chains, thereby triggering smooth muscle relaxation. (582)

N-CAM (neural cell adhesion molecule) Plasma membrane glycoprotein that mediates cell-cell adhesion in neurons. (233)

N-linked glycosylation Addition of sugar chains to the amino group of asparagine in protein molecules. *Also see* O-linked glycosylation. (269)

N-terminus End of a polypeptide chain that contains the first amino acid to be incorporated during mRNA translation; usually retains a free amino group. (21, 487)

Na+–K+ ATPase (Na+–K+ pump) *See* sodium-potassium pump.

NAD+ and NADH Oxidized and reduced forms of nicotinamide adenine dinucleotide, a coenzyme involved in numerous cellular oxidation-reduction reactions, including key steps in glycolysis, the Krebs cycle, and electron transport. (57, 311)

NADP+ and NADPH Oxidized and reduced forms of nicotinamide adenine dinucleotide phosphate, a coenzyme that functions as the final electron acceptor in the light reactions of photosynthesis. (58, 371)

NADP+ reductase Flavoprotein in the photosynthetic electron transfer chain that transfers electrons from ferredoxin to NADP+. (370)

natural killer (NK) cell Lymphocyte that can destroy tumor cells or virus-infected cells in the absence of either antibody formation or a typical cell-mediated immune response. (718)

negative control Inhibition of transcription by a DNA-binding regulatory protein. *Also see* positive control. (436)

negative cooperativity Property of enzymes possessing multiple binding sites in which the binding of substrate to one active site causes other active sites to bind substrate with decreased affinity. *Also see* positive cooperativity. (53)

negative staining Technique in which an unstained specimen is visualized microscopically against a darkly stained background. (123)

neoplasm Growing tissue mass created by the loss of normal growth control; also called a tumor. (778)

Nernst equation Equation for calculating the membrane potential (equilibrium potential) that will exactly counterbalance the tendency of an ion to diffuse down its concentration gradient. (197)

Nernst potential *See* equilibrium potential.

nerve growth factor (NGF) Protein growth factor that maintains the viability of neurons and stimulates neurite outgrowth. (772)

nerve impulse Any change in membrane potential transmitted along nerve cells. (755)

neurites Thin cytoplasmic extensions that emerge from immature neurons and develop into axons and dendrites. (771)

neuroblastoma Tumor of immature nerve cells that can be maintained in culture and induced to differentiate. (774)

neurofilament Type of intermediate filament found in high concentration in nerve cell axons. (749)

neuromuscular junction Synapse between a nerve cell axon and a skeletal muscle cell. (573)

neuron Nerve cell consisting of a cell body, axons, and dendrites. (748)

neurosecretion Secretion of hormones into the bloodstream by neurons. (770)

neurotransmitter Substance released by a neuron that transmits nerve impulses across a synapse. (205, 757)

neurotransmitter-gated ion channel Ion channel whose permeability is regulated by the binding of a neurotransmitter. (210)

NGF *See* nerve growth factor.

nicotinamide adenine dinucleotide *See* NAD⁺.

nicotinamide adenine dinucleotide phosphate *See* NADP⁺.

nicotinic acetylcholine receptor Plasma membrane receptor whose permeability to Na⁺ increases upon binding acetylcholine; the resulting diffusion of Na⁺ into the cell depolarizes the plasma membrane. *Also see* muscarinic acetylcholine receptor. (208, 762)

nitric oxide (NO) Gas molecule produced by neurons (and certain other cell types) that transmits signals to neighboring cells by stimulating guanylyl cyclase. (766)

nitrogenous bases Nitrogen-containing ring compounds used in the construction of nucleotides and nucleic acids; include the purines adenine and guanine, and the pyrimidines cytosine, thymine, and uracil. (19)

NK cell *See* natural killer cell.

NLS *See* nuclear localization sequence.

nodes of Ranvier Periodic interruptions in the myelin sheath that expose the plasma membrane of the underlying axon. (751)

Nomarski interference microscope *See* differential interference microscope.

noncompetitive inhibitor Inhibitor of enzyme activity that binds reversibly to an enzyme at some location other than the active site. (49)

noncyclic photophosphorylation ATP synthesis coupled to the light-induced flow of electrons from water to NADP⁺ via the photosynthetic electron transfer chain. (375)

nonhistone proteins Proteins other than histones that bind to the nuclear DNA of eukaryotes; include HMG proteins, transcriptions factors, structural proteins, and enzymes. (410)

nonpolar Poorly soluble in water. *Also see* hydrophobic. (156)

nonreceptor protein-tyrosine kinase A protein-tyrosine kinase lacking a receptor domain that regulates enzyme activity. *Also see* receptor protein-tyrosine kinase. (798)

Northern blotting Technique for hybridizing electrophoretically separated RNA molecules to a radioactive DNA probe. *Also see* Southern blotting. (98)

NPC *See* nuclear pore complex.

NSF (N-ethylmaleimide-sensitive fusion protein) Soluble cytoplasmic protein that acts in conjunction with several NSF attachment proteins (SNAPs) to facilitate the fusion of various kinds of membranes. (286)

nuclear envelope Double-membrane envelope that encloses the eukaryotic nucleus; characterized by the presence of numerous nuclear pore complexes. (8, 403)

nuclear fraction One of the fractions collected by differential centrifugation during subcellular fractionation; enriched in nuclei, but may be contaminated by unbroken cells and, in the case of plants, chloroplasts. (138)

nuclear lamina Thin layer of intermediate filaments located directly beneath the inner nuclear membrane; composed of protein subunits called lamins. (403)

nuclear localization sequence (NLS) Short amino acid sequence that targets proteins for uptake into the nucleus. (405, 511)

nuclear matrix Filament network that provides a supporting framework for the nucleus. (408)

nuclear pore complex (NPC) Channel through which molecules enter and exit the nucleus, both by nonselective diffusion of smaller molecules and selective transport of larger molecules. (403)

nucleic acid Polymer composed of nucleotides joined by phosphodiester bonds. *Also see* DNA and RNA. (19)

nucleic acid hybridization Family of techniques in which single-stranded nucleic acids are allowed to bind to each other by complementary base-pairing; used for assessing whether two nucleic acids contain similar base sequences. (96)

nucleoid Region of a bacterial cell containing the chromosomal DNA; lacks an enclosing membrane. (8, 415)

nucleolus (plural, *nucleoli*) Spherical structure located in the nucleus that contains granules and fibrils involved in ribosome formation. (8, 407, 476)

nucleoplasm Fluidlike material containing the soluble components of the nucleus; comparable to the cytosol of the cytoplasm. (407)

nucleosome Repeating structural unit of eukaryotic chromatin; consists of an octamer of histones associated with about 200 base pairs of DNA. (412)

nucleotide Molecule consisting of a purine or pyrimidine base joined to a five-carbon sugar (ribose or deoxyribose) containing an attached phosphate group(s); the fundamental building block of nucleic acids. (14, 19)

nucleus Large, double-membrane-enclosed organelle that contains the chromosomal DNA of eukaryotic cells. (8, 401)

nude mice Mice born without a thymus and therefore lacking T lymphocytes; thus they are unable to carry out cell-mediated immunity or produce antibodies. (735, 783)

nurse cells Accessory cells connected to invertebrate oocytes by cytoplasmic bridges that allow nutrients to be directly transferred to the oocyte. (677)

O-linked glycosylation Addition of sugar chains to the hydroxyl groups of serine, threonine, and hydroxylysine in protein molecules. *Also see* N-linked glycosylation. (279)

Okazaki fragments Short fragments of newly synthesized, lagging-strand DNA that are joined together by DNA ligase during DNA replication. (71)

oligodendrocyte Type of glial cell that forms myelin sheaths in the brain and spinal cord. (749)

oncogenes Genes whose presence can cause a cell to become malignant; arise by mutation from normal cellular genes called proto-oncogenes. (794)

oncogenic virus Virus that can cause cancer. (792)

oogenesis Process by which diploid oogonia are converted into haploid egg cells. (674)

operator DNA site to which a repressor protein binds. (433)

opiate receptors Plasma membrane receptors for enkephalins and endorphins; occur mainly in brain neurons that mediate the perception of pain. (765)

organelle Any discrete intracellular structure that is specialized for carrying out a particular function. (5)

organic molecule Molecule containing covalently linked carbon atoms. (14)

osmosis Movement of water through a semipermeable membrane driven by a difference in solute concentration. (176)

outer membrane Outermost of the two membranes enclosing a mitochondrion, chloroplast, or nucleus; also, the membrane enriched in lipopolysaccharides that lies outside the murein wall of Gram-negative bacteria. (250, 316, 359, 403)

ovary Organ that produces egg cells in both animals and flowering plants. (674, 683)

oxidation Process of giving up electrons during a chemical reaction. (57, 310)

β-oxidation pathway Cyclic set of reactions for converting fatty acids to acetyl CoA. (324)

oxidative phosphorylation ATP synthesis coupled to the transport of electrons through the respiratory chain; mediated by an electrochemical proton gradient established during electron transfer. (315, 327, 333)

P face (protoplasmic face) During freeze-fracturing, the fracture face derived from the half of a membrane that lies adjacent to the cytoplasm. *Also see* E face. (124)

P site *See* peptidyl site.

p39^{mos} Protein kinase that inhibits the breakdown of cyclin. (678)

p53 Tumor suppressor gene that codes for a transcription factor involved in preventing damaged cells from proliferating. (800)

P680 Reaction-center chlorophyll of photosystem II. (371)

P700 Reaction-center chlorophyll of photosystem I. (371)

pachytene Stage during prophase I of meiosis when crossing over between homologous chromosomes takes place. (549)

packing ratio Ratio obtained by dividing the length of a DNA molecule by the length of the particle or fiber in which the DNA is packaged. (414)

palindrome DNA sequence in which the base sequence read in the 5' → 3' direction in each strand is the same. (99, 436)

Pap smear Screening procedure for uterine cancer in which cells present in vaginal secretions are examined under the microscope. (783)

paper chromatography Type of chromatography in which the molecules being separated flow across a sheet of filter paper. (148)

paracrine Type of signaling in which a secreted molecule acts on nearby cells. (205)

Pasteur effect Decrease in the rate of glycolysis that is observed when anaerobic cells are exposed to oxygen. (343)

patch clamp technique Technique in which a tiny micropipette placed on the surface of a cell is used to measure the movement of ions through individual ion channels. (201)

PCR *See* polymerase chain reaction.

PDGF *See* platelet-derived growth factor.

pectins Polysaccharides rich in galacturonic acid found in plant cell walls, where they form a matrix in which cellulose microfibrils are embedded. (244)

penicillin Drug that blocks the transpeptidation reaction involved in murein synthesis, thereby inhibiting formation of the bacterial cell wall. (251)

pentose Any sugar with five carbon atoms—for example, ribose and deoxyribose. (19)

PEP carboxylase Enzyme that catalyzes the initial CO_2-fixing reaction in C_4 plants; joins CO_2 with phosphoenolpyruvate to form the C_4 acid, oxaloacetate. (393)

peptide bond Covalent linkage between the carboxyl group ($-COO^-$) of one amino acid and the amino group ($-NH_3^+$) of another amino acid. (21)

peptidoglycan *See* murein.

peptidyl site (P site) Ribosomal site that contains the growing polypeptide chain at the beginning of each elongation cycle. *Also see* aminoacyl site. (492)

peptidyl transferase Ribosomal component that catalyzes peptide bond formation; in bacterial ribosomes, tentatively identified as 23S rRNA. (495)

perforin Protein secreted by killer T cells that destroys target cells by creating pores in the target cell plasma membrane. (737)

peripheral protein Hydrophilic protein bound by ionic bonds to a membrane surface. (169)

periplasmic space Space between the plasma membrane and outer membrane of Gram-negative bacteria. (251)

peroxisome Family of membrane-enclosed cytoplasmic organelles containing enzymes involved in the metabolism of hydrogen peroxide, lipids, and carbohydrates. (8, 347)

petites Class of yeast mutants characterized by slow growth and defective mitochondria. (642)

phage *See* bacteriophage.

phagocytosis Uptake of particulate matter into cells via membrane vesicles that pinch off from the plasma membrane. (293)

phagosome Membrane vesicle that has been pinched off from the plasma membrane by the process of phagocytosis. (293)

phalloidin Drug that promotes actin polymerization. (584)

phase-contrast microscopy Type of light microscopy that combines refracted and unrefracted light waves, thereby allowing differences in thickness and refractive index to be visualized as differences in brightness. (119)

pheophytin Molecule of chlorophyll *a* containing two hydrogen atoms in place of the central Mg^{2+}. (373)

phorbol esters Family of tumor promoters that mimic the ability of diacylglycerol to activate protein kinase C. (221, 790)

phosphocreatine Energy-storing molecule that replenishes ATP during muscle contraction by donating its phosphate group to ADP. (578)

phosphodiester bond Covalent linkage in which two parts of a molecule are joined through a phosphate group. (19)

phosphodiesterase Enzyme that catalyzes the cleavage of phosphodiester bonds, especially in cyclic AMP. (211)

phosphoenolpyruvate carboxylase *See* PEP carboxylase.

phosphoinositide signaling pathway Pathway by which extracellular signaling molecules activate phospholipase C, which in turn catalyzes formation of the second messengers inositol trisphosphate and diacylglycerol. (218)

phospholipase C Enzyme that cleaves phosphatidylinositol bisphosphate into inositol trisphosphate and diacylglycerol, which act as second messengers in the phosphoinositide signaling pathway. (218)

phospholipid Lipid possessing a covalently attached phosphate group and therefore exhibiting both hydrophilic and hydrophobic properties; main component of the lipid bilayer that forms the structural backbone of all cell membranes. (16, 161)

phospholipid transfer protein Cytoplasmic protein that removes a specific type of phospholipid from one membrane and inserts it into another membrane. (302)

phospholipid translocator Membrane protein that catalyzes the flip-flop of membrane lipids from one side of a lipid bilayer to the other; also called flippase. (261, 302)

phosphorylase kinase Protein kinase that catalyzes the phosphorylation and resulting activation of glycogen phosphorylase, thereby promoting glycogen breakdown. (216)

phosphorylation Addition of a phosphate group. (54)

phosphotransferase pathway System for transporting sugars across the bacterial plasma membrane that involves the transfer of a phosphate group from phosphoenolpyruvate to the sugar being transported. (188)

photon Fundamental particle of light energy. (369)

photophosphorylation ATP synthesis coupled to the light-induced flow of electrons through the photosynthetic electron transfer chain; mediated by an electrochemical proton gradient established during electron transfer. (375)

photorespiration Light-dependent pathway that decreases the efficiency of photosynthesis by oxidizing reduced carbon compounds without capturing the released energy; occurs when O_2 substitutes for CO_2 in the reaction catalyzed by rubisco. (391)

photosynthesis Process by which plants and certain bacteria capture light energy and use it to drive the fixation of CO_2, yielding carbohydrates and oxygen. (355)

photosynthetic electron transfer chain Series of electron carriers that transfers electrons from water to $NADP^+$, driven by energy derived from light. (370)

photosystem Mixture of chlorophyll, accessory pigments, and proteins that function together in thylakoid membranes as a light-absorbing unit for photosynthesis. (370)

photosystem I Chloroplast photosystem that transfers electrons from plastocyanin to ferredoxin; site of the reaction-center chlorophyll, P700. (373)

photosystem II Chloroplast photosystem that transfers electrons from water to plastoquinone; site of the reaction-center chlorophyll, P680. (372)

phragmoplast Layer of membrane vesicles that forms across the spindle equator at the beginning of cytokinesis in plant cells. (546)

phycobilins Accessory photosynthetic pigments that give red algae and cyanobacteria their characteristic red and blue colors. (367)

phycobilisome granules Granules attached to the thylakoid membranes of red algae and cyanobacteria; consist of phycobilins bound to protein. (368)

phytochrome Light-absorbing protein that controls a variety of light-induced events in plants. (663)

P_i A free phosphate group. (56)

pigment granules Pigment-containing vesicles localized near the periphery of egg cells, especially at the animal pole. (677)

pinocytosis Nonselective uptake of extracellular fluid into membrane vesicles that pinch off from the plasma membrane. (294)

plaque Clear zone produced when bacterial cells in a small region of a culture dish are destroyed by infection with bacteriophage. For another use of the same term, *see* plaque-bearing junction. (103)

plaque-bearing junction Type of cell junction that is connected to either intermediate filaments or actin filaments via a dense fibrous structure called a plaque; examples include desmosomes, hemidesmosomes, and adherens junctions (adhesion belts and focal adhesions). (238)

plasma cell Antibody-secreting cell that arises from the proliferation of activated B lymphocytes. (732)

plasma membrane Lipid bilayer membrane (with associated proteins) that encloses all cells, regulating the passage of material into and out of the cell. (156)

plasmids Small circular DNA molecules that can replicate independent of chromosomal DNA; useful as cloning vectors. (101)

plasmodesmata (singular, *plasmodesma*) Narrow channels in the cell walls of adjacent plant cells that allow direct cytoplasmic continuity between adjacent cells. (246)

plasmolysis Shrinkage of a plant cell that causes the cytoplasm to pull away from the cell wall; occurs when cells are placed in a hypertonic solution. (177)

plastids Family of plant cytoplasmic organelles derived from proplastids; include chloroplasts, etioplasts, chromoplasts, amyloplasts, proteinoplasts, and elaioplasts. (661)

plastocyanin Copper-containing protein that transfers electrons to photosystem I during the light reactions of photosynthesis. (370)

plastoquinone Lipid component of thylakoid membranes that transfers electrons from photosystem II to the cytochrome b_6f complex during the light reactions of photosynthesis. (370)

platelet-derived growth factor (PDGF) Protein growth factor that stimulates the proliferation of connective tissue and smooth muscle cells by binding to a receptor that exhibits protein-tyrosine kinase activity. (526)

plus end Fast-growing end of a microtubule or actin filament. (568, 598)

pmf *See* proton motive force.

polar Readily soluble in water; exhibiting one region displaying a partial positive charge and another region displaying a partial negative charge. *Also see* hydrophilic. (11, 156)

polar bodies Tiny haploid cells that are produced during oogenesis but eventually disappear. (678)

polar lobe Transient cytoplasmic protrusion formed at the vegetal pole during the first few cleavages of a mollusk egg. (696)

polar microtubules Spindle microtubules that span the area between the two spindle poles. (539)

polar nuclei Nuclei generated by mitotic division of haploid megaspores during female gametogenesis in flowering plants; destined to fuse with a sperm nucleus to produce endosperm. (686)

polarization microscopy Type of light microscopy that uses polarized light to detect structures that are birefringent—that is, capable of rotating polarized light. (120)

pollen grain Haploid microspore of a flowering plant encased in a thick spore coat; gives rise via mitotic divisions to a tube cell and two sperm cells. (685)

pollen tube Cytoplasmic projection that emerges from the tube cell when a pollen grain lands on a stigma. (694)

poly-A tail Stretch of about 50 to 250 adenine nucleotides added to the 3' end of most eukaryotic mRNAs after transcription is completed. (428)

polycistronic mRNA Messenger RNA molecule coding for more than one polypeptide chain. (425)

polymer Large molecule constructed by joining together smaller molecules called monomers. (14)

polymerase chain reaction (PCR) Procedure that uses DNA polymerase to produce large quantities of a DNA sequence that is initially present in small amounts; requires that part of the sequence of the gene being copied be known. (105)

polypeptide Polymer of amino acids linked together by peptide bonds. (21)

polyprotein Polypeptide chain that can be cleaved into more than one protein product. (502)

polysaccharide Polymer of monosaccharides linked together by glycosidic bonds. (14)

polysome Cluster of ribosomes simultaneously translating the same mRNA molecule. (497)

polyspermy Fertilization of an egg by more than one sperm cell. (688)

polytene chromosome Giant chromosome containing multiple copies of the same DNA molecule. (446)

porins Transmembrane proteins found in the outer membranes of Gram-negative bacteria, mitochondria, and chloroplasts; form pores that permit the passive diffusion of small hydrophilic molecules. (251, 320, 364)

positive control Activation of transcription by a DNA-binding regulatory protein. *Also see* negative control. (436)

positive cooperativity Property of enzymes possessing multiple binding sites in which the binding of substrate to one active site causes other active sites to bind substrate with increased affinity. *Also see* negative cooperativity. (53)

positive staining Technique in which a stained specimen is viewed microscopically against an unstained background. (124)

potocytosis Transport of small molecules and ions into cells by caveolae that do not pinch off completely from the plasma membrane. (297)

PP1 *See* protein phosphatase-1.

ppGpp and **pppGpp (guanosine tetraphosphate** and **guanosine pentaphosphate)** Nucleotides that accumulate in bacteria and inhibit rRNA synthesis when protein synthesis has been curtailed by an inadequate supply of amino acids. (504)

pre-mRNA RNA molecule that is cleaved and processed to yield a mature messenger RNA. (427)

pre-rRNA RNA molecule that is cleaved and processed to yield mature ribosomal RNAs. (480)

prenylation Addition of an isoprenoid lipid group to a protein molecule; a step in the formation of lipid-anchored membrane proteins. (305)

primary cell wall Flexible portion of the plant cell wall formed while cell growth is still occurring; contains a loosely organized network of cellulose microfibrils. (245)

primary culture Eukaryotic cell population that is capable of dividing only a limited number of times in culture. (116)

primary embryonic induction Induction that leads to the formation of an entire new embryo. (701)

primary lysosome Lysosome that has not yet carried out any digestive activity. (292)

primary structure Linear sequence of amino acids in a protein molecule. (21)

primordial germ cells Diploid cells that undergo meiosis to produce either sperm or egg cells. (674)

primosome Protein complex containing DNA helicase and DNA primase that moves along the lagging strand during DNA replication, unwinding the double helix and synthesizing RNA primers as it proceeds. (74)

procollagen Triple helix comprised of three collagen precursor chains containing extra amino acids at both ends. (228)

proenzyme Enzyme precursor that is converted to an active enzyme by cleavage of the polypeptide chain. (54, 284)

prohormone Hormone precursor that is converted to an active hormone by cleavage of the polypeptide chain. (284)

prokaryotic cells Single-celled organisms that lack a membrane-enclosed nucleus; comprised of subgroups called eubacteria and archaebacteria. (8)

prolamellar body Highly ordered array of membrane tubules that develops in chloroplasts deprived of light. (662)

promitochondria Small, double-membrane enclosed vesicles found in anaerobic yeast; differentiate into mitochondria when oxygen is introduced. (660)

promoter site DNA region to which RNA polymerase binds before initiating transcription. (419)

promotion stage of carcinogenesis Gradual, partially reversible process by which preneoplastic cells are converted into cancer cells by agents that stimulate cell division. (789)

proofreading Correction of mismatched base pairs during DNA replication by the 3'-exonuclease activity of DNA polymerase. (72)

prophase Initial phase of mitosis, characterized by chromosome condensation, spindle assembly, and nuclear envelope breakdown; prophase I of meiosis is more complex, consisting of stages called leptotene, zygotene, pachytene, diplotene, and diakinesis. (536, 549)

proplastid Double-membrane-enclosed, plant cytoplasmic organelle that develops into several kinds of plastids, including chloroplasts. (661)

prosthetic group Small, tightly bound organic molecule that is required for the activity of an enzyme or other protein. (47)

protamines Small, extremely basic proteins that bind to DNA in place of histones in the sperm cells of certain animals. (410)

proteasome Multiprotein complex that catalyzes the ATP-dependent breakdown of proteins linked to ubiquitin. (510)

protein Macromolecule consisting of one or more polypeptide chains, each containing from a few dozen to hundreds or even thousands of amino acids. (21)

protein disulfide isomerase Enzyme that catalyzes the formation and breakage of disulfide bonds, thereby facilitating protein folding. (272)

protein kinases Large family of enzymes that catalyze the transfer of a phosphate group from ATP to an amino acid in a protein molecule; include protein-serine/threonine kinases and protein-tyrosine kinases. (214)

protein kinase A A protein-serine/threonine kinase that is activated by the second messenger, cyclic AMP. (214)

protein kinase C A protein-serine/threonine kinase that is activated by the second messenger, diacylglycerol. (221)

protein kinase G A protein-serine/threonine kinase that is activated by the second messenger, cyclic GMP. (222)

protein phosphatase inhibitor-1 Protein that inactivates protein phosphatase-1. (216)

protein phosphatase-1 (PP1) Enzyme that removes phosphate groups from, and thereby regulates the catalytic activity of, enzymes involved in glycogen metabolism. (216)

protein-serine/threonine kinases Enzymes that catalyze phosphorylation of the amino acids serine and threonine in protein molecules. (221, 798)

protein-translocating channel ER membrane channel through which newly forming polypeptides pass as they are synthesized on membrane-bound ribosomes. (266)

protein-tyrosine kinases Enzymes that catalyze phosphorylation of the amino acid tyrosine in protein molecules. (221, 527, 797)

proteinoids Polymers formed when amino acids are heated in the absence of water. (635)

proteoglycans Complexes between proteins and glycosaminoglycans found in the extracellular matrix of animal tissues. (225)

protoeukaryote Hypothetical evolutionary ancestor of present-day eukaryotic cells; postulated to contain an enveloped nucleus, but no mitochondria or chloroplasts. (640)

proton gradient *See* electrochemical proton gradient.

proton motive force (pmf) Force exerted by an electrochemical proton gradient that tends to drive protons back down the gradient. (337)

proto-oncogenes Normal cellular genes that can converted into oncogenes by mutation, amplification, or DNA rearrangement. (794)

protoplast A plant or bacterial cell whose cell wall has been removed. (5, 249, 362, 703)

provirus DNA molecule that is copied from retroviral RNA by reverse transcriptase and then integrated into the host chromosomal DNA. (794)

pseudopodia (singular, *pseudopodium*) Large, blunt-ended cytoplasmic protrusions involved in cell crawling by amoebae. (586)

puff Uncoiled region of a polytene chromosome that corresponds to a site of active gene transcription. *Also see* Balbiani ring. (447)

pulse-chase experiment Procedure in which cells are briefly exposed to a radioactive compound and the fate of the incorporated radioactivity is then monitored. (142)

pulsed-field gel electrophoresis Type of electrophoresis in which the direction of the electric field is rapidly altered, thereby improving the resolution obtained with large DNA fragments. (145)

pumps Membrane transport proteins that carry out active transport. (188)

purine Nitrogen-containing double-ring compound whose derivatives include adenine and guanine. (19)

puromycin Inhibitor of protein synthesis whose structure mimics the end of a tRNA molecule. (500)

purple membrane Regions of the plasma membrane of *Halobacterium* that contain the light-absorbing protein, bacteriorhodopsin. (189)

pyrenoids Granular deposits of crystallized rubisco found in the stroma of algal chloroplasts. (364)

pyrimidine Nitrogen-containing ring compound whose derivatives include cytosine, thymine, and uracil. (19)

quaternary structure Formation of a multisubunit protein from more than one polypeptide chain. (26)

R looping Technique in which single-stranded RNA is hybridized to double-stranded DNA under conditions that favor the formation of hybrids between RNA and DNA. (429)

R sequences Short complementary regions at the ends of V, D, and J gene segments that promote recombination. (724)

Rab proteins Family of GTP-binding proteins that have been implicated in a variety of membrane budding and fusion events. (285)

radiation Forms of energy that travel through space. (791)

radioactive Spontaneous disintegration of a chemical isotope into other kinds of atoms accompanied by the emission of radiation. (141)

Ras Plasma membrane G protein that transmits signals from external growth factors to the cell interior; commonly mutated in human cancers. (527, 798)

reaction-center complex Site within a photosystem that receives energy from the light-harvesting complex; contains a pair of chlorophyll *a* molecules that pass an excited electron to the photosynthetic electron transfer chain. (370)

receptor Any protein that contains a binding site for a specific signaling molecule. (203)

receptor protein-tyrosine kinase Transmembrane protein containing a growth factor receptor domain at the outer plasma membrane surface and a protein-tyrosine kinase domain at the inner membrane surface. *Also see* nonreceptor protein-tyrosine kinase. (798)

receptor-mediated endocytosis Binding of an external ligand to a plasma membrane receptor followed by endocytosis of coated vesicles containing the receptor-ligand complex. (295)

recombinant DNA DNA molecule created by joining DNA fragments derived from two or more sources. (100)

recombination Exchange of genetic information between two different DNA molecules, such as that occurring between homologous chromosomes during prophase I of meiosis. (548, 553)

redox couple Oxidized and reduced forms of the same molecule or ion. (329)

redox potential *See* standard redox potential.

reduction Process of gaining electrons during a chemical reaction. (57, 310)

regulated secretion Exocytosis of materials stored in secretory vesicles in response to a triggering stimulus. *Also see* constitutive secretion. (283)

regulatory subunit Subunit that contains the allosteric site of an allosterically controlled enzyme or other protein. (52)

relaxed DNA DNA that is not supercoiled. (415)

release factors Proteins that recognize stop codons and trigger the release of newly completed polypeptide chains during the process of protein synthesis. (496)

repeated DNA sequences DNA sequences present in multiple copies within an organism's chromosomal DNA; in eukaryotes, most repeated DNA sequences do not code for polypeptides. (443)

replication fork Y-shaped structure located where the two strands of a DNA double helix have become unwound and are being replicated. (71)

replication origin Site within a DNA molecule where replication is initiated. (528)

replicon Total length of DNA replicated from a single replication origin; an average eukaryotic chromosome contains about a thousand replicons. (528)

repressor protein Protein that inhibits the transcription of specific genes by binding to DNA; the opposite of an activator protein. (433)

resact Small peptide in the jelly coat of sea urchin eggs that attracts sperm cells. (686)

residual body Secondary lysosome containing accumulated undigested materials. (295)

resolving power Ability of a microscope to distinguish adjacent objects as separate entities. (118)

respiration *See* cellular respiration.

respiratory chain Series of electron carriers that transfer electrons in a stepwise fashion from NADH or $FADH_2$ to oxygen; also called electron transport chain. (327)

respiratory complex I Multiprotein complex of the inner mitochondrial membrane that catalyzes the transfer of electrons from NADH to ubiquinone. (332)

respiratory complex II Multiprotein complex of the inner mitochondrial membrane that catalyzes the transfer of electrons from succinate to ubiquinone. (332)

respiratory complex III Multiprotein complex of the inner mitochondrial membrane that catalyzes the transfer of electrons from ubiquinone to cytochrome *c*. (332)

respiratory complex IV Multiprotein complex of the inner mitochondrial membrane that catalyzes the transfer of electrons from cytochrome *c* to oxygen. (332)

respiratory control Regulatory mechanism that slows the flow of electrons through the respiratory chain when ADP is not available. (344)

response element DNA sequence that regulates the transcription of adjacent genes by binding to a gene-specific transcription factor. (422, 458)

restriction endonucleases (restriction enzymes) Enzymes that cut DNA at specific restriction sites four to eight bases long. (98)

restriction fragment length polymorphisms (RFLPs) Differences in restriction maps between individuals caused by mutations that alter restriction sites. (108)

restriction map Map of a DNA molecule indicating the location of restriction sites for various restriction endonucleases. (99)

restriction point A point late during G_1 phase of the eukaryotic cell cycle when cells become committed to proceeding through the rest of the cell cycle and dividing; also called START. (523)

restriction site DNA base sequence four to eight bases long that is cleaved by a specific restriction endonuclease. (98)

retinal Derivative of vitamin A that functions as the light-absorbing prosthetic group of bacteriorhodopsin and rhodopsin. (189)

retinoic acid Derivative of vitamin A that functions as a morphogen; guides the process of limb formation during embryonic development. (707)

retrovirus Any RNA virus that uses reverse transcriptase to make a DNA copy of its RNA for insertion into the host cell chromosomal DNA. (74, 794)

reverse transcriptase Enzyme that synthesizes double-stranded DNA using RNA as a template. (75, 794)

RFLP *See* restriction fragment length polymorphisms.

rho Bacterial protein that binds to the 3' end of newly forming RNA molecules, triggering the termination of transcription. (424)

rhodopsin Light-absorbing pigment of vertebrate photoreceptor cells that consists of a protein joined to retinal. (768)

ribonucleic acid *See* RNA.

ribose Five-carbon sugar employed in the construction of nucleotides and RNA. (19)

ribosomal proteins Proteins that bind to ribosomal RNAs to form ribosomes. (471)

ribosomal RNA (rRNA) RNA components of the ribosome; prokaryotic ribosomes contain 23S, 16S, and 5S rRNAs, while eukaryotic cytoplasmic ribosomes contain 28S, 18S, 5.8S, and 5S rRNAs. (471)

ribosome Small particle constructed from rRNA and protein that functions as the site of protein synthesis in the cytoplasm of prokaryotes and in the cytoplasm, mitochondria, and chloroplasts of eukaryotes; composed of large and small subunits. (8, 470, 654)

ribozyme RNA molecule exhibiting catalytic activity. (38, 482)

ribulose 1,5-bisphosphate carboxylase *See* rubisco.

RNA (ribonucleic acid) Polymer constructed from nucleotides consisting of ribose phosphate linked to either cytosine, guanine, adenine, or uracil. *Also see* messenger RNA, ribosomal RNA, and transfer RNA. (19)

RNA editing Altering the base sequence of an mRNA molecule by the insertion, removal, or modification of nucleotides; observed mainly in mitochondrial and chloroplast mRNAs. (653)

RNA polymerase Enzyme that catalyzes RNA synthesis in the 5' → 3' direction using DNA as a template. (81, 417)

RNA primer Short RNA fragment, synthesized by DNA primase, that serves as an initiation site for DNA synthesis. (73)

RNA processing Conversion of an initial RNA transcript into a mature mRNA, rRNA, or tRNA by the removal, addition, and/or chemical modification of nucleotide sequences. (426)

RNA splicing Removal of introns from an initial RNA transcript, thereby generating the mature form of the RNA molecule. (84, 430, 482)

rods Photoreceptor cells of the vertebrate retina that are more sensitive to dim light than cones but cannot distinguish colors. (768)

rolling circle replication Mechanism for replicating circular DNA in which the 3' end of one strand is extended by a replication process that displaces the 5' end of the same strand from the circle. (478)

rosettes Cellulose-synthesizing enzyme complexes located in the plasma membrane of plant cells. (245)

rough ER Endoplasmic reticulum containing attached ribosomes. (258)

rRNA *See* ribosomal RNA.

rubisco Enzyme that catalyzes the CO_2-fixing step of the Calvin cycle; joins CO_2 to ribulose 1,5-bisphosphate to form two molecules of 3-phosphoglycerate. (390)

ruffling Fluttering or undulating movements observed at the leading edge of a crawling cell. (589)

S phase Stage during the eukaryotic cell cycle when the chromosomal DNA is replicated. (523)

saltatory movements Sudden rapid movements of cell organelles interspersed with periods of immobility. (614)

sarcolemma Plasma membrane of a skeletal muscle cell. (564)

sarcoma Any cancer arising in supporting tissues of mesodermal origin, such as bone, fat, cartilage, and muscle. (778)

sarcomere Region extending from one Z disk to the next in skeletal muscle myofibrils. (564)

sarcoplasmic reticulum System of flattened membrane channels that cover the surface of skeletal muscle myofibrils and store the calcium ions that trigger contraction; comparable to the endoplasmic reticulum of other cell types. (574)

saturated fatty acid Fatty acid whose carbon chain does not contain any double (or triple) bonds. (15)

saturation Point at which further increases in substrate concentration cause no further increase in reaction rate for an enzyme-catalyzed reaction. (41)

scanning electron microscope (SEM) Microscope that produces a three-dimensional image from electrons that are deflected and emitted from the surface of a specimen being scanned by an electron beam. (125)

scanning probe microscopes Instruments containing a tiny probe that moves over the surface of a specimen, allowing the surface features of individual molecules to be visualized. (127)

Schwann cell Type of glial cell that forms the myelin sheaths that cover the axons of peripheral nerves. (751)

SDS polyacrylamide gel electrophoresis Gel electrophoresis of proteins in the presence of the detergent sodium dodecyl sulfate; allows individual polypeptide chains to be separated based solely on differences in size. (145)

second law of thermodynamics Law stating that all events proceed in the direction that causes the entropy of the universe to increase. (34)

second messengers Molecules that transmit signals from extracellular signaling ligands to the cell interior; examples include cyclic AMP, inositol trisphosphate, and diacylglycerol. (212)

secondary cell wall Rigid portion of the plant cell wall that develops beneath the primary cell wall after cell growth has ceased; contains densely packed, highly organized bundles of cellulose microfibrils. (245)

secondary inductions Interactions during embryonic development in which one group of cells induces a neighboring group of cells to differentiate into a particular structure. (701)

secondary lysosome Vesicle formed by the fusion of a primary lysosome with a phagosome. (293)

secondary structure Folding of a polypeptide chain produced by hydrogen bonding between the —C=O group of one peptide bond and the —NH group of another. *Also see* alpha helix and beta sheet. (23)

secretory vesicle Membrane-enclosed cytoplasmic vesicle containing material destined to be discharged from the cell by exocytosis. (283, 756)

sedimentation coefficient Measure of the rate at which a particle (or molecule) moves in a centrifugal field; provides a rough estimate of the particle's size and shape. (133)

self-assembly Spontaneous assembly of macromolecules into multimolecular structures without the need for external components that guide or catalyze the process. (632)

self-splicing Intron removal that proceeds in the absence of snRNPs or any other protein-containing catalyst. (432, 482)

SEM *See* scanning electron microscope.

semiconservative replication Model of DNA replication in which each newly formed DNA molecule consists of one old strand and one newly synthesized strand. (67)

sensory cells Cells specialized for detecting stimuli such as light, pressure, movement, and temperature. (768)

Sertoli cells Cells that provide protection and nutritional support for developing sperm. (680)

shadow casting Technique in which a specimen is sprayed with heavy metal vapor prior to electron microscopic examination. (124)

Shine-Dalgarno sequence Purine-rich stretch of nucleotides located prior to the start codon of bacterial mRNAs; binds mRNA to the ribosome by forming complementary base pairs with the 3' end of 16S rRNA. (490)

shuttle vector DNA cloning vector that is capable of replicating in two or more cell types. (101)

signal peptidase Enzyme bound to ER membranes that removes the signal sequence of a newly forming polypeptide chain as it emerges into the lumen of the ER. (266)

signal recognition particle (SRP) RNA-protein complex that binds to the signal sequence located at the N-terminus of a newly forming polypeptide chain; involved in attaching the polypeptide and its associated ribosome to the ER membrane. (265)

signal sequence Short stretch of amino acids at the N-terminus of newly forming polypeptide chains that causes the polypeptide and its associated ribosome to bind to ER membranes, where the polypeptide is then translocated into the ER lumen. (265)

signal transduction Process by which the binding of a signaling molecule to the cell surface triggers changes within the cell. (206)

silencers DNA sequences that inhibit transcription when placed either upstream or downstream from a promoter site. (423)

simple diffusion Unaided net movement of a substance down its electrochemical gradient. (175)

single-stranded DNA-binding proteins Proteins that stabilize the single-stranded state of DNA at the replication fork; also called helix-destabilizing proteins. (74)

sliding filament model Well-supported model postulating that muscle contraction is caused by sliding of the thin actin filaments relative to the thick myosin filaments. (570)

slow axonal transport Movement of proteins and cytoskeletal filaments down a nerve cell axon at rates of 1–5 mm per day. *Also see* fast axonal transport. (616, 751)

slow chemical transmission Synaptic transmission mediated by neurotransmitters that bind to receptors linked to G proteins. (761)

smooth ER Endoplasmic reticulum lacking attached ribosomes. (258)

SNAPs (NSF attachment proteins) Cytoplasmic proteins that act in conjunction with NSF (N-ethylmaleimide-sensitive fusion protein) to facilitate the fusion of various kinds of membranes. (286)

SNAREs (SNAP receptors) Membrane proteins that act as attachment sites for SNAPs during membrane fusion. (286)

snRNPs (snurps) Small nuclear ribonucleoprotein particles involved in RNA splicing. *Also see* spliceosome. (431)

snurps *See* snRNPs.

sodium-potassium pump (Na⁺–K⁺ ATPase) ATP-driven membrane transport protein that couples the active transport of Na⁺ out of cells to the active transport of K⁺ into cells. (187, 188)

solenoid Spiral coil of nucleosomes that forms the 30-nm chromatin fiber. (414)

solute Any dissolved molecule or ion. (175)

Southern blotting Technique for hybridizing electrophoretically separated DNA molecules to a radioactive DNA probe. *Also see* Northern blotting. (97)

spectrin Protein crosslinked by short actin filaments to form a meshwork associated with the cytoplasmic surface of the red cell plasma membrane. (174, 592)

sperm cell The male haploid gamete. (680)

spermatid Haploid cell produced by meiotic division of primordial germ cells (spermatogonia) in male animals; subsequently differentiates into a functional sperm cell. (680)

spermatogenesis Process by which diploid spermatogonia are converted into haploid sperm cells. (680)

spermiogenesis Differentiation of spermatids into mature sperm cells. (681)

spherosomes Small membrane–enclosed cytoplasmic organelles in plant cells that contain lysosomal enzymes and large amounts of lipid. (289)

spindle Structure composed of microtubules that separates the two sets of chromosomes during mitosis; a comparable structure separates chromosomes during meiosis. (537)

spirilla (singular, *spirillum*) Spirally shaped bacteria. (248)

spliceosome Splicing complex constructed from a group of snRNPs bound to both ends of an intron in pre-mRNA. (431)

splicing *See* RNA splicing.

SRP *See* signal recognition particle.

stacked thylakoids Thylakoid membranes stacked on each other like a pile of coins; also called grana. (359)

stain Substance that binds to cell components and alters their color or electron-scattering properties prior to microscopic examination. (119, 123)

standard free-energy change (ΔG°′) Amount of free energy released during the conversion of reactants to products at standard conditions—that is, 25°C, 1 atmosphere pressure, pH 7.0, and all reactants and products maintained at concentrations of 1.0 *M*. (35)

standard redox potential Measure of the tendency of a redox couple to donate electrons; determined by measuring the voltage generated when a redox couple is placed on one side of an electrical cell and a mixture of H⁺ and H₂ is placed on the other side under standard conditions—that is, 25°C, 1 atmosphere pressure, pH 7.0, and 1.0 *M* concentration of reactants. (329)

starch Polysaccharide composed of glucose molecules joined by α(1 → 4) glycosidic bonds; main storage form of glucose in plants. (14)

START *See* restriction point.

start codon The codon AUG in mRNA when it functions as the initiation point for protein synthesis. (87)

stereo electron microscopy Technique in which a three-dimensional image is created by photographing the same specimen at two slightly different angles. (123)

stereocilia Cell-surface projections resembling large microvilli observed on sensory epithelial cells of the inner ear; supported by actin filaments and hence unrelated to true cilia in either structure or function. (591)

steroids Lipid derivatives of the four-membered phenanthrene ring; function as hormones and as constituents of cell membranes. (16, 161)

stigma Sticky projection of the ovary of flowering plants that is specialized for trapping pollen. (692)

stomata (singular, *stoma*) Pores on the surface of a plant leaf that can be opened or closed to control gas and water exchange. (393)

stop codon The mRNA codons UAG, UAA, and UGA, which signal the termination of protein synthesis. (87)

stop-transfer sequence Stretch of amino acids that halts the translocation of a newly forming polypeptide chain through a protein-translocating channel; employed for anchoring newly forming membrane proteins within the lipid bilayer. (268)

streptomycin Inhibitor of protein synthesis that interferes with the binding of tRNAs to the ribosome; also causes misreading of the genetic code. (501)

stress fibers Bundles of contractile filaments resembling tiny myofibrils that reside in the cytoplasm of cultured fibroblasts; contain actin, myosin, and other cytoskeletal proteins. (584)

stringent response Decrease in the rate of protein, rRNA, and tRNA synthesis observed in bacterial cells grown in the presence of an inadequate supply of amino acids. (504)

stroma Space enclosed by the chloroplast envelope; site of the Calvin cycle. (359)

stroma center Array of tightly packed fibrils containing the enzyme rubisco seen in the stroma of higher plant chloroplasts. (364)

subcellular fractionation Family of techniques for isolating organelles from cell homogenates using various types of centrifugation. (133)

substrate Substance that is acted upon by an enzyme. (37)

substrate-level phosphorylation ATP synthesis that is part of a chemical reaction in which a substrate is converted to a product. (342)

substratum Solid surface over which a cell moves or upon which a cell grows. (586)

sugars Small organic molecules characterized by a 1:2:1 ratio of carbon, hydrogen, and oxygen. (14)

supercoiling Coiling of a DNA double helix upon itself, either in a circular DNA molecule or in a DNA loop anchored at both ends. (415)

supernatant The material that remains in solution after centrifugation. (135)

suppressor mutation Mutation that alters the anticodon of a tRNA molecule, causing it to insert an amino acid where a stop codon generated by another mutation would otherwise have caused premature termination of protein synthesis. (496)

suppressor T cell Type of T lymphocyte that inhibits the ability of other lymphocytes to respond to antigen stimulation. (741)

Svedberg unit (S) A sedimentation coefficient of 10⁻¹³ seconds. (134)

symport Membrane protein that transports two different ions or molecules in the same direction across a membrane. (180)

synapse Cell junction specialized for transmitting a chemical or electrical signal between a nerve cell and another nerve cell, muscle cell, or secretory cell. (205, 755)

synapsin Membrane protein that links synaptic vesicles to the cytoskeleton. (759)

synapsis Close pairing between homologous chromosomes during the zygotene phase of prophase I of meiosis. (549)

synaptic cleft Space separating the plasma membranes of presynaptic and postsynaptic cells in the region of a synapse. (755)

synaptic vesicles Small neurotransmitter-filled vesicles located directly beneath the plasma membrane at the axon tip. (756)

synaptonemal complex Zipperlike, protein-containing structure that runs lengthwise between homologous chromosomes that have become closely paired during prophase I of meiosis. (558)

synergid cells One of the cell types produced by mitotic division of haploid megaspores in flowering plants; function in transporting nutrients to the egg. (686)

T cell *See* T lymphocyte.

T-cell antigen receptor *See* TCR.

T lymphocyte (T cell) Type of lymphocyte that develops in the thymus and carries TCR molecules on its surface. *Also see* helper, killer, memory, and suppressor T cells. (717)

T tubules (transverse tubules) Membrane channels that transmit action potentials from the surface of skeletal muscle cells to the cell interior, where T tubules make close contact with the sarcoplasmic reticulum. (574)

talin Protein that links fibronectin receptors in the plasma membrane to the protein vinculin, which in turn attaches to actin filaments; also found in the plaques of focal adhesions. (231, 240)

targeting signals Amino acid sequences that direct newly formed polypeptide chains to their proper destination. (264)

TATA box Eukaryotic promoter sequence located about 30 bases upstream from the transcriptional start site of most genes transcribed by RNA polymerase II. (420)

tautomers Alternate chemical forms of the same molecule that differ in their hydrogen-bonding properties. (91)

taxol Substance isolated from the yew tree that stabilizes microtubules. (602)

TCR (T-cell antigen receptor) Plasma membrane proteins of helper T cells and killer T cells that bind to antigen–MHC II and antigen–MHC I complexes, respectively. (729)

teichoic acids Negatively charged polymers constructed from glycerol or ribitol derivatives; linked to murein in the cell wall of Gram-positive bacteria. (250)

TEL sequence Short, repeated DNA sequence found at the ends (telomeres) of linear chromosomes. (533)

telomerase Enzyme that synthesizes additional copies of a *TEL* sequence. (534)

telomere Either end of a linear chromosome. (533)

telophase Stage during mitosis or meiosis when the chromosomes arrive at the spindle poles and uncoil, accompanied by spindle breakdown and formation of new nuclear envelopes. (536, 543, 552)

TEM *See* transmission electron microscope.

temperature-sensitive mutant Cells that exhibit a mutant trait only when grown at a slightly elevated temperature. (70, 519)

template Any molecule that specifies the pattern for the synthesis of a second molecule of complementary structure; usually refers to a nucleic acid whose base sequence directs the synthesis of another nucleic acid by complementary base-pairing. (67)

teratocarcinoma Malignant tumor of primordial germ cells that is capable of differentiating into a variety of specialized cell types. (802)

terminal web Network of filaments located at the base of microvilli; contains actin, myosin, and several other cytoskeletal proteins. (591)

terpenes Lipids constructed from the five-carbon compound, isoprene; examples include vitamin A and ubiquinone. (16)

tertiary structure Folding of a polypeptide chain into domains caused by interactions between amino acid side chains. (24)

tetrad Tightly bound pair of homologous chromosomes formed during prophase I of meiosis; consists of four chromatids. (551)

thermodynamics Study of energy transformations. (33)

thick filaments Myosin-containing filaments found in the myofibrils of skeletal muscle cells. (564)

thin filaments Actin-containing filaments found in the myofibrils of skeletal muscle cells. (564)

thin sections Tissue slices 50 to 100 nm in thickness prepared for examination by electron microscopy. (123)

thin-layer chromatography (TLC) Type of chromatography in which the matrix phase is spread out as a thin layer on glass or plastic. (148)

threshold potential Value of the membrane potential that must be reached before an action potential is triggered. (754)

thylakoid lumen Internal space bounded by thylakoid membranes. (359)

thylakoid membranes System of flattened membrane sacs located within the chloroplast stroma, arranged as both stacked and unstacked thylakoids; site of the photosynthetic electron transfer chain and photophosphorylation. (359)

thymine (T) Pyrimidine base found in DNA (in place of uracil in RNA); forms a complementary base pair with adenine by hydrogen bonding. (19)

Ti plasmid DNA molecule that causes crown gall tumors when transferred into plants by bacteria; used as a cloning vector for introducing foreign genes into plant cells. (107, 802)

tight junction Type of cell junction in which the adjacent plasma membranes of neighboring cells are tightly sealed, thereby preventing molecules from diffusing from one side of an epithelial cell layer to the other by passing through the spaces between adjoining cells. (236)

tissue-specific genes Genes that are only expressed in certain cell types because they code for proteins that carry out specialized functions. *Also see* housekeeping genes. (704)

TLC *See* thin-layer chromatography.

T_m *See* melting temperature.

tolerance State in which the immune system fails to respond to a particular antigen, usually employed to protect individuals from attacking their own antigens. (739)

topoisomerases Enzymes that alter DNA supercoiling by introducing transient breaks in one or both DNA strands. (74, 415)

***trans* face** Region of the Golgi complex that is opposite from the *cis* face; usually oriented toward the cell surface. (274)

***trans* Golgi network** Network of Golgi channels and vesicles located adjacent to the *trans* face of the Golgi complex. (274)

transcription Process by which RNA polymerase utilizes one DNA strand as a template for guiding the synthesis of a complementary RNA molecule. (83)

transcription factors Accessory proteins required by RNA polymerase for binding to promoter sequences and initiating RNA synthesis. (422)

transcription unit DNA region bounded by a single transcription initiation site and a single transcription termination site. (480, 676)

transcytosis Endocytosis of material into vesicles that move to the opposite side of the cell and fuse with the plasma membrane, releasing the material into the extracellular space. (296)

transduction Transfer of bacterial DNA sequences from one bacterium to another by a bacteriophage. (553)

transfer RNA (tRNA) Family of small RNA molecules, each binding a specific amino acid and possessing an anticodon that recognizes a specific codon in mRNA. (88)

transferrin receptor Plasma membrane protein involved in transporting iron into cells. (506)

transformation *See* genetic transformation.

transformed cells Cells that have acquired an unlimited capacity to divide in culture, usually by exposure to cancer-causing agents. (116)

transgenic Carrying a gene that has been experimentally introduced from another organism. (110)

transition state Intermediate stage in a chemical reaction, characterized by a higher free energy than that of the initial reactants. (39)

translation Process by which the base sequence of an mRNA molecule guides the sequence of amino acids incorporated into a polypeptide chain; occurs on ribosomes. (83)

translational repressor Protein that binds to and inhibits the translation of specific mRNAs. (504)

translocation Movement of an mRNA molecule across a ribosome by a distance of three nucleotides, bringing the next codon into position for translation; the term *translocation* can also refer to movement of a chromosome segment to a new chromosomal location or to the movement of a newly forming polypeptide chain through a protein-translocating channel. (266, 495, 797)

transmission electron microscope (TEM) Microscope that produces an image from electrons transmitted through a specimen. (122)

transport proteins Membrane proteins that transport specific substances across membranes much faster than would be possible by simple diffusion. *Also see* facilitated diffusion and active transport. (179)

transport vesicles Membrane-enclosed cytoplasmic vesicles that rapidly transport materials to the cell surface as part of the process of constitutive secretion. (283)

transposable elements Mobile DNA sequences that migrate from one location to another within DNA; grouped into two categories called insertion sequences and transposons. (95)

transposase Protein that catalyzes the removal of a transposable genetic element from one region in a DNA molecule and its insertion somewhere else. (95)

transposon Large transposable genetic element that contains genes in addition to those required for moving the transposon itself. (95)

transverse diffusion *See* flip-flop.

transverse tubules *See* T tubules.

treadmilling Process in which the addition of actin monomers to the plus end of an actin filament is balanced by the loss of actin monomers from the minus end; a comparable phenomenon may also occur with microtubules. (588, 598)

triacylglycerols Family of lipids formed by joining various fatty acids to glycerol; also called triglycerides or fats. (15)

triad Region where a T tubule passes between the terminal cisternae of the sarcoplasmic reticulum in skeletal muscle. (575)

triglycerides *See* triacylglycerols.

triplet code A reference to the genetic code, which is read from mRNA in units of three bases called codons. (85)

tRNA *See* transfer RNA.

tropomyosin Protein associated with actin filaments that blocks the interaction between actin and myosin in the absence of Ca^{2+}. (576)

troponin Ca^{2+}-binding protein associated with actin filaments that displaces tropomyosin in the presence of Ca^{2+}, thereby activating contraction. (576)

***trp* operon (tryptophan operon)** Group of adjoining bacterial genes that code for enzymes involved in tryptophan metabolism and whose transcription is inhibited by the *trp* repressor. (438)

tube cell Haploid cell within a pollen grain that gives rise to the pollen tube, which penetrates the stigma and enters the ovary. (685)

tubulin Protein constructed from α- and β-tubulin that polymerizes into microtubules. (595)

tumor *See* neoplasm.

tumor suppressor genes Genes whose absence (or loss of function) can cause a cell to become malignant. (799)

ubiquinone Lipid component of mitochondrial inner membranes that transfers electrons from respiratory complexes I and II to respiratory complex III. (328)

ubiquitin Small protein that is linked to other proteins as a way of marking the targeted protein for degradation. (510)

ultracentrifuge Instrument capable of generating centrifugal forces that are large enough to sediment cell organelles and macromolecules. (133)

ultraviolet radiation Mutagenic type of radiation present in sunlight that triggers the formation of thymine dimers in DNA. (92)

uncoupling agent Substance that abolishes the ATP synthesis that normally accompanies electron transfer through the respiratory chain or the photosynthetic electron transfer chain; makes membranes permeable to protons, thereby dissipating the electrochemical proton gradient. (338, 381)

uniport Membrane protein that transports a single ion or molecule from one side of a membrane to the other. (180)

unique-sequence DNA DNA sequences that are present in one copy per haploid set of chromosomes. (443)

unsaturated fatty acid Fatty acid whose carbon chain contains double (or triple) bonds. (15)

unstacked thylakoids Membrane-bound channels that pass from one stack of thylakoid membranes to another. (359)

upstream Located toward the 5' end of the DNA strand lying opposite the strand that serves as the template for transcription. (420)

uracil (U) Pyrimidine base found in RNA (in place of thymine in DNA); forms a complementary base pair with adenine by hydrogen bonding. (19)

urkaryote Hypothetical ancestral eukaryote that first diverged from archaebacteria several billion years ago; later developed into a nucleus-containing cell called a protoeukaryote. (640)

V gene segment DNA segment coding for most of the variable region of an antibody light or heavy chain; the remainder of the variable region is encoded by a relatively short J gene segment and, in the case of heavy chains, a D gene segment. (723)

V_{max} *See* maximum velocity.

vacuoles Large membrane-bound vesicles that function as storage compartments and in maintaining water balance; especially prominent in plant cells. (8)

van der Waals force Weak attractive force between atoms that is optimal when atoms are separated by a distance of 0.3 to 0.4 nm. (25)

variable number of tandem repeats *See* VNTRs.

variable region Region of an antibody or TCR molecule whose amino acid sequence varies between antibody or TCR molecules that bind to different antigens. (720)

vegetal pole Region of an egg cell opposite to the animal pole; enriched in yolk platelets, when present. (677)

velocity centrifugation Centrifugation under conditions in which the density of the components being separated is greater than the density of the solution, thereby allowing molecules and organelles to be separated based on differences in size. (135)

villin Protein that crosslinks the actin filament bundles of microvilli. *Also see* fimbrin. (591)

vinculin Protein that links actin cytoskeletal filaments to the protein talin, which in turn attaches to fibronectin receptors in the plasma membrane; also found in the plaques of adherens junctions. (232, 240)

viral cloning vector Viral DNA molecule that is combined with foreign DNA to create a recombinant virus that can be used for gene cloning. (101)

virulent virus Virus that immediately directs an infected cell to make more virus particles. (29)

virus Infectious agent consisting of DNA or RNA surrounded by a protein coat; can reproduce only inside living cells. (28)

vitelline envelope Polysaccharide-rich external coat that surrounds nonmammalian egg cells; in mammals, the comparable structure is called the zona pellucida. (678)

vitellogenin Protein complex synthesized in the liver and transported to the ovary, where it is taken up by growing oocytes and stored in yolk platelets. (677)

VNTRs (variable number of tandem repeats) Short repeated DNA sequences whose variability between individuals forms the basis for DNA fingerprinting. (108)

voltage-gated channel Ion channel whose permeability is altered by changes in the membrane potential. (201)

water-oxidizing clock Four-step cycle used by photosystem II for removing electrons from water and releasing oxygen. (373)

Western blotting Technique for identifying electrophoretically separated proteins using specific antibodies. (146)

wobble hypothesis Flexibility in base-pairing between the third base of a codon and the corresponding base in its anticodon. (494)

X-ray diffraction Technique for determining the three-dimensional structure of macromolecules based on the pattern produced when a narrow beam of X-rays is passed through a sample, usually a crystal or fiber. (149)

xenograft Tissue or organ transplanted between members of different species. (738)

yolk platelets Cytoplasmic vesicles in oocytes that store nutrients for use by the embryo after fertilization. (677)

Z disk Thin, dense structure that runs down the middle of each I band in a skeletal muscle myofibril; structure into which actin thin filaments are inserted. (564)

Z scheme Diagram of the photosynthetic electron transfer chain in which the electron carriers are arranged according to their free-energy levels. (372)

zinc finger DNA-binding motif found in many proteins that regulate gene transcription; consists of a loop that binds a single atom of zinc; in cysteine-histidine zinc fingers, two cysteines and two histidines bind the zinc atom, while in cysteine-cysteine zinc fingers, four cysteines bind the zinc atom. (459, 463)

zona pellucida Polysaccharide-rich external coat that surrounds mammalian egg cells; in nonmammalian eggs, the comparable structure is called the vitelline envelope. (678)

ZPA (zone of polarizing activity) Small region at the posterior end of a developing limb bud that produces the morphogen retinoic acid, forming a gradient that guides limb formation. (707)

zygote Diploid cell produced by the fusion of a sperm cell with an egg cell. (688)

zygotene Stage during prophase I of meiosis when homologous chromosomes become closely paired by the process of synapsis. (549)

Photograph
and
Illustration
Credits

PHOTOGRAPH CREDITS

Unless otherwise acknowledged, all photographs are the property of Scott, Foresman and Company.

Part 1 Opener: *The Body Victorius,* copyright Lennart Nilsson from Bonnier Fakta.

Figure 1-1 (Top): Reproduction of a plate from Robert Hook's book *Microphagia,* published 1665.

Figure 1-1 (Bottom): Courtesy of B. J. Ford.

Figure 1-2: Courtesy of D. Fawcett.

Figure 1-3: Biophoto Associates/Photo Researchers.

Figure 1-4: Courtesy of G. J. Brewer.

Figure 1-13 (Top left): D. W. Fawcett, M. D./Photo Researchers.

Figure 1-13 (Top right): Courtesy of B. J. Ford.

Figure 1-13 (Bottom): Courtesy of K. Mühlenthaler.

Figure 1-27 (Left, center left, and center right): Courtesy of R. C. Williams.

Figure 1-27 (Right): Courtesy of W. G. Laver.

Figure 1-28: Courtesy of A. K. Kleinschmidt.

Figure 2-19: Courtesy of R. C. Williams.

Figure 3-2: L. D. Simon/Photo Researchers.

Figure 3-17 (Left): Bill Longcore/SS/Photo Researchers.

Figure 3-17 (Right): Jackie Lewin/Royal Free Hospital/SS/Photo Researchers.

Figure 3-20: Courtesy of T. O. Caspersson.

Figure 3-22: Courtesy of D. M. Prescott.

Figure 3-43: SPL/SS/Photo Researchers.

Figure 3-51: Courtesy of R. T. Fraley.

Figure 3-54: Courtesy of R. L. Brinster and R. E. Hammer, School of Veterinary Medicine, University of Pennsylvania.

Figure 4-1: Courtesy of K. K. Sanford.

Figure 4-2: Biophoto Associates/Photo Researchers.

Figure 4-3: Courtesy of Sigma Chemical Company.

Figure 4-4: Courtesy of K. K. Sanford.

Figure 4-7 (Top): 1988 Jim Solliday/Biological Photo Service.

Figure 4-7 (Top center): M. I.Walker/Photo Researchers.

Figure 4-7 (Center): 1989 Jim Solliday/Biological Photo Service.

Figure 4-7 (Bottom center): 1989 Jim Solliday/Biological Photo Service.

Figure 4-7 (Bottom): Ray Simons Photoresearchers.

Figure 4-9: Courtesy of R. Sommers and D. Pisdon.

Figure 4-10 (Left): Michael Abbey/Photo Researchers.

Figure 4-10 (Right): Courtesy of H. Cheng and C. P. Leblond.

Figure 4-11: Courtesy of P. Favard, N. Carosso, and R. McIntosh

Figure 4-12 (Left): Courtesy of Dr. Michael F. Moody.

Figure 4-12 (Center): Courtesy of J. Woodhead-Galloway.

Figure 4-12 (Right): Omikron/Photo Researchers.

Figure 4-13: Courtesy of R. L. Roberts, R. G. Kessel, and H.-N. Tung

Figure 4-14 (Top): Courtesy of K. D. Allen.

Figure 4-14 (Bottom): Courtesy of T. H. Giddings, Jr.

Figure 4-15: D. W. Fawcett/John Heuser/Photo Researchers.

Figure 4-17: Courtesy of K. R. Porter.

Figure 4-18: Courtesy of P. Arscott.

Figure 4-19: Courtesy of L. Lavia.

Figure 4-20: Courtesy of S. H. Blose.

Figure 4-21: Courtesy of S. L. Erlandsen.

Figure 4-22: Courtesy of J. R. Maddock.

Figure 4-23 (Top): Courtesy of J. C. Roland.

Figure 4-23 (Bottom): Courtesy of D. F. Bainton.

Figure 4-24 (Top): Courtesy of I. S. Edelman.

Figure 4-24 (Bottom): Courtesy of M. M. Salpeter.

Figure 4-29 (Center and right): Courtesy of P. Baudhuin.

Figure 4-29 (Left): Courtesy of D. J. Morré.

Figure 4-35: BioRad Laboratories, Life Sciences Group, Genetic Systems Division, Hercules, LA.

Figure 4-36: Courtesy of S. H. Blose.

Figure 4-38: Courtesy of P. O'Farrell.

Figure 4-40: Courtesy of R. Douce.

Figure 4-41 (Top): Courtesy of J. C. Kendrew.

Figure 4-41 (Center): Courtesy of M. H. F. Wilkins.

Figure 4-41 (Bottom): Courtesy of P. Luger.

Part 2 Opener: Conly L. Rieder, *The Journal of Cell Biology*, Volume 122, Number 2, July 1993, pp. 361–372.

Figure 5-2: Don Fawcett, M.D./Photo Researchers.

Figure 5-3: Dr. Cedric S. Raine.

Figure 5-4 (Top): Dr. Daniel Branton.

Figure 5-4 (Bottom): Courtesy of R. B. Park.

Figure 5-8: Courtesy of P. M. Frederik.

Figure 5-12: Courtesy of T. Steck.

Figure 5-20: Courtesy of I. Yahara.

Figure 5-21: Courtesy of A. E. Sowers.

Figure 5-23: Courtesy of T. Steck.

Figure 5-25: Courtesy of Dr. Daniel Branton.

Figure 6-5: Courtesy of B. Sakmann.

Figure 6-12 (Top): Courtesy of M. M. Salpeter.

Figure 6-12 (Bottom): Courtesy of F. Hucho.

Figure 6-24: Courtesy of R. D. Burgoyne.

Figure 6-28: Courtesy of Professors Richard G. Kessel and Randy Kardon, University of Iowa.

Figure 6-30: Reproduced from Rosenberg, L., W. Hellmann, and A. K. Kleinschmidt (1975). *Journal of Biological Chemistry* 250:1877–1883 by copyright permission of the American Society for Biochemistry and Molecular Biology, Bethesda, MD.

Figure 6-31 (Left): Courtesy of D. R. Keene.

Figure 6-31 (Right): Courtesy of N. Simonescu.

Figure 6-33: Biophoto Associates/SS/Photo Researchers.

Figure 6-36: Courtesy of R. O. Hynes.

Figure 6-37: G. W. Willis/Biological Photo Service.

Figure 6-40: Courtesy of J. E. Michaels.

Figure 6-41: Courtesy of G. M. Edelman.

Figure 6-42: Courtesy of M. Takeichi.

Figure 6-45: Courtesy of D. S. Friend.

Figure 6-46 (Top): Courtesy of R. S. Decker.

Figure 6-46 (Bottom): Courtesy of L. A. Staehelin.

Figure 6-47: Courtesy of R. S. Decker.

Figure 6-48: Courtesy of D. E. Kelly.

Figure 6-49: Courtesy of R. L. Roberts and Richard G. Kessel.

Figure 6-51: Courtesy of W. R. Loewenstein.

Figure 6-52 (Top): Courtesy of R. S. Decker.

Figure 6-52 (Bottom): Courtesy of E. L. Benedetti.

Figure 6-53: Courtesy of W. R. Loewenstein.

Figure 6-54: Biophoto Associates/Photo Researchers.

Figure 6-57: Courtesy of K. Mühlenthaler.

Figure 6-58: Courtesy of T. H. Giddings, Jr.

Figure 6-59: Courtesy of H. H. Mollenhauer.

Figure 6-60 (Left and center): Courtesy of K. Amako.

Figure 6-60 (Right): Courtesy of S. E. Erlandsen.

Figure 6-61: Courtesy of D. Abram.

Figure 6-62: Courtesy of J. W. Costerson.

Figure 6-65: Courtesy of R. E. Levin.

Figure 6-66: Courtesy of R. N. Goodman.

Figure 7-1: Courtesy of K. R. Porter.

Figure 7-2 (Top left): Courtesy of Don W. Fawcett.

Figure 7-2 (Top right): Courtesy of M. Bielinska.

Figure 7-2 (Bottom left and right): Courtesy of K. Tanaka.

Figure 7-5: Courtesy of C. de Duve.

Figure 7-8: Courtesy of S. Orrenius.

Figure 7-18: Courtesy of Richard G. Kessel.

Figure 7-19: Courtesy of M. J. Wynne.

Figure 7-20: Courtesy of D. S. Friend.

Figure 7-21 (Left): Courtesy of H. H. Mollenhauer.

Figure 7-21 (Right): Courtesy of J. M. Sturgess.

Figure 7-22: Courtesy of J. D. Jamieson.

Figure 7-24: Courtesy of D. S. Friend.

Figure 7-26: Courtesy of C. P. LeBlond.

Figure 7-28: Courtesy of Dr. Daniel Branton.

Figure 7-30: Courtesy of L. Orci.

Figure 7-34: Courtesy of P. Baudhuin.

Figure 7-35 (Top): Courtesy of Dr. Daniel Branton.

Figure 7-35 (Bottom): Courtesy of P. Baudhuin.

Figure 7-37: Courtesy of N. Poux.

Figure 7-39: Courtesy of M. Locke.

Figure 7-41: Courtesy of D. Zucker-Franklin.

Figure 7-42 (Top): Courtesy of Dr. Daniel Branton.

Figure 7-42 (Bottom): R. N. Band and H. S. Pankratz, Michigan State University/Biological Photo Service.

Figure 7-44: Courtesy of R. M. Steinman.

Figure 7-45: Courtesy of E. Essner.

Figure 7-46: Courtesy of M. C. Willingham.

Figure 7-47 (Left and center): J. E. Heuser.

Figure 7-47 (Right): N. Hirokawa and J. E. Heuser from D. Fawcett.

Figure 7-50: Courtesy of H. Loeb.

Figure 8-5: Courtesy of B. Tandler.

Figure 8-6 (Left): Courtesy of N. B. Rewcastle and A. P. Anzil.

Figure 8-6 (Right): Courtesy of T. Samorajski, J. M. Ordy, and J. R. Keefe.

Figure 8-7: Courtesy of H. Fernández–Morán.

Figure 8-8 (Left): Courtesy of B. Sacktor.

Figure 8-8 (Right): Courtesy of J. D. Jamieson.

Figure 8-9: Courtesy of M. L. Vorbeck.

Figure 8-10: Courtesy of Don W. Fawcett.

Figure 8-11: Courtesy of C. R. Hackenbrock.

Figure 8-13: Courtesy of B. Chance.

Figure 8-15: Courtesy of W. A. Anderson.

Figure 8-25: Courtesy of C. R. Hackenbrock.

Figure 8-26: Courtesy of E. Racker.

Figure 8-34: Courtesy of C. R. Hackenbrock.

Figure 8-35: Courtesy of A. F. Brodie.

Figure 8-39: Courtesy of J. M. Barrett.

Figure 8-40: Courtesy of S. Goldfischer.

Figure 9-4: Courtesy of W. J. Hayden.

Figure 9-5: Courtesy of J. Rosado–Alberio.

Figure 9-6: Courtesy of M. P. Garber.

Figure 9-7: Courtesy of J. Rosado–Alberio.

Figure 9-8: Courtesy of E. Gantt.

Figure 9-10 (Top left, bottom left): Courtesy of A. A. Benson.

Figure 9-10 (Top right, bottom right): Courtesy of R. B. Park.

Figure 9-11: Courtesy of J. Rosado–Alberio.

Figure 9-12: Courtesy of J. V. Possingham.

Figure 9-13: Courtesy of B. E. S. Gunning.

Figure 9-16: Courtesy of E. Gantt.

Figure 9-24: Courtesy of M. P. Garber.

Figure 9-31: Courtesy of K. D. Allen.

Figure 9-34: Courtesy of A. Melis.

Figure 9-35: Courtesy of M. Calvin.

Figure 9-41: Courtesy of E. H. Newcomb.

Figure 9-45: Courtesy of J. O. Berry.

Figure 9-46: Courtesy of N. J. Lang and B. A. Whitton.

Figure 9-47: Courtesy of A. R. Varga.

Figure 10-2: Courtesy of E. W. Daniels.

Figure 10-4: Courtesy of M. W. Goldberg.

Figure 10-5: Courtesy of C. M. Feldherr.

Figure 10-6: Courtesy of B. J. Stevens.

Figure 10-7: Courtesy of Richard G. Kessel.

Figure 10-9: Courtesy of R. Berezney.

Figure 10-10: Courtesy of R. Chalkley.

Figure 10-12: Courtesy of E. H. McConkey.

Figure 10-13: Courtesy of Jack Griffith.

Figure 10-15: Courtesy of R. D. Kornberg.

Figure 10-17: Courtesy of Barbara Hamkalo.

Figure 10-19: Courtesy of U. K. Laemmli.

Figure 10-20: Reproduced from Kobayashi, H., K. Kobayashi, and Y. Kobayashi (1977). *Journal of Bacteriology* 132: 262–269 by copyright permission of the American Society for Microbiology.

Figure 10-21: Courtesy of J. C. Wang.

Figure 10-23: Dr. Gopal/SPL/Photo Researchers.

Figure 10-24: Courtesy of V. G. Allfrey.

Figure 10-37: Courtesy of G. P. Georgiev.

Figure 10-40: Courtesy of P. A. Sharp.

Figure 10-42: Courtesy of P. Leder.

Figure 10-44 (Left): Courtesy of A. L. Beyer.

Figure 10-44 (Right): Courtesy of Jack Griffith.

Figure 10-58: Courtesy of E. Sidebottom.

Figure 10-59: Courtesy of J. Gall.

Figure 10-60: Courtesy of V. Sorsa.

Figure 10-62 (Left): Courtesy of J. M. Amabis.

Figure 10-62 (Right): Courtesy of B. Daneholt.

Figure 10-63: Courtesy of W. Beerman.

Figure 10-64: Courtesy of Th. K. H. Holt.

Figure 10-65: Courtesy of M. Ashburner.

Figure 10-66: Courtesy of G. R. Wilson.

Figure 10-69: Courtesy of O. L. Miller, Jr.

Figure 10-70: Reproduced from Saragosti, S., G. Moyne, and M. Yaniv (1980). *Cell* 20: 65-73 by copyright permission of Cell Press, Cambridge, MA.

Figure 10-72: Courtesy of A. Rich.

Figure 10-75: Courtesy of T. C. Kaufman.

Figure 11-2: Courtesy of J. A. Lake.

Figure 11-8: Courtesy of J. A. Lake.

Figure 11-10: Courtesy of U. Scheer.

Figure 11-11: Courtesy of R. P. Perry.

Figure 11-13: Courtesy of M. L. Parduc.

Figure 11-15: Courtesy of O. L. Miller, Jr.

Figure 11-21: Courtesy of M. Alfert.

Figure 11-22: Courtesy of O. L. Miller, Jr.

Figure 11-35: Courtesy of A. Rich.

Figure 11-38: Courtesy of A. Rich.

Figure 11-39: Courtesy of O. Behnke.

Figure 11-40: Courtesy of O. L. Miller, Jr.

Figure 11-51: Courtesy of W. Baumeister.

Figure 12-2: Courtesy of J. Cairns.

Figure 12-3: Courtesy of D. M. Prescott.

Figure 12-5: Courtesy of M. L. Higgins and L. Daneo-Moore.

Figure 12-6: Courtesy of H. R. Hohl.

Figure 12-10: Courtesy of J. H. Taylor.

Figure 12-11 (Top, bottom): Courtesy of J. A. Huberman.

Figure 12-11 (Center): Courtesy of D. R. Wolstenholme.

Figure 12-14: Courtesy of R. A. Laskey.

Figure 12-15: Courtesy of E. Stubblefield.

Figure 12-18: Courtesy of R. Moyzis.

Figure 12-19: Courtesy of Dr. Andrew S. Bajer.

Figure 12-21: Courtesy of M. Schibler.

Figure 12-24 (Left): J. F. Gennaro/Photo Researchers.

Figure 12-24 (Right): CNRI/SPL/Photo Researchers.

Figure 12-25: Courtesy of Christine J. Harrison, PhD., Patterson Institute for Cancer Research, Christie Hospital and Holt Radium Institute, Manchester, England.

Figure 12-29: Courtesy of K. L. McDonald.

Figure 12-31: Courtesy of T. E. Schroeder.

Figure 12-32: Courtesy of R. C. Buck.

Figure 12-33: Courtesy of Dr. Andrew Bajer.

Figure 12-37: Courtesy of B. John.

Figure 12-39: Courtesy of B. John.

Figure 12-46: Courtesy of R. B. Inman.

Figure 12-47: Courtesy of D. E. Comings.

Figure 12-48: Courtesy of T. F. Roth.

Figure 13-1: Courtesy of M. H. Ross.

Figure 13-3: Courtesy of F. A. Pepe.

Figure 13-4: Courtesy of H. S. Sugi.

Figure 13-5 (Top): Courtesy of Dr. John Heuser, Washington University.

Figure 13-5 (Bottom): Courtesy of Dr. John Heuser, Washington University.

Figure 13-7: Courtesy of H. E. Huxley.

Figure 13-8: Courtesy of H. E. Huxley.

Figure 13-9: Courtesy of H. E. Huxley.

Figure 13-11: Courtesy of H. E. Huxley.

Figure 13-14: Courtesy of E. G. Gray.

Figure 13-17: Courtesy H. E. Huxley.

Figure 13-19: Courtesy of L. D. Peachey.

Figure 13-21: Courtesy of C. Franzini-Armstrong.

Figure 13-22: Courtesy of J. R. Sommer.

Figure 13-23: Courtesy of G. Gabella.

Figure 13-24: Courtesy of P. Cooke.

Figure 13-26: Courtesy of R. D. Goldman.

Figure 13-28 (Left, middle): Courtesy of E. Lazarides.

Figure 13-28 (Right): Courtesy of T. D. Pollard.

Figure 13-30: Courtesy of N. K. Wessells.

Figure 13-31: Courtesy of R. D. Allen.

Figure 13-34: Courtesy of J. Hartwig.

Figure 13-36: Courtesy of G. Albrecht-Buehler.

Figure 13-37 (Left): Courtesy of M. S. Mooseker.

Figure 13-37 (Right): Courtesy of J. Heuser.

Figure 13-38: Courtesy of Professors Richard Kessel and Randy Kardon, University of Iowa.

Figure 13-40: Courtesy of R. Norberg.

Figure 13-41: Courtesy of R. Rajaraman.

Figure 13-42: Courtesy of L. D. Russell.

Figure 13-43 (Left): Courtesy of A. Klug.

Figure 13-43 (Right): Courtesy of P. R. Burton.

Figure 13-46: Courtesy of J. L. Rosenbaum.

Figure 13-48: Courtesy of B. R. Brinkley.

Figure 13-49: Courtesy of R. W. Weisenberg.

Figure 13-50: Courtesy of A. T. Mariassy.

Figure 13-52: Courtesy of W. L. Dentler.

Figure 13-54: Courtesy of R. W. Linck.

Figure 13-55: Courtesy of I. R. Gibbons.

Figure 13-56: Courtesy of M. Holwill.

Figure 13-57: Courtesy of P. Satir.

Figure 13-58: Courtesy of I. R. Gibbons.

Figure 13-60: Courtesy of J. L. Rosenbaum.

Figure 13-61: Courtesy of F. D. Warner.

Figure 13-62: Courtesy of M. McGill.

Figure 13-64: Courtesy of S. P. Sorokin.

Figure 13-65 (Top left and top right): Courtesy of L. E. Roth.

Figure 13-65 (Bottom left): Courtesy of L. G. Tilney.

Figure 13-65 (Bottom right): Courtesy of J. B. Tucker.

Figure 13-67: Courtesy of N. Hirokawa.

Figure 13-68: Courtesy of D. G. Weiss.

Figure 13-70: Courtesy of B. R. Brinkley.

Figure 13-72: Courtesy of W. W. Franke.

Figure 13-73: Courtesy of P. C. Bridgeman.

Figure 13-74: Courtesy of S. Penman.

Figure 13-75: Courtesy of J. Adler.

Figure 13-79: Courtesy of M. Simon.

Figure 13-80: Dr. Kwand Shing Kim/Peter Arnold, Inc.

Figure 14-1: Courtesy of J. W. Schopf.

Figure 14-2: Roger Ressmeyer/Starlight.

Figure 14-4: Courtesy of S. Fox.

Figure 14-5: Courtesy of S. Fox.

Figure 14-9: Courtesy of E. H. Coe.

Figure 14-10: Courtesy of Y. Yotsuyanagi.

Figure 14-11: Courtesy of P. R. Burton.

Figure 14-12: Courtesy of M. K. Nass.

Figure 14-15: Courtesy of M. K. Nass.

Figure 14-20 (Left): Courtesy of J. V. Possingham.

Figure 14-20 (Right): Courtesy of R. Charret and J. André.

Figure 14-22 (Left): Courtesy of D. L. Robberson.

Figure 14-22 (Right): Courtesy of K. Koike and D. R. Wolstenholme.

Figure 14-24: Courtesy of N. Davidson.

Figure 14-27: Courtesy of H. Falk.

Figure 14-28: Courtesy of W. J. Larsen.

Figure 14-30 (Top): Courtesy of J. V. Possingham.

Figure 14-30 (Bottom): Courtesy of R. M. Leech.

Figure 14-31: Micrograph by W. P. Wergin, Courtesy of E. H. Newcomb.

Figure 14-33: Courtesy of T. E. Weier.

Figure 14-34: Courtesy of L. Bogorad.

Figure 14-37: Courtesy of W. T. Hall.

Part 3 Opener: Courtesy of Christina M. Alberini, Mirella Ghirardi, Richard Metz, and Eric R. Krandel, *Cell* 78:1099–1114, March 25, 1994. Copyright permission of Cell Press, Cambridge, MA.

Figure 15-2 (Top): Courtesy of J. G. Gall.

Figure 15-2 (Bottom): Courtesy of O. L. Miller, Jr.

Figure 15-4: Courtesy of Richard G. Kessel.

Figure 15-5: Courtesy of P. Bagavandoss.

Figure 15-6: Courtesy of Richard G. Kessel.

Figure 15-7: Courtesy of J. N. Dumont.

Figure 15-9: Courtesy of E. Anderson.

Figure 15-11: D. W. Fawcett/SS/Photo Researchers.

Figure 15-12: Courtesy of W. A. Anderson.

Figure 15-13: Courtesy of D. M. Phillips.

Figure 15-14: Courtesy of D. S. Friend.

Figure 15-15: Courtesy of S. A. Pratt.

Figure 15-18: Courtesy of R. Yanagemachi.

Figure 15-19 (Left): Courtesy of G. Schatten.

Figure 15-19 (Right): Courtesy of E. Anderson.

Figure 15-20: Courtesy of M. Tegner.

Figure 15-21: Courtesy of R. D. Grey.

Figure 15-22: Courtesy of Y. Hiramoto.

Figure 15-23: Courtesy of E. Anderson.

Figure 15-29: Courtesy of Richard G. Kessel.

Figure 15-33: Courtesy of J. Wartiovaara.

Figure 15-34: Courtesy of J. B. Gurdon.

Figure 15-35: Courtesy of D. A. Evans.

Figure 15-36: Courtesy of M. S. Steinberg.

Figure 15-40: Courtesy of L. Hayflick.

Figure 16-4: Courtesy of D. W. Metzger.

Figure 16-16: Courtesy of Don W. Fawcett.

Figure 16-22 (Left): Courtesy of E. R. Podack.

Figure 16-22 (Right): Courtesy of G. Kaplan.

Figure 16-28: Courtesy of A. Fauci.

Figure 17-1: Courtesy of A. Peters.

Figure 17-3: Courtesy of Dr. Nobutaka Hirokawa.

Figure 17-10: Courtesy of Dr. Cedric S. Raine.

Figure 17-12: Courtesy of S. G. Waxman.

Figure 17-13: Courtesy of S. G. Waxman.

Figure 17-14: Courtesy of J. E. Heuser.

Figure 17-16: Courtesy of A. B. Harris.

Figure 17-17: Courtesy of J. E. Heuser.

Figure 17-23: Courtesy of S. H. Snyder.

Figure 17-25: Courtesy of E. R. Lewis, Y. Y. Zeevi and F. S. Werblin.

Figure 17-28: Courtesy of S. M. Rothman.

Figure 17-29: Courtesy of R. Levi-Montalcini.

Figure 18-1: Courtesy of G. D. Abrams.

Figure 18-5: Courtesy of K. M. Robinson.

Figure 18-7: Courtesy of G. D. Abrams.

Figure 18-9: Courtesy of L. Sachs.

Figure 18-10: Courtesy of W. R. Loewenstein.

Figure 18-11: Courtesy of P. C. Nowell.

Figure 18-14: Courtesy of M. A. Epstein.

Figure 18-19: Courtesy of L. C. Stevens.

Figure 18-20: Courtesy of B. Mintz and K. Illmensee.

TEXT AND LINE ILLUSTRATION CREDITS

Figure 3-3: Adapted from A. D. Hershey and M. Chase (1952). *J. Gen. Physiol.* 36:39–56.

Figure 3-9: Adapted from R. T. Okazaki, K. Okazaki, K. Sakabe, K. Sugimoto, and A. Sugino (1968). *Proc. Natl. Acad. Sci. USA* 59:598–602.

Figure 3-16: Adapted from G. W. Beadle and E. L. Tatum (1941). *Proc. Natl. Acad. Sci. USA* 27:499–506.

Figure 3-21: Adapted from P. Siekevitz (1952). *J. Biol. Chem.* 195:549–565, and from J. W. Littlefield, E. B. Keller, J. Gross, and P. C. Zamecnik (1955). *J. Biol. Chem.* 217:111–124.

Figure 3-25: Adapted from J. Marmur, C. M. Greenspan, E. Palecek, F. M. Kahan, J. Levine, and M. Mandel (1962). *Cold Spr. Harbor Symp. Quant. Biol.* 28:191–199.

Figure 3-28: Adapted from M. B. Hoagland, M. L. Stephenson, J. F. Scott, L. I. Hect, and P. C. Zamecnik (1958). *J. Biol. Chem.* 231:241–257.

Figure 3-48: Adapted from A. M. Maxam and W. Gilbert (1977). *Proc. Natl. Acad. Sci. USA* 74:560–564.

Figure 3-49: Adapted from F. Sanger, S. Nicklen, and A. R. Coulson (1977). *Proc. Natl. Acad. Sci. USA* 74:5463–5467.

Figure 5-34: Adapted from T. Krasne, R. Eisenmann, and S. Szabo (1971). *Science* 174: 412–415.

Figure 5-35: Adapted from R. LeFevre (1961). *Pharmacol. Rev.* 13:39–49.

Figure 5-39: Adapted from J. C. Skow (1957). *Biochim. Biophys. Acta* 23: 394–401.

Figure 5-43: Adapted from I. Bihler and R. K. Crane (1962). *Biochim. Biophys. Acta* 59:79–83 and from *Membranes and Ion Transport*, E. E. Bittar, ed., Vol. 1, copyright 1970, Wiley-Interscience, New York.

Figure 6-4: Adapted from A. L. Hodgkin and B. Katz (1949). *J. Physiol.* 108:37–77.

Figure 6-8: Adapted from B. H. Ginsberg in *Biochemical Actions of Hormones*, G. Litwack, ed., Vol. IV, pp. 313–344, copyright 1977, Academic Press, Inc., Orlando, FL.

Figure 6-20: Adapted from W. Y. Ling and J. M. Marsh (1977). *Endocrinology* 100:1571–1578.

Figure 6-34: Figure from *Molecular Biology of the Cell*, 3rd ed., by Alberts et al., 1994. Reprinted by permission of Garland Publishing, Inc.

Figure 7-9: Adapted from L. Ernster and S. Orrenius (1973). *Drug Metab. Dispos.* 1:66–73 and from R. Kuntzman, W. Levin, M. Jacobson, and A. H. Cooney (1973). *Life Sci.* 7:215–224.

Figure 7-11: Adapted from C. M. Redman and D. D. Sabatini (1966). *Proc. Natl. Acad. Sci. USA* 56: 608–615.

Figure 7-23: Adapted from J. D. Jamieson (1967). *J. Cell Biol.* 34:597–615 and from J. D. Castle, J. D. Jamieson, and G. E. Palade (1971). *J. Cell Biol.* 53:290–311.

Figure 7-32: Adapted from J. Berthet, L. Berthet, F. Appelmans, and C. deDuve (1951). *Biochem. J.* 50:182–189.

Figure 7-33: Adapted from C. deDuve, in *Subcellular Particles*, T. Hayashi, ed., pp. 128–159, copyright 1959, Academic Press, Inc., Orlando, FL.

Figure 7-36: Adapted from J. P. Milson and C. H. Wynn (1973). *Biochem. J.* 132:493–500.

Figure 7-52: Adapted from J. E. Rothman and E. P. Kennedy (1977). *Proc. Natl. Acad. Sci. USA* 74:1821–1825.

Figure 8-37: Adapted from C. deDuve (1975). *Science* 189:186–194.

Figure 8-38: Adapted from C. deDuve (1975). *Science* 189: 186–194.

Figure 8-42: Adapted from C. deDuve (1969). *Proc. Roy. Soc. B* 173:71–83.

Figure 9-1: Adapted from R. Emerson and W. Arnold (1932). *J. Gen. Physiol.* 15:391–420.

Figure 9-3: Adapted from R. B. Park (1966). *Int. Rev. Cytol.* 20:67–95.

Figure 9-17: Adapted from R. Emerson and W. Arnold (1932). *J. Gen. Physiol.* 16:191–205.

Figure 9-19: Adapted from R. Emerson and C. M. Lewis (1943). *Amer. J. Bot.* 30: 165–178 and from R. Emerson, R. Chalmers, and C. Cederstrand (1957). *Proc. Natl. Acad. Sci. USA* 43:133–143.

Figure 9-25: Adapted from J. Neumann and A. T. Jagendorf (1984). *Arch. Biochem. Biophys.* 107:109–119.

Figure 9-30: Adapted from G. Hind and A. T. Jagendorf (1963). *Proc. Natl. Acad. Sci. USA* 49:715–719 and from A. T. Jagendorf and J. Neumann (1965). *J. Biol. Chem.* 240:3210–3215.

Figure 9-37: Adapted from A. T. Wilson and M. Calvin (1955). *J. Amer. Chem. Soc.* 77:5948–5957, and from J. A. Bassham, K. Shibata, K. Steenberg, J. Bourdon, and M. Calvin (1956). *J. Amer. Chem. Soc.* 78:4120–4124.

Figure 9-42: Adapted from H. S. Johnson and M. D. Hatch (1968). *Biochem. J.* 114:127–134.

Figure 10-3: Reproduced from "Architecture of the Xenopus Nuclear Pore Complex" by C. W. Akey and M. Radermacher in *Journal of Cell Biology*, 1993, vol. 122, p. 15 by copyright permission of The Rockefeller University Press.

Figure 10-25: Adapted from R. G. Roeder and W. J. Rutter (1970). *Proc. Natl. Acad. Sci. USA* 65:675—692, and from S. P. Blatti, C. J. Ingles, T. J. Lindell, P. W. Morris, R. F. Weaver, F. Weinberg, and W. J. Rutter (1970). *Cold Spr. Harbor Symp. Quant. Biol.* 35:649–657.

Figure 10-26: Adapted from R. Weinmann and R. G. Roeder (1974). *Proc. Natl. Acad. Sci. USA* 71:1790–1794.

Figure 10-34: Adapted from H. Harris (1959). *Biochem. J.* 73:362–369.

Figure 10-35: Adapted from K. Scherrer, H. Latham, and J. E. Darnell (1963). *Proc. Natl. Acad. Sci. USA* 49:240–248.

Figure 10-47: Adapted from W. Gilbert and B. Mueller-Hill (1967). *Proc. Natl. Acad. Sci. USA* 58:2415–2421.

Figure 10-55: Adapted from R. J. Britten and D. E. Kohne (1968). *Science* 161:529—540.

Figure 10-71: Adapted from A. H.-J. Wang, G. J. Quigley, F. J. Kolpak, J. L. Crawford, J. H. vanBoom, G. vanderMarel, and A. Rich (1979). *Nature* 282:680–686.

Figure 11-5: Figure "Transfer RNA Shields Specific Nucleotides in 16S Ribosomal RNA from Attack by Chemical Probes" by Danesh Moazed and Harry F. Noller in *Cell*, December 26, 1986. Copyright © 1986 by Cell Press. Reprinted by permission of Cell Press.

Figure 11-7: Figure "The ribosome cycle" from *The Ribosome Returns* by Peter B. Moore in *Nature*, January 21, 1988. Reprinted with permission. Copyright © 1988 by Macmillan Magazines Limited.

Figure 11-9: Adapted from J. A. Lake (1981). *Sci. Amer.* 245 (Aug):84–97.

Figure 11-12: Adapted from M. Birnstiel (1966). *Nat. Cancer Inst. Monogr.* 23:431–444.

Figure 11-16 (Left): Adapted from J. E. Darnell (1968). *Bacteriol. Rev.* 32:262–281, and from K. Scherrer, H. Latham, and J. E. Darnell (1963). *Proc. Natl. Acad. Sci. USA* 49:240–248.

Figure 11-16 (Right): Adapted from J. R. Warner and R. Soeiro (1967). *Proc. Natl. Acad. Sci. USA* 58:1984–1990.

Figure 11-24: Adapted from H. M. Dintzis (1961). *Proc. Natl. Acad. Sci. USA* 47:247–261.

Figure 11-25: Adapted from R. Kaempfer (1968). *Proc. Natl. Acad. Sci. USA* 61:106–113.

Figure 11-34: Adapted from J. Warner, P. M. Knopf, and A. Rich (1963). *Proc. Natl. Acad. Sci. USA* 49:122–129.

Figure 11-37: Adapted from H. Noll, T. Staehelin, and F. O. Wettstein (1963). *Nature* 198:632–638.

Figure 11-45: Adapted from G. Huez, G. Marbaix, E. Hubert, M. Leclercq, U. Nudel, H. Soreq, R. Salomon, B. Lebleu, M.

Revel, and U. Z. Littauer (1974). *Proc. Natl. Acad. Sci. USA* 71:3143–3146.

Figure 11-47: Adapted from J. M. Gilbert and W. F. Anderson (1970). *J. Biol. Chem.* 245:2342–2349.

Figure 12-12: *Genetics*, 3rd ed., by Peter J. Russell. Copyright c 1992 by Peter J. Russell. Reprinted by permission of Harper-Collins College Publishers.

Figure 12-42: From *Biochemistry* by C. K. Mathews and K. E. vanHolde, p. 877. Copyright © 1990 by The Benjamin/Cummings Publishing Company. Reprinted by permission.

Figure 12-43: Adapted from J. A. Taylor (1965). *J. Cell Biol.* 25:57–67.

Figure 12-45: From *Biochemistry* by C. K. Mathews and K. E. vanHolde, p. 878. Copyright © 1990 by The Benjamin/Cummings Publishing Company. Reprinted by permission.

Figure 13-12: Adapted from A. M. Gordon, A. F. Huxley, and F. J. Julian (1966). *J. Physiol.* 184:170–192.

Figure 13-16: Adapted from L. D. Peachey (1965). *J. Cell Biol.* 25:209–231.

Figure 13-18: Adapted from C. C. Ashley and E. B. Ridgeway (1970). *J. Physiol.* 209:105—130.

Figure 13-20: Adapted from *Muscle Physiology*, F. D. Carlson and D. R. Wilkie, copyright © 1974 by Prentice-Hall, New York.

Figure 13-53: Adapted from P. Satir (1974). *Sci. Amer.* 231 (Oct):45–52.

Figure 13-61: Adapted from H. Pitelka in *Cilia and Flagella*, M. A. Sleigh, ed., pp. 437–469, copyright ©1974 by Academic Press, Inc., Orlando, FL.

Figure 13-76: Adapted from M. L. DePamphilis and J. Adler (1971). *J. Bacteriol.* 105:384–385.

Figure 13-77: Model depicting the arrangement of flagellin subunits in this type of flagellum A. Lowry and L. Hanson, *Journal of Molecular Biology*, Vol. 11, page 293, 1965. Reprinted by permission of Academic Press Ltd., London, England.

Figure 13-78: Adapted from T. Iino (1969). *J. Gen. Microbiol.* 56:227–239.

Figure 14-13: Adapted from D. J. L. Luck and E. Reich (1964). *Proc. Natl. Acad. Sci. USA* 52:931–938.

Figure 14-17: Adapted from J. C. Mounolou, H. Jakob, and P. P. Slonimski (1966). *Biochem. Biophys. Res. Commun.* 24:218–224.

Figure 14-18: Adapted from E. S. Goldring, L. I. Grossman, D. Krupnick, D. R. Cryer, and J. Marmur (1970). *J. Mol. Biol.* 52:323–335.

Figure 14-21: Adapted from K.-S. Chiang and N. Sueoka (1967). *Proc. Natl. Acad. Sci. USA* 57:1506–1513.

Figure 14-29: Adapted from D. J. L. Luck (1965). *J. Cell Biol.* 24: 461–470.

Figure 15-25: Adapted from D. Epel (1977). *Sci. Amer.* 237 (Nov):129–139.

Figure 15-26: Adapted from D. Epel (1967). *Proc. Natl. Acad. Sci. USA* 57:899–906.

Figure 15-37: Adapted from J. E. Morris and A. A. Moscona (1970). *Science* 167:1736–1738.

Figure 15-41: Adapted from L. Hayflick (1975). *BioScience* 25:629–637.

Figure 17-4: Adapted from *Medical Neurobiology*, W. D. Willis, Jr. and R. G. Grossman, 3rd. ed., copyright 1981, C. V. Mosby Company, St. Louis.

Figure 17-5: Adapted from J. H. Schwartz (1980). *Sci. Amer.* 242 (Apr):152–157.

Figure 17-6: Adapted from H. L. Fernandez, P. R. Burton, and F. E. Samson (1971). *J. Cell Biol.* 51:176–192.

Figure 17-9: Adapted from T. Rushton (1951). *J. Physiol.* 115:101–122.

Figure 17-17: Diagram modified from J. E. Heuser and T. S. Reese (1973). *J. Cell Biol.* 57:315–334, by copyright permission of The Rockefeller University Press, New York.

Figure 17-22: Adapted from E. Ramon-Moliner, in *The Structure and Function of Nervous Tissue*, G. H. Bourne, ed., Vol. I, pp. 205–267, copyright 1968, Academic Press, Orlando, FL.

Figure 17-27: Adapted from *Principles of Neural Science*, E. R. Kandel and J. H. Schwartz, 2nd ed., copyright 1985, Elsevier, New York.

Figure 17-29: Adapted from N. K. Wesells, B. S. Spooner, J. F. Ash, M. O. Bradley, M. A. Luduena, E. L. Taylor, J. T. Wrenn, and K. M. Yamada (1971). *Science* 171:135–143.

Figure 18-8: Adapted from S. A. Aaronson and G. J. Todaro (1968). *Science* 162:1024–1026.

Figure 18-12: Adapted from R. K. Boutwell (1964). *Prog. Exp. Tumor Res.* 4:207–250.

Figure 18-13: Adapted from P. Brooks and P. D. Lawley (1964). *Nature* 202:781–784.

Figure 18-21: Adapted from P. Buell and J. E. Dunn, Jr. (1965). *Cancer* 18:656–664.

Figure 18-24: Adapted from S. Meselson and L. Russell in *Origins of Human Cancers: A Book*, H. H. Hiatt, J. D. Watson, and J. A. Winsten, eds., copyright 1977, pp. 1473–1482, Cold Spring Harbor Laboratory, Cold Spring Harbor, NY.

Figure 18-25: Adapted from B. Armstrong and R. Doll (1975). *Int. J. Cancer* 15:617–631.

Index

Note: **Boldface** page numbers refer to pages where the term appears in boldface in the text.